MANAGERIAL STATISTICS

8E

GERALD KELLER

Wilfrid Laurier University

and

Joseph L. Rotman School of Management

University of Toronto

Australia • Brazil • Japan • Korea • Mexico • Singapore • Spain • United Kingdom • United States

Managerial Statistics, Eighth Edition
Gerald Keller

VP/Editorial Director: Jack W. Calhoun
VP/Editor-in-Chief: Alex von Rosenberg
Acquisitions Editor: Charles McCormick, Jr.
Developmental Editor: Elizabeth Lowry
Editorial Assistant: Bryn Lathrop
Sr. Marketing Communications Manager: Larry Qualls
Marketing Communications Manager: Libby Shipp
Senior Marketing Coordinator: Angela Glassmeyer
Content Project Manager: Jacquelyn K Featherly
Technology Project Manager: John Rich
Senior Manufacturing Coordinator: Diane Gibbons
Production House/Compositor: ICC Macmillan Inc.
Art Director: Stacy Jenkins Shirley
Internal Designer: Ke Design, Mason OH
Photography Manager: Deanna Ettinger
Photo Researcher: Terri Miller

Cengage Learning–Australia/New Zealand
www.cengage.com.au
tel: (61) 3 9685 4111

Cengage Learning–UK/Europe/Middle East/Africa
www.cengage.co.uk
tel: (44) 207 067 2500

Cengage Learning–Asia
www.cengageasia.com
tel: (65) 6410 1200

Cengage Learning–India
www.cengage.co.in
tel: (91) 11 30484837/38

Cengage Learning–Latin America
www.cengage.com.mx
tel: +52 (55) 1500 6000

Cengage Learning–Brazil
www.cengage.com.br
tel: (011) 3665-9900

Represented in Canada by Nelson Education, Ltd.
www.nelson.com
tel: (416) 752 9100 / (800) 668 0671

Library of Congress Control Number: 2007936976
Student Edition Package 13: 978-0-324-56955-1
Student Edition Package 10: 0-324-56955-6
Student Edition ISBN 13: 978-0-324-56954-4
Student Edition ISBN 10: 0-324-56954-8

South-Western Cengage Learning
5191 Natorp Boulevard
Mason, OH 45040
USA

Cengage Learning products are represented in Canada by Nelson Education, Ltd.

For your course and learning solutions, visit **academic.cengage.com.**

Printed in China
1 2 3 4 5 6 7 11 10 09 08

BRIEF CONTENTS

CONTENTS

PREFACE

Businesses are increasingly using statistical techniques to convert data into information. For students preparing for the business world, it is not enough merely to focus on mastering a diverse set of statistical techniques and calculations. A course and its attendant textbook must provide a complete picture of statistical concepts and their applications to the real world. *Managerial Statistics* is designed to demonstrate that statistics methods are vital tools for today's managers and economists.

To fulfill this objective requires several features that I have built into this book. First, I have included data-driven examples, exercises, and cases that demonstrate statistical applications that are and can be used by marketing managers, financial analysts, accountants, economists, operations managers, and others. Many are accompanied by large and real or realistic data sets. Second, I reinforce the applied nature of the discipline by teaching students how to choose the correct statistical technique. Third, I teach students the concepts that are essential to interpreting the statistical results.

Why I Wrote This Book

Business is complex and requires effective management to succeed. Managing complexity requires many skills. There are more competitors, more places to sell products, and more places to locate workers. As a consequence, effective decision making is more crucial than ever before. On the other hand, managers have more access to larger and more detailed data that are potential sources of information. However, to achieve this potential requires that managers know how to convert data into information. This knowledge extends well beyond the arithmetic of calculating statistics. Unfortunately, this is what most textbooks offer—a series of unconnected techniques illustrated mostly using manual calculations. This continues a pattern that goes back many years. What is required is a complete approach to applying statistical techniques.

When I started teaching statistics in 1971, books demonstrated how to calculate statistics and, in some cases, how various formulas were derived. One reason for doing so was the belief that by doing calculations by hand, students would be able to understand the techniques and concepts. When the first edition of this book was published in 1988, an important goal was to teach students to identify the correct technique. Through the next seven editions, I refined my approach to emphasize interpretation and decision making equally. I divide the solution of statistical problems into three stages and include them in every appropriate example: (1) *identify* the technique, (2) *compute* the statistics, and (3) *interpret* the results. The *compute* stage can be completed in any or all of three ways: manually (with the aid of a calculator), using Excel, and using Minitab. For those courses that wish to use the computer extensively, manual calculations can be played down or omitted completely. Conversely, those that wish to emphasize manual calculations may easily do so, and the computer solutions can be selectively introduced or skipped entirely. This approach is designed to provide maximum flexibility and leaves to the instructor the decision of if and when to introduce the computer.

I believe that my approach offers several advantages.

- Emphasis on identification and interpretation provides students with practical skills they can apply to real problems they will face whether a course uses manual or computer calculations.
- Students learn that statistics is a method of converting data into information. With 866 data files and corresponding problems that ask students to interpret statistical results, students are provided ample opportunities to practice data analysis and decision making.
- The optional use of the computer allows for larger and more realistic exercises and examples.

Placing calculations in the context of a larger problem allows instructors to focus on more important aspects of the decision problem. For example, more attention needs to be devoted to interpreting statistical results. To properly interpret statistical results requires an understanding of the probability and statistical concepts that underlie the techniques and an understanding of the context of the problems. An essential aspect of my approach is teaching students the concepts. I do so in two ways:

- First, there are 19 Java applets that allow students to see for themselves how statistical techniques are derived without going through the sometimes complicated mathematical derivations.
- Second, I have created a number of Excel worksheets that allow students to perform "what-if" analyses. Students can easily see the effect of changing the components of a statistical technique, such as the effect of increasing the sample size.

Efforts to teach statistics as a valuable and necessary tool in business and economics are made more difficult by the positioning of the statistics course in most curricula. The required statistics course in most undergraduate programs appears in the first or second year. In many graduate programs, the statistics course is offered in the first semester of a three-semester program and the first year of a two-year program. Accounting, economics, finance, human resource management, marketing, and operations management are usually taught after the statistics course. Consequently, most students will not be able to understand the general context of the statistical application. This deficiency is addressed in this book by "Applications in . . ." sections, subsections, and boxes. Illustrations of statistical applications in business with which students are unfamiliar are preceded by an explanation of the background material.

- For example, to illustrate graphical techniques, we use an example that compares the histograms of the returns on two different investments. To explain what financial analysts look for in the histograms requires an understanding that risk is measured by the amount of variation in the returns. The example is preceded by an "Applications in Finance" box that discusses how return on investment is computed and used.
- Later when I present the normal distribution, I feature another "Applications in Finance" box to show why the standard deviation of the returns measures the risk of that investment.
- Thirty-six application boxes are scattered throughout the book.

Some applications are so large that I devote an entire section or subsection to the topic. For example, in the chapter that introduces the confidence interval estimator of a proportion, I also present market segmentation. In that section, I show how the confidence interval estimate of a population proportion can yield estimates of the sizes of market segments. In other chapters, I illustrate various statistical techniques by showing how marketing managers can apply these techniques to determine the differences that exist between market segments. There are seven such sections and two subsections in this book. The "Applications in . . ." segments provide great motivation to the student who asks, How will I ever use this technique?

New in This Edition

In the first seven editions of this book, we offered two review chapters. The first reviewed inference about one and two populations of interval and nominal data. This was originally designed to be a pre-midterm test review. The second appeared at the end of the book and was used to review all the inferential material before the final exam. I decided that in this edition two reviews were not enough. Consequently, I have six review appendixes. These appear at the ends of Chapters 13, 14, 15, 16, 17, and 19, and each provides a list of the techniques covered to that point, a flowchart, exercises, and cases.

Nonparametric statistical techniques (Chapter 19) are presented immediately after the chapter on multiple regression model building.

Appendix 19 provides a complete list of all the statistical techniques (not including forecasting and statistical process control, a flowchart, review exercises, and cases).

The last chapter is now Chapter 23, which is a brief summary of the book and a list of the 12 statistical concepts needed by students after the final exam.

Chapters 2 and 4 now feature more real data. These include the following:

1. The question of global warming (monthly temperature anomalies from three sources dating back to 1880 and carbon-dioxide readings)
2. Updated team payrolls and the number of team wins in baseball, football, basketball, and hockey
3. The actual prices of gasoline and oil, allowing students to see whether real prices have risen and the relationship between the price of oil and the price of gasoline
4. The market model has been moved from Chapter 17 (in the 7th edition) to Chapter 4 with actual data from the NYSE, NASDAQ, and the TSE

I've created many new examples and exercises. Here are the numbers for the 8th edition: textbook: 153 solved examples, 1,768 exercises, 34 cases, 841 data sets (with code names and permanent names); 35 CD appendixes: 37 solved examples, 98 exercises, and 25 data sets for a grand total of 190 worked examples, 1,866 exercises, 34 cases, and 866 data sets.

Guided Book Tour

Data-Driven: The Big Picture

Solving statistical problems begins with a problem and data. The ability to select the right method by problem objective and data type is **a valuable tool for business.** Since business decisions are driven by data, students will leave this course equipped with the tools they need to make effective, informed decisions in all areas of the business world.

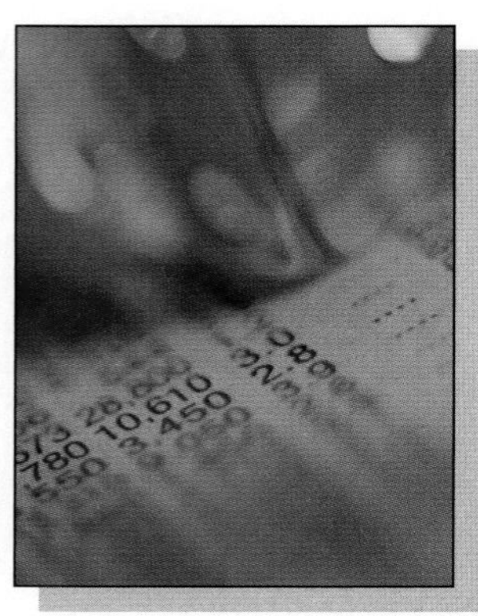

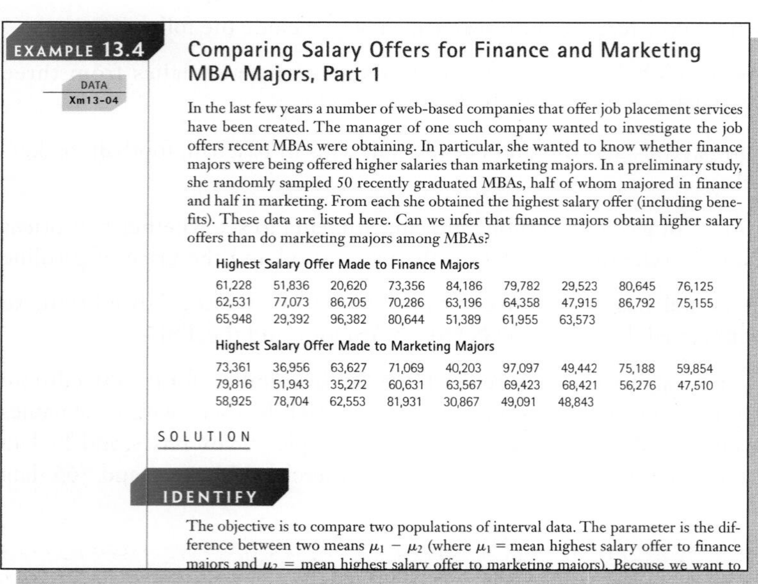

EXAMPLE 13.4

DATA Xm13-04

Comparing Salary Offers for Finance and Marketing MBA Majors, Part 1

In the last few years a number of web-based companies that offer job placement services have been created. The manager of one such company wanted to investigate the job offers recent MBAs were obtaining. In particular, she wanted to know whether finance majors were being offered higher salaries than marketing majors. In a preliminary study, she randomly sampled 50 recently graduated MBAs, half of whom majored in finance and half in marketing. From each she obtained the highest salary offer (including benefits). These data are listed here. Can we infer that finance majors obtain higher salary offers than do marketing majors among MBAs?

Highest Salary Offer Made to Finance Majors

61,228	51,836	20,620	73,356	84,186	79,782	29,523	80,645	76,125
62,531	77,073	86,705	70,286	63,196	64,358	47,915	86,792	75,155
65,948	29,392	96,382	80,644	51,389	61,955	63,573		

Highest Salary Offer Made to Marketing Majors

73,361	36,956	63,627	71,069	40,203	97,097	49,442	75,188	59,854
79,816	51,943	35,272	60,631	63,567	69,423	68,421	56,276	47,510
58,925	78,704	62,553	81,931	30,867	49,091	48,843		

SOLUTION

IDENTIFY

The objective is to compare two populations of interval data. The parameter is the difference between two means $\mu_1 - \mu_2$ (where μ_1 = mean highest salary offer to finance majors and μ_2 = mean highest salary offer to marketing majors). Because we want to

Identify the Correct Technique

Examples introduce the first crucial step in this three-step (Identify-Compute-Interpret) approach. Every example's solution begins by examining the data type and problem objective and then identifying the right technique to solve the problem.

Factors That Identify the t-Test and Estimator of μ_D

1. **Problem objective:** Compare two populations
2. **Data type:** Interval
3. **Descriptive measurement:** Central location
4. **Experimental design:** Matched pairs

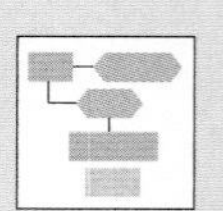

Factors That Identify . . . boxes are found in each chapter after a technique or concept has been introduced. These boxes allow students to see a technique's essential requirements and give them a way to easily review their understanding. These essential requirements are revisited in the review appendixes, where they are illustrated in flowcharts.

APPENDIX 14 / REVIEW OF CHAPTERS 12 TO 14

The number of techniques introduced in Chapters 12 to 14 is up to 23. As we did in Appendix 13, we provide a table of the techniques with formulas and required conditions, a flowchart to help you identify the correct technique, and 18 exercises to give you practice in how to choose the appropriate method. The table and the flowchart have been amended to include the three analysis of variance techniques introduced in this chapter and the three multiple comparison methods.

TABLE **A14.1** **Summary of Statistical Techniques in Chapters 12 to 14**

- t-test of μ
- Estimator of μ (including small population estimator of μ and large and small population estimators of $N\mu$)
- χ^2-test of σ^2
- Estimator of σ^2
- z-test of p
- Estimator of p (including small population estimator of p and large and small population estimators of Np)
- Equal-variances t-test of $\mu_1 - \mu_2$
- Equal-variances estimator of $\mu_1 - \mu_2$
- Unequal-variances t-test of $\mu_1 - \mu_2$
- Unequal-variances estimator of $\mu_1 - \mu_2$
- t-test of μ_D
- Estimator of μ_D
- F-test of σ_1^2/σ_2^2
- Estimator of σ_1^2/σ_2^2
- z-test of $p_1 - p_2$ (Case 1)
- z-test of $p_1 - p_2$ (Case 2)
- Estimator of $p_1 - p_2$
- One-way analysis of variance (including multiple comparisons)
- Two-way (randomized blocks) analysis of variance
- Two-factor analysis of variance

Review of Descriptive Techniques shows how the different types of data can be described graphically. Exercises on the CD-ROM let students practice what they've learned.

A GUIDE TO STATISTICAL TECHNIQUES

Problem Objectives

DATA TYPES	Describe a Population	Compare Two Populations	Compare Two or More Populations	Analyze Relationship between Two Variables
Interval	Histogram **Section 2.3** Ogive **Section 2.3** Stem-and-leaf **Section 2.3** Box plot **Section 4.3** Mean, median, and mode **Section 4.1** Range, variance, and standard deviation **Section 4.2** Percentiles and quartiles **Section 4.3** t-test and estimator of a mean **Section 12.1** Chi-squared test and estimator of a variance **Section 12.2**	Equal-variances t-test and estimator of the difference between two means: independent samples **Section 13.1** Unequal-variances t-test and estimator of the difference between two means: independent samples **Section 13.1** t-test and estimator of mean difference **Section 13.3** F-test and estimator of ratio of two variances **Section 13.4** Wilcoxon rank sum test **Section 19.1** Wilcoxon signed rank sum test **Section 19.2**	One-way analysis of variance **Section 14.1** LSD multiple comparison method **Section 14.2** Tukey's multiple comparison method **Section 14.2** Two-way analysis of variance **Section 14.4** Two-factor analysis of variance **Section 14.5** Kruskal-Wallis test **Section 19.3** Friedman test **Section 19.4**	Scatter diagram **Section 2.6** Covariance **Section 4.4** Coefficient of correlation **Section 4.4** Coefficient of determination **Section 4.4** Least squares line **Section 4.4** Simple linear regression and correlation **Chapter 16** Spearman rank correlation **Section 19.5**
Nominal	Frequency distribution **Section 2.2** Bar chart **Section 2.2** Pie chart **Section 2.2** Line chart **Section 2.4** z-test and estimator of a proportion **Section 12.3** Chi-squared goodness-of-fit test **Section 15.1**	z-test and estimator of the difference between two proportions **Section 13.5** Chi-squared test of a contingency table **Section 15.2**	Chi-squared test of a contingency table **Section 15.2**	Cross-classification table **Section 2.5** Chi-squared test of a contingency table **Section 15.2**
Ordinal	Box plot **Section 4.3** Median **Section 4.1** Percentiles and quartiles **Section 4.3**	Wilcoxon rank sum test **Section 19.1** Sign test **Section 19.2**	Kruskal-Wallis test **Section 19.3** Friedman test **Section 19.4**	Spearman rank correlation **Section 19.5**

A Guide to Statistical Techniques, found on the inside front cover of the text, pulls everything together into one useful table that helps students identify which technique to perform based on the problem objective and data type.

More Data Sets

A total of 866 data sets available on the CD-ROM provide ample practice. These data sets often contain real or realistic data, are typically large, and are formatted for Excel, Minitab, SPSS, SAS, JMP IN, and ASCII.

Prevalent use of data in examples, exercises, and cases is highlighted by the accompanying data icon, which alerts students to go to the CD.

11.40 Xr11-40 A highway patrol officer believes that the average speed of cars traveling over a certain stretch of highway exceeds the posted limit of 55 mph. The speeds of a random sample of 200 cars were recorded. Do these data provide sufficient evidence at the 1% significance level to support the officer's belief? What is the p-value of the test? (Assume that the standard deviation is known to be 5.)

11.41 Xr11-41 An automotive expert claims that the large number of self-serve gasoline stations has resulted in poor automobile maintenance, and that the average tire pressure is more than 4 pounds per square inch (psi) below its manufacturer's specification. As a quick test, 50 tires are examined, and the number of psi each tire is below specification is recorded. If we assume that tire pressure is normally distributed with $\sigma = 1.5$ psi, can we infer at the 10% significance level that the expert is correct? What is the p-value?

11.42 Xr11-42 For the past few years, the number of customers of a drive-up bank in New York has averaged 20 per hour, with a standard deviation of 3 per hour. This year, another bank 1 mile away opened a drive-up window. The manager of the first bank believes that this will result in a decrease in the number of customers. The number of customers who arrived during 36 randomly selected hours was recorded. Can we conclude at the 5% significance level that the manager is correct? What is the p-value?

11.43 Xr11-43 A fast-food franchiser is considering building a restaurant at a certain location. Based on financial analyses, a site is acceptable only if the number of pedestrians passing the location averages more than 100 per hour. The number of pedestrians observed for each of 40 hours was recorded. Assuming that the population standard deviation is known to be 16, can we conclude at the 1% significance level that the site is acceptable?

11.44 Xr11-44 Many Alpine ski centers base their projections of revenues and profits on the assumption that the average Alpine skier skis four times per year. To investigate the validity of this assumption, a random sample of 63 skiers is drawn and each is asked to report the nu[...] skied the previous year. If w[...] standard deviation is 2, can w[...] significance level that the ass[...]

11.45 Xr11-45 The golf professional [...] claims that members who h[...] from him lowered their hand[...] five strokes. The club manager[...] claim by randomly sampling[...] have had lessons and asking ea[...] duction in handicap where a negative number indicates an increase in the handicap. Assuming that the reduction in handicap is approximately normally distributed with a standard deviation of two strokes, test the golf professional's claim using a 10% significance level.

11.46 Xr11-46 The current no-smoking regulations in office buildings require workers who smoke to take breaks and leave the building in order to satisfy their habits. A study indicates that such workers average 32 minutes per day taking smoking breaks. The standard deviation is 8 minutes. To help reduce the average break, rooms with powerful exhausts were installed in the buildings. To see whether these rooms serve their designed purpose, a ra[...] taken. The to[...] desks was me[...] whether there[...] away from th[...] interpret it [...] Type II error[...]

11.47 Xr11-47 A low[...] brand golf ba[...] 230 yards an[...] Nike has jus[...] been endorse[...] the ball will [...] the claim, th[...] ball and meas[...] determine wh[...] nificance leve[...]

EXAMPLE 13.9 Test Marketing of Package Designs, Part 1

DATA Xm13-09

The General Products Company produces and sells a variety of household products. Because of stiff competition, one of its products, a bath soap, is not selling well. Hoping to improve sales, General Products decided to introduce more attractive packaging. The company's advertising agency developed two new designs. The first design features several bright colors to distinguish it from other brands. The second design is light green in color with just the company's logo on it. As a test to determine which design is better, the marketing manager selected two supermarkets. In one supermarket the soap was

CASE 13.1 Do Banks Discriminate against Women Business Owners? Part 1*

DATA C13-01

© Bruce Ayres/Stone/Getty Images

Increasingly, more women are becoming owners of small businesses. However, questions concerning how they are treated by banks and other financial institutions have been raised by women's groups. Banks are particularly important to small businesses, since studies show that bank financing represents about one-quarter of total debt, and that for medium-size businesses the proportion rises to approximately one-half. If women's requests for loans are rejected more frequently than are men's requests, or if women must pay higher interest charges than men do, women have cause for complaint. Banks might then be subject to criminal as well as civil suits. To examine this issue, a research project was launched.

The researchers surveyed a total of 1,165 business owners, of whom 115 were women. The percentage of women in the sample, 9.9%, compares favorably with other sources that indicate that women own about 10% of established small businesses at the time. The survey asked a series of questions to men and women business owners who applied for loans during the previous month. It also determined the nature of the business, its size, and its age. Additionally, the owners were asked about their experiences in dealing with banks. The questions asked in the survey included the following:

1. What is the gender of the owner?
   1. female 2. male
2. Was the loan approved?
   1. no 2. yes
3. If it was approved, what interest rate did you get? How much above the prime rate was your rate?

Of the 115 women who asked for a loan, 14 were turned down. A total of 98 men who asked for a loan were rejected. The rates above prime for all loans that were granted were recorded. What do these data disclose about possible gender bias by the banks?

*Adapted from A. L. Riding and C. S. Swift, "Giving Credit Where It's Due: Women Business Owners and Canadian Financial Institutions," Carleton University Working Paper, Series WPS 89-07, 1989.

Flexible to Use

Although many texts today incorporate the use of the computer, ***Managerial Statistics*** is designed for maximum flexibility and ease of use for both instructors and students. To this end, parallel illustration of both manual and computer printouts is provided throughout the text. This approach **allows you to choose** which, if any, computer program to use. Regardless of the method or software you choose, the output and instructions that you need are provided! Also, instructions for both SPSS and JMP IN can be found on the Keller Online Book Companion Website at international.cengage.com.

Compute the Statistics

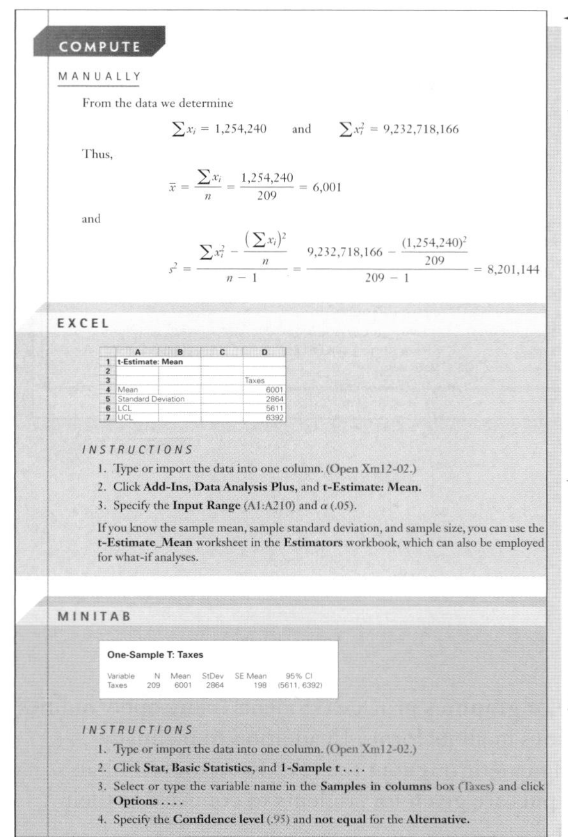

COMPUTE

MANUALLY

From the data we determine

$$\sum x_i = 1{,}254{,}240 \quad \text{and} \quad \sum x_i^2 = 9{,}232{,}718{,}166$$

Thus,

$$\bar{x} = \frac{\sum x_i}{n} = \frac{1{,}254{,}240}{209} = 6{,}001$$

and

$$s^2 = \frac{\sum x_i^2 - \frac{\left(\sum x_i\right)^2}{n}}{n-1} = \frac{9{,}232{,}718{,}166 - \frac{(1{,}254{,}240)^2}{209}}{209-1} = 8{,}201{,}144$$

EXCEL

	A	B	C	D
1	t-Estimate: Mean			
2				
3				Taxes
4	Mean			6001
5	Standard Deviation			2864
6	LCL			5611
7	UCL			6392

INSTRUCTIONS

1. Type or import the data into one column. (Open Xm12-02.)
2. Click **Add-Ins, Data Analysis Plus,** and **t-Estimate: Mean.**
3. Specify the **Input Range** (A1:A210) and α (.05).

If you know the sample mean, sample standard deviation, and sample size, you can use the **t-Estimate_Mean** worksheet in the **Estimators** workbook, which can also be employed for what-if analyses.

MINITAB

One-Sample T: Taxes

Variable	N	Mean	StDev	SE Mean	95% CI
Taxes	209	6001	2864	198	(5611, 6392)

INSTRUCTIONS

1. Type or import the data into one column. (Open Xm12-02.)
2. Click **Stat, Basic Statistics,** and **1-Sample t . . . .**
3. Select or type the variable name in the **Samples in columns** box (Taxes) and click **Options . . . .**
4. Specify the **Confidence level** (.95) and **not equal** for the **Alternative.**

Once the correct technique has been identified, examples take students to the next level within the solution by asking them to compute the statistics.

Manual calculation of the problem is presented first in each "Compute" section of the examples.

Step-by-step instructions in the use of **Excel** and **Minitab** immediately follow the manual presentation. Instructions appear in the book with the printouts—there's no need to incur the extra expense for separate software manuals. SPSS and JMP IN are also available at no cost on the Keller companion Website.

Appendix A provides summary statistics that allow students to solve applied exercises with data files by hand. Offering unparalleled flexibility, this feature allows virtually *all* exercises to be solved by hand!

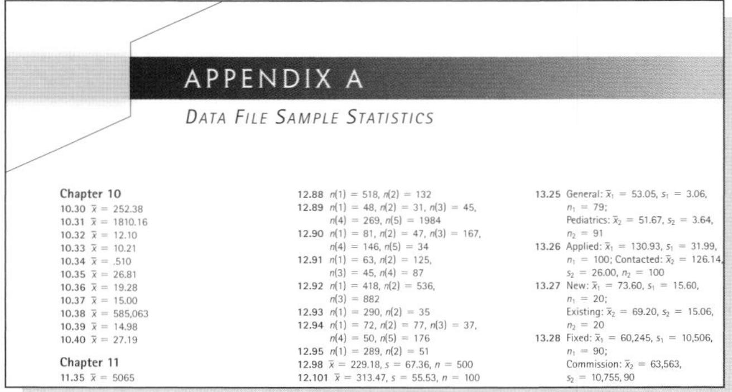

APPENDIX A

DATA FILE SAMPLE STATISTICS

Chapter 10

10.30 $\bar{x} = 252.38$
10.31 $\bar{x} = 1810.16$
10.32 $\bar{x} = 12.10$
10.33 $\bar{x} = 10.21$
10.34 $\bar{x} = .510$
10.35 $\bar{x} = 26.81$
10.36 $\bar{x} = 19.28$
10.37 $\bar{x} = 15.00$
10.38 $\bar{x} = 585{,}063$
10.39 $\bar{x} = 14.98$
10.40 $\bar{x} = 27.19$

Chapter 11

11.35 $\bar{x} = 5065$

12.88 $n(1) = 518$, $n(2) = 132$
12.89 $n(1) = 48$, $n(2) = 31$, $n(3) = 45$, $n(4) = 269$, $n(5) = 1984$
12.90 $n(1) = 81$, $n(2) = 47$, $n(3) = 167$, $n(4) = 146$, $n(5) = 34$
12.91 $n(1) = 63$, $n(2) = 125$, $n(3) = 45$, $n(4) = 87$
12.92 $n(1) = 418$, $n(2) = 536$, $n(3) = 882$
12.93 $n(1) = 290$, $n(2) = 35$
12.94 $n(1) = 72$, $n(2) = 77$, $n(3) = 37$, $n(4) = 50$, $n(5) = 176$
12.95 $n(1) = 289$, $n(2) = 51$
12.98 $\bar{x} = 229.18$, $s = 67.36$, $n = 500$
12.101 $\bar{x} = 313.47$, $s = 55.53$, $n = 100$

13.25 General: $\bar{x}_1 = 53.05$, $s_1 = 3.06$, $n_1 = 79$; Pediatrics: $\bar{x}_2 = 51.67$, $s_2 = 3.64$, $n_2 = 91$
13.26 Applied: $\bar{x}_1 = 130.93$, $s_1 = 31.99$, $n_1 = 100$; Contacted: $\bar{x}_2 = 126.14$, $s_2 = 26.00$, $n_2 = 100$
13.27 New: $\bar{x}_1 = 73.60$, $s_1 = 15.60$, $n_1 = 20$; Existing: $\bar{x}_2 = 69.20$, $s_2 = 15.06$, $n_2 = 20$
13.28 Fixed: $\bar{x}_1 = 60{,}245$, $s_1 = 10{,}506$, $n_1 = 90$; Commission: $\bar{x}_2 = 63{,}563$, $s_2 = 10{,}755, 90$

CD APPENDIX F / HYPERGEOMETRIC DISTRIBUTION

A hypergeometric experiment is an experiment where a sample of n items is taken without replacement from a finite population of N items, each of which is classified as a success or a failure. (If the sampling is done with replacement the experiment is binomial.) Let k = number of successes and $(N-k)$ is the number of failures in the population.

> **Hypergeometric Random Variable**
> The **hypergeometric random variable** is the number of success in a hypergeometric experiment.

A hypergeometric random variable is a discrete random variable that can take on any one of the values $0, 1, 2, \ldots, n$. The hypergeometric probability distribution can be derived using the multiplication, addition, and complement rules or more easily by applying a probability tree.

In addition, **CD Appendixes** are included on the CD-ROM. There are 35 appendixes covering such topics as the hypergeometric distribution, index numbers, and more detailed instructions for Excel and Minitab.

Flexible Learning

For visual learners, the **Seeing Statistics** feature refers to online Java applets developed by Gary McClelland of the University of Colorado, which use the interactive nature of the Web to illustrate key statistical concepts. With 19 applets and 82 follow-up exercises, students can explore and interpret statistical concepts, leading them to greater intuitive understanding. All Seeing Statistics applets can be found on the accompanying CD-ROM.

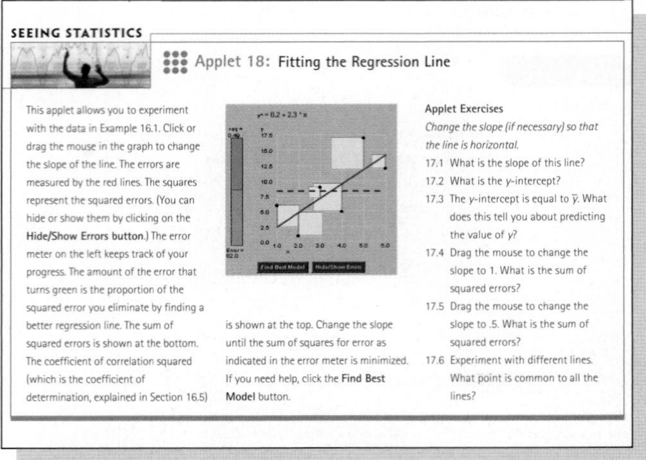

SEEING STATISTICS

Applet 18: **Fitting the Regression Line**

This applet allows you to experiment with the data in Example 16.1. Click or drag the mouse in the graph to change the slope of the line. The errors are measured by the red lines. The squares represent the squared errors. (You can hide or show them by clicking on the **Hide/Show Errors button**.) The error meter on the left keeps track of your progress. The amount of the error that turns green is the proportion of the squared error you eliminate by finding a better regression line. The sum of squared errors is shown at the bottom. The coefficient of correlation squared (which is the coefficient of determination, explained in Section 16.5) is shown at the top. Change the slope until the sum of squares for error as indicated in the error meter is minimized. If you need help, click the **Find Best Model** button.

Applet Exercises

Change the slope (if necessary) so that the line is horizontal.

17.1 What is the slope of this line?

17.2 What is the y-intercept?

17.3 The y-intercept is equal to $\bar{y}$. What does this tell you about predicting the value of y?

17.4 Drag the mouse to change the slope to 1. What is the sum of squared errors?

17.5 Drag the mouse to change the slope to .5. What is the sum of squared errors?

17.6 Experiment with different lines. What point is common to all the lines?

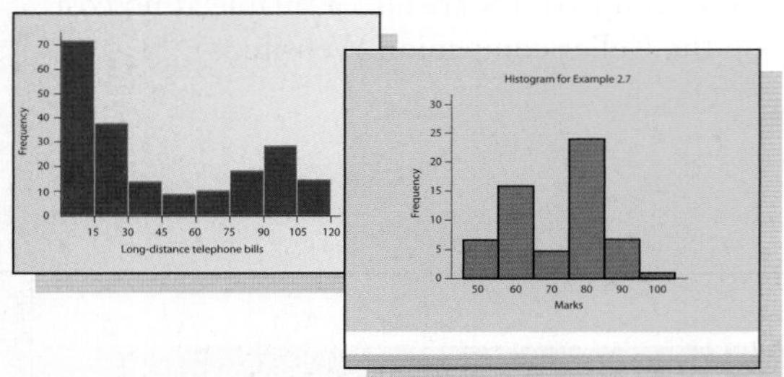

Ample use of graphics provides students many opportunities to see statistics in all its forms. In addition to manually presented figures throughout the text, Excel and Minitab graphic outputs are given for students to compare to their own results.

Applied: Bridging The Gap

In the real world, it is not enough to know *how* to generate the statistics. To be truly effective, a business person must also know how to **interpret and articulate** the results. Furthermore, students need a framework to understand and apply statistics **within a realistic setting** by using realistic data in exercises, examples, and case studies.

Interpret the Results

Examples round out the final component of the Identify-Compute-Interpret approach by asking students to interpret the results in the context of a business-related decision. This final step motivates and shows how statistics is used in everyday business situations.

3.3 PRESENTING STATISTICS: WRITTEN REPORTS AND ORAL PRESENTATIONS

Throughout this book we present a variety of statistical techniques. Our emphasis is on applying the correct statistical technique and the proper interpretation of the resulting statistics. However, the ability to communicate your findings both orally and in writing is a critical skill. In this section we provide general guidelines for both forms of

Writing a Report

Just as there are many ways to write a statistics textbook, there are also many different ways to write a report. Here is our suggested method. Reports should contain the following steps:

1. **State your objective.** In Chapter 2 we introduced statistical techniques by describing the type of information needed and the type of data produced by the experiment. For example, there are many studies that t... service is better than one or more other similar prod... simply state the purpose of the statistical analysis and ... of the experiment and the decisions that may follow.

2. **Describe the experiment.** It is important to know yo... dividuals who have little knowledge of statistics, you ... However, it is likely that some members of your audi... edge of the subject and thus will want to know how t... particular, they will want you to assure them that th... conducted.

New coverage of writing reports and creating presentations in Chapter 3 sets up exercises that ask students to articulate their findings to nonstatisticians.

The following exercises require the use of a computer and software. The answers may be calculated manually. See Appendix A for the sample statistics. ***Use a 5% significance level unless specified otherwise.***

12.76 Xr12-76* There is a looming crisis in universities and colleges across North America. In most places enrollments are increasing, requiring more instructors. However, there are not enough Ph.D.'s to fill the vacancies now. Moreover, among current professors, a large proportion are nearing retirement age. On top of these problems, some universities allow professors over the age of 60 to retire early. To help devise a plan to deal with the crisis, a consultant surveyed 521 55- to 64-year-old professors and asked each whether he or she intended to retire before 65. The responses are 1 = No and 2 = Yes.

a. Estimate with 95% confidence the proportion of professors who plan on early retirement.

b. Write a report for the university president describing your statistical analysis.

12.77 Refer to Exercise 12.76. If the number of professors between the ages of 55 and 64 is 75,000, estimate the total number of such professors who plan to retire early.

12.78 Xr12-78 To determine how many Americans smoke, annual surveys are conducted by the U.S. National Center for Health Statistics. The survey asks a random sample of Americans whether they smoke on some days. The responses are 1 = No and 2 = Yes. Estimate with 95% con-

12.81 Xr12-81* An important decision faces Christmas holiday celebrators: buy a real or artificial tree? A sample of 1,508 male and female respondents 18 years of age and over was interviewed. Respondents were asked whether they preferred a real (1) or artificial (2) tree. If there are 6 million Canadian households that buy Christmas trees, estimate with 95% confidence the total number of Canadian households that would prefer artificial Christmas trees. (*Toronto Star* November 29, 2006)

12.82 Xr12-82* Because television audiences of newscasts tend to be older (and because older people suffer from a variety of medical ailments) pharmaceutical companies' advertising often appears on national news in the three networks (ABC, CBS, and NBC). The ads concern prescription drugs such as those to treat heartburn. To determine how effective the ads are, a survey was undertaken. Adults over 50 who regularly watch network newscasts were asked whether they had contacted their physician to ask about one of the prescription drugs advertised during the newscast. The responses (1 = No and 2 = Yes) were recorded.

a. Estimate with 95% confidence the fraction of adults over 50 who have contacted their physician to inquire about a prescription drug.

b. Prepare a presentation to the executives of a pharmaceutical company that discusses your analysis.

12.83 Xr12-83 A professor of business statistics recently

An Applied Approach

With **Applications in . . .** sections and boxes, ***Managerial Statistics*** now includes 45 **applications** (in finance, marketing, operations management, human resources, economics, and accounting) highlighting how statistics is used in those professions. For example, "Applications in Accounting: Auditing" shows how statistics are used to estimate several parameters in auditing and uses a real application (GAO). An optional section, "Applications in Professional Sports: Baseball" contains a subsection on the success of the Oakland Athletics.

In addition to sections and boxes, **Applications in . . . exercises** can be found within the exercise sections to further reinforce the big picture.

4.5 (OPTIONAL) APPLICATIONS IN PROFESSIONAL SPORTS: BASEBALL

In the chapter-opening example we provided the payrolls and the number of wins from the 2006 season. We discovered that there is a weak positive linear relationship between number of wins and payroll. The strength of the linear relationship tells us that some teams with large payrolls are not successful on the field, whereas some teams with small payrolls win a large number of games. It would appear that while the amount of money teams spend is a factor, another factor is *how* teams spend their money. In this section we will analyze the five seasons between 2002 and 2006 to see how small-payroll teams succeed.

Professional sports in North America is a multibillion dollar business. The cost of a new franchise in baseball, football, basketball, and hockey is often in the hundreds of millions of dollars. Although some teams are financially successful during losing sea-

APPLICATIONS in HUMAN RESOURC...

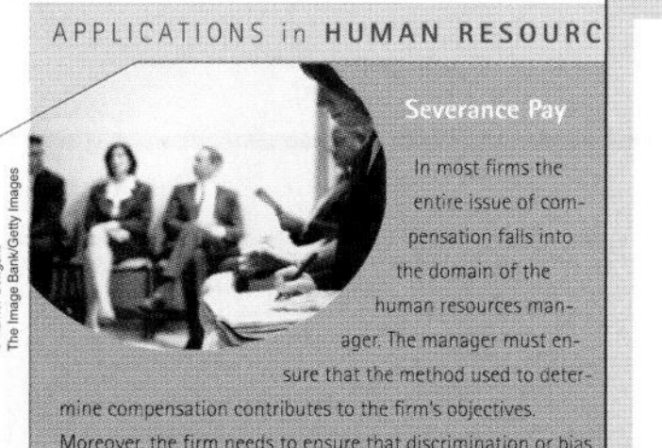

Severance Pay

In most firms the entire issue of compensation falls into the domain of the human resources manager. The manager must ensure that the method used to determine compensation contributes to the firm's objectives. Moreover, the firm needs to ensure that discrimination or bias of any kind is not a factor. Another function of the personnel manager is to develop severance packages for employees whose services are no longer needed because of downsizing or merger. The size and nature of severance is rarely pa... any working agreement and must be determined by a v... of factors. Regression analysis is often useful in this are...

17.5 Xr17-05 When one company buys another... pany, it is not unusual that some workers a... minated. The severance benefits offered ... laid-off workers are often the subject of di... Suppose that the Laurier Company re... bought the Western Company and subseq... terminated 20 of Western's employees. As ... the buyout agreement, it was promised th...

APPLICATIONS in MARKETING

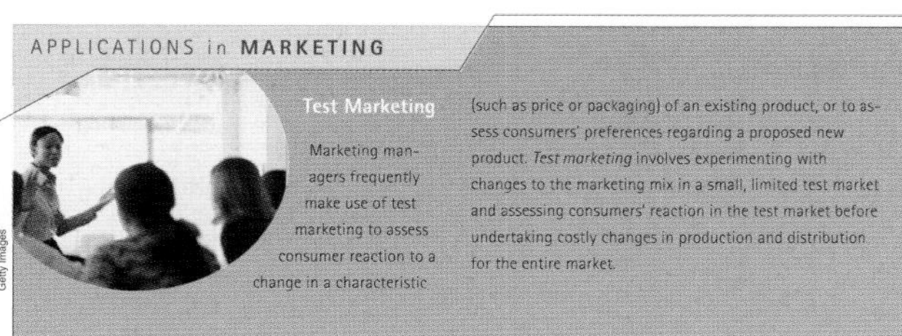

Test Marketing

Marketing managers frequently make use of test marketing to assess consumer reaction to a change in a characteristic (such as price or packaging) of an existing product, or to assess consumers' preferences regarding a proposed new product. *Test marketing* involves experimenting with changes to the marketing mix in a small, limited test market and assessing consumers' reaction in the test market before undertaking costly changes in production and distribution for the entire market.

SSA Envelope Plan

DATA Xm11-00

Federal Express (FedEx) sends invoices to customers requesting payment within 30 days. The bill lists an address and customers are expected to use their own envelopes to return their payments. Currently the mean and standard deviation of the amount of time taken to pay bills are 24 days and 6 days, respectively. The chief financial officer (CFO) believes that including a stamped self-addressed (SSA) envelope would decrease the amount of time. She calculates that the improved cash flow from a 2-day decrease in the payment period would pay for the costs of the envelopes and stamps. Any further decrease in the payment period would generate a profit. To test her belief, she randomly selects 220 customers and includes a stamped self-addressed envelope with their invoices. The numbers of days until payment is received were recorded. Can the CFO conclude that the plan will be profitable?

© Mark Richards/PhotoEdit

After we've introduced the required tools, we'll return to this question and answer it. **(See page 358).**

Chapter-opening examples and solutions present compelling discussions of how the techniques and concepts introduced in that chapter are applied to real-world problems. These examples are then revisited with a solution as each chapter unfolds, applying the methodologies introduced in the chapter.

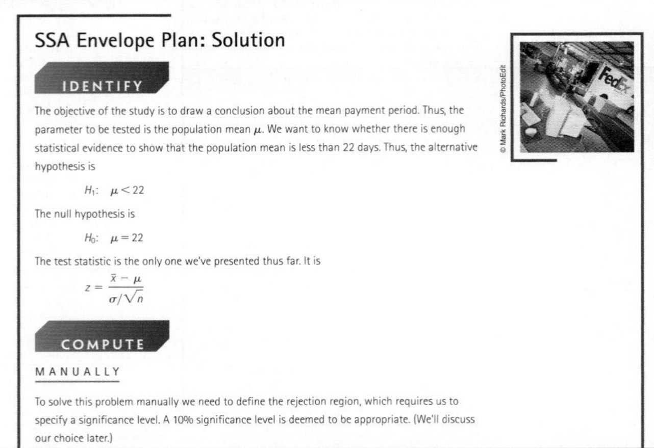

SSA Envelope Plan: Solution

IDENTIFY

The objective of the study is to draw a conclusion about the mean payment period. Thus, the parameter to be tested is the population mean μ. We want to know whether there is enough statistical evidence to show that the population mean is less than 22 days. Thus, the alternative hypothesis is

$$H_1: \quad \mu < 22$$

The null hypothesis is

$$H_0: \quad \mu = 22$$

The test statistic is the only one we've presented thus far. It is

$$z = \frac{\bar{x} - \mu}{\sigma/\sqrt{n}}$$

COMPUTE

MANUALLY

To solve this problem manually we need to define the rejection region, which requires us to specify a significance level. A 10% significance level is deemed to be appropriate. (We'll discuss our choice later.)

© Mark Richards/PhotoEdit

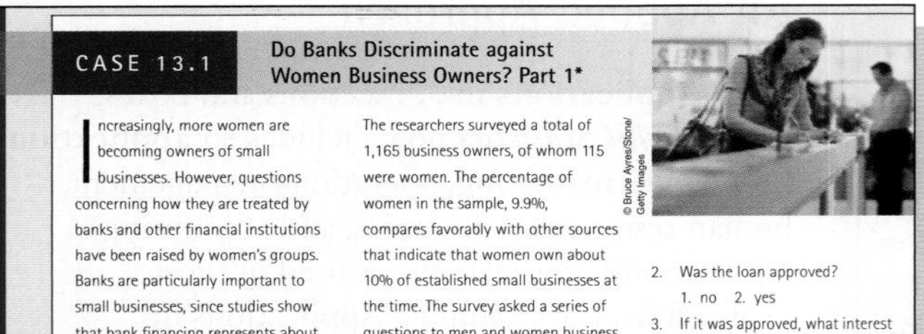

CASE 13.1 Do Banks Discriminate against Women Business Owners? Part 1*

Increasingly, more women are becoming owners of small businesses. However, questions concerning how they are treated by banks and other financial institutions have been raised by women's groups. Banks are particularly important to small businesses, since studies show that bank financing represents about one-quarter of total debt, and that for

The researchers surveyed a total of 1,165 business owners, of whom 115 were women. The percentage of women in the sample, 9.9%, compares favorably with other sources that indicate that women own about 10% of established small businesses at the time. The survey asked a series of questions to men and women business owners who applied for loans during

2. Was the loan approved?
   1. no 2. yes
3. If it was approved, what interest rate did you get? How much above

© Bruce Ayres/Stone/Getty Images

Many of the **examples, exercises, and cases are based on actual studies** performed by statisticians and published in journals, newspapers, and magazines, or presented at conferences. Many data files were re-created to produce the original results.

A total of 1,866 exercises, many of them new or updated, offer ample practice for students to use statistics in an applied context.

CHAPTER SUMMARY

The analysis of variance allows us to test for differences between populations when the data are interval. The analyses of the results of three different experimental designs were presented in this chapter. They were the one-way analysis of variance. The second experimental design also defines the treatments on the basis of one factor. However, the randomized block design uses data gathered by observing the results of a matched or blocked experiment (two-way analysis of variance). The third design is the two-factor experiment wherein the treatments are defined as the combinations of the levels of two factors. All the analyses of variance are based on partitioning the total sum of squares into sources of variation from which the mean squares and F-statistics are computed.

Additionally, we introduced three multiple comparison methods, which allow us to determine which means differ in the one-way analysis of variance.

Finally, we described an important application in operations management that employs the analysis of variance.

IMPORTANT TERMS

Analysis of variance 514
Treatment means 514
One-way analysis of variance 514
Response variable 516
Responses 516
Experimental units 516
Factor 516
Level 516
Between-treatments variation 516
Sum of squares for treatments (SST) 516
Within-treatments variation 517
Sum of squares for error (SSE) 517
Mean squares 518
Mean square for treatments 518
Mean square for error 518
F-Statistic 519
Analysis of variance (ANOVA) table 520
Total variation 520
SS(Total) 520
Completely randomized design 522
Multiple comparisons 530
Least Significance Difference 532
Bonferroni adjustment 533
Tukey's multiple comparison method 534
Multifactor experiment 539
Randomized block design 539
Repeated measures 539
Two-way analysis of variance 540
Fixed effects analysis of variance 540
Random effects analysis of variance 540
Sum of squares for blocks 540
Factorial experiment 549
Interactions 551
Complete factorial experiment 551
Replicate 551
Balanced 551

Resources

Learning Resources

Student's Suite CD-ROM (ISBN 0-324-56956-4). Included with every new copy of the text, this learning tool includes interactive concept simulation exercises from *Seeing Statistics*, *Data Analysis Plus* add-in, as well as a new Treeplan add-in, 866 data sets, optional topics, and 35 CD appendixes.

Companion Website accessible at international.cengage.com. View a host of resources, including SPSS and JMP software instruction and data sets, relevant links and resources, and more.

Student Solutions Manual. Students can check their understanding with this manual, which includes worked solutions of even-numbered exercises from the text. The Student Solutions Manual can be found at international.cengage.com.

Teaching Resources

For a complete listing of our extensive instructor resources, please go to international.cengage.com or contact your local Cengage sales representative.

Acknowledgments

Although there is only one name on the cover of this book, the number of people who made contributions is large. I would like to acknowledge the work of all of them, with particular emphasis on the following.

Curt Hinrichs was the editor of the fourth, fifth, sixth, and seventh editions of this book. His knowledge, guidance, and enthusiasm helped make this book a success. I will always be grateful to him.

Paul Baum, California State University, Northridge, and John Lawrence, California State University, Fullerton, reviewed page proofs for several editions and found and corrected various mistakes. Along the way they made numerous suggestions and recommendations that improved the book tremendously.

Deborah Rumsey, Ohio State University, produced the test bank stored on the Instructor's Suite CD-ROM. Mohammed El-Saidi, Ferris State University, created the test bank for several earlier editions.

Trent Tucker, Wilfrid Laurier University, and Zvi Goldstein, California State University, Fullerton, each produced a set of PowerPoint slides.

The following individuals played important roles in the production of the book.

Senior Acquisitions Editor:
Charles McCormick, Jr.

Developmental Editor:
Elizabeth Lowry

Content Project Manager:
Jacquelyn K Featherly

The author extends thanks also to the survey participants and reviewers of the previous editions: Paul Baum, California State University, Northridge; Nagraj Balakrishnan, Clemson University; Howard Clayton, Auburn University; Philip Cross, Georgetown University; Barry Cuffe, Wingate University; Ernest Demba, Washington

Balakrishnan, Clemson University; Howard Clayton, Auburn University; Philip Cross, Georgetown University; Barry Cuffe, Wingate University; Ernest Demba, Washington University-St. Louis; Neal Duffy, State University of New York, Plattsburgh; John Dutton, North Carolina State University; Erick Elder, University of Arkansas; Mohammed El-Saidi, Ferris State University; Grace Esimai, University of Texas at Arlington; Abe Feinberg, California State University, Northridge; Samuel Graves, Boston College; Robert Gould, UCLA; John Hebert, Virginia Tech; James Hightower, California State University, Fullerton; Bo Honore, Princeton University; Onisforos Iordanou, Hunter College; Gordon Johnson, California State University, Northridge; Hilke Kayser, Hamilton College; Kenneth Klassen, California State University, Northridge; Roger Kleckner, Bowling Green State University-Firelands; Harry Kypraios, Rollins College; John Lawrence, California State University, Fullerton; Dennis Lin, Pennsylvania State University; Neal Long, Stetson University; George Marcoulides, California State University, Fullerton; Paul Mason, University of North Florida; Walter Mayer, University of Mississippi; John McDonald, Flinders University; Richard McGowan, Boston College; Richard McGrath, Bowling Green State University; Amy Miko, St. Francis College; Janis Miller, Clemson University; Glenn Milligan, Ohio State University; James Moran, Oregon State University; Patricia Mullins, University of Wisconsin; Kevin Murphy, Oakland University; Pin Ng, University of Illinois; Des Nicholls, Australian National University; Andrew Paizis, Queens College; David Pentico, Duquesne University; Ira Perelle, Mercy College; Nelson Perera, University of Wollongong; Amy Puelz, Southern Methodist University; Lawrence Ries, University of Missouri; Colleen Quinn, Seneca College; Tony Quon, University of Ottawa; Madhu Rao, Bowling Green State University; Phil Roth, Clemson University; Farhad Saboori, Albright College; Don St. Jean, George Brown College; Hedayeh Samavati, Indiana–Purdue University; Sandy Shroeder, Ohio Northern University; Jineshwar Singh, George Brown College; Natalia Smirnova, Queens College; Eric Sowey, University of New South Wales; Cyrus Stanier, Virginia Tech; Stan Stephenson, Southwest Texas State University; Arnold Stromberg, University of Kentucky; Steve Thorpe, University of Northern Iowa; Sheldon Vernon, Houston Baptist University; and W. F. Younkin, University of Miami.

1

© SuperStock/Jupiterimages

WHAT IS STATISTICS?

Introduction

Statistics is a way to get information from data. That's it! Most of this textbook is devoted to describing how, when, and why managers and statistics practitioners* conduct statistical procedures. You may ask, "If that's all there is to statistics, why is this book (and most other statistics books) so large?" The answer is that there are different kinds of information and data to which students of applied statistics should be exposed. We demonstrate some of these with a case and two examples that are featured later in this book. The first may be of particular interest to you.

---

*The term *statistician* is used to describe so many different kinds of occupations that it has ceased to have any meaning. It is used, for example, to describe both a person who calculates baseball statistics and an individual educated in statistical principles. We will describe the former as a *statistics practitioner* and the

(continued)

Example 2.6 Business Statistics Marks (see Chapter 2)

A student enrolled in a business program is attending his first class of the required statistics course. The student is somewhat apprehensive because he believes the myth that the course is difficult. To alleviate his anxiety the student asks the professor about last year's marks. Because, like all other statistics professors this one is friendly and helpful, he obliges the student and provides a list of the final marks, which are composed of term work plus the final exam. What information can the student obtain from the list?

This is a typical statistics problem. The student has the data (marks) and needs to apply statistical techniques to get the information he requires. This is a function of **descriptive statistics.**

Descriptive Statistics

Descriptive statistics deals with methods of organizing, summarizing, and presenting data in a convenient and informative way. One form of descriptive statistics uses graphical techniques, which allow statistics practitioners to present data in ways that make it easy for the reader to extract useful information. In Chapter 2 we will present a variety of graphical methods.

Another form of descriptive statistics uses numerical techniques to summarize data. One such method that you have already used frequently calculates the average or mean. In the same way that you calculate the average age of the employees of a company, we can compute the mean mark of last year's statistics course. Chapter 4 introduces several numerical statistical measures that describe different features of the data.

The actual technique we use depends on what specific information we would like to extract. In this example, we can see at least three important pieces of information. The first is the "typical" mark. We call this a *measure of central location.* The average is one such measure. In Chapter 4 we will introduce another useful measure of central location, the median. Suppose the student was told that the average mark last year was 67. Is this enough information to reduce his anxiety? The student would likely respond "no" because he would like to know whether most of the marks were close to 67 or were scattered far below and above the average. He needs a *measure of variability.* The simplest such measure is the *range,* which is calculated by subtracting the smallest number from the largest. Suppose the largest mark is 96 and the smallest is 24. Unfortunately, this provides little information. We need other measures, which will be introduced in Chapter 4. Moreover, the student must determine more about the marks. In particular he needs to know how the marks are distributed between 24 and 96. The best way to do this is to use a graphical technique, the histogram, to be introduced in Chapter 2.

---

latter as a *statistician*. A statistics practitioner is a person who uses statistical techniques properly. Examples of statistics practitioners include the following:

1. A financial analyst who develops stock portfolios based on historical rates of return
2. An economist who uses statistical models to help explain and predict variables such as inflation rate, unemployment rate, and changes in the gross domestic product
3. A market researcher who surveys consumers and converts the responses into useful information. Our goal in this book is to convert you into one such capable individual.

The term *statistician* refers to an individual who works with the mathematics of statistics. His or her work involves research that develops techniques and concepts that in the future may help the statistics practitioner. Statisticians are also statistics practitioners, frequently conducting empirical research and consulting. The author of this book is a statistician. If you're taking a statistics course, your instructor is probably a statistician.

Case 12.1 Pepsi's Exclusivity Agreement with a University (see Chapter 12)

In the last few years, colleges and universities have signed exclusivity agreements with a variety of private companies. These agreements bind the university to sell that company's products exclusively on the campus. Many of the agreements involve food and beverage firms.

A large university with a total enrollment of about 50,000 students has offered Pepsi-Cola an exclusivity agreement that would give Pepsi exclusive rights to sell its products at all university facilities for the next year with an option for future years. In return, the university would receive 35% of the on-campus revenues and an additional lump sum of $200,000 per year. Pepsi has been given 2 weeks to respond.

The management at Pepsi quickly reviews what they know. The market for soft drinks is measured in terms of 12-ounce cans. Pepsi currently sells an average of 22,000 cans per week (over the 40 weeks of the year that the university operates). The cans sell for an average of 75 cents each. The costs including labor amount to 20 cents per can. Pepsi is unsure of its market share but suspects it is considerably less than 50%. A quick analysis reveals that if its current market share were 25%, then, with an exclusivity agreement, Pepsi would sell 88,000 (22,000 is 25% of 88,000) cans per week or 3,520,000 cans per year. The gross revenue would be computed as follows:*

$$\text{Gross revenue} = 3{,}520{,}000 \times \$.75/\text{can} = \$2{,}640{,}000$$

This figure must be multiplied by 65% because the university would rake in 35% of the gross. Thus,

$$\text{Gross revenue after deducting 35\% university take}$$
$$= 65\% \times \$2{,}640{,}000 = \$1{,}716{,}000$$

The total cost of 20 cents per can (or $704,000) and the annual payment to the university of $200,000 are subtracted to obtain the net profit:

$$\text{Net profit} = \$1{,}716{,}000 - \$704{,}000 - \$200{,}000 = \$812{,}000$$

Pepsi's current annual profit is

$$40 \text{ weeks} \times 22{,}000 \text{ cans/week} \times \$.55 = \$484{,}000$$

If the current market share is 25%, the potential gain from the agreement is

$$\$812{,}000 - \$484{,}000 = \$328{,}000$$

The only problem with this analysis is that Pepsi does not know how many soft drinks are sold weekly at the university. Coke is not likely to supply Pepsi with information about its sales, which together with Pepsi's line of products constitute virtually the entire market.

Pepsi assigned a recent university graduate to survey the university's students to supply the missing information. Accordingly, she organizes a survey that asks 500 students to keep track of the number of soft drinks they purchase in the next 7 days. The responses are stored in a file on the disk that accompanies this book.

Inferential Statistics

The information we would like to acquire in Case 12.1 is an estimate of annual profits from the exclusivity agreement. The data are the numbers of cans of soft drinks consumed in 7 days by the 500 students in the sample. We can use descriptive techniques to

---

*We have created an Excel spreadsheet that does the calculations for this case. To access it, click **Excel Workbooks** and **Case 12.1.** The only cell you may alter is cell C3, which contains the average number of soft drinks sold per week per student, assuming a total of 88,000 drinks sold per year.

learn more about the data. In this case, however, we are not so much interested in what the 500 students are reporting as we are in knowing the mean number of soft drinks consumed by all 50,000 students on campus. To accomplish this goal we need another branch of statistics—**inferential statistics.**

Inferential statistics is a body of methods used to draw conclusions or inferences about characteristics of populations based on sample data. The population in question in this case is the soft drink consumption of the university's 50,000 students. The cost of interviewing each student would be prohibitive and extremely time consuming. Statistical techniques make such endeavors unnecessary. Instead, we can sample a much smaller number of students (the sample size is 500) and infer from the data the number of soft drinks consumed by all 50,000 students. We can then estimate annual profits for Pepsi.

Example 12.5 Exit Polls (see Chapter 12)

When an election for political office takes place, the television networks cancel regular programming and provide election coverage instead. When the ballots are counted, the results are reported. However, for important offices such as president or senator in large states, the networks actively compete to see which will be the first to predict a winner. This is done through *exit polls*, wherein a random sample of voters who exit the polling booth is asked for whom they voted. From the data the sample proportion of voters supporting the candidates is computed. A statistical technique is applied to determine whether there is enough evidence to infer that the leading candidate will garner enough votes to win. Suppose that the exit poll results from the state of Florida during the 2000 year elections were recorded. Although there were a number of candidates running for president, the exit pollsters recorded only the votes of the two candidates who had any chance of winning, the Republican candidate George W. Bush and the Democrat Albert Gore. The results (765 people who voted for either Bush or Gore) were stored on a file on the disk. The network analysts would like to know whether they can conclude that George W. Bush will win the state of Florida.

Example 12.5 describes a very common application of statistical inference. The population the television networks wanted to make inferences about is the approximately 5 million Floridians who voted for Bush or Gore for president. The sample consisted of the 765 people randomly selected by the polling company who voted for either of the two main candidates. The characteristic of the population that we would like to know is the proportion of the Florida total electorate that voted for Bush. Specifically, we would like to know whether more than 50% of the electorate voted for Bush (counting only those who voted for either the Republican or Democratic candidate). It must be made clear that, because we will not ask every one of the 5 million actual voters for whom they voted, we cannot predict the outcome with 100% certainty. This is a fact that statistics practitioners and even students of statistics must understand. A sample that is only a small fraction of the size of the population can lead to correct inferences only a certain percentage of the time. You will find that statistics practitioners can control that fraction and usually set it between 90% and 99%.

Incidentally, on the night of the United States election in November 2000, the networks goofed badly. Using exit polls as well as the results of previous elections, all four networks concluded at about 8:00 P.M. that Al Gore would win the state of Florida. Shortly after 10:00 P.M. with a large percentage of the actual vote having been counted, the networks reversed course and declared that George W. Bush would win the state of Florida. By 2:00 A.M. another verdict was declared: The result was too close to call. In the future, this experience will likely be used by statistics instructors when teaching how *not* to use statistics.

Notice that, contrary to what you probably believed, data are not necessarily numbers. The marks in Example 2.6 and the number of soft drinks consumed in a week in Case 12.1, of course, are numbers; however, the votes in Example 12.5 are not. In Chapter 2, we will discuss the different types of data you will encounter in statistical applications and how to deal with them.

1.1 KEY STATISTICAL CONCEPTS

Statistical inference problems involve three key concepts: the population, the sample, and the statistical inference. We now discuss each of these concepts in more detail.

Population

A **population** is the group of all items of interest to a statistics practitioner. It is frequently very large and may, in fact, be infinitely large. In the language of statistics, *population* does not necessarily refer to a group of people. It may, for example, refer to the population of diameters of ball bearings produced at a large plant. In Case 12.1, the population of interest consists of the 50,000 students on campus. In Example 12.5 the population consists of the Floridians who voted for Bush or Gore.

A descriptive measure of a population is called a **parameter.** The parameter of interest in Case 12.1 is the mean number of soft drinks consumed by all the students at the university. The parameter in Example 12.5 is the proportion of the 5 million Florida voters who voted for Bush. In most applications of inferential statistics the parameter represents the information we need.

Sample

A **sample** is a set of data drawn from the population. A descriptive measure of a sample is called a **statistic.** We use statistics to make inferences about parameters. In Case 12.1, the statistic we would compute is the mean number of soft drinks consumed in the last week by the 500 students in the sample. We would then use the sample mean to infer the value of the population mean, which is the parameter of interest in this problem. In Example 12.5, we compute the proportion of the sample of 765 Floridians who voted for Bush. The sample statistic is then used to make inferences about the population of all 5 million votes. That is, we predict the election results even before the actual count.

Statistical Inference

Statistical inference is the process of making an estimate, prediction, or decision about a population based on sample data. Because populations are almost always very large, investigating each member of the population would be impractical and expensive. It is far easier and cheaper to take a sample from the population of interest and draw conclusions or make estimates about the population on the basis of information provided by the sample. However, such conclusions and estimates are not always going to be correct. For this reason, we build into the statistical inference a measure of reliability. There are two such measures, the **confidence level** and the **significance level.** The *confidence level* is the proportion of times that an estimating procedure will be correct. For example, in Case 12.1, we will produce an estimate of the average number of soft drinks to be consumed by all 50,000 students that has a confidence level of 95%. In other words, in the long run, estimates based on this form of statistical inference will be correct 95% of the time. When the purpose of the statistical inference is to draw a conclusion about a population, the *significance level* measures how frequently the conclusion will be wrong in the long run. For example, suppose that as a result of the analysis in Example 12.5, we conclude that more than 50% of the electorate will vote for George W. Bush, and thus he will win the state of Florida. A 5% significance level means that, in the long run, this type of conclusion will be wrong 5% of the time.

1.2 Statistical Applications in Business

An important function of statistics courses in business and economics programs is to demonstrate that statistical analysis plays an important role in virtually all aspects of the business and economics. We intend to do so through examples, exercises, and cases. However, we assume that most students taking their first statistics course have not taken courses in most of the other subjects in management programs. To understand fully how statistics is used in these and other subjects, it is necessary to know something about them. To provide sufficient background to understand the statistical application we introduce applications in accounting, economics, finance, human resources management, marketing, and operations management. We will provide readers with some background to these applications by describing their functions in two ways.

Application Sections and Subsections

We feature six sections that describe statistical applications in the functional areas of business. For example, in Section 7.3 we show an application in finance which describes a financial analyst's use of probability and statistics to construct portfolios that decrease risk. In Section 12.5 we describe an application in accounting that uses statistical techniques in auditing to produce important information.

One section and two subsections demonstrate the uses of probability and statistics in specific industries. Section 4.5 introduces an interesting application of statistics in professional baseball. Subsections in Sections 6.4 and 18.4 present applications in medical testing (useful in the medical insurance industry) and banking, respectively.

Application Boxes

For other topics that require less detailed description we provide application boxes with a relatively brief description of the background followed by examples or exercises. These boxes are scattered throughout the book. For example, in Chapter 2 we discuss a job a marketing manager may need to undertake to determine the appropriate price for a product. To understand the context we need to provide a description of marketing management. The statistical application will follow.

1.3 Statistics and the Computer

In virtually all applications of statistics, the statistics practitioner must deal with large amounts of data. For example, Case 12.1 (Pepsi-Cola) involves 500 observations. To estimate annual profits, the statistics practitioner would have to perform computations on the data; although the calculations do not require any great mathematical skill, the sheer amount of arithmetic makes this aspect of the statistical method time-consuming and tedious.

Fortunately, numerous commercially prepared computer programs are available to perform the arithmetic. We have chosen to use Microsoft Excel, which is a spreadsheet program, and Minitab, which is a statistical software package. (We use the latest versions of both software, Office 2007 and Minitab 15.) We chose Excel because we believe that it is and will continue to be the most popular spreadsheet package. One of its drawbacks is that it does not offer a complete set of the statistical techniques we introduce in this book. Consequently, we created add-ins that can be loaded onto your computer to enable you to use Excel for all statistical procedures introduced in this book. The add-ins are stored on the CD that accompanies the book and, when installed, will appear as *Data Analysis Plus©* on Excel's menu.

Also available on the CD ROM are introductions to Excel and Minitab, and detailed instructions for both software packages.

Appendix 1 describes the contents of the CD ROM that accompanies this book and provides instructions on how to extract the various components. The CD ROM also has a README file with more details.

A large proportion of the examples, exercises, and cases feature large data sets also stored on the CD ROM. These are denoted with the file name in the margin. We demonstrate the solution to the statistical examples in three ways: manually, by employing Excel, and by using Minitab. Moreover, we will provide detailed instructions for all techniques.

The files contain the data needed to produce the solution. However, in many real applications of statistics additional data are collected. For instance, in Example 12.5, the pollster often records the gender and asks the voter for other information including race, religion, education, and income. Many other data sets are similarly constructed. In later chapters we will return to these files and require other statistical techniques to extract the needed information. (Files that contain additional data are denoted by an asterisk on the file name.)

The approach we prefer to take is to minimize the time spent on manual computations and to focus instead on selecting the appropriate method for dealing with a problem and on interpreting the output after the computer has performed the necessary computations. In this way, we hope to demonstrate that statistics can be as interesting and practical as any other subject in your curriculum.

Applets and Spreadsheets

Books written for statistics courses taken by mathematics or statistics majors are considerably different from this one. Not surprisingly, such courses feature mathematical proofs of theorems and derivations of most procedures. When the material is covered in this way, the underlying concepts that support statistical inference are exposed and relatively easy to see. However, this book was created for an applied course in statistics. Consequently, we do not address directly the mathematical principles of statistics. However, as we pointed out previously, one of the most important functions of statistics practitioners is to properly interpret statistical results, whether produced manually or by computer. And, to correctly interpret statistics, students require an understanding of the principles of statistics.

To help students understand the basic foundation, we offer two approaches. First, we have created several Excel spreadsheets that allow for what-if analyses. By changing some of the input value, students can see for themselves how statistics works. (The name derives from, *What* happens to the statistics *if* I change this value?) Second, we have created applets, which are computer programs that perform similar what-if analyses or simulations. The applets and the spreadsheet applications appear in a number of chapters, where they are explained in greater detail.

1.4 World Wide Web and Learning Center

To assist students in the various aspects of using the computer to learn statistics, we have created a web page. It offers useful information, including additional exercises and cases, corrections to the different printings and supplements, and updates on the data sets and macros. Additionally, you can e-mail the author to make comments and ask questions about the installation of the files stored on the disks. The site can be accessed from the publisher's home page at http://www.academic.cengage.com.

IMPORTANT TERMS

Descriptive statistics 2
Inferential statistics 4
Population 5
Parameter 5
Sample 5
Statistic 5
Statistical inference 5
Confidence level 5
Significance level 5

Chapter Exercises

1.1 Suppose you believe that, in general, graduates who have majored in *your* subject are offered higher salaries upon graduating than are graduates of other programs. Describe a statistical experiment that could help test your belief.

1.2 A politician who is running for the office of mayor of a city with 25,000 registered voters commissions a survey. In the survey, 48% of the 200 registered voters interviewed say they plan to vote for her.

a. What is the population of interest?
b. What is the sample?
c. Is the value 48% a parameter or a statistic? Explain.

1.3 A manufacturer of computer chips claims that less than 10% of his products are defective. When 1,000 chips were drawn from a large production, 7.5% were found to be defective.

a. What is the population of interest?
b. What is the sample?
c. What is the parameter?
d. What is the statistic?
e. Does the value 10% refer to the parameter or to the statistic?
f. Is the value 7.5% a parameter or a statistic?
g. Explain briefly how the statistic can be used to make inferences about the parameter to test the claim.

1.4 Briefly describe the difference between descriptive statistics and inferential statistics.

1.5 In your own words, define and give an example of each of the following statistical terms:

a. population
b. sample
c. parameter
d. statistic
e. statistical inference

1.6 Xm01-06 The owner of a large fleet of taxis is trying to estimate his costs for next year's operations. One major cost is fuel purchases. To estimate fuel purchases, the owner needs to know the total distance his taxis will travel next year, the cost of a gallon of fuel, and the fuel mileage of his taxis. The owner has been provided with the first two figures (distance estimate and cost of a gallon of fuel). However, because of the high cost of gasoline, the owner has recently converted his taxis to operate on propane. He has measured and recorded the propane mileage (in miles per gallon) for 50 taxis.

a. What is the population of interest?
b. What is the parameter the owner needs?
c. What is the sample?
d. What is the statistic?
e. Describe briefly how the statistic will produce the kind of information the owner wants.

1.7 You are shown a coin that its owner says is fair in the sense that it will produce the same number of heads and tails when flipped a very large number of times.

a. Describe an experiment to test this claim.
b. What is the population in your experiment?
c. What is the sample?
d. What is the parameter?
e. What is the statistic?
f. Describe briefly how statistical inference can be used to test the claim.

1.8 Suppose that in Exercise 1.7 you decide to flip the coin 100 times.

a. What conclusion would you be likely to draw if you observed 95 heads?
b. What conclusion would you be likely to draw if you observed 55 heads?
c. Do you believe that, if you flip a perfectly fair coin 100 times, you will always observe exactly 50 heads? If you answered "no," what numbers do you think are possible? If you answered "yes," how many heads would you observe if you flipped the coin twice? Try it several times, and report the results.

APPENDIX 1 / INSTRUCTIONS FOR THE CD ROM

The CD ROM that accompanies this book contains the following features:

Data Analysis Plus 7.0 (a statistical software add-in for Excel, which is consistent with Office 2007)

A help file for Data Analysis Plus 7.0

Data Analysis Plus 5.1, which works with earlier versions of Excel (Office 1997, 2000, XP, 2003)

A help file for Data Analysis Plus 5.1

Data files in the following formats: ASCII, Excel, JMP, Minitab, SAS, and SPSS

Excel workbooks

TreePlan (a decision analysis software add-in for Excel)

Seeing Statistics (Java applets that teach a number of important statistical concepts)

CD appendixes (35 additional topics that are not covered in the book)

Formula card listing every formula in the book

Installation Instructions

Insert the CD into the appropriate drive. The CD should automatically start. If it does not, double-click the Keller8 CD ROM icon.

Double-click the "Install Data Analysis Plus 7.0" or "Install Data Analysis Plus 5.1" button and it will be installed into the XlSTART folder of the most recent version of Excel on your computer. If properly installed, Data Analysis Plus will be a menu item in Excel. The help file for Data Analysis Plus will be stored directly in your computer's My Documents folder. It will appear when you click the Help button or when you make a mistake when using Data Analysis plus.

Double-click the "Install Data Sets" button, and the data sets will also be installed.

The Excel workbooks, Seeing Statistics Applets, and CD Appendixes will be accessed by the buttons on the CD ROM. (You will have to insert the CD ROM each time you wish to access any of these features.) Alternatively, you can store the Excel workbooks and CD Appendixes on your hard drive.

To manually install Data Analysis Plus and the data files, open the CD ROM, navigate the manual install folder, and click the README file. It will contain the instructions you need.

For technical support, please call 1-800-423-0563 or fax 859-647-5045 or go to http://www.academic.cengage.com/support/tech_support_form.html. Refer to *Managerial Statistics* eighth edition, by Gerald Keller (ISBN 0324569556).

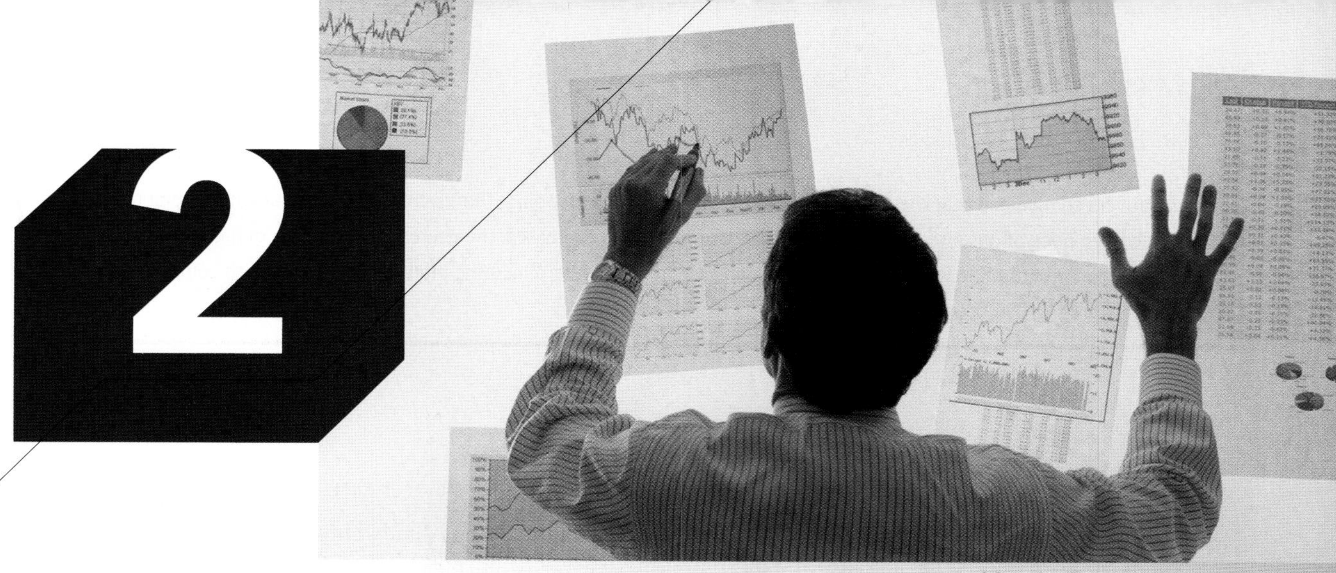

© Steve Cole/Digital Vision/Getty Images

2 GRAPHICAL AND TABULAR DESCRIPTIVE TECHNIQUES

Were Oil Companies Gouging Customers, 1999–2006?

DATA Xm02-00

The price of oil has been increasing for several reasons. First, oil is a finite resource; the world will eventually run out. In 2005 the world was consuming more than 80 million barrels per day or more than 30 billion barrels per year. (See Exercise 2.13.) The total proven world reserves of oil is 1,227 billion barrels. (See Exercise 2.11.) At today's consumption levels the proven reserves will be exhausted in 40 years. Second, China and India's industries are rapidly increasing, which requires ever-increasing amounts of oil. Third, over the last 8 years hurricanes have threatened the oil rigs in the Gulf of Mexico.

The result of the price increases in oil is reflected in the price of gasoline. In December 1998 the average retail price of gasoline in the United States was $1.046 per U.S. gallon (a U.S. gallon is equal to 3.79 liters) and the price of oil (West Texas intermediate crude) was $11.28 per barrel (a barrel is equal to 42 U.S. gallons). Over the next 8 years the price of both substantially increased.

© Comstock Images/Jupiterimages

On page 71 you will find our answer.

Many drivers complained that the oil companies were guilty of price gouging. That is, they believed that when the price of oil increased, the price of gas also increased, but when the price of oil decreased, the decrease in the price of gasoline seemed to lag behind. To determine whether this perception is accurate we determined the monthly figures for both commodities. These are stored in Ch02:\Xm02-00. Were oil and gas prices related?

INTRODUCTION

In Chapter 1, we pointed out that statistics is divided into two basic areas: descriptive statistics and inferential statistics. The purpose of this chapter, together with the next two, is to present the principal methods that fall under the heading of descriptive statistics. In this chapter we introduce graphical and tabular statistical methods that allow managers to summarize data visually to produce useful information, often used in decision making. Chapter 3 discusses ways to use the techniques introduced in this chapter in an effective and accurate way. Another class of descriptive techniques, numerical methods, is introduced in Chapter 4.

Managers frequently have access to large masses of potentially useful data. But before the data can be used to support a decision, they must be organized and summarized. Consider, for example, the problems faced by managers who have access to the databases created by the use of debit cards. The database consists of the personal information supplied by the customer when he or she applied for the debit card. This information includes age, gender, residence, and income of the cardholder. In addition each time the card is used the database grows to include a history of the timing, price, and brand of each product so purchased. Using the appropriate statistical technique, managers can determine which segments of the market are buying their company's brands. Specialized marketing campaigns, including telemarketing, can be developed. Both descriptive and inferential statistics would likely be employed in the analysis.

Descriptive statistics involves arranging, summarizing, and presenting a set of data in such a way that useful information is produced. Its methods make use of graphical techniques and numerical descriptive measures (such as averages) to summarize and present the data, allowing managers to make decisions based on the information generated. Although descriptive statistical methods are quite straightforward, their importance should not be underestimated. Most management, business, and economics students will encounter numerous opportunities to make valuable use of graphical and numerical descriptive techniques when preparing reports and presentations in the workplace. According to a Wharton Business School study, top managers reach a consensus 25% more quickly when responding to a presentation in which graphics are used.

In Chapter 1 we introduced the distinction between a population and a sample. Recall that a *population* is the entire set of observations under study, whereas a *sample* is a subset of a population. The descriptive methods presented in this chapter and in Chapter 4 apply to both a set of data constituting a population and a set of data constituting a sample.

In both the preface and Chapter 1 we pointed out that a critical part of your education as statistics practitioners includes an understanding not only of *how* to draw graphs and calculate statistics (manually or by computer) but *when* to use each technique that we cover. The two most important factors that determine the appropriate method to use are the type of data and the information that is needed. Both are discussed next.

2.1 TYPES OF DATA AND INFORMATION

The objective of statistics is to extract information from data. There are different types of data and information. To help explain this important principle, we need to define some terms.

A **variable** is some characteristic of a population or sample. For example, the mark on a statistics exam is a characteristic of statistics exams that is certainly of interest to readers of this book. Not all students achieve the same mark. The marks will vary from student to student, thus the name *variable*. The price of a stock is another variable. The prices of most stocks vary daily. We usually represent the name of the variable using upper case letters such as X, Y, and Z.

The **values** of the variable are the possible observations of the variable. The values of statistics exam marks are the integers between 0 and 100 (assuming the exam is marked out of 100). The values of a stock price are real numbers that are usually measured in dollars and cents (sometimes in fractions of a cent). The values range from 0 to hundreds of dollars.

Data* are the observed values of a variable. For example, suppose that we observe the midterm test marks of 10 students, which are

67 74 71 83 93 55 48 82 68 62

These are the data from which we will extract the information we seek. Incidentally, data is plural for **datum.** The mark of one student is a datum.

When most people think of data they think of sets of numbers. However, there are three types of data. They are interval, nominal, and ordinal data.†

Interval data are real numbers, such as heights, weights, incomes, and distances. We also refer to this type of data as **quantitative** or **numerical.**

The values of **nominal** data are categories. For example, responses to questions about marital status produce nominal data. The values of this variable are single, married, divorced, and widowed. Notice that the values are not numbers but instead are words describing the categories. We often record nominal data by arbitrarily assigning a number to each category. For example, we could record marital status using the following codes:

Single	1
Married	2
Divorced	3
Widowed	4

However, any other numbering system is valid provided that each category has a different number assigned to it. Here is another coding system that is as valid as the previous one.

Single	7
Married	4
Divorced	13
Widowed	1

Nominal data are also called **qualitative** or **categorical.**

---

*Unfortunately, the term *data*, like the term *statistician*, has taken on a number of different meanings. For example, dictionaries define data as facts, information, or statistics. In the language of computers, data may refer to any piece of information such as this textbook or an essay you have written. Such definitions make it difficult for us to present *statistics* as a method of converting *data* into *information*. In this book we carefully distinguish among the three terms.

†There are actually four types of data, the fourth being ratio data. However, for statistical purposes there is no difference between ratio and interval data. Consequently, we combine the two types.

The third type of data is ordinal. **Ordinal** data appear to be nominal, but their values are in order. For example, at the completion of most college and university courses, students are asked to evaluate the course. The variables are the ratings of various aspects of the course, including the professor. Suppose that in a particular college the values are

poor, fair, good, very good, and excellent

The difference between nominal and ordinal types of data is that the values of the latter are in order. Consequently, when assigning codes to the values, we should maintain the order of the values. For example, we can record the students' evaluations as

Poor	1
Fair	2
Good	3
Very good	4
Excellent	5

Because the only constraint that we impose on our choice of codes is that the order must be maintained, we can use any set of codes that are in order. For example, we can also assign the following codes:

Poor	6
Fair	18
Good	23
Very good	45
Excellent	88

As we discuss in Chapter 19, which introduces statistical inference techniques for ordinal data, the use of any code that preserves the order of the data will produce exactly the same result.

Students often have difficulty distinguishing between ordinal and interval data. The critical difference between them is that the intervals or differences between values of interval data are consistent and meaningful (that's why this type of data is called interval). For example, the difference between marks of 85 and 80 is the same five-mark difference that exists between 75 and 70. That is, we can calculate the difference and interpret the results.

Because the codes representing ordinal data are arbitrarily assigned except for the order, we cannot calculate and interpret differences. For example, using a 1-2-3-4-5 coding system to represent poor, fair, good, very good, and excellent, we note that the difference between excellent and very good is identical to the difference between good and fair. With a 6-18-23-45-88 coding, the difference between excellent and very good is 43, and the difference between good and fair is 5. Because both coding systems are valid we cannot use either system to compute and interpret differences.

Here is another example. Suppose that you are given the following list of the most active stocks traded on the NASDAQ in descending order of magnitude:

Order	Most Active Stocks
1	Microsoft
2	Cisco Systems
3	Dell Computer
4	Sun Microsystems
5	JDS Uniphase

Does this information allow you to conclude that the difference between the number of stocks traded in Microsoft and Cisco Systems is the same as the difference in the number of stocks traded between Dell Computer and Sun Microsystems? The answer is no because we have information only about the order of the numbers of trades, which are ordinal, and not the numbers of trades themselves, which are interval. That is, the difference between 1 and 2 is not necessarily the same as the difference between 3 and 4.

Calculations for Types of Data

Interval Data

All calculations are permitted on interval data. We often describe a set of interval data by calculating the average. For example, the average of the 10 marks listed on page 13 is 70.3. As you will discover, there are several other important statistics that we will introduce.

Nominal Data

Because the codes of nominal data are completely arbitrary, we cannot perform any calculations on these codes. To understand why, consider a survey that asks people to report their marital status. Suppose that the first 10 people surveyed gave the following responses:

single, married, married, married, widowed, single, married, married, single, divorced

Using the codes

Single	1
Married	2
Divorced	3
Widowed	4

we would record these responses as

1 2 2 2 4 1 2 2 1 3

The average of these numerical codes is 2.0. Does this mean that the average person is married? Now suppose four more persons were interviewed, of whom three are widowed and one is divorced. The data are given here:

1 2 2 2 4 1 2 2 1 3 4 4 4 3

The average of these 14 codes is 2.5. Does this mean that the average person is married, but halfway to getting divorced? The answer to both questions is an emphatic "no." This example illustrates a fundamental truth about nominal data: Calculations based on the codes used to store this type of data are meaningless. All that we are permitted to do with nominal data is count the occurrences of each category. Thus, we would describe the 14 observations by counting the number of each marital status category and reporting the frequency as shown in the following table:

Category	Code	Frequency
Single	1	3
Married	2	5
Divorced	3	2
Widowed	4	4

Ordinal Data

The most important aspect of ordinal data is the order of the values. As a result, the only permissible calculations are ones involving a ranking process. For example, we can place all the data in order and select the code that lies in the middle. As we discuss in Chapter 4, this descriptive measurement is called the *median*.

Hierarchy of Data

The data types can be placed in order of the permissible calculations. At the top of the list we place the interval data type because virtually *all* computations are allowed. The nominal data type is at the bottom because *no* calculations other than determining frequencies are permitted. (We are permitted to perform calculations using the frequencies of codes. However, this differs from performing calculations on the codes themselves.) In between interval and nominal data lies the ordinal data type. Permissible calculations are ones that rank the data.

Higher-level data types may be treated as lower-level ones. For example, in universities and colleges we convert the marks in a course, which are interval, to letter grades, which are ordinal. Some graduate courses feature only a pass or fail designation. In this case, the interval data are converted to nominal. It is important to point out that when we convert higher-level data as lower-level we lose information. For example a mark of 83 on an accounting course exam gives far more information about the performance of that student than does a letter grade of A, which is the letter grade for marks between 80 and 90. As a result we do not convert data unless it is necessary to do so. We will discuss this later.

It is also important to note that we cannot treat lower-level data types as higher-level types.

The definitions and hierarchy are summarized in the following box.

Types of Data

Interval

Values are real numbers.

All calculations are valid.

Data may be treated as ordinal or nominal.

Ordinal

Values must represent the ranked order of the data.

Calculations based on an ordering process are valid.

Data may be treated as nominal but not as interval.

Nominal

Values are the arbitrary numbers that represent categories.

Only calculations based on the frequencies of occurrence are valid.

Data may not be treated as ordinal or interval.

Interval, Ordinal, and Nominal Variables

The variables whose observations constitute our data will be given the same name as the type of data. Thus, for example, interval data are the observations of an interval variable.

Problem Objectives and Information

In presenting the different types of data, we introduced a critical factor in deciding which statistical procedure to use. A second factor is the type of information we need to produce from our data. We discuss the different types of information in greater detail in Section 11.4 when we introduce *problem objectives*. However, in this part of the book

(Chapters 2–5) we will use statistical techniques to describe a set of data, compare two or more sets of data, and to describe the relationship between two variables. In Section 2.2 we introduce graphical and tabular techniques employed to describe a set of nominal data. Section 2.3 introduces graphical methods to describe a set of interval data. In Section 2.4 we present a graphical technique to describe a time series. Section 2.5 shows how to describe the relationship between two nominal variables and to compare two or more sets of nominal data. Section 2.6 presents methods to describe the relationship between two interval variables.

EXERCISES

2.1 Provide two examples each of nominal, ordinal, and interval data.

2.2 For each of the following examples of data, determine the type.
a. the number of miles joggers run per week
b. the starting salaries of graduates of MBA programs
c. the months in which a firm's employees choose to take their vacations
d. the final letter grades received by students in a statistics course

2.3 For each of the following examples of data, determine the type.
a. the weekly closing price of the stock of Amazon.com
b. the month of highest vacancy rate at a La Quinta motel
c. the size of soft drink (small, medium, or large) ordered by a sample of McDonald's customers
d. the number of Toyotas imported monthly by the United States over the last 5 years
e. the marks achieved by the students in a statistics course final exam marked out of 100

2.4 The placement office at a university regularly surveys the graduates 1 year after graduation and asks for the following information. For each, determine the type of data.
a. What is your occupation?
b. What is your income?
c. What degree did you obtain?
d. What is the amount of your student loan?
e. How would you rate the quality of instruction? (excellent, very good, good, fair, poor)

2.5 Residents of condominiums were recently surveyed and asked a series of questions. Identify the type of data for each question.
a. What is your age?
b. On what floor is your condominium?
c. Do you own or rent?
d. How large is your condominium (in square feet)?
e. Does your condominium have a pool?

2.6 A sample of shoppers at a mall was asked the following questions. Identify the type of data each question would produce.
a. What is your age?
b. How much did you spend?
c. What is your marital status?
d. Rate the availability of parking: excellent, good, fair, or poor
e. How many stores did you enter?

2.7 Information about a magazine's readers is of interest to both the publisher and the magazine's advertisers. A survey of readers asked respondents to complete the following:
a. Age
b. Gender
c. Marital status
d. Number of magazine subscriptions
e. Annual income
f. Rate the quality of our magazine: excellent, good, fair, or poor

For each item identify the resulting data type.

2.8 Baseball fans are regularly asked to offer their opinions about various aspects of the sport. A survey asked the following questions. Identify the type of data.
a. How many games do you attend annually?
b. How would you rate the quality of entertainment? (excellent, very good, good, fair, poor)
c. Do you have season tickets?
d. How would you rate the quality of the food? (edible, barely edible, horrible)

2.9 A survey of golfers asked the following questions. Identify the type of data each question produces.
a. How many rounds of golf do you play annually?
b. Are you a member of a private club?
c. What brand of clubs do you own?

2.10 At the end of the term, university and college students often complete questionnaires about their courses. Suppose that in one university, students were asked the following.
a. Rate the course. (highly relevant, relevant, irrelevant)
b. Rate the professor. (very effective, effective, not too effective, not at all effective)
c. What was your midterm grade (A, B, C, D, F)?

Determine the type of data each question produces.

2.2 Graphical and Tabular Techniques to Describe Nominal Data

As we discussed in Section 2.1, the only allowable calculation on nominal data is to count the frequency of each value of the variable. We can summarize the data in a table that presents the categories and their counts called a **frequency distribution.** A *relative frequency distribution* lists the categories and the proportion with which each occurs. We can use graphical techniques to present a picture of the data. There are two graphical methods we can use: the *bar chart* and the *pie chart.*

EXAMPLE 2.1

DATA
Xm02-01*

Light Beer Preference Survey

In 2006 total light beer sales in the United States was approximately 3 million gallons (*Source:* http://Adage.com). With this large a market breweries often need to know more about who is buying their product. The marketing manager of a major brewery wanted to analyze the light beer sales among college and university students who do drink light beer. A random sample of 285 graduating students was asked to report which of the following is their favorite light beer:

1. Bud Light
2. Busch Light
3. Coors Light
4. Michelob Light
5. Miller Lite
6. Natural Light
7. Other brands

The responses were recorded using the codes 1, 2, 3, 4, 5, 6, and 7, respectively. The data are listed here, and the entire data set is stored on the CD that accompanies this book. The name of the file is listed in the margin. The file also contains each graduate's identification number and gender. The additional data are not needed in this example but will be used later in this book to produce other information for the manager. (Examples and exercises with additional data are indicated with an asterisk next to the file name.)

Construct a frequency and relative frequency distribution for these data and graphically summarize the data by producing a bar chart and a pie chart.

Light Beer Preferences

1	1	1	1	2	4	3	5	1	3	1	3	7	5	1
1	5	2	1	5	1	3	3	3	1	1	5	3	1	5
5	1	1	3	3	5	5	6	3	5	3	5	5	5	1
1	2	1	1	5	5	3	2	1	6	1	1	4	5	1
3	3	5	4	7	6	6	4	4	6	5	2	1	1	5
3	3	1	3	5	3	3	7	3	7	2	1	5	7	
3	6	2	6	3	6	6	6	5	6	1	1	6	3	
7	1	1	1	5	1	3	1	3	7	7	2	1	1	
2	5	3	1	1	3	1	1	7	5	3	2	1	1	
6	5	7	1	3	2	1	3	1	1	7	5	5	6	
1	4	6	1	3	1	1	5	5	5	5	1	5	5	
6	1	3	3	1	3	7	1	1	1	2	4	1	1	
3	3	7	5	5	1	1	3	5	1	5	4	5	3	
4	1	4	5	3	1	5	3	3	3	1	1	5	3	
5	6	4	3	5	6	4	6	5	5	5	5	3	1	
2	3	2	7	5	1	6	6	2	3	3	3	1	1	
5	1	4	6	3	5	1	1	2	1	5	6	1	1	
5	1	3	5	1	1	1	3	7	3	1	6	3	1	
2	2	5	1	3	5	5	2	3	1	1	3	6	1	
1	1	1	7	3	1	5	3	3	3	5	3	1	7	

SOLUTION

Scan the data. Have you learned anything about the choices of these 285 students? Unless you have special skills you have probably learned little about the numbers. To extract useful information requires the application of a statistical or graphical technique. To choose the appropriate technique we must first identify the type of data. In this example the data are nominal because the numbers represent categories. The only calculation permitted on nominal data is to count the number of occurrences of each category. Hence we count the number of 1's, 2's, 3's, 4's, 5's, 6's, and 7's. The list of the categories and their counts constitute the frequency distribution. The relative frequency distribution is produced by converting the frequencies into proportions. The frequency and relative frequency distributions are combined in Table 2.1.

TABLE **2.1** Frequency and Relative Frequency Distributions for Example 2.1

Light Beer Brand	Frequency	Relative Frequency
Bud Light	90	31.6%
Busch Light	19	6.7
Coors Light	62	21.8
Michelob Light	13	4.6
Miller Lite	59	20.7
Natural Light	25	8.8
Other brands	17	6.0
Total	285	100

As we promised in Chapter 1 (and the preface) we demonstrate the solution of all examples in this book using three approaches (where feasible): manually, using Excel, and using Minitab. For Excel and Minitab we provide not only the printout, but also instructions to produce them. (SPSS and SAS are available on our web site. See page 8. General and specific instructions are provided in the CD appendixes.)

EXCEL

INSTRUCTIONS

(Specific commands for this example are highlighted.)

1. Type or import the data into one or more columns. (Open Xm02-01.)
2. Activate any empty cell and type

=COUNTIF ([Input range], [Criteria])

Input range are the cells containing the data. In this example the range is B1:B286. The criteria are the codes you want to count (1) (2) (3) (4) (5) (6) (7). To count the number of 1's (Bud Light) type

=COUNTIF (B1:B286, 1)

And the frequency will appear in the dialog box. Change the criteria to produce the frequency of the other categories.

MINITAB

Tally for Discrete Variables: Brand

Brand	Count	Percent
1	90	31.58
2	19	6.67
3	62	21.75
4	13	4.56
5	59	20.70
6	25	8.77
7	17	5.96
N =	285	

INSTRUCTIONS

(Specific commands for this example are highlighted.)

1. Type or import the data into one column. (Open Xm02-01.)
2. Click **Stat, Tables,** and **Tally Individual Variables.**
3. Type or use the **Select** button to specify the name of the variable or the column where the data are stored in the **Variables** box (Brand). Under **Display,** click **Counts** and **Percents.**

INTERPRET

Budweiser, Coors, and Miller are by far the most popular light beers among college and university seniors.

Bar and Pie Charts

The information contained in the data is summarized well in the table. However, graphical techniques generally catch a reader's eye more quickly than does a table of numbers. Two graphical techniques can be used to display the results shown in the table. A **bar chart** is often used to display frequencies; a **pie chart** graphically shows relative frequencies.

The bar chart is created by drawing a rectangle representing each category. The height of the rectangle represents the frequency. The base is arbitrary. Figure 2.1 depicts the manually drawn bar chart for Example 2.1.

FIGURE **2.1** Bar Chart for Example 2.1

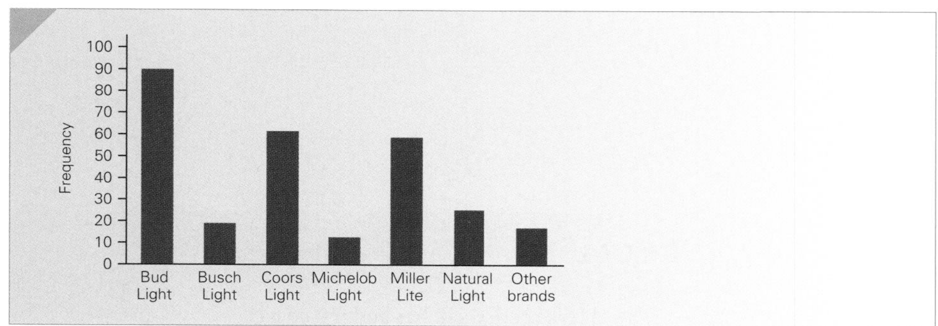

If we wish to emphasize the relative frequencies instead of drawing the bar chart, we draw the pie chart. A pie chart is simply a circle subdivided into slices that represent the categories. It is drawn so that the size of each slice is proportional to the percentage corresponding to that category. For example, since the entire circle is composed of 360 degrees, a category that contains 25% of the observations is represented by a slice of the pie that contains 25% of 360 degrees, which is equal to 90 degrees. The number of degrees for each category in Example 2.1 is shown in Table 2.2.

TABLE **2.2** Proportion in Each Category in Example 2.1

Light Beer Brand	Relative Frequency	Slice of the Pie (Degrees)
Bud Light	31.6%	113.7°
Busch Light	6.7	24.0
Coors Light	21.8	78.3
Michelob Light	4.6	16.4
Miller Lite	20.7	74.5
Natural Light	8.8	31.6
Other brands	6.0	21.5
Total	100.00	360

Figure 2.2 was drawn from these results.

FIGURE **2.2** Pie Chart for Example 2.1

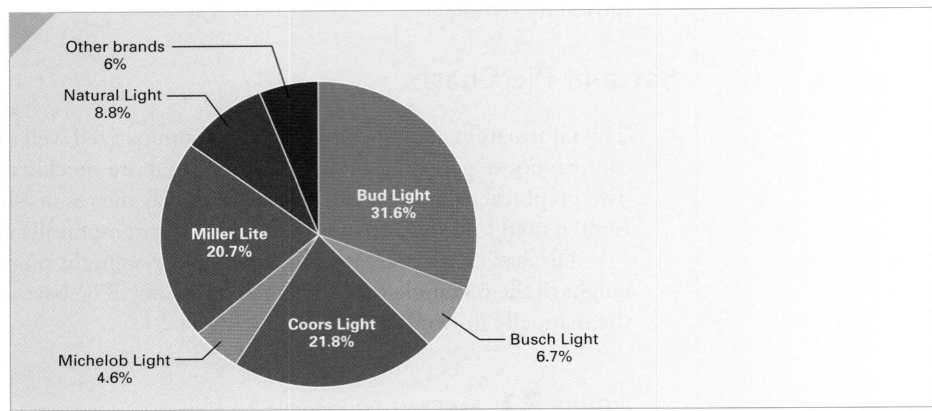

EXCEL

Here are Excel's bar and pie charts.

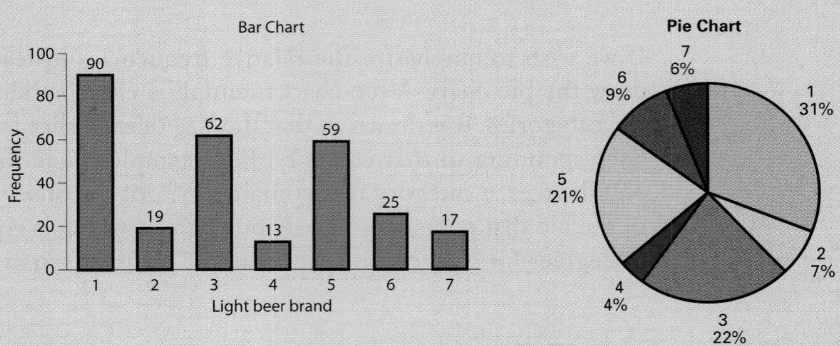

INSTRUCTIONS

1. After creating the frequency distribution, highlight the column of frequencies.
2. For a bar chart click **Insert, Column,** and the first **2-D Column.**
3. Click **Chart Tools** (if it does not appear, click inside the box containing the bar chart) and **Layout.** This will allow you to make changes to the chart. We removed the **Gridlines,** the **Legend,** and clicked the **Data Labels** to create the titles.
4. For a pie chart, click **Pie** and **Chart Tools** to edit the graph.

MINITAB

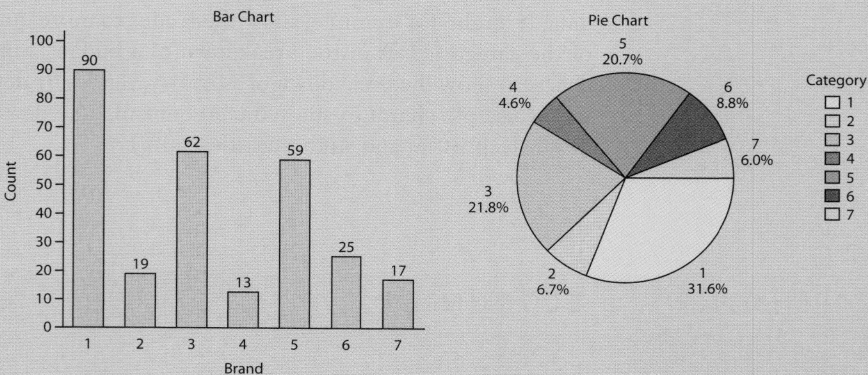

INSTRUCTIONS

1. Type or import the data into one column. (Open Xm02-01.)

For a bar chart:

2. Click **Graph** and **Bar Chart.**
3. In the **Bars represent** box, click **Counts of unique values** and select **Simple.**
4. Type or use the **Select** button to specify the variable in the **Variables** box (Brand).

We clicked **Labels** and added the title and clicked **Data Labels** and **use y-value labels** to display the frequencies at the top of the columns.

For a pie chart:

2. Click **Graph** and **Pie Chart.**
3. Click **Chart Counts of unique values** and in the **Categorical variables** box type or use the **Select** button to specify the variable (Brand).

We clicked **Labels** and added the title. We clicked **Slice Labels** and clicked **Category name** and **Percent.**

INTERPRET

The bar chart focuses on the frequencies. As you can see, Bud Light is the most popular light beer with 90 college and university seniors selecting it as their favorite. Coors Light and Miller Lite are the second and third most popular light beers.

The pie chart focuses on the proportions. Bud Light is the choice of almost one-third of college seniors.

Other Applications of Pie Charts and Bar Charts

Pie and bar charts are used widely in newspapers, magazines, and business and government reports. One of the reasons for this appeal is that they are eye-catching and can attract the reader's interest whereas a table of numbers might not. Perhaps no one understands this better than the newspaper *USA Today*, which typically has a colored graph on the front page and others inside. Pie and bar charts are frequently used to simply present numbers associated with categories. The only reason to use a bar or pie chart in such a situation would be to enhance the reader's ability to grasp the substance of the data. It might, for example, allow the reader to more quickly recognize the relative sizes of the categories, as in the breakdown of a budget. Similarly, treasurers might use pie charts to show the breakdown of a firm's revenues by department, or university students might use pie charts to show the amount of time devoted to daily activities (e.g., eat, 10%; sleep, 30%; and study statistics, 60%).

APPLICATIONS in ECONOMICS

Macroeconomics

Macroeconomics is a major branch of economics that deals with the behavior of the economy as a whole. Macroeconomists develop mathematical models that predict variables such as gross domestic product, unemployment rates, and inflation. These are used by governments and corporations to help develop strategies. For example, central banks attempt to control inflation by lowering or raising interest rates. To do this requires economists to determine the effect of a variety of variables, including the supply and demand for energy.

APPLICATIONS in ECONOMICS

Energy Economics

One variable that has had a large influence on the economies of virtually every country is energy. The 1973 oil crisis, wherein the price of oil quadrupled over a short period of time, is generally considered to be one of the largest financial shocks to our economy. In fact, economists often refer to two different economies: before the 1973 oil crisis and after.

Unfortunately the world will be facing more shocks to our economy because of energy for two primary reasons. The first is the depletion of nonrenewable sources of energy and the resulting price increases. The second is the possibility that burning fossil fuels and the creation of carbon dioxide may be the cause of global warming. One economist predicted that the cost of global warming will be calculated in the trillions of dollars. Statistics can play an important role by determining whether the earth's temperature has been increasing and if so, whether carbon dioxide is the cause. (See Cases 2.1 and 4.1.)

In this chapter you will encounter other examples and exercises involving the issue of energy.

EXAMPLE 2.2 Energy Consumption in the United States

DATA Xm02-02*

Table 2.3 lists the total energy consumption of the United States from all sources in 2005. To make it easier to see the details, the table measures the heat content in metric tons (1,000 kilograms) of oil equivalent. For example, the United States burned an amount of coal and coal products equivalent to 545,258 metric tons of oil. Use an appropriate graphical technique to depict these figures.

TABLE **2.3** Energy Consumption in the United States by Source, 2005

	Heat Content
Nonrenewable Energy Sources	
Coal and coal products	545,258
Oil	903,440
Natural gas	517,881
Nuclear	209,890
Renewable Energy Sources	
Hydroelectric	18,251
Solid biomass	52,473
Other (liquid biomass, geothermal, solar, wind, and tide, wave, and ocean)	20,533
Total	2,267,726

Source: International Energy Association.

SOLUTION

We're interested in describing the proportion of total energy consumption for each source. Thus, the appropriate technique is the pie chart. Figure 2.3 depicts the manually drawn pie charts. Excel's and Minitab's are virtually identical.

FIGURE **2.3** Pie Chart for Example 2.2

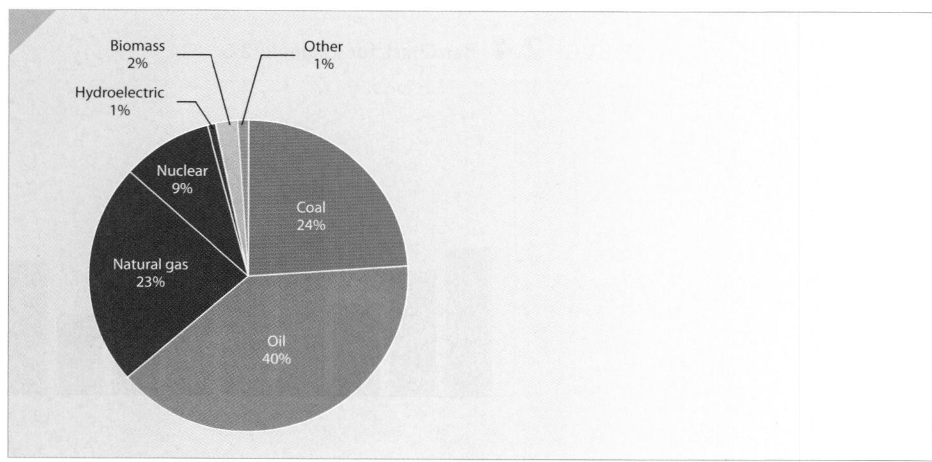

INTERPRET

The United States is very dependent on oil, coal, and natural gas. Almost 90% of energy use is based on these sources. See Exercises 2.11 to 2.15 for more information on the subject.

EXAMPLE 2.3

DATA
Xm02-03

Per Capita Beer Consumption (Ten Selected Countries)

Table 2.4 lists the per capita beer consumption for each of ten countries around the world and graphically presents these numbers.

TABLE **2.4** Per Capita Beer Consumption 2004

Country	Consumption (liters/year)
Australia	109.9
Belgium	93.0
Canada	68.3
Czech Republic	156.9
Germany	115.8
Ireland	131.1
New Zealand	77.0
Russia	58.9
United Kingdom	99.0
United States	81.6

Source: Kirin Brewery Company.

SOLUTION

In this example we're primarily interested in the numbers. There is no use in presenting proportions here. Figure 2.4 shows the bar chart that was manually drawn. The computer versions are the same.

FIGURE **2.4** Bar Chart for Example 2.3

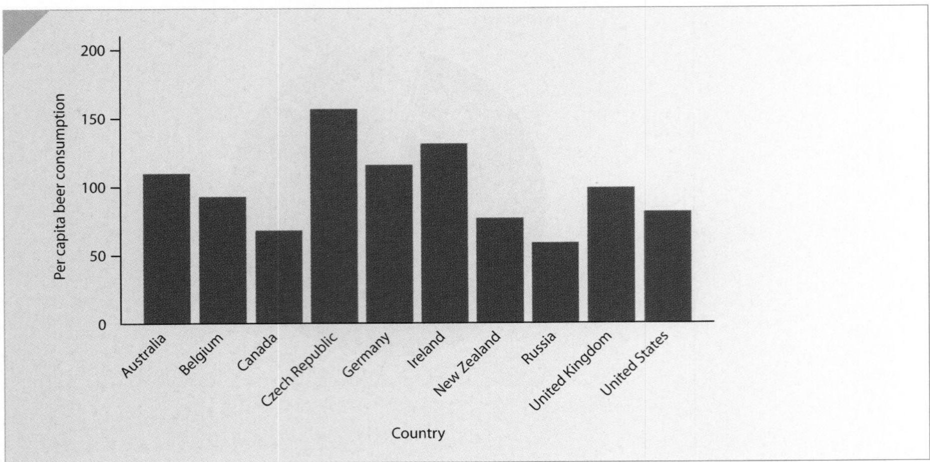

INTERPRET

The Czech Republic is the number one consumer of beer followed by Ireland, Germany, and Australia. Both the United States and Canada rank far lower. Surprised?

We complete this section by describing when bar and pie charts are used to summarize and present data.

Factors That Identify When to Use Frequency and Relative Frequency Tables, Bar and Pie Charts

1. **Objective:** Describe a single set of data
2. **Data type:** Nominal

EXERCISES

2.11 Xr02-11 When will the world run out of oil? One way to judge is to determine the oil reserves of the countries around the world. The next table displays the known oil reserves of the top 15 countries. Graphically describe the figures.

Country	Reserves
Angola	25,000,000,000
Canada	178,900,000,000
China	18,260,000,000
Iran	133,300,000,000
Iraq	112,500,000,000
Kazakhstan	26,000,000,000
Kuwait	96,500,000,000
Libya	40,000,000,000
Mexico	33,310,000,000
Nigeria	36,000,000,000
Russia	69,000,000,000
Saudi Arabia	262,700,000,000
United Arab Emirates	97,800,000,000
United States	22,450,000,000
Venezuela	75,590,000,000

Source: CIA World Factbook, July 28, 2005.

2.12 Refer to Exercise 2.11. The total reserves in the world are 1,349,417,153,000 barrels. The total reserves of the top 15 countries are 1,227,310,000,000 barrels. Use a graphical technique that emphasizes the percentage breakdown of the top 15 countries plus others.

2.13 Xr02-13 The following table lists the average oil consumption per day for 15 countries. Use a graphical technique to present these figures.

Country	Consumption (barrels per day)
Brazil	2,100,000
Canada	2,193,000
China	6,391,000
France	2,060,000
Germany	2,677,000
India	2,320,000
Italy	1,874,000
Japan	5,578,000
South Korea	2,168,000
Mexico	1,752,000
Russia	2,800,000
Saudi Arabia	1,775,000
Spain	1,544,000
United Kingdom	1,722,000
United States	20,030,000

Source: CIA World Factbook, January 10, 2005.

2.14 Xr02-14 As pointed out in the chapter-opening example, there are 42 gallons in a barrel of oil. The number of products produced and the proportion of the total are listed in the following table. Draw a graph to depict these numbers.

Product	Percent of Total
Gasoline	51.4%
Distillate fuel oil	15.3
Jet fuel	12.6
Still gas	5.4
Marketable coke	5.0
Residual fuel oil	3.3
Liquefied refinery gas	2.8
Asphalt and road oil	1.9
Lubricants	.9
Other	1.5

Source: California Energy Commission, based on 2004 data.

2.15 Xr02-15* The following table displays the energy consumption pattern of Australia. The figures measure the heat content in metric tons (1,000 kilograms) of oil equivalent. Draw a graph that depicts these numbers.

	Australia
Nonrenewable Energy Sources	
Coal and coal products	55,385
Oil	33,185
Natural gas	20,350
Nuclear	0
Renewable Energy Sources	
Hydroelectric	1,388
Solid biomass	4,741
Other (liquid biomass, geothermal, solar, wind, and tide, wave, and ocean)	347
Total	115,396

Source: International Energy Association.

2.16 Xr02-16 The planet may be threatened by global warming, possibly caused by burning fossil fuels (petroleum, natural gas, and coal) that produce carbon dioxide (CO_2). The following table lists the top 15 producers of CO_2 and the annual amounts (million of metric tons) from fossil fuels. Graphically depict these figures.

Country	CO_2	Country	CO_2
Australia	103	Japan	329
Canada	164	Mexico	110
China	966	Russia	438
France	112	South Africa	112
Germany	230	South Korea	128
India	279	United Kingdom	154
Iran	101	United States	1,582
Italy	127		

Source: Statistical Abstract of the United States, 2006, Table 1325.

2.17 Xr02-17 The production of steel has often been used as a measure of the economic strength of a country. The following table lists the steel production in the 14 largest steel-producing nations in a recent year. (The units are millions of metric tons.) Use a graphical technique to display these figures.

Country	Steel	Country	Steel
Brazil	26.7	Japan	102.9
Canada	15.1	Russia	57.6
China	143.3	South Korea	43.9
France	19.4	Spain	16.7
Germany	44.8	Taiwan	17.1
India	27.3	Ukraine	33.1
Italy	26.5	United States	90.1

Source: National Post Business, July 2004.

2.18 Xr02-18 In 2003 (latest figures available) the United States generated 236.2 million tons of garbage. The following table lists the amounts by source. Use one or more graphical techniques to present these figures.

Source	Amount (millions of tons)
Paper and paperboard	83.1
Glass	12.5
Metals	18.9
Plastics, rubber, leather, and textiles	44.1
Wood	13.7
Food wastes	27.6
Yard and other wastes	36.2

Source: Statistical Abstract of the United States, 2006, Table 363.

2.19 In Canada, liquor is sold only in stores owned and operated by the provincial government. The component costs of a bottle of liquor are listed below. Draw a graph that exhibits these numbers.

Distiller's selling price (includes corporate, municipal, and employee taxes)	$3.85
Provincial markup in provincial government store	$10.27
Provincial sales tax	$2.09
Federal GST (goods and sales tax 7%)	$1.22
Federal excise duty	$3.32
Total	$20.75

Source: Association of Canadian Distillers.

2.20 Xr02-20 In 1997 there were 5,419,720 women-owned firms in the United States. Each was categorized by industry group. The following table lists the number of women-owned firms in each industry group.

Industry	Frequency
Agriculture	74,444
Mining	20,030
Construction	157,173
Manufacturing	121,108
Transportation and public utilities	128,999
Wholesale trade	125,645
Retail trade	919,990
Finance, insurance, and real estate	479,469
Services	2,981,266
Other	411,596

Source: Statistical Abstract of the United States, 2003, Table 751.

a. Draw a bar chart.
b. Draw a pie chart.
c. What information is conveyed by each chart?

The following exercises require a computer and software.

2.21 Xr02-21 What are the most important characteristics of colleges and universities? This question was asked of a sample of college-bound high school seniors. The responses are

1. Location
2. Majors
3. Academic reputation
4. Career focus
5. Community
6. Number of students.

The results are stored using the codes. Use a graphical technique to summarize and present the data.

2.22 Xr02-22 Where do consumers get information about cars? A sample of recent car buyers was asked to identify the most useful source of information about the cars they purchased. The responses are

1. Consumer guide
2. Dealership
3. Word of mouth
4. Internet

The responses were stored using the codes. Graphically depict the responses. (*Source: Automotive Retailing Today*, The Gallup Organization.)

2.23 Xr02-23 A survey asked 392 homeowners which area of the home they would most like to renovate. The responses and frequencies are shown next. Use a graphical technique to present these results. Briefly summarize your findings

Area	Code
Basement	1
Bathroom	2
Bedroom	3
Kitchen	4
Living/dining room	5

Source: Toronto Star, November 23, 2004.

2.24 Xr02-24 Subway train riders frequently pass the time by reading a newspaper. New York City has a subway and four newspapers. A sample of 360 subway riders who regularly read a newspaper was asked to identify that newspaper. The responses are

1. *New York Daily News*
2. *New York Post*
3. *New York Times*
4. *Wall Street Journal*

The responses were recorded using the numerical codes shown.

a. Produce a frequency distribution and a relative frequency distribution.
b. Draw an appropriate graph to summarize the data. What does the graph tell you?

2.25 Xr02-25 Who applies to MBA programs? To help determine the background of the applicants, a sample of 230 applicants to a university's business school was asked to report their undergraduate degree. The degrees were recorded using these codes.

1. BA
2. BBA
3. BEng
4. BSc
5. Other

a. Determine the frequency distribution.
b. Draw a bar chart.
c. Draw a pie chart.
d. What do the charts tell you about the sample of MBA applicants?

2.26 Xr02-26 Each day a restaurant lists on its menu five daily specials. Today's specials are

1. Fried chicken
2. Meat loaf
3. Turkey pot pie
4. Fillet of sole
5. Lasagna

A sample of customers' choices was recorded (using the numerical codes). Draw a graph that describes the most important aspect of these data.

2.27 Xr02-27 Most universities have several different kinds of residences on campus. To help long-term planning, one university surveyed a sample of graduate students and asked them to report their marital status or relationship. The possible responses are

1. Single
2. Married
3. Divorced
4. Other

The responses for a sample of 250 students were recorded and saved. Draw a graph that summarizes the information that you deem necessary.

2.28 Xr02-28 A number of business and economics courses requires the use of computers. As a result many students buy their own computer. A survey asks students to identify which computer brand they have purchased. The responses are

1. IBM
2. Compaq
3. Dell
4. Other

a. Use a graphical technique that depicts the frequencies.
b. Graphically depict the proportions.
c. What do the charts tell you about the brands of computers used by the students?

2.29 Xr02-29 An increasing number of statistics courses use a computer and software rather than manual calculations. A survey of statistics instructors asked each to report the software his or her course uses. The responses are

1. Excel
2. Minitab
3. SAS
4. SPSS
5. Other

a. Produce a frequency distribution.
b. Graphically summarize the data so that the proportions are depicted.
c. What do the charts tell you about the software choices?

2.30 Xr02-30 Opinions about the economy are important measures because they can become self-fulfilling prophecies. Annual surveys are conducted to determine the level of confidence in the future prospects of the economy. A sample of 1,000 adults was asked, Compared with last year, do you think this coming year will be

1. better?
2. the same?
3. worse?

Use a suitable graphical technique to summarize these data. Describe what you have learned.

2.3 Graphical Techniques to Describe Interval Data

In this section, we introduce several graphical methods that are used when the data are interval. The most important of these graphical methods is the histogram. As you will see, the histogram is not only a powerful graphical technique used to summarize interval data, but it is also used to help explain an important aspect of probability (see Chapter 8).

APPLICATIONS in MARKETING

© Digital Vision/Getty Images

Pricing

Traditionally, marketing has been defined in terms of the four P's: product, price, promotion, and place. *Marketing management* is the functional area of business that focuses on the development of a product, together with its pricing, promotion, and distribution. Decisions are made in these four areas with a view to satisfying the wants and needs of consumers, while also satisfying the firm's objective.

The pricing decision must be addressed both for a new product, and from time to time, for an existing product. Anyone buying a product such as a personal computer has been confronted with a wide variety of prices, accompanied by a correspondingly wide variety of features. From a vendor's standpoint, establishing the appropriate price and corresponding set of attributes for a product is complicated and must be done in the context of the overall marketing plan for the product.

EXAMPLE 2.4

DATA
Xm02-04

Analysis of Long-Distance Telephone Bills

Following deregulation of telephone service, several new companies were created to compete in the business of providing long-distance telephone service. In almost all cases these companies competed on price since the service each offered is similar. Pricing a service or product in the face of stiff competition is very difficult. Factors to be considered include supply, demand, price elasticity, and the actions of competitors. Long-distance packages may employ per minute charges, a flat monthly rate, or some combination of the two. Determining the appropriate rate structure is facilitated by acquiring information about the behaviors of customers and in particular the size of monthly long-distance bills.

As part of a larger study, a long-distance company wanted to acquire information about the monthly bills of new subscribers in the first month after signing with the company. The company's marketing manager conducted a survey of 200 new residential subscribers wherein the first month's bills were recorded. These data are listed here. The general manager planned to present his findings to senior executives. What information can be extracted from these data?

Long-Distance Telephone Bills

42.19	39.21	75.71	8.37	1.62	28.77	35.32	13.90	114.67	15.30
38.45	48.54	88.62	7.18	91.10	9.12	117.69	9.22	27.57	75.49
29.23	93.31	99.50	11.07	10.88	118.75	106.84	109.94	64.78	68.69
89.35	104.88	85.00	1.47	30.62	0	8.40	10.70	45.81	35.00
118.04	30.61	0	26.40	100.05	13.95	90.04	0	56.04	9.12
110.46	22.57	8.41	13.26	26.97	14.34	3.85	11.27	20.39	18.49
0	63.70	70.48	21.13	15.43	79.52	91.56	72.02	31.77	84.12
72.88	104.84	92.88	95.03	29.25	2.72	10.13	7.74	94.67	13.68
83.05	6.45	3.20	29.04	1.88	9.63	5.72	5.04	44.32	20.84
95.73	16.47	115.50	5.42	16.44	21.34	33.69	33.40	3.69	100.04
103.15	89.50	2.42	77.21	109.08	104.40	115.78	6.95	19.34	112.94
94.52	13.36	1.08	72.47	2.45	2.88	0.98	6.48	13.54	20.12
26.84	44.16	76.69	0	21.97	65.90	19.45	11.64	18.89	53.21
93.93	92.97	13.62	5.64	17.12	20.55	0	83.26	1.57	15.30
90.26	99.56	88.51	6.48	19.70	3.43	27.21	15.42	0	49.24
72.78	92.62	55.99	6.95	6.93	10.44	89.27	24.49	5.20	9.44
101.36	78.89	12.24	19.60	10.05	21.36	14.49	89.13	2.80	2.67
104.80	87.71	119.63	8.11	99.03	24.42	92.17	111.14	5.10	4.69
74.01	93.57	23.31	9.01	29.24	95.52	21.00	92.64	3.03	41.38
56.01	0	11.05	84.77	15.21	6.72	106.59	53.90	9.16	45.77

SOLUTION

There is little information developed by casually reading through the 200 observations. The manager can probably see that most of the bills are under \$100, but that is likely to be the extent of the information garnered from browsing through the data. If he examines the data more carefully, he may discover that the smallest bill is \$0 and the largest is \$119.63. He has now developed some information. However, his presentation to senior executives will be most unimpressive if no other information is produced. For example, someone is likely to ask how the numbers are distributed between 0 and 119.63. Are there many small bills and few large bills? What is the "typical" bill? Are the bills somewhat similar or do they vary considerably?

To help answer these questions and others like them, the marketing manager can construct a frequency distribution from which a histogram can be drawn. In the previous section a frequency distribution was created by counting the number of times each category of the nominal variable occurred. We create a frequency distribution for interval data by counting the number of observations that fall into each of a series of intervals, called **classes,** that cover the complete range of observations. We discuss how to decide the number of classes and the upper and lower limits of the intervals later. We have chosen eight classes defined in such a way that each observation falls into one and only one class. These classes are defined as follows:

Classes

Amounts that are less than or equal to 15

Amounts that are more than 15 but less than or equal to 30

Amounts that are more than 30 but less than or equal to 45

Amounts that are more than 45 but less than or equal to 60

Amounts that are more than 60 but less than or equal to 75

Amounts that are more than 75 but less than or equal to 90

Amounts that are more than 90 but less than or equal to 105

Amounts that are more than 105 but less than or equal to 120

Notice that the intervals do not overlap, so that there is no uncertainty about which interval to assign to any observation. Moreover, because the smallest number is 0 and the largest is 119.63, every observation will be assigned to a class. Finally, the intervals are equally wide. Although this is not essential, it makes the task of reading and interpreting the graph easier.

To create the frequency distribution manually, we count the number of observations that fall into each interval. Table 2.5 presents the frequency distribution.

TABLE **2.5** Frequency Distribution of the Long-Distance Bills in Example 2.4

Class Limits	Frequency
0 to 15*	71
15 to 30	37
30 to 45	13
45 to 60	9
60 to 75	10
75 to 90	18
90 to 105	28
105 to 120	14
Total	200

*Classes contain observations greater than their lower limits (except for the first class) and less than or equal to their upper limits.

Although the frequency distribution provides information about how the numbers are distributed, the information is more easily understood and imparted by drawing a picture or graph. The graph is called a **histogram.** A histogram is created by drawing rectangles whose bases are the intervals and whose heights are the frequencies. Figure 2.5 exhibits the histogram that was drawn by hand.

FIGURE **2.5** Histogram for Example 2.4

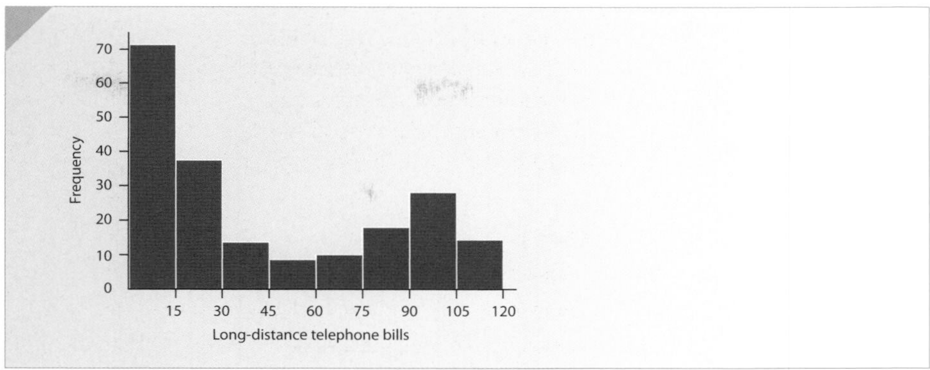

EXCEL

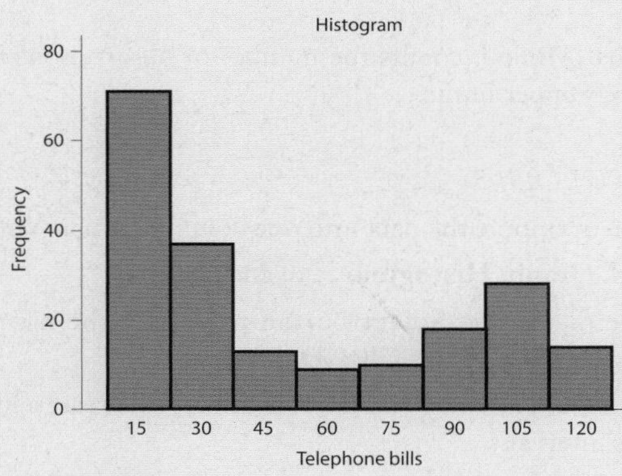

INSTRUCTIONS

1. Type or import the data into one column. (Open Xm02-04.) In another column type the upper limits of the class intervals. Excel calls them *bins*. (You can put any name in the first row; we typed "Telephone bills.")
2. Click **Data, Data Analysis,** and **Histogram.** If Data Analysis does not appear in the menu box, see CD Appendix A1.
3. Specify the **Input Range** (A1:A201) and the **Bin Range** (B1:B9). Click **Chart Output.** Click **Labels** if the first row contains names.
4. To remove the gaps, place the cursor over one of the rectangles and click the right button of the mouse. Click (with the left button) **Format Data Series. . . .** Move the pointer to **Gap Width** and use the slider to change the number from 150 to 0.

Except for the first class, Excel counts the number of observations in each class that are greater than the lower limit and less than or equal to the upper limit.

Note that the numbers along the horizontal axis represent the upper limits of each class although they appear to be placed in the centers. If you wish, you can replace these numbers with the actual midpoints by making changes to the frequency distribution in cells A1:B14 (change 15 to 7.5, 30 to 22.5, . . . , and 120 to 112.5).

You can also convert the histogram to list relative frequencies instead of frequencies. To do so, change the frequencies to relative frequencies by dividing each frequency by 200. That is replace 71 by .355, 37 by .185, . . . , and 14 by .07.

If you have difficulty with this technique turn to CD Appendix A2 or A3, which provides step-by-step instructions for Excel and provides troubleshooting tips.

MINITAB

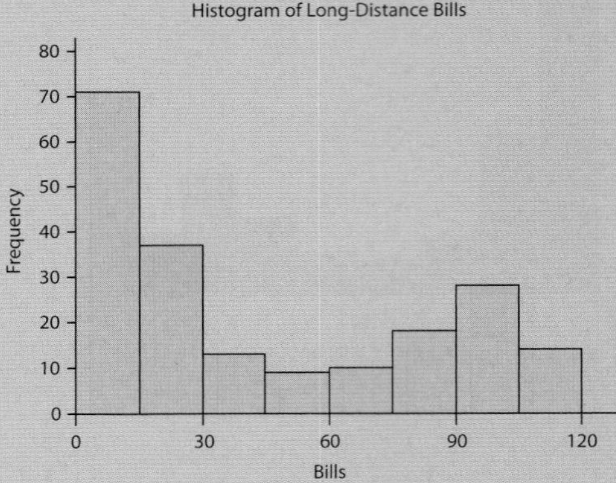

Note that Minitab counts the number of observations in each class that are strictly less than their upper limits.

INSTRUCTIONS

1. Type or import the data into one column. (Open Xm02-04.)
2. Click **Graph, Histogram . . . ,** and **Simple.**
3. Type or use the **Select** button to specify the name of the variable in the **Graph variables** box (Bills). Click **Data View.**
4. Click **Data Display** and **Bars.** Minitab will create a histogram using its own choices of class intervals.
5. To choose your own classes, double-click the horizontal axis. Click **Binning.**
6. Under **Interval Type** choose **Cutpoint.** Under **Interval Definition,** choose **Midpoint/Cutpoint positions** and type in your choices (0 15 30 45 60 75 90 105 120) to produce the histogram shown here.

INTERPRET

The histogram gives us a clear view of the way the bills are distributed. About half the monthly bills are small ($0 to $30), a few bills are in the middle range ($30 to $75), and a relatively large number of long-distance bills are at the high end of the range. It would appear from this sample of first-month long-distance bills that the company's customers are split unevenly between light and heavy users of long-distance telephone service. If the company assumes that this pattern will continue, it must address a number of pricing issues. For example, customers who incurred large monthly bills may be targets of competitors who offer flat rates for 15-minute or 30-minute calls. The company needs to know more about these customers. With the additional information, the marketing manager may suggest an alteration of its pricing.

Determining the Number of Class Intervals

The number of class intervals we select depends entirely on the number of observations in the data set. The more observations we have, the larger the number of class intervals we need to use to draw a useful histogram. Table 2.6 provides guidelines on choosing the number of classes. In Example 2.4 we had 200 observations. The table tells us to use 7, 8, 9, or 10 classes.

TABLE **2.6** **Approximate Number of Classes in Frequency Distributions**

Number of Observations	Number of Classes
Less than 50	5–7
50–200	7–9
200–500	9–10
500–1,000	10–11
1,000–5,000	11–13
5,000–50,000	13–17
More than 50,000	17–20

An alternative to the guidelines listed in Table 2.6 is to use Sturges's formula, which recommends that the number of class intervals be determined by the following:

$$\text{Number of class intervals} = 1 + 3.3 \log(n)$$

For example, if $n = 50$ Sturges's formula becomes

$$\text{Number of class intervals} = 1 + 3.3 \log(50) = 1 + 3.3(1.7) = 6.6$$

which we round to 7.

Class Interval Widths

We determine the approximate width of the classes by subtracting the smallest observation from the largest and dividing the difference by the number of classes. Thus,

$$\text{Class width} = \frac{\text{Largest observation} - \text{Smallest observation}}{\text{Number of classes}}$$

In Example 2.4 we calculated

$$\text{Class width} = \frac{119.63 - 0}{8} = 14.95$$

We often round the result to some convenient value. We then define our class limits by selecting a lower limit for the first class from which all other limits are determined. The only condition we apply is that the first class interval must contain the smallest observation. In Example 2.4, we rounded the class width to 15 and set the lower limit of the first class to 0. Thus, the first class is defined as "Amounts that are greater than or equal to 0 but less than or equal to 15." (Minitab users should remember that the classes are defined as the number of observations that are *strictly less* than their upper limits.)

Table 2.6 and Sturges's formula are guidelines only. It is more important to choose classes that are easy to interpret. For example, suppose that we have recorded the marks on an exam of the 100 students registered in the course where the highest mark is 94 and the lowest is 48. Table 2.6 suggests that we use 7, 8, or 9 classes and Sturges's formula computes the approximate number of classes as

$$\text{Number of class intervals} = 1 + 3.3 \log(100) = 1 + 3.3(2) = 7.6$$

which we round to 8. Thus,

$$\text{Class width} = \frac{94 - 48}{8} = 5.75$$

which we would round to 6. We could then produce a histogram whose upper limits of the class intervals are 50, 56, 62, . . . , 98. Because of the rounding and the way in which we defined the class limits, the number of classes is 9. However, a histogram that is easier to interpret would be produced using classes whose widths are 5. That is, the upper limits would be 50, 55, 60, . . . , 95. The number of classes in this case would be 10.

Shapes of Histograms

The purpose of drawing histograms, like that of all other statistical techniques, is to acquire information. Once we have the information, we frequently need to describe what we've learned to others. We describe the shape of histograms on the basis of the following characteristics.

Symmetry

A histogram is said to be **symmetric** if, when we draw a vertical line down the center of the histogram, the two sides are identical in shape and size. Figure 2.6 depicts three symmetric histograms.

FIGURE **2.6** Three Symmetric Histograms

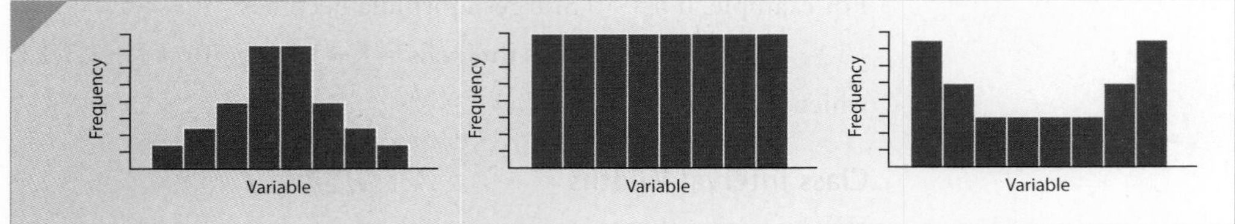

Skewness

A skewed histogram is one with a long tail extending to either the right or the left. The former is called **positively skewed,** and the latter is called **negatively skewed.** Figure 2.7 shows examples of both. Incomes of employees in large firms tend to be positively skewed, because there is a large number of relatively low-paid workers and a small number of well-paid executives. The time taken by students to write exams is frequently negatively skewed because few students hand in their exams early; most prefer to reread their papers and hand them in near the end of the scheduled test period.

FIGURE **2.7** Positively and Negatively Skewed Histograms

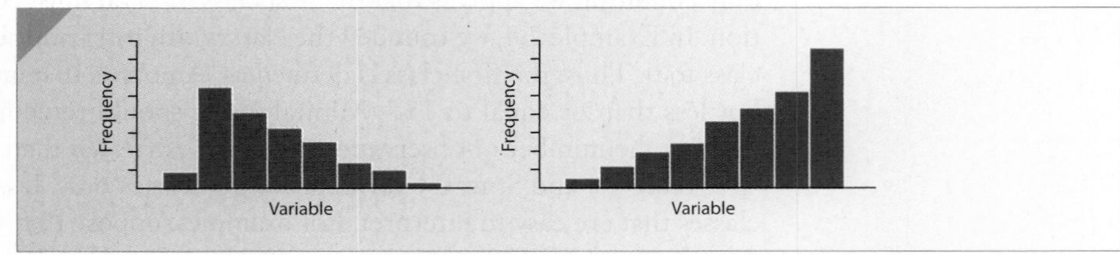

Number of Modal Classes

As we discuss in Chapter 4, a *mode* is the observation that occurs with the greatest frequency. A **modal class** is the class with the largest number of observations. A **unimodal histogram** is one with a single peak. The histogram in Figure 2.8 is unimodal. A **bimodal**

FIGURE **2.8**
A Unimodal Histogram

histogram is one with two peaks, not necessarily equal in height. Bimodal histograms often indicate that two different distributions are present. (See Example 2.7.) Figure 2.9 depicts bimodal histograms.

FIGURE **2.9**
Bimodal Histograms

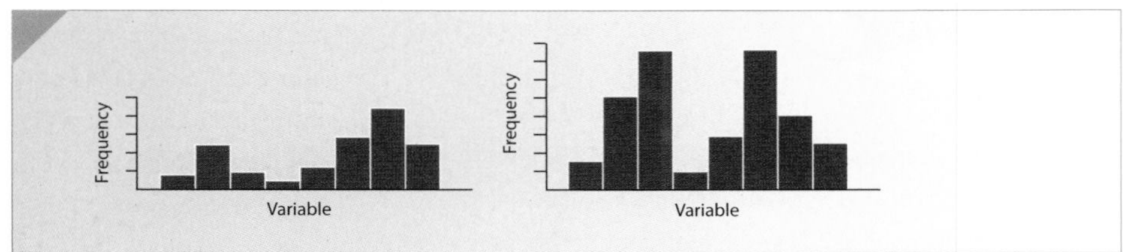

Bell Shape

A special type of symmetric unimodal histogram is one that is bell shaped. In Chapter 8 we will explain why this type of histogram is important. Figure 2.10 exhibits a bell-shaped histogram.

FIGURE **2.10**
Bell-Shaped Histogram

Now that we know what to look for, let's examine some examples of histograms and see what we can discover.

APPLICATIONS in **FINANCE**

© Michele Westmorland/Stone/ Getty Images

Stock and Bond Valuation

A basic understanding of how financial assets, such as stocks and bonds, are valued is critical to good financial management. Understanding the basics of valuation is necessary for capital budgeting and capital structure decisions. Moreover, understanding the basics of valuing investments such as stocks and bonds is at the heart of the huge and growing discipline known as *investment management.*

It is important for a financial manager to be familiar with the main characteristics of the capital markets where long-term financial assets such as stocks and bonds trade. A well-functioning capital market provides managers with useful information concerning the appropriate prices and rates of return that are required for a variety of financial securities with differing levels of risk. Statistical methods can be used to analyze capital markets and summarize their characteristics, such as the shape of the distribution of stock or bond returns.

APPLICATIONS in FINANCE

© George Doyle/Stockbyte/ Getty Images

Return on Investment

The return on an investment is calculated by dividing the gain (or loss) by the value of the investment. For example, a $100 investment that is worth $106 after 1 year has a 6% rate of return. A $100 investment that loses $20 has a −20% rate of return. For many investments, including individual stocks and stock portfolios (combinations of various stocks), the rate of return is a variable. That is, the investor does not know in advance what the rate of return will be. It could be positive number in which case the investor makes money or negative and the investor loses money.

Investors are torn between two goals. The first is to maximize the rate of return on investment. The second goal is to reduce risk. If we draw a histogram of the returns for a certain investment, the location of the center of the histogram gives us some information about the return one might expect from that investment. The spread or variation of the histogram provides us with guidance about the risk. If there is little variation an investor can be quite confident in predicting what his or her rate of return will be. If there is a great deal of variation, the return becomes much less predictable and thus riskier. Minimizing the risk becomes an important goal for investors and financial analysts.

EXAMPLE 2.5 Comparing Returns on Two Investments

DATA Xm02-05

Suppose that you are facing a decision about where to invest that small fortune that remains after you have deducted the anticipated expenses for the next year from the earnings from your summer job. A friend has suggested two types of investment, and to help make the decision you acquire some rates of return from each type. You would like to know what you can expect by way of the return on your investment, as well as other types of information, such as whether the rates are spread out over a wide range (making the investment risky) or are grouped tightly together (indicating relatively low risk). Do the data indicate that it is possible that you can do extremely well with little likelihood of a large loss? Is it likely that you could lose money (negative rate of return)?

The returns for the two types of investments are listed here. Draw histograms for each set of returns and report on your findings. Which investment would you choose and why?

Returns on Investment A

30.00	6.93	13.77	−8.55
−2.13	−13.24	22.42	−5.29
4.30	−18.95	34.40	−7.04
25.00	9.43	49.87	−12.11
12.89	1.21	22.92	12.89
−20.24	31.76	20.95	63.00
1.20	11.07	43.71	−19.27
−2.59	8.47	−12.83	−9.22
33.00	36.08	0.52	−17.00
14.26	−21.95	61.00	17.30
−15.83	10.33	−11.96	52.00
0.63	12.68	1.94	
38.00	13.09	28.45	

Returns on Investment B

30.33	−34.75	30.31	24.30
−30.37	54.19	6.06	−10.01
−5.61	44.00	14.73	35.24
29.00	−20.23	36.13	40.70
−26.01	4.16	1.53	22.18
0.46	10.03	17.61	3.24
2.07	10.51	1.20	25.10
29.44	39.04	9.94	−24.24
11.00	24.76	−33.39	−38.47
−25.93	15.28	58.67	13.44
8.29	34.21	0.25	68.00
61.00	52.00	5.23	
−20.44	−32.17	66.00	

SOLUTION

We draw the histograms of the returns on the two investments. We'll use Excel and Minitab to do the work.

EXCEL

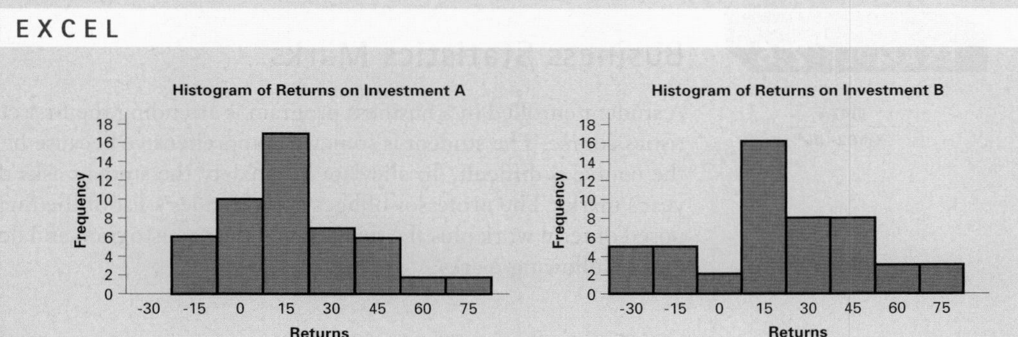

MINITAB

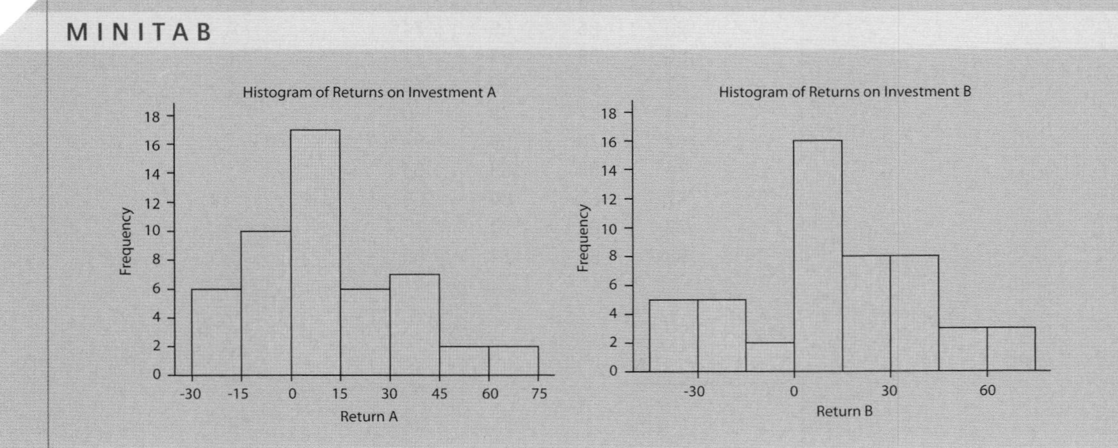

INTERPRET

Comparing the two histograms, we can extract the following information:

1. The center of the histogram of the returns of investment A is slightly lower than that for investment B.
2. The spread of returns for investment A is considerably less than that for investment B.
3. Both histograms are slightly positively skewed.

These findings suggest that investment A is superior. Although the returns for A are slightly less than those for B, the wider spread for B makes it unappealing to most investors. Both investments allow for the possibility of a relatively large return.

The interpretation of the histograms is somewhat subjective. Other viewers may not concur with our conclusion. In such cases numerical techniques provide the detail and precision lacking in most graphs. We will redo this example in Chapter 4 to illustrate how numerical techniques compare to graphical ones.

EXAMPLE 2.6

DATA
Xm02-06*

Business Statistics Marks

A student enrolled in a business program is attending the first class of the required statistics course. The student is somewhat apprehensive because he believes the myth that the course is difficult. To alleviate his anxiety the student asks the professor about last year's marks. The professor obliges and provides a list of the final marks, which is composed of term work plus the final exam. Draw a histogram and describe the result, based on the following marks:

65	81	72	59
71	53	85	66
66	70	72	71
79	76	77	68
65	73	64	72
82	73	77	75
80	85	89	74
86	83	87	77
67	80	78	69
64	67	79	60
62	78	59	92
74	68	63	69
67	67	84	69
72	62	74	73
68	83	74	65

SOLUTION

EXCEL

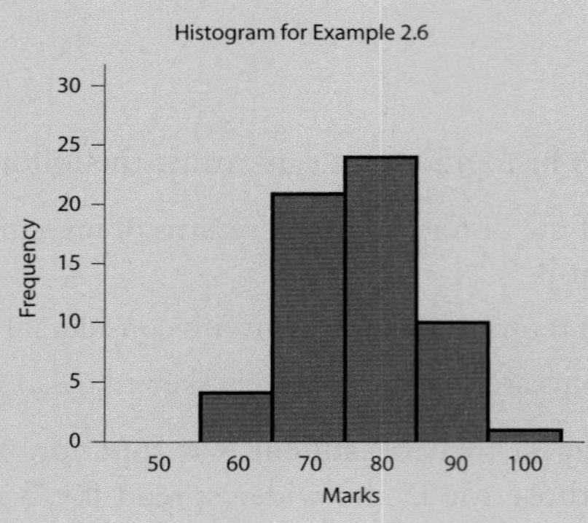

MINITAB

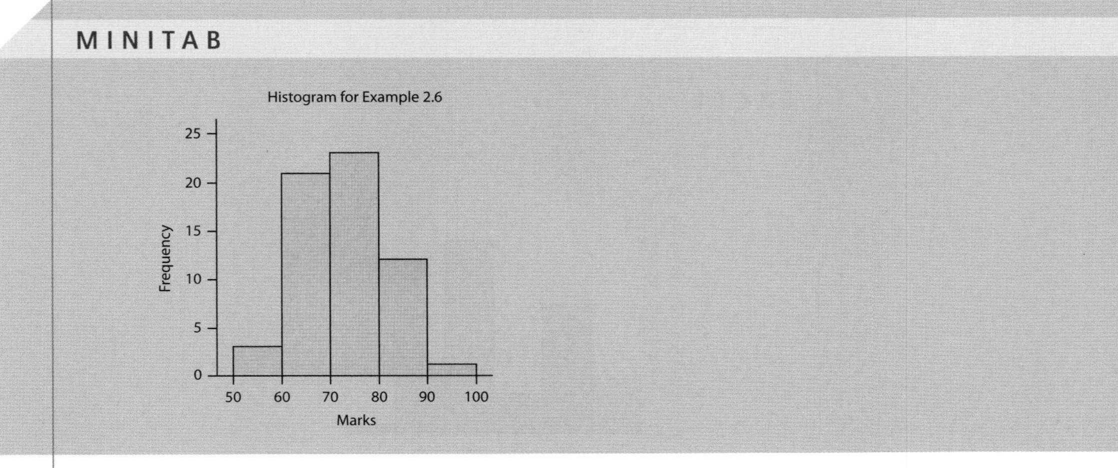

INTERPRET

The histogram is unimodal and approximately symmetric. There are no marks below 50, with the great majority of marks between 60 and 90. The modal class is 70 to 80, and the center of the distribution is approximately 75.

EXAMPLE 2.7

DATA Xm02-07*

Mathematical Statistics Marks

Suppose the student in Example 2.6 obtained a list of last year's marks in a mathematical statistics course. This course emphasizes derivations and proofs of theorems. Use the accompanying data to draw a histogram and compare it to the one produced in Example 2.6. What does this histogram tell you?

77	67	53	54
74	82	75	44
75	55	76	54
75	73	59	60
67	92	82	50
72	75	82	52
81	75	70	47
76	52	71	46
79	72	75	50
73	78	74	51
59	83	53	44
83	81	49	52
77	73	56	53
74	72	61	56
78	71	61	53

SOLUTION

EXCEL

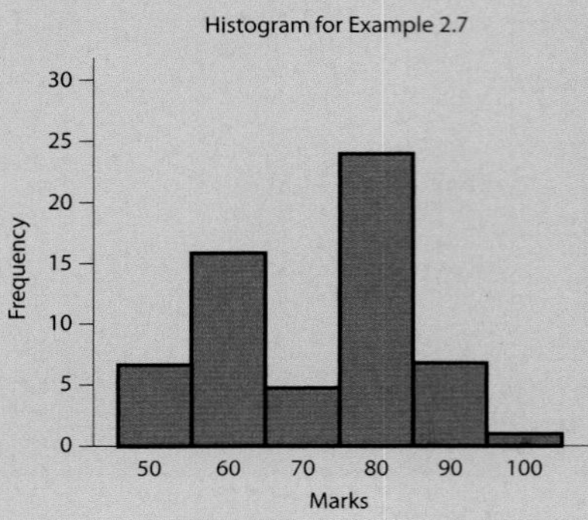

MINITAB

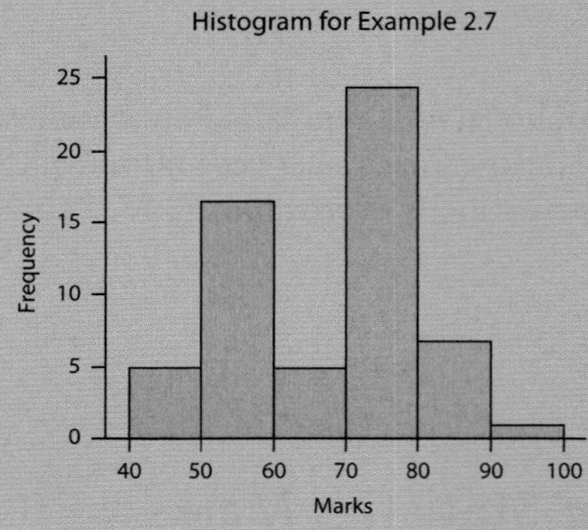

INTERPRET

The histogram is bimodal. The larger modal class is composed of the marks in the 70s. The smaller modal class includes the marks that are in the 50s. There appear to be few marks in the 60s. This histogram suggests that there are two groups of students. Because of the emphasis on mathematics in the course, one may conclude that those who performed poorly in the course are weaker mathematically than those who performed well. The histograms in this example and in Example 2.6 suggest that the courses are quite different from one another and have a completely different distribution of marks.

Stem-and-Leaf Display

One of the drawbacks of the histogram is that we lose potentially useful information by classifying the observations. In Example 2.4, we learned that there are 71 observations that fall between 0 and 15. By classifying the observations we did acquire useful information. However, the histogram focuses our attention on the frequency of each class and by doing so sacrifices whatever information was contained in the actual observations. A statistician named John Tukey introduced the **stem-and-leaf display,** which is a method that to some extent overcomes this loss.

The first step in developing a stem-and-leaf display is to split each observation into two parts, a stem and a leaf. There are several different ways of doing this. For example, the number 12.3 can be split so that the stem is 12 and the leaf is 3. In this definition the stem consists of the digits to the left of the decimal and the leaf is the digit to the right of the decimal. Another method can define the stem as 1 and the leaf as 2 (ignoring the 3). In this definition the stem is the number of tens and the leaf is the number of ones. We'll use this definition to create a stem-and-leaf display for Example 2.4.

The first observation is 42.19. Thus, the stem is 4 and the leaf is 2. The second observation is 38.45, which has a stem of 3 and a leaf of 8. We continue converting each number in this way. The stem-and-leaf display consists of listing the stems 0, 1, 2, . . . , 11. After each stem we list that stem's leaves, usually in ascending order. Figure 2.11 depicts the manually created stem-and-leaf display.

FIGURE **2.11** Stem-and-Leaf Display for Example 2.4

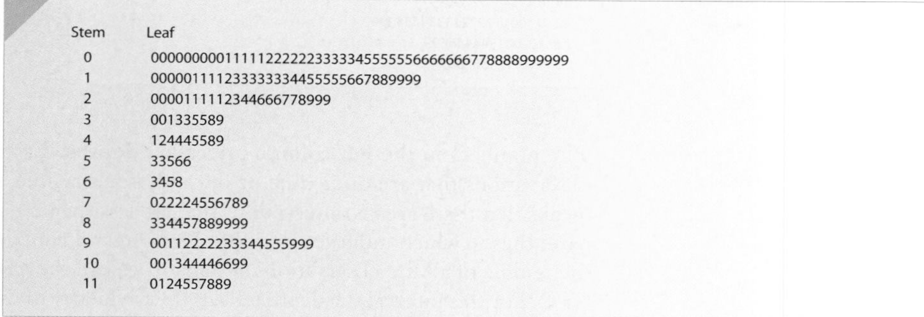

Stem	Leaf
0	00000000011111222222333334555555666666677888899999
1	000001111233333334455555667889999
2	00001111123446667789999
3	001335589
4	124445589
5	33566
6	3458
7	022224556789
8	334457889999
9	00112222233344555999
10	001344446699
11	0124557889

As you can see the stem-and-leaf display is similar to a histogram turned on its side. The length of each line represents the frequency in the class interval defined by the stems. The advantage of the stem-and-leaf display over the histogram is that we can see the actual observations.

EXCEL

	A	B	C	D	E	F	G
1	Stem & Leaf Display						
2							
3	Stems	Leaves					
4	0	->00000000011111222222333334555555666666677888899999					
5	1	->000001111233333334455555667889999					
6	2	->00001111123446667789999					
7	3	->001335589					
8	4	->12445589					
9	5	->33566					
10	6	->3458					
11	7	->022224556789					
12	8	->334457889999					
13	9	->00112222233344555999					
14	10	->001344446699					
15	11	->0124557889					
16							
17							

INSTRUCTIONS

1. Type or import the data into one column. (Open Xm02-04.)
2. Click **Add-ins, Data Analysis Plus,** and **Stem-and-leaf Display.**
3. Specify the **Input Range** (A1:A201). Click one of the values of **Increment** (the increment is the difference between stems) (10).

This the first use of the add-ins that accompany this book. If you're using Office 2007 note that Data Analysis Plus appears twice on the Add-Ins menu. This is the result of the way Excel stores add-ins. Click either one.

MINITAB

```
Stem-and-Leaf Display: Bills

Stem-and-leaf of Bills  N = 200
Leaf Unit = 1.0

 52   0  0000000001111122222233333455555566666667788889999999
 85   1  000001111233333334455555667889999
(23)  2  00001111123446667789999
 92   3  001335589
 83   4  12445589
 75   5  33566
 70   6  3458
 66   7  022224556789
 54   8  334457889999
 42   9  00112222233344555999
 22  10  001344446699
 10  11  0124557889
```

The numbers in the left column are called **depths.** Each depth counts the number of observations that are on its line or beyond. For example, the second depth is 85, which means that there are 85 observations that are less than 20. The third depth is displayed in parentheses, which indicates that the third interval contains the observation that falls in the middle of all the observations, a statistic we call the *median* (to be presented in Chapter 4). For this interval the depth tells us the frequency of the interval. That is, there are 23 observations that are greater than or equal to 20, but less than 30. The fourth depth is 92, which tells us that there are 92 observations that are greater than or equal to 30. Notice that for classes below the median, the depth reports the number of observations that are less than the upper limit of that class. For classes that are above the median, the depth reports the number of observations that are greater than or equal to the lower limit of that class.

INSTRUCTIONS

1. Type or import the data into one column. (Open Xm02-04.)
2. Click **Graph** and **Stem-and-Leaf . . . .**
3. Type or use the **Select** button to specify the variable in the **Variables** box (Bills). Type the increment in the **Increment** box (10).

Ogive

The frequency distribution lists the number of observations that fall into each class interval. We can also create a **relative frequency distribution** by dividing the frequencies by the number of observations. Table 2.7 displays the relative frequency distribution for Example 2.4.

TABLE **2.7**
Relative Frequency Distribution for Example 2.4

Class Limits	Relative Frequency
0 to 15	71/200 = .355
15 to 30	37/200 = .185
30 to 45	13/200 = .065
45 to 60	9/200 = .045
60 to 75	10/200 = .050
75 to 90	18/200 = .090
90 to 105	28/200 = .140
105 to 120	14/200 = .070
Total	200/200 = 1.0

As you can see, the relative frequency distribution highlights the proportion of the observations that fall into each class. In some situations we may wish to highlight the proportion of observations that lie below each of the class limits. In such cases we create the **cumulative relative frequency distribution.** Table 2.8 displays this type of distribution for Example 2.4.

TABLE **2.8**
Cumulative Relative Frequency Distribution for Example 2.4

Class Limits	Relative Frequency	Cumulative Relative Frequency
0 to 15	71/200 = .355	71/200 = .355
15 to 30	37/200 = .185	108/200 = .540
30 to 45	13/200 = .065	121/200 = .605
45 to 60	9/200 = .045	130/200 = .650
60 to 75	10/200 = .05	140/200 = .700
75 to 90	18/200 = .09	158/200 = .790
90 to 105	28/200 = .14	186/200 = .930
105 to 120	14/200 = .07	200/200 = 1.00

From Table 2.8, you can see that, for example, 54% of the bills were less than or equal to $30 and that 79% of the bills were less than or equal to $90.

Another way of presenting this information is the **ogive,** which is a graphical representation of the cumulative relative frequencies. Figure 2.12 is the manually drawn ogive for Example 2.4.

FIGURE **2.12**
Ogive for Example 2.4

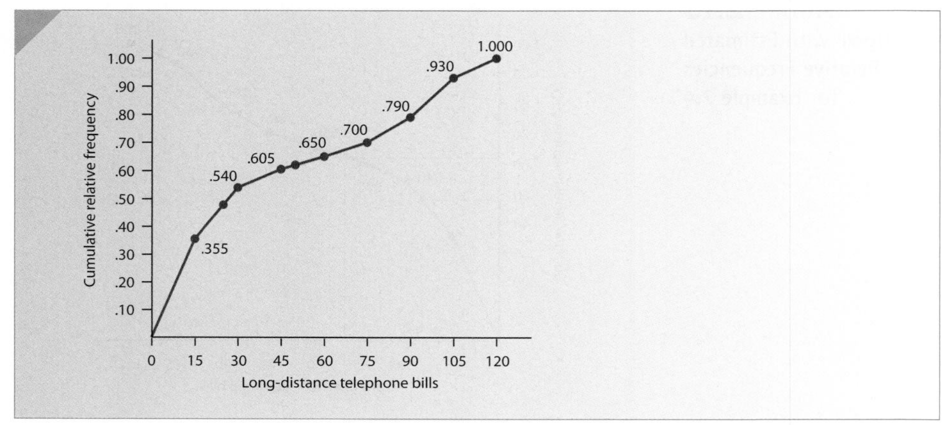

EXCEL

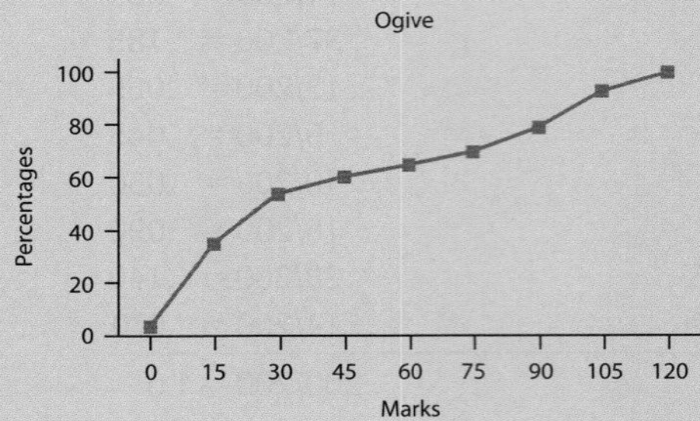

INSTRUCTIONS

Follow instructions to create a histogram. Make the first bin's upper limit a number that is slightly smaller than the smallest number in the data set. Move the cursor to **Chart Output** and click. Do the same for **Cumulative Percentage.** Remove the "More" category. Click on any of the rectangles and click **Delete. Change the Scale,** if necessary. (Right-click the vertical or horizontal axis, click **Format Axis . . .** , and change the **Maximum** value of **Y** equal to 1.0.)

MINITAB

Minitab does not draw ogives.

We can use the ogive to estimate the cumulative relative frequencies of other values. For example, we estimate that about 62% of the bills lie below \$50 and that about 48% lie below \$25. (See Figure 2.13.)

FIGURE **2.13** Ogive with Estimated Relative Frequencies for Example 2.4

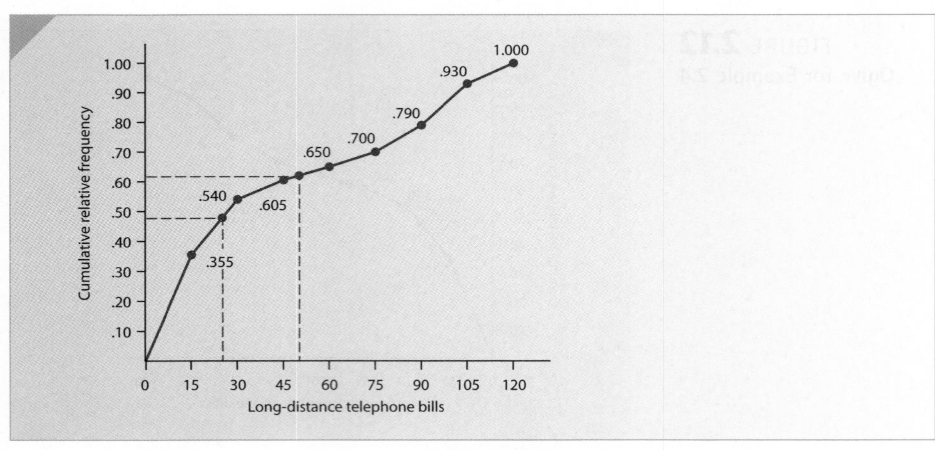

Here is a summary of this section's techniques

Factors That Identify When to Use a Histogram, Ogive, or Stem-and-Leaf Display

1. **Objective:** Describe a single set of data
2. **Data type:** Interval

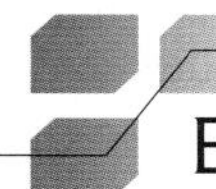

EXERCISES

2.31 How many classes should a histogram contain if the number of observations is 250?

2.32 Determine the number of classes of a histogram for 700 observations.

2.33 A data set consists of 125 observations that range between 37 and 188.

a. What is an appropriate number of classes to have in the histogram?
b. What class intervals would you suggest?

2.34 A statistics practitioner would like to draw a histogram of 62 observations that range from 5.2 to 6.1.

a. What is an appropriate number of class intervals?
b. Define the upper limits of the classes you would use.

2.35 Xr02-35 The number of items rejected daily by a manufacturer because of defects was recorded for the past 30 days. The results are as follows.

4	9	13	7	5	8	12	15	5	7	3
8	15	17	19	6	4	10	8	22	16	9
5	3	9	19	14	13	18	7			

a. Construct a histogram.
b. Construct an ogive.
c. Describe the shape of the histogram.

2.36 Xr02-36 The final exam in a third-year organizational behavior course requires students to write several essay-style answers. The numbers of pages for a sample of 25 exams were recorded. These data are shown here.

5	8	9	3	12	8	5	7	3	8	9	5	2
7	12	9	6	3	8	7	10	9	12	7	3	

a. Draw a histogram.
b. Draw an ogive.
c. Describe what you've learned from the answers to Parts a and b.

2.37 Xr02-37 A large investment firm on Wall Street wants to review the distribution of ages of its stockbrokers. The firm believes that this information can be useful in developing plans to recruit new brokers. The ages of a sample of 40 brokers are shown here.

46	28	51	34	29	40	38	33	41	52
53	40	50	33	36	41	25	38	37	41
36	50	46	33	61	48	32	28	30	49
41	37	26	39	35	39	46	26	31	35

a. Draw a stem-and-leaf display.
b. Draw a histogram.
c. Draw an ogive.
d. Describe what you have learned.

2.38 Xr02-38 The numbers of weekly sales calls by a sample of 30 telemarketers are listed here. Draw a histogram of these data and describe it.

14	8	6	12	21	4	9	3	25	17
9	5	8	18	16	3	17	19	10	15
5	20	17	14	19	7	10	15	10	8

2.39 Xr02-39 The amount of time (in seconds) needed to complete a critical task on an assembly line was measured for a sample of 50 assemblies. These data are as follows:

30.3	34.5	31.1	30.9	33.7
31.9	33.1	31.1	30.0	32.7
34.4	30.1	34.6	31.6	32.4
32.8	31.0	30.2	30.2	32.8
31.1	30.7	33.1	34.4	31.0
32.2	30.9	32.1	34.2	30.7
30.7	30.7	30.6	30.2	33.4
36.8	30.2	31.5	30.1	35.7
30.5	30.6	30.2	31.4	30.7
30.6	37.9	30.3	34.1	30.4

a. Draw a stem-and-leaf display.
b. Draw a histogram.
c. Describe the histogram.

2.40 Xr02-40 A survey of individuals in a mall asked 60 people how many stores they will enter during this visit to the mall. The responses are listed here.

3	2	4	3	3	9
2	4	3	6	2	2
8	7	6	4	5	1
5	2	3	1	1	7
3	4	1	1	4	8
0	2	5	4	4	4
6	2	2	5	3	8
4	3	1	6	9	1
4	4	1	0	4	6
5	5	5	1	4	3

a. Draw a histogram.
b. Draw an ogive.
c. Describe your findings.

2.41 Xr02-41 A survey of 50 baseball fans asked each to report the number of games they attended last year. The results are listed here. Use an appropriate graphical technique to present these data and describe what you have learned.

5	15	14	7	8
16	26	6	15	23
11	15	6	4	7
8	19	16	9	9
8	7	10	5	8
8	6	6	21	10
5	24	5	28	9
11	20	24	5	13
14	9	25	10	24
10	18	22	12	17

2.42 Xr02-42 To help determine the need for more golf courses, a survey was undertaken. A sample of 75 self-declared golfers was asked how many rounds of golf they played last year. These data are as follows:

18	26	16	35	30
15	18	15	18	29
25	30	35	14	20
18	24	21	25	18
29	23	15	19	27
28	9	17	28	25
23	20	24	28	36
20	30	26	12	31
13	26	22	30	29
26	17	32	36	24
29	18	38	31	36
24	30	20	13	23
3	28	5	14	24
13	18	10	14	16
28	19	10	42	22

a. Draw a histogram.
b. Draw a stem-and-leaf display.
c. Draw an ogive.
d. Describe what you have learned.

The following exercises require a computer and statistical software.

2.43 Xr02-43 The annual incomes for a sample of 200 first-year accountants were recorded. Summarize these data using a graphical method. Describe your results.

2.44 Xr02-44 The real estate board in a wealthy suburb of Los Angeles wanted to investigate the distribution of the prices (in $thousands) of homes sold during the past year.
a. Draw a histogram.
b. Draw an ogive.
c. Draw a stem-and-leaf display (if your software allows it).
d. Describe what you have learned.

2.45 Xr02-45 The number of customers entering a bank in the first hour of operation for each of the last 200 days was recorded. Use a graphical technique to extract information. Describe your findings.

2.46 Xr02-46 The lengths of time (in minutes) to serve 420 customers at a local restaurant were recorded.
a. How many bins should a histogram of these data contain?
b. Draw a histogram using the number of bins specified in Part a.
c. Is the histogram symmetric or skewed?
d. Is the histogram bell shaped?

2.47 Xr02-47 The marks of 320 students on an economics midterm test were recorded. Use a graphical technique to summarize these data. What does the graph tell you?

2.48 Xr02-48 The lengths (in inches) of 150 newborn babies were recorded. Use whichever graphical technique you judge suitable to describe these data. What have you learned from the graph?

2.49 Xr02-49 The number of copies made by an office copier was recorded for each of the past 75 days. Graph the data using a suitable technique. Describe what the graph tells you.

2.50 Xr02-50 Each of a sample of 240 tomatoes grown with a new type of fertilizer was weighed (in ounces) and recorded. Draw a histogram and describe your findings.

2.51 Xr02-51 The volume of water used by each of a sample of 350 households was measured (in gallons) and recorded. Use a suitable graphical statistical method to summarize the data. What does the graph tell you?

2.52 Xr02-52 The number of books shipped out daily by Amazon.com was recorded for 100 days. Draw a histogram and describe your findings.

APPLICATIONS in **BANKING**

© Comstock Images/Jupiterimages

Credit Scorecards

Credit scorecards are used by banks and financial institutions to determine whether applicants will receive loans. The scorecard is the product of a statistical technique that converts questions about income, residence, and other variables into a score. The higher the score, the higher the probability is that the applicant will repay. The scorecard is a formula produced by a statistical technique called logistic regression, which will be introduced in Chapter 18. For example, a scorecard may score age categories in the following way:

Less than 25	20 points
25 to 39	24
40 to 55	30
Over 55	38

Other variables would be scored similarly. The sum for all variables would be the applicant's score. A cutoff score would be used to predict those who will repay and those who will default. Because no scorecard is perfect, it is possible to make two types of error: granting credit to those who will default and not lending money to those who would have repaid.

2.53 Xr02-53 A small bank that heretofore did not use a scorecard wanted to determine whether a scorecard would be advantageous. The bank manager took a random sample of 300 loans that were granted and scored each on a scorecard borrowed from a similar bank. This scorecard is based on the responses supplied by the applicants to questions such as age, marital status, and household income. The cutoff is 650, which means that those scoring below are predicted to default and those scoring above are predicted to repay. Two hundred twenty of the loans were repaid, the rest were not. The scores of those who repaid and the scores of those who defaulted were recorded.

a. Use a graphical technique to present the scores of those who repaid.
b. Use a graphical technique to present the scores of those who defaulted.
c. What have you learned about the scorecard?

2.54 Xr02-54 Refer to Exercise 2.53. The bank decided to try another scorecard, this one based not on the responses of the applicants, but on credit bureau reports which list problems such as late payments and previous defaults. The scores using the new scorecard of those who repaid and the scores of those who did not repay were recorded. The cutoff score is 650.

a. Use a graphical technique to present the scores of those who repaid.
b. Use a graphical technique to present the scores of those who defaulted.
c. What have you learned about the scorecard?
d. Compare the results of this exercise with those of Exercise 2.53. Which scorecard appears to be better?

2.4 DESCRIBING TIME-SERIES DATA

Besides classifying data by type, we can also classify them according to whether the observations are measured at the same time or whether they represent measurements at successive points in time. The former are called **cross-sectional data** and the latter, **time-series data.**

The techniques described in Sections 2.2 and 2.3 are applied to cross-sectional data. In Example 2.1 the survey asked 285 graduating seniors to identify their favorite light beer. The likely time period to acquire these data was 2 or 3 days. All the data for Example 2.4 were probably determined within the same day. We can probably say the same thing for Examples 2.5 to 2.7.

To give another example, consider a real estate consultant who feels that the selling price of a house is a function of its size, age, and lot size. To estimate the specific form of the function, she samples say, 100 homes recently sold and records the price, size, age, and lot size for each home. These data are cross-sectional, in that they all are observations at the same point in time. The real estate consultant is also working on a separate project to forecast the monthly housing starts in the northeastern United States over the next year. To do so, she collects the monthly housing starts in this region for each of the past 5 years. These 60 values (housing starts) represent time-series data, because they are observations taken over time.

Note that the original data may be interval or nominal. All the illustrations above deal with interval data. A time series can also list the frequencies and relative frequencies of a nominal variable over a number of time periods. For example, a brand-preference survey asks consumers to identify their favorite brand. These data are nominal. If we repeat the survey once a month for several years, the proportion of consumers who prefer a certain company's product each month would constitute a time series.

Line Chart

Time-series data are often graphically depicted on a **line chart,** which is a plot of the variable over time. It is created by plotting the value of the variable on the vertical axis and the time periods on the horizontal axis.

The chapter-opening example addresses the issue of the relationship between the price of gasoline and the price of oil. We will introduce the technique we need to answer the question in Section 2.6. Another question that arises is, is the recent price of gasoline high compared to the past prices?

EXAMPLE 2.8 Price of Gasoline

DATA Xm02-08

We recorded the monthly average retail price of gasoline since 1978. Some of these data are displayed below. Draw a line chart to describe these data and briefly describe the results.

Year	Month	Price
1978	1	63.1
1978	2	62.9
1978	3	62.9
1978	4	63.1
1978	5	63.7
1978	6	64.5
1978	7	65.5
1978	8	66.3
1978	9	66.9
1978	10	67.1
1978	11	67.6
1978	12	68.5
⋮	⋮	⋮
2006	1	235.9
2006	2	235.4
2006	3	244.4
2006	4	280.1
2006	5	299.3
2006	6	296.3

SOLUTION

Here is the line chart for the average monthly gasoline prices.

FIGURE **2.14** Line Chart for Example 2.8

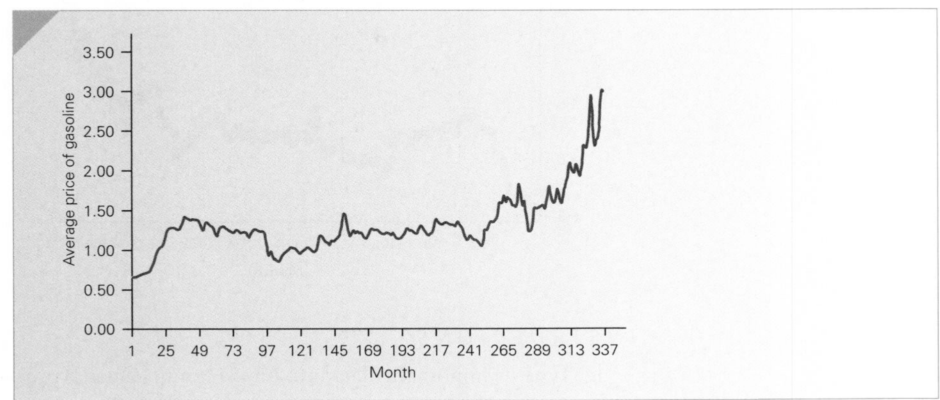

EXCEL

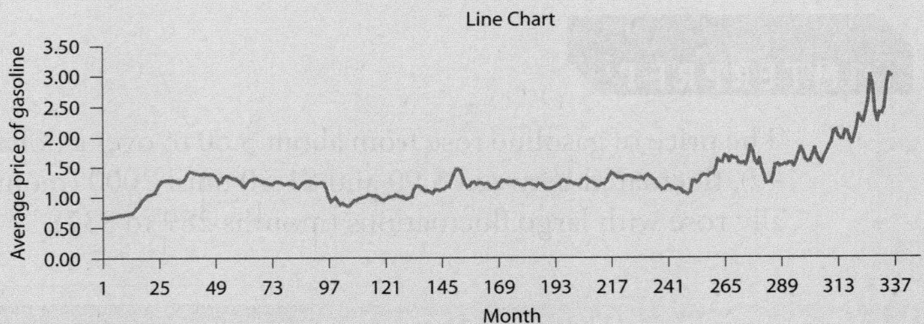

INSTRUCTIONS

1. Type or import the data into one column. (Open Xm02-08.)
2. Highlight the column of data. Click **Insert, Line,** and the first **2-D Line.** Click **Chart Tools** and **Layout** to make whatever changes you wish.

You can draw two or more line charts (for two or more variables) by highlighting all columns of data you wish to graph.

MINITAB

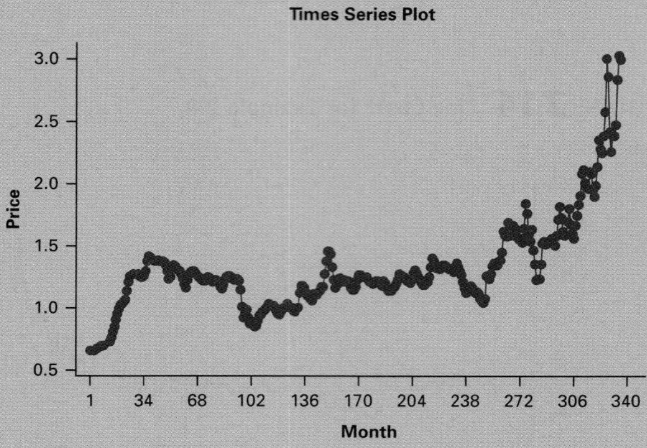

INSTRUCTIONS

1. Type or import the data into one column. (Open Xm02-08.)
2. Click **Graph** and **Time Series Plot . . . .** Click **Simple.**
3. In the **series** box type or use the **Select** button to specify the variable (Price). Click **Time/Scale.**
4. Click the **Time** tab and under **Time Scale** click **Index.**

INTERPRET

The price of gasoline rose from about \$.60 to over a dollar in the late 1970s (months 1 to 49), fluctuated between \$.90 and \$1.50 until 2000 (months 49 to 289), and then generally rose with large fluctuations (months 289 to 337).

APPLICATIONS in ECONOMICS

Measuring Inflation: Consumer Price Index*

Inflation is the increase in the prices for goods and services. In most countries inflation is measured using the consumer price index (CPI). The consumer price index works with a basket of some 300 goods and services in the United States (and a similar number in other countries), including such diverse items as food, housing, clothing, transportation, health, and recreation. The basket is defined for the "typical" or "average" middle-income family, and the set of items and their weights are revised periodically (every 10 years in the United States and every 7 years in Canada).

Prices for each item in this basket are computed on a monthly basis and the CPI is computed from these prices. Here is how it works. We start by setting a period of time as the base. In the United States the base is the years 1982–1984. Suppose that the basket of goods and services cost \$1,000 during this period. Thus, the base is \$1,000 and the CPI is set at 100. Suppose that in the next month (January 1985) the price increases to \$1,010. The CPI for January 1985

*CD Appendix AC describes index numbers and how they are calculated.

(continued)

is calculated in the following way:

$$\text{CPI (January 1985)} = \frac{1{,}010}{1{,}000} \times 100 = 101$$

If the price increases to 1,050 in the next month the CPI is

$$\text{CPI (February 1985)} = \frac{1{,}050}{1{,}000} \times 100 = 105$$

The CPI, despite never really being intended to serve as the official measure of inflation, has come to be interpreted in this way by the general public. Pension-plan payments, old-age social security, and some labor contracts are automatically linked to the CPI and automatically indexed (so it is claimed) to the level of inflation. Despite its flaws, the consumer price index is used in numerous applications. One application involves adjusting prices by removing the effect of inflation, making it possible to track the "real" changes in a time series of prices.

In Example 2.8 the figures shown are the actual prices measured in what are called *current* dollars. To remove the effect of inflation we divide the monthly prices by the CPI for that month and multiply by 100. These prices are then measured in *constant* 1982–1984 dollars. This makes it easier to see what has happened to the prices of the goods and services of interest.

We created two data sets to help you calculate prices in constant 1982–1984 dollars. File Ch02:\\CPI-Annual and Ch02:\\CPI-Monthly list the values of the CPI where 1982–1984 is set at 100 for annual values and monthly values, respectively.

EXAMPLE 2.9 Price of Gasoline in 1982–1984 Constant Dollars

DATA Xm02-09

Remove the effect of inflation in Example 2.8 to determine whether gasoline prices are higher than they have been in the past.

SOLUTION

Here are the 1978 to 2006 average monthly prices of gasoline, the CPI, and the adjusted prices. The adjusted figures for all months were used in the line chart. The Excel and Minitab charts are similar.

Year	Month	Price	CPI	Adjusted Price
1978	1	0.631	62.5	1.010
1978	2	0.629	62.9	1.000
1978	3	0.629	63.4	0.992
1978	4	0.631	63.9	0.987
1978	5	0.637	64.5	0.988
1978	6	0.645	65.2	0.989
1978	7	0.655	65.7	0.997
1978	8	0.663	66.0	1.005
1978	9	0.669	66.5	1.006
1978	10	0.671	67.1	1.000
1978	11	0.676	67.4	1.003
1978	12	0.685	67.7	1.012
⋮	⋮	⋮	⋮	⋮
2006	1	2.359	198.3	1.190
2006	2	2.354	198.7	1.185
2006	3	2.444	199.8	1.223
2006	4	2.801	201.5	1.390
2006	5	2.993	202.5	1.478
2006	6	2.963	202.9	1.460

FIGURE **2.15** Line Chart for Example 2.9

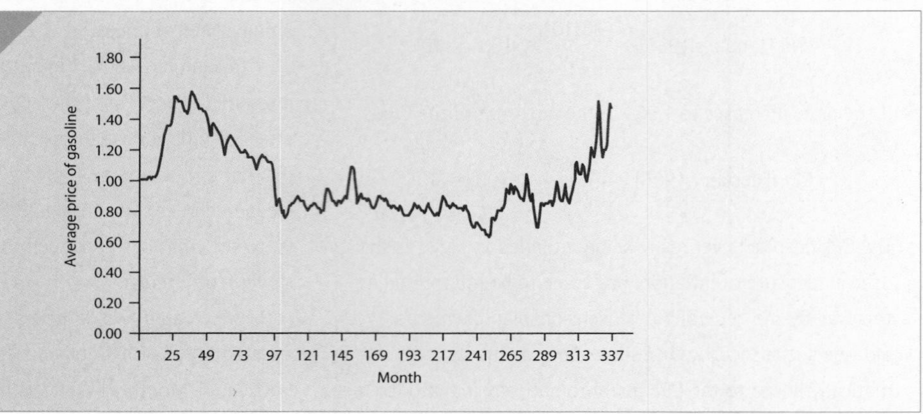

INTERPRET

Using constant 1982–1984 dollars, we can see that the average price of a gallon of gasoline hit its peak around 1980 (month 25). From there it generally (but not steadily) decreased and then started rising at the beginning of 2002 (month 289). However, the adjusted price is still lower than it was at the beginning of 1980.

There are two more factors to consider in judging whether the price of gasoline is high. The first is distance traveled and the second is fuel consumption. Exercise 2.67 deals with this issue.

EXERCISES

2.55 Xr02-55 The major networks compete to broadcast the summer and winter Olympic games. The first winter games were broadcasted in 1960 from Squaw Valley. The following table lists the amounts spent by the winning network.

Year	Site	Network	Price ($millions)
1960	Squaw Valley	CBS	0.39
1964	Innsbruck	ABC	0.60
1968	Grenoble	ABC	2.5
1972	Sapporo	NBC	6.4
1976	Innsbruck	ABC	10.0
1980	Lake Placid	ABC	15.5
1984	Sarajevo	ABC	91.5
1988	Calgary	ABC	309
1992	Albertville	CBS	243
1994	Lillehammer	CBS	300
1998	Nagano	CBS	375
2002	Salt Lake City	NBC	545
2006	Turin	NBC	613

Draw a chart to describe these prices paid by the networks.

2.56 Xr02-56 Every month for the past 2 years a poll was taken to measure the perceptions of people who rate the president of the United States in his job performance. The percentages who judge that performance as satisfactory are listed here. Draw a line chart, and describe what the chart tells you.

63	67	69	66	57	59	64	65	62	68	66	59
54	51	54	57	60	64	61	58	54	53	50	54

2.57 Xr02-57 The United States spends more money on health care than any other country. To gauge how fast costs are increasing the following table was produced, which lists the total health care expenditures in the United States annually for 1993 to 2003 (costs are in $billions).

a. Graphically present these data.
b. Use the data in CPI-Annual to remove the effect of inflation. Graph the results and describe your findings.

Year	Total National Health Expenditure	Year	Total National Health Expenditure
1993	899	1999	1220
1994	948	2000	1310
1995	993	2001	1425
1996	1039	2002	1559
1997	1093	2003	1679
1998	1150		

Source: U.S. Centers for Medicare and Medical Services, *Statistical Abstract of the United States*, 2006, Table 119.

2.58 Xr02-58 The number of earned degrees (thousands) for males and females is listed below for the years 1987 to 2003. Graph both sets of data. What do the graphs tell you?

Year	Female	Male	Year	Female	Male
1987	941	882	1996	1254	994
1988	954	881	1997	1290	998
1989	987	886	1998	1305	993
1990	1036	904	1999	1331	992
1991	1098	927	2000	1369	1016
1992	1147	961	2001	1392	1024
1993	1181	986	2002	1442	1052
1994	1211	995	2003	1518	1103
1995	1222	996			

Source: U.S. National Center for Education Statistics, *Statistical Abstract of the United States*, 2006, Table 286.

2.59 Xr02-59 The number of property crimes (burglary, larceny, theft, car theft) (in thousands) for the years 1988 to 2003 are listed next. Draw a line chart and interpret the results.

Year	Number	Year	Number
1988	12,357	1996	11,805
1989	12,605	1997	11,558
1990	12,655	1998	10,952
1991	12,961	1999	10,208
1992	12,506	2000	10,183
1993	12,219	2001	10,437
1994	12,132	2002	10,455
1995	12,604	2003	10,436

Source: U.S. Federal Bureau of Investigation, *Statistical Abstract of the United States*, 2006, Table 293.

2.60 Xr02-60 Refer to Exercise 2.59. Another way of measuring the number of property crimes is to calculate the number of crimes per 100,000 of population. This allows us to remove the effect of the increasing population. Graph these data and interpret your findings.

Year	Number	Year	Number
1988	5054	1996	4451
1989	5107	1997	4316
1990	5073	1998	4053
1991	5140	1999	3744
1992	4904	2000	3618
1993	4740	2001	3658
1994	4660	2002	3631
1995	4591	2003	3588

Source: U.S. National Center for Education Statistics, *Statistical Abstract of the United States*, 2006, Table 286.

2.61 Xr02-61 The gross domestic product (GDP) is the sum total of the economic output of a country. It is an important measure of the wealth of a country. The following table lists the year and the GDP in billions of current dollars.

a. Graph the GDP. What have you learned?
b. Use the data in CPI-Annual to compute the per capita GDP in constant 1982–1984 dollars. Graph the results and describe your findings.

1980	1981	1982	1983	1984	1985	1986
2,790	3,128	3,255	3,537	3,933	4,220	4,463
1987	**1988**	**1989**	**1990**	**1991**	**1992**	**1993**
4,739	5,104	5,484	5,803	5,996	6,338	6,657
1994	**1995**	**1996**	**1997**	**1998**	**1999**	**2000**
7,072	7,398	7,817	8,304	8,747	9,268	9,817
2001	**2002**	**2003**	**2004**	**2005**	**2006**	
10,128	10,470	10,961	11,712	12,456	13,247	

Source: Statistical Abstract of the United States, 2006, Table 657.

2.62 Xr02-62 The average daily U.S. oil consumption and production (thousands of barrels) are shown for the years 1973 to 2005. Use a graphical technique to describe these figures. What does the graph tell you?

Year	Consumption	Production
1973	17,318	9209
1974	16,655	8776
1975	16,323	8376
1976	17,460	8132
1977	18,443	8245
1978	18,857	8706
1979	18,527	8551
1980	17,060	8597
1981	16,061	8572
1982	15,301	8649
1983	15,228	8689
1984	15,722	8879

(continued)

Year	Consumption	Production
1985	15,726	8972
1986	16,277	8683
1987	16,666	8349
1988	17,284	8140
1989	17,327	7615
1990	16,988	7356
1991	16,710	7418
1992	17,031	7172
1993	17,328	6847
1994	17,721	6662
1995	17,730	6561
1996	18,308	6465
1997	18,618	6452
1998	18,913	6253
1999	19,515	5882
2000	19,699	5822
2001	19,647	5801
2002	19,758	5746
2003	20,034	5682
2004	20,731	5419
2005	20,799	5179

Source: U.S. Department of Energy.

2.63 Xr02-63 Has housing been a hedge against inflation in the last 20 years? To answer this question we produced the following table, which lists the average selling price of one-family homes in all of the United States, the Northeast, Midwest, South, and West for the years 1987 to 2004, as well as the annual CPI. For the entire country and for each area, use a graphical technique to determine whether housing prices stayed ahead of inflation?

Year	All	Northeast	Midwest	South	West	CPI
1987	85,600	133,300	66,000	80,400	113,200	113.6
1988	89,300	143,000	68,400	82,200	124,900	118.3
1989	89,500	127,700	71,800	84,400	127,100	124.0
1990	92,000	126,400	75,300	85,100	129,600	130.7
1991	97,100	129,100	79,500	88,500	135,300	136.2
1992	99,700	128,900	83,000	91,500	131,500	140.3
1993	103,100	129,100	86,000	94,300	132,500	144.5
1994	107,200	129,100	89,300	95,700	139,400	148.2
1995	110,500	126,700	94,800	97,700	141,000	152.4
1996	115,800	127,800	101,000	103,400	147,100	156.9
1997	121,800	131,800	107,000	109,600	155,200	160.5
1998	128,400	135,900	114,300	116,200	164,800	163.0
1999	133,300	139,000	119,600	120,300	173,900	166.6
2000	139,000	139,400	123,600	128,300	183,000	172.2
2001	147,800	146,500	130,200	137,400	194,500	177.1
2002	158,100	164,300	136,000	147,300	215,400	179.9
2003	170,000	190,500	141,300	157,100	234,200	184.0
2004	184,100	220,000	149,000	169,000	265,800	188.9

Source: Statistical Abstract of the United States, 2006, Table 940.

2.64 Xr02-64 How has the size of government changed? To help answer this question we recorded the U.S. federal budget receipts and outlays (billions of current dollars) for the years 1980 to 2005.

a. Use a graphical technique to describe the size of the budget.
b. Calculate the budget deficits and graph these figures.
c. Briefly describe what you have learned.

Year	Receipts	Outlays	Year	Receipts	Outlays
1980	517.1	590.9	1993	1154.4	1409.5
1981	599.3	678.2	1994	1258.6	1461.9
1982	617.8	745.7	1995	1351.8	1515.8
1983	600.6	808.4	1996	1453.1	1560.5
1984	666.5	851.9	1997	1579.3	1601.3
1985	734.1	946.4	1998	1721.8	1652.6
1986	769.2	990.4	1999	1827.5	1701.9
1987	854.4	1004.1	2000	2025.2	1789.1
1988	909.3	1064.5	2001	1991.2	1863.9
1989	991.2	1143.6	2002	1853.2	2011.0
1990	1032.0	1253.2	2003	1782.3	2159.9
1991	1055.0	1324.4	2004	1880.1	2292.2
1992	1091.3	1381.7	2005	2052.8	2479.4

Source: U.S. Office of Management and Budget.

2.65 Refer to Exercise 2.64. Another way of judging the size of budget surplus/deficits is to calculate the deficit as a percentage of GDP. Use the data in Exercises 2.61 and 2.64 to calculate this variable and use a graphical technique to display the results.

2.66 Repeat Exercise 2.65 using the CPI-Annual file to convert all amounts to constant 1982–1984 dollars. Draw a line chart to show these data.

2.67 Xr02-67 Refer to Example 2.9. The following table lists the average gasoline consumption in miles per gallon (mpg) and the average distance (thousands of miles) driven by cars in each of the years 1980 to 2003. [The file contains the average price in December, the annual CPI, fuel consumption, and distance (thousands.)] For each year calculate the inflation-adjusted cost per year of driving. Use a graphical technique to present the results.

Year	Mpg	Distance	Year	Mpg	Distance
1980	13.3	9.5	1989	15.9	10.9
1981	13.6	9.5	1990	16.4	11.1
1982	14.1	9.6	1991	16.9	11.3
1983	14.2	9.8	1992	16.9	11.6
1984	14.5	10.0	1993	16.7	11.6
1985	14.6	10.0	1994	16.7	11.7
1986	14.7	10.1	1995	16.8	11.8
1987	15.1	10.5	1996	16.9	11.8
1988	15.6	10.7	1997	17.0	12.1

(continued)

1998	16.9	12.2	2001	17.1	11.9
1999	17.7	12.2	2002	16.9	12.2
2000	16.9	12.2	2003	17.0	12.2

The following exercises require a computer and software.

2.68 Xr02-68 The monthly value of U.S. exports to Canada (in $millions) and imports from Canada from 1974 to 2006 were recorded. (*Source:* Federal Reserve Economic Data)

a. Draw a line chart of U.S. exports to Canada.
b. Draw a line chart of U.S. imports from Canada.
c. Calculate the trade balance and draw a line chart.
d. What do all the charts reveal?

2.69 Xr02-69 The monthly value of U.S. exports to Japan (in $millions) and imports from Japan from 1974 to 2006 were recorded. (*Source:* Federal Reserve Economic Data)

a. Draw a line chart of U.S. exports to Japan.
b. Draw a line chart of U.S. imports from Japan.
c. Calculate the trade balance and draw a line chart.
d. What do all the charts reveal?

2.70 Xr02-70 The value of the Canadian dollar in U.S. dollars was recorded monthly for the period 1971 to 2006. Draw a graph of these figures and interpret your findings.

2.71 Xr02-71 The value of the Japanese yen in U.S. dollars was recorded monthly for the period 1971 to 2006. Draw a graph of these figures and interpret your findings.

2.72 Xr02-72 The Dow Jones Industrial Average was recorded monthly for the years 1950 to 2006. Use a graph to describe these numbers. (*Source: The Wall Street Journal*)

2.73 Refer to Exercise 2.72. Use the CPI-monthly file to measure the Dow Jones Industrial Average in 1982–1984 constant dollars. What have you learned?

2.5 Describing the Relationship between Two Nominal Variables and Comparing Two or More Nominal Data Sets

In Section 2.2 we presented graphical and tabular techniques used to summarize a set of nominal data. Techniques applied to single sets of data are called **univariate.** There are many situations where we wish to depict the relationship between variables; in such cases **bivariate** methods are required. A **cross-classification table** (also called a **cross-tabulation table**) is used to describe the relationship between two nominal variables. A variation of the bar chart introduced in Section 2.2 is employed to graphically describe the relationship. The same technique is used to compare two or more sets of nominal data.

Tabular Method of Describing the Relationship between Two Nominal Variables

To describe the relationship between two nominal variables, we must remember that we are permitted only to determine the frequency of the values. As a first step we need to produce a cross-classification table, which lists the frequency of each combination of the values of the two variables.

EXAMPLE 2.10

DATA
Xm02-10

Newspaper Readership Survey

In a major North American city there are four competing newspapers: the *Globe and Mail* (*G&M*), *Post*, *Sun*, and *Star*. To help design advertising campaigns, the advertising managers of the newspapers need to know which segments of the newspaper market are reading their papers. A survey was conducted to analyze the relationship between newspapers read and occupation. A sample of newspaper readers was asked to report which newspaper they read: *Globe and Mail* (1) *Post* (2), *Star* (3), *Sun* (4), and to indicate whether they were blue-collar worker (1), white-collar worker (2), or professional (3). The responses are stored on the CD using the codes. Some of the data are listed here.

Reader	Occupation	Newspaper
1	2	2
2	1	4
3	2	1
⋮	⋮	⋮
352	3	2
353	1	3
354	2	3

Determine whether the two nominal variables are related.

SOLUTION

By counting the number of times each of the 12 combinations occurs, we produced Table 2.9.

TABLE **2.9** Cross-Classification Table of Frequencies for Example 2.10

	Newspaper				
Occupation	*G&M*	*Post*	*Star*	*Sun*	Total
Blue collar	27	18	38	37	120
White collar	29	43	21	15	108
Professional	33	51	22	20	126
Total	89	112	81	72	354

If occupation and newspaper are related, there will be differences in the newspapers read among the occupations. An easy way to see this is to convert the frequencies in each row (or column) to relative frequencies in each row (or column). That is, compute the row (or column) totals and divide each frequency by its row (or column) total, as shown in Table 2.10. Totals may not equal 1 because of rounding.

TABLE **2.10** Row Relative Frequencies for Example 2.10

	Newspaper				
Occupation	*G&M*	*Post*	*Star*	*Sun*	Total
Blue collar	.23	.15	.32	.31	1.00
White collar	.27	.40	.19	.14	1.00
Professional	.26	.40	.17	.16	1.00
Total	.25	.32	.23	.20	1.00

EXCEL

There are several methods by which Excel can produce the cross-classification table. We will use and describe the PivotTable in two ways: first, to create the contingency table featuring the counts and second, to produce a table showing the row relative frequencies.

	A	B	C	D	E	F
3	Count of Reader	Newspaper				
4	Occupation	1	2	3	4	Grand Total
5	1	27	18	38	37	120
6	2	29	43	21	15	108
7	3	33	51	22	20	126
8	Grand Total	89	112	81	72	354

	A	B	C	D	E	F
3	Count of Reader	Newspaper				
4	Occupation	1	2	3	4	Grand Total
5	1	0.23	0.15	0.32	0.31	1.00
6	2	0.27	0.40	0.19	0.14	1.00
7	3	0.26	0.40	0.17	0.16	1.00
8	Grand Total	0.25	0.32	0.23	0.20	1.00

INSTRUCTIONS

The data must be stored in (at least) three columns as we've done in Xm02-10. Put the cursor somewhere in the data range.

1. Click **Insert** and **PivotTable.**
2. Make sure that the Table/Range is correct.
3. Drag the Occupation button to the **ROW** section of the box. Drag the Newspaper button to the **COLUMN** section. Drag the Reader button to the **DATA** field. Right-click any number in the table, click **Summarize Data By,** and check **Counts.** To convert to row percentages right-click any number, click **Summarize Data By, More options . . . ,** and **Show values as.** Scroll down and click **% of rows.** (We then formatted the data into decimals.)

MINITAB

Tabulated statistics: Occupation, Newspaper

Rows: Occupation Columns: Newspaper

	1	2	3	4	All
1	27	18	38	37	120
	22.50	15.00	31.67	30.83	100.00
2	29	43	21	15	108
	26.85	39.81	19.44	13.89	100.00
3	33	51	22	20	126
	26.19	40.48	17.46	15.87	100.00
All	89	112	81	72	354
	25.14	31.64	22.88	20.34	100.00

Cell Contents: Count
% of Row

INSTRUCTIONS

1. Type or import the data into two columns. (Open Xm02-10.)
2. Click **Stat, Tables,** and **Cross Tabulation and Chi-square.**
3. Type or use the **Select** button to specify the **Categorical variables: For rows** (Occupation) and **For columns** (Newspaper).
4. Under **Display,** click **Counts** and **Row percents** (or any you wish).

INTERPRET

Notice that the relative frequencies in the second and third rows are similar and that there are large differences between row 1 and rows 2 and 3. This tells us that blue-collar workers tend to read different newspapers from both white-collar workers and professionals and that white-collar workers and professionals are quite similar in their newspaper choice.

Graphing the Relationship between Two Nominal Variables

We have chosen to draw three bar charts, one for each occupation depicting the four newspapers. We'll use Excel and Minitab for this purpose. The manually drawn charts are identical.

EXCEL

There are several ways to graphically display the relationship between two nominal variables. We have chosen two dimensional bar charts for each of the three occupations. The charts can be created from the output of the PivotTable (either counts as we have done) or row proportions.

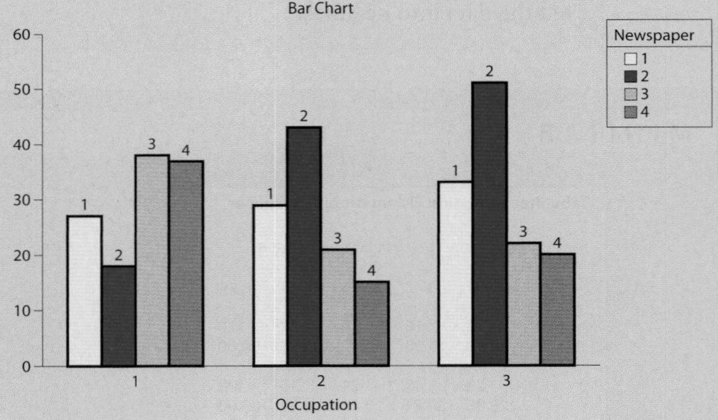

INSTRUCTIONS

From the cross-classification table, click **Insert** and **Column.** You can do the same from any completed cross-classification table.

MINITAB

Minitab can draw bar charts from the raw data.

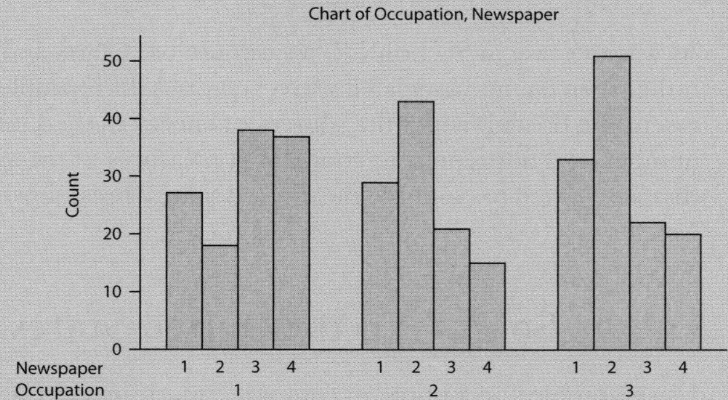

INSTRUCTIONS

1. Click **Graph** and **Bar Chart.**
2. In the **Bars represent** box, specify **Counts of unique values.** Select **Cluster.**
3. In the **Categorical variables** box, type or select the two variables (Newspaper Occupation).

If you or someone else has created the cross-classification table, Minitab can draw bar charts directly from the table.

INSTRUCTIONS

1. Start with a completed cross-classification table such as Table 2.9.
2. Click **Graph** and **Bar Chart.**
3. In the **Bars represent** box, click **Values from a table.** Choose **Two-way table Cluster.**
4. In the **Graph variables** box, **Select** the columns of numbers in the table. In the **Row labels** box, **Select** the column with the categories.

INTERPRET

If the two variables are unrelated, the patterns exhibited in the bar charts should be approximately the same. If some relationship exists, then some bar charts will differ from others.

The graphs tell us the same story as did the table. The shapes of the bar charts for occupations 2 and 3 (White collar and Professional) are very similar. Both differ considerably from the bar chart for occupation 1 (Blue-collar).

Comparing Two or More Sets of Nominal Data

We can interpret the results of the cross-classification table of the bar charts in a different way. In Example 2.10 we can consider the three occupations as defining three different populations. If differences exist between the columns of the frequency distributions (or between the bar charts), we can conclude that differences exist among the three populations. Alternatively, we can consider the readership of the four newspapers as four different populations. If differences exist between the frequencies or the bar charts, we conclude that there are differences among the four populations.

Other Applications of Bar Charts and Pie Charts

As was the case in Section 2.2, we can use bar charts and pie charts to present numbers rather than counts associated with categories. In Example 2.2 we used a pie chart to represent the breakdown of the sources of energy in the United States. In that instance the numbers did not represent counts, but measures of the amount of energy. Suppose now that we wanted to compare the United States' breakdown of energy sources with that of Canada's.

EXAMPLE 2.11

DATA Xm02-02*

Energy Sources in the United States and Canada

Use a graphical technique to compare American and Canadian sources of energy.

	United States	Canada
Nonrenewable Energy Sources		
Coal and coal products	545,258	30,775
Oil	903,440	88,850
Natural gas	517,881	71,477
Nuclear	209,890	20,103
Renewable Energy Sources		
Hydroelectric	18,251	28,541
Solid biomass	52,473	10,424
Other (liquid biomass, geothermal, solar, wind, and tide, wave, and ocean)	20,533	25
Total	2,267,726	250,195

Source: International Energy Association.

SOLUTION

To compare the United States's sources of energy with Canada's, we draw two pie charts as follows.

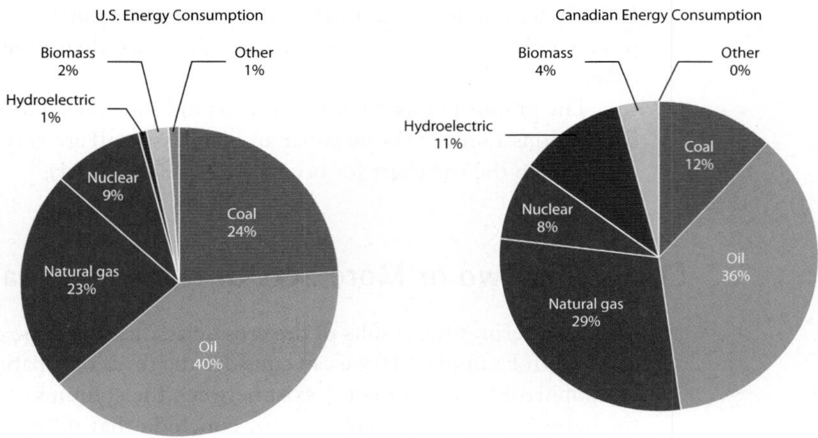

INTERPRET

Both countries are similarly very dependent on oil. The United States consumes 40% of its total energy using oil. Canada's proportion is 36%. The figures for nuclear energy are also similar: the United States 9%, Canada 8%. However, there are substantial differences in the remaining sources. Coal and natural gas make up 24% and 23%, respectively, of American needs and hydroelectric is minimal (1%). Canada uses more natural gas (29%), more hydroelectric (11%), and less coal (12%).

Notice that the same information displayed in the pie charts is available from the table. However, it is often easier to understand the information when it appears in graphical form.

Data Formats

There are several ways to store the data to be used in this section to produce a table and/or a bar or pie chart.

1. The data are in two columns where the first column represents the categories of the first nominal variable and the second column stores the categories for the second variable. Each row represents one observation of the two variables. The number of observations in each column must be the same. Excel and Minitab can produce a cross-classification table from these data. (Additionally, to use Excel's PivotTable there must be a third variable representing the observation number.) This is the way the data for Example 2.10 were stored.

2. The data are stored in two or more columns where each column represents the same variable in a different sample or population. For example, the variable may be the type of undergraduate degree of applicants to an MBA program, and there may be five universities we wish to compare. To produce a cross-classification table, we would have to count the number of observations of each category (undergraduate degree) in each column.

3. The table representing counts in a cross-classification table may have already been created. Alternatively, the statistics practitioner may have a table of numbers rather than counts as was the case with Example 2.11.

We complete this section with the factors that identify the use of the techniques introduced here.

Factors That Identify When to Use a Cross-Classification Table

1. **Objective:** Describe the relationship between two variables; compare two or more sets of data
2. **Data type:** Nominal

EXERCISES

2.74 Xr02-74 Has the educational level of adults changed over the past 15 years? To help answer this question the Bureau of Labor Statistics compiled the following table, which lists the number (1,000) of adults 25 years of age and older who are employed. Use a graphical technique to present these figures.

Educational Level	1992	1995	2000	2004
Less than high school diploma	13,418	11,972	12,486	12,513
High school graduate	37,910	36,692	37,699	37,790
Less than a bachelor's degree	27,048	30,927	33,257	34,412
College/university graduate	28,113	31,149	36,619	40,418

Source: Statistical Abstract of the United States, 2006, Table 580.

2.75 Xr02-75 How do governments spend the tax dollars they collect and has this changed over the past 15 years? The following table displays the amounts spent by the federal, state, and local governments on consumption expenditures and gross investments ($billions). Consumption expenditures are services (such as education). Gross investments consist of expenditures on fixed assets (such as roads, bridges, and highways). Use a graphical technique to present these figures. Have the ways governments spend money changed over the past 15 years?

Level of Government and Type	1990	1995	2000	2004
Federal national defense				
Consumption	308.1	297.3	321.5	477.5
Gross	65.9	51.4	48.8	70.4
Federal nondefense				
Consumption	111.7	143.2	177.8	227.0
Gross	22.6	27.3	30.7	35.0
State and local				
Consumption	544.6	696.1	917.8	1099.7
Gross	127.2	154.0	225.0	274.3

Source: Statistical Abstract of the United States, 2006, Table 419.

2.76 Xr02-15* Refer to Exercise 2.15. The following table displays the energy consumption patterns of Australia and New Zealand. The figures measure the heat content in metric tons (1,000 kilograms) of oil equivalent. Use a graphical technique to display the differences between the sources of energy for the two countries.

	Australia	New Zealand
Nonrenewable Energy Sources		
Coal and coal products	55,385	1,281
Oil	33,185	6,275
Natural gas	20,350	5,324
Nuclear	0	0
Renewable Energy Sources		
Hydroelectric	1,388	1,848
Solid biomass	4,741	805
Other (liquid biomass, geothermal, solar, wind, and tide, wave, and ocean)	347	2,761
Total	115,396	18,294

Source: International Energy Association.

The following exercises require a computer and software.

2.77 Xm02-01* Refer to Example 2.1. The gender (1 = male and 2 = female) of the respondents was also recorded. Use a graphical technique to determine whether the choice of light beers differs between genders.

2.78 Xr02-78 The average loss due to robbery in the United States in 2003 was $1,244 (*Source:* U.S. Federal Bureau of Investigation). Suppose that government agency wanted to know whether the type of robbery differs between 1990 and 2003. A random sample of robbery reports was taken from each of these years and the types recorded using the codes below. Determine whether there are differences in the types of robbery between 1990 and 2003. (Adapted from *Statistical Abstract of the United States*, 2006, Table 307.)

1. Street or highway
2. Gas station
3. Convenience store
4. Residence
5. Bank
6. Other

2.79 Xr02-79 The associate dean of a business school was looking for ways to improve the quality of the applicants to its MBA program. In particular she wanted to know whether the undergraduate degree of applicants differed among her school and the three nearby universities with MBA programs. She sampled 100 applicants of her program and an equal number from each of the other universities. She recorded their undergraduate degree (1 = BA, 2 = BEng, 3 = BBA, 4 = other) as well the university (codes 1, 2, 3, and 4). Use a graphical technique to determine whether the undergraduate degree and the university each person applied to appear to be related.

2.80 Xr02-80 Is there brand loyalty among car owners in their purchases of gasoline? To help answer the question a random sample of car owners was asked to record the brand of gasoline in their last two purchases (1 = Exxon, 2 = Amoco, 3 = Texaco, 4 = Other). Use a graphical technique to formulate your answer.

2.81 Xr02-81 The costs of smoking for individuals, companies for whom they work, and society in general is in the many billions of dollars. In an effort to reduce smoking, various government and nongovernment organizations have undertaken information campaigns about the dangers of smoking. Most of these have been directed at young people. This raises the question: Are you more likely to smoke if your parents smoke? To shed light on the issue a sample of 20- to 40-year-old people were asked whether they smoked and whether their parents smoked. The results are stored the following way:

Column 1: 1 = do not smoke, 2 = smoke
Column 2: 1 = neither parent smoked,
2 = father smoked, 3 = mother smoked,
4 = both parents smoked

Use a graphical technique to produce the information you need.

2.82 Xr02-82 In 2004, 4,456,000 men and 3,694,000 women were unemployed at some time during the year (*Source:* Bureau of Labor Statistics). A statistics practitioner wanted to investigate the reason for that unemployment status and whether the reasons differed by gender. A random sample of people 16 years of age and older was drawn. The reasons given for their status are

1. Lost job
2. Left job
3. Reentrants
4. New entrants

Determine whether there are differences between unemployed men and women in terms of the reasons for unemployment. (Adapted from *Statistical Abstract of the United States*, 2006, Table 612.)

2.83 Xr02-83 In 2004 the total number of prescriptions sold in the United States was 3,274,000,000 (*Source:* National Association of Drug Store Chains). The sales manager of a chain of drug stores wanted to determine whether changes in where the prescriptions were filled had changed. A survey of prescriptions was undertaken in 1995, 2000, and 2005. The year and type of each prescription were recorded using the codes below. Determine whether there are differences between the years. (Adapted from the *Statistical Abstract of the United States*, 2006, Table 126.)

1. Traditional chain store
2. Independent drug store
3. Mass merchant
4. Supermarket
5. Mail order

2.6 Describing the Relationship between Two Interval Variables

Statistics practitioners frequently need to know how two interval variables are related. For example, financial analysts need to understand how the returns of individual stocks and the returns of the entire market are related. Marketing managers need to understand the relationship between sales and advertising. Economists develop statistical techniques to describe the relationship between such variables as unemployment rates and inflation. The technique is called a **scatter diagram.**

To draw a scatter diagram we need data for two variables. In applications where one variable depends to some degree on the other variable, we label the dependent variable Y and the other, called the independent variable, X. For example, an individual's income depends somewhat on the number of years of education. Accordingly, we identify income as the dependent variable and label it Y, and we identify years of education as the independent variable and label it X. In other cases where there is no dependency evident, we label the variables arbitrarily.

EXAMPLE 2.12 Analyzing the Relationship between Price and Size of Houses

A real estate agent wanted to know to what extent the selling price of a home is related to its size. To acquire this information he took a sample of 12 homes that had recently sold, recording the price in thousands of dollars and the size in hundreds of square feet. These data are listed in the accompanying table. Use a graphical technique to describe the relationship between size and price.

Size	23	18	26	20	22	14	33	28	23	20	27	18
Price	315	229	355	261	234	216	308	306	289	204	265	195

SOLUTION

Using the guideline just stated, we label the price of the house Y (dependent variable) and the size X (independent variable). Figure 2.16 depicts the scatter diagram.

FIGURE **2.16** Scatter Diagram for Example 2.12

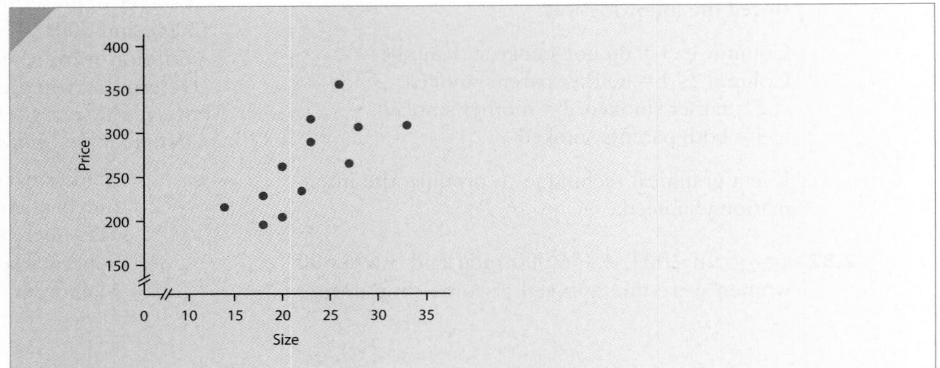

EXCEL

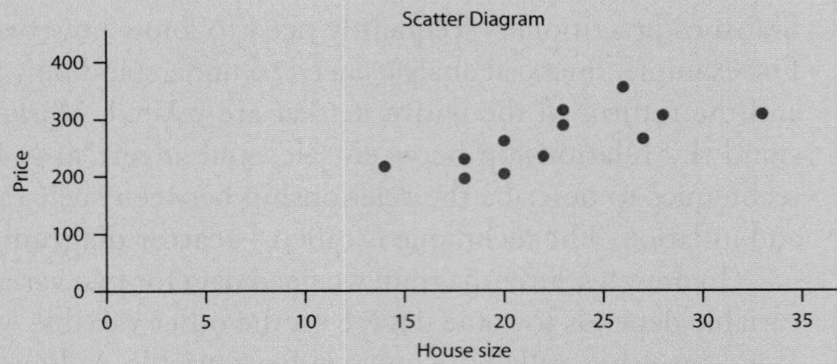

(continued)

INSTRUCTIONS

1. Type or import the data into two adjacent columns. Store variable X in the first column and variable Y in the next column. (Open Xm02-12.)
2. Click **Insert** and **Scatter.**
3. To make cosmetic changes, click **Chart Tools** and **Layout.** (We chose to add titles and remove the gridlines.) If you wish to change the scale, click **Axes, Primary Horizontal axis** or **Primary Vertical axis, More primary Horizontal** or **Vertical Axis Options . . .** and make the changes you want.

MINITAB

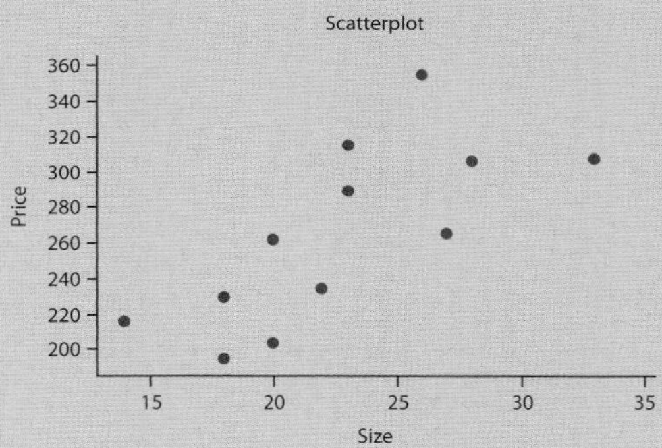

INSTRUCTIONS

1. Type or import the data into two columns. (Open Xm02-12.)
2. Click **Graph** and **Scatterplot . . . .**
3. Click **Simple.**
4. Type or use the **Select** button to specify the variable to appear on the Y-axis (Price) and the X-axis (Size).

INTERPRET

The scatter diagram reveals that, in general, the greater the size of the house, the greater the price. However, there are other variables that determine price. Further analysis may reveal what these other variables are.

Patterns of Scatter Diagrams

As was the case with histograms, we frequently need to describe verbally how two variables are related. The two most important characteristics are the strength and direction of the linear relationship.

Linearity

To determine the strength of the linear relationship we draw a straight line through the points in such a way that the line represents the relationship. If most of the points fall close to the line we say that there is a **linear relationship.** If most of the points appear to be scattered randomly with only a semblance of a straight line, there is no, or at best, a weak linear relationship. Figure 2.17 depicts several scatter diagrams that exhibit various levels of linearity.

In drawing the line freehand we would attempt to draw it so that it passes through the middle of the data. Unfortunately, different people drawing a straight line through the same set of data will produce somewhat different lines. Fortunately, statisticians have produced an objective way to draw the straight line. The method is called the *least squares method,* and it will be presented in Chapter 4 and employed in Chapters 16, 17, and 18.

FIGURE **2.17**
Scatter Diagrams Depicting Linearity

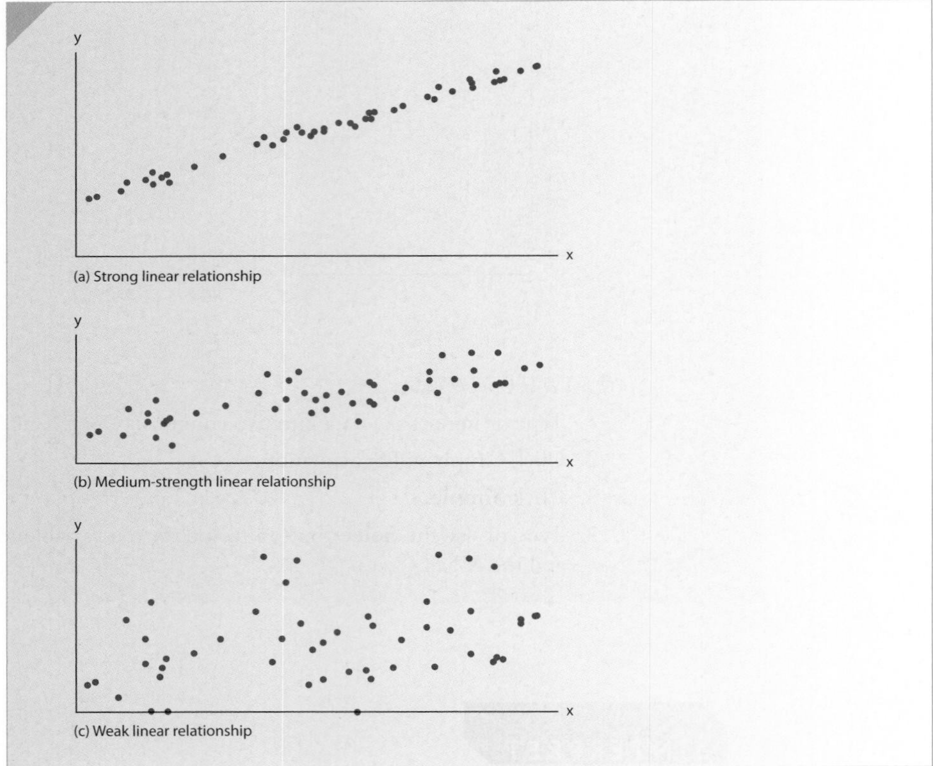

Note that there may well be some other type of relationship, such as a quadratic or exponential one.

Direction

If, in general, when one variable increases and the other does also, we say that there is a **positive linear relationship.** When the two variables tend to move in opposite directions, we describe the nature of their association as a **negative linear relationship.**

(The terms *positive* and *negative* will be explained in Chapter 4.) See Figure 2.18 for examples of scatter diagrams depicting a positive linear relationship, a negative linear relationship, no relationship, and a nonlinear relationship.

FIGURE **2.18**
Scatter Diagrams Describing Direction

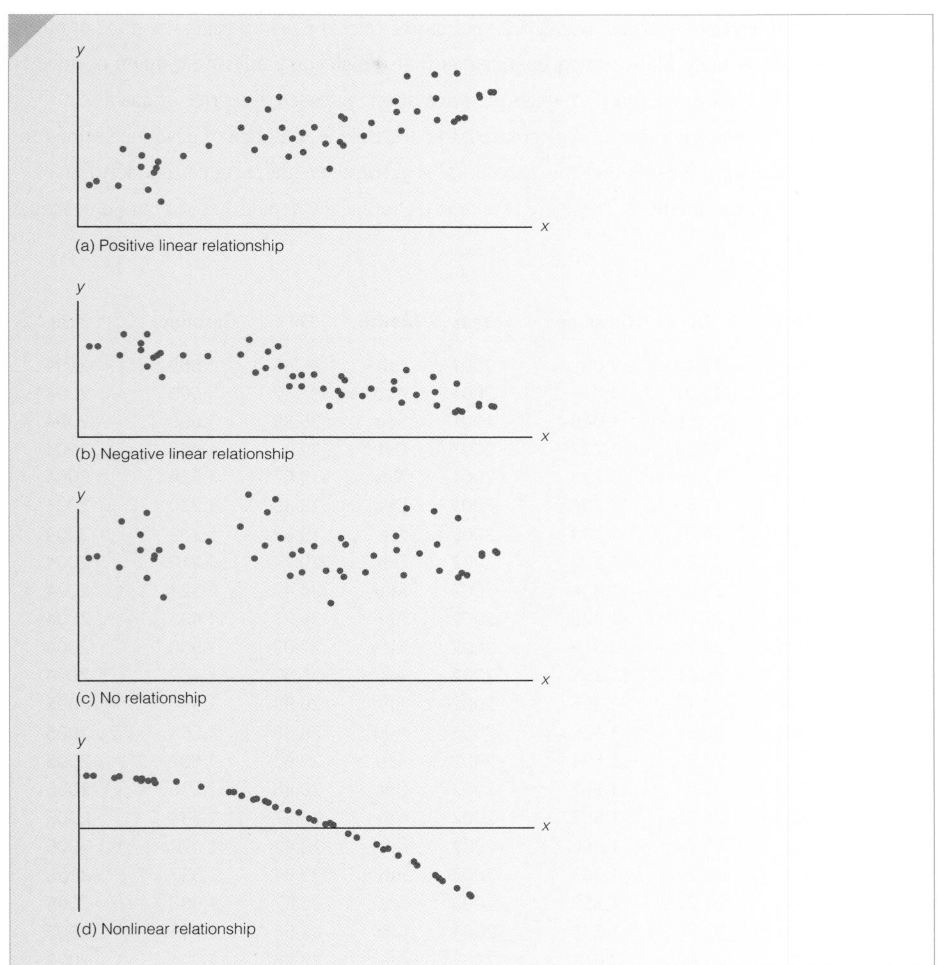

Interpreting a Strong Linear Relationship

In interpreting the results of a scatter diagram it is important to understand that if two variables are linearly related, it does not mean that one is causing the other. In fact, we can never conclude that one variable causes another variable. We can express this more eloquently as

Correlation is not causation.

Now that we know what to look for we can answer the chapter-opening example.

Were Oil Companies Gouging Customers, 1999–2006?: Solution

In December 1998 the average retail price of gasoline was \$0.913 per gallon and the price of oil (West Texas intermediate crude) was \$11.28 per barrel. Over the next 8 years the price of both increased substantially. Many drivers complained that the oil companies were guilty of price gouging. That is, they believed that when the price of oil increased, the price of gas also increased, but when the price of oil decreased, the decrease in the price of gasoline seemed to lag behind. To determine whether this perception is accurate we determined the monthly figures for both commodities. These are listed next. Graphically depict these data and describe the findings.

Year	Month	Oil	Gasoline	Year	Month	Oil	Gasoline	Year	Month	Oil	Gasoline
1999	Jan	12.47	1.031	2001	Jul	26.45	1.565	2004	Jan	34.27	1.635
1999	Feb	12.01	1.014	2001	Aug	27.47	1.509	2004	Feb	34.74	1.715
1999	Mar	14.66	1.048	2001	Sep	25.88	1.609	2004	Mar	36.76	1.809
1999	Apr	17.34	1.232	2001	Oct	22.21	1.442	2004	Apr	36.69	1.875
1999	May	17.75	1.233	2001	Nov	19.67	1.324	2004	May	40.28	2.050
1999	Jun	17.89	1.204	2001	Dec	19.33	1.200	2004	Jun	38.02	2.083
1999	Jul	20.07	1.244	2002	Jan	19.67	1.209	2004	Jul	40.69	1.982
1999	Aug	21.26	1.309	2002	Feb	20.74	1.210	2004	Aug	44.94	1.941
1999	Sep	23.88	1.334	2002	Mar	24.42	1.324	2004	Sep	45.95	1.934
1999	Oct	22.64	1.329	2002	Apr	26.27	1.493	2004	Oct	53.13	2.072
1999	Nov	24.97	1.319	2002	May	27.02	1.508	2004	Nov	48.46	2.053
1999	Dec	26.08	1.353	2002	Jun	25.52	1.489	2004	Dec	43.33	1.926
2000	Jan	27.18	1.356	2002	Jul	26.94	1.496	2005	Jan	46.84	1.866
2000	Feb	29.35	1.422	2002	Aug	28.38	1.508	2005	Feb	47.97	1.960
2000	Mar	29.89	1.594	2002	Sep	29.67	1.507	2005	Mar	54.31	2.107
2000	Apr	25.74	1.561	2002	Oct	28.85	1.535	2005	Apr	53.04	2.325
2000	May	28.78	1.552	2002	Nov	26.27	1.534	2005	May	49.83	2.257
2000	Jun	31.83	1.666	2002	Dec	29.42	1.477	2005	Jun	56.26	2.218
2000	Jul	29.77	1.642	2003	Jan	32.94	1.557	2005	Jul	58.70	2.357
2000	Aug	31.22	1.559	2003	Feb	35.87	1.686	2005	Aug	64.97	2.548
2000	Sep	33.88	1.635	2003	Mar	33.55	1.791	2005	Sep	65.57	2.969
2000	Oct	33.08	1.613	2003	Apr	28.25	1.704	2005	Oct	62.37	2.830
2000	Nov	34.40	1.608	2003	May	28.14	1.587	2005	Nov	58.30	2.387
2000	Dec	28.46	1.544	2003	Jun	30.72	1.558	2005	Dec	59.43	2.230
2001	Jan	29.58	1.525	2003	Jul	30.76	1.567	2006	Jan	65.51	2.359
2001	Feb	29.61	1.538	2003	Aug	31.59	1.671	2006	Feb	61.63	2.354
2001	Mar	27.24	1.503	2003	Sep	28.29	1.771	2006	Mar	62.90	2.444
2001	Apr	27.41	1.617	2003	Oct	30.33	1.646	2006	Apr	69.69	2.801
2001	May	28.64	1.812	2003	Nov	31.09	1.578	2006	May	70.94	2.993
2001	Jun	27.60	1.731	2003	Dec	32.15	1.538	2006	Jun	70.96	2.963

Source: U.S. Department of Energy.

We label the price of gasoline Y and the price of oil X. Figure 2.19 displays the scatter diagram.

FIGURE **2.19** Scatter Diagram for Chapter-Opening Example

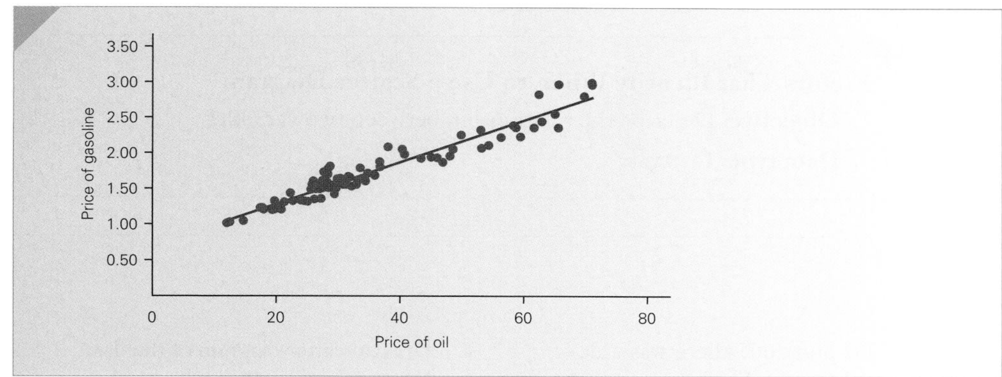

EXCEL

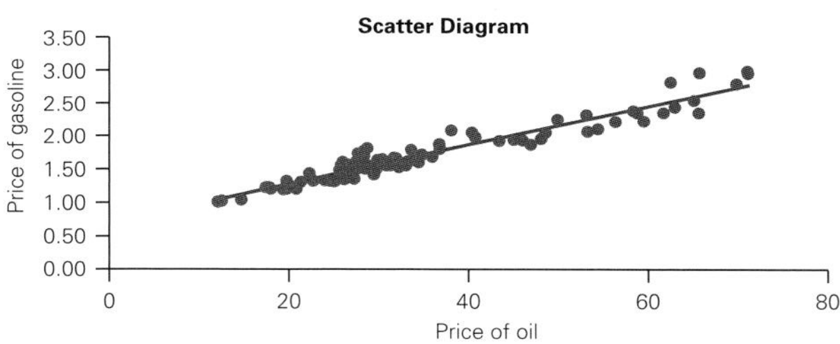

MINITAB

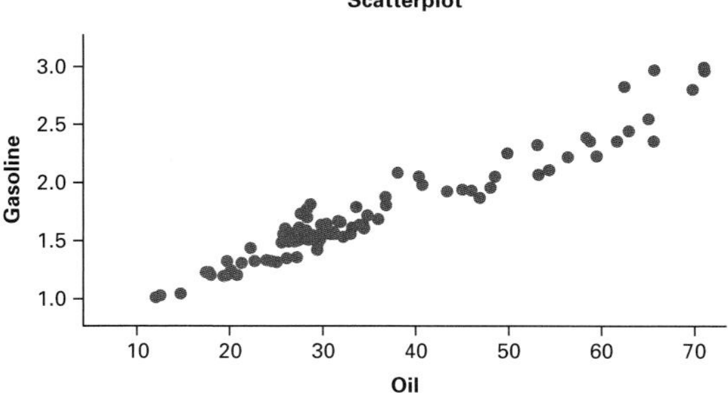

INTERPRET

The scatter diagram reveals that the two prices are strongly related linearly. When the price of oil was below $40, the relationship between the two was stronger than when the price of oil exceeded $40.

We close this section by reviewing the factors that identify the use of the scatter diagram.

Factors That Identify When to Use a Scatter Diagram

1. **Objective:** Describe the relationship between two variables
2. **Data type:** Interval

EXERCISES

2.84 Xr02-84 Between 2002 and 2005 there was a decrease in movie attendance. There are several reasons for this decline. One of them may be the increase in DVD sales. The percentage of U.S. homes with DVD players and the movie attendance (billions) in the United States for the years 2000 to 2005 are shown next. Use a graphical technique to describe the relationship between these variables.

Year	2000	2001	2002	2003	2004	2005
DVD percentage	12	23	37	42	59	74
Movie attendance	1.41	1.49	1.63	1.58	1.53	1.40

Sources: Northern Technology & Telecom Research and Motion Picture Association.

2.85 Xr02-85 Because inflation reduces the purchasing power of the dollar, investors seek investments that will provide higher returns when inflation is higher. It is frequently stated that common stocks provide just such a hedge against inflation. The annual percentage rates of return on common stock and annual inflation rates for a recent 10-year period are listed here.

Year	1	2	3	4	5	6	7	8	9	10
Returns	25	8	6	11	21	−15	12	−1	33	0
Inflation	4.4	4.2	4.1	4.0	5.2	5.0	3.8	2.1	1.7	0.2

a. Use a graphical technique to depict the relationship between the two variables.
b. Does it appear that the returns on common stocks and inflation are linearly related?

2.86 Xr02-86 In a university where calculus is a prerequisite for the statistics course, a sample of 15 students was drawn. The marks for calculus and statistics were recorded for each student. The data are as follows:

Calculus	65	58	93	68	74	81	58	85
Statistics	74	72	84	71	68	85	63	73

Calculus	88	75	63	79	80	54	72
Statistics	79	65	62	71	74	68	73

a. Draw a scatter diagram of the data.
b. What does the graph tell you about the relationship between the marks in calculus and statistics?

2.87 Xr02-87 The cost of repairing cars involved in accidents is one reason that insurance premiums are so high. In an experiment, 10 cars were driven into a wall. The speeds were varied between 2 and 20 mph. The costs of repair were estimated and are listed here. Draw an appropriate graph to analyze the relationship between the two variables. What does the graph tell you?

Speed	2	4	6	8	10	12
Cost of Repair ($)	88	124	358	519	699	816

Speed	14	16	18	20
Cost of Repair ($)	905	1521	1888	2201

2.88 Xr02-88 The growing interest in and use of the Internet have forced many companies into considering ways to sell their products on the Web. Therefore, it is of interest to these companies to determine who is using the Web. A statistics practitioner undertook a study to determine how education and Internet use are connected. She took a random sample of 15 adults (20 years of age and older) and asked each to report the years of education they had completed and the number of hours of Internet use in the previous week. These data follow.

Education	11	11	8	13	17	11	11	11
Internet use	10	5	0	14	24	0	15	12

Education	19	13	15	9	15	15	11
Internet use	20	10	5	8	12	15	0

a. Employ a suitable graph to depict the data.
b. Does it appear that there is a linear relationship between the two variables? If so, describe it.

2.89 Xr02-89 A statistics professor formed the opinion that students who handed in quiz and exams early outperformed students who handed in their papers later. To develop data to decide whether her opinion is valid she recorded the amount of time (in minutes) taken by students to submit their midterm tests (time limit 90 minutes) and the subsequent mark for a sample of 12 students.

Time	90	73	86	85	80	87	90	78	84	71	72	88
Mark	68	65	58	94	76	91	62	81	75	83	85	74

The following exercises require a computer and software.

2.90 Xr02-90 In an attempt to determine the factors that affect the amount of energy used, 200 households were analyzed. In each the number of occupants and the amount of electricity used were measured.
a. Draw a graph of the data.
b. What have you learned from the graph?

2.91 Xr02-91 Many downhill skiers eagerly look forward to the winter months and fresh snowfalls. However, winter also entails cold days. How does the temperature affect skiers' desire? To answer this question, a local ski resort recorded the temperature for 50 randomly selected days and the number of lift tickets they sold. Use a graphical technique to describe the data and interpret your results.

2.92 Xr02-92 One general belief held by observers of the business world is that taller men earn more money than shorter men. In a University of Pittsburgh study, 250 MBA graduates, all about 30 years old, were polled and asked to report their height (in inches) and their annual income (to the nearest $1,000).
a. Draw a scatter diagram of the data.
b. What have you learned from the scatter diagram?

2.93 Xr02-93 Do chief executive officers (CEOs) of publicly traded companies earn their compensation? Every year the *National Post Business* magazine attempts to answer the question by reporting the CEO's annual compensation ($1,000), the profit (or loss) ($1,000), and the three-year share return (%) for the top 50 Canadian companies. Use a graphical technique to answer the question.

2.94 Xr02-94 Are younger workers less likely to stay with their jobs? To help answer this question a random sample of workers was selected. Each was asked to report their age and how many months they have been employed with their current employer. Use a graphical technique to summarize these data. (Adapted from *Statistical Abstract of the United States*, 2006, Table 599.)

2.95 Xr02-95 A very large contribution to profits for a movie theater is the sale of popcorn, soft drinks, and candy. A movie theater manager speculated that the longer the time between showings of a movie, the greater the sales of concession items. To acquire more information the manager conducted an experiment. For a month he varied the amount of time between movie showings and calculated the sales. Use a graphical technique to help the manager determine whether longer time gaps produce higher concession stand sales.

CHAPTER SUMMARY

Descriptive statistical methods are used to summarize data sets so that we can extract the relevant information. In this chapter we presented graphical techniques.

Bar charts, pie charts, and frequency distributions are employed to summarize single sets of nominal data. Because of the restrictions applied to this type of data, all that we can show is the frequency and proportion of each category.

Histograms are used to describe a single set of interval data. Statistics practitioners examine several aspects of the shapes of histograms. These are symmetry, number of modes, and its resemblance to a bell shape.

We described the difference between time-series data and cross-sectional data. Time series are graphed by line charts.

To describe the relationship between two nominal variables, we produce cross-classification tables and bar charts. To analyze the relationship between two interval variables, we draw a scatter diagram. We look for the direction and strength of the linear relationship.

IMPORTANT TERMS

Variable 13
Values 13
Data 13
Datum 13
Interval 13
Quantitative 13
Numerical 13
Nominal 13
Qualitative 13
Categorical 13
Ordinal 14
Frequency distribution 18
Bar chart 21
Pie chart 21
Classes 32
Histogram 32
Symmetric 36
Positively skewed 36
Negatively skewed 36
Modal class 36
Unimodal histogram 36
Bimodal histogram 36
Stem-and-leaf display 42
Depths 44
Relative frequency distribution 44
Cumulative relative frequency distribution 45
Ogive 45
Credit scorecard 49
Cross-sectional data 49
Time-series data 49
Line chart 50
Univariate 57
Bivariate 57
Cross-classification table 57
Cross-tabulation table 57
Scatter diagram 65
Linear relationship 68
Positive linear relationship 68
Negative linear relationship 68

COMPUTER OUTPUT AND INSTRUCTIONS

Graphical Technique	Excel	Minitab
Bar chart	22	23
Pie chart	22	23
Histogram	33	34
Stem-and-leaf display	43	44
Ogive	46	N/A
Line chart	51	52
Scatter diagram	66	67

CHAPTER EXERCISES

The following exercises require a computer and software.

2.96 Xr02.96 Several years ago the Barnes Exhibit toured major cities all over the world, with millions of people flocking to see it. Dr. Albert Barnes was a wealthy art collector who accumulated a large number of impressionist masterpieces; the total exceeds 800 paintings. When Dr. Barnes died in 1951, he stated in his will that his collection was not to be allowed to tour. However, because of the deterioration of the exhibit's home near Philadelphia, a judge ruled that the collection could go on tour to raise enough money to renovate the building. Because of the size and value of the collection, it was predicted (correctly) that in each city a large number of people would come to view the paintings. Because space was limited, most galleries had to sell tickets that were valid at one time (much like a play). In this way, they were able to control the number of visitors at any one time. To judge how many people to let in at any time, it was necessary to know the length of time people would spend at the exhibit; longer times would dictate smaller audiences; shorter times would allow for the sale of more tickets. The manager of a gallery that will host the exhibit realized her facility can comfortably and safely hold about 250 people at any one time. Although the demand will vary throughout the day and from weekday to weekend, she believes that the demand will not drop below 500 at any time. To help make a decision about how many tickets to sell she acquired the amount of time a sample of 400 people spent at the exhibit from another city. What ticket procedure should the museum management institute?

2.97 Xr02-97 Do better golfers play faster than poorer ones? To determine whether a relationship exists, a sample of 125 foursomes was selected. Their total scores and the amount of time taken to complete the 18 holes were recorded. Graphically depict the data and describe what they tell you about the relationship between score and time.

2.98 Xr02-98 Most car rental companies keep their cars for about a year and then sell them to used car dealerships. Suppose one company decided to sell the used cars themselves. Since most used car buyers make their decision on what to buy and how much to spend based on the car's odometer reading, this would be an important issue for the car rental company. To develop information about the mileage shown on the company's rental cars the general manager took a random sample of 658 customers and recorded the average number of miles driven per day. Use a graphical technique to display these data.

2.99 Xr02-99 The monthly values of one British pound measured in American dollars since 1987 were recorded. Produce a graph that shows how the exchange rate has varied over the past 20 years. (*Source:* Federal Reserve Economic Data)

2.100 Xr02-100 How do income taxes vary from country to country? A statistics practitioner recorded the income tax rate for those individuals in the highest income group in several countries. (The figures refer to the federal income tax only in the United States and in Switzerland.) Use a graphical technique to present these numbers. (*Source:* Pricewaterhousecoopers)

2.101 Xr02-101 Which Internet search engines are the most popular? A survey undertaken by the *Financial Post* (May 14, 2004) asked random samples of Americans and Canadians that question. The responses were

1. Google
2. Microsoft (MSN)
3. Yahoo
4. Other

Use a graphical technique that compares the proportions of Americans' and Canadians' use of search engines.

2.102 Xr02-102 Is airline travel becoming safer? To help answer this question, a student recorded the number of fatal accidents and the number of deaths that occurred in the years 1987 to 2004 for scheduled airlines. Use a graphical method to answer the question. (*Source: Statistical Abstract of the United States*, 2006, Table 1056)

2.103 Xr02-103 In Chapters 16, 17, and 18 we introduce regression analysis, which addresses the relationships among variables. One of the first applications of regression analysis was to analyze the relationship between the heights of fathers and sons. Suppose that a sample of 80 fathers and sons was drawn. The heights of the fathers and of the adult sons were measured.

a. Draw a scatter diagram of the data. Draw a straight line that describes the relationship.
b. What is the direction of the line?
c. Does it appear that there is a linear relationship between the two variables?

2.104 Xr02-104 Most publicly traded companies have boards of directors. The rate of pay varies considerably. A survey was undertaken by the *Globe and Mail* (February 19, 2001) wherein 100 companies were surveyed and asked to report how much their directors were paid annually. Use a graphical technique to present these data.

2.105 Xr02-105 There are several ways to teach applied statistics. The most popular approaches are

1. Emphasize manual calculations.
2. Use a computer combined with manual calculations.
3. Use a computer exclusively with no manual calculations.

A survey of 100 statistics instructors asked each to report his or her approach. Use a graphical method to extract the most useful information about the teaching approaches.

2.106 Xr02-106 Refer to exercise 2.104. In addition to reporting the annual payment per director, the survey recorded the number of meetings last year. Use a graphical technique to summarize and present these data.

2.107 Xr02-107 One hundred students who had reported that they use their computers for at least 20 hours per week were asked to keep track of the number of crashes their computers incurred during a 12-week period. Using an appropriate statistical method, summarize the data. Describe your findings.

2.108 Xr02-108 The Red Lobster Restaurant chain conducts regular surveys of its customers to monitor the performance of individual restaurants. One of the questions asks customers to rate the overall quality of their last visit. The listed responses are Poor (1), Fair (2), Good (3), Very good (4), Excellent (5). The survey also asks respondents whether their children accompanied them (1 = yes, 2 = no) to the restaurant. Graphically depict these data and describe your findings.

2.109 Xr02-109 Studies of twins may reveal more about the "nature or nurture" debate. The issue being debated is whether nature or the environment has more of an effect on individual traits such as intelligence. Suppose that a sample of identical twins was selected and their IQs measured. Use a suitable graphical technique to depict the data, and describe what it tells you about the relationship between the IQs of identical twins.

2.110 Xr02-110 An increasing number of consumers prefer to use debit cards in place of cash or credit cards. To analyze the relationship between the amounts of purchases made with debit and credit cards, 240 people were interviewed and asked to report the amount of money spent on purchases using debit cards and the amount spent using credit cards during the last month. Draw a graph of the data and summarize your findings.

2.111 Xr02-111 The monthly values of one Australian dollar measured in American dollars since 1991 were recorded. Convert the data into the values of one American dollar measured in Australian dollars, and produce a graph that shows how the exchange rate has varied over the past 16 years. (*Source:* Federal Reserve Economic Data)

2.112 Xr02-112 The value of monthly U.S. exports to Mexico and imports from Mexico (in $millions) since 1988 were recorded. (*Source:* Federal Reserve Economic Data)

a. Draw a chart that depicts exports.
b. Draw a chart that exhibits imports.
c. Compute the trade balance and graph these data.
d. What do these charts tell you?

2.113 Xr02-113 A sample of 200 people who had purchased food at the concession stand at Yankee Stadium was asked to rate the quality of the food. The responses are

1. Poor
2. Fair
3. Good
4. Very good
5. Excellent

Draw a graph that describes the data. What does the graph tell you?

2.114 Xr02-114 The Wilfrid Laurier University bookstore conducts annual surveys of its customers. One question asks respondents to rate the prices of textbooks. The wording is, "The bookstore's prices of textbooks are reasonable." The responses are

1. Strongly disagree
2. Disagree
3. Neither agree nor disagree
4. Agree
5. Strongly agree

The responses for a group of 115 students were recorded. Graphically summarize these data and report your findings.

2.115 Xr02-115 Gold and precious metals have traditionally been considered a hedge against inflation. If this is true, we would expect that a fund made up of precious metals (gold, silver, platinum, and others) would have a strong positive relationship with the inflation rate. To see whether this is true, a statistics practitioner computed the annual inflation rate and the annual return on a precious metal fund (the fund is based on the precious metals subindex) for the years 1976 to 2003. Use a graphical technique to determine the nature of the relationship. What does the graph tell you? (*Source:* U.S. Treasury and Bridge Commodity Research Bureau)

The following exercises are based on data sets that include additional data referenced in previously presented examples and exercises.

2.116 In addition to the data discussed below in Examples 2.6 and 2.7 the professor listed the midterm mark. Conduct an analysis of the relationship between final exam mark and midterm mark in each course. What does this analysis tell you?

2.117 Examples 2.6 and 2.7 listed the final marks in the business statistics course and the mathematical statistics course. The professor also provided the final marks in the first-year required calculus course. Graphically describe the relationship between calculus and statistics marks. What information were you able to extract?

CASE 2.1 The Question of Global Warming

DATA
C02-01a
C02-01b
C02-01c
C02-01d

In the last part of the twentieth century scientists developed the theory that the planet was warming and the primary cause was the increasing amounts of carbon dioxide (CO_2), which is the product of burning oil, natural gas, and coal (fossil fuels). Although many climatologists believe in the so-called greenhouse effect, there are many others who do not subscribe to this theory. There are three critical questions that need to be answered in order to resolve the issue.

1. Is the earth actually warming? To answer this question we need accurate temperature measurements over a large number of years. But how do we measure the temperature before the invention of accurate thermometers? Moreover, how do we go about measuring the earth's temperature even with accurate thermometers?
2. If the planet is warming is there a man-made cause or is it natural fluctuation? The temperature of earth has increased and decreased many times in its long history. We've had higher temperatures and we've had lower temperatures including various ice ages. In fact, a period called the "little ice age" ended around the middle to end of the nineteenth century. Then the temperature rose until about 1940, at which point it decreased until 1975. In fact, an April 28, 1975 *Newsweek* article discussed the possibility of global cooling, which seemed to be the consensus among scientists.
3. If the planet is warming, is CO_2 the cause? There are greenhouse gases in the atmosphere, without which the earth would be considerably colder. These gases include methane, water vapor, and carbon dioxide. All occur naturally in nature. Carbon dioxide is vital to our life on earth because it is necessary for growing plants. The amount of CO_2 produced by fossil fuels is a relatively small proportion of all the CO_2 in the atmosphere.

The generally accepted procedure is to record monthly temperature anomalies. To do so, we calculate the average for each month over many years. We then calculate any deviations between the latest month's temperature reading and its average. A positive anomaly would represent a month's temperature that is above the average. A negative anomaly indicates a month where the temperature is less than the average. One key question is how we measure the temperature.

There are several ways to measure temperature. One method places thermometers at various places around the globe and then averages them. The problem is where do you place the thermometers? A complicating problem is that, on average, large cities have higher temperatures than the surrounding areas because large cities tend to trap the heat in pavements and buildings. Another somewhat more objective source of temperature is through satellite readings. Of course, we can only go back to the late 1970s when using satellites for temperature measurements began.

We've chosen to supply you with three sets of temperatures. C02-01a is the satellite-measured temperature anomalies. C02-01b stores the temperature anomalies from 1880 to 2006 produced by the National Climate Data Center (NCDC), which is affiliated with National Oceanic and Atmospheric Administration (NOAA) and C02-01c stores the data produced by Goddard Institute for Space Studies (GISS), which is part of the National Aeronautics and Space Administration (NASA).

The best measures of CO_2 levels in the atmosphere come from the Mauna Loa Observatory in Hawaii, which started measuring this variable in December 1958 and continues to do so. However, attempts to estimate CO_2 levels prior to 1958 are as controversial as the methods used to estimate temperatures. These techniques include taking ice-core samples from the arctic and measuring the amount of carbon dioxide trapped in the ice from which estimates of atmospheric CO_2 are produced. To avoid this controversy we will use the Mauna Loa Observatory numbers only. These data are stored in file C02-01d. (Note that some of the data are missing.)

a. Use all three temperature anomalies and whichever techniques you wish to determine whether there is global warming.
b. Use a graphical technique to determine whether there is a relationship between temperature anomalies and CO_2 levels.

CASE 2.2 Survey of Graduates

DATA C02-02

A survey of the business school graduates undertaken by a university placement office asked, among other questions, in which area each person was employed. The areas of employment are

1. Accounting
2. Finance
3. General management
4. Marketing/Sales
5. Other

Additional questions were asked and the responses were recorded in the following way:

Column	Variable
1	Identification number
2	Area
3	Gender (1 = female, 2 = male)
4	Job satisfaction (4 = very, 3 = quite, 2 = little, 1 = none)
5	Number of weeks job searching
6	Salary ($thousands)

The placement office wants to know the following:

a. Do female and male graduates differ in their areas of employment? If so, how?

b. Are area of employment and job satisfaction related?

c. Are salary and number of weeks needed to land the job related?

3

© John Foxx/Stockbyte/Getty Images

ART AND SCIENCE OF GRAPHICAL PRESENTATIONS

Napoleon's Invasion of and Retreat from Russia

On June 21, 1812, the French Army led by Napoleon Bonaparte invaded Russia. The military campaign was a disaster; the army was virtually annihilated. A time-series chart created by Charles Joseph Minard (1781–1870), a French engineer, is considered to be one of the best graphs ever created. The chart is effective because it depicts five variables clearly and succinctly. The variables are

1. Size of the army invading
2. Size of the army retreating
3. Location on the map
4. Temperature
5. Dates

© Ernest Meissonier/Erich Lessing/ Art Resource, NY

On page 85 we provide a copy of this chart.

INTRODUCTION

In Chapter 2, we introduced a number of graphical techniques. The emphasis was on how to construct each one manually and how to command the computer to draw them. In this chapter we discuss how to use graphical techniques effectively. We introduce the concept of **graphical excellence,** which is a term we apply to techniques that are informative and concise and that impart information clearly to their viewers. Section 3.1 discusses how to achieve graphical excellence using the methods introduced in Chapter 2. In Section 3.2, we discuss an equally important concept, graphical integrity. We demonstrate how some people use graphs and charts to purposely or inadvertently mislead readers.

3.1 GRAPHICAL EXCELLENCE

Graphical excellence is achieved when the following characteristics apply.

1. **The graph presents large data sets concisely and coherently.** Graphical techniques were created to summarize and describe large data sets. Small data sets are easily summarized with a table. One or two numbers can best be presented in a sentence.
2. **The ideas and concepts the statistics practitioner wants to deliver are clearly understood by the viewer.** The chart is designed to describe what would otherwise be described in words. An excellent chart is one that can replace a thousand words and still be clearly comprehended by its readers.
3. **The graph encourages the viewer to compare two or more variables.** Graphs displaying only one variable provide very little information. Graphs are often best used to depict relationships between two or more variables or to explain how and why the observed results occurred.
4. **The display induces the viewer to address the substance of the data and not the form of the graph.** The form of the graph is supposed to help present the substance. If the form replaces the substance, the chart is not performing its function.
5. **There is no distortion of what the data reveal.** You cannot make statistical techniques say whatever you like. A knowledgeable reader will easily see through distortions and deception. This is such an important topic that we devote Section 3.2 to its discussion.

Edward Tufte, professor of statistics at Yale University, summarized graphical excellence this way:

1. *Graphical excellence is the well-designed presentation of interesting data—a matter of substance, of statistics, and of design.*
2. *Graphical excellence is that which gives the viewer the greatest number of ideas in the shortest time with the least ink in the smallest space.*
3. *Graphical excellence is nearly always multivariate.*
4. *And graphical excellence requires telling the truth about the data.*

In attempting to demonstrate what constitutes excellence, we searched through newspapers and magazines. Unfortunately, many of the charts that appear in newspapers tend to be quite poor. However, it must be understood that the purpose of the chart is to catch the reader's eye and, to a lesser degree, to amuse. An excellent source of such charts is *USA Today.* Here are some examples.

Examples

Graphical techniques should be used when there is a large amount of data. In general, small data sets can be presented in tabular form. Examine Figure 3.1, which is a bar chart depicting the number of visitors in 1994 to Disney theme parks around the world. Does the chart provide the reader with any more information than Table 3.1 does? From both you can see that Tokyo Disneyland drew the most visitors—about 16 million—whereas the others drew between 8 and 11.2 million. The bar chart is completely unnecessary for two reasons. First, there are only six numbers represented; a data set this small does not need a graphical display. Second, there is no analysis associated with the attendance figures to explain why Tokyo Disneyland outdrew the others or how these figures are related to other variables, such as profits or sales.

FIGURE **3.1**

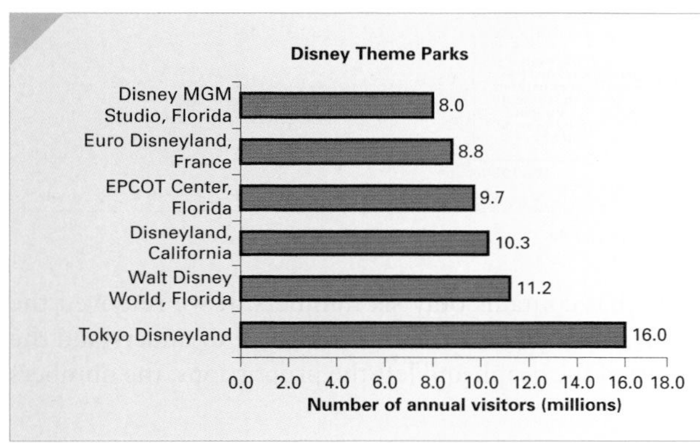

Source: USA TODAY, 1995.

TABLE **3.1** Number of People Visiting Disney Theme Parks in 1994

Park	Number of Visitors (in millions)
Tokyo Disneyland	16.0
Walt Disney World, Florida	11.2
Disneyland, California	10.3
EPCOT Center, Florida	9.7
Euro Disneyland, France	8.8
Disney-MGM Studios, Florida	8.0

Figure 3.2 displays a pie chart that contains only three numbers. The artist attempted to make the chart more interesting by modifying it to look like back-to-back CEOs. It undoubtedly caught the reader's eye but provided no information that the reader could use.

FIGURE **3.2**

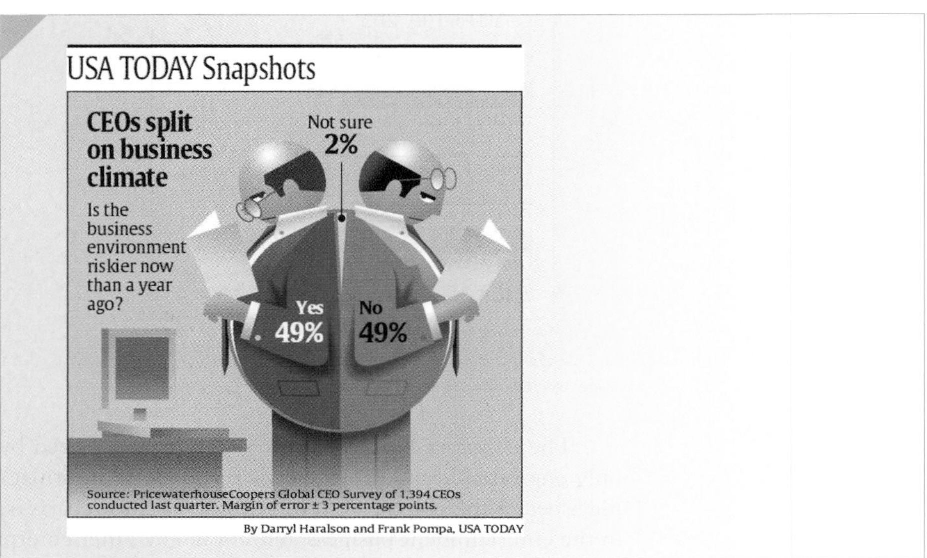

Source: USA TODAY. Copyright March 4, 2004. Reprinted with permission.

Figure 3.3 is a chart that contains only seven numbers. A table would easily suffice.

FIGURE **3.3**

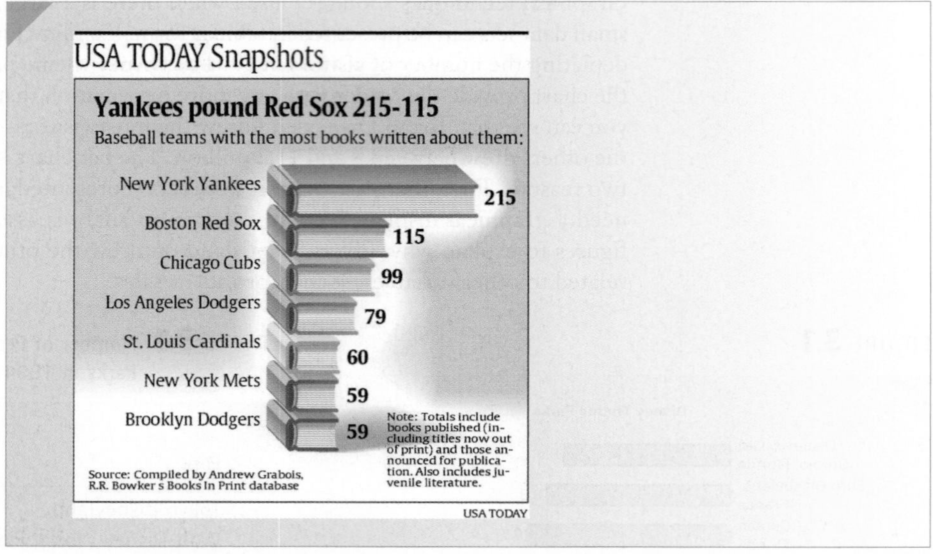

Source: USA TODAY. Copyright March 3, 2004. Reprinted with permission.

Figure 3.4 displays a bar chart that contains only six numbers. If we removed the numbers and left the bar chart alone, it would be virtually impossible to understand the chart. On the other hand, if we omitted the graph and left the proportions, the numbers would speak for themselves.

FIGURE **3.4**

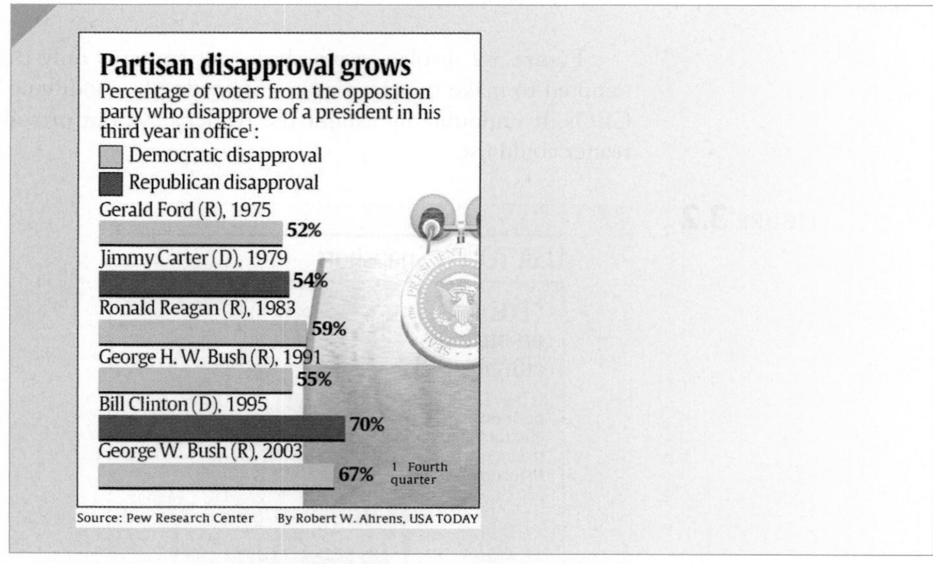

Source: USA TODAY. Copyright March 3, 2004. Reprinted with permission.

The charts in Figures 3.1 to 3.4 are characterized by small data sets that present only one variable and provide little or no useful information. In fact, the only interesting aspect is the graphic design. The goal of these charts is merely to entertain, laudable in the entertainment business, but not in any other enterprise. Although you will likely be exposed to thousands of these charts, you should not think of them as the standard form to be used in reports and presentations.

FIGURE **3.5**

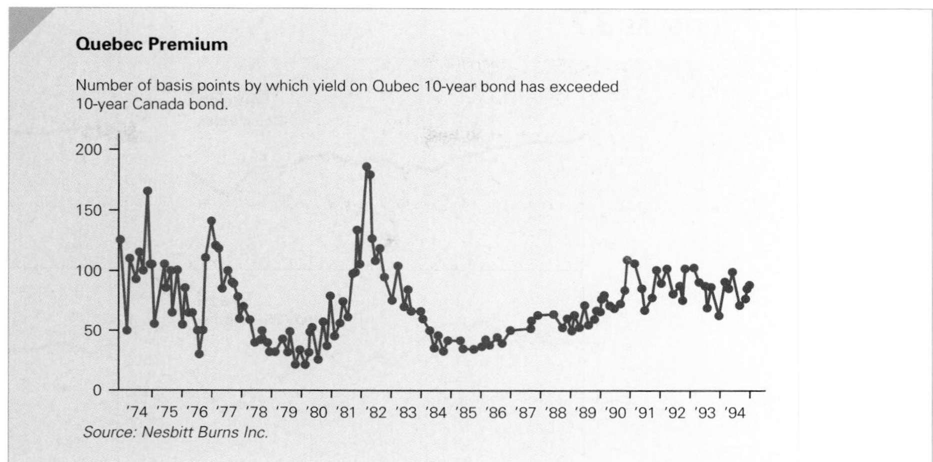

Line charts representing time-series data are often seen in the financial sections of newspapers. However, they are often devoid of substance. For example, Figure 3.5 is a time line of the number of basis points by which Quebec 10-year bonds exceed 10-year Canadian bonds. Since the late 1960s, the province of Quebec has threatened to secede from Canada, which has made many investors nervous about investing in Quebec. The chart depicts the degree of concern among investors—the greater the concern, the greater the premium that must be offered by the Quebec government. In its present form, the graph tells us very little. It could be improved by adding the line chart of one or more related variables. Examples of related variables include Quebec's budget deficits, unemployment rates, and survey results showing support for separation. The authors could also indicate the dates of Quebec's provincial elections. They could also attempt to explain the spikes in 1975, 1977, and 1982 and the steady increase since 1984.

Contrast Figure 3.5 with Figure 3.6, which plots a consumer sentiment index in the United States from 1950 to 1994. The index measures how people feel about their financial prospects. The score in 1966 was arbitrarily set equal to 100. In addition to the scores, we also see the periods during which the U.S. economy underwent recessions. The years in which a new president was inaugurated as well as other key events also appear on the chart. Examining the chart provides rich details about the factors that affect Americans' perceptions of their financial circumstances. For example, recessionary periods mostly coincide with downturns in the index. From the early 1960s to 1980, there was a general downward drift. Historians would agree that this was a troubled time in the United States. The period started with the assassination of President John Kennedy,

FIGURE **3.6**

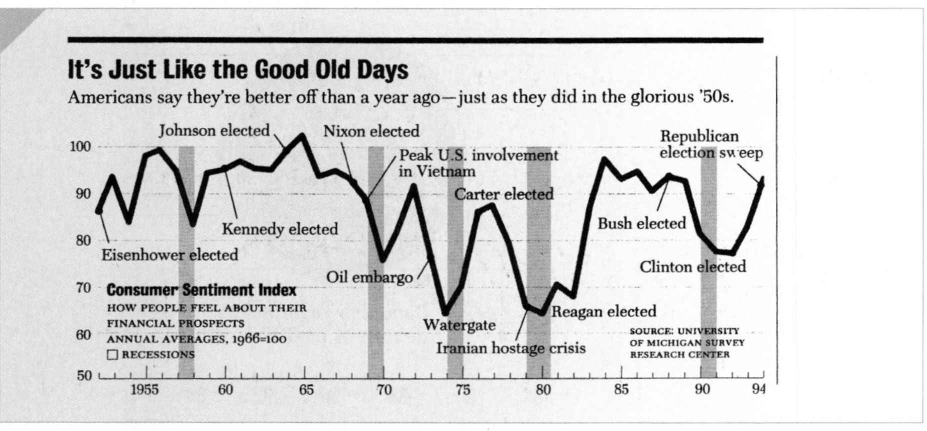

Source: © 1995 Newsweek, Inc. All rights reserved. Reprinted by permission.

FIGURE **3.7**

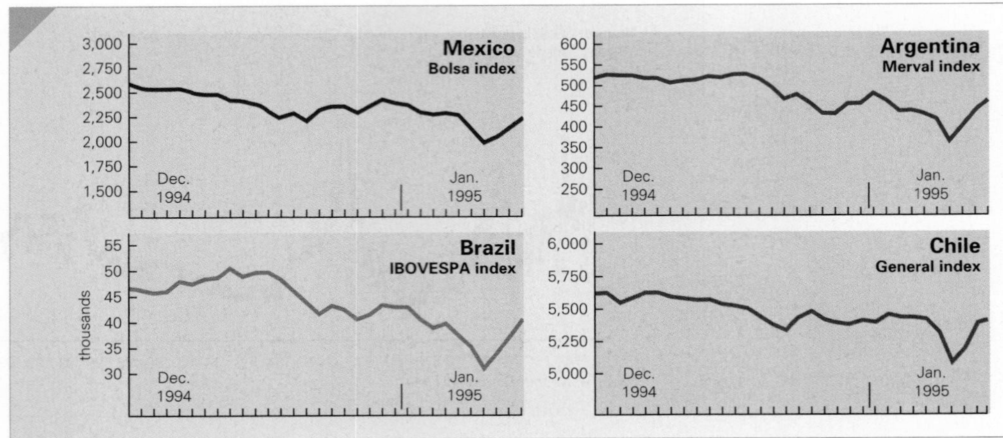

Source: © 1995 TIME Magazine.

followed by the Vietnam War, the rapid increase in the price of oil, gasoline shortages, the Watergate scandal, and the Iranian hostage crisis. The inauguration of Ronald Reagan as president in 1980 marked the end of the decline, and the sharpest increase in the index took place during his 8-year stint in office, during which the greatest boom in U.S. history occurred. This graph is more than just a graph: It's a short story.

In January 1995, a financial crisis in Mexico caused many foreign investors to sell their Mexican holdings. As a result, the Mexican stock market (as measured by the Bolsa Index) fell by 6.6% on one day (January 9). In a story about the widespread effect of this event, *Time* magazine published the graphs shown in Figure 3.7, which are line charts for the stock market indexes in Mexico, Argentina, Brazil, and Chile. The story is summarized concisely and clearly by the graphs. The shock to the Mexican stock market reverberated across South America with equally disastrous consequences.

In the same section, *Time* also published an article about the fall of the Canadian dollar. Figure 3.8 depicts the line chart showing the value of the Canadian dollar in U.S.

FIGURE **3.8**

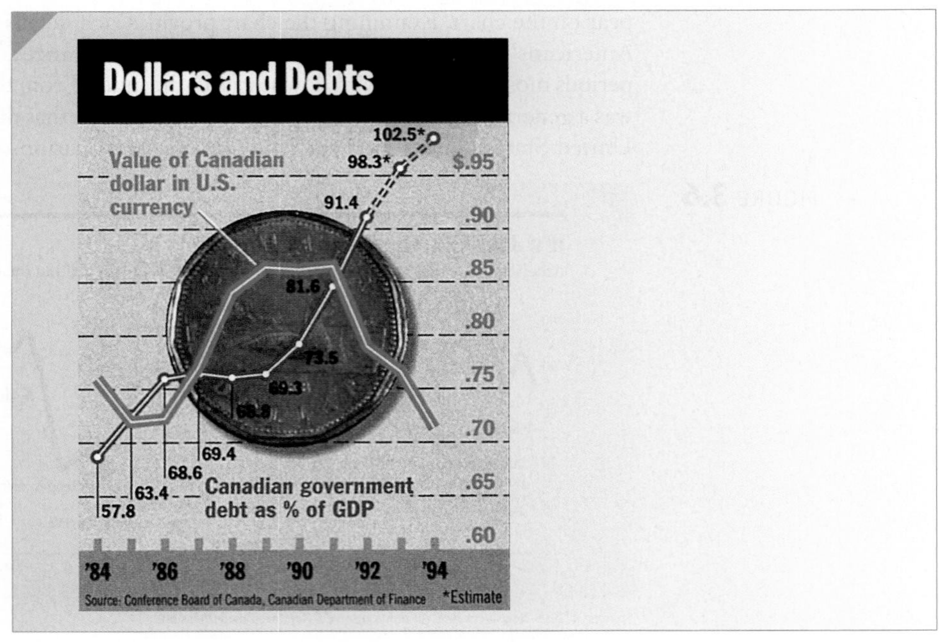

dollars. It shows that in 1990 the Canadian dollar was worth about $.86 U.S. but had dropped to about $.71 in 1994. The article discussed several reasons for this decrease in value but pointed to Canada's rapidly increasing debt as the chief cause. The chart also included Canada's debt as a percentage of gross domestic product (GDP). As you can easily see, the debt/GDP ratio was constant between 1986 and 1989 but started a sharp increase in 1990. The argument is clear: The increasing debts produced a much lower value for the dollar.

Now let's examine the chart that has been acclaimed the best chart ever drawn.

Napoleon's Invasion of and Retreat from Russia: Solution

Figure 3.9 depicts Minard's graph. The striped band is a time series depicting the size of the army at various places on the map, which is also part of the chart. When Napoleon invaded Russia by crossing the Niemen River on June 21, 1812, there were 422,000 soldiers. By the time the army reached Moscow the number had dwindled to 100,000. At that point the army started its retreat. The black band represents the army in retreat. At the bottom of the chart we see the dates starting with October 1813. Just above the dates Minard drew another time series, this one showing the temperature. It was bitterly cold during the fall and many soldiers died of exposure. As you can, see the temperature dipped to −30 on December 6. The chart is effective because it depicts five variables clearly and succinctly.

FIGURE **3.9**

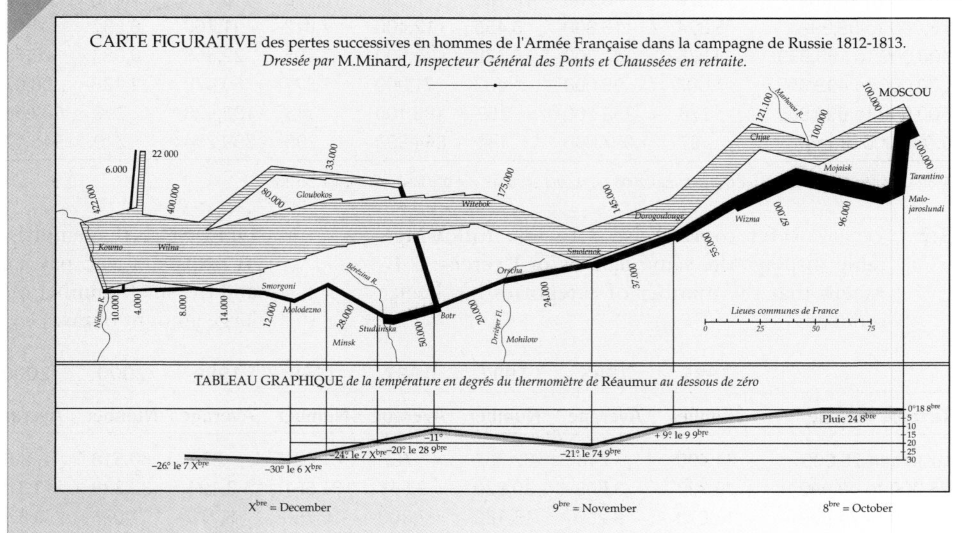

Source: Edward Tufte, *The Visual Display of Quantitative Information* (Cheshire, CT: Graphics Press, 1983), p. 41.

Figures 3.6 to 3.9 illustrate the determinants of graphical excellence. The graphs are well designed, presenting interesting data. They impart ideas concisely because they present several variables at the same time. And finally, they do not distort the data in any way.

EXERCISES

3.1 Xr03-01 The accompanying table lists the number of returns (thousands) and the average income tax paid in 1995, 1997, 1999, 2000, and 2002 in a number of taxable income categories.

a. What information do you think is the most interesting in these data? Explain.

b. A statistics practitioner would like to emphasize that for each year a small percentage of taxpayers pay a large amount of tax. Construct a chart that displays the relevant information.

c. Is your graph clear? How can you improve it?

	1995	1995	1997	1997	1999	1999	2000	2000	2002	2002
Taxable Income	Returns	Average	Returns	Average	Returns	Average	Returns	Average	Returns	Average
<1,000	3,204	200	2,931	400	2,880	518	2,966	648	3,565	1,416
1000 to 2,999	6,526	100	6,108	100	5,922	132	5,385	134	4,833	94
3,000 to 4,999	5,860	100	5,816	100	5,614	171	5,599	179	5,164	84
5,000 to 6,999	5,680	200	5,467	300	5,220	301	5,183	297	5,019	145
7,000 to 8,999	5,593	300	5,199	300	5,102	250	4,972	331	4,977	224
9,000 to 10,999	5,372	500	5,424	500	5,069	464	5,089	470	5,062	259
11,000 to 12,999	5,555	800	5,248	700	4,957	599	4,859	704	4,748	441
13,000 to 14,999	5,344	900	5,250	900	4,907	680	4,810	883	4,808	611
15,000 to 16,999	4,837	1,100	4,781	1,100	5,023	781	4,785	1,052	4,632	747
17,000 to 18,999	4,402	1,300	4,456	1,300	4,609	923	4,633	1,279	4,509	948
19,000 to 21,999	6,507	1,500	6,451	1,500	6,245	1,154	6,502	1,565	6,523	1,214
22,000 to 24,999	5,610	1,900	5,651	1,900	5,873	1,513	5,735	1,815	5,650	1,548
25,000 to 29,999	7,848	2,400	8,065	2,400	8,393	2,049	8,369	2,248	8,575	1,886
30,000 to 39,999	12,380	3,500	12,967	3,300	13,288	3,007	13,548	3,094	13,980	2,622
40,000 to 49,999	9,099	4,800	9,788	4,800	9,870	4,418	10,412	4,462	10,550	3,800
50,000 to 74,999	13,679	8,200	15,180	7,300	16,756	6,770	17,076	6,824	17,397	5,931
75,000 to 99,999	5,374	16,000	6,455	12,400	7,812	11,760	8,597	11,631	9,248	10,169
100,000 to 199,999	4,075	31,300	5,378	23,500	7,105	22,858	8,083	22,783	8,423	20,831
200,000 to 499,999	1,007	95,000	1,402	71,000	1,877	69,479	2,136	68,628	1,908	65,452
500,000 to 999,999	178	258,200	262	196,100	348	192,428	396	192,092	337	188,463
1,000,000 or more	87	1,077,000	144	844,800	205	889,234	240	945,172	169	805,212

Source: *Statistical Abstract of the United States*, 2000, 2003, and 2006 and the U.S. Internal Revenue Service.

3.2 Xr03-02 Refer to Exercise 3.1. The following table displays the same data as in Exercise 3.1 except that the number of categories has been reduced.

a. Draw charts that emphasize the large number of taxpayers who pay a small amount of tax and the small number of taxpayers who pay a large amount of tax.

	1995	1995	1997	1997	1999	1999	2000	2000	2002	2002
Taxable Income	Number	Average	Number	Average	Number	Average	Number	Average	Number	Average
Less than 25,000	64,490	746	62,782	767	61,421	635	60,518	800	59,490	648
25,000 to 49,999	29,327	3,609	30,820	3,541	31,551	3,194	32,329	3,316	33,105	2,807
50,000 to 74,999	13,679	8,200	15,180	7,300	16,756	6,770	17,076	6,824	17,397	5,931
75,000 to 100,000	5,374	16,000	6,455	12,400	7,812	11,760	8,597	11,631	9,248	10,169
Over 100,000	5,347	67,864	7,186	55,518	9,535	56,851	10,855	58,374	10,837	46,132

3.3 Refer to Exercise 3.2 Draw a chart or charts that show that the relationship between the number of taxpayers in each category and the average amount of tax is unchanged from 1995, 1997, 1999, 2000, and 2002.

3.4 Xr03-04 A computer company has diversified its operations into financial services, construction, manufacturing, and hotels. In a recent annual report, the following tables were provided. Create charts to present these data so that the differences between last year and the previous year are clear. (*Note:* It may be necessary to draw the charts manually.)

Region	Sales (millions of dollars) by Region Last Year	Previous Year
United States	67.3	40.4
Canada	20.9	18.9
Europe	37.9	35.5
Australasia	26.2	10.3
Total	152.2	105.1

Division	Sales (millions of dollars) by Division Last Year	Previous Year
Customer service	54.6	43.8
Library systems	49.3	30.5
Construction/property management	17.5	7.7
Manufacturing and distribution	15.4	8.9
Financial systems	9.4	10.9
Hotels and clubs	5.9	3.4

3.5 Xr03-05 The following table lists the number (in thousands) of violent crimes and property crimes committed annually in 1985 to 2003 (last year data available).

a. Draw a chart that displays both sets of data.
b. Does it appear that crime rates are decreasing? Explain.
c. Is there another variable that should be included to show the trends in crime rates?

Year	Violent Crimes	Property Crimes
1985	1,328	11,103
1986	1,489	11,723
1987	1,484	12,025
1988	1,566	12,357
1989	1,646	12,605
1990	1,820	12,655
1991	1,912	12,961
1992	1,932	12,506
1993	1,926	12,219
1994	1,858	12,132
1995	1,799	12,064
1996	1,689	11,805
1997	1,636	11,558
1998	1,534	10,952
1999	1,426	10,208
2000	1,425	10,183
2001	1,439	10,437
2002	1,424	10,455
2003	1,381	10,436

Source: Statistical Abstract of the United States, 2006, Table 293.

3.6 Xm03-06 Refer to Exercise 3.5. We've added the United States population.

a. Incorporate this variable into your charts to show crime rate trends.
b. Summarize tour findings.
c. Can you think of another demographic variable that may explain crime rate trends?

3.7 Xm03-07 Refer to Exercises 3.5 and 3.6. We've included the number of Americans aged 15 to 24.

a. What is the significance of adding the populations aged 15 to 24?
b. Include these data in your analysis. What have you discovered?

3.8 Xr03-08 Because Canada and the United States are the largest trading partners in the world, the exchange rate is vitally important. It is often the case that the difference between the two countries' prime rate is a determinant in the exchange rate. An economist has recorded the monthly Canadian and American prime interest rates and the exchange rate (the value of one U.S. dollar in Canadian dollars) from 2000 to 2007.

a. Draw a line chart of all three time series.
b. What have you learned?
c. Can you improve the chart so that more information is derived? Explain.

3.9 Xr03-09 To determine premiums for automobile insurance, companies must have an understanding of the variables that affect whether a driver will have an accident. The age of the driver may top the list of variables. The following table lists the number of drivers in the United States, the number of fatal accidents, and the number of total accidents in each age group in 2002.

a. Calculate the accident rate (per driver) and the fatal accident rate (per 1,000 drivers) for each age group.
b. Graphically depict the relationship between the ages of drivers, their accident rates, and their fatal accident rates (per 1,000 drivers).
c. Briefly describe what you have learned.

Age Group	Number of Drivers (1,000s)	Number of Accidents (1,000s)	Number of Fatal Accidents
Under 20	9,508	3,543	6,118
20–24	16,768	2,901	5,907
25–34	33,734	7,061	10,288
35–44	41,040	6,665	10,309
45–54	38,711	5,136	8,274
55–64	25,609	2,775	5,322
65–74	15,812	1,498	2,793
Over 74	12,118	1,121	3,689
Total	193,300	30,700	52,700

Source: National Safety Council.

3.10 Xr03-10 During 2002 in the state of Florida, a total of 365,474 drivers were involved in car accidents. The accompanying table breaks down this number by the age group of the driver and whether the driver was injured or killed. (There were actually 371,877 accidents; however, in 6,413 of these, the driver's age was not recorded.)

a. Calculate the injury rate (per 100 accidents) and the death rate (per accident) for each age group.
b. Graphically depict the relationship between the ages of drivers, their injury rate (per 100 accidents), and their death rate.
c. Briefly describe what you have learned from these graphs.
d. What is the difference between the information extracted from Exercise 3.9 and this one?

Age Group	Number of Accidents	Drivers Injured	Drivers Killed
20 or less	52,313	21,762	217
21–24	38,449	16,016	185
25–34	78,703	31,503	324
35–44	76,152	30,542	389
45–54	54,699	22,638	260
55–64	31,985	13,210	167
65–74	18,896	7,892	133
75–85	11,526	5,106	138
85 or more	2,751	1,223	65
Total	365,474	149,892	1,878

Source: Florida Department of Highway Safety and Motor Vehicles.

3.2 Graphical Deception

The use of graphs and charts is pervasive in newspapers, magazines, business and economic reports, and seminars, in large part due to the increasing availability of computers and software that allow the storage, retrieval, manipulation, and summary of large masses of raw data. It is therefore more important than ever to be able to evaluate critically the information presented by means of graphical techniques. In the final analysis, graphical techniques merely create a visual impression, which is easy to distort. In fact, distortion is so easy and commonplace that in 1992 the Canadian Institute of Chartered Accountants found it necessary to begin setting guidelines for financial graphics, after a study of hundreds of the annual reports of major corporations found that 8% contained at least one misleading graph that covered up bad results. Although the heading for this section mentions deception, it is quite possible for an inexperienced person inadvertently to create distorted impressions with graphs. In any event, you should be aware of possible methods of **graphical deception.** This section illustrates a few of them.

The first thing to watch for is a graph without a scale on one axis. The line chart of a firm's sales in Figure 3.10 might represent a growth rate of 100% or 1% over the 5 years depicted, depending on the vertical scale. It is best simply to ignore such graphs.

FIGURE **3.10**

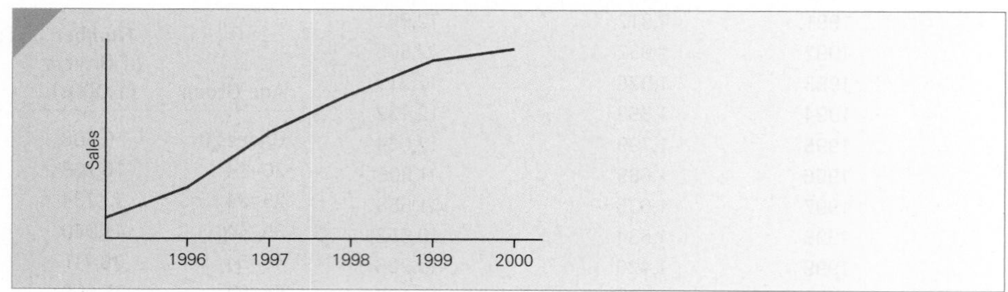

A second trap to avoid is being influenced by a graph's caption. Your impression of the trend in interest rates might be different depending on whether you read a newspaper carrying caption (a) or caption (b) in Figure 3.11.

FIGURE **3.11**

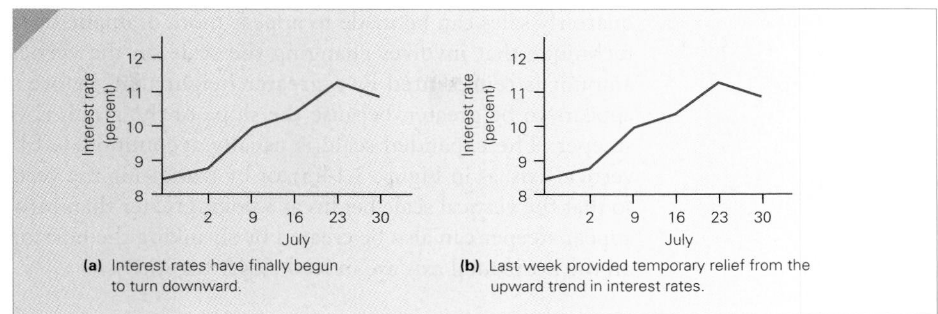

(a) Interest rates have finally begun to turn downward. **(b)** Last week provided temporary relief from the upward trend in interest rates.

Perspective is often distorted if only absolute changes in value, rather than percentage changes, are reported. A $1 drop in the price of your $2 stock is relatively more distressing than a $1 drop in the price of your $100 stock. On January 9, 1986, newspapers throughout North America displayed graphs similar to the one shown in Figure 3.12 and reported that the stock market, as measured by the Dow Jones Industrial Average (DJIA), had suffered its worst 1-day loss ever on the previous day. The loss was 39 points exceeding even the loss of Black Tuesday—October 28, 1929. While the loss was indeed a large one, many news reports failed to mention that the 1986 level of the DJIA was much higher than the 1929 level. A better perspective on the situation could be gained by noticing that the loss on January 8, 1986, represented a 2.5% decline, whereas the decline in 1929 was 12.8%. As a point of interest, we note that the stock market was 12% higher within 2 months of this historic drop and 40% higher 1 year later. The worst 1-day loss ever, 22%, occurred on October 19, 1987.

FIGURE **3.12**

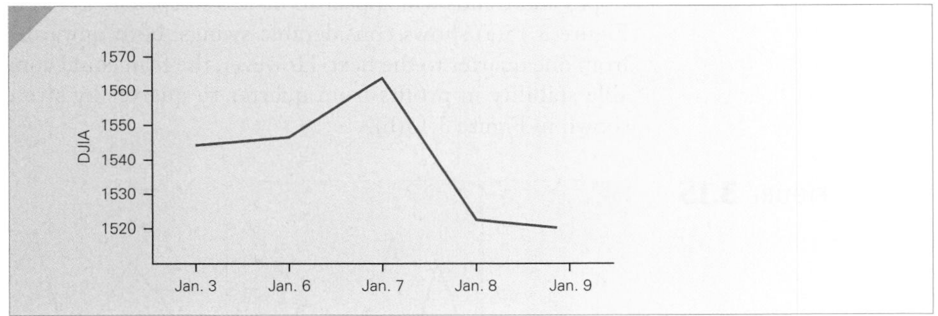

We now turn to some rather subtle methods of creating distorted impressions with graphs. Consider the graph in Figure 3.13 which depicts the growth in a firm's quarterly sales during the past year, from $100 million to $110 million. This 10% growth in

FIGURE **3.13**

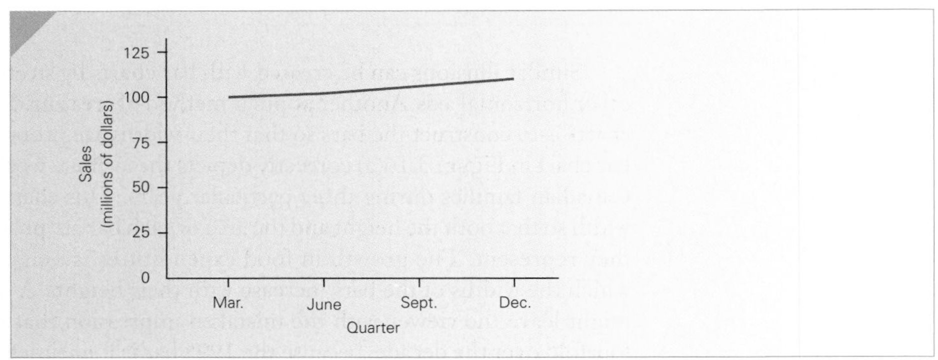

quarterly sales can be made to appear more dramatic by stretching the vertical axis—a technique that involves changing the scale on the vertical axis so that a given dollar amount is represented by a greater height than before. As a result, the rise in sales appears to be greater, because the slope of the graph is visually (but not numerically) steeper. The expanded scale is usually accommodated by employing a break in the vertical axis, as in Figure 3.14(a), or by truncating the vertical axis, as in Figure 3.14(b), so that the vertical scale begins at a point greater than zero. The effect of making slopes appear steeper can also be created by shrinking the horizontal axis, in which case points on the horizontal axis are moved closer together.

FIGURE **3.14**

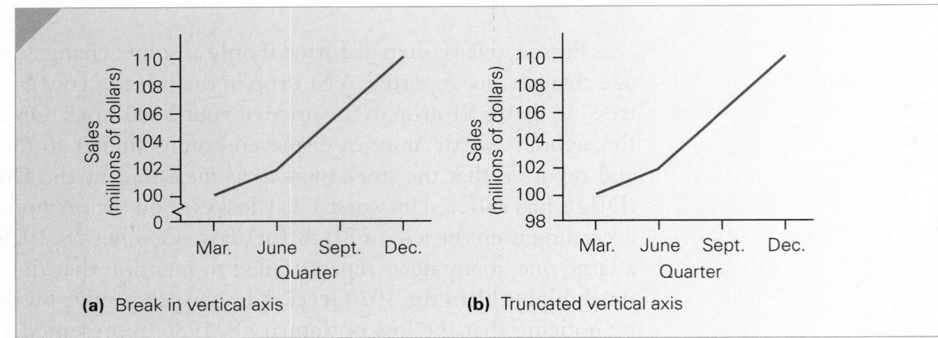

(a) Break in vertical axis **(b)** Truncated vertical axis

Just the opposite effect is obtained by stretching the horizontal axis, that is, spreading out the points on the horizontal axis to increase the distance between them so that slopes and trends will appear to be less steep. The graph of a firm's profits presented in Figure 3.15(a) shows considerable swings, both upward and downward in the profits from one quarter to the next. However, the firm could convey the impression of reasonable stability in profits from quarter to quarter by stretching the horizontal axis, as shown in Figure 3.15(b).

FIGURE **3.15**

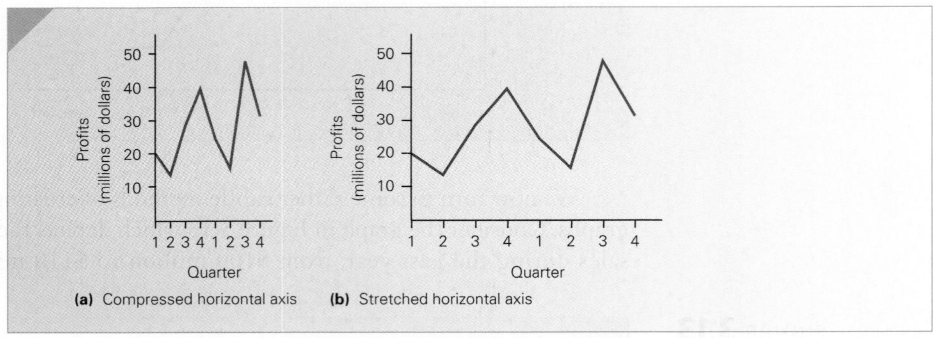

(a) Compressed horizontal axis **(b)** Stretched horizontal axis

Similar illusions can be created with bar charts by stretching or shrinking the vertical or horizontal axis. Another popular method of creating distorted impressions with bar charts is to construct the bars so that their widths are proportional to their heights. The bar chart in Figure 3.16(a) correctly depicts the average weekly amount spent on food by Canadian families during three particular years. This chart correctly uses bars of equal width so that both the height and the area of each bar are proportional to the expenditures they represent. The growth in food expenditures is exaggerated in Figure 3.16(b), in which the widths of the bars increase with their heights. A quick glance at this bar chart might leave the viewer with the mistaken impression that food expenditures increased fourfold over the decade, because the 1995 bar is four times the size of the 1985 bar.

FIGURE **3.16**

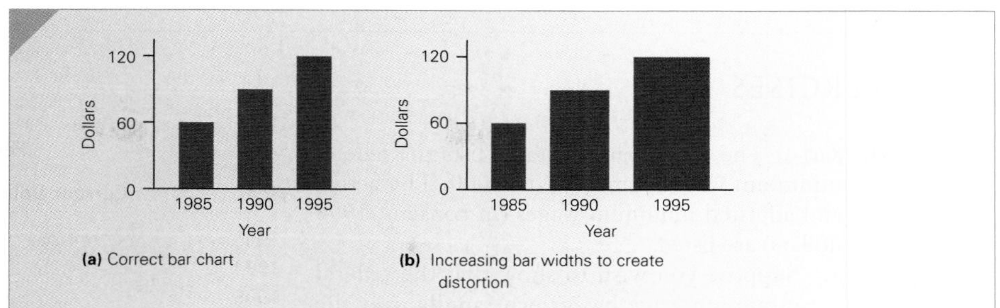

(a) Correct bar chart

(b) Increasing bar widths to create distortion

You should be on the lookout for size distortions, particularly in pictograms, which replace the bars with pictures of objects (such as bags of money, people, or animals) to enhance the visual appeal. Figure 3.17 displays the misuse of a pictogram—the snowman grows in width as well as height. The proper use of a pictogram is shown in Figure 3.18, which effectively uses pictures of Coca-Cola bottles.

FIGURE **3.17**

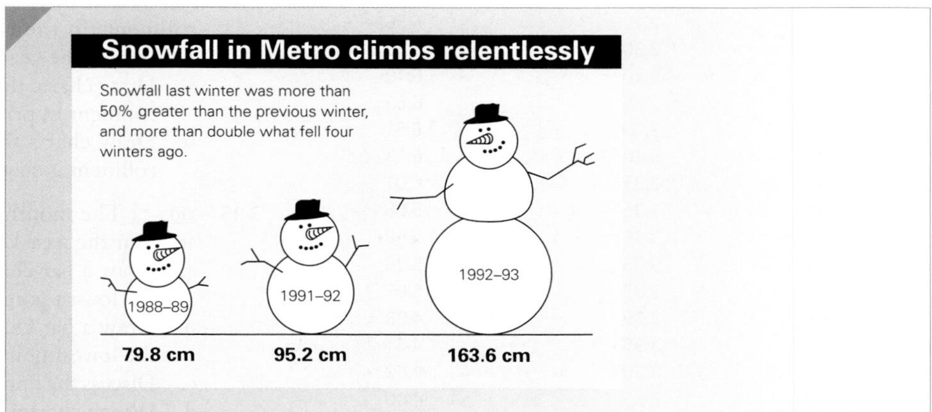

FIGURE **3.18**

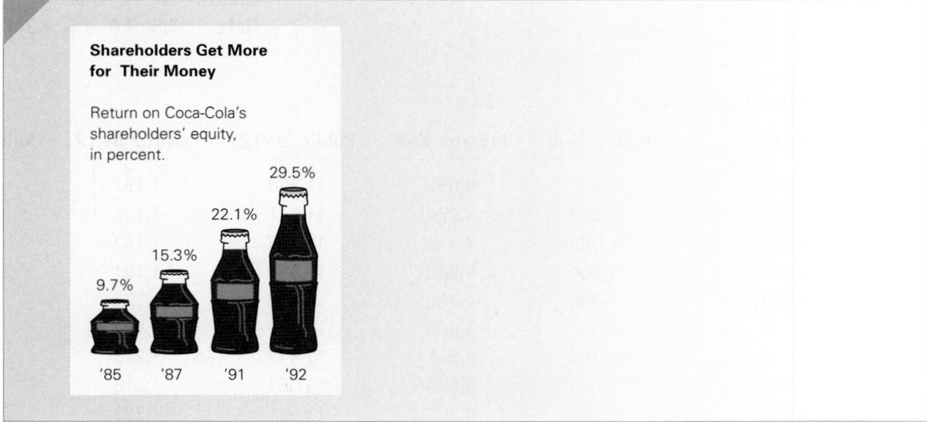

The preceding examples of creating a distorted impression using graphs are not exhaustive, but they include some of the more popular methods. They should also serve to make the point that graphical techniques are used to create a visual impression, and the impression you obtain may be a distorted one unless you examine the graph with care. You are less likely to be misled if you focus your attention on the numerical values that the graph represents. Begin by carefully noting the scales on both axes; graphs with unmarked axes should be ignored completely.

EXERCISES

3.11 Xr03-11 The accompanying table lists the federal minimum wage from 1975 to 2000. The actual and adjusted minimum wages (in constant 1998 dollars) are listed.

a. Suppose you wish to show that the federal minimum wage has grown rapidly over the years. Draw an appropriate chart.
b. Draw a chart to display the actual changes in the federal minimum wage.

Year	Federal Minimum Wage: Current Dollars	Federal Minimum Wage: Constant (1998) Dollars
1975	2.10	6.36
1976	2.30	6.59
1977	2.30	6.19
1978	2.65	6.63
1979	2.90	6.51
1980	3.10	6.13
1981	3.35	6.01
1982	3.35	5.66
1983	3.35	5.48
1984	3.35	5.26
1985	3.35	5.07
1986	3.35	4.98
1987	3.35	4.81
1988	3.35	4.62
1989	3.35	4.40
1990	3.80	4.74
1991	4.25	5.09
1992	4.25	4.94
1993	4.25	4.79
1994	4.25	4.67
1995	4.25	4.55
1996	4.75	4.93
1997	5.15	5.23
1998	5.15	5.15
1999	5.15	5.15
2000	5.15	5.15

Source: U.S. Employment Standards Administration.

3.12 Xr03-12 The table at the bottom shows school enrollment (in thousands) for public and private schools for the years 1990 to 2005.

a. Draw charts that allow you to claim that enrollment in private schools is "skyrocketing."
b. Draw charts that "prove" public school enrollment is stagnant.

3.13 Xr03-13 The monthly unemployment rate in one state for the past 12 months is listed here.

a. Draw a bar chart of these data with 6.0% as the lowest point on the vertical axis.
b. Draw a bar chart of these data with 0.0% as the lowest point on the vertical axis.
c. Discuss the impression given by the two charts.
d. Which chart would you use? Explain.

Month	1	2	3	4	5	6	7	8	9	10	11	12
Rate	7.5	7.6	7.5	7.3	7.2	7.1	7.0	6.7	6.4	6.5	6.3	6.0

Data for Exercise 3.12

Year	Public_K–8	Private_K–8	Public_9–12	Private_9–12	College–Public	College–Private
1990	29,878	4,095	11,338	1,137	10,845	2,974
1991	30,506	4,074	11,541	1,125	11,310	3,049
1992	31,088	4,212	11,735	1,163	11,385	3,103
1993	31,504	4,280	11,961	1,191	11,189	3,116
1994	31,898	4,360	12,213	1,236	11,134	3,145
1995	32,341	4,465	12,500	1,197	11,092	3,169
1996	32,764	4,551	12,847	1,213	11,102	3,247
1997	33,073	4,623	13,054	1,218	11,196	3,306
1998	33,344	4,702	13,191	1,235	11,138	3,369
1999	33,488	4,765	13,369	1,254	11,309	3,482
2000	33,709	4,678	13,514	1,266	11,753	3,560
2001	33,854	4,668	13,722	1,276	11,864	3,578
2002	33,756	4,660	13,857	1,292	11,986	3,622
2003	33,677	4,644	14,069	1,310	12,101	3,655
2004	33,500	4,620	14,346	1,334	12,247	3,699
2005	33,315	4,603	14,597	1,351	12,388	3,746

Source: U.S. National Center for Education Statistics.

3.14 Xr03-14 The following table lists the percentage of single and married women in the United States who had jobs outside the home during the period 1985 to 2004.

a. Construct a chart that shows that the percentage of married women who are working outside the home has not changed much in the past 19 years.

b. Use a chart to show that the percentage of single women in the workforce has increased "dramatically."

Year	Single	Married
1985	66.6	53.8
1986	67.2	54.9
1987	67.4	55.9
1988	67.7	56.7
1989	68.0	57.8
1990	66.7	58.4
1991	66.2	58.5
1992	66.2	59.3
1993	66.2	59.4
1994	66.7	60.7
1995	66.8	61.0
1996	67.1	61.2
1997	67.9	61.6
1998	68.5	61.2
1999	68.7	61.2
2000	68.9	61.1
2001	68.1	61.2
2002	67.4	61.0
2003	66.2	61.0
2004	65.9	60.5

Source: U.S. Bureau of Labor Statistics.

3.15 Xr03-15 The accompanying table lists the average test scores in the Scholastic Assessment Test (SAT) for the years 1967, 1970, 1975, 1980, 1985, 1990, 1995, 1997 to 2004.

Year	Verbal All	Verbal Male	Verbal Female	Math All	Math Male	Math Female
1967	543	540	545	516	535	495
1970	537	536	538	512	531	493
1975	512	515	509	498	518	479
1980	502	506	498	492	515	473
1985	509	514	503	500	522	480
1990	500	505	496	501	521	483
1995	504	505	502	506	525	490
1997	505	507	503	511	530	494
1998	505	509	502	512	531	496
1999	505	509	502	511	531	495
2000	505	507	504	514	533	498
2001	506	509	502	514	533	498
2002	504	507	502	516	534	500
2003	507	512	503	519	537	503
2004	508	512	504	518	537	501

Source: Statistical Abstract of the United States, 2003 and 2006.

Draw a chart for each of the following.

a. You wish to show that both verbal and mathematics test scores for all students have not changed much over the years.

b. The exact opposite of Part a.

c. You want to claim that there are no differences between genders.

d. You want to "prove" that differences between genders exist.

3.3 PRESENTING STATISTICS: WRITTEN REPORTS AND ORAL PRESENTATIONS

Throughout this book we present a variety of statistical techniques. Our emphasis is on applying the correct statistical technique and the proper interpretation of the resulting statistics. However, the ability to communicate your findings both orally and in writing is a critical skill. In this section we provide general guidelines for both forms of communication. Some of the exercises and cases that appear later in this book ask you not only to employ a statistical technique but to prepare a written report or design an oral presentation.

Writing a Report

Just as there are many ways to write a statistics textbook, there are also many different ways to write a report. Here is our suggested method. Reports should contain the following steps:

1. **State your objective.** In Chapter 2 we introduced statistical techniques by describing the type of information needed and the type of data produced by the experiment. For example, there are many studies that test whether one product or service is better than one or more other similar products or services. You should simply state the purpose of the statistical analysis and include the possible results of the experiment and the decisions that may follow.

2. **Describe the experiment.** It is important to know your audience. If there are individuals who have little knowledge of statistics, you must keep this very simple. However, it is likely that some members of your audience will have some knowledge of the subject and thus will want to know how the experiment was done. In particular, they will want you to assure them that the experiment was properly conducted.

3. **Describe your results.** If possible, include charts. In addition to replacing hundreds of words, a well-designed chart exhibits clarity and brevity. Of course, do not include graphs and charts that can more easily be replaced by a table or a sentence. Moreover, place graphs to appear in the body of the report and not in the appendixes. Be precise, which requires that you fully understand the statistical technique and the proper way to interpret the statistics.

   Be honest. Some statisticians describe this component as an ethical issue. That is, it is a violation of ethical standards to purposely mislead readers of the report. In fact, it is an issue of practicality. For example, what is the purpose of a statistics practitioner lying or exaggerating the effects of a new drug? If, as a result of the statistical report the company proceeds with the new drug and it turns out to be ineffective, or worse dangerous, what are the consequences for the statistics practitioner and the company?

4. **Discuss limitations of the statistical techniques.** A statistical analysis will rarely be definitive. Your report should discuss possible problems with the analysis, including violations of the required conditions which are described throughout this book. It is not necessary to include computer printouts. If you do, be selective. Include only those that are critical to your report.

Making an Oral Presentation

The general guidelines for making presentations are similar to those for writing reports.

1. **Know your audience.** Take the time to determine who is in your audience and what kind of information they will be expecting from you. Additionally, determine the level of statistical knowledge. If it is low, avoid the use of statistical terms without defining them.

2. **Restrict your main points to the objectives of the study.** Few listeners will be interested in the details of your statistical analysis. Your audience will be most interested in your conclusions and recommendations.

3. **Stay within your time limits.** If your presentation is expected to exceed your time limit, stick to the main points. Hand out additional written information if necessary.

4. **Use graphs.** It is often much easier to explain even relatively complex ideas if you provide excellent graphs.

5. **Provide handouts.** Handouts with copies of your graphs make it easier for your audience to follow your explanations.

CHAPTER SUMMARY

This chapter completes our discussion of graphical techniques, which began in Chapter 2. In Chapter 2, we showed how and when to construct the graphs. In this chapter, we provided guidelines for the application of graphical methods. We illustrated graphical excellence and graphical deception, and in so doing, we showed you what to do and what not to do.

IMPORTANT TERMS

Graphical excellence 80

Graphical deception 88

© Dave Greenwood/Photonica/Getty Images

NUMERICAL DESCRIPTIVE TECHNIQUES

The Cost of One More Win in Major League Baseball

DATA Xm04-00

In the era of free agency professional sports teams must compete for the services of the best players. It is generally believed that only teams whose salaries place them in the top quarter have a chance of winning the championship. Efforts have been made to provide balance by establishing salary caps or some form of equalization. To examine the problem we gathered data from the 2006 baseball season. For each team in major league baseball, we recorded the number of wins and the team payroll.

To make informed decisions, we need to know how the number of wins and the team payroll are related. After the statistical technique is presented, we return to this problem and solve it.

© Ezra Shaw/Allsport/Getty Images

See pages 135–136.

INTRODUCTION

In Chapter 2 we presented several graphical techniques that describe data. In this chapter we introduce numerical descriptive techniques, which allow the statistics practitioner to be more precise in describing various characteristics of a sample or population. These techniques are critical to the development of statistical inference.

As we pointed out in Chapter 2, arithmetic calculations can be applied to interval data only. Consequently, most of the techniques introduced here may be used only to numerically describe interval data. However, some of the techniques can be used for ordinal data, and one of the techniques can be employed for nominal data.

When we introduced the histogram, we commented that there are several bits of information that we look for. The first is the location of the center of the data. In Section 4.1 we will present measures of central location. Another important characteristic that we seek from a histogram is the spread of the data. The spread will be measured more precisely by measures of variability, which we present in Section 4.2. Section 4.3 introduces measures of relative standing and another graphical technique, the box plot.

In Section 2.6 we introduced the scatter diagram, which is a graphical method that we use to analyze the relationship between two interval variables. The numerical counterparts to the scatter diagram are called measures of linear relationship, and they are presented in Section 4.4.

Sections 4.5 and 4.6 feature statistical applications in baseball and finance, respectively. In Section 4.7 we compare the information provided by graphical and numerical techniques. Finally, we complete this chapter by providing guidelines on how to explore data and retrieve information.

Sample Statistic or Population Parameter

Recall the terms introduced in Chapter 1: population, sample, parameter, and statistic. A parameter is a descriptive measurement about a population, and a statistic is a descriptive measurement about a sample. In this chapter we introduce a dozen descriptive measurements. For each we describe how to calculate both the population parameter and the sample statistic. However, in most realistic applications, populations are very large—in fact, virtually infinite. The formulas describing the calculation of parameters are not practical and are seldom used. They are provided here primarily to teach the concept and the notation. In Chapter 7 we introduce probability distributions, which describe populations. At that time we show how parameters are calculated from probability distributions. In general, small data sets of the type we feature in this book are samples.

4.1 / MEASURES OF CENTRAL LOCATION

Arithmetic Mean

There are three different measures that we use to describe the center of a set of data. The first is the best known, the **arithmetic mean,** which we'll refer to simply as the *mean*. Students may be more familiar with its other name, the *average*. The mean is computed by summing the observations and dividing by the number of observations. We label the observations in a sample $x_1, x_2, \dots, x_n$, where x_1 is the first observation, x_2 is the second, and so on until x_n, where n is the sample size. As a result the sample mean is denoted $\bar{x}$. In a population the number of observations is labeled N and the population mean is denoted by μ (Greek letter *mu*).

Mean

$$\text{Population mean: } \mu = \frac{\sum_{i=1}^{N} x_i}{N}$$

$$\text{Sample mean: } \bar{x} = \frac{\sum_{i=1}^{n} x_i}{n}$$

EXAMPLE 4.1

Mean Time Spent on the Internet

A sample of 10 adults was asked to report the number of hours they spent on the Internet the previous month. The results are listed here. Manually calculate the sample mean.

0 7 12 5 33 14 8 0 9 22

SOLUTION

Using our notation we have $x_1 = 0$, $x_2 = 7, \ldots, x_{10} = 22$, and $n = 10$. The sample mean is

$$\bar{x} = \frac{\sum_{i=1}^{n} x_i}{n} = \frac{0 + 7 + 12 + 5 + 33 + 14 + 8 + 0 + 9 + 22}{10} = \frac{110}{10} = 11.0$$

EXAMPLE 4.2

DATA
Xm02-04

Mean Long-Distance Telephone Bill

Refer to Example 2.4. Find the mean long-distance telephone bill.

SOLUTION

To calculate the mean we add the observations and divide the sum by the size of the sample. Thus,

$$\bar{x} = \frac{\sum_{i=1}^{n} x_i}{n} = \frac{42.19 + 38.45 + \cdots + 45.77}{200} = \frac{8717.52}{200} = 43.59$$

USING THE COMPUTER

There are several ways to command Excel and Minitab to compute the mean. If we simply want to compute the mean and no other statistics, we can proceed as follows.

EXCEL

INSTRUCTIONS

Type or import the data into one or more columns. (Open Xm02-04.) Type into any empty cell

=AVERAGE([Input range])

For Example 4.2, we would type into any cell

=AVERAGE(A1:A201)

The active cell would store the mean as 43.5876.

MINITAB

INSTRUCTIONS

1. Type or import the data into one column. (Open Xm02-04.)
2. Click **Calc** and **Column Statistics . . . .** Specify **Mean** in the **Statistic** box. Type or use the **Select** button to specify the **Input variable** and click **OK.** The sample mean is outputted in the session window as 43.5876.

Median

The second most popular measure of central location is the *median*.

Median

The **median** is calculated by placing all the observations in order (ascending or descending). The observation that falls in the middle is the median. The sample and population medians are computed in the same way.

When there is an even number of observations, the median is determined by averaging the two observations in the middle.

EXAMPLE 4.3

Median Time Spent on Internet

Find the median for the data in Example 4.1.

SOLUTION

When placed in ascending order, the data appear as follows:

0 0 5 7 8 9 12 14 22 33

The median is the average of the fifth and sixth observations (the middle two), which are 8 and 9, respectively. Thus, the median is 8.5.

EXAMPLE 4.4

Median Long-Distance Telephone Bill

DATA
Xm02-04

Find the median of the 200 observations in Example 2.4.

SOLUTION

All the observations were placed in order. We observed that the 100th and 101st observations are 26.84 and 26.97, respectively. Thus, the median is the average of these two numbers.

$$\text{Median} = \frac{26.84 + 26.97}{2} = 26.905$$

EXCEL

INSTRUCTIONS

To calculate the median, substitute **MEDIAN** in place of **AVERAGE** in the instructions for the mean (page 99). The median is reported as 26.905.

MINITAB

INSTRUCTIONS

Follow the instructions for the mean (page 99) to compute the mean except click **Median** instead of **Mean.** The median is outputted as 26.905 in the session window.

INTERPRET

Half the observations are below 26.905, and half the observations are above 26.905.

Mode

The third and last measure of central location that we present here is the **mode.**

Mode
The mode is defined as the observation (or observations) that occurs with the greatest frequency. Both the statistic and parameter are computed in the same way.

For populations and large samples, it is preferable to report the modal class, which we defined in Chapter 2.

There are several problems with using the mode as a measure of central location. First, in a small sample it may not be a very good measure. Second, it may not be unique.

EXAMPLE 4.5

Mode Time Spent on Internet

Find the mode for the data in Example 4.1.

SOLUTION

All observations except 0 occur once. There are two 0s. Thus, the mode is 0. As you can see, this is a poor measure of central location. It is nowhere near the center of the data. Compare this with the mean 11.0 and median 8.5 and you can appreciate that in this example the mean and median are superior measures.

EXAMPLE 4.6

Mode of Long-Distance Bill

DATA
Xm02-04

Determine the mode for Example 2.4.

SOLUTION

An examination of the 200 observations reveals that, except for 0, it appears that each number is unique. However, there are 8 zeroes, which indicate that the mode is 0.

EXCEL

INSTRUCTIONS

To compute the mode substitute **MODE** in place of **AVERAGE** in the previous instructions. Note that if there is more than one mode, Excel prints only the smallest one, without indicating whether there are other modes. In this example Excel reports that the mode is 0.

MINITAB

Minitab does not compute the mode directly. However, it will provide a count of the frequency of each number using the **Tally** command. (See page 20.)

Excel and Minitab: Printing All the Measures of Central Location Plus Other Statistics

Both Excel and Minitab can produce the measures of central location plus a variety of others that we will introduce in later sections.

EXCEL

Excel Output for Examples 4.2, 4.4, and 4.6

	A	B
1	*Bills*	
2		
3	Mean	43.59
4	Standard Error	2.76
5	Median	26.91
6	Mode	0
7	Standard Deviation	38.97
8	Sample Variance	1518.64
9	Kurtosis	-1.29
10	Skewness	0.54
11	Range	119.63
12	Minimum	0
13	Maximum	119.63
14	Sum	8717.5
15	Count	200

Excel reports the mean, median, and mode as the same values we obtained previously. Most of the other statistics will be discussed later.

INSTRUCTIONS

1. Type or import the data into one column. (Open Xm02-04.)
2. Click **Data, Data Analysis,** and **Descriptive Statistics.**
3. Specify the **Input Range** (A1:A201) and click **Summary statistics.**

MINITAB

Minitab Output for Examples 4.2 and 4.4

Descriptive Statistics: Bills

Variable	N	N*	Mean	StDev	Variance	Minimum	Q1	Median	Q3
Bills	200	0	43.59	38.97	1518.64	0.000000000	9.28	26.91	84.94

Variable	Maximum
Bills	119.63

INSTRUCTIONS

1. Type or import the data into one column. (Open Xm02-04.)
2. Click **Stat, Basic Statistics,** and **Display Descriptive Statistics . . . .**
3. Type or use **Select** to identify the name of the variable or column (Bills). Click **Statistics . . .** to add or delete particular statistics.

Mean, Median, Mode: Which Is Best?

With three measures from which to choose, which one should we use? There are several factors to consider when making our choice of measure of central location. The mean is generally our first selection. However, there are several circumstances when the median is better. The mode is seldom the best measure of central location. One advantage the median holds is that it not as sensitive to extreme values as is the mean. To illustrate, consider the data in Example 4.1. The mean was 11.0 and the median was 8.5. Now suppose that the respondent who reported 33 hours actually reported 133 hours (obviously an Internet addict). The mean becomes

$$\bar{x} = \frac{\sum_{i=1}^{n} x_i}{n} = \frac{0 + 7 + 12 + 5 + 133 + 14 + 8 + 0 + 9 + 22}{10} = \frac{210}{10} = 21.0$$

This value is only exceeded by only two of the ten observations in the sample, making this statistic a poor measure of *central* location. The median stays the same. When there is a relatively small number of extreme observations (either very small or very large, but not both), the median usually produces a better measure of the center of the data.

To see another advantage of the median over the mean, suppose you and your classmates have written a statistics test and the instructor is returning the graded tests. What piece of information is most important to you? The answer, of course, is *your* mark. What is the next important bit of information? The answer is how well you performed relative to the class. Most students ask their instructor for the class mean. This is the wrong statistic to request. You want the *median* because it divides the class into two halves. This information allows you to identify which half of the class your mark falls into. The median provides this information; the mean does not. Nevertheless, the mean can also be useful in this scenario. If there are several sections of the course, the section means can be compared to determine whose class performed best (or worst).

Measures of Central Location for Ordinal and Nominal Data

When the data are interval, we can use any of the three measures of central location. However, for ordinal and nominal data the calculation of the mean is not valid. Because the calculation of the median begins by placing the data in order, this statistic is appropriate for ordinal data. The mode, which is determined by counting the frequency of each observation, is appropriate for nominal data. However, nominal data do not have a "center," so we cannot interpret the mode of nominal data in that way. It is generally pointless to compute the mode of nominal data.

APPLICATIONS in **FINANCE**

Image State Royalty-free

Geometric Mean

The arithmetic mean is the single most popular and useful measure of central location. We noted certain situations where the median is a better measure of central location. However, there is another circumstance where neither the mean nor the median is the best measure. When the variable is a growth rate or rate of change, such as the value of an investment over periods of time, we need another measure. This will become apparent from the following illustration.

Suppose you make a 2-year investment of $1,000 and it grows by 100% to $2,000 during the first year. During the second year, however, the investment suffers a 50% loss, from $2,000 back to $1,000. The rates of return for years 1 and 2 are $R_1 = 100\%$ and $R_2 = -50\%$, respectively. The arithmetic mean (and the median) is computed as

$$\bar{R} = \frac{R_1 + R_2}{2} = \frac{100 + (-50)}{2} = 25\%$$

But this figure is misleading. Because there was no change in the value of the investment from the beginning to the end of the 2-year period, the "average" compounded rate of return is 0%. As you will see, this is the value of the *geometric mean.*

(continued)

Geometric Mean

Let R_i denote the rate of return (in decimal form) in period i ($i = 1, 2, \ldots, n$). The **geometric mean** R_g of the returns $R_1, R_2, \ldots, R_n$ is defined such that

$$(1 + R_g)^n = (1 + R_1)(1 + R_2) \cdots (1 + R_n)$$

Solving for R_g we produce the following formula.

$$R_g = \sqrt[n]{(1 + R_1)(1 + R_2) \cdots (1 + R_n)} - 1$$

The geometric mean of our investment illustration is

$$R_g = \sqrt[n]{(1 + R_1)(1 + R_2) \cdots (1 + R_n)} - 1$$

$$= \sqrt[2]{(1 + 1)(1 + [-.50])} - 1 = 1 - 1 = 0$$

The geometric mean is therefore 0%. This is the single "average" return that allows us to compute the value of the investment at the end of the investment period from the beginning value. Thus, using the formula for compound interest with the rate = 0%, we find

Value at the end of the investment period

$$= 1{,}000(1 + R_g)^2 = 1{,}000(1 + 0)^2 = 1{,}000$$

The geometric mean is used whenever we wish to find the "average" growth rate, or rate of change, in a variable *over time*. However, the arithmetic mean of n returns (or growth rates) is the appropriate mean to calculate if you wish to estimate the mean rate of return (or growth rate) for any *single* period in the future. That is, in the illustration above if we wanted to estimate the rate of return in year 3, we would use the arithmetic mean of the two annual rates of return, which we found to be 25%.

EXCEL

INSTRUCTIONS

1. Type or import the values of $1 + R_1$ into a column.
2. Follow the instructions to produce the mean (page 99) except substitute **GEOMEAN** in place of **AVERAGE.**
3. To determine the geometric mean, subtract 1 from the number produced.

MINITAB

Minitab does not compute the geometric mean.

Here is a summary of the numerical techniques introduced in this section and when to use them.

Factors That Identify When to Compute the Mean

1. **Objective:** Describe a single set of data
2. **Type of data:** Interval
3. **Descriptive measurement:** Central location

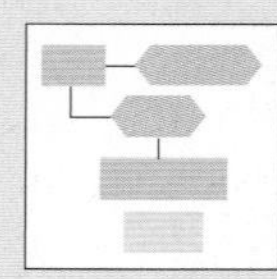

Factors That Identify When to Compute the Median

1. **Objective:** Describe a single set of data
2. **Type of data:** Ordinal or interval (with extreme observations)
3. **Descriptive measurement:** Central location

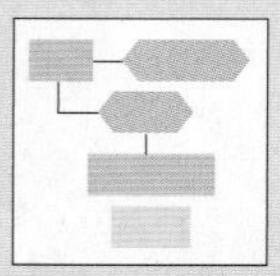

Factors That Identify When to Compute the Mode

1. **Objective:** Describe a single set of data
2. **Type of data:** Nominal, ordinal, interval

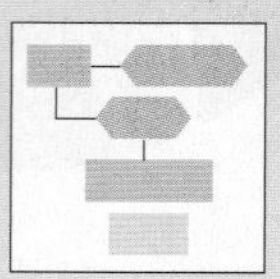

Factors That Identify When to Compute the Geometric Mean

1. **Objective:** Describe a single set of data
2. **Type of data:** Interval; growth rates

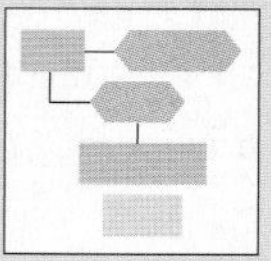

EXERCISES

4.1 A sample of 12 people was asked how much change they had in their pockets and wallets. The responses (in cents) are

52 25 15 0 104 44 60 30 33 81 40 5

Determine the mean, median, and mode for these data.

4.2 The number of sick days due to colds and flu last year was recorded by a sample of 15 adults. The data are

5 7 0 3 15 6 5 9 3 8 10 5 2 0 12

Compute the mean, median, and mode.

4.3 A random sample of 12 joggers was asked to keep track and report the number of miles they ran last week. The responses are

5.5 7.2 1.6 22.0 8.7 2.8
5.3 3.4 12.5 18.6 8.3 6.6

a. Compute the three statistics that measure central location.
b. Briefly describe what each statistic tells you.

4.4 The midterm test for a statistics course has a time limit of 1 hour. However, like most statistics exams this one was quite easy. To assess how easy, the professor recorded the amount of time taken by a sample of nine students to hand in their test papers. The times (rounded to the nearest minute) are

33 29 45 60 42 19 52 38 36

a. Compute the mean, median, and mode.
b. What have you learned from the three statistics calculated in Part a?

4.5 The professors at Wilfrid Laurier University are required to submit their final exams to the registrar's office 10 days before the end of the semester. The exam coordinator sampled 20 professors and recorded the number of days before the final exam that each submitted his or her exam. The results are

14 8 3 2 6 4 9 13 10 12
7 4 9 13 15 8 11 12 4 0

a. Compute the mean, median, and mode.
b. Briefly describe what each statistic tells you.

4.6 Compute the geometric mean of the following rates of return.

.25 −.10 .50

4.7 What is the geometric mean of the following rates of return?

.50 .30 −.50 −.25

4.8 The following returns were realized on an investment over a 5-year period.

Year	1	2	3	4	5
Rate of Return	.10	.22	.06	−.05	.20

a. Compute the mean and median of the returns.
b. Compute the geometric mean.
c. Which one of the three statistics computed in Parts a and b best describes the return over the 5-year period? Explain.

4.9 An investment you made 5 years ago has realized the following rates of return:

Year	1	2	3	4	5
Rate of Return	−.15	−.20	.15	−.08	.50

a. Compute the mean and median of the rates of return.
b. Compute the geometric mean.
c. Which one of the three statistics computed in Parts a and b best describes the return over the 5-year period? Explain.

4.10 An investment of $1,000 you made 4 years ago was worth $1,200 after the first year, $1,200 after the second year, $1,500 after the third year, and $2,000 today.
a. Compute the annual rates of return.
b. Compute the mean and median of the rates of return.
c. Compute the geometric mean.
d. Discuss whether the mean, median, or geometric mean is the best measure of the performance of the investment.

4.11 Suppose that you bought a stock 6 years ago at $12. The stock's price at the end of each year is shown here.

Year	1	2	3	4	5	6
Price	10	14	15	22	30	25

a. Compute the rate of return for each year.
b. Compute the mean and median of the rates of return.
c. Compute the geometric mean of the rates of return.
d. Explain why the best statistic to use to describe what happened to the price of the stock over the 6-year period is the geometric mean.

The following exercises require the use of a computer and software.

4.12 Xr04-12 The starting salaries of a sample of 125 recent MBA graduates are recorded.

a. Determine the mean and median of these data.

b. What do these two statistics tell you about the starting salaries of MBA graduates?

4.13 Xr04-13 To determine whether changing the color of their invoices would improve the speed of payment a company selected 200 customers at random and sent their invoices on blue paper. The number of days until the bills were paid was recorded. Calculate the mean and median of these data. Report what you have discovered.

4.14 Xr04-14 A survey undertaken by the U.S. Bureau of Labor Statistics, Annual Consumer Expenditure, asks American adults to report the amount of money spent on reading material in 2003. (Adapted from *Statistical Abstract of the United States*, 2006, Table 669)

a. Compute the mean and median of the sample.

b. What do the statistics computed in Part a tell you about the reading materials expenditures?

4.15 Xr04-14 A survey of 225 workers in Los Angeles and 260 workers in New York asked each to report the average amount of time spent commuting to work. (Adapted from *Statistical Abstract of the United States*, 2006, Table 1083)

a. Compute the mean and median of the commuting times for workers in Los Angeles.

b. Repeat Part a for New York workers.

c. Summarize your findings.

4.16 Xr04-16 Employee training and education have become important factors in the success of many firms. An annual survey undertaken by the U.S. Bureau of Labor Statistics attempts to measure the amount of training. A survey of employers with more than 500 employees asked the personnel manager to report the number of hours of employee training for the 6-month period May to October. (Adapted from *Statistical Abstract of the United States*, 1999, Table 691)

a. Compute the mean and median.

b. Interpret the statistics you computed.

4.17 Xr04-17 In an effort to slow drivers, traffic engineers painted a solid line 3 feet from the curb over the entire length of a road and filled the space with diagonal lines. The lines made the road look narrower. A sample of car speeds was taken after the lines were drawn.

a. Compute the mean, median, and mode of these data.

b. Briefly describe the information you acquired from each statistic calculated in Part a.

4.18 Xr04-18 How much do Americans spend on various food groups? A sample of 350 American families were surveyed and asked to report the amount of money spent annually on fruits and vegetables. Compute the mean and median of these data and interpret the results. (Adapted from *Statistical Abstract of the United States*, 2006, Table 667)

4.2 MEASURES OF VARIABILITY

The statistics introduced in Section 4.1 serve to provide information about the central location of the data. However, as we have already discussed in Chapter 2, there are other characteristics of data that are of interest to practitioners of statistics. One such characteristic is the spread or variability of the data. In this section we introduce four **measures of variability.** We begin with the simplest.

Range

Range

Range = Largest observation − Smallest observation

The advantage of the range is its simplicity. The disadvantage is also its simplicity. Because the range is calculated from only two observations, it tells us nothing about the other observations. Consider the following two sets of data:

Set 1: 4 4 4 4 4 50

Set 2: 4 8 15 24 39 50

The range of both sets is 46. The two sets of data are completely different and yet their ranges are the same. To measure variability, we need other statistics that incorporate all the data and not just two observations.

Variance

The **variance** and its related measure, the standard deviation, are arguably the most important statistics. They are used to measure variability, but, as you will discover, they play a vital role in almost all statistical inference procedures.

Variance

Population variance: $$\sigma^2 = \frac{\sum_{i=1}^{N}(x_i - \mu)^2}{N}$$

Sample variance:* $$s^2 = \frac{\sum_{i=1}^{n}(x_i - \bar{x})^2}{n - 1}$$

The population variance is represented by σ^2 (Greek letter *sigma* squared).

Examine the formula for the sample variance s^2. It may appear to be illogical that in calculating s^2 we divide the sum of squared deviations by $n - 1$ rather than by n. However, we do so for the following reason. Population parameters in practical settings are seldom known. One of the objectives of statistical inference is to estimate the parameter from the statistic. For example, we estimate the population mean μ from the sample mean $\bar{x}$. Although it is not obviously logical, the statistic created by dividing $\sum(x_i - \bar{x})^2$ by $n - 1$ is a better estimator than is the one created by dividing by n. We will discuss this issue in greater detail in Section 10.1.

To compute the sample variance s^2, we begin by calculating the sample mean $\bar{x}$. Next we compute the difference (also call the *deviation*) between each observation and the mean. We square the deviations and sum. Finally, we divide the sum of squared deviations by $n - 1$.

We'll illustrate with a simple example. Suppose that we have the following observations of the numbers of hours five students spent studying statistics last week:

8 4 9 11 3

*Technically, the variance of the sample is calculated by dividing the sum of squared deviations by n. The statistic computed by dividing the sum of squared deviations by $n - 1$ is called the *sample variance corrected for the mean*. Because this statistic is used extensively, we will shorten its name to *sample variance*.

The mean is

$$\bar{x} = \frac{8 + 4 + 9 + 11 + 3}{5} = \frac{35}{5} = 7$$

For each observation we determine its deviation from the mean. The deviation is squared and the sum of squares determined as shown in Table 4.1.

TABLE **4.1** Calculation of Sample Variance

x_i	$(x_i - \bar{x})$	$(x_i - \bar{x})^2$
8	(8–7) = 1	$(1)^2 = 1$
4	(4–7) = –3	$(-3)^2 = 9$
9	(9–7) = 2	$(2)^2 = 4$
11	(11–7) = 4	$(4)^2 = 16$
3	(3–7) = –4	$(-4)^2 = 16$
	$\sum_{i=1}^{5}(x_i - \bar{x}) = 0$	$\sum_{i=1}^{5}(x_i - \bar{x})^2 = 46$

The sample variance is

$$s^2 = \frac{\sum_{i=1}^{n}(x_i - \bar{x})^2}{n - 1} = \frac{46}{5 - 1} = 11.5$$

The calculation of this statistic raises several questions. Why do we square the deviations before averaging? If you examine the deviations you will see that some of the deviations are positive and some are negative. When you add them together, the sum is 0. This will always be the case because the sum of the positive deviations will always equal the sum of the negative deviations. Consequently, we square the deviations to avoid the "canceling effect."

Is it possible to avoid the canceling effect without squaring? We could average the *absolute* value of the deviations. In fact, such a statistic has already been invented. It is called the **mean absolute deviation** or MAD. However, this statistic has limited utility and is seldom calculated.

What is the unit of measurement of the variance? Because we squared the deviations, we also squared the units. In this illustration the units were hours (of study). Thus, the sample variance is 11.5 hours^2.

EXAMPLE 4.7 Summer Jobs

The following are the number of summer jobs a sample of six students applied for. Find the mean and variance of these data.

17 15 23 7 9 13

SOLUTION

The mean of the six observations is

$$\bar{x} = \frac{17 + 15 + 23 + 7 + 9 + 13}{6} = \frac{84}{6} = 14 \text{ jobs}$$

The sample variance is

$$s^2 = \frac{\sum_{i=1}^{n}(x_i - \bar{x})^2}{n-1}$$

$$= \frac{(17-14)^2 + (15-14)^2 + (23-14)^2 + (7-14)^2 + (9-14)^2 + (13-14)^2}{6-1}$$

$$= \frac{9+1+81+49+25+1}{5} = \frac{166}{5} = 33.2 \text{ jobs}^2$$

(Optional) Shortcut Method for Variance

The calculations for larger data sets are quite time-consuming. The following shortcut for the sample variance may help lighten the load.

Shortcut for Sample Variance

$$s^2 = \frac{1}{n-1}\left[\sum_{i=1}^{n} x_i^2 - \frac{\left(\sum_{i=1}^{n} x_i\right)^2}{n}\right]$$

To illustrate we'll do Example 4.7 again.

$$\sum_{i=1}^{n} x_i^2 = 17^2 + 15^2 + 23^2 + 7^2 + 9^2 + 13^2 = 1{,}342$$

$$\sum_{i=1}^{n} x_i = 17 + 15 + 23 + 7 + 9 + 13 = 84$$

$$\left(\sum_{i=1}^{n} x_i\right)^2 = 84^2 = 7{,}056$$

$$s^2 = \frac{1}{n-1}\left[\sum_{i=1}^{n} x_i^2 - \frac{\left(\sum_{i=1}^{n} x_i\right)^2}{n}\right] = \frac{1}{6-1}\left[1{,}342 - \frac{7{,}056}{6}\right] = 33.2 \text{ jobs}^2$$

Notice that we produced the same exact answer.

EXCEL

INSTRUCTIONS

Follow the instructions to compute the mean (page 99) except type **VAR** instead of **AVERAGE.**

MINITAB

Minitab does not compute the variance directly.

Interpreting the Variance

We calculated the variance in Example 4.7 to be 33.2 $jobs^2$. What does this statistic tell us? Unfortunately, the variance provides us with only a rough idea about the amount of variation in the data. However, this statistic is useful when comparing two or more sets of data of the same type of variable. If the variance of one data set is larger than that of a second data set, we interpret that to mean that the observations in the first set display more variation than the observations in the second set.

The problem of interpretation is caused by the way the variance is computed. Because we squared the deviations from the mean, the unit attached to the variance is the square of the unit attached to the original observations. That is, in Example 4.7 the unit of the data is jobs; the unit of the variance is jobs squared. This contributes to the problem of interpretation. We resolve this difficulty by calculating another related measure of variability.

Standard Deviation

Standard Deviation

Population standard deviation: $\sigma = \sqrt{\sigma^2}$

Sample standard deviation: $s = \sqrt{s^2}$

The **standard deviation** is simply the positive square root of the variance. Thus, in Example 4.7 the sample standard deviation is

$$s = \sqrt{s^2} = \sqrt{33.2} = 5.76 \text{ jobs}$$

Notice that the unit associated with the standard deviation is the unit of the original data set.

EXAMPLE 4.8 Comparing the Consistency of Two Types of Golf Clubs

DATA Xm04-08

Consistency is the hallmark of a good golfer. Golf equipment manufacturers are constantly seeking ways to improve their products. Suppose that a recent innovation is designed to improve the consistency of its users. As a test a golfer was asked to hit 150 shots using a 7-iron, 75 of which were hit with his current club and 75 with the new innovative 7-iron. The distances were measured and recorded. Which 7-iron is more consistent?

SOLUTION

To gauge the consistency, we must determine the standard deviations. (We could also compute the variances, but as we just pointed out, the standard deviation is easier to interpret.) We can get Excel and Minitab to print the sample standard deviations. Alternatively, we can calculate all the descriptive statistics, a course of action we recommend because we often need several statistics. The printouts for both 7-irons are shown here.

EXCEL

	A	B	C	D	E
1	*Current*			*Innovation*	
2					
3	Mean	150.55		Mean	150.15
4	Standard Error	0.67		Standard Error	0.36
5	Median	151		Median	150
6	Mode	150		Mode	149
7	Standard Deviation	5.79		Standard Deviation	3.09
8	Sample Variance	33.55		Sample Variance	9.56
9	Kurtosis	0.13		Kurtosis	-0.89
10	Skewness	-0.43		Skewness	0.18
11	Range	28		Range	12
12	Minimum	134		Minimum	144
13	Maximum	162		Maximum	156
14	Sum	11291		Sum	11261
15	Count	75		Count	75

MINITAB

Descriptive Statistics: Current, Innovation

Variable	N	N*	Mean	StDev	Variance	Minimum	Q1	Median	Q3
Current	75	0	150.55	5.79	33.55	134.00	148.00	151.00	155.00
Innovation	75	0	150.15	3.09	9.56	144.00	148.00	150.00	152.00

Variable	Maximum
Current	162.00
Innovation	156.00

INTERPRET

The standard deviation of the distances of the current 7-iron is 5.79 yards whereas that of the innovative 7-iron is 3.09 yards. Based on this sample, the innovative club is more consistent. Because the mean distances are similar it would appear that the new club is indeed superior.

Interpreting the Standard Deviation

Knowing the mean and standard deviation allows the statistics practitioner to extract useful bits of information. The information depends on the shape of the histogram. If the histogram is bell shaped, we can use the **Empirical Rule.**

Empirical Rule

1. Approximately 68% of all observations fall within one standard deviation of the mean.
2. Approximately 95% of all observations fall within two standard deviations of the mean.
3. Approximately 99.7% of all observations fall within three standard deviations of the mean.

EXAMPLE 4.9

Using the Empirical Rule to Interpret Standard Deviation

After an analysis of the returns on an investment, a statistics practitioner discovered that the histogram is bell shaped and that the mean and standard deviation are 10% and 8%, respectively. What can you say about the way the returns are distributed?

SOLUTION

Because the histogram is bell shaped, we can apply the Empirical Rule. Thus,

1. Approximately 68% of the returns lie between 2% (the mean minus one standard deviation = 10 − 8) and 18% (the mean plus one standard deviation =10 + 8).
2. Approximately 95% of the returns lie between −6% [the mean minus two standard deviations = 10 − 2(8)] and 26% [the mean plus two standard deviations = 10 + 2(8)].
3. Approximately 99.7% of the returns lie between −14% [the mean minus three standard deviations = 10 − 3(8)] and 34% [the mean plus three standard deviations = 10 + 3(8)].

A more general interpretation of the standard deviation is derived from **Chebysheff's Theorem,** which applies to all shapes of histograms.

Chebysheff's Theorem

The proportion of observations in any sample or population that lie within k standard deviations of the mean is at least

$$1 - \frac{1}{k^2} \quad \text{for } k > 1$$

When $k = 2$, Chebysheff's Theorem states that at least three-quarters (75%) of all observations lie within two standard deviations of the mean. With $k = 3$, Chebysheff's Theorem states that at least eight-ninths (88.9%) of all observations lie within three standard deviations of the mean.

Note that the Empirical Rule provides approximate proportions, whereas Chebysheff's Theorem provides lower bounds on the proportions contained in the intervals.

EXAMPLE 4.10

Using Chebysheff's Theorem to Interpret Standard Deviation

The annual salaries of the employees of a chain of computer stores produced a positively skewed histogram. The mean and standard deviation are $28,000 and $3,000, respectively. What can you say about the salaries at this chain?

SOLUTION

Because the histogram is not bell shaped, we cannot use the Empirical Rule. We must employ Chebysheff's Theorem instead.

The intervals created by adding and subtracting two and three standard deviations to and from the mean are as follows:

1. At least 75% of the salaries lie between \$22,000 [the mean minus two standard deviations = 28,000 − 2(3,000)] and \$34,000 [the mean plus two standard deviations = 28,000 + 2(3,000)].
2. At least 88.9% of the salaries lie between \$19,000 [the mean minus three standard deviations = 28,000 − 3(3,000)] and \$37,000 [the mean plus three standard deviations = 28,000 + 3(3,000)].

Coefficient of Variation

Is a standard deviation of 10 a large number indicating great variability or is it a small number indicating little variability? The answer depends somewhat on the magnitude of the observations in the data set. If the observations are in the millions, a standard deviation of 10 will probably be considered a small number. On the other hand, if the observations are less than 50, the standard deviation of 10 would be seen as a large number. This logic lies behind yet another measure of variability, the *coefficient of variation*.

Coefficient of Variation

The **coefficient of variation** of a set of observations is the standard deviation of the observations divided by their mean.

$$\text{Population coefficient of variation: } CV = \frac{\sigma}{\mu}$$

$$\text{Sample coefficient of variation: } cv = \frac{s}{\bar{x}}$$

Measures of Variability for Ordinal and Nominal Data

The measures of variability introduced in this section can be used only for interval data. The next section will feature a measure that can be used to describe the variability of ordinal data. There are no measures of variability for nominal data.

Approximating the Mean and Variance from Grouped Data

The statistical methods presented in this chapter are used to compute descriptive statistics from data. However, in some circumstances the statistics practitioner does not have the raw data but instead has a frequency distribution. This is often the case when data are supplied by government organizations. In CD Appendix C we provide the formulas used to approximate the sample mean and variance.

We complete this section by reviewing the factors that identify the use of measures of variability.

Factors That Identify When to Compute the Range, Variance, Standard Deviation, and Coefficient of Variation

1. **Objective:** Describe a single set of data
2. **Type of Data:** Interval
3. **Descriptive measurement:** Variability

EXERCISES

4.19 Calculate the variance of the following data.

9 3 7 4 1 7 5 4

4.20 Calculate the variance of the following data.

4 5 3 6 5 6 5 6

4.21 Determine the variance and standard deviation of the following sample.

12 6 22 31 23 13 15 17 21

4.22 Find the variance and standard deviation of the following sample.

0 −5 −3 6 4 −4 1 −5 0 3

4.23 Examine the three samples listed here. Without performing any calculations, indicate which sample has the largest amount of variation and which sample has the smallest amount of variation. Explain how you produced your answer.
a. 17 29 12 16 11
b. 22 18 23 20 17
c. 24 37 6 39 29

4.24 Refer to Exercise 4.23. Calculate the variance for each part. Was your answer in Exercise 4.23 correct?

4.25 A friend calculates a variance and reports that it is −25.0. How do you know that he has made a serious calculation error?

4.26 Create a sample of five numbers whose mean is 6 and whose standard deviation is 0.

4.27 A set of data whose histogram is bell shaped yields a mean and standard deviation of 50 and 4, respectively. Approximately what proportion of observations
a. are between 46 and 54?
b. are between 42 and 58?
c. are between 38 and 62?

4.28 Refer to Exercise 4.27. Approximately what proportion of observations
a. are less than 46?
b. are less than 58?
c. are greater than 54?

4.29 A set of data whose histogram is extremely skewed yields a mean and standard deviation of 70 and 12, respectively. What is the minimum proportion of observations that
a. are between 46 and 94?
b. are between 34 and 106?

4.30 A statistics practitioner determined that the mean and standard deviation of a data set were 120 and 30, respectively. What can you say about the proportions of observations that lie between each of the following intervals?
a. 90 and 150
b. 60 and 180
c. 30 and 210

The following exercises require a computer and software.

4.31 Xr04-31 There has been much media coverage of the high cost of medicinal drugs in the United States. One concern is the large variation from pharmacy to pharmacy. To investigate, a consumer advocacy group took a random sample of 100 pharmacies around the country and recorded the price (in dollars per 100 pills) of Prozac. Compute the range, variance, and standard deviation of the prices. Discuss what these statistics tell you.

4.32 Xr04-32 Many traffic experts argue that the most important factor in accidents is not the average speed of cars, but the amount of variation. Suppose that the speeds of a sample of 200 cars were taken over a stretch of highway that has seen numerous accidents. Compute the variance and standard deviation of the speeds, and interpret the results.

4.33 Xr04-33 Three men were trying to make the football team as punters. The coach had each of them punt the ball 50 times and the distances were recorded.

a. Compute the variance and standard deviation for each punter.
b. What do these statistics tell you about the punters?

4.34 Xr04-34 Variance is often used to measure quality in production-line products. Suppose that a sample of steel rods that are supposed to be exactly 100 cm long is taken. The length of each is determined, and the results are recorded. Calculate the variance and the standard deviation. Briefly describe what these statistics tell you.

4.35 Xr04-35 To learn more about the size of withdrawals at a banking machine, the proprietor took a sample of 75 withdrawals and recorded the amounts. Determine the mean and standard deviation of these data, and describe what these two statistics tell you about the withdrawal amounts.

4.36 Xr04-36 Everyone is familiar with waiting lines or queues. For example, people wait in line at a supermarket to go through the checkout counter. There are two factors that determine how long the queue becomes. One is the speed of service. The other is the number of arrivals at the checkout counter. The mean number of arrivals is an important number, but so is the standard deviation. Suppose that a consultant for the supermarket counts the number of arrivals per hour during a sample of 150 hours.

a. Compute the standard deviation of the number of arrivals.
b. Assuming that the histogram is bell shaped, interpret the standard deviation.

4.3 Measures of Relative Standing and Box Plots

Measures of relative standing are designed to provide information about the position of particular values relative to the entire data set. We've already presented one measure of relative standing, the median, which is also a measure of central location. Recall that the median divides the data set into halves, allowing the statistics practitioner to determine which half of the data set each observation lies in. The statistics we're about to introduce will give you much more detailed information.

Percentile

The *P*th **percentile** is the value for which *P* percent are less than that value and $(100 - P)\%$ are greater than that value.

The scores and the percentiles of the SAT (Scholastic Achievement Test) and the GMAT (Graduate Management Admission Test), as well as various other admissions tests, are reported to students taking them. Suppose for example, that your SAT score is reported to be at the 60th percentile. This means that 60% of all the other marks are below yours and 40% are above it. You now know exactly where you stand relative to the population of SAT scores.

We have special names for the 25th, 50th, and 75th percentiles. Because these three statistics divide the set of data into quarters, these measures of relative standing are also called **quartiles.** The *first* or *lower quartile* is labeled Q_1. It is equal to the 25th percentile. The *second quartile*, Q_2 is equal to the 50th percentile, which is also the median. The *third* or *upper quartile*, Q_3 is equal to the 75th percentile. Incidentally, many people confuse the terms *quartile* and *quarter*. A common error is to state that someone is in the lower *quartile* of a group when they actually mean that someone is in the lower *quarter* of a group.

Besides quartiles, we can also convert percentiles into quintiles and deciles. *Quintiles* divide the data into fifths, and *deciles* divide the data into tenths.

Locating Percentiles

The following formula allows us to approximate the location of any percentile.

Location of a Percentile

$$L_P = (n + 1)\frac{P}{100}$$

where L_P is the location of the Pth percentile.

EXAMPLE 4.11

Percentiles of Time Spent on Internet

Calculate the 25th, 50th, and 75th percentiles (first, second, and third quartiles) of the data in Example 4.1.

SOLUTION

Placing the 10 observations in ascending order we get

0 0 5 7 8 9 12 14 22 33

The location of the 25th percentile is

$$L_{25} = (10 + 1)\frac{25}{100} = (11)(.25) = 2.75$$

The 25th percentile is three-quarters of the distance between the second (which is 0) and the third (which is 5) observations. Three-quarters of the distance is

$$(.75)(5 - 0) = 3.75$$

Because the second observation is 0, the 25th percentile is $0 + 3.75 = 3.75$.

To locate the 50th percentile, we substitute $P = 50$ into the formula and produce

$$L_{50} = (10 + 1)\frac{50}{100} = (11)(.5) = 5.5$$

which means that the 50th percentile is halfway between the fifth and sixth observations. The fifth and sixth observations are 8 and 9, respectively. The 50th percentile is 8.5. This is the median calculated in Example 4.3.

The 75th percentile's location is

$$L_{75} = (10 + 1)\frac{75}{100} = (11)(.75) = 8.25$$

Thus, it is located one-quarter of the distance between the eighth and the ninth observations, which are 14 and 22, respectively. One-quarter of the distance is

$$(.25)(22 - 14) = 2$$

which means that the 75th percentile is

$$14 + 2 = 16$$

EXAMPLE 4.12

Quartiles of Long-Distance Telephone Bills

DATA Xm02-04

Determine the quartiles for Example 2.4.

SOLUTION

EXCEL

	A	B
1	Bills	
2		
3	Mean	43.59
4	Standard Error	2.76
5	Median	26.91
6	Mode	0
7	Standard Deviation	38.97
8	Sample Variance	1518.64
9	Kurtosis	-1.29
10	Skewness	0.54
11	Range	119.63
12	Minimum	0
13	Maximum	119.63
14	Sum	8717.52
15	Count	200
16	Largest(50)	85
17	Smallest(50)	9.22

INSTRUCTIONS

Follow the instructions for **Descriptive Statistics** (page 102). In the dialog box, click **Kth Largest** and type in the integer closest to $n/4$. Repeat for **Kth Smallest,** typing in the integer closest to $n/4$.

Excel approximates the third and first percentiles in the following way. The **Largest(50)** is 85, which is the number such that 150 numbers are below it and 49 numbers are above it. The **Smallest(50)** is 9.22, which is the number such that 49 numbers are below it and 150 numbers are above it. The median is 26.91, a statistic we discussed in Example 4.4.

MINITAB

Descriptive Statistics: Bills

Variable	N	N*	Mean	StDev	Variance	Minimum	Q1	Median	Q3
Bills	200	0	43.59	38.97	1518.64	0.000000000	9.28	26.91	84.94

Variable	Maximum
Bills	119.63

Minitab outputs the first and third quartiles as Q1 (9.28) and Q3 (84.94), respectively. (See page 102.)

We can often get an idea of the shape of the histogram from the quartiles. For example, if the first and second quartiles are closer to each other than are the second and third quartiles, the histogram is positively skewed. If the first and second quartiles are farther apart than the second and third quartiles, the histogram is negatively skewed. If the difference between the first and second quartiles is approximately equal to the difference between the second and third quartiles, the histogram is approximately symmetric. The box plot described subsequently is particularly useful in this regard.

Interquartile Range

The quartiles can be used to create another measure of variability, the **interquartile range,** which is defined as follows.

Interquartile Range

$$\text{Interquartile range} = Q_3 - Q_1$$

The interquartile range measures the spread of the middle 50% of the observations. Large values of this statistic mean that the first and third quartiles are far apart, indicating a high level of variability.

EXAMPLE 4.13

Interquartile Range of Long-Distance Telephone Bills

Determine the interquartile range for Example 2.4.

SOLUTION

Using Excel's approximations of the first and third quartiles, we find

$$\text{Interquartile range} = Q_3 - Q_1 = 85 - 9.22 = 75.78$$

Box Plots

Now, that we have introduced quartiles we can present one more graphical technique, the **box plot.** This technique graphs five statistics, the minimum and maximum observations, and the first, second, and third quartiles. It also depicts other features of a set of data. Figure 4.1 exhibits the box plot of the data in Example 4.1.

FIGURE **4.1** Box Plot for Example 4.1

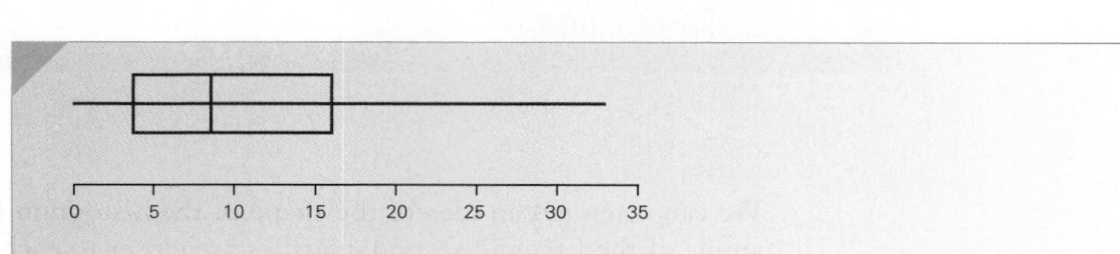

The three vertical lines of the box are the first, second, and third quartiles. The lines extending to the left and right are called *whiskers*. Any points that lie outside the whiskers are called *outliers*. The whiskers extend outward to the smaller of 1.5 times the interquartile range or to the most extreme point that is not an outlier.

Outliers

Outliers are unusually large or small observations. Because an outlier is considerably removed from the main body of the data set, its validity is suspect. Consequently, outliers should be checked to determine that they are not the result of an error in recording their values. Outliers can also represent unusual observations that should be investigated. For example, if a salesperson's performance is an outlier on the high end of the distribution, the company could profit by determining what sets that salesperson apart from the others.

EXAMPLE 4.14

Box Plot of Long-Distance Telephone Bills

DATA Xm02-04

Draw the box plot for Example 2.4.

SOLUTION

EXCEL

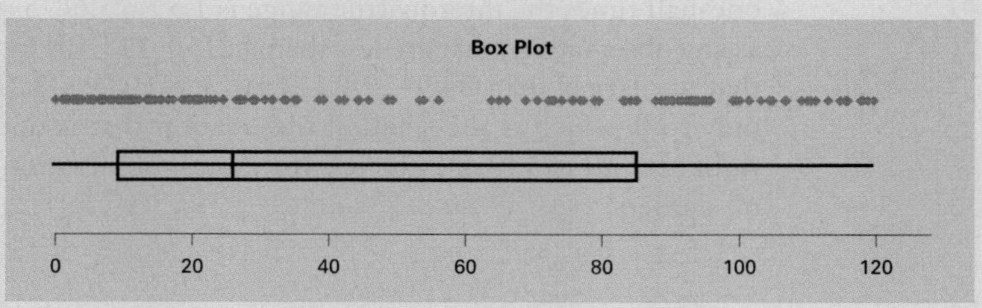

INSTRUCTIONS

1. Type or import the data into one column or two or more adjacent columns. (Open Xm02-04.)
2. Click **Add-Ins, Data Analysis Plus,** and **Box Plot.**
3. Specify the **Input Range** (A1:A201).

A box plot will be created for each column of data that you have specified or highlighted.

Notice that the quartiles produced in the **Box Plot** are not exactly the same as those produced by **Descriptive Statistics.** The **Box Plot** command uses a slightly different method than the **Descriptive Methods** command. Note that Excel in Office 2007 takes an unusually long time to draw box plots.

MINITAB

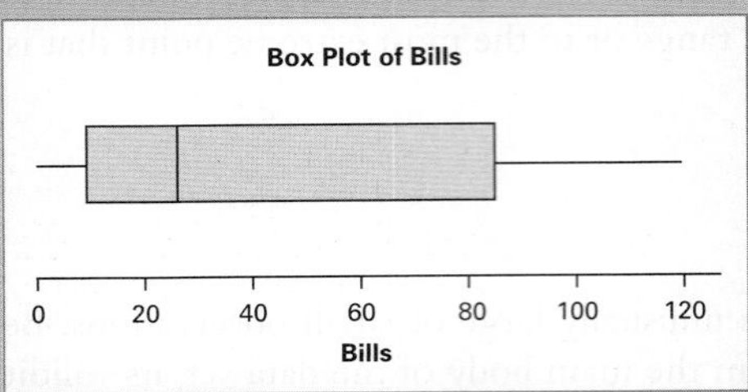

INSTRUCTIONS

1. Type or import the data into one column or more columns. (Open Xm02-04.)
2. Click **Graph** and **Box Plot . . . .**
3. Click **Simple** if there is only one column of data or **Multiple Y's** if there are two or more columns.
4. Type or **Select** the variable or variables in the **Graph variables** box (Bills).
5. The box plot will be drawn so that the values (Bills) will appear on the vertical axis. To turn the box plot on its side, click **Scale . . . , Axes and Ticks,** and **Transpose value and category scales.**

INTERPRET

The smallest value is 0 and the largest is 119.63. The first, second, and third quartiles are 9.275, 26.905, and 84.9425, respectively. The interquartile range is 75.6675. One and one-half times the interquartile range is $1.5 \times 75.6675 = 113.5013$. Outliers are defined as any observations that are less than $9.275 - 113.5013 = -104.226$ and any observations that are larger than $84.9425 + 113.5013 = 198.4438$. The whisker to the left extends only to 0, which is the smallest observation that is not an outlier. The whisker to the right extends to 119.63, which is the largest observation that is not an outlier. There are no outliers.

The box plot is particularly useful when comparing two or more data sets.

EXAMPLE 4.15 Comparing Service Times of Fast-Food Restaurants' Drive Throughs

DATA Xm04-15

A large number of fast-food restaurants with drive-through windows offer drivers and their passengers the advantages of quick service. To measure how good the service is, an organization called QSR planned a study in which the amount of time taken by a sample of drive-through customers at each of five restaurants was recorded. Compare the five sets of data using a box plot and interpret the results.

SOLUTION

We use the computer and our software to produce the box plots.

EXCEL

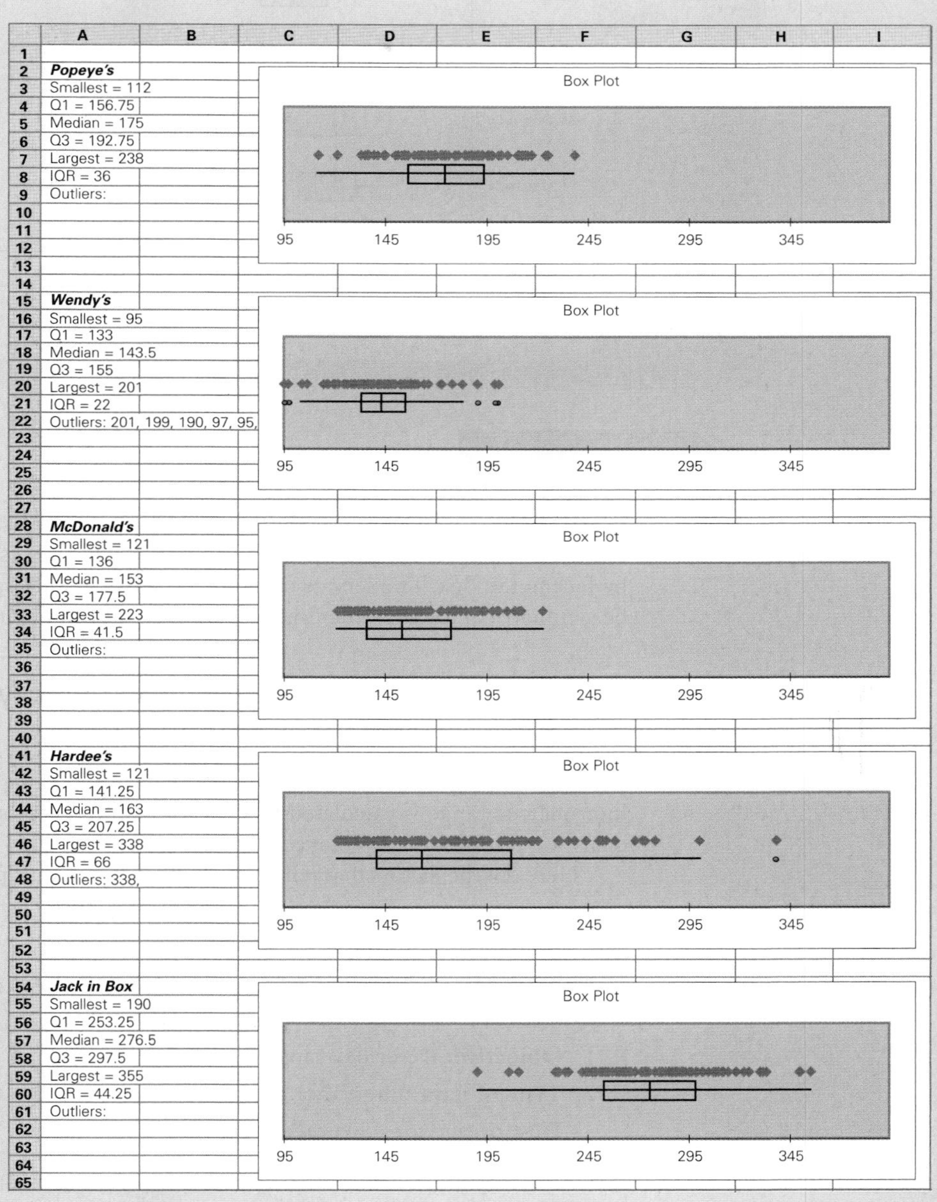

MINITAB

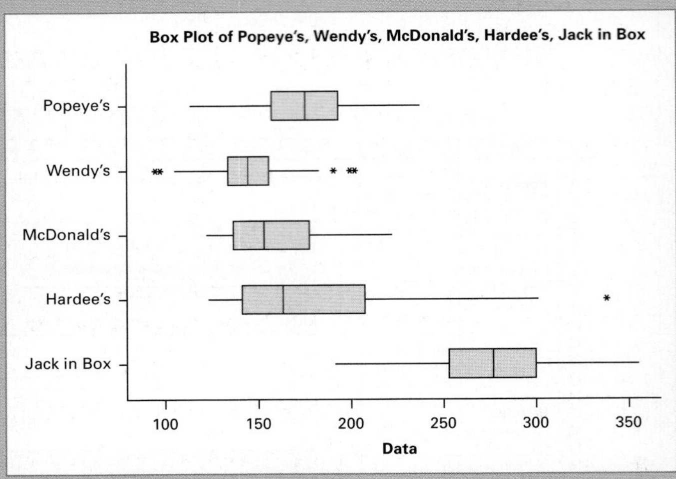

INTERPRET

Wendy's times appear to be the lowest and most consistent. The service times for Hardee's display considerably more variability. The slowest service times are provided by Jack in the Box. The service times for Popeye's, Wendy's, and Jack in the Box seem to be symmetric. However, the times for McDonald's and Hardee's are positively skewed.

Measures of Relative Standing and Variability for Ordinal Data

Because the measures of relative standing are computed by ordering the data, these statistics are appropriate for ordinal as well as for interval data. Furthermore, because the interquartile range is calculated by taking the difference between the upper and lower quartiles, it too can be employed to measure the variability of ordinal data.

Here are the factors that tell us when to use the techniques presented in this section.

Factors That Identify When to Compute Percentiles and Quartiles

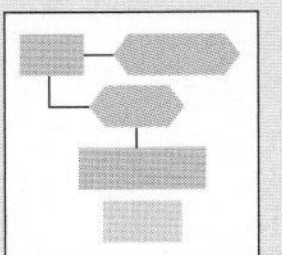

1. **Objective:** Describe a single set of data
2. **Type of data:** Interval or ordinal
3. **Descriptive measurement:** Relative standing

Factors That Identify When to Compute the Interquartile Range

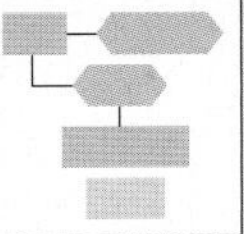

1. **Objective:** Describe a single set of data
2. **Type of data:** Interval or ordinal
3. **Descriptive measurement:** Variability

EXERCISES

4.37 Calculate the first, second, and third quartiles of the following sample.

5 8 2 9 5 3 7 4 2 7 4 10 4 3 5

4.38 Find the third and eighth deciles (30th and 80th percentiles) of the following data set.

26 23 29 31 24 22 15 31 30 20

4.39 Find the first and second quintiles (20th and 40th percentiles) of the data shown here.

52 61 88 43 64 71 39 73 51 60

4.40 Determine the first, second, and third quartiles of the following data.

10.5 14.7 15.3 17.7 15.9 12.2 10.0
14.1 13.9 18.5 13.9 15.1 14.7

4.41 Calculate the 3rd and 6th deciles of the accompanying data.

7 18 12 17 29 18 4 27 30 2
4 10 21 5 8

4.42 Refer to Exercise 4.40. Determine the interquartile range.

4.43 Refer to Exercise 4.37. Determine the interquartile range.

4.44 Compute the interquartile range from the following data.

5 8 14 6 21 11 9 10 18 2

4.45 Draw the box plot of the following set of data.

9 28 15 21 12 22 29 20
23 31 11 19 24 16 13

4.46 Given this data, draw a box plot.

65 80 39 22 74 61 63 46 72 34
30 34 69 31 46 39 57 79 89 41

The following exercises require a computer and software.

4.47 Xr04-47 Accountemps, a company that supplies temporary workers, sponsored a survey of 100 executives. Each was asked to report the number of minutes they spend screening each job resume they receive.
a. Compute the quartiles.
b. What information did you derive from the quartiles? What does this suggest about writing your resume?

4.48 Xr04-48 How much do pets cost? A random sample of dog and cat owners was asked to compute the amounts of money spent on their pets (exclusive of pet food). Draw a box plot for each data set and describe your findings.

4.49 Xr04-49 The Travel Industry Association of America sponsored a poll that asked a random sample of people how much they spent in preparation for pleasure travel. Determine the quartiles and describe what they tell you.

4.50 Xr04-50 The career counseling center at a university wanted to learn more about the starting salaries of the university's graduates. They asked each graduate to report the highest salary offer received. The survey also asked each graduate to report the degree and starting salary (column 1 = BA, column 2 = BSc, column 3 = BBA, column 4 = other). Draw box plots to compare the four groups of starting salaries. Report your findings.

4.51 Xr04-51 A random sample of Boston Marathon runners was drawn and the times to complete the race were recorded.
a. Draw the box plot.
b. What are the quartiles?
c. Identify outliers.
d. What information does the box plot deliver?

4.52 Xr04-52 Do golfers who are members of private courses play faster than players on a public course? The amount of time taken for a sample of private-course and public-course golfers was recorded.
a. Draw box plots for each sample.
b. What do the box plots tell you?

4.53 Xr04-53 For many restaurants the amount of time customers linger over coffee and dessert negatively affects profits. To learn more about this variable, a sample of 200 restaurant groups was observed and the amount of time customers spent in the restaurant was recorded.
a. Calculate the quartiles of these data.
b. What do these statistics tell you about the amount of time spent in this restaurant?

4.54 Xr04-54 Homeowners were surveyed and asked to report the size of their mortgage payments. Compute the quartiles and describe what they tell you. (Adapted from *Statistical Abstract of the United States*, 2000, Table 1209)

4.4 Measures of Linear Relationship

In Chapter 2 we introduced the scatter diagram, a graphical technique that describes the relationship between two interval variables. At that time we pointed out that we were particularly interested in the direction and strength of the linear relationship. We now present three numerical **measures of linear relationship** that provide this information. They are the *covariance*, *coefficient of correlation*, and *coefficient of determination*. Later in this section we discuss another related numerical technique, the *least squares line*.

Covariance

As we did in Chapter 2, we label one variable X and the other Y.

Covariance

Population covariance: $$\sigma_{xy} = \frac{\sum_{i=1}^{N}(x_i - \mu_x)(y_i - \mu_y)}{N}$$

Sample covariance: $$s_{xy} = \frac{\sum_{i=1}^{n}(x_i - \bar{x})(y_i - \bar{y})}{n - 1}$$

The denominator in the calculation of the sample **covariance** is $n - 1$, not the more logical n for the same reason we divide by $n - 1$ to calculate the sample variance (see page 107). If you plan to compute the sample covariance manually, here is a shortcut calculation.

Shortcut for Sample Covariance

$$s_{xy} = \frac{1}{n-1}\left[\sum_{i=1}^{n} x_i y_i - \frac{\sum_{i=1}^{n} x_i \sum_{i=1}^{n} y_i}{n}\right]$$

To illustrate how covariance measures the linear relationship, examine the following three sets of data.

Set 1

x_i	y_i	$(x_i - \bar{x})$	$(y_i - \bar{y})$	$(x_i - \bar{x})(y_i - \bar{y})$
2	13	−3	−7	21
6	20	1	0	0
7	27	2	7	14
$\bar{x} = 5$	$\bar{y} = 20$			$s_{xy} = 35/2 = 17.5$

Set 2

x_i	y_i	$(x_i - \bar{x})$	$(y_i - \bar{y})$	$(x_i - \bar{x})(y_i - \bar{y})$
2	27	−3	7	−21
6	20	1	0	0
7	13	2	−7	−14
$\bar{x} = 5$	$\bar{y} = 20$			$s_{xy} = -35/2 = -17.5$

Set 3

x_i	y_i	$(x_i - \bar{x})$	$(y_i - \bar{y})$	$(x_i - \bar{x})(y_i - \bar{y})$
2	20	−3	0	0
6	27	1	7	7
7	13	2	−7	−14
$\bar{x} = 5$	$\bar{y} = 20$			$s_{xy} = -7/2 = -3.5$

Notice that the values of x are the same in all three sets and that the values of y are also the same. The only difference is the order of the values of y.

In set 1, as x increases so does y. When x is larger than its mean, y is at least as large as its mean. Thus $(x_i - \bar{x})$ and $(y_i - \bar{y})$ have the same sign or 0. Their product is also positive or 0. Consequently, the covariance is a positive number. Generally, when two variables move in the same direction (both increase or both decrease), the covariance will be a large positive number.

If you examine set 2, you will discover that as x increases, y decreases. When x is larger than its mean, y is less than or equal to its mean. As a result when $(x_i - \bar{x})$ is positive, $(y_i - \bar{y})$ is negative or 0. Their products are either negative or 0. It follows that the covariance is a negative number. In general, when two variables move in opposite directions, the covariance is a large negative number.

In set 3, as x increases, y does not exhibit any particular direction. One of the products $(x_i - \bar{x})(y_i - \bar{y})$ is 0, one is positive, and one is negative. The resulting covariance is a small number. In general, when there is no particular pattern, the covariance is a small number.

We would like to extract two pieces of information. The first is the sign of the covariance, which tells us the nature of the relationship. The second is the magnitude, which describes the strength of the association. Unfortunately, the magnitude may be difficult to judge. For example, if you're told that the covariance between two variables is 500, does this mean that there is a strong linear relationship? The answer is that it is impossible to judge without additional statistics. Fortunately, we can improve on the information provided by this statistic by creating another one.

Coefficient of Correlation

The **coefficient of correlation** is defined as the covariance divided by the standard deviations of the variables.

Coefficient of Correlation

Population coefficient of correlation: $\rho = \dfrac{\sigma_{xy}}{\sigma_x \sigma_y}$

Sample coefficient of correlation: $r = \dfrac{s_{xy}}{s_x s_y}$

The population parameter is denoted by the Greek letter *rho*.

The advantage that the coefficient of correlation has over the covariance is that the former has a set lower and upper limit. The limits are −1 and +1, respectively. That is,

$$-1 \le r \le +1 \quad \text{and} \quad -1 \le \rho \le +1$$

When the coefficient of correlation equals −1, there is a negative linear relationship and the scatter diagram exhibits a straight line. When the coefficient of correlation equals +1, there is a perfect positive relationship. When the coefficient of correlation equals 0, there is no linear relationship. All other values of correlation are judged in relation to these three values. The drawback to the coefficient of correlation is that except for the three values −1, 0, and +1, we cannot interpret the correlation. For example, suppose that we calculated the coefficient of correlation to be −.4. What does this tell us? It tells us two things. The minus sign tells us the relationship is negative and since .4 is closer to 0 than to 1, we judge that the linear relationship is weak. In many applications we need a better interpretation than the "linear relationship is weak." Fortunately, there is yet another measure of the strength of a linear relationship, which gives us more information. It is the *coefficient of determination,* which we introduce later in this section.

EXAMPLE 4.16 Calculating the Coefficient of Correlation

Calculate the coefficient of correlation for the three sets of data on pages 124–125.

SOLUTION

Because we've already calculated the covariances, we need to compute only the standard deviations of X and Y.

$$\bar{x} = \frac{2 + 6 + 7}{3} = 5.0$$

$$\bar{y} = \frac{13 + 20 + 27}{3} = 20.0$$

$$s_x^2 = \frac{(2-5)^2 + (6-5)^2 + (7-5)^2}{3-1} = \frac{9 + 1 + 4}{2} = 7.0$$

$$s_y^2 = \frac{(13-20)^2 + (20-20)^2 + (27-20)^2}{3-1}$$

$$= \frac{49 + 0 + 49}{2} = 49.0$$

The standard deviations are

$$s_x = \sqrt{7.0} = 2.65$$

$$s_y = \sqrt{49.0} = 7.00$$

The coefficients of correlation are

$$\text{Set 1:} \quad r = \frac{s_{xy}}{s_x s_y} = \frac{17.5}{(2.65)(7.0)} = .943$$

$$\text{Set 2:} \quad r = \frac{s_{xy}}{s_x s_y} = \frac{-17.5}{(2.65)(7.0)} = -.943$$

$$\text{Set 3:} \quad r = \frac{s_{xy}}{s_x s_y} = \frac{-3.5}{(2.65)(7.0)} = -.189$$

It is now easier to see the strength of the linear relationship between X and Y.

Comparing the Scatter Diagram, Covariance, and Coefficient of Correlation

The scatter diagram depicts relationships graphically; the covariance and the coefficient of correlation describe the linear relationship numerically. Figures 4.2, 4.3, and 4.4 depict three scatter diagrams. To show how the graphical and numerical techniques compare, we calculated the covariance and the coefficient of correlation for each. (The data are stored in files Fig04-02, Fig04-03, and Fig04-04.) As you can see Figure 4.2 depicts a strong positive relationship between the two variables. The covariance is 36.87 and the coefficient of correlation is .9641. The variables in Figure 4.3 produced a relatively strong negative linear relationship; the covariance and coefficient of correlation are −34.18 and −.8791, respectively. The covariance and coefficient of correlation for the data in Figure 4.4 are 2.07 and .1206, respectively. There is no apparent linear relationship in this figure.

FIGURE **4.2** Strong Positive Linear Relationship

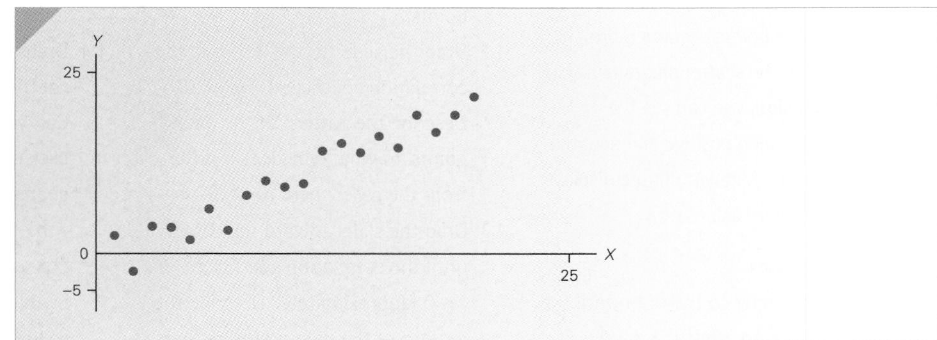

FIGURE **4.3** Strong Negative Linear Relationship

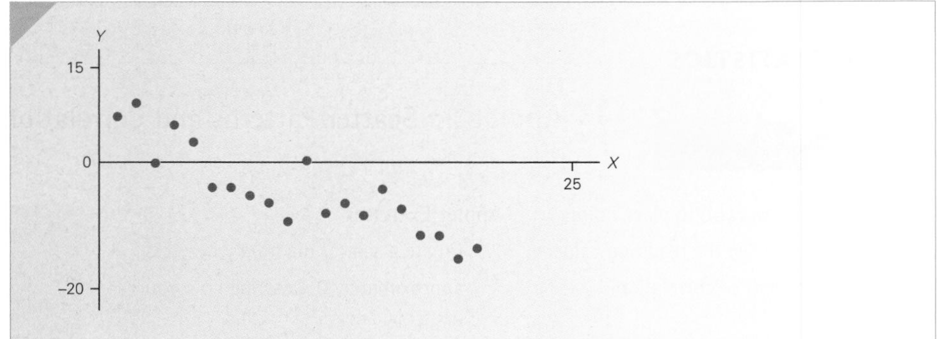

FIGURE **4.4** No Linear Relationship

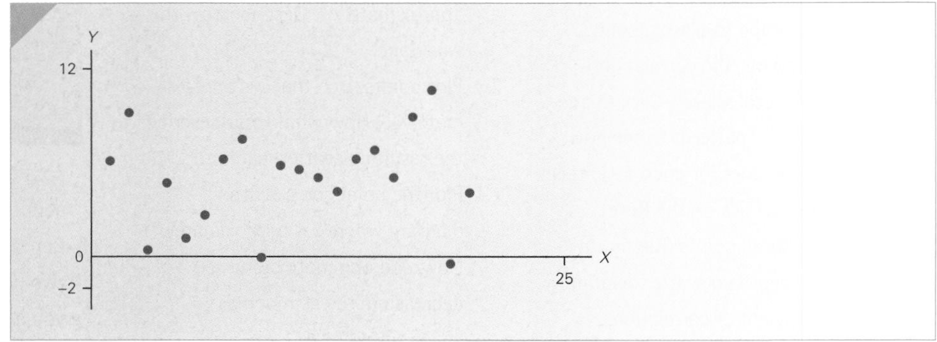

SEEING STATISTICS

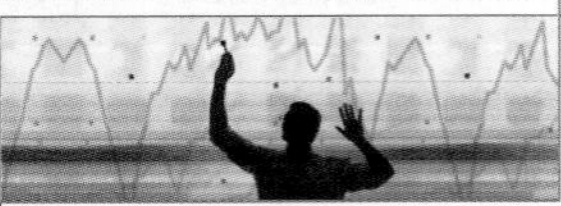

Applet 1: Scatter Diagrams and Correlation

In Section 1.3 we introduced applets as a method to allow students of applied statistics to see how statistical techniques work and to gain insights into the underlying principles. The applets are stored on the CD that accompanies this book. See the README file for instructions on how to use them.

Instructions for Applet 1

Use your mouse to move the slider in the graph. As you move the slider, observe how the coefficient of correlation changes as the points become more "organized" in the scatter diagram. If you click **Switch sign**, you can see the difference between positive and negative coefficients. The following figure displays the applet for two values of *r*.

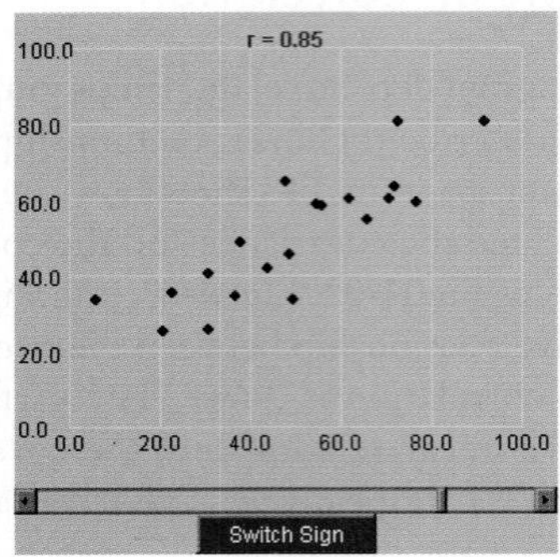

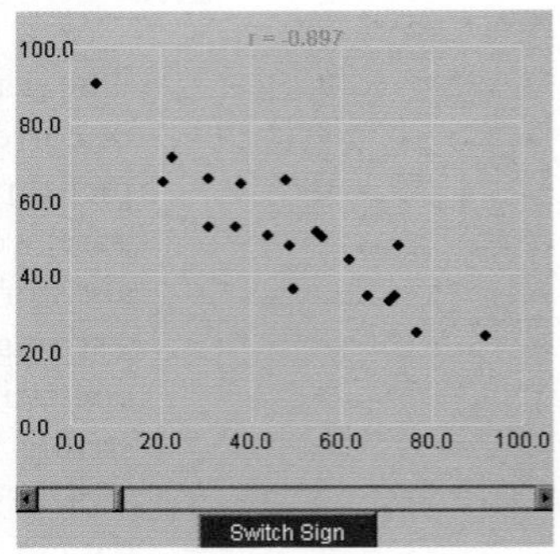

Applet Exercises

1.1 Drag the slider to the right until the correlation coefficient *r* is 1.0. Describe the pattern of the data points.

1.2 Drag the slider to the left until the correlation coefficient *r* is −1.0. Describe the pattern of the data points. In what way does it differ from the case where $r = 1.0$?

1.3 Drag the slider toward the center until the correlation coefficient *r* is 0 (approximately). Describe the pattern of the data points. Is there a pattern? Or do the points appear to be scattered randomly?

1.4 Drag the slider until the correlation coefficient *r* is .5 (approximately). Can you detect a pattern? Now click on the "Switch Sign" button to change the correlation coefficient *r* to −.5. How does the pattern change when the sign switches? Switch back and forth several times so you can see the changes.

SEEING STATISTICS

Applet 2: Scatter Patterns and Correlation

This applet allows you to place points on a graph and see the resulting value of the coefficient of correlation.

Instructions

Click on the graph to place a point. As you add points, the correlation coefficient is recalculated. Click to add points in various patterns to see how the correlation does (or does not) reflect those patterns. Click on the **Reset** button to clear all points. The figure shown here depicts a scatter diagram and its coefficient of correlation.

Applet Exercises

2.1 Create a scatter diagram where *r* is approximately 0. Describe how you did it.

2.2 Create a scatter diagram where *r* is approximately 1. Describe how this was done.

2.3 Plot points such that *r* is approximately .5. How would you describe the resulting scatter diagram?

2.4 Plot the points on a scatter diagram where *r* is approximately 1. Now, add one more point, decreasing *r* by as much as possible. What does this tell you about extreme points?

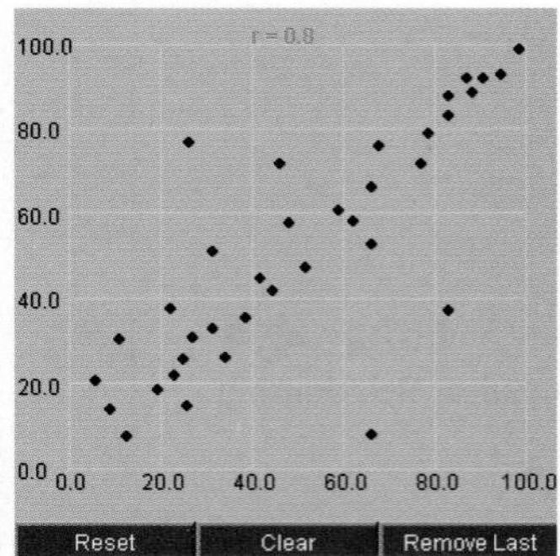

2.5 Repeat Applet Exercise 2.4 adding two points. How close to $r = 0$ did you get?

Least Squares Method

When we presented the scatter diagram in Section 2.6, we pointed out that we were interested in measuring the strength and direction of the linear relationship. Both can be more easily judged by drawing a straight line through the data. However, if different people draw a line through the same data set, it is likely that each person's line will differ from all the others. Moreover, we often need to know the equation of the line. Consequently, we need an objective method of producing a straight line. Such a method has been developed; it is called the **least squares method.**

The least squares method produces a straight line drawn through the points so that the sum of squared deviations between the points and the line is minimized. The line is represented by the equation

$$\hat{y} = b_0 + b_1 x$$

where b_0 is the y-intercept (where the line intercepts the y-axis), and b_1 is the slope (defined as rise/run), and $\hat{y}$ is the value of y determined by the line. The coefficients b_0 and b_1 are derived using calculus so that we minimize the sum of squared deviations:

$$\sum_{i=1}^{n}(y_i - \hat{y}_i)^2$$

Least Squares Line Coefficients

$$b_1 = \frac{s_{xy}}{s_x^2}$$

$$b_0 = \bar{y} - b_1\bar{x}$$

APPLICATIONS in **ACCOUNTING**

Breakeven Analysis

Breakeven analysis is an extremely important business tool, one that you will likely encounter repeatedly in your course of studies. It can be used to determine how much sales volume your business needs to start making a profit.

In Chapter 2 (page 30) we briefly introduced the four P's of marketing and illustrated the problem of pricing with Example 2.4. Breakeven analysis is especially useful when managers are attempting to determine the appropriate price for the company's products and services.

A company's profit can be calculated simply as

Profit = (Price per unit − variable cost per unit) × (Number of units sold) − Fixed costs

The breakeven point is the number of units sold such that the profit is 0. Thus, the breakeven point is calculated as

Number of units sold = Fixed cost/(Price − Variable cost)

Managers can use the formula to help determine the price that will produce a profit. However, to do so requires knowledge of the fixed and variable costs. For example, suppose that a bakery sells only loaves of bread. The bread sells for $1.20, the variable cost is $0.40, and the fixed annual costs are $10,000. The breakeven point is

Number of units sold = 10,000/(1.20 − 0.40) = 12,500

The bakery must sell more than 12,500 loaves per year to make a profit.

In the next application box we discuss fixed and variable costs.

APPLICATIONS in ACCOUNTING

© Steve Allen/Brand X Pictures/ Jupiter images

Fixed and Variable Costs

Fixed costs are costs that must be paid whether or not any units are produced. These costs are "fixed" over a specified period of time or range of production. Variable costs are costs that vary directly with the number of products produced. For the previous bakery example, the fixed costs would include rent and maintenance of the shop, wages paid to employees, advertising costs, telephone, and any other costs that are not related to the number of loaves baked. The variable cost is primarily the cost of ingredients, which rises in relation to the number of loaves baked.

There are some expenses that are mixed. For the bakery example one such cost is the cost of electricity. Electricity is needed for lights, which is considered a fixed cost, but also for the ovens and other equipment, which are variable costs.

There are several ways to break the mixed costs into fixed and variable components. One such method is the least squares line. That is, we express the total costs of some component as

$$y = b_0 + b_1x$$

where y = total mixed cost, b_0 = fixed cost, b_1 = variable cost, and x is the number of units.

EXAMPLE 4.17 Estimating Fixed and Variable Costs

DATA Xm04-17

A tool and die maker operates out of a small shop making specialized tools. He is considering increasing the size of his business and needs to know more about his costs. One such cost is electricity, which he needs to operate his machines and lights. (Some jobs require that he turn on extra bright lights to illuminate his work.) He keeps track of his daily electricity costs and the number of tools that he made that day. These data are listed next. Determine the fixed and variable electricity costs.

Day	1	2	3	4	5	6	7	8	9	10
Number of Tools	7	3	2	5	8	11	5	15	3	6
Electricity Cost	23.80	11.89	15.98	26.11	31.79	39.93	12.27	40.06	21.38	18.65

SOLUTION

The dependent variable is the daily cost of electricity and the independent variable is the number of tools. To calculate the coefficients of the least squares line and other statistics (calculated below) we need the sum of X, Y, XY, X^2, and Y^2.

Day	X	Y	XY	X^2	Y^2
1	7	23.80	166.6	49	566.44
2	3	11.89	35.67	9	141.37
3	2	15.98	31.96	4	255.36
4	5	26.11	130.55	25	681.73
5	8	31.79	254.32	64	1010.60
6	11	39.93	439.23	121	1594.40
7	5	12.27	61.35	25	150.55
8	15	40.06	600.90	225	1604.80
9	3	21.38	64.14	9	457.10
10	6	18.65	111.90	36	347.82
Total	65	241.86	1,896.62	567	6,810.20

Covariance

$$s_{xy} = \frac{1}{n-1}\left[\sum_{i=1}^{n} x_i y_i - \frac{\sum_{i=1}^{n} x_i \sum_{i=1}^{n} y_i}{n}\right] = \frac{1}{10-1}\left[1{,}896.62 - \frac{(65)(241.86)}{10}\right] = 36.06$$

Variance of X

$$s_x^2 = \frac{1}{n-1}\left[\sum_{i=1}^{n} x_i^2 - \frac{\left(\sum_{i=1}^{n} x_i\right)^2}{n}\right] = \frac{1}{10-1}\left[567 - \frac{(65)^2}{10}\right] = 16.06$$

Sample means

$$\bar{x} = \frac{\sum x_i}{n} = \frac{65}{10} = 6.5$$

$$\bar{y} = \frac{\sum y_i}{n} = \frac{241.86}{10} = 24.19$$

The coefficients of the least squares line are

Slope

$$b_1 = \frac{s_{xy}}{s_x^2} = \frac{36.06}{16.06} = 2.25$$

y-intercept

$$b_0 = \bar{y} - b_1\bar{x} = 24.19 - (2.25)(6.5) = 9.57$$

The least squares line is

$$\hat{y} = 9.57 + 2.25x$$

EXCEL

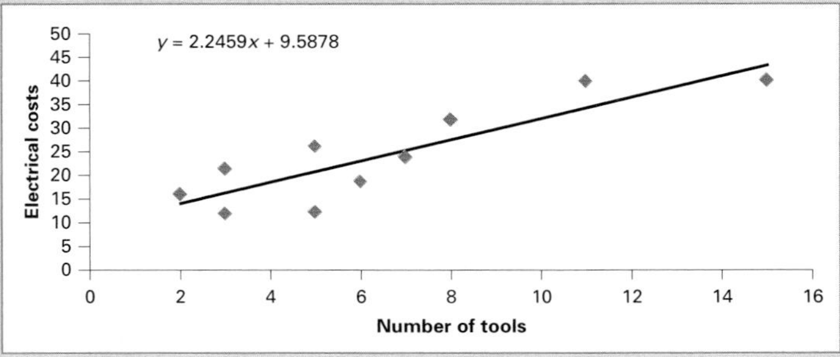

INSTRUCTIONS

1. Type or import the data into two columns where the first column stores X and the second stores Y. (Open Xm04-17.) Highlight the columns containing the variables. Follow the instructions to draw a scatter diagram (page 67).
2. In the **Chart Tools** and **Layout** menu, click **Trendline** and **Linear Trendline.**
3. Click **Trendline** and **More Trendline Options . . . .** Click **Display Equation on Chart.**

MINITAB

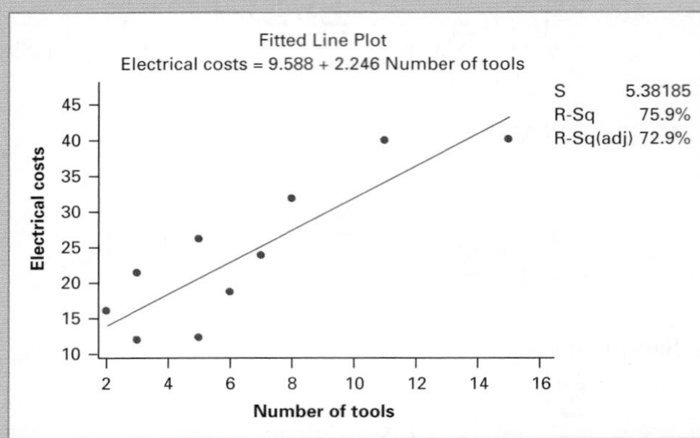

INSTRUCTIONS

1. Type or import the data into two columns. (Open Xm04-17.)
2. Click **Stat, Regression,** and **Fitted Line Plot.**
3. Specify the **Response [Y]** (Electrical cost) and the **Predictor [X]** (Number of Tools) variables. Specify **Linear.**

INTERPRET

The slope is defined as rise/run, which means that it is the change in y (rise) for a 1-unit increase in x (run). Put less mathematically, the slope measures the *marginal* rate of change in the dependent variable. The marginal rate of change refers to the effect of increasing the independent variable by one additional unit. In this example the slope is 2.25, which means that in this sample, for each 1-unit increase in the number of tools, the marginal increase in the electricity cost is 2.25. Thus, the estimated variable cost is \$2.25 per tool.

The y-intercept is 9.57. That is, the line strikes the y-axis at 9.57. This is simply the value of $\hat{y}$ when $x = 0$. However, when $x = 0$, we are producing no tools and hence the estimated fixed cost of electricity is \$9.57 per day.

Because the costs are estimates based on a straight line we often need to know how well the line fits the data.

EXAMPLE 4.18 Measuring the Strength of the Linear Relationship

DATA Xm04-17

Calculate the coefficient of correlation for Example 4.17.

SOLUTION

To calculate the coefficient of correlation we need the covariance and the standard deviations of both variables. The covariance and the variance of X were calculated in Example 4.17. The covariance is

$$s_{xy} = 36.06$$

and the variance of X is

$$s_x^2 = 16.06$$

Standard deviation of X is

$$s_x = \sqrt{s_x^2} = \sqrt{16.06} = 4.01$$

All we need is the standard deviation of Y.

$$s_y^2 = \frac{1}{n-1}\left[\sum_{i=1}^{n} y_i^2 - \frac{\left(\sum_{i=1}^{n} y_i\right)^2}{n}\right] = \frac{1}{10-1}\left[6{,}810.20 - \frac{(241.86)^2}{10}\right] = 106.73$$

$$s_y = \sqrt{s_y^2} = \sqrt{106.73} = 10.33$$

The coefficient of correlation is

$$r = \frac{s_{xy}}{s_x\, s_y} = \frac{36.06}{(4.01)(10.33)} = .8705$$

EXCEL

As was the case with the other statistics introduced in this chapter there is more than one way to calculate the coefficient of correlation and the covariance. Here are the instructions for both.

INSTRUCTIONS

1. Type or import the data into two columns. (Open Xm04-17.) Type the following into any empty cell.

   =**CORREL**[Input range of one variable], [Input range of second variable])

   In this example we would enter

   =**CORREL** (B1:B11, C1:C11)

To calculate the covariance, replace **CORREL** with **COVAR.**

Another method, which is also useful if you have more than two variables and you would like to compute the coefficient of correlation and/or the covariance for each pair of variables, is to produce the correlation matrix and the variance-covariance matrix. We do the correlation matrix first.

	A	B	C
1		*Number of tools*	*Electrical costs*
2	Number of tools	1	
3	Electrical costs	0.8711	1

INSTRUCTIONS

1. Type or import the data into adjacent columns. (Open Xm04-17.)
2. Click **Data, Data Analysis,** and **Correlation.**
3. Specify the **Input Range** (B1:C11).

The coefficient of correlation between number of tools and electrical costs is .8711 (slightly different from the manually calculated value). (The two 1's on the diagonal of the matrix are the coefficients of number of tools and number of tools, and electrical costs and electrical costs, telling you the obvious.)

Incidentally, the formulas for the population parameter ρ (Greek letter *rho*) and for the sample statistic r produce exactly the same value.

The variance-covariance matrix is shown next.

	A	B	C
1		*Number of tools*	*Electrical costs*
2	Number of tools	14.45	
3	Electrical costs	32.45	96.06

INSTRUCTIONS

1. Type or import the data into adjacent columns. (Open Xm04-17.)
2. Click **Data, Data Analysis,** and **Covariance.**
3. Specify the **Input Range** (B1:C11).

Unfortunately, Excel computes the population parameters. That is, the variance of the number of tools is $\sigma_x^2 = 14.45$ and the variance of the electrical costs is $\sigma_y^2 = 96.06$, and the covariance $\sigma_{xy} = 32.45$. You can convert these parameters to statistics by multiplying each by $n/(n-1)$.

	D	E	F
1		*Number of tools*	*Electrical costs*
2	Number of tools	16.06	
3	Electrical costs	36.06	106.73

MINITAB

Correlations: Number of tools, Electrical costs

Pearson correlation of Number of tools and Electrical costs = 0.871

INSTRUCTIONS

1. Type or import the data into two columns. (Open Xm04-17.)
2. Click **Calc, Basic Statistics** and **Correlation . . . .**
3. In the **Variables** box, type **Select** the variables **(Number of Tools, Electrical Costs).**

Covariances: Number of tools, Electrical costs

	Number of tools	Electrical costs
Number of tools	16.0556	
Electrical costs	36.0589	106.7301

INSTRUCTIONS

Click **Covariance . . .** instead of **Correlation . . .** in step 2 above.

INTERPRET

The coefficient of correlation is .8711, which tells us that there is a positive linear relationship between the number of tools and the electricity cost. The coefficient of correlation tells us that the linear relationship is quite strong and thus the estimates of the fixed and variable costs should be good.

Coefficient of Determination

When we introduced the coefficient of correlation (page 125) we pointed out that except for −1, 0, and +1 we cannot precisely interpret its meaning. We can judge the coefficient of correlation in relation to its proximity to only −1, 0, and +1. Fortunately, we have another measure that can be precisely interpreted. It is the coefficient of determination, which is calculated by squaring the coefficient of correlation. For this reason we denote it R^2.

The **coefficient of determination** measures the amount of variation in the dependent variable that is explained by the variation in the independent variable. For example, if the coefficient of correlation is −1 or +1, a scatter diagram would display all the points lining up in a straight line. The coefficient of determination is 1, which we interpret to mean that 100% of the variation in the dependent variable Y is explained by the variation in the independent variable X. If the coefficient of correlation is 0, then there is no linear relationship between the two variables, $R^2 = 0$, and none of the variation in Y is explained by the variation in X. In Example 4.18 the coefficient of correlation was calculated to be $r = .8711$. Thus, the coefficient of determination is

$$r^2 = (.8711)^2 = .7588$$

This tells us that 75.88% of the variation in electrical costs is explained by the number of tools. The remaining 24.12% is unexplained.

Using the Computer

EXCEL

You can use Excel to calculate the coefficient of correlation and then square the result. Alternatively, use Excel to draw the least squares line. After doing so, click **Trendline,** More **Trendline Options . . .**, and **Display R-squared value on chart.**

MINITAB

Minitab automatically prints the coefficient of determination.

The concept of explained variation is an extremely important one in statistics. We return to this idea repeatedly in Chapters 13, 14, 16, 17, and 18. In Chapter 16 we explain why we interpret the coefficient of determination in the way that we do.

Cost of One More Win: Solution

© Ezra Shaw/Allsport/Getty Images

To determine the cost of an additional win, we must describe the relationship between two variables. To do so, we use the least squares method to produce a straight line through the data. Because we believe that the number of games a baseball teams wins depends to some extent on its team payroll, we label Wins as the dependent variable and Payroll as the independent variable.

EXCEL

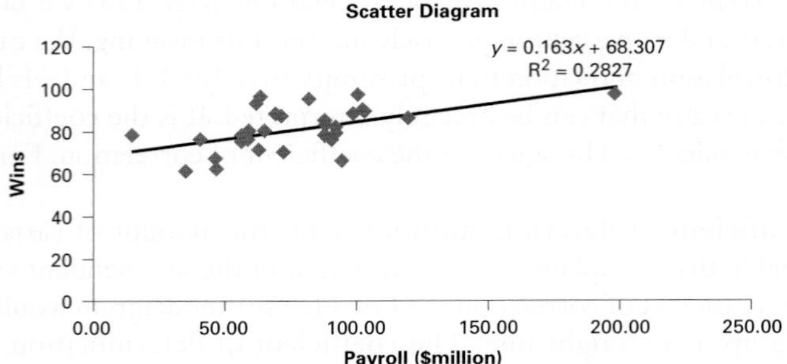

As you can see Excel output the least squares line and the coefficient of determination.

MINITAB

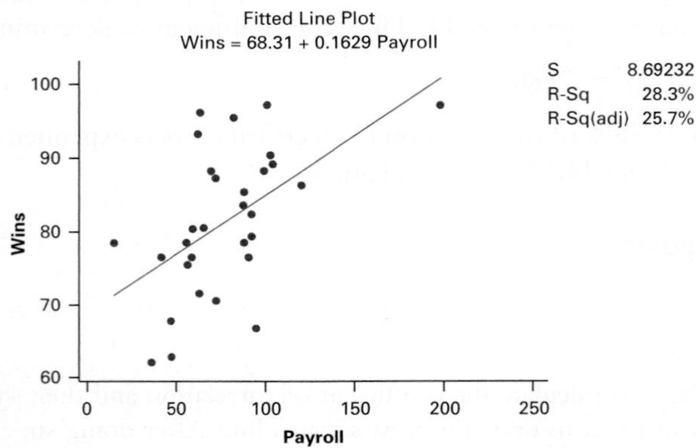

INTERPRET

The least squares line is

$$\hat{y} = 68.31 + .1629x$$

The slope is equal to .1629, which is the marginal rate of change in games won for each 1-unit increase in payroll. Because payroll is measured in millions of dollars, we estimate that for each one-million-dollar increase in the payroll the number of games won increases on average by .1629. Thus, to win one more game requires on average an additional expenditure of an incredible $6,138,735 (calculated as 1 million/.1629).

Besides analyzing the least squares line, we should determine the strength of the linear relationship. The coefficient of determination is .282, which means that the variation in the teams' payroll explains 28.2% of the variation in the teams number of games won. This suggests that there are some teams that win a small number of games with large payrolls, whereas others win a large number of games with small payrolls. In the next section we will return to this issue and examine why some teams perform better than predicted by the least squares line.

Interpreting Correlation

Because of its importance we remind you about the correct interpretation of the analysis of the relationship between two interval variables that we discussed in Chapter 2. That is, if two variables are linearly related, it does not mean that *X* is causing *Y*. It may mean that another variable is causing both *X* and *Y* or that *Y* is causing *X*. Remember,

Correlation is not causation.

We complete this section with a review of when to use the techniques introduced in this section.

Factors That Identify When to Compute Covariance, Coefficient of Correlation, Coefficient of Determination, and Least Squares Line

1. **Objective:** Describe the relationship between two variables
2. **Type of data:** Interval

EXERCISES

4.55 The covariance of two variables has been calculated to be −150. What does the statistic tell you about the relationship between the two variables?

4.56 Refer to Exercise 4.55. You've now learned that the two sample standard deviations are 16 and 12.

a. Calculate the coefficient of correlation. What does this statistic tell you about the relationship between the two variables?
b. Calculate the coefficient of determination and describe what this says about the relationship between the two variables.

4.57 Xr04-57 A retailer wanted to estimate the monthly fixed and variable selling expenses. As a first step she collected data from the past 8 months. The total selling expenses (in $thousands) and the total sales (in $thousands) were recorded and listed below.

Total Sales	Selling Expenses
20	14
40	16
60	18
50	17
50	18
55	18
60	18
70	20

a. Compute the covariance and the coefficient of determination and describe what these statistics tell you.
b. Determine the least squares line and use it to produce the estimates the retailer wants.

4.58 Xr04-58 Are the marks one receives in a course related to the amount of time spent studying the subject? To analyze this mysterious possibility a student took a random sample of 10 students who had enrolled in an accounting class last semester. He asked each to report his or her mark in the course and the total number of hours spent studying accounting. These data are listed here.

Marks	77	63	79	86	51	78	83	90	65	47
Time Spent Studying	40	42	37	47	25	44	41	48	35	28

a. Calculate the covariance.
b. Calculate the coefficient of correlation.
c. Calculate the coefficient of determination.
d. Determine the least squares line.
e. What do the statistics calculated above tell you about the relationship between marks and study time?

4.59 Xr04-59 Students who apply to MBA programs must take the Graduate Management Admission Test (GMAT). University admissions committees use the GMAT score as one of the critical indicators of how well a student is likely to perform in the MBA program. However, the GMAT may not be a very strong indicator for all MBA programs. Suppose that an MBA program designed for middle managers who wish to upgrade their skills was launched 3 years ago. To judge how well the GMAT score predicts MBA performance, a sample of 12 graduates was taken. Their grade point average in the MBA program (values from 0 to 12) and the GMAT score (values range from 200 to 800) are listed here. Compute the covariance and the coefficient of determination and interpret your findings.

GMAT and GPA Scores for 12 MBA Students

GMAT	599	689	584	631	594	643
MBA GPA	9.6	8.8	7.4	10.0	7.8	9.2
GMAT	656	594	710	611	593	683
MBA GPA	9.6	8.4	11.2	7.6	8.8	8.0

The following exercises require a computer and software.

4.60 Xr04-60 The unemployment rate is an important measure of the economic health of a country. The unemployment rate measures the percentage of people who are looking for work and who are without jobs. Another way of measuring this economic variable is to calculate the employment rate, which is the percentage of adults who are employed. Here are the unemployment rates and employment rates of 19 countries.

Country	Unemployment Rate	Employment Rate
Australia	6.7	70.7
Austria	3.6	74.8
Belgium	6.6	59.9
Canada	7.2	72.0
Denmark	4.3	77.0
Finland	9.1	68.1
France	8.6	63.2
Germany	7.9	69.0
Hungary	5.8	55.4
Ireland	3.8	67.3
Japan	5.0	74.3
Netherlands	2.4	65.4
New Zealand	5.3	62.3
Poland	18.2	53.5
Portugal	4.1	72.2
Spain	13.0	57.5
Sweden	5.1	73.0
United Kingdom	5.0	72.2
United States	4.8	73.1

Source: National Post Business.

Calculate the coefficient of determination and describe what you have learned.

4.61 Xr04-61 All Canadians have government-funded health insurance, which pays for any medical care they require. However, when traveling out of the country, Canadians usually acquire supplementary health insurance to cover the difference between the costs incurred for emergency treatment and what the government program pays. In the United States this cost differential can be prohibitive. Until recently, private insurance companies (such as Blue Cross/Blue Shield) charged everyone the same weekly rate, regardless of age. However, because of rising costs and the realization that older people frequently incur greater medical emergency expenses, insurers had to change their premium plans. They decided to offer rates that depend on the age of the customer. To help determine the new rates, one insurance company gathered data concerning the age and mean daily medical expenses of a random sample of 1,348 Canadians during the previous 12-month period.

a. Calculate the coefficient of determination.
b. What does the statistic calculated in Part a tell you?
c. Determine the least squares line.
d. Interpret the coefficients.
e. What rate plan would you suggest?

4.62 Xr04-62 A real estate developer of single-family dwellings across the country is in the process of developing plans for the next several years. An analyst for the company believes that interest rates are likely to increase but remain at low levels. To help make decisions about the number of homes to build, the developer acquired the monthly bank prime rate and the number of new single family homes sold monthly (thousands) from 1963 to 2007. (*Source:* Federal Reserve Statistics and U.S. Census Bureau)

a. Calculate the coefficient of determination. Explain what these statistics tell you about the relationship between the prime bank rate and the number of single-family homes sold.
b. Write a report to the company president describing your findings. (See Section 3.3 for guidelines on writing reports.)

4.63 Xr04-63 When the price of crude oil increases, do oil companies drill more oil wells? To determine the strength and nature of the relationship an economist recorded the price of a barrel of domestic crude oil (West Texas Crude) and the number of exploratory oil wells drilled for each month from 1973 to 2006. Analyze the data and explain what you have discovered. (*Source:* U.S. Department of Energy)

4.64 Xr04-64 One way of measuring the extent of unemployment is through the help wanted index, which measures the number of want ads in the nation's newspapers. The higher the index, the greater is the demand for workers. Another measure is the unemployment rate among insured workers. An economist wanted to know whether these two variables are related, and if so, how. He acquired the help wanted index and unemployment rates for each quarter between 1951 and 2006. Determine the strength and direction of the relationship. (*Source:* U.S. Department of Labor Statistics)

4.65 Xr04-65 A manufacturing firm produces its products in batches using sophisticated machines and equipment. The general manager wanted to investigate the relationship between direct labor costs and the number of units produced per batch. He recorded the data from the last 30 batches. Determine the fixed and variable labor costs.

4.66 [Xr04-66] A manufacturer has recorded its cost of electricity and the total number of hours of machine time for each of 52 weeks. Estimate the fixed and variable electricity costs.

4.67 Xr04-67 The chapter-opening example showed that there is a linear relationship between a baseball team's payroll and the number of wins. This raises the question: Are success on the field and attendance related? If the answer is no, then profit-driven owners may not be inclined to spend money to improve their teams. The statistics practitioner recorded the number of wins and the average home attendance for the 2006 baseball season.

a. Calculate whichever parameters you wish to help guide baseball owners.
b. Estimate the marginal number of tickets sold for each additional game won.

4.68 Xr04-68 Refer to Exercise 4.67. The practitioner also recorded the average away attendance for each team. Since visiting teams take a share of the gate, owners should be interested in this analysis.

a. Is visiting team attendance related to number of wins?
b. Estimate the marginal number of tickets sold for each additional game won.

4.69 Xr04-69 The number of wins and payrolls for the each team in the National Basketball Association in the 2005–2006 season were recorded.

a. Determine the marginal cost of one more win.
b. Calculate the coefficient of determination and describe what this number tells you.

4.70 Xr04-70 The number of wins and payrolls for each team in the National Football League the 2005–2006 season were recorded.

a. Determine the marginal cost of one more win.
b. Calculate the coefficient of determination and describe what this number tells you.

4.5 (Optional) Applications in Professional Sports: Baseball

In the chapter-opening example we provided the payrolls and the number of wins from the 2006 season. We discovered that there is a weak positive linear relationship between number of wins and payroll. The strength of the linear relationship tells us that some teams with large payrolls are not successful on the field, whereas some teams with small payrolls win a large number of games. It would appear that while the amount of money teams spend is a factor, another factor is *how* teams spend their money. In this section we will analyze the five seasons between 2002 and 2006 to see how small-payroll teams succeed.

Professional sports in North America is a multibillion dollar business. The cost of a new franchise in baseball, football, basketball, and hockey is often in the hundreds of millions of dollars. Although some teams are financially successful during losing seasons, success on the field is often related to financial success. (Exercises 4.67 and 4.68 reveal that there is a financial incentive to win more games.)

It is obvious that winning teams have better players. But how does a team get better players? There are three ways wherein teams acquire new players:

1. They can draft players from high school and college.
2. They can sign free agents on their team or other teams.
3. They can trade with other teams.

Drafting Players

Every year high school and university players are drafted by major league baseball teams. The order of the draft is in reverse order of the winning percentage the previous season. Teams that rank low in the standings rank high in the draft. A team that drafts and signs a player owns the rights to that player for his first 7 years in the minor leagues and his first 6 years in the major leagues. The decision on whom to draft and in what order is made by the general manager and a group of scouts who travel the country watching high school and college games. Young players are often invited to a camp where variables such as running speed, home run power, and, for pitchers, velocity are measured. They are often judged by whether a young man "looks" like a player. That is, taller more athletic men are judged to be better than shorter heavier ones.

Free Agency

For the first 3 years in the major leagues the team can pay a player the minimum, which in 2003 was $300,000 per year. After 3 years the player is eligible for arbitration. A successful player can usually increase his salary to $2 or $3 million through arbitration. After 6 years the player can become a free agent and can sign with any major league team. The top free agents such as Alex Rodriguez or Jason Giambi can make well in excess of $10,000,000 per year in a multiyear contract.

Trading

Teams will often trade with each other in the hope that the player they acquire will help the team more than the player they traded away. Many trades produce little improvement in both teams. However, in the history of baseball there have been some very one sided trades that resulted in great improvement in one team and the weakening of the other.

As you can see from the solved example "Cost of One More Win," on page 135 there is a great variation in team payrolls. In 2006 the New York Yankees spent $199 million, while the Tampa Bay Devil Rays spent $35 million (the amounts listed are payrolls at the beginning of the season). To a very large extent the ability to finance expensive teams is a function of the city in which the team plays. For example, New York, Los Angeles, Atlanta, and Arlington, Texas, are large-market teams. Tampa Bay, Oakland, and Minnesota are small-market teams. Large-market teams can afford higher salaries because of the higher gate receipts and higher fees for local television. This means that small-market teams cannot compete for the services of the top free agency players, and thus are more likely to be less successful than large market teams.

The question arises: Can small-market teams be successful on the field, and if so, how? The answer lies in how players are assessed for the draft and for trades. The decisions about whom to draft, whom to trade for, and whom to give in return are made by the team's general manager with the assistance of his assistants and the team's scouts. Because scouts are usually former major league and minor league players who were trained by other former minor league and major league players, they tend to generally agree on the value of the players in the draft. Similarly, teams making trades often trade players of "equal" value. As a result most teams evaluate players in the same way, yielding little differences in a player's worth. This raises the question: How can a team get the edge on other teams? The answer lies in statistics.

You won't be surprised to learn that the two most important variables in determining the number of wins is the number of runs the team scores and the number of runs the team allows. The number of runs allowed is a function of the quality of the team's pitchers and, to a lesser extent, the defense. Most major league teams evaluate pitchers on the velocity of their fastball. Velocities in the 90- to 100-mile per hour range get the scouts' attention. High school and college pitchers whose fastball speed is in the 80s are seldom drafted in the early rounds even when they appear to be successful in allowing a small number of opposing team's runs.

Scouts also seek out high school and college players with high batting averages and who hit home runs in high school and college.

The only way that small-budget teams can succeed is for them to evaluate players more accurately. In practice this means that they need to judge players differently from the other teams. In the following analysis we concentrate on the number of runs a team scores and the statistics that are related to this variable.

If the scouts are correct in their method of evaluating young players, the variables that would be most strongly related to the number of runs a team scores are batting average (BA) and the number of home runs (HR). (A player's batting average is computed by calculating the number of times the player hits divided by the number of at bats less bases on balls.) The coefficients of correlation for seasons 2002 to 2006 are listed here.

	Year				
Coefficients of Correlation	**2002**	**2003**	**2004**	**2005**	**2006**
Number of runs and batting average	.828	.889	.803	.780	.672
Number of runs and home runs	.682	.747	.765	.713	.559

Are there better statistics? That is, are there other team statistics that correlate more highly with the number of runs a team scores? There are two candidates. The first is the teams' on-base average (OBA); the second is the slugging percentage (SLG). The OBA is the number of hits plus bases on balls plus hit by pitcher divided by the number of at bats. The SLG is calculated by dividing the total number of bases (single = 1, double = 2, triple = 3, and home run = 4) by the number of at bats minus bases on balls and hit by pitcher. The coefficients of correlation are listed here.

	Year				
Coefficients of Correlation	**2002**	**2003**	**2004**	**2005**	**2006**
Number of runs and on-base average	.835	.916	.875	.783	.800
Number of runs and slugging percentage	.913	.951	.929	.790	.854

As you can see, for all five seasons the OBA had a higher correlation with runs than did the BA.

Comparing the coefficients of correlation of runs with HR and SLG, we can see that in all five seasons SLG was more strongly correlated than was HR.

As we've pointed out previously, we cannot definitively conclude a causal relationship from these statistics. However, because most decisions are based on the BA and HR, these statistics suggest that general managers should place much greater weight on batters' ability to get on base instead of simply reading the batting averages.

The Oakland Athletics (and a Statistics) Success Story*

In the past five years no team has been as successful as the Oakland Athletics in converting a small payroll into a large number of wins. In 2002 Oakland's payroll was $40 million and the team won 103 games. In the same season the New York Yankees spent $126 million and won the same number of games. In 2003 Oakland won 96 games, second to New York's 101 games. Oakland's payroll was $50 million whereas New York's was $153 million. In 2004 Oakland won 91 games with a payroll of $59 million and the Yankees won 101 games with a payroll of $184 million. Oakland won 88 games in 2005 and the Yankees won 95 games. Payrolls were Oakland $55 million; Yankees $208 million. In 2006 the team payrolls were Oakland $62 million; Yankees $199 million. Team wins were Oakland 93, Yankees 97.

The Athletics owe their success to general managers who were willing to rethink how teams win. The first of these general managers was Sandy Alderson, who was hired

*The Oakland success story is described in the book *Moneyball: The Art of Winning an Unfair Game* by Michael Lewis (New York, London: W. W. Norton).

by Oakland in 1993. He was a complete outsider with no baseball experience. This was an unusual hire in an organization in which managers and general managers are either former players or individuals who worked their way up the organization after years of service in a variety of jobs. At the time Oakland was a high-payroll team that was successful on the field. It was in the World Series in 1988, 1989, and 1990, and had the highest payroll in 1991. The team was owned by Walter A. Haas, Jr, who was willing to spend whatever was necessary to win. When he died in 1995 the new owners decided that the payroll was too large and restricted it. This forced Alderson to rethink strategy.

Sandy Alderson was a lawyer and a former marine. Because he was an outsider, he approached the job in a completely different way. He examined each aspect of the game and, among other things, concluded that before three outs everything was possible, but that after three outs nothing was possible. This led him to the conclusion that the way to score runs is to minimize each player's probability of making an out. Rather than judge a player by his batting average, which is the way every other general manager assessed players, it would make more sense to judge the player on his on-base average. The on-base average (explained previously) is the probability of *not* making an out. Thus was born something quite rare in baseball—a new idea. Alderson's replacement is Billy Beane, who continued and extended Alderson's thinking, including hiring a Harvard graduate to help manage the statistics.

The success of the Oakland Athletics raises the question: Why don't other teams do the same? The answer is that some teams have learned from Oakland. The Boston Red Sox and the Toronto Blue Jays have hired young general managers who share Beane's philosophy. Because virtually all the other teams' general managers are "old" baseball types, one assumes that they are unwilling to change their philosophy.

4.6 (Optional) Applications in Finance: Market Model

In the Applications in Finance boxes on pages 37 and 38 we introduced the terms *return on investment* and *risk*. We described two goals of investing. The first is to maximize the expected or mean return and the second is to minimize the risk. Financial analysts use a variety of statistical techniques to achieve these goals. Most investors are risk-averse, which means that for them minimizing risk is of paramount importance. In Section 4.2 we pointed out that variance and standard deviation are used to measure the risk associated with investments.

APPLICATIONS in FINANCE

Stock Market Indexes

Stock markets like the New York Stock Exchange (NYSE), NASDAQ, Toronto Stock Exchange (TSE), and many others around the world calculate indexes to provide information about the prices of stocks on their exchanges. A stock market index is composed of a number of stocks that more or less represent the entire market. For example, the Dow Jones Industrial Average (DJIA) is the average price of a group of 30 NYSE stocks of large publicly traded companies. The Standard and Poor's 500 (S&P) is the average price of 500 NYSE stocks. These indexes represent their stock exchanges and give readers a quick view of how well the exchange is doing as well the economy of the country as a whole. The NASDAQ 100 is the average price of the 100 largest non-financial companies on the NASDAQ exchange. The S&P/TSX Composite Index is composed of the largest companies on the TSE.

In this section we describe one of the most important applications of the use of a least squares line. It is the well-known and often applied *market model*. This model assumes that the rate of return on a stock is linearly related to the rate of return on the stock market index. The return on the index is calculated in the same way the return on a single stock is computed. For example, if the index at the end of last year was 10,000 and the value at the end of this year is 11,000, the market index annual return is 10%. The return on the stock is the dependent variable Y and the return on the index is the independent variable X.

We use the least squares line to represent the linear relationship between X and Y. The coefficient b_1 is called the stock's *beta coefficient*, which measures how sensitive the stock's rate of return is to changes in the level of the overall market. For example, if b_1 is greater than 1, the stock's rate of return is more sensitive to changes in the level of the overall market than is the average stock. To illustrate, suppose that $b_1 = 2$. Then a 1% increase in the index results in an average increase of 2% in the stock's return. A 1% decrease in the index produces an average 2% decrease in the stock's return. Thus, a stock with a beta coefficient greater than 1 will tend to be more volatile than the market.

EXAMPLE 4.19

Market Model for Research in Motion

DATA
Xm04-19

The monthly rates of return for Research in Motion (makers of the Blackberry) and the NASDAQ index (a measure of the overall NASDAQ stock market) were recorded for each month between January 2001 and December 2006. Some of these data are shown below. Estimate the market model and analyze the results.

Date	Index	RIM
Jan. 2001	0.122326	−0.17725
Feb. 2001	−0.22393	−0.41234
Mar. 2001	−0.14479	−0.43226
Apr. 2001	0.149968	0.544627
May 2001	−0.00272	−0.03833
June 2001	0.023715	−0.01165
Jul. 2006	−0.03712	−0.05934
Aug. 2006	0.044122	0.257047
Sep. 2006	0.034198	0.244242
Oct. 2006	0.047945	0.144472
Nov. 2006	0.02749	0.181733
Dec. 2006	−0.00678	−0.07959

SOLUTION

Excel's scatter diagram and least squares line are shown below. (Minitab produces a similar result.) We added the equation and the coefficient of determination to the scatter diagram.

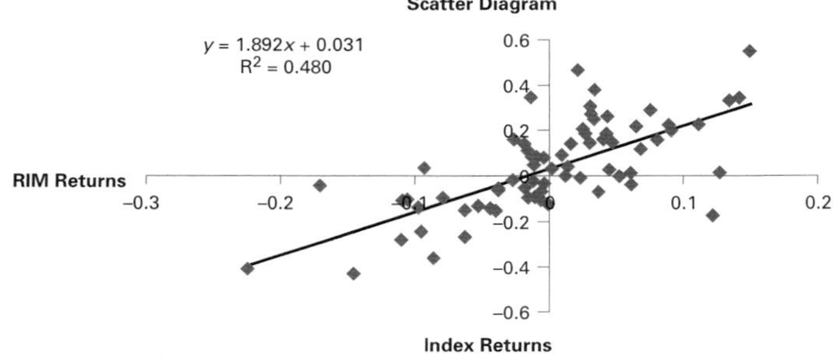

We note that the slope coefficient for RIM is 1.892. We interpret this to mean that in this sample for each 1% increase in the NADAQ Index return, the average increase in RIM's return is 1.892%. Because b_1 is greater than 1, we conclude that the return on investing in Research in Motion is more volatile and therefore riskier than the entire market.

Systematic and Firm-Specific Risk

The slope coefficient b_1 is a measure of the stock's *market-related* (or *systematic*) *risk* because it measures the volatility of the stock price that is related to the overall market volatility. The slope coefficient only informs us about the nature of the relationship between the two sets of returns. It tells us nothing about the *strength* of the linear relationship.

The coefficient of determination measures the proportion of the total risk that is market related. In this case we see that 48.0% of RIM's total risk is market related. That is, 48.0% of the variation in RIM's returns are explained by the variation in the NASDAQ Index's returns. The remaining 52.0% is the proportion of the risk that is associated with events specific to RIM, rather than the market. Financial analysts (and most everyone else) call this the *firm-specific* (or *nonsystematic*) *risk*. The firm-specific risk is attributable to variables and events not included in the market model, such as the effectiveness of RIM's sales force and managers. This is the part of the risk that can be "diversified away" by creating a portfolio of stocks as discussed in Section 7.3. We cannot however, diversify away the part of the risk that is market related.

When a portfolio has been created, we can estimate its beta by averaging the betas of the stocks that compose the portfolio. If an investor believes that the market is likely to rise, a portfolio with a beta coefficient greater than 1 is desirable. Risk-averse investors or ones who believe that the market will fall will seek out portfolios with betas less than 1.

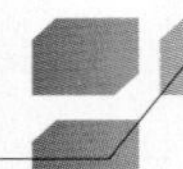

EXERCISES

The following exercises require the use of a computer and software.

4.71 Xr04-71 We have recorded the monthly returns for the S&P 500 index and the following six stocks listed on the New York Stock Exchange for the period January 2001 to December 2006.

Coca-Cola
Genentech
General Electric
General Motors
McDonald's
Motorola

Calculate the beta coefficient for each stock and briefly describe what it means.

(Excel users: Note that to use the scatter diagram to compute the beta coefficient, the data must be stored in two adjacent columns. The first must contain the returns on the index and the second stores the returns for whichever stock whose coefficient you wish to calculate.)

4.72 Xm04-72 Monthly returns for the Toronto Stock Exchange Index and the following stocks on the Toronto Stock Exchange were recorded for the years 2001 to 2006.

Barrick Gold
Bell Canada Enterprises (BCE)
Bank of Montreal (BMO)
MDS Laboratories
Petro-Canada
Research in Motion (RIM)

Calculate the beta coefficient for each stock and discuss what you have learned about each stock.

4.73 Xm04-73 We calculated the returns on the NASDAQ Index and the following stocks on the NASDAQ Exchange for the period January 2001 to December 2006.

Amgen
Ballard Power Systems
Cisco Systems
Intel
Microsoft

Calculate the beta coefficient for each stock and briefly describe what it means.

4.7 COMPARING GRAPHICAL AND NUMERICAL TECHNIQUES

As we mentioned before, graphical techniques are useful in producing a quick picture of the data. For example, you learn something about the location, spread, and shape of a set of interval data when you examine its histogram. Numerical techniques provide the same approximate information. We have measures of central location, measures of variability, and measures of relative standing that do what the histogram does. The scatter diagram graphically describes the relationship between two interval variables. But so do the numerical measures covariance, coefficient of correlation, coefficient of determination, and least squares line. Why then do we need to learn both categories of techniques? The answer is that they differ in the information each provides. We illustrate the difference between graphical and numerical methods by redoing four examples we used to illustrate graphical techniques in Chapter 2.

EXAMPLE 2.5

Comparing Returns on Two Investments

In Example 2.5 we wanted to judge which investment appeared to be better. As we discussed in the Applications in Finance: Return on Investment (page 38), we judge investments in terms of the return we can expect and its risk. We drew histograms and attempted to interpret them. The centers of the histograms provided us with information about the expected return, and their spreads gauged the risk. However, the histograms were not clear. Fortunately, we can use numerical measures. The mean and median provide us with information about the return we can expect, and the variance or standard deviation tell us about the risk associated with each investment.

Here are the descriptive statistics produced by Excel. Minitab's are similar. (We combined the output into one worksheet.)

Microsoft Excel Output for Example 2.5

	A	B	C	D	E
1	*Return A*			*Return B*	
2					
3	Mean	10.95		Mean	12.76
4	Standard Error	3.10		Standard Error	3.97
5	Median	9.88		Median	10.76
6	Mode	12.89		Mode	#N/A
7	Standard Deviation	21.89		Standard Deviation	28.05
8	Sample Variance	479.35		Sample Variance	786.62
9	Kurtosis	-0.32		Kurtosis	-0.62
10	Skewness	0.54		Skewness	0.01
11	Range	84.95		Range	106.47
12	Minimum	-21.95		Minimum	-38.47
13	Maximum	63		Maximum	68
14	Sum	547.27		Sum	638.01
15	Count	50		Count	50

We can now see that investment B has a larger mean and median, but that investment A has a smaller variance and standard deviation. If an investor were interested in low-risk investments, he or she would choose investment A. If you reexamine the histograms from Example 2.5 (page 39), you will see that the precision provided by the numerical techniques (mean, median, and standard deviation) provides more useful information than did the histograms.

EXAMPLES 2.6 AND 2.7

Business Statistics Marks; Mathematical Statistical Marks

In these examples we wanted to see what differences existed between the marks in the two statistics classes. Here are the descriptive statistics. (We combined the two printouts in one worksheet.)

Microsoft Excel Output for Examples 2.6 and 2.7

	A	B	C	D	E
1	*Marks (Example 2.6)*			*Marks (Example 2.7)*	
2					
3	Mean	72.67		Mean	66.40
4	Standard Error	1.07		Standard Error	1.61
5	Median	72		Median	71.50
6	Mode	67		Mode	75
7	Standard Deviation	8.29		Standard Deviation	12.47
8	Sample Variance	68.77		Sample Variance	155.50
9	Kurtosis	-0.36		Kurtosis	-1.24
10	Skewness	0.16		Skewness	-0.22
11	Range	39		Range	48
12	Minimum	53		Minimum	44
13	Maximum	92		Maximum	92
14	Sum	4360		Sum	3984
15	Count	60		Count	60
16	Largest(15)	79		Largest(15)	76
17	Smallest(15)	67		Smallest(15)	53

The statistics tell us that the mean and median of the marks in the business statistics course (Example 2.6) are higher than in the mathematical statistics course (Example 2.7). We found that the histogram of the mathematical statistics marks was bimodal, which we interpreted to mean that this type of approach created differences between students. The unimodal histogram of the business statistics marks informed us that this approach eliminated those differences.

Chapter 2 Opening Example

In this example we wanted to know whether the prices of gasoline and oil were related. The scatter diagram did reveal a strong positive linear relationship. We can improve upon the quality of this information by computing the coefficient of correlation and drawing the least squares line.

Excel Output for Chapter 2 Opening Example: Coefficient of Correlation

	A	B	C
1		*Oil*	*Gasoline*
2	Oil	1	
3	Gasoline	0.8574	1

The coefficient of correlation seems to confirm what we learned from the scatter diagram. That is, there is a moderately strong positive linear relationship between the two variables.

Excel Output for Chapter 2 Opening Example: Least Squares Line

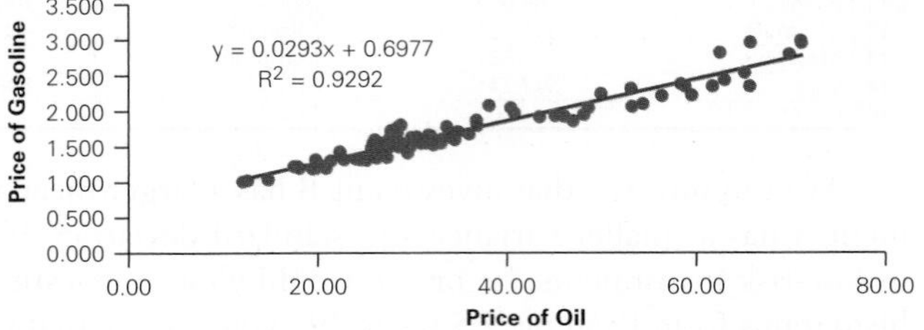

The slope coefficient tells us that for each dollar increase in the price of a barrel of oil, the price of a (U.S.) gallon of gasoline increases on average by 2.9 cents. However, because there are 42 gallons per barrel, we would expect a dollar increase in a barrel of oil to yield a 2.4* cents per gallon (calculated as \$1.00/42) increase. It does appear that the oil companies are taking some small advantage by adding an extra half cent per gallon. The coefficient of determination is .929, which indicates that 92.9% of the variation in gasoline prices is explained by the variation in oil prices.

*This is a simplification. In fact, a barrel of oil yields a variety of other profitable products. See Exercise 2.14.

EXERCISES

The following exercises require a computer and statistical software.

4.74 Refer to Exercise 2.54.

a. Calculate the mean, median, and standard deviation of the scores of those who repaid and of those who defaulted.
b. Do these statistics produce more useful information than the histograms?

4.75 Refer to Exercise 2.53.

a. Draw box plots of the scores of those who repaid and of those who defaulted.
b. Compare the information gleaned from the histograms to that contained in the box plots. Which are better?

4.76 Calculate the coefficient of determination for Exercise 2.86. Is this more informative than the scatter diagram?

4.77 Refer to Exercise 2.87. Compute the coefficients of the least squares line and compare your results with the scatter diagram.

4.78 Compute the coefficient of determination and the least squares line for Exercise 2.92. Compare this information with that developed by the scatter diagram alone.

4.79 Refer to Exercise 2.95. Calculate the coefficient of determination and the least squares line. Is this more informative than the scatter diagram?

4.80 a. Calculate the coefficients of the least squares line for the data in Example 2.12.
b. Interpret the coefficients.
c. Is this information more useful than the information extracted from the scatter diagram?

4.81 In Exercise 4.50 you drew box plots. Draw histograms instead and compare the results.

4.82 Refer to Exercise 4.52. Draw histograms of the data. What have you learned?

4.8 GENERAL GUIDELINES FOR EXPLORING DATA

The purpose of applying graphical and numerical techniques is to describe and summarize data. Statisticians usually apply graphical techniques as a first step because we need to know the shape of the distribution. The shape of the distribution helps answer the following questions:

1. Where is the approximate center of the distribution?
2. Are the observations close to one another, or are they widely dispersed?
3. Is the distribution unimodal, bimodal, or multimodal? If there is more than one mode, where are the peaks, and where are the valleys?
4. Is the distribution symmetric? If not, is it skewed? If symmetric, is it bell shaped?

Histograms and box plots provide most of the answers. We can frequently make several inferences about the nature of the data from the shape. For example, we can assess the relative risk of investments by noting their spreads. We can attempt to improve the teaching of a course by examining whether the distribution of final grades is bimodal or skewed.

The shape can also provide some guidance on which numerical techniques to use. As we noted in this chapter, the central location of highly skewed data may be more appropriately measured by the median. We may also choose to use the interquartile range instead of the standard deviation to describe the spread of skewed data.

When we have an understanding of the structure of the data, we may do additional analysis. For example, we often want to determine how one variable, or several variables, affect another. Scatter diagrams, covariance, the coefficient of correlation, and the coefficient of determination are useful techniques for detecting relationships between variables. A number of techniques to be introduced later in this book will help uncover the nature of these associations.

Chapter Summary

This chapter extended our discussion of descriptive statistics, which deals with methods of summarizing and presenting the essential information contained in a set of data. After constructing a frequency distribution to obtain a general idea about the distribution of a data set, we can use numerical measures to describe the central location and variability of interval data. Three popular measures of central location, or averages, are the mean, the median, and the mode. Taken by themselves, these measures provide an inadequate description of the data because they say nothing about the extent to which the data vary. Information regarding the variability of interval data is conveyed by such numerical measures as the range, variance, and standard deviation.

For the special case in which a sample of measurements has a mound-shaped distribution, the Empirical Rule provides a good approximation of the percentages of measurements that fall within one, two, and three standard deviations of the mean. Chebysheff's Theorem applies to all sets of data no matter the shape of the histogram.

Measures of relative standing that were presented in this chapter are percentiles and quartiles. The box plot graphically depicts these measures as well as several others. The linear relationship between two interval variables is measured by the covariance, the coefficient of correlation, the coefficient of determination, and the least squares line.

IMPORTANT TERMS

SYMBOLS

Symbol	Pronounced	Represents
μ	*mu*	Population mean
σ^2	*sigma-squared*	Population variance
σ	*sigma*	Population standard deviation
ρ	*rho*	Population coefficient of correlation
$\sum$	*Sum of*	Summation

Symbol	Pronounced	Represents
$\sum_{i=1}^{n} x_i$	*Sum of x_i from 1 to n*	Summation of n numbers
$\hat{y}$	*y-hat*	Fitted or calculated value of y
b_0	*b-zero*	y-intercept
b_1	*b-one*	Slope coefficient

FORMULAS

Population mean

$$\mu = \frac{\sum_{i=1}^{N} x_i}{N}$$

Sample mean

$$\bar{x} = \frac{\sum_{i=1}^{n} x_i}{n}$$

Range

Largest observation − Smallest observation

Population variance

$$\sigma^2 = \frac{\sum_{i=1}^{N} (x_i - \mu)^2}{N}$$

Sample variance

$$S^2 = \frac{\sum_{i=1}^{n} (x_i - \bar{x})^2}{n - 1}$$

Population standard deviation

$$\sigma = \sqrt{\sigma^2}$$

Sample standard deviation

$$s = \sqrt{s^2}$$

Population covariance

$$\sigma_{xy} = \frac{\sum_{i=1}^{N} (x_i - \mu_x)(y_i - \mu_y)}{N}$$

Sample covariance

$$s_{xy} = \frac{\sum_{i=1}^{n} (x_i - \bar{x})(y_i - \bar{y})}{n - 1}$$

Population coefficient of correlation

$$\rho = \frac{\sigma_{xy}}{\sigma_x \sigma_y}$$

Sample coefficient of correlation

$$r = \frac{s_{xy}}{s_x s_y}$$

Coefficient of determination

$$R^2 = r^2$$

Slope coefficient

$$b_1 = \frac{s_{xy}}{s_x^2}$$

y-intercept

$$b_0 = \bar{y} - b_1 \bar{x}$$

COMPUTER OUTPUT AND INSTRUCTIONS

Technique	Excel	Minitab
Mean	99	99
Median	100	100
Mode	101	101
Variance	109	N/A
Standard deviation	111	111
Descriptive statistics	117	117
Box plot	119	120
Least squares line	131	132
Covariance	133	134
Correlation	133	134
Coefficient of determination	135	135

CHAPTER EXERCISES

4.83 Xr04-83* Increasing tuition has resulted in some students being saddled with large debts upon graduation. To examine this issue, a random sample of recent graduates was asked to report whether they had student loans, and, if so, how much was the debt at graduation.

a. Compute all three measures of central location.
b. What do these statistics reveal about student loan debt at graduation?

4.84 Xr04-84* The Internet is growing rapidly with an increasing number of regular users. However, among people older than 50, Internet use is still relatively low. To learn more about this issue, a sample of 250 men and women older than 50 who had used the Internet at least once were selected. The number of hours on the Internet during the past month was recorded.

a. Calculate the mean and median.
b. Calculate the variance and standard deviation.
c. Draw a box plot.
d. Briefly describe what you have learned from the statistics you calculated.

4.85 Xr04-85* A sample was drawn of 1-acre plots of land planted with corn. The crop yields were recorded. Calculate the descriptive statistics you judge to be useful. Interpret these statistics.

4.86 Xr04-86* Chris Golfnut loves the game of golf. Chris also loves statistics. Combining both passions, Chris records a sample of 100 scores.

a. What statistics should Chris compute to describe the scores?
b. Calculate the mean and standard deviation of the scores.
c. Briefly describe what the statistics computed in Part b divulge.

4.87 Refer to Exercise 4.85. For each plot the amounts of rainfall were also recorded.

a. Compute the coefficient of determination.
b. Determine the coefficients of the least squares line.
c. Describe what these statistics tell you about the relationship between crop yield and rainfall.
d. Discuss the information obtained here and in Exercise 4.86.

4.88 Xr04-88* The temperature in December in Buffalo, New York, is often below 40 degrees Fahrenheit (4 degrees Celsius). Not surprisingly, when the National Football League Buffalo Bills play at home in December, coffee is a popular item at the concession stand. The concession manager would like to acquire more information so that he can manage inventories more efficiently. The number of cups of coffee sold during 50 games played in December in Buffalo were recorded.

a. Determine the mean and median.
b. Determine the variance and standard deviation.
c. Draw a box plot.
d. Briefly describe what you have learned from your statistical analysis.

4.89 Refer to Exercise 4.85. For each plot the amounts of fertilizer were recorded.

a. Compute the coefficient of determination.
b. Determine the coefficients of the least squares line.
c. Describe what these statistics tell you about the relationship between crop yield and the amount of fertilizer.
d. Discuss the information obtained here and in Exercise 4.86.

4.90 Xr04-90* Osteoporosis is a condition in which bone density decreases, often resulting in broken bones. Bone density usually peaks at age 30 and decreases thereafter. To understand more about the condition a random sample of women aged 50 and over were recruited. Each woman's bone density loss was recorded.

a. Compute the mean and median of these data.
b. Compute the standard deviation of the bone density losses.
c. Describe what you have learned from the statistics.

4.91 Refer to Exercise 4.84. In addition to Internet use, the numbers of years of education were recorded.

a. Compute the coefficient of determination.
b. Determine the coefficients of the least squares line.
c. Describe what these statistics tell you about the relationship between Internet use and education.
d. Discuss the information obtained here and in Exercise 4.92.

4.92 Refer to Exercise 4.88. Suppose that in addition to recording the coffee sales, the manager also recorded the average temperature (measured in degrees Fahrenheit) during the game. These data together with the number of cups of coffee sold were recorded.

a. Compute the coefficient of determination.
b. Determine the coefficients of the least squares line.
c. What have you learned from the statistics calculated in Parts a and b about the relationship between the number of cups of coffee sold and the temperature?
d. Discuss the information obtained here and in Exercise 4.80. Which is more useful to the manager?

4.93 Refer to Exercise 4.86. For each score Chris also recorded the number of putts as well as his scores. Conduct an analysis of both sets of data. What conclusions can be achieved from the statistics?

4.94 Refer to Exercise 4.90. In addition to the bone density losses, the ages of the women were also recorded. Compute the coefficient of determination and describe what this statistic tells you.

CASE 4.1 Return to the Global Warming Question

DATA
C04-01a
C04-01b
C04-01c
C04-01d

Now that we have presented techniques that allow us to conduct more precise analyses we'll return to Case 2.1. Recall that there are two issues in this discussion. First, is there global warming and second if so, is carbon dioxide the cause? The only tools available at the end of Chapter 2 were graphical techniques including line charts and scatter diagrams. You are now invited to apply the more precise techniques in this chapter to answer the same questions.

Here are the data sets you can work with.

C04-01a: Column 1: Months numbered 1 to 313
Column 2: Temperature anomalies measured by satellite

C04-01b: Column 1: Months numbered 1 to 1,525
Column 2: Temperature anomalies produced by the National Climate Data Center

C04-01c: Column 1: Months numbered 1 to 1,524
Column 2: Temperature anomalies determined by GISS

C04-01d: Monthly carbon dioxide levels measured by the Mauna Loa Observatory

a. Use the least squares method to estimate monthly temperature anomalies for all three data sets. Estimate the total increase in temperature anomalies for each data source.
b. Calculate the least squares line and the coefficient of correlation between CO_2 levels and temperature anomalies and describe your findings.

CASE 4.2 The Effect of the Players' Strike in the 2004–2005 Hockey Season

DATA
C04-02a
C04-02b

The 2004–2005 hockey season was cancelled because of a player strike. The key issue in this labor dispute was a "salary cap." The team owners wanted a salary cap to cut their costs. The owners of small-market teams wanted the cap to help their teams be competitive. Of course, caps on salaries would lower the salaries of most players and as a result the players association fought against it. The team owners prevailed and the collective bargaining agreement specified a salary cap of $39 million and a floor of $21.5 million for the 2005–2006 season.

Conduct an analysis of the 2003–2004 season (C04-02a) and the 2005–2006 season (C04-02b). For each season,

a. Estimate how much on average a team needs to spend to win one more game.
b. Measure the strength of the linear relationship.
c. Discuss the differences between the two seasons.

CASE 4.3 Cadillac's Lagging Sales

DATA
C04-03

© Eyewire Collection/Getty Images

For many years the top-selling luxury car in North America was the Cadillac. It reigned from 1950 to 1998. In 2000 it sank to sixth, behind Lexus, BMW, and Mercedes. Cadillac's sales during 2000 were half those of 1978, Cadillac's peak year. The problem is that Cadillac appears to appeal mainly to older males. Younger people seeking luxury cars shop elsewhere. Although Cadillac made $700 million in 2000, if the company doesn't pick up sales among younger shoppers, profits will go the way of fins. To put Cadillac back on top, it is necessary to understand who is buying. A survey of luxury car buyers was undertaken. The following information was gathered from random samples of recent buyers of the luxury cars.

Column	Luxury Car
1	BMW
2	Cadillac
3	Lexus
4	Lincoln
5	Mercedes-Benz

In the data set columns 1 to 5 contain the ages of the owners of the five luxury cars, columns 6 to 10 show their household incomes, and columns 11 to 15 store their years of education. (*Source*: Company reports and *Newsweek* magazine, May 28, 2001)

a. Use statistics to describe the ages, household incomes, and education of the five groups of car buyers.

b. Use box plots to compare the ages, household incomes, and education of the five groups of car buyers.

c. Write a brief report describing your findings.

CASE 4.4 Quebec Referendum Vote: Was There Electoral Fraud?*

© Bettman/Corbis

DATA C04-04

Since the 1960s Quebecois have been debating whether to separate from Canada and form an independent nation. A referendum was held on October 30, 1995, in which the people of Quebec voted not to separate. The vote was extremely close with the "no" side winning by only 52,448 votes. A large number of "no" votes was cast by the non-Francophone (Non-French speaking) people of Quebec, who make up about 20% of the population and who very much want to remain Canadians. The remaining 80% are Francophones, a majority of whom voted "yes."

After the votes were counted, it became clear that the tallied vote was much closer than it should have been. Supporters of the "no" side charged that poll scrutineers, all of whom were appointed by the pro-separatist provincial government, rejected a disproportionate number of ballots in ridings where the percentage of "yes" votes was low and where there are large numbers of Allophone (people whose first language is neither English nor French) and Anglophone (English-speaking) residents. (Electoral laws require the rejection of ballots that do not appear to be properly marked.) They were outraged that in a strong democracy like Canada, votes would be rigged much as they are in many nondemocratic countries around the world.

If, in ridings where there was a low percentage of "yes" votes there was a high percentage of rejected ballots, this would be evidence of electoral fraud. Moreover, if, in ridings where there were large percentages of Allophone and/or Anglophone voters, there were high percentages of rejected ballots, this too would constitute evidence of fraud on the part of the scrutineers and possibly the government.

To determine the veracity of the charges, the following variables were recorded for each riding.

Percentage of rejected ballots in referendum

Percentage of "yes" votes

Percentage of Allophones

Percentage of Anglophones

Conduct a statistical analysis of these data to determine whether there are indications that electoral fraud took place.

*This case is based on Jason Cawley and Paul Sommers, "Voting Irregularities in the 1995 Referendum on Quebec Sovereignty," *Chance*, Vol. 9, No. 4 (Fall 1996). We are grateful to Dr. Paul Sommers of Middlebury College, for his assistance in writing this case.

APPENDIX 4 / REVIEW OF DESCRIPTIVE TECHNIQUES

Here is a list of the statistical techniques introduced in Chapters 2 and 4. This is followed by a flowchart designed to help you select the most appropriate method to use to address any problem requiring a descriptive method.

To provide practice in identifying the correct descriptive method to use, we have created a number of review exercises. These are in CD Appendix D.

Graphical Techniques

Histogram
Stem-and-leaf display
Ogive
Bar chart
Pie chart
Scatter diagram
Line chart (time series)
Box plot

Numerical Techniques

Measures of Central Location

Mean
Median
Mode
Geometric mean (growth rates)

Measures of Variability

Range
Variance
Standard deviation
Coefficient of variation
Interquartile range

Measures of Relative Standing

Percentiles
Quartiles

Measures of Linear Relationship

Covariance
Coefficient of correlation
Coefficient of determination
Least squares line

Flowchart: Graphical and Numerical Techniques

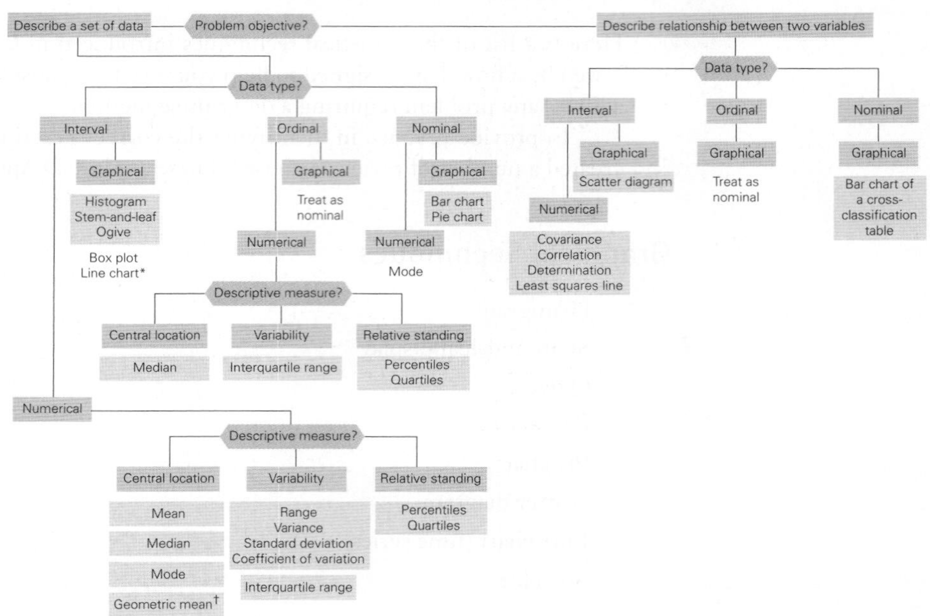

*Time-series data
†Growth rates

© Richard Iwasaki/Stock Connection/Jupiterimages

5 DATA COLLECTION AND SAMPLING

Sampling and the Census

© Spencer Grant/PhotoEdit

See page 165.

The census, which is conducted every 10 years in the United States, serves an important function. It is the basis for deciding how many congressional representatives and how many votes in the electoral college each state will have. Businesses often use the information derived from the census to help make decisions about products, advertising, and plant locations.

One of the problems with the census is the issue of undercounting, which occurs when some people are not included. For example, the 1990 census reported that 12.05% of adults were African American; the true value was 12.41%. To address undercounting, the Census Bureau adjusts the numbers it gets from the census. The adjustment is based on another survey. The mechanism is called the Accuracy and Coverage Evaluation. Using sampling methods described in this chapter, the Census Bureau is able to adjust the numbers in American subgroups. For example, the Bureau may discover that the number of Hispanics has been undercounted or that the number of people living in California has not been accurately counted.

Later in this chapter we'll discuss how the sampling is conducted and how the adjustments are made.

INTRODUCTION

In Chapter 1, we briefly introduced the concept of statistical inference—the process of inferring information about a population from a sample. Because information about populations can usually be described by parameters, the statistical technique used generally deals with drawing inferences about population parameters from sample statistics. (Recall that a parameter is a measurement about a population, and a statistic is a measurement about a sample.)

Working within the covers of a statistics textbook, we can assume that population parameters are known. In real life, however, calculating parameters is virtually impossible because populations tend to be very large. As a result, most population parameters are not only unknown, but also unknowable. The problem that motivates the subject of statistical inference is that we often need information about the value of parameters in order to make decisions. For example, to make decisions about whether to expand a line of clothing, we may need to know the mean annual expenditure on clothing by North American adults. Because the size of this population is approximately two hundred million, determining the mean is prohibitive. However, if we are willing to accept less than 100% accuracy, we can use statistical inference to obtain an estimate. Rather than investigating the entire population, we select a sample of people, determine the annual expenditures on clothing in this group, and calculate the sample mean. Although the probability that the sample mean will equal the population mean is very small, we would expect them to be close. For many decisions we need to know how close. We postpone that discussion until Chapters 10 and 11. In this chapter we will discuss the basic concepts and techniques of sampling itself. But first we take a look at various sources for collecting data.

5.1 METHODS OF COLLECTING DATA

Most of this book addresses the problem of converting data into information. The question arises: Where do data come from? The answer is that there are a large number of methods that produce data. Before we proceed, however, we'll remind you of the definition of data introduced in Section 2.1. Data are the observed values of a variable. That is, we define a variable or variables that are of interest to us and then proceed to collect observations of those variables.

Direct Observation

The simplest method of obtaining data is by direct observation. When data are gathered in this way, they are said to be **observational.** For example, suppose that a researcher for a pharmaceutical company wants to determine whether aspirin does reduce the incidence of heart attacks. Observational data may be gathered by selecting a sample of men and women and asking each whether he or she has taken aspirin regularly over the past 2 years. Each person would be asked whether he or she had suffered a heart attack over the same period. The proportions reporting heart attacks would be compared and a statistical technique that is introduced in Chapter 13 would be used to determine whether aspirin is effective in reducing the likelihood of heart attacks. There are many drawbacks to this method. One of the most critical is that it is difficult to produce useful information in this way. For example, if the statistics practitioner concludes that people who take aspirin suffer fewer heart attacks, can we conclude that aspirin is effective? It may be that people who take aspirin tend to be more health conscious, and health conscious people tend to have fewer heart attacks. The one advantage to direct observation is that it is relatively inexpensive.

Experiments

A more expensive but better way to produce data is through experiments. Data produced in this manner are called **experimental.** In the aspirin illustration, a statistics practitioner can randomly select men and women. The sample would be divided into two groups. One group would take aspirin regularly and the other would not. After 2 years, the statistics practitioner would determine the proportion of people in each group who had suffered a heart attack and again statistical methods would be used to determine whether aspirin works. If we find that the aspirin group suffered fewer heart attacks, we may more confidently conclude that taking aspirin regularly is a healthy decision.

Surveys

One of the most familiar methods of collecting data is the survey, which solicits information from people concerning such things as their income, family size, and opinions on various issues. We're all familiar, for example, with opinion polls that accompany each political election. The Gallup Poll and the Harris Survey are two well-known surveys of public opinion whose results are often reported by the media. But the majority of surveys are conducted for private use. Private surveys are used extensively by market researchers to determine the preferences and attitudes of consumers and voters. The results can be used for a variety of purposes, from helping to determine the target market for an advertising campaign to modifying a candidate's platform in an election campaign. As an illustration, consider a television network that has hired a market research firm to provide the network with a profile of owners of luxury automobiles, including what they watch on television and at what times. The network could then use this information to develop a package of recommended time slots for Cadillac commercials, including costs, which it would present to General Motors. It is quite likely that many students reading this book will one day be marketing executives who will "live and die" by such market research data.

An important aspect of surveys is the **response rate.** The response rate is the proportion of all people who were selected who complete the survey. As we discuss in the next section, a low response rate can destroy the validity of any conclusion resulting from the statistical analysis. Statistics practitioners need to ensure that data are reliable.

Personal Interview

Many researchers feel that the best way to survey people is by means of a personal interview, which involves an interviewer soliciting information from a respondent by asking prepared questions. A personal interview has the advantage of having a higher expected response rate than other methods of data collection. In addition, there will probably be fewer incorrect responses resulting from respondents misunderstanding some questions, because the interviewer can clarify misunderstandings when asked to. But the interviewer must also be careful not to say too much, for fear of biasing the response. To avoid introducing such biases, as well as to reap the potential benefits of a personal interview, the interviewer must be well trained in proper interviewing techniques and well informed on the purpose of the study. The main disadvantage of personal interviews is that they are expensive, especially when travel is involved.

Telephone Interview

A telephone interview is usually less expensive, but it is also less personal and has a lower expected response rate. Unless the issue is of interest, many people will refuse to respond to telephone surveys. This problem is exacerbated by telemarketers trying to sell something.

Self-Administered Survey

A third popular method of data collection is the self-administered questionnaire, which is usually mailed to a sample of people. This is an inexpensive method of conducting a survey and is therefore attractive when the number of people to be surveyed is large. But self-administered questionnaires usually have a low response rate and may have a relatively high number of incorrect responses due to respondents misunderstanding some questions.

Questionnaire Design

Whether a questionnaire is self-administered or completed by an interviewer, it must be well designed. Proper questionnaire design takes knowledge, experience, time, and money. Some basic points to consider regarding questionnaire design follow.

1. First and foremost, the questionnaire should be kept as short as possible to encourage respondents to complete it. Most people are unwilling to spend much time filling out a questionnaire.
2. The questions themselves should also be short, as well as simply and clearly worded, to enable respondents to answer quickly, correctly, and without ambiguity. Even familiar terms, such as "unemployed" and "family," must be defined carefully because several interpretations are possible.
3. Questionnaires often begin with simple demographic questions to help respondents get started and become comfortable quickly.
4. Dichotomous questions (questions with only two possible responses, such as "yes" and "no," and multiple-choice questions) are useful and popular because of their simplicity, but they, too, have possible shortcomings. For example, a respondent's choice of yes or no to a question may depend on certain assumptions not stated in the question. In the case of a multiple-choice question, a respondent may feel that none of the choices offered is suitable.
5. Open-ended questions provide an opportunity for respondents to express opinions more fully, but they are time consuming and more difficult to tabulate and analyze.
6. Avoid using leading questions, such as "Wouldn't you agree that the statistics exam was too difficult?" These types of questions tend to lead the respondent to a particular answer.
7. Time permitting, it is useful to pretest a questionnaire on a small number of people in order to uncover potential problems, such as ambiguous wording.
8. Finally, when preparing the questions, think about how you intend to tabulate and analyze the responses. First determine whether you are soliciting values (i.e., responses) for an interval variable or a nominal variable. Then consider which type of statistical techniques—descriptive or inferential—you intend to apply to the data to be collected, and note the requirements of the specific techniques to be used. Thinking about these questions will help ensure that the questionnaire is designed to collect the data you need.

Whatever method is used to collect primary data, we need to know something about sampling, the subject of the next section.

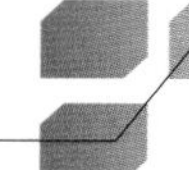

EXERCISES

5.1 Briefly describe the difference between observational and experimental data.

5.2 A soft-drink manufacturer has been supplying its cola drink in bottles to grocery stores and in cans to small convenience stores. The company is analyzing sales of this cola drink to determine which type of packaging is preferred by consumers.
a. Is this study observational or experimental? Explain your answer.
b. Outline a better method for determining whether a store will be supplied with cola in bottles or in cans so that future sales data will be more helpful in assessing the preferred type of packaging.

5.3 a. Briefly describe how you might design a study to investigate the relationship between smoking and lung cancer.
b. Is your study in Part a observational or experimental? Explain why.

5.4 a. List three methods of conducting a survey of people.
b. Give an important advantage and disadvantage of each of the methods listed in Part a.

5.5 List five important points to consider when designing a questionnaire.

5.2 SAMPLING

The chief motive for examining a sample rather than a population is cost. Statistical inference permits us to draw conclusions about a population parameter based on a sample that is quite small in comparison to the size of the population. For example, television executives want to know the proportion of television viewers who watch a network's programs. Because 100 million people may be watching television in the United States on a given evening, determining the actual proportion of the population that is watching certain programs is impractical and prohibitively expensive. The Nielsen ratings provide approximations of the desired information by observing what is watched by a sample of 5,000 television viewers. The proportion of households watching a particular program can be calculated for the households in the Nielsen sample. This sample proportion is then used as an **estimate** of the proportion of all households (the population proportion) that watched the program.

Another illustration of sampling can be taken from the field of quality management. To ensure that a production process is operating properly, the operations manager needs to know what proportion of items being produced is defective. If the quality technician must destroy the item to determine whether it is defective, then there is no alternative to sampling: A complete inspection of the product population would destroy the entire output of the production process.

We know that the sample proportion of television viewers or of defective items is probably not exactly equal to the population proportion we want to estimate. Nonetheless, the sample statistic can come quite close to the parameter it is designed to estimate if the **target population** (the population about which we want to draw inferences) and the **sampled population** (the actual population from which the sample has been taken) are the same. In practice, these may not be the same. One of statistics' most famous failures illustrates this phenomenon.

The *Literary Digest* was a popular magazine of the 1920s and 1930s that had correctly predicted the outcomes of several presidential elections. In 1936, the *Digest* predicted that the Republican candidate, Alfred Landon, would defeat the Democratic incumbent, Franklin D. Roosevelt, by a 3 to 2 margin. But in that election, Roosevelt defeated Landon in a landslide victory, garnering the support of 62% of the electorate. The source of this blunder was the sampling procedure, and there were two distinct mistakes.* First, the *Digest* sent out 10 million sample ballots to prospective voters. However, most of the names of these people were taken from the *Digest*'s subscription list and from telephone directories. Subscribers to the magazine and people who owned telephones tended to be wealthier than average and such people then, as today, tended to vote Republican. Additionally, only 2.3 million ballots were returned resulting in a self-selected sample.

Self-selected samples are almost always biased, because the individuals who participate in them are more keenly interested in the issue than are the other members of the population. You often find similar surveys conducted today when radio and television stations ask people to call and give their opinion on an issue of interest. Again only listeners who are concerned about the topic and have enough patience to get through to the station will be included in the sample. Hence, the sampled population is composed entirely of people who are interested in the issue, whereas the target population is made up of all the people within the listening radius of the radio station. As a result, the conclusions drawn from such surveys are frequently wrong.

An excellent example of this phenomenon occurred on ABC's *Nightline* in 1984. Viewers were given a 900 telephone number (cost: 50 cents) and asked to phone in their responses to the question of whether the United Nations should continue to be located in the United States. More than 186,000 people called, with 67% responding "no." At the same time, a (more scientific) market research poll of 500 people revealed that 72% wanted the United Nations to remain in the United States. In general, because the true value of the parameter being estimated is never known, these surveys give the impression of providing useful information. In fact, the results of such surveys are likely to be no more accurate than the results of the 1936 *Literary Digest* poll or *Nightline*'s phone-in show. Statisticians have coined two terms to describe these polls: SLOP (self-selected opinion poll) and *oy vey* (from the Yiddish lament), both of which convey the contempt that statisticians have for such data-gathering processes.

*Many statisticians ascribe the *Literary Digest*'s statistical debacle to the wrong causes. For an understanding of what really happened, read Maurice C. Bryson, "The Literary Digest Poll: Making of a Statistical Myth," *American Statistician* 30(4) (November 1976): 184–185.

EXERCISES

5.6 For each of the following sampling plans, indicate why the target population and the sampled population are not the same.

a. To determine the opinions and attitudes of customers who regularly shop at a particular mall, a surveyor stands outside a large department store in the mall and randomly selects people to participate in the survey.

b. A library wants to estimate the proportion of its books that have been damaged. The librarians decide to select one book per shelf as a sample by measuring 12 inches from the left edge of each shelf and selecting the book in that location.

c. Political surveyors visit 200 residences during one afternoon to ask eligible voters present in the house at the time whom they intend to vote for.

5.7 a. Describe why the *Literary Digest* poll of 1936 has become infamous.

b. What caused this poll to be so wrong?

5.8 a. What is meant by a self-selected sample?
b. Give an example of a recent poll that involved a self-selected sample.
c. Why are self-selected samples not desirable?

5.9 A regular feature in a newspaper asks readers to respond via e-mail to a survey that requires a yes or no response. In the following day's newspaper, the percentage of yes and no responses are reported. Discuss why we should ignore these statistics.

5.10 Suppose your statistics professor distributes a questionnaire about the course. One of the questions asks, "Would you recommend this course to a friend?" Can the professor use the results to infer something about all statistics courses? Explain.

5.3 Sampling Plans

Our objective in this section is to introduce three different sampling plans: simple random sampling, stratified random sampling, and cluster sampling. We begin our presentation with the most basic design.

Simple Random Sampling

Simple Random Sample

A **simple random sample** is a sample selected in such a way that every possible sample with the same number of observations is equally likely to be chosen.

One way to conduct a simple random sample is to assign a number to each element in the population, write these numbers on individual slips of paper, toss them into a hat, and draw the required number of slips (the sample size, n) from the hat. This is the kind of procedure that occurs in raffles, when all the ticket stubs go into a large rotating drum from which the winners are selected.

Sometimes the elements of the population are already numbered. For example, virtually all adults have Social Security numbers (in the United States) or Social Insurance numbers (in Canada); all employees of large corporations have employee numbers; many people have driver's license numbers, medical plan numbers, student numbers, and so on. In such cases, choosing which sampling procedure to use is simply a matter of deciding how to select from among these numbers.

In other cases, the existing form of numbering has built-in flaws that make it inappropriate as a source of samples. Not everyone has a phone number, for example, so the telephone book does not list all the people in a given area. Many households have two (or more) adults, but only one phone listing. Couples often list the phone number under the man's name, so telephone listings are likely to be disproportionately male. Some people do not have phones, some have unlisted phone numbers, and some have more than one phone; these differences mean that each element of the population does not have an equal probability of being selected.

After each element of the chosen population has been assigned a unique number, sample numbers can be selected at random. A random number table can be used to select these sample numbers. (See, for example, *CRC Standard Management Tables*, W. H. Beyer, ed., Boca Raton FL: CRC Press.) Alternatively, we can use Excel to perform this function.

EXAMPLE 5.1 Random Sample of Income Tax Returns

A government income tax auditor has been given responsibility for 1,000 tax returns. A computer is used to check the arithmetic of each return. However, to determine whether the returns have been completed honestly, the auditor must check each entry and confirm its veracity. Because it takes, on average, 1 hour to completely audit a return and she has only 1 week to complete the task, the auditor has decided to randomly select 40 returns. The returns are numbered from 1 to 1,000. Use a computer random number generator to select the sample for the auditor.

SOLUTION

We generated 50 numbers between 1 and 1,000 even though we needed only 40 numbers. We did so because it is likely that there will be some duplicates. We will use the first 40 unique random numbers to select our sample. The following numbers were generated by Excel. The instructions for both Excel and Minitab are provided here. [Notice that the 24th and 36th (counting down the columns) numbers generated were the same—467.]

Computer-Generated Random Numbers

383	246	372	952	75
101	46	356	54	199
597	33	911	706	65
900	165	467	817	359
885	220	427	973	488
959	18	304	467	512
15	286	976	301	374
408	344	807	751	986
864	554	992	352	41
139	358	257	776	231

EXCEL

INSTRUCTIONS

1. Click **Data, Data Analysis,** and **Random Number Generation.**
2. Specify the **Number of Variables** (1) and the **Number of Random Numbers** (50).
3. Select Uniform Distribution.
4. Specify the range of the uniform distribution (**Parameters**) (0 and 1). Click **OK.** Column A will fill with 50 numbers that range between 0 and 1.
5. Multiply column A by 1,000 and store the products in column B.
6. Make cell C1 active, and click f_x, **Math & Trig, ROUNDUP,** and **OK.**
7. Specify the first number to be rounded (B1).
8. Type the **number of digits** (decimal places) (0). Click **OK.**
9. Complete column C.

The first four steps command Excel to generate 50 uniformly distributed random numbers between 0 and 1 to be stored in column A. Steps 5 through 9 convert these random numbers to integers between 1 and 1,000. Each tax return has the same probability (1/1,000 = .001) of being selected. Thus, each member of the population is equally likely to be included in the sample.

MINITAB

INSTRUCTIONS

1. Click **Calc**, **Random Data,** and **Integer . . . .**
2. Type the number of random numbers you wish (50).
3. Specify where the numbers are to be stored (C1).
4. Specify the **Minimum value** (1).
5. Specify the **Maximum value** (1000). Click **OK.**

INTERPRET

The auditor would examine the tax returns selected by the computer. She would pick returns numbered 383, 101, 597, . . . , 352, 776, and 75 (the first 40 unique numbers). Each of these returns would be audited to determine whether they are fraudulent. If the objective is to audit these 40 returns, no statistical procedure would be employed. However, if the objective is to estimate the proportion of all 1,000 returns that are dishonest, she would use one of the inferential techniques presented later in this book.

Stratified Random Sampling

In making inferences about a population, we attempt to extract as much information as possible from a sample. The basic sampling plan, simple random sampling, often accomplishes this goal at low cost. Other methods, however, can be used to increase the amount of information about the population. One such procedure is *stratified random sampling*.

Stratified Random Sample

A **stratified random sample** is obtained by separating the population into mutually exclusive sets, or strata, and then drawing simple random samples from each stratum.

Examples of criteria for separating a population into strata (and of the strata themselves) follow.

1. Gender
  male
  female

2. Age
  under 20
  20–30
  31–40
  41–50
  51–60
  over 60

3. Occupation
   professional
   clerical
   blue-collar
   other
4. Household income
   under $25,000
   $25,000–$49,999
   $50,000–$74,999
   over $75,000

To illustrate, suppose a public opinion survey is to be conducted to determine how many people favor a tax increase. A stratified random sample could be obtained by selecting a random sample of people from each of the four income groups we just described. We usually stratify in a way that enables us to obtain particular kinds of information. In this example, we would like to know whether people in the different income categories differ in their opinions about the proposed tax increase, because the tax increase will affect the strata differently. We avoid stratifying when there is no connection between the survey and the strata. For example, little purpose is served in trying to determine whether people within religious strata have divergent opinions about the tax increase.

One advantage of stratification is that, besides acquiring information about the entire population, we can also make inferences within each stratum or compare strata. For instance, we can estimate what proportion of the lowest income group favors the tax increase, or we can compare the highest and lowest income groups to determine whether they differ in their support of the tax increase.

Any stratification must be done in such a way that the strata are mutually exclusive: Each member of the population must be assigned to exactly one stratum. After the population has been stratified in this way, we can use simple random sampling to generate the complete sample. There are several ways to do this. For example, we can draw random samples from each of the four income groups according to their proportions in the population. Thus, if in the population the relative frequencies of the four groups are as listed here, our sample will be stratified in the same proportions. If a total sample of 1,000 is to be drawn, we will randomly select 250 from stratum 1, 400 from stratum 2, 300 from stratum 3, and 50 from stratum 4.

Stratum	Income Categories	Population Proportions
1	under $25,000	25%
2	25,000–49,999	40
3	50,000–74,999	30
4	over 75,000	5

The problem with this approach, however, is that if we want to make inferences about the last stratum, a sample of 50 may be too small to produce useful information. In such cases, we usually increase the sample size of the smallest stratum to ensure that the sample data provide enough information for our purposes. An adjustment must then be made before we attempt to draw inferences about the entire population. The required procedure is beyond the level of this book. We recommend that anyone planning such a survey consult an expert statistician or a reference book on the subject. Better still, become an expert statistician yourself by taking additional statistics courses.

Cluster Sampling

Cluster Sample
A **cluster sample** is a simple random sample of groups or clusters of elements.

Cluster sampling is particularly useful when it is difficult or costly to develop a complete list of the population members (making it difficult and costly to generate a simple random sample). It is also useful whenever the population elements are widely dispersed geographically. For example, suppose we wanted to estimate the average annual household income in a large city. To use simple random sampling, we would need a complete list of households in the city from which to sample. To use stratified random sampling, we would need the list of households, and we would also need to have each household categorized by some other variable (such as age of household head) in order to develop the strata. A less expensive alternative would be to let each block within the city represent a cluster. A sample of clusters could then be randomly selected, and every household within these clusters could be questioned to determine income. By reducing the distances the surveyor must cover to gather data, cluster sampling reduces the cost.

But cluster sampling also increases sampling error (see Section 5.4), because households belonging to the same cluster are likely to be similar in many respects, including household income. This can be partially offset by using some of the cost savings to choose a larger sample than would be used for a simple random sample.

Sample Size

Whichever type of sampling plan you select, you still have to decide what size sample to use. Determining the appropriate sample size will be addressed in detail in Chapters 10 and 12. Until then, we can rely on our intuition, which tells us that the larger the sample size is, the more accurate we can expect the sample estimates to be.

Sampling and the Census

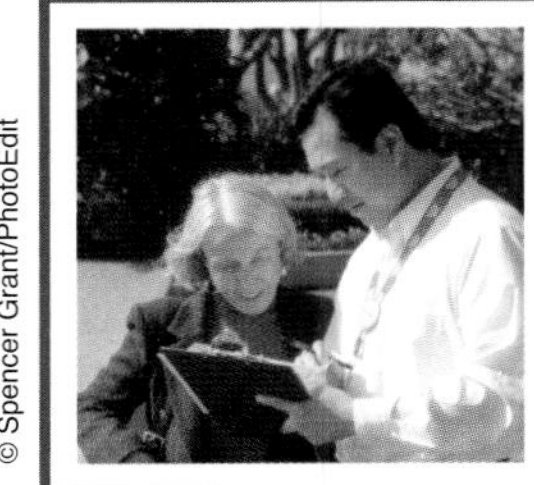
© Spencer Grant/PhotoEdit

To adjust for undercounting, the Census Bureau conducts cluster sampling. The clusters are geographic blocks. For the year 2000 census, the Bureau randomly sampled 11,800 blocks, which contained 314,000 housing units. Each unit was intensively revisited to ensure that all residents were counted. From the results of this survey, the Census Bureau estimated the number of people missed by the first census in various subgroups, defined by several variables including gender, race, and age. Because of the importance of determining state populations, adjustments were made to state totals. For example, by comparing the results of the census and of the sampling, the Bureau determined that the undercount in the state of Texas was 1.7087%. The official census produced a state population of 20,851,820. Taking 1.7087% of this total produced an adjustment of 356,295. Using this method changed the population of the state of Texas to 21,208,115.

It should be noted that this process is contentious. The controversy centers on the way in which subgroups are defined. Changing the definition alters the undercounts, making this statistical technique subject to politicking.

EXERCISES

5.11 A statistics practitioner would like to conduct a survey to ask people their views on a proposed new shopping mall in their community. According to the latest census, there are 800 households in the community. The statistician has numbered each household (from 1 to 800), and she would like to randomly select 25 of these households to participate in the study. Use Excel or Minitab to generate the sample.

5.12 A safety expert wants to determine the proportion of cars in his state with worn tire treads. The state license plate contains six digits. Use Excel or Minitab to generate a sample of 20 cars to be examined.

5.13 A large university campus has 60,000 students. The president of the students' association wants to conduct a survey of the students to determine their views on an increase in the student activity fee. She would like to acquire information about all the students but would also like to compare the school of business, the faculty of arts and sciences, and the graduate school. Describe a sampling plan that accomplishes these goals.

5.14 A telemarketing firm has recorded the households that have purchased one or more of the company's products. These number in the millions. They would like to conduct a survey of purchasers to acquire information about their attitude concerning the timing of the telephone calls. The president of the company would like to know the views of all purchasers but would also like to compare the attitudes of people in the West, South, North, and East. Describe a suitable sampling plan.

5.15 The operations manager of a large plant with four departments wants to estimate the person-hours lost per month due to accidents. Describe a sampling plan that would be suitable for estimating the plantwide loss and for comparing departments.

5.16 A statistics practitioner wants to estimate the mean age of children in his city. Unfortunately, he does not have a complete list of households. Describe a sampling plan that would be suitable for his purposes.

5.4 SAMPLING AND NONSAMPLING ERRORS

Two major types of error can arise when a sample of observations is taken from a population: *sampling error* and *nonsampling error*. Anyone reviewing the results of sample surveys and studies, as well as statistics practitioners conducting surveys and applying statistical techniques, should understand the sources of these errors.

Sampling Error

Sampling error refers to differences between the sample and the population that exists only because of the observations that happened to be selected for the sample. Sampling error is an error that we expect to occur when we make a statement about a population that is based only on the observations contained in a sample taken from the population. To illustrate, suppose that we wish to determine the mean annual income of North American blue-collar workers. To determine this parameter we would have to ask each North American blue-collar worker what his or her income is and then calculate the mean of all the responses. Because the size of this population is several million, the task is both expensive and impractical. We can use statistical inference to estimate the mean income μ of the population if we are willing to accept less than 100% accuracy. We record the incomes of a sample of the workers and find the mean $\bar{x}$ of this sample of incomes. This sample mean is an estimate of the desired population mean. But the value of the sample mean will deviate from the population mean simply by chance, because

SEEING STATISTICS

Applet 3: Sampling

When you select this applet, you will see 100 circles. Imagine that each of the circles represents a household. You want to estimate the proportion of households having high-speed Internet access (DSL, cable modem, etc.). You may collect data from a sample of 10 households by clicking on a household's circle. If the circle turns red, then the household has high-speed Internet access. If the circle turns green, then the household does not have high-speed access. After collecting your sample and obtaining your estimate, click on the **Show All** button to see information for all the households. How well did your sample estimate the true proportion? Click the **Reset** button to try again. (*Note:* This page uses a randomly determined base proportion each time this page is loaded/reloaded.)

Applet Exercises

3.1 Run the applet 25 times. How many times did the sample proportion equal the population proportion?

3.2 Run the applet 20 times. For each simulation, record the sample proportion of homes with high-speed Internet access as well as the population proportion. Compute the average sampling error.

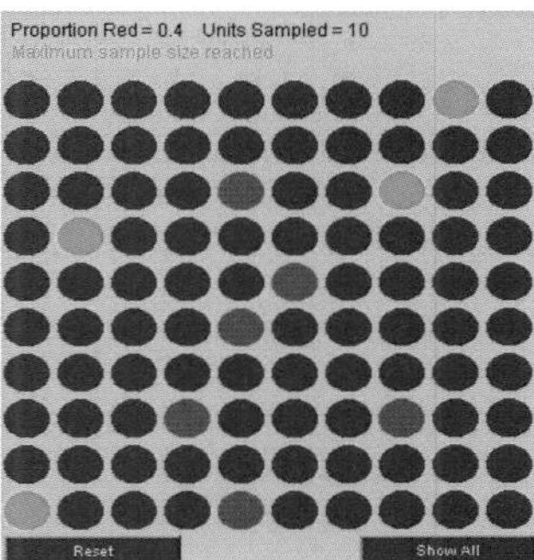

the value of the sample mean depends on which incomes just happened to be selected for the sample. The difference between the true (unknown) value of the population mean and its estimate, the sample mean, is the sampling error. The size of this deviation may be large simply due to bad luck—bad luck that a particularly unrepresentative sample happened to be selected. The only way we can reduce the expected size of this error is to take a larger sample.

Given a fixed sample size, the best we can do is to state the probability that the sampling error is less than a certain amount (as we will discuss in Chapter 10). It is common today for such a statement to accompany the results of an opinion poll. If an opinion poll states that, based on sample results, the incumbent candidate for mayor has the support of 54% of eligible voters in an upcoming election, the statement may be accompanied by the following explanatory note: This percentage is correct to within three percentage points, 19 times out of 20. This statement means that we estimate that the actual level of support for the candidate is between 51% and 57%, and that in the long run this type of procedure is correct 95% of the time.

Nonsampling Error

Nonsampling error is more serious than sampling error, because taking a larger sample won't diminish the size, or the possibility of occurrence, of this error. Even a census can (and probably will) contain nonsampling errors. **Nonsampling errors** are due to mistakes made in the acquisition of data or due to the sample observations being selected improperly.

1. *Errors in data acquisition.* This type of error arises from the recording of incorrect responses. Incorrect responses may be the result of incorrect measurements being taken because of faulty equipment, mistakes made during transcription from

primary sources, inaccurate recording of data due to misinterpretation of terms, or inaccurate responses to questions concerning sensitive issues such as sexual activity or possible tax evasion.

2. *Nonresponse error.* **Nonresponse error** refers to error (or **bias**) introduced when responses are not obtained from some members of the sample. When this happens, the sample observations that are collected may not be representative of the target population, resulting in biased results (as was discussed in Section 5.2). Nonresponse can occur for a number of reasons. An interviewer may be unable to contact a person listed in the sample, or the sampled person may refuse to respond for some reason. In either case, responses are not obtained from a sampled person, and bias is introduced. The problem of nonresponse is even greater when self-administered questionnaires are used rather than an interviewer, who can attempt to reduce the nonresponse rate by means of callbacks. As noted previously, the *Literary Digest* fiasco was largely due to a high nonresponse rate, resulting in a biased, self-selected sample.

3. *Selection bias.* **Selection bias** occurs when the sampling plan is such that some members of the target population cannot possibly be selected for inclusion in the sample. Together with nonresponse error, selection bias played a role in the *Literary Digest* poll being so wrong, as voters without telephones or without a subscription to *Literary Digest* were excluded from possible inclusion in the sample taken.

EXERCISES

5.17 Is it possible for a sample to yield better results than a census? Explain.

5.18 a. Explain the difference between sampling error and nonsampling error.

b. Which type of error in Part a is more serious? Why?

5.19 Briefly describe three types of nonsampling error.

CHAPTER SUMMARY

Because most populations are very large, it is extremely costly and impractical to investigate each member of the population to determine the values of the parameters. As a practical alternative, we take a sample from the population and use the sample statistics to draw inferences about the parameters. Care must be taken to ensure that the sampled population is the same as the target population.

We can choose from among several different sampling plans, including simple random sampling, stratified random sampling, and cluster sampling. Whatever sampling plan is used, it is important to realize that both sampling error and nonsampling error will occur and to understand what the sources of these errors are.

IMPORTANT TERMS

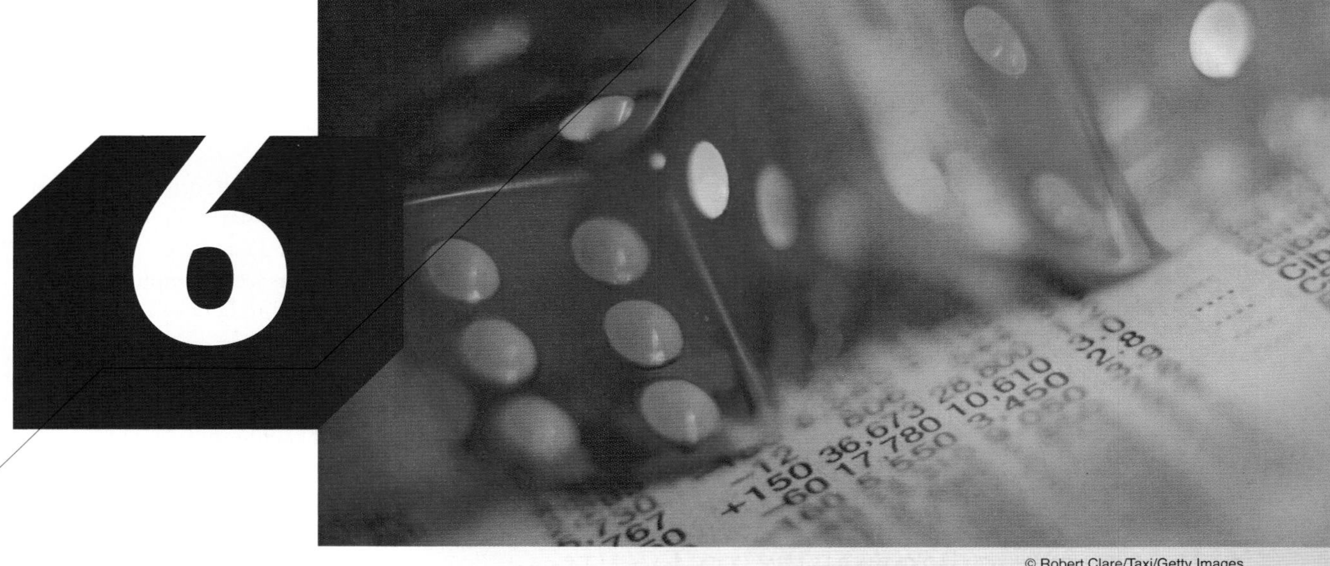

© Robert Clare/Taxi/Getty Images

6 PROBABILITY

Auditing Tax Returns

Government auditors routinely check tax returns to determine whether calculation errors were made. They also attempt to detect fraudulent returns. There are several methods that dishonest taxpayers use to evade income tax. One method is not to declare various sources of income. Auditors have several detection methods, including spending patterns. Another form of tax fraud is to invent deductions that are not real. After analyzing the returns of thousands of self-employed taxpayers, an auditor has determined that 45% of fraudulent returns contain two suspicious deductions, 28% contain one suspicious deduction, and the rest no suspicious deductions. Among honest returns the rates are 11% for two deductions, 18% for one deduction, and 71% for no deductions. The auditor believes that 5% of the returns of self-employed individuals contain significant fraud. The auditor has just received a tax return for a self-employed individual that contains one suspicious expense deduction. What is the probability that this tax return contains significant fraud?

© Gary Buss/Taxi/Getty Images

See page 196 for the answer.

INTRODUCTION

In Chapters 2, 3, and 4 we introduced graphical and numerical descriptive methods. Although the methods are useful on their own, we are particularly interested in developing statistical inference. As we pointed out in Chapter 1, statistical inference is the process by which we acquire information about populations from samples. A critical component of inference is *probability* because it provides the link between the population and the sample.

Our primary objective in this and the following two chapters is to develop the probability-based tools that are at the basis of statistical inference. However, probability can also play a critical role in decision making, a subject we explore in Chapter 22.

6.1 ASSIGNING PROBABILITY TO EVENTS

To introduce probability, we must first define a *random experiment*.

Random Experiment

A **random experiment** is an action or process that leads to one of several possible outcomes.

Here are six illustrations of random experiments and their outcomes.

Illustration 1. *Experiment*: Flip a coin.
Outcomes: Heads and tails

Illustration 2. *Experiment*: Record marks on a statistics test (out of 100).
Outcomes: Numbers between 0 and 100

Illustration 3. *Experiment*: Record grade on a statistics test.
Outcomes: A, B, C, D, and F

Illustration 4. *Experiment*: Record student evaluations of a course.
Outcomes: Poor, fair, good, very good, and excellent

Illustration 5. *Experiment*: Measure the time to assemble a computer.
Outcomes: Number whose smallest possible value is 0 seconds with no predefined upper limit

Illustration 6. *Experiment*: Record the party that a voter will vote for in an upcoming election.
Outcomes: Party A, Party B, . . .

The first step in assigning probabilities is to produce a list of the outcomes. The listed outcomes must be **exhaustive,** which means that all possible outcomes must be included. Additionally, the outcomes must be **mutually exclusive,** which means that no two outcomes can occur at the same time.

To illustrate the concept of exhaustive outcomes consider this list of the outcomes of the toss of a die:

1 2 3 4 5

This list is not exhaustive, because we have omitted 6.

The concept of mutual exclusiveness can be seen by listing the following outcomes in illustration 2:

0–50 50–60 60–70 70–80 80–100

If these intervals include both the lower and upper limits, these outcomes are not mutually exclusive because two outcomes can occur for any student. For example, if a student receives a mark of 70, both the third and fourth outcomes occur.

It should be noted that we could produce more than one list of exhaustive and mutually exclusive outcomes. For example, here is another list of outcomes for illustration 3:

Pass and fail

A list of exhaustive and mutually exclusive outcomes is called a *sample space* and is denoted by S. The outcomes are denoted by $O_1, O_2, \ldots, O_k$.

Sample Space

A **sample space** of a random experiment is a list of all possible outcomes of the experiment. The outcomes must be exhaustive and mutually exclusive.

Using set notation we represent the sample space and its outcomes as

$$S = \{O_1, O_2, \ldots, O_k\}$$

Once a sample space has been prepared we begin the task of assigning probabilities to the outcomes. There are three ways to assign probability to outcomes. However it is done, there are two rules governing probabilities as stated in the next box.

Requirements of Probabilities

Given a sample space $S = \{O_1, O_2, \ldots, O_k\}$, the probabilities assigned to the outcomes must satisfy two requirements.

1. The probability of any outcome must lie between 0 and 1. That is,

$$0 \leq P(O_i) \leq 1 \quad \text{for each } i$$

[**Note:** $P(O_i)$ is the notation we use to represent the probability of outcome i.]

2. The sum of the probabilities of all the outcomes in a sample space must be 1. That is,

$$\sum_{i=1}^{k} P(O_i) = 1$$

Three Approaches to Assigning Probabilities

The **classical approach** is used by mathematicians to help determine probability associated with games of chance. For example, the classical approach specifies that the probabilities of heads and tails in the flip of a balanced coin are equal to each other. Because the sum of the probabilities must be 1, the probability of heads and the probability of tails are both 50%. Similarly, the six possible outcomes of the toss of a balanced die have the same probability; each is assigned a probability of 1/6. In some experiments it is necessary to develop mathematical ways to count the number of outcomes. For example, to determine the probability of winning a lottery we need to determine the number of possible combinations. For details on how to count events see CD Appendix E.

The **relative frequency approach** defines probability as the long-run relative frequency with which an outcome occurs. For example, suppose that we know that of the last

1,000 students who took the statistics course you're now taking, 200 received a grade of A. The relative frequency of A's is then 200/1000 or 20%. This figure represents an estimate of the probability of obtaining a grade of A in the course. It is only an estimate because the relative frequency approach defines probability as the "long-run" relative frequency. One thousand students do not constitute the long run. The larger the number of students whose grades we have observed, the better the estimate becomes. In theory we would have to observe an infinite number of grades to determine the exact probability.

When it is not reasonable to use the classical approach and there is no history of the outcomes, we have no alternative but to employ the **subjective approach.** In the subjective approach we define probability as the degree of belief that we hold in the occurrence of an event. An excellent example is derived from the field of investment. An investor would like to know the probability that a particular stock will increase in value. Using the subjective approach, the investor would analyze a number of factors associated with the stock and the stock market in general and, using his or her judgment, assign a probability to the outcomes of interest.

Defining Events

An individual outcome of a sample space is called a *simple event*. All other events are composed of the simple events in a sample space.

Event

An **event** is a collection or set of one or more simple events in a sample space.

In illustration 2 we can define the event, achieve a grade of A, as the set of numbers that lie between 80 and 100, inclusive. Using set notation we have

$$A = \{80, 81, 82, \ldots, 99, 100\}$$

Similarly,

$$F = \{0, 1, 2, \ldots, 48, 49\}$$

Probability of Events

We can now define the probability of any event.

Probability of an Event

The probability of an event is the sum of the probabilities of the simple events that constitute the event.

For example, suppose that in illustration 3, we employed the relative frequency approach to assign probabilities to the simple events as follows:

$$P(A) = .20$$
$$P(B) = .30$$
$$P(C) = .25$$
$$P(D) = .15$$
$$P(F) = .10$$

The probability of the event, pass the course, is

$$P(\text{Pass the course}) = P(A) + P(B) + P(C) + P(D) = .20 + .30 + .25 + .15 = .90$$

Interpreting Probability

No matter what method was used to assign probability, we interpret it using the relative frequency approach for an infinite number of experiments. For example, an investor may have used the subjective approach to determine that there is a 65% probability that a particular stock's price will increase over the next month. However, we interpret the 65% figure to mean that if we had an infinite number of stocks with exactly the same economic and market characteristics as the one the investor will buy, 65% of them will increase in price over the next month. Similarly, we can determine that the probability of throwing a 5 with a balanced die is 1/6. We may have used the classical approach to determine this probability. However, we interpret the number as the proportion of times that a 5 is observed on a balanced die thrown an infinite number of times.

This relative frequency approach is useful to interpret probability statements such as those heard from weather forecasters or scientists. You will also discover that this is the way we link the population and the sample in statistical inference.

EXERCISES

6.1 The weather forecaster reports that the probability of rain tomorrow is 10%.
a. Which approach was used to arrive at this number?
b. How do you interpret the probability?

6.2 A sportscaster states that he believes that the probability that the New York Yankees will win the World Series this year is 25%.
a. Which method was used to assign that probability?
b. How would you interpret the probability?

6.3 A quiz contains a multiple-choice questions with five possible answers, only one of which is correct. A student plans to guess the answer because he knows absolutely nothing about the subject.
a. Produce the sample space for each question.
b. Assign probabilities to the simple events in the sample space you produced.
c. Which approach did you use to answer Part b?
d. Interpret the probabilities you assigned in Part b.

6.4 An investor tells you that in her estimation there is a 60% probability that the Dow Jones Industrial Index will increase tomorrow.
a. Which approach was used to produce this figure?
b. Interpret the 60% probability.

6.5 The sample space of the toss of a fair die is

$$S = \{1, 2, 3, 4, 5, 6\}$$

If the die is balanced each simple event has the same probability. Find the probability of the following events.
a. An even number
b. A number less than or equal to 4
c. A number greater than or equal to 5

6.6 Four candidates are running for mayor. The four candidates are Adams, Brown, Collins, and Dalton. Determine the sample space of the results of the election.

6.7 Refer to Exercise 6.6. Employing the subjective approach, a political scientist has assigned the following probabilities:

$P(\text{Adams wins}) = .42$

$P(\text{Brown wins}) = .09$

$P(\text{Collins wins}) = .27$

$P(\text{Dalton wins}) = .22$

Determine the probabilities of the following events.
a. Adams loses.
b. Either Brown or Dalton wins.
c. Either Adams, Brown, or Collins wins.

6.8 The manager of a computer store has kept track of the number of computers sold per day. On the basis of this information, the manager produced the following list of the number of daily sales.

Number of Computers Sold	Probability
0	.08
1	.17
2	.26
3	.21
4	.18
5	.10

a. If we define the experiment as observing the number of computers sold tomorrow, determine the sample space.
b. Use set notation to define the event, sell more than 3 computers.
c. What is the probability of selling 5 computers?
d. What is the probability of selling 2, 3, or 4 computers?
e. What is the probability of selling 6 computers?

6.9 Three contractors (call them contractors 1, 2, and 3) bid on a project to build a new bridge. What is the sample space?

6.10 Refer to Exercise 6.9. Suppose that you believe that contractor 1 is twice as likely to win as contractor 3 and that contractor 2 is three times as likely to win as contractor 3. What are the probabilities of winning for each contractor?

6.11 Shoppers can pay for their purchases with cash, a credit card, or a debit card. Suppose that the proprietor of a shop determines that 60% of her customers use a credit card, 30% pay with cash, and the rest use a debit card.
a. Determine the sample space for this experiment.
b. Assign probabilities to the simple events.
c. Which method did you use in Part b?

6.12 Refer to Exercise 6.11.
a. What is the probability that a customer does not use a credit card?
b. What is the probability that a customer pays in cash or with a credit card?

6.13 A survey asks adults to report their marital status. The sample space is S = {single, married, divorced, widowed}. Use set notation to represent the event the adult is not married.

6.14 Refer to Exercise 6.13. Suppose that in the city in which the survey is conducted, 50% of adults are married, 15% are single, 25% are divorced, and 10% are widowed.
a. Assign probabilities to each simple event in the sample space.
b. Which approach did you use in Part a?

6.15 Refer to Exercises 6.13 and 6.14. Find the probability of each of the following events.
a. The adult is single.
b. The adult is not divorced
c. The adult is either widowed or divorced.

6.2 Joint, Marginal, and Conditional Probability

In the previous section we described how to produce a sample space and assign probabilities to the simple events in the sample space. Although this method of determining probability is useful, we need to develop more sophisticated methods. In this section we discuss how to calculate the probability of more complicated events from the probability of related events. Here is an illustration of the process.

The sample space for the toss of a die is

$$S = \{1, 2, 3, 4, 5, 6\}$$

If the die is balanced, the probability of each simple event is 1/6. In most parlor games and casinos, players toss two dice. To determine playing and wagering strategies, players need to compute the probabilities of various totals of the two dice. For example, the probability of tossing a total of 3 with two dice is 2/36. This probability was derived by creating combinations of the simple events. There are several different types of combinations. One of the most important types is the **intersection** of two events.

Intersection

Intersection of Events *A* and *B*

The **intersection** of events A and B is the event that occurs when both A and B occur. It is denoted as

$$A \text{ and } B$$

The probability of the intersection is called the **joint probability.**

For example, one way to toss a 3 with two dice is to toss a 1 on the first die *and* a 2 on the second die, which is the intersection of two simple events. Incidentally, to compute the probability of a total of 3, we need to combine this intersection with another intersection, namely, a 2 on the first die and a 1 on the second die. This type of combination is called a *union* of two events and it will be described later in this section. Here is another illustration.

APPLICATIONS in **FINANCE**

Mutual Funds

A mutual fund is a pool of investments made on behalf of people who share similar objectives. In most cases a professional manager who has been educated in finance and statistics manages the fund. He or she makes decisions to buy and sell individual stocks and bonds in accordance with a specified investment philosophy. For example, there are funds that concentrate on other publicly traded mutual fund companies. Other mutual funds specialize in Internet (so-called "dot-coms") stocks whereas others buy stocks of biotech firms. Surprisingly, most mutual funds do not outperform the market. That is, the increase in the net asset value (NAV) of the mutual fund is often less than the increase in the value of stock indexes that represent their stock markets. One reason for this is the management expense ratio (MER) which is a measure of the costs charged to the fund by the manager to cover expenses, including the salary and bonus of the managers. The MERs for most funds range from .5% to more than 4%. The ultimate success of the fund depends on the skill and knowledge of the fund manager. This raises the question: Which managers do best?

EXAMPLE 6.1

Determinants of Success among Mutual Fund Managers—Part 1*

Why are some mutual fund managers more successful than others? One possible factor is where the manager earned his or her MBA. Suppose that a potential investor examined the relationship between how well the mutual fund performs and where the fund manager earned his or her MBA. After the analysis, Table 6.1, a table of joint probabilities, was developed. Analyze these probabilities and interpret the results.

*This example is adapted from Judith Chevalier and Glenn Ellison, "Are Some Mutual Fund Managers Better than Others? Cross-Sectional Patterns in Behavior and Performance," Working paper 5852, National Bureau of Economic Research.

TABLE **6.1** Determinants of Success among Mutual Fund Managers, Part 1

	Mutual Fund Outperforms Market	Mutual Fund Does Not Outperform Market
Top-20 MBA program	.11	.29
Not top-20 MBA program	.06	.54

Table 6.1 tells us that the joint probability that a mutual fund outperforms the market *and* that its manager graduated from a top-20 MBA program is .11. That is, 11% of all mutual funds outperform the market and their managers graduated from a top-20 MBA program. The other three joint probabilities are defined similarly. That is,

The probability that a mutual fund outperforms the market and its manager did not graduate from a top-20 MBA program is .06.

The probability that a mutual fund does not outperform the market and its manager graduated from a top-20 MBA program is .29.

The probability that a mutual fund does not outperform the market and its manager did not graduate from a top-20 MBA program is .54.

To help make our task easier we'll use notation to represent the events. Let

A_1 = Fund manager graduated from a top-20 MBA program

A_2 = Fund manager did not graduate from a top-20 MBA program

B_1 = Fund outperforms the market

B_2 = Fund does not outperform the market

Thus,

$$P(A_1 \text{ and } B_1) = .11$$
$$P(A_2 \text{ and } B_1) = .06$$
$$P(A_1 \text{ and } B_2) = .29$$
$$P(A_2 \text{ and } B_2) = .54$$

Marginal Probability

The joint probabilities in Table 6.1 allow us to compute various probabilities. **Marginal probabilities,** computed by adding across rows or down columns, are so named because they are calculated in the margins of the table.

Adding across the first row produces

$$P(A_1 \text{ and } B_1) + P(A_1 \text{ and } B_2) = .11 + .29 = .40$$

Notice that both intersections state that the manager graduated from a top-20 MBA program (represented by A_1). Thus, when randomly selecting mutual funds, the probability that its manager graduated from a top-20 MBA program is .40. Expressed as relative frequency, 40% of all mutual fund managers graduated from a top-20 MBA program.

Adding across the second row,

$$P(A_2 \text{ and } B_1) + P(A_2 \text{ and } B_2) = .06 + .54 = .60$$

This probability tells us that 60% of all mutual fund managers did not graduate from a top-20 MBA program (represented by A_2). Notice that the probability that a mutual fund manager graduated from a top-20 MBA program and the probability that the manager did not graduate from a top-20 MBA program add to 1.

Adding down the columns produces the following marginal probabilities.

Column 1:

$$P(A_1 \text{ and } B_1) + P(A_2 \text{ and } B_1) = .11 + .06 = .17$$

Column 2:

$$P(A_1 \text{ and } B_2) + P(A_2 \text{ and } B_2) = .29 + .54 = .83$$

These marginal probabilities tell us that 17% of all mutual funds outperform the market and that 83% of mutual funds do not outperform the market.

Table 6.2 lists all the joint and marginal probabilities.

TABLE **6.2** Joint and Marginal Probabilities

	Mutual Fund Outperforms Market	Mutual Fund Does Not Outperform Market	Totals
Top-20 MBA program	$P(A_1$ and $B_1) = .11$	$P(A_1$ and $B_2) = .29$	$P(A_1) = .40$
Not top-20 MBA program	$P(A_2$ and $B_1) = .06$	$P(A_2$ and $B_2) = .54$	$P(A_2) = .60$
Totals	$P(B_1) = .17$	$P(B_2) = .83$	1.00

Conditional Probability

We frequently need to know how two events are related. In particular, we would like to know the probability of one event given the occurrence of another related event. For example, we would certainly like to know the probability that a fund managed by a graduate of a top-20 MBA program will outperform the market. Such a probability will allow us to make an informed decision about where to invest our money. This probability is called a **conditional probability** because we want to know the probability that a fund will outperform the market *given* the condition that the manager graduated from a top-20 MBA program. The conditional probability that we seek is represented by

$$P(B_1 \mid A_1)$$

where the "|" represents the word *given*. Here is how we compute this conditional probability.

The marginal probability that a manager graduated from a top-20 MBA program is .40, which is made up of two joint probabilities. They are the probability that the mutual fund outperforms the market and the manager graduated from a top-20 MBA program [$P(A_1$ and $B_1)$] and the probability that the fund does not outperform the market and the manager graduated from a top-20 MBA program [$P(A_1$ and $B_2)$]. Their joint probabilities are .11 and .29, respectively. We can interpret these numbers in the following way. On average for every 100 mutual funds, 40 will be managed by a graduate of a top-20 MBA program. Of these 40 managers, on average 11 of them will manage a mutual fund that will outperform the market. Thus, the conditional probability is $11/40 = .275$. Notice that this ratio is the same as the ratio of the joint probability to the marginal probability $.11/.40$. All conditional probabilities can be computed this way.

Conditional Probability

The probability of event A given event B is

$$P(A \mid B) = \frac{P(A \text{ and } B)}{P(B)}$$

The probability of event B given event A is

$$P(B \mid A) = \frac{P(A \text{ and } B)}{P(A)}$$

EXAMPLE 6.2

Determinants of Success among Mutual Fund Managers—Part 2

Suppose that in Example 6.1 we select one mutual fund at random and discover that it did not outperform the market. What is the probability that a graduate of a top-20 MBA program manages it?

SOLUTION

We wish to find a conditional probability. The condition is that the fund did not outperform the market (event B_2), and the event whose probability we seek is that the fund is managed by a graduate of a top-20 MBA program (event A_1). Thus, we want to compute the following probability:

$$P(A_1 \mid B_2)$$

Using the conditional probability formula, we find

$$P(A_1 \mid B_2) = P(A_1 \text{ and } B_2)/P(B_2) = .29/.83 = .3494$$

Thus, 34.94% of all mutual funds that do not outperform the market are managed by top-20 MBA program graduates.

The calculation of conditional probabilities raises the question of whether the two events, the fund outperformed the market and the manager graduated from a top-20 MBA program, are related, a subject we tackle next.

Independence

One of the objectives of calculating conditional probability is to determine whether two events are related. In particular, we would like to know whether they are **independent events.**

Independent Events

Two events A and B are said to be independent if

$$P(A \mid B) = P(A)$$

or

$$P(B \mid A) = P(B)$$

Put another way, two events are independent if the probability of one event is not affected by the occurrence of the other event.

EXAMPLE 6.3

Determinants of Success among Mutual Fund Managers—Part 3

Determine whether the event that the manager graduated from a top-20 MBA program and the event the fund outperforms the market are independent events.

SOLUTION

We wish to determine whether A_1 and B_1 are independent. To do so we must calculate the probability of A_1 given B_1. That is,

$$P(A_1|B_1) = \frac{P(A_1 \text{ and } B_1)}{P(B_1)} = \frac{.11}{.17} = .647$$

The marginal probability that a manager graduated from a top-20 MBA program is

$$P(A_1) = .40$$

Since the two probabilities are not equal, we conclude that the two events are dependent. Incidentally, we could have made the decision by calculating $P(B_1 \mid A_1) = .275$ and observing that it is not equal to $P(B_1) = .17$.

Note that there are three other combinations of events in this problem. They are (A_1 and B_2), (A_2 and B_1), (A_2 and B_2) [ignoring mutually exclusive combinations (A_1 and A_2) and (B_1 and B_2), which are dependent]. In each combination the two events are dependent. In this type of problem where there are only four combinations, if one combination is dependent, all four will be dependent. Similarly, if one combination is independent, all four will be independent. This rule does not apply to any other situation.

Union

Another event that is the combination of other events is the *union*.

Union of Events A and B

The **union** of events A and B is the event that occurs when either A or B or both occur. It is denoted as

$$A \text{ or } B$$

EXAMPLE 6.4

Determinants of Success among Mutual Fund Managers—Part 4

Determine the probability that a randomly selected fund outperforms the market or the manager graduated from a top-20 MBA program.

SOLUTION

We want to compute the probability of the union of two events

$$P(A_1 \text{ or } B_1)$$

The union A_1 or B_1 consists of three events. That is, the union occurs whenever any of the following joint events occurs:

1. Fund outperforms the market and the manager graduated from a top-20 MBA program
2. Fund outperforms the market and the manager did not graduate from a top-20 MBA program
3. Fund does not outperform the market and the manager graduated from a top-20 MBA program

Their probabilities are

$$P(A_1 \text{ and } B_1) = .11$$
$$P(A_2 \text{ and } B_1) = .06$$
$$P(A_1 \text{ and } B_2) = .29$$

Thus, the probability of the union, the fund outperforms the market or the manager graduated from a top-20 MBA program, is the sum of the three probabilities. That is,

$$P(A_1 \text{ or } B_1) = P(A_1 \text{ and } B_1) + P(A_2 \text{ and } B_1) + P(A_1 \text{ and } B_2) = .11 + .06 + .29 = .46$$

Notice that there is another way to produce this probability. Of the four probabilities in Table 6.1, the only one representing an event that is not part of the union is the probability of the event the fund does not outperform the market and the manager did not graduate from a top-20 MBA program. That probability is

$$P(A_2 \text{ and } B_2) = .54$$

which is the probability that the union *does not* occur. Thus, the probability of the union is

$$P(A_1 \text{ or } B_1) = 1 - P(A_2 \text{ and } B_2) = 1 - .54 = .46$$

Thus, we determined that 46% of mutual funds either outperform the market or are managed by a top-20 MBA program graduate or have both characteristics.

EXERCISES

6.16 Given the following table of joint probabilities, calculate the marginal probabilities.

	A_1	A_2	A_3
B_1	.1	.3	.2
B_2	.2	.1	.1

6.17 Calculate the marginal probabilities from the following table of joint probabilities.

	A_1	A_2
B_1	.4	.3
B_2	.2	.1

6.18 Refer to Exercise 6.17.
a. Determine $P(A_1 | B_1)$.
b. Determine $P(A_2 | B_1)$.
c. Did your answers to Parts a and b sum to 1? Is this a coincidence? Explain.

6.19 Refer to Exercise 6.17. Calculate the following probabilities.
a. $P(A_1 | B_2)$
b. $P(B_2 | A_1)$
c. Did you expect the answers to Parts a and b to be reciprocals? That is, did you expect that $P(A_1 | B_2) = 1/P(B_2 | A_1)$? Why is this impossible (unless both probabilities are 1)?

6.20 Are the events in Exercise 6.17 independent? Explain.

6.21 Refer to Exercise 6.17. Compute the following.
a. $P(A_1 \text{ or } B_1)$
b. $P(A_1 \text{ or } B_2)$
c. $P(A_1 \text{ or } A_2)$

6.22 Suppose that you have been given the following joint probabilities. Are the events independent? Explain.

	A_1	A_2
B_1	.20	.60
B_2	.05	.15

6.23 Determine whether the events are independent from the following joint probabilities.

	A_1	A_2
B_1	.20	.15
B_2	.60	.05

6.24 Suppose we have the following joint probabilities.

	A_1	A_2	A_3
B_1	.15	.20	.10
B_2	.25	.25	.05

Compute the marginal probabilities.

6.25 Refer to Exercise 6.24.
a. Compute $P(A_2 | B_2)$.
b. Compute $P(B_2 | A_2)$.
c. Compute $P(B_1 | A_2)$.

6.26 Refer to Exercise 6.24.
a. Compute $P(A_1 \text{ or } A_2)$.
b. Compute $P(A_2 \text{ or } B_2)$.
c. Compute $P(A_3 \text{ or } B_1)$.

6.27 Discrimination in the workplace is illegal and companies that do so are often sued. The female instructors at a large university recently lodged a complaint about the most recent round of promotions from assistant professor to associate professor. An analysis of the relationship between gender and promotion produced the following joint probabilities.

	Promoted	Not Promoted
Female	.03	.12
Male	.17	.68

a. What is the rate of promotion among female assistant professors?
b. What is the rate of promotion among male assistant professors?
c. Is it reasonable to accuse the university of gender bias?

6.28 A department store analyzed its most recent sales and determined the relationship between the way the customer paid for the item and the price category of the item. The joint probabilities in the following table were calculated.

	Cash	Credit Card	Debit Card
Under $20	.09	.03	.04
$20–$100	.05	.21	.18
Over $100	.03	.23	.14

a. What proportion of purchases was paid by debit card?
b. Find the probability that a credit card purchase was over $100.
c. Determine the proportion of purchases made by credit card or by debit card.

6.29 The following table lists the probabilities of unemployed females and males and their educational attainment.

	Female	Male
Less than high school	.083	.109
High school graduate	.153	.190
Some college/university—no degree	.132	.133
College/university graduate	.091	.108

Source: Statistical Abstract of the United States, 2006, Table 615

a. If one unemployed person is selected at random, what is the probability that he or she did not finish high school?
b. If an unemployed female is selected at random, what is the probability that she has a college or university degree?
c. If an unemployed high school graduate is selected at random, what is the probability that he is a male?

6.30 The costs of medical care in North America are increasing faster than inflation, and with the baby boom generation soon to need health care, it becomes imperative that countries find ways to reduce both costs and demand. The following table lists the joint probabilities associated with smoking and lung disease among 60- to 65-year-old men.

	He Is a Smoker	He Is a Nonsmoker
He has lung disease	.12	.03
He does not have lung disease	.19	.66

One 60- to 65-year-old man is selected at random. What is the probability of the following events?
a. He is a smoker.
b. He does not have lung disease.

c. He has lung disease given that he is a smoker.
d. He has lung disease given that he does not smoke.

6.31 Refer to Exercise 6.30. Are smoking and lung disease among 60- to 65-year-old men related?

6.32 The method of instruction in college and university applied statistics courses is changing. Historically, most courses were taught with an emphasis on manual calculation. The alternative is to employ a computer and a software package to perform the calculations. An analysis of applied statistics courses investigated whether the instructor's educational background is primarily mathematics (or statistics) or some other field. The result of this analysis is the accompanying table of joint probabilities.

	Statistics Course Emphasizes Manual Calculations	Statistics Course Employs Computer and Software
Mathematics or statistics education	.23	.36
Other education	.11	.30

a. What is the probability that a randomly selected applied statistics course instructor whose education was in statistics emphasizes manual calculations?
b. What proportion of applied statistics courses employ a computer and software?
c. Are the educational background of the instructor and the way his or her course is taught independent?

6.33 A restaurant chain routinely surveys customers and among other questions asks each customer whether he or she would return and to rate the quality of food. Summarizing hundreds of thousands of questionnaires produced this table of joint probabilities.

Rating	Customer Will Return	Customer Will Not Return
Poor	.02	.10
Fair	.08	.09
Good	.35	.14
Excellent	.20	.02

a. What proportion of customers say that they will return and rate the restaurant's food as good?
b. What proportion of customers who say that they will return rate the restaurant's food as good?
c. What proportion of customers who rate the restaurant's food as good say that they will return?
d. Discuss the differences in your answers to Parts a, b, and c.

6.34 To determine whether drinking alcoholic beverages has an effect on the bacteria that cause ulcers, researchers developed the following table of joint probabilities.

Number of Alcoholic Drinks per Day	Ulcer	No Ulcer
None	.01	.22
One	.03	.19
Two	.03	.32
More than two	.04	.16

a. What proportion of people have ulcers?
b. What is the probability that a teetotaler (no alcoholic beverages) develops an ulcer?
c. What is the probability that someone who has an ulcer does not drink alcohol?
d. What is the probability that someone who has an ulcer drinks alcohol?

6.35 An analysis of fired or laid-off workers, their age, and the reason for their departure produced the following table of joint probabilities.

	Age Category			
Reason for Job Loss	20–24	25–54	55–64	65 and Over
Plant or company closed or moved	.015	.324	.075	.017
Insufficient work	.008	.225	.037	.014
Position or shift abolished	.005	.219	.054	.008

Source: Statistical Abstract of the United States, 2006, Table 601

a. What is the probability that a 25–54-year-old employee was laid off or fired because of insufficient work?
b. What proportion of laid-off or fired workers is 65 and over?
c. What is the probability that a laid-off or fired worker because the plant or company closed is 65 and over?

6.36 Many critics of television claim that there is too much violence and that it has a negative impact on society. However, there may also be a negative effect on advertisers. To examine this issue, researchers developed two versions of a cops-and-robbers made-for-television movie. One version depicted several violent crimes, and the other removed these scenes. In the middle of the movie one 60-second commercial was shown advertising a new product and brand name. At the end of the movie, viewers were asked to name the brand. After observing the results, the researchers produced the following table of joint probabilities.

	Watch Violent Movie	Watch Nonviolent Movie
Remember the brand name	.15	.18
Do not remember the brand name	.35	.32

a. What proportion of viewers remember the brand name?
b. What proportion of viewers who watch the violent movie remember the brand name?
c. Does watching a violent movie affect whether the viewer will remember the brand name? Explain.

6.37 Is there a relationship between the male hormone testosterone and criminal behavior? To answer this question, medical researchers measured the testosterone level of penitentiary inmates and recorded whether they were convicted of murder. After analyzing the results they produced the following table of joint probabilities.

Testosterone Level	Murderer	Other Felon
Above average	.27	.24
Below average	.21	.28

a. What proportion of murderers have above average testosterone levels?
b. Are levels of testosterone and the crime committed independent? Explain.

6.38 According to the U.S. National Center for Education Statistics, more than 72 million American workers 18 years and over use computers at work. From this study, which was conducted in 1994 and 1998, the following table of joint probabilities was developed.

	Computer Usage	
Gender	Uses a Spreadsheet or Database	Does Not Use Spreadsheet of Database
Female	.311	.209
Male	.312	.168

Source: Statistical Abstract of the United States, 2003, Table 634

If one worker is selected at random, find the following probabilities.
a. *P*(Worker uses a spreadsheet or database)
b. *P*(Male worker uses a spreadsheet or database)
c. *P*(Female worker uses a spreadsheet or database)

6.39 Refer to Exercise 6.38. Are gender and use of a spreadsheet or database independent events?

6.40 The annual report of the U.S. Bureau of Labor Statistics lists the number of employees who receive various types of benefits. From the information in the January 1999 report, the following probabilities were derived.

	Dental Care	
Type of Worker	Provided by Employer	Not Provided by Employer
Professional/technical	.166	.094
Clerical/sales	.195	.135
Blue collar/services	.230	.180

Source: Statistical Abstract of the United States, 2000, Table 703

a. What proportion of workers are provided with dental care?
b. What proportion of professional/technical workers are provided with dental care?
c. What proportion of blue collar workers have dental care?

6.41 A firm has classified its customers in two ways: (1) according to whether the account is overdue and(2) whether the account is new (less than 12 months) or old. An analysis of the firm's records provided the input for the following table of joint probabilities.

	Overdue	Not Overdue
New	.06	.13
Old	.52	.29

One account is randomly selected.
a. If the account is overdue, what is the probability that it is new?
b. If the account is new, what is the probability that it is overdue?
c. Is the age of the account related to whether it is overdue? Explain.

6.42 How are the size of a firm (measured in terms of the number of employees) and the type of firm related? To help answer the question, an analyst referred to the U.S. Census and developed the following table of joint probabilities.

Number of Employees	Industry		
	Construction	Manufacturing	Retail
Under 20	.2307	.0993	.5009
20 to 99	.0189	.0347	.0876
100 or more	.0019	.0147	.0113

Source: Statistical Abstract of the United States, 2000, Table 868

If one firm is selected at random, find the probability of the following events.

a. The firm employs fewer than 20 employees.
b. The firm is in the retail industry.
c. A firm in the construction industry employs between 20 and 99 workers.

6.43 Credit scorecards are used by financial institutions to help decide to whom loans should be granted (see the Applications in Banking: Credit Scorecards summary on page 49). An analysis of the records of one bank produced the following probabilities.

Loan Performance	Score	
	Under 400	400 or More
Fully repaid	.19	.64
Defaulted	.13	.04

a. What proportion of loans are fully repaid?
b. What proportion of loans given to scorers of less than 400 fully repay?
c. What proportion of loans given to scorers of 400 or more fully repay?
d. Are score and whether the loan is fully repaid independent? Explain.

6.44 A retail outlet wanted to know whether its weekly advertisement in the daily newspaper works. To acquire this critical information, the store manager surveyed the people who entered the store and determined whether each individual saw the ad and whether a purchase was made. From the information developed, the manager produced the following table of joint probabilities. Are the ads effective? Explain.

	Purchase	No Purchase
See ad	.18	.42
Do not see ad	.12	.28

6.45 To gauge the relationship between education and unemployment an economist turned to the U.S. Census, from which the following table was produced.

Education	Employed	Unemployed
Not a high school graduate	.0975	.0080
High school graduate	.3108	.0128
Some college, no degree	.1785	.0062
Associate's degree	.0849	.0023
Bachelor's degree	.1959	.0041
Advanced degree	.0975	.0015

Source: Statistical Abstract of the United States, 2000, Table 251

a. What is the probability that a high school graduate is unemployed?
b. Determine the probability that a randomly selected individual is employed.
c. Find the probability that an unemployed person possesses an advanced degree.
d. What is the probability that a randomly selected person did not finish high school?

6.46 The decision about where to build a new plant is a major one for most companies. One of the factors that is often considered is the education level of the location's residents. Census information may be useful in this regard. After analyzing a recent census, a company produced the following joint probabilities.

Education	Region			
	Northwest	Midwest	South	West
Not a high school graduate	.0301	.0318	.0683	.0359
High school graduate	.0711	.0843	.1174	.0608
Some college, no degree	.0262	.0410	.0605	.0456
Associate's degree	.0143	.0180	.0248	.0181
Bachelor's degree	.0350	.0368	.0559	.0418
Advanced degree	.0190	.0184	.0269	.0180

Source: Statistical Abstract of the United States, 2000, Table 251

a. Determine the probability that a person living in the West has a bachelor's degree.
b. Find the probability that a high school graduate lives in the Northwest.
c. What is the probability that a person selected at random lives in the South?
d. What is the probability that a person selected at random does not live in the South?

6.3 Probability Rules and Trees

In Section 6.2 we introduced intersection and union and described how to determine the probability of the intersection and the union of two events. In this section we present other methods of determining these probabilities. We introduce three rules that enable us to calculate the probability of more complex events from the probability of simpler events.

Complement Rule

The **complement** of event A is the event that occurs when event A does not occur. The complement of event A is denoted by A^C. The **complement rule** defined here derives from the fact that the probability of an event and the probability of the event's complement must sum to 1.

Complement Rule

$$P(A^C) = 1 - P(A)$$

for any event A.

We will demonstrate the use of this rule after we introduce the next rule.

Multiplication Rule

The **multiplication rule** is used to calculate the joint probability of two events. It is based on the formula for conditional probability supplied in the previous section. That is, from the following formula

$$P(A \mid B) = \frac{P(A \text{ and } B)}{P(B)}$$

we derive the multiplication rule simply by multiplying both sides by $P(B)$.

Multiplication Rule

The joint probability of any two events A and B is

$$P(A \text{ and } B) = P(B)P(A \mid B)$$

or altering the notation

$$P(A \text{ and } B) = P(A)P(B \mid A)$$

If A and B are independent events, $P(A \mid B) = P(A)$ and $P(B \mid A) = P(B)$. It follows that the joint probability of two independent events is simply the product of the probabilities of the two events. We can express this as a special form of the multiplication rule.

Multiplication Rule for Independent Events

The joint probability of any two independent events A and B is

$$P(A \text{ and } B) = P(A)P(B)$$

EXAMPLE 6.5* Selecting Two Students without Replacement

A graduate statistics course has seven male and three female students. The professor wants to select two students at random to help her conduct a research project. What is the probability that the two students chosen are female?

SOLUTION

Let A represent the event that the first student chosen is female and B represent the event that the second student chosen is also female. We want the joint probability $P(A \text{ and } B)$. Consequently, we apply the multiplication rule:

$$P(A \text{ and } B) = P(A)P(B \mid A)$$

Because there are three female students in a class of ten, the probability that the first student chosen is female is

$$P(A) = 3/10$$

After the first student is chosen, there are only nine students left. Given that the first student chosen was female, there are only two female students left. It follows that

$$P(B \mid A) = 2/9$$

Thus the joint probability is

$$P(A \text{ and } B) = P(A)P(B \mid A) = \left(\frac{3}{10}\right)\left(\frac{2}{9}\right) = \frac{6}{90} = .067$$

EXAMPLE 6.6 Selecting Two Students with Replacement

Refer to Example 6.5. The professor who teaches the course is suffering from the flu and will be unavailable for two classes. The professor's replacement will teach the next two classes. His style is to select one student at random and pick on him or her to answer questions during that class. What is the probability that the two students chosen are female?

SOLUTION

The form of the question is the same as in Example 6.5: We wish to compute the probability of choosing two female students. However, the experiment is slightly different. It is now possible to choose the *same* student in each of the two classes the replacement teaches. Thus A and B are independent events, and we apply the multiplication rule for independent events:

$$P(A \text{ and } B) = P(A)P(B)$$

The probability of choosing a female student in each of the two classes is the same. That is,

$$P(A) = 3/10 \text{ and } P(B) = 3/10$$

Hence,

$$P(A \text{ and } B) = P(A)P(B) = \left(\frac{3}{10}\right)\left(\frac{3}{10}\right) = \frac{9}{100} = .09$$

*This example can be solved using the hypergeometric distribution, which is described in CD Appendix F.

Addition Rule

The **addition rule** enables us to calculate the probability of the union of two events.

Addition Rule

The probability that event A, or event B, or both occur is

$$P(A \text{ or } B) = P(A) + P(B) - P(A \text{ and } B)$$

If you're like most students, you're wondering why we subtract the joint probability from the sum of the probabilities of A and B. To understand why this is necessary, examine Table 6.2 (page 177), which we have reproduced here as Table 6.3.

TABLE **6.3**
Joint and Marginal Probabilities

	B_1	B_2	Totals
A_1	$P(A_1 \text{ and } B_1) = .11$	$P(A_1 \text{ and } B_2) = .29$	$P(A_1) = .40$
A_2	$P(A_2 \text{ and } B_1) = .06$	$P(A_2 \text{ and } B_2) = .54$	$P(A_2) = .60$
Totals	$P(B_1) = .17$	$P(B_2) = .83$	1.00

This table summarizes how the marginal probabilities were computed. For example, the marginal probability of A_1 and the marginal probability of B_1 were calculated as

$$P(A_1) = P(A_1 \text{ and } B_1) + P(A_1 \text{ and } B_2) = .11 + .29 = .40$$

$$P(B_1) = P(A_1 \text{ and } B_1) + P(A_2 \text{ and } B_1) = .11 + .06 = .17$$

If we now attempt to calculate the probability of the union of A_1 and B_1 by summing their probabilities, we find

$$P(A_1) + P(B_1) = .11 + .29 + .11 + .06$$

Notice that we added the joint probability of A_1 and B_1 (which is .11) twice. To correct the double counting, we subtract the joint probability from the sum of the probabilities of A_1 and B_1. Thus,

$$\begin{aligned} P(A_1 \text{ or } B_1) &= P(A_1) + P(B_1) - P(A_1 \text{ and } B_1) \\ &= [.11 + .29] + [.11 + .06] - .11 \\ &= .40 + .17 - .11 = .46 \end{aligned}$$

This is the probability of the union of A_1 and B_1, which we calculated in Example 6.4 (page 179).

As was the case with the multiplication rule, there is a special form of the addition rule. When two events are mutually exclusive (which means that the two events cannot occur together), their joint probability is 0.

Addition Rule for Mutually Exclusive Events

The probability of the union of two mutually exclusive events A and B is

$$P(A \text{ or } B) = P(A) + P(B)$$

EXAMPLE 6.7 Applying the Addition Rule

In a large city, two newspapers are published, the *Sun* and the *Post*. The circulation departments report that 22% of the city's households have a subscription to the *Sun* and 35% subscribe to the *Post*. A survey reveals that 6% of all households subscribe to both newspapers. What proportion of the city's households subscribe to either newspaper?

SOLUTION

We can express this question as, What is the probability of selecting a household at random that subscribes to the *Sun*, the *Post*, or both? Another way of asking the question is, What is the probability that a randomly selected household subscribes to *at least one* of the newspapers? It is now clear that we seek the probability of the union, and we must apply the addition rule. Let A = the household subscribes to the *Sun* and B = the household subscribes to the *Post*. We perform the following calculation:

$$P(A \text{ or } B) = P(A) + P(B) - P(A \text{ and } B) = .22 + .35 - .06 = .51$$

The probability that a randomly selected household subscribes to either newspaper is .51. Expressed as relative frequency, 51% of the city's households subscribe to either newspaper.

Probability Trees

An effective and simpler method of applying the probability rules is the probability tree, wherein the events in an experiment are represented by lines. The resulting figure resembles a tree, hence the name. We will illustrate the probability tree with several examples, including two that we addressed using the probability rules alone.

In Example 6.5 we wanted to find the probability of choosing two female students, where the two choices had to be different. The tree diagram in Figure 6.1 describes this experiment. Notice that the first two branches represent the two possibilities, female and male students, on the first choice. The second set of branches represents the two possibilities on the second choice. The probabilities of female and male student chosen first are 3/10 and 7/10, respectively. The probabilities for the second set of branches are conditional probabilities based on the choice of the first student selected.

We calculate the joint probabilities by multiplying the probabilities on the linked branches. Thus, the probability of choosing two female students is $P(F \text{ and } F) = (3/10)(2/9) = 6/90$. The remaining joint probabilities are computed similarly.

FIGURE 6.1 Probability Tree for Example 6.5

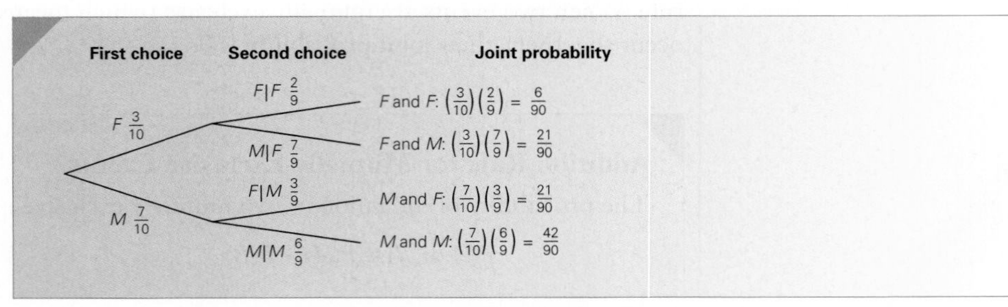

In Example 6.6, the experiment was similar to that of Example 6.5. However, the student selected on the first choice was returned to the pool of students and was eligible to be chosen again. Thus, the probabilities on the second set of branches remain the same as the probabilities on the first set, and the probability tree is drawn with these changes, as shown in Figure 6.2.

FIGURE **6.2** Probability Tree for Example 6.6

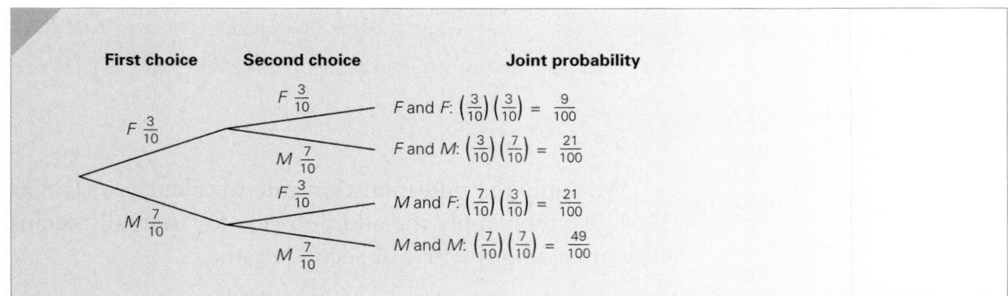

The advantage of a probability tree on this type of problem is that it restrains its users from making the wrong calculation. Once the tree is drawn and the probabilities of the branches inserted, virtually the only allowable calculation is the multiplication of the probabilities of linked branches. An easy check on those calculations is available. The joint probabilities at the ends of the branches must sum to 1, because all possible events are listed. In both figures notice that the joint probabilities do indeed sum to 1.

The special form of the addition rule for mutually exclusive events can be applied to the joint probabilities. In both probability trees, we can compute the probability that one student chosen is female and one is male simply by adding the joint probabilities. For the tree in Example 6.5, we have

$$P(F \text{ and } M) + P(M \text{ and } F) = 21/90 + 21/90 = 42/90$$

In the probability tree in Example 6.6, we find

$$P(F \text{ and } M) + P(M \text{ and } F) = 21/100 + 21/100 = 42/100$$

EXAMPLE 6.8

Probability of Passing the Bar Exam

Students who graduate from law schools must still pass a bar exam before becoming lawyers. Suppose that in a particular jurisdiction the pass rate for first-time test takers is 72%. Candidates who fail the first exam may take it again several months later. Of those who fail their first test, 88% pass their second attempt. Find the probability that a randomly selected law school graduate becomes a lawyer. Assume that candidates cannot take the exam more than twice.

SOLUTION

The probability tree in Figure 6.3 is employed to describe the experiment. Note that we use the complement rule to determine the probability of failing each exam.

FIGURE **6.3** Probability Tree for Example 6.8

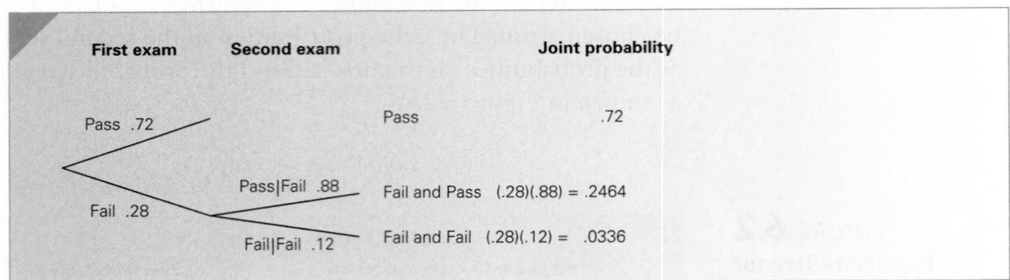

We apply the multiplication rule to calculate *P*(Fail and Pass), which we find to be .2464. We then apply the addition rule for mutually exclusive events to find the probability of passing the first or second exam:

$$P(\text{Pass [on first exam]}) + P(\text{Fail [on first exam] and Pass [on second exam]})$$
$$= .72 + .2464 = .9664$$

Thus, 96.64% of applicants become lawyers by passing the first or second exam.

EXERCISES

6.47 Given the following probabilities, compute all joint probabilities.

$P(A) = .9$ $\quad P(A^C) = .1$
$P(B|A) = .4$ $\quad P(B|A^C) = .7$

6.48 Determine all joint probabilities from the following.

$P(A) = .8$ $\quad P(A^C) = .2$
$P(B|A) = .4$ $\quad P(B|A^C) = .7$

6.49 Draw a probability tree to compute the joint probabilities from the following probabilities.

$P(A) = .5$ $\quad P(A^C) = .5$
$P(B|A) = .4$ $\quad P(B|A^C) = .7$

6.50 Given the following probabilities, draw a probability tree to compute the joint probabilities.

$P(A) = .8$ $\quad P(A^C) = .2$
$P(B|A) = .3$ $\quad P(B|A^C) = .3$

6.51 Given the following probabilities, find the joint probability *P*(*A* and *B*).

$P(A) = .7$ $\quad P(B|A) = .3$

6.52 Approximately 10% of people are left-handed. If two people are selected at random, what is the probability of the following events?

a. Both are right-handed.
b. Both are left-handed.
c. One is right-handed and the other is left-handed.
d. At least one is right-handed.

6.53 Refer to Exercise 6.52. Suppose that three people are selected at random.

a. Draw a probability tree to depict the experiment.
b. If we use the notation *RRR* to describe the selection of three right-handed people, similarly describe the remaining seven events. (Use *L* for left-hander.)
c. How many of the events yield no right-handers, one right-hander, two right-handers, three right-handers?
d. Find the probability of no right-handers, one right-hander, two right-handers, three right-handers.

6.54 Suppose that there are 100 students in your accounting class, of whom 10 are left-handed. Two students are selected at random.

a. Draw a probability tree and insert the probabilities for each branch.

What is the probability of the following events?

b. Both are right-handed.
c. Both are left-handed.
d. One is right-handed and the other is left-handed.
e. At least one is right-handed.

6.55 Refer to Exercise 6.54. Suppose that three people are selected at random.

a. Draw a probability tree and insert the probabilities of each branch.
b. What is the probability of no right-handers, one right-hander, two right-handers, three right-handers?

6.56 An aerospace company has submitted bids on two separate federal government defense contracts. The company president believes that there is a 40% probability of winning the first contract. If they win the first contract, the probability of winning the second is 70%. However, if they lose the first contract, the president thinks that the probability of winning the second contract decreases to 50%.

a. What is the probability that they win both contracts?
b. What is the probability that they lose both contracts?
c. What is the probability that they win only one contract?

6.57 A telemarketer calls people and tries to sell them a subscription to a daily newspaper. On 20% of her calls, there is no answer or the line is busy. She sells subscriptions to 5% of the remaining calls. For what proportion of calls does she make a sale?

6.58 A foreman for an injection-molding firm admits that on 10% of his shifts, he forgets to shut off the injection machine on his line. This causes the machine to overheat, increasing the probability from 2% to 20% that a defective molding will be produced during the early morning run. What proportion of moldings from the early morning run is defective?

6.59 A study undertaken by the Miami-Dade Supervisor of Elections in 2002 revealed that 44% of registered voters are Democrats, 37% are Republicans, and 19% are others. If two registered voters are selected at random, what is the probability that both of them have the same party affiliation? (*Source: Miami Herald*, April 11, 2002)

6.60 In early 2001 the United States Census Bureau started releasing the results of the latest census. Among many other pieces of information, the bureau recorded the race or ethnicity of the residents of every county in every state. From these results the bureau calculated a "diversity index," which measures the probability that two people chosen at random are of different races or ethnicities. Suppose that the census determined that in a county in Wisconsin 80% of its residents are White, 15% are Black, and 5% are Asian. Calculate the diversity index for this county.

6.61 A survey of middle-age men reveals that 28% of them are balding at the crown of their heads. Moreover, it is known that such men have an 18% probability of suffering a heart attack in the next 10 years. Men who are not balding in this way have an 11% probability of a heart attack. Find the probability that a middle-age man will suffer a heart attack sometime in the next 10 years.

6.62 The chartered financial analyst (CFA) is a designation earned after taking three annual exams (CFA I, II, and III). The exams are taken in early June. Candidates who pass an exam are eligible to take the exam for the next level in the following year. The pass rates for levels I, II, and III are .57, .73, and .85, respectively. Suppose that 3,000 candidates take the level I exam, 2,500 take the level II exam, and 2,000 take the level III exam. Suppose that one student is selected at random. What is the probability that he or she has passed the exam? (*Source:* Institute of Financial Analysts)

6.63 The Nickels restaurant chain regularly conducts surveys of its customers. Respondents are asked to assess food quality, service, and price. The responses are

Excellent Good Fair

They are also asked whether they would come back. After analyzing the responses, an expert in probability determined that 87% of customers say that they will return. Of those who so indicate, 57% rate the restaurant as excellent, 36% rate it as good, and the remainder rate it as fair. Of those who say that they won't return, the probabilities are 14%, 32%, and 54%, respectively. What proportion of customers rate the restaurant as good?

6.64 Researchers at the University of Pennsylvania School of Medicine have determined that children under 2 years old who sleep with the lights on have a 36% chance of becoming myopic before they are 16. Children who sleep in darkness have a 21% probability of becoming myopic. A survey indicates that 28% of children under 2 sleep with some light on. Find the probability that a child under 16 is myopic.

6.65 All printed circuit boards (PCBs) that are manufactured at a certain plant are inspected. An analysis of the company's records indicates that 22% of all PCBs are flawed in some way. Of those that are flawed, 84% are reparable and the rest must be discarded. If a newly produced PCB is randomly selected, what is the probability that it does not have to be discarded?

6.66 A financial analyst has determined that there is a 22% probability that a mutual fund will outperform the market over a 1-year period provided that it outperformed the market the previous year. If only 15% of mutual funds outperform the market during any year, what is the probability that a mutual fund will outperform the market 2 years in a row?

6.67 An investor believes that on a day when the Dow Jones Industrial Average (DJIA) increases, the probability that the NASDAQ also increases is 77%. If the investor believes that there is a 60% probability that the DJIA will increase tomorrow, what is the probability that the NASDAQ will increase as well?

6.68 The controls of an airplane have several backup systems or redundancies, so that if one fails the plane will continue to operate. Suppose that the mechanism that controls the flaps has two backups. If the probability that the main control fails is .0001 and the probability that each backup will fail is .01, what is the probability that all three fail to operate?

6.69 According to *TNS Intersearch*, 69% of wireless web users use it primarily for receiving and sending e-mail. Suppose that three wireless web users are selected at random. What is the probability that all of them use it primarily for e-mail?

6.70 A financial analyst estimates that the probability that the economy will experience a recession in the next 12 months is 25%. She also believes that if the economy encounters a recession, the probability that her mutual fund will increase in value is 20%. If there is no recession, the probability that the mutual fund will increase in value is 75%. Find the probability that the mutual fund's value will increase.

6.4 BAYES'S LAW

Conditional probability is often used to gauge the relationship between two events. In many of the examples and exercises you've already encountered, conditional probability measures the probability that an event occurs given that a possible cause of the event has occurred. In Example 6.2 we calculated the probability that a mutual fund outperforms the market (the effect) given that the fund manager graduated from a top-20 MBA program (the possible cause). There are situations, however, where we witness a particular event and we need to compute the probability of one of its possible causes. **Bayes's Law** is the technique we use.

EXAMPLE 6.9 Should an MBA Applicant Take a Preparatory Course?

The Graduate Management Admission Test (GMAT) is a requirement for all applicants of MBA programs. There are a variety of preparatory courses designed to help improve GMAT scores, which range from 200 to 800. Suppose that a survey of MBA students reveals that among GMAT scorers above 650, 52% took a preparatory course, whereas among GMAT scorers of less than 650 only 23% took a preparatory course. An applicant to an MBA program has determined that he needs a score of more than 650 to get into a certain MBA program, but he feels that his probability of getting that high a score is quite low—10%. He is considering taking a preparatory course that costs $500. He is willing to do so only if his probability of achieving 650 or more doubles. What should he do?

SOLUTION

The easiest way to address this problem is to draw a tree diagram. The following notation will be used:

A = GMAT score is 650 or more

A^C = GMAT score less than 650

B = Take preparatory course

B^C = Do not take preparatory course

The probability of scoring 650 or more is

$$P(A) = .10$$

The complement rule gives us

$$P(A^C) = 1 - .10 = .90$$

Conditional probabilities are

$$P(B \mid A) = .52$$

and

$$P(B \mid A^C) = .23$$

Again using the complement rule, we find the following conditional probabilities:

$$P(B^C \mid A) = 1 - .52 = .48$$

and

$$P(B^C \mid A^C) = 1 - .23 = .77$$

We would like to determine the probability that he achieves a GMAT score of 650 or more given that he takes the preparatory course. That is, we need to compute

$$P(A \mid B)$$

Using the definition of conditional probability (page 178), we have

$$P(A \mid B) = \frac{P(A \text{ and } B)}{P(B)}$$

Neither the numerator nor the denominator is known. The probability tree (Figure 6.4) will provide us with the probabilities.

FIGURE **6.4** **Probability Tree for Example 6.9**

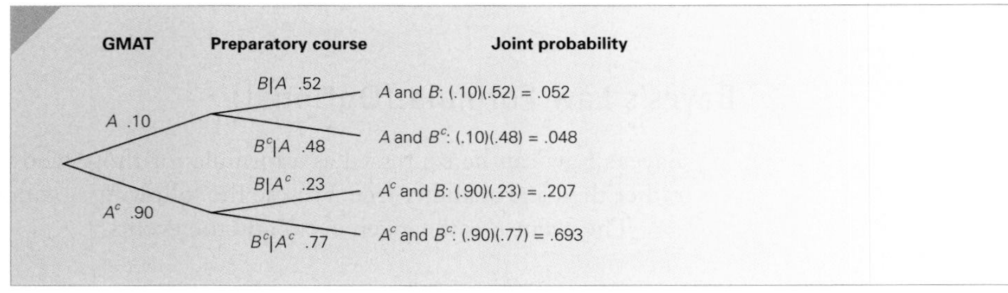

As you can see,

$$P(A \text{ and } B) = (.10)(.52) = .052$$
$$P(A^C \text{ and } B) = (.90)(.23) = .207$$

and

$$P(B) = P(A \text{ and } B) + P(A^C \text{ and } B) = .052 + .207 = .259$$

Thus,

$$P(A \mid B) = \frac{P(A \text{ and } B)}{P(B)} = \frac{.052}{.259} = .201$$

The probability of scoring 650 or more on the GMAT doubles when the preparatory course is taken.

Thomas Bayes first employed the calculation of conditional probability, as shown in Example 6.9, during the eighteenth century. Accordingly, it is called Bayes's Law.

The probabilities $P(A)$ and $P(A^C)$ are called **prior probabilities** because they are determined *prior* to the decision about taking the preparatory course. The conditional probabilities are called **likelihood probabilities** for reasons that are beyond the mathematics in this book. Finally, the conditional probability $P(A \mid B)$ and similar conditional probabilities $P(A^C \mid B)$, $P(A \mid B^C)$, and $P(A^C \mid B^C)$ are called **posterior probabilities** or **revised probabilities,** because the prior probabilities are revised *after* the decision about taking the preparatory course.

You may be wondering why we did not get $P(A \mid B)$ directly. That is, why not survey people who took the preparatory course and ask whether they received a score of 650 or more? The answer is that using the likelihood probabilities and using Bayes's Law allows individuals to set their own prior probabilities, which can then be revised. For example, another MBA applicant may assess her probability of scoring 650 or more as .40. Inputting the new prior probabilities produces the following probabilities:

$$P(A \text{ and } B) = (.40)(.52) = .208$$
$$P(A^C \text{ and } B) = (.60)(.23) = .138$$
$$P(B) = P(A \text{ and } B) + P(A^C \text{ and } B) = .208 + .138 = .346$$
$$P(A \mid B) = \frac{P(A \text{ and } B)}{P(B)} = \frac{.208}{.346} = .601$$

The probability of achieving a GMAT score of 650 or more increases by a more modest 50% (from .40 to .601).

Bayes's Law Formula (Optional)

Bayes's Law can be expressed as a formula for those who prefer an algebraic approach rather than a probability tree. We use the following notation.

The event B is the given event and the events

$$A_1, A_2, \ldots, A_k$$

are the events for which prior probabilities are known. That is,

$$P(A_1), P(A_2), \ldots, P(A_k)$$

are the prior probabilities.

The likelihood probabilities are

$$P(B \mid A_1), P(B \mid A_2), \ldots, P(B \mid A_k)$$

and

$$P(A_1 \mid B), P(A_2 \mid B), \ldots, P(A_k \mid B)$$

are the posterior probabilities, which represent the probabilities we seek.

Bayes' Law Formula

$$P(A_i \mid B) = \frac{P(A_i)P(B \mid A_i)}{P(A_1)P(B \mid A_1) + P(A_2)P(B \mid A_2) + \cdots + P(A_k)P(B \mid A_k)}$$

To illustrate the use of the formula, we'll redo Example 6.9. We begin by defining the events.

A_1 = GMAT score is 650 or more

A_2 = GMAT score less than 650

B = Take preparatory course

The probabilities are

$$P(A_1) = .10$$

The complement rule gives us

$$P(A_2) = 1 - .10 = .90$$

Conditional probabilities are

$$P(B \mid A_1) = .52$$

and

$$P(B \mid A_2) = .23$$

Substituting the prior and likelihood probabilities into the Bayes's Law formula yields the following:

$$P(A_1 \mid B) = \frac{P(A_1)P(B \mid A_1)}{P(A_1)P(B \mid A_1) + P(A_2)P(B \mid A_2)} = \frac{(.10)(.52)}{(.10)(.52) + (.90)(.23)}$$

$$= \frac{.052}{.052 + .207} = \frac{.052}{.259} = .201$$

As you can see, the calculation of the Bayes's Law formula produces the same results as the probability tree.

Auditing Tax Returns: Solution

We need to revise the prior probability that this return contains significant fraud. The tree shown in Figure 6.5 details the calculation.

F = Tax return is fraudulent

F^C = Tax return is honest

E_0 = Tax return contains no expense deductions

E_1 = Tax return contains expense deduction

E_2 = tax return contains 2 expense deductions

$$P(E_1) = P(F \text{ and } E_1) + P(F^C \text{ and } E_1) = .0140 + .1710 = .1850$$

$$P(F | E_1) = P(F \text{ and } E_1)/P(E_1) = .0140/.1850 = .0757$$

The probability that this return is fraudulent is .0757.

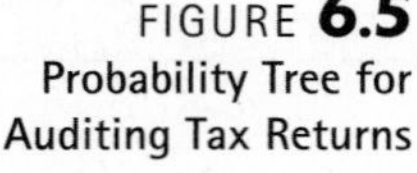

FIGURE **6.5**
Probability Tree for Auditing Tax Returns

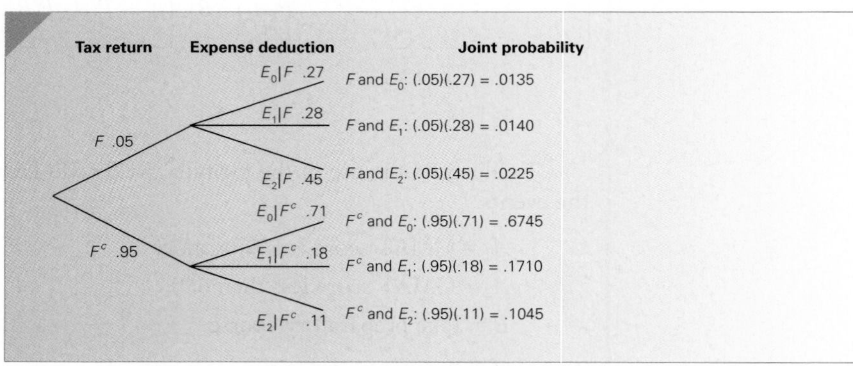

Applications in Medicine and Medical Insurance (Optional)

Physicians routinely perform medical tests, called *screenings*, on their patients. Screening tests are conducted for all patients in a particular age and gender group, regardless of their symptoms. For example, men in their 50s are advised to take a PSA test to determine whether there is evidence of prostate cancer. Women undergo a Pap test for cervical cancer. Unfortunately, few of these tests are 100% accurate. Most can produce *false-positive* and *false-negative* results. A false-positive result is one in which the patient does not have the disease, but the test shows positive. A false-negative result is one in which the patient does have the disease, but the test produces a negative result. The consequences of each test are serious and costly. A false-negative test results in not detecting a disease in a patient, therefore postponing, perhaps indefinitely, treatment. A false-positive test leads to apprehension and fear for the patient. In most cases the patient is required to undergo further testing such as a biopsy. The unnecessary follow-up procedure can pose medical risks.

False-positive test results have financial repercussions. That is, the cost of the follow-up procedure is usually far more expensive than the screening test. Medical insurance companies as well as government-funded plans are all adversely affected by false-positive test results. Compounding the problem is that physicians and patients are incapable of properly interpreting the results. A correct analysis can save both lives and money.

Bayes's Law is the vehicle we use to determine the true probabilities associated with screening tests. Applying the complement rule to the false-positive and false-negative rates produces the conditional probabilities that represent correct conclusions. Prior probabilities are usually derived by looking at the overall proportion of people with the diseases. In some cases the prior probabilities may themselves have been revised because of heredity or demographic variables such as age or race. Bayes's Law allows us to revise the prior probability after the test result is positive or negative.

Example 6.10 is based on the actual false-positive and false-negative rates. Note, however, that different sources provide somewhat different probabilities. The differences may be due to the way positive and negative results are defined or to the way technicians conduct the tests. Students who are affected by the diseases described in the example and exercises should seek clarification from their physicians.

EXAMPLE 6.10

Probability of Prostate Cancer

Prostate cancer is the most common form of cancer found in men. The probability of developing prostate cancer over a lifetime is 16%. (This figure may be higher since many prostate cancers go undetected.) Many physicians routinely perform a PSA test, particularly for men over age 50. Prostate specific antigen (PSA) is a protein produced only by the prostate gland and thus is fairly easy to detect. Normally men have PSA levels between 0 and 4 mg/ml. Readings above 4 may be considered high and potentially indicative of cancer. However, PSA levels tend to rise with age even among men who are cancer-free. Studies have shown that the test is not very accurate. In fact, the probability of having an elevated PSA level given that the man does not have cancer (false-positive) is .135. If the man does have cancer, the probability of a normal PSA level (false-negative) is almost .300. (This figure may vary by age and by the definition of *high* PSA level.) If a physician concludes that the PSA is high, a biopsy is performed. Besides the concerns and health needs of the men, there are also financial costs. The cost of the blood test is low (approximately $50). However, the cost of the biopsy is considerably higher (approximately $1,000). A false-positive PSA test will lead to an unnecessary biopsy. Because the PSA test is so inaccurate, some private and public medical plans do not pay for it. Suppose you are a manager in a medical insurance company and must decide on guidelines for whom should be routinely screened for prostate cancer. An analysis of prostate cancer incidence and age produces the following table of probabilities. (The probability of a man under 40 developing prostate cancer is less than .0001, small enough to treat as 0.)

Age	Probability of Developing Prostate Cancer
40 up to, but not including, 50	.010
50 up to, but not including, 60	.022
60 up to, but not including, 70	.046
Over 70	.079

Assume that a man in each of the age categories undergoes a PSA test with a positive result. Calculate the probability that each man actually has prostate cancer and the probability that he does not. Perform a cost-benefit analysis to determine the cost per cancer detected.

SOLUTION

As we did in Example 6.9 and the chapter-opening example we'll draw a probability tree (Figure 6.6). The notation is

$C =$ Has prostate cancer

$C^C =$ Does not have prostate cancer

$PT =$ Positive test result

$NT =$ Negative test result

Starting with a man between 40 and 50 years old, we have the following probabilities

Prior

$P(C) = .010$

$P(C^C) = 1 - .010 = .990$

Likelihood probabilities

False-negative: $P(NT \mid C) = .300$

True-positive: $P(PT \mid C) = 1 - .300 = .700$

False-positive: $P(PT \mid C^C) = .135$

True-negative: $P(NT \mid C^C) = 1 - .135 = .865$

FIGURE **6.6** **Probability Tree for Example 6.10**

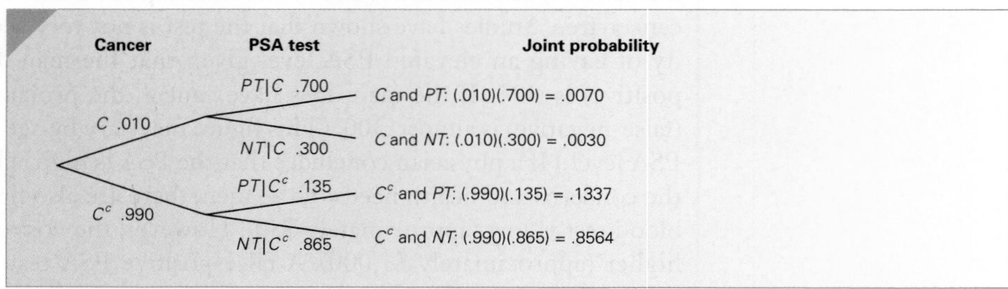

The tree allows you to determine the probability of obtaining a positive test result. It is

$$P(PT) = P(C \text{ and } PT) + P(C^C \text{ and } PT) = .0070 + .1337 = .1407$$

We can now compute the probability that the man has prostate cancer given a positive test result:

$$P(C \mid PT) = \frac{P(C \text{ and } PT)}{P(PT)} = \frac{.0070}{.1407} = .0498$$

The probability that he does not have prostate cancer is

$$P(C^C \mid PT) = 1 - P(C \mid PT) = 1 - .0498 = .9502$$

We can repeat the process for the other age categories. Here are the results.

	Probabilities Given a Positive PSA Test	
Age	Has Prostate Cancer	Does Not Have Prostate Cancer
40 to 50	.0498	.9502
50 to 60	.1045	.8955
60 to 70	.2000	.8000
Over 70	.3078	.6922

The following table lists the proportion of each age category wherein the PSA test is positive $[P(PT)]$

Age	Proportion of Tests That Are Positive	Number of Biopsies Performed per Million	Number of Cancers Detected	Number of Biopsies per Cancer Detected
40 to 50	.1407	140,700	.0498(140,700) = 7,007	20.10
50 to 60	.1474	147,400	.1045(147,400) = 15,403	9.57
60 to 70	.1610	161,000	.2000(161,000) = 32,200	5.00
Over 70	.1796	179,600	.3078(179,600) = 55,281	3.25

If we assume a cost of $1,000 per biopsy, the cost per cancer detected is $20,100 for 40 to 50, $9,570 for 50 to 60, $5,000 for 60 to 70, and $3,250 for over 70.

We have created an Excel spreadsheet to help you perform the calculations in Example 6.10. Open the **Excel Workbooks** folder and select **Medical screening.** There are three cells that you may alter. In cell B5 enter a new prior probability for prostate cancer. Its complement will be calculated in cell B15. In cells D6 and D15, type new values for the false-negative and false-positive rates, respectively. Excel will do the rest. We will use this spreadsheet to demonstrate some terminology standard in medical testing.

Terminology

We will illustrate the terms using the probabilities calculated for the 40 to 50 age category.

The false negative rate is .300. Its complement is the likelihood probability $P(PT \mid C)$, called the **sensitivity.** It is equal to $1 - .300 = .700$. Among men with prostate cancer, this is the proportion of men who will get a positive test result.

The complement of the false-positive rate (.135) is $P(NT \mid C^C)$, which is called the **specificity.** This likelihood probability is $1 - .135 = .865$

The posterior probability that someone has prostate cancer given a positive test result $[P(C \mid PT) = .0498]$ is called the **positive predictive value.** Using Bayes's Law, we can compute the other three posterior probabilities.

Probability that the patient does not have prostate cancer given a positive test result is

$$P(C^C \mid PT) = .9502$$

Probability that the patient has prostate cancer given a negative test result is

$$P(C \mid NT) = .0035$$

Probability that the patient does not have prostate cancer given a negative test result is

$$P(C^C \mid NT) = .9965$$

This revised probability is called the **negative predictive value.**

If you review the computations made previously, you'll realize that the prior probabilities are as important as the probabilities associated with the test results (the likelihood probabilities) in determining the posterior probabilities. The following table shows the prior probabilities and the revised probabilities.

Age	Prior Probabilities for Prostate Cancer	Posterior Probabilities Given a Positive PSA Test
40 to 50	.010	.0498
50 to 60	.022	.1045
60 to 70	.046	.2000
Over 70	.079	.3078

As you can see if the prior probability is low, then unless the screening test is quite accurate, the revised probability will still be quite low.

To see the effects of different likelihood probabilities, suppose the PSA test is a perfect predictor. That is, the false-positive and false-negative rates are 0. Figure 6.7 displays the probability tree.

FIGURE **6.7** Probability Tree for Example 6.10 with a Perfect Predictor Test

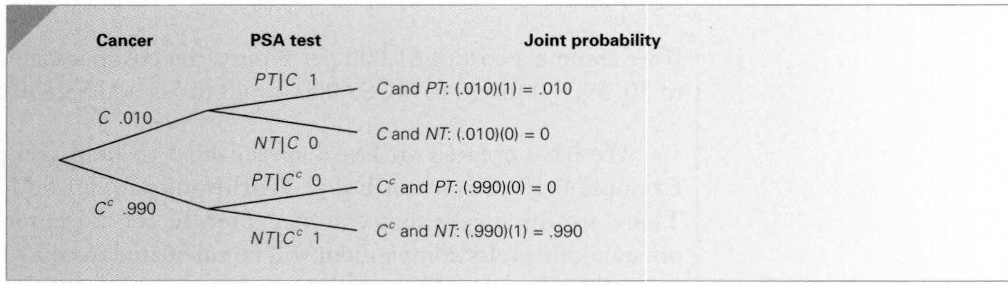

We find

$$P(PT) = P(C \text{ and } PT) + P(C^C \text{ and } PT) = .010 + 0 = .010$$

$$P(C|PT) = \frac{P(C \text{ and } PT)}{P(PT)} = \frac{.010}{.010} = 1.00$$

Now we calculate the probability of prostate cancer when the test is negative.

$$P(NT) = P(C \text{ and } NT) + P(C^C \text{ and } NT) = 0 + .990 = .990$$

$$P(C|NT) = \frac{P(C \text{ and } NT)}{P(NT)} = \frac{0}{.990} = 0$$

Thus, if the test is a perfect predictor and a man has a positive test, then, as expected, the probability that he has prostate cancer is 1.0. The probability that he does not have cancer when the test is negative is 0.

Now suppose that the test is always wrong. That is, the false-positive and false-negative rates are 100%. The probability tree is shown in Figure 6.8.

FIGURE **6.8** Probability Tree for Example 6.10 with a Test That Is Always Wrong

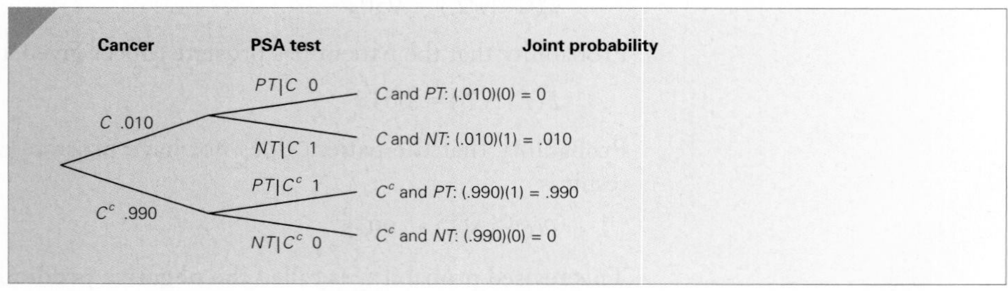

$$P(PT) = P(C \text{ and } PT) + P(C^C \text{ and } PT) = 0 + .990 = .990$$

$$P(C|PT) = \frac{P(C \text{ and } PT)}{P(PT)} = \frac{0}{.990} = 0$$

and

$$P(NT) = P(C \text{ and } NT) + P(C^C \text{ and } NT) = .010 + 0 = .010$$

$$P(C|NT) = \frac{P(C \text{ and } NT)}{P(NT)} = \frac{.010}{.010} = 1.00$$

Notice we have another perfect predictor except that it is reversed. The probability of prostate cancer given a positive test result is 0, but the probability becomes 1.00 when the test is negative.

Finally we consider the situation when the set of likelihood probabilities are the same. Figure 6.9 depicts the probability tree for a 40- to 50-year-old male and the probability of a positive test is (say) .3 and the probability of a negative test is .7.

FIGURE 6.9 Probability Tree for Example 6.10 with Identical Likelihood Probabilities

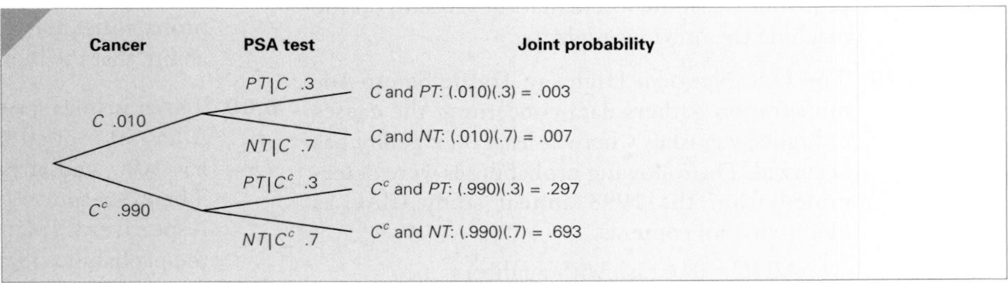

$$P(PT) = P(C \text{ and } PT) + P(C^C \text{ and } PT) = .003 + .297 = .300$$

$$P(C|PT) = \frac{P(C \text{ and } PT)}{P(PT)} = \frac{.003}{.300} = .01$$

and

$$P(NT) = P(C \text{ and } NT) + P(C^C \text{ and } NT) = .007 + .693 = .700$$

$$P(C|NT) = \frac{P(C \text{ and } NT)}{P(NT)} = \frac{.007}{.700} = .01$$

As you can see, the posterior and prior probabilities are the same. That is, the PSA test does not change the prior probabilities. Obviously, the test is useless.

We could have used any probability for the false-positive and false-negative rates, including .5. If we had used .5, then one way of performing this PSA test is to flip a fair coin. One side would be interpreted as positive and the other side as negative. It is clear that such a test has no predictive power.

The exercises and Case 6.4 offer the probabilities for several other screening tests.

EXERCISES

6.71 Refer to Exercise 6.47. Determine $P(A \mid B)$.

6.72 Refer to Exercise 6.48. Find the following.
a. $P(A \mid B)$
b. $P(A^C \mid B)$
c. $P(A \mid B^C)$
d. $P(A^C \mid B^C)$

6.73 Refer to Example 6.9. An MBA applicant believes that the probability of scoring more than 650 on the GMAT without the preparatory course is .95. What is the probability of attaining that level after taking the preparatory course?

6.74 Refer to Exercise 6.58. The plant manager randomly selects a molding from the early morning run and discovers it is defective. What is the probability that the foreman forgot to shut off the machine the previous night?

6.75 The U.S. National Highway Traffic Safety Administration gathers data concerning the causes of highway crashes where at least one fatality has occurred. The following probabilities were determined from the 1998 annual study (BAC is blood-alcohol content).

$P(BAC = 0 \mid \text{Crash with fatality}) = .616$

$P(BAC \text{ is between } .01 \text{ and } .09 \mid \text{Crash with fatality}) = .300$

$P(BAC \text{ is greater than } .09 \mid \text{Crash with fatality}) = .084$

Source: Statistical Abstract of the United States, 2000 Table 1042.

Suppose that over a certain stretch of highway during a 1-year period, the probability of being involved in a crash that results in at least one fatality is .01. It has been estimated that 12% of the drivers on this highway drive while their BAC is greater than .09. Determine the probability of a crash with at least one fatality if a driver drives while legally intoxicated (BAC greater than .09).

6.76 Refer to Exercise 6.62. A randomly selected candidate who took a CFA exam tells you that he has passed the exam. What is the probability that he took the CFA I exam?

6.77 Bad gums may mean a bad heart. Researchers discovered that 85% of people who have suffered a heart attack had periodontal disease, an inflammation of the gums. Only 29% of healthy people have this disease. Suppose that in a certain community heart attacks are quite rare, occurring with only 10% probability. If someone has periodontal disease, what is the probability that he or she will have a heart attack?

6.78 Refer to Exercise 6.77. If 40% of the people in a community will have a heart attack, what is the probability that a person with periodontal disease will have a heart attack?

6.79 Data from the Office on Smoking and Health, Centers for Disease Control and Prevention, indicate that 40% of adults who did not finish high school, 34% of high school graduates, 24% of adults who completed some college, and 14% of college graduates smoke. Suppose that one individual is selected at random and it is discovered that the individual smokes. What is the probability that the individual is a college graduate? Use the probabilities in Exercise 6.45 to calculate the probability that the individual is a college graduate.

6.80 Three airlines serve a small town in Ohio. Airline A has 50% of all the scheduled flights, airline B has 30%, and airline C has the remaining 20%. Their on-time rates are 80%, 65%, and 40% respectively. A plane has just left on time. What is the probability that it was airline A?

6.81 Your favorite team is in the final playoffs. You have assigned a probability of 60% that they will win the championship. Past records indicate that when teams win the championship, they win the first game of the series 70% of the time. When they lose the series, they win the first game 25% of the time. The first game is over; your team has lost. What is the probability that they will win the series?

The following exercises are based on the Applications in Medical Screening and Medical Insurance subsection.

6.82 Transplant operations have become routine. One common transplant operation is for kidneys. The most dangerous aspect of the procedure is the possibility that the body may reject the new organ. There are several new drugs available for such circumstances, and the earlier the drug is administered, the higher the probability of averting rejection. The *New England Journal of Medicine* recently reported the development of a new urine test to detect early warning signs that the body is rejecting a transplanted kidney. However, like most other tests, the new test is not perfect. When the test is conducted on someone whose kidney will be rejected, approximately one out of five tests will be negative (i.e., the test is wrong). When the test is conducted on a person whose kidney will not be rejected, 8% will show a positive test result (i.e., another incorrect result). Physicians know that in about 35% of kidney transplants the body tries to reject the organ. Suppose that the test was performed and the test is positive (indicating early warning of rejection). What is the probability that the body is attempting to reject the kidney?

6.83 The Rapid Test is used to determine whether someone has HIV (the virus that causes AIDS). The false-positive and false-negative rates are .027 and .080, respectively. A physician has just received the Rapid Test report that his patient tested positive. Before receiving the result the physician assigned his patient to the low-risk group (defined on the basis of several variables) with only a 0.5% probability of having HIV. What is the probability that the patient actually has HIV?

6.84 What are the sensitivity, specificity, positive predictive value, and negative predictive value in the previous exercise?

6.85 The Pap smear is the standard test for cervical cancer. The false-positive rate is .636; the false-negative rate is .180. Family history and age are factors that must be considered when assigning a probability of cervical cancer. Suppose that after determining a medical history, a physician determines that the proportion of women his patient's age and with her family history that have cervical cancer is 2%. Determine the effects a positive and a negative Pap smear test have on the probability that the patient has cervical cancer.

6.5 Identifying the Correct Method

As we've previously pointed out, the emphasis in this book will be on identifying the correct statistical technique to use. In Chapters 2 and 4 we showed how to summarize data by first identifying the appropriate method to use. Although it is difficult to offer strict rules on which probability method to use, nevertheless we can provide some general guidelines.

In the examples and exercises in this text (and most other introductory statistics books), the key issue is whether joint probabilities are provided or are required.

Joint Probabilities Are Given

In Section 6.2 we addressed problems where the joint probabilities were given. In these problems, we can compute marginal probabilities by adding across rows and down columns. We can use the joint and marginal probabilities to compute conditional probabilities, for which a formula is available. This allows us to determine whether the events described by the table are independent or dependent.

We can also apply the addition rule to compute the probability that either of two events occurs.

Joint Probabilities Are Required

The previous section introduced three probability rules and probability trees. We need to apply some or all of these rules in circumstances where one or more joint probabilities are required. We apply the multiplication rule (either by formula or through a probability tree) to calculate the probability of intersections. In some problems we're interested in adding these joint probabilities. We're actually applying the addition rule for mutually exclusive events here. We also frequently use the complement rule. In addition, we can also calculate new conditional probabilities using Bayes's Law.

CHAPTER SUMMARY

The first step in assigning probability is to create an exhaustive and mutually exclusive list of outcomes. The second step is to use the classical, relative frequency, or subjective approach and assign probability to the outcomes. There are a variety of methods available to compute the probability of other events. These methods include probability rules and trees.

An important application of these rules is Bayes's Law, which allows us to compute conditional probabilities from other forms of probability.

IMPORTANT TERMS

FORMULAS

Conditional probability

$P(A \mid B) = P(A \text{ and } B)/P(B)$

Complement rule

$P(A^C) = 1 - P(A)$

Multiplication rule

$P(A \text{ and } B) = P(A \mid B)P(B)$

Addition rule

$P(A \text{ or } B) = P(A) + P(B) - P(A \text{ and } B)$

CHAPTER EXERCISES

6.86 In a class on probability a statistics professor flips two balanced coins. Both fall to the floor and roll under his desk. A student in the first row informs the professor that he can see both coins. He reports that at least one of them shows tails. What is the probability that the other coin is also tails? (Beware the obvious.)

6.87 In Canada criminals are entitled to parole after serving only one-third of their sentence. Virtually all prisoners, with several exceptions including murderers, are released after serving two-thirds of their sentence. The government has proposed a new law that would create a special category of inmates based on whether they had committed crimes involving violence or drugs. Such criminals would be subject to additional detention if the Correction Services judges them highly likely to reoffend. Currently, 27% of prisoners who are released commit another crime within 2 years of release. Among those who have reoffended, 41% would have been detained under the new law, whereas 31% of those who have not reoffended would have been detained.

a. What is the probability that a prisoner who would have been detained under the new law does commit another crime within 2 years?

b. What is the probability that a prisoner who would not have been detained under the new law does commit another crime within 2 years?

6.88 Refer to Exercise 6.86. Suppose the student informs the professor that he can see only one coin and it shows tails. What is the probability that the other coin is also tails?

6.89 A statistics professor believes that there is a relationship between the number of missed classes and the grade on his midterm test. After examining his records, he produced the following table of joint probabilities.

	Student Fails the Test	Student Passes the Test
Student misses fewer than 5 classes	.02	.86
Student misses 5 or more classes	.09	.03

a. What is the pass rate on the midterm test?
b. What proportion of students who miss 5 or more classes passes the midterm test?
c. What proportion of students who miss fewer than 5 classes passes the midterm test?
d. Are the events independent?

6.90 A union's executive conducted a survey of its members to determine what the membership felt were the important issues to be resolved during upcoming negotiations with management. The results indicate that 74% felt that job security was an important issue, whereas 65% identified pension benefits as an important issue. Of those who felt that pension benefits were important, 60% also felt that job security was an important issue. One member is selected at random.
a. What is the probability that he or she felt that both job security and pension benefits were important?
b. What is the probability that the member felt that at least one of these two issues was important?

6.91 A telemarketer sells magazine subscriptions over the telephone. The probability of a busy signal or no answer is 65%. If the telemarketer does make contact, the probability of 0, 1, 2, or 3 magazine subscriptions is .5, .25, .20, and .05, respectively. Find the probability that in one call she sells no magazines.

6.92 Researchers have developed statistical models based on financial ratios that predict whether a company will go bankrupt over the next 12 months. In a test of one such model, the model correctly predicted the bankruptcy of 85% of firms that did in fact fail, and it correctly predicted nonbankruptcy for 74% of firms that did not fail. Suppose that we expect 8% of the firms in a particular city to fail over the next year. Suppose that the model predicts bankruptcy for a firm that you own. What is the probability that your firm will fail within the next 12 months?

6.93 In a four-cylinder engine there are four spark plugs. If any one of them malfunctions, the car will idle roughly and power will be lost. Suppose that for a certain brand of spark plugs the probability that a spark plug will function properly after 5,000 miles is .90. Assuming that the spark plugs operate independently, what is the probability that the car will idle roughly after 5,000 miles?

6.94 The owner of an appliance store is interested in the relationship between the price at which an item is sold (regular or sale price) and the customer's decision on whether to purchase an extended warranty. After analyzing her records she produced the following joint probabilities.

	Purchased Extended Warranty	Did Not Purchase Extended Warranty
Regular price	.21	.57
Sale price	.14	.08

a. What is the probability that a customer who bought an item at the regular price purchased the extended warranty?
b. What proportion of customers buy an extended warranty?
c. Are the events independent? Explain.

6.95 The effect of an antidepressant drug varies from person to person. Suppose that the drug is effective on 80% of women and 65% of men. It is known that 66% of the people who take the drug are women. What is the probability that the drug is effective?

6.96 A statistics professor and his wife are planning to take a 2-week vacation in Hawaii, but they can't decide whether to spend 1 week on each of the islands of Maui and Oahu, 2 weeks on Maui, or 2 weeks on Oahu. Placing their faith in random chance, they insert two Maui brochures in one envelope, two Oahu brochures in a second envelope, and one brochure from each island in a third envelope. The professor's wife will select one envelope at random, and their vacation schedule will be based on the brochures of the islands so selected. After his wife randomly selects an envelope, the professor removes one brochure from the envelope (without looking at the second brochure) and observes that it is a Maui brochure. What is the probability that the other brochure in the envelope is a Maui brochure? (Proceed with caution; the problem is more difficult than it appears.)

6.97 Refer to Exercise 6.95. Suppose that you are told that the drug is effective. What is the probability that the drug-taker is a man?

6.98 How does level of affluence affect health care? To address one dimension of the problem a group of heart attack victims was selected. Each was categorized as a low-, medium-, or high-income earner. Each was also categorized as having survived or died. A demographer notes that in our society 21% fall into the low-income group, 49% are in the medium-income group and 30% are in the high-income group. Furthermore, an analysis of heart attack victims reveals that 12% of low-income people, 9% of medium-income people, and 7% of high-income people die of heart attacks. Find the probability that a survivor of a heart attack is in the low-income group.

6.99 Laser surgery to fix short-sightedness is becoming more popular. However, for some people, a second procedure is necessary. The following table lists the joint probabilities of needing a second procedure and whether the patient has a corrective lens with a factor (diopter) of minus 8 or less.

	Vision Corrective Factor of More Than Minus 8	Vision Corrective Factor of Minus 8 or Less
First procedure is successful	.66	.15
Second procedure is required	.05	.14

a. Find the probability that a second procedure is required.
b. Determine the probability that someone whose corrective lens factor is minus 8 or less does not require a second procedure.
c. Are the events independent? Explain your answer.

6.100 A customer service supervisor regularly conducts a survey of customer satisfaction. The results of the latest survey indicate that 8% of customers were not satisfied with the service they received at their last visit to the store. Of those who are not satisfied, only 22% return to the store within a year. Of those who are satisfied, 64% return within a year. A customer has just entered the store. In response to your question he informs you that it is less than 1 year since his last visit to the store. What is the probability that he was satisfied with the service he received?

6.101 A construction company has bid on two contracts. The probability of winning contract A is .3. If the company wins contract A, the probability of winning contract B is .4. If the company loses contract A, the probability of winning contract B decreases to .2. Find the probability of the following events.
a. Winning both contracts
b. Winning exactly one contract
c. Winning at least one contract

6.102 Casino Windsor conducts surveys to determine the opinions of its customers. Among other questions respondents are asked to give their opinion about "Your overall impression of Casino Windsor." The responses are

Excellent Good Average Poor

Additionally, the gender of the respondent is noted. After analyzing the results the following table of joint probabilities was produced.

Rating	Women	Men
Excellent	.27	.22
Good	.14	.10
Average	.06	.12
Poor	.03	.06

a. What proportion of customers rate Casino Windsor as excellent?
b. Determine the probability that a male customer rates Casino Windsor as excellent.
c. Find the probability that a customer who rates Casino Windsor as excellent is a man.
d. Are gender and rating independent? Explain your answer.

6.103 The following table lists the joint probabilities of achieving grades of A and not achieving A's in two MBA courses.

	Achieves a Grade of A in Marketing	Does Not Achieve a Grade of A in Marketing
Achieves a grade of A in statistics	.053	.130
Does not achieve a grade of A in statistics	.237	.580

a. What is the probability that a student achieves a grade of A in marketing?
b. What is the probability that a student achieves a grade of A in marketing, given that he or she does not achieve a grade of A in statistics?
c. Are achieving grades of A in marketing and statistics independent events? Explain.

CASE 6.1 Let's Make a Deal

© Michael Newman/PhotoEdit

A number of years ago, there was a popular television game show called *Let's Make a Deal*. The host, Monty Hall, would randomly select contestants from the audience and, as the title suggests, he would make deals for prizes. Contestants would be given relatively modest prizes and would then be offered the opportunity to risk those prizes to win better ones.

Suppose that you are a contestant on this show. Monty has just given you a free trip touring toxic waste sites around the country. He now offers you a trade: Give up the trip in exchange for a gamble. On the stage are three curtains, A, B, and C. Behind one of them is a brand new car worth $20,000. Behind the other two curtains, the stage is empty. You decide to gamble and select curtain A. In an attempt to make things more interesting, Monty then exposes an empty stage by opening curtain C (he knows there is nothing behind curtain C). He then offers you the free trip again if you quit now or, if you like, propose another deal (i.e., you can keep your choice of curtain A or perhaps switch to curtain B). What do you do?

To help you answer that question, try first answering these questions.

1. Before Monty shows you what's behind curtain C, what is the probability that the car is behind curtain A? What is the probability that the car is behind curtain B?
2. After Monty shows you what's behind curtain C, what is the probability that the car is behind curtain A? What is the probability that the car is behind curtain B?

CASE 6.2 To Bunt or Not to Bunt, That Is the Question

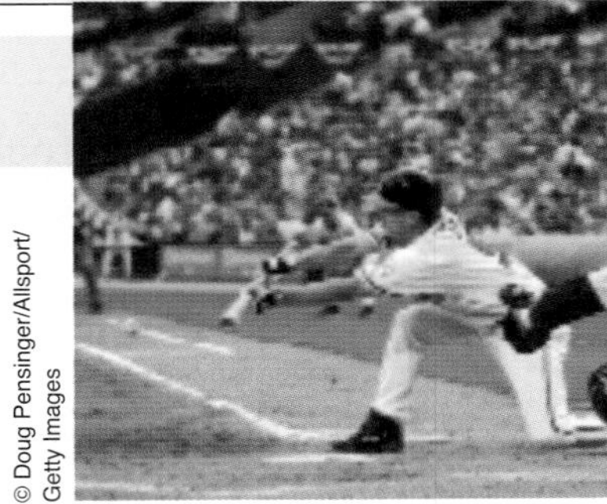
© Doug Pensinger/Allsport/Getty Images

No sport generates as many statistics as baseball. Reporters, managers, and fans argue and discuss strategies on the basis of these statistics. An article in *Chance* ("A Statistician Reads the Sports Page," Hal S. Stern, Vol. 1, Winter 1997) offers baseball lovers another opportunity to analyze numbers associated with the game. Table 1 lists the probabilities of scoring at least one run in situations that are defined by the number of outs and the bases occupied. For example, the probability of scoring at least one run when there are no outs and a man is on first base is .39. If the bases are loaded with one out, the probability of scoring any runs is .67.

TABLE 1 Probability of Scoring Any Runs

Bases Occupied	0 Outs	1 Out	2 Outs
Bases empty	.26	.16	.07
First base	.39	.26	.13
Second base	.57	.42	.24
Third base	.72	.55	.28
First base and second base	.59	.45	.24
First base and third base	.76	.61	.37
Second base and third base	.83	.74	.37
Bases loaded	.81	.67	.43

(Probabilities are based on results from the American League during the 1989 season. The results for the National League are also shown in the article and are similar.) Table 1 allows us to determine the best strategy in a variety of circumstances. This case will concentrate on the strategy of the sacrifice bunt. The purpose of the sacrifice bunt is to sacrifice the batter to move base runners to the next base. It can be employed when there are fewer than two outs and men on base. Ignoring the suicide squeeze, any of four outcomes can occur:

1. The bunt is successful. The runner (or runners) advances one base, and the batter is out.

2. The batter is out but fails to advance the runner.
3. The batter bunts into a double play.
4. The batter is safe (hit or error), and the runner advances.

Suppose that you are an American League manager. The game is tied in the middle innings of a game, and there is a runner on first base with no one out. Given the following probabilities of the four outcomes of a bunt for the batter at the plate, should you signal the batter to sacrifice bunt?

$P(\text{Outcome } 1) = .75$
$P(\text{Outcome } 2) = .10$
$P(\text{Outcome } 3) = .10$
$P(\text{Outcome } 4) = .05$

Assume for simplicity that after the hit or error in outcome 4, there will be men on first and second base and no one out.

CASE 6.3 Should He Attempt to Steal a Base?

© Reuters/Corbis

Refer to Case 6.2. Another baseball strategy is to attempt to steal second base. Historically, the probability of a successful steal of second base is approximately 68%. The probability of being thrown out is 32% (We'll ignore the relatively rare event wherein the catcher throws the ball into center field allowing the base runner to advance to third base.) Suppose there is a runner on first base. For each of the possible number of outs (0, 1, or 2), determine whether it is advisable to have the runner attempt to steal second base.

CASE 6.4 Maternal Serum Screening Test for Down Syndrome

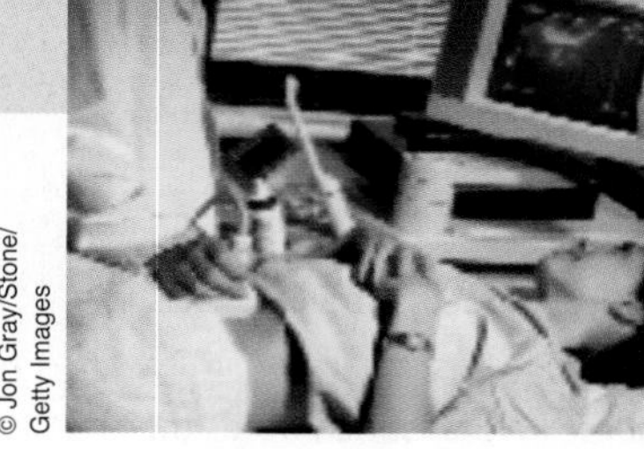
© Jon Gray/Stone/ Getty Images

Pregnant women are screened for a birth defect called Down syndrome. Down syndrome babies are mentally and physically challenged. Some mothers choose to abort the fetus when they are certain that their baby will be born with the syndrome. The most common screening is maternal serum screening, a blood test that looks for markers in the blood to indicate whether the birth defect may occur. The false-positive and false-negative rates vary according to the age of the mother.

Mother's Age	False-Positive Rate	False-Negative Rate
Under 30	.04	.376
30–34	.082	.290
35–37	.178	.269
Over 38	.343	.029

The probability that a baby has Down syndrome is primarily a function of the mother's age. The probabilities are listed here.

Age	Probability of Down Syndrome
25	1/1300
30	1/900
35	1/350
40	1/100
45	1/25
49	1/12

a. For each of the ages 25, 30, 35, 40, 45, and 49, determine the probability of Down syndrome if the maternity serum screening produces a positive result.

b. Repeat for a negative result.

© Comstock Images/Jupiterimages

7 RANDOM VARIABLES AND DISCRETE PROBABILITY DISTRIBUTIONS

Investing to Maximize Returns and Minimize Risk

DATA Xm07-00

An investor has $100,000 to invest in the stock market. She is interested in developing a stock portfolio made up of General Electric, General Motors, McDonald's, and Motorola. However, she doesn't know how much to invest in each one. She wants to maximize her return, but she would also like to minimize the risk. She has computed the monthly returns for all four stocks during a 60-month period (January 2001 to December 2006). After some consideration, she narrowed her choices down to the following three. What should she do?

1. $25,000 in each stock
2. General Electric: $10,000, General Motors: $20,000, McDonald's: $30,000, Motorola: $40,000
3. General Electric: $10,000, General Motors: $50,000, McDonald's: $30,000, Motorola: $10,000

© Terry Vine/Blend Images/ Jupiterimages

We will provide our answer after we've developed the necessary tools in Section 7.3.

INTRODUCTION

In this chapter, we extend the concepts and techniques of probability introduced in Chapter 6. We present random variables and probability distributions, which are essential in the development of statistical inference.

Here is a brief glimpse into the wonderful world of statistical inference. Suppose that you flip a coin 100 times and count the number of heads. The objective is to determine whether we can infer from the count that the coin is not balanced. It is reasonable to believe that observing a large number of heads (say, 90) or a small number (say, 15) would be a statistical indication of an unbalanced coin. However, where do we draw the line? At 75 or 65 or 55? Without knowing the probability of the frequency of the number of heads from a balanced coin, we cannot draw any conclusions from the sample of 100 coin flips.

The concepts and techniques of probability introduced in this chapter will allow us to calculate the probability we seek. As a first step we introduce random variables and probability distributions.

7.1 RANDOM VARIABLES AND PROBABILITY DISTRIBUTIONS

Consider an experiment where we flip two balanced coins and observe the results. We can represent the events as

Heads on the first coin and heads on the second coin
Heads on the first coin and tails on the second coin
Tails on the first coin and heads on the second coin
Tails on the first coin and tails on the second coin

However, we can list the events in a different way. Instead of defining the events by describing the outcome of each coin, we can count the number of heads (or, if we wish, the number of tails). Thus, the events are now

2 heads
1 heads
1 heads
0 heads

The number of heads is called the **random variable.** We often label the random variable X, and we're interested in the probability of each value of X. Thus, in this illustration the values of X are 0, 1, and 2.

Here is another example. In many parlor games as well as in the game of craps played in casinos, the player tosses two dice. One way of listing the events is to describe the number on the first die and the number on the second die as follows.

1, 1	1, 2	1, 3	1, 4	1, 5	1, 6
2, 1	2, 2	2, 3	2, 4	2, 5	2, 6
3, 1	3, 2	3, 3	3, 4	3, 5	3, 6
4, 1	4, 2	4, 3	4, 4	4, 5	4, 6
5, 1	5, 2	5, 3	5, 4	5, 5	5, 6
6, 1	6, 2	6, 3	6, 4	6, 5	6, 6

However, in almost all games the player is primarily interested in the total. Accordingly, we can list the totals of the two dice instead of the individual numbers.

2	3	4	5	6	7
3	4	5	6	7	8
4	5	6	7	8	9
5	6	7	8	9	10
6	7	8	9	10	11
7	8	9	10	11	12

If we define the random variable X as the total of the two dice, X can equal 2, 3, 4, 5, 6, 7, 8, 9, 10, 11, and 12.

Random Variable

A **random variable** is a function or rule that assigns a number to each outcome of an experiment.

In some experiments the outcomes are numbers. For example, when we observe the return on an investment or measure the amount of time to assemble a computer, the experiment produces events that are numbers. Simply stated, the value of a random variable is a numerical event.

There are two types of random variables, discrete and continuous. A **discrete random variable** is one that can take on a countable number of values. For example, if we define X as the number of heads observed in an experiment that flips a coin 10 times, the values of X are 0, 1, 2, . . . , 10. The variable X can assume a total of 11 values. Obviously, we counted the number of values; hence X is discrete.

A **continuous random variable** is one whose values are uncountable. An excellent example of a continuous random variable is the amount of time to complete a task. For example, let X = time to write a statistics exam in a university where the time limit is 3 hours and students cannot leave before 30 minutes. The smallest value of X is 30 minutes. If we attempt to count the number of values that X can take on, we need to identify the next value. Is it 30.1 minutes? 30.01 minutes? 30.001 minutes? None of these is the second possible value of X because there exist numbers larger than 30 and smaller than 30.001. It becomes clear that we cannot identify the second, or third, or any other values of X (except for the largest value 180 minutes). Thus, we cannot count the number of values and X is continuous.

A **probability distribution** is a table, formula, or graph that describes the values of a random variable and the probability associated with these values. We will address discrete probability distributions in the rest of this chapter and cover continuous distributions in Chapter 8.

As we noted above, an uppercase letter will represent the *name* of the random variable, usually X. Its lowercase counterpart will represent the value of the random variable. Thus, we represent the probability that the random variable X will equal x as

$$P(X = x)$$

or more simply

$$P(x)$$

Discrete Probability Distributions

The probabilities of the values of a discrete random variable may be derived by means of probability tools such as tree diagrams or by applying one of the definitions of probability. However, two fundamental requirements apply as stated in the box.

Requirements for a Distribution of a Discrete Random Variable

1. $0 \leq P(x) \leq 1$ for all x
2. $\sum_{\text{all } x} P(x) = 1$

where the random variable can assume values x and $P(x)$ is the probability that the random variable is equal to x.

These requirements are equivalent to the rules of probability provided in Chapter 6. To illustrate, consider the following example.

EXAMPLE 7.1

Probability Distribution of the Number of Color Televisions

The *Statistical Abstract of the United States* is published annually. It contains a wide variety of information based on the census as well as other sources. The objective is to provide information about a variety of different aspects of the lives of the country's residents. One of the questions asks households to report the number of color televisions in the household. The following table summarizes the data. Develop the probability distribution of the random variable defined as the number of color televisions per household.

Number of Color Televisions	Number of Households (thousands)
0	1,218
1	32,379
2	37,961
3	19,387
4	7,714
5	2,842
Total	101,501

Source: Statistical Abstract of the United States, 2000, Table 1221

SOLUTION

The probability of each value of X, the number of color televisions per household, is computed as the relative frequency. We divide the frequency for each value of X by the total number of households, producing the following probability distribution.

x	$P(x)$
0	1,218/101,501 = .012
1	32,379/101,501 = .319
2	37,961/101,501 = .374
3	19,387/101,501 = .191
4	7,714/101,501 = .076
5	2,842/101,501 = .028
Total	1.000

As you can see, the requirements are satisfied. Each probability lies between 0 and 1 and the total is 1.

We interpret the probabilities in the same way we did in Chapter 6. For example, if we select one household at random, the probability that it owns three color televisions is

$$P(3) = .191$$

We can also apply the addition rule for mutually exclusive events. (The values of X are mutually exclusive; a household can own 0, 1, 2, 3, 4, or 5 color televisions.) The probability that a randomly selected household owns two or more color televisions is

$$P(X \geq 2) = P(2) + P(3) + P(4) + P(5) = .374 + .191 + .076 + .028 = .669$$

In Example 7.1, we calculated the probabilities using census information about the entire population. The next example illustrates the use of the techniques introduced in Chapter 6 to develop a probability distribution.

EXAMPLE 7.2

Probability Distribution of the Number of Sales

A mutual fund salesperson has arranged to call on three people tomorrow. Based on past experience, the salesperson knows that there is a 20% chance of closing a sale on each call. Determine the probability distribution of the number of sales the salesperson will make.

SOLUTION

We can use the probability rules and trees introduced in Section 6.3. Figure 7.1 displays the probability tree for this example. Let X = the number of sales.

FIGURE **7.1**

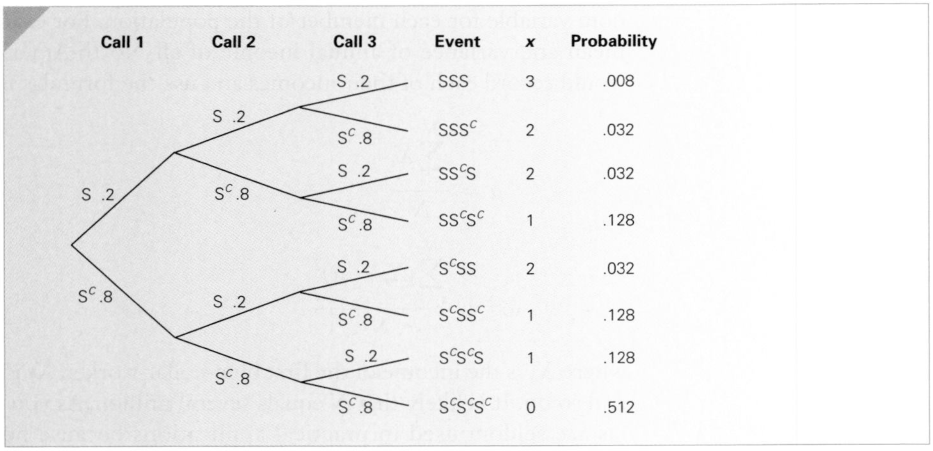

Call 1	Call 2	Call 3	Event	x	Probability
S .2	S .2	S .2	SSS	3	.008
		S^C .8	SSS^C	2	.032
	S^C .8	S .2	SS^CS	2	.032
		S^C .8	SS^CS^C	1	.128
S^C .8	S .2	S .2	S^CSS	2	.032
		S^C .8	S^CSS^C	1	.128
	S^C .8	S .2	S^CS^CS	1	.128
		S^C .8	$S^CS^CS^C$	0	.512

The tree exhibits each of the eight possible outcomes and their probabilities. We see that there is 1 outcome that represents no sales and its probability is $P(0) = .512$. There are three outcomes representing one sale, each with probability .128, so we add these probabilities. Thus,

$$P(1) = .128 + .128 + .128 = 3(.128) = .384$$

The probability of two sales is computed similarly:

$$P(2) = 3(.032) = .096$$

There is one outcome where there are three sales:

$$P(3) = .008$$

The probability distribution of X is listed in Table 7.1.

TABLE **7.1** Probability Distribution of the Number of Sales in Example 7.2

x	$P(x)$
0	.512
1	.384
2	.096
3	.008

Probability Distributions and Populations

The importance of probability distributions derives from their use as representatives of populations. In Example 7.1 the distribution provided us with information about the population of numbers of color televisions per household. In Example 7.2 the population was the number of sales made in three calls by the salesperson. And as we noted before, statistical inference deals with inference about populations.

Describing the Population/Probability Distribution

In Chapter 4, we showed how to calculate the mean, variance, and standard deviation of a population. The formulas we provided were based on knowing the value of the random variable for each member of the population. For example, if we want to know the mean and variance of annual income of all North American blue-collar workers, we would record each of their incomes and use the formulas introduced in Chapter 4:

$$\mu = \frac{\sum_{i=1}^{N} X_i}{N}$$

$$\sigma^2 = \frac{\sum_{i=1}^{N} (X_i - \mu)^2}{N}$$

where X_1 is the income of the first blue-collar worker, X_2 is the second worker's income, and so on. It is likely that N equals several million. As you can appreciate, these formulas are seldom used in practical applications because populations are so large. It is unlikely that we would be able to record all the incomes in the population of North American blue-collar workers. However, probability distributions often represent populations. Rather than record each of the many observations in a population, we list the values and their associated probabilities as we did in deriving the probability distribution of the number of color televisions per household in Example 7.1 and the number of successes in three calls by the mutual fund salesperson. These can be used to compute the mean and variance of the population.

The population mean is the weighted average of all of its values. The weights are the probabilities. This parameter is also called the **expected value** of X and is represented by $E(X)$.

Population Mean

$$E(X) = \mu = \sum_{\text{all } x} xP(x)$$

The population variance is calculated similarly. It is the weighted average of the squared deviations from the mean.

Population Variance

$$V(X) = \sigma^2 = \sum_{\text{all } x} (x - \mu)^2 P(x)$$

There is a shortcut calculation that simplifies the calculations for the population variance. This formula is not an approximation; it will yield the same value as the formula above.

Shortcut Calculation for Population Variance

$$V(X) = \sigma^2 = \sum_{\text{all } x} x^2 P(x) - \mu^2$$

The standard deviation is defined as in Chapter 4.

Population Standard Deviation

$$\sigma = \sqrt{\sigma^2}$$

EXAMPLE 7.3

Describing the Population of the Number of Color Televisions

Find the mean, variance, and standard deviation for the population of the number of color televisions per household in Example 7.1.

SOLUTION

The mean of X is

$$\begin{aligned} E(X) = \mu &= \sum_{\text{all } x} xP(x) = 0P(0) + 1P(1) + 2P(2) + 3P(3) + 4P(4) + 5P(5) \\ &= 0(.012) + 1(.319) + 2(.374) + 3(.191) + 4(.076) + 5(.028) \\ &= 2.084 \end{aligned}$$

Notice that the random variable can assume integer values only, yet the mean is 2.084.

The variance of X is

$$V(X) = \sigma^2 = \sum_{\text{all } x}(x - \mu)^2 P(x)$$

$$= (0 - 2.084)^2(.012) + (1 - 2.084)^2(.319) + (2 - 2.084)^2(.374)$$

$$+ (3 - 2.084)^2(.191) + (4 - 2.084)^2(.076) + (5 - 2.084)^2(.028)$$

$$= 1.107$$

To demonstrate the shortcut method we'll use it to recompute the variance:

$$\sum_{\text{all } x} x^2 P(x) = 0^2(.012) + 1^2(.319) + 2^2(.374) + 3^2(.191) + 4^2(.076) + 5^2(.028)$$

$$= 5.450$$

and

$$\mu = 2.084$$

Thus,

$$\sigma^2 = \sum_{\text{all } x} x^2 P(x) - \mu^2 = 5.450 - (2.084)^2 = 1.107$$

The standard deviation is

$$\sigma = \sqrt{\sigma^2} = \sqrt{1.107} = 1.052$$

These parameters tell us that the mean and standard deviation of the number of color televisions per household are 2.084 and 1.052, respectively.

Laws of Expected Value and Variance

As you will discover, we often create new variables that are functions of other random variables. The formulas given in the next two boxes allow us to quickly determine the expected value and variance of these new variables. In the notation used here, X is the random variable and c is a constant.

Laws of Expected Value

1. $E(c) = c$
2. $E(X + c) = E(X) + c$
3. $E(cX) = cE(X)$

Laws of Variance

1. $V(c) = 0$
2. $V(X + c) = V(X)$
3. $V(cX) = c^2V(X)$

EXAMPLE 7.4

Describing the Population of Monthly Profits

The monthly sales at a computer store have a mean of $25,000 and a standard deviation of $4,000. Profits are calculated by multiplying sales by 30% and subtracting fixed costs of $6,000. Find the mean and standard deviation of monthly profits.

SOLUTION

We can describe the relationship between profits and sales by the following equation:

$$\text{Profit} = .30(\text{Sales}) - 6{,}000$$

The expected or mean profit is

$$E(\text{Profit}) = E[.30(\text{Sales}) - 6{,}000]$$

Applying the second law of expected value, we produce

$$E(\text{Profit}) = E[.30(\text{Sales})] - 6{,}000$$

Applying law 3 yields

$$E(\text{Profit}) = .30E(\text{Sales}) - 6{,}000 = .30(25{,}000) - 6{,}000 = 1{,}500$$

Thus, the mean monthly profit is $1,500.

The variance is

$$V(\text{Profit}) = V[.30(\text{Sales}) - 6{,}000]$$

The second law of variance states that

$$V(\text{Profit}) = V[.30(\text{Sales})]$$

and law 3 yields

$$V(\text{Profit}) = (.30)^2\, V(\text{Sales}) = .09(4{,}000)^2 = 1{,}440{,}000$$

Thus, the standard deviation of monthly profits is

$$\sigma_{\text{Profit}} = \sqrt{1{,}440{,}000} = \$1{,}200$$

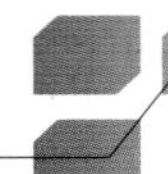

EXERCISES

7.1 The number of accidents that occur on a busy stretch of highway is a random variable.
a. What are the possible values of this random variable?
b. Are the values countable? Explain.
c. Is there a finite number of values? Explain.
d. Is the random variable discrete or continuous? Explain.

7.2 The distance a car travels on a tank of gasoline is a random variable.
a. What are the possible values of this random variable?
b. Are the values countable? Explain.
c. Is there a finite number of values? Explain.
d. Is the random variable discrete or continuous? Explain.

7.3 The amount of money students earn on their summer jobs is a random variable.
a. What are the possible values of this random variable?
b. Are the values countable? Explain.
c. Is there a finite number of values? Explain.
d. Is the random variable discrete or continuous? Explain.

7.4 The mark on a statistics exam that consists of 100 multiple-choice questions is a random variable.

a. What are the possible values of this random variable?
b. Are the values countable? Explain.
c. Is there a finite number of values? Explain.
d. Is the random variable discrete or continuous? Explain.

7.5 Determine whether each of the following is a valid probability distribution.

a.

x	0	1	2	3
$P(x)$	.1	.3	.4	.1

b.

x	5	−6	10	0
$P(x)$	.01	.01	.01	.97

c.

x	14	12	−7	13
$P(x)$	.25	.46	.04	.24

7.6 Let X be the random variable designating the number of spots that turn up when a balanced die is rolled. What is the probability distribution of X?

7.7 The recent census in a large county revealed the following probability distribution for the number of children under 18 per household.

Number of Children	0	1	2
Number of Households	24,750	37,950	59,400

Number of Children	3	4	5
Number of Households	29,700	9,900	3,300

a. Develop the probability distribution of X, the number of children under 18 per household.
b. Determine the following probabilities.

$P(X \le 2)$
$P(X > 2)$
$P(X \ge 4)$

7.8 Using historical records, the personnel manager of a plant has determined the probability distribution of X, the number of employees absent per day. It is

x	0	1	2	3	4	5	6	7
$P(x)$	.005	.025	.310	.340	.220	.080	.019	.001

a. Find the following probabilities.

$P(2 \le X \le 5)$
$P(X > 5)$
$P(X < 4)$

b. Calculate the mean of the population.
c. Calculate the standard deviation of the population.

7.9 Second-year business students at many universities are required to take 10 one-semester courses. The number of courses that result in a grade of A is a discrete random variable. Suppose that each value of this random variable has the same probability. Determine the probability distribution.

7.10 The random variable X has the following probability distribution.

x	−3	2	6	8
$P(x)$	.2	.3	.4	.1

Find the following probabilities.
a. $P(X > 0)$
b. $P(X \ge 1)$
c. $P(X \ge 2)$
d. $P(2 \le X \le 5)$

7.11 An Internet pharmacy advertises that it will deliver the over-the-counter products that customers purchase in 3 to 6 days. The manager of the company wanted to be more precise in its advertising. Accordingly, she recorded the number of days it took to deliver to customers. From the data the following probability distribution was developed.

Number of Days	0	1	2	3	4	5	6	7	8
Probability	0	0	.01	.04	.28	.42	.21	.02	.02

a. What is the probability that a delivery will be made within the advertised 3- to 6-day period?
b. What is the probability that a delivery will be late?
c. What is the probability that a delivery will be early?

7.12 A gambler believes that a strategy called "doubling up" is an effective way to gamble. The method requires the gambler to double the stake after each loss. Thus, if the initial bet is \$1, after losing he will double the bet until he wins. After a win, he resorts back to a \$1 bet. The result is that he will net \$1 for every win. The problem, however, is that he will eventually run out of money or bump up against the table limit. Suppose that for a certain game the probability of winning is .5 and that losing 6 in a row will result in bankrupting the gambler. Find the probability of losing six times in a row.

7.13 The probability that a university graduate will be offered no jobs within a month of graduation is estimated to be 5%. The probability of receiving one, two, and three job offers has similarly been estimated to be 43%, 31%, and 21%, respectively. Determine the following probabilities.
a. A graduate is offered fewer than two jobs.
b. A graduate is offered more than one job.

7.14 Use a probability tree to compute the probability of the following events when flipping two fair coins.

a. Heads on the first coin and heads on the second coin
b. Heads on the first coin and tails on the second coin
c. Tails on the first coin and heads on the second coin
d. Tails on the first coin and tails on the second coin

7.15 Refer to Exercise 7.14. Find the following probabilities.

a. No heads
b. One head
c. Two heads
d. At least one head

7.16 Draw a probability tree to describe the flipping of three fair coins.

7.17 Refer to Exercise 7.16. Find the following probabilities.

a. Two heads
b. One head
c. At least one head
d. At least two heads

7.18 The random variable X has the following distribution.

x	−2	5	7	8
$P(x)$	.59	.15	.25	.01

a. Find the mean and variance for the probability distribution below.
b. Determine the probability distribution of Y where $Y = 5X$.
c. Use the probability distribution in Part b to compute the mean and variance of Y.
d. Use the laws of expected value and variance to find the expected value and variance of Y from the parameters of X.

7.19 We are given the following probability distribution.

x	0	1	2	3
$P(x)$	.4	.3	.2	.1

a. Calculate the mean, variance, and standard deviation.
b. Suppose that $Y = 3X + 2$. For each value of X, determine the value of Y. What is the probability distribution of Y?
c. Calculate the mean, variance, and standard deviation from the probability distribution of Y.
d. Use the laws of expected value and variance to calculate the mean, variance, and standard deviation of Y from the mean, variance, and standard deviation of X. Compare your answers in Parts c and d. Are they the same (except for rounding)?

7.20 The number of pizzas delivered to university students each month is a random variable with the following probability distribution.

x	0	1	2	3
$P(x)$	.1	.3	.4	.2

a. Find the probability that a student has received delivery of two or more pizzas this month.
b. Determine the mean and variance of the number of pizzas delivered to students each month.

7.21 Refer to Exercise 7.20. If the pizzeria makes a profit of $3 per pizza, determine the mean and variance of the profits per student.

7.22 After watching a number of children playing games at a video arcade, a statistics practitioner estimated the following probability distribution of X, the number of games per visit.

x	1	2	3	4	5	6	7
$P(x)$	.05	.15	.15	.25	.20	.10	.10

a. What is the probability that a child will play more than four games?
b. What is the probability that a child will play at least two games?

7.23 Refer to Exercise 7.22. Determine the mean and variance of the number of games played.

7.24 Refer to Exercise 7.23. Suppose that each game costs the player 25 cents. Use the laws of expected value and variance to determine the expected value and variance of the amount of money the arcade takes in.

7.25 Refer to Exercise 7.22.

a. Determine the probability distribution of the amount of money the arcade takes in per child.
b. Use the probability distribution to calculate the mean and variance of the amount of money the arcade takes in.
c. Compare the answers in Part b with those of Exercise 7.24. Are they identical (except for rounding errors)?

7.26 A survey of Amazon.com shoppers reveals the following probability distribution of the number of books purchased per hit.

x	0	1	2	3	4	5	6	7
P(x)	.35	.25	.20	.08	.06	.03	.02	.01

a. What is the probability that an Amazon.com visitor will buy four books?
b. What is the probability that an Amazon.com visitor will buy eight books?
c. What is the probability that an Amazon.com visitor will not buy any books?
d. What is the probability that an Amazon.com visitor will buy at least one book?

7.27 A university librarian produced the following probability distribution of the number of times a student walks into the library over the period of a semester.

x	0	5	10	15	20	25	30	40	50	75	100
P(x)	.22	.29	.12	.09	.08	.05	.04	.04	.03	.03	.01

Find the following probabilities.
a. $P(X \geq 20)$
b. $P(X = 60)$
c. $P(X > 50)$
d. $P(X > 100)$

7.28 After analyzing the frequency with which cross-country skiers participate in their sport, a sportswriter created the following probability distribution for X = number of times per year cross-country skiers ski.

x	0	1	2	3	4	5	6	7	8
P(x)	.04	.09	.19	.21	.16	.12	.08	.06	.05

Find the following.
a. $P(3)$
b. $P(X \geq 5)$
c. $P(5 \leq X \leq 7)$

7.29 The natural remedy echinacea is reputed to boost the immune system, which will reduce the number of flu and colds. A 6-month study was undertaken to determine whether the remedy works. From this study, the following probability distribution of the number of respiratory infections per year (X) for echinacea users was produced.

x	0	1	2	3	4
P(x)	.45	.31	.17	.06	.01

Find the following probabilities.
a. An echinacea user has more than one infection per year.
b. An echinacea user has no infections per year.
c. An echinacea user has between one and three (inclusive) infections per year.

7.30 A shopping mall estimates the probability distribution of the number of stores mall customers actually enter, as shown in the table.

x	0	1	2	3	4	5	6
P(x)	.04	.19	.22	.28	.12	.09	.06

Find the mean and standard deviation of the number of stores entered.

7.31 Refer to Exercise 7.30. Suppose that, on average, customers spend 10 minutes in each store they enter. Find the mean and standard deviation of the total amount of time customers spend in stores.

7.32 When parking a car in a downtown parking lot, drivers pay according to the number of hours or parts thereof. The probability distribution of the number of hours cars are parked has been estimated as follows.

x	1	2	3	4	5	6	7	8
P(x)	.24	.18	.13	.10	.07	.04	.04	.20

Find the mean and standard deviation of the number of hours cars are parked in the lot.

7.33 Refer to Exercise 7.32. The cost of parking is \$2.50 per hour. Calculate the mean and standard deviation of the amount of revenue each car generates.

7.34 You have been given the choice of receiving \$500 in cash or receiving a gold coin that has a face value of \$100. However, the actual value of the gold coin depends on its gold content. You are told that the coin has a 40% probability of being worth \$400, a 30% probability of being worth \$900 and a 30% probability of being worth its face value. Basing your decision on expected value, should you choose the coin?

7.35 The manager of a bookstore recorded the number of customers who arrive at a checkout counter every 5 minutes from which the following distribution was calculated. Calculate the mean and standard deviation of the random variable.

x	0	1	2	3	4
P(x)	.10	.20	.25	.25	.20

7.36 The owner of a small firm has just purchased a personal computer, which she expects will serve her for the next 2 years. The owner has been told that she "must" buy a surge suppressor to provide protection for her new hardware against possible surges or variations in the electrical current, which have the capacity to damage the computer. The amount of damage to the computer depends on the strength of the surge. It has been estimated that there is a 1% chance of incurring \$400 damage, a 2% chance of incurring \$200 damage, and 10% chance of \$100 damage. An inexpensive suppressor, which would provide protection for only one surge can be purchased. How much should the owner be willing to pay if she makes decisions on the basis of expected value?

7.37 It cost one dollar to buy a lottery ticket, which has five prizes. The prizes and the probability that a player wins the prize are listed here. Calculate the expected value of the payoff.

Prize	\$1 million	\$200,000	\$50,000
Probability	1/10 million	1/1 million	1/500,000

Prize	\$10,000	\$1,000
Probability	1/50,000	1/10,000

7.38 After an analysis of incoming faxes, the manager of an accounting firm determined the probability distribution of the number of pages per facsimile as follows:

x	1	2	3	4	5	6	7
P(*x*)	.05	.12	.20	.30	.15	.10	.08

Compute the mean and variance of the number of pages per fax.

7.39 Refer to Exercise 7.38. Further analysis by the manager (in Exercise 7.38) revealed that the cost of processing each page of a fax is \$.25. Determine the mean and variance of the cost per fax.

7.40 To examine the effectiveness of its four annual advertising promotions, a mail-order company has sent a questionnaire to each of its customers, asking how many of the previous year's promotions prompted orders that would not otherwise have been made. The table lists the probabilities that were derived from the questionnaire, where X is the random variable representing the number of promotions that prompted orders. If we assume that overall customer behavior next year will be the same as last year, what is the expected number of promotions that each customer will take advantage of next year by ordering goods that otherwise would not be purchased?

x	0	1	2	3	4
P(*x*)	.10	.25	.40	.20	.05

7.41 Refer to exercise 7.40. A previous analysis of historical records found that the mean value of orders for promotional goods is \$20, with the company earning a gross profit of 20% on each order. Calculate the expected value of the profit contribution next year.

7.42 Refer to Exercises 7.40 and 7.41. The fixed cost of conducting the four promotions is estimated to be \$15,000 with a variable cost of \$3.00 per customer for mailing and handling costs. How large a customer base does the company need to cover the cost of promotions?

7.2 Bivariate Distributions

Thus far, we have dealt with the distribution of a *single* variable. However, there are circumstances where we need to know about the relationship between two variables. Recall that we have addressed this problem statistically in Chapter 2 by drawing the scatter diagram and in Chapter 4 by calculating the covariance and the coefficient of correlation. In this section we present the **bivariate distribution,** which provides probabilities of combinations of two variables. Incidentally, when we need to distinguish between the bivariate distributions and the distributions of one variable, we'll refer to the latter as univariate distributions.

The joint probability that two variables will assume the values x and y is denoted $P(x, y)$. A bivariate (or joint) probability distribution of X and Y is a table or formula that lists the joint probabilities for all pairs of values of x and y. As was the case with univariate distributions, the joint probability must satisfy two requirements.

Requirements for a Discrete Bivariate Distribution

1. $0 \le P(x, y) \le 1$ for all pairs of values (x, y)
2. $\sum_{\text{all } x} \sum_{\text{all } y} P(x, y) = 1$

EXAMPLE 7.5 Bivariate Distribution of the Number of House Sales

Xavier and Yvette are real estate agents. Let X denote the number of houses that Xavier will sell in a month and let Y denote the number of houses Yvette will sell in a month. An analysis of their past monthly performances has the following joint probabilities.

Bivariate Probability Distribution

		X		
		0	1	2
Y	0	.12	.42	.06
	1	.21	.06	.03
	2	.07	.02	.01

We interpret these joint probabilities in the same way we did in Chapter 6. For example, the probability that Xavier sells 0 houses and Yvette sells 1 house in the month is $P(0, 1) = .21$.

Marginal Probabilities

As we did in Chapter 6, we can calculate the marginal probabilities by summing across rows or down columns.

Marginal Probability Distribution of X in Example 7.5

$$P(X = 0) = P(0, 0) + P(0, 1) + P(0, 2) = .12 + .21 + .07 = .4$$
$$P(X = 1) = P(1, 0) + P(1, 1) + P(1, 2) = .42 + .06 + .02 = .5$$
$$P(X = 2) = P(2, 0) + P(2, 1) + P(2, 2) = .06 + .03 + .01 = .1$$

The marginal probability distribution of X is

x	P(x)
0	.4
1	.5
2	.1

Marginal Probability Distribution of Y in Example 7.5

$$P(Y = 0) = P(0, 0) + P(1, 0) + P(2, 0) = .12 + .42 + .06 = .6$$
$$P(Y = 1) = P(0, 1) + P(1, 1) + P(2, 1) = .21 + .06 + .03 = .3$$
$$P(Y = 2) = P(0, 2) + P(1, 2) + P(2, 2) = .07 + .02 + .01 = .1$$

The marginal probability distribution of Y is

y	P(y)
0	.6
1	.3
2	.1

Notice that both marginal probability distributions meet the requirements; the probabilities are between 0 and 1, and they add to 1.

Describing the Bivariate Distribution

As we did with the univariate distribution, we often describe the bivariate distribution by computing the mean, variance, and standard deviation of each variable. We do so by utilizing the marginal probabilities.

Expected Value, Variance, and Standard Deviation of *X* in Example 7.5

$$E(X) = \mu_X = \sum xP(x) = 0(.4) + 1(.5) + 2(.1) = .7$$

$$V(X) = \sigma_X^2 = \sum (x - \mu_X)^2 P(x) = (0 - .7)^2(.4) + (1 - .7)^2(.5) + (2 - .7)^2(.1) = .41$$

$$\sigma_X = \sqrt{\sigma_X^2} = \sqrt{.41} = .64$$

Expected Value, Variance, and Standard Deviation of *Y* in Example 7.5

$$E(Y) = \mu_Y = \sum yP(y) = 0(.6) + 1(.3) + 2(.1) = .5$$

$$V(Y) = \sigma_Y^2 = \sum (y - \mu_Y)^2 P(y) = (0 - .5)^2(.6) + (1 - .5)^2(.3) + (2 - .5)^2(.1) = .45$$

$$\sigma_Y = \sqrt{\sigma_Y^2} = \sqrt{.45} = .67$$

There are two more parameters we can and need to compute. Both deal with the relationship between the two variables. They are the covariance and the coefficient of correlation. Recall that both were introduced in Chapter 4, where the formulas were based on the assumption that we knew each of the N observations of the population. In this chapter we compute parameters like the covariance and the coefficient of correlation from the bivariate distribution.

Covariance

The covariance of two discrete variables is defined as

$$\text{COV}(X, Y) = \sigma_{xy} = \sum_{\text{all } x} \sum_{\text{all } y} (x - \mu_X)(y - \mu_Y)P(x, y)$$

Notice that we multiply the deviations from the mean for both X and Y and then multiply by the joint probability.

The calculations are simplified by the following shortcut method.

Shortcut Calculation for Covariance

$$\text{COV}(X, Y) = \sigma_{xy} = \sum_{\text{all } x} \sum_{\text{all } y} xyP(x, y) - \mu_X \mu_Y$$

The coefficient of correlation is calculated in the same way as in Chapter 4.

Coefficient of Correlation

$$\rho = \frac{\sigma_{xy}}{\sigma_x \sigma_y}$$

EXAMPLE 7.6 Describing the Bivariate Distribution

Compute the covariance and the coefficient of correlation between the numbers of houses sold by the two agents in Example 7.5.

SOLUTION

We start by computing the covariance.

$$\begin{aligned}\sigma_{xy} &= \sum_{\text{all } x}\sum_{\text{all } y}(x-\mu_X)(y-\mu_Y)P(x,y)\\ &= (0-.7)(0-.5)(.12)+(1-.7)(0-.5)(.42)+(2-.7)(0-.5)(.06)\\ &\quad +(0-.7)(1-.5)(.21)+(1-.7)(1-.5)(.06)+(2-.7)(1-.5)(03)\\ &\quad +(0-.7)(2-.5)(.07)+(1-.7)(2-.5)(.02)+(2-.7)(2-.5)(.01)\\ &= -.15\end{aligned}$$

As we did with the shortcut method for the variance, we'll recalculate the covariance using its shortcut method:

$$\begin{aligned}\sum_{\text{all } x}\sum_{\text{all } y} xyP(x,y) &= (0)(0)(.12)+(1)(0)(.42)+(2)(0)(.06)+(0)(1)(.21)+(1)(1)(.06)\\ &\quad +(2)(1)(.03)+(0)(2)(.07)+(1)(2)(.02)+(2)(2)(.01)\\ &= .2\end{aligned}$$

Using the expected values computed above we find

$$\sigma_{xy} = \sum_{\text{all } x}\sum_{\text{all } y} xyP(x,y) - \mu_X\mu_Y = .2 - (.7)(.5) = -.15$$

We also computed the standard deviations above. Thus, the coefficient of correlation is

$$\rho = \frac{\sigma_{xy}}{\sigma_X\sigma_Y} = \frac{-.15}{(.64)(.67)} = -.35$$

There is a weak negative relationship between the two variables, the number of houses Xavier will sell in a month (X) and the number of houses Yvette will sell in a month (Y).

Sum of Two Variables

The bivariate distribution allows us to develop the probability distribution of any combination of the two variables. Of particular interest to us is the sum of two variables. The analysis of this type of distribution leads to an important statistical application in finance, which we present in the next section.

To demonstrate how to develop the probability distribution of the sum of two variables from their bivariate distribution return to Example 7.5. The sum of the two variables X and Y is the total number of houses sold per month. The possible values of $X + Y$ are 0, 1, 2, 3, and 4. The probability that $X + Y = 2$, for example is obtained by summing the joint probabilities of all pairs of values of X and Y that sum to 2:

$$P(X+Y=2) = P(0,2) + P(1,1) + P(2,0) = .07 + .06 + .06 = .19$$

We calculate the probabilities of the other values of $X + Y$ similarly, producing the following table.

Probability Distribution of $X + Y$ in Example 7.5

$x + y$	0	1	2	3	4
$P(x + y)$	.12	.63	.19	.05	.01

We can compute the expected value, variance, and standard deviation of $X + Y$ in the usual way.

$$E(X+Y) = 0(.12) + 1(.63) + 2(.19) + 3(.05) + 4(.01) = 1.2$$

$$V(X+Y) = \sigma^2_{X+Y} = (0 - 1.2)^2(.12) + (1 - 1.2)^2(.63) + (2 - 1.2)^2(.19) + (3 - 1.2)^2(.05) + (4 - 1.2)^2(.01) = .56$$

$$\sigma_{X+Y} = \sqrt{.56} = .75$$

We can derive a number of laws that enable us to compute the expected value and variance of the sum of two variables.

Laws of Expected Value and Variance of the Sum of Two Variables

1. $E(X+Y) = E(X) + E(Y)$
2. $V(X+Y) = V(X) + V(Y) + 2\text{COV}(X, Y)$

If X and Y are independent, $\text{COV}(X, Y) = 0$ and thus $V(X+Y) = V(X) + V(Y)$

EXAMPLE 7.7

Describing the Population of the Total Number of House Sales

Use the rules of expected value and variance of the sum of two variables to calculate the mean and variance of the total number of houses sold per month in Example 7.5.

SOLUTION

Using law 1, we compute the expected value of $X + Y$:

$$E(X+Y) = E(X) + E(Y) = .7 + .5 = 1.2$$

which is the same value we produced directly from the probability distribution of $X + Y$.

We apply law 3 to determine the variance:

$$V(X+Y) = V(X) + V(Y) + 2\text{COV}(X, Y) = .41 + .45 + 2(-.15) = .56$$

This is the same value we obtained from the probability distribution of $X + Y$.

We will encounter several applications where we need the laws of expected value and variance for the sum of two variables. Additionally, we will demonstrate an important application in operations management where we need the formulas for the expected value and variance of the sum of more than two variables. See Exercises 7.57–7.60.

EXERCISES

7.43 The following table lists the bivariate distribution of X and Y.

y \ x	1	2
1	.5	.1
2	.1	.3

a. Find the marginal probability distribution of X.
b. Find the marginal probability distribution of Y.
c. Compute the mean and variance of X.
d. Compute the mean and variance of Y.

7.44 Refer to Exercise 7.43. Compute the covariance and the coefficient of correlation.

7.45 Refer to Exercise 7.43. Use the laws of expected value and variance of the sum of two variables to compute the mean and variance of $X + Y$.

7.46 Refer to Exercise 7.43.
a. Determine the distribution of the $X + Y$.
b. Determine the mean and variance of $X + Y$.
c. Does your answer to Part b equal the answer to Exercise 7.45?

7.47 The bivariate distribution of X and Y is described here.

y \ x	1	2
1	.28	.42
2	.12	.18

a. Find the marginal probability distribution of X.
b. Find the marginal probability distribution of Y.
c. Compute the mean and variance of X.
d. Compute the mean and variance of Y.

7.48 Refer to Exercise 7.47. Compute the covariance and the coefficient of correlation.

7.49 Refer to Exercise 7.47. Use the laws of expected value and variance of the sum of two variables to compute the mean and variance of $X + Y$.

7.50 Refer to Exercise 7.47.
a. Determine the distribution of $X + Y$.
b. Determine the mean and variance of $X + Y$.
c. Does your answer to Part b equal the answer to Exercise 7.49?

7.51 The joint probability distribution of X and Y is shown in the following table.

y \ x	1	2	3
1	.42	.12	.06
2	.28	.08	.04

a. Determine the marginal distributions of X and Y.
b. Compute the covariance and coefficient of correlation between X and Y.
c. Develop the probability distribution of $X + Y$.

7.52 The following distributions of X and of Y have been developed. If X and Y are independent, determine the joint probability distribution of X and Y.

x	0	1	2
$p(x)$	.6	.3	.1

y	1	2
$p(y)$	.7	.3

7.53 The distributions of X and of Y are described here. If X and Y are independent, determine the joint probability distribution of X and Y.

x	0	1
$P(x)$	.2	.8

y	1	2	3
$P(y)$	.2	.4	.4

7.54 After analyzing several months of sales data, the owner of an appliance store produced the following joint probability distribution of the number of refrigerators and stoves sold daily.

	Refrigerators		
Stoves	**0**	**1**	**2**
0	.08	.14	.12
1	.09	.17	.13
2	.05	.18	.04

a. Find the marginal probability distribution of the number of refrigerators sold daily.
b. Find the marginal probability distribution of the number of stoves sold daily.
c. Compute the mean and variance of the number of refrigerators sold daily.
d. Compute the mean and variance of the number of stoves sold daily.
e. Compute the covariance and the coefficient of correlation.

7.55 Canadians who visit the United States often buy liquor and cigarettes, which are much cheaper in the United States. However, there are limitations. Canadians visiting in the United States for more than 2 days are allowed to bring into Canada one bottle of liquor and one carton of cigarettes. A Canada customs agent has produced the following joint probability distribution of the number of bottles of liquor and the number of cartons of cigarettes imported by Canadians who have visited the United States for 2 or more days.

	Bottles of Liquor	
Cartons of Cigarettes	0	1
0	.63	.18
1	.09	.10

a. Find the marginal probability distribution of the number of bottles imported.
b. Find the marginal probability distribution of the number of cigarette cartons imported.
c. Compute the mean and variance of the number of bottles imported.
d. Compute the mean and variance of the number of cigarette cartons imported.
e. Compute the covariance and the coefficient of correlation.

7.56 Refer to Exercise 7.54. Find the following conditional probabilities.
a. P(1 refrigerator | 0 stoves)
b. P(0 stoves | 1 refrigerator)
c. P(2 refrigerators | 2 stoves)

APPLICATIONS in OPERATIONS MANAGEMENT

© BananaStock/Jupiterimages

PERT/CPM

PERT (Project Evaluation and Review Technique) and CPM (Critical Path Method) are related management science techniques that help operations managers control the activities and the amount of time it takes to complete a project. Both techniques are based on the order in which the activities must be performed. For example, in building a house, the excavation of the foundation must precede the pouring of the foundation, which in turn precedes the framing. A **path** is defined as a sequence of related activities that leads from the starting point to the completion of a project. In most projects there are several paths with differing amounts of time needed for their completion. The longest path is called the **critical path** because any delay in the activities along this path will result in a delay in the completion of the project. In some versions of PERT/CPM, the activity completion times are fixed and the chief task of the operations manager is to determine the critical path. In other versions each activity's completion time is considered to be a random variable, where the mean and variance can be estimated. By extending the laws of expected value and variance for the sum of two variables to more than two variables, we produce the following, where $X_1, X_2, \ldots, X_k$ are the times for the completion of activities $1, 2, \ldots, k$, respectively. These times are independent random variables.

Laws of Expected Value and Variance for the Sum of More Than Two Independent Variables

1. $E(X_1 + X_2 + \cdots + X_k) = E(X_1) + E(X_2) + \cdots + E(X_k)$
2. $V(X_1 + X_2 + \cdots + X_k) = V(X_1) + V(X_2) + \cdots + V(X_k)$

Using these laws we can then produce the expected value and variance for the complete project. Exercises 7.57–7.60 address this problem.

7.57 There are four activities along the critical path for a project. The expected values and variances of the completion times of the activities are listed here. Determine the expected value and variance of the completion time of the project.

Activity	Expected Completion Time (Days)	Variance
1	18	8
2	12	5
3	27	6
4	8	2

7.58 The operations manager of a large plant wishes to overhaul a machine. After conducting a PERT/CPM analysis he has developed the following critical path.
1. Disassemble machine
2. Determine parts that need replacing

(continued)

3. Find needed parts in inventory
4. Reassemble machine
5. Test machine

He has estimated the mean (in minutes) and variances of the completion times as follows.

Activity	Mean	Variance
1	35	8
2	20	5
3	20	4
4	50	12
5	20	2

Determine the mean and standard deviation of the completion time of the project.

7.59 In preparing to launch a new product, a marketing manager has determined the critical path for her department. The activities and the mean and variance of the completion time for each activity along the critical path are shown in the accompanying table. Determine the mean and variance of the completion time of the project.

Activity	Expected Completion Time (Days)	Variance
Develop survey questionnaire	8	2
Pretest the questionnaire	14	5
Revise the questionnaire	5	1
Hire survey company	3	1
Conduct survey	30	8
Analyze data	30	10
Prepare report	10	3

7.60 A professor of business statistics is about to begin work on a new research project. Because his time is quite limited, he has developed a PERT/CPM critical path, which consists of the following activities:

1. Conduct a search for relevant research articles
2. Write proposal for a research grant
3. Perform the analysis
4. Write the article and send to journal
5. Wait for reviews
6. Revise on the basis of the reviews and re-submit

The mean and standard deviation (in days) of the completion times are as follows

Activity	Mean	Variance
1	10	9
2	3	0
3	30	100
4	5	1
5	100	400
6	20	64

Compute the mean and standard deviation of the completion time of the entire project.

7.3 (Optional) Applications in Finance: Portfolio Diversification and Asset Allocation

In this section we introduce an important application in finance that is based on the previous section.

In Chapters 2 (page 37) and 4 (page 145) we described how the variance or standard deviation can be used to measure the risk associated with an investment. Most investors tend to be risk averse, which means that they prefer to have lower risk associated with their investments. One of the ways in which financial analysts lower the risk that is associated with the stock market is through diversification. This strategy was first mathematically developed by Harry Markowitz in 1952. His model paved the way for the development of modern portfolio theory (MPT), which is the concept underlying mutual funds (see page 175).

To illustrate the basics of portfolio diversification, consider an investor who forms a portfolio, consisting of only two stocks, by investing $4,000 in one stock and $6,000 in a second stock. Suppose that the results after 1 year are as listed here. (We've previously defined return on investment. See Applications in Finance: Return on Investment on page 38.)

One-Year Results

Stock	Initial Investment	Value of Investment After One Year	Rate of Return on Investment
1	\$4,000	\$5,000	$R_1 = .25(25\%)$
2	\$6,000	\$5,400	$R_2 = -.10(-10\%)$
Total	\$10,000	\$10,400	$R_p = .04(4\%)$

Another way of calculating the portfolio return R_p is to compute the weighted average of the individual stock returns R_1 and R_2, where the weights w_1 and w_2 are the proportions of the initial \$10,000 invested in stocks 1 and 2, respectively. In this illustration, $w_1 = .4$ and $w_2 = .6$. (Note that w_1 and w_2 must always sum to 1 because the two stocks constitute the entire portfolio.) The weighted average of the two returns is

$$R_p = w_1R_1 + w_2R_2$$
$$= (.4)(.25) + (.6)(-.10) = .04$$

This is how portfolio returns are calculated. However, when the initial investments are made, the investor does not know what the returns will be. In fact, the returns are random variables. We are interested in determining the expected value and variance of the portfolio. The formulas in the box were derived from the laws of expected value and variance introduced in the two previous sections.

Mean and Variance of a Portfolio of Two Stocks

$$E(R_p) = w_1E(R_1) + w_2E(R_2)$$
$$V(R_p) = w_1^2V(R_1) + w_2^2V(R_2) + 2w_1w_2\text{COV}(R_1, R_2)$$
$$= w_1^2\sigma_1^2 + w_2^2\sigma_2^2 + 2w_1w_2\rho\sigma_1\sigma_2$$

where w_1 and w_2 are the proportions or weights of investments 1 and 2, $E(R_1)$ and $E(R_2)$ are their expected values, σ_1 and σ_2 are their standard deviations, $\text{COV}(R_1, R_2)$ is the covariance, and ρ is the coefficient of correlation. (Recall that $\rho = \frac{\text{COV}(R_1, R_2)}{\sigma_1\sigma_2}$, which means that $\text{COV}(R_1, R_2) = \rho\sigma_1\sigma_2$.)

EXAMPLE 7.8

Describing the Population of the Returns on a Portfolio

An investor has decided to form a portfolio by putting 25% of his money into McDonald's stock and 75% into Cisco Systems stock. The investor assumes that the expected returns will be 8% and 15%, respectively, and that the standard deviations will be 12% and 22%, respectively.

a. Find the expected return on the portfolio.
b. Compute the standard deviation of the returns on the portfolio assuming that
   i. the two stocks' returns are perfectly positively correlated.
   ii. the coefficient of correlation is .5.
   iii. the two stocks' returns are uncorrelated.

SOLUTION

a. The expected values of the two stocks are

$$E(R_1) = .08 \qquad \text{and} \qquad E(R_2) = .15$$

The weights are $w_1 = .25$ and $w_2 = .75$.
Thus,

$$\begin{aligned} E(R_p) &= w_1 E(R_1) + w_2 E(R_2) \\ &= .25(.08) + .75(.15) = .1325 \end{aligned}$$

b. The standard deviations are

$$\sigma_1 = .12 \qquad \text{and} \qquad \sigma_2 = .22$$

Thus,

$$\begin{aligned} V(R_p) &= w_1^2\sigma_1^2 + w_2^2\sigma_2^2 + 2w_1w_2\rho\sigma_1\sigma_2 \\ &= (.25^2)(.12^2) + (.75^2)(.22^2) + 2(.25)(.75)\,\rho\,(.12)(.22) \\ &= .0281 + .0099\rho \end{aligned}$$

When $\rho = 1$,

$$V(R_p) = .0281 + .0099(1) = .0380$$

$$\text{Standard deviation} = \sqrt{V(R_p)} = \sqrt{.0380} = .1949$$

When $\rho = .5$,

$$V(R_p) = .0281 + .0099(.5) = .0331$$

$$\text{Standard deviation} = \sqrt{V(R_p)} = \sqrt{.0331} = .1819$$

When $\rho = 0$,

$$V(R_p) = .0281 + .0099(0) = .0281$$

$$\text{Standard deviation} = \sqrt{V(R_p)} = \sqrt{.0281} = .1676$$

Notice that the variance and standard deviation of the portfolio returns decrease as the coefficient of correlation decreases.

Portfolio Diversification in Practice

The formulas introduced in this section require that we know the expected values, variances, and covariance (or coefficient of correlation) of the investments we're interested in. The question arises: How do we determine these parameters? (Incidentally, this question is rarely addressed in finance textbooks!) The most common procedure is to estimate the parameters from historical data, using sample statistics.

Portfolios with More Than Two Stocks

We can extend the formulas that describe the mean and variance of the returns of a portfolio of two stocks to a portfolio of any number of stocks.

Mean and Variance of a Portfolio of k Stocks

$$E(R_p) = \sum_{i=1}^{k} w_i E(R_i)$$

$$V(R_p) = \sum_{i=1}^{k} w_i^2 \sigma_i^2 + 2\sum_{i=1}^{k} \sum_{j=i+1}^{k} w_i w_j \text{COV}(R_i, R_j)$$

Where R_i is the return of the ith stock, w_i is the proportion of the portfolio invested in stock i, and k is the number of stocks in the portfolio.

When k is greater than 2, the calculations can be tedious and time-consuming. For example, when $k = 3$, we need to know the values of the three weights, three expected values, three variances, and three covariances. When $k = 4$, there are four expected values, four variances, and six covariances. [The number of covariances required in general is $k(k - 1)/2$.] To assist you, we have created an Excel worksheet to perform the computations when $k = 2$, 3, or 4. To demonstrate we'll return to the problem described in this chapter's introduction.

Investing to Maximize Returns and Minimize Risk: Solution

© Terry Vine/Blend Images/Jupiterimages

Because of the large number of calculations, we will solve this problem using only Excel. From the file we compute the means of each stock's returns.

Excel Means

	A	B	C	D
74	0.000305	0.002339	0.00791	0.007997

Next we compute the variance-covariance matrix. (The commands are the same as those described in Chapter 4—simply include all the columns of the returns of the investments you wish to include in the portfolio.)

Excel Variance–Covariance Matrix

	A	B	C	D	E
1		**GE**	**GM**	**McDonald's**	**Motorola**
2	GE	0.003493			
3	GM	0.001076	0.011016		
4	McDonalds	0.001528	0.001989	0.005409	
5	Motorola	0.000933	0.004131	0.002515	0.010277

Notice that the variances of the returns are listed on the diagonal. Thus, for example, the variance of the 60 monthly returns of General Electric is .003493. The covariances appear below the diagonal. The covariance between the returns of McDonald's and Motorola is .002515.

The means and the variance-covariance matrix are copied to the spreadsheet using the commands described here. The weights are typed producing the accompanying output.

Excel Worksheet: Portfolio Diversification Plan 1

	A	B	C	D	E	F
1	**Portfolio of 4 Stocks**					
2			**GE**	**GM**	**McDonald's**	**Motorola**
3	**Variance-Covariance Matrix**	**GE**	0.003493			
4		**GM**	0.001076	0.011016		
5		**McDonald's**	0.001528	0.001989	0.005409	
6		**Motorola**	0.000933	0.004131	0.002515	0.010277
7						
8	**Expected Returns**		0.000305	0.002339	0.007910	0.007997
9						
10	**Weights**		0.2500	0.2500	0.2500	0.2500
11						
12	**Portfolio Return**					
13	**Expected Value**	**0.0046**				
14	**Variance**	**0.0034**				
15	**Standard Deviation**	**0.0584**				

The expected return on the portfolio is .0046 and the variance is .0034.

INSTRUCTIONS

1. Open the file containing the returns. In this example open file **Xm07-00.**
2. Compute the means of the columns containing the returns of the stocks in the portfolio.
3. Using the commands described in Chapter 4 (page 133), compute the variance-covariance matrix.
4. Open the **Portfolio Diversification** workbook. Use the tab to select the **4 Stocks** worksheet. DO NOT CHANGE ANY CELLS THAT APPEAR IN BOLD PRINT. DO NOT SAVE ANY WORKSHEETS.
5. Copy the means into cells C8 to F8. (Use **Copy, Paste Special** with **Values and number formats.)**
6. Copy the variance-covariance matrix (including row and column labels) into columns B, C, D, E, and F.
7. Type the weights into cells C10 to F10.

The mean, variance, and standard deviation of the portfolio will be printed. Use similar commands for 2-stock and 3-stock portfolios.

The results for plan 2 are

	A	B
12	**Portfolio Return**	
13	**Expected Value**	**0.0061**
14	**Variance**	**0.0043**
15	**Standard Deviation**	**0.0657**

Plan 3

	A	B
12	**Portfolio Return**	
13	**Expected Value**	**0.0044**
14	**Variance**	**0.0048**
15	**Standard Deviation**	**0.0690**

Plan 3 has the smallest expected value and the largest variance, making it the worst of the three plans. Plan 2 has the largest expected value, whereas plan 1 has the smallest variance. If the investor is like most investors, she would select Plan 1 because of its lower risk. Other more daring investors may choose plan 2 to take advantage of its higher expected value.

In this example we showed how to compute the expected return, variance, and standard deviation from a sample of returns on the investments for any combination of weights. (We illustrated the process with three sets of weights.) It is possible to determine the "optimal" weights that minimize risk for a given expected value or maximize expected return for a given standard deviation. This is an extremely important function of financial analysts and investment advisors. Solutions can be determined using a management science technique called *linear programming*, a subject taught by most schools of business and faculties of management.

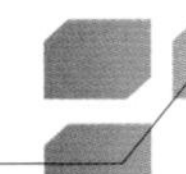

EXERCISES

7.61 Describe what happens to the expected value and standard deviation of the portfolio returns when the coefficient of correlation decreases.

7.62 A portfolio is composed of two stocks. The proportion of each stock, their expected values, and standard deviations are listed next.

Stock	1	2
Proportion of Portfolio	.30	.70
Mean	.12	.25
Standard Deviation	.02	.15

For each of the following coefficients of correlation, calculate the expected value and standard deviation of the portfolio.

a. $\rho = .5$
b. $\rho = .2$
c. $\rho = 0$

7.63 An investor is given the following information about the returns on two stocks.

Stock	1	2
Mean	.09	.13
Standard Deviation	.15	.21

a. If he is most interested in maximizing his returns, which stock should he choose?
b. If he is most interested in minimizing his risk, which stock should he choose?

7.64 Refer to Exercise 7.63. Compute the expected value and standard deviation of the portfolio composed of 60% stock 1 and 40% stock 2. The coefficient of correlation is .4.

7.65 Refer to Exercise 7.63. Compute the expected value and standard deviation of the portfolio composed of 30% stock 1 and 70% stock 2.

The following exercises require the use of a computer.

Xr07-66 *The monthly returns for the following stocks on the New York Stock Exchange were recorded: Coca-Cola, Genentech, General Electric, General Motors, McDonald's, Motorola.*

The next seven exercises are based on this set of data.

7.66 a. Calculate the mean and variance of the monthly return for each stock.
b. Determine the variance-covariance matrix.

7.67 Select the two stocks with the largest means and construct a portfolio consisting of equal amounts of both. Determine the expected value and standard deviation of the portfolio.

7.68 Select the two stocks with the smallest variances and construct a portfolio consisting of equal amounts of both. Determine the expected value and standard deviation of the portfolio.

7.69 Describe the results of Exercises 7.66 to 7.68.

7.70 An investor wants to develop a portfolio composed of shares of Coca-Cola, Genentech, and General electric. Calculate the expected value and standard deviation of the returns for a portfolio with equal proportions of all three stocks.

7.71 Suppose you want a portfolio composed of General Electric, General Motors, McDonald's, and Motorola. Find the expected value and standard deviation of the returns for the following portfolio.

General Electric	30%
General Motors	20%
McDonald's	10%
Motorola	40%

7.72 Repeat Exercise 7.71 using the following proportions. Compare your results with those of Exercise 7.71.

General Electric	30%
General Motors	10%
McDonald's	40%
Motorola	20%

The following six exercises are directed at Canadian students.

Xr07-73 *The monthly returns for the following stocks on the Toronto Stock Exchange were recorded: Barrick Gold, Bell Canada Enterprises (BCE), Bank of Montreal (BMO), MDS Laboratories, Petro-Canada, Research in Motion (RIM).*

The next seven exercises are based on this set of data.

7.73 a. Calculate the mean and variance of the monthly return for each stock.
b. Determine the correlation matrix.

7.74 Select the two stocks with the largest means and construct a portfolio consisting of equal amounts of both. Determine the expected value and standard deviation of the portfolio.

7.75 Select the two stocks with the smallest variances and construct a portfolio consisting of equal amounts of both. Determine the expected value and standard deviation of the portfolio.

7.76 Describe the results of Exercises 7.73 to 7.75.

7.77 An investor wants to develop a portfolio composed of shares of Barrick Gold, Bell Canada

Enterprises, and Bank of Montreal. Calculate the expected value and standard deviation of the returns for a portfolio with the following proportions.

Barrick Gold	30%
Bell Canada Enterprises	20%
Bank of Montreal	50%

7.78 Suppose you want a portfolio composed of Bank of Montreal, MDS Laboratories, Petro-Canada, and Research in Motion. Find the expected value and standard deviation of the returns for the following portfolio.

Bank of Montreal	20%
MDS Laboratories	30%
Petro-Canada	30%
Research in Motion	20%

7.79 Repeat Exercise 7.78 using the following proportions. Compare your results with those of Exercise 7.78.

Bank of Montreal	50%
MDS Laboratories	10%
Petro-Canada	30%
Research in Motion	10%

Xr07-80 *The monthly returns for the following stocks on the NASDAQ Stock Exchange were recorded: Amgen, Ballard Power Systems, Cisco Systems, Intel, Microsoft, and Research in Motion.*

The next four exercises are based on this set of data.

7.80 a. Calculate the mean and variance of the monthly return for each stock.

b. Determine which four stocks you would include in your portfolio if you wanted a large expected value.

c. Determine which four stocks you would include in your portfolio if you wanted a small variance.

7.81 Suppose you want a portfolio composed of Cisco Systems, Intel, Microsoft, and Research in Motion. Find the expected value and standard deviation of the returns for the following portfolio.

Cisco Systems	30%
Intel	15%
Microsoft	25%
Research in Motion	30%

7.82 An investor wants to acquire a portfolio composed of Cisco Systems, Intel, Microsoft, and Research in Motion. Moreover he wants the expected value to be at least 1%. Try several sets of proportions (remember they must add to 1.0) to see if you can find the portfolio with the smallest variance.

7.83 Refer to Exercise 7.81.

a. Compute the expected value and variance of the portfolio described next.

Cisco Systems	11.35%
Intel	3.52%
Microsoft	81.33%
Research in Motion	3.79%

b. Can you do better? That is, can you find a portfolio whose expected value is greater than or equal to 1% and whose variance is less than the one you calculated in Part a? (*Hint:* Don't spend too much time at this. You won't be able to do better.)

c. If you want to learn how we produced the portfolio above, take a course that teaches linear and nonlinear programming.

7.4 BINOMIAL DISTRIBUTION

Now that we've introduced probability distributions in general, we need to introduce several specific probability distributions. In this section we present the *binomial distribution*.

The binomial distribution is the result of a *binomial experiment*, which has the following properties.

Binomial Experiment

1. The **binomial experiment** consists of a fixed number of trials. We represent the number of trials by n.
2. On each trial there are two possible outcomes. We label one outcome a *success*, and the other a *failure*.
3. The probability of success is p. The probability of failure is $1 - p$.
4. The trials are independent, which means that the outcome of one trial does not affect the outcomes of any other trials.

If properties 2, 3, and 4 are satisfied, we say that each trial is a **Bernoulli process.** Adding property 1 yields the binomial experiment. The random variable of a binomial experiment is defined as the number of successes in the n trials. It is called the **binomial random variable.** Here are several examples of binomial experiments.

1. Flip a coin 10 times. The two outcomes per trial are heads and tails. The terms *success* and *failure* are arbitrary. We can label either outcome success. However, generally, we call success anything we're looking for. For example, if we were betting on heads, we would label heads a success. If the coin is fair, the probability of heads is 50%. Thus, $p = .5$. Finally, we can see that the trials are independent, because the outcome of one coin flip cannot possibly affect the outcomes of other flips.

2. Draw five cards out of a shuffled deck. We can label as success whatever card we seek. For example, if we wish to know the probability of receiving five clubs, a club is labeled a success. On the first draw, the probability of a club is $13/52 = .25$. However, if we draw a second card without replacing the first card and shuffling, the trials are not independent. To see why, suppose that the first draw is a club. If we draw again without replacement, the probability of drawing a second club is $12/51$, which is not .25. In this experiment, the trials are *not* independent.* Hence, this is not a binomial experiment. However, if we replace the card and shuffle before drawing again, the experiment is binomial. Note that in most card games, we do not replace the card, and as a result the experiment is not binomial.

3. A political survey asks 1,500 voters for whom they intend to vote in an approaching election. In most elections in the United States, there are only two candidates, the Republican and Democratic nominees. Thus, we have two outcomes per trial. The trials are independent, because the choice of one voter does not affect the choice of other voters. In Canada, and in other countries with a parliamentary system of government, there are usually several candidates in the race. However, we can label a vote for our favored candidate (or the party that is paying us to do the survey) a success and all the others are failures.

As you will discover, the third example is a very common application of statistical inference. The actual value of p is unknown, and the job of the statistics practitioner is to estimate its value. By understanding the probability distribution that uses p, we will be able to develop the statistical tools to estimate p.

Binomial Random Variable

The binomial random variable is the number of successes in the experiment's n trials. It can take on values $0, 1, 2, \ldots, n$. Thus, the random variable is discrete. To proceed we must be capable of calculating the probability associated with each value.

Using a probability tree, we draw a series of branches as depicted in Figure 7.2. The stages represent the outcomes for each of the n trials. At each stage there are two branches representing success and failure. To calculate the probability that there are X successes in n trials, we note that for each success in the sequence we must multiply by p. And, if there are X successes, there must be $n - X$ failures. For each failure in the sequence we multiply by $1 - p$. Thus, the probability for each sequence of branches that represent x successes and $n - x$ failures has probability

$$p^x (1 - p)^{n-x}$$

*The hypergeometric distribution (see CD Appendix F) is used to calculate probabilities in such cases.

There are a number of branches that yield x successes and $n - x$ failures. For example, there are two ways to produce exactly one success and one failure in two trials—SF and FS. To count the number of branch sequences that produce x successes and $n - x$ failures, we use the combinatorial formula

$$C_x^n = \frac{n!}{x!(n-x)!}$$

where $n! = n(n-1)(n-2)\cdots(2)(1)$. For example, $3! = 3(2)(1) = 6$. Incidentally, although it may not appear to be logical $0! = 1$.

FIGURE **7.2** Probability Tree for a Binomial Experiment

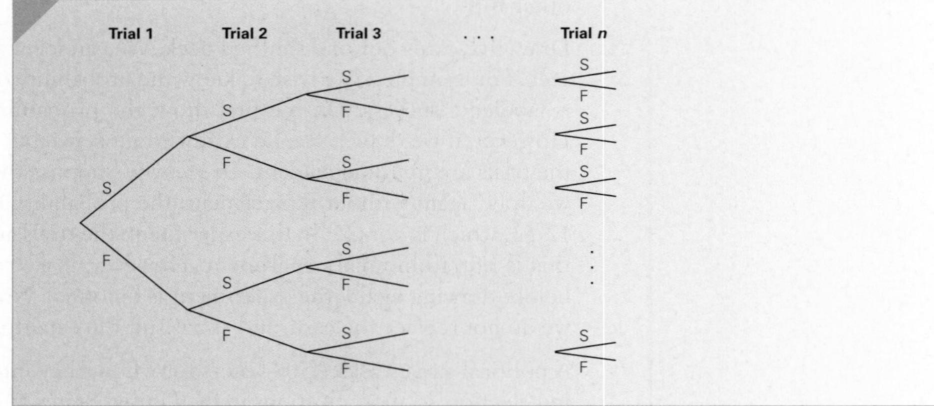

Pulling together the two components of the probability distribution yields the following.

Binomial Probability Distribution

The probability of x successes in a binomial experiment with n trials and probability of success $= p$ is

$$P(x) = \frac{n!}{x!(n-x)!}\, p^x(1-p)^{n-x} \quad \text{for } x = 0, 1, 2, \ldots, n$$

EXAMPLE 7.9 Pat Statsdud and the Statistics Quiz

Pat Statsdud is a student taking a statistics course. Unfortunately, Pat is not a good student. Pat does not read the textbook before class, does not do homework, and regularly misses class. Pat intends to rely on luck to pass the next quiz. The quiz consists of 10 multiple-choice questions. Each question has five possible answers, only one of which is correct. Pat plans to guess the answer to each question.

a. What is the probability that Pat gets no answers correct?
b. What is the probability that Pat gets two answers correct?

SOLUTION

The experiment consists of 10 identical trials, each with two possible outcomes and where success is defined as a correct answer. Because Pat intends to guess, the probability of success is 1/5 or .2. Finally, the trials are independent because the outcome of any of the questions does not affect the outcomes of any other questions. These four properties tell us that the experiment is binomial with $n = 10$ and $p = .2$.

a. From

$$P(x) = \frac{n!}{x!(n-x)!} p^x(1-p)^{n-x}$$

we produce the probability of no successes by letting $n = 10$, $p = .2$, and $x = 0$. Hence,

$$P(0) = \frac{10!}{0!(10-0)!}(.2)^0(1-.2)^{10-0}$$

The combinatorial part of the formula is $\frac{10!}{0!10!}$ which is 1. This is the number of ways to get 0 correct and 10 incorrect. Obviously there is only one way to produce $X = 0$. And because $(.2)^0 = 1$,

$$P(0) = 1(1)(.8)^{10} = .1074$$

b. The probability of two correct answers is computed similarly by substituting $n = 10$, $p = .2$, and $x = 2$:

$$P(x) = \frac{n!}{x!(n-x)!} p^x(1-p)^{n-x}$$

$$P(2) = \frac{10!}{2!(10-2)!}(.2)^2(1-.2)^{10-2}$$

$$= \frac{(10)(9)(8)(7)(6)(5)(4)(3)(2)(1)}{(2)(1)(8)(7)(6)(5)(4)(3)(2)(1)}(.04)(.1678)$$

$$= 45(.006712) = .3020$$

In this calculation we discovered that there are 45 ways to get exactly two correct and eight incorrect answers, and that each such outcome has probability .006712. Multiplying the two numbers produces a probability of .3020.

Cumulative Probability

The formula of the binomial distribution allows us to determine the probability that X equals individual values. In Example 7.9, the values of interest were 0 and 2. There are many circumstances where we wish to find the probability that a random variable is less than or equal to a value. That is, we want to determine $P(X \leq x)$, where x is that value. Such a probability is called a **cumulative probability.**

EXAMPLE 7.10 Will Pat Fail the Quiz?

Find the probability that Pat fails the quiz. A mark is considered a failure if it is less than 50%.

SOLUTION

In this quiz, a mark of less than 5 is a failure. Because the marks must be integers, a mark of 4 or less is a failure. We wish to determine $P(X \leq 4)$. So,

$$P(X \leq 4) = P(0) + P(1) + P(2) + P(3) + P(4)$$

From Example 7.9, we know $P(0) = .1074$ and $P(2) = .3020$. Using the binomial formula, we find $P(1) = .2684$, $P(3) = .2013$, and $P(4) = .0881$. Thus,

$$P(X \leq 4) = .1074 + .2684 + .3020 + .2013 + .0881 = .9672$$

There is a 96.72% probability that Pat will fail the quiz by guessing the answer for each question.

Binomial Table

There is another way to determine binomial probabilities. Table 1 in Appendix B provides cumulative binomial probabilities for selected values of n and p. We can use this table to answer the question in Example 7.10, where we need $P(X \leq 4)$. Refer to Table 1, find $n = 10$, and in that table find $p = .20$. The values in that column are $P(X \leq x)$ for $x = 0, 1, 2, \ldots, 10$, which are shown in Table 7.2.

TABLE **7.2** Cumulative Binomial Probabilities with $n = 10$ and $p = .2$

x	$P(X \leq x)$
0	.1074
1	.3758
2	.6778
3	.8791
4	.9672
5	.9636
6	.9991
7	.9999
8	1.000
9	1.000
10	1.000

The first cumulative probability is $P(X \leq 0)$, which is $P(0) = .1074$. The probability we need for Example 7.10 is $P(X \leq 4) = .9672$, which is the same value we obtained manually.

We can use the table and the complement rule to determine probabilities of the type $P(X \geq x)$. For example, to find the probability that Pat will pass the quiz, we note that

$$P(X \leq 4) + P(X \geq 5) = 1$$

Thus,

$$P(X \geq 5) = 1 - P(X \leq 4) = 1 - .9672 = .0328$$

Using Table 1 to Find the Binomial Probability $P(X \geq x)$

$$P(X \geq x) = 1 - P(X \leq [x - 1])$$

The table is also useful in determining the probability of an individual value of X. For example, to find the probability that Pat will get exactly two right answers we note that

$$P(X \leq 2) = P(0) + P(1) + P(2)$$

and

$$P(X \leq 1) = P(0) + P(1)$$

The difference between these two cumulative probabilities is $P(2)$. Thus

$$P(2) = P(X \leq 2) - P(X \leq 1) = .6778 - .3758 = .3020$$

Using Table 1 to Find the Binomial Probability $P(X = x)$

$$P(x) = P(X \leq x) - P(X \leq [x - 1])$$

Using the Computer

EXCEL

INSTRUCTIONS

Type the following into any empty cell.

=BINOMDIST([x], [n], [p], [True] or [False])

Typing "True" calculates a cumulative probability and typing "False" computes the probability of an individual value of X. For Example 7.9a, type

=BINOMDIST(0, 10, .2, False)

For Example 7.10, enter

=BINOMDIST(4, 10, .2, True)

MINITAB

INSTRUCTIONS

This is the first of seven probability distributions for which we provide instructions. All work in the same way. Click **Calc**, **Probability Distributions,** and the specific distribution whose probability you wish to compute. In this case select **Binomial . . . .** Check either **Probability** or **Cumulative probability.** If you wish to make a probability statement about one value of x, specify **Input constant** and type the value of x. If you wish to make probability statements about several values of x from the same binomial distribution, type the values of x into a column before checking **Calc.** Choose **Input column** and type the name of the column. Finally, enter the components of the distribution. For the binomial, enter the **Number of trials** n and the **Event probability** p.

For the other six distributions we list the distribution (here it is **Binomial**) and the components only (for this distribution it is n and p).

Mean and Variance of a Binomial Distribution

Statisticians have developed general formulas for the mean, variance, and standard deviation of a binomial random variable. They are

$$\mu = np$$

$$\sigma^2 = np(1 - p)$$

$$\sigma = \sqrt{np(1 - p)}$$

EXAMPLE 7.11 Pat Statsdud Has Been Cloned!

Suppose that a professor has a class full of students like Pat (a nightmare!). What is the mean mark? What is the standard deviation?

SOLUTION

The mean mark for a class of Pat Statsduds is

$$\mu = np = 10(.2) = 2$$

The standard deviation is

$$\sigma = \sqrt{np(1 - p)} = \sqrt{10(.2)(1 - .2)} = 1.26$$

EXERCISES

7.84 Given a binomial random variable with $n = 10$ and $p = .3$, use the formula to find the following probabilities.
a. $P(X = 3)$
b. $P(X = 5)$
c. $P(X = 8)$

7.85 Repeat Exercise 7.84 using Table 1 in Appendix B.

7.86 Repeat Exercise 7.84 using Excel or Minitab.

7.87 Given a binomial random variable with $n = 6$ and $p = .2$, use the formula to find the following probabilities.
a. $P(X = 2)$
b. $P(X = 3)$
c. $P(X = 5)$

7.88 Repeat Exercise 7.87 using Table 1 in Appendix B.

7.89 Repeat Exercise 7.87 using Excel or Minitab.

7.90 Suppose X is a binomial random variable with $n = 25$ and $p = .7$. Use Table 1 to find the following.
a. $P(X = 18)$
b. $P(X = 15)$
c. $P(X \leq 20)$
d. $P(X \geq 16)$

7.91 Repeat Exercise 7.90 using Excel or Minitab.

7.92 A sign on the gas pumps of a chain of gasoline stations encourages customers to have their oil checked, claiming that one out of four cars needs to have oil added. If this is true, what is the probability of the following events?
a. One out of the next four cars needs oil
b. Two out of the next eight cars need oil
c. Three out of the next twelve cars need oil

7.93 The leading brand of dishwasher detergent has a 30% market share. A sample of 25 dishwasher detergent customers was taken. What is the probability that 10 or fewer customers chose the leading brand?

7.94 A certain type of tomato seed germinates 90% of the time. A backyard farmer planted 25 seeds.
a. What is the probability that exactly 20 germinate?
b. What is the probability that 20 or more germinate?

c. What is the probability that 24 or fewer germinate?
d. What is the expected number of seeds that germinate?

7.95 According to the American Academy of Cosmetic Dentistry, 75% of adults believe that an unattractive smile hurts career success. Suppose that 25 adults are randomly selected. What is the probability that 15 or more of them would agree with the claim?

7.96 A student majoring in accounting is trying to decide on the number of firms to which he should apply. Given his work experience and grades, he can expect to receive a job offer from 70% of the firms to which he applies. The student decides to apply to only four firms. What is the probability that he receives no job offers?

7.97 In the United States, voters who are neither Democrat nor Republican are called Independent. It is believed that 10% of all voters are Independent. A survey asked 25 people to identify themselves as Democrat, Republican, or Independent.
a. What is the probability that none of the people are Independent?
b. What is the probability that fewer than five people are Independent?
c. What is the probability that more than two people are Independent?

7.98 Most Internet service providers (ISPs) attempt to provide a large enough service so that customers seldom encounter a busy signal. Suppose that the customers of one ISP encounter busy signals 8% of the time. During the week a customer of this ISP called 25 times. What is the probability that she did not encounter any busy signals?

7.99 Major software manufacturers offer a help line that allows customers to call and receive assistance in solving their problems. However, because of the volume of calls, customers frequently are put on hold. One software manufacturer claims that only 20% of callers are put on hold. Suppose that 100 customers call. What is the probability that more than 25 of them are put on hold?

7.100 A statistics practitioner working for major league baseball determined the probability that the hitter will be out on ground balls is .75. In a game where there are 20 ground balls, find the probability that all of them were outs.

The following exercises are best solved with a computer.

7.101 The probability of winning a game of craps (a dice-throwing game played in casinos) is 244/495.
a. What is the probability of winning 5 or more times in 10 games?
b. What is the expected number of wins in 100 games?

7.102 In the game of blackjack as played in casinos in Las Vegas, Atlantic City, Niagara Falls, as well as many other cities, the dealer has the advantage. Most players do not play very well. As a result, the probability that the average player wins a hand is about 45%. Find the probability that an average player wins.
a. Twice in 5 hands.
b. Ten or more times in 25 hands.

7.103 There are several books that teach blackjack players the "basic strategy," which increases the probability of winning any hand to 50%. Repeat Exercise 7.102, assuming the player plays the basic strategy.

7.104 The best way of winning at blackjack is to "case the deck," which involves counting tens, nontens, and aces. For card counters, the probability of winning a hand may increase to 52%. Repeat Exercise 7.102 for a card counter.

7.105 In the game of roulette, a steel ball is rolled onto a wheel that contains 18 red, 18 black, and 2 green slots. If the ball is rolled 25 times, find the probabilities of the following events.
a. The ball falls into the green slots two or more times.
b. The ball does not fall into any green slots.
c. The ball falls into black slots 15 or more times.
d. The ball falls into red slots 10 or fewer times.

7.106 According to a Gallup Poll conducted March 5–7, 2001, 52% of American adults think that protecting the environment should be given priority over developing U.S. energy supplies. Thirty-six percent think that developing energy supplies is more important, and 6% believe the two are equally important. The rest had no opinion. Suppose that a sample of 100 American adults is quizzed on the subject. What is the probability of the following events?
a. Fifty or more think that protecting the environment should be given priority.
b. Thirty or fewer think that developing energy supplies is more important.
c. Five or fewer have no opinion.

7.107 In a *Bon Appetit* poll, 38% of people said that chocolate was their favorite flavor of ice cream. A sample of 20 people was asked to name their favorite flavor of ice cream. What is the probability that half or more of them prefer chocolate?

7.108 The statistics practitioner in Exercise 7.100 also determined that if a batter hits a line drive, the probability of an out is .23. Determine the following probabilities.

a. In a game with 10 line drives, at least 5 are outs.
b. In a game with 25 line drives, there are 5 outs or less.

7.109 According to the last census, 45% of working women held full-time jobs in 2002. If a random sample of 50 working women is drawn, what is the probability that 19 or more hold full-time jobs?

7.5 Poisson Distribution

Another useful discrete probability distribution is the **Poisson probability distribution,** named after its French creator. Like the binomial random variable, the *Poisson random variable* is the number of occurrences of events, which we'll continue to call successes. The difference between the two random variables is that a binomial random variable is the number of successes in a set number of trials whereas a Poisson random variable is the number of successes in an interval of time or specific region of space. Here are several examples of Poisson random variables.

1. The number of cars arriving at a service station in 1 hour. (The interval of time is 1 hour.)
2. The number of flaws in a bolt of cloth. (The specific region is a bolt of cloth.)
3. The number of accidents in 1 day on a particular stretch of highway. (The interval is defined by both time, 1 day, and space, the particular stretch of highway.)

The Poisson experiment is described in the box.

Poisson Experiment

A **Poisson experiment** is characterized by the following properties:

1. The number of successes that occur in any interval is independent of the number of successes that occur in any other interval.
2. The probability of a success in an interval is the same for all equal-size intervals.
3. The probability of a success in an interval is proportional to the size of the interval.
4. The probability of more than one success in an interval approaches 0 as the interval becomes smaller.

Poisson Random Variable

The **Poisson random variable** is the number of successes that occur in a period of time or an interval of space in a Poisson experiment.

There are several ways to derive the probability distribution of a Poisson random variable. However, all are beyond the mathematical level of this book. We simply provide the formula and illustrate how it is used.

Poisson Probability Distribution

The probability that a Poisson random variable assumes a value of x in a specific interval is

$$P(x) = \frac{e^{-\mu}\mu^x}{x!} \quad \text{for } x = 0, 1, 2, \ldots$$

where μ is the mean number of successes in the interval or region and e is the base of the natural logarithm (approximately 2.71828). Incidentally, the variance of a Poisson random variable is equal to its mean. That is, $\sigma^2 = \mu$.

EXAMPLE 7.12

Probability of the Number of Typographical Errors in Textbooks

A statistics instructor has observed that the number of typographical errors in new editions of textbooks varies considerably from book to book. After some analysis, he concludes that the number of errors is Poisson distributed with a mean of 1.5 per 100 pages. The instructor randomly selects 100 pages of a new book. What is the probability that there are no typographical errors?

SOLUTION

We want to determine the probability that a Poisson random variable with a mean of 1.5 is equal to 0. Using the formula

$$P(x) = \frac{e^{-\mu}\mu^x}{x!}$$

and substituting $x = 0$ and $\mu = 1.5$, we get

$$P(0) = \frac{e^{-1.5}1.5^0}{0!} = \frac{(2.71828)^{-1.5}(1)}{1} = .2231$$

The probability that in the 100 pages selected there are no errors is .2231.

Notice that in Example 7.12 we wanted to find the probability of 0 typographical errors in 100 pages given a mean of 1.5 typos in 100 pages. The next example illustrates how we calculate the probability of events where the intervals or regions do not match.

EXAMPLE 7.13

Probability of the Number of Typographical Errors in 400 Pages

Refer to Example 7.12. Suppose that the instructor has just received a copy of a new statistics book. He notices that there are 400 pages.

a. What is the probability that there are no typos?

b. What is the probability that there are five or fewer typos?

SOLUTION

The specific region that we're interested in is 400 pages. To calculate Poisson probabilities associated with this region, we must determine the mean number of typos per 400 pages. Because the mean is specified as 1.5 per 100 pages, we multiply this figure by 4 to convert to 400 pages. Thus, $\mu = 6$ typos per 400 pages.

a. The probability of no typos is

$$P(0) = \frac{e^{-6}6^0}{0!} = \frac{(2.71828)^{-6}(1)}{1} = .002479$$

b. We want to determine the probability that a Poisson random variable with a mean of 6 is 5 or less. That is, we want to calculate

$$P(X \le 5) = P(0) + P(1) + P(2) + P(3) + P(4) + P(5)$$

To produce this probability we need to compute the six probabilities in the summation.

$$P(0) = .002479$$

$$P(1) = \frac{e^{-\mu}\mu^x}{x!} = \frac{e^{-6}6^1}{1!} = \frac{(2.71828)^{-6}(6)}{1} = .01487$$

$$P(2) = \frac{e^{-\mu}\mu^x}{x!} = \frac{e^{-6}6^2}{2!} = \frac{(2.71828)^{-6}(36)}{2} = .04462$$

$$P(3) = \frac{e^{-\mu}\mu^x}{x!} = \frac{e^{-6}6^3}{3!} = \frac{(2.71828)^{-6}(216)}{6} = .08924$$

$$P(4) = \frac{e^{-\mu}\mu^x}{x!} = \frac{e^{-6}6^4}{4!} = \frac{(2.71828)^{-6}(1296)}{24} = .1339$$

$$P(5) = \frac{e^{-\mu}\mu^x}{x!} = \frac{e^{-6}6^5}{5!} = \frac{(2.71828)^{-6}(7776)}{120} = .1606$$

Thus,

$$P(X \le 5) = .002479 + .01487 + .04462 + .08924 + .1339 + .1606 = .4457$$

The probability of observing 5 or fewer typos in this book is .4457.

Poisson Table

As was the case with the binomial distribution, a table is available that makes it easier to compute Poisson probabilities of individual values of x as well as cumulative and related probabilities.

Table 2 in Appendix B provides cumulative Poisson probabilities for selected values of μ. This table makes it easy to find cumulative probabilities like those in Example 7.13, Part b, where we found $P(X \le 5)$. To do so, find $\mu = 6$ in Table 2. The values in that column are $P(X \le x)$ for $x = 0, 1, 2, \ldots$, which are shown in Table 7.3.

TABLE **7.3**
Cumulative Poisson Probabilities for $\mu = 6$

x	$P(X \le x)$
0	.0025
1	.0174
2	.0620
3	.1512
4	.2851
5	.4457
6	.6063
7	.7440
8	.8472
9	.9161
10	.9574
11	.9799
12	.9912
13	.9964
14	.9986
15	.9995
16	.9998
17	.9999
18	1.0000

Theoretically, a Poisson random variable has no upper limit. The table provides cumulative probabilities until the sum is 1.0000 (using four decimal places).

The first cumulative probability is $P(X \le 0)$, which is $P(0) = .0025$. The probability we need for Example 7.13, Part b, is $P(X \le 5) = .4457$, which is the same value we obtained manually.

Like Table 1 for binomial probabilities, Table 2 can be used to determine probabilities of the type $P(X \ge x)$. For example, to find the probability that in Example 7.13 there are 6 or more typos, we note that $P(X \le 5) + P(X \ge 6) = 1$. Thus,

$$P(X \ge 6) = 1 - P(X \le 5) = 1 - .4457 = .5543$$

Using Table 2 to Find the Poisson Probability $P(X \ge x)$

$$P(X \ge x) = 1 - P(X \le [x - 1])$$

We can also use the table to determine the probability of one individual value of X. For example, to find the probability that the book contains exactly 10 typos, we note that

$$P(X \le 10) = P(0) + P(1) + \cdots + P(9) + P(10)$$

and

$$P(X \le 9) = P(0) + P(1) + \cdots + P(9)$$

The difference between these two cumulative probabilities is $P(10)$. Thus,

$$P(10) = P(X \le 10) - P(X \le 9) = .9574 - .9161 = .0413$$

Using Table 2 to Find the Poisson Probability $P(X = x)$

$$P(x) = P(X \leq x) - P(X \leq [x - 1])$$

Using the Computer

EXCEL

INSTRUCTIONS

Type the following into any empty cell:

=**POISSON**([x], [μ], [True] or [False])

We calculate the probability in Example 7.12 by typing

=**POISSON**(0, 1.5, False)

For Example 7.13, we type

=**POISSON**(5, 6, True)

MINITAB

INSTRUCTIONS

Click **Calc, Probability Distributions,** and **Poisson . . .** . and type the mean.

EXERCISES

7.110 Given a Poisson random variable with $\mu = 2$, use the formula to find the following probabilities.
a. $P(X = 0)$
b. $P(X = 3)$
c. $P(X = 5)$

7.111 Given that X is a Poisson random variable with $\mu = .5$, use the formula to determine the following probabilities.
a. $P(X = 0)$
b. $P(X = 1)$
c. $P(X = 2)$

7.112 The number of accidents that occur at a busy intersection is Poisson distributed with a mean of 3.5 per week. Find the probability of the following events.
a. No accidents in one week
b. Five or more accidents in one week
c. One accident today

7.113 Snowfalls occur randomly and independently over the course of winter in a Minnesota city. The average is one snowfall every 3 days.
a. What is the probability of five snowfalls in 2 weeks?
b. Find the probability of a snowfall today.

7.114 The number of students who seek assistance with their statistics assignments is Poisson distributed with a mean of two per day.
a. What is the probability that no students seek assistance tomorrow?
b. Find the probability that 10 students seek assistance in a week.

7.115 Hits on a personal website occur quite infrequently. They occur randomly and independently with an average of five per week.
a. Find the probability that the site gets 10 or more hits in a week.
b. Determine the probability that the site gets 20 or more hits in 2 weeks.

7.116 In older cities across North America, infrastructure is deteriorating. One of the areas is water lines that supply homes and businesses. In a report to the Toronto city council, it was reported that there are on average 30 water line breaks per 100 kilometers per year in the city of Toronto. Outside of Toronto, the average number of breaks is 15 per 100 kilometers per year.

a. Find the probability that, in a stretch of 100 kilometers in Toronto, there will be 35 or more breaks next year.
b. Find the probability that there will be 12 or fewer breaks in a stretch of 100 kilometers outside of Toronto next year.

7.117 The number of bank robberies that occur in a large North American city is Poisson distributed with a mean of 1.8 per day. Find the probabilities of the following events.

a. Three or more bank robberies in a day
b. Between 10 and 15 (inclusive) robberies during a 5-day period

7.118 Flaws in a carpet tend to occur randomly and independently at a rate of one every 200 square feet. What is the probability that a carpet that is 8 feet by 10 feet contains no flaws?

7.119 Complaints about an Internet brokerage firm occur at a rate of five per day. The number of complaints appears to be Poisson distributed.

a. Find the probability that the firm receives 10 or more complaints in a day.
b. Find the probability that the firm receives 25 or more complaints in a 5-day period.

APPLICATIONS in OPERATIONS MANAGEMENT

© Ghislain and Marie David de Lossy/ The Image Bank/Getty Images

Waiting Lines

Everyone is familiar with waiting lines. We wait in line at banks, groceries, and fast-food restaurants. There are also waiting lines in firms where trucks wait to load and unload and on assembly lines where stations wait for new parts. Management scientists have developed mathematical models that allow managers to determine the operating characteristics of waiting lines. Some of the operating characteristics are

The probability that there are no units in the system
The average number of units in the waiting line
The average time a unit spends in the waiting line
The probability that an arriving unit must wait for service

The Poisson probability distribution is used extensively in waiting line (also called queuing) models. Many models assume that the arrival of units for service is Poisson distributed with a specific value of μ. In the next chapter we will discuss the operating characteristics of waiting lines. Exercises 7.120–7.122 require the calculation of the probability of a number of arrivals.

7.120 The number of trucks crossing at the Ambassador Bridge connecting Detroit, Michigan, and Windsor, Ontario, is Poisson distributed with a mean of 1.5 per minute.

a. What is the probability that in any 1-minute time span two or more trucks will cross the bridge?
b. What is the probability that fewer than four trucks will cross the bridge over the next 4 minutes?

7.121 Cars arriving for gasoline at a particular gas station follow a Poisson distribution with a mean of 5 per hour.

a. Determine the probability that over the next hour only one car will arrive.
b. Compute the probability that in the next 3 hours more than 20 cars will arrive.

7.122 The number of users of an automatic banking machine is Poisson distributed. The mean number of users per 5-minute interval is 1.5. Find the probability of the following events.

a. No users in the next 5 minutes
b. Five or fewer users in the next 15 minutes
c. Three or more users in the next 10 minutes

CHAPTER SUMMARY

There are two types of random variable. A discrete random variable is one whose values are countable. A continuous random variable can assume an uncountable number of values. In this chapter we discussed discrete random variables and their probability distributions. We defined the expected value, variance, and standard deviation of a population represented by a discrete probability distribution. Also introduced in this chapter were bivariate discrete distributions on which an important application in finance was based. Finally, the two most important discrete distributions, the binomial and the Poisson, were presented.

IMPORTANT TERMS

Random variable 210
Discrete random variable 211
Continuous random variable 211
Probability distribution 211
Expected value 215
Bivariate distribution 221
Binomial experiment 234
Bernoulli process 235
Binomial random variable 235
Binomial probability distribution 236
Cumulative probability 237
Poisson probability distribution 242
Poisson experiment 242
Poisson random variable 242

SYMBOLS

Symbol	Pronounced	Represents
$\sum_{\text{all } x} x$	*Sum of x for all values of x*	Summation
C_x^n	*n-choose x*	Number of combinations
$n!$	*n-factorial*	$n(n-1)(n-2)\cdots(3)(2)(1)$
e		2.71828 . . .

FORMULAS

Expected value (mean)

$$E(X) = \mu = \sum_{\text{all } x} xP(x)$$

Variance

$$V(x) = \sigma^2 = \sum_{\text{all } x} (x - \mu)^2 P(x)$$

Standard deviation

$$\sigma = \sqrt{\sigma^2}$$

Covariance

$$COV(X, Y) = \sigma_{xy} = \sum (x - \mu_x)(y - \mu_y)P(x, y)$$

Coefficient of correlation

$$\rho = \frac{COV(X, Y)}{\sigma_x \sigma_y} = \frac{\sigma_{xy}}{\sigma_x \sigma_y}$$

Laws of expected value

1. $E(c) = c$
2. $E(X + c) = E(X) + c$
3. $E(cX) = cE(X)$

Laws of variance

1. $V(c) = 0$
2. $V(X + c) = V(X)$
3. $V(cX) = c^2 V(X)$

Laws of expected value and variance of the sum of two variables

1. $E(X + Y) = E(X) + E(Y)$
2. $V(X + Y) = V(X) + V(Y) + 2COV(X, Y)$

Laws of expected value and variance for the sum of k variables, where $k \geq 2$

1. $E(X_1 + X_2 + \cdots + X_k) = E(X_1) + E(X_2) + \cdots + E(X_K)$
2. $V(X_1 + X_2 + \cdots + X_k) = V(X_1) + V(X_2) + \cdots + V(X_K)$

if the variables are independent

Mean and variance of a portfolio of two stocks

$$E(R_p) = w_1 E(R_1) + w_2 E(R_2)$$

$$V(R_p) = w_1^2 V(R_1) + w_2^2 V(R_2) + 2w_1 w_2 COV(R_1, R_2) = w_1^2 \sigma_1^2 + w_2^2 \sigma_2^2 + 2w_1 w_2 \rho \sigma_1 \sigma_2$$

Mean and variance of a portfolio of k stocks

$$E(R_p) = \sum_{i=1}^{k} w_i E(R_i)$$

$$V(R_p) = \sum_{i=1}^{k} w_i^2 \sigma_i^2 + 2 \sum_{i=1}^{k} \sum_{j=i+1}^{k} w_i w_j \text{COV}(R_i, R_j)$$

Binomial probability

$$P(X = x) = \frac{n!}{x!(n-x)!} p^x (1-p)^{n-x}$$

$$\mu = np$$

$$\sigma^2 = np(1-p)$$

$$\sigma = \sqrt{np(1-p)}$$

Poisson probability

$$P(X = x) = \frac{e^{-\mu}\mu^x}{x!}$$

COMPUTER INSTRUCTIONS

Probability Distribution	Excel	Minitab
Binomial	239	239
Poisson	246	246

CHAPTER EXERCISES

7.123 When Earth traveled through the storm of meteorites trailing the comet Tempel-Tuttle on November 17, 1998, the storm was 1,000 times as intense as the average meteor storm. Before the comet arrived, telecommunication companies worried about the potential damage that might be inflicted on the approximately 650 satellites in orbit. It was estimated that each satellite had a 1% chance of being hit, causing damage to the satellite's electronic system. One company had five satellites in orbit at the time. Determine the probability distribution of the number of the company's satellites that would be damaged.

7.124 Shutouts in the National Hockey League occur randomly and independently at a rate of 1 every 20 games. Calculate the probability of the following events.

a. 2 shutouts in the next 10 games
b. 25 shutouts in 400 games
c. a shutout in tonight's game

7.125 In a recent election the mayor received 60% of the vote. Last week a survey was undertaken that asked 100 people whether they would vote for the mayor. Assuming that her popularity has not changed, what is the probability that more than 50 people in the sample would vote for the mayor?

7.126 An auditor is preparing for a physical count of inventory as a means of verifying its value. Items counted are reconciled with a list prepared by the storeroom supervisor. In one particular firm, 20% of the items counted cannot be reconciled without reviewing invoices. The auditor selects 10 items. Find the probability that 6 or more items cannot be reconciled.

7.127 It is recommended that women over 40 have a mammogram annually. A recent report indicated that if a woman has annual mammograms over a 10-year period, there is a 60% probability that there will be at least one false-positive result. (A false-positive mammogram test result is one that indicates the presence of cancer when in fact there is no cancer.) If the annual test results are independent, what is the probability that in any one year a mammogram will produce a false-positive result? (*Hint:* Find the value of p such that the probability that a binomial random variable with $n = 10$ is greater than or equal to 1 is .60.)

7.128 The number of 60-minute cassettes that can be played on a Walkman before the battery expires is a variable. The distribution is shown here.

Number of 60-Minutes Cassettes on a Walkman Before Battery Expires	5	6	7	8	9	10
Probability	.05	.16	.41	.27	.07	.04

a. Calculate the mean number of cassettes.
b. Find the standard deviation.

7.129 According to the U.S. census one-third of all businesses are owned by women. If we select 25 businesses at random, what is the probability that 10 or more of them are owned by women?

7.130 Lotteries are an important income source for various governments around the world. However, the availability of lotteries and other forms of gambling have created a social problem, gambling addicts. A critic of government-controlled gambling contends that 30% of people who regularly buy lottery tickets are gambling addicts. If we randomly select 10 people among those who report that they regularly buy lottery tickets, what is the probability that more than five of them are addicts?

7.131 Advertising researchers have developed a theory that states that commercials that appear in violent television shows are less likely to be remembered and will thus be less effective. After examining samples of viewers who watch violent and nonviolent programs and asking them a series of five questions about the commercials, the researchers produced the following probability distributions of the number of correct answers.

Viewers of Violent Shows

x	0	1	2	3	4	5
$P(x)$	.36	.22	.20	.09	.08	.05

Viewers of Nonviolent Shows

x	0	1	2	3	4	5
$P(x)$	.15	.18	.23	.26	.10	.08

a. Calculate the mean and standard deviation of the number of correct answers among viewers of violent television programs.
b. Calculate the mean and standard deviation of the number of correct answers among viewers of nonviolent television programs.

7.132 The percentage of customers who enter a restaurant and ask to be seated in a smoking section is 15%. Suppose that 100 people enter the restaurant.
a. What is the expected number of people who request a smoking table?
b. What is the standard deviation of the number of requests for a smoking table?
c. What is the probability that 20 or more people request a smoking table?

7.133 A pharmaceutical researcher working on a cure for baldness noticed that middle-age men who are balding at the crown of their head have a 45% probability of suffering a heart attack over the next decade. In a sample of 100 middle-age balding men, what are the following probabilities?
a. More than 50 will suffer a heart attack in the next decade.
b. Fewer than 44 will suffer a heart attack in the next decade.
c. Exactly 45 will suffer a heart attack in the next decade.

7.134 The number of arrivals at a car wash is Poisson distributed with a mean of eight per hour.
a. What is the probability that 10 cars will arrive in the next hour?
b. What is the probability that more than 5 cars will arrive in the next hour?
c. What is the probability that fewer than 12 cars will arrive in the next hour?

7.135 Researchers at the University of Pennsylvania School of Medicine theorized that children under 2 years old who sleep in rooms with the light on have a 40% probability of becoming myopic by age 16. Suppose that researchers found 25 children who slept with the light on before they were 2.
a. What is the probability that 10 of them will become myopic before the age of 16?
b. What is the probability that fewer than 5 of them will become myopic before the age of 16?
c. What is the probability that more than 15 of them will become myopic before the age of 16?

7.136 The number of magazine subscriptions per household is represented by the following probability distribution.

Magazine Subscriptions per Household	0	1	2	3	4
Probability	.48	.35	.08	.05	.04

a. Calculate the mean number of magazine subscriptions per household.
b. Find the standard deviation.

7.137 According to climatologists, the long-term average for Atlantic storms is 9.6 per season (June 1 to November 30) with 6 becoming hurricanes and 2.3 becoming intense hurricanes. Find the probability of the following events. (*Source: Globe and Mail*, December 3, 2004)
a. Ten or more Atlantic storms
b. Five or fewer hurricanes
c. Three or more intense hurricanes

7.138 The final exam in a one-term statistics course is taken in the December exam period. Students who are sick or have other legitimate reasons for missing the exam are allowed to write a deferred exam scheduled for the first week in January. A statistics professor has observed that only 2% of all students legitimately miss the December final exam. Suppose that the professor has 40 students registered this term.
a. How many students can the professor expect to miss the December exam?
b. What is the probability that the professor will not have to create a deferred exam?

7.139 Most Miami Beach restaurants offer "early-bird" specials. These are lower-priced meals that are available only from 4:00 to 6:00. However, not all customers who arrive between 4:00 and 6:00 order the special. In fact, only 70% do.

a. Find the probability that of 80 customers between 4:00 and 6:00, more than 65 order the special.
b. What is the expected number of customers who order the special?
c. What is the standard deviation?

7.140 In 2000 Northwest Airlines boasted that 77.4% of its flights were on time. If we select five Northwest flights at random, what is the probability that all five are on time? (*Source:* Department of Transportation).

CASE 7.1 To Bunt or Not to Bunt, That Is the Question—Part 2

© Tom Pigeon/Allsport/ Getty Images

In Case 6.2 we presented the probabilities of scoring at least one run and asked you to determine whether the manager should signal for the batter to sacrifice bunt. The decision was made on the basis of comparing the probability of scoring at least one run when the manager signaled for the bunt and when he signaled the batter to swing away. Another factor that should be incorporated into the decision is the *number* of runs the manager expects his team to score. In the same article referred to in Case 6.2, the author also computed the expected number of runs scored for each situation. Table 1 lists the expected number of runs in situations that are defined by the number of outs and the bases occupied.

Table 1 Expected Number of Runs Scored

Bases Occupied	0 Out	1 Out	2 Outs
Bases empty	.49	.27	.10
First base	.85	.52	.23
Second base	1.06	.69	.34
Third base	1.21	.82	.38
First base and second base	1.46	1.00	.48
First base and third base	1.65	1.10	.51
Second base and third base	1.94	1.50	.62
Bases loaded	2.31	1.62	.82

Assume that the manager wishes to score as many runs as possible. Using the same probabilities of the four outcomes of a bunt listed in Case 6.2, determine whether the manager should signal the batter to sacrifice bunt.

© Bob Llewellyn/ImageState/Jupiterimages

CONTINUOUS PROBABILITY DISTRIBUTIONS

Minimum GMAT Score to Enter Executive MBA Program

A university has just approved a new Executive MBA Program. The new director believes that to maintain the prestigious image of the business school, the new program must be seen as having high standards. Accordingly, the Faculty Council decides that one of the entrance requirements will be that applicants must score in the top 1% of GMAT (Graduate Management Admission Test) scores. The director knows that GMAT scores are normally distributed with a mean of 490 and a standard deviation of 61. The only thing she doesn't know is what the minimum GMAT score for admission should be.

After introducing the normal distribution, we will return to this question and answer it.

© F64/Digital Vision/Getty Images

See pages 270 and 271.

INTRODUCTION

This chapter completes our presentation of probability by introducing continuous random variables and their distributions. In Chapter 7 we introduced discrete probability distributions that are employed to calculate the probability associated with discrete random variables. In Section 7.4 we introduced the binomial distribution, which allows us to determine the probability that the random variable equals a particular value (the number of successes). In this way we connected the population represented by the probability distribution with a sample of nominal data. In this chapter we introduce continuous probability distributions, which are used to calculate the probability associated with an interval variable. By doing so, we develop the link between a population and a sample of interval data.

Section 8.1 introduces probability density functions and demonstrates with the uniform density function how probability is calculated. In Section 8.2 we focus on the normal distribution, one of the most important distributions because of its role in the development of statistical inference. Section 8.3 introduces the exponential distribution, a distribution that has proven to be useful in various management science applications. Finally, in Section 8.4 we introduce three additional continuous distributions. They will be used in statistical inference throughout the book.

8.1 PROBABILITY DENSITY FUNCTIONS

A continuous random variable is one that can assume an uncountable number of values. Because this type of random variable is so different from a discrete variable, we need to treat it completely differently. First, we cannot list the possible values because there is an infinite number of them. Second, because there is an infinite number of values, the probability of each individual value is virtually 0. Consequently, we can determine the probability of only a range of values. To illustrate how this is done, consider the histogram we created for the long-distance telephone bills (Example 2.4), which is depicted in Figure 8.1.

FIGURE **8.1** Histogram for Example 2.4

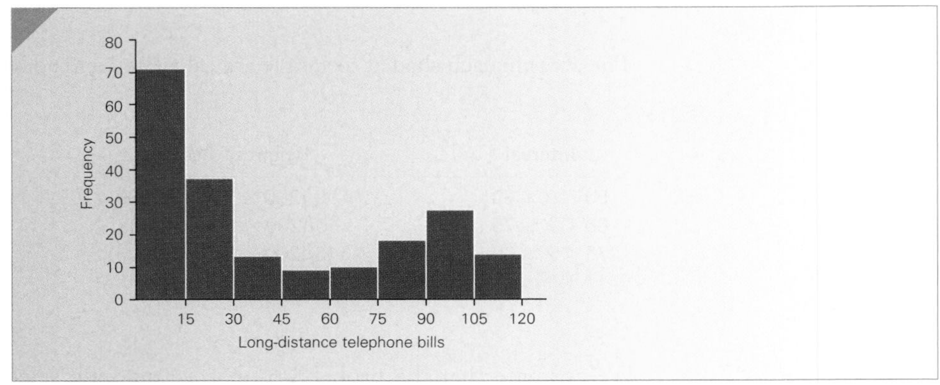

We found, for example that the relative frequency of the interval 15 to 30 was 37/200. Using the relative frequency approach, we estimate that the probability that a randomly selected long-distance bill will fall between \$15 and \$30 is 37/200 = .185. We can similarly estimate the probabilities of the other intervals in the histogram.

Interval	Relative Frequency
$0 \le X \le 15$	71/200
$15 < X \le 30$	37/200
$30 < X \le 45$	13/200
$45 < X \le 60$	9/200
$60 < X \le 75$	10/200
$75 < X \le 90$	18/200
$90 < X \le 105$	28/200
$105 < X \le 120$	14/200

Notice that the sum of the probabilities equals 1. To proceed, we set the values along the vertical axis so that the *area* in all the rectangles together adds to 1. We accomplish this by dividing each relative frequency by the width of the interval, which is 15. The result is a rectangle over each interval whose *area* equals the probability that the random variable will fall into that interval.

To determine probabilities of ranges other than the ones created when we drew the histogram, we apply the same approach. For example, the probability that a long-distance bill will fall between \$50 and \$80 is equal to the area between 50 and 80 as shown in Figure 8.2.

FIGURE **8.2** Histogram for Example 2.4: Relative Frequencies Divided by Interval Width

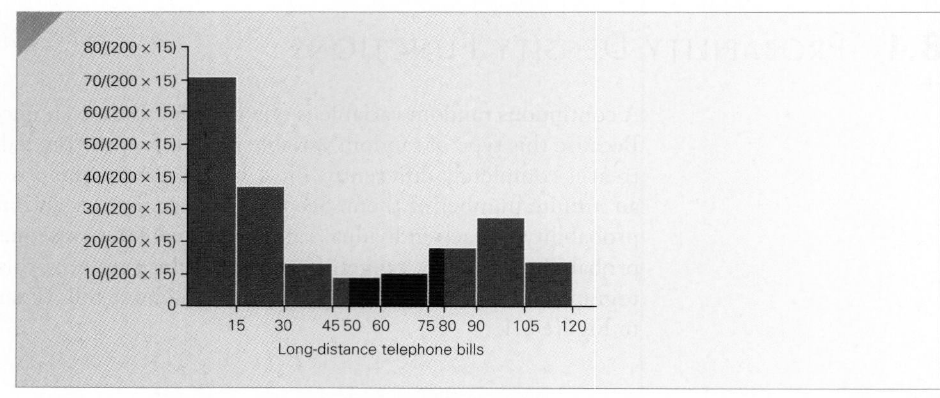

The areas in each shaded rectangle are calculated and added together as follows:

Interval	Height of Rectangle	Base Multiplied by Height
$50 < X \le 60$	$9/(200 \times 15) = .00300$	$(60 - 50) \times .00300 = .030$
$60 < X \le 75$	$10/(200 \times 15) = .00333$	$(75 - 60) \times .00333 = .050$
$75 < X \le 80$	$18/(200 \times 15) = .00600$	$(80 - 75) \times .00600 = .030$
		Total = .110

We estimate that the probability that a randomly selected long-distance bill falls between \$50 and \$80 is .11.

If the histogram is drawn with a large number of small intervals, we can smooth the edges of the rectangles to produce a smooth curve as shown in Figure 8.3. In many cases it is possible to determine a function $f(x)$ that approximates the curve. The function is called a **probability density function.** Its requirements are stated in the box.

Requirements for a Probability Density Function

The following requirements apply to a probability density function $f(x)$ whose range is $a \le x \le b$.

1. $f(x) \ge 0$ for all x between a and b.
2. The total area under the curve between a and b is 1.0.

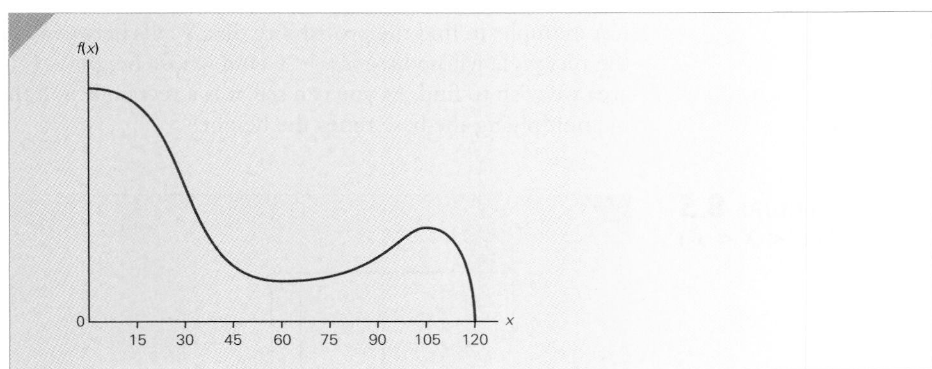

FIGURE **8.3**
Density Function for Example 2.4

Integral calculus* can often be used to calculate the area under a curve. Fortunately, the probabilities corresponding to continuous probability distributions that we deal with do not require this mathematical tool. The distributions will be either simple or too complex for calculus. Let's start with the simplest continuous distribution.

Uniform Distribution

To illustrate how we find the area under the curve that describes a probability density function, consider the **uniform probability distribution,** also called the **rectangular probability distribution.**

Uniform Probability Density Function

The uniform distribution is described by the function

$$f(x) = \frac{1}{b - a} \quad \text{where } a \le x \le b$$

The function is graphed in Figure 8.4. You can see why the distribution is called *rectangular.*

*CD Appendix G demonstrates how to use integral calculus to determine probabilities and parameters for continuous random variables.

FIGURE **8.4**
Uniform Distribution

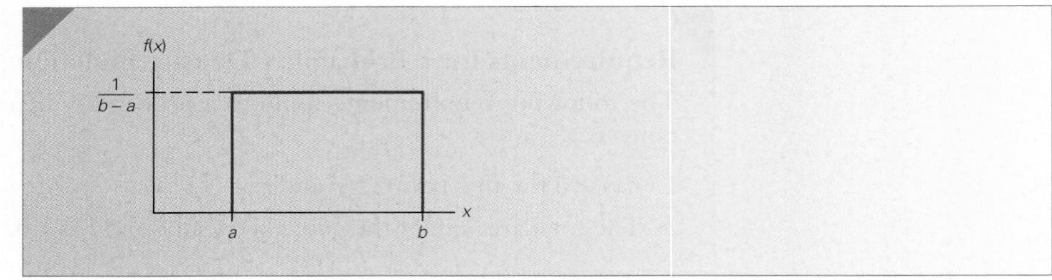

To calculate the probability of any interval, simply find the area under the curve. For example, to find the probability that X falls between x_1 and x_2, determine the area in the rectangle whose base is $x_2 - x_1$ and whose height is $1/(b - a)$. Figure 8.5 depicts the area we wish to find. As you can see, it is a rectangle and the area of a rectangle is found by multiplying the base times the height.

FIGURE **8.5**
$P(x_1 < X < x_2)$

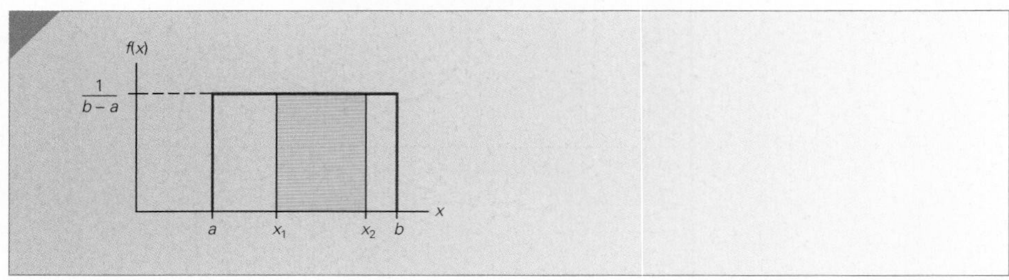

Thus,

$$P(x_1 < X < x_2) = \text{Base} \times \text{Height} = (x_2 - x_1) \times \frac{1}{b - a}$$

EXAMPLE 8.1

Uniformly Distributed Gasoline Sales

The amount of gasoline sold daily at a service station is uniformly distributed with a minimum of 2,000 gallons and a maximum of 5,000 gallons.

a. Find the probability that daily sales will fall between 2,500 and 3,000 gallons.
b. What is the probability that the service station will sell at least 4,000 gallons?
c. What is the probability that the station will sell exactly 2,500 gallons?

SOLUTION

The probability density function is

$$f(x) = \frac{1}{5{,}000 - 2{,}000} = \frac{1}{3{,}000} \qquad 2{,}000 \le x \le 5{,}000$$

a. The probability that X falls between 2,500 and 3,000 is the area under the curve between 2,500 and 3,000 as depicted in Figure 8.6a. The area of a rectangle is the base times the height. Thus,

$$P(2{,}500 \le X \le 3{,}000) = (3{,}000 - 2{,}500) \times \left(\frac{1}{3{,}000}\right) = .1667$$

b. $P(X \geq 4{,}000) = (5{,}000 - 4{,}000) \times \left(\frac{1}{3{,}000}\right) = .3333$ [See Figure 8.6(b).]

c. $P(X = 2{,}500) = 0$

Because there is an uncountable infinite number of values of X, the probability of each individual value is zero. Moreover, as you can see from Figure 8.6c, the area of a line is 0.

Because the probability that a continuous random variable equals any individual value is 0, there is no difference between $P(2{,}500 \leq X \leq 3{,}000)$ and $P(2{,}500 < X < 3{,}000)$. Of course, we cannot say the same thing about discrete random variables.

FIGURE **8.6** **Density Functions for Example 8.1**

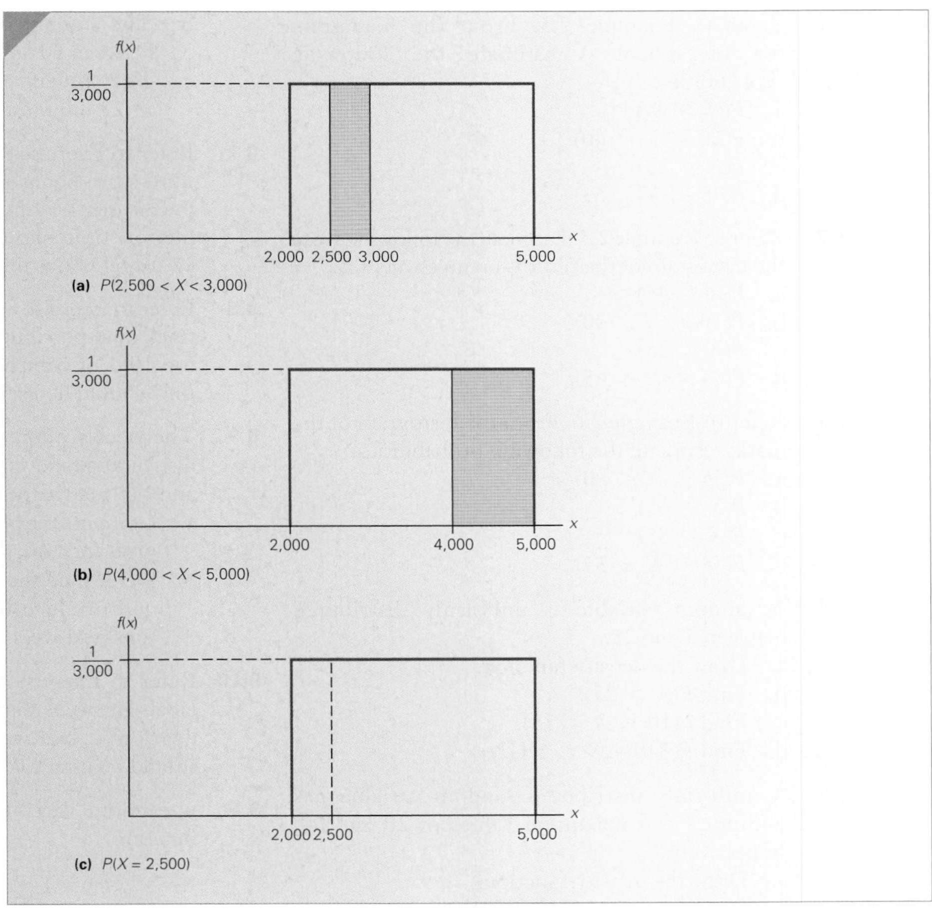

Using a Continuous Distribution to Approximate a Discrete Distribution

In our definition of discrete and continuous random variables, we distinguish between them by noting whether the number of possible values is countable or uncountable. However, in practice, we frequently use a continuous distribution to approximate a discrete one when the number of values the variable can assume is countable but large.

For example, the number of possible values of weekly income is countable. The values of weekly income expressed in dollars are 0, .01, .02, ... Although there is no set upper limit we can easily identify (and thus, count) all the possible values. Consequently, weekly income is a discrete random variable. However, because it can assume such a large number of values, we prefer to employ a continuous probability distribution to determine the probability associated with such variables. In the next section we introduce the normal distribution, which is often used to describe discrete random variables that can assume a large number of values.

EXERCISES

8.1 Refer to Example 2.5. From the histogram for Investment A estimate the following probabilities.

a. $P(X > 45)$
b. $P(10 < X < 40)$
c. $P(X < 25)$
d. $P(35 < X < 65)$

8.2 Refer to Example 2.5. Estimate the following from the histogram of the returns on investment B.

a. $P(X > 45)$
b. $P(10 < X < 40)$
c. $P(X < 25)$
d. $P(35 < X < 65)$

8.3 Refer to Example 2.6. From the histogram of the marks, estimate the following probabilities.

a. $P(55 < X < 80)$
b. $P(X > 65)$
c. $P(X < 85)$
d. $P(75 < X < 85)$

8.4 A random variable is uniformly distributed between 5 and 25.

a. Draw the density function.
b. Find $P(X > 25)$.
c. Find $P(10 < X < 15)$.
d. Find $P(5.0 < X < 5.1)$.

8.5 A uniformly distributed random variable has minimum and maximum values of 20 and 60, respectively.

a. Draw the density function.
b. Determine $P(35 < X < 45)$.
c. Draw the density function including the calculation of the probability in Part b.

8.6 The amount of time it takes for a student to complete a statistics quiz is uniformly distributed between 30 and 60 minutes. One student is selected at random. Find the probability of the following events.

a. The student requires more than 55 minutes to complete the quiz.
b. The student completes the quiz in a time between 30 and 40 minutes.
c. The student completes the quiz in exactly 37.23 minutes.

8.7 Refer to Exercise 8.6. The professor wants to reward (with bonus marks) students who are in the lowest quarter of completion times. What completion time should he use for the cutoff for awarding bonus marks?

8.8 Refer to Exercise 8.6. The professor would like to track (and possibly help) students who are in the top 10% of completion times. What completion time should he use?

8.9 The weekly output of a steel mill is a uniformly distributed random variable that lies between 110 and 175 metric tons.

a. Compute the probability that the steel mill will produce more than 150 metric tons next week.
b. Determine the probability that the steel mill will produce between 120 and 160 metric tons next week.

8.10 Refer to Exercise 8.9. The operations manager labels any week that is in the bottom 20% of production a "bad week." How many metric tons should be used to define a bad week?

8.11 A random variable has the following density function.

$$f(x) = 1 - .5x \quad 0 < x < 2$$

a. Graph the density function.
b. Verify that $f(x)$ is a density function.
c. Find $P(X > 1)$.
d. Find $P(X < .5)$.
e. Find $P(X = 1.5)$.

8.12 The following function is the density function for the random variable X:

$$f(x) = \frac{x - 1}{8} \quad 1 < x < 5$$

a. Graph the density function.
b. Find the probability that X lies between 2 and 4.
c. What is the probability that X is less than 3?

8.13 The following density function describes the random variable X.

$$f(x) = \begin{cases} \dfrac{x}{25} & 0 < x < 5 \\ \dfrac{10 - x}{25} & 5 < x < 10 \end{cases}$$

a. Graph the density function.
b. Find the probability that X lies between 1 and 3.
c. What is the probability that X lies between 4 and 8?
d. Compute the probability that X is less than 7.
e. Find the probability that X is greater than 3.

8.14 The following is a graph of a density function.

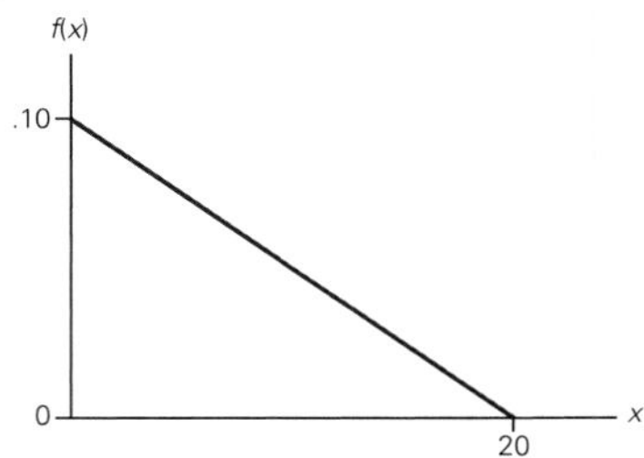

a. Determine the density function.
b. Find the probability that X is greater than 10.
c. Find the probability that X lies between 6 and 12.

8.2 Normal Distribution

The **normal distribution** is the most important of all probability distributions because of its crucial role in statistical inference.

Normal Density Function

The probability density function of a **normal random variable** is

$$f(x) = \frac{1}{\sigma\sqrt{2\pi}} e^{-\frac{1}{2}\left(\frac{x-\mu}{\sigma}\right)^2} \quad -\infty < x < \infty$$

where $e = 2.71828\ldots$ and $\pi = 3.14159\ldots$

Figure 8.7 depicts a normal distribution. Notice that the curve is symmetric about its mean and the random variable ranges between $-\infty$ and $+\infty$.

FIGURE **8.7**
Normal Distribution

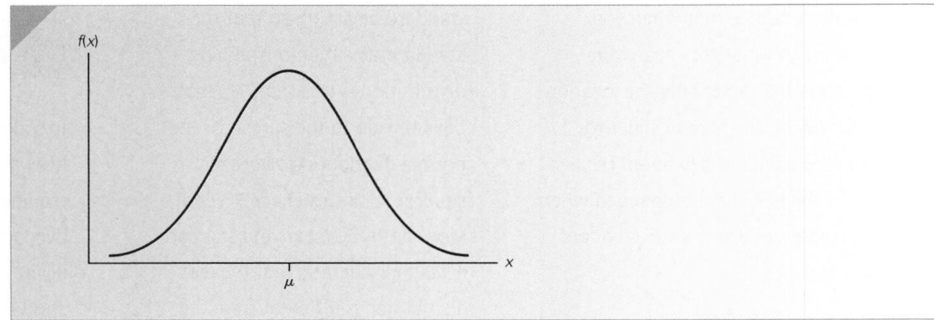

The normal distribution is described by two parameters, the mean μ and the standard deviation σ. In Figure 8.8, we demonstrate the effect of changing the value of μ. Obviously, increasing μ shifts the curve to the right and decreasing μ shifts it to the left.

FIGURE 8.8
Normal Distributions with the Same Variance but Different Means

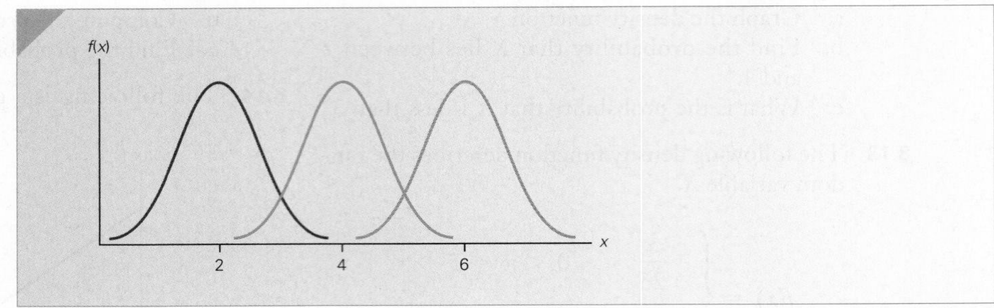

Figure 8.9 describes the effect of σ. Larger values of σ widen the curve and smaller ones narrow it.

FIGURE 8.9
Normal Distributions with the Same Means but Different Variances

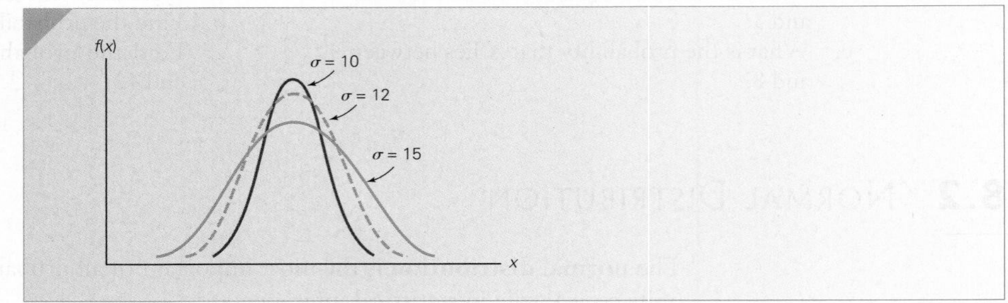

SEEING STATISTICS

Applet 4: Normal Distribution Parameters

This applet can be used to see the effect of changing the values of the mean and standard deviation of a normal distribution.

Move the top slider left or right to decrease or increase the mean of the distribution. Notice that when you change the value of the mean, the shape stays the same; only the location changes. Move the second slider to change the standard deviation. The shape of the bell curve is changed when you increase or decrease the standard deviation.

Applet Exercises

4.1 Move the slider bar for the standard deviation so that the standard deviation of the red distribution is greater than 1. What does this do to the spread of the normal distribution? Does it squeeze it or stretch it?

4.2 Move the slider bar for the standard deviation so that the standard deviation of the red distribution is less than 1. What does this do to the spread of the normal distribution? Does it squeeze it or stretch it?

4.3 Move both the mean and standard deviation sliders so that the red distribution is different from the blue distribution. What would you need to subtract from the red values to slide the red distribution back (forward) so that the centers of the red and blue distributions would overlap? By what would you need to divide the red values to squeeze or stretch the red distribution so that it would have the same spread as the blue distribution?

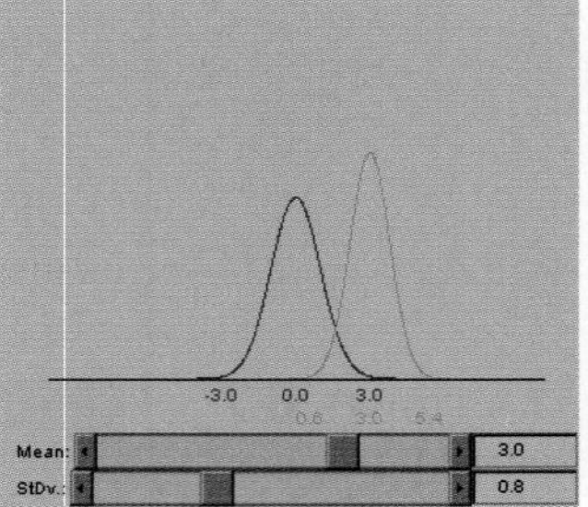

Calculating Normal Probabilities

To calculate the probability that a normal random variable falls into any interval, we must compute the area in the interval under the curve. Unfortunately, the function is not as simple as the uniform precluding the use of simple mathematics or even integral calculus. Instead we will resort to using a probability table similar to Tables 1 and 2 in Appendix B used to calculate binomial and Poisson probabilities, respectively. Recall that to determine binomial probabilities from Table 1 we needed probabilities for values of n and a separate column for selected values of p. Similarly, to find Poisson probabilities, we needed a separate column for each value of μ that we chose to include in Table 2. It would appear then that we will need a separate table for normal probabilities for a selected set of values of μ and σ. Fortunately, this won't be necessary. Instead, we reduce the number of tables needed to one by standardizing the random variable. We standardize a random variable by subtracting its mean and dividing by its standard deviation. When the variable is normal, the transformed variable is called a **standard normal random variable** and is denoted by Z. That is,

$$Z = \frac{X - \mu}{\sigma}$$

The probability statement about X is transformed by this formula into a statement about Z. To illustrate how we proceed consider the following example.

EXAMPLE 8.2

Normally Distributed Gasoline Sales

Suppose that at another gas station the daily demand for regular gasoline is normally distributed with a mean of 1,000 gallons and a standard deviation of 100 gallons. The station manager has just opened the station for business and notes that there is exactly 1,100 gallons of regular gasoline in storage. The next delivery is scheduled later today at the close of business. The manager would like to know the probability that he will have enough regular gasoline to satisfy today's demands.

SOLUTION

The amount of gasoline on hand will be sufficient to satisfy demand if the demand is less than the supply. We label the demand for regular gasoline as X, and we want to find the probability

$$P(X \leq 1{,}100)$$

Note that since X is a continuous random variable, we can also express the probability as

$$P(X < 1{,}100)$$

because the area for $X = 1{,}100$ is 0.

Figure 8.10 describes a normal curve with mean of 1,000 and standard deviation of 100, and the area we want to find.

FIGURE **8.10** $P(X < 1{,}100)$

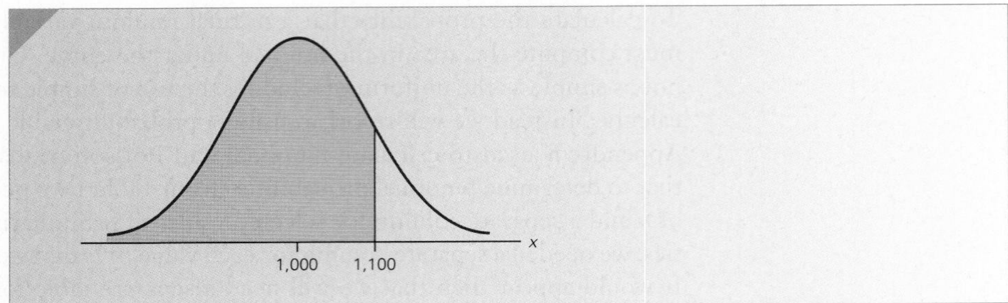

The first step is to standardize *X*. However, if we perform any operations on *X*, we must perform the same operations on 1,100. Thus,

$$P(X < 1{,}100) = P\left(\frac{X - \mu}{\sigma} < \frac{1{,}100 - 1{,}000}{100}\right) = P(Z < 1.00)$$

Figure 8.11 describes the transformation that has taken place. Notice that the variable *X* was transformed into *Z*, and 1,100 was transformed into 1.00. However, the area has not changed. That is, the probability that we wish to compute $P(X < 1{,}100)$ is identical to $P(Z < 1.00)$.

FIGURE **8.11** $P(Z < 1.00)$

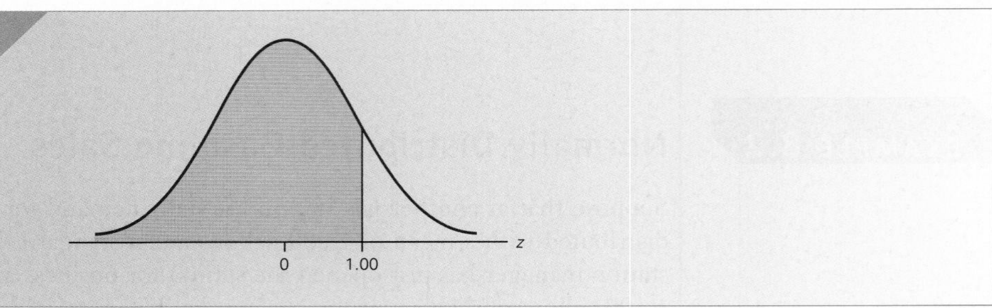

The values of *Z* specify the location of the corresponding value of *X*. A value of $Z = 1$ corresponds to a value of *X* that is 1 standard deviation above the mean. Notice as well that the mean of *Z*, which is 0, corresponds to the mean of *X*.

If we know the mean and standard deviation of a normally distributed random variable, we can always transform the probability statement about *X* into a probability statement about *Z*. Consequently, we need only one table, Table 3 in Appendix B, the standard normal probability table, which is reproduced here as Table 8.1.*

This table is similar to the ones we used for the binomial and Poisson distributions. That is, this table lists cumulative probabilities

$$P(Z < z)$$

for values of *z* ranging from −3.09 to +3.09.

To use the table we simply find the value of *z* and read the probability. For example, the probability $P(Z < 2.00)$ is found by finding 2.0 in the left margin, and under the heading .00, finding .9772. The probability $P(Z < 2.01)$ is found in the same row, but under the heading .01. It is .9778.

---

*In previous editions we have used another table, one that lists $P(0 < Z < z)$ for $z = 0, 0.1, 0.2, \ldots 3.09$. CD Appendix H provides instructions and examples using this table.

TABLE **8.1** Normal Probabilities $P(Z < z)$

Z	0.00	0.01	0.02	0.03	0.04	0.05	0.06	0.07	0.08	0.09
−3.0	0.0013	0.0013	0.0013	0.0012	0.0012	0.0011	0.0011	0.0011	0.0010	0.0010
−2.9	0.0019	0.0018	0.0018	0.0017	0.0016	0.0016	0.0015	0.0015	0.0014	0.0014
−2.8	0.0026	0.0025	0.0024	0.0023	0.0023	0.0022	0.0021	0.0021	0.0020	0.0019
−2.7	0.0035	0.0034	0.0033	0.0032	0.0031	0.0030	0.0029	0.0028	0.0027	0.0026
−2.6	0.0047	0.0045	0.0044	0.0043	0.0041	0.0040	0.0039	0.0038	0.0037	0.0036
−2.5	0.0062	0.0060	0.0059	0.0057	0.0055	0.0054	0.0052	0.0051	0.0049	0.0048
−2.4	0.0082	0.0080	0.0078	0.0075	0.0073	0.0071	0.0069	0.0068	0.0066	0.0064
−2.3	0.0107	0.0104	0.0102	0.0099	0.0096	0.0094	0.0091	0.0089	0.0087	0.0084
−2.2	0.0139	0.0136	0.0132	0.0129	0.0125	0.0122	0.0119	0.0116	0.0113	0.0110
−2.1	0.0179	0.0174	0.0170	0.0166	0.0162	0.0158	0.0154	0.0150	0.0146	0.0143
−2.0	0.0228	0.0222	0.0217	0.0212	0.0207	0.0202	0.0197	0.0192	0.0188	0.0183
−1.9	0.0287	0.0281	0.0274	0.0268	0.0262	0.0256	0.0250	0.0244	0.0239	0.0233
−1.8	0.0359	0.0351	0.0344	0.0336	0.0329	0.0322	0.0314	0.0307	0.0301	0.0294
−1.7	0.0446	0.0436	0.0427	0.0418	0.0409	0.0401	0.0392	0.0384	0.0375	0.0367
−1.6	0.0548	0.0537	0.0526	0.0516	0.0505	0.0495	0.0485	0.0475	0.0465	0.0455
−1.5	0.0668	0.0655	0.0643	0.0630	0.0618	0.0606	0.0594	0.0582	0.0571	0.0559
−1.4	0.0808	0.0793	0.0778	0.0764	0.0749	0.0735	0.0721	0.0708	0.0694	0.0681
−1.3	0.0968	0.0951	0.0934	0.0918	0.0901	0.0885	0.0869	0.0853	0.0838	0.0823
−1.2	0.1151	0.1131	0.1112	0.1093	0.1075	0.1056	0.1038	0.1020	0.1003	0.0985
−1.1	0.1357	0.1335	0.1314	0.1292	0.1271	0.1251	0.1230	0.1210	0.1190	0.1170
−1.0	0.1587	0.1562	0.1539	0.1515	0.1492	0.1469	0.1446	0.1423	0.1401	0.1379
−0.9	0.1841	0.1814	0.1788	0.1762	0.1736	0.1711	0.1685	0.1660	0.1635	0.1611
−0.8	0.2119	0.2090	0.2061	0.2033	0.2005	0.1977	0.1949	0.1922	0.1894	0.1867
−0.7	0.2420	0.2389	0.2358	0.2327	0.2296	0.2266	0.2236	0.2206	0.2177	0.2148
−0.6	0.2743	0.2709	0.2676	0.2643	0.2611	0.2578	0.2546	0.2514	0.2483	0.2451
−0.5	0.3085	0.3050	0.3015	0.2981	0.2946	0.2912	0.2877	0.2843	0.2810	0.2776
−0.4	0.3446	0.3409	0.3372	0.3336	0.3300	0.3264	0.3228	0.3192	0.3156	0.3121
−0.3	0.3821	0.3783	0.3745	0.3707	0.3669	0.3632	0.3594	0.3557	0.3520	0.3483
−0.2	0.4207	0.4168	0.4129	0.4090	0.4052	0.4013	0.3974	0.3936	0.3897	0.3859
−0.1	0.4602	0.4562	0.4522	0.4483	0.4443	0.4404	0.4364	0.4325	0.4286	0.4247
−0.0	0.5000	0.4960	0.4920	0.4880	0.4840	0.4801	0.4761	0.4721	0.4681	0.4641
0.0	0.5000	0.5040	0.5080	0.5120	0.5160	0.5199	0.5239	0.5279	0.5319	0.5359
0.1	0.5398	0.5438	0.5478	0.5517	0.5557	0.5596	0.5636	0.5675	0.5714	0.5753
0.2	0.5793	0.5832	0.5871	0.5910	0.5948	0.5987	0.6026	0.6064	0.6103	0.6141
0.3	0.6179	0.6217	0.6255	0.6293	0.6331	0.6368	0.6406	0.6443	0.6480	0.6517
0.4	0.6554	0.6591	0.6628	0.6664	0.6700	0.6736	0.6772	0.6808	0.6844	0.6879
0.5	0.6915	0.6950	0.6985	0.7019	0.7054	0.7088	0.7123	0.7157	0.7190	0.7224
0.6	0.7257	0.7291	0.7324	0.7357	0.7389	0.7422	0.7454	0.7486	0.7517	0.7549
0.7	0.7580	0.7611	0.7642	0.7673	0.7704	0.7734	0.7764	0.7794	0.7823	0.7852
0.8	0.7881	0.7910	0.7939	0.7967	0.7995	0.8023	0.8051	0.8078	0.8106	0.8133
0.9	0.8159	0.8186	0.8212	0.8238	0.8264	0.8289	0.8315	0.8340	0.8365	0.8389
1.0	0.8413	0.8438	0.8461	0.8485	0.8508	0.8531	0.8554	0.8577	0.8599	0.8621
1.1	0.8643	0.8665	0.8686	0.8708	0.8729	0.8749	0.8770	0.8790	0.8810	0.8830
1.2	0.8849	0.8869	0.8888	0.8907	0.8925	0.8944	0.8962	0.8980	0.8997	0.9015
1.3	0.9032	0.9049	0.9066	0.9082	0.9099	0.9115	0.9131	0.9147	0.9162	0.9177
1.4	0.9192	0.9207	0.9222	0.9236	0.9251	0.9265	0.9279	0.9292	0.9306	0.9319
1.5	0.9332	0.9345	0.9357	0.9370	0.9382	0.9394	0.9406	0.9418	0.9429	0.9441
1.6	0.9452	0.9463	0.9474	0.9484	0.9495	0.9505	0.9515	0.9525	0.9535	0.9545
1.7	0.9554	0.9564	0.9573	0.9582	0.9591	0.9599	0.9608	0.9616	0.9625	0.9633
1.8	0.9641	0.9649	0.9656	0.9664	0.9671	0.9678	0.9686	0.9693	0.9699	0.9706
1.9	0.9713	0.9719	0.9726	0.9732	0.9738	0.9744	0.9750	0.9756	0.9761	0.9767
2.0	0.9772	0.9778	0.9783	0.9788	0.9793	0.9798	0.9803	0.9808	0.9812	0.9817
2.1	0.9821	0.9826	0.9830	0.9834	0.9838	0.9842	0.9846	0.9850	0.9854	0.9857
2.2	0.9861	0.9864	0.9868	0.9871	0.9875	0.9878	0.9881	0.9884	0.9887	0.9890
2.3	0.9893	0.9896	0.9898	0.9901	0.9904	0.9906	0.9909	0.9911	0.9913	0.9916
2.4	0.9918	0.9920	0.9922	0.9925	0.9927	0.9929	0.9931	0.9932	0.9934	0.9936
2.5	0.9938	0.9940	0.9941	0.9943	0.9945	0.9946	0.9948	0.9949	0.9951	0.9952
2.6	0.9953	0.9955	0.9956	0.9957	0.9959	0.9960	0.9961	0.9962	0.9963	0.9964
2.7	0.9965	0.9966	0.9967	0.9968	0.9969	0.9970	0.9971	0.9972	0.9973	0.9974
2.8	0.9974	0.9975	0.9976	0.9977	0.9977	0.9978	0.9979	0.9979	0.9980	0.9981
2.9	0.9981	0.9982	0.9982	0.9983	0.9984	0.9984	0.9985	0.9985	0.9986	0.9986
3.0	0.9987	0.9987	0.9987	0.9988	0.9988	0.9989	0.9989	0.9989	0.9990	0.9990

Returning to Example 8.2, the probability we seek is found in Table 8.1 by finding 1.0 in the left margin. The number to its right under the heading .00 is .8413. See Figure 8.12.

FIGURE **8.12** $P(Z < 1.00)$

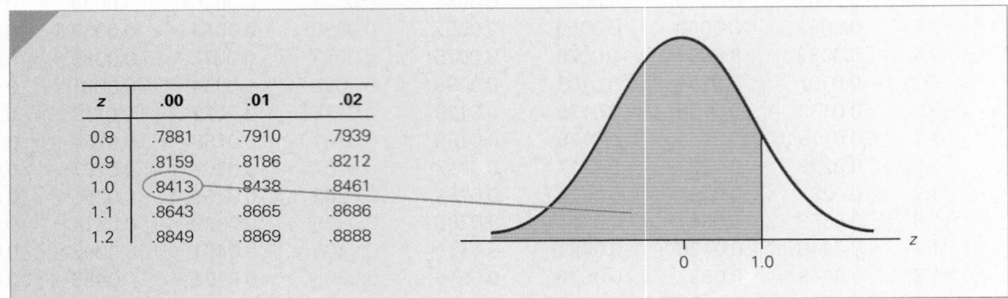

z .00 .01 .02
0.8 .7881 .7910 .7939
0.9 .8159 .8186 .8212
1.0 .8413 .8438 .8461
1.1 .8643 .8665 .8686
1.2 .8849 .8869 .8888
z
0
1.0

As was the case with Tables 1 and 2, we can also determine the probability that the standard normal random variable is greater than some value of z. For example, we find the probability that Z is greater than 1.80 by determining the probability that Z is less than 1.80 and subtracting that value from 1. Applying the complement rule, we get

$$P(Z > 1.80) = 1 - P(Z < 1.80) = 1 - .9641 = .0359$$

See Figure 8.13.

FIGURE **8.13** $P(Z > 1.80)$

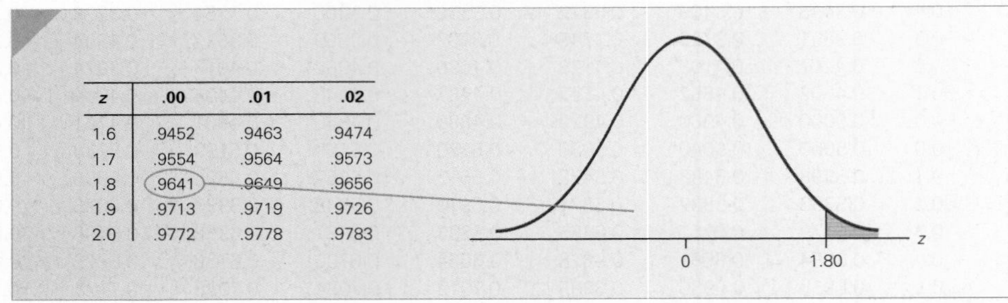

z .00 .01 .02
1.6 .9452 .9463 .9474
1.7 .9554 .9564 .9573
1.8 .9641 .9649 .9656
1.9 .9713 .9719 .9726
2.0 .9772 .9778 .9783
z
0
1.80

We can also easily determine the probability that a standard normal random variable lies between 2 values of z. For example, we find the probability

$$P(-0.71 < Z < 0.92)$$

by finding the 2 cumulative probabilities and calculating their difference. That is,

$$P(Z < -0.71) = .2389$$

and

$$P(Z < 0.92) = .8212$$

Hence,

$$P(-0.71 < Z < 0.92) = P(Z < 0.92) - P(Z < -0.71) = .8212 - .2389 = .5823$$

Figure 8.14 depicts this calculation.

Notice that the largest value of z in the table is 3.09, and that $P(Z < 3.09) = .9990$. This means that

$$P(Z > 3.09) = 1 - .9990 = .0010$$

However, because the table lists no values beyond 3.09, we approximate any area beyond 3.10 as 0. That is,

$$P(Z > 3.10) = P(Z < -3.10) \approx 0$$

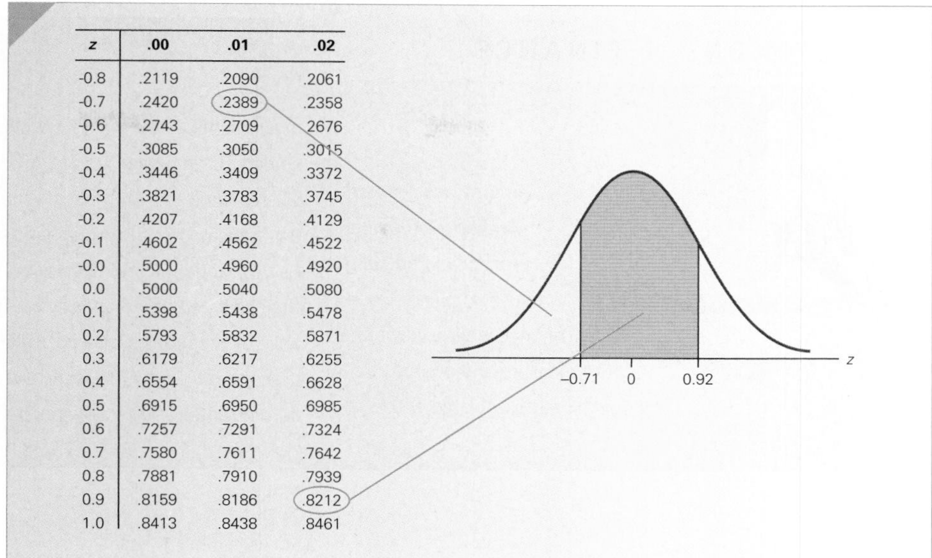

z	.00	.01	.02
-0.8	.2119	.2090	.2061
-0.7	.2420	.2389	.2358
-0.6	.2743	.2709	.2676
-0.5	.3085	.3050	.3015
-0.4	.3446	.3409	.3372
-0.3	.3821	.3783	.3745
-0.2	.4207	.4168	.4129
-0.1	.4602	.4562	.4522
-0.0	.5000	.4960	.4920
0.0	.5000	.5040	.5080
0.1	.5398	.5438	.5478
0.2	.5793	.5832	.5871
0.3	.6179	.6217	.6255
0.4	.6554	.6591	.6628
0.5	.6915	.6950	.6985
0.6	.7257	.7291	.7324
0.7	.7580	.7611	.7642
0.8	.7881	.7910	.7939
0.9	.8159	.8186	.8212
1.0	.8413	.8438	.8461

FIGURE **8.14**
$P(-0.71 < Z < 0.92)$

Recall that in Tables 1 and 2 we were able to use the table to find the probability that X is *equal* to some value of x, but we won't do the same with the normal table. Remember that the normal random variable is continuous, and the probability that a continuous random variable is equal to any single value is 0.

SEEING STATISTICS

Applet 5: Normal Distribution Areas

This applet can be used to show the calculation of the probability of any interval for any values of μ and σ. Click or drag anywhere in the graph to move the nearest end to that point. Adjust the ends to correspond to either z-scores or actual scores. The area under the normal curve between the two endpoints is highlighted in red. The size of this area corresponds to the probability of obtaining a score between the two endpoints. You can change the mean and standard deviation of the actual scores by changing the numbers in the text boxes. After changing a number, press the **Enter** or **Return** key to update the graph. When this page first loads, the mean and standard deviation correspond to a mean of 50 and a standard deviation of 10.

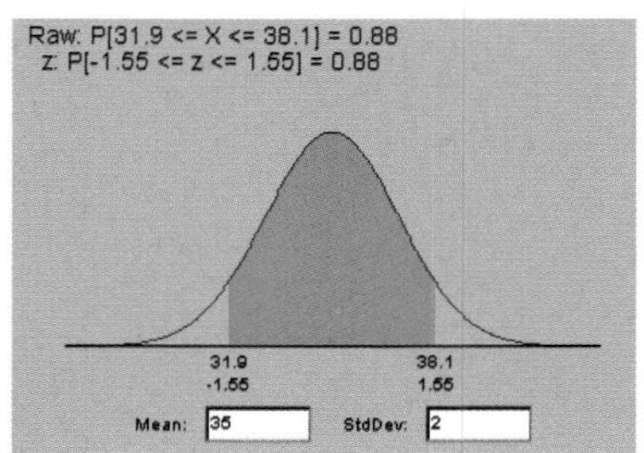

Applet Exercises

The graph is initially set with mean of 50 and standard deviation of 10. Change it so that it represents the distribution of IQs, which are normally distributed with a mean of 100 and a standard deviation of 16.

5.1 About what proportion of people have IQ scores equal to or less than 116?

5.2 About what proportion of people have IQ scores between 100 and 116?

5.3 About what proportion have IQ scores greater than 120?

5.4 About what proportion of the scores are within one standard deviation of the mean?

5.5 About what proportion of the scores are within two standard deviations of the mean?

5.6 About what proportion of the scores are within three standard deviations of the mean?

APPLICATIONS in **FINANCE**

Measuring Risk

In previous chapters we discussed several probability and statistical applications in finance where we wanted to measure and perhaps reduce the risk associated with investments. In Example 2.5 we drew histograms to gauge the spread of the histogram of the returns on two investments. We repeated this example in Chapter 4 where we computed the standard deviation and variance as numerical measures of risk. In Section 7.3 we developed an important application in finance where the emphasis was placed on reducing the variance of the returns on a portfolio. However, we have not demonstrated why risk is measured by the variance and standard deviation. The following example corrects this deficiency.

EXAMPLE 8.3

Probability of a Negative Return on Investment

Consider an investment whose return is normally distributed with a mean of 10% and a standard deviation of 5%.

a. Determine the probability of losing money.

b. Find the probability of losing money when the standard deviation is equal to 10%.

SOLUTION

a. The investment loses money when the return is negative. Thus, we wish to determine

$$P(X < 0)$$

The first step is to standardize both X and 0 in the probability statement:

$$P(X < 0) = P\left(\frac{X - \mu}{\sigma} < \frac{0 - 10}{5}\right) = P(Z < -2.00) = .0228$$

Therefore, the probability of losing money is .0228. Figure 8.15a depicts this calculation.

b. If we increase the standard deviation to 10%, the probability of suffering a loss becomes

$$P(X < 0) = P\left(\frac{X - \mu}{\sigma} < \frac{0 - 10}{10}\right) = P(Z < -1.00) = .1587$$

As you can see, increasing the standard deviation increases the probability of losing money [see Figure 8.15(b)]. It should be noted that increasing the standard deviation will also increase the probability that the return will exceed some relatively large amount. However, because investors tend to be risk averse, we emphasize the increased probability of negative returns when discussing the effect of increasing the standard deviation.

FIGURE **8.15a** $P(x < 0)$ with $\sigma = 5$

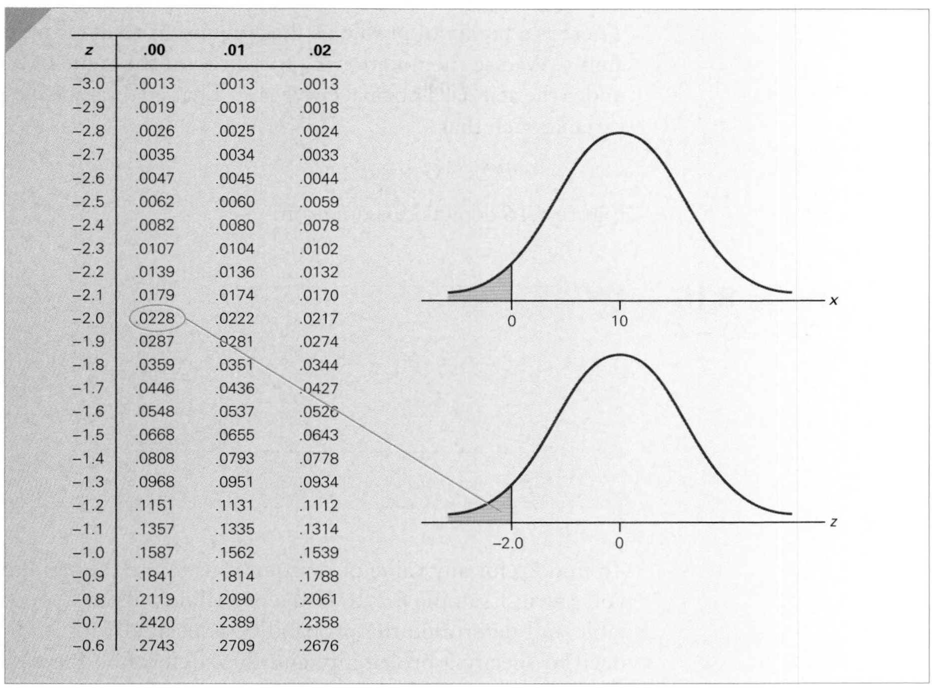

z	.00	.01	.02
−3.0	.0013	.0013	.0013
−2.9	.0019	.0018	.0018
−2.8	.0026	.0025	.0024
−2.7	.0035	.0034	.0033
−2.6	.0047	.0045	.0044
−2.5	.0062	.0060	.0059
−2.4	.0082	.0080	.0078
−2.3	.0107	.0104	.0102
−2.2	.0139	.0136	.0132
−2.1	.0179	.0174	.0170
−2.0	.0228	.0222	.0217
−1.9	.0287	.0281	.0274
−1.8	.0359	.0351	.0344
−1.7	.0446	.0436	.0427
−1.6	.0548	.0537	.0526
−1.5	.0668	.0655	.0643
−1.4	.0808	.0793	.0778
−1.3	.0968	.0951	.0934
−1.2	.1151	.1131	.1112
−1.1	.1357	.1335	.1314
−1.0	.1587	.1562	.1539
−0.9	.1841	.1814	.1788
−0.8	.2119	.2090	.2061
−0.7	.2420	.2389	.2358
−0.6	.2743	.2709	.2676

FIGURE **8.15b** $P(x < 0)$ with $\sigma = 10$

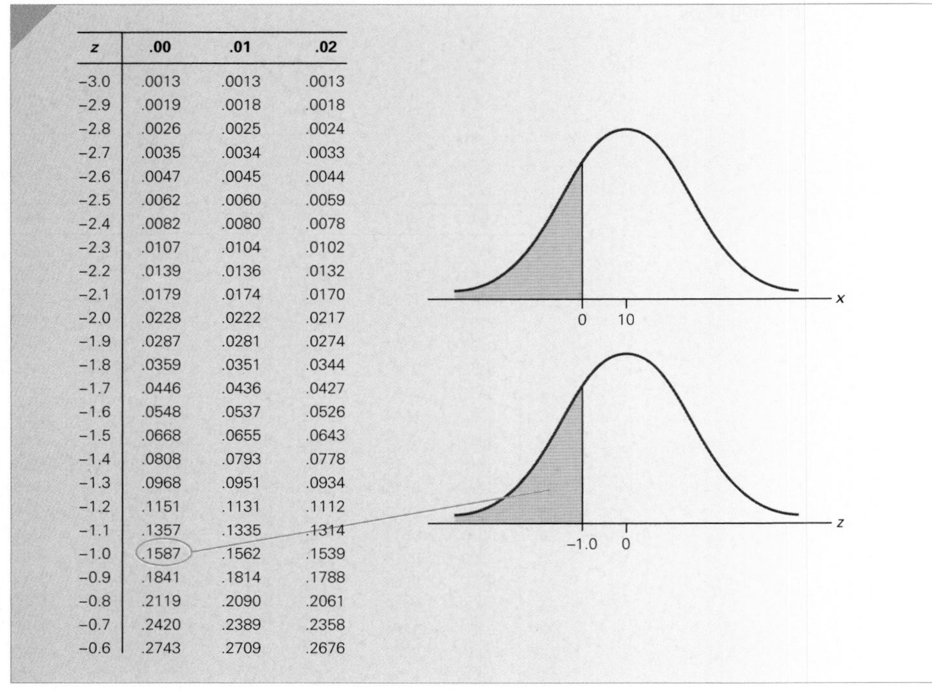

z	.00	.01	.02
−3.0	.0013	.0013	.0013
−2.9	.0019	.0018	.0018
−2.8	.0026	.0025	.0024
−2.7	.0035	.0034	.0033
−2.6	.0047	.0045	.0044
−2.5	.0062	.0060	.0059
−2.4	.0082	.0080	.0078
−2.3	.0107	.0104	.0102
−2.2	.0139	.0136	.0132
−2.1	.0179	.0174	.0170
−2.0	.0228	.0222	.0217
−1.9	.0287	.0281	.0274
−1.8	.0359	.0351	.0344
−1.7	.0446	.0436	.0427
−1.6	.0548	.0537	.0526
−1.5	.0668	.0655	.0643
−1.4	.0808	.0793	.0778
−1.3	.0968	.0951	.0934
−1.2	.1151	.1131	.1112
−1.1	.1357	.1335	.1314
−1.0	.1587	.1562	.1539
−0.9	.1841	.1814	.1788
−0.8	.2119	.2090	.2061
−0.7	.2420	.2389	.2358
−0.6	.2743	.2709	.2676

Finding Values of Z

There is a family of problems that require us to determine the value of Z given a probability. We use the notation Z_A to represent the value of z such that the area to its right under the standard normal curve is A. That is, Z_A is a value of a standard normal random variable such that

$$P(Z > Z_A) = A$$

Figure 8.16 depicts this notation.

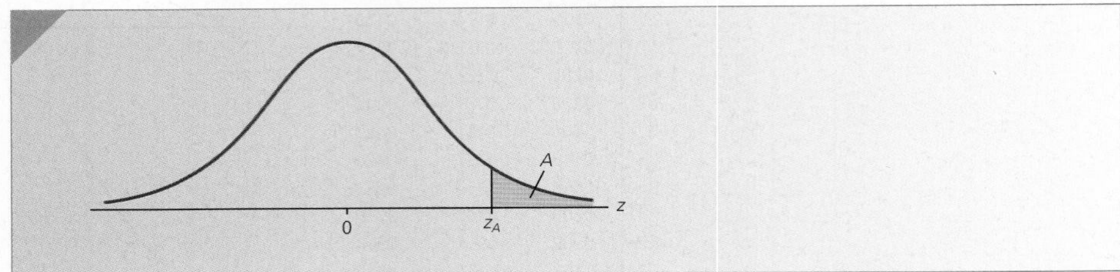

FIGURE 8.16 $P(Z > Z_A) = A$

To find Z_A for any value of A requires us to use the standard normal table backward. As you saw in Example 8.2, to find a probability about Z we must find the value of z in the table and determine the probability associated with it. To use the table backward, we need to specify a probability and then determine the z-value associated with it. We'll demonstrate by finding $Z_{.025}$. Figure 8.17 depicts the standard normal curve and $Z_{.025}$.

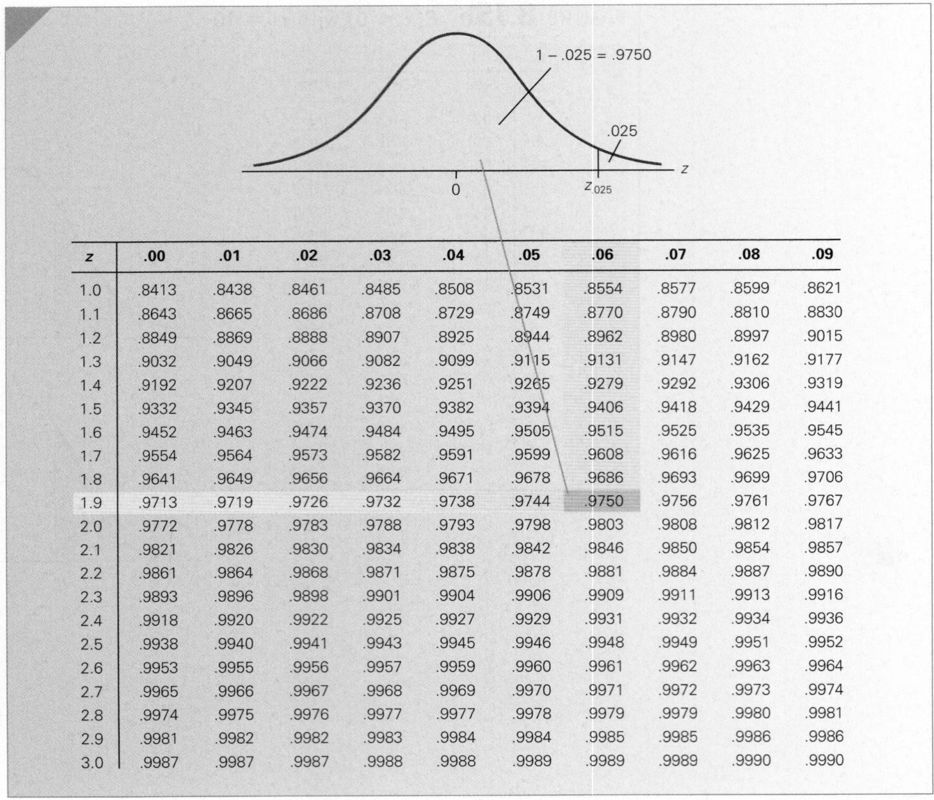

z	.00	.01	.02	.03	.04	.05	.06	.07	.08	.09
1.0	.8413	.8438	.8461	.8485	.8508	.8531	.8554	.8577	.8599	.8621
1.1	.8643	.8665	.8686	.8708	.8729	.8749	.8770	.8790	.8810	.8830
1.2	.8849	.8869	.8888	.8907	.8925	.8944	.8962	.8980	.8997	.9015
1.3	.9032	.9049	.9066	.9082	.9099	.9115	.9131	.9147	.9162	.9177
1.4	.9192	.9207	.9222	.9236	.9251	.9265	.9279	.9292	.9306	.9319
1.5	.9332	.9345	.9357	.9370	.9382	.9394	.9406	.9418	.9429	.9441
1.6	.9452	.9463	.9474	.9484	.9495	.9505	.9515	.9525	.9535	.9545
1.7	.9554	.9564	.9573	.9582	.9591	.9599	.9608	.9616	.9625	.9633
1.8	.9641	.9649	.9656	.9664	.9671	.9678	.9686	.9693	.9699	.9706
1.9	.9713	.9719	.9726	.9732	.9738	.9744	.9750	.9756	.9761	.9767
2.0	.9772	.9778	.9783	.9788	.9793	.9798	.9803	.9808	.9812	.9817
2.1	.9821	.9826	.9830	.9834	.9838	.9842	.9846	.9850	.9854	.9857
2.2	.9861	.9864	.9868	.9871	.9875	.9878	.9881	.9884	.9887	.9890
2.3	.9893	.9896	.9898	.9901	.9904	.9906	.9909	.9911	.9913	.9916
2.4	.9918	.9920	.9922	.9925	.9927	.9929	.9931	.9932	.9934	.9936
2.5	.9938	.9940	.9941	.9943	.9945	.9946	.9948	.9949	.9951	.9952
2.6	.9953	.9955	.9956	.9957	.9959	.9960	.9961	.9962	.9963	.9964
2.7	.9965	.9966	.9967	.9968	.9969	.9970	.9971	.9972	.9973	.9974
2.8	.9974	.9975	.9976	.9977	.9977	.9978	.9979	.9979	.9980	.9981
2.9	.9981	.9982	.9982	.9983	.9984	.9984	.9985	.9985	.9986	.9986
3.0	.9987	.9987	.9987	.9988	.9988	.9989	.9989	.9989	.9990	.9990

FIGURE 8.17 Finding $Z_{.025}$

Because of the format of the standard normal table, we begin by determining the area *less than* $Z_{.025}$, which is $1 - .025 = .9750$. (Notice that we expressed this probability with four decimal places to make it easier for you to see what you need to do.) We now search through the probability part of the table looking for .9750. When we locate it, we see that the z-value associated with it is 1.96.

Thus, $Z_{.025} = 1.96$, which means that $P(Z > 1.96) = .025$.

EXAMPLE 8.4

Finding $Z_{.05}$

Find the value of a standard normal random variable such that the probability that the random variable is greater than it is 5%.

SOLUTION

We wish to determine $Z_{.05}$. Figure 8.18 depicts the normal curve and $Z_{.05}$. If .05 is the area in the tail, then the probability less than $Z_{.05}$ must be $1 - .05 = .9500$. To find $Z_{.05}$, we search the table looking for the probability .9500. We don't find this probability, but we find two values that are equally close: .9495 and .9505. The Z-values associated with these probabilities are 1.64 and 1.65, respectively. The average is taken as $Z_{.05}$. Thus, $Z_{.05} = 1.645$.

FIGURE **8.18** Finding $Z_{.05}$

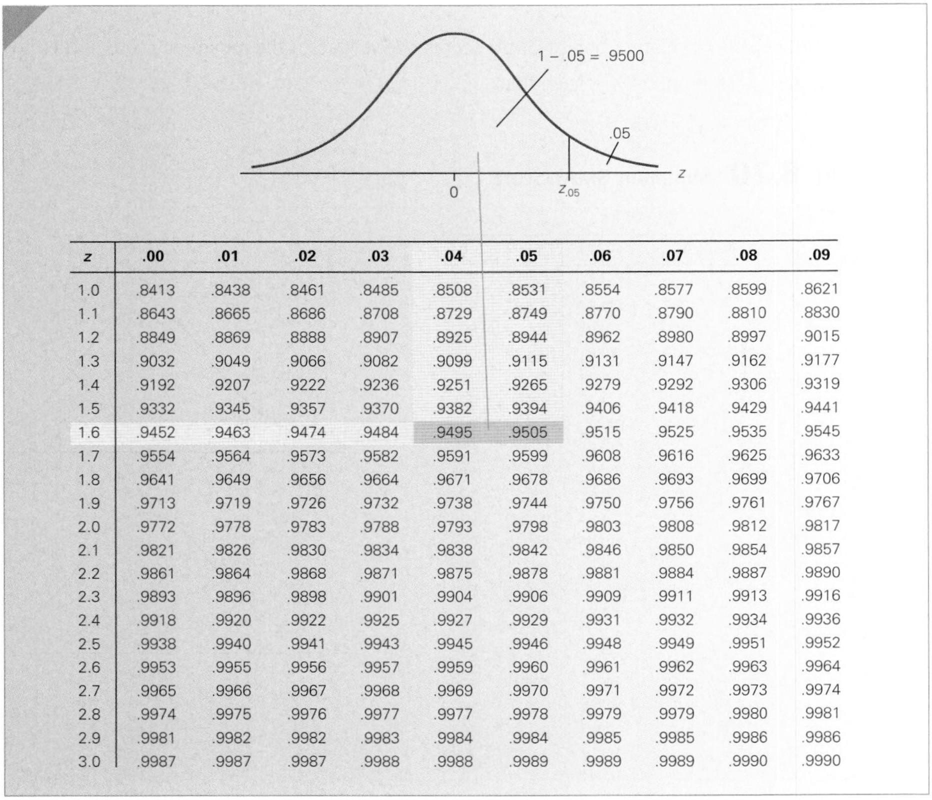

z	.00	.01	.02	.03	.04	.05	.06	.07	.08	.09
1.0	.8413	.8438	.8461	.8485	.8508	.8531	.8554	.8577	.8599	.8621
1.1	.8643	.8665	.8686	.8708	.8729	.8749	.8770	.8790	.8810	.8830
1.2	.8849	.8869	.8888	.8907	.8925	.8944	.8962	.8980	.8997	.9015
1.3	.9032	.9049	.9066	.9082	.9099	.9115	.9131	.9147	.9162	.9177
1.4	.9192	.9207	.9222	.9236	.9251	.9265	.9279	.9292	.9306	.9319
1.5	.9332	.9345	.9357	.9370	.9382	.9394	.9406	.9418	.9429	.9441
1.6	.9452	.9463	.9474	.9484	.9495	.9505	.9515	.9525	.9535	.9545
1.7	.9554	.9564	.9573	.9582	.9591	.9599	.9608	.9616	.9625	.9633
1.8	.9641	.9649	.9656	.9664	.9671	.9678	.9686	.9693	.9699	.9706
1.9	.9713	.9719	.9726	.9732	.9738	.9744	.9750	.9756	.9761	.9767
2.0	.9772	.9778	.9783	.9788	.9793	.9798	.9803	.9808	.9812	.9817
2.1	.9821	.9826	.9830	.9834	.9838	.9842	.9846	.9850	.9854	.9857
2.2	.9861	.9864	.9868	.9871	.9875	.9878	.9881	.9884	.9887	.9890
2.3	.9893	.9896	.9898	.9901	.9904	.9906	.9909	.9911	.9913	.9916
2.4	.9918	.9920	.9922	.9925	.9927	.9929	.9931	.9932	.9934	.9936
2.5	.9938	.9940	.9941	.9943	.9945	.9946	.9948	.9949	.9951	.9952
2.6	.9953	.9955	.9956	.9957	.9959	.9960	.9961	.9962	.9963	.9964
2.7	.9965	.9966	.9967	.9968	.9969	.9970	.9971	.9972	.9973	.9974
2.8	.9974	.9975	.9976	.9977	.9977	.9978	.9979	.9979	.9980	.9981
2.9	.9981	.9982	.9982	.9983	.9984	.9984	.9985	.9985	.9986	.9986
3.0	.9987	.9987	.9987	.9988	.9988	.9989	.9989	.9989	.9990	.9990

EXAMPLE 8.5 Finding $-Z_{.05}$

Find the value of a standard normal random variable such that the probability that the random variable is less than it is 5%.

SOLUTION

Because the standard normal curve is symmetric about 0, we wish to find $-Z_{.05}$. In Example 8.4 we found $Z_{.05} = 1.645$. Thus, $-Z_{.05} = -1.645$. See Figure 8.19.

FIGURE **8.19** $-Z_{.05}$

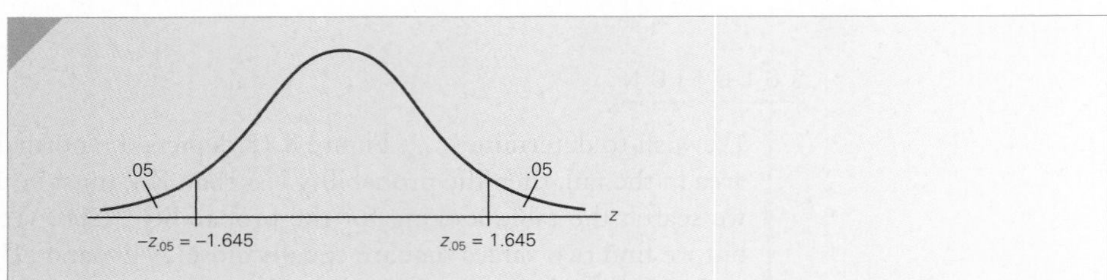

Minimum GMAT Score to Enter Executive MBA Program: Solution

Figure 8.20 depicts the distribution of GMAT scores. We've labeled the minimum score needed to enter the new MBA Program $X_{.01}$, such that

$$P(X > X_{.01}) = .01$$

FIGURE **8.20** Minimum GMAT Score

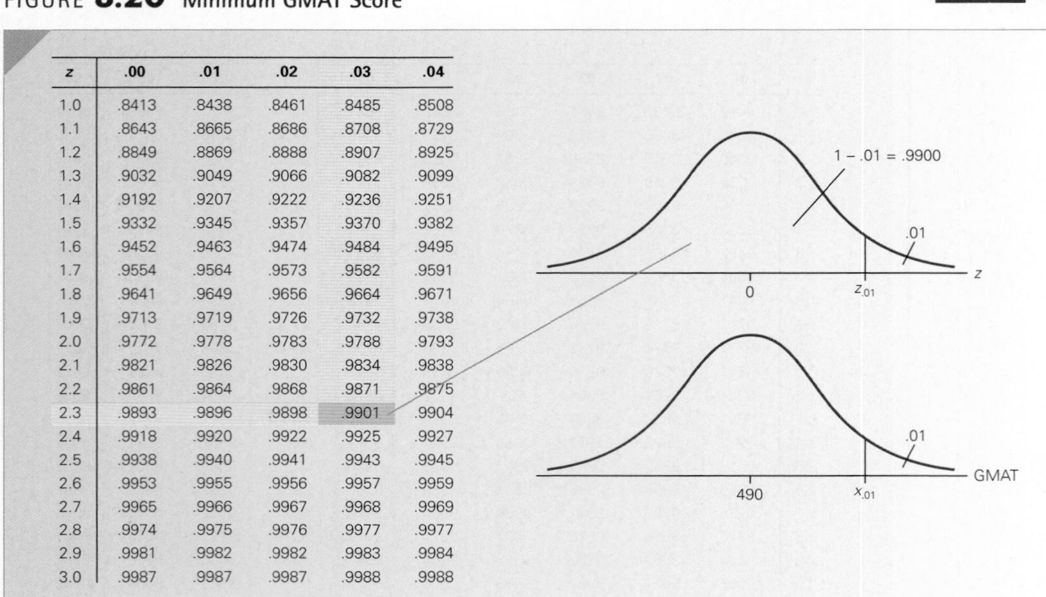

z	.00	.01	.02	.03	.04
1.0	.8413	.8438	.8461	.8485	.8508
1.1	.8643	.8665	.8686	.8708	.8729
1.2	.8849	.8869	.8888	.8907	.8925
1.3	.9032	.9049	.9066	.9082	.9099
1.4	.9192	.9207	.9222	.9236	.9251
1.5	.9332	.9345	.9357	.9370	.9382
1.6	.9452	.9463	.9474	.9484	.9495
1.7	.9554	.9564	.9573	.9582	.9591
1.8	.9641	.9649	.9656	.9664	.9671
1.9	.9713	.9719	.9726	.9732	.9738
2.0	.9772	.9778	.9783	.9788	.9793
2.1	.9821	.9826	.9830	.9834	.9838
2.2	.9861	.9864	.9868	.9871	.9875
2.3	.9893	.9896	.9898	.9901	.9904
2.4	.9918	.9920	.9922	.9925	.9927
2.5	.9938	.9940	.9941	.9943	.9945
2.6	.9953	.9955	.9956	.9957	.9959
2.7	.9965	.9966	.9967	.9968	.9969
2.8	.9974	.9975	.9976	.9977	.9977
2.9	.9981	.9982	.9982	.9983	.9984
3.0	.9987	.9987	.9987	.9988	.9988

Below the normal curve, we depict the standard normal curve and $Z_{.01}$. We can determine the value of $Z_{.01}$ as we did Example 8.4. In the standard normal table we find $1 - .01 = .9900$ (its closest value in the table is .9901) and the Z-value 2.33. Thus, the standardized value of $X_{.01}$ is $Z_{.01} = 2.33$. To find $X_{.01}$, we must ***unstandardize*** $Z_{.01}$. We do so by solving for $X_{.01}$ in the equation

$$Z_{.01} = \frac{X_{.01} - \mu}{\sigma}$$

Substituting $Z_{.01} = 2.33$, $\mu = 490$, and $\sigma = 61$, we find

$$2.33 = \frac{X_{.01} - 490}{61}$$

Solving, we get

$$X_{.01} = 2.33(61) + 490 = 632.13$$

Rounding up (GMAT scores are integers), we find that the minimum GMAT score to enter the Executive MBA Program is 633.

Z_A and Percentiles

In Chapter 4 we introduced percentiles, which are measures of relative standing. The values of Z_A are the $100(1 - A)$th percentiles of a standard normal random variable. For example, $Z_{.05} = 1.645$, which means that 1.645 is the 95th percentile; 95% of all values of Z are below it and 5% are above it. We interpret other values of Z_A similarly.

Using the Computer

EXCEL

INSTRUCTIONS

We can use Excel to compute probabilities as well as values of X and Z. To compute cumulative normal probabilities $P(X < x)$, type (in any cell)

=**NORMDIST**([X],[μ],[σ],True)

(Typing "True" yields a cumulative probability. Typing "False" will produce the value of the normal density function, a number with little meaning.)

If you type 0 for μ and 1 for σ, you will obtain standard normal probabilities. Alternatively, type

NORMSDIST instead of **NORMDIST** and enter the value of z.

In Example 8.2, we found $P(X < 1{,}100) = P(Z < 1.00) = .8413$. To instruct Excel to calculate this probability, we enter

=NORMDIST(1100, 1000, 100, True)

or

=NORMSDIST(1.00)

To calculate a value for Z_A, type

=NORMSINV([1 − A])

In Example 8.4, we would type

=NORMSINV(.95)

and produce 1.6449. We calculated $Z_{.05} = 1.645$.

To calculate a value of x given the probability $P(X > x) = A$, enter

=NORMINV([1 − A], μ, σ)

The chapter-opening example would be solved by typing

=NORMINV(.99, 490, 61)

which yields 632.

MINITAB

INSTRUCTIONS

We can use Minitab to compute probabilities as well as values of X and Z.

Check **Calc, Probability Distributions,** and **Normal . . .** and either **Cumulative probability** [to determine $P(X < x)$] or **Inverse cumulative probability** to find the value of x. Specify the **Mean** and **Standard deviation.**

APPLICATIONS in OPERATIONS MANAGEMENT

© John Zoiner/Workbook Stock/ Jupiterimages

Inventory Management

Every organization maintains some inventory, which is defined as a stock of items. For example, grocery stores hold inventories of almost all the products they sell. When the total number of products drops to a specified level, the manager arranges for the delivery of more products. An automobile repair shop keeps an inventory of a large number of replacement parts. A school keeps stock of items that it uses regularly, including chalk, pens, envelopes, file folders, and paper clips. There are costs associated with inventories.

These include the cost of capital, losses (theft and obsolescence), and warehouse space, as well as maintenance and record keeping. Management scientists have developed many models to help determine the optimum inventory level that balances the cost of inventory with the cost of shortages and the cost of making many small orders. Several of these models are deterministic—that is, they assume that the demand for the product is constant. However, in most realistic situations the demand is a random variable. One commonly applied probabilistic model assumes that the demand during lead time is a normally distributed random variable. *Lead time* is defined as the amount of time between when the order is placed and when it is delivered.

The quantity ordered is usually calculated by attempting to minimize the total costs, including the cost of ordering and the cost of maintaining inventory. (This topic is discussed in most management science courses.) Another critical decision involves the *reorder point*, which is the level of inventory at which an order is issued to its supplier. If the reorder point is too low, the company will run out of product, suffering the loss of sales and potentially customers who will go to a competitor. If the reorder point is too high, the company will be carrying too much inventory, which costs money to buy and store. In some companies inventory has a tendency to walk out the back door or become obsolete. As a result, managers create a *safety stock*, which is the extra amount of inventory to reduce the times when the company has a shortage. They do so by setting a service level, which is the probability that the company will not experience a shortage. The method used to determine the reorder point is be demonstrated with Example 8.6.

EXAMPLE 8.6 Determining the Reorder Point

During the spring the demand for electric fans at a large home improvement store is quite strong. The company tracks inventory using a computer system so that it knows how many fans are in the inventory at any time. The policy is to order a new shipment of 250 fans when the inventory level falls to the reorder point, which is 150. However, this policy has resulted in frequent shortages resulting in lost sales because both lead time and demand are highly variable. The manager would like to reduce the incidence of shortages so that only 5% of orders will arrive after inventory drops to 0 (resulting in a shortage). This policy is expressed as a 95% service level. From previous periods the company has determined that demand during lead time is normally distributed with a mean of 200 and a standard deviation of 50. Find the reorder point.

SOLUTION

The reorder point is set so that the probability that demand during lead time exceeds this quantity is 5%. Figure 8.21 depicts demand during lead time and the reorder point. As we did in the solution to the chapter-opening example, we find the standard normal value such that the area to its right is .05. The standardized value of the reorder point is $Z_{.05} = 1.645$. To find the reorder point (ROP) we must unstandardize $Z_{.05}$.

$$Z_{.05} = \frac{\text{ROP} - \mu}{\sigma}$$

$$1.645 = \frac{\text{ROP} - 200}{50}$$

$$\text{ROP} = 50(1.645) + 200 = 282.25$$

which we round up to 283. The policy is to order a new batch of fans when there are 283 fans left in inventory.

FIGURE **8.21** Distribution of Demand during Lead Time

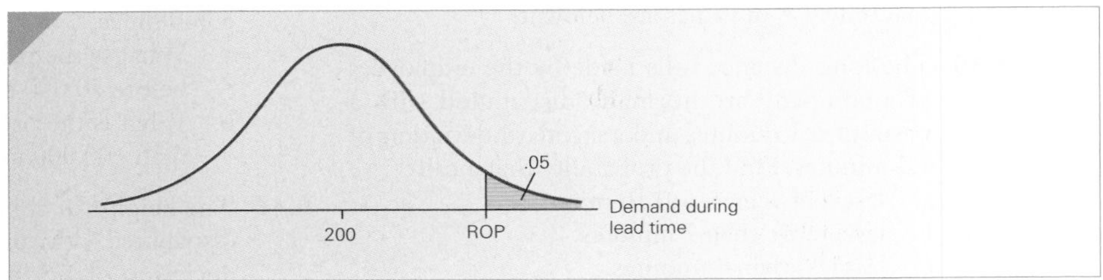

EXERCISES

In Exercises 8.15 to 8.30 find the following probabilities.

8.15 $P(Z < 1.50)$

8.16 $P(Z < 1.51)$

8.17 $P(Z < 1.55)$

8.18 $P(Z < -1.59)$

8.19 $P(Z < -1.60)$

8.20 $P(Z < -2.30)$

8.21 $P(-1.40 < Z < .60)$

8.22 $P(Z > -1.44)$

8.23 $P(Z < 2.03)$

8.24 $P(Z > 1.67)$

8.25 $P(Z < 2.84)$

8.26 $P(1.14 < Z < 2.43)$

8.27 $P(-0.91 < Z < -0.33)$

8.28 $P(Z > 3.09)$

8.29 $P(Z > 0)$

8.30 $P(Z > 4.0)$

8.31 Find $z_{.02}$.

8.32 Find $z_{.045}$.

8.33 Find $z_{.20}$.

8.34 X is normally distributed with a mean of 100 and a standard deviation of 20. What is the probability that X is greater than 145?

8.35 X is normally distributed with a mean of 250 and a standard deviation of 40. What value of X does only the top 15% exceed?

8.36 X is normally distributed with a mean of 1,000 and a standard deviation of 250. What is the probability that X lies between 800 and 1,100?

8.37 X is normally distributed with a mean of 50 and a standard deviation of 8. What value of X is such that only 8% of values are below it?

8.38 The long-distance calls made by the employees of a company are normally distributed with a mean of 6.3 minutes and a standard deviation of 2.2 minutes. Find the probability that a call

a. lasts between 5 and 10 minutes.
b. lasts more than 7 minutes.
c. last less than 4 minutes.

8.39 Refer to Exercise 8.38. How long do the longest 10% of calls last?

8.40 The lifetimes of light bulbs that are advertised to last for 5,000 hours are normally distributed with a mean of 5,100 hours and a standard deviation of 200 hours. What is the probability that a bulb lasts longer than the advertised figure?

8.41 Refer to Exercise 8.40. If we wanted to be sure that 98% of all bulbs last longer than the advertised figure, what figure should be advertised?

8.42 Travelbyus is an Internet-based travel agency wherein customers can see videos of the cities they plan to visit. The number of hits daily is a normally distributed random variable with a mean of 10,000 and a standard deviation of 2,400.

a. What is the probability of getting more than 12,000 hits?
b. What is the probability of getting fewer than 9,000 hits?

8.43 Refer to Exercise 8.42. Some Internet sites have bandwidths that are not sufficient to handle all their traffic, often causing the system to crash. Bandwidth can be measured by the number of hits it can handle. How large a bandwidth should Travelbyus have in order to handle 99.9% of daily traffic?

8.44 A new car that is a gas- and electric-powered hybrid has recently hit the market. The distance traveled on 1 gallon of fuel is normally distributed with a mean of 65 miles and a standard deviation of 4 miles. Find the probability of the following events.

a. The car travels more than 70 miles per gallon.
b. The car travels less than 60 miles per gallon.
c. The car travels between 55 and 70 miles per gallon.

8.45 The top-selling Red and Voss tire is rated 70,000 miles, which means nothing. In fact, the distance the tires can run until they wear out is a normally distributed random variable with a mean of 82,000 miles and a standard deviation of 6,400 miles.

a. What is the probability that a tire wears out before 70,000 miles?
b. What is the probability that a tire lasts more than 100,000 miles?

8.46 The heights of children 2 years old are normally distributed with a mean of 32 inches and a standard deviation of 1.5 inches. Pediatricians regularly measure the heights of toddlers to determine whether there is a problem. There may be a problem when a child is in the top or bottom 5% of heights. Determine the heights of 2-year-old children that could be a problem.

8.47 Refer to Exercise 8.46. Find the probability of these events.
a. A 2-year-old child is taller than 36 inches.
b. A 2-year-old child is shorter than 34 inches.
c. A 2-year-old child is between 30 and 33 inches tall.

8.48 University and college students average 7.2 hours of sleep per night with a standard deviation of 40 minutes. If the amount of sleep is normally distributed, what proportion of university and college students sleep for more than 8 hours?

8.49 Refer to Exercise 8.48. Find the amount of sleep that is exceeded by only 25% of students.

8.50 The amount of time devoted to studying statistics each week by students who achieve a grade of A in the course is a normally distributed random variable with a mean of 7.5 hours and a standard deviation of 2.1 hours.
a. What proportion of A students study for more than 10 hours per week?
b. Find the probability that an A student spends between 7 and 9 hours studying.
c. What proportion of A students spend less than 3 hours studying?
d. What is the amount of time below which only 5% of all A students spend studying?

8.51 The number of pages printed before replacing the cartridge in a laser printer is normally distributed with a mean of 11,500 pages and a standard deviation of 800 pages. A new cartridge has just been installed.
a. What is the probability that the printer produces more than 12,000 pages before this cartridge must be replaced?
b. What is the probability that the printer produces fewer than 10,000 pages?

8.52 Refer to Exercise 8.51. The manufacturer wants to provide guidelines to potential customers advising them of the minimum number of pages they can expect from each cartridge. How many pages should it advertise if the company wants to be correct 99% of the time?

8.53 Battery manufacturers compete on the basis of the amount of time their products last in cameras and toys. A manufacturer of alkaline batteries has observed that its batteries last for an average of 26 hours when used in a toy racing car. The amount of time is normally distributed with a standard deviation of 2.5 hours.
a. What is the probability that the battery lasts between 24 and 28 hours?
b. What is the probability that the battery lasts longer than 28 hours?
c. What is the probability that the battery lasts less than 24 hours?

8.54 Because of the relatively high interest rates, most consumers attempt to pay off their credit card bills promptly. However, this is not always possible. An analysis of the amount of interest paid monthly by a bank's Visa cardholders reveals that the amount is normally distributed with a mean of $27 and a standard deviation of $7.
a. What proportion of the bank's Visa cardholders pay more than $30 in interest?
b. What proportion of the bank's Visa cardholders pay more than $40 in interest?
c. What proportion of the bank's Visa cardholders pay less than $15 in interest?
d. What interest payment is exceeded by only 20% of the bank's Visa cardholders?

8.55 It is said that sufferers of a cold virus experience symptoms for 7 days. However, the amount of time is actually a normally distributed random variable whose mean is 7.5 days and whose standard deviation is 1.2 days.
a. What proportion of cold sufferers experiences less than 4 days of symptoms?
b. What proportion of cold sufferers experiences symptoms for between 7 and 10 days?

8.56 How much money does a typical family of four spend at McDonald's restaurants per visit? The amount is a normally distributed random variable whose mean is $16.40 and whose standard deviation is $2.75.
a. Find the probability that a family of four spends less than $10.
b. What is the amount below which only 10% of families of four spend at McDonald's?

8.57 The final marks in a statistics course are normally distributed with a mean of 70 and a standard deviation of 10. The professor must convert all marks to letter grades. She decides that she wants 10% A's, 30% B's, 40% C's, 15% D's, and 5% F's. Determine the cutoffs for each letter grade.

8.58 Mensa is an organization whose members possess IQs that are in the top 2% of the population. It is known that IQs are normally distributed with a mean of 100 and a standard deviation of 16. Find the minimum IQ needed to be a Mensa member.

8.59 According to the 2001 Canadian census, university-educated Canadians earned a mean income of \$61,823. The standard deviation is \$17,301. If incomes are normally distributed, what is the probability that a randomly selected university-educated Canadian earns more than \$70,000?

8.60 The census referred to in the previous exercise also reported that college-educated Canadians earn on average \$41,825. Suppose that incomes are normally distributed with a standard deviation of \$13,444. Find the probability that a randomly selected college-educated Canadian earns less than \$45,000.

8.61 The lifetimes of televisions produced by the Hishobi Company are normally distributed with a mean of 75 months and a standard deviation of 8 months. If the manufacturer wants to have to replace only 1% of its televisions, what should its warranty be?

8.62 According to the *Statistical Abstract of the United States* (2000, Table 764) the mean family net worth of families whose head is between 35 and 44 years old is approximately \$99,700. If family net worth is normally distributed with a standard deviation of \$30,000, find the probability that a randomly selected family whose head is between 35 and 44 years old has a net worth greater than \$150,000.

8.63 A retailer of computing products sells a variety of computer-related products. One of his most popular products is an HP Laser Printer. The average weekly demand is 200. Lead time for a new order from the manufacturer to arrive is 1 week. If the demand for printers were constant, the retailer would reorder when there were exactly 200 printers in inventory. However, the demand is a random variable. An analysis of previous weeks reveals that the weekly demand standard deviation is 30. The retailer knows that if a customer wants to buy an HP Laser Printer but he has none available he will lose that sale plus possibly additional sales. He wants the probability of running short in any week to be no more than 6%. How many HP Laser Printers should he have in stock when he reorders from the manufacturer?

8.64 The demand for a daily newspaper at a newsstand at a busy intersection is known to be normally distributed with a mean of 150 and a standard deviation of 25. How many newspapers should the newsstand operator order to ensure that he runs short on no more than 20% of days?

8.65 Every day a bakery prepares its famous marble rye. A statistically savvy customer determined that daily demand is normally distributed with a mean of 850 and a standard deviation of 90. How many loaves should the bakery make if it wants the probability of running short on any day to be no more than 30%?

8.66 Refer to Exercise 8.65. Any marble ryes that are unsold at the end of the day are marked down and sold for half price. How many loaves should the bakery prepare so that the proportion of days that result in unsold loaves is no more than 60%?

APPLICATIONS in OPERATIONS MANAGEMENT

© Banana Stock/ Jupiterimages

PERT/CPM

In the Application box on page 227 we introduced PERT/CPM. The purpose of this powerful management science procedure is to determine the critical path of a project. The expected value and variance of the completion time of the project are based on the expected values and variances of the completion times of the activities on the critical path. Once we have the expected value and variance of the completion time of the project, we can use these figures to determine the probability that the project will be completed by a certain date. Statisticians have established that the completion time of the project is approximately normally distributed, enabling us to compute the needed probabilities.

8.67 Refer to Exercise 7.57. Find the probability that the project will take more than 60 days to complete.

8.68 The mean and variance of the time to complete the project in Exercise 7.58 was 145 minutes and 31 minutes^2. What is the probability that it will take less than 2.5 hours to overhaul the machine?

8.69 The annual rate of return on a mutual fund is normally distributed with a mean of 14% and a standard deviation of 18%.
a. What is the probability that the fund returns more than 25% next year?
b. What is the probability that the fund loses money next year?

8.70 In Exercise 7.64 we discovered that the expected return is .1060 and the standard deviation is .1456. Working with the assumption that returns are normally distributed, determine the probability of the following events.
a. The portfolio loses money.
b. The return on the portfolio is greater than 20%.

8.3 (OPTIONAL) EXPONENTIAL DISTRIBUTION

Another important continuous distribution is the **exponential distribution.**

Exponential Probability Density Function

A random variable X is exponentially distributed if its probability density function is given by

$$f(x) = \lambda e^{-\lambda x}, \quad x \geq 0$$

where $e = 2.71828\ldots$ and λ is the parameter of the distribution.

Statisticians have shown that the mean and standard deviation of an exponential random variable are equal to each other:

$$\mu = \sigma = 1/\lambda$$

Recall that the normal distribution is a two-parameter distribution. The distribution is completely specified once the values of the two parameters μ and σ are known. In contrast, the exponential distribution is a one-parameter distribution. The distribution is completely specified once the value of the parameter λ is known. Figure 8.22 depicts three exponential distributions, corresponding to three different values of the parameter λ. Notice that for any exponential density function $f(X)$, $f(0) = \lambda$ and $f(x)$ approaches 0 as x approaches infinity.

FIGURE **8.22**
Exponential Distributions

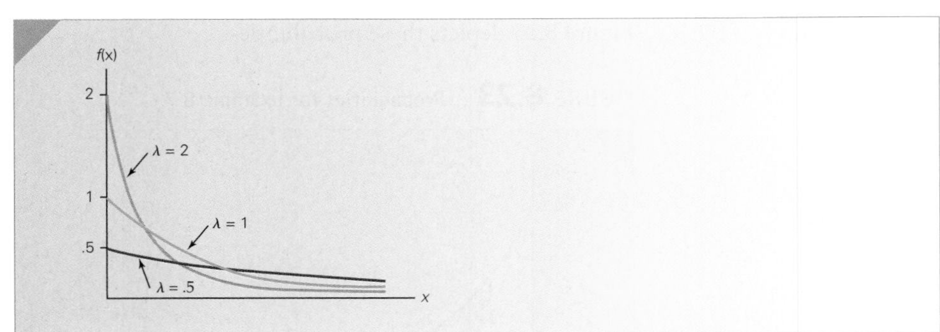

The exponential density function is easier to work with than the normal; as a result, we can develop formulas for the calculation of the probability of any ranges of values. Using integral calculus, we can determine the following probability statements.

Probability Associated with an Exponential Random Variable

If X is an exponential random variable,

$$P(X > x) = e^{-\lambda x}$$
$$P(X < x) = 1 - e^{-\lambda x}$$
$$P(x_1 < X < x_2) = P(X < x_2) - P(X < x_1) = e^{-\lambda x_1} - e^{-\lambda x_2}$$

The value of $e^{-\lambda x}$ can be obtained with the aid of a calculator.

EXAMPLE 8.7

Lifetimes of Alkaline Batteries

The lifetime of an alkaline battery (measured in hours) is exponentially distributed with $\lambda = .05$.

a. What is the mean and standard deviation of the battery's lifetime?
b. Find the probability that a battery will last between 10 and 15 hours.
c. What is the probability that a battery will last for more than 20 hours?

SOLUTION

a. The mean and standard deviation are equal to $1/\lambda$. Thus,

$$\mu = \sigma = 1/\lambda = 1/.05 = 20 \text{ hours}$$

b. Let X denote the lifetime of a battery. The required probability is

$$\begin{aligned} P(10 < X < 15) &= e^{-.05(10)} - e^{-.05(15)} \\ &= e^{-.5} - e^{-.75} \\ &= .6065 - .4724 \\ &= .1341 \end{aligned}$$

c.
$$\begin{aligned} P(X > 20) &= e^{-.05(20)} \\ &= e^{-1} \\ &= .3679 \end{aligned}$$

Figure 8.23 depicts these probabilities.

FIGURE **8.23** **Probabilities for Example 8.7**

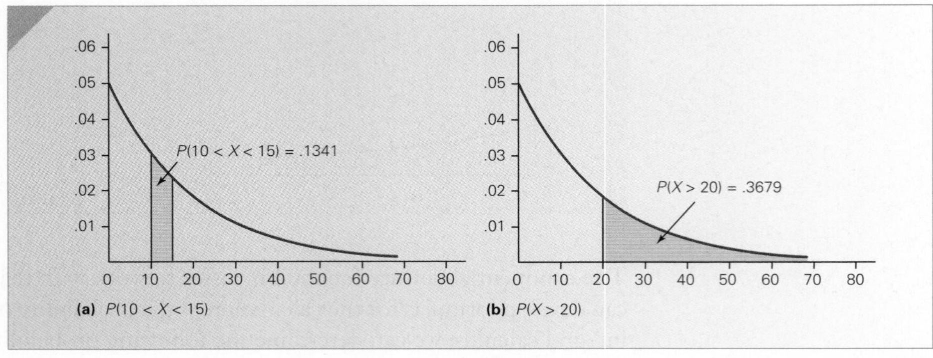

Using the Computer

EXCEL

INSTRUCTIONS

Type (in any cell)

=**EXPONDIST** ([X],[λ], True)

To produce the answer for Example 8.7c we would find $P(X < 20)$ and subtract it from 1. To find $P(X < 20)$ type

=**EXPONDIST**(20,.05, True)

which outputs .6321 and hence $P(X > 20) = 1 - .6321 = .3679$, which is exactly the number we produced manually.

MINITAB

INSTRUCTIONS

Click **Calc, Probability Distributions, Exponential** . . . and specify **Cumulative probability.** In the **Scale** box type the mean, which is $1/\lambda$. In the **Threshold** box type 0.

APPLICATIONS in OPERATIONS MANAGEMENT

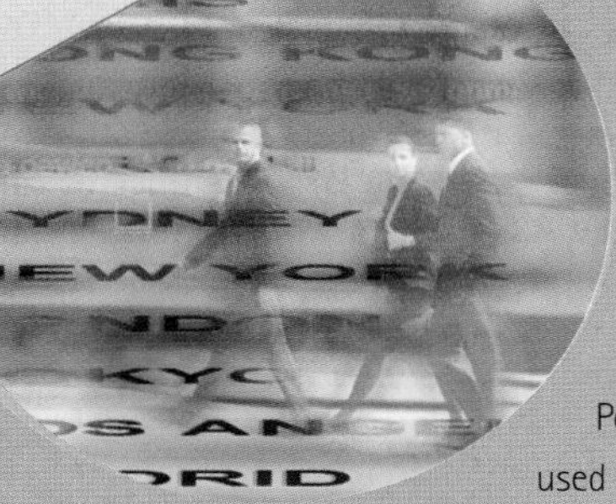

© SuperStock

Waiting Lines

In Section 7.5 we described waiting line models and described how the Poisson distribution is used to calculate the probabilities of the number of arrivals per time period. In order to calculate the operating characteristics of waiting lines, management scientists often assume that the times to complete a service are exponentially distributed. In this application the parameter λ is the service rate, which is defined as the mean number of service completions per time period. For example, if service times are exponentially distributed with λ = 5/hour, this tells us that the service rate is 5 units per hour or 5 per 60 minutes. Recall that the mean of an exponential distribution is $\mu = 1/\lambda$. In this case the service facility can complete a service in an average of 12 minutes. This was calculated as

$$\mu = \frac{1}{\lambda} = \frac{1}{5/\text{hr.}} = \frac{1}{5/60 \text{ minutes}} = \frac{60 \text{ minutes}}{5}$$
$$= 12 \text{ minutes}$$

We can use this distribution to make a variety of probability statements.

EXAMPLE 8.8 Supermarket Checkout Counter

A checkout counter at a supermarket completes the process according to an exponential distribution with a service rate of 6 per hour. A customer arrives at the checkout counter. Find the probability of the following events.

a. The service is completed in less than 5 minutes.
b. The customer leaves the checkout counter more than 10 minutes after arriving.
c. The service is completed in a time between 5 and 8 minutes.

SOLUTION

One way to solve this problem is to convert the service rate so that the time period is 1 minute. (Alternatively, we can solve by converting the probability statements so that the time periods are measured in fractions of an hour.) Let the service rate = $\lambda = .1$/minute.

a. $P(X < 5) = 1 - e^{-\lambda x} = 1 - e^{-.1(5)} = 1 - e^{-.5} = 1 - .6065 = .3935$
b. $P(X > 10) = e^{-\lambda x} = e^{-.1(10)} = e^{-1} = .3679$
c. $P(5 < X < 8) = e^{-.1(5)} = e^{-.1(8)} = e^{-.5} - e^{-.8} = .6065 - .4493 = .1572$

EXERCISES

8.71 The manager of a supermarket tracked the amount of time needed for customers to be served by the cashier. After checking with his statistics professor, he concluded that the checkout times are exponentially distributed with a mean of 6 minutes. What proportion of customers require more than 10 minutes to check out?

8.72 The time between breakdowns of aging machines is known to be exponentially distributed with a mean of 25 hours. The machine has just been repaired. Determine the probability that the next breakdown occurs more than 50 hours from now.

8.73 Because automatic banking machine (ABM) customers can perform a number of transactions, the times to complete them can be quite variable. A banking consultant has noted that the times are exponentially distributed with a mean of 125 seconds. What proportion of the ABM customers take more than 3 minutes to do their banking?

8.74 The production of a complex chemical needed for anticancer drugs is exponentially distributed with $\lambda = 6$ kilograms per hour. What is the probability that the production process requires more than 15 minutes to produce the next kilogram of drugs?

8.75 The manager of a gas station has observed that the times required by drivers to fill their car's tank and pay are quite variable. In fact, the times are exponentially distributed with a mean of 7.5 minutes. What is the probability that a car can complete the transaction in less than 5 minutes?

8.76 X is an exponential random variable with $\lambda = .3$. Find the following probabilities.

a. $P(X > 2)$
b. $P(X < 4)$
c. $P(1 < X < 2)$
d. $P(X = 3)$

8.77 Toll booths on the New York Thruway are often congested because of the large number of cars waiting to pay. A consultant working for the state concluded that if service times are measured from the time a car stops in line until it leaves, service times are exponentially distributed with a mean of 2.7 minutes. What proportion of cars can get through the toll booth in less than 3 minutes?

8.78 Let X be an exponential random variable with $\lambda = .5$. Find the following probabilities.

a. $P(X > 1)$
b. $P(X > .4)$
c. $P(X < .5)$
d. $P(X < 2)$

8.79 A bank wishing to increase its customer base advertises that it has the fastest service and that virtually all of its customers are served in less than 10 minutes. A management scientist has studied the service times and concluded that service times are exponentially distributed with a mean of 5 minutes. Determine what the bank means when it claims "virtually all" its customers are served in under 10 minutes.

8.80 X is an exponential random variable with $\lambda = .25$. Sketch the graph of the distribution of X by plotting and connecting the points representing $f(x)$ for $x = 0, 2, 4, 6, 8, 10, 15, 20$.

8.81 When trucks arrive at the Ambassador Bridge, each truck must be checked by customs agents. The times are exponentially distributed with a service rate of 10 per hour. What is the probability that a truck requires more than 15 minutes to be checked?

8.82 The random variable X is exponentially distributed with $\lambda = 3$. Sketch the graph of the distribution of X by plotting and connecting the points representing $f(x)$ for $x = 0, .5, 1, 1.5$, and 2.

8.4 Other Continuous Distributions

In this section we introduce three more continuous distributions which are used extensively in statistical inference.

Student *t* Distribution

The Student t distribution was first derived by William S. Gosset in 1908. (Gosset published his findings under the pseudonym "Student" and used the letter t to represent the random variable, hence the **Student *t* distribution**—also called the Student's t distribution.) It is very commonly used in statistical inference, and we will introduce its applications in Chapters 12, 13, 14, 16, 17, and 18.

Student *t* Density Function

The density function of the Student t distribution is as follows:

$$f(t) = \frac{\Gamma[(\nu + 1)/2]}{\sqrt{\nu\pi}\,\Gamma(\nu/2)}\left[1 + \frac{t^2}{\nu}\right]^{-(\nu+1)/2}$$

where ν (Greek letter *nu*) is the parameter of the Student t distribution called the **degrees of freedom,** $\pi = 3.14159$ (approximately), and Γ is the gamma function whose definition is not needed here.

The mean and variance of a Student t random variable are

$$E(t) = 0$$

and

$$V(t) = \frac{\nu}{\nu - 2} \quad \text{for } \nu > 2$$

Figure 8.24 depicts the Student t distribution. As you can see, this distribution is similar to the standard normal distribution. Both are symmetrical about 0. (Both random variables have a mean of 0.) We describe the Student t distribution as mound shaped, whereas the normal distribution is bell shaped.

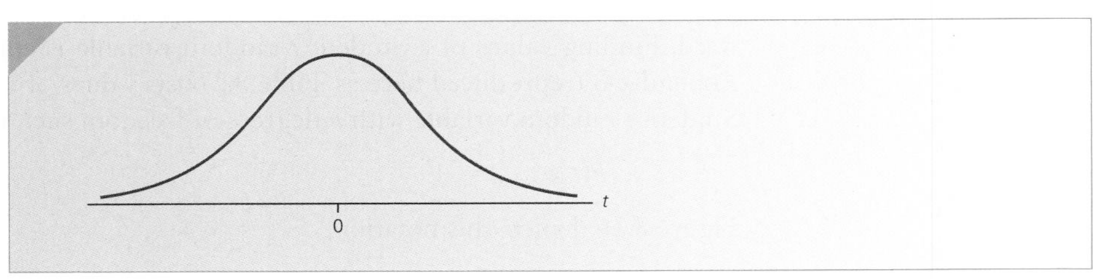

FIGURE **8.24** Student *t* Distribution

Figure 8.25 shows both a Student t and the standard normal distributions. The former is more widely spread out than the latter. [The variance of a standard normal random variable is 1, whereas the variance of a Student t random variable is $\nu/(\nu - 2)$, which is greater than 1 for all ν.]

FIGURE **8.25**
Student *t* and Normal Distributions

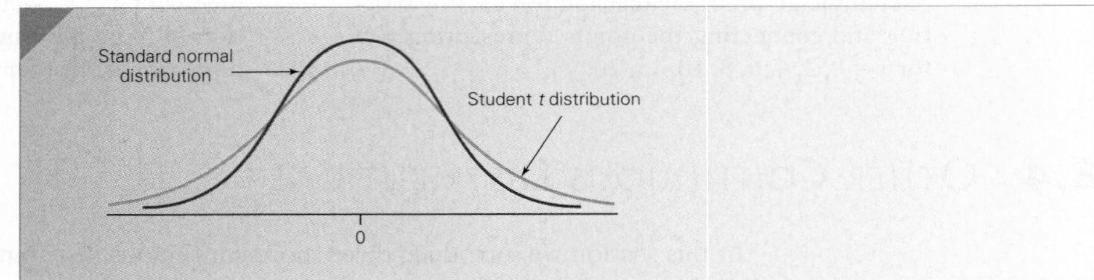

Figure 8.26 depicts Student t distributions with several different degrees of freedom. Notice that for larger degrees of freedom the Student t distribution's dispersion is smaller. For example, when $\nu = 10$, $V(t) = 1.25$; when $\nu = 50$, $V(t) = 1.042$; and when $\nu = 200$, $V(t) = 1.010$. As ν grows larger the Student t distribution approaches the standard normal distribution.

FIGURE **8.26**
Student *t* Distribution with t_A

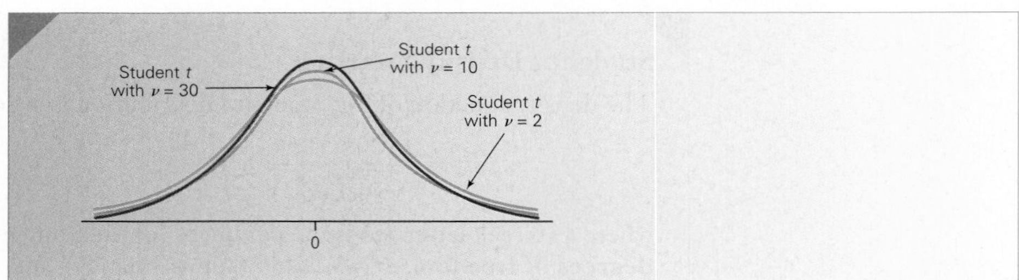

Student *t* Probabilities

For each value of ν (the number of degrees of freedom), there is a different Student t distribution. If we wanted to calculate probabilities of the Student t random variable manually as we did for the normal random variable, we would need a different table for each ν, which is not practical. Alternatively, we can use Microsoft Excel or Minitab. The instructions are given later in this section.

Determining Student *t* Values

As you will discover later in this book, the Student t distribution is used extensively in statistical inference. And for inferential methods we often need to find values of the random variable. To determine values of a normal random variable we used Table 3 backward. Finding values of a Student t random variable is considerably easier. Table 4 in Appendix B (reproduced here as Table 8.2) lists values of $t_{A,\nu}$, which are the values of a Student t random variable with ν degrees of freedom such that

$$P(t > t_{A,\nu}) = A$$

Figure 8.26 depicts this notation.

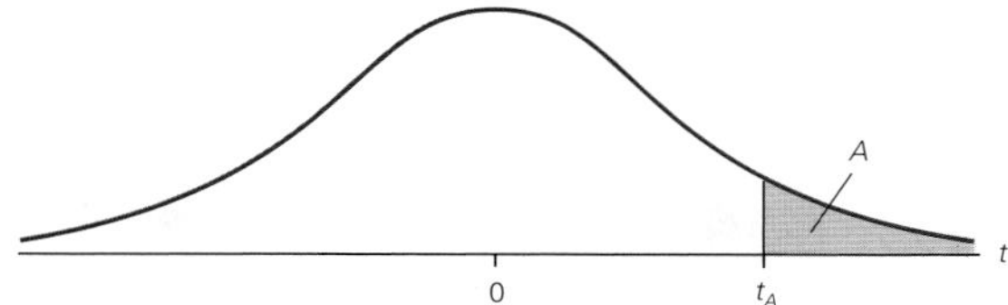

TABLE **8.2** Critical Values of t

ν	$t_{.100}$	$t_{.050}$	$t_{.025}$	$t_{.010}$	$t_{.005}$	ν	$t_{.100}$	$t_{.050}$	$t_{.025}$	$t_{.010}$	$t_{.005}$
1	3.078	6.314	12.71	31.82	63.66	29	1.311	1.699	2.045	2.462	2.756
2	1.886	2.920	4.303	6.965	9.925	30	1.310	1.697	2.042	2.457	2.750
3	1.638	2.353	3.182	4.541	5.841	35	1.306	1.690	2.030	2.438	2.724
4	1.533	2.132	2.776	3.747	4.604	40	1.303	1.684	2.021	2.423	2.704
5	1.476	2.015	2.571	3.365	4.032	45	1.301	1.679	2.014	2.412	2.690
6	1.440	1.943	2.447	3.143	3.707	50	1.299	1.676	2.009	2.403	2.678
7	1.415	1.895	2.365	2.998	3.499	55	1.297	1.673	2.004	2.396	2.668
8	1.397	1.860	2.306	2.896	3.355	60	1.296	1.671	2.000	2.390	2.660
9	1.383	1.833	2.262	2.821	3.250	65	1.295	1.669	1.997	2.385	2.654
10	1.372	1.812	2.228	2.764	3.169	70	1.294	1.667	1.994	2.381	2.648
11	1.363	1.796	2.201	2.718	3.106	75	1.293	1.665	1.992	2.377	2.643
12	1.356	1.782	2.179	2.681	3.055	80	1.292	1.664	1.990	2.374	2.639
13	1.350	1.771	2.160	2.650	3.012	85	1.292	1.663	1.988	2.371	2.635
14	1.345	1.761	2.145	2.624	2.977	90	1.291	1.662	1.987	2.368	2.632
15	1.341	1.753	2.131	2.602	2.947	95	1.291	1.661	1.985	2.366	2.629
16	1.337	1.746	2.120	2.583	2.921	100	1.290	1.660	1.984	2.364	2.626
17	1.333	1.740	2.110	2.567	2.898	110	1.289	1.659	1.982	2.361	2.621
18	1.330	1.734	2.101	2.552	2.878	120	1.289	1.658	1.980	2.358	2.617
19	1.328	1.729	2.093	2.539	2.861	130	1.288	1.657	1.978	2.355	2.614
20	1.325	1.725	2.086	2.528	2.845	140	1.288	1.656	1.977	2.353	2.611
21	1.323	1.721	2.080	2.518	2.831	150	1.287	1.655	1.976	2.351	2.609
22	1.321	1.717	2.074	2.508	2.819	160	1.287	1.654	1.975	2.350	2.607
23	1.319	1.714	2.069	2.500	2.807	170	1.287	1.654	1.974	2.348	2.605
24	1.318	1.711	2.064	2.492	2.797	180	1.286	1.653	1.973	2.347	2.603
25	1.316	1.708	2.060	2.485	2.787	190	1.286	1.653	1.973	2.346	2.602
26	1.315	1.706	2.056	2.479	2.779	200	1.286	1.653	1.972	2.345	2.601
27	1.314	1.703	2.052	2.473	2.771	∞	1.282	1.645	1.960	2.326	2.576
28	1.313	1.701	2.048	2.467	2.763						

Observe that $t_{A,\nu}$ is provided for degrees of freedom ranging from 1 to 200 and ∞. To read this table, simply identify the degrees of freedom and find that value or the closest number to it if it is not listed. Then locate the column representing the t_A value you wish. For example, if we want the value of t with 10 degrees of freedom such that the area under the Student t curve is .05, we locate 10 in the first column and move across this row until we locate the number under the heading $t_{.05}$. From Table 8.3, we find

$$t_{.05,10} = 1.812$$

If the number of degrees of freedom is not shown, find its closest value. For example, suppose we wanted to find $t_{.025,32}$. Because 32 degrees of freedom is not listed, we find the closest number of degrees of freedom, which is 30 and use $t_{.025,32} = 2.042$ as an approximation.

Because the Student t distribution is symmetric about 0, the value of t such that the area to its *left* is A is $-t_{A,\nu}$. For example, the value of t with 10 degrees of freedom such that the area to its left is .05 is

$$-t_{.05,10} = -1.812$$

TABLE **8.3**
Finding $t_{.05,10}$

Degrees of Freedom	$t_{.10}$	$t_{.05}$	$t_{.025}$	$t_{.01}$	$t_{.005}$
1	3.078	6.314	12.706	31.821	63.657
2	1.886	2.920	4.303	6.965	9.925
3	1.638	2.353	3.182	4.541	5.841
4	1.533	2.132	2.776	3.747	4.604
5	1.476	2.015	2.571	3.365	4.032
6	1.440	1.943	2.447	3.143	3.707
7	1.415	1.895	2.365	2.998	3.499
8	1.397	1.860	2.306	2.896	3.355
9	1.383	1.833	2.262	2.821	3.250
10	1.372	1.812	2.228	2.764	3.169
11	1.363	1.796	2.201	2.718	3.106
12	1.356	1.782	2.179	2.681	3.055

Notice the last row in the Student t table. The number of degrees of freedom is infinite and the t values are identical (except for the number of decimal places) to the values of z. For example,

$$t_{.10,\infty} = 1.282$$
$$t_{.05,\infty} = 1.645$$
$$t_{.025,\infty} = 1.960$$
$$t_{.01,\infty} = 2.326$$
$$t_{.005,\infty} = 2.576$$

In the previous section we showed (or showed how we determine) that

$$Z_{.10} = 1.28$$
$$Z_{.05} = 1.645$$
$$Z_{.025} = 1.96$$
$$Z_{.01} = 2.33$$
$$Z_{.005} = 2.575$$

Using the Computer

EXCEL

INSTRUCTIONS

To compute Student t probabilities type

=TDIST([x], [ν], [Tails])

where x must be positive, ν is the number of degrees of freedom, and Tails is 1 or 2. Typing 1 for Tails produces the area to the right of x. Typing 2 for Tails produces the area to the right of x plus the area to the left of $-x$. For example,

=TDIST(2,50,1) = .02547

and

=TDIST(2,50,2) = .05095

To determine t_A, type

=TINV([2A], [ν])

For example, to find $t_{.05,200}$, enter

=TINV(.10,200)

yielding 1.6525.

MINITAB

INSTRUCTIONS

Click **Calc, Probability Distributions,** and t . . . and type the **Degrees of freedom.**

SEEING STATISTICS

Applet 6: Student *t* Distribution

The Student *t* distribution applet allows you to see for yourself the shape of the distribution, how the degrees of freedom change the shape, and its resemblance to the standard normal curve. The first graph shows the comparison of the normal distribution (red curve) to Student *t* distribution (blue curve). Use the right slider to change the degrees of freedom for the *t* distribution. Use the text boxes to change either the value of *t* or the two-tail probability. Remember to press the **Return** key in the text box to record the change.

The second graph is the same as the one above except the comparison to the normal distribution has been removed. This graph is a little easier to use to find critical values of *t* or to find the probability of specific values of *t*.

Applet Exercises

The following exercises refer to Graph 1.

6.1 Set the degrees of freedom equal to 2. For values (on the horizontal axis) near 0, which curve is higher? The higher curve is more likely to have observations in that region.

6.2 Again for $df = 2$, for values around either +4 or −4, which curve is higher? In other words, which distribution is more likely to have extreme values—the normal (red) or Student *t* (blue) distribution?

The following exercises refer to Graph 2.

6.3 As you use the scrollbar to increase (slowly) the degrees of freedom, what happens to the value of $t_{.025}$ and $-t_{.025}$?

6.4 When the degrees of freedom = 100, is there still a small difference between the critical values of $t_{.025}$ and $z_{.025}$? How large do you think the degrees of freedom would have to be before the two sets of critical values were identical?

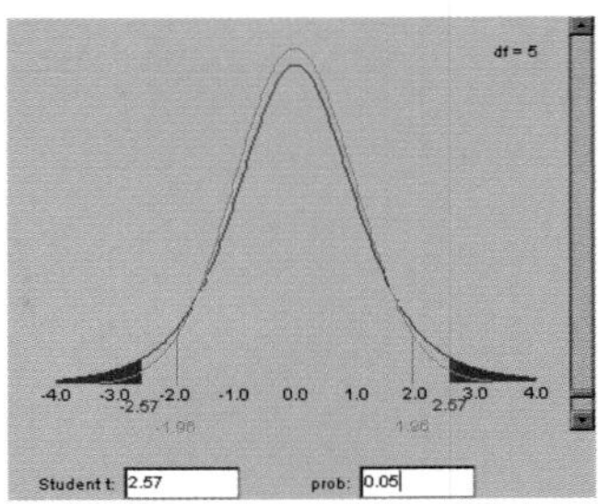

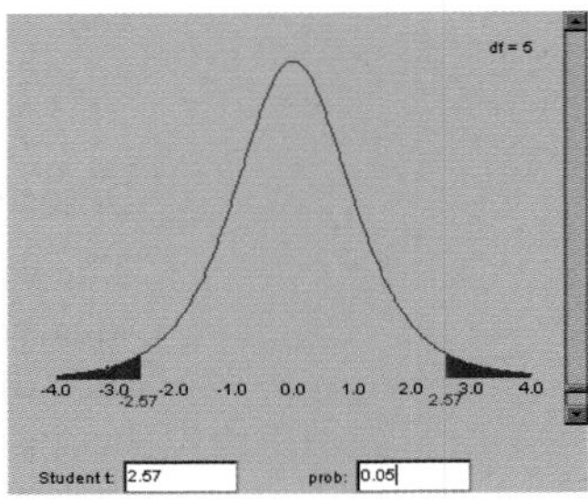

Chi-Squared Distribution

The density function of another very useful random variable is exhibited next.

Chi-Squared Density Function

The chi-squared density function is

$$f(\chi^2) = \frac{1}{\Gamma(\nu/2)} \frac{1}{2^{\nu/2}} (\chi^2)^{(\nu/2)-1} e^{-\chi^2/2} \qquad \chi^2 > 0$$

The parameter ν is the number of degrees of freedom, which, like the degrees of freedom of the Student *t* distribution, affects the shape.

Figure 8.27 depicts a **chi-squared distribution.** As you can see, it is positively skewed ranging between 0 and ∞. Like that of the Student *t* distribution, its shape depends on its number of degrees of freedom. The effect of increasing the degrees of freedom is seen in Figure 8.28.

FIGURE **8.27**

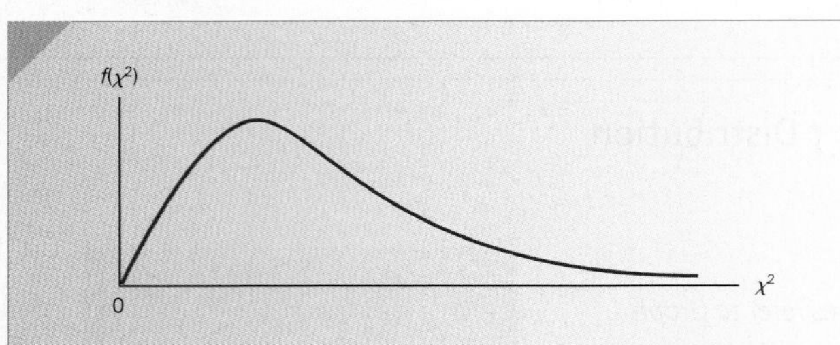

FIGURE **8.28**

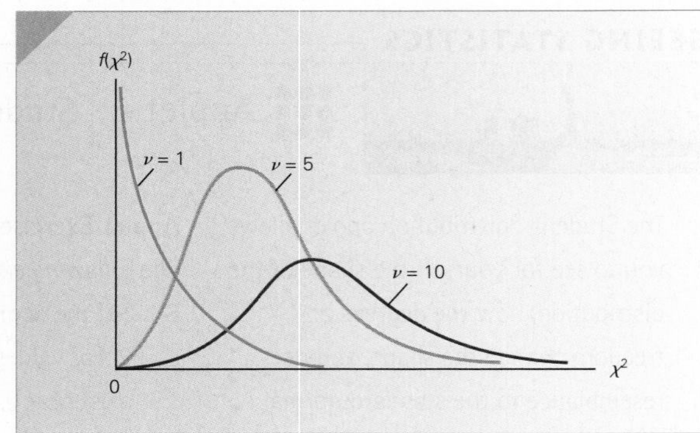

The mean and variance of a chi-squared random variable are

$$E(\chi^2) = \nu$$

and

$$V(\chi^2) = 2\nu$$

Determining Chi-Squared Values

The value of χ^2 with ν degrees of freedom such that the area to its right under the chi-squared curve is equal to A is denoted $\chi^2_{A,\nu}$. We cannot use $-\chi^2_{A,\nu}$ to represent the point such that the area to its *left* is A (as we did with the standard normal and Student *t* values) because χ^2 is always greater than 0. To represent left-tail critical values, we note that if the area to the left of a point is A, the area to its right must be $1 - A$ because the entire area under the chi-squared curve (as well as all continuous distributions) must equal 1. Thus $\chi^2_{1-A,\nu}$ denotes the point such that the area to its left is A. See Figure 8.29.

FIGURE **8.29**
χ^2_A and χ^2_{1-A}

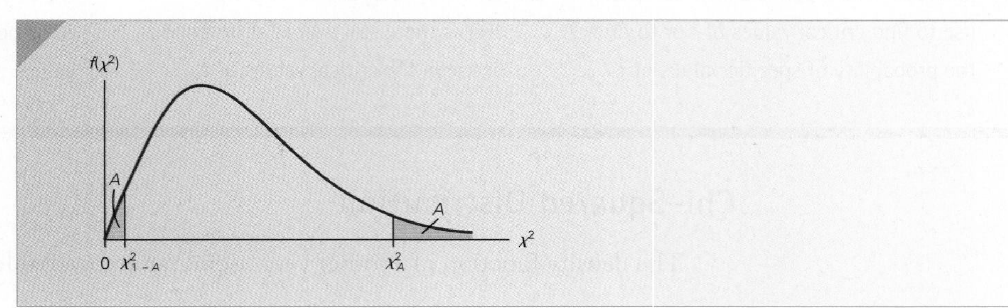

Table 5 in Appendix B (reproduced here as Table 8.4) lists critical values of the chi-squared distribution for degrees of freedom equal to 1 to 30, 40, 50, 60, 70, 80, 90, and 100. For example, to find the point in a chi-squared distribution with 8 degrees of freedom such that the area to its right is .05, locate 8 degrees of freedom in the left column and $\chi^2_{.050}$ across the top. The intersection of the row and column contains the number we seek, as shown in Table 8.5. That is,

$$\chi^2_{.050,8} = 15.5$$

To find the point in the same distribution such that the area to its *left* is .05, find the point such that the area to its *right* is .95. Locate $\chi^2_{.950}$ across the top row and 8 degrees of freedom down the left column (also shown in Table 8.5). You should see that

$$\chi^2_{.950,8} = 2.73$$

TABLE **8.4** Critical Values of χ^2

ν	$\chi^2_{.995}$	$\chi^2_{.990}$	$\chi^2_{.975}$	$\chi^2_{.950}$	$\chi^2_{.900}$	$\chi^2_{.100}$	$\chi^2_{.050}$	$\chi^2_{.025}$	$\chi^2_{.010}$	$\chi^2_{.005}$
1	0.000039	0.000157	0.000982	0.00393	0.0158	2.71	3.84	5.02	6.63	7.88
2	0.0100	0.0201	0.0506	0.103	0.211	4.61	5.99	7.38	9.21	10.6
3	0.072	0.115	0.216	0.352	0.584	6.25	7.81	9.35	11.3	12.8
4	0.207	0.297	0.484	0.711	1.06	7.78	9.49	11.1	13.3	14.9
5	0.412	0.554	0.831	1.15	1.61	9.24	11.1	12.8	15.1	16.7
6	0.676	0.872	1.24	1.64	2.20	10.6	12.6	14.4	16.8	18.5
7	0.989	1.24	1.69	2.17	2.83	12.0	14.1	16.0	18.5	20.3
8	1.34	1.65	2.18	2.73	3.49	13.4	15.5	17.5	20.1	22.0
9	1.73	2.09	2.70	3.33	4.17	14.7	16.9	19.0	21.7	23.6
10	2.16	2.56	3.25	3.94	4.87	16.0	18.3	20.5	23.2	25.2
11	2.60	3.05	3.82	4.57	5.58	17.3	19.7	21.9	24.7	26.8
12	3.07	3.57	4.40	5.23	6.30	18.5	21.0	23.3	26.2	28.3
13	3.57	4.11	5.01	5.89	7.04	19.8	22.4	24.7	27.7	29.8
14	4.07	4.66	5.63	6.57	7.79	21.1	23.7	26.1	29.1	31.3
15	4.60	5.23	6.26	7.26	8.55	22.3	25.0	27.5	30.6	32.8
16	5.14	5.81	6.91	7.96	9.31	23.5	26.3	28.8	32.0	34.3
17	5.70	6.41	7.56	8.67	10.09	24.8	27.6	30.2	33.4	35.7
18	6.26	7.01	8.23	9.39	10.86	26.0	28.9	31.5	34.8	37.2
19	6.84	7.63	8.91	10.12	11.65	27.2	30.1	32.9	36.2	38.6
20	7.43	8.26	9.59	10.85	12.44	28.4	31.4	34.2	37.6	40.0
21	8.03	8.90	10.28	11.59	13.24	29.6	32.7	35.5	38.9	41.4
22	8.64	9.54	10.98	12.34	14.04	30.8	33.9	36.8	40.3	42.8
23	9.26	10.20	11.69	13.09	14.85	32.0	35.2	38.1	41.6	44.2
24	9.89	10.86	12.40	13.85	15.66	33.2	36.4	39.4	43.0	45.6
25	10.52	11.52	13.12	14.61	16.47	34.4	37.7	40.6	44.3	46.9
26	11.16	12.20	13.84	15.38	17.29	35.6	38.9	41.9	45.6	48.3
27	11.81	12.88	14.57	16.15	18.11	36.7	40.1	43.2	47.0	49.6
28	12.46	13.56	15.31	16.93	18.94	37.9	41.3	44.5	48.3	51.0
29	13.12	14.26	16.05	17.71	19.77	39.1	42.6	45.7	49.6	52.3
30	13.79	14.95	16.79	18.49	20.60	40.3	43.8	47.0	50.9	53.7
40	20.71	22.16	24.43	26.51	29.05	51.8	55.8	59.3	63.7	66.8
50	27.99	29.71	32.36	34.76	37.69	63.2	67.5	71.4	76.2	79.5
60	35.53	37.48	40.48	43.19	46.46	74.4	79.1	83.3	88.4	92.0
70	43.28	45.44	48.76	51.74	55.33	85.5	90.5	95.0	100	104
80	51.17	53.54	57.15	60.39	64.28	96.6	102	107	112	116
90	59.20	61.75	65.65	69.13	73.29	108	113	118	124	128
100	67.33	70.06	74.22	77.93	82.36	118	124	130	136	140

TABLE **8.5** Finding $\chi^2_{.05,8}$ and $\chi^2_{.950,8}$

Degrees of Freedom	$\chi^2_{.995}$	$\chi^2_{.990}$	$\chi^2_{.975}$	$\chi^2_{.950}$	$\chi^2_{.900}$	$\chi^2_{.100}$	$\chi^2_{.050}$	$\chi^2_{.025}$	$\chi^2_{.010}$	$\chi^2_{.005}$
1	0.000039	0.000157	0.000982	0.00393	0.0158	2.71	3.84	5.02	6.63	7.88
2	0.0100	0.0201	0.0506	0.103	0.211	4.61	5.99	7.38	9.21	10.6
3	0.072	0.115	0.216	0.352	0.584	6.25	7.81	9.35	11.3	12.8
4	0.207	0.297	0.484	0.711	1.06	7.78	9.49	11.1	13.3	14.9
5	0.412	0.554	0.831	1.15	1.61	9.24	11.1	12.8	15.1	16.7
6	0.676	0.872	1.24	1.64	2.20	10.6	12.6	14.4	16.8	18.5
7	0.989	1.24	1.69	2.17	2.83	12.0	14.1	16.0	18.5	20.3
8	1.34	1.65	2.18	2.73	3.49	13.4	15.5	17.5	20.1	22.0
9	1.73	2.09	2.70	3.33	4.17	14.7	16.9	19.0	21.7	23.6
10	2.16	2.56	3.25	3.94	4.87	16.0	18.3	20.5	23.2	25.2
11	2.60	3.05	3.82	4.57	5.58	17.3	19.7	21.9	24.7	26.8

For values of degrees of freedom greater than 100, the chi-squared distribution can be approximated by a normal distribution with $\mu = \nu$ and $\sigma = \sqrt{2\nu}$.

Using the Computer

EXCEL

INSTRUCTIONS

To calculate $P(\chi^2 > x)$, type into any cell

=CHIDIST([x], [ν])

For example, **CHIDIST**(6.25139,3) = .1000.

To determine $\chi_{A,\nu}$, type

=CHIINV([A], [ν])

For example, **CHIINV**(.10,3) = 6.25139.

MINITAB

INSTRUCTIONS

Click **Calc, Probability Distributions**, and **Chi-square** . . . . Specify the **Degrees of freedom.**

SEEING STATISTICS

Applet 7: Chi-Squared Distribution

Like the Student *t* applet, this applet allows you to see how the degrees of freedom affect the shape of the chi-squared distribution. Additionally, you can use the applet to determine probabilities and values of the chi-squared random variable.

Use the right slider to change the degrees of freedom. Use the text boxes to change either the value of ChiSq or the probability. Remember to press the **Return** key in the text box to record the change.

Applet Exercises

7.1 What happens to the shape of the chi-squared distribution as the degrees of freedom increase?

7.2 Describe what happens to $\chi^2_{.05}$ when the degrees of freedom increase.

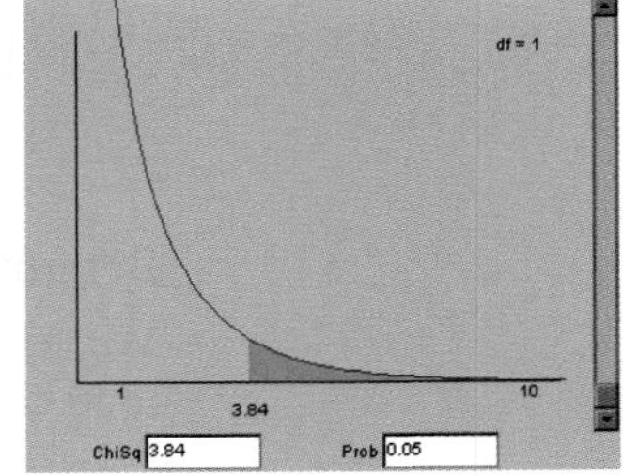

7.3 Describe what happens to $\chi^2_{.95}$ when the degrees of freedom increase.

F Distribution

The density function of the ***F* distribution** is given in the box.

F Density Function

$$f(F) = \frac{\Gamma\left(\frac{\nu_1 + \nu_2}{2}\right)}{\Gamma\left(\frac{\nu_1}{2}\right)\Gamma\left(\frac{\nu_2}{2}\right)} \left(\frac{\nu_1}{\nu_2}\right)^{\frac{\nu_1}{2}} \frac{F^{\frac{\nu_1 - 2}{2}}}{\left(1 + \frac{\nu_1 F}{\nu_2}\right)^{\frac{\nu_1 + \nu_2}{2}}} \qquad F > 0$$

where F ranges from 0 to ∞ and ν_1 and ν_2 are the parameters of the distribution called degrees of freedom. For reasons that are clearer in Chapter 13, we call ν_1 the *numerator degrees of freedom* and ν_2 the *denominator degrees of freedom.*

The mean and variance of an F random variable are

$$E(F) = \frac{\nu_2}{\nu_2 - 2} \qquad \nu_2 > 2$$

and

$$V(F) = \frac{2\nu_2^2(\nu_1 + \nu_2 - 2)}{\nu_1(\nu_2 - 2)^2(\nu_2 - 4)} \qquad \nu_2 > 4$$

Notice that the mean depends only on the denominator degrees of freedom and that for large ν_2, the mean of the F distribution is approximately 1. Figure 8.30 describes the density function when it is graphed. As you can see, the F distribution is positively skewed. Its actual shape depends on the two numbers of degrees of freedom.

FIGURE **8.30**
F Distribution

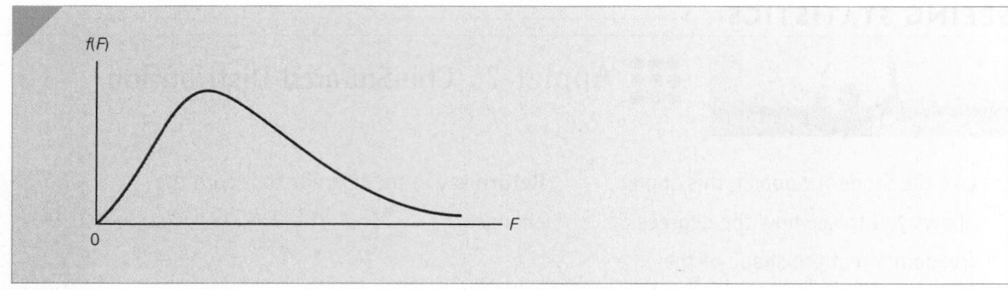

Determining Values of *F*

We define F_{A,ν_1,ν_2} as the value of F with ν_1 and ν_2 degrees of freedom such that the area to its right under the curve is A. That is,

$$P(F > F_{A,\nu_1,\nu_2}) = A$$

Because the F random variable like the chi-squared can equal only positive values, we define F_{1-A,ν_1,ν_2} as the value such that the area to its left is A. Figure 8.31 depicts this notation. Table 6 in Appendix B provides values of F_{A,ν_1,ν_2} for $A = .05, .025, .01,$ and $.005$. Part of Table 6 is reproduced here as Table 8.6.

FIGURE **8.31**
F Distribution

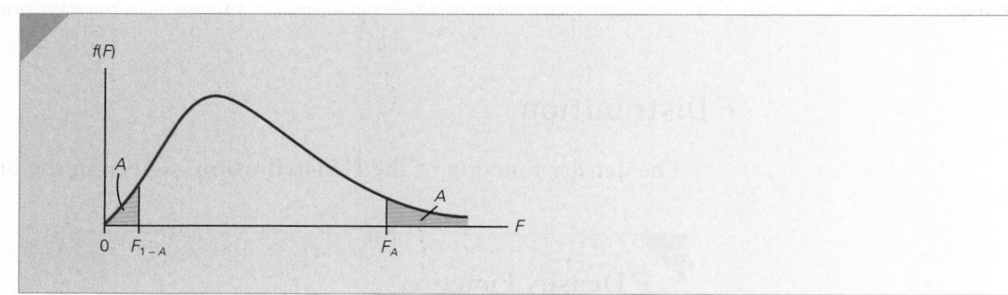

TABLE **8.6** Critical Values of F_A for $A = .05$

	ν_2									
ν_1	1	2	3	4	5	6	7	8	9	10
1	161	199	216	225	230	234	237	239	241	242
2	18.5	19.0	19.2	19.2	19.3	19.3	19.4	19.4	19.4	19.4
3	10.1	9.55	9.28	9.12	9.01	8.94	8.89	8.85	8.81	8.79
4	7.71	6.94	6.59	6.39	6.26	6.16	6.09	6.04	6.00	5.96
5	6.61	5.79	5.41	5.19	5.05	4.95	4.88	4.82	4.77	4.74
6	5.99	5.14	4.76	4.53	4.39	4.28	4.21	4.15	4.10	4.06
7	5.59	4.74	4.35	4.12	3.97	3.87	3.79	3.73	3.68	3.64
8	5.32	4.46	4.07	3.84	3.69	3.58	3.50	3.44	3.39	3.35
9	5.12	4.26	3.86	3.63	3.48	3.37	3.29	3.23	3.18	3.14
10	4.96	4.10	3.71	3.48	3.33	3.22	3.14	3.07	3.02	2.98
11	4.84	3.98	3.59	3.36	3.20	3.09	3.01	2.95	2.90	2.85
12	4.75	3.89	3.49	3.26	3.11	3.00	2.91	2.85	2.80	2.75
13	4.67	3.81	3.41	3.18	3.03	2.92	2.83	2.77	2.71	2.67
14	4.60	3.74	3.34	3.11	2.96	2.85	2.76	2.70	2.65	2.60
15	4.54	3.68	3.29	3.06	2.90	2.79	2.71	2.64	2.59	2.54
16	4.49	3.63	3.24	3.01	2.85	2.74	2.66	2.59	2.54	2.49
17	4.45	3.59	3.20	2.96	2.81	2.70	2.61	2.55	2.49	2.45
18	4.41	3.55	3.16	2.93	2.77	2.66	2.58	2.51	2.46	2.41
19	4.38	3.52	3.13	2.90	2.74	2.63	2.54	2.48	2.42	2.38

TABLE **8.6** *(Continued)*

ν_1	ν_2 1	2	3	4	5	6	7	8	9	10
20	4.35	3.49	3.10	2.87	2.71	2.60	2.51	2.45	2.39	2.35
22	4.30	3.44	3.05	2.82	2.66	2.55	2.46	2.40	2.34	2.30
24	4.26	3.40	3.01	2.78	2.62	2.51	2.42	2.36	2.30	2.25
26	4.23	3.37	2.98	2.74	2.59	2.47	2.39	2.32	2.27	2.22
28	4.20	3.34	2.95	2.71	2.56	2.45	2.36	2.29	2.24	2.19
30	4.17	3.32	2.92	2.69	2.53	2.42	2.33	2.27	2.21	2.16
35	4.12	3.27	2.87	2.64	2.49	2.37	2.29	2.22	2.16	2.11
40	4.08	3.23	2.84	2.61	2.45	2.34	2.25	2.18	2.12	2.08
45	4.06	3.20	2.81	2.58	2.42	2.31	2.22	2.15	2.10	2.05
50	4.03	3.18	2.79	2.56	2.40	2.29	2.20	2.13	2.07	2.03
60	4.00	3.15	2.76	2.53	2.37	2.25	2.17	2.10	2.04	1.99
70	3.98	3.13	2.74	2.50	2.35	2.23	2.14	2.07	2.02	1.97
80	3.96	3.11	2.72	2.49	2.33	2.21	2.13	2.06	2.00	1.95
90	3.95	3.10	2.71	2.47	2.32	2.20	2.11	2.04	1.99	1.94
100	3.94	3.09	2.70	2.46	2.31	2.19	2.10	2.03	1.97	1.93
120	3.92	3.07	2.68	2.45	2.29	2.18	2.09	2.02	1.96	1.91
140	3.91	3.06	2.67	2.44	2.28	2.16	2.08	2.01	1.95	1.90
160	3.90	3.05	2.66	2.43	2.27	2.16	2.07	2.00	1.94	1.89
180	3.89	3.05	2.65	2.42	2.26	2.15	2.06	1.99	1.93	1.88
200	3.89	3.04	2.65	2.42	2.26	2.14	2.06	1.98	1.93	1.88
∞	3.84	3.00	2.61	2.37	2.21	2.10	2.01	1.94	1.88	1.83

Values of F_{1-A,ν_1,ν_2} are unavailable. However, we do not need them because we can determine F_{1-A,ν_1,ν_2} from F_{A,ν_1,ν_2}. That is, statisticians can show that

$$F_{1-A,\nu_1,\nu_2} = \frac{1}{F_{A,\nu_2,\nu_1}}$$

To determine any critical value, find the numerator degrees of freedom ν_1 across the top of Table 6 and the denominator degrees of freedom ν_2 down the left column. The intersection of the row and column contains the number we seek. To illustrate, suppose that we want to find $F_{.05,5,7}$. Table 8.7 shows how this point is found. Locate the numerator degrees of freedom, 5, across the top and the denominator degrees of freedom, 7, down the left column. The intersection is 3.97. Thus, $F_{.05,5,7} = 3.97$.

TABLE **8.7**
Finding $F_{.05,5,7}$

Denominator Degrees of Freedom

ν_2 \ ν_1	Numerator Degrees of Freedom 1	2	3	4	5	6	7	8	9
1	161	199	216	225	230	234	237	239	241
2	18.5	19.0	19.2	19.2	19.3	19.3	19.4	19.4	19.4
3	10.1	9.55	9.28	9.12	9.01	8.94	8.89	8.85	8.81
4	7.71	6.94	6.59	6.39	6.26	6.16	6.09	6.04	6.00
5	6.61	5.79	5.41	5.19	5.05	4.95	4.88	4.82	4.77
6	5.99	5.14	4.76	4.53	4.39	4.28	4.21	4.15	4.10
7	5.59	4.74	4.35	4.12	3.97	3.87	3.79	3.73	3.68
8	5.32	4.46	4.07	3.84	3.69	3.58	3.50	3.44	3.39
9	5.12	4.26	3.86	3.63	3.48	3.37	3.29	3.23	3.18
10	4.96	4.10	3.71	3.48	3.33	3.22	3.14	3.07	3.02

Note that the order in which the degrees of freedom appear is important. To find $F_{.05,7,5}$ (numerator degrees of freedom = 7 and denominator degrees of freedom = 5), we locate 7 across the top and 5 down the side. The intersection is $F_{.05,7,5} = 4.88$.

Suppose that we want to determine the point in an F distribution with $\nu_1 = 4$ and $\nu_2 = 8$ such that the area to its right is .95. Thus,

$$F_{.95,4,8} = \frac{1}{F_{.05,8,4}} = \frac{1}{6.04} = .166$$

Using the Computer

EXCEL

INSTRUCTIONS

For probabilities, type

=**FDIST**([X], [ν_1],[ν_2])

For example, **FDIST**(3.97,5,7) = .05.

To determine F_{A,ν_1,ν_2} type

=**FINV**([A], [ν_1],[ν_2])

For example, **FINV**(.05,5,7) = 3.97.

MINITAB

INSTRUCTIONS

Click **Calc, Probability Distributions,** and **F . . . .** Specify the **Numerator degrees of freedom** and the **Denominator degrees of freedom.**

SEEING STATISTICS

Applet 8: *F* Distribution

The graph shows the *F* distribution. Use the left and right sliders to change the numerator and denominator degrees of freedom, respectively. Use the text boxes to change either the value of *F* or the probability. Remember to press the **Return** key in the text box to record the change.

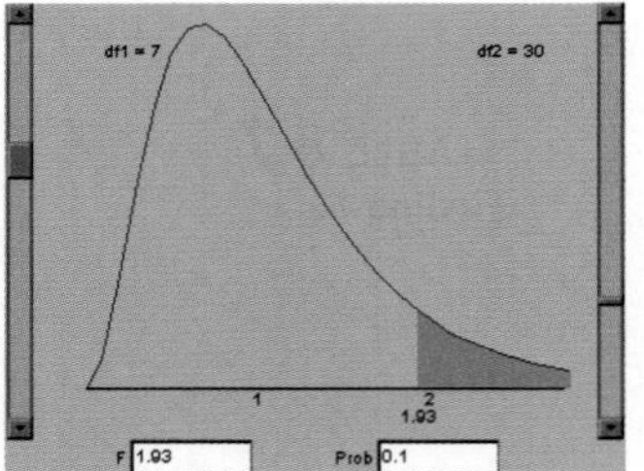

Applet Exercises

8.1 Set the numerator degrees of freedom equal to 1. What happens to the shape of the *F* distribution as the denominator degrees of freedom increase?

8.2 Set the numerator degrees of freedom equal to 10. What happens to the shape of the *F* distribution as the denominator degrees of freedom increase?

8.3 Describe what happens to $F_{.05}$ when either the numerator or the denominator degrees of freedom increase.

8.4 Describe what happens to $F_{.95}$ when either the numerator or the denominator degrees of freedom increase.

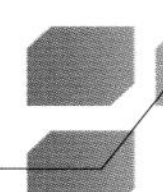

EXERCISES

Some of the following exercises require the use of a computer and software.

8.83 Use the t table (Table 4) to find the following values of t.
a. $t_{.10,15}$ b. $t_{.10,23}$ c. $t_{.025,83}$ d. $t_{.05,195}$

8.84 Use the t table (Table 4) to find the following values of t.
a. $t_{.005,33}$ b. $t_{.10,600}$ c. $t_{.05,4}$ d. $t_{.01,20}$

8.85 Use a computer to find the following values of t.
a. $t_{.10,15}$ b. $t_{.10,23}$ c. $t_{.025,83}$ d. $t_{.05,195}$

8.86 Use a computer to find the following values of t.
a. $t_{.05,143}$ b. $t_{.01,12}$ c. $t_{.025,\infty}$ d. $t_{.05,100}$

8.87 Use a computer to find the following probabilities.
a. $P(t_{64} > 2.12)$ b. $P(t_{27} > 1.90)$
c. $P(t_{159} > 1.33)$ d. $P(t_{550} > 1.85)$

8.88 Use a computer to find the following probabilities.
a. $P(t_{141} > .94)$ b. $P(t_{421} > 2.00)$
c. $P(t_{1000} > 1.96)$ d. $P(t_{82} > 1.96)$

8.89 Use the χ^2 table (Table 5) to find the following values of χ^2.
a. $\chi^2_{.10,5}$ b. $\chi^2_{.01,100}$ c. $\chi^2_{.95,18}$ d. $\chi^2_{.99,60}$

8.90 Use the χ^2 table (Table 5) to find the following values of χ^2.
a. $\chi^2_{.90,26}$ b. $\chi^2_{.01,30}$ c. $\chi^2_{.10,1}$ d. $\chi^2_{.99,80}$

8.91 Use a computer to find the following values of χ^2.
a. $\chi^2_{.25,66}$ b. $\chi^2_{.40,100}$ c. $\chi^2_{.50,17}$ d. $\chi^2_{.10,17}$

8.92 Use a computer to find the following values of χ^2.
a. $\chi^2_{.99,55}$ b. $\chi^2_{.05,800}$ c. $\chi^2_{.99,43}$ d. $\chi^2_{.10,233}$

8.93 Use a computer to find the following probabilities.
a. $P(\chi^2_{73} > 80)$ b. $P(\chi^2_{200} > 125)$
c. $P(\chi^2_{88} > 60)$ d. $P(\chi^2_{1000} > 450)$

8.94 Use a computer to find the following probabilities.
a. $P(\chi^2_{250} > 250)$ b. $P(\chi^2_{36} > 25)$
c. $P(\chi^2_{600} > 500)$ d. $P(\chi^2_{120} > 100)$

8.95 Use the F table (Table 6) to find the following values of F.
a. $F_{.05,3,7}$ b. $F_{.05,7,3}$ c. $F_{.025,5,20}$ d. $F_{.01,12,60}$

8.96 Use the F table (Table 6) to find the following values of F.
a. $F_{.025,8,22}$ b. $F_{.05,20,30}$
c. $F_{.01,9,18}$ d. $F_{.025,24,10}$

8.97 Use a computer to find the following values of F.
a. $F_{.05,70,70}$ b. $F_{.01,45,100}$
c. $F_{.025,36,50}$ d. $F_{.05,500,500}$

8.98 Use a computer to find the following values of F.
a. $F_{.01,100,150}$ b. $F_{.05,25,125}$
c. $F_{.01,11,33}$ d. $F_{.05,300,800}$

8.99 Use a computer to find the following probabilities.
a. $P(F_{7,20} > 2.5)$ b. $P(F_{18,63} > 1.4)$
c. $P(F_{34,62} > 1.8)$ d. $P(F_{200,400} > 1.1)$

8.100 Use a computer to find the following probabilities.
a. $P(F_{600,800} > 1.1)$ b. $P(F_{35,100} > 1.3)$
c. $P(F_{66,148} > 2.1)$ d. $P(F_{17,37} > 2.8)$

CHAPTER SUMMARY

This chapter dealt with continuous random variables and their distributions. Because a continuous random variable can assume an infinite number of values, the probability that the random variable equals any single value is 0. Consequently, we address the problem of computing the probability of a range of values. We showed that the probability of any interval is the area in the interval under the curve representing the density function.

We introduced the most important distribution in statistics and showed how to compute the probability that a normal random variable falls into any interval. Additionally, we demonstrated how to use the normal table backward to find values of a normal random variable given a probability. Next we introduced the exponential distribution, a distribution that is particularly useful in several management science applications. Finally we presented three more continuous random variables and their probability density functions. The Student t, chi-squared, and F distributions will be used extensively in statistical inference.

IMPORTANT TERMS

SYMBOLS

Symbol	Pronounced	Represents
π	*pi*	3.14159 . . .
Z_A	*Z-sub-A* or *Z-A*	Value of Z such that area to its right is A
ν	*nu*	Degrees of freedom
t_A	*t-sub-A* or *t-A*	Value of t such that area to its right is A
χ^2_A	*chi-squared-sub-A* or *chi-squared-A*	Value of chi-squared such that area to its right is A
F_A	*F-sub-A* or *F-A*	Value of F such that area to its right is A
ν_1	*nu-sub-one* or *nu-one*	Numerator degrees of freedom
ν_2	*nu-sub-two* or *nu-two*	Denominator degrees of freedom

COMPUTER OUTPUT AND INSTRUCTIONS

Probability/Random Variable	Excel	Minitab
Normal probability	271	272
Normal random variable	272	272
Exponential probability	279	279
Exponential random variable	279	279
Student *t* probability	284	285
Student *t* random variable	285	285
Chi-squared probability	288	288
Chi-squared random variable	288	288
F probability	292	292
F random variable	292	292

© Stockbyte/Getty Images

9 SAMPLING DISTRIBUTIONS

Salaries of a Business School's Graduates

Deans and other faculty members in professional schools often monitor how well the graduates of their programs fare in the job market. Information about the types of jobs and their salaries may provide useful information about the success of the program.

In the advertisements for a large university, the dean of the School of Business claims that the average salary of the school's graduates one year after graduation is $800 per week with a standard deviation of $100. A second-year student in the business school who has just completed his statistics course would like to check whether the claim about the mean is correct. He does a survey of 25 people who graduated one year ago and determines their weekly salary. He discovers the sample mean to be $750. To interpret his finding he needs to calculate the probability that a sample of 25 graduates would have a mean of $750 or less when the population mean is $800 and the standard deviation is $100. After calculating the probability, he needs to draw some conclusion.

© PhotoDisc/Getty Images

See page 305 for the answer.

INTRODUCTION

This chapter introduces the *sampling distribution*, a fundamental element in statistical inference. We remind you that statistical inference is the process of converting data into information. Here are the parts of the process we have thus far discussed:

1. Parameters describe populations.
2. Parameters are almost always unknown.
3. We take a random sample of a population to obtain the necessary data.
4. We calculate one or more statistics from the data.

For example, to estimate a population mean, we compute the sample mean. Although there is very little chance that the sample mean and the population mean are identical, we would expect them to be quite close. However, for the purposes of statistical inference, we need to be able to measure how close. The sampling distribution provides this service. It plays a crucial role in the process, because the measure of proximity it provides is the key to statistical inference.

9.1 SAMPLING DISTRIBUTION OF THE MEAN

A **sampling distribution** is created by, as the name suggests, sampling. There are two ways to create a sampling distribution. The first is to actually draw samples of the same size from a population, calculate the statistic of interest, and then use descriptive techniques to learn more about the sampling distribution. The second method relies on the rules of probability and the laws of expected value and variance to derive the sampling distribution. We'll demonstrate the latter approach by developing the sampling distribution of the mean of two dice.

Sampling Distribution of the Mean of Two Dice

The population is created by throwing a fair die infinitely many times, with the random variable X indicating the number of spots showing on any one throw. The probability distribution of the random variable X is as follows:

x	1	2	3	4	5	6
$p(x)$	1/6	1/6	1/6	1/6	1/6	1/6

The population is infinitely large, because we can throw the die infinitely many times (or at least imagine doing so). From the definitions of expected value and variance presented in Section 7.1, we calculate the population mean, variance, and standard deviation.

Population mean:

$$\begin{aligned}\mu &= \sum xP(x) \\ &= 1(1/6) + 2(1/6) + 3(1/6) + 4(1/6) + 5(1/6) + 6(1/6) \\ &= 3.5\end{aligned}$$

Population variance:

$$\sigma^2 = \sum(x - \mu)^2 P(x)$$
$$= (1 - 3.5)^2(1/6) + (2 - 3.5)^2(1/6) + (3 - 3.5)^2(1/6) + (4 - 3.5)^2(1/6)$$
$$+ (5 - 3.5)^2(1/6) + (6 - 3.5)^2(1/6)$$
$$= 2.92$$

Population standard deviation:

$$\sigma = \sqrt{\sigma^2} = \sqrt{2.92} = 1.71$$

The sampling distribution is created by drawing samples of size 2 from the population. In other words, we toss two dice. Figure 9.1 depicts this process where we compute the mean for each sample. Because the value of the sample mean varies randomly from sample to sample, we can regard $\overline{X}$ as a new random variable created by sampling. Table 9.1 lists all the possible samples and their corresponding values of $\overline{x}$.

FIGURE **9.1** Drawing Samples of Size 2 from a Population

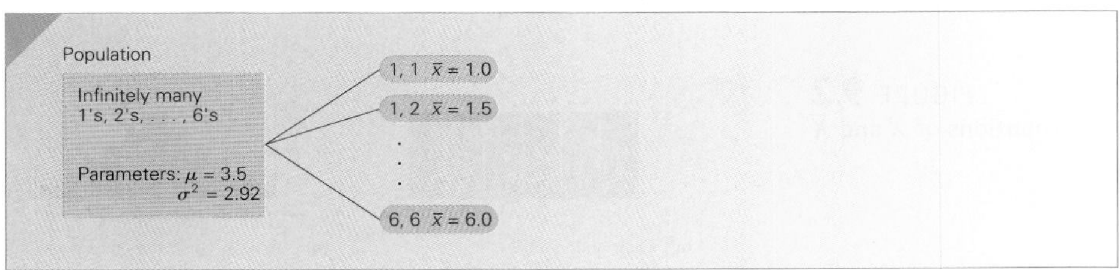

TABLE **9.1** All Samples of Size 2 and Their Means

Sample	$\overline{x}$	Sample	$\overline{x}$	Sample	$\overline{x}$
1, 1	1.0	3, 1	2.0	5, 1	3.0
1, 2	1.5	3, 2	2.5	5, 2	3.5
1, 3	2.0	3, 3	3.0	5, 3	4.0
1, 4	2.5	3, 4	3.5	5, 4	4.5
1, 5	3.0	3, 5	4.0	5, 5	5.0
1, 6	3.5	3, 6	4.5	5, 6	5.5
2, 1	1.5	4, 1	2.5	6, 1	3.5
2, 2	2.0	4, 2	3.0	6, 2	4.0
2, 3	2.5	4, 3	3.5	6, 3	4.5
2, 4	3.0	4, 4	4.0	6, 4	5.0
2, 5	3.5	4, 5	4.5	6, 5	5.5
2, 6	4.0	4, 6	5.0	6, 6	6.0

There are 36 different possible samples of size 2; because each sample is equally likely, the probability of any one sample being selected is 1/36. However, $\overline{x}$ can assume only 11 different possible values: 1.0, 1.5, 2.0,…, 6.0, with certain values of $\overline{x}$ occurring more frequently than others. The value $\overline{x} = 1.0$ occurs only once, so its probability is 1/36. The value $\overline{x} = 1.5$ can occur in two ways, (1, 2) and (2, 1), each having the same probability 1/36. Thus, $P(\overline{x} = 1.5) = 2/36$. The probabilities of the other values of $\overline{x}$ are determined in similar fashion, and the resulting sampling distribution of $\overline{X}$ is shown in Table 9.2.

TABLE 9.2 Sampling Distribution of $\overline{X}$

$\bar{x}$	$P(\bar{x})$
1.0	1/36
1.5	2/36
2.0	3/36
2.5	4/36
3.0	5/36
3.5	6/36
4.0	5/36
4.5	4/36
5.0	3/36
5.5	2/36
6.0	1/36

The most interesting aspect of the sampling distribution of $\overline{X}$ is how different it is from the distribution of X, as can be seen in Figure 9.2.

FIGURE 9.2 Distributions of X and $\overline{X}$

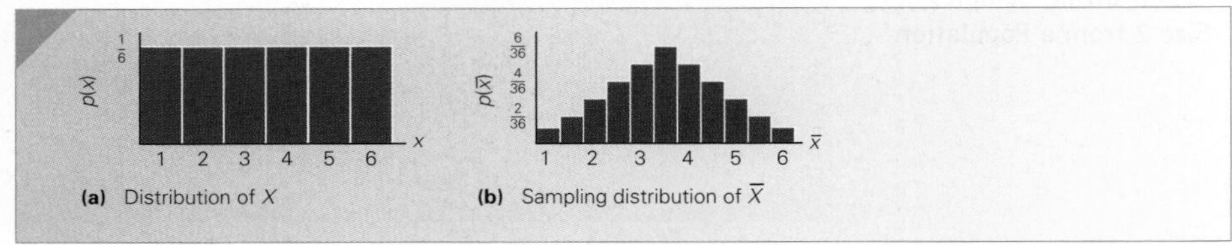

We can also compute the mean, variance, and standard deviation of the sampling distribution. Once again using the definitions of expected value and variance, we determine the following parameters of the sampling distribution.

Mean of the sampling distribution of $\overline{X}$:

$$\begin{aligned}\mu_{\bar{x}} &= \sum \bar{x}P(\bar{x}) \\ &= 1.0(1/36) + 1.5(2/36) + \cdots + 6.0(1/36) \\ &= 3.5\end{aligned}$$

Notice that the mean of the sampling distribution of $\overline{X}$ is equal to the mean of the population of the toss of a die computed previously.

Variance of the sampling distribution of $\overline{X}$:

$$\begin{aligned}\sigma^2_{\bar{x}} &= \sum (\bar{x} - \mu_{\bar{x}})^2 P(\bar{x}) \\ &= (1.0 - 3.5)^2(1/36) + (1.5 - 3.5)^2(2/36) + \cdots + (6.0 - 3.5)^2(1/36) \\ &= 1.46\end{aligned}$$

It is no coincidence that the variance of the sampling distribution of $\overline{X}$ is exactly half of the variance of the population of the toss of a die (computed previously as $\sigma^2 = 2.92$).

Standard deviation of the sampling distribution of $\overline{X}$:

$$\sigma_{\bar{x}} = \sqrt{\sigma^2_{\bar{x}}} = \sqrt{1.46} = 1.21$$

It is important to recognize that the distribution of $\overline{X}$ is different from the distribution of X as depicted in Figure 9.2. However, the two random variables are related. Their means are the same ($\mu_{\bar{x}} = \mu = 3.5$) and their variances are related ($\sigma^2_{\bar{x}} = \sigma^2/2$).

Don't get lost in the terminology and notation. Remember that μ and σ^2 are the parameters of the population of X. To create the sampling distribution of $\overline{X}$, we repeatedly drew samples of size $n = 2$ from the population and calculated $\overline{x}$ for each sample. Thus, we treat $\overline{X}$ as a brand-new random variable, with its own distribution, mean, and variance. The mean is denoted $\mu_{\overline{x}}$ and the variance is denoted $\sigma^2_{\overline{x}}$.

If we now repeat the sampling process with the same population but with other values of n, we produce somewhat different sampling distributions of $\overline{X}$. Figure 9.3 shows the sampling distributions of $\overline{X}$ when $n = 5$, 10, and 25.

FIGURE **9.3** Sampling Distributions of $\overline{X}$ for $n = 5$, 10, and 25

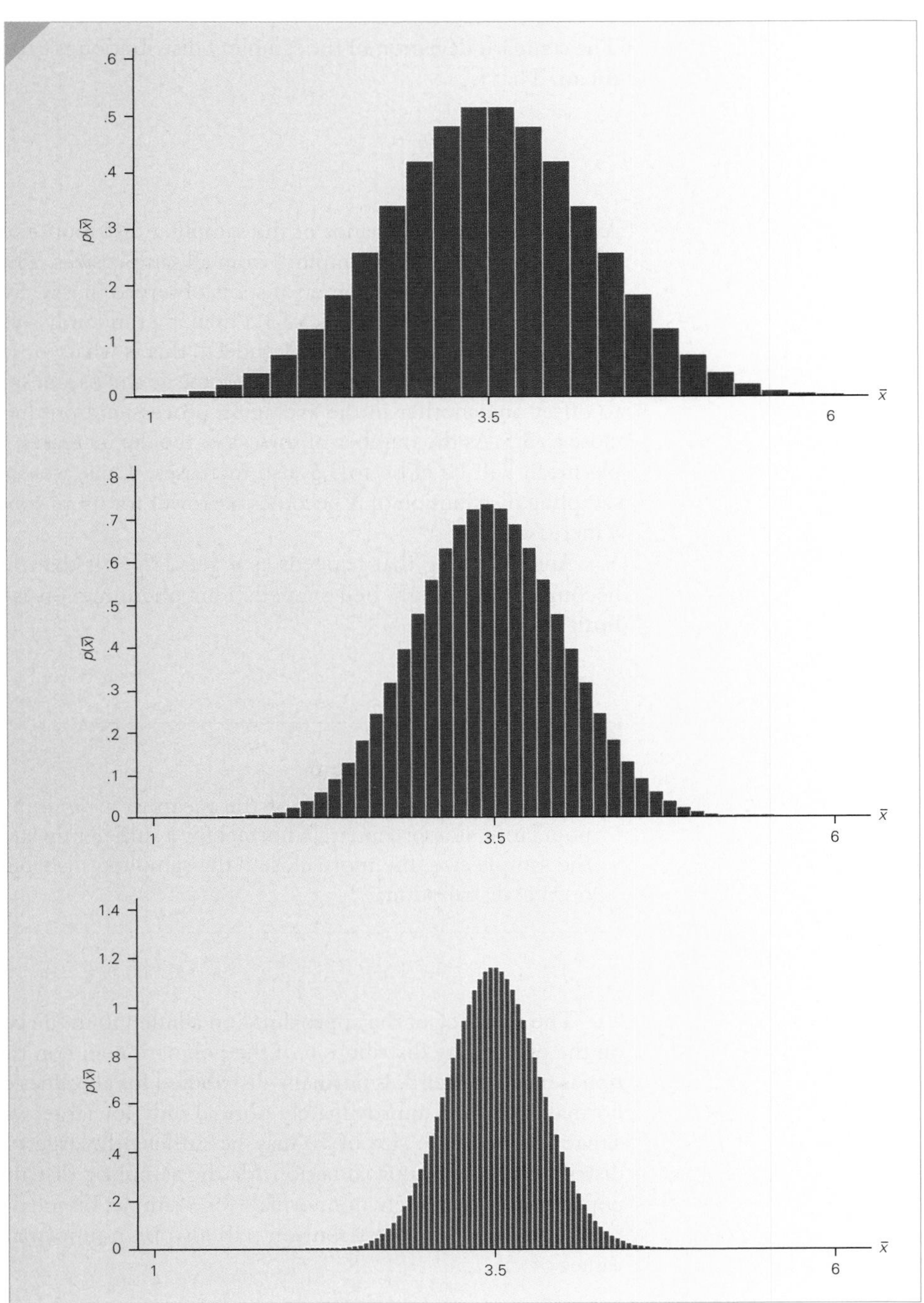

For each value of n, the mean of the sampling distribution of $\overline{X}$ is the mean of the population from which we're sampling. That is,

$$\mu_{\overline{x}} = \mu = 3.5$$

The variance of the sampling distribution of the sample mean is the variance of the population divided by the sample size. That is,

$$\sigma_{\overline{x}}^2 = \frac{\sigma^2}{n}$$

The standard deviation of the sampling distribution is called the **standard error of the mean.** That is,

$$\sigma_{\overline{x}} = \frac{\sigma}{\sqrt{n}}$$

As you can see, the variance of the sampling distribution of $\overline{X}$ is less than the variance of the population we're sampling from all sample sizes. Thus, a randomly selected value of $\overline{X}$ (the mean of the number of spots observed in, say, five throws of the die) is likely to be closer to the mean value of 3.5 than is a randomly selected value of X (the number of spots observed in one throw). Indeed, this is what you would expect, because in five throws of the die you are likely to get some 5s and 6s and some 1s and 2s, which will tend to offset one another in the averaging process and produce a sample mean reasonably close to 3.5. As the number of throws of the die increases, the probability that the sample mean will be close to 3.5 also increases. Thus, we observe in Figure 9.3 that the sampling distribution of $\overline{X}$ becomes narrower (or more concentrated about the mean) as n increases.

Another thing that happens as n gets larger is that the sampling distribution of $\overline{x}$ becomes increasingly bell shaped. This phenomenon is summarized in the **central limit theorem.**

Central Limit Theorem

The sampling distribution of the mean of a random sample drawn from any population is approximately normal for a sufficiently large sample size. The larger the sample size, the more closely the sampling distribution of $\overline{X}$ will resemble a normal distribution.

The accuracy of the approximation alluded to in the central limit theorem depends on the probability distribution of the population and on the sample size. If the population is normal, then $\overline{X}$ is normally distributed for all values of n. If the population is nonnormal, then $\overline{X}$ is approximately normal only for larger values of n. In many practical situations, a sample size of 30 may be sufficiently large to allow us to use the normal distribution as an approximation for the sampling distribution of $\overline{X}$. However, if the population is extremely nonnormal (for example, bimodal and highly skewed distributions), the sampling distribution will also be nonnormal even for moderately large values of n.

Sampling Distribution of the Mean of Any Population

We can extend the discoveries we've made to all infinitely large populations. Statisticians have shown that the mean of the sampling distribution is always equal to the mean of the population and that the standard error is equal to $\sigma/\sqrt{n}$ for infinitely large populations. (In CD Appendix I we describe how to mathematically prove that $\mu_{\overline{x}} = \mu$ and $\sigma^2_{\overline{x}} = \sigma^2/n$.) However, if the population is finite the standard error is

$$\sigma_{\overline{x}} = \frac{\sigma}{\sqrt{n}}\sqrt{\frac{N-n}{N-1}}$$

where N is the population size and $\sqrt{\frac{N-n}{N-1}}$ is called the **finite population correction factor.** The source of the correction factor is provided in CD Appendix F where we introduce the hypergeometric distribution. An analysis (see Exercises 9.13 and 9.14) reveals that if the population size is large relative to the sample size, the finite population correction factor is close to 1 and can be ignored. As a rule of thumb, we will treat any population that is at least 20 times larger than the sample size as large. In practice, most applications involve populations that qualify as large because if the population is small, it may be possible to investigate each member of the population, and in so doing, calculate the parameters precisely. As a consequence, the finite population correction factor is usually omitted. In Section 12.5 we will discuss an application where the population can be small.

We can now summarize what we know about the **sampling distribution of the sample mean** for large populations.

Sampling Distribution of the Sample Mean

1. $\mu_{\overline{x}} = \mu$
2. $\sigma^2_{\overline{x}} = \sigma^2/n$ and $\sigma_{\overline{x}} = \sigma/\sqrt{n}$
3. If X is normal, $\overline{X}$ is normal. If X is nonnormal, $\overline{X}$ is approximately normal for sufficiently large sample sizes. The definition of "sufficiently large" depends on the extent of nonnormality of X.

Creating the Sampling Distribution Empirically

In the previous analysis, we created the sampling distribution of the mean theoretically. We did so by listing all the possible samples of size 2 and their probabilities. (They were all equally likely with probability 1/36.) From this distribution, we produced the sampling distribution. We could also create the distribution empirically by actually tossing two fair dice repeatedly, calculating the sample mean for each sample, counting the number of times each value of $\overline{X}$ occurs, and computing the relative frequencies to estimate the theoretical probabilities. If we toss the two dice a large enough number of times, the relative frequencies and theoretical probabilities (computed previously) will be similar. Try it yourself. Toss two dice 500 times, calculate the mean of the two tosses, count the number of times each sample mean occurs, and construct the histogram representing the sampling distribution. Obviously this approach is far from ideal because of the excessive amount of time required to toss the dice enough times to make the relative frequencies good approximations for the theoretical probabilities. However, we can use the computer to quickly simulate tossing dice many times.

SEEING STATISTICS

Applet 9: Fair Dice 1

This applet has two parts. The first part simulates the tossing of one fair die. You can toss 1 at a time, 10 at a time, or 100 at a time. The histogram of the cumulative results is shown. The second part allows you to simulate tossing 2 dice one set at a time, 10 sets a time, or 100 sets a time. The histogram of the means of the cumulative results is exhibited. To start again, click **Refresh** or **Reload** on the browser menu. The value N represents the number of sets. The larger the value of N, the closer the histogram approximates the theoretical distribution.

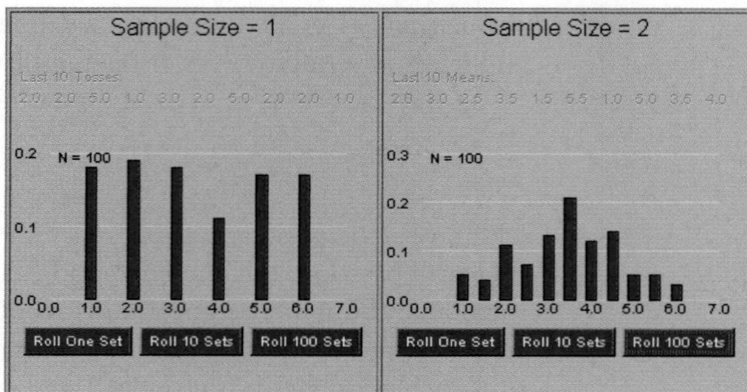

Applet Exercises

Simulate 2,500 tosses of one fair die and 2,500 tosses of two fair dice.

9.1 Does the simulated probability distribution of one die look like the theoretical distribution displayed in Figure 9.2? Discuss the reason for the deviations.

9.2 Does the simulated sampling distribution of the mean of two dice look like the theoretical distribution displayed in Figure 9.2? Discuss the reason for the deviations.

9.3 Do the distribution of one die and the sampling distribution of the mean of two dice have the same or different shapes? How would you characterize the difference?

9.4 Do the centers of the distribution of one die and the sampling distribution of the mean of two dice appear to be about the same?

9.5 Do the spreads of the distribution of one die and the sampling distribution of the mean of two dice appear to be about the same? Which one has the smaller spread?

SEEING STATISTICS

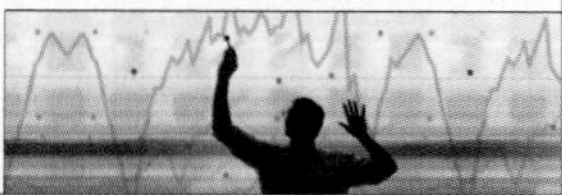

Applet 10: Fair Dice 2

This applet allows you to simulate tossing 12 fair dice and drawing the sampling distribution of the mean. As was the case with the previous applet, you can toss 1 set, 10 sets, or 100 sets. To start again, click **Refresh** or **Reload** on the browser menu.

Applet Exercises

Simulate 2,500 tosses of 12 fair dice.

10.1 Does the simulated sampling distribution of $\overline{X}$ appear to be bell shaped?

10.2 Does it appear that the simulated sampling distribution of the mean of 12 fair dice is narrower than that of 2 fair dice? Explain why this is so.

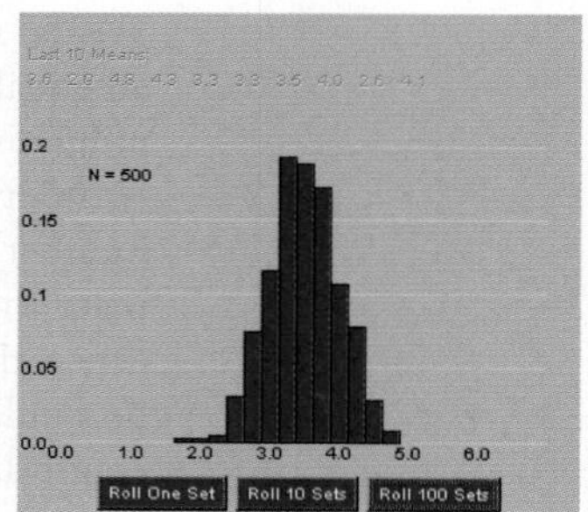

SEEING STATISTICS

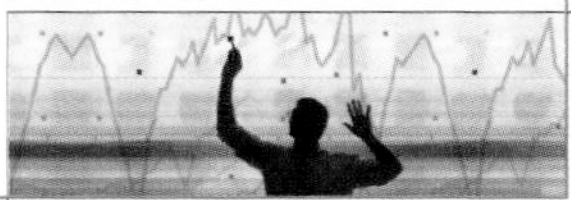

Applet 11: Loaded Dice

This applet has two parts. The first part simulates the tossing of a loaded die. "Loaded" refers to the inequality of the probabilities of the six outcomes. You can toss 1 at a time, 10 at a time, or 100 at a time. The second part allows you to simulate tossing 12 loaded dice 1 set at a time, 10 sets a time, or 100 sets at a time.

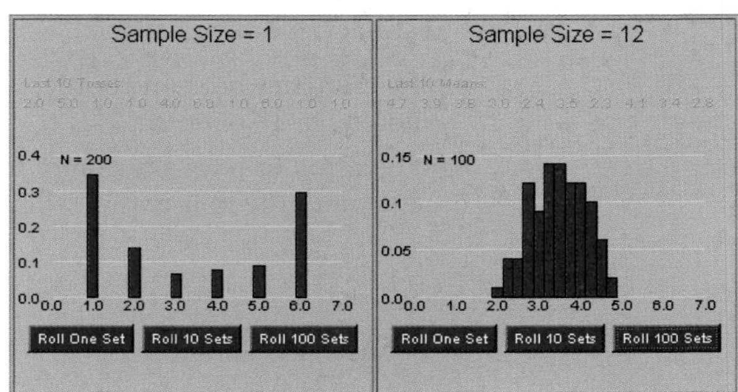

Applet Exercises

Simulate 2,500 tosses of one loaded die.

11.1 Estimate the probability of each value of X.

11.2 Use the estimated probabilities to compute the expected value, variance, and standard deviation of X.

Simulate 2,500 tosses of 12 loaded dice.

11.3 Does it appear that the mean of the simulated sampling distribution of $\overline{X}$ is equal to 3.5?

11.4 Does it appear that the standard deviation of the simulated sampling distribution of the mean of 12 loaded dice is greater than that for 12 fair dice? Explain why this is so.

11.5 Does the simulated sampling distribution of the mean of 12 loaded dice appear to be bell shaped? Explain why this is so.

SEEING STATISTICS

Applet 12: Skewed Dice

This applet has two parts. The first part simulates the tossing of a skewed die. You can toss it 1 at a time, 10 at a time, or 100 at a time. The second part allows you to simulate tossing 2 dice 1 set at a time, 10 sets a time, or 100 sets at a time.

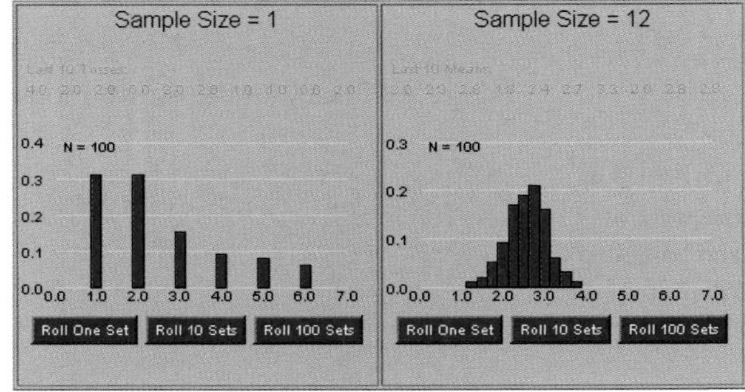

Applet Exercises

Simulate 2,500 tosses of one skewed die.

12.1 Estimate the probability of each value of X.

12.2 Use the estimated probabilities to compute the expected value, variance, and standard deviation of X.

Simulate 2,500 tosses of 12 skewed dice.

12.3 Does it appear that the mean of the simulated sampling distribution of $\overline{X}$ is less than 3.5?

12.4 Does the simulated sampling distribution of the mean of 12 skewed dice appear to be bell shaped? Explain why this is so.

EXAMPLE 9.1 Contents of a 32-Ounce Bottle

The foreman of a bottling plant has observed that the amount of soda in each 32-ounce bottle is actually a normally distributed random variable, with a mean of 32.2 ounces and a standard deviation of .3 ounce.

a. If a customer buys one bottle, what is the probability that the bottle will contain more than 32 ounces?

b. If a customer buys a carton of four bottles, what is the probability that the mean amount of the four bottles will be greater than 32 ounces?

SOLUTION

a. Because the random variable is the amount of soda in one bottle, we want to find $P(X > 32)$, where X is normally distributed, $\mu = 32.2$, and $\sigma = .3$. Hence,

$$P(X > 32) = P\left(\frac{X - \mu}{\sigma} > \frac{32 - 32.2}{.3}\right)$$
$$= P(Z > -.67)$$
$$= 1 - P(Z < -.67)$$
$$= 1 - .2514 = .7486$$

b. Now we want to find the probability that the mean amount of four filled bottles exceeds 32 ounces. That is, we want $P(\overline{X} > 32)$. From our previous analysis and from the central limit theorem, we know the following:

1. $\overline{X}$ is normally distributed.
2. $\mu_{\overline{x}} = \mu = 32.2$
3. $\sigma_{\overline{x}} = \sigma/\sqrt{n} = .3/\sqrt{4} = .15$

Hence,

$$P(\overline{X} > 32) = P\left(\frac{\overline{X} - \mu_{\overline{x}}}{\sigma_{\overline{x}}} > \frac{32 - 32.2}{.15}\right)$$
$$= P(Z > -1.33) = 1 - P(Z < -1.33) = 1 - .0918 = .9082$$

Figure 9.4 illustrates the distributions used in this example.

FIGURE **9.4** Distribution of X and Sampling Distribution of $\overline{X}$

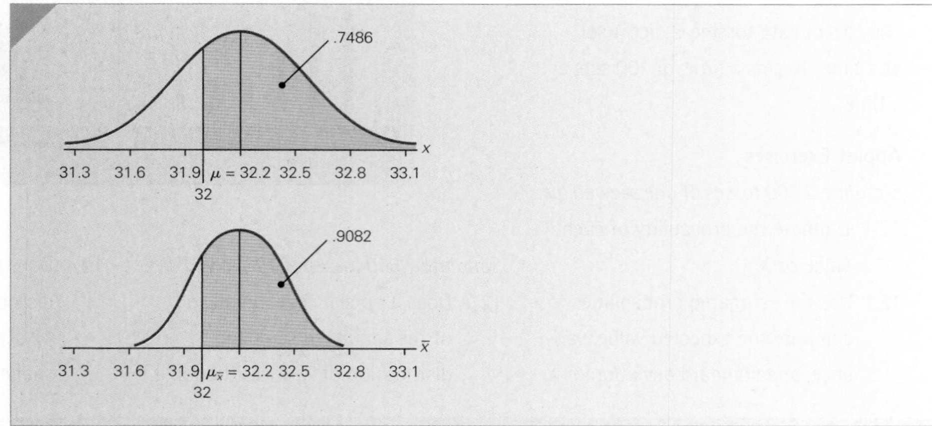

In Example 9.1(b), we began with the assumption that both μ and σ were known. Then, using the sampling distribution, we made a probability statement about $\overline{X}$. Unfortunately, the values of μ and σ are not usually known, so an analysis such as that in Example 9.1 cannot usually be conducted. However, we can use the sampling distribution to infer something about an unknown value of μ on the basis of a sample mean.

Salaries of a Business School's Graduates: Solution

© PhotoDisc/Getty Images

We want to find the probability that the sample mean is less than $750. Thus, we seek

$$P(\overline{X} < 750)$$

The distribution of X, the weekly income, is likely to be positively skewed, but not sufficiently so to make the distribution of $\overline{X}$ nonnormal. As a result, we may assume that $\overline{X}$ is normal with mean $\mu_{\overline{x}} = \mu = 800$ and standard deviation $\sigma_{\overline{x}} = \sigma/\sqrt{n} = 100/\sqrt{25} = 20$. Thus,

$$P(\overline{X} < 750) = P\left(\frac{\overline{X} - \mu_{\overline{x}}}{\sigma_{\overline{x}}} < \frac{750 - 800}{20}\right) = P(Z < -2.5) = .0062$$

Figure 9.5 illustrates the distribution.

FIGURE **9.5** $P(\overline{X} < 750)$

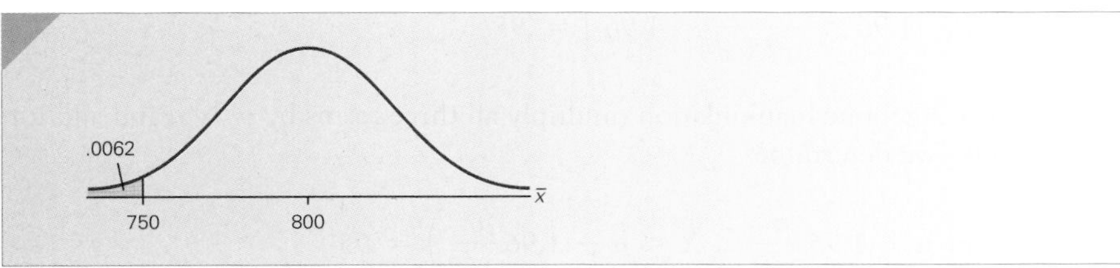

The probability of observing a sample mean as low as $750 when the population mean is $800 is extremely small. Because this event is quite unlikely, we would have to conclude that the dean's claim is not justified.

Using the Sampling Distribution for Inference

Our conclusion in the chapter-opening example illustrates how the sampling distribution can be used to make inferences about population parameters. The first form of inference is estimation, which we introduce in the next chapter. In preparation for this momentous occasion, we'll present another way of expressing the probability associated with the sampling distribution.

Recall the notation introduced in Section 8.2 (see page 268). We defined z_A to be the value of z such that the area to the right of z_A under the standard normal curve is equal to A. We also showed that $z_{.025} = 1.96$. Because the standard normal distribution is symmetric about 0, the area to the left of -1.96 is also .025. The area between -1.96

and 1.96 is .95. Figure 9.6 depicts this notation. We can express the notation algebraically as

$$P(-1.96 < Z < 1.96) = .95$$

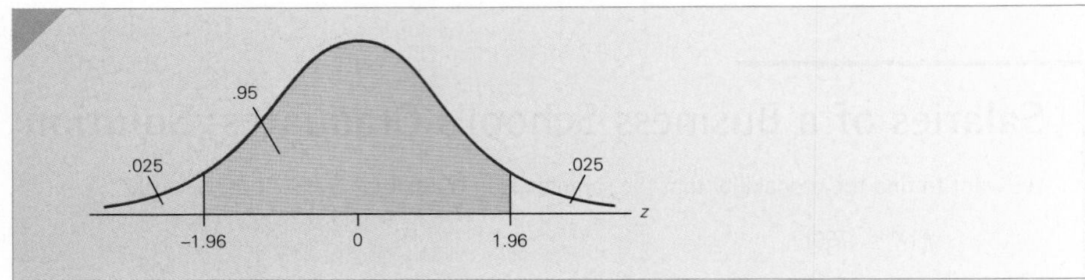

FIGURE 9.6 $P(-1.96 < Z < 1.96) = .05$

In this section we established that

$$Z = \frac{\overline{X} - \mu}{\sigma/\sqrt{n}}$$

is standard normally distributed. Substituting this form of Z into the previous probability statement, we produce

$$P\left(-1.96 < \frac{\overline{X} - \mu}{\sigma/\sqrt{n}} < 1.96\right) = .95$$

With a little algebraic manipulation (multiply all three terms by $\sigma/\sqrt{n}$ and add μ to all three terms), we determine

$$P\left(\mu - 1.96\frac{\sigma}{\sqrt{n}} < \overline{X} < \mu + 1.96\frac{\sigma}{\sqrt{n}}\right) = .95$$

Returning to the chapter-opening example where $\mu = 800$, $\sigma = 100$, and $n = 25$, we compute

$$P\left(800 - 1.96\frac{100}{\sqrt{25}} < \overline{X} < 800 + 1.96\frac{100}{\sqrt{25}}\right) = .95$$

Thus, we can say that

$$P(760.8 < \overline{X} < 839.2) = .95$$

This tells us that there is a 95% probability that a sample mean will fall between 760.8 and 839.2. Because the sample mean was computed to be $750, we would have to conclude that the dean's claim is not supported by the statistic.

Changing the probability from .95 to .90 changes the probability statement to

$$P\left(\mu - 1.645\frac{\sigma}{\sqrt{n}} < \overline{X} < \mu + 1.645\frac{\sigma}{\sqrt{n}}\right) = .90$$

We can also produce a general form of this statement:

$$P\left(\mu - z_{\alpha/2}\frac{\sigma}{\sqrt{n}} < \overline{X} < \mu + z_{\alpha/2}\frac{\sigma}{\sqrt{n}}\right) = 1 - \alpha$$

In this formula α (Greek letter *alpha*) is the probability that $\overline{X}$ does not fall into the interval. To apply this formula, all we need to do is substitute the values for μ, σ, n, and α. For example, with $\mu = 800$, $\sigma = 100$, $n = 25$ and $\alpha = .01$, we produce

$$P\left(\mu - z_{.005}\frac{\sigma}{\sqrt{n}} < \overline{X} < \mu + z_{.005}\frac{\sigma}{\sqrt{n}}\right) = 1 - .01$$

$$P\left(800 - 2.575\frac{100}{\sqrt{25}} < \overline{X} < 800 + 2.575\frac{100}{\sqrt{25}}\right) = .99$$

$$P(748.5 < \overline{X} < 851.5) = .99$$

which is another probability statement about $\overline{X}$. In Section 10.2, we will use a similar type of probability statement to derive the first statistical inference technique.

EXERCISES

9.1 Let X represent the result of the toss of a fair die. Find the following probabilities.
a. $P(X = 1)$
b. $P(X = 6)$

9.2 Let $\overline{X}$ represent the mean of the toss of two fair dice. Use the probabilities listed in Table 9.2 to determine the following probabilities.
a. $P(\overline{X} = 1)$
b. $P(\overline{X} = 6)$

9.3 An experiment consists of tossing five balanced dice. Find the following probabilities. (Determine the exact probabilities as we did in Tables 9.1 and 9.2 for two dice.)
a. $P(\overline{X} = 1)$
b. $P(\overline{X} = 6)$

9.4 Refer to Exercises 9.1 to 9.3. What do the probabilities tell you about the variances of X and $\overline{X}$?

9.5 A normally distributed population has a mean of 40 and a standard deviation of 12. What does the central limit theorem say about the sampling distribution of the mean if samples of size 100 are drawn from this population?

9.6 Refer to Exercise 9.5. Suppose that the population is not normally distributed. Does this change your answer? Explain.

9.7 A sample of $n = 16$ observations is drawn from a normal population with $\mu = 1{,}000$ and $\sigma = 200$. Find the following.
a. $P(\overline{X} > 1{,}050)$
b. $P(\overline{X} < 960)$
c. $P(\overline{X} > 1{,}100)$

9.8 Repeat Exercise 9.7 with $n = 25$.

9.9 Repeat Exercise 9.7 with $n = 100$.

9.10 Given a normal population whose mean is 50 and whose standard deviation is 5,
a. find the probability that a random sample of 4 has a mean between 49 and 52.
b. find the probability that a random sample of 16 has a mean between 49 and 52.
c. find the probability that a random sample of 25 has a mean between 49 and 52.

9.11 Repeat Exercise 9.10 for a standard deviation of 10.

9.12 Repeat Exercise 9.10 for a standard deviation of 20.

9.13 a. Calculate the finite population correction factor when the population size is $N = 1{,}000$ and the sample size is $n = 100$.
b. Repeat Part a when $N = 3{,}000$.
c. Repeat Part a when $N = 5{,}000$.
d. What have you learned about the finite population correction factor when N is large relative to n?

9.14 a. Suppose that the standard deviation of a population with $N = 10{,}000$ members is 500. Determine the standard error of the sampling distribution of the mean when the sample size is 1,000.
b. Repeat Part a when $n = 500$.
c. Repeat Part a $n = 100$.

9.15 The heights of North American women are normally distributed with a mean of 64 inches and a standard deviation of 2 inches.
a. What is the probability that a randomly selected woman is taller than 66 inches?

b. A random sample of four women is selected. What is the probability that the sample mean height is greater than 66 inches?
c. What is the probability that the mean height of a random sample of 100 women is greater than 66 inches?

9.16 Refer to Exercise 9.15. If the population of women's heights is not normally distributed, which if any, of the questions can you answer? Explain.

9.17 An automatic machine in a manufacturing process is operating properly if the lengths of an important subcomponent are normally distributed with mean = 117 cm and standard deviation = 5.2 cm.
a. Find the probability that one selected subcomponent is longer than 120 cm.
b. Find the probability that if four subcomponents are randomly selected, their mean length exceeds 120 cm.
c. Find the probability that if four subcomponents are randomly selected, all four have lengths that exceed 120 cm.

9.18 The amount of time the university professors devote to their jobs per week is normally distributed with a mean of 52 hours and a standard deviation of 6 hours.
a. What is the probability that a professor works for more than 60 hours per week?
b. Find the probability that the mean amount of work per week for three randomly selected professors is more than 60 hours.
c. Find the probability that if three professors are randomly selected, all three work for more than 60 hours per week.

9.19 The number of pizzas consumed per month by university students is normally distributed with a mean of 10 and a standard deviation of 3.
a. What proportion of students consume more than 12 pizzas per month?
b. What is the probability that in a random sample of 25 students more than 275 pizzas are consumed? (*Hint:* What is the mean number of pizzas consumed by the sample of 25 students?)

9.20 The marks on a statistics midterm test are normally distributed with a mean of 78 and a standard deviation of 6.
a. What proportion of the class has a midterm mark of less than 75?
b. What is the probability that a class of 50 has an average midterm mark that is less than 75?

9.21 The amount of time spent by North American adults watching television per day is normally distributed with a mean of 6 hours and a standard deviation of 1.5 hours.
a. What is the probability that a randomly selected North American adult watches television for more than 7 hours per day?
b. What is the probability that the average time watching television by a random sample of five North American adults is more than 7 hours?
c. What is the probability that, in a random sample of five North American adults, all watch television for more than 7 hours per day?

9.22 The manufacturer of cans of salmon that are supposed to have a net weight of 6 ounces tells you that the net weight is actually a normal random variable with a mean of 6.05 ounces and a standard deviation of .18 ounces. Suppose that you draw a random sample of 36 cans.
a. Find the probability that the mean weight of the sample is less than 5.97 ounces.
b. Suppose your random sample of 36 cans of salmon produced a mean weight that is less than 5.97 ounces. Comment on the statement made by the manufacturer.

9.23 The number of customers who enter a supermarket each hour is normally distributed with a mean of 600 and a standard deviation of 200. The supermarket is open 16 hours per day. What is the probability that the total number of customers who enter the supermarket in one day is greater than 10,000? (*Hint:* Calculate the average hourly number of customers necessary to exceed 10,000 in one 16-hour day.)

9.24 The sign on the elevator in the Peters Building, which houses the School of Business and Economics at Wilfrid Laurier University, states, "Maximum Capacity 1,140 kilograms (2500 pounds) or 16 Persons." A professor of statistics wonders what the probability is that 16 persons would weigh more than 1,140 kilograms. Discuss what the professor needs (besides the ability to perform the calculations) in order to satisfy his curiosity.

9.25 Refer to Exercise 9.24. Suppose that the professor discovers that the weights of people who use the elevator are normally distributed with an average of 75 kilograms and a standard deviation of 10 kilograms. Calculate the probability that the professor seeks.

9.26 The time it takes for a statistics professor to mark his midterm test is normally distributed with a mean of 4.8 minutes and a standard deviation of 1.3 minutes. There are 60 students in the professor's class. What is the probability that he needs more than 5 hours to mark all the midterm tests? (The 60 midterm tests of the students in this year's class can be considered a random sample of the many thousands of midterm tests the professor has marked and will mark.)

9.27 Refer to Exercise 9.26. Does your answer change if you discover that the times needed to mark a midterm test are not normally distributed?

9.28 The restaurant in a large commercial building provides coffee for the building's occupants. The restaurateur has determined that the mean number of cups of coffee consumed in a day by all the occupants is 2.0 with a standard deviation of .6. A new tenant of the building intends to have a total of 125 new employees. What is the probability that the new employees will consume more than 240 cups per day?

9.29 The number of pages produced by a fax machine in a busy office is normally distributed with a mean of 275 and a standard deviation of 75. Determine the probability that in 1 week (5 days) more than 1,500 faxes will be received?

9.2 Sampling Distribution of a Proportion

In Section 7.4 we introduced the binomial distribution whose parameter is p, the probability of success in any trial. In order to compute binomial probabilities, we assumed that p was known. However, in the real world p is unknown, requiring the statistics practitioner to estimate its value from a sample. The estimator of a population proportion of successes is the sample proportion. That is, we count the number of successes in a sample and compute

$$\hat{P} = \frac{X}{n}$$

($\hat{P}$ is read as *p-hat*) where X is the number of successes and n is the sample size. When we take a sample of size n, we're actually conducting a binomial experiment and as a result X is binomially distributed. Thus, the probability of any value of $\hat{P}$ can be calculated from its value of X. For example, suppose that we have a binomial experiment with $n = 10$ and $p = .4$. To find the probability that the sample proportion $\hat{P}$ is less than or equal to .50, we find the probability that X is less than or equal to 5 (because $5/10 = .50$). From Table 1 in Appendix B we find with $n = 10$ and $p = .4$

$$P(\hat{P} \le .50) = P(X \le 5) = .8338$$

We can calculate the probability associated with other values of $\hat{P}$ similarly.

Discrete distributions such as the binomial do not lend themselves easily to the kinds of calculation needed for inference. And inference is the reason we need sampling distributions. Fortunately, we can approximate the binomial distribution by a normal distribution.

What follows is an explanation of how and why the normal distribution can be used to approximate a binomial distribution. Disinterested readers can skip to page 313 where we present the approximate **sampling distribution of a sample proportion.**

(Optional) Normal Approximation to the Binomial Distribution

Recall how we introduced continuous probability distributions in Chapter 8. We developed the density function by converting a histogram so that the total area in the rectangles equaled 1. We can do the same for a binomial distribution. To illustrate, let X be a binomial random variable with $n = 20$ and $p = .5$. We can easily determine the probability of each value of X, where $X = 0, 1, 2, \ldots, 19, 20$. A rectangle representing a value of x is drawn so that its area equals the probability. We accomplish this by letting the height of the rectangle equal the probability and the base of the rectangle equal 1. Thus the base of each rectangle for x is the interval $x - .5$ to $x + .5$. Figure 9.7 depicts this graph. As you can see, the rectangle representing $x = 10$ is the rectangle whose base is the interval 9.5 to 10.5 and whose height is $P(X = 10) = .1762$.

FIGURE **9.7**
Binomial Distribution with $n = 20$ and $p = .5$

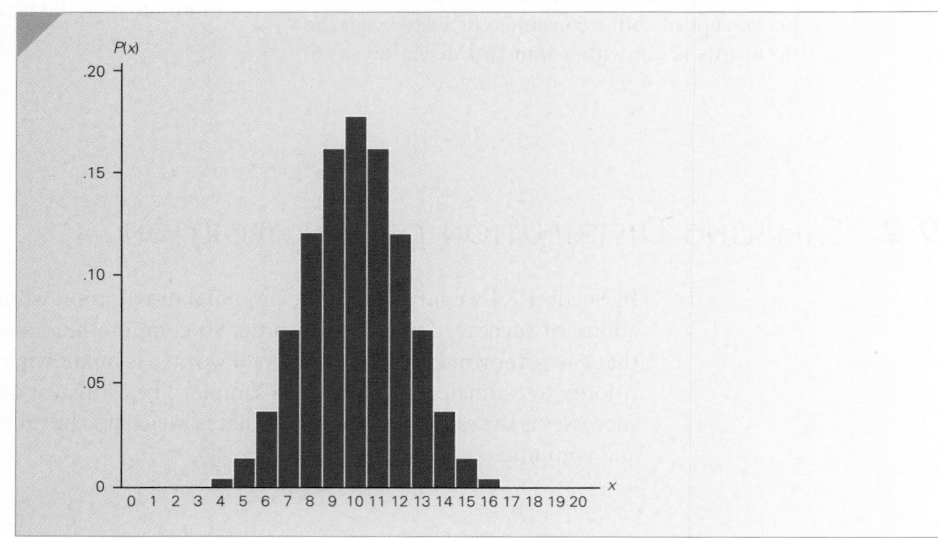

If we now smooth the ends of the rectangles, we produce a bell-shaped curve as seen in Figure 9.8. Thus, to use the normal approximation, all we need do is find the area under the *normal* curve between 9.5 and 10.5. To find normal probabilities requires us

FIGURE **9.8**
Binomial Distribution with $n = 20$ and $p = .5$ and Normal Approximation

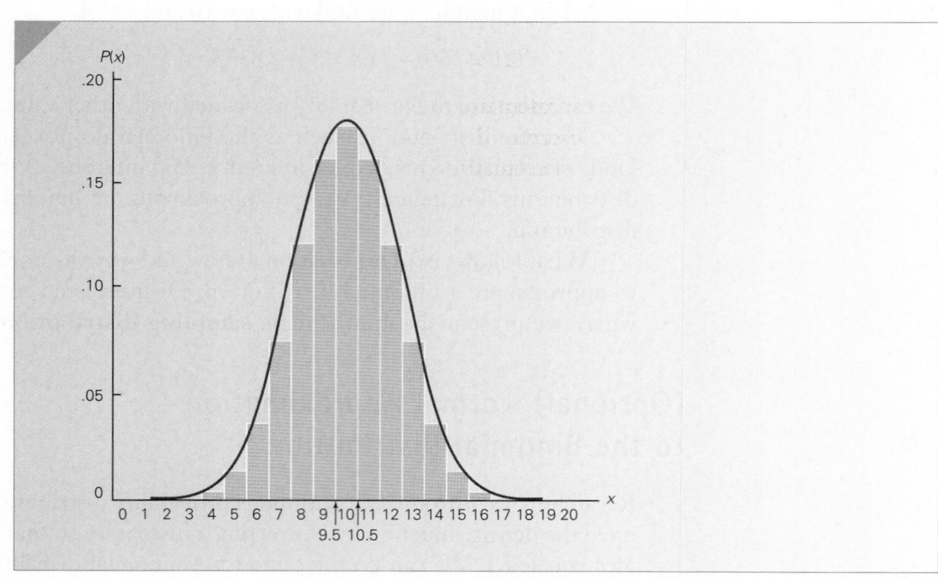

to first standardize x by subtracting the mean and dividing by the standard deviation. The values for μ and σ are derived from the binomial distribution being approximated. In Section 7.4 we pointed out that

$$\mu = np$$

and

$$\sigma = \sqrt{np(1-p)}$$

For $n = 20$ and $p = .5$, we have

$$\mu = np = 20(.5) = 10$$

and

$$\sigma = \sqrt{np(1-p)} = \sqrt{20(.5)(1-.5)} = 2.24$$

To calculate the probability that $X = 10$ using the normal distribution requires that we find the area under the normal curve between 9.5 and 10.5. That is,

$$P(X = 10) \approx P(9.5 < Y < 10.5)$$

where Y is a normal random variable approximating the binomial random variable X. We standardize Y and use Table 3 of Appendix B to find

$$P(9.5 < Y < 10.5) = P\left(\frac{9.5-10}{2.24} < \frac{Y-\mu}{\sigma} < \frac{10.5-10}{2.24}\right)$$
$$= P(-.22 < Z < .22) = (Z < .22) - P(Z < -.22)$$
$$= .5871 - .4129 = .1742$$

The actual probability that X equals 10 is

$$P(X = 10) = .1762$$

As you can see, the approximation is quite good.

Notice that to draw a binomial distribution, which is discrete, it was necessary to draw rectangles whose bases were constructed by adding and subtracting .5 to the values of X. The .5 is called the **continuity correction factor.**

The approximation for any other value of X would proceed in the same manner. In general, the binomial probability $P(X = x)$ is approximated by the area under a normal curve between $x - .5$ and $x + .5$. To find the binomial probability $P(X \leq x)$, we calculate the area under the normal curve to the left of $x + .5$. For the same binomial random variable, the probability that its value is less than or equal to 8 is $P(X \leq 8) = .2517$. The normal approximation is

$$P(X \leq 8) \approx P(Y < 8.5) = P\left(\frac{Y-\mu}{\sigma} < \frac{8.5-10}{2.24}\right) = P(Z < -.67) = .2514$$

We find the area under the normal curve to the right of $x - .5$ to determine the binomial probability $P(X \geq x)$. To illustrate, the probability that the binomial random variable (with $n = 20$ and $p = .5$) is greater than or equal to 14 is $P(X \geq 14) = .0577$. The normal approximation is

$$P(X \geq 14) \approx P(Y > 13.5) = P\left(\frac{Y-\mu}{\sigma} > \frac{13.5-10}{2.24}\right)$$
$$= P(Z > 1.56) = .0594$$

SEEING STATISTICS

Applet 13: Normal Approximation to Binomial Probabilities

This applet shows how well the normal distribution approximates the binomial distribution. Select values for n and p, which will specify a binomial distribution. Then set a value for k. The applet calculates and graphs both the binomial and normal probabilities for $P(X \le k)$.

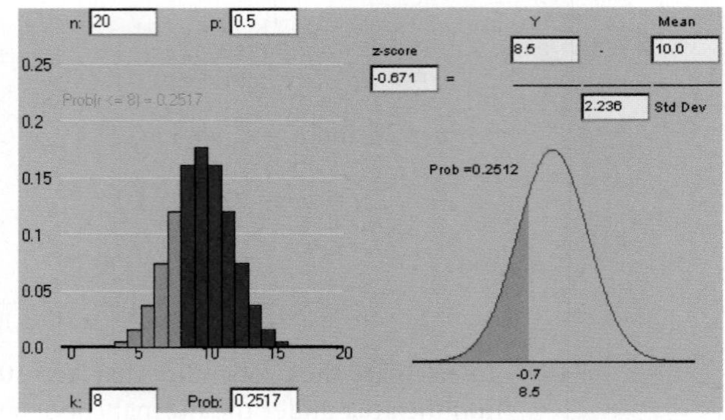

Applet Exercises

13.1 Given a binomial distribution with $n = 5$ and $p = .2$, use the applet to compute the actual and normal approximations of the following.

a. $P(X \le 0)$
b. $P(X \le 1)$
c. $P(X \le 2)$
d. $P(X \le 3)$

Describe how well the normal distribution approximates the binomial when n is small and when p is small.

13.2 Repeat Exercise 13.1 with $p = .5$. Describe how well the normal distribution approximates the binomial when n is small and when p is .5.

13.3 Suppose that X is a binomial random variable with $n = 10$ and $p = .2$. Use the applet to calculate the actual and normal approximations of the following.

a. $P(X \le 2)$
b. $P(X \le 3)$
c. $P(X \le 4)$
d. $P(X \le 5)$

Describe how well the normal distribution approximates the binomial when $n = 10$ and when p is small.

13.4 Repeat Exercise 13.3 with $p = .5$. Describe how well the normal distribution approximates the binomial when $n = 10$ and when p is .5.

13.5 Describe the effect on the normal approximation to the binomial as n increases.

Omitting the Correction Factor for Continuity

When calculating the probability of *individual* values of X as we did when we computed the probability that X equals 10 above, the correction factor *must* be used. If we don't, we are left with finding the area in a line, which is 0. When computing the probability of a *range* of values of X, we can omit the correction factor. However, the omission of the correction factor will decrease the accuracy of the approximation. For example, if we approximate $P(X \le 8)$ as we did previously except without the correction factor, we find

$$P(X \le 8) \approx P(Y < 8) = P\left(\frac{Y - \mu}{\sigma} < \frac{8 - 10}{2.24}\right) = P(Z < -.89) = .1867$$

The absolute size of the error between the actual cumulative binomial probability and its normal approximation is quite small when the values of x are in the tail regions of the distribution. For example, the probability that a binomial random variable with $n = 20$ and $p = .5$ is less than or equal to 3 is

$$P(X \le 3) = .0013$$

The normal approximation with the correction factor is

$$P(X \leq 3) \approx P(Y < 3.5) = P\left(\frac{Y - \mu}{\sigma} < \frac{3.5 - 10}{2.24}\right) = P(Z < -2.90) = .0019$$

The normal approximation without the correction factor is (using Excel)

$$P(X \leq 3) \approx P(Y < 3) = P\left(\frac{Y - \mu}{\sigma} < \frac{3 - 10}{2.24}\right) = P(Z < -3.13) = .0009$$

For larger values of n, the differences between the normal approximation with and without the correction factor are small even for values of X near the center of the distribution. For example, the probability that a binomial random variable with $n = 1{,}000$ and $p = .3$ is less than or equal to 260 is

$P(X \leq 260) = .0029$ (using Excel)

The normal approximation with the correction factor is

$$P(X \leq 260) \approx P(Y < 260.5) = P\left(\frac{Y - \mu}{\sigma} < \frac{260.5 - 300}{14.49}\right) = P(Z < -2.73) = .0032$$

The normal approximation without the correction factor is

$$P(X \leq 260) \approx P(Y < 260) = P\left(\frac{Y - \mu}{\sigma} < \frac{260 - 300}{14.49}\right) = P(Z < -2.76) = .0029$$

As we pointed out, the normal approximation of the binomial distribution is made necessary by the needs of statistical inference. As you will discover, statistical inference generally involves the use of large values of n and the part of the sampling distribution that is of greatest interest lies in the tail regions. The correction factor was a temporary tool that allowed us to convince you that a binomial distribution can be approximated by a normal distribution. Now that we have done so, we will use the normal approximation of the binomial distribution to approximate the sampling distribution of a sample proportion and in such applications the correction factor will be omitted.

Approximate Sampling Distribution of a Sample Proportion

Using the laws of expected value and variance (see CD Appendix I), we can determine the mean, variance, and standard deviation of $\hat{P}$. We will summarize what we have learned.

Sampling Distribution of a Sample Proportion

1. $\hat{P}$ is approximately normally distributed provided that np and $n(1 - p)$ are greater than or equal to 5.
2. The expected value: $E(\hat{P}) = p$
3. The variance: $V(\hat{P}) = \sigma^2_{\hat{p}} = \dfrac{p(1 - p)}{n}$*
4. The standard deviation: $\sigma_{\hat{p}} = \sqrt{p(1 - p)/n}$

(The standard deviation of $\hat{P}$ is called the **standard error of the proportion.**)

*As was the case with the standard error of the mean (page 301), the standard error of a proportion is $\sqrt{p(1 - p)/n}$ when sampling from infinitely large populations. When the population is finite the standard error of the proportion must include the finite population correction factor, which can be omitted when the population is large relative to the sample size, a very common occurrence in practice.

The sample size requirement is theoretical because in practice much larger sample sizes are needed for the normal approximation to be useful.

EXAMPLE 9.2 Political Survey

In the last election a state representative received 52% of the votes cast. One year after the election the representative organized a survey that asked a random sample of 300 people whether they would vote for him in the next election. If we assume that his popularity has not changed, what is the probability that more than half of the sample would vote for him?

SOLUTION

The number of respondents who would vote for the representative is a binomial random variable with $n = 300$ and $p = .52$. We want to determine the probability that the sample proportion is greater than 50%. That is, we want to find $P(\hat{P} > .50)$.

We now know that the sample proportion $\hat{P}$ is approximately normally distributed with mean $p = .52$ and standard deviation $= \sqrt{p(1 - p)/n} = \sqrt{(.52)(.48)/300} = .0288$.

Thus, we calculate

$$P(\hat{P} > .50) = P\left(\frac{\hat{P} - p}{\sqrt{p(1 - p)/n}} > \frac{.50 - .52}{.0288}\right)$$

$$= P(Z > -.69) = 1 - P(Z < -.69) = 1 - .2451 = .7549$$

If we assume that the level of support remains at 52%, the probability that more than half the sample of 300 people would vote for the representative is .7549.

EXERCISES

Use the normal approximation without the correction factor to find the probabilities in the following exercises.

9.30 a. In a binomial experiment with $n = 300$ and $p = .5$, find the probability that $\hat{P}$ is greater than 60%.

b. Repeat Part a with $p = .55$.

c. Repeat Part a with $p = .6$.

9.31 a. The probability of success on any trial of a binomial experiment is 25%. Find the probability that the proportion of successes in a sample of 500 is less than 22%.

b. Repeat Part a with $n = 800$.

c. Repeat Part a with $n = 1{,}000$.

9.32 Determine the probability that in a sample of 100 the sample proportion is less than .75 if $p = .80$.

9.33 A binomial experiment where $p = .4$ is conducted. Find the probability that in a sample of 60 the proportion of successes exceeds .35.

9.34 The proportion of eligible voters in the next election who will vote for the incumbent is assumed to be 55%. What is the probability that in a random sample of 500 voters less than 49% say they will vote for the incumbent?

9.35 The assembly line that produces an electronic component of a missile system has historically resulted in a 2% defective rate. A random sample of 800 components is drawn. What is the probability that the defective rate is greater than 4%? Suppose that in the random sample the defective rate is 4%. What does that suggest about the defective rate on the assembly line?

9.36 a. The manufacturer of aspirin claims that the proportion of headache sufferers who get relief with just two aspirins is 53%. What is the probability that in a random sample of 400 headache sufferers, less than 50% obtain relief? If 50% of the sample actually obtained relief, what does this suggest about the manufacturer's claim?

b. Repeat Part a using a sample of 1,000.

9.37 The manager of a restaurant in a commercial building has determined that the proportion of customers who drink tea is 14%. What is the probability that in the next 100 customers at least 10% will be tea drinkers?

9.38 A commercial for a manufacturer of household appliances claims that 3% of all its products require a service call in the first year. A consumer protection association wants to check the claim by surveying 400 households that recently purchased one of the company's appliances. What is the probability that more than 5% require a service call within the first year? What would you say about the commercial's honesty if in a random sample of 400 households 5% report at least one service call?

9.39 The Laurier Company's brand has a market share of 30%. Suppose that in a survey 1,000 consumers of the product are asked which brand they prefer. What is the probability that more than 32% of the respondents say they prefer the Laurier brand?

9.40 A university bookstore claims that 50% of its customers are satisfied with the service and prices.

a. If this claim is true what is the probability that in a random sample of 600 customers less than 45% are satisfied?
b. Suppose that in a random sample of 600 customers, 270 express satisfaction with the bookstore. What does this tell you about the bookstore's claim?

9.41 A psychologist believes that 80% of male drivers when lost continue to drive hoping to find the location they seek rather than ask directions. To examine this belief, he took a random sample of 350 male drivers and asked each what they did when lost. If the belief is true, determine the probability that less than 75% said they continue driving.

9.42 The Red Lobster restaurant chain regularly surveys its customers. On the basis of these surveys, the management of the chain claims that 75% of its customers rate the food as excellent. A consumer testing service wants to examine the claim by asking 460 customers to rate the food. What is the probability that less than 70% rate the food as excellent?

9.43 An accounting professor claims that no more than one-quarter of undergraduate business students will major in accounting. What is the probability that in a random sample of 1,200 undergraduate business students, 336 or more will major in accounting?

9.44 Refer to Exercise 9.43. A survey of a random sample of 1,200 undergraduate business students indicates that there are 336 students who plan to major in accounting. What does this tell you about the professor's claim?

9.3 Sampling Distribution of the Difference between Two Means

Another sampling distribution that you will soon encounter is that of the **difference between two sample means.** The sampling plan calls for independent random samples drawn from each of two normal populations. The samples are said to be independent if the selection of the members of one sample is independent of the selection of the members of the second sample. We will expand upon this discussion in Chapter 13. We are interested in the sampling distribution of the difference between the two sample means.

In Section 9.1 we introduced the central limit theorem, which states that in repeated sampling from a normal population whose mean is μ and whose standard deviation is σ, the sampling distribution of the sample mean is normal with mean μ and standard deviation $\sigma/\sqrt{n}$. Statisticians have shown that the difference between two independent normal random variables is also normally distributed. Thus, the difference between two sample means $\overline{X}_1 - \overline{X}_2$ is normally distributed if both populations are normal.

Through the use of the laws of expected value and variance (see CD Appendix I), we derive the expected value and variance of the sampling distribution of $\overline{X}_1 - \overline{X}_2$:

$$\mu_{\bar{x}_1-\bar{x}_2} = \mu_1 - \mu_2$$

and

$$\sigma^2_{\bar{x}_1-\bar{x}_2} = \frac{\sigma_1^2}{n_1} + \frac{\sigma_2^2}{n_2}$$

Thus, it follows that in repeated independent sampling from two populations with means μ_1 and μ_2 and standard deviations σ_1 and σ_2, respectively, the sampling distribution of $\overline{X}_1 - \overline{X}_2$ is normal with mean

$$\mu_{\bar{x}_1-\bar{x}_2} = \mu_1 - \mu_2$$

and standard deviation (which is the **standard error of the difference between two means**)

$$\sigma_{\bar{x}_1-\bar{x}_2} = \sqrt{\frac{\sigma_1^2}{n_1} + \frac{\sigma_2^2}{n_2}}$$

If the populations are nonnormal, then the sampling distribution is only approximately normal for large sample sizes. The required sample sizes depend on the extent of nonnormality. However, for most populations, sample sizes of 30 or more are sufficient.

Figure 9.9 depicts the sampling distribution of the difference between two means.

FIGURE **9.9** Sampling Distribution of $\overline{X}_1 - \overline{X}_2$

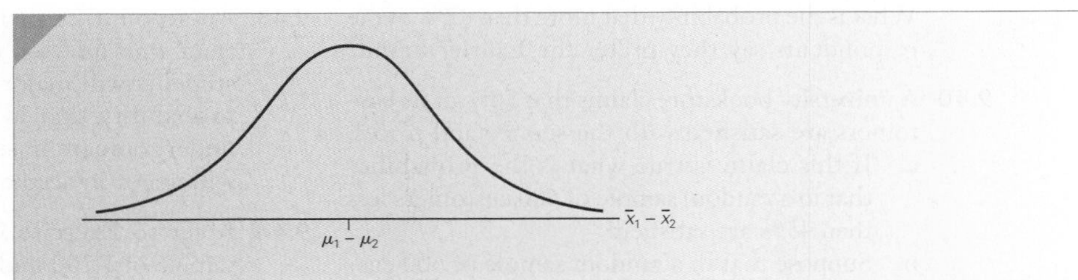

EXAMPLE 9.3

Starting Salaries of MBAs

Suppose that the starting salaries of MBAs at Wilfrid Laurier University (WLU) are normally distributed with a mean of \$62,000 and a standard deviation of \$14,500. The starting salaries of MBAs at the University of Western Ontario (UWO) are normally distributed with a mean of \$60,000 and a standard deviation of \$18,300. If a random sample of 50 WLU MBAs and a random sample of 60 UWO MBAs are selected, what is the probability that the sample mean starting salary of WLU graduates will exceed that of the UWO graduates?

SOLUTION

We want to determine $P(\overline{X}_1 - \overline{X}_2 > 0)$. We know that $\overline{X}_1 - \overline{X}_2$ is normally distributed with mean $\mu_1 - \mu_2 = 62{,}000 - 60{,}000 = 2{,}000$ and standard deviation

$$\sqrt{\frac{\sigma_1^2}{n_1} + \frac{\sigma_2^2}{n_2}} = \sqrt{\frac{14{,}500^2}{50} + \frac{18{,}300^2}{60}} = 3{,}128$$

We can standardize the variable and refer to Table 3 of Appendix B:

$$P(\overline{X}_1 - \overline{X}_2 > 0) = P\left(\frac{(\overline{X}_1 - \overline{X}_2) - (\mu_1 - \mu_2)}{\sqrt{\frac{\sigma_1^2}{n_1} + \frac{\sigma_2^2}{n_2}}} > \frac{0 - 2{,}000}{3{,}128}\right)$$

$$= P(Z > -.64) = 1 - P(Z < -.64) = 1 - .2611 = .7389$$

There is a .7389 probability that for a sample of size 50 from the WLU graduates and a sample of size 60 from the UWO graduates, the sample mean starting salary of WLU graduates will exceed the sample mean of UWO graduates.

SEEING STATISTICS

Applet 14: Distribution of the Differences between Means

The first part of this applet depicts two graphs. The first graph shows the distribution of the random variable of two populations. Moving the top slider shifts the first distribution left or right. The right slider controls the value of the population standard deviations, which are assumed to be equal. By moving each slider, you can see the relationship between the two populations.

The second graph describes the sampling distribution of the mean of each population in the first graph. Moving the right slider increases or decreases the sample size, which is the same for both samples.

The second part of the applet has three graphs. The first two graphs are identical to the graphs in the first part. The third graph depicts the sampling distribution of the difference between the two sample means from the populations described previously.

Moving the sliders allows you to see the effect on the sampling distribution of $\bar{x}_1 - \bar{x}_2$ of changing the relationship among the two population means, the common population standard deviation, and the sample size.

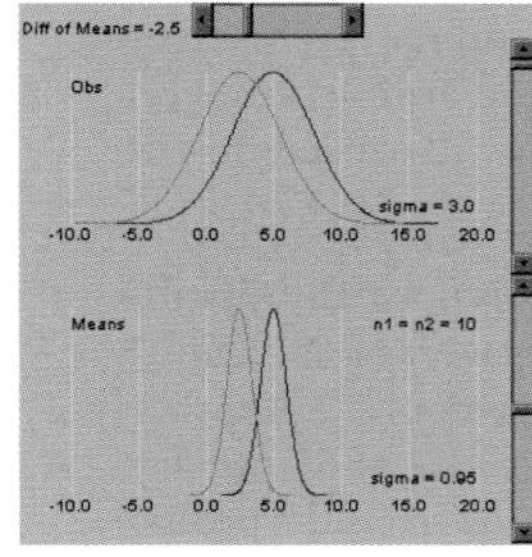

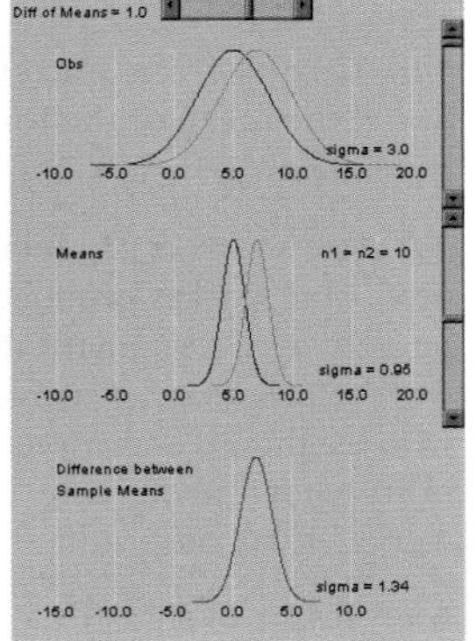

Applet Exercises

14.1 Describe the effect of changing the difference between population means from −5.0 to 4.5 on the population random variables, the sampling distribution of $\bar{x}_1$, the sampling distribution of $\bar{x}_2$, and the sampling distribution of $\bar{x}_1 - \bar{x}_2$. Describe what happened.

14.2 Describe the effect of changing the standard deviations from $\sigma_1 = \sigma_2 = 1.1$ to $\sigma_1 = \sigma_2 = 3.0$ on the population random variables, the sampling distribution of $\bar{x}_1$, the sampling distribution of $\bar{x}_2$, and the sampling distribution of $\bar{x}_1 - \bar{x}_2$. Describe what happened.

14.3 Describe the effect of changing the sample sizes from $n_1 = n_2 = 2$ to $n_1 = n_2 = 20$ on the sampling distribution of $\bar{x}_1$, the sampling distribution of $\bar{x}_2$, and the sampling distribution of $\bar{x}_1 - \bar{x}_2$. Describe the effect.

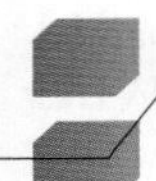

EXERCISES

9.45 The manager of a restaurant believes that waiters and waitresses who introduce themselves by telling customers their names will get larger tips than those who don't. In fact, she claims that the average tip for the former group is 18% whereas that of the latter is only 15%. If tips are normally distributed with a standard deviation of 3%, what is the probability that in a random sample of 10 tips recorded from waiters and waitresses who introduce themselves and 10 tips from waiters and waitresses who don't, the mean of the former will exceed that of the latter?

9.46 A factory's worker productivity is normally distributed. One worker produces an average of 75 units per day with a standard deviation of 20. Another worker produces at an average rate of 65 per day with a standard deviation of 21. What is the probability that during one week (5 working days) worker 1 will outproduce worker 2?

9.47 A professor of statistics noticed that the marks in his course are normally distributed. He has also noticed that his morning classes average 73% with a standard deviation of 12% on their final exams. His afternoon classes average 77% with a standard deviation of 10%. What is the probability that the mean mark of four randomly selected students from a morning class is greater than the average mark of four randomly selected students from an afternoon class?

9.48 The average North American loses an average of 15 days per year due to colds and flu. The natural remedy echinacea is reputed to boost the immune system. One manufacturer of echinacea pills claims that consumers of its product will reduce the number of days lost to colds and flu by one-third. To test the claim, a random sample of 50 people was drawn. Half took echinacea, and the other half took placebos. If we assume that the standard deviation of the number of days lost to colds and flu with and without echinacea is 3 days, find the probability that the mean number of days lost for echinacea users is less than that for nonusers.

9.49 Suppose that we have two normal populations with the means and standard deviations listed here. If random samples of size 25 are drawn from each population, what is the probability that the mean of sample 1 is greater than the mean of sample 2?

Population 1: $\mu = 40, \sigma = 6$

Population 2: $\mu = 38, \sigma = 8$

9.50 Repeat Exercise 9.49 assuming that the standard deviations are 12 and 16, respectively.

9.51 Repeat Exercise 9.49 assuming that the means are 140 and 138, respectively.

9.52 Independent random samples of 10 observations each are drawn from normal populations. The parameters of these populations are

Population 1: $\mu = 280, \sigma = 25$

Population 2: $\mu = 270, \sigma = 30$

Find the probability that the mean of sample 1 is greater than the mean of sample 2 by more than 25.

9.53 Repeat Exercise 9.52 with samples of size 50.

9.54 Repeat Exercise 9.52 with samples of size 100.

9.4 FROM HERE TO INFERENCE

The primary function of the sampling distribution is statistical inference. To see how the sampling distribution contributes to the development of inferential methods, we need to briefly review how we got to this point.

In Chapters 7 and 8 we introduced probability distributions, which allowed us to make probability statements about values of the random variable. A prerequisite of this calculation is knowledge of the distribution and the relevant parameters. In Example 7.9, we needed to know that the probability that Pat Statsdud guesses the correct answer is 20% ($p = .2$) and that the number of correct answers (successes) in 10 questions (trials) is a binomial random variable. We could then compute the probability of any number of successes. In Example 8.3, we needed to know that the return on investment is normally distributed with a mean of 10% and a standard deviation of 5%. These three bits of information allowed us to calculate the probability of various values of the random variable.

Figure 9.10 symbolically represents the use of probability distributions. Simply put, knowledge of the population and its parameter(s) allows us to use the probability distribution to make probability statements about individual members of the population. The direction of the arrows indicates the direction of the flow of information.

FIGURE **9.10**
Probability Distribution

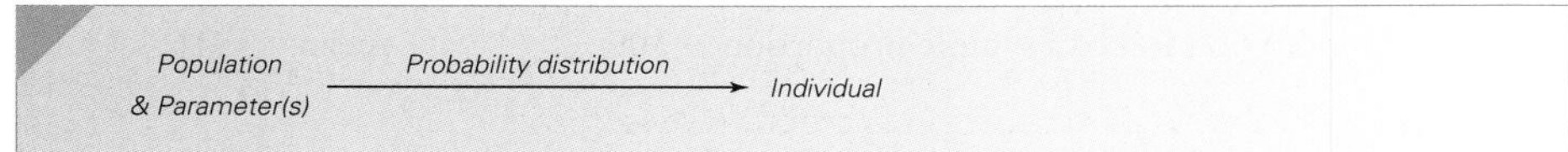

In this chapter we developed the sampling distribution, wherein knowledge of the parameter(s) and some information about the distribution allow us to make probability statements about a sample statistic. In Example 9.1b, knowing the population mean and standard deviation and assuming that the population is not extremely nonnormal enabled us to calculate a probability statement about a sample mean. Figure 9.11 describes the application of sampling distributions.

FIGURE **9.11**
Sampling Distribution

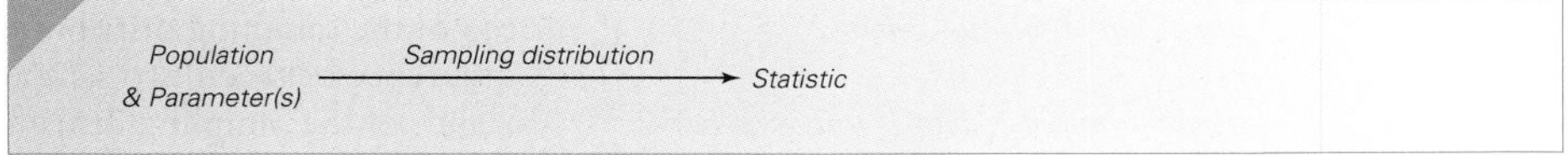

Notice that in applying both probability distributions and sampling distributions, we must know the value of the relevant parameters, a highly unlikely circumstance. In the real world, parameters are almost always unknown because they represent descriptive measurements about extremely large populations. Statistical inference addresses this problem. It does so by reversing the direction of the flow of knowledge in Figure 9.11. In Figure 9.12 we display the character of statistical inference. Starting in Chapter 10, we will assume that most population parameters are unknown. The statistics practitioner will sample from the population and compute the required statistic. The sampling distribution of that statistic will enable us to draw inferences about the parameter.

FIGURE **9.12**
Sampling Distribution in Inference

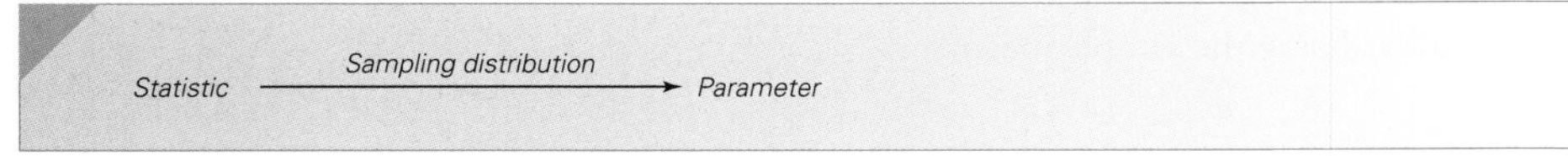

You may be surprised to learn that, by and large, that is all we do in the remainder of this book. Why then do we need another 14 chapters? They are necessary because there are many more parameter and sampling distribution combinations that define the inferential procedures to be presented in an introductory statistics course. However, they all work in the same way. If you understand how one procedure is developed, you will likely understand all of them. Our task in the next two chapters is to ensure that you understand the first inferential method. Your job is identical.

Chapter Summary

The sampling distribution of a statistic is created by repeated sampling from one population. In this chapter we introduced the sampling distribution of the mean, the proportion, and the difference between two means. We described how these distributions are created theoretically and empirically.

Important Terms

Sampling distribution 296
Standard error of the mean 300
Central limit theorem 300
Finite population correction factor 301
Sampling distribution of the sample mean 301
Sampling distribution of a sample proportion 309
Continuity correction factor 311
Standard error of the proportion 313
Sampling distribution of the difference between two sample means 315
Standard error of the difference between two means 316

Symbols

Symbol	Pronounced	Represents
$\mu_{\bar{x}}$	*mu-x-bar*	Mean of the sampling distribution of the sample mean
$\sigma^2_{\bar{x}}$	*sigma-squared-x-bar*	Variance of the sampling distribution of the sample mean
$\sigma_{\bar{x}}$	*sigma-x-bar*	Standard deviation (standard error) of the sampling distribution of the sample mean
α	*alpha*	Probability
$\hat{P}$	*p-hat*	Sample proportion
$\sigma^2_{\hat{p}}$	*sigma-squared-p-hat*	Variance of the sampling distribution of the sample proportion
$\sigma_{\hat{p}}$	*sigma-p-hat*	Standard deviation (standard error) of the sampling distribution of the sample proportion
$\mu_{\bar{x}_1-\bar{x}_2}$	*mu-x-bar-1-minus-x-bar-2*	Mean of the sampling distribution of the difference between two sample means
$\sigma^2_{\bar{x}_1-\bar{x}_2}$	*sigma-squared-x-bar-1-minus-x-bar-2*	Variance of the sampling distribution of the difference between two sample means
$\sigma_{\bar{x}_1-\bar{x}_2}$	*sigma-x-bar-1-minus-x-bar-2*	Standard deviation (standard error) of the sampling distribution of the difference between two sample means

Formulas

Expected value of the sample mean

$$E(\bar{X}) = \mu_{\bar{x}} = \mu$$

Variance of the sample mean

$$V(\bar{X}) = \sigma^2_{\bar{x}} = \frac{\sigma^2}{n}$$

Standard error of the sample mean

$$\sigma_{\bar{x}} = \frac{\sigma}{\sqrt{n}}$$

Standardizing the sample mean

$$Z = \frac{\bar{X} - \mu}{\sigma/\sqrt{n}}$$

Expected value of the sample proportion

$$E(\hat{P}) = \mu_{\hat{p}} = p$$

Variance of the sample proportion

$$V(\hat{P}) = \sigma^2_{\hat{p}} = \frac{p(1-p)}{n}$$

Standard error of the sample proportion

$$\sigma_{\hat{p}} = \sqrt{\frac{p(1-p)}{n}}$$

Standardizing the sample proportion

$$Z = \frac{\hat{P} - p}{\sqrt{p(1-p)/n}}$$

Expected value of the difference between two means

$$E(\bar{X}_1 - \bar{X}_2) = \mu_{\bar{x}_1-\bar{x}_2} = \mu_1 - \mu_2$$

Variance of the difference between two means

$$V(\bar{X}_1 - \bar{X}_2) = \sigma^2_{\bar{x}_1-\bar{x}_2} = \frac{\sigma_1^2}{n_1} + \frac{\sigma_2^2}{n_2}$$

Standard error of the difference between two means

$$\sigma_{\bar{x}_1-\bar{x}_2} = \sqrt{\frac{\sigma_1^2}{n_1} + \frac{\sigma_2^2}{n_2}}$$

Standardizing the difference between two sample means

$$Z = \frac{(\bar{X}_1 - \bar{X}_2) - (\mu_1 - \mu_2)}{\sqrt{\frac{\sigma_1^2}{n_1} + \frac{\sigma_2^2}{n_2}}}$$

10

© age fotostock/Super Stock

INTRODUCTION TO ESTIMATION

Determining the Sample Size to Estimate the Mean Tree Diameter

A lumber company has just acquired the rights to a large tract of land containing thousands of trees. Lumber companies need to be able to estimate the amount of lumber that they can harvest in a tract of land to determine whether the effort will be profitable. To do so, they must estimate the mean diameter of the trees. It has been decided to estimate that parameter to within 1 inch with 90% confidence. A forester familiar with the territory guesses that the diameters of the trees are normally distributed with a standard deviation of 6 inches. Using the formula on page 340, he determines that he should sample 98 trees. After sampling 98 trees, the forester calculates the sample mean to be 25 inches. Suppose that after he has completed his sampling and calculations, he discovers that the actual standard deviation is 12 inches. Will he be satisfied with the result?

© Randy Mayor/Botanical/Jupiterimages

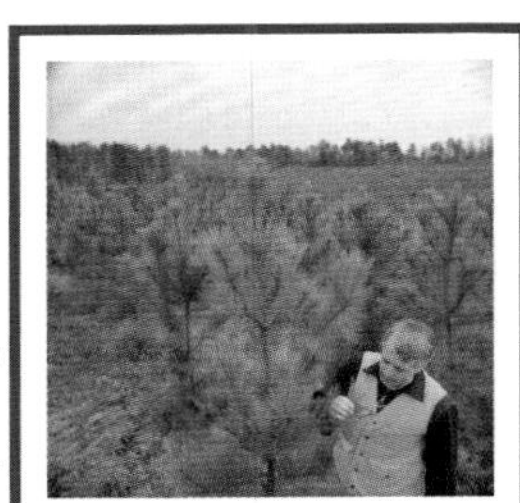

See page 340 for the solution.

INTRODUCTION

Having discussed descriptive statistics (Chapter 4), probability distributions (Chapters 7 and 8), and sampling distributions (Chapter 9), we are ready to tackle statistical inference. As we explained in Chapter 1, *statistical inference* is the process by which we acquire information and draw conclusions about populations from samples. There are two general procedures for making inferences about populations: *estimation* and *hypothesis testing*. In this chapter, we introduce the concepts and foundations of estimation and demonstrate them with simple examples. In Chapter 11, we describe the fundamentals of hypothesis testing. Because most of what we do in the remainder of this book applies the concepts of estimation and hypothesis testing, understanding Chapters 10 and 11 is vital to your development as a statistics practitioner.

10.1 CONCEPTS OF ESTIMATION

As its name suggests, the objective of estimation is to determine the approximate value of a population parameter on the basis of a sample statistic. For example, the sample mean is employed to estimate the population mean. We refer to the sample mean as the *estimator* of the population mean. Once the sample mean has been computed, its value is called the *estimate*. In this chapter we will introduce the statistical process whereby we estimate a population mean using sample data. In the rest of the book, we use the concepts and techniques introduced here for other parameters.

Point and Interval Estimators

We can use sample data to estimate a population parameter in two ways. First, we can compute the value of the estimator and consider that value as the estimate of the parameter. Such an estimator is called a *point estimator*.

Point Estimator

A **point estimator** draws inferences about a population by estimating the value of an unknown parameter using a single value or point.

There are three drawbacks to using point estimators. First, it is virtually certain that the estimate will be wrong. (The probability that a continuous random variable will equal a specific value is 0. That is, the probability that $\bar{x}$ will exactly equal μ is 0.) Second, we often need to know how close the estimator is to the parameter. Third, in drawing inferences about a population, it is intuitively reasonable to expect that a large sample will produce more accurate results, because it contains more information than a smaller sample does. But point estimators don't have the capacity to reflect the effects of larger sample sizes. As a consequence we use the second method of estimating a population parameter, the *interval estimator*.

Interval Estimator

An **interval estimator** draws inferences about a population by estimating the value of an unknown parameter using an interval.

As you will see, the interval estimator is affected by the sample size; because it possesses this feature, we will deal mostly with interval estimators in this text.

To illustrate the difference between point and interval estimators, suppose that a statistics professor wants to estimate the mean summer income of his second-year business students. Selecting 25 students at random, he calculates the sample mean weekly income to be \$400. The point estimate is the sample mean. That is, he estimates the mean weekly summer income of all second-year business students to be \$400. Using the technique described subsequently, he may instead use an interval estimate; he estimates that the mean weekly summer income of second-year business students to lie between \$380 and \$420.

Numerous applications of estimation occur in the real world. For example, television network executives want to know the proportion of television viewers who are tuned in to their networks; an economist wants to know the mean income of university graduates and, a medical researcher wishes to estimate the recovery rate of heart attack victims treated with a new drug. In each of these cases, to accomplish the objective exactly, the statistics practitioner would have to examine each member of the population and then calculate the parameter of interest. For instance, network executives would have to ask each person in the country what he or she is watching to determine the proportion of people who are watching their shows. Since there are millions of television viewers, the task is both impractical and prohibitively expensive. An alternative would be to take a random sample from this population, calculate the sample proportion, and use that as an estimator of the population proportion. The use of the sample proportion to estimate the population proportion seems logical. The selection of the sample statistic to be used as an estimator, however, depends on the characteristics of that statistic. Naturally, we want to use the statistic with the most desirable qualities for our purposes.

One desirable quality of an estimator is *unbiasedness.*

Unbiased Estimator

An **unbiased estimator** of a population parameter is an estimator whose expected value is equal to that parameter.

This means that, if you were to take an infinite number of samples and calculate the value of the estimator in each sample, the average value of the estimators would equal the parameter. This amounts to saying that, on average, the sample statistic is equal to the parameter.

We know that the sample mean $\overline{X}$ is an unbiased estimator of the population mean μ. In presenting the sampling distribution of $\overline{X}$ in Section 9.1, we stated that $E(\overline{X}) = \mu$. We also know that the sample proportion is an unbiased estimator of the population proportion because $E(\hat{P}) = p$ and that the difference between two sample means is an unbiased estimator of the difference between two population means because $E(\overline{X}_1 - \overline{X}_2) = \mu_1 - \mu_2$.

Recall that in Chapter 4 we defined the sample variance as

$$s^2 = \frac{\sum (x_i - \bar{x})^2}{n - 1}$$

At the time, it seemed odd that we divided by $n - 1$ rather than by n. The reason for choosing $n - 1$ was to make $E(s^2) = \sigma^2$ so that this definition makes the sample variance an unbiased estimator of the population variance. (The proof of this statement requires about a page of algebraic manipulation, which is more than we would be comfortable presenting here.) Had we defined the sample variance using n in the denominator, the resulting statistic would be a biased estimator of the population variance, one whose expected value is less than the parameter.

Knowing that an estimator is unbiased only assures us that its expected value equals the parameter; it does not tell us how close the estimator is to the parameter. Another desirable quality is that as the sample size grows larger, the sample statistic should come closer to the population parameter. This quality is called *consistency.*

Consistency

An unbiased estimator is said to be **consistent** if the difference between the estimator and the parameter grows smaller as the sample size grows larger.

The measure we use to gauge closeness is the variance (or the standard deviation). Thus, $\bar{X}$ is a consistent estimator of μ, because the variance of $\bar{X}$ is σ^2/n. This implies that as n grows larger, the variance of $\bar{X}$ grows smaller. As a consequence, an increasing proportion of sample means falls close to μ.

Figure 10.1 depicts two sampling distributions of $\bar{X}$ when samples are drawn from a population whose mean is 0 and whose standard deviation is 10. One sampling distribution is based on samples of size 25, and the other is based on samples of size 100. The former is more spread out than the latter.

FIGURE **10.1** Sampling Distribution of $\bar{X}$ with $n = 25$ and $n = 100$

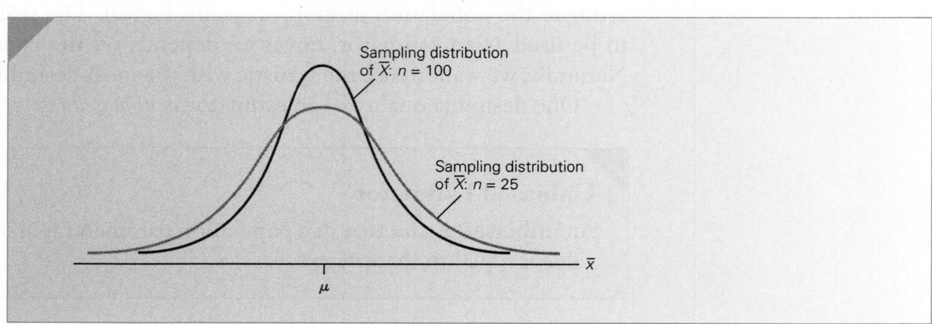

Similarly, $\hat{P}$ is a consistent estimator of p because it is unbiased and the variance of $\hat{P}$ is $p(1-p)/n$, which grows smaller as n grows larger.

A third desirable quality is *relative efficiency,* which compares two unbiased estimators of a parameter.

Relative Efficiency

If there are two unbiased estimators of a parameter, the one whose variance is smaller is said to be **relatively more efficient.**

We have already seen that the sample mean is an unbiased estimator of the population mean and that its variance is σ^2/n. In the next section we will discuss the use of the sample median as an estimator of the population mean. Statisticians have established that the sample median is an unbiased estimator but that its variance is greater than that of the sample mean (when the population is normal). As a consequence, the sample mean is relatively more efficient than the sample median when estimating the population mean.

In the remaining chapters of this book, we will present the statistical inference of a number of different population parameters. In each case, we will select a sample statistic that is unbiased and consistent. When there is more than one such statistic, we will choose the one that is relatively efficient to serve as the estimator.

Developing an Understanding of Statistical Concepts

In this section we described three desirable characteristics of estimators: unbiasedness, consistency, and relative efficiency. An understanding of statistics requires that you know that there are several potential estimators for each parameter, but that we choose the estimators used in this book because they possess these characteristics.

EXERCISES

10.1 How do point estimators and interval estimators differ?

10.2 Define unbiasedness.

10.3 Draw a sampling distribution of an unbiased estimator.

10.4 Draw a sampling distribution of a biased estimator.

10.5 Define consistency.

10.6 Draw diagrams representing what happens to the sampling distribution of a consistent estimator when the sample size increases.

10.7 Define relative efficiency.

10.8 Draw a diagram that shows the sampling distribution representing two unbiased estimators, one of which is relatively efficient.

10.2 ESTIMATING THE POPULATION MEAN WHEN THE POPULATION STANDARD DEVIATION IS KNOWN

We now describe how an interval estimator is produced from a sampling distribution. We choose to demonstrate estimation with an example that is unrealistic. However, this liability is offset by the example's simplicity. When you understand more about estimation, you will be able to apply the technique to more realistic situations.

Suppose we have a population with mean μ and standard deviation σ. The population mean is assumed to be unknown, and our task is to estimate its value. As we just discussed, the estimation procedure requires the statistics practitioner to draw a random sample of size n and calculate the sample mean $\bar{x}$.

The central limit theorem presented in Section 9.1 stated that $\bar{X}$ is normally distributed if X is normally distributed, or approximately normally distributed if X is nonnormal and n is sufficiently large. This means that the variable

$$Z = \frac{\bar{X} - \mu}{\sigma/\sqrt{n}}$$

is standard normally distributed (or approximately so). In Section 9.1 (page 306) we developed the following probability statement associated with the sampling distribution of the mean:

$$P\left(\mu - Z_{\alpha/2}\frac{\sigma}{\sqrt{n}} < \bar{X} < \mu + Z_{\alpha/2}\frac{\sigma}{\sqrt{n}}\right) = 1 - \alpha$$

which was derived from

$$P\left(-Z_{\alpha/2} < \frac{\overline{X} - \mu}{\sigma/\sqrt{n}} < Z_{\alpha/2}\right) = 1 - \alpha$$

Using a similar algebraic manipulation, we can express the probability in a slightly different form:

$$P\left(\overline{X} - Z_{\alpha/2}\frac{\sigma}{\sqrt{n}} < \mu < \overline{X} + Z_{\alpha/2}\frac{\sigma}{\sqrt{n}}\right) = 1 - \alpha$$

Notice that in this form the population mean is in the center of the interval created by adding and subtracting $Z_{\alpha/2}$ standard errors to and from the sample mean. It is important for you to understand that this is merely another form of probability statement about the sample mean. This equation says that, with repeated sampling from this population, the proportion of values of $\overline{X}$ for which the interval

$$\overline{X} - Z_{\alpha/2}\frac{\sigma}{\sqrt{n}}, \quad \overline{X} + Z_{\alpha/2}\frac{\sigma}{\sqrt{n}}$$

includes the population mean μ is equal to $1 - \alpha$. This form of probability statement is very useful to us because it is the **confidence interval estimator of μ.**

Confidence Interval Estimator of μ*

$$\bar{x} - z_{\alpha/2}\frac{\sigma}{\sqrt{n}}, \quad \bar{x} + z_{\alpha/2}\frac{\sigma}{\sqrt{n}}$$

The probability $1 - \alpha$ is called the **confidence level.**

$\bar{x} - z_{\alpha/2}\frac{\sigma}{\sqrt{n}}$ is called the **lower confidence limit (LCL).**

$\bar{x} + z_{\alpha/2}\frac{\sigma}{\sqrt{n}}$ is called the **upper confidence limit (UCL).**

We often represent the confidence interval estimator as

$$\bar{x} \pm z_{\alpha/2}\frac{\sigma}{\sqrt{n}}$$

where the minus sign defines the lower confidence limit and the plus sign defines the upper confidence limit.

To apply this formula we specify the confidence level $1 - \alpha$, from which we determine α, $\alpha/2$, $z_{\alpha/2}$ (from Table 3 in Appendix B). Because the confidence level is the

*Since Chapter 7 we've been using the convention whereby an uppercase letter (usually X) represents a random variable and a lowercase letter (usually x) represents one of its values. However, in the formulas used in statistical inference, the distinction between the variable and its value becomes blurred. Accordingly, we will discontinue the notational convention and simply use lowercase letters except when we wish to make a probability statement.

probability that the interval includes the actual value of μ, we generally set $1 - \alpha$ close to 1 (usually between .90 and .99).

In Table 10.1, we list four commonly used confidence levels and their associated values of $z_{\alpha/2}$. For example, if the confidence level is $1 - \alpha = .95$, $\alpha = .05$, $\alpha/2 = .025$, and $z_{\alpha/2} = z_{.025} = 1.96$. The resulting confidence interval estimator is then called the 95% confidence interval estimator of μ.

TABLE **10.1**
Four Commonly Used Confidence Levels and $z_{\alpha/2}$

$1 - \alpha$	α	$\alpha/2$	$z_{\alpha/2}$
.90	.10	.05	$z_{.05} = 1.645$
.95	.05	.025	$z_{.025} = 1.96$
.98	.02	.01	$z_{.01} = 2.33$
.99	.01	.005	$z_{.005} = 2.575$

The following example illustrates how statistical techniques are applied. It also illustrates how we intend to solve problems in the rest of this book. The solution process that we advocate and use throughout this book is by and large the same one that statistics practitioners use to apply their skills in the real world. The process is divided into three stages. Simply stated, the stages are (1) the activities we perform before the calculations, (2) the calculations, and (3) the activities we perform after the calculations.

In stage 1 we determine the appropriate statistical technique to employ. Of course, for this example you will have no difficulty identifying the technique, since at this point you know only one. (In practice, stage 1 also addresses the problem of *how* to gather the data. The methods used in the examples, exercises, and cases are described in the problem.)

In the second stage we calculate the statistics. We will do this in three ways.* To illustrate how the computations are completed, we will do the arithmetic manually with the assistance of a calculator. Solving problems by hand often provides insights into the statistical inference technique. Additionally, we will use the computer in two ways. First, in Excel we will use the Analysis ToolPak (**Data** menu item **Data Analysis**) or the add-ins we created for this book (**Add-Ins** menu item **Data Analysis Plus**). (Additionally, we will create, where possible, a spreadsheet that uses built-in statistical functions. Some will be exhibited in this book but most will be available in CD Appendix J only.) Finally, we will use Minitab, one of the easiest software packages to use.

In the third and last stage of the solution, we intend to interpret the results and deal with the question presented in the problem. To be capable of properly interpreting statistical results, one needs to have an understanding of the fundamental principles underlying statistical inference.

*We anticipate that students in most statistics classes will use only one of the three methods of computing statistics, the choice made by the instructor. If such is the case, readers are directed to ignore the other two.

APPLICATIONS in OPERATIONS MANAGEMENT

© Comstock Images/Jupiterimages

Inventory Management

Operations managers use inventory models to determine the stock level that minimizes total costs. In Section 8.2 we showed how the probabilistic model is used to make the inventory level decision (see page 272). One component of that model is the mean demand during lead time. Recall that lead time refers to the interval between the time an order is made and when it is delivered. Demand during lead time is a random variable that is often assumed to be normally distributed. There are several ways to determine mean demand during lead time, but the simplest is to estimate that quantity from a sample.

EXAMPLE 10.1 Doll Computer Company

DATA Xm10-01

The Doll Computer Company makes its own computers and delivers them directly to customers who order them via the Internet. Doll competes primarily on price and speed of delivery. To achieve its objective of speed, Doll makes each of its five most popular computers and transports them to warehouses across the country. The computers are stored in the warehouses from which it generally takes 1 day to deliver a computer to the customer. This strategy requires high levels of inventory that add considerably to the cost. To lower these costs, the operations manager wants to use an inventory model. He notes that both daily demand and lead time are random variables. He concludes that demand during lead time is normally distributed and he needs to know the mean to compute the optimum inventory level. He observes 25 lead time periods and records the demand during each period. These data are listed here. The manager would like a 95% confidence interval estimate of the mean demand during lead time. From long experience the manager knows that the standard deviation is 75 computers.

Demand During Lead Time

235	374	309	499	253
421	361	514	462	369
394	439	348	344	330
261	374	302	466	535
386	316	296	332	334

SOLUTION

IDENTIFY

To ultimately determine the optimum inventory level, the manager must know the mean demand during lead time. Thus, the parameter to be estimated is μ. At this point, we have described only one estimator. Thus, the confidence interval estimator that we intend to use is

$$\bar{x} \pm z_{\alpha/2}\frac{\sigma}{\sqrt{n}}$$

The next step is to perform the calculations. As we discussed previously, we will perform the calculations in three ways: manually, using Excel, and using Minitab.

COMPUTE

MANUALLY

We need four values to construct the confidence interval estimate of μ. They are

$$\bar{x} \quad z_{\alpha/2} \quad \sigma \quad n$$

Using a calculator, we determine the summation $\sum x_i = 9{,}254$. From this we find

$$\bar{x} = \frac{\sum x_i}{n} = \frac{9{,}254}{25} = 370.16$$

The confidence level is set at 95%; thus $1 - \alpha = .95$, $\alpha = 1 - .95 = .05$, and $\alpha/2 = .025$. From Table 3 in Appendix B or from Table 10.1, we find

$$z_{\alpha/2} = z_{.025} = 1.96$$

The population standard deviation is $\sigma = 75$, and the sample size is 25. Substituting $\bar{x}$, $z_{\alpha/2}$, σ, and n into the confidence interval estimator, we find

$$\bar{x} \pm z_{\alpha/2}\frac{\sigma}{\sqrt{n}} = 370.16 \pm z_{.025}\frac{75}{\sqrt{25}} = 370.16 \pm 1.96\frac{75}{\sqrt{25}} = 370.16 \pm 29.40$$

The lower and upper confidence limits are LCL = 340.76 and UCL = 399.56, respectively.

EXCEL

	A	B	C
1	*z*-Estimate: Mean		
2			
3			*Demand*
4	Mean		370.16
5	Standard Deviation		80.783
6	Observations		25
7	SIGMA		75
8	LCL		340.76
9	UCL		399.56

INSTRUCTIONS

1. Type or import the data into one column. (Open Xm10-01.)
2. Click **Add-Ins, Data Analysis Plus,** and **Z-Estimate: Mean.**
3. Fill in the dialog box: **Input Range** (A1:A26), type the value for the **Standard Deviation** (75), click **Labels** if the first row contains the name of the variable, and specify the confidence level by typing the value of α (.05).

There is another way to produce the interval estimate for this problem. If you have already calculated the sample mean and know the sample size and population standard deviation, you need not employ the data set and **Data Analysis Plus** described here. Instead, open the **Estimators** workbook (it will be in the same directory as the chapter directories).

(continued)

This workbook contains eight spreadsheets, each showing a confidence interval estimator that is presented in this book. Use the tabs to find and click **Z-Estimate_Mean.** The worksheet that will be opened represents the solution to Example 10.1. We typed the values of $\bar{x}$ (370.16), σ (75), and n (25) in cells B3, B4, and B5, respectively, and the confidence level (.95) in cell B6. The confidence interval estimator is automatically computed. The completed worksheet is shown here. As you can see we produce the same result.

Estimators Workbook

	A	B	C	D	E
1	z-Estimate of a Mean				
2					
3	Sample mean	370.16	Confidence Interval Estimate		
4	Population standard deviation	75	370.16	±	29.40
5	Sample size	25	Lower confidence limit		340.76
6	Confidence level	0.95	Upper confidence limit		399.56

In addition to providing another method of using Excel, this spreadsheet allows you to perform a "what-if" analysis. That is, this worksheet provides you the opportunity to learn how changing some of the inputs affects the estimate. For example, type 0.99 in cell B6 to see what happens to the size of the interval when you increase the confidence level. Type 1000 in cell B5 to examine the effect of increasing the sample size. Type 10 in cell B4 to see what happens when the population standard deviation is smaller.

There are eight spreadsheets in the **Estimators** workbook. Each is associated with another confidence interval estimator that will be introduced in Chapters 12 and 13.

MINITAB

One-Sample Z: Demand

The assumed standard deviation = 75

```
Variable   N     Mean   StDev  SE Mean          95% CI
Demand    25  370.160  80.783   15.000  (340.761, 399.559)
```

The output includes the sample standard deviation (**StDev** = 80.783), which is not needed for this interval estimate. Also printed is the standard error (**SE Mean** = $\sigma/\sqrt{n}$ = 15.0), and last, but not least, the 95% confidence interval estimate of the population mean.

INSTRUCTIONS

1. Type or import that data into one column. (Open Xm10-01.)
2. Click **Stat, Basic Statistics,** and **1-Sample Z . . . .**
3. Type or use the **Select** button to specify the name of the variable or the column it is stored in, in the **Samples in columns** box (Demand), type the value of the population standard deviation (75), and click **Options . . . .**
4. Type the value for the confidence level (.95) and in the **Alternative** box select **not equal.**

INTERPRET

The operations manager estimates that the mean demand during lead time lies between 340.76 and 399.56. He can use this estimate as an input in developing an inventory policy. The model discussed in Section 8.2 computes the reorder point assuming a particular value of the mean demand during lead time. In this example he could have used the sample mean as a point estimator of the mean demand, from which the inventory policy could be determined. However, the use of the confidence interval estimator allows the manager to use both the lower and upper limits so that he can understand the possible outcomes.

Interpreting the Confidence Interval Estimate

Some people erroneously interpret the confidence interval estimate in Example 10.1 to mean that there is a 95% probability that the population mean lies between 340.76 and 399.56. This interpretation is wrong because it implies that the population mean is a variable about which we can make probability statements. In fact, the population mean is a fixed, but unknown quantity. Consequently, we cannot interpret the confidence interval estimate of μ as a probability statement about μ. To translate the confidence interval estimate properly, we must remember that the confidence interval estimator was derived from the sampling distribution of the sample mean. In Section 9.1, we used the sampling distribution to make probability statements about the sample mean. Although the form has changed, the confidence interval estimator is also a probability statement about the sample mean. It states that there is $1 - \alpha$ probability that the sample mean will be equal to a value such that the interval $\bar{x} - z_{\alpha/2}\sigma/\sqrt{n}$ to $\bar{x} + z_{\alpha/2}\sigma/\sqrt{n}$ will include the population mean. Once the sample mean is computed, the interval acts as the lower and upper limits of the interval estimate of the population mean.

As an illustration, suppose we want to estimate the mean value of the distribution resulting from the throw of a fair die. Because we know the distribution, we also know that $\mu = 3.5$ and $\sigma = 1.71$. Pretend now that we know only that $\sigma = 1.71$, that μ is unknown, and that we want to estimate its value. To estimate μ, we draw a sample of size $n = 100$ and calculate $\bar{x}$. The confidence interval estimator of μ is

$$\bar{x} \pm z_{\alpha/2}\frac{\sigma}{\sqrt{n}}$$

The 90% confidence interval estimator is

$$\bar{x} \pm z_{\alpha/2}\frac{\sigma}{\sqrt{n}} = \bar{x} \pm 1.645\frac{1.71}{\sqrt{100}} = \bar{x} \pm .281$$

This notation means that, if we repeatedly draw samples of size 100 from this population, 90% of the values of $\bar{x}$ will be such that μ would lie somewhere between $\bar{x} - .281$ and $\bar{x} + .281$, and 10% of the values of $\bar{x}$ will produce intervals that would not include μ. Now, imagine that we draw 40 samples of 100 observations each. The values of $\bar{x}$ and the resulting confidence interval estimates of μ are shown in Table 10.2. Notice that not all the intervals include the true value of the parameter. Samples 5, 16, 22, and 34 produce values of $\bar{x}$ that in turn produce intervals that exclude μ.

Students often react to this situation by asking, What went wrong with samples 5, 16, 22, and 34? The answer is nothing. Statistics does not promise 100% certainty. In fact, in this illustration, we expected 90% of the intervals to include μ and 10% to exclude μ. Since we produced 40 intervals, we expected that 4.0 (10% of 40) intervals would not contain $\mu = 3.5$.* It is important to understand that, even when the statistics practitioner performs experiments properly, a certain proportion (in this example, 10%) of the experiments will produce incorrect estimates by random chance.

TABLE **10.2**
90% Confidence Interval Estimates of μ

Sample	$\bar{x}$	LCL = $\bar{x}$ − .281	UCL = $\bar{x}$ + .281	Does Interval Include μ = 3.5?
1	3.550	3.269	3.831	Yes
2	3.610	3.329	3.891	Yes
3	3.470	3.189	3.751	Yes
4	3.480	3.199	3.761	Yes
5	3.800	3.519	4.081	No
6	3.370	3.089	3.651	Yes
7	3.480	3.199	3.761	Yes
8	3.520	3.239	3.801	Yes
9	3.740	3.459	4.021	Yes
10	3.510	3.229	3.791	Yes
11	3.230	2.949	3.511	Yes
12	3.450	3.169	3.731	Yes
13	3.570	3.289	3.851	Yes
14	3.770	3.489	4.051	Yes
15	3.310	3.029	3.591	Yes
16	3.100	2.819	3.381	No
17	3.500	3.219	3.781	Yes
18	3.550	3.269	3.831	Yes
19	3.650	3.369	3.931	Yes
20	3.280	2.999	3.561	Yes
21	3.400	3.119	3.681	Yes
22	3.880	3.599	4.161	No
23	3.760	3.479	4.041	Yes
24	3.400	3.119	3.681	Yes
25	3.340	3.059	3.621	Yes
26	3.650	3.369	3.931	Yes
27	3.450	3.169	3.731	Yes
28	3.470	3.189	3.751	Yes
29	3.580	3.299	3.861	Yes
30	3.360	3.079	3.641	Yes
31	3.710	3.429	3.991	Yes
32	3.510	3.229	3.791	Yes
33	3.420	3.139	3.701	Yes
34	3.110	2.829	3.391	No
35	3.290	3.009	3.571	Yes
36	3.640	3.359	3.921	Yes
37	3.390	3.109	3.671	Yes
38	3.750	3.469	4.031	Yes
39	3.260	2.979	3.541	Yes
40	3.540	3.259	3.821	Yes

*In this illustration, exactly 10% of the sample means produced interval estimates that excluded the value of μ, but this will not always be the case. Remember, we expect 10% of the sample means in the long run to result in intervals excluding μ. This group of 40 sample means does not constitute "the long run."

We can improve the confidence associated with the interval estimate. If we let the confidence level $1 - \alpha$ equal .95, the 95% confidence interval estimator is

$$\bar{x} \pm z_{\alpha/2}\frac{\sigma}{\sqrt{n}} = \bar{x} \pm 1.96\frac{1.71}{\sqrt{100}} = \bar{x} \pm .335$$

Because this interval is wider, it is more likely to include the value of μ. If you redo Table 10.2, this time using a 95% confidence interval estimator, only samples 16, 22, and 34 will produce intervals that do not include μ. (Notice that we expected 5% of the intervals to exclude μ and that we actually observed $3/40 = 7.5\%$.) The 99% confidence interval estimator is

$$\bar{x} \pm z_{\alpha/2}\frac{\sigma}{\sqrt{n}} = \bar{x} \pm 2.575\frac{1.71}{\sqrt{100}} = \bar{x} \pm .440$$

Applying this interval estimate to the sample means listed in Table 10.2 would result in having all 40 interval estimates include the population mean $\mu = 3.5$. (We expected 1% of the intervals to exclude μ; we observed $0/40 = 0\%$.)

SEEING STATISTICS

Applet 15: Confidence Interval Estimates of a Mean

The simulations used in the applets introduced in Chapter 9 can be used here to demonstrate how confidence interval estimates are interpreted. This applet generates samples of size 100 from the population of the toss of a die. We know that the population mean is $\mu = 3.5$ and that the standard deviation is 1.71. The 95% confidence interval estimator is

$$\bar{x} \pm z_{\alpha/2}\frac{\sigma}{\sqrt{n}} = \bar{x} \pm 1.96\frac{1.71}{\sqrt{100}} = \bar{x} \pm .335$$

The applet will generate one sample, 10 samples, or 100 samples at a time. The resulting confidence interval is displayed as a horizontal line between the upper and lower ends of the confidence interval. The true mean of 3.5 is the green vertical line. If the confidence interval includes the true population mean of 3.5 (i.e., if the confidence interval line overlaps the green vertical line), it is displayed in blue. If the confidence interval does not include the true mean, it is displayed in red.

After you understand the basics, click the Sample 10 button a few times to see 10 confidence intervals (but not their calculations) at once. Then click on the Sample 100 button to generate 100 samples and confidence intervals.

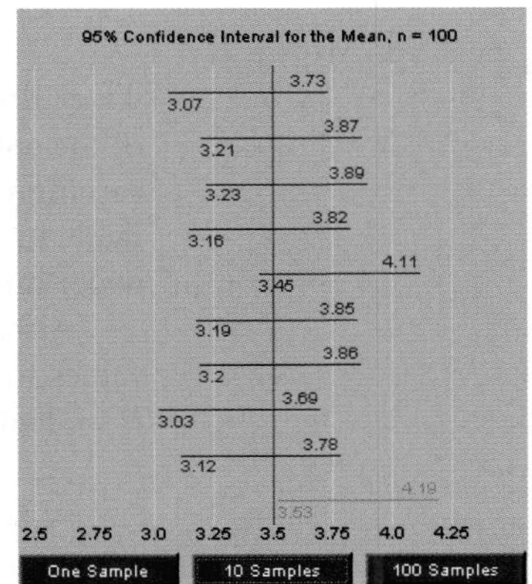

Applet Exercises

Simulate 100 samples.

15.1 Are all the confidence interval estimates identical?

15.2 Count the number of confidence interval estimates that include the true value of the mean.

15.3 How many intervals did you expect to see that correctly included the mean?

15.4 What do these exercises tell you about the proper interpretation of a confidence interval estimate?

In actual practice only one sample will be drawn and thus only one value of $\bar{x}$ will be calculated. The resulting interval estimate will either correctly include the parameter or incorrectly exclude it. Unfortunately, statistics practitioners do not know whether they are correct in each case; they know only that, in the long run, they will incorrectly estimate the parameter some of the time. Statistics practitioners accept that as a fact of life.

We summarize our calculations in Example 10.1 as follows. We estimate that the mean demand during lead time falls between 340.76 and 399.56, and this type of estimator is correct 95% of the time. Thus, the confidence level applies to our estimation procedure and not to any one interval. Incidentally, the media often refer to the 95% figure as "19 times out of 20," which emphasizes the long-run aspect of the confidence level.

Information and the Width of the Interval

Interval estimation, like all other statistical techniques, is designed to convert data into information. However, a wide interval provides little information. For example, suppose that as a result of a statistical study we estimate with 95% confidence that the average starting salary of an accountant lies between \$15,000 and \$100,000. This interval is so wide that very little information was derived from the data. Suppose, however, that the interval estimate was \$52,000 to \$55,000. This interval is much narrower, providing accounting students more precise information about the mean starting salary.

The width of the confidence interval estimate is a function of the population standard deviation, the confidence level, and the sample size. Consider Example 10.1, where σ was assumed to be 75. The interval estimate was 370.16 ± 29.40. If σ equaled 150, the 95% confidence interval estimate would become

$$\bar{x} \pm z_{\alpha/2}\frac{\sigma}{\sqrt{n}} = 370.16 \pm z_{.025}\frac{150}{\sqrt{25}} = 370.16 \pm 1.96\frac{150}{\sqrt{25}} = 370.16 \pm 58.80$$

Thus, doubling the population standard deviation has the effect of doubling the width of the confidence interval estimate. This result is quite logical. If there is a great deal of variation in the random variable (measured by a large standard deviation), it is more difficult to accurately estimate the population mean. That difficulty is translated into a wider interval.

Although we have no control over the value of σ, we do have the power to select values for the other two elements. In Example 10.1 we chose a 95% confidence level. If we had chosen 90% instead, the interval estimate would have been

$$\bar{x} \pm z_{\alpha/2}\frac{\sigma}{\sqrt{n}} = 370.16 \pm z_{.05}\frac{75}{\sqrt{25}} = 370.16 \pm 1.645\frac{75}{\sqrt{25}} = 370.16 \pm 24.68$$

A 99% confidence level results in this interval estimate:

$$\bar{x} \pm z_{\alpha/2}\frac{\sigma}{\sqrt{n}} = 370.16 \pm z_{.005}\frac{75}{\sqrt{25}} = 370.16 \pm 2.575\frac{75}{\sqrt{25}} = 370.16 \pm 38.63$$

As you can see, decreasing the confidence level narrows the interval; increasing it widens the interval. However, a large confidence level is generally desirable since that means a larger proportion of confidence interval estimates that will be correct in the long run. There is a direct relationship between the width of the interval and the confidence level. This is because in order to be more confident in the estimate we need to widen the interval. (The analogy is that to be more likely to capture a butterfly, we need a larger butterfly net.) The trade-off between increased confidence and the resulting wider confidence interval estimates must be resolved by the statistics practitioner. As a general rule, however, 95% confidence is considered "standard."

The third element is the sample size. Had the sample size been 100 instead of 25, the confidence interval estimate would become

$$\bar{x} \pm z_{\alpha/2}\frac{\sigma}{\sqrt{n}} = 370.16 \pm z_{.025}\frac{75}{\sqrt{100}} = 370.16 \pm 1.96\frac{75}{\sqrt{100}} = 370.16 \pm 14.70$$

Increasing the sample size fourfold decreases the width of the interval by half. A larger sample size provides more potential information. The increased amount of information is reflected in a narrower interval. However, there is another trade-off: Increasing the sample size increases the sampling cost. We will discuss these issues when we present sample size selection in Section 10.3.

(Optional) Estimating the Population Mean Using the Sample Median

To understand why the sample mean is most often used to estimate a population mean, let's examine the properties of the sampling distribution of the sample median (denoted here as m). The sampling distribution of a sample median is normally distributed provided that the population is normal. Its mean and standard deviation are

$$\mu_m = \mu$$

and

$$\sigma_m = \frac{1.2533\sigma}{\sqrt{n}}$$

Using the same algebraic steps that we used above, we derive the confidence interval estimator of a population mean using the sample median

$$m \pm z_{\alpha/2}\frac{1.2533\sigma}{\sqrt{n}}$$

To illustrate, suppose that we have drawn the following random sample from a normal population whose standard deviation is 2.

1 1 1 3 4 5 6 7 8

The sample mean is $\bar{x} = 4$ and the median is $m = 4$.
The 95% confidence interval estimates using the sample mean and the sample median are

$$\bar{x} \pm z_{\alpha/2}\frac{\sigma}{\sqrt{n}} = 4.0 \pm 1.96\frac{2}{\sqrt{9}} = 4 \pm 1.307$$

$$m \pm z_{\alpha/2}\frac{1.2533\sigma}{\sqrt{n}} = 4.0 \pm 1.96\frac{(1.2533)(2)}{\sqrt{9}} = 4 \pm 1.638$$

As you can see, the interval based on the sample mean is narrower; as we pointed out previously narrower intervals provide more precise information. To understand why the sample mean produces better estimators than the sample median, recall how the median is calculated. We simply put the data in order and select the observation that falls in the middle. Thus, as far as the median is concerned the data appear as

1 2 3 4 5 6 7 8 9

By ignoring the actual observations and using their ranks instead, we lose information. With less information we have less precision in the interval estimators and so ultimately make poorer decisions.

EXERCISES

Developing an Understanding of Statistical Concepts

Exercises 10.9 to 10.16 are "what-if analyses" designed to determine what happens to the interval estimate when the confidence level, sample size, and standard deviation change. These problems can be solved manually or using Excel's ***Z-Estimate_Mean*** *worksheet in the* ***Estimators*** *workbook.*

10.9 a. A statistics practitioner took a random sample of 50 observations from a population whose standard deviation is 25 and computed the sample mean to be 100. Estimate the population mean with 90% confidence.

b. Repeat Part a using a 95% confidence level.

c. Repeat Part a using a 99% confidence level.

d. Describe the effect on the confidence interval estimate of increasing the confidence level.

10.10 a. The mean of a random sample of 25 observations from a normal population whose standard deviation is 50 is 200. Estimate the population mean with 95% confidence

b. Repeat Part a changing the population standard deviation to 25.

c. Repeat Part a changing the population standard deviation to 10.

d. Describe what happens to the confidence interval estimate when the standard deviation is decreased.

10.11 a. A random sample of 25 was drawn from a normal distribution whose standard deviation is 5. The sample mean is 80. Determine the 95% confidence interval estimate of the population mean.

b. Repeat Part a with a sample size of 100.

c. Repeat Part a with a sample size of 400.

d. Describe what happens to the confidence interval estimate when the sample size increases.

10.12 a. Given the following information, determine the 98% confidence interval estimate of the population mean:

$$\bar{x} = 500 \quad \sigma = 12 \quad n = 50$$

b. Repeat Part a using a 95% confidence level.

c. Repeat Part a using a 90% confidence level.

d. Review Parts a–c and discuss the effect on the confidence interval estimator of decreasing the confidence level.

10.13 a. The mean of a sample of 25 was calculated as $\bar{x} = 500$. The sample was randomly drawn from a population whose standard deviation is 15. Estimate the population mean with 99% confidence.

b. Repeat Part a changing the population standard deviation to 30.

c. Repeat Part a changing the population standard deviation to 60.

d. Describe what happens to the confidence interval estimate when the standard deviation is increased.

10.14 a. A statistics practitioner randomly sampled 100 observations from a population whose standard deviation is 5 and found that $\bar{x}$ is 10. Estimate the population mean with 90% confidence.

b. Repeat Part a with a sample size of 25.

c. Repeat Part a with a sample size of 10.

d. Describe what happens to the confidence interval estimate when the sample size decreases.

10.15 a. From the information given here, determine the 95% confidence interval estimate of the population mean.

$$\bar{x} = 100 \quad \sigma = 20 \quad n = 25$$

b. Repeat Part a with $\bar{x} = 200$.

c. Repeat Part a with $\bar{x} = 500$.

d. Describe what happens to the width of the confidence interval estimate when the sample mean increases.

10.16 a. A random sample of 100 observations was randomly drawn from a population whose standard deviation is 5. The sample mean was calculated as $\bar{x} = 400$. Estimate the population mean with 99% confidence.

b. Repeat Part a with $\bar{x} = 200$.

c. Repeat Part a with $\bar{x} = 100$.

d. Describe what happens to the width of the confidence interval estimate when the sample mean decreases.

Exercises 10.17 to 10.20 are based on the optional subsection "Estimating the Population Mean Using the Sample Median." All exercises assume that the population is normal.

10.17 Is the sample median an unbiased estimator of the population mean? Explain.

10.18 Is the sample median a consistent estimator of the population mean? Explain.

10.19 Show that the sample mean is relatively more efficient than the sample median when estimating the population mean.

10.20 a. Given the following information, determine the 90% confidence interval estimate of the population mean using the sample median.

Sample median = 500, $\sigma = 12$, and $n = 50$

b. Compare your answer in Part a to that produced in Part c of Exercise 10.12. Why is the confidence interval estimate based on the sample median wider than that based on the sample mean?

Applications

The following exercises may be answered manually or with the assistance of a computer. The names of the files containing the data are shown.

10.21 Xr10-21 The following data represent a random sample of 9 marks (out of 10) on a statistics quiz. The marks are normally distributed with a standard deviation of 2. Estimate the population mean with 90% confidence.

7 9 7 5 4 8 3 10 9

10.22 Xr10-22 The following observations are the ages of a random sample of 8 men in a bar. It is known that the ages are normally distributed with a standard deviation of 10. Determine the 95% confidence interval estimate of the population mean. Interpret the interval estimate.

52 68 22 35 30 56 39 48

10.23 Xr10-23 How many rounds of golf do physicians (who play golf) play per year? A survey of 12 physicians revealed the following numbers:

3 41 17 1 33 37 18 15 17 12 29 51

Estimate with 95% confidence the mean number of rounds per year played by physicians, assuming that the number of rounds is normally distributed with a standard deviation of 12.

10.24 Xr10-24 Among the most exciting aspects of a university professor's life are the departmental meetings where such critical issues as the color the walls will be painted and who gets a new desk are decided. A sample of 20 professors was asked how many hours per year are devoted to these meetings. The responses are listed here. Assuming that the variable is normally distributed with a standard deviation of 8 hours, estimate the mean number of hours spent at departmental meetings by all professors. Use a confidence level of 90%.

14 17 3 6 17 3 8 4 20 15
7 9 0 5 11 15 18 13 8 4

10.25 Xr10-25 The number of cars sold annually by used car salespeople is normally distributed with a standard deviation of 15. A random sample of 15 salespeople was taken and the number of cars each sold is listed here. Find the 95% confidence interval estimate of the population mean. Interpret the interval estimate.

79 43 58 66 101 63 79 33 58
71 60 101 74 55 88

10.26 Xr10-26 It is known that the amount of time needed to change the oil on a car is normally distributed with a standard deviation of 5 minutes. The amount of time to complete a random sample of 10 oil changes was recorded and listed here. Compute the 99% confidence interval estimate of the mean of the population.

11 10 16 15 18 12 25 20 18 24

10.27 Xr10-27 Suppose that the amount of time teenagers spend weekly working at part-time jobs is normally distributed with a standard deviation of 40 minutes. A random sample of 15 teenagers was drawn and each reported the amount of time spent at part-time jobs (in minutes). These are listed here. Determine the 95% confidence interval estimate of the population mean.

180 130 150 165 90 130 120 60 200
180 80 240 210 150 125

10.28 Xr10-28 One of the few negative side effects of quitting smoking is weight gain. Suppose that the weight gain in the 12 months following a cessation in smoking is normally distributed with a standard deviation of 6 pounds. To estimate the mean weight gain, a random sample of 13 quitters was drawn and their weights recorded and listed here. Determine the 90% confidence interval estimate of the mean 12-month weight gain for all quitters.

16 23 8 2 14 22 18 11 10 19 5 8 15

10.29 Xr10-29 Because of different sales ability, experience, and devotion, the incomes of real estate agents vary considerably. Suppose that in a large city the annual income is normally distributed with a standard deviation of $15,000. A random sample of 16 real estate agents was asked to report their annual income (in $1,000). The responses are listed here. Determine the 99% confidence interval estimate of the mean annual income of all real estate agents in the city.

65 94 57 111 83 61 50 73 68 80
93 84 113 41 60 77

The following exercises require the use of a computer and software. The answers may be calculated manually. See Appendix A for the sample statistics.

10.30 Xr10-30 A survey of 400 statistics professors was undertaken. Each professor was asked how much time was devoted to teaching graphical techniques. We believe that the times are normally distributed with a standard deviation of 30 minutes. Estimate the population mean with 95% confidence.

10.31 Xr10-31 In a survey conducted to determine, among other things, the cost of vacations, 64 individuals were randomly sampled. Each person was asked to compute the cost of her or his most recent vacation. Assuming that the standard deviation is \$400, estimate with 95% confidence the average cost of all vacations.

10.32 Xr10-32 In an article about *disinflation*, various investments were examined. The investments included stocks, bonds, and real estate. Suppose that a random sample of 200 rates of return on real estate investments was computed and recorded. Assuming that the standard deviation of all rates of return on real estate investments is 2.1%, estimate the mean rate of return on all real estate investments with 90% confidence. Interpret the estimate.

10.33 Xr10-33 A statistics professor is in the process of investigating how many classes university students miss each semester. To help answer this question, she took a random sample of 100 university students and asked each to report how many classes he or she had missed in the previous semester. Estimate the mean number of classes missed by all students at the university. Use a 99% confidence level and assume that the population standard deviation is known to be 2.2 classes.

10.34 Xr10-34 As part of a project to develop better lawn fertilizers, a research chemist wanted to determine the mean weekly growth rate of Kentucky bluegrass, a common type of grass. A sample of 250 blades of grass was measured, and the amount of growth in 1 week was recorded. Assuming that weekly growth is normally distributed with a standard deviation of .10 inch, estimate with 99% confidence the mean weekly growth of Kentucky bluegrass. Briefly describe what the interval estimate tells you about the growth of Kentucky bluegrass.

10.35 Xr10-35 A time study of a large production facility was undertaken to determine the mean time required to assemble a cell phone. A random sample of the times to assemble 50 cell phones was recorded. An analysis of the assembly times reveals that they are normally distributed with a standard deviation of 1.3 minutes. Estimate with 95% confidence the mean assembly time for all cell phones. What do your results tell you about the assembly times?

10.36 Xr10-36 The image of the Japanese manager is that of a workaholic with little or no leisure time. In a survey, a random sample of 250 Japanese middle managers was asked how many hours per week they spent in leisure activities (e.g., sports, movies, television). The results of the survey were recorded. Assuming that the population standard deviation is 6 hours, estimate with 90% confidence the mean leisure time per week for all Japanese middle managers. What do these results tell you?

10.37 Xr10-37 One measure of physical fitness is the amount of time it takes for the pulse rate to return to normal after exercise. A random sample of 100 women age 40 to 50 exercised on stationary bicycles for 30 minutes. The amount of time it took for their pulse rates to return to pre-exercise levels was measured and recorded. If the times are normally distributed with a standard deviation of 2.3 minutes, estimate with 99% confidence the true mean pulse-recovery time for all 40- to 50-year-old women. Interpret the results.

10.38 Xr10-38 A survey of 80 randomly selected companies asked them to report the annual income of their presidents. Assuming that incomes are normally distributed with a standard deviation of \$30,000, determine the 90% confidence interval estimate of the mean annual income of all company presidents. Interpret the statistical results.

10.39 Xr10-39 To help make a decision about expansion plans, the president of a music company needs to know how many compact discs teenagers buy annually. Accordingly, he commissions a survey of 250 teenagers. Each is asked to report how many CDs he or she purchased in the previous 12 months. Estimate with 90% confidence the mean annual number of CDs purchased by all teenagers. Assume that the population standard deviation is three CDs.

Statistical Applications in Marketing: Advertising

© Jonathan Nourok/PhotoEdit

One of the major tools in the promotion mix is advertising. An important decision to be made by the advertising manager is how to allocate the company's total advertising budget among the various competing media types, including television, radio, and newspapers. Ultimately, the manager wants to know, for example, which television programs are most watched by potential customers, and how effective it is to sponsor these programs through advertising. But first the manager must assess the size of the audience, which involves estimating the amount of exposure potential customers have to the various media types, such as television.

10.40 Xr10-40 The sponsors of television shows targeted at the children's market wanted to know the amount of time children spend watching television, since the types and number of programs and commercials are greatly influenced by this information. As a result, it was decided to survey 100 North American children and ask them to keep track of the number of hours of television they watch each week. From past experience, it is known that the population standard deviation of the weekly amount of television watched is $\sigma = 8.0$ hours. The television sponsors want an estimate of the amount of television watched by the average North American child. A confidence level of 95% is judged to be appropriate.

10.3 SELECTING THE SAMPLE SIZE

As we discussed in the previous section, if the interval estimate is too wide, it provides little information. In Example 10.1 the interval estimate was 340.76 to 399.56. If the manager is to use this estimate as input for an inventory model, he needs greater precision. Fortunately, statistics practitioners can control the width of the interval by determining the sample size necessary to produce narrow intervals.

To understand how and why we can determine the sample size, we discuss the sampling error.

Error of Estimation

In Chapter 5 we pointed out that sampling error is the difference between the sample and the population that exists only because of the observations that happened to be selected for the sample. Now that we have discussed estimation we can define the sampling error as the difference between an estimator and a parameter. We can also define this difference as the **error of estimation.** In this chapter this can be expressed as the difference between $\overline{X}$ and μ. In our derivation of the confidence interval estimator of μ (see page 306), we expressed the following probability:

$$P\left(-Z_{\alpha/2} < \frac{\overline{X} - \mu}{\sigma/\sqrt{n}} < Z_{\alpha/2}\right) = 1 - \alpha$$

which can also be expressed as

$$P\left(-Z_{\alpha/2}\frac{\sigma}{\sqrt{n}} < \overline{X} - \mu < +Z_{\alpha/2}\frac{\sigma}{\sqrt{n}}\right) = 1 - \alpha$$

This tells us that the difference between $\overline{X}$ and μ lies between $-Z_{\alpha/2}\,\sigma/\sqrt{n}$ and $+Z_{\alpha/2}\,\sigma/\sqrt{n}$ with probability $1 - \alpha$. Expressed another way, we have with probability $1 - \alpha$,

$$|\overline{X} - \mu| < Z_{\alpha/2}\frac{\sigma}{\sqrt{n}}$$

In other words the error of estimation is less than $Z_{\alpha/2}\,\sigma/\sqrt{n}$. We interpret this to mean that $Z_{\alpha/2}\,\sigma/\sqrt{n}$ is the maximum error of estimation that we are willing to tolerate. We label this value B, which stands for the **bound on the error of estimation.** That is,

$$B = Z_{\alpha/2}\frac{\sigma}{\sqrt{n}}$$

Determining the Sample Size

We can solve the equation for n if the population standard deviation σ, the confidence level $1 - \alpha$, and the bound on the error of estimation B are known. Solving for n we produce

Sample Size to Estimate a Mean

$$n = \left(\frac{z_{\alpha/2}\,\sigma}{B}\right)^2$$

To illustrate, suppose that in Example 10.1 before gathering the data, the manager had decided that he needed to estimate the mean demand during lead time to within 16 units, which is the bound on the error of estimation. We also have $1 - \alpha = .95$ and $\sigma = 75$. We calculate

$$n = \left(\frac{z_{\alpha/2}\sigma}{B}\right)^2 = \left(\frac{(1.96)(75)}{16}\right)^2 = 84.41$$

Because n must be an integer and because we want the bound on the error of estimation to be *no more* than 16, any noninteger value must be rounded up. Thus, the value of n is rounded to 85, which means that to be 95% confident that the error of estimation will be no larger than 16, we need to randomly sample 85 lead time intervals.

Determining the Sample Size to Estimate the Mean Tree Diameter: Solution

© Randy Mayor/Botanical/Jupiterimages

Before the sample was taken, the forester determined the sample size as follows. The bound on the error of estimation is $B = 1$. The confidence level is 90% ($1 - \alpha = .90$). Thus $\alpha = .10$ and $\alpha/2 = .05$. It follows that $z_{\alpha/2} = 1.645$. The population standard deviation is assumed to be $\sigma = 6$. Thus,

$$n = \left(\frac{z_{\alpha/2}\sigma}{B}\right)^2 = \left(\frac{1.645 \times 6}{1}\right)^2 = 97.42$$

which is rounded to 98.

However, after the sample is taken, the forester discovered that $\sigma = 12$. The 90% confidence interval estimate is

$$\bar{x} \pm z_{\alpha/2}\frac{\sigma}{\sqrt{n}} = 25 \pm z_{.05}\frac{12}{\sqrt{98}} = 25 \pm 1.645\frac{12}{\sqrt{98}} = 25 \pm 2$$

As you can see the bound on the error of estimation is 2 and not 1. The interval is twice as wide as it was designed to be. The resulting estimate will not be as precise as needed.

In this chapter we have assumed that we know the value of the population standard deviation. In practice this is seldom the case. (In Chapter 12 we introduce a more realistic confidence interval estimator of the population mean.) It is frequently necessary to "guesstimate" the value of σ to calculate the sample size. That is, we must use our knowledge of the variable with which we're dealing to assign some value to σ. Unfortunately, we cannot be very precise in this guess. However, in guesstimating the value of σ, we prefer to err on the high side. For the chapter-opening example, if the forester had determined the sample size using $\sigma = 12$, he would have computed

$$n = \left(\frac{z_{\alpha/2}\sigma}{B}\right)^2 = \left(\frac{(1.645)(12)}{1}\right)^2 = 389.67 \text{ (rounded to 390)}$$

Using $n = 390$ (assuming that the sample mean is again 25), the 90% confidence interval estimate is

$$\bar{x} \pm z_{\alpha/2}\frac{\sigma}{\sqrt{n}} = 25 \pm 1.645\frac{12}{\sqrt{390}} = 25 \pm 1$$

This interval is as narrow as the forester wanted.

What happens if the standard deviation is *smaller* than assumed? If we discover that the standard deviation is less than we assumed when we determined the sample size, the confidence interval estimator will be narrower, and therefore more precise. Suppose that after the sample of 98 trees was taken (assuming again that $\sigma = 6$), the forester discovers that $\sigma = 3$. The confidence interval estimate is

$$\bar{x} \pm z_{\alpha/2}\frac{\sigma}{\sqrt{n}} = 25 \pm 1.645\frac{3}{\sqrt{98}} = 25 \pm 0.5$$

which is narrower than the forester wanted. Although this means that he would have sampled more trees than needed, the additional cost is relatively low when compared to the value of the information derived.

EXERCISES

Developing an Understanding of Statistical Concepts

10.41 The operations manager of a plant making cellular telephones has proposed rearranging the production process to be more efficient. She wants to estimate the time to assemble the telephone using the new arrangement. She believes that the population standard deviation is 15 seconds. How large a sample of workers should she take to estimate the mean assembly time to within 2 seconds with 95% confidence?

10.42 The label on 1-gallon cans of paint states that the amount of paint in the can is sufficient to paint 400 square feet. However, this number is quite variable. In fact, the amount of coverage is known to be approximately normally distributed with standard deviation of 25 square feet. How large a sample should be taken to estimate the true mean coverage of all 1-gallon cans to within 5 square feet with 95% confidence?

10.43 A medical researcher wants to investigate the amount of time it takes for patients' headache pain to be relieved after taking a new prescription painkiller. She plans to use statistical methods to estimate the mean of the population of relief times. She believes that the population is normally distributed with a standard deviation of 20 minutes. How large a sample should she take to estimate the mean time to within 1 minute with 90% confidence?

10.44 A statistics professor wants to compare today's students with those 25 years ago. All his current students' marks are stored on a computer so that he can easily determine the population mean. However, the marks 25 years ago reside only in his musty files. He does not want to retrieve all the marks and will be satisfied with a 95% confidence interval estimate of the mean mark 25 years ago. If he assumes that the population standard deviation is 12, how large a sample should he take to estimate the mean to within 2 marks?

10.45 The operations manager of a large production plant would like to estimate the average amount of time workers take to assemble a new electronic component. After observing a number of workers assembling similar devices, she guesses that the standard deviation is 6 minutes. How large a sample of workers should she take if she wishes to estimate the mean assembly time to within 20 seconds? Assume that the confidence level is to be 99%.

10.46 a. A statistics practitioner would like to estimate a population mean to within 10 units. The confidence level has been set at 95% and $\sigma = 200$. Determine the sample size.

b. Suppose that the sample mean was calculated as 500. Estimate the population mean with 95% confidence.

10.47 a. Repeat Part b of Exercise 10.46 after discovering that the population standard deviation is actually 100.

b. Repeat Part b of Exercise 10.46 after discovering that the population standard deviation is actually 400.

10.48 Review Exercises 10.46 and 10.47. Describe what happens to the confidence interval estimate when

a. the standard deviation is equal to the value used to determine the sample size
b. the standard deviation is smaller than the one used to determine the sample size
c. the standard deviation is larger than the one used to determine the sample size

10.49 A medical statistician wants to estimate the average weight loss of people who are on a new diet plan. In a preliminary study, he guesses that the standard deviation of the population of weight losses is about 10 pounds. How large a sample should he take to estimate the mean weight loss to within 2 pounds, with 90% confidence?

10.50 a. Determine the sample size necessary to estimate a population mean to within 1 with 90% confidence given that the population standard deviation is 10.

b. Suppose that the sample mean was calculated as 150. Estimate the population mean with 90% confidence.

10.51 a. Repeat Exercise 10.50b after discovering that the population standard deviation is actually 5.

b. Repeat Exercise 10.50b after discovering that the population standard deviation is actually 20.

10.52 Review Exercises 10.50 and 10.51. Describe what happens to the confidence interval estimate when

a. the standard deviation is equal to the value used to determine the sample size
b. the standard deviation is smaller than the one used to determine the sample size
c. the standard deviation is larger than the one used to determine the sample size

10.53 a. A statistics practitioner would like to estimate a population mean to within 50 units with 99% confidence given that the population standard deviation is 250. What sample size should be used?

b. Redo Part a changing the standard deviation to 50.
c. Redo Part a using a 95% confidence level.
d. Redo Part a wherein we wish to estimate the population mean to within 10 units.

10.54 Review the results of Exercise 10.53. Describe what happens to the sample size when

a. the population standard deviation decreases
b. the confidence level decreases
c. the bound on the error of estimation decreases

10.55 a. Determine the sample size required to estimate a population mean to within 10 units given that the population standard deviation is 50. A confidence level of 90% is judged to be appropriate.

b. Repeat Part a changing the standard deviation to 100.
c. Redo Part a using a 95% confidence level.
d. Repeat Part a wherein we wish to estimate the population mean to within 20 units.

10.56 Review Exercise 10.55. Describe what happens to the sample size when

a. the population standard deviation increases
b. the confidence level increases
c. the bound on the error of estimation increases

CHAPTER SUMMARY

This chapter introduced the concepts of estimation and the estimator of a population mean when the population variance is known. It also presented a formula to calculate the sample size necessary to estimate a population mean.

IMPORTANT TERMS

Point estimator 322
Interval estimator 322
Unbiased estimator 323
Consistency 324
Relative efficiency 324
Confidence interval estimator of μ 326
Confidence level 326
Lower confidence limit (LCL) 326
Upper confidence limit (UCL) 326
Error of estimation 339
Bound on the error of estimation 339

SYMBOLS

Symbol	Pronounced	Represents
$1-\alpha$	*One-minus-alpha*	Confidence level
B		Bound on the error of estimation
$z_{\alpha/2}$	*z-alpha-by-2*	Value of Z such that the area to its right is equal to $\alpha/2$

FORMULAS

Confidence interval estimator of μ with σ known

$$\bar{x} \pm z_{\alpha/2}\frac{\sigma}{\sqrt{n}}$$

Sample size to estimate μ

$$n = \left(\frac{z_{\alpha/2}\sigma}{B}\right)^2$$

COMPUTER OUTPUT AND INSTRUCTIONS

Technique	Excel	Minitab
Confidence interval estimate of μ	329	330

11

© Steve Hathaway/Workbook Stock/Jupiterimages

INTRODUCTION TO HYPOTHESIS TESTING

SSA Envelope Plan

DATA Xm11-00

Federal Express (FedEx) sends invoices to customers requesting payment within 30 days. The bill lists an address and customers are expected to use their own envelopes to return their payments. Currently the mean and standard deviation of the amount of time taken to pay bills are 24 days and 6 days, respectively. The chief financial officer (CFO) believes that including a stamped self-addressed (SSA) envelope would decrease the amount of time. She calculates that the improved cash flow from a 2-day decrease in the payment period would pay for the costs of the envelopes and stamps. Any further decrease in the payment period would generate a profit. To test her belief, she randomly selects 220 customers and includes a stamped self-addressed envelope with their invoices. The numbers of days until payment is received were recorded. Can the CFO conclude that the plan will be profitable?

© Mark Richards/PhotoEdit

After we've introduced the required tools, we'll return to this question and answer it. (See page 358).

Introduction

In Chapter 10, we introduced estimation and showed how it is used. Now we're going to present the second general procedure of making inferences about a population—hypothesis testing. The purpose of this type of inference is to determine whether enough statistical evidence exists to enable us to conclude that a belief or hypothesis about a parameter is supported by the data. You will discover that hypothesis testing has a wide variety of applications in business and economics, as well as many other fields. This chapter will lay the foundation upon which the rest of the book is based. As such it represents a critical contribution to your development as a statistics practitioner.

In the next section, we will introduce the concepts of hypothesis testing, and in Section 11.2 we will develop the method employed to test a hypothesis about a population mean when the population standard deviation is known. The rest of the chapter deals with related topics.

11.1 Concepts of Hypothesis Testing

The term **hypothesis testing** is likely new to most readers, but the concepts underlying hypothesis testing are quite familiar. There are a variety of nonstatistical applications of hypothesis testing, the best known of which is a criminal trial.

When a person is accused of a crime, he or she faces a trial. The prosecution presents its case and a jury must make a decision on the basis of the evidence presented. In fact, the jury conducts a test of hypothesis. There are actually two hypotheses that are tested. The first is called the **null hypothesis** and is represented by H_0 (pronounced *H-nought: nought* is a British term for zero). It is

H_0: The defendant is innocent.

The second is called the **alternative** or **research hypothesis** and is denoted H_1. In a criminal trial it is

H_1: The defendant is guilty.

Of course, the jury does not know which hypothesis is correct. They must make a decision on the basis of the evidence presented by both the prosecution and the defense. There are only two possible decisions. Convict or acquit the defendant. In statistical parlance, convicting the defendant is equivalent to *rejecting the null hypothesis in favor of the alternative*. That is, the jury is saying that there was enough evidence to conclude that the defendant was guilty. Acquitting a defendant is phrased as *not rejecting the null hypothesis in favor of the alternative*, which means that the jury decided that there was not enough evidence to conclude that the defendant was guilty. Notice that we do not say that we accept the null hypothesis. In a criminal trial, that would be interpreted as finding the defendant innocent. Our justice system does not allow this decision.

There are two possible errors. A **Type I error** occurs when we reject a true null hypothesis. A **Type II error** is defined as not rejecting a false null hypothesis. In the criminal trial, a Type I error is made when an innocent person is wrongly convicted. A Type II error occurs when a guilty defendant is acquitted. The probability of a Type I error is denoted by α, which is also called the **significance level.** The probability of a Type II error is denoted by β (Greek letter *beta*). The error probabilities α and β are inversely related, meaning that any attempt to reduce one will increase the other. Table 11.1 summarizes the terminology and the concepts.

TABLE **11.1** Terminology of Hypothesis Testing

Decision	H_0 Is True (Defendant Is Innocent)	H_0 Is False (Defendant Is Guilty)
REJECT H_0 Convict defendant	Type I error P(Type II error) $= \alpha$	Correct decision
DO NOT REJECT H_0 Acquit defendant	Correct decision	Type II error P(Type II error) $= \beta$

In our justice system, Type I errors are regarded as more serious. As a consequence, the system is set up so that the probability of a Type I error is small. This is arranged by placing the burden of proof on the prosecution (the prosecution must prove guilt—the defense need not prove anything) and by having judges instruct the jury to find the defendant guilty only if there is "evidence beyond a reasonable doubt." In the absence of enough evidence, the jury must acquit even though there may be some evidence of guilt. The consequence of this arrangement is that the probability of acquitting guilty people is relatively large. Oliver Wendell Holmes, a United States Supreme Court justice, once phrased the relationship between the probabilities of Type I and Type II errors in the following way: "Better to acquit 100 guilty men than convict one innocent one." In Justice Holmes's opinion, the probability of a Type I error should be 1/100 of the probability of a Type II error.

The critical concepts in hypothesis testing follow.

1. There are two hypotheses. One is called the null hypothesis and the other the alternative or research hypothesis.
2. The testing procedure begins with the assumption that the null hypothesis is true.
3. The goal of the process is to determine whether there is enough evidence to infer that the alternative hypothesis is true.
4. There are two possible decisions:

   Conclude that there is enough evidence to support the alternative hypothesis

   Conclude that there is not enough evidence to support the alternative hypothesis
5. Two possible errors can be made in any test. A Type I error occurs when we reject a true null hypothesis and a Type II error occurs when we don't reject a false null hypothesis. The probabilities of Type I and Type II errors are

   $P(\text{Type I error}) = \alpha$

   $P(\text{Type II error}) = \beta$

Let's extend these concepts to statistical hypothesis testing.

In statistics we frequently test hypotheses about parameters. The hypotheses we test are generated by questions that managers need to answer. To illustrate, suppose that in Example 10.1 (page 328) the operations manager did not want to estimate the mean demand during lead time but instead wanted to know whether the mean is different from 350, which may be the point at which the current inventory policy needs to be altered. That is, the manager wants to determine whether he can infer that μ is not equal to 350. We can rephrase the question so that it now reads, Is there enough evidence to conclude that μ is not equal to 350? This wording is analogous to the criminal trial wherein the jury is asked to determine whether there is enough evidence to conclude that the defendant is guilty. Thus, the alternative (research) hypothesis is

$$H_1: \quad \mu \neq 350$$

In a criminal trial the process begins with the assumption that the defendant is innocent. In a similar fashion we start with the assumption that the parameter equals the value we're testing. Consequently, the operations manager would assume that $\mu = 350$ and the null hypothesis is expressed as

$$H_0: \quad \mu = 350$$

When we state the hypotheses we list the null first followed by the alternative hypothesis. To determine whether the mean is different from 350 we test

$$H_0: \quad \mu = 350$$
$$H_1: \quad \mu \neq 350$$

Now suppose that in this illustration the current inventory policy is based on an analysis that revealed that the actual mean demand during lead time is 350. After a vigorous advertising campaign, the manager suspects that there has been an increase in demand and thus an increase in mean demand during lead time. To test whether there is evidence of an increase, the manager would specify the alternative hypothesis as

$$H_1: \quad \mu > 350$$

Because the manager knew that the mean was (and maybe still is) 350, the null hypothesis would state

$$H_0: \quad \mu = 350$$

Further suppose that the manager does not know the actual mean demand during lead time, but the current inventory policy is based on the assumption that the mean is *less than or equal to* 350. If the advertising campaign increases the mean to a quantity larger than 350, a new inventory plan will have to be instituted. In this scenario the hypotheses become

$$H_0: \quad \mu \leq 350$$
$$H_1: \quad \mu > 350$$

Notice that in both illustrations the alternative hypothesis is designed to determine whether there is enough evidence to conclude that the mean is greater than 350. Although the two null hypotheses are different (one states that the mean is equal to 350 and the other states that the mean is less than or equal to 350), when the test is conducted the process begins by assuming that the mean is *equal to* 350, In other words, no matter the form of the null hypothesis, we use the equal sign in the null hypothesis. Here is the reason. If there is enough evidence to conclude that the alternative hypothesis (the mean is greater than 350) is true when we assume that the mean is *equal to* 350, we would certainly draw the same conclusion when we assume that the mean is a value that is *less than* 350. As a result, the null hypothesis will always state that the parameter equals the value specified in the alternative hypothesis.

To emphasize this point, suppose the manager now wanted to determine whether there has been a decrease in the mean demand during lead time. We express the null and alternative hypotheses as

$$H_0: \quad \mu = 350$$
$$H_1: \quad \mu < 350$$

The hypotheses are often set up to reflect a manager's decision problem wherein the null hypothesis represents the *status quo*. Often this takes the form of some course of action such as maintaining a particular inventory policy. If there is evidence of an increase or decrease in the value of the parameter, a new course of action will be taken. Examples include deciding to produce a new product, switching to a better drug to treat an illness, or sentencing a defendant to prison.

The next element in the procedure is to randomly sample the population and calculate the sample mean. This is called the **test statistic.** The test statistic is the criterion upon which we base our decision about the hypotheses. (In the criminal trial analogy, this is equivalent to the evidence presented in the case.) The test statistic is based on the best estimator of the parameter. In Chapter 10 we stated that the best estimator of a population mean is the sample mean.

If the test statistic's value is inconsistent with the null hypothesis, we reject the null hypothesis and infer that the alternative hypothesis is true. For example, if we're trying to decide whether the mean is greater than 350, a large value of $\bar{x}$ (say, 600) would provide enough evidence. If $\bar{x}$ is close to 350 (say, 355), we would say that this does not provide much evidence to infer that the mean is greater than 350. In the absence of sufficient evidence, we do not reject the null hypothesis in favor of the alternative. (In the absence of sufficient evidence of guilt, a jury finds the defendant not guilty.)

In a criminal trial "sufficient evidence" is defined as "evidence beyond a reasonable doubt." In statistics we need to use the test statistic's sampling distribution to define "sufficient evidence." We will do so in the next section.

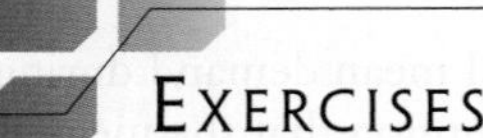

EXERCISES

Eexercises 11.1–11.5 feature nonstatistical applications of hypothesis testing. For each, identify the hypotheses, define Type I and Type II errors, and discuss the consequences of each error. In setting up the hypotheses, you will have to consider where to place the "burden of proof."

11.1 It is the responsibility of the federal government to judge the safety and effectiveness of new drugs. There are two possible decisions: approve the drug or disapprove the drug.

11.2 You are contemplating a Ph.D. in business or economics. If you succeed, a life of fame, fortune, and happiness awaits you. If you fail, you've wasted 5 years of your life. Should you go for it?

11.3 You are the center fielder of the New York Yankees. It is the bottom of the ninth inning of the seventh game of the World Series. The Yanks lead by 2 with 2 outs and men on second and third. The batter is known to hit for high average and runs very well, but with mediocre power. A single will tie the game and a hit over your head will likely result in the Yanks losing. Do you play shallow?

11.4 You are faced with two investments. One is very risky, but the potential returns are high. The other is safe, but the potential is quite limited. Pick one.

11.5 You are the pilot of a jumbo jet. You smell smoke in the cockpit. The nearest airport is less than 5 minutes away. Should you land the plane immediately?

11.6 Several years ago in a high-profile case, a defendant was acquitted in a double-murder trial but was subsequently found responsible for the deaths in a civil trial. (Guess the name of the defendant—the answer is in Appendix C.) In a civil trial the plaintiff (the victims' relatives) are required only to show that the preponderance of evidence points to the guilt of the defendant. Aside from the other issues in the cases, discuss why these results are logical.

11.2 TESTING THE POPULATION MEAN WHEN THE POPULATION STANDARD DEVIATION IS KNOWN

To illustrate the process, consider the following example.

EXAMPLE 11.1

Department Store's New Billing System

DATA
Xm011-01

The manager of a department store is thinking about establishing a new billing system for the store's credit customers. After a thorough financial analysis, she determines that the new system will be cost-effective only if the mean monthly account is more than \$170. A random sample of 400 monthly accounts is drawn, for which the sample mean is \$178. The manager knows that the accounts are approximately normally distributed with a standard deviation of \$65. Can the manager conclude from this that the new system will be cost-effective?

SOLUTION

IDENTIFY

This example deals with the population of the credit accounts at the store. To conclude that the system will be cost-effective requires the manager to show that the mean account for all customers is greater than \$170. Consequently, we set up the alternative hypothesis to express this circumstance:

$$H_1: \quad \mu > 170 \qquad \text{(Install new system)}$$

If the mean is less than or equal to 170, the system will not be cost-effective. The null hypothesis can be expressed as

$$H_0: \quad \mu \leq 170 \qquad \text{(Do not install new system)}$$

However, as was discussed in Section 11.1, we will actually test $\mu = 170$, which is how we specify the null hypothesis:

$$H_0: \quad \mu = 170$$

As we previously pointed out, the test statistic is the best estimator of the parameter. In Chapter 10, we used the sample mean to estimate the population mean. To conduct this test, we ask and answer the following question: Is a sample mean of 178 sufficiently greater than 170 to allow us to confidently infer that the population mean is greater than 170?

There are two approaches to answering this question. The first is called the *rejection region method*. It can be used in conjunction with the computer, but it is mandatory for those computing statistics manually. The second is the *p-value approach*, which in general can be employed only in conjunction with a computer and statistical software. We recommend, however, that users of statistical software be familiar with both approaches.

Rejection Region

It seems reasonable to reject the null hypothesis in favor of the alternative if the value of the sample mean is large relative to 170. If we had calculated the sample mean to be say, 500, it would be quite apparent that the null hypothesis is false and we would reject it.

On the other hand, values of $\bar{x}$ close to 170, such as 171, do not allow us to reject the null hypothesis because it is entirely possible to observe a sample mean of 171 from a population whose mean is 170. Unfortunately, the decision is not always so obvious. In this example, the sample mean was calculated to be 178, a value apparently neither very far away from nor very close to 170. To make a decision about this sample mean, we set up the *rejection region*.

Rejection Region

The **rejection region** is a range of values such that if the test statistic falls into that range, we decide to reject the null hypothesis in favor of the alternative hypothesis.

Suppose we define the value of the sample mean that is just large enough to reject the null hypothesis as $\bar{x}_L$. The rejection region is

$$\bar{x} > \bar{x}_L$$

Since a Type I error is defined as rejecting a true null hypothesis, and the probability of committing a Type I error is α, it follows that

$$\alpha = P(\text{rejecting } H_0 \text{ given that } H_0 \text{ is true})$$

$$= P(\bar{x} > \bar{x}_L \text{ given that } H_0 \text{ is true})$$

Figure 11.1 depicts the sampling distribution and the rejection region.

FIGURE **11.1**
Sampling Distribution for Example 11.1

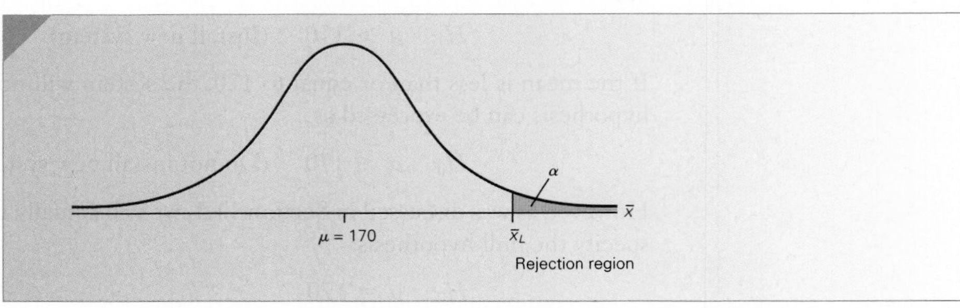

From Section 9.1, we know that the sampling distribution of $\bar{x}$ is normal or approximately normal, with mean μ and standard deviation $\sigma/\sqrt{n}$. As a result, we can standardize $\bar{x}$ and obtain the following probability:

$$P\left(\frac{\bar{x} - \mu}{\sigma/\sqrt{n}} > \frac{\bar{x}_L - \mu}{\sigma/\sqrt{n}}\right) = P\left(Z > \frac{\bar{x}_L - \mu}{\sigma/\sqrt{n}}\right) = \alpha$$

From Section 8.2, we defined z_α to be the value of a standard normal random variable such that

$$P(Z > z_\alpha) = \alpha$$

Since both probability statements involve the same distribution (standard normal) and the same probability (α), it follows that the limits are identical. Thus,

$$\frac{\bar{x}_L - \mu}{\sigma/\sqrt{n}} = z_\alpha$$

We know that $\sigma = 65$ and $n = 400$. Because the probabilities defined above are conditional upon the null hypothesis being true, we have $\mu = 170$. To calculate the rejection region, we need a value of α, the significance level. Suppose that the manager chose α to be 5%. It follows that $z_\alpha = z_{.05} = 1.645$. We can now calculate the value of $\bar{x}_L$:

$$\frac{\bar{x}_L - \mu}{\sigma/\sqrt{n}} = z_\alpha$$

$$\frac{\bar{x}_L - 170}{65/\sqrt{400}} = 1.645$$

$$\bar{x}_L = 175.34$$

Therefore the rejection region is

$$\bar{x} > 175.34$$

The sample mean was computed to be 178. Because the test statistic (sample mean) is in the rejection region (it is greater than 175.34), we reject the null hypothesis. Thus, there is sufficient evidence to infer that the mean monthly account is greater than \$170.

Our calculations determined that any value of $\bar{x}$ above 175.34 represents an event that is quite unlikely when sampling (with $n = 400$) from a population whose mean is 170 (and whose standard deviation is 65). This suggests that the assumption that the null hypothesis is true is incorrect, and consequently we reject the null hypothesis in favor of the alternative hypothesis.

Standardized Test Statistic

The preceding test used the test statistic $\bar{x}$; as a result, the rejection region had to be set up in terms of $\bar{x}$. An easier method specifies that the test statistic be the standardized value of $\bar{x}$. That is, we use the **standardized test statistic**

$$z = \frac{\bar{x} - \mu}{\sigma/\sqrt{n}}$$

and the rejection region consists of all values of z that are greater than z_α. Algebraically, the rejection region is

$$z > z_\alpha$$

We can redo Example 11.1 using the standardized test statistic.

The rejection region is

$$z > z_\alpha = z_{.05} = 1.645$$

The value of the test statistic is calculated next:

$$z = \frac{\bar{x} - \mu}{\sigma/\sqrt{n}} = \frac{178 - 170}{65/\sqrt{400}} = 2.46$$

Because 2.46 is greater than 1.645, reject the null hypothesis and conclude that there is enough evidence to infer that the mean monthly account is greater than \$170.

As you can see, the conclusions we draw from using the test statistic $\bar{x}$ and the standardized test statistic z are identical. Figures 11.2 and 11.3 depict the two sampling distributions, highlighting the equivalence of the two tests.

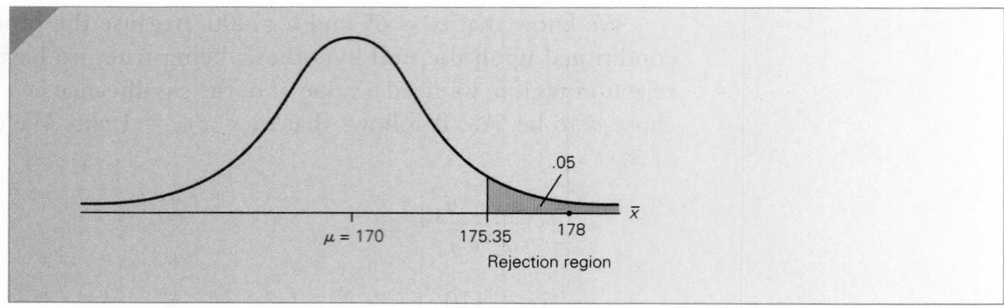

FIGURE **11.2**
Sampling Distribution of $\bar{X}$ for Example 11.1

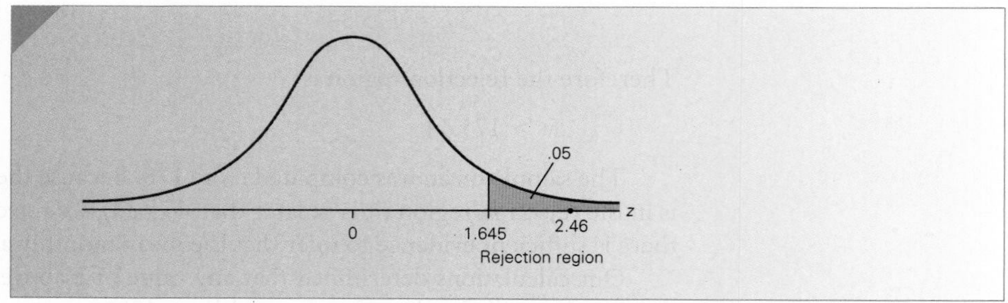

FIGURE **11.3**
Sampling Distribution of Z for Example 11.1

Because it is convenient and because statistical software packages employ it, the standardized test statistic will be used throughout this book. For simplicity we will refer to the *standardized test statistic* simply as the *test statistic.*

Incidentally, when a null hypothesis is rejected, the test is said to be **statistically significant** at whatever significance level the test was conducted. Summarizing Example 11.1, we would say that the test was significant at the 5% significance level.

p-Value

There are several drawbacks to the rejection region method. Foremost among them is the type of information provided by the result of the test. The rejection region method produces a yes or no response to the question, Is there sufficient statistical evidence to infer that the alternative hypothesis is true? The implication is that the result of the test of hypothesis will be converted automatically into one of two possible courses of action: one action as a result of rejecting the null hypothesis in favor of the alternative and another as a result of not rejecting the null hypothesis in favor of the alternative. In Example 11.1 the rejection of the null hypothesis seems to imply that the new billing system will be installed.

In fact, this is not the way in which the result of a statistical analysis is utilized. The statistical procedure is only one of several factors considered by a manager when making a decision. In Example 11.1 the manager discovered that there was enough statistical evidence to conclude that the mean monthly account is greater than $170. However, before taking any action, the manager would like to consider a number of factors including the cost and feasibility of restructuring the billing system and the possibility of making an error, in this case a Type I error.

What is needed to take full advantage of the information available from the test result and make a better decision is a measure of the amount of statistical evidence supporting the alternative hypothesis so that it can be weighed in relation to the other factors, especially the financial ones. The *p-value of a test* provides this measure.

p-Value

The **p-value of a test** is the probability of observing a test statistic at least as extreme as the one computed given that the null hypothesis is true.

In Example 11.1 the p-value is the probability of observing a sample mean at least as large as 178 when the population mean is 170. Thus,

$$p\text{-value} = P(\overline{X} > 178) = P\left(\frac{\overline{X} - \mu}{\sigma/\sqrt{n}} > \frac{178 - 170}{65/\sqrt{400}}\right) = P(Z > 2.46)$$

$$= 1 - P(Z < 2.46) = 1 - .9931 = .0069$$

Figure 11.4 describes this calculation.

FIGURE **11.4** p-Value for Example 11.1

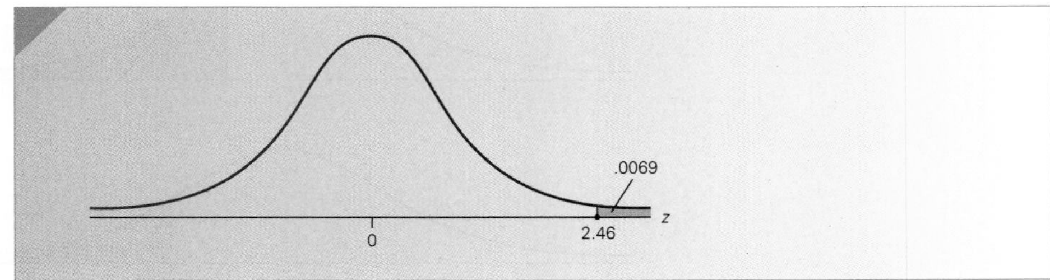

Interpreting the p-Value

To properly interpret the results of an inferential procedure, you must remember that the technique is based on the sampling distribution. The sampling distribution allows us to make probability statements about a sample statistic assuming knowledge of the population parameter. Thus, the probability of observing a sample mean at least as large as 178 from a population whose mean is 170 is .0069, which is very small. In other words, we have just observed an unlikely event, an event so unlikely that we seriously doubt the assumption that began the process, that the null hypothesis is true. Consequently, we have reason to reject the null hypothesis and support the alternative.

Students may be tempted to simplify the interpretation by stating that the p-value is the probability that the null hypothesis is true. Don't! As was the case with interpreting the confidence interval estimator, you cannot make a probability statement about a parameter. It is not a random variable.

The p-value of a test provides valuable information because it is a measure of the amount of statistical evidence that supports the alternative hypothesis. To understand this interpretation fully, refer to Table 11.2 where we list several values of $\bar{x}$, their z-statistics, and p-values for Example 11.1. Notice that the closer $\bar{x}$ is to the hypothesized mean, 170, the larger the p-value is. The farther $\bar{x}$ is above 170, the smaller the p-value is. Values of $\bar{x}$ far above 170 tend to indicate that the alternative hypothesis is true. Thus, the smaller the p-value, the more the statistical evidence supports the alternative hypothesis. Figure 11.5 graphically depicts the information in Table 11.2.

TABLE **11.2**
Test Statistics and p-Values for Example 11.1

Sample Mean $\bar{x}$	Test Statistic $z = \frac{\bar{x} - \mu}{\sigma/\sqrt{n}} = \frac{\bar{x} - 170}{65/\sqrt{400}}$	p-Value
170	0	.5000
172	0.62	.2676
174	1.23	.1093
176	1.85	.0322
178	2.46	.0069
180	3.08	.0010

FIGURE **11.5**
p-Values for Example 11.1

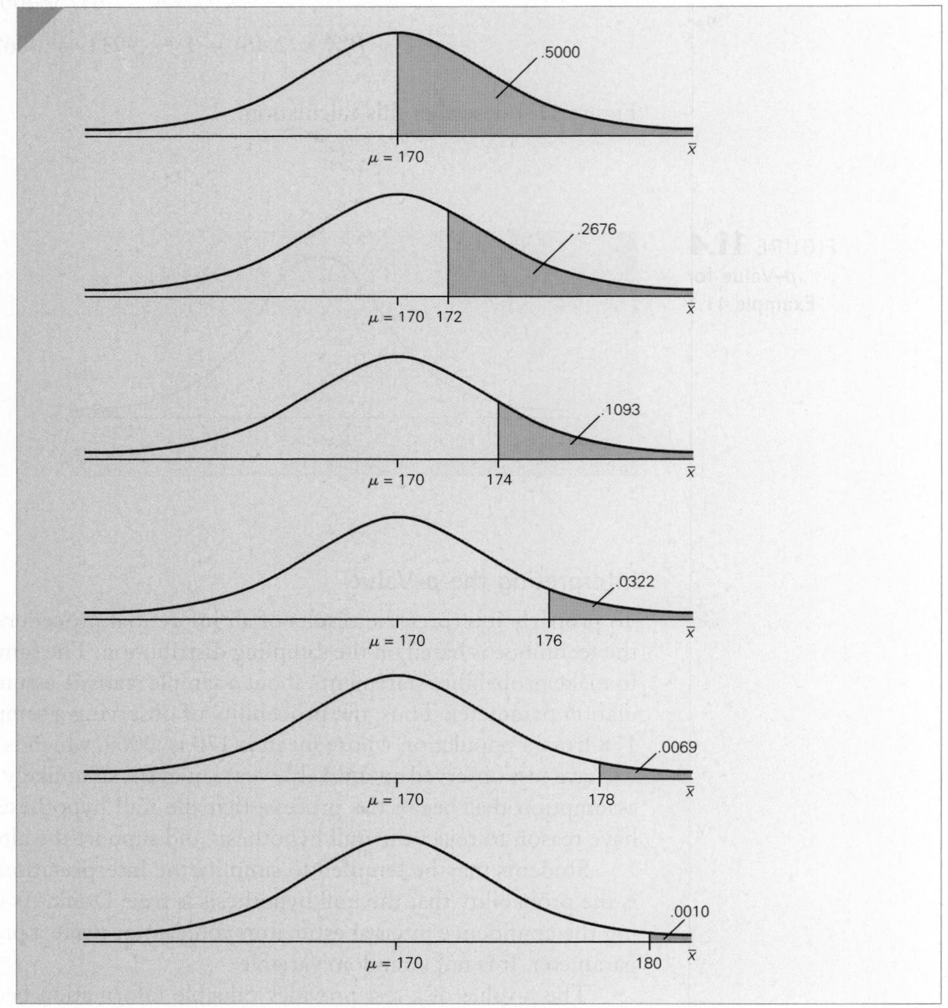

This raises the question, How small does the p-value have to be to infer that the alternative hypothesis is true? In general, the answer depends on a number of factors, including the costs of making Type I and Type II errors. In Example 11.1, a Type I error would occur if the manager adopts the new billing system when it is not cost-effective. If the cost of this error is high, we attempt to minimize its probability. In the rejection region method, we do so by setting the significance level quite low, say 1%. Using the p-value method, we would insist that the p-value be quite small, providing sufficient evidence to infer that the mean monthly account is greater than \$170 before proceeding with the new billing system.

Describing the p-Value

Statistics practitioners can translate p-values using the following descriptive terms:

If the p-value is less than .01, we say that there is *overwhelming* evidence to infer that the alternative hypothesis is true. We also say that the test is **highly significant.**

If the p-value lies between .01 and .05, there is *strong* evidence to infer that the alternative hypothesis is true. The result is deemed to be **significant.**

If the p-value is between .05 and .10, we say that there is *weak* evidence to indicate that the alternative hypothesis is true. When the p-value is greater than 5%, we say that the result is **not statistically significant.**

When the p-value exceeds .10, we say that there is no evidence to infer that the alternative hypothesis is true.

Figure 11.6 summarizes these terms.

FIGURE **11.6** Test Statistics and p-Values for Example 11.1

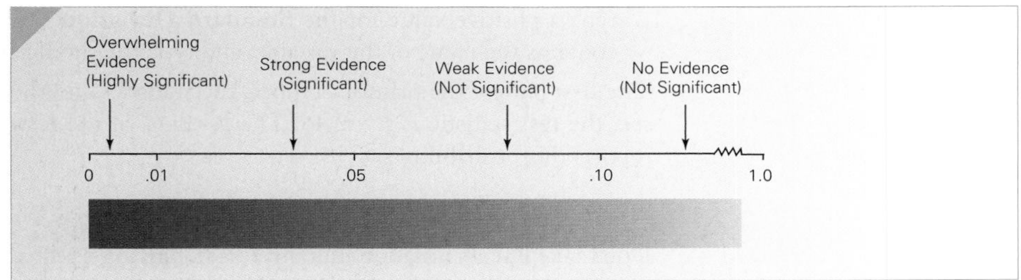

The p-Value and Rejection Region Methods

If we so choose, we can use the p-value to make the same type of decisions we make in the rejection region method. The rejection region method requires the decision maker to select a significance level from which the rejection region is constructed. We then decide to reject or not reject the null hypothesis. Another way of making that type of decision is to compare the p-value with the selected value of the significance level. If the p-value is less than α, we judge the p-value to be small enough to reject the null hypothesis. If the p-value is greater than α, we do not reject the null hypothesis.

Solving Manually, Using Excel, and Using Minitab

As you have already seen, we offer three ways to solve statistical problems. When we perform the calculations manually, we will use the rejection region approach. We will set up the rejection region using the test statistic's sampling distribution and associated table (in Appendix B). The calculations will be performed manually and a reject/do not reject decision will be made. In this chapter it is possible to compute the p-value of the test manually. However, in later chapters we will be using test statistics that are not normally distributed, making it impossible to calculate the p-values manually. In these instances manual calculations require the decision to be made via the rejection region method only.

Most software packages that compute statistics, including Excel and Minitab, print the p-value of the test. When we employ the computer, we will not set up the rejection region. Instead we will focus on the interpretation of the p-value.

EXCEL

	A	B	C	D
1	Z-Test: Mean			
2				
3				Accounts
4	Mean			178.00
5	Standard Deviation			68.37
6	Observations			400
7	Hypothesized Mean			170
8	SIGMA			65
9	z Stat			2.46
10	P(Z<=z) one-tail			0.0069
11	z Critical one-tail			1.6449
12	P(Z<=z) two-tail			0.0138
13	z Critical two-tail			1.96

INSTRUCTIONS

1. Type or import the data into one column. (Open Xm11-01.)
2. Click **Add-Ins, Data Analysis Plus,** and **Z-Test: Mean.**
3. Fill in the dialog box: **Input Range** (A1:A401), type the **Hypothesized Mean** (170), type a positive value for the **Standard Deviation** (65), click **Labels** if the first row contains the name of the variable, and type the significance level α (.05).

The first part of the printout reports the statistics and the details of the test. As you can see, the test statistic is $z = 2.46$. The *p*-value* of the test is $P(Z > 2.46) = .0069$. Excel reports this probability as

$$P(Z <= z) \text{ one-tail}$$

Don't take Excel's notation literally. It is not giving us the probability that Z is less than or equal to the value of the z-statistic. Also printed is the critical value of the rejection region shown as

$$Z \text{ Critical one-tail}$$

The printout shown here was produced from the raw data. That is, we input the 400 observations in the data set and the computer calculated the value of the test statistic and the *p*-value. As was the case with estimation, if you have calculated the value of the sample mean, you can use an Excel worksheet we created for this book.

Open the **Test Statistics** workbook and find the worksheet ***z*-Test_Mean.** Input the components of the test $\bar{x}$, σ, n and the hypothesized value of μ. Excel then computes the required statistics. This spreadsheet also allows you to conduct a what-if analysis. Try changing any of the inputs to discover their effects.

Test Statistics Workbook

	A	B	C	D
1	**z-Test of a Mean**			
2				
3	**Sample mean**	178	**z Stat**	**2.46**
4	**Population standard deviation**	65	**P(Z<=z) one-tail**	**0.0069**
5	**Sample size**	400	**z Critical one-tail**	**1.6449**
6	**Hypothesized mean**	170	**P(Z<=z) two-tail**	**0.0138**
7	**Alpha**	0.05	**z Critical two-tail**	**1.9600**

*Excel provides two probabilities in its printout. The way in which we determine the *p*-value of the test from the printout is somewhat more complicated. Interested students are advised to read CD Appendix K.

MINITAB

One-Sample Z: Accounts

Test of mu = 170 vs > 170
The assumed standard deviation = 65

Variable	N	Mean	StDev	SE Mean	95% Lower Bound	Z	P
Accounts	400	177.997	68.367	3.250	172.651	2.46	0.007

INSTRUCTIONS

1. Type or import the data into one column. (Open Xm11-01.)
2. Click **Stat, Basic Statistics,** and **1-Sample Z . . . .**
3. Type or use the **Select** button to specify the name of the variable or the column in the **Samples in Columns** box (Accounts). Type the value of the **Standard deviation** (65), check the **Perform hypothesis test** box, and type the value of μ under the null hypothesis in the **Hypothesized mean** box (170).
4. Click **Options . . .** and specify the form of the alternative hypothesis in the **Alternative** box (greater than).

Interpreting the Results of a Test

In Example 11.1, we rejected the null hypothesis. Does this prove that the alternative hypothesis is true? The answer is no; because our conclusion is based on sample data (and not on the entire population), we can never *prove* anything by using statistical inference. Consequently, we summarize the test by stating that there is enough statistical evidence to infer that the null hypothesis is false and that the alternative hypothesis is true.

Now suppose that $\bar{x}$ had equaled 174 instead of 178. We would then have calculated $z = 1.23$ (p-value $= .1093$), which is not in the rejection region. Could we conclude on this basis that there is enough statistical evidence to infer that the null hypothesis is true and hence that $\mu = 170$? Again the answer is "no" because it is absurd to suggest that a sample mean of 174 provides enough evidence to infer that the population mean is 170. (If it proved anything, it would prove that the population mean is 174.) Because we're testing a single value of the parameter under the null hypothesis, we can never have enough statistical evidence to establish that the null hypothesis is true (unless we sample the entire population). (The same argument is valid if you set up the null hypothesis as H_0: $\mu \leq 170$. It would be illogical to conclude that a sample mean of 174 provides enough evidence to conclude that the population mean is *less than or equal to 170*.)

Consequently, if the value of the test statistic does not fall into the rejection region (or the p-value is large), rather than say we accept the null hypothesis (which implies that we're stating that the null hypothesis is true), we state that we do not reject the null hypothesis, and we conclude that not enough evidence exists to show that the alternative hypothesis is true. Although it may appear to be the case, we are not being overly technical. Your ability to set up tests of hypotheses properly and to interpret their results correctly very much depends on your understanding of this point. The point is that the conclusion is based on the alternative hypothesis. In the final analysis, there are only two possible conclusions of a test of hypothesis.

Conclusions of a Test of Hypothesis

If we reject the null hypothesis, we conclude that there is enough statistical evidence to infer that the alternative hypothesis is true.

If we do not reject the null hypothesis, we conclude that there is not enough statistical evidence to infer that the alternative hypothesis is true.

Observe that the alternative hypothesis is the focus of the conclusion. It represents what we are investigating. That is why it is also called the research hypothesis. Whatever you're trying to show statistically must be represented by the alternative hypothesis (bearing in mind that you have only three choices for the alternative hypothesis—the parameter is greater than, less than, or not equal to the value specified in the null hypothesis).

When we introduced statistical inference in Chapter 10, we pointed out that the first step in the solution is to identify the technique. When the problem involves hypothesis testing, part of this process is the specification of the hypotheses. Because the alternative hypothesis represents the condition we're researching, we will identify it first. The null hypothesis automatically follows because the null hypothesis must specify equality. However, by tradition, when we list the two hypotheses, the null hypothesis comes first, followed by the alternative hypothesis. All examples in this book will follow that format.

SSA Envelope Plan: Solution

© Mark Richards/PhotoEdit

IDENTIFY

The objective of the study is to draw a conclusion about the mean payment period. Thus, the parameter to be tested is the population mean μ. We want to know whether there is enough statistical evidence to show that the population mean is less than 22 days. Thus, the alternative hypothesis is

$$H_1: \quad \mu < 22$$

The null hypothesis is

$$H_0: \quad \mu = 22$$

The test statistic is the only one we've presented thus far. It is

$$z = \frac{\bar{x} - \mu}{\sigma / \sqrt{n}}$$

COMPUTE

MANUALLY

To solve this problem manually we need to define the rejection region, which requires us to specify a significance level. A 10% significance level is deemed to be appropriate. (We'll discuss our choice later.)

We wish to reject the null hypothesis in favor of the alternative only if the sample mean and hence the value of the test statistic is small enough. As a result we locate the rejection region in the left tail of the sampling distribution. To understand why, remember that we're trying to decide whether there is enough statistical evidence to infer that the mean is less than 22 (which is the alternative hypothesis). If we observe a large sample mean (and hence a large value of z), do we want to reject the null hypothesis in favor of the alternative? The answer is an emphatic "no." It is illogical to think that if the sample mean is, say, 30, there is enough evidence to conclude that the mean payment period for all customers would be less than 22. Consequently, we want to reject the null hypothesis only if the sample mean (and hence the value of the test statistic z) is small. How small is small enough? The answer is determined by the significance level and the rejection region. Thus, we set up the rejection region as

$$z < -z_{\alpha} = -z_{.10} = -1.28$$

Note that the direction of the inequality in the rejection region ($z < -z_{\alpha}$) matches the direction of the inequality in the alternative hypothesis ($\mu < 22$). Also note that we use the negative sign, because the rejection region is in the left tail (containing values of z less than 0) of the sampling distribution.

From the data, we compute the sum and the sample mean. They are

$$\sum x_i = 4{,}759$$

$$\bar{x} = \frac{\sum x_i}{220} = \frac{4{,}759}{220} = 21.63$$

We will assume that the standard deviation of the payment periods for the SSA plan is unchanged from its current value of $\sigma = 6$. The sample size is $n = 220$, and the value of μ is hypothesized to be 22. We compute the value of the test statistic as

$$z = \frac{\bar{x} - \mu}{\sigma/\sqrt{n}} = \frac{21.63 - 22}{6/\sqrt{220}} = -.91$$

Because the value of the test statistic, $z = -.91$, is not less than -1.28, we do not reject the null hypothesis and we do not conclude that the alternative hypothesis is true. There is insufficient evidence to infer that the mean is less than 22 days.

We can determine the p-value of the test as follows:

$$p\text{-value} = P(Z < -.91) = .1814$$

In this type of one-tail (left-tail) test of hypothesis, we calculate the p-value as $P(Z < z)$ where z is the actual value of the test statistic. Figure 11.7 depicts the sampling distribution, rejection region, and p-value.

EXCEL

	A	B	C	D
1	**Z-Test: Mean**			
2				
3				Payment
4	Mean			21.63
5	Standard Deviation			5.84
6	Observations			220
7	Hypothesized Mean			22
8	SIGMA			6
9	z Stat			-0.91
10	P(Z<=z) one-tail			0.1814
11	z Critical one-tail			1.6449
12	P(Z<=z) two-tail			0.3628
13	z Critical two-tail			1.96

MINITAB

One-Sample Z: Payment

Test of mu = 22 vs < 22
The assumed standard deviation = 6

Variable	N	Mean	StDev	SE Mean	95% Upper Bound	Z	P
Payment	220	21.6318	5.8353	0.4045	22.2972	-0.91	0.181

INTERPRET

The value of the test statistic is −.91 and its *p*-value is .1814, a figure that does not allow us to reject the null hypothesis. Because we were not able to reject the null hypothesis, we say that there is not enough evidence to infer that the mean payment period is less than 22 days. Note that there was some evidence to indicate that the mean of the entire population of payment periods is less than 22 days. We did calculate the sample mean to be 21.63. However, to reject the null hypothesis we need *enough* statistical evidence, and in this case we simply did not have enough reason to reject the null hypothesis in favor of the alternative. In the absence of evidence to show that the mean payment period for all customers sent a stamped self-addressed envelope would be less than 22 days, we cannot infer that the plan would be profitable.

A Type I error occurs when we conclude that the plan works when it actually does not. The cost of this mistake is not high. A Type II error occurs when we don't adopt the SSA envelope plan when it would reduce costs. The cost of this mistake can be high. As a consequence, we would like to minimize the probability of a Type II error. Thus, we chose a large value for the probability of a Type I error; we set $\alpha = .10$.

Figure 11.7 exhibits the sampling distribution for this example.

FIGURE **11.7**
Sampling Distribution for SSA Envelope Example

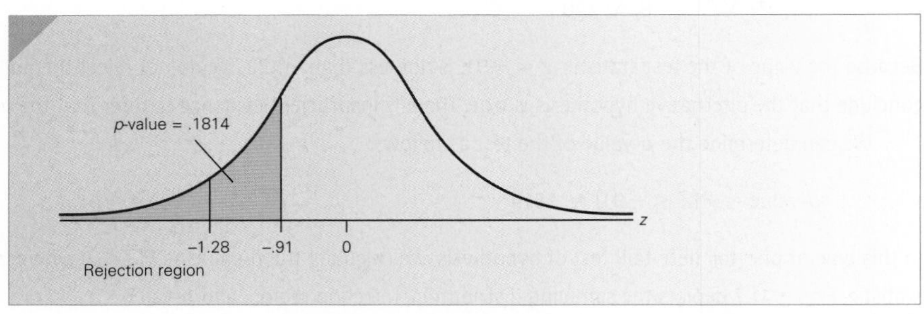

One- and Two-Tail Tests

The statistical tests conducted in Example 11.1 and the SSA Envelope example are called **one-tail tests** because the rejection region is located in only one tail of the sampling distribution. The *p*-value is also computed by finding the area in one tail of the sampling distribution. The right tail in Example 11.1 is the important one because the alternative hypothesis specifies that the mean is *greater than* 170. In the SSA envelope example, the left tail is emphasized because the alternative hypothesis specifies that the mean is *less than* 22.

We now present an example that requires a **two-tail test.**

EXAMPLE 11.2

DATA Xm11-02

Comparison of AT&T and Its Competitor

In recent years, a number of companies have been formed that offer competition to AT&T in long-distance calls. All advertise that their rates are lower than AT&T's, and as a result their bills will be lower. AT&T has responded by arguing that for the average consumer there will be no difference in billing. Suppose that a statistics practitioner working for AT&T determines that the mean and standard deviation of monthly long-distance bills for all its residential customers are \$17.09 and \$3.87, respectively. He then takes a random sample of 100 customers and recalculates their last month's bill using the rates quoted by a leading competitor. Assuming that the standard deviation of this population is the same as for AT&T, can we conclude at the 5% significance level that there is a difference between the average AT&T bill and that of the leading competitor?

SOLUTION

IDENTIFY

In this problem we want to know whether the mean monthly long-distance bill is different from \$17.09. Consequently, we set up the alternative hypothesis to express this condition:

$$H_1: \quad \mu \neq 17.09$$

The null hypothesis specifies that the mean is equal to the value specified under the alternative hypothesis. Hence,

$$H_0: \quad \mu = 17.09$$

COMPUTE

MANUALLY

To set up the rejection region, we need to realize that we can reject the null hypothesis when the test statistic is large or when it is small. That is, we must set up a *two-tail rejection region*. Because the total area in the rejection region must be α, we divide this probability by 2. Thus, the rejection region* is

$$z < -z_{\alpha/2} \quad \text{or} \quad z > z_{\alpha/2}$$

For $\alpha = .05$, $\alpha/2 = .025$, and $z_{\alpha/2} = z_{.025} = 1.96$,

$$z < -1.96 \quad \text{or} \quad z > 1.96$$

*Statistics practitioners often represent this rejection region as $|z| > z_{\alpha/2}$, which reads, "the *absolute* value of z is greater than $z_{\alpha/2}$." We prefer our method because it is clear that we are performing a two-tail test.

From the data we compute

$$\sum x_i = 1{,}754.99$$

$$\bar{x} = \frac{\sum x_i}{n} = \frac{1{,}754.99}{100} = 17.55$$

The value of the test statistic is

$$z = \frac{\bar{x} - \mu}{\sigma/\sqrt{n}} = \frac{17.55 - 17.09}{3.87/\sqrt{100}} = 1.19$$

Because 1.19 is neither greater than 1.96 nor less than −1.96, we cannot reject the null hypothesis.

We can also calculate the p-value of the test. Because it is a two-tail test, we determine the p-value by finding the area in both tails. That is,

$$p\text{-value} = P(Z < -1.19) + P(Z > 1.19) = .1170 + .1170 = .2340$$

Or, more simply multiply the probability in one tail by 2.

In general, the p-value in a two-tail test is determined by

$$p\text{-value} = 2P(Z > |z|)$$

where z is the actual value of the test statistic and $|z|$ is its absolute value.

EXCEL

	A	B	C	D
1	**Z-Test: Mean**			
2				
3				*Bills*
4	Mean			17.55
5	Standard Deviation			3.94
6	Observations			100
7	Hypothesized Mean			17.09
8	SIGMA			3.87
9	z Stat			1.19
10	P(Z<=z) one-tail			0.1173
11	z Critical one-tail			1.6449
12	P(Z<=z) two-tail			0.2346
13	z Critical two-tail			1.96

MINITAB

One-Sample Z: Bills

Test of mu = 17.09 vs not = 17.09
The assumed standard deviation = 3.87

Variable	N	Mean	StDev	SE Mean	95% CI	Z	P
Bills	100	17.5499	3.9382	0.3870	(16.7914, 18.3084)	1.19	0.235

INTERPRET

There is not enough evidence to infer that the mean long-distance bill is different from AT&T's mean of $17.09. Figure 11.8 depicts the sampling distribution for this example.

FIGURE **11.8** Sampling Distribution for Example 11.2

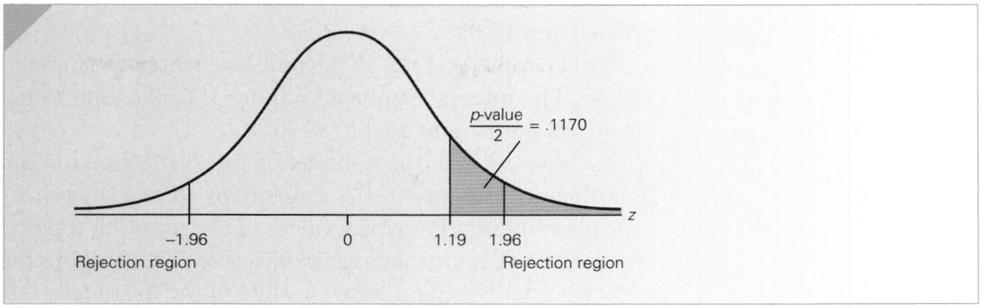

When Do We Conduct One- and Two-Tail Tests?

A two-tail test is conducted whenever the alternative hypothesis specifies that the mean is *not equal* to the value stated in the null hypothesis, that is, when the hypotheses assume the following form:

$$H_0: \quad \mu = \mu_0$$
$$H_1: \quad \mu \neq \mu_0$$

There are two one-tail tests. We conduct a one-tail test that focuses on the right tail of the sampling distribution whenever we want to know whether there is enough evidence to infer that the mean is greater than the quantity specified by the null hypothesis, that is, when the hypotheses are

$$H_0: \quad \mu = \mu_0$$
$$H_1: \quad \mu > \mu_0$$

The second one-tail test involves the left tail of the sampling distribution. It is used when the statistics practitioner wants to determine whether there is enough evidence to infer that the mean is less than the value of the mean stated in the null hypothesis. The resulting hypotheses appear in this form:

$$H_0: \quad \mu = \mu_0$$
$$H_1: \quad \mu < \mu_0$$

The techniques introduced in Chapters 12, 13, 16, 17, 18, and 19 require you to decide which of the three forms of the test to employ. Make your decision in the same way as we described the process.

Testing Hypotheses and Confidence Interval Estimators

As you've seen, the test statistic and the confidence interval estimator are both derived from the sampling distribution. It shouldn't be a surprise then that we can use the confidence interval estimator to test hypotheses. To illustrate, consider Example 11.2. The 95% confidence interval estimate of the population mean is

$$\bar{x} \pm z_{\alpha/2}\frac{\sigma}{\sqrt{n}} = 17.55 \pm 1.96\frac{3.87}{\sqrt{100}} = 17.55 \pm .76$$

$$LCL = 16.79 \quad \text{and} \quad UCL = 18.31$$

We estimate that μ lies between \$16.79 and \$18.31. Because this interval includes 17.09, we cannot conclude that there is sufficient evidence to infer that the population mean differs from 17.09.

In Example 11.1 the 95% confidence interval estimate is LCL = 171.63 and UCL = 184.37. The interval estimate excludes 170, allowing us to conclude that the population mean account is not equal to \$170.

As you can see, the confidence interval estimator can be used to conduct tests of hypotheses. This process is equivalent to the rejection region approach. However, instead of finding the critical values of the rejection region and determining whether the test statistic falls into the rejection region, we compute the interval estimate and determine whether the hypothesized value of the mean falls into the interval.

Using the interval estimator to test hypotheses has the advantage of simplicity. Apparently, we don't need the formula for the test statistic; we need only the interval estimator. However, there are two serious drawbacks.

First, when conducting a one-tail test, our conclusion may not answer the original question. In Example 11.1 we wanted to know whether there was enough evidence to infer that the mean is *greater than* 170. The estimate concludes that the mean *differs from* 170. You may be tempted to say that since the entire interval is greater than 170, there is enough statistical evidence to infer that the population mean is greater than 170. However, in attempting to draw this conclusion, we run into the problem of determining the procedure's significance level. Is it 5% or is it 2.5%? We may be able to overcome this problem through the use of **one-sided confidence interval estimators.** However, if the purpose of using confidence interval estimators instead of test statistics is simplicity, one-sided estimators are a contradiction.

Second, the confidence interval estimator does not yield a *p*-value, which we have argued is the better way to draw inferences about a parameter. Using the confidence interval estimator to test hypotheses forces the decision maker into making a reject/don't reject decision rather than providing information about how much statistical evidence exists to be judged with other factors in the decision process. Furthermore, we only postpone the point in time when a test of hypothesis must be used. In later chapters we will present problems where only a test produces the information we need to make decisions.

Developing an Understanding of Statistical Concepts 1

As is the case with the confidence interval estimator, the test of hypothesis is based on the sampling distribution of the sample statistic. The result of a test of hypothesis is a probability statement about the sample statistic. We assume that the population mean is specified by the null hypothesis. We then compute the test statistic and determine how likely it is to observe this large (or small) a value when the null hypothesis is true. If the probability is small, we conclude that the assumption that the null hypothesis is true is unfounded and we reject it.

Developing an Understanding of Statistical Concepts 2

When we (or the computer) calculate the value of the test statistic

$$z = \frac{\bar{x} - \mu}{\sigma/\sqrt{n}}$$

we're also measuring the difference between the sample statistic $\bar{x}$ and the hypothesized value of the parameter μ in terms of the standard error $\sigma/\sqrt{n}$. In Example 11.2 we found that the value of the test statistic was $z = 1.19$. This means that the sample mean was 1.19 standard errors above the hypothesized value of μ. The standard normal probability table told us that this value is not considered unlikely. As a result we did not reject the null hypothesis.

The concept of measuring the difference between the sample statistic and the hypothesized value of the parameter in terms of the standard errors is one that will be used throughout this book.

EXERCISES

Developing an Understanding of Statistical Concepts

In Exercises 11.7—11.12, calculate the value of the test statistic, set up the rejection region, determine the p-value, interpret the result, and draw the sampling distribution.

11.7 H_0: $\mu = 1{,}000$
H_1: $\mu \neq 1{,}000$
$\sigma = 200, \quad n = 100, \quad \bar{x} = 980, \quad \alpha = .01$

11.8 H_0: $\mu = 50$
H_1: $\mu > 50$
$\sigma = 5, \quad n = 9, \quad \bar{x} = 51, \quad \alpha = .03$

11.9 H_0: $\mu = 15$
H_1: $\mu < 15$
$\sigma = 2, \quad n = 25, \quad \bar{x} = 14.3, \quad \alpha = .10$

11.10 H_0: $\mu = 100$
H_1: $\mu \neq 100$
$\sigma = 10, \quad n = 100, \quad \bar{x} = 100, \quad \alpha = .05$

11.11 H_0: $\mu = 70$
H_1: $\mu > 70$
$\sigma = 20, \quad n = 100, \quad \bar{x} = 80, \quad \alpha = .01$

11.12 H_0: $\mu = 50$
H_1: $\mu < 50$
$\sigma = 15, \quad n = 100, \quad \bar{x} = 48, \quad \alpha = .05$

Exercises 11.13 to 11.27 are "what-if analyses" designed to determine what happens to the test statistic and p-value when the sample size, standard deviation, and sample mean change. These problems can be solved manually or using the Test Statistics Workbook.

11.13 a. Compute the p-value in order to test the following hypotheses given that $\bar{x} = 52$, $n = 9$, $\sigma = 5$.

H_0: $\mu = 50$
H_1: $\mu > 50$

b. Repeat Part a with $n = 25$.
c. Repeat Part a with $n = 100$.
d. Describe what happens to the value of the test statistic and its p-value when the sample size increases.

11.14 a. A statistics practitioner formulated the following hypotheses.

H_0: $\mu = 200$
H_1: $\mu < 200$

and learned that $\bar{x} = 190$, $n = 9$, and $\sigma = 50$. Compute the p-value of the test.
b. Repeat Part a with $\sigma = 30$.
c. Repeat Part a with $\sigma = 10$.
d. Discuss what happens to the value of the test statistic and its p-value when the standard deviation decreases.

11.15 a. Given the following hypotheses, determine the p-value when $\bar{x} = 21$, $n = 25$, and $\sigma = 5$.

H_0: $\mu = 20$
H_1: $\mu \neq 20$

b. Repeat Part a with $\bar{x} = 22$.
c. Repeat Part a with $\bar{x} = 23$.
d. Describe what happens to the value of the test statistic and its p-value when the value of $\bar{x}$ increases.

11.16 a. Test these hypotheses by calculating the p-value given that $\bar{x} = 99$, $n = 100$, and $\sigma = 8$.

H_0: $\mu = 100$
H_1: $\mu \neq 100$

b. Repeat Part a with $n = 50$.
c. Repeat Part a with $n = 20$.
d. What is the effect on the value of the test statistic and the p-value of the test when the sample size decreases?

11.17 a. Find the p-value of the following test given that $\bar{x} = 990$, $n = 100$, and $\sigma = 25$.

H_0: $\mu = 1{,}000$
H_1: $\mu < 1{,}000$

b. Repeat Part a with $\sigma = 50$.
c. Repeat Part a with $\sigma = 100$.
d. Describe what happens to the value of the test statistic and its p-value when the standard deviation increases.

11.18 a. Calculate the p-value of the test described here.

$$H_0: \mu = 60$$
$$H_1: \mu > 60$$
$$\bar{x} = 72, \quad n = 25, \quad \sigma = 20$$

b. Repeat Part a with $\bar{x} = 68$.
c. Repeat Part a with $\bar{x} = 64$.
d. Describe the effect on the test statistic and the p-value of the test when the value of $\bar{x}$ decreases.

11.19 Redo Example 11.1 with
a. $n = 200$
b. $n = 100$
c. Describe the effect on the test statistic and the p-value when n increases.

11.20 Redo Example 11.1 with
a. $\sigma = 35$
b. $\sigma = 100$
c. Describe the effect on the test statistic and the p-value when σ increases.

11.21 Perform a what-if analysis to calculate the p-values in Table 11.2.

11.22 Redo the SSA example with
a. $n = 100$
b. $n = 500$
c. What is the effect on the test statistic and the p-value when n increases?

11.23 Redo the SSA example with
a. $\sigma = 3$
b. $\sigma = 12$
c. Discuss the effect on the test statistic and the p-value when σ increases.

11.24 For the SSA example, create a table that shows the effect on the test statistic and the p-value of decreasing the value of the sample mean. Use $\bar{x}$ = 22.0, 21.8, 21.6, 21.4, 21.2, 21.0, 20.8, 20.6, and 20.4.

11.25 Redo Example 11.2 with
a. $n = 50$
b. $n = 400$
c. Briefly describe the effect on the test statistic and the p-value when n increases.

11.26 Redo Example 11.2 with
a. $\sigma = 2$
b. $\sigma = 10$
c. What happens to the test statistic and the p-value when σ increases?

11.27 Refer to Example 11.2. Create a table that shows the effect on the test statistic and the p-value of changing the value of the sample mean. Use $\bar{x}$ = 15.0, 15.5, 16.0, 16.5, 17.0, 17.5, 18.0, 18.5, and 19.0

Applications

The following exercises may be answered manually or with the assistance of a computer. The files containing the data are given.

11.28 Xr11-28 A business student claims that on average an MBA student is required to prepare more than five cases per week. To examine the claim, a statistics professor asks a random sample of 10 MBA students to report the number of cases they prepare weekly. The results are exhibited here. Can the professor conclude at the 5% significance level that the claim is true, assuming that the number of cases is normally distributed with a standard deviation of 1.5?

2 7 4 8 9 5 11 3 7 4

11.29 Xr11-29 A random sample of 18 young adult men (20–30 years old) was sampled. Each person was asked how many minutes of sports they watched on television daily. The responses are listed here. It is known that $\sigma = 10$. Test to determine at the 5% significance level whether there is enough statistical evidence to infer that the mean amount of television watched daily by all young adult men is greater than 50 minutes.

50 48 65 74 66 37 45 68 64
65 58 55 52 63 59 57 74 65

11.30 Xr11-30 The club professional at a difficult public course boasts that his course is so tough that the average golfer loses a dozen or more golf balls during a round of golf. A dubious golfer sets out to show that the pro is fibbing. He asks a random sample of 15 golfers who just completed their rounds to report the number of golf balls each lost. Assuming that the number of golf balls lost is normally distributed with a standard deviation of 3, can we infer at the 10% significance level that the average number of golf balls lost is less than 12?

1 14 8 15 17 10 12 6
14 21 15 9 11 4 8

11.31 Xr11-31 A random sample of 12 second-year university students enrolled in a business statistics course was drawn. At the course's completion, each student was asked how many hours he or she spent doing homework in statistics. The data are listed here. It is known that the population standard deviation is $\sigma = 8.0$. The instructor has recommended that students devote 3 hours per week for the duration of the 12-week semester, for a total of 36 hours. Test to determine whether there is evidence that the average student spent less than the recommended amount of time. Compute the p-value of the test.

31 40 26 30 36 38 29 40 38 30 35 38

11.32 Xr11-32 The owner of a public golf course is concerned about slow play, which clogs the course and results in selling fewer rounds. She believes the problem lies in the amount of time taken to sink putts on the green. To investigate the problem, she randomly samples 10 foursomes and measures the amount of time they spend on the 18th green. The data are listed here. Assuming that the times are normally distributed with a standard deviation of 2 minutes, test to determine whether the owner can infer at the 5% significance level that the mean amount of time spent putting on the 18th green is greater than 6 minutes.

8 11 5 6 7 8 6 4 8 3

11.33 Xr11-33 A machine that produces ball bearings is set so that the average diameter is .50 inch. A sample of 10 ball bearings was measured with the results shown here. Assuming that the standard deviation is .05 inch, can we conclude at the 5% significance level that the mean diameter is not .50 inch?

.48 .50 .49 .52 .53 .48 .49 .47 .46 .51

11.34 Xr11-34 Spam e-mail has become a serious and costly nuisance. An office manager believes that the average amount of time spent by office workers reading and deleting spam exceeds 25 minutes per day. To test this belief, he takes a random sample of 18 workers and measures the amount of time each spends reading and deleting spam. The results are listed here. If the population of times is normal with a standard deviation of 12 minutes, can the manager infer at the 1% significance level that he is correct?

35 48 29 44 17 21 32 28 34
23 13 9 11 30 42 37 43 38

The following exercises require the use of a computer and software. The answers may be calculated manually. See Appendix A for the sample statistics.

11.35 [Xr11-35] A manufacturer of light bulbs advertises that, on average, its long-life bulb will last more than 5,000 hours. To test the claim, a statistician took a random sample of 100 bulbs and measured the amount of time until each bulb burned out. If we assume that the lifetime of this type of bulb has a standard deviation of 400 hours, can we conclude at the 5% significance level that the claim is true?

11.36 Xr11-36 In the midst of labor-management negotiations, the president of a company argues that the company's blue-collar workers, who are paid an average of $30,000 per year, are well paid because the mean annual income of all blue-collar workers in the country is less than $30,000. That figure is disputed by the union, which does not believe that the mean blue-collar income is less than $30,000. To test the company president's belief, an arbitrator draws a random sample of 350 blue-collar workers from across the country and asks each to report his or her annual income. If the arbitrator assumes that the blue-collar incomes are normally distributed with a standard deviation of $8,000, can it be inferred at the 5% significance level that the company president is correct?

11.37 Xr11-37 A dean of a business school claims that the GMAT scores of applicants to the school's MBA program have increased during the past 5 years. Five years ago, the mean and standard deviation of GMAT scores of MBA applicants were 560 and 50, respectively. Twenty applications for this year's program were randomly selected and the GMAT scores recorded. If we assume that the distribution of GMAT scores of this year's applicants is the same as that of 5 years ago, with the possible exception of the mean, can we conclude at the 5% significance level that the dean's claim is true?

11.38 Xr11-38 Past experience indicates that the monthly long-distance telephone bill is normally distributed with a mean of $17.85 and a standard deviation of $3.87. After an advertising campaign aimed at increasing long-distance telephone usage, a random sample of 25 household bills was taken.

a. Do the data allow us to infer at the 10% significance level that the campaign was successful?
b. What assumption must you make to answer Part a?

11.39 Xr11-39 In an attempt to reduce the number of person-hours lost as a result of industrial accidents, a large production plant installed new safety equipment. In a test of the effectiveness of the equipment, a random sample of 50 departments was chosen. The number of person-hours lost in the month prior to and the month after the installation of the safety equipment was recorded. The percentage change was calculated and recorded. Assume that the population standard deviation is $\sigma = 6$. Can we infer at the 10% significance level that the new safety equipment is effective?

11.40 Xr11-40 A highway patrol officer believes that the average speed of cars traveling over a certain stretch of highway exceeds the posted limit of 55 mph. The speeds of a random sample of 200 cars were recorded. Do these data provide sufficient evidence at the 1% significance level to support the officer's belief? What is the p-value of the test? (Assume that the standard deviation is known to be 5.)

11.41 Xr11-41 An automotive expert claims that the large number of self-serve gasoline stations has resulted in poor automobile maintenance, and that the average tire pressure is more than 4 pounds per square inch (psi) below its manufacturer's specification. As a quick test, 50 tires are examined, and the number of psi each tire is below specification is recorded. If we assume that tire pressure is normally distributed with $\sigma = 1.5$ psi, can we infer at the 10% significance level that the expert is correct? What is the p-value?

11.42 Xr11-42 For the past few years, the number of customers of a drive-up bank in New York has averaged 20 per hour, with a standard deviation of 3 per hour. This year, another bank 1 mile away opened a drive-up window. The manager of the first bank believes that this will result in a decrease in the number of customers. The number of customers who arrived during 36 randomly selected hours was recorded. Can we conclude at the 5% significance level that the manager is correct? What is the p-value?

11.43 Xr11-43 A fast-food franchiser is considering building a restaurant at a certain location. Based on financial analyses, a site is acceptable only if the number of pedestrians passing the location averages more than 100 per hour. The number of pedestrians observed for each of 40 hours was recorded. Assuming that the population standard deviation is known to be 16, can we conclude at the 1% significance level that the site is acceptable?

11.44 Xr11-44 Many Alpine ski centers base their projections of revenues and profits on the assumption that the average Alpine skier skis four times per year. To investigate the validity of this assumption, a random sample of 63 skiers is drawn and each is asked to report the number of times they skied the previous year. If we assume that the standard deviation is 2, can we infer at the 10% significance level that the assumption is wrong?

11.45 Xr11-45 The golf professional at a private course claims that members who have taken lessons from him lowered their handicap by more than five strokes. The club manager decides to test the claim by randomly sampling 25 members who have had lessons and asking each to report the reduction in handicap where a negative number indicates an increase in the handicap. Assuming that the reduction in handicap is approximately normally distributed with a standard deviation of two strokes, test the golf professional's claim using a 10% significance level.

11.46 Xr11-46 The current no-smoking regulations in office buildings require workers who smoke to take breaks and leave the building in order to satisfy their habits. A study indicates that such workers average 32 minutes per day taking smoking breaks. The standard deviation is 8 minutes. To help reduce the average break, rooms with powerful exhausts were installed in the buildings. To see whether these rooms serve their designed purpose, a random sample of 110 smokers was taken. The total amount of time away from their desks was measured for 1 day. Test to determine whether there has been a decrease in the mean time away from their desks. Compute the p-value and interpret it relative to the costs of Type I and Type II errors.

11.47 Xr11-47 A low-handicap golfer who uses Titleist brand golf balls observed that his average drive is 230 yards and the standard deviation is 10 yards. Nike has just introduced a new ball, which has been endorsed by Tiger Woods. Nike claims that the ball will travel farther than Titleist. To test the claim, the golfer hits 100 drives with a Nike ball and measures the distances. Conduct a test to determine whether Nike is correct. Use a 5% significance level.

11.3 CALCULATING THE PROBABILITY OF A TYPE II ERROR

To properly interpret the results of a test of hypothesis requires that you be able to specify an appropriate significance level or to judge the p-value of a test. However, it also requires that you have an understanding of the relationship between Type I and Type II errors. In this section, we describe how the probability of a Type II error is computed and interpreted.

Recall Example 11.1, where we conducted the test using the sample mean as the test statistic and we computed the rejection region (with $\alpha = .05$) as

$$\bar{x} > 175.34$$

A Type II error occurs when a false null hypothesis is not rejected. In Example 11.1, if $\bar{x}$ is less than 175.34, we will not reject the null hypothesis. If we do not reject the null hypothesis, we will not install the new billing system. Thus, the consequence of a Type II error in this example is that we will not install the new system when it would be cost-effective. The probability of this occurring is the probability of a Type II error. It is defined as

$$\beta = P(\overline{X} < 175.34, \text{ given that the null hypothesis is false})$$

The condition that the null hypothesis is false tells us only that the mean is not equal to 170. If we want to compute β, we need to specify a value for μ. Suppose that when the mean account is at least \$180, the new billing system's savings become so attractive that the manager would hate to make the mistake of not installing the system. As a result, she would like to determine the probability of not installing the new system when it would produce large cost savings. Because calculating probability from an approximately normal sampling distribution requires a value of μ (as well as σ and n), we will calculate the probability of not installing the new system when μ is *equal* to 180:

$$\beta = P(\overline{X} < 175.34, \text{ given that } \mu = 180)$$

We know that $\bar{x}$ is approximately normally distributed with mean μ and standard deviation $\sigma/\sqrt{n}$. To proceed, we standardize $\bar{x}$ and use the standard normal table (Table 3 in Appendix B):

$$\beta = P\left(\frac{\overline{X} - \mu}{\sigma/\sqrt{n}} < \frac{175.34 - 180}{65/\sqrt{400}}\right) = P(Z < -1.43) = .0764$$

This tells us that when the mean account is actually \$180, the probability of incorrectly not rejecting the null hypothesis is .0764. Figure 11.9 graphically depicts how the calculation was performed. Notice that to calculate the probability of a Type II error, we had to express the rejection region in terms of the unstandardized test statistic $\bar{x}$, and we had to specify a value for μ other than the one shown in the null hypothesis. In this illustration, the value of μ used was based on a financial analysis indicating that when μ is at least \$180, the cost savings would be very attractive.

FIGURE 11.9 Calculating β for $\mu = 180$, $\alpha = .05$, and $n = 400$

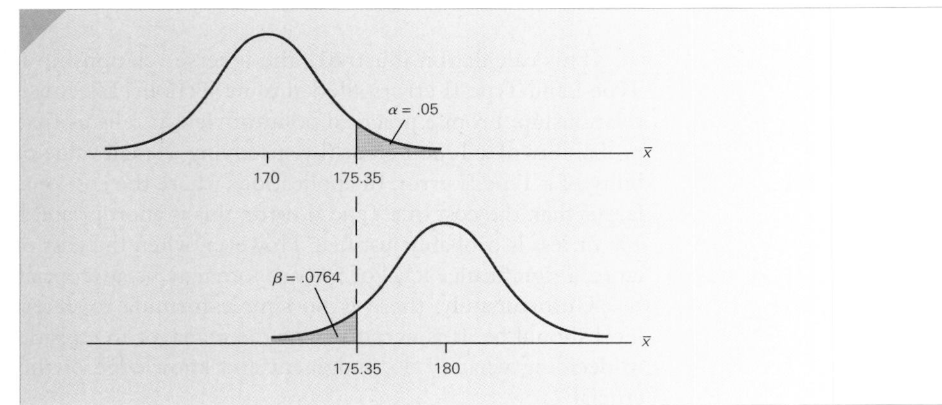

Effect on β of Changing α

Suppose that in the previous illustration we had used a significance level of 1% instead of 5%. The rejection region expressed in terms of the standardized test statistic would be

$$z > z_{.01} = 2.33$$

or

$$\frac{\bar{x} - 170}{65/\sqrt{400}} > 2.33$$

Solving for $\bar{x}$, we find the rejection region in terms of the unstandardized test statistic:

$$\bar{x} > 177.57$$

The probability of a Type II error when $\mu = 180$ is

$$\beta = P\left(\frac{\bar{x} - \mu}{\sigma/\sqrt{n}} < \frac{177.57 - 180}{65/\sqrt{400}}\right) = P(Z < -.75) = .2266$$

Figure 11.10 depicts this calculation. Compare this figure with Figure 11.9. As you can see, by decreasing the significance level from 5% to 1%, we have shifted the critical value of the rejection region to the right and thus enlarged the area where the null hypothesis is not rejected. The probability of a Type II error increases from .0764 to .2266.

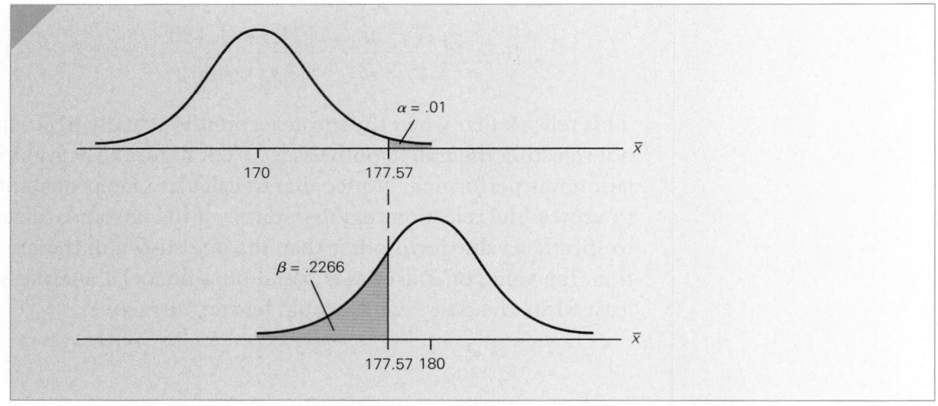

FIGURE **11.10** Calculating β for $\mu = 180$, $\alpha = .01$, and $n = 400$

This calculation illustrates the inverse relationship between the probabilities of Type I and Type II errors alluded to in Section 11.2. It is important to understand this relationship. From a practical point of view, it tells us that if you want to decrease the probability of a Type I error (by specifying a small value of α), you increase the probability of a Type II error. In applications where the cost of a Type I error is considerably larger than the cost of a Type II error, this is appropriate. In fact, a significance level of 1% or less is probably justified. However, when the cost of a Type II error is relatively large, a significance level of 5% or more may be appropriate.

Unfortunately, there is no simple formula to determine what the significance level should be. It is necessary for the manager to consider the costs of both mistakes in deciding what to do. Judgment and knowledge of the factors in the decision are crucial.

Judging the Test

There is another important concept to be derived from this section. A statistical test of hypothesis is effectively defined by the significance level and the sample size, both of which are selected by the statistics practitioner. We can judge how well the test functions by calculating the probability of a Type II error at some value of the parameter. To illustrate, in Example 11.1 the manager chose a sample size of 400 and a 5% significance level on which to base her decision. With those selections, we found β to be .0764 when the actual mean is 180. If we believe that the cost of a Type II error is high and thus that the probability is too large, we have two ways to reduce the probability. We can increase the value of α; however, this would result in an increase in the chance of making a Type I error, which is very costly.

Alternatively, we can increase the sample size. Suppose that the manager chose a sample size of 1,000. We'll now recalculate β with $n = 1{,}000$ (and $\alpha = .05$). The rejection region is

$$z > z_{.05} = 1.645$$

or

$$\frac{\bar{x} - 170}{65/\sqrt{1{,}000}} > 1.645$$

which yields

$$\bar{x} > 173.38$$

The probability of a Type II error is

$$\beta = P\left(\frac{\bar{x} - \mu}{\sigma/\sqrt{n}} < \frac{173.38 - 180}{65/\sqrt{1{,}000}}\right) = P(Z < -3.22) = 0 \text{ (approximately)}$$

In this case we maintained the same value of α (.05), but we reduced the probability of not installing the system when the actual mean account is \$180 to virtually 0.

Developing an Understanding of Statistical Concepts: Larger Sample Size Equals More Information Equals Better Decisions

Figure 11.11 displays the previous calculation. When compared with Figure 11.9, we can see that the sampling distribution of the mean is narrower because the standard error of the mean $\sigma/\sqrt{n}$ becomes smaller as n increases. Narrower distributions represent more

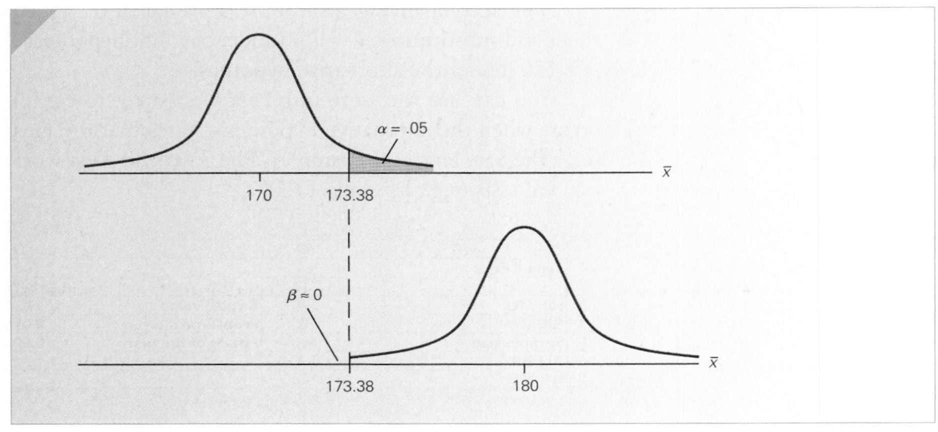

FIGURE **11.11** Calculating β for $\mu = 180$, $\alpha = .05$, and $n = 1{,}000$

information. The increased information is reflected in a smaller probability of a Type II error.

The calculation of the probability of a Type II error for $n = 400$ and for $n = 1{,}000$ illustrates a concept whose importance cannot be overstated. By increasing the sample size, we reduce the probability of a Type II error. By reducing the probability of a Type II error, we make this type of error less frequently. Hence, larger sample sizes allow us to make better decisions in the long run. This finding lies at the heart of applied statistical analysis and reinforces the book's first sentence, "Statistics is a way to get information from data."

Throughout this book we introduce a variety of applications in accounting, finance, marketing, operations management, human resources management, and economics. In all such applications the statistics practitioner must make a decision, which involves converting data into information. The more information, the better the decision. Without such information, decisions must be based on guesswork, instinct, and luck. A famous statistician, W. Edwards Deming said it best: "Without data you're just another person with an opinion."

Power of a Test

Another way of expressing how well a test performs is to report its *power:* the probability of its leading us to reject the null hypothesis when it is false. Thus, the power of a test is $1 - \beta$.

When more than one test can be performed in a given situation, we would naturally prefer to use the test that is correct more frequently. If (given the same alternative hypothesis, sample size, and significance level) one test has a higher power than a second test, the first test is said to be more powerful.

Using the Computer

EXCEL

We have made it possible to utilize Excel to calculate β for any test of hypothesis.

Open the **Beta-mean** workbook. There are three worksheets. They are **Right-tail test, Left-tail test,** and **Two-tail test.** Find the appropriate worksheet for the test of hypothesis you are analyzing and type values for μ (under the null hypothesis), σ, n, α, and μ (actual value under the alternative hypothesis).

The accompanying printout was produced by selecting the **Right-tail Test** worksheet and substituting $\mu = 170$ (under the null hypothesis), $\sigma = 65$, $n = 400$, $\alpha = .05$, and $\mu = 180$ (under the alternative hypothesis).

You can use the **Left-tail Test** worksheet to compute the probability of Type II errors when the alternative hypothesis states that the mean is less than a specified value (e.g., the SSA envelope example). The **Two-tail Test** worksheet is used to compute β for two tail tests (e.g., Example 11.2).

	A	B	C	D
1	Type II Error			
2				
3	H0: MU	170	Critical value	175.35
4	SIGMA	65	Prob(Type II error)	0.0761
5	Sample size	400	Power of the test	0.9239
6	ALPHA	0.05		
7	H1: MU	180		

MINITAB

Minitab computes the power of the test.

```
Power and Sample Size

1-Sample Z Test

Testing mean = null (versus > null)
Calculating power for mean = null + difference
Alpha = 0.05  Assumed standard deviation = 65


              Sample
Difference     Size      Power
        10      400   0.923938
```

INSTRUCTIONS

1. Click **Stat, Power and Sample Size,** and **1-Sample Z . . . .**
2. Specify the sample size in the **Sample sizes** box. (You can specify more than one value of n. Minitab will compute the power for each value.) Type the difference between the actual value of μ and the value of μ under the null hypothesis. (You can specify more than one value.) Type the value of the standard deviation in the **Standard deviation** box.
3. Click **Options . . .** and specify the **Alternative Hypothesis** and the **Significance level.**

For Example 11.1 we typed **400** to select the **Sample sizes,** the **Differences** was **10** (=180 − 170), **Standard deviation** was **65,** the **Alternative Hypothesis** was **Greater than**, and the **Significance level** was **0.05.**

Operating Characteristic Curve

To compute the probability of a Type II error, we must specify the significance level, the sample size, and an alternative value of the population mean. One way to keep track of all these components is to draw the **operating characteristic (OC) curve,** which plots the values of β versus the values of μ. Because of the time-consuming nature of these calculations, the computer is a virtual necessity. To illustrate, we'll draw the OC curve for Example 11.1. We used Excel (we could have used Minitab instead) to compute the probability of Type II error in Example 11.1 for $\mu = 170, 171, \ldots, 185$, with $n = 400$. Figure 11.12

FIGURE **11.12**
Operating Characteristic Curve for Example 11.1

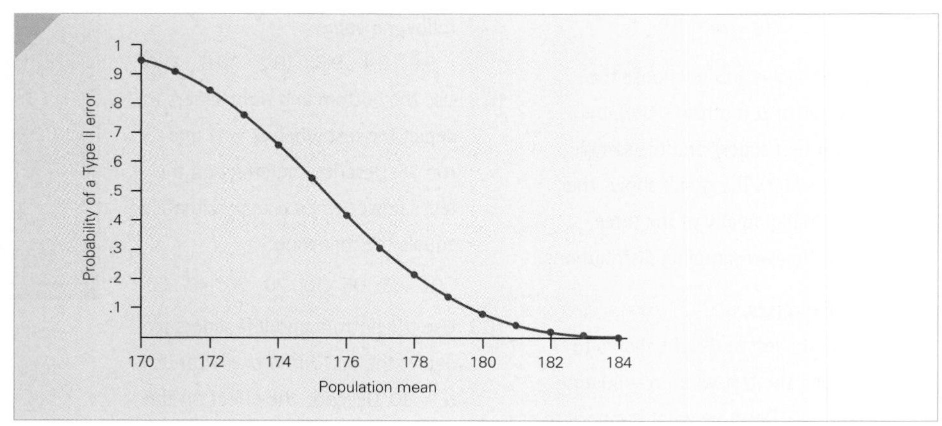

depicts this curve. Notice as the alternative value of μ increases, the value of β decreases. This tells us that as the alternative value of μ moves farther from the value of μ under the null hypothesis, the probability of a Type II error decreases. In other words, it becomes easier to distinguish between $\mu = 170$ and other values of μ when μ is farther from 170. Note that when $\mu = 170$ (the hypothesized value of μ), $\beta = 1 - \alpha$.

The OC curve can also be useful in selecting a sample size. Figure 11.13 shows the OC curve for Example 11.1 with $n = 100, 400, 1{,}000$, and 2,000. An examination of this chart sheds some light concerning the effect that increasing the sample size has on how well the test performs at different values of μ. For example, we can see that smaller sample sizes will work well to distinguish between 170 and values of μ larger than 180. However, to distinguish between 170 and smaller values of μ requires larger sample sizes. Although the information is imprecise, it does allow us to select a sample size that is suitable for our purposes.

FIGURE 11.13
Operating Characteristic Curve for Example 11.1 for n = 100, 400, 1,000, and 2,000

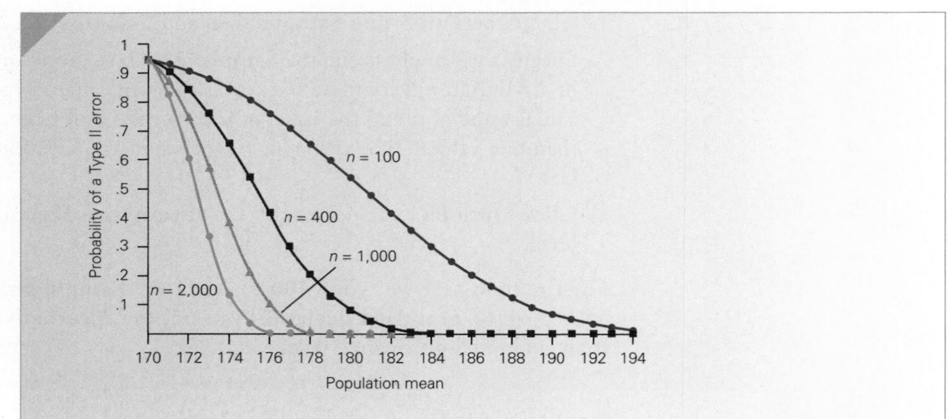

SEEING STATISTICS

Applet 16: Power of a *z*-Test

We are given the following hypotheses to test:

H_0: $\mu = 10$
H_1: $\mu \neq 10$

The applet allows you to choose the actual value of μ (bottom slider), the value of α (left slider), and the sample size (right slider). The graph shows the effect of changing any of the three values on the two sampling distributions.

Applet Exercises

16.1 Use the left and right sliders to depict the test when $n = 50$ and $\alpha = .10$. Describe what happens to the power of the test (Power = $1 - \beta$) when the actual value of μ approximately equals the following values:

9.0 9.4 9.8 10.2 10.6 11.0

16.2 Use the bottom and right sliders to depict the test when $\mu = 11$ and $n = 25$. Describe the effect on the test's power when α approximately equals the following:

.01 .03 .05 .10 .20 .30 .40 .50

16.3 Use the bottom and left sliders to depict the test when $\mu = 11$ and $\alpha = .10$. Describe the effect on the test's power when n equals the following:

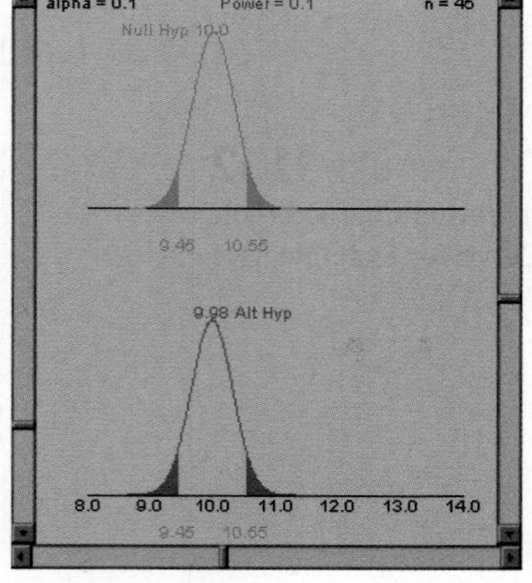

2 5 10 25 50 75 100

Determining the Alternative Hypothesis to Define Type I and Type II Errors

We've already discussed how the alternative hypothesis is determined. It represents the condition we're investigating. In Example 11.1 we wanted to know whether there was sufficient statistical evidence to infer that the new billing system would be cost-effective, that is, whether the mean monthly account is greater than \$170. In this textbook you will encounter many problems using similar phraseology. Your job will be to conduct the test that answers the question.

In real life, however, the manager (that's you 5 years from now) will be asking and answering the question. In general, you will find that the question can be posed in two ways. In Example 11.1 we asked whether there was evidence to conclude that the new system would be cost-effective. Another way of investigating the issue is to determine whether there is sufficient evidence to infer that the new system would *not* be cost-effective. We remind you of the criminal trial analogy. In a criminal trial the burden of proof falls on the prosecution to prove that the defendant is guilty. In other countries with less emphasis on individual rights, the defendant is required to prove his or her innocence. In the United States and Canada (and in other countries), we chose the former because we consider the conviction of an innocent defendant to be the greater error. Thus, the test is set up with the null and alternative hypotheses as described in Section 11.1.

In a statistical test where we are responsible for asking the question, as well as answering it, we must ask the question so that we directly control the error that is more costly. As you have already seen, we control the probability of a Type I error by specifying its value (the significance level). Consider Example 11.1 once again. There are two possible errors: conclude that the billing system is cost-effective, when it isn't and conclude that the system is not cost-effective when it is. If the manager concludes that the billing plan is cost-effective, the company will install the new system. If, in reality, the system is not cost-effective, the company will incur a loss. On the other hand, if the manager concludes that the billing plan is not going to be cost-effective, the company will not install the system. However, if the system is actually cost-effective the company will lose the potential gain from installing it. Which cost is greater?

Suppose we believe that the cost of installing a system that is not cost-effective is higher than the potential loss of not installing an effective system. The error we wish to avoid is the erroneous conclusion that the system is cost-effective. We define this as a Type I error. As a result, the burden of proof is placed on the system to deliver sufficient statistical evidence that the mean account is greater than \$170. The null and alternative hypotheses are as formulated previously:

$$H_0: \quad \mu = 170$$
$$H_1: \quad \mu > 170$$

However, if we believe that the potential loss of not installing the new system when it would be cost-effective is the larger cost, we would place the burden of proof on the manager to infer that the mean monthly account is less than \$170. Consequently, the hypotheses would be

$$H_0: \quad \mu = 170$$
$$H_1: \quad \mu < 170$$

This discussion emphasizes the need in practice to examine the costs of making both types of error before setting up the hypotheses. However, it is important for readers to understand that the questions posed in exercises throughout this book have already taken these costs into consideration. Accordingly, your task is to set up the hypotheses to answer the questions.

EXERCISES

Developing an Understanding of Statistical Concepts

11.48 a. Find the probability of a Type II error for the following test of hypothesis, given that $\mu = 196$:

H_0: $\mu = 200$

H_1: $\mu < 200$

The significance level is 10%, the population standard deviation is 30, and the sample size is 25.

b. Repeat Part a with $n = 100$.

c. Describe the effect on β of increasing n.

11.49 Suppose that in Example 11.1 we wanted to determine whether there was sufficient evidence to conclude that the new system would *not* be cost-effective. Set up the null and alternative hypotheses and discuss the consequences of Type I and Type II errors. Conduct the test. Is your conclusion the same as the one reached in Example 11.1? Explain.

11.50 A statistics practitioner wants to test the following hypotheses with $\sigma = 20$ and $n = 100$:

H_0: $\mu = 100$

H_1: $\mu > 100$

a. Using $\alpha = .10$ find the probability of a Type II error when $\mu = 102$.

b. Repeat Part a with $\alpha = .02$.

c. Describe the effect on β of decreasing α.

11.51 Draw the operating characteristic curve for $n =$ 10, 50, and 100 for the following test:

H_0: $\mu = 400$

H_1: $\mu > 400$

$\alpha = .05, \quad \sigma = 50$

11.52 a. Calculate the probability of a Type II error for the following hypotheses when $\mu = 37$:

H_0: $\mu = 40$

H_1: $\mu < 40$

The significance level is 5%, the population standard deviation is 5, and the sample size is 25.

b. Repeat Part a with $\alpha = 15\%$.

c. Describe the effect on β of increasing α.

11.53 Draw the figures of the sampling distributions for Exercises 11.50 and 11.52.

11.54 For each of Exercises 11.56–11.58, draw the sampling distributions similar to Figure 11.9.

11.55 For the test of hypothesis

H_0: $\mu = 1{,}000$

H_1: $\mu \neq 1{,}000$

$\alpha = .05, \quad \sigma = 200$

draw the operating characteristic curve for $n =$ 25, 100, and 200.

11.56 Determine β for the following test of hypothesis, given that $\mu = 48$.

H_0: $\mu = 50$

H_1: $\mu < 50$

$\alpha = .05, \quad \sigma = 10, \quad n = 40$

11.57 Calculate the probability of a Type II error for the following test of hypothesis, given that $\mu = 203$.

H_0: $\mu = 200$

H_1: $\mu \neq 200$

$\alpha = .05, \quad \sigma = 10, \quad n = 100$

11.58 Find the probability of a Type II error for the following test of hypothesis, given that $\mu = 1{,}050$.

H_0: $\mu = 1{,}000$

H_1: $\mu > 1{,}000$

$\alpha = .01, \quad \sigma = 50, \quad n = 25$

11.59 a. Determine β for the following test of hypothesis, given that $\mu = 310$:

H_0: $\mu = 300$

H_1: $\mu > 300$

The statistics practitioner knows that the population standard deviation is 50, the significance level is 5%, and the sample size is 81.

b. Repeat Part a with $n = 36$.

c. Describe the effect on β of decreasing n.

11.60 For Exercises 11.48 and 11.59, draw the sampling distributions similar to Figure 11.9.

Applications

11.61 The feasibility of constructing a profitable electricity-producing windmill depends on the mean velocity of the wind. For a certain type of windmill, the mean would have to exceed 20 miles per hour in order for its construction to be warranted. The determination of a site's feasibility is a two-stage process. In the first stage, readings of the wind velocity are taken and the mean is calculated. The test is designed to answer the question, Is the site feasible? In other words, is there sufficient evidence to conclude that the mean wind velocity exceeds 20 mph? If there is enough evidence, further testing is conducted. If there is not enough evidence, the site is removed from consideration. Discuss the consequences and potential costs of Type I and Type II errors.

11.62 The number of potential sites for the first-stage test in Exercise 11.61 is quite large and the readings can be expensive. Accordingly, the test is conducted with a sample of 25 observations. Because the second-stage cost is high, the significance level is set at 1%. A financial analysis of the potential profits and costs reveals that if the mean wind velocity is as high as 25 mph, the windmill would be extremely profitable. Calculate the probability that the first-stage test will not conclude that the site is feasible when the actual mean wind velocity is 25 mph. (Assume that σ is 8.) Discuss how the process can be improved.

11.63 A school-board administrator believes that the average number of days absent per year among students is less than 10 days. From past experience, he knows that the population standard deviation is 3 days. In testing to determine whether his belief is true, he could use one of the following plans:

i. $n = 100, \quad \alpha = .01$
ii. $n = 75, \quad \alpha = .05$
iii. $n = 50, \quad \alpha = .10$

Which plan has the lowest probability of a Type II error, given that the true population average is 9 days?

11.64 The fast-food franchiser in Exercise 11.43 was unable to provide enough evidence that the site is acceptable. She is concerned that she may be missing an opportunity to locate the restaurant in a profitable location. She feels that if the actual mean is 104, the restaurant is likely to be very successful. Determine the probability of a Type II error when the mean is 104. Suggest ways to improve this probability.

11.65 The test of hypothesis in the SSA example concluded that there was not enough evidence to infer that the plan would be profitable. The company would hate to not institute the plan if the actual reduction was as little as 3 days (i.e., $\mu = 21$). Calculate the relevant probability and describe how the company should use this information.

11.66 Refer to Exercise 11.46. A financial analyst has determined that a 2-minute reduction in the average break would increase productivity. As a result, the company would hate to lose this opportunity. Calculate the probability of erroneously concluding that the renovation would not be successful when the average break is 30 minutes. If this probability is high, describe how it can be reduced.

11.67 In Exercise 11.39 we tested to determine whether the installation of safety equipment was effective in reducing person-hours lost to industrial accidents. The null and alternative hypotheses were

H_0: $\mu = 0$
H_1: $\mu < 0$

with $\sigma = 6$, $\alpha = .10$, $n = 50$, and μ = the mean percentage change. The test failed to indicate that the new safety equipment is effective. The manager is concerned that the test was not sensitive enough to detect small, but important changes. In particular, he worries that if the true reduction in time lost to accidents is actually 2% (i.e., $\mu = -2$), the firm may miss the opportunity to install very effective equipment. Find the probability that the test with $\sigma = 6$, $\alpha = .10$, and $n = 50$ will fail to conclude that such equipment is effective. Discuss ways to decrease this probability.

11.4 The Road Ahead

We had two principal goals to accomplish in Chapters 10 and 11. First, we wanted to present the concepts of estimation and hypothesis testing. Second, we wanted to show how to produce confidence interval estimates and conduct tests of hypotheses. The importance of both of these goals should not be underestimated. Almost everything that follows this chapter will involve either estimating a parameter or testing a set of hypotheses. Consequently, Sections 10.2 and 11.2 set the pattern for the way in which statistical techniques are applied. It is no exaggeration to state that if you understand how to produce and use confidence interval estimates and how to conduct and interpret hypothesis tests, then you are well on your way to the ultimate goal of being competent at analyzing, interpreting, and presenting data. It is fair for you to ask what more you must accomplish to achieve this goal. The answer, simply put, is much more of the same.

In the chapters that follow, we plan to present about three dozen different statistical techniques that can be (and frequently are) employed by statistics practitioners. To calculate the value of test statistics or confidence interval estimates requires nothing more than the ability to add, subtract, multiply, divide, and compute square roots. If you intend to use the computer, all you need to know are the commands. The key, then, to applying statistics is knowing which formula to calculate or which set of commands to issue. Thus, the real challenge of the subject lies in being able to define the problem and identify which statistical method is the most appropriate one to use.

Most students have some difficulty recognizing the particular kind of statistical problem they are addressing unless, of course, the problem appears among the exercises at the end of a section that just introduced the technique needed. Unfortunately, in practice, statistical problems do not appear already so identified. Consequently, we have adopted an approach to teaching statistics that is designed to help identify the statistical technique.

A number of factors determine which statistical method should be used, but two are especially important: the type of data and the purpose of the statistical inference. In Chapter 2, we pointed out that there are effectively three types of data—interval, ordinal, and nominal. Recall that nominal data represent categories such as marital status, occupation, and gender. Statistics practitioners often record nominal data by assigning numbers to the responses (e.g., 1 = single; 2 = married; 3 = divorced; 4 = widowed). Because these numbers are assigned completely arbitrarily, any calculations performed on them are meaningless. All that we can do with nominal data is count the number of times each category is observed. Ordinal data are obtained from questions whose answers represent a rating or ranking system. For example, if students are asked to rate a university professor, the responses may be excellent, good, fair, or poor. To draw inferences about such data, we convert the responses to numbers. Any numbering system is valid as long as the order of the responses is preserved. Thus, "4 = excellent; 3 = good; 2 = fair; 1 = poor" is just as valid as "15 = excellent; 8 = good; 5 = fair; 2 = poor." Because of this feature, the most appropriate statistical procedures for ordinal data are ones based on a ranking process.

Interval data are real numbers, such as those representing income, age, height, weight, and volume. Computation of means and variances is permissible.

The second key factor in determining the statistical technique is the purpose of doing the work. Every statistical method has some specific objective. We address five such objectives in this book.

Problem Objectives

1. **Describe a population.** Our objective here is to describe some property of a population of interest. The decision about which property to describe is generally dictated by the type of data. For example, suppose the population of interest consists of all purchasers of home computers. If we are interested in the purchasers' incomes (for which the data are interval), we may calculate the mean or the variance to describe that aspect of the population. But if we are interested in the brand of computer that has been bought (for which the data are nominal), all we can do is compute the proportion of the population that purchases each brand.

2. **Compare two populations.** In this case, our goal is to compare a property of one population with a corresponding property of a second population. For example, suppose the populations of interest are male and female purchasers of computers. We could compare the means of their incomes, or we could compare the proportion of each population that purchases a certain brand. Once again, the data type generally determines what kinds of properties we compare.

3. **Compare two or more populations.** We might want to compare the average income in each of several locations in order (for example) to decide where to build a new shopping center. Or we might want to compare the proportions of defective items in a number of production lines in order to determine which line is the best. In each case, the problem objective involves comparing two or more populations.

4. **Analyze the relationship between two variables.** There are numerous situations in which we want to know how one variable is related to another. Governments need to know what effect rising interest rates have on the unemployment rate. Companies want to investigate how the sizes of their advertising budgets influence sales volume. In most of the problems in this introductory text, the two variables to be analyzed will be of the same type; we will not attempt to cover the fairly large body of statistical techniques that has been developed to deal with two variables of different types.

5. **Analyze the relationship among two or more variables.** Our objective here is usually to forecast one variable (called the dependent variable) on the basis of several other variables (called independent variables). We will deal with this problem only in situations in which all variables are interval.

Table 11.3 lists the types of data and the five problem objectives. For each combination, the table specifies the chapter and/or section where the appropriate statistical technique is presented. For your convenience, a more detailed version of this table is reproduced inside the front cover of this book.

TABLE **11.3** **Guide to Statistical Inference Showing Where Each Technique Is Introduced**

	Data Type		
Problem Objective	**Nominal**	**Ordinal**	**Interval**
Describe a population	Sections 12.3, 15.1	Not covered	Sections 12.1, 12.2
Compare two populations	Sections 13.5, 15.2	Sections 19.1, 19.2	Sections 13.1, 13.3, 13.4, 19.1, 19.2
Compare two or more populations	Section 15.2	Sections 19.3, 19.4	Chapter 14 Sections 19.3, 19.4
Analyze the relationship between two variables	Section 15.2	Section 19.5	Chapter 16
Analyze the relationship among two or more variables	Not covered	Not covered	Chapters 17, 18

Derivations

Because this book is about statistical applications, we assume that our readers have little interest in the mathematical derivations of the techniques described. However, it might be helpful for you to have some understanding about the process that produces the formulas.

As described previously, factors such as the problem objective and the type of data determine the parameter to be estimated and tested. For each parameter, statisticians have determined which statistic to use. That statistic has a sampling distribution that can usually be expressed as a formula. For example, in this chapter, the parameter of interest was the population mean μ, whose best estimator is the sample mean $\bar{x}$. Assuming that the population standard deviation σ is known, the sampling distribution of $\overline{X}$ is normal (or approximately so) with mean μ and standard deviation $\sigma/\sqrt{n}$. The sampling distribution can be described by the formula

$$z = \frac{\overline{X} - \mu}{\sigma/\sqrt{n}}$$

This formula also describes the test statistic for μ with σ known. With a little algebra, we were able to derive (in Section 10.2) the confidence interval estimator of μ.

In future chapters, we will repeat this process, which in several cases involves the introduction of a new sampling distribution. Although its shape and formula will differ from the sampling distribution used in this chapter, the pattern will be the same. In general, the formula that expresses the sampling distribution will describe the test statistic. Then some algebraic manipulation (which we will not show) produces the interval estimator. Consequently, we will reverse the order of presentation of the two techniques. That is, we will present the test of hypothesis first, followed by the confidence interval estimator.

CHAPTER SUMMARY

In this chapter, we introduced the concepts of hypothesis testing and applied them to testing hypotheses about a population mean. We showed how to specify the null and alternative hypotheses, set up the rejection region, compute the value of the test statistic, and finally, to make a decision. Equally as important, we discussed how to interpret the test results. This chapter also demonstrated another way to make decisions; by calculating and using the *p*-value of the test. To help interpret test results, we showed how to calculate the probability of a Type II error. Finally, we provided a road map of how we plan to present statistical techniques.

IMPORTANT TERMS

Hypothesis testing 345
Null hypothesis 345
Alternative or research hypothesis 345
Type I error 345
Type II error 345
Significance level 345
Test statistic 348
Rejection region 350
Standardized test statistic 351
Statistically significant 352
p-value of a test 353
Highly significant 355
Significant 355
Not significant 355
One-tail test 360
Two-tail test 360
One-sided confidence interval estimator 364
Operating characteristic curve 373

SYMBOLS

Symbol	Pronounced	Represents
H_0	*H-nought*	Null hypothesis
H_1	*H-one*	Alternative (research) hypothesis
α	*alpha*	Probability of a Type I error
β	*beta*	Probability of a Type II error
$\bar{x}_L$	*x-bar-sub L or x-bar-L*	Value of $\bar{x}$ large enough to reject H_0
$\lvert z \rvert$	*Absolute z*	Absolute value of z

FORMULA

Test statistic for μ

$$z = \frac{\bar{x} - \mu}{\sigma/\sqrt{n}}$$

COMPUTER OUTPUT AND INSTRUCTIONS

Technique	Excel	Minitab
Test of μ	356	357
Probability of a Type II error (and power)	372	373

© Chris Ryan/OJO Images/Getty Images

12 INFERENCE ABOUT A POPULATION

Nielsen Ratings

DATA Xm12-00*

Statistical techniques play a vital role in helping advertisers determine how many viewers watch the shows that they sponsor. Although several companies sample television viewers to determine what shows they watch, the best known is the A. C. Nielsen firm. The Nielsen ratings are based on a random sample of approximately 5,000 of the 110 million households in the United States with at least one television (in 2007). A meter attached to the televisions in the selected households keeps track of when the televisions are turned on and what channels they are tuned to. The data are sent to the Nielsen's computer every night from which Nielsen computes the rating and sponsors can determine the number of viewers and the potential value of any commercials.

© Brand X Pictures/Jupiterimages

On page 410 we provide a solution to this problem.

The results from Sunday, April 1, 2007, for the time slot 9:00 to 9:30 P.M. have been recorded using the following codes:

Network	Show	Code
ABC	*Desperate Housewives*	1
CBS	*The Amazing Race 11*	2
NBC	*Deal or No Deal*	3
Fox	*Family Guy*	4
Television turned off or watched some other channel		5

Source: Televisionweek, April 7, 2007.

NBC would like to use the data to estimate how many of the households were tuned to its program *Deal or No Deal.*

INTRODUCTION

In the previous two chapters, we introduced the concepts of statistical inference and showed how to estimate and test a population mean. However, the illustration we chose is unrealistic because the techniques require us to use the population standard deviation σ, which, in general, is unknown. The purpose, then, of Chapters 10 and 11 was to set the pattern for the way in which we plan to present other statistical techniques. That is, we will begin by identifying the parameter to be estimated or tested. We will then specify the parameter's estimator (each parameter has an estimator chosen because of the characteristics we discussed at the beginning of Chapter 10) and its sampling distribution. Using simple mathematics, statisticians have derived the interval estimator and the test statistic. This pattern will be used repeatedly as we introduce new techniques.

In Section 11.4, we described the five problem objectives addressed in this book, and we laid out the order of presentation of the statistical methods. In this chapter, we will present techniques employed when the problem objective is to describe a population. When the data are interval, the parameters of interest are the population mean μ and the population variance σ^2. In Section 12.1, we describe how to make inferences about the population mean under the more realistic assumption that the population standard deviation is unknown. In Section 12.2, we continue to deal with interval data, but our parameter of interest becomes the population variance.

In Chapter 2 and in Section 11.4, we pointed out that when the data are nominal, the only computation that makes sense is determining the proportion of times each value occurs. Section 12.3 discusses inference about the proportion p. In Section 12.4 we present an important application in marketing, market segmentation, and in Section 12.5 we discuss how the statistical techniques introduced in this chapter are used in auditing.

12.1 INFERENCE ABOUT A POPULATION MEAN WHEN THE STANDARD DEVIATION IS UNKNOWN

In Sections 10.2 and 11.2, we demonstrated how to estimate and test the population mean when the population standard deviation is known. The confidence interval estimator and the test statistic were derived from the sampling distribution of the sample mean with σ known, expressed as

$$z = \frac{\bar{x} - \mu}{\sigma/\sqrt{n}}$$

In this section, we take a more realistic approach by acknowledging that if the population mean is unknown, so is the population standard deviation. Consequently, the previous sampling distribution cannot be used. Instead, we substitute the sample standard deviation s in place of the unknown population standard deviation σ. The result is called a ***t*-statistic** because that is what mathematician William S. Gosset called it. In 1908, Gosset showed that the t-statistic defined as

$$t = \frac{\bar{x} - \mu}{s/\sqrt{n}}$$

is Student t distributed when the sampled population is normal. (Gosset published his findings under the pseudonym "Student," hence the **Student *t* distribution.**) Recall that we introduced the Student t distribution in Section 8.4.

With exactly the same logic used to develop the test statistic in Section 11.2 and the confidence interval estimator in Section 10.2, we derive the following inferential methods.

Test Statistic for μ When σ Is Unknown

When the population standard deviation is unknown and the population is normal, the test statistic for testing hypotheses about μ is

$$t = \frac{\bar{x} - \mu}{s/\sqrt{n}}$$

which is Student t distributed with $\nu = n - 1$ degrees of freedom.

Confidence Interval Estimator of μ When σ Is Unknown

$$\bar{x} \pm t_{\alpha/2}\frac{s}{\sqrt{n}} \qquad \nu = n - 1$$

These formulas now make obsolete the test statistic and interval estimator employed in Chapters 10 and 11 to estimate and test a population mean. Although we continue to use the concepts developed in Chapters 10 and 11 (as well as all the other chapters), we will no longer use the z-statistic and the z-estimator of μ. All future inferential problems involving a population mean will be solved using the t-statistic and t-estimator of μ shown in the preceding boxes.

EXAMPLE 12.1

DATA Xm12-01*

Newspaper Recycling Plant

It is likely that in the near future nations will have to do more to save the environment. Possible actions include reducing energy use and recycling. Currently (2007), most products manufactured from recycled material are considerably more expensive than those manufactured from material found in the earth. For example, it is approximately three times as expensive to produce glass bottles from recycled glass as from silica sand, soda ash, and limestone, all plentiful materials mined in numerous countries. It is more expensive to manufacture aluminum cans from recycled cans than from bauxite. Newspapers are an exception. It can be profitable to recycle newspaper. A major expense is the collection from homes. In recent years a number of companies have gone into the business of collecting used newspapers from households and recycling them. A financial analyst for one such company has recently computed that the firm would make a profit if the mean weekly newspaper collection from each household exceeded 2.0 pounds. In a study to determine the feasibility of a recycling plant, a random sample of 148 households was drawn from a large community, and the weekly weight of newspapers discarded for recycling for each household was recorded and is listed next. Do these data provide sufficient evidence to allow the analyst to conclude that a recycling plant would be profitable?

Weights of Discarded Newspapers

2.5	0.7	3.4	1.8	1.9	2.0	1.3	1.2	2.2	0.9	2.7	2.9	1.5	1.5	2.2
3.2	0.7	2.3	3.1	1.3	4.2	3.4	1.5	2.1	1.0	2.4	1.8	0.9	1.3	2.6
3.6	0.8	3.0	2.8	3.6	3.1	2.4	3.2	4.4	4.1	1.5	1.9	3.2	1.9	1.6
3.0	3.7	1.7	3.1	2.4	3.0	1.5	3.1	2.4	2.1	2.1	2.3	0.7	0.9	2.7
1.2	2.2	1.3	3.0	3.0	2.2	1.5	2.7	0.9	2.5	3.2	3.7	1.9	2.0	3.7
2.3	0.6	0.0	1.0	1.4	0.9	2.6	2.1	3.4	0.5	4.1	2.2	3.4	3.3	0.0
2.2	4.2	1.1	2.3	3.1	1.7	2.8	2.5	1.8	1.7	0.6	3.6	1.4	2.2	2.2
1.3	1.7	3.0	0.8	1.6	1.8	1.4	3.0	1.9	2.7	0.8	3.3	2.5	1.5	2.2
2.6	3.2	1.0	3.2	1.6	3.4	1.7	2.3	2.6	1.4	3.3	1.3	2.4	2.0	
1.3	1.8	3.3	2.2	1.4	3.2	4.3	0.0	2.0	1.8	0.0	1.7	2.6	3.1	

SOLUTION

IDENTIFY

The problem objective is to describe the population of the amounts of newspaper discarded by each household in the population. The data are interval, indicating that the parameter to be tested is the population mean. Because the financial analyst needs to determine whether the mean is greater than 2.0 pounds, the alternative hypothesis is

$$H_1: \ \mu > 2.0$$

As usual, the null hypothesis states that the mean is equal to the value listed in the alternative hypothesis:

$$H_0: \ \mu = 2.0$$

The test statistic is

$$t = \frac{\bar{x} - \mu}{s/\sqrt{n}} \qquad \nu = n - 1$$

COMPUTE

MANUALLY

The manager believes that the cost of a Type I error (concluding that the mean is greater than 2 when it isn't) is quite high. Consequently, he sets the significance level at 1%. The rejection region is

$$t > t_{\alpha,\nu} = t_{.01,148} \approx t_{.01,150} = 2.351$$

To calculate the value of the test statistic, we need to calculate the sample mean $\bar{x}$ and the sample standard deviation s. From the data we determine

$$\sum x_i = 322.7 \qquad \text{and} \qquad \sum x_i^2 = 845.1$$

Thus,

$$\bar{x} = \frac{\sum x_i}{n} = \frac{322.7}{148} = 2.18$$

$$s^2 = \frac{\sum x_i^2 - \frac{\left(\sum x_i\right)^2}{n}}{n-1} = \frac{845.1 - \frac{(322.7)^2}{148}}{148 - 1} = .962$$

and

$$s = \sqrt{s^2} = \sqrt{.962} = .981$$

The value of μ is to be found in the null hypothesis. It is 2.0. The value of the test statistic is

$$t = \frac{\bar{x} - \mu}{s/\sqrt{n}} = \frac{2.18 - 2.0}{.981/\sqrt{148}} = 2.23$$

Because 2.23 is not greater than 2.351, we cannot reject the null hypothesis in favor of the alternative. (Students performing the calculations manually can approximate the p-value. CD Appendix L describes how.)

EXCEL

	A	B	C	D
1	**t-Test: Mean**			
2				
3				Newspaper
4	Mean			2.18
5	Standard Deviation			0.98
6	Hypothesized Mean			2
7	df			147
8	t Stat			2.24
9	P(T<=t) one-tail			0.0134
10	t Critical one-tail			2.3520
11	P(T<=t) two-tail			0.0268
12	t Critical two-tail			2.6097

INSTRUCTIONS

1. Type or import the data into one column. (Open Xm12-01.)
2. Click **Add-Ins, Data Analysis Plus**, and **t-Test: Mean.**
3. Specify the **Input Range** (A1:A149), the **Hypothesized Mean** (2), and α (.01).

To conduct this test from statistics or to perform a what-if analysis, open the **Test Statistics** workbook and find the **t-Test_Mean** worksheet. Substitute values for the sample mean, sample standard deviation, sample size, and hypothesized value of the population mean. (See CD Appendix J for the spreadsheet for this example.)

MINITAB

One-Sample T: Newspaper

Test of mu = 2 vs > 2

Variable	N	Mean	StDev	SE Mean	95% Lower Bound	T	P
Newspaper	148	2.1804	0.9812	0.0807	2.0469	2.24	0.013

INSTRUCTIONS

1. Type or import the data into one column. (Open Xm12-01.)
2. Click **Stat, Basic Statistics,** and **1-Sample t . . . .**
3. Type or use the **Select** button to specify the name of the variable or the column in the **Samples in columns** box (Newspaper), choose **Perform hypothesis test** and type the value of μ in the **Hypothesized mean** box (2), and click **Options . . . .**
4. Select one of **less than, not equal,** or **greater than** in the **Alternative** box (greater than).

INTERPRET

The value of the test statistic is $t = 2.24$ and its p-value is .0134. There is not enough evidence to infer that the mean weight of discarded newspapers is greater than 2.0. Note that there is some evidence; the p-value is .0134. However, because we wanted the probability of a Type I error to be small, we insisted on a 1% significance level. Thus, we cannot conclude that the recycling plant would be profitable.

Figure 12.1 exhibits the sampling distribution for this example.

FIGURE **12.1** Sampling Distribution for Example 12.1

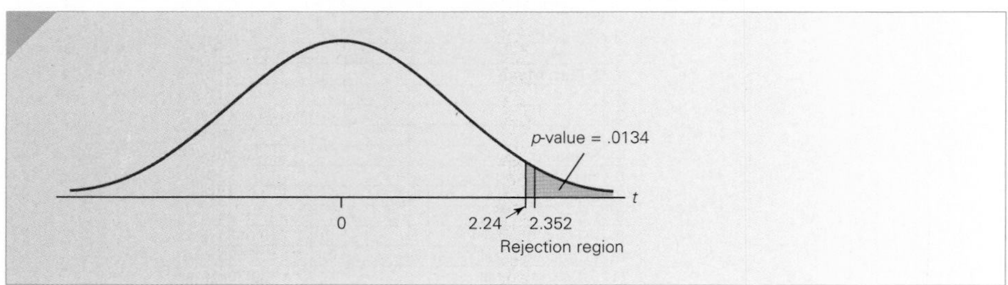

EXAMPLE 12.2

DATA
Xm12-02

Tax Collected from Audited Returns

In 2004 (the latest year reported), 130,134,000 tax returns were filed in the United States. The Internal Revenue Service (IRS) examined 0.77% or 1,008,000 of them to determine if they were correctly done. To determine how well the auditors are performing, a random sample of these returns was drawn and the additional tax was reported, which is listed next. Estimate with 95% confidence the mean additional income tax collected from the 1,008,000 files audited.

*Sources: *Statistical Abstract of the United States*, 2006, Table 471 and U.S. Internal Revenue Service, *IRS Data Book*, annual, Publication 55B.

Additional Income Tax

6,039	4,119	5,637	5,616	3,718	6,294	356	4,166	3,765	4,877	7,726	13,226
5,147	4,268	7,462	5,759	5,259	4,332	7,865	5,931	3,924	5,022	1,753	16,916
4,384	8,155	6,715	2,167	515	8,415	2,962	6,622	7,121	6,244	5,477	10,513
3,790	4,737	4,280	6,147	5,490	6,139	5,093	0	4,894	5,694	6,509	10,249
5,713	4,082	6,069	5,233	7,855	0	4,214	1,353	9,533	6,171	5,502	11,993
4,818	4,367	6,856	5,528	6,233	6,234	6,755	6,824	7,697	2,679	10,322	11,195
7,798	4,134	5,567	0	7,273	7,087	8,659	5,126	8,206	2,292	11,657	10,783
6,687	3,896	4,582	5,539	4,101	5,577	3,284	7,118	3,651	4,749	7,942	10,591
6,511	1,477	2,357	3,991	8,363	5,507	5,862	3,527	3,412	6,778	11,373	8,293
1,600	1,507	5,241	6,991	5,711	6,333	6,971	4,324	5,589	7,204	6,076	8,610
6,766	5,209	5,200	6,559	5,650	7,671	6,766	4,912	6,687	3,135	9,974	15,226
4,817	4,770	3,342	5,824	5,984	5,396	9,090	3,568	6,003	2,006	11,427	
3,189	5,469	3,269	5,682	4,024	6,165	6,429	8,169	7,856	6,363	14,545	
6,637	4,558	4,115	8,592	7,208	6,665	3,635	2,966	7,554	6,750	10,626	
5,888	2,276	9,112	5,104	5,622	6,897	2,862	1,156	6,081	5,572	13,251	
4,034	7,404	5,476	1,983	4,934	5,531	252	8,457	3,315	7,252	17,994	
9,023	5,074	7,013	6,173	7,598	4,886	4,710	4,637	7,932	7,274	10,275	
7,792	4,457	7,794	5,023	9,384	4,352	5,241	0	5,531	5,437	9,111	

SOLUTION

IDENTIFY

The problem objective is to describe the population of additional income tax. The data are interval and hence, the parameter is the population mean μ. The question asks us to estimate this parameter. The confidence interval estimator is

$$\bar{x} \pm t_{\alpha/2} \frac{s}{\sqrt{n}}$$

COMPUTE

MANUALLY

From the data we determine

$$\sum x_i = 1{,}254{,}240 \qquad \text{and} \qquad \sum x_i^2 = 9{,}232{,}718{,}166$$

Thus,

$$\bar{x} = \frac{\sum x_i}{n} = \frac{1{,}254{,}240}{209} = 6{,}001$$

and

$$s^2 = \frac{\sum x_i^2 - \frac{\left(\sum x_i\right)^2}{n}}{n - 1} = \frac{9{,}232{,}718{,}166 - \frac{(1{,}254{,}240)^2}{209}}{209 - 1} = 8{,}201{,}144$$

Thus,

$$s = \sqrt{s^2} = \sqrt{8{,}201{,}144} = 2{,}864$$

Because we want a 95% confidence interval estimate, $1 - \alpha = .95, \alpha = .05, \alpha/2 = .025$, and $t_{\alpha/2,\nu} = t_{.025,208} \approx t_{.025,200} = 1.972$. Thus, the 95% confidence interval estimate of μ is

$$\bar{x} \pm t_{\alpha/2}\frac{s}{\sqrt{n}} = 6{,}001 \pm 1.972\frac{2{,}864}{\sqrt{209}} = 6{,}001 \pm 391$$

or

$$\text{LCL} = \$5{,}610 \qquad \text{UCL} = \$6{,}392$$

EXCEL

	A	B	C	D
1	t-Estimate: Mean			
2				
3				Taxes
4	Mean			6001
5	Standard Deviation			2864
6	LCL			5611
7	UCL			6392

INSTRUCTIONS

1. Type or import the data into one column. (Open Xm12-02.)
2. Click **Add-Ins, Data Analysis Plus,** and **t-Estimate: Mean.**
3. Specify the **Input Range** (A1:A210) and α (.05).

If you know the sample mean, sample standard deviation, and sample size, you can use the **t-Estimate_Mean** worksheet in the **Estimators** workbook, which can also be employed for what-if analyses.

MINITAB

One-Sample T: Taxes

Variable	N	Mean	StDev	SE Mean	95% CI
Taxes	209	6001	2864	198	(5611, 6392)

INSTRUCTIONS

1. Type or import the data into one column. (Open Xm12-02.)
2. Click **Stat, Basic Statistics,** and **1-Sample t . . . .**
3. Select or type the variable name in the **Samples in columns** box (Taxes) and click **Options . . . .**
4. Specify the **Confidence level** (.95) and **not equal** for the **Alternative.**

INTERPRET

We estimate that the mean additional tax collected lies between $5,611 and $6,392 (Excel's figures). We can use this estimate to help decide whether the IRS is auditing the individuals who should be audited.

Checking the Required Conditions

When we introduced the Student t distribution, we pointed out that the t-statistic is Student t distributed if the population from which we've sampled is normal. However, statisticians have shown that the mathematical process that derived the Student t distribution is **robust,** which means that if the population is nonnormal, the results of the t-test and confidence interval estimate are still valid provided that the population is not *extremely* nonnormal.* To check this requirement, we draw the histogram and determine whether it is far from bell shaped. Figures 12.2 and 12.3 depict the Excel histograms for Examples 12.1 and 12.2, respectively. (The Minitab histograms are similar.) Both histograms suggest that the variables are not extremely nonnormal.

FIGURE **12.2** Histogram for Example 12.1

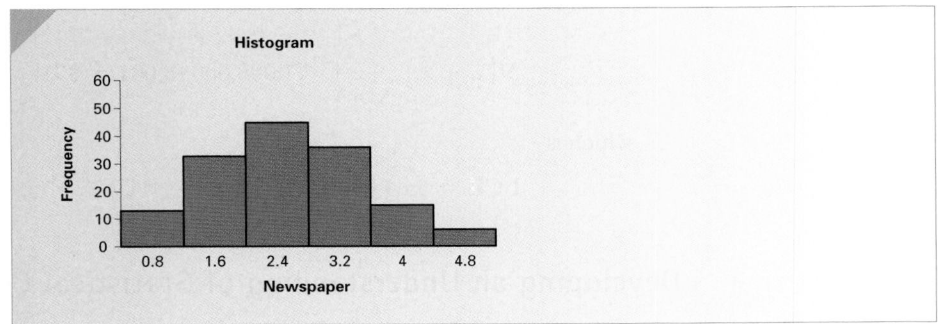

FIGURE **12.3** Histogram for Example 12.2

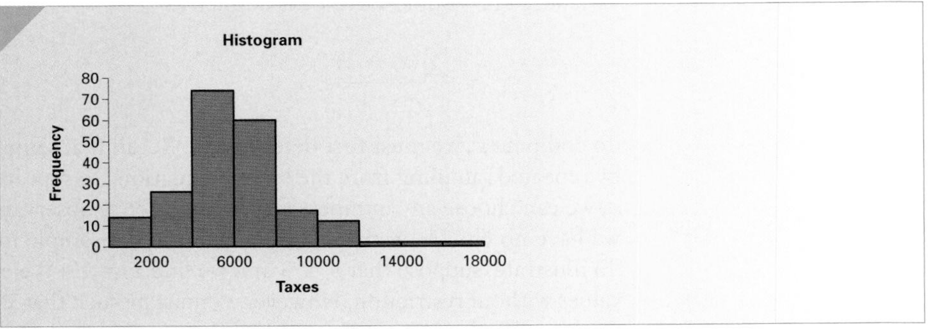

Estimating the Totals of Finite Populations

The inferential techniques introduced thus far were derived by assuming infinitely large populations. In practice, however, most populations are finite. (Infinite populations are usually the result of some endlessly repeatable process, such as flipping a coin or selecting items with replacement.) When the population is small, we must adjust the test statistic and interval estimator using the finite population correction factor introduced in

*Statisticians have shown that when the sample size is large, the results of a t-test and estimator of a mean are valid even when the population is extermely nonnormal. The sample size required depends on the extent of nonnormality.

Chapter 9 (page 301). (In Section 12.5 we feature an application in accounting that requires the use of the correction factor.) However, in populations that are large relative to the sample size, we can ignore the correction factor. Large populations are defined as populations that are at least 20 times the sample size.

Finite populations allow us to use the confidence interval estimator of a mean to produce a confidence interval estimator of the population total. To estimate the total, we multiply the lower and upper confidence limits of the estimate of the mean by the population size. Thus, the confidence interval estimator of the total is

$$N\left(\bar{x} \pm t_{\alpha/2}\frac{s}{\sqrt{n}}\right)$$

For example, suppose that we wish to estimate the total amount of additional income tax collected from the 1,008,000 returns that were examined. The 95% confidence interval estimate of the total is

$$N\left(\bar{x} \pm t_{\alpha/2}\frac{s}{\sqrt{n}}\right) = 1{,}008{,}000\,(6{,}001 \pm 391)$$

which is

$$\text{LCL} = 5{,}654{,}880{,}000 \qquad \text{and} \qquad \text{UCL} = 6{,}443{,}136{,}000$$

Developing an Understanding of Statistical Concepts 1

This section introduced the term *degrees of freedom*. We will encounter this term many times in this book, so a brief discussion of its meaning is warranted. The Student t distribution is based on using the sample variance to estimate the unknown population variance. The sample variance is defined as

$$s^2 = \frac{\sum(x_i - \bar{x})^2}{n - 1}$$

To compute s^2, we must first determine $\bar{x}$. Recall that sampling distributions are derived by repeated sampling from the same population. To repeatedly take samples to compute s^2, we can choose any numbers for the first $n - 1$ observations in the sample. However, we have no choice on the nth value because the sample mean must be calculated first. To illustrate, suppose that $n = 3$ and we find $\bar{x} = 10$. We can have x_1 and x_2 assume any values without restriction. However, x_3 must be such that $\bar{x} = 10$. For example, if $x_1 = 6$ and $x_2 = 8$, then x_3 must equal 16. Therefore, there are only 2 degrees of freedom in our selection of the sample. We say that we lose 1 degree of freedom because we had to calculate $\bar{x}$.

Notice that the denominator in the calculation of s^2 is equal to the number of degrees of freedom. This is not a coincidence and will be repeated throughout this book.

Developing an Understanding of Statistical Concepts 2

The t-statistic like the z-statistic measures the difference between the sample mean $\bar{x}$ and the hypothesized value of μ in terms of the number of standard errors. However, when the population standard deviation σ is unknown, we estimate the standard error by $s/\sqrt{n}$.

Developing an Understanding of Statistical Concepts 3

When we introduced the Student t distribution in Section 8.4, we pointed out that it is more widely spread out than the standard normal. This circumstance is logical. The only variable in the z-statistic is the sample mean $\bar{x}$, which will vary from sample to sample. The t-statistic has two variables: the sample mean $\bar{x}$ and the sample standard deviation s, both of which will vary from sample to sample. Because of the greater uncertainty, the t-statistic will display greater variability. Exercises 12.13–12.20 address this concept.

We complete this section with a review of how we identify the techniques introduced in this section.

Factors That Identify the t-Test and Estimator of μ

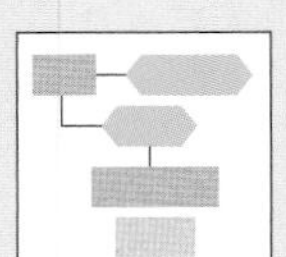

1. **Problem objective:** Describe a population
2. **Data type:** Interval
3. **Type of descriptive measurement:** Central location

EXERCISES

Developing an Understanding of Statistical Concepts

The following exercises are "what-if analyses" designed to determine what happens to the test statistics and interval estimates when elements of the statistical inference change. These problems can be solved manually or using Excel's ***Test Statistics*** *or* ***Estimators*** *workbooks.*

12.1 a. A random sample of 25 was drawn from a population. The sample mean and standard deviation are $\bar{x} = 510$ and $s = 125$. Estimate μ with 95% confidence.
b. Repeat Part a with $n = 50$.
c. Repeat Part a with $n = 100$.
d. Describe what happens to the confidence interval estimate when the sample size increases.

12.2 a. The mean and standard deviation of a sample of 100 is $\bar{x} = 1{,}500$ and $s = 300$. Estimate the population mean with 95% confidence.
b. Repeat Part a with $s = 200$.
c. Repeat Part a with $s = 100$.
d. Discuss the effect on the confidence interval estimate of decreasing the standard deviation s.

12.3 a. A statistics practitioner drew a random sample of 400 observations and found that $\bar{x} = 700$ and $s = 100$. Estimate the population mean with 90% confidence.
b. Repeat Part a with a 95% confidence level.
c. Repeat Part a with a 99% confidence level.
d. What is the effect on the confidence interval estimate of increasing the confidence level?

12.4 a. The mean and standard deviation of a sample of 100 are

$$\bar{x} = 10 \quad \text{and} \quad s = 1$$

Estimate the population mean with 95% confidence.
b. Repeat Part a with $s = 4$.
c. Repeat Part a with $s = 10$.
d. Discuss the effect on the confidence interval estimate of increasing the standard deviation s.

12.5 a. A statistics practitioner calculated the mean and standard deviation from a sample of 51. They are $\bar{x} = 120$ and $s = 15$. Estimate the population mean with 95% confidence.
b. Repeat Part a with a 90% confidence level.
c. Repeat Part a with an 80% confidence level.
d. What is the effect on the confidence interval estimate of decreasing the confidence level?

12.6 a. The sample mean and standard deviation from a sample of 81 observations are $\bar{x} = 63$ and $s = 8$. Estimate μ with 95% confidence.
b. Repeat Part a with $n = 64$.
c. Repeat Part a with $n = 36$.
d. Describe what happens to the confidence interval estimate when the sample size decreases.

12.7 a. The sample mean and standard deviation from a random sample of 10 observations from a normal population were computed as $\bar{x} = 23$ and $s = 9$. Calculate the value of the test statistic (and for Excel users, the p-value) of the test required to determine whether there is enough evidence to infer at the 5% significance level that the population mean is greater than 20.

b. Repeat Part a with $n = 30$.

c. Repeat Part a with $n = 50$.

d. Describe the effect on the t-statistic (and for Excel users, the p-value) of increasing the sample size.

12.8 a. A statistics practitioner is in the process of testing to determine whether there is enough evidence to infer that the population mean is different from 180. She calculated the mean and standard deviation of a sample of 200 observations as $\bar{x} = 175$ and $s = 22$. Calculate the value of the test statistic (and for Excel users, the p-value) of the test required to determine whether there is enough evidence at the 5% significance level.

b. Repeat Part a with $s = 45$.

c. Repeat Part a with $s = 60$.

d. Discuss what happens to the t-statistic (and for Excel users, the p-value) when the standard deviation increases.

12.9 a. Calculate the test statistic (and for Excel users, the p-value) when is $\bar{x} = 145$, $s = 50$, and $n = 100$. Use a 5% significance level.

$$H_0: \mu = 150$$
$$H_1: \mu < 150$$

b. Repeat Part a with $\bar{x} = 140$.

c. Repeat Part a with $\bar{x} = 135$.

d. What happens to the t-statistic (and for Excel users, the p-value) when the sample mean decreases?

12.10 a. A random sample of 25 observations was drawn from a normal population. The sample mean and sample standard deviation are $\bar{x} = 52$ and $s = 15$. Calculate the test statistic (and for Excel users, the p-value) of a test to determine if there is enough evidence at the 10% significance level to infer that the population mean is not equal to 50.

b. Repeat Part a with $n = 15$.

c. Repeat Part a with $n = 5$.

d. Discuss what happens to the t-statistic (and for Excel users, the p-value) when the sample size decreases.

12.11 a. A statistics practitioner wishes to test the following hypotheses:

$$H_0: \mu = 600$$
$$H_1: \mu < 600$$

A sample of 50 observations yielded the statistics $\bar{x} = 585$ and $s = 45$. Calculate the test statistic (and for Excel users, the p-value) of a test to determine whether there is enough evidence at the 10% significance level to infer that the alternative hypothesis is true.

b. Repeat Part a with $\bar{x} = 590$.

c. Repeat Part a with $\bar{x} = 595$.

d. Describe the effect of decreasing the sample mean.

12.12 a. To test the following hypotheses, a statistics practitioner randomly sampled 100 observations and found $\bar{x} = 106$ and $s = 35$. Calculate the test statistic (and for Excel users, the p-value) of a test to determine whether there is enough evidence at the 1% significance level to infer that the alternative hypothesis is true.

$$H_0: \mu = 100$$
$$H_1: \mu > 100$$

b. Repeat Part a with $s = 25$.

c. Repeat Part a with $s = 15$.

d. Discuss what happens to the t-statistic (and for Excel users, the p-value) when the standard deviation decreases.

12.13 A random sample of 8 observations was drawn from a normal population. The sample mean and sample standard deviation are $\bar{x} = 40$ and $s = 10$.

a. Estimate the population mean with 95% confidence.

b. Repeat Part a assuming that you know that the population standard deviation is $\sigma = 10$.

c. Explain why the interval estimate produced in Part b is narrower than that in Part a.

12.14 a. Estimate the population mean with 90% confidence given the following:

$$\bar{x} = 175, \quad s = 30, \quad \text{and} \quad n = 5$$

b. Repeat Part a assuming that you know that the population standard deviation is $\sigma = 30$.

c. Explain why the interval estimate produced in Part b is narrower than that in Part a.

12.15 a. After sampling 1,000 members of a normal population, you find $\bar{x} = 15{,}500$ and $s = 9{,}950$. Estimate the population mean with 90% confidence.

b. Repeat Part a assuming that you know that the population standard deviation is $\sigma = 9{,}950$.

c. Explain why the interval estimates were virtually identical.

12.16 a. In a random sample of 500 observations drawn from a normal population, the sample mean and sample standard deviation were calculated as $\bar{x} = 350$ and $s = 100$. Estimate the population mean with 99% confidence.

b. Repeat Part a assuming that you know that the population standard deviation is $\sigma = 100$.

c. Explain why the interval estimates were virtually identical.

12.17 a. A random sample of 11 observations was taken from a normal population. The sample mean and standard deviation are $\bar{x} = 74.5$ and $s = 9$. Can we infer at the 5% significance level that the population mean is greater than 70?

b. Repeat Part a assuming that you know that the population standard deviation is $\sigma = 90$.

c. Explain why the conclusions produced in Parts a and b differ.

12.18 a. A statistics practitioner randomly sampled 10 observations and found $\bar{x} = 103$ and $s = 17$. Is there sufficient evidence at the 10% significance level to conclude that the population mean is less than 110?

b. Repeat Part a assuming that you know that the population standard deviation is $\sigma = 17$.

c. Explain why the conclusions produced in Parts a and b differ.

12.19 a. A statistics practitioner randomly sampled 1,500 observations and found $\bar{x} = 14$ and $s = 25$. Test to determine whether there is enough evidence at the 5% significance level to infer that the population mean is less than 15.

b. Repeat Part a assuming that you know that the population standard deviation is $\sigma = 25$.

c. Explain why the conclusions produced in Parts a and b are virtually identical.

12.20 a. Test the following hypotheses with $\alpha = .05$ given that $\bar{x} = 405$, $s = 100$, and $n = 1{,}000$:

$$H_0: \quad \mu = 400$$
$$H_1: \quad \mu > 400$$

b. Repeat Part a assuming that you know that the population standard deviation is $\sigma = 100$.

c. Explain why the conclusions produced in Parts a and b are virtually identical.

Applications

The following exercises may be answered manually or with the assistance of a computer. The data are stored in files. ***Assume that the random variable is normally distributed.***

12.21 Xr12-21 A courier service advertises that its average delivery time is less than 6 hours for local deliveries. A random sample of times for 12 deliveries to an address across town was recorded. These data are shown here. Is this sufficient evidence to support the courier's advertisement, at the 5% level of significance?

3.03	6.33	6.50	5.22	3.56	6.76
7.98	4.82	7.96	4.54	5.09	6.46

12.22 Xr12-22 How much money do winners go home with from the television quiz show *Jeopardy*? To determine an answer, a random sample of winners was drawn and the amount of money each won was recorded and is listed here. Estimate with 95% confidence the mean winnings for all the show's players.

26,650	6,060	52,820	8,490	13,660
25,840	49,840	23,790	51,480	18,960
990	11,450	41,810	21,060	7,860

12.23 Xr12-23 A diet doctor claims that the average North American is more than 20 pounds overweight. To test his claim, a random sample of 20 North Americans was weighed, and the difference between their actual weight and their ideal weight was calculated. The data are listed here. Do these data allow us to infer at the 5% significance level that the doctor's claim is true?

16	23	18	41	22	18	23	19	22	15
18	35	16	15	17	19	23	15	16	26

12.24 Xr12-24 A federal agency responsible for enforcing laws governing weights and measures routinely inspects packages to determine whether the weight of the contents is at least as great as that advertised on the package. A random sample of 18 containers whose packaging states that the contents weigh 8 ounces was drawn. The contents were weighed and the results follow. Can we conclude at the 1% significance level that on average the containers are mislabeled?

7.80	7.91	7.93	7.99	7.94	7.75
7.97	7.95	7.79	8.06	7.82	7.89
7.92	7.87	7.92	7.98	8.05	7.91

12.25 Xr12-25 A parking control officer is conducting an analysis of the amount of time left on parking meters. A quick survey of 15 cars that have just left their metered parking spaces produced the following times (in minutes). Estimate with 95% confidence the mean amount of time left for all the city's meters.

22	15	1	14	0	9	17	31
18	26	23	15	33	28	20	

12.26 Xr12-26 Part of a university professors' job is to publish his or her research. This task often entails reading a variety of journal articles to keep up to date. To help determine faculty standards, a dean of a business school surveyed a random sample of

12 professors across the country and asked them to count the number of journal articles they read in a typical month. These data are listed here. Estimate with 90% confidence the mean number of journal articles read monthly by professors.

9 17 4 23 56 30 41 45 21 10 44 20

12.27 Xr12-27 Most owners of digital cameras store their pictures on the camera. Some will eventually download these to a computer or print them using their own printers or a commercial printer. A film-processing company wanted to know how many pictures were stored on computers. A random sample of 10 digital camera owners produced the data given here. Estimate with 95% confidence the mean number of pictures stored on digital cameras.

25 6 22 26 31 18 13 20 14 2

12.28 Xr12-28 University bookstores order books that instructors adopt for their courses. The number of copies ordered matches the projected demand. However, at the end of the semester the bookstore has too many copies on hand and must return them to the publisher. A bookstore has a policy that the proportion of books returned should be kept as small as possible. The average is supposed to be less than 10%. To see whether the policy is working, a random sample of book titles was drawn and the fraction of the total originally ordered that are returned is recorded and listed here. Can we infer at the 10% significance level that the mean proportion of returns is less than 10%?

4 15 11 7 5 9 4 3 5 8

The following exercises require the use of a computer and software. The answers may be calculated manually. See Appendix A for the sample statistics. ***Use a 5% significance level unless specified otherwise.***

12.29 Xr12-29* A growing concern for educators in the United States is the number of teenagers who have part-time jobs while they attend high school. It is generally believed that the amount of time teenagers spend working is deducted from the amount of time devoted to schoolwork. To investigate this problem, a school guidance counselor took a random sample of 200 15-year-old high school students and asked how many hours per week each worked at a part-time job. Estimate with 95% confidence the mean amount of time all 15-year-old high school students devote per week to part-time jobs.

12.30 Xr12-30 A company that produces universal remote controls wanted to determine the number of remote control devices American homes contain. The company hired a statistician to survey 240 randomly selected homes and determine the number of remote controls. If there are 100 million households, estimate with 99% confidence the total number of remote controls in the United States.

12.31 Xr12-31 How much time do executives spend each day reading and sending e-mail. A survey was conducted by Accountemps (reported in *USA Today*, March 12, 2001). The responses (in minutes) were recorded. Can we infer from these data that the mean amount of time spent by all executives reading and sending e-mail daily differs from 60 minutes?

12.32 Xr12-32 Bankers and economists watch for signs that the economy is slowing. One statistic they monitor is consumer debt, particularly credit card debt. The Federal Reserve conducts surveys of consumer finances every 3 years. In the last survey, the survey determined that 23.8% of American households have no credit cards and another 31.2% of the households paid off their most recent credit card bills. The remainder, approximately 50 million households, did not pay their credit card bills in the previous month. A random sample of these households was drawn. Each household in the sample reported how much credit card debt they currently carry. The Federal Reserve would like an estimate (with 95% confidence) of the total credit card debt in the United States.

12.33 Xr12-33* OfficeMax, a chain that sells a wide variety of office equipment, often features sales of products whose prices are reduced because of rebates. Some rebates are so large that the effective price becomes $0. The goal is to lure customers into the store to buy other nonsale items. A secondary objective is to acquire addresses and telephone numbers to sell to telemarketers and other mass marketers. During one week in January, OfficeMax offered a 100-pack of CD-ROMs (regular price $29.99 minus $10 instant rebate, $12 manufacturer's rebate, and $8 OfficeMax mail-in rebate). The number of packages was limited and no rain checks were issued. In all the OfficeMax stores there were 2,800 packages in stock. All were sold. A random sample of 122 buyers was undertaken. Each was asked to report the total value of the other purchases made that day.

a. Estimate with 95% the total spent on other products purchased by those who bought the CD-ROMs.
b. Write a report to the president of OfficeMax describing your analysis. (See Section 3.3 for guidelines.)

12.34 Xr12-34 An increasing number of North Americans regularly take vitamin or herbal remedies daily. To gauge this phenomenon, a random sample of Americans was asked to report the number of vitamin and herbal supplements they take daily.

Estimate with 95% confidence the mean number of vitamin and herbal supplements Americans take daily.

12.35 Xr12-35 Generic drug sales make up about half of all prescriptions sold in the United States. The marketing manager for a pharmaceutical company wanted to acquire more information about the sales of generic prescription drugs. To do so, she randomly sampled 900 customers who recently filled prescriptions for generic drugs and recorded the cost of each prescription. Estimate with 95% confidence the mean cost of all generic prescription drugs. (Adapted from the *Statistical Abstract of the United States*, 2006, Table 126)

12.36 Xr12-36 Traffic congestion seems to worsen each year. This raises the question, How much does roadway congestion cost the United States annually? The Federal Highway Administration's Highway Performance Monitoring System conducts an analysis to produce an estimate of the total cost. Drivers in the 73 most congested areas in the United States were sampled and for each driver the congestion cost in time and gasoline was recorded. The total number of drivers in these 73 areas was 128,000,000. Estimate with 95% confidence the total cost of congestion in the 73 areas. (Adapted from the *Statistical Abstract of the United States*, 2006, Table 1082)

12.37 Xr12-37 To help estimate the size of the disposable razor market, a random sample of men was asked to count the number of shaves they used each razor for. Assume that each razor is used once per day. Estimate with 95% confidence the number of days a pack of 10 razors will last.

12.38 Xr12-38 Because of the enormity of the viewing audience, firms that advertise during the Super Bowl create special commercials that tend to be quite entertaining. Thirty-second commercials cost $2.3 million during the 2001 Super Bowl game. A random sample of people who watched the game was asked how many commercials they watched in their entirety. Do these data allow us to infer that the mean number of commercials watched is greater than 15?

12.39 Xr12-39 On a per capita basis, the United States spends far more on health than any other country. To help assess the costs, annual surveys are undertaken. One such survey asks a sample of Americans to report the number of times they visited a health care professional in the year. The data for 1998 were recorded. In 1998 the United States population was 270,509,000. Estimate with 95% confidence the total number of visits to a health care professional. (Adapted from U.S. National Center for Health Statistics, United States, 2000, and the *Statistical Abstract of the United States*, 2000, Table 189)

12.40 Xr12-40 Companies that sell groceries over the Internet are called e-grocers. Customers enter their orders, pay by credit card, and receive delivery by truck. A potential e-grocer analyzed the market and determined that to be profitable the average order would have to exceed $85. To determine whether an e-grocery would be profitable in one large city, she offered the service and recorded the size of the order for a random sample of customers.

a. Can we infer from these data that an e-grocery will be profitable in this city?
b. Prepare a presentation to the investors who wish to put money into this company. (See Section 3.3 for guidelines in making presentations.)

12.41 Xr12-41 During the last decade, a number of institutions dedicated to improving the quality of products and services in the United States have been formed. Many of these groups annually give awards to companies that produce high-quality goods and services. An investor believes that publicly traded companies that win awards are likely to outperform companies that do not win such awards. To help determine his return on investment in such companies, he took a random sample of 83 firms that won quality awards the previous year and computed the annual return he would have received had he invested. The investor would like an estimate of the returns he can expect. A 95% confidence level is deemed appropriate.

12.42 Xr12-42 Couriers such as UPS and FedEx compete on service and price. One way to reduce costs is to keep labor costs low by hiring and laying off workers to meet demand. This strategy requires managers to hire and train new workers. But newly hired and trained workers are not as productive as more experienced ones. Thus, determining the number of workers required and the work schedule is difficult. The current work schedule is based on the belief that trainees will achieve more than 90% of the level of experienced workers within 1 week of hiring. To determine the accuracy of this number, an operations manager conducted an experiment. Fifty trainees were observed for 1 hour and the numbers of packages processed and routed were recorded. It is known that experienced workers process an average of 500 packages per hour. The manager is concerned that if he concludes that the mean is greater than 450 when it isn't, the result will be some late deliveries. Can the manager conclude from the data that the belief is correct?

12.2 INFERENCE ABOUT A POPULATION VARIANCE

In Section 12.1, where we presented the inferential methods about a population mean, we were interested in acquiring information about the central location of the population. As a result, we tested and estimated the population mean. If we are interested instead in drawing inferences about a population's variability, the parameter we need to investigate is the population variance σ^2. Inference about the variance can be used to make decisions in a variety of problems. In an example illustrating the use of the normal distribution in Section 8.2, we showed why variance is a measure of risk. In Section 7.3, we described an important application in finance wherein stock diversification was shown to reduce the variance of a portfolio and in so doing, reduce the risk associated with that portfolio. In both sections we assumed that the population variances were known. In this section we take a more realistic approach and acknowledge that we need to use statistical techniques to draw inferences about a population variance.

Another application of the use of variance comes from operations management. Quality technicians attempt to ensure that their company's products consistently meet specifications. One way of judging the consistency of a production process is to compute the variance of the size, weight, or volume of the product; that is, if the variation in product size, weight, or volume is large, it is likely that an unsatisfactorily large number of products will lie outside the specifications for that product. We will return to this subject later in this book. In Section 14.6, we discuss how operations managers search for and reduce the variation in production processes.

The task of deriving the test statistic and the interval estimator provides us with another opportunity to show how statistical techniques in general are developed. We begin by identifying the best estimator. That estimator has a sampling distribution, from which we produce the test statistic and the interval estimator.

Statistic and Sampling Distribution

The estimator of σ^2 is the sample variance introduced in Section 4.2. The statistic s^2 has the desirable characteristics presented in Section 10.1; that is, s^2 is an unbiased, consistent estimator of σ^2.

Statisticians have shown that the sum of squared deviations from the mean $\sum(x_i - \bar{x})^2$ [which is equal to $(n - 1)s^2$] divided by the population variance is chi-squared distributed with $\nu = n - 1$ degrees of freedom provided that the sampled population is normal. The statistic

$$\chi^2 = \frac{(n-1)s^2}{\sigma^2}$$

is called the **chi-squared statistic** (χ^2-statistic). The chi-squared distribution was introduced in Section 8.4.

Testing and Estimating a Population Variance

As we discussed in Section 11.4, the formula that describes the sampling distribution is the formula of the test statistic.

Test Statistic for σ^2

The test statistic used to test hypotheses about σ^2 is

$$\chi^2 = \frac{(n-1)s^2}{\sigma^2}$$

which is chi-squared distributed with $\nu = n - 1$ degrees of freedom when the population random variable is normally distributed with variance equal to σ^2.

Using the notation introduced in Section 8.4, we can make the following probability statement:

$$P\left(\chi^2_{1-\alpha/2} < \chi^2 < \chi^2_{\alpha/2}\right) = 1 - \alpha$$

Substituting

$$\chi^2 = \frac{(n-1)s^2}{\sigma^2}$$

and with some algebraic manipulation, we derive the confidence interval estimator of a population variance.

Confidence Interval Estimator of σ^2

$$\text{Lower confidence limit (LCL)} = \frac{(n-1)s^2}{\chi^2_{\alpha/2}}$$

$$\text{Upper confidence limit (UCL)} = \frac{(n-1)s^2}{\chi^2_{1-\alpha/2}}$$

APPLICATIONS in OPERATIONS MANAGEMENT

Quality

A critical aspect of production is quality. The quality of a final product is a function of the quality of the product's components. If the components don't fit, the product will not function as planned and it will likely cease functioning before its customers expect it to. For example, if a car door is not made to its specifications, it will not fit. As a result, the door will leak both water and air.

Operations managers attempt to maintain and improve the quality of products by ensuring that all components are made so that there is as little variation as possible. As you have already seen, statisticians measure variation by computing the variance.

Incidentally, an entire chapter (Chapter 21) is devoted to the topic of quality.

EXAMPLE 12.3

Consistency of a Container-Filling Machine, Part 1

DATA
Xm12-03

Container-filling machines are used to package a variety of liquids, including milk, soft drinks, and paint. Ideally, the amount of liquid should vary only slightly, since large variations will cause some containers to be underfilled (cheating the customer) and some to be overfilled (resulting in costly waste). The president of a company that developed a new type of machine boasts that this machine can fill 1-liter (1,000 cubic centimeters) containers so consistently that the variance of the fills will be less than 1 cubic centimeter. To examine the veracity of the claim, a random sample of 25 l-liter fills was taken and the results recorded. These data are listed here. Do these data allow the president to make this claim at the 5% significance level?

Fills

999.6	1000.7	999.3	1000.1	999.5
1000.5	999.7	999.6	999.1	997.8
1001.3	1000.7	999.4	1000.0	998.3
999.5	1000.1	998.3	999.2	999.2
1000.4	1000.1	1000.1	999.6	999.9

SOLUTION

IDENTIFY

The problem objective is to describe the population of l-liter fills from this machine. The data are interval, and we're interested in the variability of the fills. It follows that the parameter of interest is the population variance. Because we want to determine whether there is enough evidence to support the claim, the alternative hypothesis is

$$H_1: \quad \sigma^2 < 1$$

The null hypothesis is

$$H_0: \quad \sigma^2 = 1$$

and the test statistic we will use is

$$\chi^2 = \frac{(n - 1)s^2}{\sigma^2}$$

COMPUTE

MANUALLY

Using a calculator, we find

$$\sum x_i = 24{,}992.0 \qquad \text{and} \qquad \sum x_i^2 = 24{,}984{,}017.76$$

Thus,

$$s^2 = \frac{\sum x_i^2 - \frac{\left(\sum x_i\right)^2}{n}}{n - 1} = \frac{24{,}984{,}017.76 - \frac{(24{,}992.0)^2}{25}}{25 - 1} = .6333$$

The value of the test statistic is

$$\chi^2 = \frac{(n-1)s^2}{\sigma^2} = \frac{(25-1)(.6333)}{1} = 15.20$$

The rejection region is

$$\chi^2 < \chi^2_{1-\alpha,n-1} = \chi^2_{1-.05,25-1} = \chi^2_{.95,24} = 13.85$$

Since 15.20 is not less than 13.85, we cannot reject the null hypothesis in favor of the alternative.

EXCEL

	A	B	C	D
1	**Chi Squared Test: Variance**			
2				
3				*Fills*
4	Sample Variance			0.6333
5	Hypothesized Variance			1
6	df			24
7	chi-squared Stat			15.20
8	P (CHI<=chi) one-tail			0.0852
9	chi-squared Critical one tail	Left-tail		13.85
10		Right-tail		36.42
11	P (CHI<=chi) two-tail			0.1705
12	chi-squared Critical two tail	Left-tail		12.40
13		Right-tail		39.36

The value of the test statistic is 15.20. $P(\text{CHI} <= \text{chi})$ one-tail is the probability $P(\chi^2 < 15.20)$, which is equal to .0852. Because this is a one-tail test, the p-value is .0852.

INSTRUCTIONS

1. Type or import the data into one column. (Open Xm12-03.)
2. Click **Add-Ins, Data Analysis Plus,** and **Chi-squared Test: Variance.**
3. Specify the **Input Range** (A1:A26), type the **Hypothesized Variance** (1) and the value of α (.05).

We can also use the worksheet **Chi-squared Test_Variance** in the **Test Statistics** workbook to perform the calculations as well as to conduct what-if analyses.

MINITAB

Test and CI for One Standard Deviation: Fills

Null hypothesis sigma = 1
Alternative hypothesis sigma = < 1

Statistics

Variable	N	StDev	Variance
Fills	25	0.796	0.633

Tests

Variable	Method	Chi-Square	DF	P-Value
Fills	Standard	15.20	24.00	0.085

INSTRUCTIONS

Some of the output has been deleted.

Besides computing the *z*-statistic and *p*-value, and because we're conducting a one-tail test, Minitab calculates a one-sided confidence interval estimate. (See page 363 for a discussion of one-sided confidence interval estimators.)

1. Type or import the data into one column. (Open Xm12-03.)
2. Click **Stat, Basic Statistics,** and **1 Variance . . . .**
3. Type or use the **Select** button to specify the name of the variable or the column in the **Samples in columns** box (Fills), check **Perform hypothesis test,** and type the value of σ in the **Hypothesized standard deviation** box (1).
4. Click **Options . . .** and select one of **less than, not equal,** or **greater than** in the **Alternative** box (less than).

INTERPRET

There is not enough evidence to infer that the claim is true. As we discussed before, the result does not say that the variance is equal to 1; it merely states that we are unable to show that the variance is less than 1. Figure 12.4 depicts the sampling distribution of the test statistic.

FIGURE **12.4** Sampling Distribution for Example 12.3

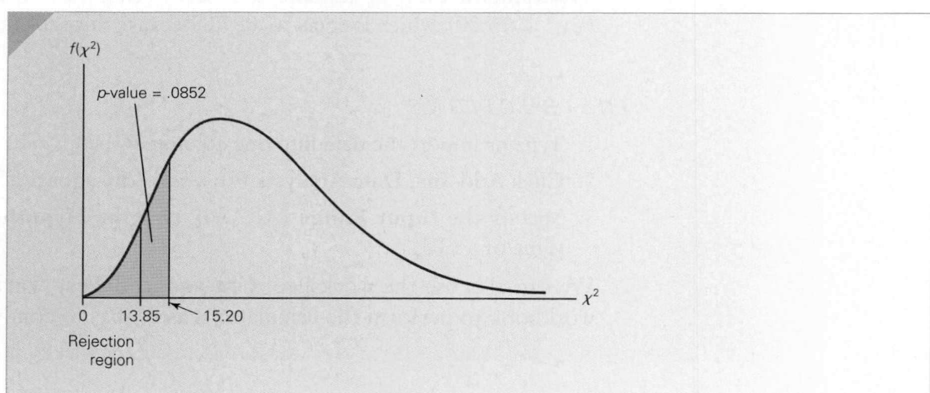

EXAMPLE 12.4

Consistency of a Container-Filling Machine, Part 2

Estimate with 99% confidence the variance of fills in Example 12.3.

SOLUTION

MANUALLY

In the solution to Example 12.3, we found $(n - 1)s^2$ to be 15.20. From Table 5 in Appendix B, we find

$$\chi^2_{\alpha/2,n-1} = \chi^2_{.005,24} = 45.56$$

$$\chi^2_{1-\alpha/2,n-1} = \chi^2_{.995,24} = 9.89$$

Thus,

$$\text{LCL} = \frac{(n-1)s^2}{\chi^2_{\alpha/2}} = \frac{15.20}{45.56} = .3336$$

$$\text{UCL} = \frac{(n-1)s^2}{\chi^2_{1-\alpha/2}} = \frac{15.20}{9.89} = 1.537$$

We estimate that the variance of fills is a number that lies between .3336 and 1.537.

EXCEL

	A	B
1	**Chi Squared Estimate: Variance**	
2		
3		*Fills*
4	Sample Variance	0.6333
5	df	24
6	LCL	0.3336
7	UCL	1.5375

INSTRUCTIONS

1. Type or import the data into one column. (Open Xm12-03.)
2. Click **Add-Ins, Data Analysis Plus,** and **Chi-squared Estimate: Variance.**
3. Specify the **Input Range** (A1:A26) and α (.01).

If you know the sample variance and sample size you can employ the **chi-squared Estimate_Variance** worksheet in the **Estimators** workbook to perform the calculations. This spreadsheet can be used to conduct what-if analyses. (See CD Appendix J.)

MINITAB

Test and CI for One Standard Deviation: Fills

Statistics

Variable	N	StDev	Variance
Fills	25	0.796	0.633

99% Confidence Intervals

Variable	Method	CI for StDev	CI for Variance
Fills	Standard	(0.578, 1.240)	(0.334, 1.537)

INSTRUCTIONS

Some of the output has been deleted.

1. Type or import the data into one column. (Open Xm12-03.)
2. Click **Stat, Basic Statistics,** and **1 Variance . . . .**

(continued)

3. Type or use the **Select** button to specify the name of the variable or the column in the **Samples in columns** box (Fills).
4. Click **Options . . .**, type the **Confidence level,** and select one of **less than, not equal,** or **greater than** in the **Alternative** box (not equal).

INTERPRET

In Example 12.3, we saw that there was not sufficient evidence to infer that the population variance is less than 1. Here we see that σ^2 is estimated to lie between .3336 and 1.5375. Part of this interval is above 1, which tells us that the variance may be larger than 1, confirming the conclusion we reached in Example 12.3. We may be able to use the estimate to predict the percentage of overfilled and underfilled bottles. This may allow us to choose among competing machines.

Checking the Required Condition

Like the t-test and estimator of μ introduced in Section 12.1, the chi-squared test and estimator of σ^2 theoretically require that the sample population be normal. In practice, however, the technique is valid as long as the population is not extremely nonnormal. We can gauge the extent of nonnormality by drawing the histogram. Figure 12.5 depicts Excel's version of this histogram. As you can see, the fills appear to be somewhat asymmetric. However the variable does not appear to be very nonnormal. We conclude that the normality requirement is not seriously violated.

FIGURE **12.5** Histogram for Examples 12.3 and 12.4

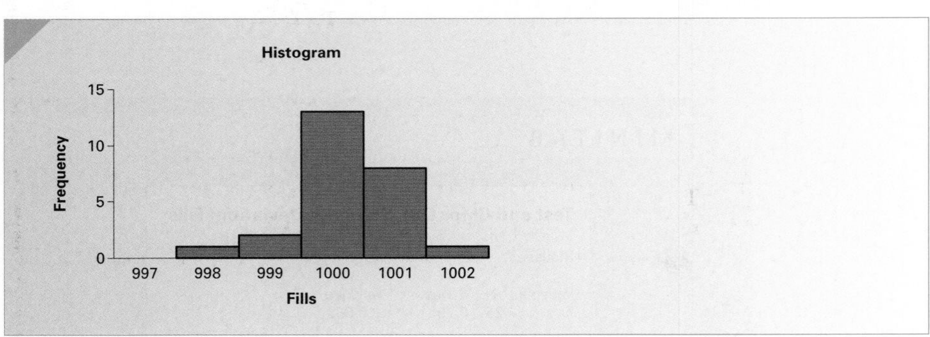

Here is how we recognize when to use the techniques introduced in this section.

Factors That Identify the Chi-Squared Test and Estimator of σ^2

1. **Problem objective:** Describe a population
2. **Data type:** Interval
3. **Type of descriptive measurement:** Variability

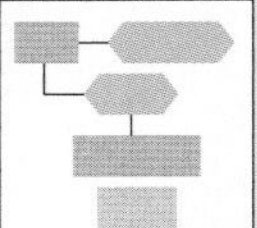

EXERCISES

Developing an Understanding of Statistical Concepts

*The following exercises are "what-if analyses" designed to determine what happens to the test statistics and interval estimates when elements of the statistical inference change. These problems can be solved manually or using Excel's **Test Statistics** or **Estimators** workbook.*

12.43 a. A random sample of 100 observations was drawn from a normal population. The sample variance was calculated to be $s^2 = 220$. Test with $\alpha = .05$ to determine whether we can infer that the population variance differs from 300.
b. Repeat Part a, changing the sample size to 50.
c. What is the effect of decreasing the sample size?

12.44 a. The sample variance of a random sample of 50 observations from a normal population was found to be $s^2 = 80$. Can we infer at the 1% significance level that σ^2 is less than 100?
b. Repeat Part a, increasing the sample size to 100.
c. What is the effect of increasing the sample size?

12.45 a. Estimate σ^2 with 90% confidence given that $n = 15$ and $s^2 = 12$.
b. Repeat Part a with $n = 30$.
c. What is the effect of increasing the sample size?

Applications

12.46 Xr12-46 The weights of a random sample of cereal boxes that are supposed to weigh 1 pound are listed here. Estimate the variance of the entire population of cereal box weights with 90% confidence.

1.05 1.03 .98 1.00 .99 .97 1.01 .96

12.47 Xr12-47 After many years of teaching, a statistics professor computed the variance of the marks on her final exam and found it to be $\sigma^2 = 250$. She recently made changes to the way in which the final exam is marked and wondered whether this would result in a reduction in the variance. A random sample of this year's final exam marks are listed here. Can the professor infer at the 10% significance level that the variance has decreased?

57 92 99 73 62 64 75 70 88 60

12.48 Xr12-48 With gasoline prices increasing, drivers are becoming more concerned with their cars' gasoline consumption. For the past 5 years, a driver has tracked the gas mileage of his car and found that the variance from fill-up to fill-up was $\sigma^2 = 23$ mpg^2. Now that his car is 5 years old, he would like to know whether the variability of gas mileage has changed. He recorded the gas mileage from his last eight fill-ups; these are listed here. Conduct a test at a 10% significance level to infer whether the variability has changed.

28 25 29 25 32 36 27 24

12.49 Xr12-49 During annual checkups, physicians routinely send their patients to medical laboratories to have various tests performed. One such test determines the cholesterol level in patients' blood. However, not all tests are conducted in the same way. To acquire more information, a man was sent to 10 laboratories and had his cholesterol level measured in each. The results are listed here. Estimate with 95% confidence the variance of these measurements.

188 193 186 184 190 195 187 190
192 196

The following exercises require the use of a computer and software. The answers may be calculated manually. See Appendix A for the sample statistics.

12.50 Xr12-50 One important factor in inventory control is the variance of the daily demand for the product. A management scientist has developed the optimal order quantity and reorder point, assuming that the variance is equal to 250. Recently, the company has experienced some inventory problems, which induced the operations manager to doubt the assumption. To examine the problem, the manager took a sample of 25 days and recorded the demand.
a. Do these data provide sufficient evidence at the 5% significance level to infer that the management scientist's assumption about the variance is wrong?
b. What is the required condition for the statistical procedure in Part a?
c. Does it appear that the required condition is not satisfied?

12.51 Xr12-51 Some traffic experts believe that the major cause of highway collisions is the differing speeds of cars. That is, when some cars are driven slowly while others are driven at speeds well in excess of the speed limit, cars tend to congregate in bunches, increasing the probability of accidents. Thus, the greater the variation in speeds, the greater will be the number of collisions that occur. Suppose that one expert believes that when the variance exceeds 18 mph^2, the number of accidents will be unacceptably high. A random sample of the speeds of 245 cars on a highway with one of the highest accident rates in the country is taken. Can we conclude at the 10% significance level that the variance in speeds exceeds 18 mph^2?

12.52 Xr12-52 The job placement service at a university observed the not unexpected result of the variance in marks and work experience of the university's graduates: Some graduates received numerous offers whereas others received far fewer. To learn more about the problem, a survey of 90 recent graduates was conducted wherein each was asked how many job offers they received. Estimate with 90% confidence the variance in the number of job offers made to the university's graduates.

12.53 Xr12-53 One problem facing the managers of maintenance departments is when to change the bulbs in streetlamps. If bulbs are changed only when they burn out, it is quite costly to send crews out to change only one bulb at a time. This method also requires someone to report the problem and, in the meantime, the light is off. If each bulb lasts approximately the same amount of time, they can all be replaced periodically, producing significant cost savings in maintenance. Suppose that a financial analysis of the lights at Yankee Stadium has concluded that it will pay to replace all of the lightbulbs at the same time if the variance of the lives of the bulbs is less than 200 hours^2. The lengths of life of the last 100 bulbs were recorded. What conclusion can be drawn from these data? Use a 5% significance level.

12.54 Xr12-54 Home blood-pressure monitors have been on the market for several years. This device allows people with high blood pressure to measure their own and determine whether additional medication is necessary. Concern has been expressed about inaccurate readings. To judge the severity of the problem, a laboratory technician measured his own blood pressure 25 times using the leading brand of monitors. Estimate the population variance with 95% confidence.

12.3 Inference about a Population Proportion

In this section we continue to address the problem of describing a population. However, we shift our attention to populations of nominal data, which means that the population consists of nominal or categorical values. For example, in a brand preference survey where the statistics practitioner asks consumers of a particular product which brand they purchase, the values of the random variable are the brands. If there are five brands, the values could be represented by their names, by letters (A, B, C, D, and E), or by numbers (1, 2, 3, 4, and 5). When numbers are used, it should be understood that the numbers only represent the name of the brand, are completely arbitrarily assigned, and cannot be treated as real numbers, that is, we cannot calculate means and variances.

Parameter

Recall the discussion of types of data in Chapter 2. When the data are nominal, all that we are permitted to do to describe the population or sample is count the number of occurrences of each value. From the counts we calculate proportions. Thus, the parameter of interest in describing a population of nominal data is the population proportion p. In Section 7.4 this parameter was used to calculate probabilities based on the binomial experiment. One of the characteristics of the binomial experiment is that there are only two possible outcomes per trial. Most practical applications of inference about p involve more than two outcomes. However, in most cases we're interested in only one outcome, which we label a "success." All other outcomes are labeled as "failures." For example, in brand-preference surveys we are interested in our company's brand. In political surveys we wish to estimate or test the proportion of voters who will vote for one particular candidate—likely the one who has paid for the survey.

Statistic and Sampling Distribution

The logical statistic used to estimate and test the population proportion is the sample proportion defined as

$$\hat{p} = \frac{x}{n}$$

where x is the number of successes in the sample and n is the sample size. In Section 9.2 we presented the approximate sampling distribution of $\hat{P}$. (The actual distribution is based on the binomial distribution, which does not lend itself to statistical inference.) The sampling distribution of $\hat{p}$ is approximately normal with mean p and standard deviation $\sqrt{p(1-p)/n}$ (provided that np and $n(1-p)$ are greater than 5). We express this sampling distribution as

$$z = \frac{\hat{P} - p}{\sqrt{p(1-p)/n}}$$

Testing and Estimating a Proportion

As you have already seen, the formula that summarizes the sampling distribution also represents the test statistic.

Test Statistic for *p*

$$z = \frac{\hat{p} - p}{\sqrt{p(1-p)/n}}$$

which is approximately normal for np and $n(1-p)$ greater than 5.

Using the same algebra employed in Sections 10.2 and 12.1, we attempt to derive the confidence interval estimator of p from the sampling distribution. The result is

$$\hat{p} \pm z_{\alpha/2}\sqrt{p(1-p)/n}$$

This formula, although technically correct, is useless. To understand why, examine the standard error of the sampling distribution $\sqrt{p(1-p)/n}$. To produce the interval estimate, we must compute the standard error, which requires us to know the value of p, the parameter we wish to estimate. This is the first of several statistical techniques where we face the same problem—how to determine the value of the standard error. In this application the problem is easily and logically solved: Simply estimate the value of p with $\hat{p}$. Thus, we estimate the standard error with $\sqrt{\hat{p}(1-\hat{p})/n}$.

Confidence Interval Estimator of p

$$\hat{p} \pm z_{\alpha/2}\sqrt{\hat{p}(1 - \hat{p})/n}$$

which is valid provided that $n\hat{p}$ and $n(1 - \hat{p})$ are greater than 5.

EXAMPLE 12.5

DATA Xm12-05*

Election Day Exit Poll

When an election for political office takes place, the television networks cancel regular programming and instead provide election coverage. When the ballots are counted, the results are reported. However, for important offices such as president or senator in large states, the networks actively compete to see which will be the first to predict a winner. This is done through exit polls,* wherein a random sample of voters who exit the polling booth is asked for whom they voted. From the data, the sample proportion of voters supporting the candidates is computed. A statistical technique is applied to determine whether there is enough evidence to infer that the leading candidate will garner enough votes to win. Suppose that in the exit poll from the state of Florida during the 2000 year elections, the pollsters recorded only the votes of the two candidates who had any chance of winning, Democrat Albert Gore (code = 1) and the Republican candidate George W. Bush (code = 2). The polls close at 8:00 P.M. Can the networks conclude from these data that the Republican candidate will win the state? Should the network announce at 8:01 P.M. that the Republican candidate will win?

SOLUTION

IDENTIFY

The problem objective is to describe the population of votes in the state. The data are nominal because the values are "Democrat" (code = 1) and "Republican" (code = 2). Thus the parameter to be tested is the proportion of votes in the entire state that are for the Republican candidate. Because we want to determine whether the network can declare the Republican to be the winner at 8:01 P.M., the alternative hypothesis is

$$H_1: \quad p > .5$$

which makes the null hypothesis

$$H_0: \quad p = .5$$

The test statistic is

$$z = \frac{\hat{p} - p}{\sqrt{p(1 - p)/n}}$$

*Warren Mitofsky is generally credited for creating the Election Day exit poll in 1967 when he worked for CBS News. Mitofsky claimed to have correctly predicted 2,500 elections and only six wrong. Exit polls are considered so accurate that when the exit poll and the actual election result differ, some newspaper and television reporters claim that the election result is wrong! In the 2004 presidential election the exit poll showed John Kerry leading. However, when the ballots were counted, George Bush won the state of Ohio. Conspiracy theorists now believe that the Ohio election was stolen by the Republicans using the exit poll as their "proof." However, Mitofsky's own analysis found that the exit poll was improperly conducted, resulting in many Republican voters refusing to participate in the poll. Blame was placed on poorly trained interviewers (*Source: Amstat News*, December 2006).

COMPUTE

MANUALLY

It appears that this is a "standard" problem, which requires a 5% significance level. Thus, the rejection region is

$$z > z_\alpha = z_{.05} = 1.645$$

From the file we count the number of "successes," which is the number of votes cast for the Republican and find $x = 407$. The sample size is 765. Hence, the sample proportion is

$$\hat{p} = \frac{x}{n} = \frac{407}{765} = .532$$

The value of the test statistic is

$$z = \frac{\hat{p} - p}{\sqrt{p(1 - p)/n}} = \frac{.532 - .5}{\sqrt{.5(1 - .5)/765}} = 1.77$$

Since the test statistic is (approximately) normally distributed, we can determine the p-value. It is

$$p\text{-value} = P(z > 1.77) = 1 - p(Z < 1.77) = 1 - .9616 = .0384$$

There is enough evidence at the 5% significance level that the Republican candidate has won.

EXCEL

	A	B	C	D
1	**z-Test: Proportion**			
2				
3				*Votes*
4	Sample Proportion			0.532
5	Observations			765
6	Hypothesized Proportion			0.5
7	z Stat			1.7716
8	P(Z<=z) one-tail			0.0382
9	z Critical one-tail			1.6449
10	P(Z<=z) two-tail			0.0764
11	z Critical two-tail			1.9600

INSTRUCTIONS

1. Type or import the data into one column. (Open Xm12-05.)
2. Click **Add-Ins, Data Analysis Plus,** and **Z-Test: Proportion.**
3. Specify the **Input Range** (A1:A766), type the **Code for Success** (2), the **Hypothesized Proportion** (.5), and a value of α (.05).

To complete the technique from the sample proportion or to conduct a what-if analysis, open the **Test Statistics** workbook and the **z-Test_Proportion** worksheet. (CD Appendix M describes how to use an Excel spreadsheet to calculate the probability of a Type II error when testing a proportion.)

MINITAB

Test and CI for One Proportion: Votes

Test of p = 0.5 vs p > 0.5

Event = 2

Variable	X	N	Sample p	95% Lower Bound	Z-Value	P-Value
Votes	407	765	0.532026	0.502352	1.77	0.038

Using the normal approximation

As was the case with the test of a variance, Minitab calculates a one-sided confidence interval estimate when we're conducting a one-tail test.

INSTRUCTIONS

The data must represent successes and failures. The codes can be numbers or text. There can be only two kinds of entries, one representing success and the other representing failure. If numbers are used, Minitab will interpret the larger one as a success.

1. Type or import the data into one column. (Open Xm12-05.)
2. Click **Stat, Basic Statistics**, and **1 Proportion . . . .**
3. Use the **Select** button or type the name of the variable or its column in the **Samples in columns** box (Votes) and check **Perform hypothesis test** and type the **Hypothesized proportion** (.5).
4. Click **Options . . .** and specify the **Alternative** hypothesis (greater than). To use the normal approximation of the binomial, click **Use test and interval based on normal approximation.**

INTERPRET

The value of the test statistic is $z = 1.77$ and the one-tail p-value $= .0382$. Using a 5% significance level, we reject the null hypothesis and conclude that there is enough evidence to infer that George W. Bush won the presidential election in the state of Florida.

One of the key issues to consider here is the cost of Type I and Type II errors. A Type I error occurs if we conclude that the Republican will win when in fact he has lost. Such an error would mean that a network would announce at 8:01 P.M. that the Republican has won and then later in the evening would have to admit to a mistake. If a particular network were the only one that made this error, it would cast doubt on their integrity and possibly affect the number of viewers.

This is exactly what happened on the evening of the U. S. presidential elections in November 2000. Shortly after the polls closed at 8:00 P.M., all the networks declared that the Democratic candidate Albert Gore would win in the state of Florida. A couple of hours later, the networks admitted that a mistake had been made and the Republican candidate George W. Bush had won. Several hours later they again admitted a mistake and finally declared the race too close to call. Fortunately for each network, all the networks made the same mistake. However, if one network had not done this, it would have developed a better track record, which could have been used in future advertisements for news shows and likely drawn more viewers.

Missing Data

In real statistical applications we occasionally find that the data set is incomplete. In some instances the statistics practitioner may have failed to properly record some observations or some data may have been lost. In other cases respondents may refuse to answer. For example, in political surveys where the statistics practitioner asks voters for whom they intend to vote in the next election, some people will answer that they haven't decided or that their vote is confidential and refuse to answer. In surveys where respondents are asked to report their income, people often refuse to divulge this information. This is a troublesome issue for statistics practitioners. We can't force people to answer our questions. However, if the number of nonresponses is high, the results of our analysis may be invalid because the sample is no longer truly random. To understand why, suppose that people who are in the top quarter of household incomes regularly refuse to answer questions about their incomes. The resulting estimate of the population household income mean will be lower than the actual value.

The issue can be complicated. There are several ways to compensate for nonresponses. The simplest method is eliminating them. To illustrate, suppose that in a political survey respondents are asked for whom they intend to vote in a two-candidate race. Surveyors record the results as 1 = Candidate A, 2 = Candidate B, 3 = "Don't know," and 4 = "Refuse to say." If we wish to infer something about the proportion of decided voters who will vote for Candidate A, we can simply omit codes 3 and 4. If we're doing the work manually, we will count the number of voters who prefer Candidate A and the number who prefer Candidate B. The sum of these two numbers is the total sample size.

In the language of statistical software, nonresponses that we wish to eliminate are collectively called "missing data." Software packages deal with missing data in different ways. CD Appendix N describes how to address the problem of missing data in Excel and in Minitab.

Estimating the Total Number of Successes in a Large Finite Population

As was the case with the inference about a mean, the techniques in this section assume infinitely large populations. When the populations are small, it is necessary to include the finite population correction factor. A population is small when it is less than 20 times the sample size. When the population is large and finite, we can estimate the total number of successes in the population.

To produce the confidence interval estimator of the total, we multiply the lower and upper confidence limits of the interval estimator of the proportion of successes by the population size. The confidence interval estimator of the total number of successes in a large finite population is

$$N\left(\hat{p} \pm z_{\alpha/2}\sqrt{\frac{\hat{p}(1-\hat{p})}{n}}\right)$$

We will use this estimator in the chapter-opening example and several of this section's exercises.

Nielsen Ratings: Solution

© Brand X Pictures/Jupiterimages

IDENTIFY

The problem objective is to describe the population of television shows watched by viewers across the country. The data are nominal. The combination of problem objective and data type make the parameter to be estimated the proportion of the entire population that watched *Deal or No Deal* (code = 3). The confidence interval estimator of the proportion is

$$\hat{p} \pm z_{\alpha/2}\sqrt{\frac{\hat{p}(1-\hat{p})}{n}}$$

COMPUTE

MANUALLY

To solve manually, we count the number of 3's in the file. We find this value to be 418. Thus,

$$\hat{p} = \frac{x}{n} = \frac{418}{5{,}000} = .0836$$

The confidence level is $1 - \alpha = .95$. It follows that $\alpha = .05$, $\alpha/2 = .025$, and $z_{\alpha/2} = z_{.025} = 1.96$. The 95% confidence interval estimate of p is

$$\hat{p} \pm z_{\alpha/2}\sqrt{\frac{\hat{p}(1-\hat{p})}{n}} = .0836 \pm 1.96\sqrt{\frac{(.0836)(1-.0836)}{5{,}000}} = .0836 \pm .0077$$

$$LCL = .0759 \qquad UCL = .0913$$

EXCEL

	A	B
1	**z-Estimate: Proportion**	
2		*Program*
3	Sample Proportion	0.0836
4	Observations	5000
5	LCL	0.0759
6	UCL	0.0913

INSTRUCTIONS

1. Type or import the data into one column. (Open Xm12-00.)
2. Click **Add-Ins, Data Analysis Plus,** and **Z-Estimate: Proportion.**
3. Specify the **Input Range** (A1:A5001), the **Code for Success** (3), and the value of α (.05).

To complete the technique from the sample proportion or to conduct a what-if analysis, open the **Estimators** workbook and the **z-Estimate_Proportion** worksheet.

MINITAB

Minitab requires that the data set contain only two values, the larger of which would be considered a success. In this example there are five values. If there are more than two codes or if the code for success is smaller than that for failure, we must recode.

Test and CI for One Proportion: Recoded Programs

Event = 3

Variable	X	N	Sample p	95% CI
Recoded Programs	418	5000	0.083600	(0.075928, 0.091272)

Using the normal approximation.

INSTRUCTIONS

Recode data

1. Click **Data, Code,** and **Numeric to Numeric . . . .**
2. In the **Code data from columns** box, type or **Select** the data you wish to recode.
3. In the **Store coded data in columns** box, type the column where the recoded data are to be stored. (We named the column "Recoded Programs.")
4. Specify the **Original values:** you wish to recode and their **New:** values.

Estimate the proportion

1. Click **Stat, Basic Statistics,** and **1 Proportion . . . .**
2. In the **Samples in columns** box, type or **Select** the data (Recoded Programs).
3. Click **Options . . . .**
4. Specify the **Confidence level:** (.95), select **Alternative: not equal,** and **Use test and interval based on normal distribution.**

INTERPRET

We estimate that between 7.59% and 9.13% of all households with televisions had tuned to *Deal or No Deal* on Sunday, April 1, 2007, at 9:00 to 9:30. If we multiply these figures by the total number of television households, 110 million, we produce an interval estimate of the number of televisions tuned to *Deal or No Deal*. Thus,

$$\text{LCL} = .0759 \times 110 \text{ million} = 8.349 \text{ million}$$

and

$$\text{UCL} = .0913 \times 110 \text{ million} = 10.043 \text{ million}$$

Sponsoring companies can then determine the value of any commercials that appeared on the show.

Selecting the Sample Size to Estimate the Proportion

When we introduced the sample size selection method to estimate a mean in Section 10.3, we pointed out that the sample size depends on the confidence level and the bound on the error of estimation that the statistics practitioner is willing to tolerate. When the parameter to be estimated is a proportion, the bound on the error of estimation is

$$B = z_{\alpha/2}\sqrt{\frac{\hat{p}(1-\hat{p})}{n}}$$

Solving for n, we produce the required sample size as indicated in the box.

Sample Size to Estimate a Proportion

$$n = \left(\frac{z_{\alpha/2}\sqrt{\hat{p}(1-\hat{p})}}{B}\right)^2$$

To illustrate the use of this formula, suppose that in a brand preference survey we want to estimate the proportion of consumers who prefer our company's brand to within .03 with 95% confidence. This means that the bound on the error of estimation is $B = .03$. Since $1 - \alpha = .95$, $\alpha = .05$, $\alpha/2 = .025$, and $z_{\alpha/2} = z_{.025} = 1.96$, therefore,

$$n = \left(\frac{1.96\sqrt{\hat{p}(1-\hat{p})}}{.03}\right)^2$$

To solve for n, we need to know $\hat{p}$. Unfortunately, this value is unknown, because the sample has not yet been taken. At this point, we can use either of two methods to solve for n.

Method 1

If we have no knowledge of even the approximate value of $\hat{p}$, we let $\hat{p} = .5$. We choose $\hat{p} = .5$ because the product $\hat{p}(1 - \hat{p})$ equals its maximum value at $\hat{p} = .5$. (Figure 12.6 illustrates this point.) This, in turn, results in a conservative value of n, and as a result, the confidence interval will be no wider than the interval $\hat{p} \pm .03$. If, when the sample is drawn, $\hat{p}$ does not equal .5, the confidence interval estimate will be better (that is, narrower) than planned. Thus,

$$n = \left(\frac{1.96\sqrt{(.5)(.5)}}{.03}\right)^2 = (32.67)^2 = 1{,}068$$

If it turns out that $\hat{p} = .5$, the interval estimate is $\hat{p} \pm .03$. If not, the interval estimate will be narrower. For instance, if it turns out that $\hat{p} = .2$, the estimate is $\hat{p} \pm .024$, which is better than we had planned.

FIGURE **12.6**
Plot of $\hat{p}$ versus $\hat{p}$ $(1 - \hat{p})$

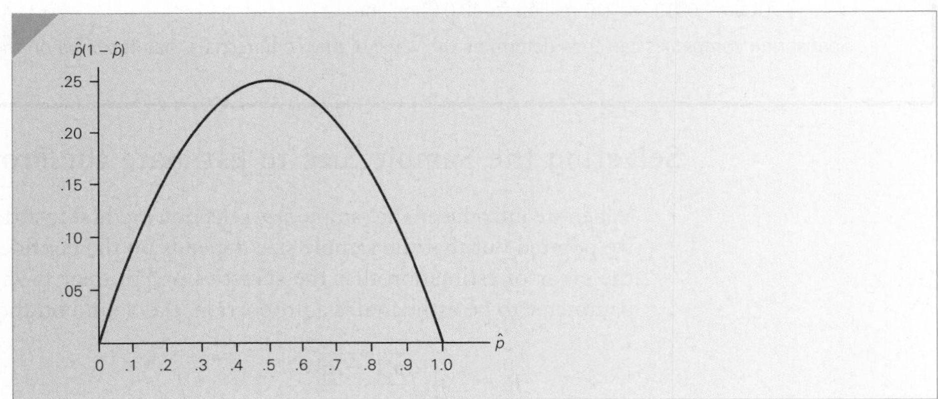

Method 2

If we have some idea about the value of $\hat{p}$, we can use that quantity to determine n. For example, if we believe that $\hat{p}$ will turn out to be approximately .2, we can solve for n as follows:

$$n = \left(\frac{1.96\sqrt{(.2)(.8)}}{.03}\right)^2 = (26.13)^2 = 683$$

Notice that this produces a smaller value of n (thus reducing sampling costs) than does method 1. If $\hat{p}$ actually lies between .2 and .8, however, the estimate will not be as good as we wanted, because the interval will be wider than desired.

Method 1 is often used to determine the sample size used in public opinion surveys reported by newspapers, magazines, television, and radio. These polls usually estimate proportions to within 3%, with 95% confidence. (The media often state the confidence level as "19 times out of 20.") If you've ever wondered why opinion polls almost always estimate proportions to within 3%, consider the sample size required to estimate a proportion to within 1%:

$$n = \left(\frac{1.96\sqrt{(.5)(.5)}}{.01}\right)^2 = (98)^2 = 9{,}604$$

The sample size 9,604 is nine times the sample size needed to estimate a proportion to within 3%. Thus, to divide the width of the interval by 3 requires multiplying the sample size by 9. The cost would also increase considerably. For most applications, the increase in accuracy (created by decreasing the width of the confidence interval estimate) does not overcome the increased cost. Confidence interval estimates with 5% or 10% bounds (sample sizes 385 and 97, respectively) are generally considered too wide to be useful. Thus, the 3% bound provides a reasonable compromise between cost and accuracy.

Wilson Estimators (Optional)

When applying the confidence interval estimator of a proportion when success is a relatively rare event, it is possible to find no successes, especially if the sample size is small. To illustrate, suppose that a sample of 100 produced $x = 0$, which means that $\hat{p} = 0$. The 95% confidence interval estimator of the proportion of successes in the population becomes

$$\hat{p} \pm z_{\alpha/2}\sqrt{\frac{\hat{p}(1-\hat{p})}{n}} = 0 \pm 1.96\sqrt{\frac{0(1-0)}{100}} = 0 \pm 0$$

This implies that if we find no successes in the sample, then there is no chance of finding a success in the population. Drawing such a conclusion from virtually any sample size is unacceptable. The remedy may be a suggestion made by Edwin Wilson in 1927. The Wilson estimate denoted $\tilde{p}$ (pronounced p-tilde) is computed by adding 2 to the number of successes in the sample and 4 to the sample size. Thus,

$$\tilde{p} = \frac{x+2}{n+4}$$

The standard error of $\tilde{p}$ is

$$\sigma_{\tilde{p}} = \sqrt{\frac{\tilde{p}(1-\tilde{p})}{n+4}}$$

Confidence Interval Estimator of p Using the Wilson Estimate

$$\tilde{p} \pm z_{\alpha/2}\sqrt{\frac{\tilde{p}(1-\tilde{p})}{n+4}}$$

Exercises 12.72–12.74 require the use of this technique.

We complete this section by reviewing the factors that tell us when to test and estimate a population proportion.

Factors That Identify the z-Test and Interval Estimator of p

1. **Problem objective:** Describe a population
2. **Data type:** Nominal

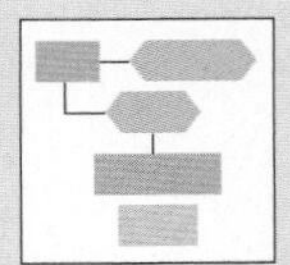

EXERCISES

Developing an Understanding of Statistical Concepts

Exercises 12.55 to 12.58 are "what-if analyses" designed to determine what happens to the test statistics and interval estimates when elements of the statistical inference change. These problems can be solved manually or using Excel's ***Test Statistics*** *or* ***Estimators*** *workbooks.*

12.55 a. In a random sample of 500 observations, we found the proportion of successes to be 48%. Estimate with 95% confidence the population proportion of successes.
b. Repeat Part a with $n = 200$.
c. Repeat Part a with $n = 1{,}000$.
d. Describe the effect on the confidence interval estimate of increasing the sample size.

12.56 a. The proportion of successes in a random sample of 400 was calculated as 50%. Estimate the population proportion with 95% confidence.
b. Repeat Part a with $\hat{p} = 33\%$.
c. Repeat Part a with $\hat{p} = 10\%$.
d. Discuss the effect on the width of the confidence interval estimate of reducing the sample proportion.

12.57 a. Calculate the p-value of the test of the following hypotheses given that $\hat{p} = .63$ and $n = 100$:

H_0: $p = .60$

H_1: $p > .60$

b. Repeat Part a with $n = 200$.
c. Repeat Part a with $n = 400$.
d. Describe the effect on the p-value of increasing the sample size.

12.58 a. A statistics practitioner wants to test the following hypotheses:

H_0: $p = .70$

H_1: $p > .70$

A random sample of 100 produced $\hat{p} = .73$. Calculate the p-value of the test.

b. Repeat Part a with $\hat{p} = .72$.
c. Repeat Part a with $\hat{p} = .71$.
d. Describe the effect on the z-statistic and its p-value of decreasing the sample proportion.

12.59 Determine the sample size necessary to estimate a population proportion to within .03 with 90% confidence assuming you have no knowledge of the approximate value of the sample proportion.

12.60 Suppose that you used the sample size calculated in Exercise 12.59 and found $\hat{p} = .5$.
a. Estimate the population proportion with 90% confidence.
b. Is this the result you expected? Explain.

12.61 Suppose that you used the sample size calculated in Exercise 12.59 and found $\hat{p} = .75$.
a. Estimate the population proportion with 90% confidence.
b. Is this the result you expected? Explain.
c. If you were hired to conduct this analysis, would the person who hired you be satisfied with the interval estimate you produced? Explain.

12.62 Redo Exercise 12.59, assuming that you know that the sample proportion will be no less than .75.

12.63 Suppose that you used the sample size calculated in Exercise 12.62 and found $\hat{p} = .75$.
a. Estimate the population proportion with 90% confidence.
b. Is this the result you expected? Explain.

12.64 Suppose that you used the sample size calculated in Exercise 12.62 and found $\hat{p} = .92$.
a. Estimate the population proportion with 90% confidence.
b. Is this the result you expected? Explain.
c. If you were hired to conduct this analysis, would the person who hired you be satisfied with the interval estimate you produced? Explain.

12.65 Suppose that you used the sample size calculated in Exercise 12.62 and found $\hat{p} = .5$.

a. Estimate the population proportion with 90% confidence.

b. Is this the result you expected? Explain.

c. If you were hired to conduct this analysis, would the person who hired you be satisfied with the interval estimate you produced? Explain.

Applications

12.66 A statistics practitioner working for major league baseball wants to supply radio and television commentators with interesting statistics. He observed several hundred games and counted the number of times a runner on first base attempted to steal second base. He found there were 373 such events of which 259 were successful. Estimate with 95% confidence the proportion of all attempted thefts of second base that are successful.

12.67 In some states the law requires drivers to turn on their headlights when driving in the rain. A highway patrol officer believes that less than one-quarter of all drivers follow this rule. As a test, he randomly samples 200 cars driving in the rain and counts the number whose headlights are turned on. He finds this number to be 41. Does the officer have enough evidence at the 10% significance level to support his belief?

12.68 A dean of a business school wanted to know whether the graduates of her school used a statistical inference technique during their first year of employment after graduation. She surveyed 314 graduates and asked about the use of statistical techniques. After tallying up the responses, she found that 204 used statistical inference within one year of graduation. Estimate with 90% confidence the proportion of all business school graduates who use their statistical education within a year of graduation.

12.69 Has the recent drop in airplane passengers resulted in better on-time performance? Before the recent downturn, one airline bragged that 92% of its flights were on time. A random sample of 165 flights completed this year reveals that 153 were on time. Can we conclude at the 5% significance level that the airline's on-time performance has improved?

12.70 What type of educational background do CEOs have? In one survey 344 CEOs of medium and large companies were asked whether they had an MBA degree. There were 97 MBAs. Estimate with 95% confidence the proportion of all CEOs of medium and large companies who have MBAs.

12.71 The GO transportation system of buses and commuter trains operates on the honor system. Train travelers are expected to buy their tickets before boarding the train. Only a small number of people will be checked on the train to see whether they bought a ticket. Suppose that a random sample of 400 train travelers was sampled and 68 of them had failed to buy a ticket. Estimate with 95% confidence the proportion of all train travelers who do not buy a ticket.

12.72 Refer to Exercise 12.71. Assuming that there are 1 million travelers per year and the fare is $3.00, estimate with 95% confidence the amount of revenue lost each year.

The following exercises require the use of the Wilson Estimator.

12.73 In Chapter 6 we discussed how an understanding of probability allows one to properly interpret the results of medical screening tests. The use of Bayes's Law requires a set of prior probabilities, which are based on historical records. Suppose that a physician wanted to estimate the probability that a woman under 35 years of age would give birth to a Down syndrome baby. She randomly sampled 200 births and discovered only one such case. Use the Wilson estimator to produce a 95% confidence interval estimate of the proportion of women under 35 who will have a Down syndrome baby.

12.74 Spam is of concern to anyone with an e-mail address. Several companies offer protection by eliminating Spam e-mails as soon as they hit an inbox. To examine one such product, a manager randomly sampled his daily e-mails for 50 days after installing spam software. A total of 374 e-mails was received, of which 3 were spam. Use the Wilson estimator to estimate with 90% confidence the proportion of spam e-mails that get through.

12.75 A management professor was in the process of investigating the relationship between education and managerial level achieved. The source of his data was a survey of 385 CEOs of medium and large companies. He discovered that there was only 1 CEO who did not have at least one university degree. Estimate (using a Wilson estimator) with 99% confidence the proportion of CEOs of medium and large companies with no university degrees.

The following exercises require the use of a computer and software. The answers may be calculated manually. See Appendix A for the sample statistics. ***Use a 5% significance level unless specified otherwise.***

12.76 Xr12-76* There is a looming crisis in universities and colleges across North America. In most places enrollments are increasing, requiring more instructors. However, there are not enough Ph.D.'s to fill the vacancies now. Moreover, among current professors, a large proportion are nearing retirement age. On top of these problems, some universities allow professors over the age of 60 to retire early. To help devise a plan to deal with the crisis, a consultant surveyed 521 55- to 64-year-old professors and asked each whether he or she intended to retire before 65. The responses are 1 = No and 2 = Yes.

a. Estimate with 95% confidence the proportion of professors who plan on early retirement.
b. Write a report for the university president describing your statistical analysis.

12.77 Refer to Exercise 12.76. If the number of professors between the ages of 55 and 64 is 75,000, estimate the total number of such professors who plan to retire early.

12.78 Xr12-78 To determine how many Americans smoke, annual surveys are conducted by the U.S. National Center for Health Statistics. The survey asks a random sample of Americans whether they smoke on some days. The responses are 1 = No and 2 = Yes. Estimate with 95% confidence the proportion of Americans who smoke. (Adapted from the *Statistical Abstract of the United States*, 2000, Table 226)

12.79 Xr12-79 The results of an annual Claimant Satisfaction Survey of policyholders who have had a claim with State Farm Insurance Company revealed a 90% satisfaction rate for claim service. To check the accuracy of this claim, a random sample of State Farm claimants was asked to rate whether they were satisfied with the quality of the service (1 = Satisfied and 2 = Unsatisfied). Can we infer that the satisfaction rate is less than 90%?

12.80 Xr12-80 An increasing number of people are giving gift certificates as Christmas presents. To measure the extent of this practice, a random sample of people was asked (survey conducted December 26–29) whether they had received a gift certificate for Christmas. The responses are recorded as 1 = No and 2 = Yes. Estimate with 95% confidence the proportion of people who received a gift certificate for Christmas.

12.81 Xr12-81* An important decision faces Christmas holiday celebrators: buy a real or artificial tree? A sample of 1,508 male and female respondents 18 years of age and over was interviewed. Respondents were asked whether they preferred a real (1) or artificial (2) tree. If there are 6 million Canadian households that buy Christmas trees, estimate with 95% confidence the total number of Canadian households that would prefer artificial Christmas trees. (*Toronto Star* November 29, 2006)

12.82 Xr12-82* Because television audiences of newscasts tend to be older (and because older people suffer from a variety of medical ailments) pharmaceutical companies' advertising often appears on national news in the three networks (ABC, CBS, and NBC). The ads concern prescription drugs such as those to treat heartburn. To determine how effective the ads are, a survey was undertaken. Adults over 50 who regularly watch network newscasts were asked whether they had contacted their physician to ask about one of the prescription drugs advertised during the newscast. The responses (1 = No and 2 = Yes) were recorded.

a. Estimate with 95% confidence the fraction of adults over 50 who have contacted their physician to inquire about a prescription drug.
b. Prepare a presentation to the executives of a pharmaceutical company that discusses your analysis.

12.83 Xr12-83 A professor of business statistics recently adopted a new textbook. At the completion of the course, 100 randomly selected students were asked to assess the book. The responses are as follows.

Excellent (1), Good (2), Adequate (3), Poor (4)

The results are stored using the codes in parentheses. Do the data allow us to conclude at the 10% significance level that more than 50% of all business students would rate the book as excellent?

12.84 Refer to Exercise 12.83. Do the data allow us to conclude at the 10% significance level that more than 90% of all business students would rate it as at least adequate?

12.85 Xm12-00* Refer to the chapter-opening example. Estimate with 95% confidence the number of television households that were tuned to the ABC program *Desperate Housewives*.

12.86 Xr12-86* In the fall of 1998, a newspaper publisher launched a new "national" newspaper in

Canada. It was believed that the new newspaper would have to capture more than 12% of the Toronto market to be financially viable. During the planning stages of this new newspaper, a market survey was conducted of a sample of 400 Toronto readers. The survey provided a brief description of the proposed newspaper and then asked whether the survey participant would subscribe to the newspaper if the cost did not exceed $20 per month. The responses are 1 = No and 2 = Yes. Can the publisher conclude that the proposed newspaper will be financially viable?

12.87 Xr12-87 Obesity is not only a health problem, but also an economic one. Obesity has been related to a wide variety of medical problems including heart attacks, strokes, and cancer. To help gauge the extent of the problem, the U.S. Center for Health Statistics took a survey of Americans and determined whether each was obese (1 = Obese and 2 = Not obese). Assuming a total population of 270 million people, estimate with 90% confidence the total number of Americans who are obese. (Adapted from the *Statistical Abstract of the United States*, 2000, Table 231)

12.88 Xr12-88 Chlorofluorocarbons (CFCs) are used in air conditioners. However, CFCs damage the ozone layer, which protects us from the sun's harmful rays. As a result, many jurisdictions have banned the production and use of CFCs. The latest jurisdiction to do so is the province of Ontario, which has banned the use of CFCs in car and truck air conditioners. However, it is not known how many vehicles will be affected by the new legislation. A survey of 650 vehicles was undertaken. Each vehicle was identified as either using CFCs (2) or not (1).

a. If there are 5 million vehicles registered in Ontario, estimate with 95% confidence the number of vehicles affected by the new law.
b. Write a report for the premier of the province describing what you have learned about the problem.

12.89 Xr12-89 Refer to the chapter-opening example. Besides tracking the total number of television households, Nielsen also tracks various age groups defined by the age of the head of the household. The results for the 18 to 49 age group on Wednesday, January 24, 2007, at 8:00 to 8:30 P.M. were recorded. Estimate with 95% confidence the number of televisions in the 18 to 49 age group that had received *American Idol 6*. There are 50 million households whose head is 18 to 49 years old.

Network	Show	Code
ABC	*George Lopez Show*	1
CBS	*Armed and Famous*	2
NBC	*Friday Night Lights*	3
Fox	*American Idol 6*	4
Television turned off or watched some other channel		5

Source: Televisionweek, February 2, 2007

12.4 (Optional) Applications in Marketing: Market Segmentation

Mass marketing refers to the mass production and marketing by a company of a single product for the entire market. Mass marketing is especially effective for commodity goods such as gasoline, which are very difficult to differentiate from the competition, except through price and convenience of availability. But generally speaking, mass marketing has given way to target marketing, which focuses on satisfying the demands of a particular segment of the entire market. For example, the Coca-Cola Company has moved from the mass marketing of a single beverage to the production of several different beverages. Among the cola products, there are Coca-Cola Classic, Diet Coke, and Caffeine-Free Diet Coke. Each product is aimed at a different market segment.

Because there is no single way to segment a market, managers must consider several different variables (or characteristics) that could be used to identify segments. Surveys of customers are used to gather data about various aspects of the market, and statistical techniques are applied to define the segments. Market segmentation separates consumers of a product into different groups in such a way that members of each group are similar to each other, and there are differences between groups. Market segmentation grew out of the realization that a single product can seldom satisfy the needs and wants of all consumers. Managers must then formulate a strategy to target these profitable segments, using the four elements of the marketing mix: product, pricing, promotion, and place.

There are many ways to segment a market. Table 12.1 lists several different segmentation variables and their market segments. For example, car manufacturers can use education levels to segment the market. It is likely that high school graduates would be quite similar to others in this group and that members of this group would differ from university graduates. We would expect those differences to include the types and brands of cars each group would choose to buy. However, it is likely that income level would differentiate more clearly between segments. Statistical techniques can be used to help determine the best way to segment the market. These statistical techniques are more advanced than this textbook. Consequently, we will focus our attention on other statistical applications.

It is important for marketing managers to know the size of the segment because the size (among other parameters) determines its profitability. Not all segments are worth pursuing. In some instances, the size of the segment is too small or the costs of satisfying it may be too high. The size can be determined in several ways. The census provides useful information. For example, we can determine the number of Americans in various age categories or the size of geographic residences. For other segments we may need to survey members of a general population and use the inferential techniques introduced in the previous section where we showed how to estimate the total number of successes.

TABLE **12.1** Market Segmentation

Segmentation Variable	Segments
Geographic	
Countries	Brazil, Canada, China, France, United States
Country regions	Midwest, Northeast, Southwest, Southeast
Demographic	
Age	Under 5, 5–12, 13–19, 20–29, 30–50, over 50
Education	Some high school, high school graduate, some college, college or university graduate
Income	Under \$20,000, \$20,000–29,999, \$30,000–49,999, over \$50,000
Marital status	Single, married, divorced, widowed
Social	
Religion	Catholic, Protestant, Jewish, Muslim, Buddhist
Class	Upper class, middle class, working class, lower class
Behavior	
Media usage	TV, Internet, newspaper, magazine
Payment method	Cash, check, Visa, Mastercard

In Section 12.3 we showed how to estimate the total number of successes in a large finite population. The confidence interval estimator is

$$N\left(\hat{p} \pm z_{\alpha/2}\sqrt{\frac{\hat{p}(1-\hat{p})}{n}}\right)$$

The following example demonstrates the use of this estimator in market segmentation.

EXAMPLE 12.6

DATA Xm12-06*

Segmenting the Breakfast Cereal Market

In segmenting the breakfast cereal market, a food manufacturer uses health and diet consciousness as the segmentation variable. Four segments are developed:

1. Concerned about eating healthy foods
2. Concerned primarily about weight
3. Concerned about health because of illness
4. Unconcerned

To distinguish between groups, surveys are conducted. On the basis of a questionnaire, people are categorized as belonging to one of these groups. A recent survey asked a random sample of 1,250 American adults (20 and over) to complete the questionnaire. The categories were recorded using the codes. The most recent census reveals that there are 194,506,000 Americans who are 20 and over. Estimate with 95% confidence the number of American adults who are concerned about eating healthy foods.

SOLUTION

IDENTIFY

The problem objective is to describe the population of American adults. The data are nominal. Consequently, the parameter we wish to estimate is the proportion p of American adults who classify themselves as concerned about eating healthy. The confidence interval estimator we need to employ is

$$\hat{p} \pm z_{\alpha/2}\sqrt{\frac{\hat{p}(1-\hat{p})}{n}}$$

from which we will produce the estimate of the size of the market segment.

COMPUTE

MANUALLY

To solve manually, we count the number of 1's in the file. We find this value to be 269. Thus,

$$\hat{p} = \frac{x}{n} = \frac{269}{1{,}250} = .2152$$

The confidence level is $1 - \alpha = .95$. It follows that $\alpha = .05$, $\alpha/2 = .025$, and $z_{\alpha/2} = z_{.025} = 1.96$. The 95% confidence interval estimate of p is

$$\hat{p} \pm z_{\alpha/2}\sqrt{\frac{\hat{p}(1-\hat{p})}{n}} = .2152 \pm 1.96\sqrt{\frac{(.2152)(1-.2152)}{1{,}250}} = .2152 \pm .0228$$

$$\text{LCL} = .1924 \qquad \text{UCL} = .2380$$

EXCEL

	A	B
1	**z-Estimate: Proportion**	
2		*Group*
3	Sample Proportion	0.2152
4	Observations	1250
5	LCL	0.1924
6	UCL	0.2380

MINITAB

Test and CI for One Proportion:

Sample	X	N	Sample p	95% CI
1	269	1250	0.215200	(0.192418, 0.237982)

Using the normal approximation.

INTERPRET

We estimate that the proportion of American adults who are in group 1 lies between .1924 and .2380. Because there are 194,506,000 adults in the population, we estimate that the number of adults who belong to group 1 falls between

$$\text{LCL} = N\left(\hat{p} - z_{\alpha/2}\sqrt{\frac{\hat{p}(1-\hat{p})}{n}}\right) = 194{,}506{,}000(.1924) = 37{,}422{,}954$$

and

$$\text{UCL} = N\left(\hat{p} + z_{\alpha/2}\sqrt{\frac{\hat{p}(1-\hat{p})}{n}}\right) = 194{,}506{,}000(.2380) = 46{,}292{,}428$$

We will return to the subject of market segmentation in other chapters where we demonstrate how statistics can be used to determine whether differences actually exist between segments.

EXERCISES

The following exercises may be solved manually. See Appendix A for the sample statistics.

12.90 Xr12-90 A new credit card company is investigating various market segments to determine whether it is profitable to direct its advertising specifically at each one. One of the market segments is composed of Hispanic people. The latest census indicates that there are 19,108,000 Hispanic people in the United States. A survey of 475 Hispanics asked each how they usually pay for products that they purchase. The responses are

1. Cash
2. Check
3. Visa
4. Mastercard
5. Other credit card

Estimate with 95% confidence the number of Hispanics in the United States who usually pay by credit card.

12.91 Xr12-91* A California university is investigating expanding its evening programs. It wants to target people between 25 and 55 years old who have completed high school but did not complete college or a university. To help determine the extent and type of offerings, the university needs to know the size of its target market. A survey of 320 California adults was drawn and each person was asked to identify his or her highest educational attainment. The responses are

1. Did not complete high school
2. Completed high school only
3. Some college or university
4. College or university graduate

The *Statistical Abstract of the United States* (2003, Table 20) indicates that there are 15,517,000 Californians between the ages of 25 and 55. Estimate with 95% confidence the number of Californians between 25 and 55 years of age who are in the market segment the university wishes to target.

12.92 Xr12-92* The JCPenney department store chain segments the market for women's apparel by its identification of values. The three segments are

1. Conservative
2. Traditional
3. Contemporary

Questionnaires about personal and family values are used to identify which segment a woman falls into. Suppose that the questionnaire was sent to a random sample of 1,836 women. Each woman was classified using the codes 1, 2, and 3. The latest census reveals that there are 107,194,000 adult women in the United States (*Statistical Abstract of the United States*, 2003, Table 11). Use a 95% confidence level.

a. Estimate the proportion of adult American women who are classified as traditional.
b. Estimate the size of the traditional market segment.

12.93 Xr12-93 Most life insurance companies are leery about offering policies to people over 64. When they do, the premiums must be high enough to overcome the predicted length of life. The president of one life insurance company was thinking about offering special discounts to Americans over 64 who held full-time jobs. The plan was based on the belief that full-time workers over 64 are likely to be in good health and would likely live well into their eighties. To help decide what to do, he organized a survey of a random sample of the 35.6 million American adults over 64 (*Statistical Abstract of the United States*, 2003, Table 11). He asked a random sample of 325 Americans over 64 whether they currently hold a full-time job (1 = No and 2 = Yes).

a. Estimate with 95% confidence the size of this market segment.
b. Write a report to the executives of an insurance company detailing your statistical analysis.

12.94 Xr12-94 An advertising company was awarded the contract to design advertising for Rolls Royce automobiles. An executive in the firm decided to pitch the product not only to the affluent in the United States, but also to those who think they are in the top 1% of income earners in the country. A survey was undertaken, which among other questions asked respondents 25 and over where their annual income ranked. The following responses were given.

1 = Top 1%
2 = Top 5% but not top 1%
3 = Top 10% but not top 5%
4 = Top 25% but not top 10%
5 = bottom 75%

Estimate with 90% confidence the number of Americans 25 and over who believe they are in the top 1% of income earners. The number of Americans over 25 is 187 million. (*Statistical Abstract of the United States*, 2003, Table 11)

12.95 Xr12-95 Suppose the survey in the previous exercise also asked those who were not in the top 1% whether they believed that within 5 years they would be in the top 1% (1 = will not be in top 1% within 5 years and 2 = will be in top 1% within 5 years). Estimate with 95% confidence the number of Americans who believe that they will be in the top 1% of income earners within 5 years.

12.5 (OPTIONAL) APPLICATIONS IN ACCOUNTING AUDITING

Introduction to Accounting

The functional areas of finance, operations, and marketing are directly involved in a company's production and delivery of goods and services to its customers. Although it is not directly involved in the financing, production, and marketing of a product, *accounting* is the functional area that collects, organizes, and provides information about a company's activities that helps managers in these other areas to make decisions. Accounting information is provided both for internal use (such as for planning, control, decision making, and performance evaluation) and for external use (such as keeping investors informed). The terms *managerial accounting* and *financial accounting* are used to distinguish between the internal and external focuses, respectively, of a company's accounting activities. For planning purposes, accountants prepare budgets, which include forecasts of sales revenues and the associated costs. The statistical forecasting methods described in Chapter 20 can be used to assist in preparing sales forecasts. Forecasting costs often make use of a cost function, which expresses the relationship between a cost and some measure of the level of activity (such as production) that creates that cost. Cost functions can be estimated using a statistical procedure we described in Chapter 4.

Financial accounting typically communicates its information in the form of financial statements. Publicly held corporations are required to obtain independent external audits of the financial statements to assess their validity. Statistical sampling plays an important role in selecting samples of units (such as accounts or invoices) for inspection by auditors and in detecting errors.

Introduction to Auditing

Both private and public companies generate financial statements. These financial statements summarize the company's financial transactions that have occurred throughout the year. The summaries are often made up of thousands or even hundreds of thousands of transactions that take place over the course of the year.

Users of these financial statements (investors, bankers, etc.) want assurance from an independent third party of the fairness of the results of the company's operations, cash flows, and financial position. These users also want to be assured that the financial statements are free of significant errors. *Auditing* is the methodology that is used to give these assurances to the users of the company's financial statements.

Certified Public Accountants in the United States and Chartered Accountants in Canada often conduct company audits. Because the number of transactions in many firms is prohibitively large, sampling and statistical techniques must be employed. (In recent years technological advances in large firms have made it possible for auditors to examine the entire population.) Some audits produce estimates of annual expenses; others estimate the total of various errors, which may be judged relative to some standard called the *materiality limit*. This term is used when an audit indicates that the financial statement contains errors large enough to be necessary to take corrective action. The source of the limit is auditing rules that we won't pursue here.

Estimating the Total of a Population

Some of the statistical information auditors will use in writing their reports will be produced by estimating the total of an interval variable. For example, auditors will often estimate the annual totals of certain types of expenses, such as amounts spent on meals and entertainment. Government tax auditors may estimate the total deductions during a year made by individuals and corporations claiming automobile expenses. For this purpose we may use the confidence interval estimator of a total introduced on page 390, that is,

$$N\left(\bar{x} \pm t_{\alpha/2}\frac{s}{\sqrt{n}}\right)$$

This technique assumes a large finite population. However, one of the characteristics of many audits is that the populations are small. Recall that we label a population small if it is less than 20 times the sample size. Although we've made reference to how we address the problem of estimating in small populations, we've yet to show details, a circumstance we now correct.

Estimating the Mean and the Total in Small Populations

When the population is small relative to the sample size, we must incorporate the finite population correction factor (FPCF) in the calculation of the standard error of estimate of the mean.

Finite Population Correction Factor

$$\text{FPCF} = \sqrt{\frac{N-n}{N-1}}$$

where N is the population size and n is the sample size.

When estimating the mean when the population is small, we must multiply the FPCF by the standard error of the mean, which yields the following.

Confidence Interval Estimator of μ When N Is Small

$$\bar{x} \pm t_{\alpha/2}\frac{s}{\sqrt{n}}\sqrt{\frac{N-n}{N-1}}$$

We estimate the total by multiplying the lower and upper limits of the estimator of the mean by the population size.

Confidence Interval Estimator of a Total When N Is Small

$$N\left(\bar{x} \pm t_{\alpha/2}\frac{s}{\sqrt{n}}\sqrt{\frac{N-n}{N-1}}\right)$$

Examples 12.7 and 12.8 illustrate the use of confidence interval estimators of population totals in auditing.

EXAMPLE 12.7* GAO Audit of the U.S. Forest Service

DATA
Xm12-07

The General Accounting Office (GAO) is responsible for producing reports concerning how various government agencies and departments spend their allotted funds. (A similar function is performed by the Auditor General in Canada.) Since 1999, GAO has designated the U.S. Forest Service's financial management as a high-risk area because of internal control and accounting weaknesses that have been identified by the GAO. In 2001 the Forest Service used purchase cards and convenience checks to make 1.1 million purchases totaling $320 million. The GAO performed an audit on these purchases to determine (among other things) the dollar amount of duplicate transactions. An analysis revealed that there were 8,659 duplicate transactions. A random sample of 125 of these transactions was drawn. The amounts are listed here. The GAO would like a 95% confidence interval estimate of the total amount of duplicate transactions.

689	794	1061	965	652	900	554	1218	1038	1230	657	152	1454
1027	381	985	921	1123	752	1219	915	514	775	838	757	293
804	644	1393	350	313	1372	614	1305	1738	988	1524	437	1254
718	1456	131	1193	1055	478	714	855	1406	202	1271	892	1738
817	1120	1166	1185	1405	1063	739	1387	996	1384	1239	1005	621
434	1062	950	1872	693	1425	765	1226	305	869	1431	420	
1451	199	911	1751	1501	792	503	696	1060	35	1051	1157	
1485	974	860	415	821	593	873	1125	2147	1001	211	1047	
331	1521	837	1169	827	829	1032	1291	681	922	967	1190	
536	771	1200	1462	1289	792	347	1274	880	336	272	506	

SOLUTION

IDENTIFY

The objective is to describe the population of the amounts of duplicate transactions (interval data). Thus, the parameter to be estimated is the population mean. The confidence interval estimator of the mean will be used to estimate the total spent. Because the population is larger than 20 times the sample size, we will use the large population technique.

COMPUTE

MANUALLY

We begin by calculating the sum and the sum of squares:

$$\sum x_1 = 116{,}209 \qquad \text{and} \qquad \sum x_i^2 = 128{,}673{,}337$$

The sample mean is

$$\bar{x} = \frac{\sum x_i}{n} = \frac{116{,}209}{125} = 929.67$$

*This example was adapted from GAO Report Number 03-786.

The sample variance and standard deviation are

$$s^2 = \frac{\sum x_i^2 - \frac{\left(\sum x_i\right)^2}{n}}{n - 1} = \frac{128{,}673{,}337 - \frac{(116{,}209)^2}{125}}{125 - 1} = 166{,}428.0$$

and

$$s = \sqrt{s^2} = \sqrt{166{,}428} = 408.0$$

Because we want a 95% confidence interval estimate, $1 - \alpha = .95$, $\alpha = .05$, $\alpha/2 = .025$, and $t_{\alpha/2,n-1} = t_{.025,125-1} = t_{.025,124} \approx t_{.025,120} = 1.980$. Thus, the 95% confidence interval estimate of μ is

$$\bar{x} \pm t_{\alpha/2}\frac{s}{\sqrt{n}} = 929.67 \pm 1.980\frac{408.0}{\sqrt{125}} = 929.67 \pm 72.25$$

or $\quad$ LCL = 857.42 $\quad$ and $\quad$ UCL = 1,001.92

The estimate of the total amount in duplicate transactions lies between

LCL = 8,659(857.42) = \$7,424,400

and

UCL = 8,659(1,001.92) = \$8,675,625

EXCEL

	A	B	C	D
1	**t-Estimate: Mean**			
2				
3				*Duplicates*
4	Mean			929.67
5	Standard Deviation			407.96
6	LCL			857.45
7	UCL			1001.89

The total amount in duplicate transactions is estimated to lie between

LCL = 8,659(857.45) = \$7,424,666

and

UCL = 8,659(1,001.89) = \$8,675,394

MINITAB

One-Sample T: Duplicates

Variable	N	Mean	StDev	SE Mean	95% CI
Duplicates	125	929.672	407.956	36.489	(857.451, 1001.893)

The total amount in duplicate transactions is estimated to lie between

LCL = 8,659(857.45) = \$7,424,666

and

UCL = 8,659(1,001.89) = \$8,675,394

INTERPRET

The General Accounting Office estimates that the total amount in duplicate transactions lies between \$7,424,666 and \$8,675,394. It is the responsibility of the department administrators to correct the problem.

EXAMPLE 12.8 Audit of Purchase Orders at a Car Dealership

DATA Xm12-08

An accountant was performing an audit for a car dealership. Working through the records of sales in the parts department, she discovered that for the month of August purchase orders were handled by temporary employees because of illness and vacations among the regular staff. She decided to randomly sample 96 of the 866 purchase orders for the month of August. The differences between the purchase orders and the records on the computer were recorded and listed here. The auditor would like to estimate the total error.

0	0	0	0	0	43.53	0	30.65	20.23	0
43.86	0	0	0	30.58	0	0	0	0	0
0	17.68	0	0	0	0	38.11	0	14.17	0
26.24	0	58.88	53.04	0	0	0	0	0	0
18.71	0	0	0	−20.4	0	0	0	0	0
0	0	0	20.10	0	13.11	0	0	0	46.69
31.59	−25.1	0	0	0	50.08	0	30.21	0	
0	−31.2	0	0	0	0	0	−27	15.65	
0	−20	0	39.26	0	0	0	0	0	
20.72	−30	54.33	26.19	0	30.1	0	0	0	

SOLUTION

IDENTIFY

The objective is to describe the population of errors, which are interval. The parameter to be estimated is the population mean, from which the total will be estimated. Because the population is less than 20 times the sample size, we will use the small-population formula.

COMPUTE

MANUALLY

The sum and sum of squares are

$$\sum x_i = 620 \qquad \text{and} \qquad \sum x_i^2 = 33{,}366$$

The sample mean, variance, and standard deviation are

$$\bar{x} = \frac{\sum x_i}{n} = \frac{620}{96} = 6.46$$

$$s^2 = \frac{\sum x_i^2 - \frac{\left(\sum x_i\right)^2}{n}}{n-1} = \frac{33{,}366 - \frac{(620)^2}{96}}{96-1} = 309.06$$

$$s = \sqrt{s^2} = \sqrt{309.06} = 17.58$$

With $n = 96$, we have $t_{\alpha/2,n-1} = t_{.025,96-1} = t_{.025,95} \approx t_{.025,100} = 1.984$. The 95% confidence interval estimate of the total is

$$N\left(\bar{x} \pm t_{\alpha/2}\frac{s}{\sqrt{n}}\sqrt{\frac{N-n}{N-1}}\right) = 866\left(6.46 \pm 1.984\frac{17.58}{\sqrt{96}}\sqrt{\frac{866-96}{866-1}}\right)$$

$$= 866(6.46 \pm 3.36)$$

$$= 5{,}594.36 \pm 2{,}909.76$$

or

$$\text{LCL} = 2{,}684.60 \quad \text{and} \quad \text{UCL} = 8{,}504.12$$

EXCEL

	A	B	C	D	E
1	t-Estimate of a Total				
2					
3	Sample mean	6.46	Confidence Interval Estimate		
4	Sample standard deviation	17.58	5594.36	±	2910.41
5	Sample size	96	Lower confidence limit		2683.95
6	Population size	866	Upper confidence limit		8504.77
7	Confidence level	0.95			

INSTRUCTIONS

Calculate the sample mean and sample standard deviation. Open the **Estimators Small Population** workbook and the **t-Estimate_Total** worksheet. Type the values of $\bar{x}$, s, n, N, and $1 - \alpha$ into cells B3 to B7, respectively.

MINITAB

Minitab does not compute confidence interval estimator of the mean of a small population.

INTERPRET

We estimate that the total error during the month of August lies between $2,683.95 and $8,504.77.

Estimating the Total Number of Successes in a Population

Another way to apply statistics in auditing is to estimate the number of items that are in error. In the previous section, we used the confidence interval estimator of a proportion to estimate the total number of successes in a large finite population. If the population is small (the population is less than 20 times the sample size), we include the finite population correction factor in the estimator of a proportion.

Confidence Interval Estimator of p in a Small Population

$$\hat{p} \pm z_{\alpha/2}\sqrt{\frac{\hat{p}(1-\hat{p})}{n}}\sqrt{\frac{N-n}{N-1}}$$

We can now estimate the total number of successes in a small population by multiplying the proportion by the population size.

Confidence Interval Estimator of the Total Number of Successes in a Small Population

$$N\left(\hat{p} \pm z_{\alpha/2}\sqrt{\frac{\hat{p}(1-\hat{p})}{n}}\sqrt{\frac{N-n}{N-1}}\right)$$

EXAMPLE 12.9

Audit of Work Orders at a Car Dealership

The auditor in Example 12.8 also reviewed the work orders to determine whether there were any deviations from the standard operating procedure. These include repairs made without customer authorization, incorrect codes used for parts, and others. Records indicate that the number of work orders for the year totaled 11,054. A random sample of 750 work orders was audited and 87 were found to have some irregularities. Estimate with 95% confidence the total number of work orders with irregularities.

SOLUTION

IDENTIFY

The objective is to describe the population of work orders, which are nominal. Thus, the parameter to be estimated is the population proportion of work orders with irregularities. The confidence interval estimator of the proportion will be used to estimate the total number of work orders for the year that have irregularities. The population is smaller than 20 times the sample size; we will use the small population technique.

COMPUTE

MANUALLY

The sample proportion is

$$\hat{p} = \frac{87}{750} = .116$$

The 95% confidence interval estimator of the total number of work orders with irregularities is

$$N\left(\hat{p} \pm z_{\alpha/2}\sqrt{\frac{\hat{p}(1-\hat{p})}{n}}\sqrt{\frac{N-n}{N-1}}\right)$$

$$= 11{,}054\left(.116 \pm 1.96\sqrt{\frac{.116(1-.116)}{750}}\sqrt{\frac{11{,}054-750}{11{,}054-1}}\right)$$

$$= 11{,}054\,(.116 \pm .022) = 1{,}282 \pm 243$$

The lower and upper confidence limits are

$$\text{LCL} = 1{,}039 \qquad \text{UCL} = 1{,}525$$

EXCEL

	A	B	C	D	E
1	z-Estimate of a Total				
2					
3	Sample proportion	0.116	Confidence Interval Estimate		
4	Sample size	750	1282	±	245
5	Population size	11054	Lower confidence limit		1038
6	Confidence level	0.95	Upper confidence limit		1527

INSTRUCTIONS

Open the **Estimators Small Population** workbook and the **z-Estimate_Total** worksheet. Type the values of $\hat{p}$, n, N, and $1 - \alpha$ into cells B3 to B6, respectively.

MINITAB

Minitab does not produce this confidence interval estimator.

INTERPRET

The total number of work orders with irregularities is estimated to lie between 1,038 and 1,527.

EXERCISES

12.96 A government tax auditor is examining the books of a large auto-wrecking yard. Auto wreckers buy cars that have been in collisions severe enough to render the car unfit. They then remove parts that are in good working order and sell them to repair facilities. The auditor believes that some expensive parts are sold for cash that is not reported on the firm's tax return. The auditor notes that the records indicate that a total of 2,453 engines have been removed from wrecks and the serial numbers recorded. She randomly samples 80 serial numbers and determines whether they are in inventory. She finds 4 are missing. She would like a 95% confidence interval estimate of the total number of engines missing.

12.97 Many employees are part of a company drug plan wherein some portion of their costs of prescription drugs is covered. However, some drugs and over-the-counter-medications are not covered. An audit of an insurance company's prescription coverage reveals that in the past year there were 1,258,401 requests for reimbursements, of which 118,653 were approved. A random sample of 559 reimbursement requests that were approved was drawn. It was determined that 29 of these should not have been approved.

a. Estimate with 99% confidence the total number of requests that were improperly approved.

b. Describe your findings in a report to the insurance company's executives.

12.98 Xr12-98 After a bank demanded payment on a loan, an auto wrecker was forced into bankruptcy. The trustee was given the task of estimating the value of all the items in inventory. Computer records indicate that there is a total of 73,544 items in inventory. A random sample of 500 items is drawn and their value recorded. Estimate with 95% confidence the value of the inventory.

12.99 When a professor is awarded a grant to help fund his or her research, the university is given the money to administer. To spend the money, the professor must authorize the university to pay for equipment or labor. The accountant for a large university was given the task of auditing the expenses. The university is currently administering a total of 2,490 grants. The accountant takes a random sample of 125 of these and determines that 14 were not properly authorized. Estimate with 90% confidence the total number of grants that contained improperly awarded funds.

12.100 Most banks use credit scoring to make decisions about loan requests. Although the number of financial institutions that allow branch managers to override the credit scorecard is decreasing, there are a number of banks that do allow the practice. One such bank with many branches wanted to determine whether this practice has led to losses. The accountant took a random sample of 85 of the 1,864 loans that were granted after the manager overrode the credit scorecard. The number of loans that resulted in a default was 5.

a. Calculate the 90% confidence interval estimate of the total number of loans granted by overriding the scorecard that resulted in default.
b. Prepare a presentation to the bank's board of directors describing your findings.

12.101 Xr12-101 The auditor in Example 12.7 was asked to report on the amounts that were split transactions, which are purchases made by using more than one purchase card. There were 1,431 split transactions, of which the auditor sampled 100. The GAO wants a 95% confidence interval estimate of the total in split transactions.

12.102 Financial institutions rely on credit bureaus to provide data for the scorecards that those institutions use to make loan decisions. A bank was recently encountering problems and wondered whether they were receiving accurate reports. With the approval of the credit bureau, the bank appointed an auditor. In the past month the bank requested information about 3,745 individuals. The auditor randomly sampled 200 of these and found 18 with errors. The auditor would like to know the total number of individual accounts with errors. A 95% confidence level is considered appropriate.

12.103 Xr12-103 Bursaries are loans made to students who need the money to support themselves while attending universities and colleges. The loans are made by banks that are guaranteed repayment by the government. A government auditor wanted to know the status of the loans made to students at one university. She randomly sampled 188 of the 2,684 loans that are outstanding. The total amounts owing were recorded. Estimate with 90% confidence the total outstanding amount of the loans made to students at the university.

12.104 In a recent municipal election, the list of "eligible" voters contained 102,412 names. However, the list may contain a substantial number of names of voters who, for one reason or another, are ineligible. To examine the problem, an auditor was hired to delve into the matter. He took a random sample of 317 names from the list and determined that 38 of them were ineligible. The municipality would like a 95% confidence interval estimate of the total number of ineligible voters.

CHAPTER SUMMARY

The inferential methods presented in this chapter address the problem of describing a single population. When the data are interval, the parameters of interest are the population mean μ and the population variance σ^2. The Student t distribution is used to test and estimate the mean when the population standard deviation is unknown. The chi-squared distribution is used to make inferences about a population variance. When the data are nominal, the parameter to be tested and estimated is the population proportion p. The sample proportion follows an approximate normal distribution, which produces the test statistic and the interval estimator. We also discussed how to determine the sample size required to estimate a population proportion. We introduced market segmentation and described how statistical techniques presented in this chapter can be used to estimate the size of a segment. Finally, we described how statistical methods are used in auditing.

IMPORTANT TERMS

t-statistic 383
Student t distribution 383
Robust 389
Chi-squared statistic 396

SYMBOLS

Symbol	Pronounced	Represents
ν	*nu*	Degrees of freedom
χ^2	*chi-squared*	Chi-squared statistic
$\hat{p}$	*p-hat*	Sample proportion
$\tilde{p}$	*p-tilde*	Wilson estimator

FORMULAS

Test statistic for μ

$$t = \frac{\bar{x} - \mu}{s/\sqrt{n}}$$

Confidence interval estimator of μ

$$\bar{x} \pm t_{\alpha/2}\frac{s}{\sqrt{n}}$$

Test statistic for σ^2

$$\chi^2 = \frac{(n-1)s^2}{\sigma^2}$$

Confidence interval estimator of σ^2

$$\text{LCL} = \frac{(n-1)s^2}{\chi^2_{\alpha/2}}$$

$$\text{UCL} = \frac{(n-1)s^2}{\chi^2_{1-\alpha/2}}$$

Test statistic for p

$$z = \frac{\hat{p} - p}{\sqrt{p(1-p)/n}}$$

Confidence interval estimator of p

$$\hat{p} \pm z_{\alpha/2}\sqrt{\hat{p}(1-\hat{p})/n}$$

Sample size to estimate p

$$n = \left(\frac{z_{\alpha/2}\sqrt{\hat{p}(1-\hat{p})}}{B}\right)^2$$

Wilson estimator

$$\tilde{p} = \frac{x+2}{n+4}$$

Confidence interval estimator of p using the Wilson estimate

$$\tilde{p} \pm z_{\alpha/2}\sqrt{\tilde{p}(1-\hat{p})/(n+4)}$$

Confidence interval estimator of the total of a large finite population

$$N\left(\bar{x} \pm t_{\alpha/2}\frac{s}{\sqrt{n}}\right)$$

Confidence interval estimator of the total number of successes in a large finite population

$$N\left(\hat{p} \pm z_{\alpha/2}\sqrt{\frac{\hat{p}(1-\hat{p}}{n}}\right)$$

Confidence interval estimator of μ when the population is small

$$\bar{x} \pm t_{\alpha/2}\frac{s}{\sqrt{n}}\sqrt{\frac{N-n}{N-1}}$$

Confidence interval estimator of the total in a small population

$$N\left(\bar{x} \pm t_{\alpha/2}\frac{s}{\sqrt{n}}\sqrt{\frac{N-n}{N-1}}\right)$$

Confidence interval estimator of p when the population is small

$$\hat{p} \pm z_{\alpha/2}\sqrt{\frac{\hat{p}(1-\hat{p})}{n}}\sqrt{\frac{N-n}{N-1}}$$

Confidence interval estimator of the total number of successes in a small population

$$N\left(\hat{p} \pm z_{\alpha/2}\sqrt{\frac{\hat{p}(1-\hat{p})}{n}}\sqrt{\frac{N-n}{N-1}}\right)$$

COMPUTER OUTPUT AND INSTRUCTIONS

Technique	Excel	Minitab
t-test of μ	385	386
t-estimator of μ	388	388
Chi-squared test of σ^2	399	399
Chi-squared estimator of σ^2	401	401
z-test of p	407	408
z-estimator of p	410	410

We present the flowchart in Figure 12.7 as part of our ongoing effort to help you identify the appropriate statistical technique. This flowchart shows the techniques introduced in this chapter only. As we add new techniques in the upcoming chapters, we will expand this flowchart until it contains all the statistical inference techniques covered in this book. Use the flowchart to select the correct method in the chapter exercises that follow.

FIGURE **12.7**
Flowchart of Techniques in Chapter 12

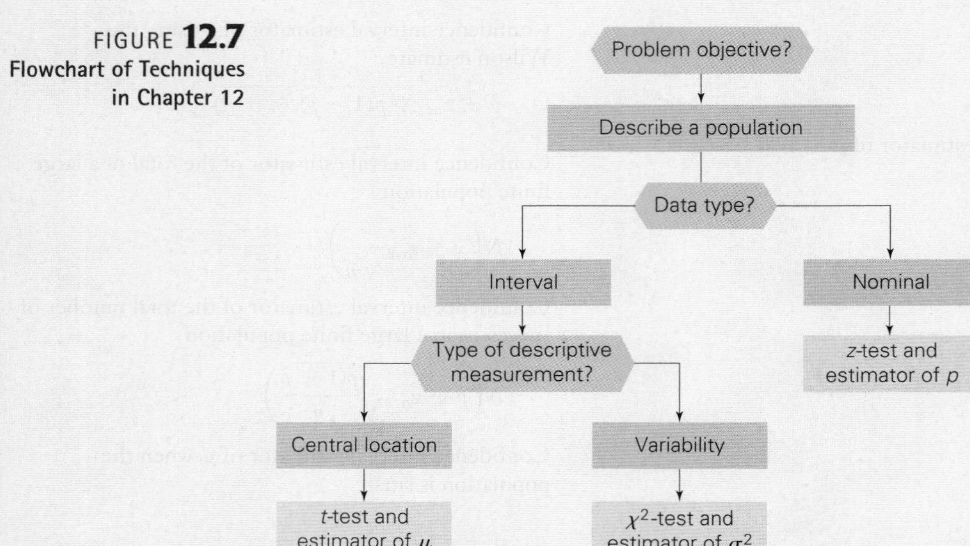

CHAPTER EXERCISES

The following exercises require the use of a computer and software. ***Use a 5% significance level unless specified otherwise.***

12.105 Xr12-105 One of the issues that came up in a recent municipal election was the high cost of housing. A candidate seeking to unseat an incumbent claimed that the average family spends more than 30% of its annual income on housing. A housing expert was asked to investigate the claim. A random sample of 125 households was drawn, and each household was asked to report the percentage of household income spent on housing costs.

a. Is there enough evidence to infer that the candidate is correct?
b. Using a confidence level of 95%, estimate the mean percentage of household income spent on housing by all households.
c. What is the required condition for the techniques used in Parts a and b? Use a graphical technique to check whether it is satisfied.

12.106 Xr12-106 The "just-in-time" policy of inventory control (developed by the Japanese) is growing in popularity. For example, General Motors recently spent \$2 billion on its Oshawa, Ontario, plant so that it will be less than 1 hour from most suppliers. Suppose that an automobile parts supplier claims to deliver parts to any manufacturer in an average time of less than 1 hour. In an effort to test the claim, a manufacturer recorded the times (in minutes) of 24 deliveries from this supplier. Can we conclude that the supplier's assertion is correct?

12.107 Xr12-107 Robots are being used with increasing frequency on production lines to perform monotonous tasks. To determine whether a robot welder should replace human welders in producing automobiles, an experiment was performed. The time for the robot to complete a series of welds was found to be 38 seconds. A random sample of 20 workers was taken, and the time for each worker to complete the welds was measured. The mean was calculated to be 38 seconds, the same as the robot's time. However, the robot's time did not vary, whereas there was variation among the workers' times. An analysis of the production line revealed that if the variance exceeds 17 seconds^2, there will be problems. Perform an analysis of the data, and determine whether problems using human welders are likely.

12.108 Xr12-108 Opinion Research International surveyed people whose household incomes exceed $50,000 and asked each for their top money-related new year's resolutions. The responses are

1. Get out of credit card debt
2. Retire before age 65
3. Die broke
4. Make do with current finances
5. Look for higher paying job

Estimate with 90% confidence the proportion of people whose household incomes exceed $50,000 whose top money-related resolution is to get out of credit card debt.

12.109 Xr12-109 Suppose that, in a large state university (with numerous campuses), the marks in an introductory statistics course are normally distributed with a mean of 68%. To determine the effect of requiring students to pass a calculus test (which at present is not a prerequisite), a random sample of 50 students who have taken calculus is given a statistics course. The marks out of 100 were recorded.

a. Estimate with 95% confidence the mean statistics mark for all students who have taken calculus.
b. Do these data provide evidence to infer that students with a calculus background would perform better in statistics than students with no calculus?

12.110 Xr12-110 Duplicate bridge is a game in which players compete for master points. When a player receives 300 master points, he or she becomes a life master. Since that title comes with a year's free subscription to the American Contract Bridge League's (ACBL) monthly bulletin, the ACBL is interested in knowing the status of nonlife masters. Suppose that a random sample of 80 nonlife masters was asked how many master points they have. The ACBL would like an estimate of the mean number of master points held by all nonlife masters. A confidence level of 90% is considered adequate in this case.

12.111 Xr12-111 A national health care system was an issue in the 1992 presidential election campaign and is likely to be a subject of debate for many years. The issue arose because of the large number of Americans who have no health insurance. Under the present system, free health care is available to poor people, whereas relatively well-off Americans buy their own health insurance. Those who are considered working poor and who are in the lower-middle-class economic stratum appear to be most unlikely to have adequate medical insurance. To investigate this problem, a statistician surveyed 250 families whose gross income last year was between $10,000 and $15,000. Family heads were asked whether they have medical insurance coverage (2 = Has medical insurance and 1 = Doesn't have medical insurance). The statistics practitioner wanted an estimate of the fraction of all families whose incomes are in the range of $10,000 to $15,000 who have medical insurance. Perform the necessary calculations to produce an interval estimate with 90% confidence.

12.112 Xr12-112 The routes of postal deliverers are carefully planned so that each deliverer works between 7 and 7.5 hours per shift. The planned routes assume an average walking speed of 2 miles per hour and no shortcuts across lawns. In an experiment to examine the amount of time deliverers actually spend completing their shifts, a random sample of 75 postal deliverers was secretly timed.

a. Estimate with 99% confidence the mean shift time for all postal deliverers.
b. Check to determine whether the required condition for this statistical inference is satisfied.
c. Is there enough evidence at the 10% significance level to conclude that postal workers are on average spending less than 7 hours per day doing their jobs?

12.113 Xr12-113 As you can easily appreciate, the number of Internet users is rapidly increasing. A recent survey reveals that there are about 50 million Internet users in North America. Suppose that a survey of 200 of these people asked them to report the number of hours they spent on the Internet last week. Estimate with 95% confidence the annual total amount of time spent by all North Americans on the Internet.

12.114 Xr12-114 The manager of a branch of a major bank wants to improve service. She is thinking about giving $1 to any customer who waits in line for a period of time that is considered excessive. (The bank ultimately decided that more than 8 minutes is excessive.) However, to get a better idea about the level of current service, she undertakes a survey of customers. A student is hired to measure the time spent waiting in line by a random sample of 50 customers. Using a stopwatch, the student determined the amount of time between the time the customer joined the line and the time he or she reached the teller. The times were recorded. Construct a 90% confidence interval estimate of the mean waiting time for the bank's customers.

12.115 Xr12-115 In an examination of consumer loyalty in the travel business, 72 first-time visitors to a tourist attraction were asked whether they planned to return. The responses were recorded where 2 = Yes and 1 = No. Estimate with 95% confidence the proportion of all first-time visitors who planned to return to the same destination.

12.116 Xr12-116 Engineers who are in charge of the production of springs used to make car seats are concerned about the variability in the length of the springs. The springs are designed to be 500 mm long. When the springs are too long, they will loosen and fall out. When they are too short, they will not fit into the frames. The springs that are too long and too short must be reworked at considerable additional cost. The engineers have calculated that a standard deviation of 2 mm will result in an acceptable number of springs that must be reworked. A random sample of 100 springs was measured. Can we infer at the 5% significance level that the number of springs requiring reworking is unacceptably large?

12.117 Xr12-117 Refer to Exercise 12.116. Suppose the engineers recoded the data so that springs that were the correct length were recorded as 1, springs that were too long were recorded as 2, and springs that were too short were recorded as 3. Can we infer at the 10% significance level that less than 90% of the springs are the correct length?

12.118 Xr12-118 An advertisement for a major home appliance manufacturer claims that its repair personnel are the loneliest in the world because its appliances require the smallest number of service calls. To examine this claim, a researcher drew a random sample of 100 owners of 5-year-old washing machines. The number of service calls made in the 5-year period were recorded. Find the 90% confidence interval estimate of the mean number of service calls for all 5-year-old washing machines.

12.119 Xr12-119 An oil company sends out monthly statements to its customers who purchased gasoline and other items using the company's credit card. Until now, the company has not included a preaddressed envelope for returning payments. The average and the standard deviation of the number of days before payment is received are 9.8 and 3.2, respectively. As an experiment to determine whether enclosing preaddressed envelopes speeds up payment, 150 customers selected at random were sent preaddressed envelopes with their bills. The numbers of days to payment were recorded.

a. Do the data provide sufficient evidence at the 10% level of significance to establish that enclosure of preaddressed envelopes improves the average speed of payments?
b. Can we conclude at the 10% significance level that the variability in payment speeds decreases when a preaddressed envelope is sent?

12.120 A rock promoter is in the process of deciding whether to book a new band for a rock concert. He knows that this band appeals almost exclusively to teenagers. According to the latest census, there are 400,000 teenagers in the area. The promoter decides to do a survey to try to estimate the proportion of teenagers who will attend the concert. How large a sample should be taken in order to estimate the proportion to within .02 with 99% confidence?

12.121 Xr12-121 In Exercise 12.120, suppose that the promoter decided to draw a sample of size 600 (because of financial considerations). Each teenager was asked whether he or she would attend the concert (2 = Yes, I will attend; 1 = No, I will not attend). Estimate with 95% confidence the number of teenagers who will attend the concert.

12.122 Xr12-122 The owner of a downtown parking lot suspects that the person he hired to run the lot is stealing some money. The receipts as provided by the employee indicate that the average number of cars parked in the lot is 125 per day and that, on average, each car is parked for 3.5 hours. To determine whether the employee is stealing, the owner watches the lot for 5 days. On those days, the numbers of cars parked are as follows:

120 130 124 127 128

The time spent on the lot for the 629 cars that the owner observed during the 5 days was recorded. Can the owner conclude at the 1% level of significance that the employee is stealing? (*Hint:* Since there are two ways to steal, two tests should be performed.)

12.123 Xr12-123 Jim Cramer hosts CNBC's *Mad Money* program. Mr. Cramer regularly makes suggestions about which stocks to buy and sell. How well have Mr. Cramer's picks performed over the past two years (2005 to 2007)? To answer the question, a random sample of Mr. Cramer's picks was selected. The name of the stock, the buy price of the stock, the current or sold price, and the percent return were recorded. (*Source:* YourMoneyWatch.com)

a. Estimate with 95% confidence the mean return for all of Mr. Cramer's selections.
b. Over the two-year period, the Standard and Poor 500 Index rose by 16%. Is there sufficient evidence to infer that Mr. Cramer's picks have done less well?

CASE 12.1 Pepsi's Exclusivity Agreement with a University

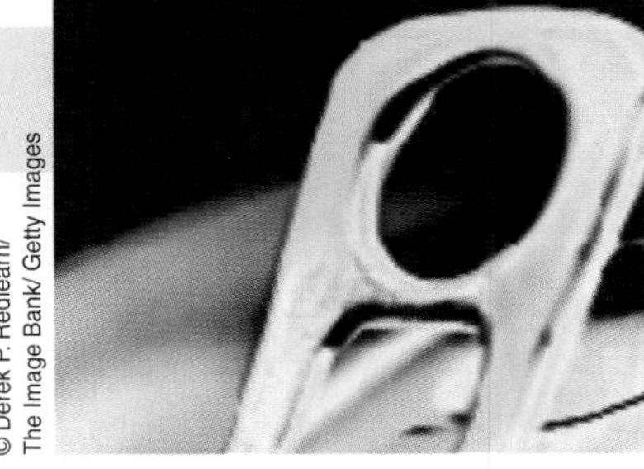

© Derek P. Redfearn/ The Image Bank/ Getty Images

DATA C12-01

In the last few years, colleges and universities have signed exclusivity agreements with a variety of private companies. These agreements bind the university to sell that company's products exclusively on the campus. Many of the agreements involve food and beverage firms.

A large university with a total enrollment of about 50,000 students has offered Pepsi-Cola an exclusivity agreement, which would give Pepsi exclusive rights to sell its products at all university facilities for the next year and an option for future years. In return the university would receive 35% of the on-campus revenues and an additional lump sum of $200,000 per year. Pepsi has been given 2 weeks to respond.

The management at Pepsi quickly reviews what it knows. The market for soft drinks is measured in terms of the equivalent of 12-ounce cans. Pepsi currently sells an average of 22,000 cans or their equivalents per week (over the 40 weeks of the year that the university operates). The cans sell for an average of $.75 each. The costs including labor amount to $.20 per can. Pepsi is unsure of its market share but suspects it is considerably less than 50%. A quick analysis reveals that if their current market share were 25%, then with an exclusivity agreement Pepsi would sell 88,000 cans per week. Thus, annual sales would be or 3,520,000 cans per year (calculated as 88,000 cans per week × 40 weeks). The gross revenue would be computed as follows:*

Gross revenue = 3,520,000 cans × $.75 revenue/can = $2,640,000

This figure must be multiplied by 65% since the university would rake in 35% of the gross. Thus,

65% × $2,640,000 = $1,716,000

The total cost of 20 cents per can (or $704,000) and the annual payment to the university of $200,000 is subtracted to obtain the net profit:

Net profit = $1,716,000 − $704,000 − $200,000 = $812,000

Their current annual profit is

Current profit = 40 weeks × 22,000 cans/week × $.55/can = $484,000

If the current market share is 25%, the potential gain from the agreement is

$812,000 − $484,000 = $328,000

The only problem with this analysis is that Pepsi does not know how many soft drinks are sold weekly at the university. In addition, Coke is not likely to supply Pepsi with information about its sales, which together with Pepsi's line of products constitutes virtually the entire market.

A recent graduate of a business program believes that a survey of the university's students can supply the needed information. Accordingly, she organizes a survey that asks 500 students to keep track of the number of soft drinks they purchase on campus over the next 7 days. Perform a statistical analysis to extract the needed information from the data. Estimate with 95% confidence the parameter that is at the core of the decision problem. Use the estimate to compute estimates of the annual profit. Assume that Coke and Pepsi drinkers would be willing to buy either product in the absence of their first choice.

a. On the basis of maximizing profits from sales of soft drinks at the university, should Pepsi agree to the exclusivity agreement?
b. Write a report to the company's executives describing your analysis.

*We have created an Excel spreadsheet that does the calculations for this case. To access it, click **Excel Workbooks** and **Case 12.1.** The only cell you may alter is cell C3, which contains the average number of soft drinks sold per week per student, assuming a total of 88,000 drinks sold per year.

CASE 12.2 Pepsi's Exclusivity Agreement with a University: The Coke Side of the Equation

© Comstock Images/Royalty-free/ Getty Images

DATA C12-01

While the executives of Pepsi Cola are trying to decide what to do, the university informs them that a similar offer has gone out to the Coca Cola Company. Furthermore, if both companies want exclusive rights, then a bidding war will take place. The executives at Pepsi would like to know how likely it is that Coke will want exclusive rights under the conditions outlined by the university.

Perform a similar analysis to the one you did in Case 12.1, but this time from Coke's point of view. Is it likely that Coke will want to conclude an exclusivity agreement with the university? Discuss the reasons for your conclusions.

CASE 12.3 Estimating Total Medical Costs

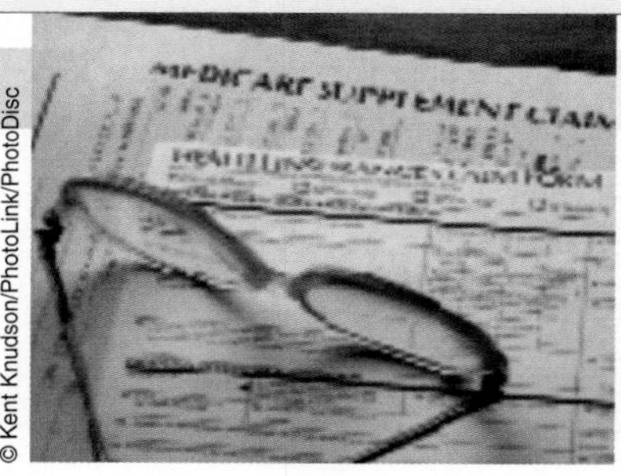
© Kent Knudson/PhotoLink/PhotoDisc

DATA C12-03

Virtually, all countries have universal government-run health care systems. The United States is one notable exception. This is an issue in every election, with some politicians pushing for the United States to adopt a program similar to Canada's.

In Canada, hospitals are financed and administered by provincial governments. Physicians are paid by the government for each patient service. As a result, Canadians pay nothing for these services. The revenues that support the system are derived through income tax, corporate taxes, and sales taxes. Despite higher taxes in Canada than those in the United States, the system is chronically underfunded, resulting in long waiting times for sometimes, critical procedures. For example, in some provinces newly diagnosed cancer victims must wait several weeks before treatments can begin. Virtually everyone agrees that more money is needed. No one can agree however, on how much is needed. Unfortunately, the problem is going to worsen. Canada, like the United States, has an aging population due to the large numbers constituting the baby boom (those born between 1946 and 1966), and because medical costs are generally higher for older people.

One of the first steps in addressing the problem is to forecast medical costs, particularly for the 20-year period starting when the first baby boomers reaches 60 years of age (in the year 2006). A statistics practitioner has been given the task of making these predictions. Accordingly, random samples of four groups of Canadians were drawn. They are

Group	Ages
1	45–64
2	65–74
3	75–84
4	85 and over

The medical expenses for the previous 12 months were recorded and stored in

Age Category	2011	2016	2021	2026	2031
45–64	9,718	10,013	10,065	9,996	10,016
65–74	2,644	3,344	3,992	4,511	4,846
75–84	1,600	1,718	2,045	2,627	3,169
85+	639	738	810	909	1,121

Source: Statistics Canada.

columns 1 to 4, respectively in C12-03. Projections for 2011, 2016, 2021, 2026, and 2031 of the numbers of Canadians (in thousands) in each age category are listed above.

a. Determine the 95% confidence interval estimates of the mean medical costs for each of the four age categories.

b. For each year listed, determine the 95% confidence interval estimates of the total medical costs for Canadians 45 years old and older.

13

© mason morfit 2002/Workbook Stock/Jupiterimages

INFERENCE ABOUT COMPARING TWO POPULATIONS

Estimating the Cost of a Life Saved

Two drugs are used to treat heart attack victims. Streptokinase, which has been available since 1959, costs about $460 per dose. The second drug is t-PA, a genetically engineered product that sells for about $2,900 per dose. Both Streptokinase and t-PA work by opening the arteries and dissolving blood clots, which are the cause of heart attacks. Several previous studies have failed to reveal any differences between the effects of the two drugs. Consequently, in many countries where health care is funded by governments, physicians are required to use the less expensive Streptokinase. However, t-PA's maker, Genentech Inc., contended that in the earlier studies showing no difference between the two drugs, t-PA was not used in the right way. Genentech decided to sponsor a more thorough experiment. The experiment was organized in 15 countries,

© Michael Agliolo/ International Stock

The solution is shown on page 492.

including the United States and Canada, and involved a total of 41,000 patients. In this study, t-PA was given to patients in 90 minutes instead of 3 hours as in previous trials. Half of the sample of 41,000 patients was treated by a rapid injection of t-PA with intravenous heparin, while the other half received Streptokinase along with heparin. The number of deaths in each sample was recorded. A total of 1,497 patients treated with Streptokinase died, while 1,292 patients who received t-PA died. Estimate the cost per life saved by using t-PA instead of Streptokinase.

INTRODUCTION

We can compare learning how to use statistical techniques to learning how to drive a car. We began by describing what you are going to do in this course (Chapter 1), and then presented the essential background material (Chapters 2–9). Learning the concepts of statistical inference and applying them the way we did in Chapters 10 and 11 is akin to driving a car in an empty parking lot. You're driving, but it's not a realistic experience. Learning Chapter 12 is like driving on a quiet side street with little traffic. The experience represents real driving, but many of the difficulties have been eliminated. In this chapter, you begin to drive for real, with many of the actual problems faced by licensed drivers, and the experience prepares you to tackle the next difficulty.

In this chapter, we present a variety of techniques whose objective is to compare two populations. In Sections 13.1 and 13.3, we deal with interval variables; the parameter of interest is the difference between two means. The difference between these two sections introduces yet another factor that determines the correct statistical method—the design of the experiment used to gather the data. In Section 13.1, the samples are independently drawn, whereas in Section 13.3, the samples are taken from a matched pairs experiment. In Section 13.2, we discuss the difference between observational and experimental data, a distinction that is critical to the way in which we interpret statistical results.

Section 13.4 presents the procedures employed to infer whether two population variances differ. The parameter is the ratio σ_1^2/σ_2^2. (When comparing two variances, we use the ratio rather than the difference because of the nature of the sampling distribution.)

Section 13.5 addresses the problem of comparing two populations of nominal data. The parameter to be tested and estimated is the difference between two proportions.

13.1 INFERENCE ABOUT THE DIFFERENCE BETWEEN TWO MEANS: INDEPENDENT SAMPLES

In order to test and estimate the difference between two population means, the statistics practitioner draws random samples from each of two populations. In this section, we discuss independent samples. In Section 13.3, where we present the matched pairs experiment, the distinction between independent samples and matched pairs will be made clear. For now, we define independent samples as samples completely unrelated to one another.

Figure 13.1 depicts the sampling process. Observe that we draw a sample of size n_1 from population 1 and a sample of size n_2 from population 2. For each sample, we compute the sample means and sample variances.

FIGURE **13.1**
Independent Samples from Two Populations

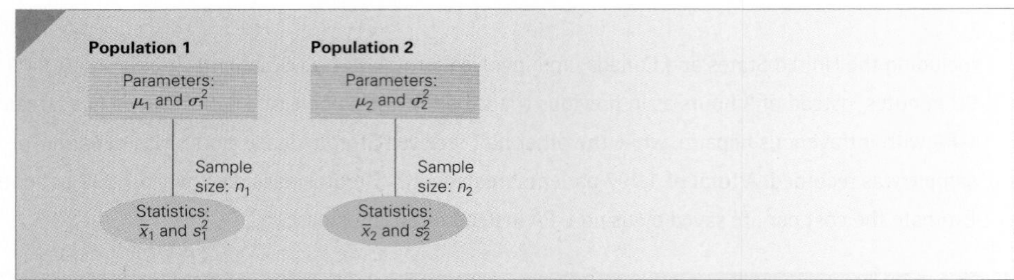

The best estimator of the difference between two population means $\mu_1 - \mu_2$ is the difference between two sample means $\bar{x}_1 - \bar{x}_2$. In Section 9.3 we presented the sampling distribution of $\bar{x}_1 - \bar{x}_2$.

Sampling Distribution of $\bar{x}_1 - \bar{x}_2$

1. $\bar{x}_1 - \bar{x}_2$ is normally distributed if the populations are normal and approximately normal if the populations are nonnormal and the sample sizes are large.
2. The expected value of $\bar{x}_1 - \bar{x}_2$ is

$$E(\bar{x}_1 - \bar{x}_2) = \mu_1 - \mu_2$$

3. The variance of $\bar{x}_1 - \bar{x}_2$ is

$$V(\bar{x}_1 - \bar{x}_2) = \frac{\sigma_1^2}{n_1} + \frac{\sigma_2^2}{n_2}$$

The standard error of $\bar{x}_1 - \bar{x}_2$ is

$$\sqrt{\frac{\sigma_1^2}{n_1} + \frac{\sigma_2^2}{n_2}}$$

Thus,

$$z = \frac{(\bar{x}_1 - \bar{x}_2) - (\mu_1 - \mu_2)}{\sqrt{\frac{\sigma_1^2}{n_1} + \frac{\sigma_2^2}{n_2}}}$$

is a standard normal (or approximately normal) random variable. It follows that the test statistic is

$$z = \frac{(\bar{x}_1 - \bar{x}_2) - (\mu_1 - \mu_2)}{\sqrt{\frac{\sigma_1^2}{n_1} + \frac{\sigma_2^2}{n_2}}}$$

The interval estimator is

$$(\bar{x}_1 - \bar{x}_2) \pm z_{\alpha/2}\sqrt{\frac{\sigma_1^2}{n_1} + \frac{\sigma_2^2}{n_2}}$$

However, these formulas are rarely used because the population variances σ_1^2 and σ_2^2 are virtually always unknown. Consequently, it is necessary to estimate the standard error of the sampling distribution. (We've encountered this problem before; when we derived the confidence interval estimator of a population proportion in Section 12.3, it was necessary to estimate the standard error of $\hat{p}$.) The way to do this depends on whether the two unknown population variances are equal. When they are equal, the test statistic is defined in the following way.

Test Statistic for $\mu_1 - \mu_2$ When $\sigma_1^2 = \sigma_2^2$

$$t = \frac{(\bar{x}_1 - \bar{x}_2) - (\mu_1 - \mu_2)}{\sqrt{s_p^2\left(\frac{1}{n_1} + \frac{1}{n_2}\right)}} \qquad \nu = n_1 + n_2 - 2$$

where

$$s_p^2 = \frac{(n_1 - 1)s_1^2 + (n_2 - 1)s_2^2}{n_1 + n_2 - 2}$$

The quantity s_p^2 is called the **pooled variance estimator.** It is the weighted average of the two sample variances with the number of degrees of freedom in each sample used as weights. The requirement that the population variances be equal makes this calculation feasible, because we need only one estimate of the common value of σ_1^2 and σ_2^2. It makes sense for us to use the pooled variance estimator because, in combining both samples, we produce a better estimate.

The test statistic is Student t distributed with $n_1 + n_2 - 2$ degrees of freedom, provided that the two populations are normal. The confidence interval estimator is derived by mathematics that by now has become routine.

Confidence Interval Estimator of $\mu_1 - \mu_2$ When $\sigma_1^2 = \sigma_2^2$

$$(\bar{x}_1 - \bar{x}_2) \pm t_{\alpha/2}\sqrt{s_p^2\left(\frac{1}{n_1} + \frac{1}{n_2}\right)} \qquad \nu = n_1 + n_2 - 2$$

We will refer to these formulas as the **equal-variances test statistic** and **confidence interval estimator,** respectively.

When the population variances are unequal, we cannot use the pooled variance estimate. Instead, we estimate each population variance with its sample variance. Unfortunately, the sampling distribution of the resulting statistic

$$\frac{(\bar{x}_1 - \bar{x}_2) - (\mu_1 - \mu_2)}{\sqrt{\frac{s_1^2}{n_1} + \frac{s_2^2}{n_2}}}$$

is neither normally nor Student t distributed. However, it can be approximated by a Student t distribution with degrees of freedom equal to

$$\nu = \frac{\left(s_1^2/n_1 + s_2^2/n_2\right)^2}{\frac{\left(s_1^2/n_1\right)^2}{n_1 - 1} + \frac{\left(s_2^2/n_2\right)^2}{n_2 - 1}}$$

(It is usually necessary to round this number to the nearest integer.) The test statistic and confidence interval estimator are easily derived from the sampling distribution.

Test Statistic for $\mu_1 - \mu_2$ When $\sigma_1^2 \neq \sigma_2^2$

$$t = \frac{(\bar{x}_1 - \bar{x}_2) - (\mu_1 - \mu_2)}{\sqrt{\left(\frac{s_1^2}{n_1} + \frac{s_2^2}{n_2}\right)}} \qquad \nu = \frac{\left(s_1^2/n_1 + s_2^2/n_2\right)^2}{\frac{\left(s_1^2/n_1\right)^2}{n_1 - 1} + \frac{\left(s_2^2/n_2\right)^2}{n_2 - 1}}$$

Confidence Interval Estimator of $\mu_1 - \mu_2$ When $\sigma_1^2 \neq \sigma_2^2$

$$(\bar{x}_1 - \bar{x}_2) \pm t_{\alpha/2}\sqrt{\left(\frac{s_1^2}{n_1} + \frac{s_2^2}{n_2}\right)} \qquad \nu = \frac{\left(s_1^2/n_1 + s_2^2/n_2\right)^2}{\frac{\left(s_1^2/n_1\right)^2}{n_1 - 1} + \frac{\left(s_2^2/n_2\right)^2}{n_2 - 1}}$$

We will refer to these formulas as the **unequal-variances test statistic** and **confidence interval estimator,** respectively.

The question naturally arises, How do we know when the population variances are equal? The answer is that since σ_1^2 and σ_2^2 are unknown, we can't know for certain whether they're equal. However, we can perform a statistical test to determine whether there is evidence to infer that the population variances differ. We conduct the F-test of the ratio of two variances, which we briefly present here and save the details for Section 13.4.

Testing the Population Variances

The hypotheses to be tested are

$$H_0: \quad \sigma_1^2/\sigma_2^2 = 1$$
$$H_1: \quad \sigma_1^2/\sigma_2^2 \neq 1$$

The test statistic is the ratio of the sample variances s_1^2/s_2^2, which is F-distributed with degrees of freedom $\nu_1 = n_1 - 1$ and $\nu_2 = n_2 - 2$. Recall that we introduced the F-distribution in Section 8.4. The required condition is the same as that for the t-test of $\mu_1 - \mu_2$, which is that both populations are normally distributed.

This is a two-tail test so that the rejection region is

$$F > F_{\alpha/2,\nu_1,\nu_2} \qquad \text{or} \qquad F < F_{1-\alpha/2,\nu_1,\nu_2}$$

Put simply, we will reject the null hypothesis that states that the population variances are equal when the ratio of the sample variances is large or if it is small. Table 6 in Appendix B, which lists the critical values of the F distribution, defines "large" and "small."

Decision Rule: Equal-Variances or Unequal-Variances *t*-Tests and Estimators

Recall that we can never have enough statistical evidence to conclude that the null hypothesis is true. This means that we can only determine whether there is enough evidence to infer that the population variances *differ*. Accordingly, we adopt the following rule: We will use the equal-variances test statistic and confidence interval estimator unless there is evidence (based on the *F*-test of the population variances) to indicate that the population variances are unequal, in which case we will apply the unequal-variances test statistic and confidence interval estimator.

EXAMPLE 13.1*

DATA Xm13-01

Direct and Broker-Purchased Mutual Funds

Millions of investors buy mutual funds (see page 175 for a description of mutual funds), choosing from thousands of possibilities. Some funds can be purchased directly from banks or other financial institutions while others must be purchased through brokers, who charge a fee for this service. This raises the question, Can investors do better by buying mutual funds directly than by purchasing mutual funds through brokers? To help answer this question, a group of researchers randomly sampled the annual returns from mutual funds that can be acquired directly and mutual funds that are bought through brokers and recorded the net annual returns, which are the returns on investment after deducting all relevant fees. These are listed next.

Direct					Broker				
9.33	4.68	4.23	14.69	10.29	3.24	3.71	16.4	4.36	9.43
6.94	3.09	10.28	−2.97	4.39	−6.76	13.15	6.39	−11.07	8.31
16.17	7.26	7.1	10.37	−2.06	12.8	11.05	−1.9	9.24	−3.99
16.97	2.05	−3.09	−0.63	7.66	11.1	−3.12	9.49	−2.67	−4.44
5.94	13.07	5.6	−0.15	10.83	2.73	8.94	6.7	8.97	8.63
12.61	0.59	5.27	0.27	14.48	−0.13	2.74	0.19	1.87	7.06
3.33	13.57	8.09	4.59	4.8	18.22	4.07	12.39	−1.53	1.57
16.13	0.35	15.05	6.38	13.12	−0.8	5.6	6.54	5.23	−8.44
11.2	2.69	13.21	−0.24	−6.54	−5.75	−0.85	10.92	6.87	−5.72
1.14	18.45	1.72	10.32	−1.06	2.59	−0.28	−2.15	−1.69	6.95

Can we conclude at the 5% significance level that directly purchased mutual funds outperform mutual funds bought through brokers?

SOLUTION

IDENTIFY

To answer the question, we need to compare the population of returns from direct and the returns from broker-bought mutual funds. The data are obviously interval (we've recorded real numbers). This problem objective–data type combination tells us that the parameter to be tested is the difference between two means $\mu_1 - \mu_2$. The hypothesis to

*Source: D. Bergstresser, J. Chalmers, and P. Tufano, "Assessing the Costs and Benefits of Brokers in the Mutual Fund Industry."

be tested is that the mean net annual return from directly purchased mutual funds (μ_1) is larger than the mean of broker-purchased funds (μ_2). Hence, the alternative hypothesis is

$$H_1: \quad (\mu_1 - \mu_2) > 0$$

As usual, the null hypothesis automatically follows:

$$H_0: \quad (\mu_1 - \mu_2) = 0$$

To decide which of the t-tests of $\mu_1 - \mu_2$ to apply, we conduct the F-test of σ_1^2/σ_2^2:

$$H_0: \quad \sigma_1^2/\sigma_2^2 = 1$$
$$H_1: \quad \sigma_1^2/\sigma_2^2 \neq 1$$

COMPUTE

MANUALLY

From the data we calculated the following statistics:

$s_1^2 = 37.49$ and $s_2^2 = 43.34$

Test statistic: $F = s_1^2/s_2^2 = 37.49/43.34 = 0.86$

Rejection region: $F > F_{\alpha/2,\nu_1,\nu_2} = F_{.025,49,49} \approx F_{.025,50,50} = 1.75$

or

$$F < F_{1-\alpha/2,\nu_1,\nu_2} = F_{.975,49,49} = 1/F_{.025,49,49} \approx 1/F_{.025,50,50} = 1/1.75 = .57$$

Because $F = .86$ is not greater than 1.75 or smaller than .57, we cannot reject the null hypothesis.

EXCEL

	A	B	C
1	F-Test: Two-Sample for Variances		
2			
3		*Direct*	*Broker*
4	Mean	6.63	3.72
5	Variance	37.49	43.34
6	Observations	50	50
7	df	49	49
8	F	0.8650	
9	P(F<=f) one-tail	0.3068	
10	F Critical one-tail	0.6222	

The value of the test statistic is $F = .8650$. Excel outputs the one-tail p-value. Because we're conducting a two-tail test, we double that value. Thus, the p-value of the test we're conducting is $2 \times .3068 = .6136$.

INSTRUCTIONS

1. Type or import the data into one column. (Open Xm13-01.)
2. Click **Data, Data Analysis,** and **F-Test Two-Sample for Variances.**
3. Specify the **Variable 1 Range** (A1:A51) and the **Variable 2 Range** (B1:B51). Type a value for α (.05).

MINITAB

Test for Equal Variances: Direct, Broker

F-Test (Normal Distribution)
Test statistic = 0.86, p-value = 0.614

INSTRUCTIONS

(Note: Some of the printout has been omitted.)

1. Type or import the data into one column. (Open Xm13-01.)
2. Click **Stat, Basic Statistics,** and **2 Variances . . . .**
3. In the **Samples in different columns** box, select the **First** (Direct) and **Second** (Broker) variables.

INTERPRET

There is not enough evidence to infer that the population variances differ. It follows that we must apply the equal-variances t-test of $\mu_1 - \mu_2$.

The hypotheses are

$$H_0: \quad (\mu_1 - \mu_2) = 0$$
$$H_1: \quad (\mu_1 - \mu_2) > 0$$

COMPUTE

MANUALLY

From the data we calculated the following statistics:

$$\bar{x}_1 = 6.63$$
$$\bar{x}_2 = 3.72$$
$$s_1^2 = 37.49$$
$$s_2^2 = 43.34$$

The pooled variance estimator is

$$s_p^2 = \frac{(n_1 - 1)s_1^2 + (n_2 - 1)s_2^2}{n_1 + n_2 - 2}$$
$$= \frac{(50 - 1)37.49 + (50 - 1)43.34}{50 + 50 - 2}$$
$$= 40.42$$

The number of degrees of freedom of the test statistic is

$$\nu = n_1 + n_2 - 2 = 50 + 50 - 2 = 98$$

The rejection region is

$$t > t_{\alpha,\nu} = t_{.05,98} \approx t_{.05,100} = 1.660$$

We determine that the value of the test statistic is

$$t = \frac{(\bar{x}_1 - \bar{x}_2) - (\mu_1 - \mu_2)}{\sqrt{s_p^2\left(\frac{1}{n_1} + \frac{1}{n_2}\right)}}$$

$$= \frac{(6.63 - 3.72) - 0}{\sqrt{40.42\left(\frac{1}{50} + \frac{1}{50}\right)}}$$

$$= 2.29$$

EXCEL

	A	B	C
1	t-Test: Two-Sample Assuming Equal Variances		
2			
3		*Direct*	*Broker*
4	Mean	6.63	3.72
5	Variance	37.49	43.34
6	Observations	50	50
7	Pooled Variance	40.41	
8	Hypothesized Mean Difference	0	
9	df	98	
10	t Stat	2.29	
11	P(T<=t) one-tail	0.0122	
12	t Critical one-tail	1.6606	
13	P(T<=t) two-tail	0.0243	
14	t Critical two-tail	1.9845	

INSTRUCTIONS

1. Type or import the data into 2 columns. (Open Xm13-01.)
2. Click **Data, Data Analysis,** and **t-Test: Two-Sample Assuming Equal Variances.**
3. Specify the **Variable 1 Range (A1:A51)** and the **Variable 2 Range** (B1:B51). Type the value of the **Hypothesized Mean Difference*** (0) and type a value for α (.05).

To conduct this test from means and variances or to perform a what-if analysis, activate the **t-Test_2 Means (Eq-Var)** worksheet in the **Test Statistics** workbook (shown in CD Appendix J).

MINITAB

Two-Sample T-Test and CI: Direct, Broker

Two-sample T for Direct vs Broker

	N	Mean	StDev	SE Mean
Direct	50	6.63	6.12	0.87
Broker	50	3.72	6.58	0.93

Difference = mu (Direct) – mu (Broker)
Estimate for difference: 2.91
95% lower bond for difference: 0.80
T-Test of difference = 0 (vs >): T-Value = 2.29 P-Value = 0.012 DF = 98
Both use Pooled StDev = 6.3572

*This term is technically incorrect. Because we're testing $\mu_1 - \mu_2$, Excel should ask for and output the "Hypothesized Difference between Means."

INSTRUCTIONS

1. Type or import the data into 2 columns. (Open Xm13-01.)
2. Click **Stat, Basic Statistics,** and **2-Sample t . . . .**
3. If the data are stacked, use the **Samples in one column** box to specify the names of the variables. If the data are unstacked (as in Example 13.1) specify the **First** and **Second** variables in the **Samples in different columns** box (Direct, Broker). (See the discussion on Data Formats on page 454 for a discussion of stacked and unstacked data.) Click **Assume equal variances.** Click **Options . . . .**
4. In the **Test difference** box, type the value of the parameter under the null hypothesis (0), and select one of **less than, not equal,** or **greater than** for the **Alternative** hypothesis (greater than).

INTERPRET

The value of the test statistic is 2.29. The one-tail p-value is .0122. We observe that the p-value of the test is small (and the test statistic falls into the rejection region). As a result, we conclude that there is sufficient evidence to infer that on average directly purchased mutual funds outperform broker-purchased mutual funds.

In addition to testing a value of the difference between two population means, we can also estimate the difference between means. Next we compute the 95% confidence interval estimate of the difference between the mean return for direct and broker mutual funds.

COMPUTE

MANUALLY

The confidence interval estimator of the difference between two means with equal population variances is

$$(\bar{x}_1 - \bar{x}_2) \pm t_{\alpha/2}\sqrt{s_p^2\left(\frac{1}{n_1} + \frac{1}{n_2}\right)}$$

The 95% confidence interval estimate of the difference between the mean return for directly purchased mutual funds and the mean return of broker-bought mutual funds is

$$(\bar{x}_1 - \bar{x}_2) \pm t_{\alpha/2}\sqrt{s_p^2\left(\frac{1}{n_1} + \frac{1}{n_2}\right)} = (6.63 - 3.72) \pm 1.984\sqrt{40.42\left(\frac{1}{50} + \frac{1}{50}\right)}$$
$$= 2.91 \pm 2.52$$

The lower and upper limits are .39 and 5.43.

EXCEL

	A	B	C	D	E	F
1	t-Estimate of the Difference Between Two Means (Equal-Variances)					
2						
3		Sample 1	Sample 2	Confidence Interval Estimate		
4	Mean	6.63	3.72	2.91	±	2.52
5	Variance	37.49	43.34	Lower confidence limit		0.38
6	Sample size	50	50	Upper confidence limit		5.43
7	Pooled variance	40.41				
8	Confidence level	0.95				

INSTRUCTIONS

Open the **t-Estimate_2 Means (Eq-Var)** worksheet in the **Estimators** workbook. Substitute the values of the sample means, sample variances, and sample sizes, as well as the confidence level. If you've already tested the means, simply copy and paste the six cells containing the statistics you need (use **Copy, Paste Special,** and **Values**).

MINITAB

Two-Sample T-Test and CI: Direct, Broker

Two-sample T for Direct vs Broker

	N	Mean	StDev	SE Mean
Direct	50	6.63	6.12	0.87
Broker	50	3.72	6.58	0.93

Difference = mu (Direct) – mu (Broker)
Estimate for difference: 2.91
95% CI for difference: (0.38, 5.43)
T-Test of difference = 0 (vs not =): T-Value = 2.29 P-Value = 0.024 DF = 98
Both use Pooled StDev = 6.3572

INSTRUCTIONS

To produce a confidence interval estimate, follow the instructions for the test, but specify **not equal** for the **Alternative.** Minitab will conduct a two-tail test and produce the confidence interval estimate.

INTERPRET

We estimate that the return on directly purchased mutual funds is on average between .38 and 5.43 percentage points larger than broker-purchased mutual funds.

EXAMPLE 13.2*

DATA Xm13-02

Effect of New CEO in Family-Run Businesses

What happens to the family-run business when the boss's son or daughter takes over? Does the business do better after the change if the new boss is the offspring of the owner or does the business do better when an outsider is made chief executive officer (CEO)? In pursuit of an answer, researchers randomly selected 140 firms between 1994 and 2002, 30% of which passed ownership to an offspring and 70% appointed an outsider as CEO. For each company the researchers calculated the operating income as a proportion of assets in the year before and the year after the new CEO took over. The change

Source: M. Bennedsen and K. Nielsen, Copenhagen Business School, and D. Wolfenzon, New York University.

(operating income after − operating income before) in this variable was recorded and is listed next. Do these data allow us to infer that the effect of making an offspring CEO is different from the effect of hiring an outsider as CEO?

Offspring					**Outsider**									
−1.95	−1.38	−0.48	−0.67	2.75	0.69	6.31	0.09	4.41	0.33	5.67	2.79	−1.32	3.2	−1.16
0	0.57	0.24	2.61	0.3	−0.95	−3.04	6.79	4.62	−5.96	−0.8	5.62	5.93	−3.07	1.04
0.56	3.05	0.79	1.55		−2.2	−0.42	1.72	4.5	−2.46	1.37	−2.69	−0.45	−4.34	1.28
1.44	2.98	−1.19	−2.67		2.65	−0.89	6.64	2.37	1.59	0.72	−2.59	−3.2	−1.16	1.74
1.5	0.91	1.89	−1.91		5.39	−1.05	4.75	2.44	−2.03	4.14	2.45	5.08	−0.51	−0.14
1.41	−2.16	−3.7	1.01		4.15	−4.23	2.84	1.07	−1.69	3.04	3.39	0.23	8.68	−0.82
−0.32	1.22	−0.31	−1.62		4.28	−0.16	−2.1	−1.11	0.55	3.33	5.89	−2.69	1.43	0
−1.7	0.67	−1.37	−5.25		2.97	2.77	2.07	0.44	0.95	3.2	−0.71	3.76	−0.37	2.68
−1.66	−0.39	−3.15	0.14		4.11	−0.96	1.58	1.36	3.06	0.55	4.22	1.05	−0.49	
−1.87	−1.43	3.27	2.12		2.66	1.01	−1.98	0.88	4.83	−1.4	0.46	0.53	−0.08	

SOLUTION

IDENTIFY

The objective is to compare two populations, and the data are interval. It follows that the parameter of interest is the difference between two population means $\mu_1 - \mu_2$ where μ_1 is the mean difference for companies where the owner's son or daughter became CEO and μ_2 is the mean difference for companies that appointed an outsider as CEO.

To determine whether to apply the equal or unequal variances t-test, we use the F-test of two variances, σ_1^2/σ_2^2:

$$H_0: \quad \sigma_1^2/\sigma_2^2 = 1$$
$$H_1: \quad \sigma_1^2/\sigma_2^2 \neq 1$$

COMPUTE

MANUALLY

From the data we calculated the following statistics:

$s_1^2 = 3.79$ and $s_2^2 = 8.03$

Test statistic: $F = s_1^2/s_2^2 = 3.79/8.03 = 0.47$

The degrees of freedom are $\nu_1 = n_1 - 1 = 42 - 1 = 41$ and $\nu_2 = n_2 - 1 = 98 - 1 = 97$

Rejection region: $F > F_{\alpha/2,\nu_1,\nu_2} = F_{.025,41,97} \approx F_{.025,40,100} = 1.64$

or

$$F < F_{1-\alpha/2,\nu_1,\nu_2} = F_{.975,41,97} = 1/F_{.025,97,41} \approx 1/F_{.025,100,40} = 1/1.74 = .57$$

Because $F = .47$ is less than .57, we reject the null hypothesis.

EXCEL

	A	B	C
1	F-Test: Two-Sample for Variances		
2			
3		*Offspring*	*Outsider*
4	Mean	-0.10	1.24
5	Variance	3.79	8.03
6	Observations	42	98
7	df	41	97
8	F	0.47	
9	P(F<=f) one-tail	0.0040	
10	F Critical one-tail	0.6314	

The value of the test statistic is $F = .47$ and the p-value $= 2 \times .0040 = .0080$.

MINITAB

Test for Equal Variances: Offspring, Outsider

F-Test (Normal Distribution)
Test statistic = 0.47, p-value = 0.008

INTERPRET

There is enough evidence to infer that the population variances differ. The appropriate technique is the unequal-variances t-test of $\mu_1 - \mu_2$.

Because we want to determine whether there is a *difference* between means, the alternative hypothesis is

$$H_1: \quad (\mu_1 - \mu_2) \neq 0$$

and the null hypothesis is

$$H_0: \quad (\mu_1 - \mu_2) = 0$$

COMPUTE

MANUALLY

From the data we calculated the following statistics:

$$\bar{x}_1 = -.10$$
$$\bar{x}_2 = 1.24$$
$$s_1^2 = 3.79$$
$$s_2^2 = 8.03$$

The number of degrees of freedom of the test statistic is

$$\nu = \frac{\left(s_1^2/n_1\right)^2 + \left(s_2^2/n_2\right)^2}{\dfrac{\left(s_1^2/n_1\right)^2}{n_1 - 1} + \dfrac{\left(s_2^2/n_2\right)^2}{n_2 - 1}}$$

$$= \frac{(3.79/42)^2 + (8.03/98)^2}{\dfrac{(3.79/42)^2}{42 - 1} + \dfrac{(8.03/98)^2}{98 - 1}}$$

$$= 110.75 \text{ rounded to } 111$$

The rejection region is

$$t < -t_{\alpha/2,\nu} = -t_{.025,111} \approx -t_{.025,110} = -1.982 \quad \text{or} \quad t > t_{\alpha/2,\nu} = t_{.025,111} \approx 1.982$$

The value of the test statistic is computed next:

$$t = \frac{(\bar{x}_1 - \bar{x}_2) - (\mu_1 - \mu_2)}{\sqrt{\left(\dfrac{s_1^2}{n_1} + \dfrac{s_2^2}{n_2}\right)}}$$

$$= \frac{(-.10 - 1.24) - (0)}{\sqrt{\left(\dfrac{3.79}{42} + \dfrac{8.03}{98}\right)}}$$

$$= -3.22$$

EXCEL

	A	B	C
1	t-Test: Two-Sample Assuming Unequal Variances		
2			
3		*Offspring*	*Outsider*
4	Mean	-0.10	1.24
5	Variance	3.79	8.03
6	Observations	42	98
7	Hypothesized Mean Difference	0	
8	df	111	
9	t Stat	-3.22	
10	P(T<=t) one-tail	0.0008	
11	t Critical one-tail	1.6587	
12	P(T<=t) two-tail	0.0017	
13	t Critical two-tail	1.9816	

INSTRUCTIONS

Follow the instructions for Example 13.1, except at step 2 click **Data, Data Analysis,** and **t-Test: Two-Sample Assuming Unequal Variances.**

Use the **t-Test_2 Means (Uneq-Var)** worksheet in the **Test Statistics** workbook to complete this test from the sample statistics and to perform a what-if analysis (CD Appendix J). (CD Appendix O describes how to use an Excel spreadsheet to calculate the probability of a Type II error when testing the difference between two means.)

MINITAB

Two-Sample T-Test and CI: Offspring, Outsider

Two-sample T for Offspring vs Outsider

	N	Mean	StDev	SE Mean
Offspring	42	–0.10	1.95	0.30
Outsider	98	1.24	2.83	0.29

Difference = mu (Offspring) – mu (Outsider)
Estimate for difference: –1.336
95% CI for difference: (–2.158, –0.514)
T-Test of difference = 0 (vs not =): T-Value = –3.22 P-Value = 0.002 DF = 110

INSTRUCTIONS

Follow the instructions for Example 13.1, except at step 3 do not click **Assume equal variances.**

INTERPRET

The t-statistic is -3.22 and its p-value is .0017. Accordingly, we conclude that there is sufficient evidence to infer that the mean changes in operating income differ.

We can also draw inferences about the difference between the two population means by calculating the confidence interval estimator. We use the unequal-variances confidence interval estimator of $\mu_1 - \mu_2$ and a 95% confidence level.

COMPUTE

MANUALLY

$$(\bar{x}_1 - \bar{x}_2) \pm t_{\alpha/2}\sqrt{\left(\frac{s_1^2}{n_1} + \frac{s_2^2}{n_2}\right)}$$

$$= (-.10 - 1.24) \pm 1.982\sqrt{\left(\frac{3.79}{42} + \frac{8.03}{98}\right)}$$

$$= -1.34 \pm .82$$

$$\text{LCL} = -2.16 \quad \text{and} \quad \text{UCL} = -.52$$

EXCEL

	A	B	C	D	E	F
1	t-Estimate of the Difference Between Two Means (Unequal-Variances)					
2						
3		Sample 1	Sample 2	Confidence Interval Estimate		
4	Mean	-0.10	1.24	-1.34	±	0.82
5	Variance	3.79	8.03	Lower confidence limit		-2.16
6	Sample size	42	98	Upper confidence limit		-0.51
7	Degrees of freedom	110.75				
8	Confidence level	0.95				

INSTRUCTIONS

Activate the **t-Estimate_2 Means (Uneq-Var)** worksheet in the **Estimators** workbook and substitute the sample statistics and confidence level.

MINITAB

Minitab prints the confidence interval estimate as part of the output of the test statistic. However, you must specify the **Alternative** hypothesis as **not equal** to produce a two-sided interval.

INTERPRET

We estimate that the mean change in operating incomes for outsiders exceeds the mean change in the operating income for offspring lies between .51 and 2.16 percentage points.

Checking the Required Condition

Both the equal-variances and unequal-variances techniques require that the populations be normally distributed.* As before, we can check to see whether the requirement is satisfied by drawing the histograms of the data.

To illustrate, we used Excel (Minitab histograms are almost identical) to create the histograms for Example 13.1 (Figures 13.2 and 13.3) and Example 13.2 (Figures 13.4 and 13.5). Although the histograms are not perfectly bell shaped, it appears that in both examples the data are at least approximately normal. Because this technique is robust, we can be confident in the validity of the results.

FIGURE **13.2** Histogram of Rates of Return for Directly Purchased Mutual Funds in Example 13.1

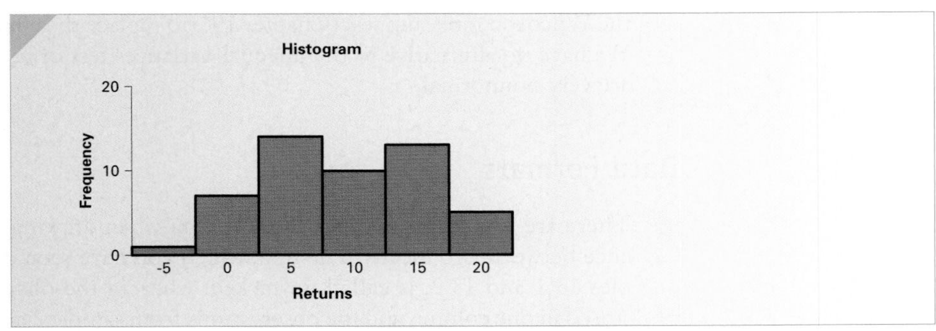

FIGURE **13.3** Histogram of Rates of Return for Broker-Purchased Mutual Funds in Example 13.1

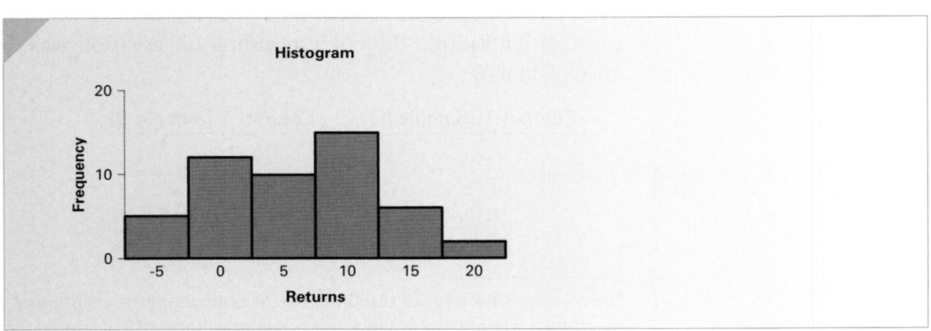

*As we pointed out in Chapter 12, large sample sizes can overcome the effects of extreme nonnormality.

FIGURE **13.4**
Histogram of Change in Operating Income for Offspring-Run Businesses in Example 13.2

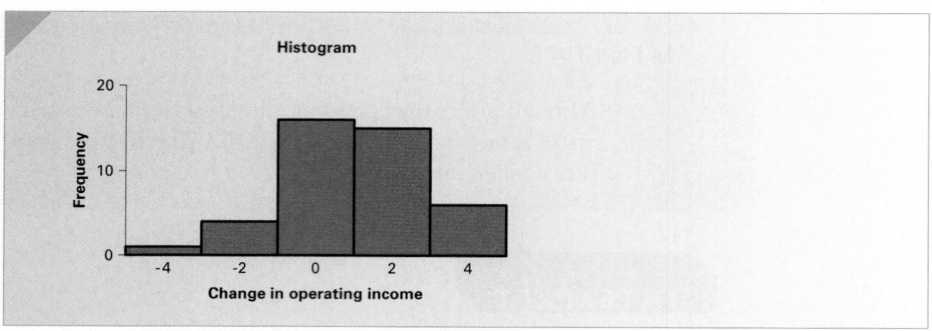

FIGURE **13.5**
Histogram of Change in Operating Income for Outsider-Run Businesses in Example 13.2

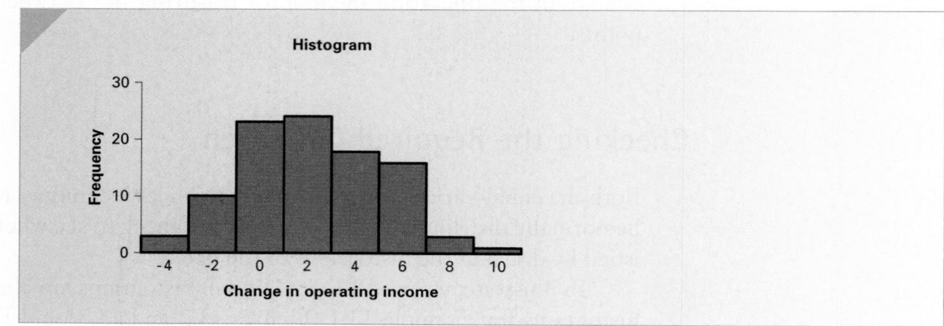

Violation of the Required Condition

When the normality requirement is unsatisfied, we can use a nonparametric technique—the Wilcoxon rank sum test (Chapter 19*) to replace the equal-variances test of $\mu_1 - \mu_2$. We have no alternative to the unequal-variances test of $\mu_1 - \mu_2$ when the populations are very nonnormal.

Data Formats

There are two formats for storing the data when drawing inferences about the difference between two means. The first, which you have seen demonstrated in both Examples 13.1 and 13.2, is called unstacked, wherein the observations from sample 1 are stored in one column and the observations from sample 2 are stored in a second column. We may also store the data in stacked format. In this format all the observations are stored in one column. A second column contains the codes, usually 1 and 2, that indicate from which sample the corresponding observation was drawn. Here is an example of unstacked data:

Column 1 (Sample 1)	Column 2 (Sample 2)
12	18
19	23
13	25

*Instructors who wish to teach the use of nonparametric techniques for testing the difference between two means when the normality requirement is not satisfied should use CD Appendix P, "Introduction to Nonparametric Techniques" and CD Appendix Q, "Wilcoxon Rank Sum Test and Wilcoxon Signed Rank Sum Test."

Here are the same data in stacked form:

Column 1	Column 2
12	1
19	1
13	1
18	2
23	2
25	2

It should be understood that the data need not be in order. Hence, they could have been stored in this way:

Column 1	Column 2
18	2
25	2
13	1
12	1
23	2
19	1

If there are two populations to compare and only one variable, it is probably better to record the data in unstacked form. However, it is frequently the case that we want to observe several variables and compare them. For example, suppose that we survey male and female MBAs and ask each to report his or her age, income, and number of years of experience. These data are usually stored in stacked form using the following format:

Column 1: Code identifying female (1) and male (2)

Column 2: Age

Column 3: Income

Column 4: Years of experience

To compare ages, we would use columns 1 and 2. Columns 1 and 3 are used to compare incomes, and columns 1 and 4 are used to compare experience levels.

Most statistical software requires one format or the other. Some but not all of Excel's techniques require unstacked data. Some of Minitab's procedures allow either format, whereas others specify only one. Fortunately, both of our software packages allow the statistics practitioner to alter the format. (See CD Appendix R for details.) We say "fortunately" because this allowed us to store the data in either form on the data disk provided with this book. In fact, we've used both forms to allow you to practice your ability to manipulate the data as necessary. You will need this ability to perform statistical techniques in this and other chapters in this book.

Developing an Understanding of Statistical Concepts 1

The formulas in this section are relatively complicated. However, conceptually both test statistics are based on the techniques we introduced in Chapter 11 and repeated in Chapter 12. That is, the value of the test statistic is the difference between the statistic $\bar{x}_1 - \bar{x}_2$ and the hypothesized value of the parameter $\mu_1 - \mu_2$ measured in terms of the standard error.

Developing an Understanding of Statistical Concepts 2

As was the case with the interval estimator of p, the standard error must be estimated from the data for all inferential procedures introduced here. The method we use to compute the standard error of $\bar{x}_1 - \bar{x}_2$ depends on whether the population variances are equal. When they are equal, we calculate and use the pooled variance estimator s_p^2. We are applying an important principle here, and we will do so again in Section 13.5 and in later chapters. The principle can be loosely stated as follows: Where possible, it is advantageous to pool sample data to estimate the standard error. In the previous application, we are able to pool because we assume that the two samples were drawn from populations with a common variance. Combining both samples increases the accuracy of the estimate. Thus, s_p^2 is a better estimator of the common variance than either s_1^2 or s_2^2 separately. When the two population variances are unequal, we cannot pool the data and produce a common estimator. We must compute s_1^2 and s_2^2 and use them to estimate σ_1^2 and σ_2^2, respectively.

Here is a summary of how we recognize the techniques presented in this section.

Factors That Identify the Equal-Variances t-Test and Estimator of $\mu_1 - \mu_2$

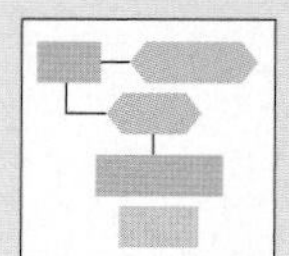

1. **Problem objective:** Compare two populations
2. **Data type:** Interval
3. **Descriptive measurement:** Central location
4. **Experimental design:** Independent samples
5. **Population variances:** Equal

Factors That Identify the Unequal-Variances t-Test and Estimator of $\mu_1 - \mu_2$

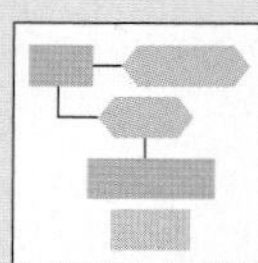

1. **Problem objective:** Compare two populations
2. **Data type:** Interval
3. **Descriptive measurement:** Central location
4. **Experimental design:** Independent samples
5. **Population variances:** Unequal

EXERCISES

Developing an Understanding of Statistical Concepts

*Exercises 13.1 to 13.6 are "what-if analyses" designed to determine what happens to the test statistics and interval estimates when elements of the statistical inference change. These problems can be solved manually or using Excel's **Test Statistics** or **Estimators** workbooks.*

13.1 In random samples of 25 from each of two normal populations, we found the following statistics:

$\bar{x}_1 = 524$ $\quad s_1 = 129$
$\bar{x}_2 = 469$ $\quad s_2 = 141$

a. Estimate the difference between the two population means with 95% confidence.
b. Repeat Part a, increasing the standard deviations to $s_1 = 255$ and $s_2 = 260$.
c. Describe what happens when the sample standard deviations get larger.
d. Repeat Part a with samples of size 100.
e. Discuss the effects of increasing the sample size.

13.2 In random samples of 12 from each two normal populations, we found the following statistics:

$\bar{x}_1 = 74$ $s_1 = 18$
$\bar{x}_2 = 71$ $s_2 = 16$

a. Test with $\alpha = .05$ to determine whether we can infer that the population means differ.
b. Repeat Part a, increasing the standard deviations to $s_1 = 210$ and $s_2 = 198$.
c. Describe what happens when the sample standard deviations get larger.
d. Repeat Part a with samples of size 150.
e. Discuss the effects of increasing the sample size.
f. Repeat Part a, changing the mean of sample 1 to $\bar{x}_1 = 76$.
g. Discuss the effect of increasing $\bar{x}_1$.

13.3 Random sampling from two normal populations produced the following results:

$\bar{x}_1 = 63$ $s_1 = 18$ $n_1 = 50$
$\bar{x}_2 = 60$ $s_2 = 7$ $n_2 = 45$

a. Estimate with 90% confidence the difference between the two population means.
b. Repeat Part a, changing the sample standard deviations to 41 and 15, respectively.
c. What happens when the sample standard deviations increase?
d. Repeat Part a, doubling the sample sizes.
e. Describe the effects of increasing the sample sizes.

13.4 Random sampling from two normal populations produced the following results:

$\bar{x}_1 = 412$ $s_1 = 128$ $n_1 = 150$
$\bar{x}_2 = 405$ $s_2 = 54$ $n_2 = 150$

a. Can we infer at the 5% significance level that μ_1 is greater than μ_2?
b. Repeat Part a, decreasing the standard deviations to $s_1 = 31$ and $s_2 = 16$.
c. Describe what happens when the sample standard deviations get smaller.
d. Repeat Part a with samples of size 20.
e. Discuss the effects of decreasing the sample size.
f. Repeat Part a, changing the mean of sample 1 to $\bar{x}_1 = 409$.
g. Discuss the effect of decreasing $\bar{x}_1$.

13.5 For each of the following, determine the number of degrees of freedom, assuming equal population variances and unequal population variances.

a. $n_1 = 15, n_2 = 15, s_1^2 = 25, s_2^2 = 15$
b. $n_1 = 10, n_2 = 16, s_1^2 = 100, s_2^2 = 15$
c. $n_1 = 50, n_2 = 50, s_1^2 = 8, s_2^2 = 14$
d. $n_1 = 60, n_2 = 45, s_1^2 = 75, s_2^2 = 10$

13.6 Refer to Exercise 13.5.

a. Confirm that in each case the number of degrees of freedom for the equal-variances test statistic and confidence interval estimator is larger than that for the unequal-variances test statistic and confidence interval estimator.
b. Try various combinations of sample sizes and sample variances to illustrate that the number of degrees of freedom for the equal-variances test statistic and confidence interval estimator is larger than that for the unequal-variances test statistic and confidence interval estimator.

Applications

13.7 Xr13-07 Every month a clothing store conducts an inventory and calculates the losses due to theft. The store would like to reduce these losses and is considering two methods. The first is to hire a security guard and the second is to install cameras. To help decide which method to choose, they hired a security guard for 6 months. During the next 6-month period, the store installed cameras. The monthly losses were recorded and are listed here. The manager decided that since the cameras were cheaper than the guard, he would install the cameras unless there was enough evidence to infer that the guard was better. What should the manager do?

Security Guard	355	284	401	398	477	254
Cameras	486	303	270	386	411	435

13.8 Xr13-08 A men's softball league is experimenting with a yellow baseball that is easier to see during night games. One way to judge the effectiveness is to count the number of errors. In a preliminary experiment, the yellow baseball was used in 10 games and the traditional white baseball was used in another 10 games. The number of errors in each game was recorded and is listed here. Can we infer that there are fewer errors on average when the yellow ball is used?

Yellow	5	2	6	7	2	5	3	8	4	9
White	7	6	8	5	9	11	8	3	6	10

13.9 Xr13-09 Many people who own digital cameras prefer to have pictures printed. In a preliminary study to determine spending patterns, a random sample of 8 digital camera owners and 8 standard camera owners were surveyed and asked how many pictures they printed in the past month. The results are presented here. Can we infer that the two groups differ in the mean number of pictures that are printed?

Digital	15	12	23	31	20	14	12	19
Standard	0	24	36	24	0	48	0	0

13.10 Xr13-10 Who spends more on their vacations, golfers or skiers? To help answer this question, a travel agency surveyed 15 customers who regularly take their spouses on either a skiing or a golfing vacation. The amounts spent on vacations last year are shown here. Can we infer that golfers and skiers differ in their vacation expenses?

Golfer	2,450	3,860	4,528	1,944	3,166	3,275
	4,490	3,685	2,950			
Skier	3,805	3,725	2,990	4,357	5,550	4,130

The following exercises require the use of a computer and software. The answers may be calculated manually. See Appendix A for the sample statistics. ***Use a 5% significance level unless specified otherwise.***

13.11 Xr13-11 The president of Tastee Inc., a baby-food producer, claims that her company's product is superior to that of her leading competitor, because babies gain weight faster with her product. (This is a good thing for babies.) To test this claim, a survey was undertaken. Mothers of newborn babies were asked which baby food they intended to feed their babies. Those who responded Tastee or the leading competitor were asked to keep track of their babies' weight gains over the next 2 months. There were 15 mothers who indicated that they would feed their babies Tastee and 25 who responded that they would feed their babies the product of the leading competitor. Each baby's weight gain (in ounces) was recorded.

a. Can we conclude, using weight gain as our criterion, that Tastee baby food is indeed superior?
b. Estimate with 95% confidence the difference between the mean weight gains of the two products.
c. Check to ensure that the required condition(s) is satisfied.

13.12 Xr13-12 Is eating oat bran an effective way to reduce cholesterol? Early studies indicated that eating oat bran daily reduces cholesterol levels by 5% to 10%. Reports of this study resulted in the introduction of many new breakfast cereals with various percentages of oat bran as an ingredient. However, an experiment performed by medical researchers in Boston, Massachusetts, cast doubt on the effectiveness of oat bran. In that study, 120 volunteers ate oat bran for breakfast, and another 120 volunteers ate another grain cereal for breakfast. At the end of 6 weeks, the percentage of cholesterol reduction was computed for both groups. Can we infer that oat bran is different from other cereals in terms of cholesterol reduction?

13.13 Xr13-13* In assessing the value of radio advertisements, sponsors consider not only the total number of listeners, but also their ages. The 18-to-34 age group is considered to spend the most money. To examine the issue, the manager of an FM station commissioned a survey. One objective was to measure the difference in listening habits between the 18-to-34 and 35-to-50 age groups. The survey asked 250 people in each age category how much time they spent listening to FM radio per day. The results (in minutes) were recorded and stored in stacked format (column 1 = Age group and column 2 = Listening times).

a. Can we conclude that a difference exists between the two groups?
b. Estimate with 95% confidence the difference in mean time listening to FM radio between the two age groups.
c. Are the required conditions satisfied for the techniques you used in Parts a and b?

13.14 Xr13-14 The cruise ship business is rapidly increasing. Although cruises have long been associated with seniors, it now appears that younger people are choosing a cruise as their vacations. To determine whether this is true, an executive for a cruise line sampled passengers 2 years ago and this year and determined their ages.

a. Do these data allow the executive to infer that cruise ships are attracting younger customers?
b. Estimate with 99% confidence the difference in ages between this year and 2 years ago.

13.15 Xr13-15* Automobile insurance companies take many factors into consideration when setting rates. These factors include age, marital status, and miles driven per year. To determine the effect of gender, a random sample of young (under 25, with at least 2 years of driving experience) male and female drivers was surveyed. Each was asked how many miles he or she had driven in the past year. The distances (in thousands of miles) are stored in stacked format (column 1 = driving distances and column 2 identifies the gender where 1 = male and code 2 = female).

a. Can we conclude that male and female drivers differ in the numbers of miles driven per year?
b. Estimate with 95% confidence the difference in mean distance driven by male and female drivers.
c. Check to ensure that the required condition(s) of the techniques used in Parts a and b is satisfied.

13.16 Xr13-16 The president of a company that manufactures automobile air conditioners is considering switching his supplier of condensers. Supplier A, the current producer of condensers for the manufacturer, prices its product 5% higher than supplier B does. Because the president wants to maintain his company's reputation for quality, he wants to be sure that supplier B's condensers last at least as long as supplier A's. After a careful analysis, the president decided to retain supplier A if there is sufficient statistical evidence that supplier A's condensers last longer on average than supplier B's condensers. In an experiment, 30 midsize cars were equipped with air conditioners using type A condensers while another 30 midsize cars were equipped with type B condensers. The number of miles (in thousands) driven by each car before the condenser broke down was recorded. Should the president retain supplier A?

13.17 Xr13-17 An important function of a firm's human resources manager is to track worker turnover. As a general rule, companies prefer to retain workers. New workers frequently need to be trained and it often takes time for new workers to learn how to perform their jobs. To investigate nationwide results, a human resources manager organized a survey wherein a random sample of men and women was asked how long they have worked for their current employer. Can we infer that men and women have different job tenures? (Adapted from the *Statistical Abstract of the United States*, 2000, Table 664)

13.18 Xr13-18 A statistics professor is about to select a statistical software package for her course. One of the most important features, according to the professor, is the ease with which students learn to use the software. She has narrowed the selection to two possibilities: software A, a menu-driven statistical package with some high-powered techniques, and software B, a spreadsheet that has the capability of performing most techniques. To help make her decision, she asks 40 statistics students selected at random to choose one of the two packages. She gives each student a statistics problem to solve by computer and the appropriate manual. The amount of time (in minutes) each student needed to complete the assignment was recorded.

a. Can the professor conclude from these data that the two software packages differ in the amount of time needed to learn how to use them? (Use a 1% significance level.)
b. Estimate with 95% confidence the difference in the mean amount of time needed to learn to use the two packages.
c. What are the required conditions for the techniques used in Parts a and b?
d. Check to see whether the required conditions are satisfied.

13.19 Xr13-19 One factor in low productivity is the amount of time wasted by workers. Wasted time includes time spent cleaning up mistakes, waiting for more material and equipment, and performing any other activity not related to production. In a project designed to examine the problem, an operations management consultant took a survey of 200 workers in companies that were classified as successful (on the basis of their latest annual profits) and another 200 workers from unsuccessful companies. The amount of time (in hours) wasted during a standard 40-hour workweek was recorded for each worker.

a. Do these data provide enough evidence at the 1% significance level to infer that the amount of time wasted in unsuccessful firms exceeds that of successful ones?
b. Estimate with 95% confidence how much more time is wasted in unsuccessful firms than in successful ones.

13.20 Xr13-20 Recent studies seem to indicate that using a cell phone while driving is dangerous. One reason for this is that a driver's reaction times may slow while he or she is talking on the phone. Researchers at Miami (Ohio) University measured the reaction times of a sample of drivers who owned a cell phone. Half the sample was tested while on the phone and the other half was tested while not on the phone. Can we conclude that reaction times are slower for drivers using cell phones?

13.21 Xr13-21 Refer to Exercise 13.20. To determine whether the type of phone usage affects reaction times, another study was launched. A group of drivers was asked to participate in a discussion. Half the group engaged in simple chitchat and the other half participated in a political discussion. Once again reaction times were measured. Can we infer that the type of telephone discussion affects reaction times?

13.22 Xr13-22 Most consumers who require someone to perform various professional services undertake research before making their selection. A random sample of people who recently selected a financial planner and a random sample of individuals who chose a stockbroker were asked to report the amount of time they spent researching before deciding. Can we infer that people spend more time researching for a financial planner than they do for a stockbroker? (*Source:* Yankelovich Partners)

13.23 Xr13-23 A recent study by researchers at North Carolina State University found thousands of errors in 12 of the most widely used high school science texts. For example, the Statue of Liberty is left-handed; volume is equal to length multiplied by depth (*Time Magazine*, February 12, 2001). The books are so bad that Philip Sadler, director of science education at the Harvard-Smithsonian Center for Astrophysics, decided to conduct a study of their effects. He recorded the physics marks of college students who had used a textbook in high school and the marks of students who did not have a high school textbook. Do these data allow us to infer that students without high school textbooks in science outperform students who used textbooks?

13.24 Xr13-24 Between Wendy's and McDonald's, which fast-food drive-through window is faster? To answer the question, a random sample of service times for each restaurant was measured. Can we infer from these data that there are differences in service times between the two chains? (*Source:* 2000 QSR Drive-Thru Time Study)

13.25 Xr13-25 The American Medical Association tracks the amount of time physicians devote to patient care per week. An insurance executive wanted to determine whether different specialties differ in the amounts of time. She acquired the results of the survey in 1997 (the most recent year available). She recorded the amount of time for physicians who are in general/family practice and those whose specialty is pediatrics. Can we infer that the two types of physicians differ in the time devoted to patient care per week? (Adapted from the American Medical Association *Socioeconomic Characteristics of Medical Practice*, 1997/1998 and the *Statistical Abstract of the United States*, 2000, Table 190)

13.26 Xr13-26 It is often useful for companies to know who their customers are and how they became customers. In a study of credit card use, a random sample of cardholders who applied for the credit card and a random sample of credit cardholders who were contacted by telemarketers or by mail were drawn. The total purchases made by each last month were recorded.

a. Can we conclude from these data that differences exist on average between the two types of customers?
b. Prepare a presentation to the bank's marketing manager describing your findings.

13.27 Xr13-27 Tire manufacturers are constantly researching ways to produce tires that last longer. Innovations are tested by professional drivers on racetracks. However, any promising inventions are also test-driven by ordinary drivers. The latter tests are closer to what the tire company's customers will actually experience. Suppose that to determine whether a new steel-belted radial tire lasts longer than the company's current model, two new-design tires were installed on the rear wheels of 20 randomly selected cars and two existing-design tires were installed on the rear wheels of another 20 cars. All drivers were told to drive in their usual way until the tires wore out. The number of miles driven by each driver was recorded. Can the company infer that the new tire will last on average longer than the existing tire?

13.28 Xr13-28 It is generally believed that salespeople who are paid on a commission basis outperform salespeople who are paid a fixed salary. Some management consultants argue, however, that in certain industries the fixed-salary salesperson may sell more because the consumer will feel less sales pressure and respond to the salesperson less as an antagonist. In an experiment to study this, a random sample of 180 salespeople from a retail clothing chain was selected. Of these, 90 salespeople were paid a fixed salary, and the remaining 90 were paid a commission on each sale. The total dollar amount of 1 month's sales for each was recorded. Can we conclude that the commission salesperson outperforms the fixed-salary salesperson?

13.29 Xr13-29 Credit scorecards were designed to be used to help financial institutions make decisions about loan applications (see page 49). However, some insurance companies have suggested that credit scores could also be used to determine insurance premiums, particularly car insurance. The Massachusetts Public Interest Research Group (MASSPIRG) has come out against this proposal. To acquire more information, an executive for a car insurance company gathered data about a random sample of the company's customers. She recorded whether the individual was involved in an accident in the last 3 years and determined the credit score. Can the executive infer that there is a difference in scores between those who have and those who do not have accidents in a 3-year period?

13.30 Xr13-30* Traditionally, wine has been sold in glass bottles with cork stoppers. The stoppers are supposed to keep air out of the bottle because oxygen is the enemy of wine, particularly red wine. Recent research appears to indicate that metal screw caps are more effective in keeping air out of the bottle. However, metal caps are perceived to be inferior and usually associated with cheaper brands of wine. To determine if this perception is wrong, a random sample of 130 people who drink at least one bottle per week on average was asked to participate in an experiment. All were given the same wine in two types of bottles. One group was given a corked bottle and the other was given a bottle with a metal cap and asked to taste the wine and indicate what they think the retail price of the wine should be. Determine whether there is enough evidence to conclude that bottles of wine with metal caps are perceived to be cheaper.

13.2 Observational and Experimental Data

As we've pointed out several times, the ability to properly interpret the results of a statistical technique is a crucial skill for students to develop. This ability is dependent on your understanding of Type I and Type II errors and the fundamental concepts that are part of statistical inference. However, there is another component that must be understood: the difference between **observational data** and **experimental data.** The difference is due to the way the data are generated. The following example will demonstrate the difference between the two types.

EXAMPLE 13.3

Dietary Effects of High-Fiber Breakfast Cereals

DATA Xm13-03

Despite some controversy, scientists generally agree that high-fiber cereals reduce the likelihood of various forms of cancer. However, one scientist claims that people who eat high-fiber cereal for breakfast will consume, on average, fewer calories for lunch than people who don't eat high-fiber cereal for breakfast. If this is true, high-fiber cereal manufacturers will be able to claim another advantage of eating their product—potential weight reduction for dieters. As a preliminary test of the claim, 150 people were randomly selected and asked what they regularly eat for breakfast and lunch. Each person was identified as either a consumer or a nonconsumer of high-fiber cereal, and the number of calories consumed at lunch was measured and recorded. These data are listed here. Can the scientist conclude at the 5% significance level that his belief is correct?

Calories Consumed at Lunch by Consumers of High-Fiber Cereal

568	646	607	555	530	714	593	647	650
498	636	529	565	566	639	551	580	629
589	739	637	568	687	693	683	532	651
681	539	617	584	694	556	667	467	
540	596	633	607	566	473	649	622	

Calories Consumed at Lunch by Nonconsumers of High-Fiber Cereal

705	754	740	569	593	637	563	421	514	536
819	741	688	547	723	553	733	812	580	833
706	628	539	710	730	620	664	547	624	644
509	537	725	679	701	679	625	643	566	594
613	748	711	674	672	599	655	693	709	596
582	663	607	505	685	566	466	624	518	750
601	526	816	527	800	484	462	549	554	582
608	541	426	679	663	739	603	726	623	788
787	462	773	830	369	717	646	645	747	
573	719	480	602	596	642	588	794	583	
428	754	632	765	758	663	476	490	573	

SOLUTION

The appropriate technique is the unequal-variances t-test of $\mu_1 - \mu_2$ where μ_1 is the mean of the number of calories for lunch by consumers of high-fiber cereal for breakfast and μ_2 is the mean of the number of calories for lunch by nonconsumers of high-fiber cereal for breakfast. [The F-test of the ratio of two variances (not shown here) yielded $F = .3845$ and p-value $= .0008$.] The hypotheses are

$$H_0: \ (\mu_1 - \mu_2) = 0$$
$$H_1: \ (\mu_1 - \mu_2) < 0$$

The Excel printout is shown next. The manually calculated and the Minitab-produced results are identical.

	A	B	C
1	t-Test: Two-Sample Assuming Unequal Variances		
2			
3		Consumers	Nonconsumers
4	Mean	604.0	633.2
5	Variance	4103	10670
6	Observations	43	107
7	Hypothesized Mean Difference	0	
8	df	123	
9	t Stat	-2.09	
10	P(T<=t) one-tail	0.0193	
11	t Critical one-tail	1.6573	
12	P(T<=t) two-tail	0.0386	
13	t Critical two-tail	1.9794	

INTERPRET

The value of the test statistic is −2.09. The one-tail p-value is .0193. We observe that the p-value of the test is small (and the test statistic falls into the rejection region). As a result, we conclude that there is sufficient evidence to infer that consumers of high-fiber cereal do eat fewer calories at lunch than do nonconsumers. From this result, we're inclined to believe that eating a high-fiber cereal at breakfast may be a way to reduce weight. However, other interpretations are plausible. For example, people who eat fewer calories are probably more health conscious, and such people are more likely to eat high-fiber cereal as part of a healthy breakfast. In this interpretation, high-fiber cereals do not necessarily lead to fewer calories at lunch. Instead another factor, general health consciousness, leads to both fewer calories at lunch and high-fiber cereal for breakfast. Notice that the conclusion of the statistical procedure is unchanged. On average, people who eat high-fiber cereal consume fewer calories at lunch. However, because of the way the data were gathered, we have more difficulty interpreting this result.

Suppose that we redo Example 13.3 using the experimental approach. We randomly select 150 people to participate in the experiment. We randomly assign 75 to eat high-fiber cereal for breakfast and the other 75 to eat something else. We then record the number of calories each person consumes at lunch. Ideally, in this experiment both groups will be similar in all other dimensions, including health consciousness. (Larger sample sizes increase the likelihood that the two groups will be similar.) If the statistical result is about the same as in Example 13.3, we may have some valid reason to believe that high-fiber cereal at breakfast leads to a decrease in caloric intake at lunch.

Experimental data are usually more expensive to obtain because of the planning required to set up the experiment; observational data usually require less work to gather. Furthermore, in many situations it is impossible to conduct a controlled experiment. For example, suppose that we want to determine whether an undergraduate degree in engineering better prepares students for an MBA than does an arts degree. In a controlled experiment, we would randomly assign some students to achieve a degree in engineering and other students to obtain an arts degree. We would then make them sign up for an MBA program where we would record their grades. Unfortunately for statistical despots (and fortunately for the rest of us), we live in a democratic society, which makes the coercion necessary to perform this controlled experiment impossible.

To answer our question about the relative performance of engineering and arts students, we have no choice but to obtain our data by observational methods. We would take a random sample of engineering students and arts students who have already entered MBA programs and record their grades. If we find that engineering students do better, we may tend to conclude that an engineering background better prepares students for an MBA program. However, it may be true that better students tend to choose engineering as their undergraduate major and that better students achieve higher grades in all programs, including the MBA program.

Although we've discussed observational and experimental data in the context of the test of the difference between two means, you should be aware that the issue of how the data are obtained is relevant to the interpretation of all the techniques that follow.

EXERCISES

13.31 Refer to Exercise 13.11. If the data are observational, describe another conclusion besides the one that infers that Tastee is better for babies.

13.32 Are the data in Exercise 13.12 observational or experimental? Explain. If the data are observational, describe a method of producing experimental data.

13.33 Are the data in Exercise 13.15 observational or experimental? Explain. If the data are observational, is it possible to collect experimental data?

13.34 Refer to Exercise 13.18.

a. Are the data observational or experimental?
b. If the data are observational, describe a method of answering the question with experimental data?
c. If the data are observational, produce another explanation for the statistical outcome.

13.35 Refer to Exercise 13.25.

a. Are the data observational or experimental? Explain.
b. If the data are observational, describe a method of answering the question with experimental data.

13.36 Suppose that you wish to test to determine whether one method of teaching statistics is better than another.

a. Describe a data-gathering process that produces observational data.
b. Describe a data-gathering process that produces experimental data.

13.37 Put yourself in place of the director of research and development for a pharmaceutical company. When a new drug is developed, it undergoes a number of tests. One of the tests is designed to determine whether the drug is safe and effective. Your company has just developed a drug that is designed to alleviate the symptoms of degenerative diseases such as multiple sclerosis. Design an experiment that tests the new drug.

13.38 You wish to determine whether MBA graduates who majored in finance attract higher starting salaries than MBA graduates who majored in marketing.

a. Describe a data-gathering process that produces observational data.
b. Describe a data-gathering process that produces experimental data.
c. If observational data indicate that finance majors attract higher salaries than do marketing majors, provide two explanations for this result.

13.39 Suppose that you are analyzing one of the hundreds of statistical studies linking smoking with lung cancer. The study analyzed thousands of randomly selected people, some of whom had lung cancer. The statistics indicate that those who have lung cancer smoked on average significantly more than those who did not have lung cancer.

a. Explain how you know that the data are observational.
b. Is there another interpretation of the statistics besides the obvious one that smoking causes lung cancer? If so, what is it? (Students who produce the best answers will be eligible for a job in the public relations department of a tobacco company.)
c. Is it possible to conduct a controlled experiment to produce data that address the question of the relationship between smoking and lung cancer? If so, describe the experiment.

13.3 Inference about the Difference between Two Means: Matched Pairs Experiment

We continue our presentation of statistical techniques that address the problem of comparing two populations of interval data. In Section 13.1, the parameter of interest was the difference between two population means, where the data were generated from independent samples. In this section, the data are gathered from a matched pairs experiment. To illustrate why matched pairs experiments are needed and how we deal with data produced in this way, consider the following example.

EXAMPLE 13.4 Comparing Salary Offers for Finance and Marketing MBA Majors, Part 1

DATA Xm13-04

In the last few years a number of web-based companies that offer job placement services have been created. The manager of one such company wanted to investigate the job offers recent MBAs were obtaining. In particular, she wanted to know whether finance majors were being offered higher salaries than marketing majors. In a preliminary study, she randomly sampled 50 recently graduated MBAs, half of whom majored in finance and half in marketing. From each she obtained the highest salary offer (including benefits). These data are listed here. Can we infer that finance majors obtain higher salary offers than do marketing majors among MBAs?

Highest Salary Offer Made to Finance Majors

61,228	51,836	20,620	73,356	84,186	79,782	29,523	80,645	76,125
62,531	77,073	86,705	70,286	63,196	64,358	47,915	86,792	75,155
65,948	29,392	96,382	80,644	51,389	61,955	63,573		

Highest Salary Offer Made to Marketing Majors

73,361	36,956	63,627	71,069	40,203	97,097	49,442	75,188	59,854
79,816	51,943	35,272	60,631	63,567	69,423	68,421	56,276	47,510
58,925	78,704	62,553	81,931	30,867	49,091	48,843		

SOLUTION

IDENTIFY

The objective is to compare two populations of interval data. The parameter is the difference between two means $\mu_1 - \mu_2$ (where μ_1 = mean highest salary offer to finance majors and μ_2 = mean highest salary offer to marketing majors). Because we want to determine whether finance majors are offered higher salaries, the alternative hypothesis will specify that μ_1 is greater than μ_2. The F-test for variances was conducted and the results indicate that there is not enough evidence to infer that the population variances differ. Hence we use the equal-variances test statistic:

$$H_0: \quad (\mu_1 - \mu_2) = 0$$

$$H_1: \quad (\mu_1 - \mu_2) > 0$$

$$\text{Test statistic:} \quad t = \frac{(\bar{x}_1 - \bar{x}_2) - (\mu_1 - \mu_2)}{\sqrt{s_p^2\left(\frac{1}{n_1} + \frac{1}{n_2}\right)}}$$

COMPUTE

MANUALLY

From the data we calculated the following statistics:

$$\bar{x}_1 = 65{,}624$$

$$\bar{x}_2 = 60{,}423$$

$$s_1^2 = 360{,}433{,}294$$

$$s_2^2 = 262{,}228{,}559$$

$$s_p^2 = \frac{(n_1 - 1)s_1^2 + (n_2 - 1)s_2^2}{n_1 + n_2 - 2}$$

$$= \frac{(25 - 1)(360{,}433{,}294) + (25 - 1)(262{,}228{,}559)}{25 + 25 - 2}$$

$$= 311{,}330{,}926$$

The value of the test statistic is computed next:

$$t = \frac{(\bar{x}_1 - \bar{x}_2) - (\mu_1 - \mu_2)}{\sqrt{s_p^2\left(\frac{1}{n_1} + \frac{1}{n_2}\right)}}$$

$$= \frac{(65{,}624 - 60{,}423) - (0)}{\sqrt{311{,}330{,}926\left(\frac{1}{25} + \frac{1}{25}\right)}}$$

$$= 1.04$$

The number of degrees of freedom of the test statistic is

$$\nu = n_1 + n_2 - 2 = 25 + 25 - 2 = 48$$

The rejection region is

$$t > t_{\alpha,\nu} = t_{.05,48} \approx 1.676$$

EXCEL

	A	B	C
1	t-Test: Two-Sample Assuming Equal Variances		
2			
3		*Finance*	*Marketing*
4	Mean	65,624	60,423
5	Variance	360,433,294	262,228,559
6	Observations	25	25
7	Pooled Variance	311,330,926	
8	Hypothesized Mean Difference	0	
9	df	48	
10	t Stat	1.04	
11	P(T<=t) one-tail	0.1513	
12	t Critical one-tail	1.6772	
13	P(T<=t) two-tail	0.3026	
14	t Critical two-tail	2.0106	

MINITAB

```
Two-Sample T-Test and CI: Finance, Marketing

Two-sample T for Finance vs Marketing

            N   Mean   StDev   SE Mean
Finance    25  65624   18985      3797
Marketing  25  60423   16193      3239


Difference = mu (Finance) – mu (Marketing)
Estimate for difference:  5201.00
95% lower bound for difference:  –3169.42
T-Test of difference = 0 (vs >): T-Value = 1.04  P-Value = 0.151  DF = 48
Both use Pooled StDev = 17644.5722
```

INTERPRET

The value of the test statistic ($t = 1.04$) and its p-value (.1513) indicate that there is very little evidence to support the hypothesis that finance majors receive higher salary offers than marketing majors.

Notice that we have some evidence to support the alternative hypothesis. The difference in sample means is

$$(\bar{x}_1 - \bar{x}_2) = (65{,}624 - 60{,}423) = 5{,}201$$

However, we judge the difference between sample means in relation to the standard error of $\bar{x}_1 - \bar{x}_2$. As we've already calculated,

$$s_p^2 = 311{,}330{,}926$$

and

$$\sqrt{s_p^2\left(\frac{1}{n_1} + \frac{1}{n_2}\right)} = 4{,}991$$

Consequently, the value of the test statistic is $t = 5{,}201/4{,}991 = 1.04$, a value that does not allow us to infer that finance majors attract higher salary offers. We can see that although the difference between the sample means was quite large, the variability of the data, as measured by s_p^2, was also large, resulting in a small test statistic value.

EXAMPLE 13.5

DATA
Xm13-05

Comparing Salary Offers for Finance and Marketing MBA Majors, Part 2

Suppose now that we redo the experiment in the following way. We examine the transcripts of finance and marketing MBA majors. We randomly select a finance and a marketing major whose grade point average (GPA) falls between 3.92 and 4 (based on a maximum of 4). We then randomly select a finance and a marketing major whose GPA is between 3.84 and 3.92. We continue this process until the 25th pair of finance and marketing majors is selected whose GPA fell between 2.0 and 2.08. (The minimum GPA required for graduation is 2.0.) As we did in Example 13.4, we recorded the highest salary offer. These data, together with the GPA group, are listed here. Can we conclude from these data that finance majors draw larger salary offers than do marketing majors?

Group	Finance	Marketing
1	95,171	89,329
2	88,009	92,705
3	98,089	99,205
4	106,322	99,003
5	74,566	74,825
6	87,089	77,038
7	88,664	78,272
8	71,200	59,462
9	69,367	51,555
10	82,618	81,591
11	69,131	68,110
12	58,187	54,970
13	64,718	68,675
14	67,716	54,110
15	49,296	46,467
16	56,625	53,559
17	63,728	46,793
18	55,425	39,984
19	37,898	30,137
20	56,244	61,965
21	51,071	47,438
22	31,235	29,662
23	32,477	33,710
24	35,274	31,989
25	45,835	38,788

SOLUTION

The experiment described in Example 13.4 is one in which the samples are independent. That is, there is no relationship between the observations in one sample and the observations in the second sample. However, in this example the experiment was designed in such a way that each observation in one sample is matched with an observation in the other sample. The matching is conducted by selecting finance and marketing majors with similar GPAs. Thus, it is logical to compare the salary offers for finance and marketing majors in each group. This type of experiment is called **matched pairs.** We now describe how we conduct the test.

For each GPA group, we calculate the matched pair difference between the salary offers for finance and marketing majors.

Group	Finance	Marketing	Difference
1	95,171	89,329	5,842
2	88,009	92,705	−4,696
3	98,089	99,205	−1,116
4	106,322	99,003	7,319
5	74,566	74,825	−259
6	87,089	77,038	10,051
7	88,664	78,272	10,392
8	71,200	59,462	11,738
9	69,367	51,555	17,812
10	82,618	81,591	1,027

(continued)

Group	Finance	Marketing	Difference
11	69,131	68,110	1,021
12	58,187	54,970	3,217
13	64,718	68,675	−3,957
14	67,716	54,110	13,606
15	49,296	46,467	2,829
16	56,625	53,559	3,066
17	63,728	46,793	16,935
18	55,425	39,984	15,441
19	37,898	30,137	7,761
20	56,244	61,965	−5,721
21	51,071	47,438	3,633
22	31,235	29,662	1,573
23	32,477	33,710	−1,233
24	35,274	31,989	3,285
25	45,835	38,788	7,047

In this experimental design the parameter of interest is the **mean of the population of differences,** which we label μ_D. Note that μ_D does in fact equal $\mu_1 - \mu_2$, but we test μ_D because of the way the experiment was designed. Hence, the hypotheses to be tested are

$$H_0: \quad \mu_D = 0$$
$$H_1: \quad \mu_D > 0$$

We have already presented inferential techniques about a population mean. Recall that in Chapter 12 we introduced the t-test of μ. Thus, to test hypotheses about μ_D, we use the following test statistic.

Test Statistic for μ_D

$$t = \frac{\bar{x}_D - \mu_D}{s_D/\sqrt{n_D}}$$

which is Student t distributed with $\nu = n_D - 1$ degrees of freedom, provided that the differences are normally distributed.

Aside from the subscript D, this test statistic is identical to the one presented in Chapter 12. We conduct the test in the usual way.

COMPUTE

MANUALLY

Using the differences computed above, we find the following statistics:

$$\bar{x}_D = 5{,}065$$
$$s_D = 6{,}647$$

from which we calculate the value of the test statistic:

$$t = \frac{\bar{x}_D - \mu_D}{s_D/\sqrt{n_D}} = \frac{5{,}065 - 0}{6{,}647/\sqrt{25}} = 3.81$$

The rejection region is

$$t > t_{\alpha,\nu} = t_{.05,24} = 1.711$$

EXCEL

	A	B	C
1	t-Test: Paired Two Sample for Means		
2			
3		*Finance*	*Marketing*
4	Mean	65,438	60,374
5	Variance	444,981,810	469,441,785
6	Observations	25	25
7	Pearson Correlation	0.9520	
8	Hypothesized Mean Difference	0	
9	df	24	
10	t Stat	3.81	
11	P(T<=t) one-tail	0.0004	
12	t Critical one-tail	1.7109	
13	P(T<=t) two-tail	0.0009	
14	t Critical two-tail	2.0639	

Excel prints the sample means, variances, and sample sizes for each sample (as well as the coefficient of correlation), which implies that the procedure uses these statistics. It doesn't. The technique is based on computing the paired differences from which the mean, variance, and sample size are determined. Excel should have printed these statistics.

INSTRUCTIONS

1. Type or import the data into two columns. (Open Xm13-05.)
2. Click **Data, Data Analysis,** and **t-Test: Paired Two- Sample for Means.**
3. Specify the **Variable 1 Range** (B1:B26) and the **Variable 2 Range** (C1:C26). Type the value of **the Hypothesized Mean Difference** (0) and specify a value for α (.05).

MINITAB

Paired T-Test and CI: Finance, Marketing

Paired T for Finance - Marketing

	N	Mean	StDev	SE Mean
Finance	25	65438.2	21094.6	4218.9
Marketing	25	60373.7	21666.6	4333.3
Difference	25	5064.52	6646.90	1329.38

95% lower bound for mean difference: 2790.11
T-Test of mean difference = 0 (vs > 0): T-Value = 3.81 P-Value = 0.000

INSTRUCTIONS

1. Type or import the data into two columns. (Open Xm13-05.)
2. Click **Stat, Basic Statistics,** and **Paired t . . . .**
3. Select the variable names of the **First sample** (Finance) and **Second sample** (Marketing). Click **Options . . . .**
4. In the **Test Mean** box, type the hypothesized mean of the paired difference (0), and specify the **Alternative** (greater than).

INTERPRET

The value of the test statistic is $t = 3.81$ with a p-value of .0004. There is now overwhelming evidence to infer that finance majors obtain higher salary offers than marketing majors. By redoing the experiment as matched pairs, we were able to extract this information from the data.

Estimating the Mean Difference

We derive the confidence interval estimator of μ_D using the usual form for the confidence interval.

Confidence Interval Estimator of μ_D

$$\bar{x}_D \pm t_{\alpha/2}\frac{s_D}{\sqrt{n_D}}$$

EXAMPLE 13.6

DATA Xm13-05

Comparing Salary Offers for Finance and Marketing MBA Majors, Part 3

Compute the 95% confidence interval estimate of the mean difference in salary offers between finance and marketing majors in Example 13.5.

SOLUTION

COMPUTE

MANUALLY

The 95% confidence interval estimate of the mean difference is

$$\bar{x}_D \pm t_{\alpha/2}\frac{s_D}{\sqrt{n_D}} = 5{,}065 \pm 2.064\frac{6{,}647}{\sqrt{25}} = 5{,}065 \pm 2{,}744$$

$$\text{LCL} = 2321 \qquad \text{and} \qquad \text{UCL} = 7809$$

EXCEL

	A	B	C	D
1	t-Estimate: Mean			
2				
3				Difference
4	Mean			5065
5	Standard Deviation			6647
6	LCL			2321
7	UCL			7808

INSTRUCTIONS

To estimate the mean difference, employ the t-estimate of the mean of the differences (see Section 12.1). You must first instruct Excel to calculate the differences.

MINITAB

Paired T-Test and CI: Finance, Marketing

Paired T for Finance - Marketing

	N	Mean	StDev	SE Mean
Finance	25	65438.2	21094.6	4218.9
Marketing	25	60373.7	21666.6	4333.3
Difference	25	5064.52	6646.90	1329.38

95% CI for mean difference: (2320.82, 7808.22)
T-Test of mean difference = 0 (vs not = 0): T-Value = 3.81 P-Value = 0.001

INSTRUCTIONS

Follow the instructions to test the paired difference. However, you must specify **not equal** for the **Alternative** hypothesis to produce the two-sided confidence interval estimate of the mean difference.

INTERPRET

We estimate that the mean salary offer to finance majors exceeds the mean salary offer to marketing majors by an amount that lies between $2,321 and $7,808 (using the computer output).

Independent Samples or Matched Pairs: Which Experimental Design Is Better?

Examples 13.4 and 13.5 demonstrated that the experimental design is an important factor in statistical inference. However, these two examples raise several questions about experimental designs.

1. Why does the matched pairs experiment result in concluding that finance majors receive higher salary offers than do marketing majors, whereas the independent samples experiment could not?

2. Should we always use the matched pairs experiment? In particular, are there disadvantages to its use?

3. How do we recognize when a matched pairs experiment has been performed?

Here are our answers.

1. The matched pairs experiment worked in Example 13.5 by reducing the variation in the data. To understand this point, examine the statistics from both examples. In Example 13.4, we found $\bar{x}_1 - \bar{x}_2 = 5{,}201$. In Example 13.5, we computed $\bar{x}_D = 5{,}065$. Thus, the numerators of the two test statistics were quite similar. However, the test statistic in Example 13.5 was much larger than the test statistic in Example 13.4 because of the standard errors. In Example 13.4, we calculated

$$s_p^2 = 311{,}330{,}926 \qquad \text{and} \qquad \sqrt{s_p^2\left(\frac{1}{n_1} + \frac{1}{n_2}\right)} = 4{,}991$$

Example 13.5 produced

$$s_D = 6{,}647 \qquad \text{and} \qquad \frac{s_D}{\sqrt{n_D}} = 1{,}329$$

As you can see, the difference in the test statistics was caused not by the numerator, but by the denominator. This raises another question: Why was the variation in the data of Example 13.4 so much greater than the variation in the data of Example 13.5? If you examine the data and statistics from Example 13.4, you will find that there was a great deal of variation *between* the salary offers in each sample. That is, some MBA graduates received high salary offers and others relatively low ones. This high level of variation, as expressed by s_p^2, made the difference between the sample means appear to be small. As a result, we could not conclude that finance majors attract higher salary offers.

Looking at the data from Example 13.5, we see that there is very little variation between the observations of the paired differences. The variation caused by different GPAs has markedly been decreased. The smaller variation causes the value of the test statistic to be larger. Consequently, we conclude that finance majors obtain higher salary offers.

2. Will the matched pairs experiment always produce a larger test statistic than the independent samples experiment? The answer is, not necessarily. Suppose that in our example we found that companies did not consider grade point averages when making decisions about how much to offer the MBA graduates. In such circumstances, the matched pairs experiment would result in no significant decrease in variation when compared to independent samples. It is possible that the matched pairs experiment may be less likely to reject the null hypothesis than the independent samples experiment. The reason can be seen by calculating the degrees of freedom. In Example 13.4, the number of degrees of freedom was 48, whereas in Example 13.5, it was 24. Even though we had the same number of observations (25 in each sample), the matched pairs experiment had half the number of degrees of freedom as the equivalent independent samples experiment. For exactly the same value of the test statistic, a smaller number of degrees of freedom in a Student t distributed test statistic yields a larger p-value. What this means is that if there is little reduction in variation to be achieved by the matched pairs experiment, the statistics practitioner should choose instead to conduct the experiment with independent samples.

3. As you've seen, in this book we deal with questions arising from experiments that have already been conducted. Consequently, one of your tasks is to determine the appropriate test statistic. In the case of comparing two populations of interval data, you must decide whether the samples are independent (in which case the parameter is $\mu_1 - \mu_2$) or matched pairs (in which case the parameter is μ_D) to select the correct test statistic. To help you do so, we suggest you ask and answer the following question: Does some natural relationship exist between each pair of observations that provides a logical reason to compare the first observation of sample 1 with the first observation of sample 2, the second observation of sample 1 with the second observation of sample 2, and so on? If so, the experiment was conducted by matched pairs. If not, it was conducted using independent samples.

Observational and Experimental Data

The points we made in Section 13.2 are also valid in this section. That is, we can design a matched pairs experiment where the data are gathered using a controlled experiment or by observation. The data in Examples 13.4 and 13.5 are observational.

As a consequence, when the statistical result provided evidence that finance majors attracted higher salary offers, it did not necessarily mean that students educated in finance are more attractive to prospective employers. It may be, for example, that better students major in finance and better students achieve higher starting salaries.

Checking the Required Condition

The validity of the results of the *t*-test and estimator of μ_D depends on the normality of the differences (or large enough sample sizes). The histogram of the differences (Figure 13.6) is positively skewed, but not enough so that the normality requirement is violated.

FIGURE **13.6**
Histogram of Differences in Example 13.5

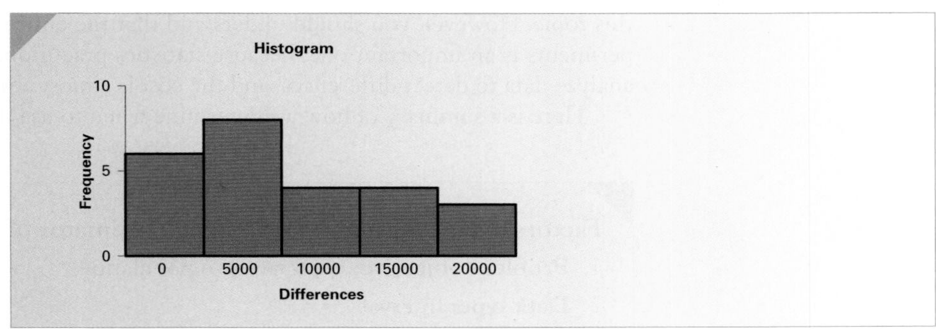

Violation of Required Condition

If the differences are very nonnormal, we cannot use the *t*-test of μ_D. We can, however, employ a nonparametric technique—the Wilcoxon signed rank sum test for matched pairs, which we present in Chapter 19.*

Developing an Understanding of Statistical Concepts 1

Two of the most important principles in statistics were applied in this section. The first is the concept of analyzing sources of variation. In Examples 13.4 and 13.5, we showed that by reducing the variation between salary offers in each sample, we were able to detect a real difference between the two majors. This was an application of the more general procedure of analyzing data and attributing some fraction of the variation to several sources. In Example 13.5, the two sources of variation were the GPA and the MBA major. However, we were not interested in the variation between graduates with differing GPAs. Instead we only wanted to eliminate that source of variation, making it easier to determine whether finance majors draw larger salary offers.

In Chapter 14, we will introduce a technique called the *analysis of variance* which does what its name suggests: It analyzes sources of variation in an attempt to detect real

*Instructors who wish to teach the use of nonparametric techniques for testing the mean difference when the normality requirement is not satisfied should use CD Appendix P, "Introduction to Nonparametric Techniques" and CD Appendix Q, "Wilcoxon Rank Sum Test and Wilcoxon Signed Rank Sum Test."

differences. In most applications of this procedure, we will be interested in each source of variation and not simply in reducing one source. We refer to the process as *explaining the variation*. The concept of explained variation is also applied in Chapters 16–18, where we introduce regression analysis.

Developing an Understanding of Statistical Concepts 2

The second principle demonstrated in this section is that statistics practitioners can design data-gathering procedures in such a way that they can analyze sources of variation. Before conducting the experiment in Example 13.5, the statistics practitioner suspected that there were large differences between graduates with different GPAs. Consequently, the experiment was organized so that the effects of those differences were mostly eliminated. It is also possible to design experiments that allow for easy detection of real differences and minimize the costs of data gathering. Unfortunately, we will not present this topic. However, you should understand that the entire subject of the design of experiments is an important one, because statistics practitioners often need to be able to analyze data to detect differences, and the cost is almost always a factor.

Here is a summary of how we determine when to use these techniques.

Factors That Identify the t-Test and Estimator of μ_D

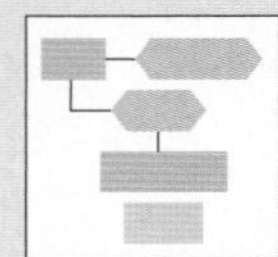

1. **Problem objective:** Compare two populations
2. **Data type:** Interval
3. **Descriptive measurement:** Central location
4. **Experimental design:** Matched pairs

EXERCISES

Applications

Conduct all tests of hypotheses at the 5% significance level.

13.40 Xr13-40 Many people use scanners to read documents and store them in a Word (or some other software) file. To help determine which brand of scanner to buy, a student conducts an experiment wherein 8 documents are scanned by each of the two scanners that he is interested in. He records the number of errors made by each. These data are listed here. Can he infer that Brand A (the more expensive scanner) is better than Brand B?

Document	1	2	3	4	5	6	7	8
Brand A	17	29	18	14	21	25	22	29
Brand B	21	38	15	19	22	30	31	37

13.41 Xr13-41 How effective are antilock brakes, which pump very rapidly rather than lock and thus avoid skids. As a test, a car buyer organized an experiment. He hit the brakes and, using a stopwatch, recorded the number of seconds it took to stop an ABS-equipped car and another identical car without ABS. The speeds when the brakes were applied and the number of seconds each took to stop on dry pavement are listed here. Can we infer that ABS is better?

Speeds	20	25	30	35	40	45	50	55
ABS	3.6	4.1	4.8	5.3	5.9	6.3	6.7	7.0
Non-ABS	3.4	4.0	5.1	5.5	6.4	6.5	6.9	7.3

13.42 Xr13-42 In a preliminary study to determine whether the installation of a camera designed to catch cars that go through red lights affects the number of violators, the number of red-light runners was recorded for each day of the week before and after the camera was installed. These data are listed here. Can we infer that the camera reduces the number of red-light runners?

Day	Sunday	Monday	Tuesday	Wednesday
Before	7	21	27	18
After	8	18	24	19

Day	Thursday	Friday	Saturday
Before	20	24	16
After	16	19	16

13.43 Xr13-43 In an effort to determine whether a new type of fertilizer is more effective than the type currently in use, researchers took 12 2-acre plots of land scattered throughout the county. Each plot was divided into two equal-size subplots, one of which was treated with the current fertilizer and the other of which was treated with the new fertilizer. Wheat was planted, and the crop yields were measured.

Plot	1	2	3	4	5	6	7	8	9	10	11	12
Current fertilizer	56	45	68	72	61	69	57	55	60	72	75	66
New fertilizer	60	49	66	73	59	67	61	60	58	75	72	68

a. Can we conclude at the 5% significance level that the new fertilizer is more effective than the current one?
b. Estimate with 95% confidence the difference in mean crop yields between the two fertilizers.
c. What is the required condition(s) for the validity of the results obtained in Parts a and b?
d. Is the required condition(s) satisfied?
e. Are these data experimental or observational? Explain.
f. How should the experiment be conducted if the researchers believed that the land throughout the county was essentially the same?

13.44 Xr13-44 The president of a large company is in the process of deciding whether to adopt a lunchtime exercise program. The purpose of such programs is to improve the health of workers and, in so doing, reduce medical expenses. To get more information, he instituted an exercise program for the employees in one office. The president knows that during the winter months medical expenses are relatively high because of the incidence of colds and flu. Consequently, he decides to use a matched pairs design by recording medical expenses for the 12 months before the program and for 12 months after the program. The "before" and "after" expenses (in thousands of dollars) are compared on a month-to-month basis and shown here.

Month	Jan.	Feb.	Mar.	Apr.	May.	Jun.
Before program	68	44	30	58	35	33
After program	59	42	20	62	25	30

Month	Jul.	Aug.	Sep.	Oct.	Nov.	Dec.
Before program	52	69	23	69	48	30
After program	56	62	25	75	40	26

a. Do the data indicate that exercise programs reduce medical expenses? (Test with $\alpha = .05$.)
b. Estimate with 95% confidence the mean savings produced by exercise programs.
c. Was it appropriate to conduct a matched pairs experiment? Explain.

The following exercises require the use of a computer and software. The answers may be calculated manually. See Appendix A for the sample statistics. ***Use a 5% significance level unless specified otherwise.***

13.45 Xr13-45 One measure of the state of the economy is the amount of money homeowners pay on their mortgage each month. To determine the extent of change between this year and 5 years ago, a random sample of 150 homeowners was drawn. The monthly mortgage payments for each homeowner for this year and for 5 years ago were recorded. (The amounts have been adjusted so that we're comparing constant dollars.) Can we infer that mortgage payments have risen over the past 5 years?

13.46 Xr13-46 Do waiters or waitresses earn larger tips? To answer this question, a restaurant consultant undertook a preliminary study. The study involved measuring the percentage of the total bill left as a tip for one randomly selected waiter and one randomly selected waitress in each of 50 restaurants during a 1-week period. What conclusions can be drawn from these data?

13.47 Xr13-47 To determine the effect of advertising in the Yellow Pages, Bell Telephone took a sample of 40 retail stores that did not advertise in the Yellow Pages last year but did so this year. The annual sales (in thousands of dollars) for each store in both years were recorded.

a. Estimate with 90% confidence the improvement in sales between the 2 years.
b. Can we infer that advertising in the Yellow Pages improves sales?
c. Check to ensure that the required condition(s) of the techniques used in Parts a and b is satisfied.
d. Would it be advantageous to perform this experiment with independent samples? Explain why or why not.

13.48 Xr13-48 Because of the high cost of energy, homeowners in northern climates need to find ways to cut their heating costs. A building contractor wanted to investigate the effect on heating costs of increasing the insulation. As an experiment, he located a large subdevelopment built around 1970 with minimal insulation. His plan was to insulate some of the houses and compare the heating costs in the insulated homes with those that remained uninsulated. However, it was clear to him that the size of the house was a critical factor in determining heating costs. Consequently, he found 16 pairs of identical-sized houses ranging from about 1,200 to 2,800 square feet. He insulated one house in each pair (levels of R20 in the walls and R32 in the attic) and left the other house unchanged. The heating cost for the following winter season was recorded for each house.

a. Do these data allow the contractor to infer at the 10% significance level that the heating cost for insulated houses is less than that for the uninsulated houses?
b. Estimate with 95% confidence the mean savings due to insulating the house.
c. What is the required condition for the use of the techniques in Parts a and b?

13.49 Xr13-49 The cost of health care is rising faster than most other items. To learn more about the problem, a survey was undertaken to determine whether differences in health care expenditures exist between men and women. The survey randomly sampled men and women aged 21, 22, . . . , 65 and determined the total amount spent on health care. Do these data allow us to infer that men and women spend different amounts on health care? (*Source:* Bureau of Labor Statistics, Consumer Expenditure Survey)

13.50 Xr13-50 The fluctuations in the stock market induce some investors to sell and move their money into more stable investments. To determine the degree to which recent fluctuations affected ownership, a random sample of 170 people who confirmed that they owned some stock was surveyed. The values of the holdings were recorded at the end of last year and at the end of the year before. Can we infer that the value of the stock holdings has decreased?

13.51 Xr13-51 Are Americans more deeply in debt this year compared to last year? To help answer this question, a statistics practitioner randomly sampled Americans this year and last year. The sampling was conducted so that the samples were matched by the age of the head of the household. For each, the ratio of debt payments to household income was recorded. Can we infer that the ratios are higher this year than last?

13.52 Xr13-52 Every April Americans and Canadians fill out their tax return forms. Many turn to tax preparation companies to do this tedious job. The question arises, Are there differences between companies? In an experiment, two of the largest companies were asked to prepare the tax returns of a sample of 55 taxpayers. The amounts of tax payable were recorded. Can we conclude that company 1's service results in higher tax payable?

13.53 Xr13-53 Refer to Exercise 13.27. Suppose now we redo the experiment in the following way. On 20 randomly selected cars, one of each type of tire is installed on the rear wheels and as before, the cars are driven until the tires wear out. The number of miles until wear-out occurred was recorded. Can we conclude from these data that the new tire is superior?

13.54 Refer to Exercises 13.27 and 13.53. Explain why the matched pairs experiment produced significant results whereas the independent samples t-test did not.

13.55 Xr13-55 Refer to Examples 13.4 and 13.5. Suppose that another experiment is conducted. Finance and marketing MBA majors were matched according to their undergraduate GPA. As in the previous examples, the highest starting salary offers were recorded. Can we infer from these data that finance majors attract higher salary offers than marketing majors?

13.56 Discuss why the experiment in Example 13.5 produced a significant test result whereas the one in Exercise 13.55 did not.

13.57 Xr13-57 Refer to Example 13.2. The actual after and before operating incomes were recorded.

a. Test to determine whether there is enough evidence to infer that for companies where an offspring takes the helm there is a decrease in operating income.
b. Is there sufficient evidence to conclude that when an outsider becomes CEO the operating income increases?

13.4 INFERENCE ABOUT THE RATIO OF TWO VARIANCES

In Sections 13.1 and 13.3, we dealt with statistical inference concerning the difference between two population means. The problem objective in each case was to compare two populations of interval data, and our interest was in comparing measures of central location. This section discusses the statistical technique to use when the problem objective and the data type are the same as in Sections 13.1 and 13.3, but our interest is in comparing variability. Here we will study the ratio of two population variances. We make inferences about the ratio because the sampling distribution is based on ratios rather than differences.

We have already encountered this technique when we used the F-test of two variances to determine which t-test and estimator of the difference between two means to use. In this section we apply the technique to other problems where our interest is in comparing the variability in two populations.

In the previous chapter, we presented the procedures used to draw inferences about a single population variance. We pointed out that variance can be used to address problems where we need to judge the consistency of a production process. We also use variance to measure the risk associated with a portfolio of investments. In this section we compare two variances, enabling us to compare the consistency of two production processes. We can also compare the relative risks of two sets of investments.

We will proceed in a manner that is probably becoming quite familiar.

Parameter

As you will see shortly, we compare two population variances by determining the ratio. Consequently, the parameter is σ_1^2/σ_2^2.

Statistic and Sampling Distribution

We have previously noted that the sample variance (defined in Chapter 4) is an unbiased and consistent estimator of the population variance. Not surprisingly, the estimator of the parameter σ_1^2/σ_2^2 is the ratio of the two sample variances drawn from their respective populations s_1^2/s_2^2.

The sampling distribution of s_1^2/s_2^2 is said to be F distributed provided that we have independently sampled from two normal populations. (The F distribution was introduced in Section 8.4.)

Statisticians have shown that the ratio of two independent chi-squared variables divided by their degrees of freedom is F distributed. The degrees of freedom of the F distribution are identical to the degrees of freedom for the two chi-squared distributions. In Section 12.2, we pointed out that $(n-1)s^2/\sigma^2$ is chi-squared distributed, provided that the sampled population is normal. If we have independent samples drawn from two normal populations, then both $(n_1-1)s_1^2/\sigma_1^2$ and $(n_2-1)s_2^2/\sigma_2^2$ are chi-squared distributed. If we divide each by their respective number of degrees of freedom and take the ratio, we produce

$$\frac{\dfrac{(n_1-1)s_1^2/\sigma_1^2}{(n_1-1)}}{\dfrac{(n_2-1)s_2^2/\sigma_2^2}{(n_2-1)}}$$

which simplifies to

$$\frac{s_1^2/\sigma_1^2}{s_2^2/\sigma_2^2}$$

This statistic is F distributed with $\nu_1 = n_1 - 1$ and $\nu_2 = n_2 - 1$ degrees of freedom. Recall that ν_1 is called the **numerator degrees of freedom** and ν_2 is called the **denominator degrees of freedom.**

Testing and Estimating a Ratio of Two Variances

In this book, our null hypothesis will always specify that the two variances are equal. As a result, the ratio will equal 1. Thus, the null hypothesis will always be expressed as

$$H_0: \quad \sigma_1^2/\sigma_2^2 = 1$$

The alternative hypothesis can state that the ratio σ_1^2/σ_2^2 is either not equal to 1, greater than 1, or less than 1. Technically, the test statistic is

$$F = \frac{s_1^2/\sigma_1^2}{s_2^2/\sigma_2^2}$$

However, under the null hypothesis, which states that $\sigma_1^2/\sigma_2^2 = 1$, the test statistic becomes as follows.

Test Statistic for σ_1^2/σ_2^2

The test statistic employed to test that σ_1^2/σ_2^2 is equal to 1 is

$$F = \frac{s_1^2}{s_2^2}$$

which is F distributed with $\nu_1 = n_1 - 1$ and $\nu_2 = n_2 - 1$ degrees of freedom provided that the populations are normal.

With the usual algebraic manipulation, we can derive the confidence interval estimator of the ratio of two population variances.

Confidence Interval Estimator of σ_1^2/σ_2^2

$$\text{LCL} = \left(\frac{s_1^2}{s_2^2}\right)\frac{1}{F_{\alpha/2,\nu_1,\nu_2}}$$

$$\text{UCL} = \left(\frac{s_1^2}{s_2^2}\right)F_{\alpha/2,\nu_2,\nu_1}$$

where $\nu_1 = n_1 - 1$ and $\nu_2 = n_2 - 1$

EXAMPLE 13.7

DATA Xm13-07

Testing the Quality of Two-Bottle Filling Machines

In Example 12.3, we applied the chi-squared test of a variance to determine whether there was sufficient evidence to conclude that the population variance was less than 1.0. Suppose that the statistics practitioner also collected data from another container-filling machine and recorded the fills of a randomly selected sample. Can we infer at the 5% significance level that the second machine is superior in its consistency?

SOLUTION

IDENTIFY

The problem objective is to compare two populations where the data are interval. Because we want information about the consistency of the two machines, the parameter we wish to test is σ_1^2/σ_2^2, where σ_1^2 is the variance of machine 1 and σ_2^2 is the variance for machine 2. We need to conduct the F-test of σ_1^2/σ_2^2 to determine whether the variance of population 2 is less than that of population 1. Expressed differently, we wish to determine whether there is enough evidence to infer that σ_1^2 is larger than σ_2^2. Hence, the hypotheses we test are

$$H_0: \quad \sigma_1^2/\sigma_2^2 = 1$$
$$H_1: \quad \sigma_1^2/\sigma_2^2 > 1$$

COMPUTE

MANUALLY

The sample variances are $s_1^2 = .6333$ and $s_2^2 = .4528$.
The value of the test statistic is

$$F = \frac{s_1^2}{s_2^2} = \frac{.6333}{.4528} = 1.40$$

The rejection region is

$$F > F_{\alpha,\nu_1,\nu_2} = F_{.05,24,24} = 1.98$$

Since the value of the test statistic is not greater than 1.98, we cannot reject the null hypothesis.

EXCEL

	A	B	C
1	F-Test Two-Sample for Variances		
2			
3		*Machine 1*	*Machine 2*
4	Mean	999.7	999.8
5	Variance	0.6333	0.4528
6	Observations	25	25
7	df	24	24
8	F	1.3988	
9	P(F<=f) one-tail	0.2085	
10	F Critical one-tail	1.9838	

The value of the test statistic is $F = 1.3988$. Excel outputs the one-tail p-value, which is .2085.

INSTRUCTIONS

1. Type or import the data into two columns. (Open Xm13-07.)
2. Click **Data, Data Analysis,** and **F-Test Two-Sample for Variances.**
3. Specify the **Variable 1 Range** (A1:A26) and the **Variable 2 Range** (B1:B26). Type a value for α (.05).

Use the **F-Test of 2 Variances** worksheet to complete this test from the sample variances and to perform a what-if analysis (CD Appendix J).

MINITAB

Test for Equal Variances: Machine 1, Machine 2

F-Test (Normal Distribution)
Test statistic = 1.40, p-value = 0.417

Note that Minitab conducts a two-tail test. Thus, the p-value = .417/2 = .2085.

INSTRUCTIONS

1. Type or import the data into two columns. (Open Xm13-07.)
2. Click **Stat, Basic Statistics,** and **2 Variances . . . .**
3. In the **Samples in different columns** box, select the **First** (Machine 1) and **Second** (Machine 2) variables.

INTERPRET

There is not enough evidence to infer that the variance of machine 2 is less than the variance of machine 1. We're confident that the normality requirement for this test is satisfied. It is the same requirement for the t-test, which we checked when we drew the histograms. (See Figures 13.2 and 13.3.)

EXAMPLE 13.8

Estimating the Ratio of the Variances in Example 13.7

DATA
Xm13-07

Determine the 95% confidence interval estimate of the ratio of the two population variances in Example 13.7.

SOLUTION

COMPUTE

MANUALLY

We find

$$F_{\alpha/2,\nu_1,\nu_2} = F_{.025,24,24} = 2.27$$

Thus,

$$\text{LCL} = \left(\frac{s_1^2}{s_2^2}\right)\frac{1}{F_{\alpha/2,\nu_1,\nu_2}} = \left(\frac{.6333}{.4528}\right)\frac{1}{2.27} = .616$$

$$\text{UCL} = \left(\frac{s_1^2}{s_2^2}\right)F_{\alpha/2,\nu_2,\nu_1} = \left(\frac{.6333}{.4528}\right)2.27 = 3.17$$

We estimate that σ_1^2/σ_2^2 lies between .616 and 3.17

EXCEL

	A	B	C	D	E
1	F-Estimate of the Ratio of Two Variances				
2					
3		Sample 1	Sample 2	Confidence Interval Estimate	
4	Sample variance	0.6333	0.4528	Lower confidence limit	0.6164
5	Sample size	25	25	Upper confidence limit	3.1741
6	Confidence level	0.95			

INSTRUCTIONS

Open the **F-Estimator_2 Variances** worksheet in the **Estimators** workbook and substitute sample variances, samples sizes, and the confidence level.

MINITAB

Minitab does not compute the estimate of the ratio of two variances.

INTERPRET

As we pointed out in Chapter 11, we can often use a confidence interval estimator to test hypotheses. In this example the interval estimate excludes the value of 1. Consequently, we can draw the same conclusion as we did in Example 13.7.

Factors That Identify the *F*-Test and Estimator of σ_1^2/σ_2^2

1. **Problem objective:** Compare two populations
2. **Data type:** Interval
3. **Descriptive measurement:** Variability

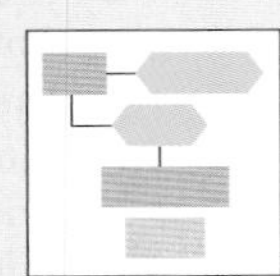

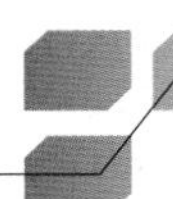

EXERCISES

Developing an Understanding of Statistical Concepts

*Exercises 13.58 and 13.59 are "what-if analyses" designed to determine what happens to the test statistics and interval estimates when elements of the statistical inference change. These problems can be solved manually or using Excel's **Test Statistics** or **Estimators** workbooks.*

13.58 Random samples from two normal populations produced the following statistics:

$$s_1^2 = 350 \quad n_1 = 30 \quad s_2^2 = 700 \quad n_2 = 30$$

a. Can we infer at the 10% significance level that the two population variances differ?

b. Repeat Part a, changing the sample sizes to $n_1 = 15$ and $n_2 = 15$.

c. Describe what happens to the test statistic when the sample sizes decrease.

13.59 Random samples from two normal populations produced the following statistics:

$$s_1^2 = 28 \quad n_1 = 10 \quad s_2^2 = 19 \quad n_2 = 10$$

a. Estimate with 95% confidence the ratio of the two population variances.

b. Repeat Part a, changing the sample sizes to $n_1 = 25$ and $n_2 = 25$.

c. Describe what happens to the width of the confidence interval estimate when the sample sizes increase.

Applications

Use a 5% significance level in all tests unless specified otherwise.

13.60 Xr13-60 The manager of a dairy is in the process of deciding which of two new carton-filling machines to use. The most important attribute is the consistency of the fills. In a preliminary study, she measured the fills in the 1-liter carton and listed them here. Can the manager infer that the two machines differ in their consistency of fills?

Machine 1 .998 .997 1.003 1.000 .999 1.000 .998 1.003 1.004 1.000

Machine 2 1.003 1.004 .997 .996 .999 1.003 1.000 1.005 1.002 1.004 .996

13.61 Xr13-61 An operations manager who supervises an assembly line has been experiencing problems with the sequencing of jobs. The problem is that bottlenecks are occurring because of the inconsistency of sequential operations. He decides to conduct an experiment wherein two different methods are used to complete the same task. He measures the times (in seconds). The data are listed here. Can he infer that the second method is more consistent than the first method?

Method 1	8.8	9.6	8.4	9.0	8.3	9.2	9.0	8.7	8.5	9.4
Method 2	9.2	9.4	8.9	9.6	9.7	8.4	8.8	8.9	9.0	9.7

13.62 Xr13-62 A statistics professor hypothesized that not only would the means vary, but also would the variances if the business statistics course was taught in two different ways but had the same final exam. He organized an experiment wherein one section of the course was taught using detailed PowerPoint® slides whereas the other required students to read the book and answer questions in class discussions. A sample of the marks was recorded and listed next. Can we infer that the variances of the marks differ between the two sections?

Class 1	64	85	80	64	48	62	75	77	50	81	90
Class 2	73	78	66	69	79	81	74	59	83	79	84

The following exercises require the use of a computer and software. The answers may be calculated manually. See Appendix A for the sample statistics.

13.63 Xr13-63 A new highway has just been completed and the government must decide on speed limits. There are several possible choices. However, on advice from police who monitor traffic, the objective was to reduce the variation in speeds, which is thought to contribute to the number of collisions. It has been acknowledged that speed contributes to the severity of collisions. It is decided to conduct an experiment to acquire more information. Signs are posted for 1 week indicating that the speed limit is 70 mph. A random sample of cars' speeds is measured. During the second week, signs are posted indicating that the maximum speed is 70 mph and that the minimum speed is 60 mph. Once again a random sample of speeds is measured. Can we infer that limiting the minimum and maximum speeds reduces the variation in speeds?

13.64 Xr13-64 In Exercise 12.52 we described the problem of whether to change all the light bulbs at Yankee Stadium or change them one by one as they burn out. There are two brands of bulbs that can be used. Because both the mean and the variance of the lengths of life are important, it was decided to test the two brands. A random sample of both brands was drawn and left on until they burned out. The times were recorded. Can the Yankee Stadium management conclude that the variances differ?

13.65 Xr13-65 In deciding where to invest her retirement fund, an investor recorded the weekly returns of two portfolios for 1 year. Can we conclude that portfolio 2 is riskier than portfolio 1?

13.66 Xr13-66 An important statistical measurement in service facilities (such as restaurants and banks) is the variability in service times. As an experiment, two bank tellers were observed, and the service times for each of 100 customers were recorded. Do these data allow us to infer at the 10% significance level that the variance in service times differs between the two tellers?

13.5 Inference about the Difference between Two Population Proportions

In this section, we present the procedures for drawing inferences about the difference between populations whose data are nominal. The number of applications of these techniques is almost limitless. For example, pharmaceutical companies test new drugs by comparing the new and old or the new versus a placebo. Marketing managers compare market shares before and after advertising campaigns. Operations managers compare defective rates between two machines. Political pollsters measure the difference in popularity before and after an election.

Parameter

When data are nominal, the only meaningful computation is to count the number of occurrences of each type of outcome and calculate proportions. Consequently, the parameter to be tested and estimated in this section is the difference between two population proportions $p_1 - p_2$.

Statistic and Sampling Distribution

To draw inferences about $p_1 - p_2$, we take a sample of size n_1 from population 1 and a sample of size n_2 from population 2 (Figure 13.7 depicts the sampling process).

FIGURE **13.7** Sampling from Two Populations of Nominal Data

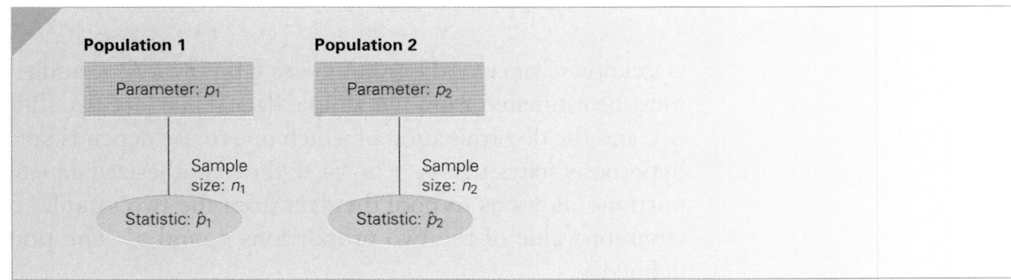

For each sample, we count the number of successes (recall that we call anything we're looking for a success), which we label x_1, and x_2, respectively. The sample proportions are then computed:

$$\hat{p}_1 = \frac{x_1}{n_1} \quad \text{and} \quad \hat{p}_2 = \frac{x_2}{n_2}$$

Statisticians have proven that the statistic $\hat{p}_1 - \hat{p}_2$ is an unbiased consistent estimator of the parameter $p_1 - p_2$. Using the same mathematics as we did in Chapter 9 to derive the sampling distribution of the sample proportion $\hat{p}$, we determine the sampling distribution of the difference between two sample proportions.

Sampling Distribution of $\hat{p}_1 - \hat{p}_2$

1. The statistic $\hat{p}_1 - \hat{p}_2$ is approximately normally distributed provided that the sample sizes are large enough so that n_1p_1, $n_1(1 - p_1)$, n_2p_2, and $n_2(1 - p_2)$ are all greater than or equal to 5. [Because p_1 and p_2 are unknown, we express the sample size requirement as $n_1\hat{p}_1$, $n_1(1 - \hat{p}_1)$, $n_2\hat{p}_2$, and $n_2(1 - \hat{p}_2)$ are greater than or equal to 5.]
2. The mean of $\hat{p}_1 - \hat{p}_2$ is
$$E(\hat{p}_1 - \hat{p}_2) = p_1 - p_2$$
3. The variance of $\hat{p}_1 - \hat{p}_2$ is
$$V(\hat{p}_1 - \hat{p}_2) = \frac{p_1(1 - p_1)}{n_1} + \frac{p_2(1 - p_2)}{n_2}$$
The standard error is
$$\sigma_{\hat{p}_1-\hat{p}_2} = \sqrt{\frac{p_1(1 - p_1)}{n_1} + \frac{p_2(1 - p_2)}{n_2}}$$

Thus, the variable

$$z = \frac{(\hat{p}_1 - \hat{p}_2) - (p_1 - p_2)}{\sqrt{\dfrac{p_1(1 - p_1)}{n_1} + \dfrac{p_2(1 - p_2)}{n_2}}}$$

is approximately standard normally distributed.

Testing and Estimating the Difference between Two Proportions

We would like to use the z-statistic just described as our test statistic; however, the standard error of $\hat{p}_1 - \hat{p}_2$, which is

$$\sigma_{\hat{p}_1 - \hat{p}_2} = \sqrt{\frac{p_1(1 - p_1)}{n_1} + \frac{p_2(1 - p_2)}{n_2}}$$

is unknown, since both p_1 and p_2 are unknown. As a result, the standard error of $\hat{p}_1 - \hat{p}_2$ must be estimated from the sample data. There are two different estimators of this quantity, and the determination of which one to use depends on the null hypothesis. If the null hypothesis states that $p_1 - p_2 = 0$, the hypothesized equality of the two population proportions allows us to pool the data from the two samples to produce an estimate of the common value of the two proportions p_1 and p_2. The **pooled proportion estimate** is defined as

$$\hat{p} = \frac{x_1 + x_2}{n_1 + n_2}$$

Thus, the estimated standard error of $\hat{p}_1 - \hat{p}_2$ is

$$\sqrt{\frac{\hat{p}(1 - \hat{p})}{n_1} + \frac{\hat{p}(1 - \hat{p})}{n_2}} = \sqrt{\hat{p}(1 - \hat{p})\left(\frac{1}{n_1} + \frac{1}{n_2}\right)}$$

The principle used in estimating the standard error of $\hat{p}_1 - \hat{p}_2$ is analogous to that applied in Section 13.1 to produce the pooled variance estimate s_p^2, which is used to test $\mu_1 - \mu_2$ with σ_1^2 and σ_2^2 unknown, but equal. The principle roughly states that, where possible, pooling data from two samples produces a better estimate of the standard error. Here, pooling is made possible by hypothesizing (under the null hypothesis) that $p_1 = p_2$. (In Section 13.1, we used the pooled variance estimate because we assumed that $\sigma_1^2 = \sigma_2^2$.) We will call this application Case 1.

Test Statistic for $p_1 - p_2$: Case 1

If the null hypothesis specifies

$$H_0: \quad (p_1 - p_2) = 0$$

the test statistic is

$$z = \frac{(\hat{p}_1 - \hat{p}_2) - (p_1 - p_2)}{\sqrt{\hat{p}(1 - \hat{p})\left(\dfrac{1}{n_1} + \dfrac{1}{n_2}\right)}}$$

Because we hypothesize that $p_1 - p_2 = 0$, we simplify the test statistic to

$$z = \frac{(\hat{p}_1 - \hat{p}_2)}{\sqrt{\hat{p}(1 - \hat{p})\left(\dfrac{1}{n_1} + \dfrac{1}{n_2}\right)}}$$

The second case applies when, under the null hypothesis, we state that $p_1 - p_2 = D$, where D is some value other than 0. Under such circumstances, we cannot pool the sample data to estimate the standard error of $\hat{p}_1 - \hat{p}_2$. The appropriate test statistic is described next as Case 2.

Test Statistic for $p_1 - p_2$: Case 2

If the null hypothesis specifies

$$H_0: \quad (p_1 - p_2) = D \qquad (D \neq 0)$$

the test statistic is

$$z = \frac{(\hat{p}_1 - \hat{p}_2) - (p_1 - p_2)}{\sqrt{\dfrac{\hat{p}_1(1 - \hat{p}_1)}{n_1} + \dfrac{\hat{p}_2(1 - \hat{p}_2)}{n_2}}}$$

which can also be expressed as

$$z = \frac{(\hat{p}_1 - \hat{p}_2) - D}{\sqrt{\dfrac{\hat{p}_1(1 - \hat{p}_1)}{n_1} + \dfrac{\hat{p}_2(1 - \hat{p}_2)}{n_2}}}$$

Notice that this test statistic is determined by simply substituting the sample statistics $\hat{p}_1$ and $\hat{p}_2$ in the standard error of $\hat{p}_1 - \hat{p}_2$.

You will find that, in most practical applications (including the exercises in this book), Case 1 applies—in most problems, we want to know whether the two population proportions differ, that is,

$$H_1: \quad (p_1 - p_2) \neq 0$$

or if one proportion exceeds the other, that is,

$$H_1: \quad (p_1 - p_2) > 0 \qquad \text{or} \qquad H_1: \quad (p_1 - p_2) < 0$$

In some other problems, however, the objective is to determine whether one proportion exceeds the other by a specific nonzero quantity. In such situations, Case 2 applies.

We derive the interval estimator of $p_1 - p_2$ in the same manner we have been using since Chapter 10.

Confidence Interval Estimator of $p_1 - p_2$

$$(\hat{p}_1 - \hat{p}_2) \pm z_{\alpha/2}\sqrt{\frac{\hat{p}_1(1 - \hat{p}_1)}{n_1} + \frac{\hat{p}_2(1 - \hat{p}_2)}{n_2}}$$

This formula is valid when $n_1\hat{p}_1$, $n_1(1 - \hat{p}_1)$, $n_2\hat{p}_2$, and $n_2(1 - \hat{p}_2)$ are greater than or equal to 5.

Notice that the standard error is estimated using the individual sample proportions rather than the pooled proportion. In this procedure we cannot assume that the population proportions are equal as we did in the Case 1 test statistic.

APPLICATIONS in MARKETING

© White Packert/The Image Bank/Getty Images

Test Marketing

Marketing managers frequently make use of test marketing to assess consumer reaction to a change in a characteristic (such as price or packaging) of an existing product, or to assess consumers' preferences regarding a proposed new product. *Test marketing* involves experimenting with changes to the marketing mix in a small, limited test market and assessing consumers' reaction in the test market before undertaking costly changes in production and distribution for the entire market.

EXAMPLE 13.9 Test Marketing of Package Designs, Part 1

DATA Xm13-09

The General Products Company produces and sells a variety of household products. Because of stiff competition, one of its products, a bath soap, is not selling well. Hoping to improve sales, General Products decided to introduce more attractive packaging. The company's advertising agency developed two new designs. The first design features several bright colors to distinguish it from other brands. The second design is light green in color with just the company's logo on it. As a test to determine which design is better, the marketing manager selected two supermarkets. In one supermarket the soap was packaged in a box using the first design and in the second supermarket the second design was used. The product scanner at each supermarket tracked every buyer of soap over a 1-week period. The supermarkets recorded the last four digits of the scanner code for each of the five brands of soap the supermarket sold. The code for the General Products brand of soap is 9077 (the other codes are 4255, 3745, 7118, and 8855). After the trial period, the scanner data were transferred to a computer file. Because the first design is more expensive, management has decided to use this design only if there is sufficient evidence to allow them to conclude that it is better. Should management switch to the brightly colored design or the simple green one?

SOLUTION

IDENTIFY

The problem objective is to compare two populations. The first is the population of soap sales in supermarket 1 and the second is the population of soap sales in supermarket 2. The data are nominal because the values are "buy General Products soap" and "buy other companies' soap." These two factors tell us that the parameter to be tested is the difference between two population proportions $p_1 - p_2$ (where p_1 and p_2 are the proportions of soap sales that are a General Products brand in supermarkets 1 and 2, respectively). Because we want to know whether there is enough evidence to adopt the brightly colored design, the alternative hypothesis is

$$H_1: \quad (p_1 - p_2) > 0$$

The null hypothesis must be

$$H_0: \quad (p_1 - p_2) = 0$$

which tells us that this is an application of Case 1. Thus, the test statistic is

$$z = \frac{(\hat{p}_1 - \hat{p}_2)}{\sqrt{\hat{p}(1 - \hat{p})\left(\frac{1}{n_1} + \frac{1}{n_2}\right)}}$$

COMPUTE

MANUALLY

To compute the test statistic manually requires the statistics practitioner to tally the number of successes in each sample, where success is represented by the code 9077. Reviewing all the sales reveals that

$x_1 = 180 \qquad n_1 = 904 \qquad x_2 = 155 \qquad n_2 = 1{,}038$

The sample proportions are

$$\hat{p}_1 = \frac{180}{904} = .1991$$

and

$$\hat{p}_2 = \frac{155}{1{,}038} = .1493$$

The pooled proportion is

$$\hat{p} = \frac{180 + 155}{904 + 1{,}038} = \frac{335}{1{,}942} = .1725$$

The value of the test statistic is

$$z = \frac{(\hat{p}_1 - \hat{p}_2)}{\sqrt{\hat{p}(1 - \hat{p})\left(\frac{1}{n_1} + \frac{1}{n_2}\right)}} = \frac{(.1991 - .1493)}{\sqrt{(.1725)(1 - .1725)\left(\frac{1}{904} + \frac{1}{1{,}038}\right)}} = 2.90$$

A 5% significance level seems to be appropriate. Thus, the rejection region is

$$z > z_\alpha = z_{.05} = 1.645$$

EXCEL

	A	B	C	D
1	**z-Test: Two Proportions**			
2				
3			*Supermarket 1*	*Supermarket 2*
4	Sample Proportions		0.1991	0.1493
5	Observations		904	1038
6	Hypothesized Difference		0	
7	z Stat		2.90	
8	P(Z<=z) one tail		0.0019	
9	z Critical one-tail		1.6449	
10	P(Z<=z) two-tail		0.0038	
11	z Critical two-tail		1.96	

INSTRUCTIONS

1. Type or import the data into two adjacent columns. (Open Xm13-09.)
2. Click **Add-Ins, Data Analysis Plus,** and **Z-Test: 2 Proportions.**
3. Specify the **Variable 1 Range** (A1:A905) and the **Variable 2 Range** (B1:B1039). Type the **Code for Success** (9077), the **Hypothesized Difference** (0), and a value for α (.05).

To conduct this procedure from the sample proportions or to perform a what-if analysis, activate the **z-Test_2 Proportions (Case 1)** worksheet in the **Test Statistics** workbook (CD Appendix J). (CD Appendix S describes how to use an Excel spreadsheet to calculate the probability of a Type II error when testing the difference between two proportions.)

MINITAB

Test and CI for Two Proportions: Supermarket 1, Supermarket 2

Event = 9077

Variable	X	N	Sample p
Supermarket 1	180	904	0.199115
Supermarket 2	155	1038	0.149326

Difference = p (Supermarket 1) – p (Supermarket 2)
Estimate for difference: 0.0497894
95% lower bound for difference: 0.0213577
Test for difference = 0 (vs > 0): Z = 2.90 P-Value = 0.002

INSTRUCTIONS

1. Type or import the data into two adjacent columns. (Open Xm13-09.) Recode the data if necessary. (Minitab requires that there be only two codes and the higher value is deemed to be a success. See CD Appendix N.)
2. Click **Stat, Basic Statistics,** and **2 Proportions . . . .**
3. In the **Samples in different columns,** specify the **First** (Supermarket 1) and **Second** (Supermarket 2) samples. Click **Options . . . .**
4. Type the value of the **Test difference** (0), specify the **Alternative** hypothesis (greater than), and click **Use pooled estimate of p for test.**

INTERPRET

The value of the test statistic is $z = 2.90$; its p-value is .0019. There is enough evidence to infer that the brightly colored design is more popular than the simple design. As a result, it is recommended that management switch to the first design.

EXAMPLE 13.10 Test Marketing of Package Designs, Part 2

DATA Xm13-09

Suppose that in Example 13.9 the additional cost of the brightly colored design requires that it outsell the simple design by more than 3%. Should management switch to the brightly colored design?

SOLUTION

IDENTIFY

The alternative hypothesis is

$$H_1: \quad (p_1 - p_2) > .03$$

and the null hypothesis follows as

$$H_0: \quad (p_1 - p_2) = .03$$

Because the null hypothesis specifies a nonzero difference, we would apply the Case 2 test statistic.

COMPUTE

MANUALLY

The value of the test statistic is

$$z = \frac{(\hat{p}_1 - \hat{p}_2) - (p_1 - p_2)}{\sqrt{\frac{\hat{p}_1(1 - \hat{p}_1)}{n_1} + \frac{\hat{p}_2(1 - \hat{p}_2)}{n_2}}} = \frac{(.1991 - .1493) - (.03)}{\sqrt{\frac{.1991(1 - .1991)}{904} + \frac{.1493(1 - .1493)}{1,038}}} = 1.15$$

EXCEL

	A	B	C	D
1	**z-Test: Two Proportions**			
2				
3			*Supermarket 1*	*Supermarket 2*
4	Sample Proportions		0.1991	0.1493
5	Observations		904	1038
6	Hypothesized Difference		0.03	
7	z Stat		1.14	
8	P(Z<=z) one tail		0.1261	
9	z Critical one-tail		1.6449	
10	P(Z<=z) two-tail		0.2522	
11	z Critical two-tail		1.96	

INSTRUCTIONS

Use the same commands we used previously, except specify that the **Hypothesized Difference** is .03. Excel will apply the Case 2 test statistic when a nonzero value is typed.

MINITAB

Test and CI for Two Proportions: Supermarket 1, Supermarket 2

Event = 9077

Variable	X	N	Sample p
Supermarket 1	180	904	0.199115
Supermarket 2	155	1038	0.149326

Difference = p (Supermarket 1) – p (Supermarket 2)
Estimate for difference: 0.0497894
95% lower bound for difference: 0.0213577
Test for difference = 0.03 (vs > 0.03): Z = 1.14 P-Value = 0.126

INSTRUCTIONS

Use the same commands detailed previously, except at step 4, specify that the **Test difference** is .03 and do not click **Use pooled estimate of p for test.**

INTERPRET

There is not enough evidence to infer that the proportion of soap customers who buy the product with the brightly colored design is more than 3% higher than the proportion of soap customers who buy the product with the simple design. In the absence of sufficient evidence, the analysis suggests that the product should be packaged using the simple design.

EXAMPLE 13.11 Test Marketing of Package Designs, Part 3

DATA Xm13-09

To help estimate the difference in profitability, the marketing manager in Examples 13.9 and 13.10 would like to estimate the difference between the two proportions. A confidence level of 95% is suggested.

SOLUTION

IDENTIFY

The parameter is $p_1 - p_2$, which is estimated by the following confidence interval estimator:

$$(\hat{p}_1 - \hat{p}_2) \pm z_{\alpha/2}\sqrt{\frac{\hat{p}_1(1 - \hat{p}_1)}{n_1} + \frac{\hat{p}_2(1 - \hat{p}_2)}{n_2}}$$

COMPUTE

MANUALLY

The sample proportions have already been computed. They are

$$\hat{p}_1 = \frac{180}{904} = .1991$$

and

$$\hat{p}_2 = \frac{155}{1,038} = .1493$$

The 95% confidence interval estimate of $p_1 - p_2$ is

$$(\hat{p}_1 - \hat{p}_2) \pm z_{\alpha/2}\sqrt{\frac{\hat{p}_1(1 - \hat{p}_1)}{n_1} + \frac{\hat{p}_2(1 - \hat{p}_2)}{n_2}}$$

$$= (.1991 - .1493) \pm 1.96\sqrt{\frac{.1991(1 - .1991)}{904} + \frac{.1493(1 - .1493)}{1,038}}$$

$$= .0498 \pm .0339$$

$$\text{LCL} = .0159 \quad \text{and} \quad \text{UCL} = .0837$$

EXCEL

	A	B	C	D
1	z-Estimate: Two Proportions			
2				
3			*Supermarket 1*	*Supermarket 2*
4	Sample Proportions		0.1991	0.1493
5	Observations		904	1038
6				
7	LCL		0.0159	
8	UCL		0.0837	

INSTRUCTIONS

1. Type or import the data into two adjacent columns. (Open Xm13-09.)
2. Click **Add-Ins, Data Analysis Plus,** and **Z-Estimate: 2 Proportions.**
3. Specify the **Variable 1 Range** (A1:A905) and the **Variable 2 Range** (B1:B1039). Specify the **Code for Success** (9077) and a value for α (.05).

To produce the confidence interval estimate of the difference between two proportions or to perform a what-if analysis, activate the **z-Estimate_2 Proportions** worksheet in the **Estimators** workbook (output shown in CD Appendix J).

MINITAB

Test and CI for Two Proportions: Supermarket 1, Supermarket 2

Event = 9077

Variable	X	N	Sample p
Supermarket 1	180	904	0.199115
Supermarket 2	155	1038	0.149326

Difference = p (Supermarket 1) – p (Supermarket 2)
Estimate for difference: 0.0497894
95% CI for difference: (0.0159109, 0.0836679)
Test for difference = 0 (vs not = 0): Z = 2.88 P-Value = 0.004

INSTRUCTIONS

Follow the commands to test hypotheses about two proportions. Specify the alternative hypothesis as **not equal** and do not click **Use pooled estimate of p for test.**

INTERPRET

We estimate that the market share for the brightly colored design is between 1.59% and 8.37% larger than the market share for the simple design.

Estimating the Cost of a Life Saved: Solution

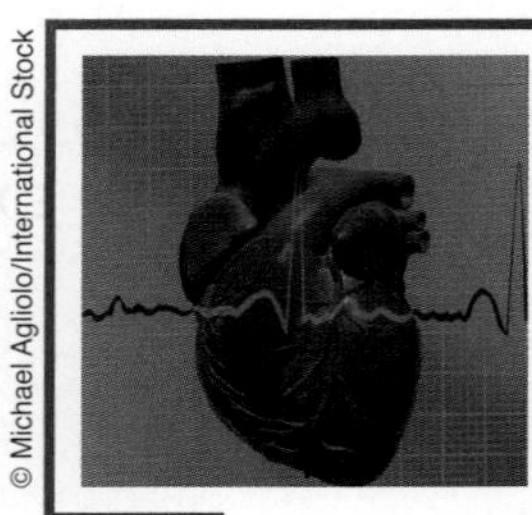
© Michael Agliolo/International Stock

IDENTIFY

The problem objective is to compare two populations: the outcomes of the treatments with Streptokinase and with t-PA. The data are nominal because we record only whether the patient lived or died. Thus, the parameter is $p_1 - p_2$, where p_1 = death rate with Streptokinase and p_2 = death rate with t-PA. Because we wish to estimate the cost per life saved, we first must estimate the difference in death rates between the two drugs. We'll use a 95% confidence level.

COMPUTE

MANUALLY

The sample proportions are

$$\hat{p}_1 = \frac{1{,}497}{20{,}500} = .0730$$

and

$$\hat{p}_2 = \frac{1{,}292}{20{,}500} = .0630$$

The 95% confidence interval estimate of the difference between death rates is

$$(\hat{p}_1 - \hat{p}_2) \pm z_{\alpha/2}\sqrt{\frac{\hat{p}_1(1-\hat{p}_1)}{n_1} + \frac{\hat{p}_2(1-\hat{p}_2)}{n_2}}$$

$$= (.0730 - .0630) \pm 1.96\sqrt{\frac{.0730(1-.0730)}{20{,}500} + \frac{.0630(1-.0630)}{20{,}500}}$$

$$= .0100 \pm .0049$$

LCL = .0051 and UCL = .0149

EXCEL

	A	B	C	D	E	F
1	z-Estimate of the Difference Between Two Proportions					
2						
3		Sample 1	Sample 2	Confidence Interval Estimate		
4	Sample proportion	0.0730	0.0630	0.0100	±	0.0049
5	Sample size	20500	20500	Lower confidence limit		0.0051
6	Confidence level	0.95		Upper confidence limit		0.0149

INSTRUCTIONS

Open the **z-Estimate_2 Proportions** worksheet in the **Estimators** workbook and substitute the sample proportions, sample sizes, and confidence level.

MINITAB

Test and CI for Two Proportions

```
Sample     X      N   Sample p
1       1497  20500   0.073024
2       1292  20500   0.063024


Difference = p (1) – p (2)
Estimate for difference:  0.01
95% CI for difference:  (0.00512657, 0.0148734)
Test for difference = 0 (vs not = 0):  Z = 4.02  P-Value = 0.000
```

INSTRUCTIONS

1. Click **Stat, Basic Statistics,** and **2 Proportions . . . .**
2. Click **Summarized data** and type the sample sizes in the **First** (20500) and **Second** (20500) **Trials** boxes and the values of x_1 and x_2 in the **First** (1497) and **Second** (1292) **Events** boxes. Click **Options . . . .**
3. Type the value of the **Confidence level** and specify the **Alternative** as **not equal.**

INTERPRET

We estimate that between .51% and 1.49% more heart attack victims will survive because of the use of t-PA instead of Streptokinase. However, the difference in cost is \$2,900 − \$460 = \$2,440. The cost per life saved by switching to t-PA is estimated to fall between

LCL = 2,440/.0149 = \$163,758

and

UCL = 2,440/.0051 = \$478,431

The factors that identify the inference about the difference between two proportions are listed below.

Factors That Identify the *z*-Test and Estimator of $p_1 - p_2$

1. **Problem objective:** Compare two populations
2. **Data type:** Nominal

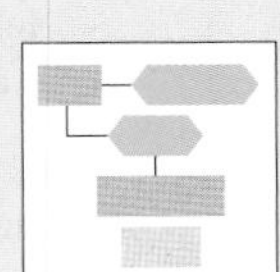

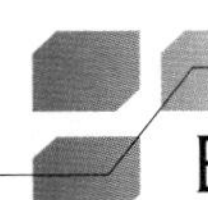

EXERCISES

Developing an Understanding of Statistical Concepts

Exercises 13.67–13.72 are "what-if analyses" designed to determine what happens to the test statistics and interval estimates when elements of the statistical inference change. These problems can be solved manually, using Excel's ***Test Statistics*** *or* ***Estimators*** *workbooks, or using Minitab.*

13.67 Random samples from two binomial populations yielded the following statistics:

$\hat{p}_1 = .45 \quad n_1 = 100 \quad \hat{p}_2 = .40 \quad n_2 = 100$

a. Calculate the *p*-value of a test to determine whether we can infer that the population proportions differ.
b. Repeat Part a, increasing the sample sizes to 400.
c. Describe what happens to the *p*-value when the sample sizes increase.

13.68 These statistics were calculated from two random samples:

$\hat{p}_1 = .60 \quad n_1 = 225 \quad \hat{p}_2 = .55 \quad n_2 = 225$

a. Calculate the *p*-value of a test to determine whether there is evidence to infer that the population proportions differ.
b. Repeat Part a with $\hat{p}_1 = .95$ and $\hat{p}_2 = .90$.
c. Describe the effect on the *p*-value of increasing the sample proportions.
d. Repeat Part a with $\hat{p}_1 = .10$ and $\hat{p}_2 = .05$.
e. Describe the effect on the *p*-value of decreasing the sample proportions.

13.69 After sampling from two binomial populations we found the following:

$\hat{p}_1 = .18 \quad n_1 = 100 \quad \hat{p}_2 = .22 \quad n_2 = 100$

a. Estimate with 90% confidence the difference in population proportions.
b. Repeat Part a, increasing the sample proportions to .48 and .52, respectively.
c. Describe the effects of increasing the sample proportions.

Applications

13.70 Many stores sell extended warranties for products they sell. These are very lucrative for store owners. To learn more about who buys these warranties, a random sample of a store's customers who recently purchased a product for which an extended warranty was available was drawn. Among other variables each respondent reported whether they paid the regular price or a sale price and whether they purchased an extended warranty.

	Regular Price	Sale Price
Sample size	229	178
Number who bought extended warranty	47	25

Can we conclude at the 10% significance level that those who paid the regular price are more likely to buy an extended warranty?

13.71 A firm has classified its customers in two ways: (1) according to whether the account is overdue and (2) whether the account is new (less than 12 months) or old. To acquire information about which customers are paying on time and which are overdue, a random sample of 292 customer accounts was drawn. Each was categorized as a new account (less than 12 months) and old, and whether the customer has paid or is overdue. The results are summarized next.

	New Account	Old Account
Sample size	83	209
Overdue account	12	49

Is there enough evidence at the 5% significance level to infer that new and old accounts are different with respect to overdue accounts?

13.72 Credit scorecards are used by financial institutions to help decide to whom loans should be granted (see the Applications in Banking: Credit Scorecards summary on page 49). An analysis of the records of a random sample of loans at one bank produced the following results:

	Score Under 600	Score 600 or More
Sample size	562	804
Number defaulted	11	7

Do these results allow us to conclude that those who score under 600 are more likely to default than those who score 600 or more? Use a 10% significance level.

13.73 Surveys have been widely used by politicians around the world as a way of monitoring the opinions of the electorate. Six months ago, a survey was undertaken to determine the degree of support for a national party leader. Of a sample of 1,100, 56% indicated that they would vote for this politician. This month, another survey of 800 voters revealed that 46% now support the leader.

a. At the 5% significance level, can we infer that the national leader's popularity has decreased?
b. At the 5% significance level, can we infer that the national leader's popularity has decreased by more than 5%?
c. Estimate with 95% confidence the decrease in percentage support between now and 6 months ago.

13.74 The process that is used to produce a complex component used in medical instruments typically results in defective rates in the 40% range. Recently, two innovative processes have been developed to replace the existing process. Process 1 appears to be more promising, but it is considerably more expensive to purchase and operate than process 2. After a thorough analysis of the costs, management decides that it will adopt process 1 only if the proportion of defective components it produces is more than 8% smaller than that produced by process 2. In a test to guide the decision, both processes were used to produce 300 components. Of the 300 components produced by process 1, 33 were found to be defective, whereas 84 out of the 300 produced by process 2 were defective. Conduct a test using a significance level of 1% to help management make a decision.

APPLICATIONS in OPERATIONS MANAGEMENT

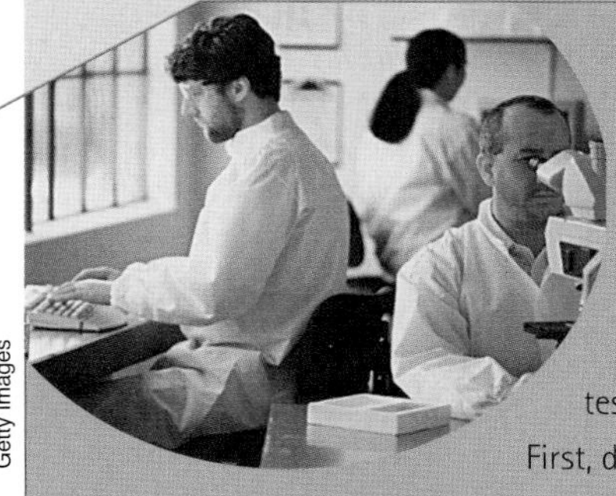

© Ryan McVay/PhotoDisc/ Getty Images

Pharmaceutical and Medical Experiments

When new products are developed, they are tested in several ways. First, does the new product work? Second, is it better than the existing product? Third, will customers buy it at a price that is profitable? Performing a customer survey or some other experiment that yields the information needed often tests the last question. This experiment is usually the domain of the marketing manager.

The other two questions are dealt with by the developers of the new product, which usually means the research department or the operations manager. When the product is a new drug, there are particular ways in which the data are gathered. The sample is divided into two groups. One group is assigned the new drug and the other is assigned a placebo, a pill that contains no medication. The experiment is often called "double-blind" because neither the subjects who take the drug nor the physician/scientist who provides the drug knows whether any individual is taking the drug or the placebo. At the end of the experiment, the data that are compiled allow statistics practitioners to do their work. Exercises 13.75–13.79 are examples of this type of statistical application. Exercise 13.80 describes a health-related problem where the use of a placebo is not possible.

13.75 Cold and allergy medicines have been available for a number of years. One serious side effect of these medications is that they cause drowsiness, which makes them dangerous for industrial workers. In recent years, a nondrowsy cold and allergy medicine has been developed. One such product, Hismanal, is claimed by its manufacturer to be the first once-a-day nondrowsy allergy medicine. The nondrowsy part of the claim is based on a clinical experiment in which 1,604 patients were given Hismanal and 1,109 patients were given a placebo. Of the first group, 7.1% reported drowsiness: of the second group, 6.4% reported drowsiness. Do these results allow us to infer at the 5% significance level that Hismanal's claim is false?

13.76 Plavix is a drug that is given to angioplasty patients to help prevent blood clots. A researcher at McMaster University organized a study that involved 12,562 patients in 482 hospitals in 28 countries. All the patients had acute coronary syndrome, which produces mild heart attacks or unstable angina, chest pain that may precede a heart attack. The patients were divided into two equal groups. Group 1 received daily Plavix pills, while group 2 received a placebo. After 1 year, 9.3% of patients on Plavix suffered a stroke or new heart attack, or had died of cardiovascular disease, compared with 11.5% of those who took the placebo.

a. Can we infer that Plavix is effective?
b. Describe your statistical analysis in a report to the marketing manager of the pharmaceutical company.

13.77 In a study that was highly publicized, doctors discovered that aspirin seems to help prevent heart attacks. The research project, which was scheduled to last for 5 years, involved 22,000 American physicians (all male). Half took an aspirin tablet three times per week, while the other half took a placebo on the same schedule. The researchers tracked each of the volunteers and updated the records regularly. Among the physicians who took aspirin, 104 suffered a heart attack; 189 physicians who took the placebo had a heart attack.

a. Determine whether these results indicate that aspirin is effective in reducing the incidence of heart attacks.
b. Write a report that describes the results of this experiment.

13.78 Exercise 13.77 described the experiment that determined that taking aspirin daily reduces one's probability of suffering a heart attack. The study was conducted in 1982 and at that time the mean age of the physicians was 50. In the years following the experiment, the physicians were monitored for other medical conditions. One of these was the incidence of cataracts. There were 1,084 cataracts in the aspirin group and 997 in the placebo group. Do these statistics allow researchers to conclude that aspirin leads to more cataracts?

13.79 According to the Canadian Cancer Society, more than 21,000 women will be diagnosed with breast cancer every year and more than 5,000 will die. (U.S. figures are more than 10 times those in Canada.) Surgery is generally considered the first method of treatment. However, many women suffer recurrences of cancer. For this reason many women are treated with Tamoxifen. But after 5 years, tumors develop a resistance to Tamoxifen. A new drug called Letrozole was developed by Novartis Pharmaceuticals to replace Tamoxifen. To determine its effectiveness, a study involving 5,187 breast cancer survivors from Canada, the United States, and Europe was undertaken. Half the sample received Letrozole and the other half a placebo. The study was to run for 5 years. However, after only 2.5 years, it was determined that 132 women receiving the placebo and 75 taking the drug had recurrences of their cancers. (The study was published in the *New England Journal of Medicine.*)

a. Do these results provide sufficient evidence to infer that Letrozole works?
b. Prepare a presentation to the board of directors of Novartis describing your analysis.

13.80 A study described in the *British Medical Journal* (January 2004) sought to determine whether exercise would help extend the lives of patients with heart failure. A sample of 801 patients with heart failure was recruited; 395 received exercise training and 406 did not. There were 88 deaths among the exercise group and 105 among those who did not exercise. Can researchers infer that exercise training reduces mortality?

The following exercises require the use of a computer and software. The answers may be calculated manually. See Appendix A for the sample statistics. ***Use a 5% significance level unless specified otherwise.***

13.81 Xr13-81 Automobile magazines often compare models and rate them in various ways. One question that is often asked of car owners, Would you buy the same model again? Suppose that a researcher for one magazine asked a random sample of Cadillac owners and a random sample of Lincoln (model LS) owners whether they plan to buy another Cadillac/Lincoln the next time they shop for a new car. The responses (1 = no and 2 = yes) were recorded. Do these data allow the researcher to infer that the two populations of car owners differ in their satisfaction levels?

13.82 Xr13-82 An insurance company is thinking about offering discounts on its life insurance policies to nonsmokers. As part of its analysis, the company randomly selects 200 men who are 60 years old and asks them whether they smoke at least one pack of cigarettes per day and if they have ever suffered from heart disease (2 = suffer from heart disease and 1 = do not suffer from heart disease).

a. Can the company conclude at the 10% significance level that smokers have a higher incidence of heart disease than nonsmokers?
b. Estimate with 90% confidence the difference in the proportions of men suffering from heart disease between smokers and nonsmokers.

13.83 Xr13-83 The impact of the accumulation of carbon dioxide in the atmosphere caused by burning fossil fuels such as oil, coal, and natural gas has been hotly debated for more than a decade. Some environmentalists and scientists have predicted that the excess carbon dioxide will increase the earth's temperature over the next 50 to 100 years with disastrous consequences. This belief is often called the "greenhouse effect." Other scientists claim that we don't know what the effect will be, and yet others believe that the earth's temperature is likely to decrease. Given the debate among scientists, it is not surprising that the general population is confused. To gauge the public's opinion on the subject, last year a random sample of 400 people was asked whether they believed in the greenhouse effect (2 = believe in greenhouse effect and 1 = do not believe in greenhouse effect). This year, 500 people were asked the same question.

a. Can we infer at the 10% significance level that there has been an increase in belief in the greenhouse effect?
b. Estimate the real change in the public's opinion about the subject. Use a 90% confidence level.

13.84 Xr13-84 Has the illicit use of drugs decreased over the past 10 years? Government agencies have undertaken surveys of Americans 12 years of age and older. Each was asked whether he or she used drugs at least once in the previous month. The results of this year's survey and the results of the survey completed 10 years ago were recorded as 1 = no and 2 = yes. Can we infer that the use of illicit drugs in the United States has increased in the past decade? (Adapted from the U.S. Substance Abuse and Mental Health Services Administration, National Household Survey on Drug Abuse)

13.85 Xr13-85 An operations manager of a computer chip maker is in the process of selecting a new machine to replace several older ones. Although technological innovations have improved the production process, it is quite common for the machines to produce defective chips. The operations manager must choose between two machines. The cost of machine A is several thousand dollars greater than the cost of machine B. After an analysis of the costs, it was determined that machine A is warranted provided that its defective rate is more than 2% less than that of machine B. To help decide, both machines are used to produce 200 chips each. Each chip was examined and whether it was defective (code = 2) or not (code = 1) was recorded. Should the operations manager select machine A?

APPLICATIONS in MARKETING

Market Segmentation

In Section 12.4, we introduced market segmentation and described how the size of market segments can be estimated. Once the segments have been defined, we can use statistical techniques to determine whether members of the segments differ in their purchases of a firm's products.

13.86 Xr13-86* The market for breakfast cereals has been divided into several segments related to health. One company identified a segment as those adults who are health conscious. The marketing manager would like to know whether this segment is more likely to purchase its Special X cereal that is pitched toward the health conscious segment. A survey of adults was undertaken. On the basis of several probing questions, each was classified as either a member of the health-conscious group (code = 1) or not (code = 2). Each respondent was also asked whether he or she buys Special X (1 = no, 2 = yes). The data were recorded in stacked format. Can we infer from these data that health-conscious adults are more likely to buy Special X?

13.87 Xr13-87* Quik Lube is a company that offers oil change service while the customer waits. Its market has been broken down into the following segments:

1. Working men and women too busy to wait at a dealer or service center
2. Spouses who work in the home
3. Retired persons
4. Other

A random sample of car owners was drawn. Each owner classified his or her market segment and also reported whether they usually use the services like Quik Lube (1 = yes and 2 = no). These data are stored in stacked format.

a. Determine whether members of segment 1 are more likely than members of segment 4 to respond that they usually use the service?
b. Can we infer that retired persons and spouses who work in the home differ in their use of services such as Quik Lube?

13.88 Xr13-88 Telemarketers obtain names and telephone numbers from several sources. To determine whether one particular source is better than a second, a random sample of names and numbers from the two different sources was obtained. For each potential customer, a statistics practitioner recorded whether that individual made a purchase (code = 2) or not (code = 1). Can we infer that differences exist between the two sources?

CHAPTER SUMMARY

In this chapter, we presented a variety of techniques that allow statistics practitioners to compare two populations. When the data are interval and we are interested in measures of central location, we encountered two more factors that must be considered when choosing the appropriate technique. When the samples are independent, we can use either the equal-variances or unequal-variances formulas. When the samples are matched pairs, we have only one set of formulas. We introduced the F-statistic, which is used to make inferences about two population variances. When the data are nominal, the parameter of interest is the difference between two proportions. For this parameter we had two test statistics and one interval estimator. Finally, we discussed observational and experimental data, important concepts in attempting to interpret statistical findings.

IMPORTANT TERMS

Pooled variance estimator 441
Equal-variances test statistic and confidence interval estimator 441
Unequal-variances test statistic and confidence interval estimator 442
Observational data 461
Experimental data 461
Matched pairs experiment 467
Mean of the population of differences 468
Numerator degrees of freedom 477
Denominator degrees of freedom 477
Pooled proportion estimator 484

SYMBOLS

Symbol	Pronounced	Represents
s_p^2	*s-sub-p-squared*	Pooled variance estimator
μ_D	*mu-sub-D* or *mu-D*	Mean of the paired differences
$\bar{x}_D$	*x-bar-sub-D* or *x-bar-D*	Sample mean of the paired differences
s_D	*s-sub-D* or *s-D*	Sample standard deviation of the paired differences
n_D	*n-sub-D* or *n-D*	Sample size of the paired differences
$\hat{p}$	*p-hat*	Pooled proportion

FORMULAS

Equal-variances t-test of $\mu_1 - \mu_2$

$$t = \frac{(\bar{x}_1 - \bar{x}_2) - (\mu_1 - \mu_2)}{\sqrt{s_p^2\left(\frac{1}{n_1} + \frac{1}{n_2}\right)}} \qquad \nu = n_1 + n_2 - 2$$

Equal-variances interval estimator of $\mu_1 - \mu_2$

$$(\bar{x}_1 - \bar{x}_2) \pm t_{\alpha/2}\sqrt{s_p^2\left(\frac{1}{n_1} + \frac{1}{n_2}\right)} \qquad \nu = n_1 + n_2 - 2$$

Unequal-variances t-test of $\mu_1 - \mu_2$

$$t = \frac{(\bar{x}_1 - \bar{x}_2) - (\mu_1 - \mu_2)}{\sqrt{\left(\frac{s_1^2}{n_1} + \frac{s_2^2}{n_2}\right)}} \qquad \nu = \frac{(s_1^2/n_1 + s_2^2/n_2)^2}{\frac{(s_1^2/n_1)^2}{n_1 - 1} + \frac{(s_2^2/n_2)^2}{n_2 - 1}}$$

Unequal-variances interval estimator of $\mu_1 - \mu_2$

$$(\bar{x}_1 - \bar{x}_2) \pm t_{\alpha/2}\sqrt{\frac{s_1^2}{n_1} + \frac{s_2^2}{n_2}} \qquad \nu = \frac{(s_1^2/n_1 + s_2^2/n_2)^2}{\frac{(s_1^2/n_1)^2}{n_1 - 1} + \frac{(s_2^2/n_2)^2}{n_2 - 1}}$$

t-test of μ_D

$$t = \frac{\bar{x}_D - \mu_D}{s_D/\sqrt{n_D}} \qquad \nu = n_D - 1$$

t-estimator of μ_D

$$\bar{x}_D \pm t_{\alpha/2}\frac{s_D}{\sqrt{n_D}} \qquad \nu = n_D - 1$$

F-test of σ_1^2/σ_2^2

$$F = \frac{s_1^2}{s_2^2} \qquad \nu_1 = n_1 - 1 \text{ and } \nu_2 = n_2 - 1$$

F-estimator of σ_1^2/σ_2^2

$$\text{LCL} = \left(\frac{s_1^2}{s_2^2}\right)\frac{1}{F_{\alpha/2,\nu_1,\nu_2}}$$

$$\text{UCL} = \left(\frac{s_1^2}{s_2^2}\right)F_{\alpha/2,\nu_2,\nu_1}$$

z-test and estimator of $p_1 - p_2$

Case 1: $$z = \frac{(\hat{p}_1 - \hat{p}_2)}{\sqrt{\hat{p}(1-\hat{p})\left(\frac{1}{n_1} + \frac{1}{n_2}\right)}}$$

Case 2: $$z = \frac{(\hat{p}_1 - \hat{p}_2) - (p_1 - p_2)}{\sqrt{\frac{\hat{p}_1(1-\hat{p}_1)}{n_1} + \frac{\hat{p}_2(1-\hat{p}_2)}{n_2}}}$$

z-estimator of $p_1 - p_2$

$$(\hat{p}_1 - \hat{p}_2) \pm z_{\alpha/2}\sqrt{\frac{\hat{p}_1(1-\hat{p}_1)}{n_1} + \frac{\hat{p}_2(1-\hat{p}_2)}{n_2}}$$

COMPUTER OUTPUT AND INSTRUCTIONS

Technique	Excel	Minitab
F-test of σ_1^2/σ_2^2	444	445
Equal-variances t-test of $\mu_1 - \mu_2$	446	446
Equal-variances estimator of $\mu_1 - \mu_2$	447	448
Unequal-variances t-test of $\mu_1 - \mu_2$	451	452
Unequal-variances estimator of $\mu_1 - \mu_2$	452	453
t-Test of μ_D	469	469
t-Estimator of μ_D	470	471
F-estimator of σ_1^2/σ_2^2	481	
z-Test of $p_1 - p_2$ (Case 1)	487	488
z-Test of $p_1 - p_2$ (Case 2)	489	489
z-Estimator of $p_1 - p_2$	491	491

CHAPTER EXERCISES

The following exercises require the use of a computer and software. ***Use a 5% significance level unless specified otherwise.***

13.89 Xr13-89 A restaurant located in an office building decides to adopt a new strategy for attracting customers to the restaurant. Every week it advertises in the city newspaper. To assess how well the advertising is working, the restaurant owner recorded the weekly gross sales for the 15 weeks after the campaign began and the weekly gross sales for the 24 weeks immediately prior to the campaign. Can the restaurateur conclude that the advertising campaign is successful?

13.90 Refer to Exercise 13.89. Assume that the profit is 20% of the gross. If the ads cost $50 per week, can the restaurateur conclude that the ads are profitable?

13.91 Xr13-91 How important to your health are regular vacations? In a study, a random sample of men and women were asked how frequently they take vacations. The men and women were divided into two groups each. The members of group 1 had suffered a heart attack; the members of group 2 had not. The number of days of vacation last year was recorded for each person. Can we infer that men and women who suffer heart attacks vacation less than those who did not suffer a heart attack?

13.92 Xr13-92 Research scientists at a pharmaceutical company have recently developed a new nonprescription sleeping pill. They decide to test its effectiveness by measuring the time it takes for people to fall asleep after taking the pill. Preliminary analysis indicates that the time to fall asleep varies considerably from one person to another. Consequently, they organize the experiment in the following way. A random sample of 100 volunteers who regularly suffer from insomnia is chosen. Each person is given one pill containing the newly developed drug and one placebo. (They do not know whether the pill they are taking is the placebo or the real thing, and the order of use is random.) Each participant is fitted with a device that measures the time until sleep occurs. Can we conclude that the new drug is effective?

13.93 Xr13-93 The city of Toronto boasts four daily newspapers. Not surprisingly, competition is keen. To help learn more about newspaper readers, an advertiser selected a random sample of people who bought their newspapers from a street vendor and people who had the newspaper delivered to their homes. Each was asked how many minutes they spent reading their newspapers. Can we infer that the amount of time reading differs between the two groups?

13.94 Xr13-94 In recent years, a number of state governments have passed mandatory seat-belt laws. Although the use of seat belts is known to save lives and reduce serious injuries, compliance with seat-belt laws is not universal. In an effort to increase the use of seat belts, a government agency sponsored a 2-year study. Among its objectives was to determine whether there was enough evidence to infer that seat-belt usage increased between last year and this year. To test this belief, random samples of drivers last year and this year were asked whether they always use their seat belts (2 = wear seat belt; 1 = do not wear seat belt). Can we infer that seat belt usage has increased over the last year?

13.95 Xr13-95 An important component of the cost of living is the amount of money spent on housing. Housing costs include rent (for tenants), mortgage payments and property tax (for home owners), heating, electricity, and water. An economist undertook a 5-year study to determine how housing costs have changed. Five years ago, he took a random sample of 200 households and recorded the percentage of total income spent on housing. This year, he took another sample of 200 households.

a. Conduct a test (with $\alpha = .10$) to determine whether the economist can infer that housing cost as a percentage of total income has increased over the last 5 years.
b. Use whatever statistical method you deem appropriate to check the required condition(s) of the test used in Part a.

13.96 Xr13-96 In designing advertising campaigns to sell magazines, it is important to know how much time each of a number of demographic groups spends reading magazines. In a preliminary study, 40 people were randomly selected. Each was asked how much time per week he or she spends reading magazines; additionally, each was categorized by gender and by income level (high or low). The data are stored in the following way: column 1 = time spent reading magazines per week in minutes for all respondents; column 2 = gender (1 = male, 2 = female); column 3 = income level (1 = low, 2 = high).

a. Is there sufficient evidence at the 10% significance level to conclude that men and women differ in the amount of time spent reading magazines?
b. Is there sufficient evidence at the 10% significance level to conclude that high-income individuals devote more time to reading magazines than low-income people?

13.97 Xr13-97 In a study to determine whether gender affects salary offers for graduating MBA students, 25 pairs of students were selected. Each pair consisted of a female and a male student who were matched according to their grade point averages, courses taken, ages, and previous work experience. The highest salary offered (in thousands of dollars) to each graduate was recorded.

a. Is there enough evidence at the 10% significance level to infer that gender is a factor in salary offers?
b. Discuss why the experiment was organized in the way it was.
c. Is the required condition for the test in Part a satisfied?

13.98 Xr13-98 Have North Americans grown to distrust television and newspaper journalists? A study was conducted this year to compare what Americans currently think of the press versus what they said 3 years ago. The survey asked respondents whether they agreed that the press tends to favor one side when reporting on political and social issues. A random sample of people was asked to participate in this year's survey. The results of a survey of another random sample taken 3 years ago are also available. The responses are 2 = agree and 1 = disagree. Can we conclude at the 10% significance level that Americans have become more distrustful of television and newspaper reporting this year than they were 3 years ago?

13.99 Xr13-99 Before deciding which of two types of stamping machines should be purchased, the plant manager of an automotive parts manufacturer wants to determine the number of units that each produces. The two machines differ in cost, reliability, and productivity. The firm's accountant has calculated that machine A must produce 25 more nondefective units per hour than machine B to warrant buying machine A. To help decide, both machines were operated for 24 hours. The total number of units and the number of defective units produced by each machine per hour were recorded. These data are stored in the following way. Column 1 = total number of units produced by machine A and column 2 = number of defectives produced by machine A; column 3 = total number of units produced by machine B; column 4 = number of defectives produced by machine B). Determine which machine should be purchased.

13.100 Refer to Exercise 13.99. Can we conclude that the defective rate differs between the two machines?

13.101 Xr13-101 The growing use of bicycles to commute to work has caused many cities to create exclusive bicycle lanes. These lanes are usually created by disallowing parking on streets that formerly allowed curbside parking. Merchants on such streets complain that the removal of parking will cause their businesses to suffer. To examine this problem, the mayor of a large city decided to launch an experiment on one busy street that had 1-hour parking meters. The meters were removed and a bicycle lane was created. The mayor asked the three businesses (a dry cleaner, a doughnut shop, and a convenience store) in one block to record daily sales for two complete weeks (Sunday to Saturday) prior to the change and two complete weeks after the change. The data are stored as follows. Column 1 = day of the week, column 2 = sales before change for dry cleaner, column 3 = sales after change for dry cleaner, column 4 = sales before change for doughnut shop, column 5 = sales after change for doughnut shop, column 6 = sales before change for convenience store, and column 7 = sales after change for convenience store). What conclusions can you draw from these data?

13.102 Xr13-102 There may be a new health concern—too much iron in our bodies. An article in the *Wall Street Journal* (January 17, 1992) reported that some scientists have implicated iron as a factor in various diseases, including cancer. Part of the problem, it is believed, is that iron builds up in the body over many years. To examine the issue, a random sample of 20-year-old men and women and 40-year-old men and women was drawn. The amount of iron in their bodies was measured and recorded. The results are stored in the following way: column 1 shows the amount of stored iron in men (in milligrams); column 2 indicates the men's ages; column 3 shows the amount of stored iron in women; column 4 lists the women's ages.

a. Conduct a test at the 10% significance level to determine whether we can infer that 40-year-old men have more iron in their bodies than do 20-year-old men.
b. Repeat Part a for women.

13.103 Xr13-103 Clinical depression is linked to several other diseases. Scientists at Johns Hopkins University undertook a study to determine whether heart disease is one of these. A group of 1,190 male medical students was tracked over a 40-year period. Of these, 132 had suffered clinically diagnosed depression. For each student the scientists recorded whether the student died of a heart attack (code = 2) or did not (code = 1).

a. Can we infer at the 1% significance level that men who are clinically depressed are more likely to die from heart diseases?
b. If the answer to Part a is "yes," can you interpret this to mean that depression causes heart disease? Explain.

13.104 Xr13-104 High blood pressure (hypertension) is a leading cause of strokes. Medical researchers are constantly seeking ways to treat patients suffering from this condition. A specialist in hypertension claims that regular aerobic exercise can reduce high blood pressure just as successfully as drugs, with none of the adverse side effects. To test the claim, 50 patients who suffer from high blood pressure were chosen to participate in an experiment. For 60 days, half the sample exercised three times per week for 1 hour and did not take medication; the other half took the standard medication. The percentage reduction in blood pressure was recorded for each individual.

a. Can we conclude at the 1% significance level that exercise is more effective than medication in reducing hypertension?
b. Estimate with 95% confidence the difference in mean percentage reduction in blood pressure between drugs and exercise programs.
c. Check to ensure that the required condition(s) of the techniques used in Parts a and b is satisfied.

13.105 Xr13-105 Most people exercise in order to lose weight. To determine better ways to lose weight, a random sample of male and female exercisers was divided into groups. The first group exercised vigorously twice a week. The second group exercised moderately four times per week. The weight loss for each individual was recorded. Can we infer that people who exercise moderately more frequently lose more weight than people who exercise vigorously?

13.106 Xr13-106 After observing the results of the test in Exercise 13.105, a statistics practitioner organized another experiment. People were matched according to gender, height, and weight. One member of each matched pair then exercised vigorously twice a week and the other member exercised moderately four times per week. The weight losses were recorded. Can we infer that people who exercise moderately lose more weight?

13.107 Xr13-107 Pass the Lotion, a long-running television commercial for Special K cereal, features a flabby sunbather who asks his wife to smear sun lotion on his back. A random sample of Special K customers and a random sample of people who do not buy Special K were asked to indicate whether they liked (code = 1) or disliked (code = 2) the ad. Can we infer that Special K buyers like the ad more than nonbuyers?

13.108 Xr13-108 Refer to Exercise 13.107. The respondents were also asked whether they thought the ad would be effective in selling the product. The responses (1 = Yes and 2 = No) were recorded. Can we infer that Special K buyers are more likely to respond yes than nonbuyers?

13.109 Xr13-109 Most English professors complain that students don't write very well. In particular, they point out that students often confuse quality and quantity. A study at the University of Texas examined this claim. In the study, undergraduate students were asked to compare the cost benefits of Japanese and American cars. All wrote their analyses on computers. Unbeknownst to the students, the computers were rigged so that some students would have to type twice as many words to fill a single page. The number of words used by each student was recorded. Can we conclude that students write in such a way as to fill the allotted space?

13.110 Xr13-110 Approximately 20 million Americans work for themselves. Most run single-person businesses out of their homes. One-quarter of these individuals use personal computers in their businesses. A market research firm, Computer Intelligence InfoCorp, wanted to know whether single-person businesses that use personal computers are more successful than those with no computer. They surveyed 150 single-person firms and recorded their annual incomes. Can we infer at the 10% significance level that single-person businesses that use a personal computer earn more than those that do not?

13.111 Xr13-111 Many small retailers advertise in their neighborhoods by sending out flyers. People deliver these to homes and are paid according to the number of flyers delivered. Each deliverer is given several streets whose homes become their responsibility. One of the ways retailers use to check the performance of deliverers is to randomly sample some of the homes and ask the home owner whether he or she received the flyer. Recently university students started a new delivery service. They have promised better service at a competitive price. A retailer wanted to know whether the new company's delivery rate is better than that of the existing firm. She had both companies deliver her flyers. Random samples of homes were drawn and each was asked whether he or she received the flyer (2 = yes and 1 = no). Can the retailer conclude that the new company is better? (Test with $\alpha = .10$.)

13.112 Xr13-112 Medical experts advocate the use of vitamin and mineral supplements to help fight infections. A study undertaken by researchers at Memorial University (reported in the British journal *Lancet*, November 1992) recruited 96 men and women age 65 and older. One-half of them received daily supplements of vitamins and minerals, whereas the other half received placebos. The supplements contained the daily recommended amounts of 18 vitamins and minerals, including vitamins B-6, B-12, C, and D, thiamine, riboflavin, niacin, calcium, copper, iodine, iron, selenium, magnesium, and zinc. The doses of vitamins A and E were slightly less than the daily requirements. The supplements included four times the amount of beta-carotene than the average person ingests daily. The number of days of illness from infections (ranging from colds to pneumonia) was recorded for each person. Can we infer that taking vitamin and mineral supplements daily increases the body's immune system?

13.113 Xr13-113 An inspector for the Atlantic City Gaming Commission suspects that a particular blackjack dealer may be cheating (in favor of the casino) when he deals at expensive tables. To test her belief, she observed 500 hands each at the \$100-limit table and the \$3,000-limit table. For each hand, she recorded whether the dealer won (code = 2) or lost (code = 1). When a tie occurs, there is no winner or loser. Can the inspector conclude at the 10% significance level that the dealer is cheating at the more expensive table?

Exercises 13.114 and 13.115 require access to the data files introduced in previous exercises.

13.114 Xr12-29* Exercise 12.29 dealt with the amount of time high school students spend per week at part-time jobs. In addition to the hours of part-time work, the school guidance counselor recorded the gender of the student surveyed (1 = female and 2 = male). Can we conclude that female and male high school students differ in the amount of time spent at part-time jobs?

13.115 Xm12-01* The company that organized the survey to determine the amount of discarded newspaper (Example 12.1) kept track of the type of neighborhood (1 = city and 2 = suburbs). Do these data allow the company management to infer that city households discard more newspaper than do suburban households?

APPLICATIONS in MARKETING

© Picturenet/Blend Images/ Jupiterimages

Market Segmentation

In Section 12.4 we introduced market segmentation. The following exercises address the problem of determining whether two market segments differ in their pattern of purchases of a particular product or service.

13.116 Xr13-116 Movie studios segment their markets by age. Two segments that are particularly important to this industry are teenagers and 20-to-30-year-olds. To assess markets and guide the making of movies, a random sample of teenagers and 20-to-30-year-olds was drawn. Each was asked to report the number of movies they saw in theaters last year. Do these data allow us to infer that teenagers see more movies than 20-to-30-year-olds?

The following exercises employ data files associated with examples and exercises seen previously in this book.

13.117 Xr12-91* In addition to asking about educational attainment, the survey conducted in Exercise 12.91 also asked whether the respondent had plans in the next 2 years to take a course (1 = no and 2 = yes). Can we conclude that Californians who did not complete high school are less likely to take a course in the university's evening program?

13.118 Xm12-06* The objective in the survey conducted in Example 12.6 was to estimate the size of the market segment of adults who are concerned about eating healthy foods. As part of the survey, each respondent was asked how much they spend on breakfast cereal in an average month. The marketing manager of a company that produces several breakfast cereals would like to know whether on average the market segment concerned about eating health foods outspends the other market segments. Write a brief report detailing your findings.

13.119 Xr12-33* In Exercise 12.33, we described how the office equipment chain OfficeMax offers rebates on some products. The goal in that exercise was to estimate the total amount spent by customers who bought the package of 100 CD-ROMS. In addition to tracking these amounts, an executive also determined the amounts spent in the store by another sample of customers who purchased a fax machine/copier (regular price $89.99 minus $40 manufacturer's rebate and $10 OfficeMax mail-in rebate). Can OfficeMax conclude that those who buy the fax/copier outspend those who buy the package of CD-ROMs? Write a brief memo to the executives of OfficeMax describing your findings and any possible recommendations.

13.120 Xr12-76* In addition to recording whether faculty members who are between 55 and 64 plan to retire before they reach 65 in Exercise 12.76, the consultant asked each to report his or her annual salary. Can the president infer that professors aged 55 to 64 who plan to retire early have higher salaries than those who don't plan to retire early?

13.121 Xr12-81* In Exercise 12.81, the statistics practitioner also recorded the gender of the respondents where 1 = female 2 = male. Can we infer that men and women differ in their choices of Christmas trees?

CASE 13.1 Do Banks Discriminate against Women Business Owners? Part 1*

© Bruce Ayres/Stone/Getty Images

DATA C13-01

Increasingly, more women are becoming owners of small businesses. However, questions concerning how they are treated by banks and other financial institutions have been raised by women's groups. Banks are particularly important to small businesses, since studies show that bank financing represents about one-quarter of total debt, and that for medium-size businesses the proportion rises to approximately one-half. If women's requests for loans are rejected more frequently than are men's requests, or if women must pay higher interest charges than men do, women have cause for complaint. Banks might then be subject to criminal as well as civil suits. To examine this issue, a research project was launched.

The researchers surveyed a total of 1,165 business owners, of whom 115 were women. The percentage of women in the sample, 9.9%, compares favorably with other sources that indicate that women own about 10% of established small businesses at the time. The survey asked a series of questions to men and women business owners who applied for loans during the previous month. It also determined the nature of the business, its size, and its age. Additionally, the owners were asked about their experiences in dealing with banks. The questions asked in the survey included the following:

1. What is the gender of the owner?
   1. female 2. male
2. Was the loan approved?
   1. no 2. yes
3. If it was approved, what interest rate did you get? How much above the prime rate was your rate?

Of the 115 women who asked for a loan, 14 were turned down. A total of 98 men who asked for a loan were rejected. The rates above prime for all loans that were granted were recorded. What do these data disclose about possible gender bias by the banks?

*Adapted from A. L. Riding and C. S. Swift, "Giving Credit Where It's Due: Women Business Owners and Canadian Financial Institutions," Carleton University Working Paper, Series WPS 89-07, 1989.

CASE 13.2 Ambulance and Fire Department Response Interval Study*

© Jeff Perkell/Index Stock Imagery/Jupiterimages

DATA C13-02

Every year, thousands of people die of heart attacks partly because of delays in waiting for emergency medical care to arrive. One form of heart attack is ventricular fibrillation rhythm, which is treated by a defibrillator. However, immediate medical attention is critical. In general, if a patient receives treatment within 8 minutes, he or she is very likely to survive. It is estimated that the probability of survival is reduced by 7% to 10% for each minute thereafter that defibrillation is delayed.

The region in the Ambulance and Fire Department Response Interval Study is composed of the three cities of Cambridge, Waterloo, and Kitchener. Each city has a fire department, and the region has a 911 emergency telephone system. When a medical-related call is received by the Police Dispatch Center, it is relayed to the Central Ambulance Communication Center (CACC). The CACC dispatches both the ambulance and the fire department to certain calls

*The authors are grateful to Bruce Jermyn for supplying this case. The data are real. However, the sample size was reduced to ease disk-storage problems.

that match one of several criteria indicating the need for fire department personnel. There are two ambulance services that cover the region: the Cambridge Memorial Hospital Ambulance Service and the Kitchener-Waterloo Regional Ambulance Service. Currently, ambulance personnel sent to the patient after a 911 call perform all defibrillation. A city counselor recently suggested that, since the fire department has more centers, it is likely that fire department personnel could arrive at the scene more quickly than ambulance personnel. A study was undertaken to determine whether fire department personnel should be trained in the use of defibrillators and sent to treat ventricular fibrillation rhythm.

Between March 1, 1994, and August 31, 1994, all calls that involved both ambulance and fire department personnel were monitored. The times for each service to arrive at the scene were recorded using the following format:

Column 1: Call number for Cambridge calls
Column 2: Time in minutes for the ambulance to arrive
Column 3: Time in minutes for fire truck to arrive
Column 4: Call number for Kitchener calls
Column 5: Time in minutes for the ambulance to arrive
Column 6: Time in minutes for fire truck to arrive
Column 7: Call number for Waterloo calls
Column 8: Time in minutes for the ambulance to arrive
Column 9: Time in minutes for fire truck to arrive

It has been decided that the training of fire department personnel is warranted only if it can be shown that a fire truck arrives at the scene on average more than 1 minute sooner than an ambulance and that the frequency of arrival within 8 minutes is greater for the fire department.

What conclusions can be drawn from the data?

CASE 13.3 How to Market Genelec

DATA C13-03

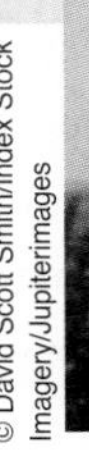

Genelec, a company that produces electric generators for houses, has made the decision to start selling in Florida. The market may be quite large because in recent years hurricanes and tornadoes have caused the power to go out sometimes for hours and other times for days. This leaves residents without electricity to run refrigerators, stoves, air conditioners, televisions, and computers. Portable generators are available that run on gasoline, but they are too small to run the whole house. Genelec produces larger generators that are run on natural gas (where available) or Liquid propane and are large enough to provide electricity for the entire house. The marketing manager is given the task of deciding how to advertise the product. A random sample of homes in South Florida is drawn and those selected are sent a brochure that describes the product and lists the prices and a short questionnaire with the following questions:

1. Would you buy a generator?
   No (1) Yes (2)
2. How long have you lived at your current address? ______ years
3. When you purchased this house did you consider the resale value?
   No (1) Yes (2)
4. Which county do you live in?
   Miami-Dade (1) Other (2)

The results of the surveys were recorded.

a. Conduct a test to determine whether the years of ownership differ between those who would and those who would not buy a generator.
b. Test to determine whether the proportion of those who would buy a generator is greater for homeowners who consider resale value than homeowners who do not consider resale value.
c. Test to determine whether the proportion of those who would buy a generator differ between Miami-Dade County and the other counties.
d. Describe how and where you would advertise.

APPENDIX 13 / REVIEW OF CHAPTERS 12 AND 13

As you may have already discovered, the ability to identify the correct statistical technique is critical, without which any calculation performed is useless. When you solved problems at the end of each section in the preceding chapters (you *have* been solving problems at the end of each section covered, haven't you?), you probably had no great difficulty identifying the correct technique to use. You used the statistical technique introduced in that section. Although those exercises provided practice in setting up hypotheses, producing computer output of tests of hypothesis and confidence interval estimators, and interpreting the results, you did not address a fundamental question faced by statistics practitioners: Which technique should I use? If you still do not appreciate the dimension of this problem, examine Table A13.1, which lists all the inferential methods covered thus far.

TABLE **A13.1** Summary of Statistical Techniques in Chapters 12 and 13

- t-test of μ
- Estimator of μ (including small population estimator of μ and large and small population estimators of $N\mu$)
- z-test of p
- Estimator of p (including small population estimator of p and large and small population estimators of Np)
- χ^2-test of σ^2
- Estimator of σ^2
- Equal-variances t-test of $\mu_1 - \mu_2$
- Equal-variances estimator of $\mu_1 - \mu_2$
- Unequal-variances t-test of $\mu_1 - \mu_2$
- Unequal-variances estimator of $\mu_1 - \mu_2$
- t-test of μ_D
- Estimator of μ_D
- F-test of σ_1^2/σ_2^2
- Estimator of σ_1^2/σ_2^2
- z-test of $p_1 - p_2$ (Case 1)
- z-test of $p_1 - p_2$ (Case 2)
- Estimator of $p_1 - p_2$

Counting tests and confidence interval estimators of a parameter as two different techniques, a total of 17 statistical procedures have been presented thus far, and there is much left to be done. Faced with statistical problems that require the use of some of these techniques (such as in real-world applications or on a quiz or midterm test), most students need some assistance in identifying the appropriate method. In this appendix and the appendixes of five more chapters you will have the opportunity to practice your decision skills; we've provided exercises and cases that require all the inferential techniques introduced in Chapters 12 and 13. Solving these problems will require you to do what statistics practitioners must do: analyze the problem, identify the technique or techniques, employ statistical software and a computer to yield the required statistics, and interpret the results.

The flowchart in Figure A13.1 represents the logical process that leads to the identification of the appropriate method. Of course, it only shows the techniques covered to this point. Chapters 14, 15, 16, 17, and 19 will include appendixes that review all the techniques introduced up to that chapter. The list and the flowchart will be expanded in each appendix and all appendixes will contain review exercises. (Some will contain cases.)

FIGURE **A13.1**
Flowchart of Techniques in Chapters 12 and 13

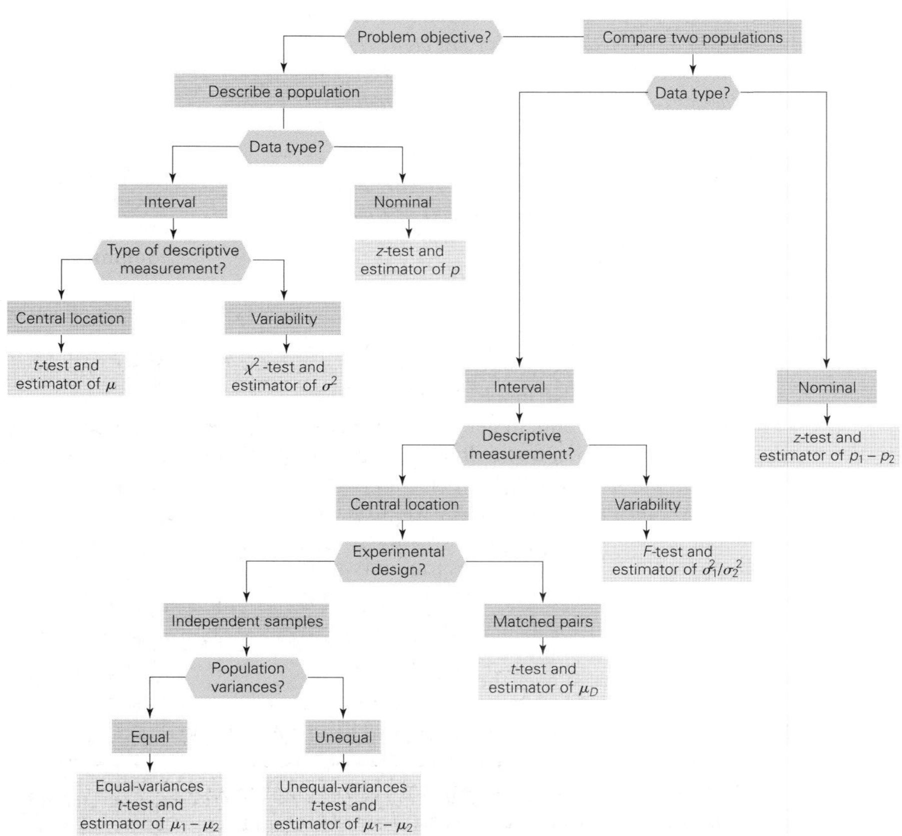

Notice that we did not specifically include the additional applications for the interval estimators of μ (estimators of totals and estimation with small populations) and of p (estimators of total number of successes and estimation with small population). We simply grouped them under the general heading.

As we pointed out in Chapter 11, the two most important factors in determining the correct statistical technique are the problem objective and the data type. In some situations, once these have been recognized, the technique automatically follows. In other cases, however, several additional factors must be identified before you can proceed. For example, when the problem objective is to compare two populations and the data are interval, three other significant issues must be addressed: the descriptive measurement (central location or variability), whether the samples are independently drawn, and, if so, whether the unknown population variances are equal.

EXERCISES

The purpose of the exercises that follow is twofold. First, the exercises provide you with practice in the critical skill of identifying the correct technique. Second, they allow you to improve your ability to determine the statistics needed to answer the question and interpret the results. We believe that the first skill is underdeveloped, because up to now you have had little practice. The exercises you've worked on have appeared at the end of sections and chapters where the correct techniques have just been presented. Determining the correct technique should not have been difficult. Because the exercises that follow were selected from the types that you have already encountered in Chapters 12 and 13, they will help you develop your technique-identification skills.

You will note that in the exercises that require a test of hypothesis, we do not specify a significance level. We have left this decision to you. After analyzing the issues raised in the exercise, use your own judgment to determine whether the p-value is small enough to reject the null hypothesis.

A13.1 XrA13-01 Shopping malls are more than places where we buy things. We go to malls to watch movies; buy breakfast, lunch, and dinner; exercise; meet friends; and, in general, to socialize. To study the trends, a sociologist took a random sample of 100 mall shoppers and asked a variety of questions. This survey was first conducted 3 years ago with another sample of 100 shoppers. In both surveys, respondents were asked to report the number of hours they spend in malls during an average week. Can we conclude that the amount of time spent at malls has decreased over the past 3 years?

A13.2 XrA13-02 It is often useful for retailers to determine why their potential customers choose to visit their store. Possible reasons include advertising, advice from a friend, or previous experience. To determine the effect of full-page advertisements in the local newspaper, the owner of an electronic-equipment store asked 200 randomly selected people who visited the store whether they had seen the ad. He also determined whether the customers had bought anything, and, if so, how much they spent. There were 113 respondents who saw the ad. Of these 49 made a purchase. Of the 87 respondents who did not see the ad, 21 made a purchase. The amounts spent were recorded.

a. Can the owner conclude that customers who see the ad are more likely to make a purchase than those who do not see the ad?
b. Can the owner conclude that customers who see the ad spend more than those who do not see the ad (among those who make a purchase)?
c. Estimate with 95% confidence the proportion of all customers who see the ad who then make a purchase.
d. Estimate with 95% confidence the mean amount spent by customers who see the ad and make a purchase.

A13.3 XrA13-03 In an attempt to reduce the number of person-hours lost as a result of industrial accidents, a large multiplant corporation installed new safety equipment in all departments and all plants. To test the effectiveness of the equipment, a random sample of 25 plants was drawn. The number of person-hours lost in the month prior to installation of the safety equipment and in the month after installation was recorded. Can we conclude that the equipment is effective?

A13.4 XrA13-04 Is the antilock braking system (ABS), now available as a standard feature on many cars, really effective? The ABS works by automatically pumping brakes extremely quickly on slippery surfaces so the brakes do not lock, avoiding an uncontrollable skid. If ABS is effective, we would expect that cars equipped with ABS would have fewer accidents, and the costs of repairs for the accidents that do occur would be smaller. To investigate the effectiveness of ABS, the Highway Loss Data Institute gathered data on a random sample of 500 General Motors cars that did not have ABS and 500 GM cars that were equipped with ABS. For each year, the institute recorded whether the car was involved in an accident and, if so, the cost of making repairs. Forty-two cars without ABS and 38 ABS-equipped cars were involved in accidents. The costs of repairs were recorded. Using frequency of accidents and cost of repairs as measures of effectiveness, can we conclude that ABS is effective? If so, estimate how much better are cars equipped with ABS compared to cars without ABS.

A13.5 XrA13-05 The electric company is considering an incentive plan to encourage its customers to pay their bills promptly. The plan is to discount the bills 1% if the customer pays within 5 days, as opposed to the usual 25 days. As an experiment, 50 customers are offered the discount on their September bill. The amount of time each takes to pay his or her bill is recorded. The amount of time a random sample of 50 customers not offered the discount take to pay their bills is also recorded. Do these data allow us to infer that the discount plan works?

A13.6 XrA13-06 Traffic experts are always looking for ways to control automobile speeds. Some communities have experimented with "traffic-calming" techniques. These include speed bumps and various obstructions that force cars to slow down to drive around them. Critics point out that the techniques are counterproductive because they cause drivers to speed on other parts of these roads. In an analysis of the effectiveness of speed bumps, a statistics practitioner organized a study over a 1-mile stretch of city road that had 10 stop signs. He then took a random sample of 100 cars and recorded their average speed (the speed limit was 30 mph) and the number of proper stops at the stop signs. He repeated the observations for another sample of 100 cars after speed bumps were placed on the road. Do these data allow the statistics practitioner to conclude that the speed bumps are effective?

A13.7 XrA13-07 The proliferation of self-serve pumps at gas stations has generally resulted in poorer automobile maintenance. One feature of poor maintenance is low tire pressure, which results in shorter tire life and higher gasoline consumption. To examine this problem, an automotive expert took a random sample of cars across the country and measured the tire pressure. The difference between the recommended tire pressure and the observed tire pressure was recorded. [A recording of 8 means that the pressure of the tire is 8 pounds per square inch (psi) less than the amount recommended by the tire manufacturer.] Suppose that for each psi below recommendation, tire life decreases by 100 miles and gasoline consumption increases by 0.1 gallon per mile. Estimate with 95% confidence the effect on tire life and gasoline consumption.

A13.8 XrA13-08 Many North American cities encourage the use of bicycles as a way to reduce pollution and traffic congestion. So many people now regularly use the bicycle to get to work and for exercise that some jurisdictions have enacted bicycle helmet laws, which specify that all bicycle riders must wear helmets to protect against head injuries. Critics of these laws complain that it is a violation of individual freedom and that helmet laws tend to discourage bicycle usage. To examine this issue, a researcher randomly sampled 50 bicycle users and asked each to record the number of miles he or she rode weekly. Several weeks later the helmet law was enacted. The number of miles each of the 50 bicycle riders rode weekly was recorded for the week after the law was passed. Can we infer from these data that the law discourages bicycle usage?

A13.9 XrA13-09 Cardizem CD is a prescription drug that is used to treat high blood pressure and angina. One common side effect of such drugs is the occurrence of headaches and dizziness. To determine whether its drug has the same side effects, the drug's manufacturer, Marion Merrell Dow, Inc., undertook a study. A random sample of 908 high blood pressure sufferers was recruited; 607 took Cardizem CD and 301 took a placebo. Each reported whether they suffered from headaches and/or dizziness (2 = yes, 1 = no). Can the pharmaceutical company scientist infer that Cardizem CD users are more likely to suffer headache and dizziness side effects than nonusers?

A13.10 XrA13-10 A fast-food franchiser is considering building a restaurant at a downtown location. Based on a financial analysis, a site is acceptable only if the number of pedestrians passing the location during the work day averages more than 200 per hour. To help decide whether to build on the site, a statistics practitioner observes the number of pedestrians who pass the site each hour over a 40-hour workweek. Should the franchiser build on this site?

A13.11 XrA13-11 Most people who quit smoking cigarettes do so for health reasons. However, some quitters find that they gain weight after quitting, and scientists estimate that the health risks of smoking two packs of cigarettes per day and of carrying 65 extra pounds of weight are about equivalent. In an attempt to learn more about the effects of quitting smoking, the U.S. Centers for Disease Control conducted a study (reported in *Time*, March 25, 1991). A sample of 1,885 smokers was taken. During the course of the experiment, some of the smokers quit their habit. The amount of weight gained by all the subjects was recorded. Do these data allow us to conclude that quitting smoking results in weight gains?

A13.12 XrA13-12 Golf equipment manufacturers compete against one another by offering a bewildering array of new products and innovations. Oversized clubs, square grooves, and graphite shafts are examples of such innovations. The effect of these new products on the average golfer is, however, much in doubt. One product, a perimeter weighted iron, was designed to increase the consistency of distance and accuracy. The most important aspect of irons is consistency, which means that ideally there should be no variation in distance from shot to shot. To examine the relative merits of two brands of perimeter-weighted irons, an average golfer used the 7-iron, hitting 100 shots using each of two brands. The distance in yards was recorded. Can the golfer conclude that brand B is superior to brand A?

A13-13 XrA13-13 Managers are frequently called upon to negotiate in a variety of settings. This calls for an ability to think logically, which requires an ability to concentrate and ignore distractions. In a study of the effect of distractions, a random sample of 208 students was drawn by psychologists at McMaster University (*National Post*, December 11, 2003). The male students were shown pictures of women of varying attractiveness. The female students were shown pictures of men of varying attractiveness. All students were then offered a choice of an immediate reward of $15 or a wait of 8 months for a reward of $75. The choices of the male and of the female students (1 = immediate reward, 2 = larger reward 8 months later) were recorded. The results are stored in the following way:

Column 1: Choices of males shown most attractive women
Column 2: Choices of males shown less attractive women
Column 3: Choices of females shown most attractive men
Column 4: Choices of females shown less attractive men

a. Can we infer that men's choices are affected by the attractiveness of women's pictures?
b. Can we infer that women's choices are affected by the attractiveness of men's pictures?

A13.14 XrA13-14 Throughout the day there are a number of exercise shows appearing on television. These usually feature attractive and fit men and women performing various exercises and urging viewers to duplicate the activity at home. Some viewers are exercisers. However, some people like to watch the shows without exercising (which explains why they use attractive people as demonstrators). Various companies sponsor the shows and there are commercial breaks. One sponsor wanted to determine whether there are differences between exercisers and nonexercisers in terms of how well they remember the sponsor's name. A random sample of viewers was selected and called after the exercise show was over. Each was asked to report whether they exercised or only watched. They were also asked to name the sponsor's brand name (2 = yes, they could, 1 = no, they couldn't). Can the sponsor conclude that exercisers are more likely to remember the sponsor's brand name than those who only watch?

A13.15 XrA13-15 According to the latest census, the number of households in a large metropolitan area is 425,000. The home delivery department of the local newspaper reports that 104,320 households receive daily home delivery. To increase home delivery sales, the marketing department launches an expensive advertising campaign. A financial analyst tells the publisher that for the campaign to be successful, home delivery sales must increase to more than 110,000 households. Anxious to see whether the campaign is working, the publisher authorizes a telephone survey of 400 households within 1 week of the beginning of the campaign and asks each household head whether he or she has the newspaper delivered. The responses were recorded where 2 = yes and 1 = no.

a. Do these data indicate that the campaign will increase home delivery sales?
b. Do these data allow the publisher to conclude that the campaign will be successful?

A13.16 XrA13-16 The Scholastic Aptitude Test (SAT), which is organized by the Educational Testing Service (ETS), is important to high school students seeking admission to colleges and universities throughout the United States. A number of companies offer courses to prepare students for the SAT. The Stanley H. Kaplan Educational Center claims that its students gain on average, more than 110 points by taking its course. ETS, however, insists that preparatory courses can improve a score by no more than 40 points. (The minimum and maximum scores of the SAT are 400 and 1,600, respectively.) Suppose a random sample of 40 students wrote the exam, then took the Kaplan preparatory course, and then took the exam again.

a. Do these data provide sufficient evidence to refute the ETS claim?
b. Do these data provide sufficient evidence to refute Kaplan's claim?

A13.17 XrA13-17 A potato chip manufacturer has contracted for the delivery of 15,000,000 kilograms of potatoes. The supplier agrees to deliver the potatoes in 15,000 equal truckloads. The manufacturer suspects that the supplier will attempt to cheat him. He has the weight of the first 50 truckloads recorded.

a. Can the manufacturer conclude from these data that the supplier is cheating him?
b. Estimate with 95% confidence the total weight of potatoes for all 15,000 truckloads.

CASE A13.1 Hormone Replacement Therapy and Its Effect on Cognitive Performance*

DATA CA13-01

For more than 20 years physicians have been treating postmenopausal women with hormone replacement therapy to reduce the various ailments that accompany menopause. A statistical study had indicated that HRT not only treated effects such as hot flashes but was also beneficial in that it reduced the incidence of breast cancer and heart attacks. Unfortunately, this study was based on observational data where causal relationships can be substantiated. In 2003, the Women's Health Initiative (WHI) published the results of another statistical analysis, which was based on experimental data gathered by a double-blind experiment that randomly assigned women to either HRT or a placebo. (See Section 13.3 for a discussion of the differences between observational and experimental data and the application box on page 495 for a short description of experiments using placebos.) The WHI study discovered that HRT actually increases the risk of breast cancer and heart disease.

Because millions of women have taken hormone replacement therapy for long periods of time, it is possible to use statistical analysis to help point the way for the investigation of other effects of HRT. To determine whether HRT helps reduce the incidence and severity of dementia (which includes Alzheimer's disease), a random sample of women over 75 was collected. Prescription records were examined to determine HRT status. Fifty-eight users and 47 nonusers of HRT participated in this study. The California Verbal Learning Test and the Logical Memory Test were used to measure cognitive performance. The data were formatted as follows:

Column 1 ID
Column 2: User (1) or nonuser (2) of HRT
Column 3: California Verbal Learning Test score (out of 100)
Column 4: Logical Memory Test score (out of 100)

Do these data present sufficient evidence that HRT helps cognitive performance in women over 75?

*This case is based on an article in the *Journal of the American Geriatrics Society*, February 2004.

CASE A13.2 Attitudes after the Women's Health Initiative Study*

DATA CA13-02

Despite extensive media coverage of the Women's Health Initiative project (Case A13.1), researchers were uncertain to what extent the important information about hormone replacement therapy (HRT) was known by both physicians and women. To measure the impact of the project, a survey was conducted. A sample of 600 women was allocated in two groups according to their socioeconomic status, high (HSES) or low (LSES). Additionally, 283 physicians were surveyed to determine their attitudes regarding HRT after the publication of the WHI report. *Note*: According to the latest census there are 836,200 physicians in the United States (*Source: Statistical Abstract of the United States*, 2003, Table 162).

*This case was adapted from a study published in *Menopause*, January 2004.

(continued)

Column 1: Physician ID
Column 2: Aware
Column 3: Modified treatment
Column 4: Women ID
Column 5: 1 = LSES, 2 = HSES
Column 6: Aware (2 = yes, 1 = no)

a. Estimate with 95% confidence the total number of physicians who are aware of the report.

b. Estimate with 95% confidence the total number of physicians who have modified their treatment of postmenopausal women as a result of the study.

c. Do the data allow for the conclusion that women in the HSES group are more aware of the report than are women in LSES group?

d. Is there enough statistical evidence to conclude that women in the HSES group are more likely to use HRT?

CASE A13.3 Quebec Separation? *Oui Ou Non*

DATA CA13-03

Since the 1960s, there has been an ongoing campaign among Quebecers to separate from Canada and form an independent nation. Should Quebec separate, the ramifications for the rest of Canada, American states that border Quebec, the North American Free Trade Agreement, and numerous multinational corporations would be enormous. In the 1993 federal election, the pro-sovereigntist *Bloc Quebecois* won 54 of Quebec's 75 seats in the House of Commons. In 1994, the separatist *Parti Quebecois* formed the provincial government in Quebec and promised to hold a referendum on separation. As with most political issues, polling plays an important role in trying to influence voters and to predict the outcome of the referendum vote. Shortly after the 1993 federal election, *The Financial Post Magazine*, in cooperation with several polling companies, conducted a survey of Quebecers.

A total of 641 adult Quebecers were interviewed. They were asked the following question. (Francophones were asked the questions in French.) The pollsters also recorded the language (English or French) in which the respondent answered.

If a referendum were held today on Quebec's sovereignty with the following question, "Do you want Quebec to separate from Canada and become an independent country?" would you vote yes or no?

1. No
2. Yes

The responses were recorded and stored in columns 1 (planned referendum vote for Francophones) and 2 (planned referendum vote for Anglophones).

Infer from the data:

a. If the referendum were held on the day of the survey, would Quebec vote to remain in Canada?
b. Estimate with 95% confidence the difference between French and English-speaking Quebecers in their support for separation.

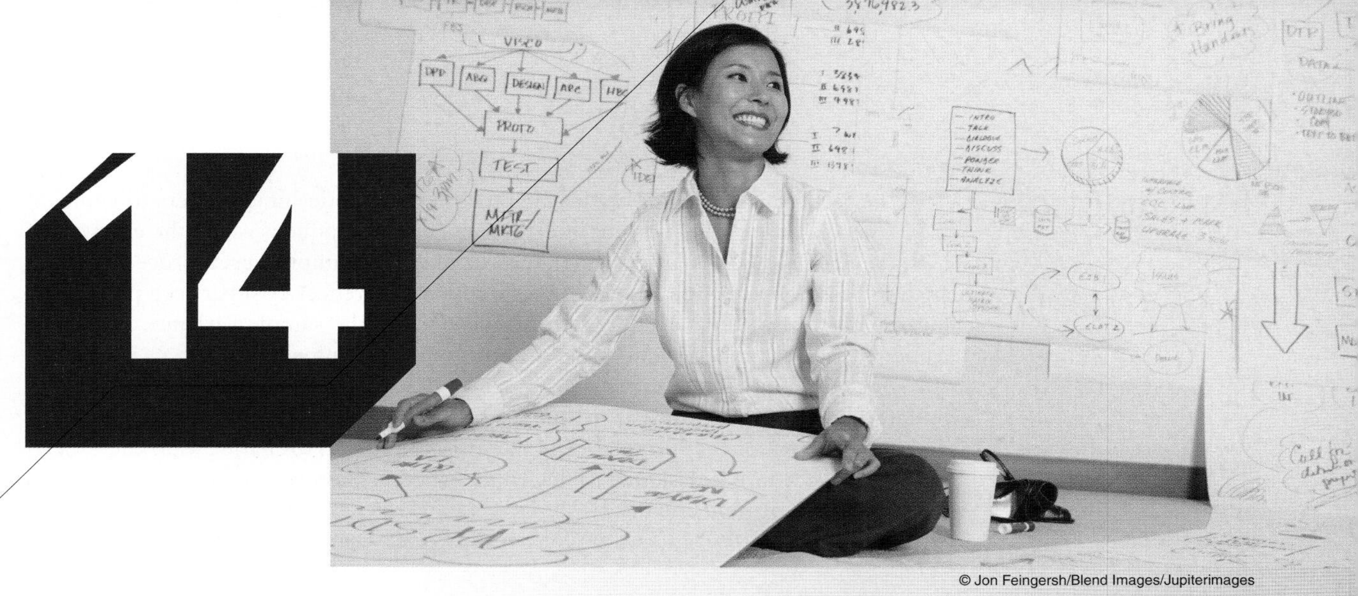

© Jon Feingersh/Blend Images/Jupiterimages

14 ANALYSIS OF VARIANCE

Causes of Variation

DATA Xm14-00

A critical component in an aircraft engine is a steel rod that must be 41.387 cm long. The operations manager has noted that there has been some variation in the lengths. In some cases the steel rods had to be discarded or reworked because they were either too short or too long. The operations manager believes that some of the variation is caused by the way the production process has been designed. Specifically, he believes that the rods vary from machine to machine and from operator to operator. To help unravel the truth, he organizes an experiment. Each of the three operators produces five rods on each of the four machines. The lengths are measured and recorded. Determine whether the machines and/or the operators are indeed sources of variation.

© Peter Poulides/Stone/Getty Images

See page 567 for the solution.

INTRODUCTION

The technique presented in this chapter allows statistics practitioners to compare two or more populations of interval data. The technique is called the **analysis of variance** and it is an extremely powerful and commonly used procedure. The analysis of variance technique determines whether differences exist between population means. Ironically, the procedure works by analyzing the sample variance, hence the name. We will examine several different forms of the technique.

One of the first applications of the analysis of variance was conducted in the 1920s to determine whether different treatments of fertilizer produced different crop yields. The terminology of that original experiment is still used. No matter what the experiment, the procedure is designed to determine whether there are significant differences between the **treatment means.**

14.1 ONE-WAY ANALYSIS OF VARIANCE

The analysis of variance is a procedure that tests to determine whether differences exist between two or more population means. The name of the technique derives from the way in which the calculations are performed. That is, the technique analyzes the variance of the data to determine whether we can infer that the population means differ. As in Chapter 13, the experimental design is a determinant in identifying the proper method to use. In this section, we describe the procedure to apply when the samples are independently drawn. The technique is called the **one-way analysis of variance.** Figure 14.1 depicts the sampling process for drawing independent samples. The mean and variance of population $j(j = 1, 2, \ldots, k)$ are labeled μ_j and σ_j^2, respectively. Both parameters are unknown. For each population, we draw independent random samples. For each sample, we can compute the mean $\bar{x}_j$ and the variance s_j^2.

FIGURE **14.1**
Sampling Scheme for Independent Samples

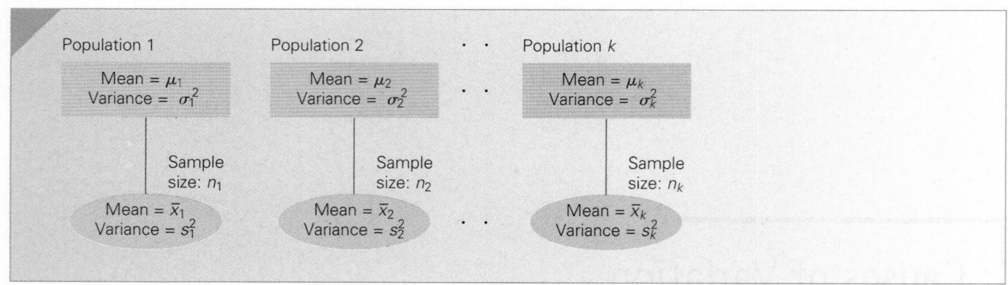

EXAMPLE 14.1*

DATA
Xm14-01

Proportion of Total Assets Invested in Stocks

In the last decade stockbrokers have drastically changed the way they do business. Internet trading has become quite common and online trades can cost as little $7. It is now easier and cheaper to invest in the stock market than ever before. What are the effects of these changes? To help answer this question, a financial analyst randomly sampled 366 American households and asked each to report the age of the head of the household

---

*Adapted from U.S. Census Bureau, "Asset Ownership of Households, May 2003," *Statistical Abstract of the United States*, 2006, Table 700.

and the proportion of their financial assets that are invested in the stock market. The age categories are

Young (under 35)
Early middle age (35 to 49)
Late middle age (50 to 65)
Senior (over 65)

The analyst was particularly interested in determining whether the ownership of stocks varied by age. Some of the data are listed next. Do these data allow the analyst to determine that there are differences in stock ownership between the four age groups?

Young	Early Middle Age	Late Middle Age	Senior
24.8	28.9	81.5	66.8
35.5	7.3	0.0	77.4
68.7	61.8	61.3	32.9
42.2	53.6	0.0	74.0
⋮	⋮	⋮	⋮

SOLUTION

You should confirm that the data are interval (percentage of total assets invested in the stock market) and that the problem objective is to compare four populations (age categories). The parameters are the four population means μ_1, μ_2, μ_3, and μ_4. The null hypothesis will state that there are no differences between the population means. Hence,

$$H_0: \quad \mu_1 = \mu_2 = \mu_3 = \mu_4$$

The analysis of variance determines whether there is enough statistical evidence to show that the null hypothesis is false. Consequently, the alternative hypothesis will always specify the following:

$$H_1: \quad \text{At least two means differ}$$

The next step is to determine the test statistic, which is somewhat more involved than the test statistics we have introduced thus far. The process of performing the analysis of variance is facilitated by the notation in Table 14.1.

TABLE **14.1** Notation for the One-Way Analysis of Variance

	Treatment						
	1	2		j		k
	x_{11}	x_{12}	...	x_{1j}	...	x_{1k}
	x_{21}	x_{22}	...	x_{2j}	...	x_{2k}
	⋮	⋮		⋮		⋮
	$x_{n_1 1}$	$x_{n_2 2}$		$x_{n_j j}$		$x_{n_k k}$
Sample size	n_1	n_2		n_j		n_k
Sample mean	$\bar{x}_1$	$\bar{x}_2$		$\bar{x}_j$		$\bar{x}_k$

x_{ij} = ith observation of the jth sample

n_j = number of observations in the sample taken from the jth population

$\bar{x}_j$ = mean of the jth sample $= \dfrac{\sum_{i=1}^{n_j} x_{ij}}{n_j}$

$\bar{\bar{x}}$ = grand mean of all the observations $= \dfrac{\sum_{j=1}^{k}\sum_{i=1}^{n_j} x_{ij}}{n}$ where $n = n_1 + n_2 + \cdots + n_k$ and k is the number of populations.

The variable X is called the **response variable** and its values are called **responses.** The unit that we measure is called an **experimental unit.** In this example, the response variable is the percentage of assets invested in stocks, and the experimental units are the heads of households sampled. The criterion by which we classify the populations is called a **factor.** Each population is called a factor **level.** The factor in Example 14.1 is the age category of the head of the household and there are four levels. Later in this chapter we'll discuss an experiment where the populations are classified using two factors. In this section we deal with single-factor experiments only.

Test Statistic

The test statistic is computed in accordance with the following rationale. If the null hypothesis is true, the population means would all be equal. We would then expect that the sample means would be close to one another. If the alternative hypothesis is true, however, there would be large differences between some of the sample means. The statistic that measures the proximity of the sample means to each other is called the **between-treatments variation** denoted **SST,** which stands for **sum of squares for treatments.**

Sum of Squares for Treatments

$$\text{SST} = \sum_{j=1}^{k} n_j(\bar{x}_j - \bar{\bar{x}})^2$$

As you can deduce from this formula, if the sample means are close to each other, all of the sample means would be close to the grand mean, and as a result, SST would be small. In fact, SST achieves its smallest value (zero) when all the sample means are equal. That is, if

$$\bar{x}_1 = \bar{x}_2 = \cdots = \bar{x}_k$$

then

$$\text{SST} = 0$$

It follows that a small value of SST supports the null hypothesis. In this example, we compute the sample means and the grand mean as

$$\bar{x}_1 = 44.40$$
$$\bar{x}_2 = 52.47$$
$$\bar{x}_3 = 51.14$$
$$\bar{x}_4 = 51.84$$
$$\bar{\bar{x}} = 50.18$$

The sample sizes are

$$n_1 = 84$$
$$n_2 = 131$$
$$n_3 = 93$$
$$n_4 = 58$$
$$n = n_1 + n_2 + n_3 + n_4 = 84 + 131 + 93 + 58 = 366$$

Then

$$\begin{aligned} \text{SST} &= \sum_{j=1}^{k} n_j(\bar{x}_j - \bar{\bar{x}})^2 \\ &= 84(44.40 - 50.18)^2 + 131(52.47 - 50.18)^2 \\ &\quad + 93(51.14 - 50.18)^2 + 58(51.84 - 50.18)^2 \\ &= 3{,}738.8 \end{aligned}$$

If large differences exist between the sample means, at least some sample means differ considerably from the grand mean, producing a large value of SST. It is then reasonable to reject the null hypothesis in favor of the alternative hypothesis. The key question to be answered in this test (as in all other statistical tests) is, How large does the statistic have to be for us to justify rejecting the null hypothesis? In our example, SST $= 3{,}738.8$. Is this value large enough to indicate that the population means differ? To answer this question, we need to know how much variation exists in the percentage of assets, which is measured by the **within-treatments variation,** which is denoted by **SSE (sum of squares for error).** The within-treatments variation provides a measure of the amount of variation in the response variable that is not caused by the treatments. In this example we are trying to determine whether the percentages of total assets invested in stocks vary by the age of the head of the household. However, there are other variables that affect the responses variable other than age. We would expect that variables such as household income, occupation, and the size of the family would play a role in determining how much money families invest in stocks. All of these (as well as others we may not even be able to identify) are sources of variation, which we would group together and call it the error. This source of variation is measured by the sum of squares for error.

Sum of Squares for Error

$$\text{SSE} = \sum_{j=1}^{k} \sum_{i=1}^{n_j} (x_{ij} - \bar{x}_j)^2$$

When SSE is partially expanded, we get

$$\text{SSE} = \sum_{i=1}^{n_1} (x_{i1} - \bar{x}_1)^2 + \sum_{i=1}^{n_2} (x_{i2} - \bar{x}_2)^2 + \cdots + \sum_{i=1}^{n_k} (x_{ik} - \bar{x}_k)^2$$

If you examine each of the k components of SSE, you'll see that each is a measure of the variability of that sample. If we divide each component by $n_j - 1$, we obtain the sample variances. We can express this by rewriting SSE as

$$\text{SSE} = (n_1 - 1)s_1^2 + (n_2 - 1)s_2^2 + \cdots + (n_k - 1)s_k^2$$

where s_j^2 is the sample variance of sample j. SSE is thus the combined or pooled variation of the k samples. This is an extension of a calculation we made in Section 13.1, where we tested and estimated the difference between two means using the pooled estimate of the common population variance (denoted s_p^2). One of the required conditions for that statistical technique is that the population variances are equal. That same condition is now necessary for us to use SSE. That is, we require that

$$\sigma_1^2 = \sigma_2^2 = \cdots = \sigma_k^2$$

Returning to our example, we calculate the sample variances as follows:

$$s_1^2 = 386.55$$

$$s_2^2 = 469.44$$

$$s_3^2 = 471.82$$

$$s_4^2 = 444.79$$

Thus,

$$\begin{aligned} \text{SSE} &= (n_1 - 1)s_1^2 + (n_2 - 1)s_2^2 + (n_3 - 1)s_3^2 + (n_4 - 1)s_4^2 \\ &= (84 - 1)(386.55) + (131 - 1)(469.44) + (93 - 1)(471.82) + (58 - 1)(444.79) \\ &= 161{,}871.3 \end{aligned}$$

The next step is to compute quantities called the **mean squares.** The **mean square for treatments** is computed by dividing SST by the number of treatments minus 1.

Mean Square for Treatments

$$\text{MST} = \frac{\text{SST}}{k - 1}$$

The **mean square for error** is determined by dividing SSE by the total sample size (labeled n) minus the number of treatments.

Mean Square for Error

$$\text{MSE} = \frac{\text{SSE}}{n - k}$$

Finally, the test statistic is defined as the ratio of the two mean squares.

Test Statistic

$$F = \frac{\text{MST}}{\text{MSE}}$$

Sampling Distribution of the Test Statistic

The test statistic is F-distributed with $k - 1$ and $n - k$ degrees of freedom provided that the response variable is normally distributed. In Section 8.4, we introduced the F distribution and in Section 13.4 we used it to test and estimate the ratio of two population variances. The test statistic in that application was the ratio of two sample variances s_1^2 and s_2^2. If you examine the definitions of SST and SSE, you will see that both measure variation similar to the numerator in the formula used to calculate the sample variance s^2 used throughout this book. When we divide SST by $k - 1$ and SSE by $n - k$ to calculate MST and MSE, respectively, we're actually computing unbiased estimators of the common population variance, assuming (as we do) that the null hypothesis is true. Thus, the ratio $F = \text{MST}/\text{MSE}$ is the ratio of two sample variances. The degrees of freedom for this application are the denominators in the mean squares, that is, $\nu_1 = k - 1$ and $\nu_2 = n - k$. For Example 14.1, the degrees of freedom are

$$\nu_1 = k - 1 = 4 - 1 = 3$$
$$\nu_2 = n - k = 366 - 4 = 362$$

In our example, we found

$$\text{MST} = \frac{\text{SST}}{k-1} = \frac{3{,}738.8}{3} = 1{,}246.27$$
$$\text{MSE} = \frac{\text{SSE}}{n-k} = \frac{161{,}871.3}{362} = 447.16$$
$$F = \frac{\text{MST}}{\text{MSE}} = \frac{1{,}246.27}{447.16} = 2.79$$

Rejection Region and *p*-Value

The purpose of calculating the ***F*-statistic** is to determine whether the value of SST is large enough to reject the null hypothesis. As you can see, if SST is large, F will be large. Hence, we reject the null hypothesis only if

$$F > F_{\alpha,k-1,n-k}$$

If we let $\alpha = .05$, the rejection region for Example 14.1 is

$$F > F_{\alpha,k-1,n-k} = F_{.05,3,362} \approx F_{.05,3,\infty} = 2.61$$

We found the value of the test statistic to be $F = 2.79$. Thus, there is enough evidence to infer that the mean percentage of total assets invested in the stock market differs between the four age groups.

The p-value of this test is

$$P(F > 2.79)$$

A computer is required to calculate this value, which is .0405.

Figure 14.2 depicts the sampling distribution for Example 14.1.

FIGURE **14.2** Sampling Distribution for Example 14.1

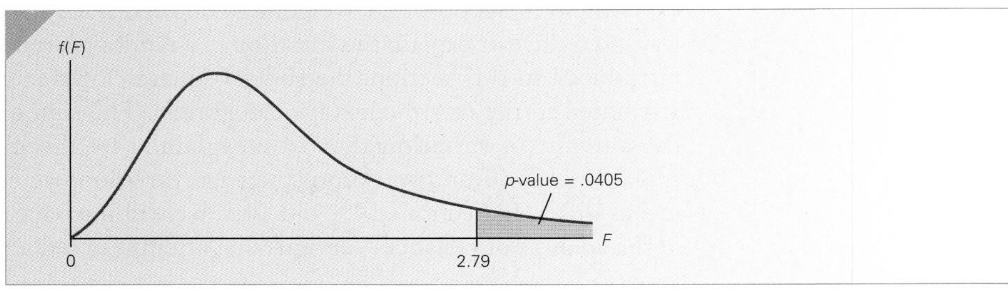

The results of the analysis of variance are usually reported in an **analysis of variance (ANOVA) table.** Table 14.2 shows the general organization of the ANOVA table, whereas Table 14.3 shows the ANOVA table for Example 14.1.

TABLE **14.2** ANOVA Table for the One-Way Analysis of Variance

Source of Variation	Degrees of Freedom	Sums of Squares	Mean Squares	*F*-Statistic
Treatments	$k - 1$	SST	MST = SST/(k − 1)	F = MST/MSE
Error	$n - k$	SSE	MSE = SSE/(n − k)	
Total	$n - 1$	SS(Total)		

TABLE **14.3** ANOVA Table for Example 14.1

Source of Variation	Degrees of Freedom	Sums of Squares	Mean Squares	*F*-Statistic
Treatments	3	3,738.8	1,246.27	2.79
Error	362	161,871.3	447.16	
Total	365	165,610.1		

The terminology used in the ANOVA table (and for that matter, in the test itself) is based on the partitioning of the sum of squares. Such partitioning is derived from the following equation (whose validity can be demonstrated by using the rules of summation):

$$\sum_{j=1}^{k}\sum_{i=1}^{n_j}(x_{ij} - \bar{\bar{x}})^2 = \sum_{j=1}^{k} n_j(\bar{x}_j - \bar{\bar{x}})^2 + \sum_{j=1}^{k}\sum_{i=1}^{n_j}(x_{ij} - \bar{x}_j)^2$$

The term on the left represents the **total variation** of all the data. This expression is denoted **SS(Total).** If we divide SS(Total) by the total sample size minus 1 (that is, by $n - 1$), we would obtain the sample variance (assuming that the null hypothesis is true). The first term on the right of the equal sign is SST, and the second term is SSE. As you can see, the total variation SS(Total) is partitioned into two sources of variation. The sum of squares for treatments (SST) is the variation attributed to the differences between the treatment means, whereas the sum of squares for error (SSE) measures the variation within the samples. The preceding equation can be restated as

$$\text{SS(Total)} = \text{SST} + \text{SSE}$$

The test is then based on the comparison of the mean squares of SST and SSE.

Recall that in discussing the advantages and disadvantages of the matched pairs experiment in Section 13.3, we pointed out that statistics practitioners frequently seek ways to reduce or explain the variation in a random variable. In the analysis of variance introduced in this section, the sum of squares for treatments explains the variation attributed to the treatments (age categories). The sum of squares for error measures the amount of variation that is unexplained by the different treatments. If SST explains a significant portion of the total variation, we conclude that the population means differ. In Sections 14.4 and 14.5, we will introduce other experimental designs of the analysis of variance—designs that attempt to reduce or explain even more of the variation.

If you've felt some appreciation of the computer and statistical software sparing you the need to manually perform the statistical techniques in earlier chapters, your appreciation should now grow, because the computer will allow you to avoid the incredibly time-consuming and boring task of performing the analysis of variance by hand. As usual, we've solved Example 14.1 using Excel and Minitab, whose outputs are shown here.

COMPUTE

EXCEL

	A	B	C	D	E	F	G
1	Anova: Single Factor						
2							
3	SUMMARY						
4	*Groups*	*Count*	*Sum*	*Average*	*Variance*		
5	Young	84	3729.5	44.40	386.55		
6	Early Middle Age	131	6873.9	52.47	469.44		
7	Late Middle Age	93	4755.9	51.14	471.82		
8	Senior	58	3006.6	51.84	444.79		
9							
10							
11	ANOVA						
12	*Source of Variation*	*SS*	*df*	*MS*	*F*	*P-value*	*F crit*
13	Between Groups	3741.4	3	1247.12	2.79	0.0405	2.6296
14	Within Groups	161871.0	362	447.16			
15							
16	Total	165612.3	365				

INSTRUCTIONS

1. Type or import the data into adjacent columns. (Open Xm14-01.)
2. Click **Data, Data Analysis,** and **Anova: Single Factor.**
3. Specify the **Input Range** (A1:D132) and a value for α (.05).

MINITAB

One-way ANOVA: Young, Early Middle Age, Late Middle Age, Senior

```
Source   DF      SS    MS     F      P
Factor    3    3741  1247  2.79  0.041
Error   362  161871   447
Total   365  165612

S = 21.15   R-Sq = 2.26%   R-Sq(adj) = 1.45%

Level               N   Mean  StDev
Young              84  44.40  19.66
Early Middle Age  131  52.47  21.67
Late Middle Age    93  51.14  21.72
Senior             58  51.84  21.09

                      Individual 95% CIs For Mean Based on Pooled StDev
Level                 --------+---------+---------+---------+---------
Young                  (-------*-------)
Early Middle Age                        (-----*-----)
Late Middle Age                      (------*------)
Senior                              (--------*--------)
                      --------+---------+---------+---------+---------
                         40.0      45.0      50.0      55.0

Pooled StDev = 21.15
```

INSTRUCTIONS

If the data are unstacked,

1. Type or import the data. (Open Xm14-01.)
2. Click **Stat, ANOVA,** and **Oneway (Unstacked) . . . .**
3. In the **Responses (in separate columns)** box, type or select the variable names of the treatments (Young, Early Middle Age, Late Middle Age, Senior).

If the data are stacked,

1. Type or import the data in two columns.
2. Click **Stat, ANOVA,** and **Oneway . . . .**
3. Type the variable name of the response variable and the name of the factor variable.

INTERPRET

The value of the test statistic is $F = 2.79$ and its p-value is .0405, which means there is evidence to infer that the percentage of total assets invested in stocks are different in at least two of the age categories.

Note that in this example the data are observational. We cannot conduct a controlled experiment. To do so would require the financial analyst to randomly assign households to each of the four age groups.

Incidentally, when the data are obtained through a controlled experiment in the one-way analysis of variance, we call the experimental design the **completely randomized design** of the analysis of variance.

Checking the Required Conditions

The F-test of the analysis of variance requires that the random variable be normally distributed with equal variances. The normality requirement is easily checked graphically by producing the histograms for each sample. From the Excel histograms in Figure 14.3, we can see that there is no reason to believe that the requirement is not satisfied.

The equality of variances is examined by printing the sample standard deviations or variances. Excel output includes the variances and Minitab calculates the standard deviations. The similarity of sample variances allows us to assume that the population variances are equal. In CD Appendix T we present Bartlett's test, a statistical procedure designed to test for the equality of variances.

Violation of the Required Conditions

If the data are not normally distributed, we can replace the one-way analysis of variance with its nonparametric counterpart, which is the Kruskal-Wallis test (see Section 19.3*). If the population variances are unequal, we can use several methods to correct the problem. However, these corrective measures are beyond the level of this book.

*Instructors who wish to teach the use of nonparametric techniques for testing the difference between two or more means when the normality requirement is not satisfied should use CD Appendix U, "Kruskal-Wallis Test and Friedman Test."

FIGURE **14.3** Histograms for Example 14.1

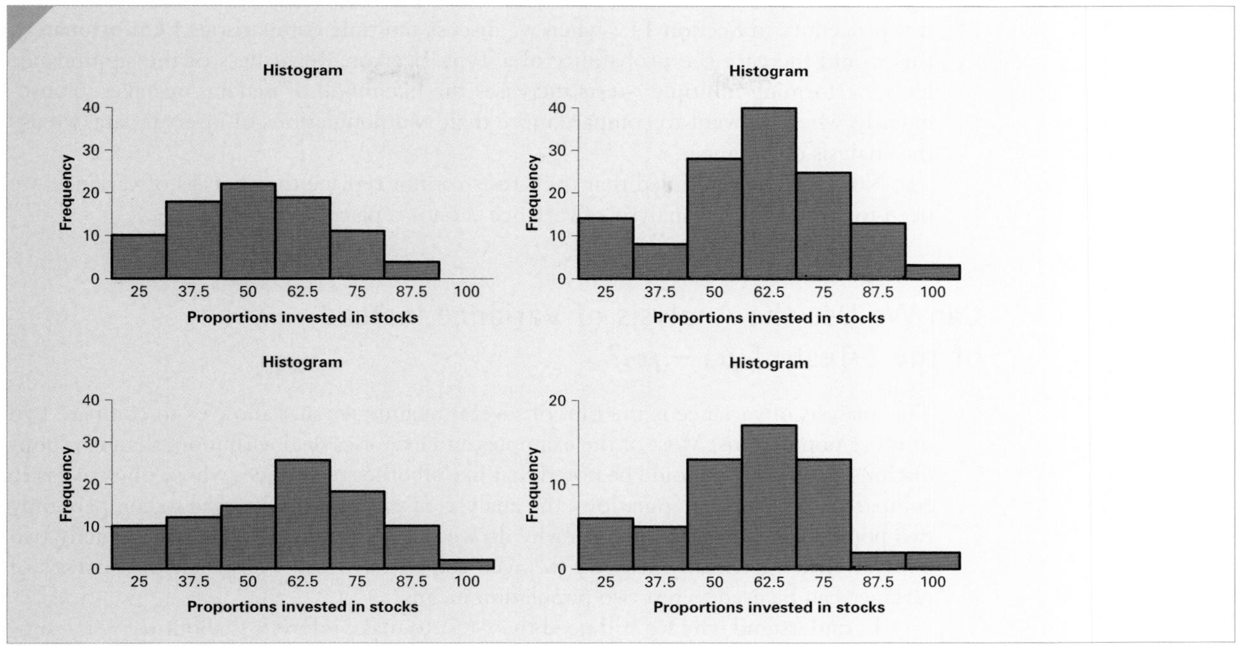

Can We Use the *t*-Test of the Difference between Two Means Instead of the Analysis of Variance?

The analysis of variance tests to determine whether there is evidence of differences between two or more population means. The t-test of $\mu_1 - \mu_2$ determines whether there is evidence of a difference between two population means. The question arises, Can we use t-tests instead of the analysis of variance? That is, instead of testing all the means in one test as in the analysis of variance, why not test each pair of means? In Example 14.1, we would test $(\mu_1 - \mu_2)$, $(\mu_1 - \mu_3)$, $(\mu_1 - \mu_4)$, $(\mu_2 - \mu_3)$, $(\mu_2 - \mu_4)$, and $(\mu_3 - \mu_4)$. If we find no evidence of a difference in each test, we would conclude that none of the means differ. If there was evidence of a difference in at least one test, we would conclude that some of the means differ.

There are two reasons why we don't use multiple t-tests instead of one F-test. First, we would have to perform many more calculations. Even with a computer, this extra work is tedious. Second, and more important, conducting multiple tests increases the probability of making Type I errors. To understand why, consider a problem where we want to compare six populations, all of which are identical. If we conduct an analysis of variance where we set the significance level at 5%, there is a 5% chance that we would reject the true null hypothesis. That is, there is a 5% chance that we would conclude that differences exist when in fact they don't.

To replace the F-test, we would perform 15 t-tests. [This number is derived from the number of combinations of pairs of means to test, which is $C_2^6 = (6 \times 5)/2 = 15$.] Each test would have a 5% probability of erroneously rejecting the null hypothesis. The probability of committing one or more Type I errors is about 54%.*

---

*The probability of committing at least one Type I error is computed from a binomial distribution with $n = 15$ and $p = .05$. Thus,

$$P(X \geq 1) = 1 - P(X = 0) = 1 - .463 = .537$$

One remedy for this problem is to decrease the significance level. In this illustration, we would perform the t-tests with $\alpha = .05/15$, which is equal to .0033. (We will use this procedure in Section 14.2 when we discuss multiple comparisons.) Unfortunately, this would increase the probability of a Type II error. Regardless of the significance level, performing multiple t-tests increases the likelihood of making mistakes. Consequently, when we want to compare more than two populations of interval data, we use the analysis of variance.

Now that we've argued that the t-tests cannot replace the analysis of variance, we need to argue that the analysis of variance cannot replace the t-test.

Can We Use the Analysis of Variance Instead of the t-Test of $\mu_1 - \mu_2$?

The analysis of variance is the first of several techniques that allow us to compare two or more populations. Most of the examples and exercises deal with more than two populations. However, it should be noted that like all other techniques whose objective is to compare two or more populations, the analysis of variance can be used to compare only two populations. If that's the case, why do we need techniques to compare exactly two populations? Specifically, why do we need the t-test of $\mu_1 - \mu_2$ when the analysis of variance can be used to test two population means?

To understand why we still need the t-test to make inferences about $\mu_1 - \mu_2$, suppose that we plan to use the analysis of variance to test two population means. The null and alternative hypotheses are

H_0: $\mu_1 = \mu_2$

H_1: At least two means differ

Of course, the alternative hypothesis specifies that $\mu_1 \neq \mu_2$. However, if we want to determine whether μ_1 is greater than μ_2 (or vice versa), we cannot use the analysis of variance because this technique allows us to test for a difference only. Thus, if we want to test to determine whether one population mean exceeds the other, we must use the t-test of $\mu_1 - \mu_2$ (with $\sigma_1^2 = \sigma_2^2$). Moreover, the analysis of variance requires that the population variances are equal. If they are not, we must use the unequal variances test statistic.

Relationship between the F-Statistic and the t-Statistic

It is probably useful for you to understand the relationship between the t-statistic and the F-statistic. The test statistic for testing hypotheses about $\mu_1 - \mu_2$ with equal variances is

$$t = \frac{(\bar{x}_1 - \bar{x}_2) - (\mu_1 - \mu_2)}{\sqrt{s_p^2\left(\frac{1}{n_1} + \frac{1}{n_2}\right)}}$$

If we square this quantity, the result is the F-statistic. That is, $F = t^2$. To illustrate this point, we'll redo the calculation of the test statistic in Example 13.1 using the analysis of variance. Recall that because we were able to assume that the population variances were equal, the value of the test statistic was as follows:

$$t = \frac{(6.63 - 3.72) - 0}{\sqrt{40.42\left(\frac{1}{50} + \frac{1}{50}\right)}}$$
$$= 2.29$$

Using the analysis of variance (the Excel output is shown here; Minitab's is similar), we find that the value of the test statistic is $F = 5.23$, which is 2.29^2. Notice though that the analysis of variance p-value is .0243, which is twice the t-test p-value, which is .0122. The reason: The analysis of variance is conducting a test to determine whether the population means *differ*. If Example 13.1 had asked to determine whether the means differ, we would have conducted a two-tail test and the p-value would be .0243, the same as the analysis of variance p-value.

Excel Analysis of Variance Output for Example 13.1

	A	B	C	D	E	F	G
1	Anova: Single Factor						
2							
3	SUMMARY						
4	*Groups*	*Count*	*Sum*	*Average*	*Variance*		
5	Direct	50	331.6	6.63	37.49		
6	Broker	50	186.2	3.72	43.34		
7							
8							
9	ANOVA						
10	*Source of Variation*	*SS*	*df*	*MS*	*F*	*P-value*	*F crit*
11	Between Groups	211.4	1	211.41	5.23	0.0243	3.9381
12	Within Groups	3960.5	98	40.41			
13							
14	Total	4172.0	99				

Developing an Understanding of Statistical Concepts

Conceptually and mathematically, the F-test of the independent samples single-factor analysis of variance is an extension of the t-test of $\mu_1 - \mu_2$. Moreover, if we simply want to determine whether a difference between two means exists, we can use the analysis of variance. The advantage of using the analysis of variance is that we can partition the total sum of squares, which enables us to measure how much variation is attributable to differences between populations and how much variation is attributable to differences within populations. As we pointed out in Section 13.3, explaining the variation is an extremely important topic, one that we will see again in other experimental designs of the analysis of variance and in regression analysis (Chapters 16, 17, and 18).

Let's review how we recognize the need to use the techniques introduced in this section.

Factors That Identify the One-Way Analysis of Variance

1. **Problem objective:** Compare two or more populations
2. **Data type:** Interval
3. **Experimental design:** Independent samples

EXERCISES

Developing an Understanding of Statistical Concepts

*Exercises 14.1–14.3 are "what-if analyses" designed to determine what happens to the test statistic when the means, variances, and sample sizes change. These problems can be solved manually or using the **ANOVA** worksheet in the **Test Statistics** workbook.*

14.1 A statistics practitioner calculated the following statistics:

	Treatment		
Statistic	1	2	3
n	5	5	5
$\bar{x}$	10	15	20
s^2	50	50	50

a. Complete the ANOVA table.
b. Repeat Part a, changing the sample sizes to 10 each.
c. Describe what happens to the F-statistic when the sample sizes increase.

14.2 You are given the following statistics:

	Treatment		
Statistic	1	2	3
n	4	4	4
$\bar{x}$	20	22	25
s^2	10	10	10

a. Complete the ANOVA table.
b. Repeat Part a, changing the variances to 25 each.
c. Describe the effect on the F-statistic of decreasing the sample variances.

14.3 The following statistics were calculated:

	Treatment			
Statistic	1	2	3	4
n	10	14	11	18
$\bar{x}$	30	35	33	40
s^2	10	10	10	10

a. Complete the ANOVA table.
b. Repeat Part a, changing the sample means to 130, 135, 133, and 140.
c. Describe the effect on the F-statistic of increasing the sample means by 100.

Applications

14.4 Xr14-04 How does an MBA major affect the number of job offers received? An MBA student randomly sampled four recent graduates in each of Finance, Marketing, and Management and asked each of them to report the number of job offers. Can we conclude at the 5% significance level that there are differences in the number of job offers between the three MBA majors?

Finance	Marketing	Management
3	1	8
1	5	5
4	3	4
1	4	6

14.5 Xr14-05 A consumer organization was concerned about the differences between the advertised sizes of containers and the actual amount of product. In a preliminary study, six packages of three different brands of margarine that are supposed to contain 500 ml were measured. The differences from 500 ml are listed here. Do these data provide sufficient evidence to conclude that differences exist between the three brands? Use $\alpha = .01$.

Brand 1	Brand 2	Brand 3
1	2	1
3	2	2
3	4	4
0	3	2
1	0	3
0	4	4

14.6 Xr14-06 Many college and university students obtain summer jobs. A statistics professor wanted to determine whether students in different degree programs earn different amounts. A random sample of 5 students in the B.A., B.Sc., and B.B.A. programs were asked to report what they earned the previous summer. The results (in $1,000s) are listed here. Can the professor infer at the 5% significance level that students in different degree programs differ in their summer earnings?

B.A.	B.Sc.	B.B.A.
3.3	3.9	4.0
2.5	5.1	6.2
4.6	3.9	6.3
5.4	6.2	5.9
3.9	4.8	6.4

14.7 Xr14-07 Spam is the price we pay to be capable of easily communicating using e-mail. Does spam affect everyone equally? In a preliminary study, university professors, administrators, and students were randomly sampled. Each person was asked to count the number of spam messages received that day. The results follow. Can we infer at the 2.5% significance level that the differing university communities differ in the amount of spam they receive in their e-mails?

Professors	Administrators	Students
7	5	12
4	9	4
0	12	5
3	16	18
18	10	15

14.8 Xr14-08 A management scientist believes that one way of judging whether a computer came equipped with enough memory is to determine the age of the computer. In a preliminary study, random samples of computer users were asked to identify the brand of computer and its age (in months). The categorized responses are shown here. Do these data provide sufficient evidence to conclude that there are differences in age between the computer brands? (Use $\alpha = .05$.)

IBM	Dell	Hewlett-Packard	Other
17	8	6	24
10	4	15	12
13	21	8	15

The following exercises require the use of a computer and software. Some answers may be calculated manually. See Appendix A for the sample statistics. ***Use a 5% significance level unless specified otherwise.***

14.9 Xr14-09 Because there are no national or regional standards, it is difficult for university admission committees to compare graduates of different high schools. University administrators have noted that an 80% average at a high school with low standards may be equivalent to a 70% average at another school with higher standards of grading. In an effort to more equitably compare applications, a pilot study was initiated. Random samples of students who were admitted the previous year from four local high schools were drawn. All the students entered the business program with averages between 70% and 80%. Their average grades in the first year at the university were computed.

a. Can the university admissions officer conclude that there are differences in grading standards between the four high schools?
b. What are the required conditions for the test conducted in Part a?
c. Does it appear that the required conditions of the test in Part a are satisfied?

14.10 Xr14-10 The friendly folks at the Internal Revenue Service (IRS) in the United States and Canada Revenue Agency (CRA) are always looking for ways to improve the wording and format of its tax return forms. Three new forms have been developed recently. To determine which, if any, are superior to the current form, 120 individuals were asked to participate in an experiment. Each of the three new forms and the currently used form were filled out by 30 different people. The amount of time (in minutes) taken by each person to complete the task was recorded.

a. What conclusions can be drawn from these data?
b. What are the required conditions for the test conducted in Part a?
c. Does it appear that the required conditions of the test in Part a are satisfied?

14.11 Xr14-11 Are proficiency test scores affected by the education of the child's parents? (Proficiency tests are administered to a sample of students in private and public schools. Test scores can range from 0 to 500.) To answer this question, a random sample of 9-year-old children was drawn. Each child's test score and the educational level of the parent with the higher level were recorded. The education categories are less than high school, high school graduate, some college, and college graduate. Can we infer that there are differences in test scores between children whose parents have different educational levels? (Adapted from the *Statistical Abstract of the United States*, 2000, Table 286)

14.12 Xr14-12 A manufacturer of outdoor brass lamps and mailboxes has received numerous complaints about premature corrosion. The manufacturer has identified the cause of the problem as the low-quality lacquer used to coat the brass. He decides to replace his current lacquer supplier with one of five possible alternatives. To judge which is best, he uses each of the five lacquers to coat 25 brass mailboxes and puts all 125 mailboxes outside. He records, for each, the number of days until the first sign of corrosion is observed.

a. Is there sufficient evidence at the 1% significance level to allow the manufacturer to conclude that differences exist between the five lacquers?
b. What are the required conditions for the test conducted in Part a?
c. Does it appear that the required conditions of the test in Part a are satisfied?

14.13 Xr14-13 In early 2001, the economy was slowing down and companies were laying off workers. A Gallup poll conducted February 9–11, 2001, asked a random sample of workers how long it would be before they had significant financial hardships if they lost their jobs and couldn't find new ones. They also classified their income. The classifications are

Over $50,000

$30,000 to $50,000

$20,000 to $30,000

Less than $20,000

Can we infer that differences exist between the four groups?

14.14 Xr14-14 In the introduction to this chapter we mentioned that the first use of the analysis of variance was in the 1920s. It was employed to determine whether different amounts of fertilizer yielded different amounts of crop. Suppose that a scientist at an agricultural college wanted to redo the original experiment using three different types of fertilizer. Accordingly, he applied fertilizer A to 20 1-acre plots of land, fertilizer B to another 20 plots, and fertilizer C to yet another 20 plots of land. At the end of the growing season, the crop yields were recorded. Can the scientist infer that differences exist between the crop yields?

14.15 Xr14-15 A study performed by a Columbia University professor (described in *Report on Business*, August 1991) counted the number of times per minute professors from three different departments said "uh" or "ah" during lectures to fill gaps between words. The data derived from observing 100 minutes from each of the three departments were recorded. If we assume that the more frequent use of "uh" and "ah" results in more boring lectures, can we conclude that some departments' professors are more boring than others?

14.16 Xr14-16 Does the level of success of publicly traded companies affect the way their board members are paid? Publicly traded companies were divided into four quarters using the rate of return in their stocks to differentiate among the companies. The annual payment (in $1,000s) to their board members was recorded. Can we infer that the amount of payment differs between the four groups of companies?

14.17 Xr14-17 In 1994 the chief executive officers of the major tobacco companies testified before a Senate subcommittee. One of the accusations made was that tobacco firms added nicotine to their cigarettes, which made them even more addictive to smokers. Company scientists argued that the amount of nicotine in cigarettes depended completely on the size of the tobacco leaf. That is, during poor growing seasons the tobacco leaves would be smaller than in normal or good growing seasons. However, since the amount of nicotine in a leaf is a fixed quantity, smaller leaves would result in cigarettes having more nicotine (since a greater fraction of the leaf would be used to make a cigarette). To examine the issue, a university chemist took random samples of tobacco leaves that were grown in greenhouses where the amount of water was allowed to vary. Three different groups of tobacco leaves were grown. Group 1 leaves were grown with about an average season's rainfall. Group 2 leaves were given about 67% of group 1's water, and group 3 leaves were given 33% of group 1's water. The size of the leaf (in grams) and the amount of nicotine in each leaf were measured.

a. Test to determine whether the leaf sizes differ between the three groups.

b. Test to determine whether the amounts of nicotine differ in the three groups.

14.18 Xr14-18 There is a bewildering number of breakfast cereals on the market. Each company produces several different products in the belief that there are distinct markets. For example, there is a market composed primarily of children, another for diet-conscious adults, and another for health-conscious adults. Each cereal the companies produce has at least one market as its target. However, consumers make their own decisions, which may or may not match the target predicted by the cereal maker. In an attempt to distinguish between consumers, a survey of adults between the ages of 25 and 65 was undertaken. Each was asked several questions including age, income, and years of education, as well as which brand of cereal they consumed most frequently. The cereal choices are

1. Sugar Smacks, a children's cereal
2. Special K, a cereal aimed at dieters
3. Fiber One, a cereal that is designed and advertised as healthy
4. Cheerios, a combination of healthy and tasty

The results of the survey were recorded using the following format:

Column 1: Cereal choice

Column 2: Age of respondent

Column 3: Annual household income

Column 4: Years of education

a. Determine whether there are differences between the ages of the consumers of the four cereals.

b. Determine whether there are differences between the incomes of the consumers of the four cereals.
c. Determine whether there are differences between the educational levels of the consumers of the four cereals.
d. Summarize your findings in Parts a through c and prepare a report describing the differences between the four groups of cereal consumers.

APPLICATIONS in **MARKETING**

Test Marketing

© Thierry Dosogne/
The Image Bank/Getty Images

In Chapter 13 we introduced test marketing, which allows us to determine whether changing some of the elements of the marketing mix yields different sales. In the next exercise we use the technique to discover the effect of different prices.

14.19 Xr14-19 A manufacturer of novelty items is undecided about the price to charge for a new product. The marketing manager knows that it should sell for about \$10 but is unsure of whether sales will vary significantly if it is priced at either \$9 or \$11. To conduct a pricing experiment, she distributes the new product to a sample of 60 stores belonging to a certain chain of variety stores. These 60 stores are all located in similar neighborhoods. The manager randomly selects 20 stores in which to sell the item at \$9, 20 stores to sell it at \$10, and the remaining 20 stores to sell it at \$11. Sales at the end of the trial period were recorded. What should the manager conclude?

APPLICATIONS in **MARKETING**

© BananaStock/Jupiterimages

Marketing Segmentation

Section 12.4 introduced market segmentation. In Chapter 13 we demonstrated how to use statistical analyses to determine whether two segments differ in their buying behavior. The next exercise requires you to apply the analysis of variance to determine whether several segments differ.

14.20 Xr14-20 After determining in Exercise 13.116 that teenagers watch more movies than do 20- to 30-year-olds, teenagers were further segmented into three age groups; 12 to 14, 15 to 16, and 17 to 19. Random samples were drawn from each segment and the number of movies each teenager saw last year was recorded. Do these data allow a marketing manager of a movie studio to conclude that differences exist between the three segments?

14.2 MULTIPLE COMPARISONS

When we conclude from the one-way analysis of variance that at least two treatment means differ, we often need to know which treatment means are responsible for these differences. For example, if an experiment is undertaken to determine whether different locations within a store produce different mean sales, the manager would be keenly interested in determining which locations result in significantly higher sales and which locations result in lower sales. Similarly, a stockbroker would like to know which one of several mutual funds outperforms the others, and a television executive would like to know which television commercials hold the viewers' attention and which are ignored.

Although it may appear that all we need to do is examine the sample means and identify the largest or the smallest to determine which population means are largest or smallest, this is not the case. To illustrate, suppose that in a five-treatment analysis of variance, we discover that differences exist and that the sample means are as follows:

$$\bar{x}_1 = 20 \quad \bar{x}_2 = 19 \quad \bar{x}_3 = 25 \quad \bar{x}_4 = 22 \quad \bar{x}_5 = 17$$

The statistics practitioner wants to know which of the following conclusions are valid:

1. μ_3 is larger than the other means.
2. μ_3 and μ_4 are larger than the other means.
3. μ_5 is smaller than the other means.
4. μ_5 and μ_2 are smaller than the other means.
5. μ_3 is larger than the other means, and μ_5 is smaller than the other means.

From the information we have, it is impossible to determine which, if any, of the statements are true. We need a statistical method to make this determination. The technique is called **multiple comparisons.**

EXAMPLE 14.2 Comparing the Costs of Repairing Car Bumpers

DATA Xm14-02

North American automobile manufacturers have become more concerned with quality because of foreign competition. One aspect of quality is the cost of repairing damage caused by accidents. A manufacturer is considering several new types of bumpers. To test how well they react to low-speed collisions, 10 bumpers of each of four different types were installed on midsize cars, which were then driven into a wall at 5 miles per hour. The cost of repairing the damage in each case was assessed. The data are shown below.

a. Is there sufficient evidence at the 5% significance level to infer that the bumpers differ in their reactions to low-speed collisions?
b. If differences exist, which bumpers differ?

Bumper 1	Bumper 2	Bumper 3	Bumper 4
610	404	599	272
354	663	426	405
234	521	429	197
399	518	621	363
278	499	426	297
358	374	414	538
379	562	332	181
548	505	460	318
196	375	494	412
444	438	637	499

SOLUTION

IDENTIFY

The problem objective is to compare four populations, the data are interval, and the samples are independent. The correct statistical method is the one-way analysis of variance, which we perform using Excel and Minitab.

COMPUTE

EXCEL

	A	B	C	D	E	F	G
1	Anova: Single Factor						
2							
3	SUMMARY						
4	*Groups*	*Count*	*Sum*	*Average*	*Variance*		
5	Bumper 1	10	3800	380.0	16924		
6	Bumper 2	10	4859	485.9	8197		
7	Bumper 3	10	4838	483.8	10426		
8	Bumper 4	10	3482	348.2	14049		
9							
10							
11	ANOVA						
12	*Source of Variation*	*SS*	*df*	*MS*	*F*	*P-value*	*F crit*
13	Between Groups	150884	3	50295	4.06	0.0139	2.8663
14	Within Groups	446368	36	12399			
15							
16	Total	597252	39				

MINITAB

```
One-way ANOVA: Bumper 1, Bumper 2, Bumper 3, Bumper 4

Source  DF      SS     MS     F      P
Factor   3  150884  50295  4.06  0.014
Error   36  446368  12399
Total   39  597252

S = 111.4   R-Sq = 25.26%   R-Sq(adj) = 19.03%

                              Individual 95% CIs For Mean Based on Pooled StDev
Level      N   Mean   StDev   ----------+-----------------+-----------------+-----------------+--------
Bumper 1  10  380.0   130.1        (---------------*--------------)
Bumper 2  10  485.9    90.5                              (---------------*---------------)
Bumper 3  10  483.8   102.1                              (--------------*---------------)
Bumper 4  10  348.2   118.5   (-----------------*--------------)
                              ----------+-----------------+-----------------+-----------------+--------
                                       320               400               480               560
```

INTERPRET

The test statistic is $F = 4.06$ and the p-value $= .0139$. There is enough statistical evidence to infer that there are differences between some of the bumpers. The question is now, Which bumpers differ?

There are several statistical inference procedures that deal with this problem. We will present three methods that allow us to determine which population means differ. All three methods apply to the one-way experiment only.

Fisher's Least Significant Difference (LSD) Method

This method was briefly introduced in Section 14.1 (page 523). To determine which population means differ, we could perform a series of t-tests of the difference between two means on all pairs of population means to determine which are significantly different. In Chapter 13 we introduced the equal-variances t-test of the difference between two means. The test statistic and confidence interval estimator are, respectively,

$$t = \frac{(\bar{x}_1 - \bar{x}_2) - (\mu_1 - \mu_2)}{\sqrt{s_p^2\left(\frac{1}{n_1} + \frac{1}{n_2}\right)}}$$

$$(\bar{x}_1 - \bar{x}_2) \pm t_{\alpha/2}\sqrt{s_p^2\left(\frac{1}{n_1} + \frac{1}{n_2}\right)}$$

with degrees of freedom $\nu = n_1 + n_2 - 2$.

Recall that s_p^2 is the pooled variance estimate, which is an unbiased estimator of the variance of the two populations. (Recall that the use of these techniques requires that the population variances be equal.) In this section we modify the test statistic and interval estimator.

Earlier in this chapter we pointed out that MSE is an unbiased estimator of the common variance of the populations we're testing. Since MSE is based on all the observations in the k samples, it will be a better estimator than s_p^2 (which is based on only two samples). Thus, we could draw inferences about every pair of means by substituting MSE for s_p^2 in the formulas for test statistic and confidence interval estimator shown previously. The number of degrees of freedom would also change to $\nu = n - k$ (where n is the total sample size). The test statistic to determine whether μ_i and μ_j differ is

$$t = \frac{(\bar{x}_i - \bar{x}_j) - (\mu_i - \mu_j)}{\sqrt{\text{MSE}\left(\frac{1}{n_i} + \frac{1}{n_j}\right)}}$$

The confidence interval estimator is

$$(\bar{x}_i - \bar{x}_j) \pm t_{\alpha/2}\sqrt{\text{MSE}\left(\frac{1}{n_i} + \frac{1}{n_j}\right)}$$

with degrees of freedom $\nu = n - k$.

We define the **least significant difference** LSD as

$$\text{LSD} = t_{\alpha/2}\sqrt{\text{MSE}\left(\frac{1}{n_i} + \frac{1}{n_j}\right)}$$

A simple way of determining whether differences exist between each pair of population means is to compare the absolute value of the difference between their two sample means and LSD. That is, we will conclude that μ_i and μ_j differ if

$$|\bar{x}_i - \bar{x}_j| > \text{LSD}$$

LSD will be the same for all pairs of means if all k sample sizes are equal. If some sample sizes differ, LSD must be calculated for each combination.

In Section 14.1 we argued that this method is flawed because it will increase the probability of committing a Type I error. That is, it is more likely than the analysis of variance to conclude that a difference exists in some of the population means when in fact none differ. On page 523 we calculated that if $k = 6$ and all population means are equal, the probability of erroneously inferring at the 5% significance level that at least two means differ is about 54%. The 5% figure is now referred to as the *comparisonwise Type I error rate*. The true probability of making at least one Type I error is called the *experimentwise Type I error rate*, denoted α_E. The experimentwise Type I error rate can be calculated as

$$\alpha_E = 1 - (1 - \alpha)^C$$

Here C is the number of pairwise comparisons, which can be calculated by $C = k(k - 1)/2$. Mathematicians have proven that

$$\alpha_E \leq C\alpha$$

which means that if we want the probability of making at least one Type I error to be no more than α_E, we simply specify $\alpha = \alpha_E/C$. The resulting procedure is called the **Bonferroni adjustment.**

Bonferroni Adjustment to LSD Method

The adjustment is made by dividing the specified experimentwise Type I error rate by the number of combinations of pairs of population means. For example, if $k = 6$, then

$$C = \frac{k(k-1)}{2} = \frac{6(5)}{2} = 15$$

If we want the true probability of a Type I error to be no more than 5%, we divide this probability by C. Thus, for each test we would use a value of α equal to

$$\alpha = \frac{\alpha_E}{C} = \frac{.05}{15} = .0033$$

We use Example 14.2 to illustrate Fisher's LSD method and the Bonferroni adjustment. The four sample means are

$$\bar{x}_1 = 380.0$$
$$\bar{x}_2 = 485.9$$
$$\bar{x}_3 = 483.8$$
$$\bar{x}_4 = 348.2$$

The pairwise absolute differences are

$$|\bar{x}_1 - \bar{x}_2| = |380.0 - 485.9| = |-105.9| = 105.9$$
$$|\bar{x}_1 - \bar{x}_3| = |380.0 - 483.8| = |-103.8| = 103.8$$
$$|\bar{x}_1 - \bar{x}_4| = |380.0 - 348.2| = |31.8| = 31.8$$
$$|\bar{x}_2 - \bar{x}_3| = |485.9 - 483.8| = |2.1| = 2.1$$
$$|\bar{x}_2 - \bar{x}_4| = |485.9 - 348.2| = |137.7| = 137.7$$
$$|\bar{x}_3 - \bar{x}_4| = |483.8 - 348.2| = |135.6| = 135.6$$

From the computer output we learn that MSE = 12,399 and $\nu = n - k = 40 - 4 = 36$. If we conduct the LSD procedure with $\alpha = .05$, we find $t_{\alpha/2,n-k} = t_{.025,36} \approx t_{.025,35} = 2.030$. Thus,

$$t_{\alpha/2}\sqrt{\text{MSE}\left(\frac{1}{n_i} + \frac{1}{n_j}\right)} = 2.030\sqrt{12{,}399\left(\frac{1}{10} + \frac{1}{10}\right)} = 101.09$$

We can see that four pairs of sample means differ by more than 101.09. That is, $|\bar{x}_1 - \bar{x}_2| = 105.9$, $|\bar{x}_1 - \bar{x}_3| = 103.8$, $|\bar{x}_2 - \bar{x}_4| = 137.7$, and $|\bar{x}_3 - \bar{x}_4| = 135.6$. Hence, μ_1 and μ_2, μ_1 and μ_3, μ_2 and μ_4, and μ_3 and μ_4 differ. The other two pairs μ_1 and μ_4, and μ_2 and μ_3 do not differ.

If we perform the LSD procedure with the Bonferroni adjustment, the number of pairwise comparisons is 6 (calculated as $C = k(k - 1)/2 = 4(3)/2$). We set $\alpha = .05/6 = .0083$. Thus, $t_{\alpha/2,36} = t_{.0042,36} = 2.794$ (available from Excel and difficult to approximate manually) and

$$t_{\alpha/2}\sqrt{\text{MSE}\left(\frac{1}{n_i} + \frac{1}{n_j}\right)} = 2.794\sqrt{12{,}399\left(\frac{1}{10} + \frac{1}{10}\right)} = 139.13$$

Now no pair of means differ since all the absolute values of the differences between sample means are less than 139.19.

The drawback to the LSD procedure is that we increase the probability of at least one Type I error. The Bonferroni adjustment corrects this problem. However, recall that the probabilities of Type I and Type II errors are inversely related. The Bonferroni adjustment uses a smaller value of α, which results in an increased probability of a Type II error. A Type II error occurs when a difference between population means exists, yet we cannot detect it. This may be the case in this example. The next multiple comparison method addresses this problem.

Tukey's Multiple Comparison Method

A more powerful test is **Tukey's multiple comparison method.** This technique determines a critical number similar to LSD for Fisher's test, denoted by ω (Greek letter *omega*) such that, if any pair of sample means has a difference greater than ω, we conclude that the pair's two corresponding population means are different.

The test is based on the Studentized range, which is defined as the variable

$$q = \frac{\bar{x}_{\text{max}} - \bar{x}_{\text{min}}}{s/\sqrt{n}}$$

where $\bar{x}_{\text{max}}$ and $\bar{x}_{\text{min}}$ are the largest and smallest sample means, respectively, assuming that there are no differences between the population means. We define ω as follows.

Critical Number ω

$$\omega = q_\alpha(k, \nu)\sqrt{\frac{\text{MSE}}{n_g}}$$

where

k = number of treatments

n = number of observations ($n = n_1 + n_2 + \cdots + n_k$)

ν = number of degrees of freedom associated with MSE ($\nu = n - k$)

n_g = number of observations in each of k samples

α = significance level

$q_\alpha(k, \nu)$ = critical value of the Studentized range

Theoretically, this procedure requires that all sample sizes be equal. However, if the sample sizes are different, we can still use this technique provided that the sample sizes are at least similar. The value of n_g used previously is the *harmonic mean* of the sample sizes. That is,

$$n_g = \frac{k}{\dfrac{1}{n_1} + \dfrac{1}{n_2} + \cdots + \dfrac{1}{n_k}}$$

Table 7 in Appendix B provides values of $q_\alpha(k, \nu)$ for a variety of values of k and ν, and for $\alpha = .01$ and .05. Applying Tukey's method to Example 14.2, we find

$$\begin{aligned} k &= 4 \\ n_1 &= n_2 = n_3 = n_4 = n_g = 10 \\ \nu &= n - k = 40 - 4 = 36 \\ \text{MSE} &= 12{,}399 \\ q_{.05}(4,37) &\approx q_{.05}(4,40) = 3.79 \end{aligned}$$

Thus,

$$\omega = q_\alpha(k,\nu)\sqrt{\frac{\text{MSE}}{n_g}} = (3.79)\sqrt{\frac{12{,}399}{10}} = 133.45$$

There are two absolute values larger than 133.45. Hence, we conclude that μ_2 and μ_4, and μ_3 and μ_4 differ. The other four pairs do not differ.

EXCEL

	A	B	C	D	E
1	**Multiple Comparisons**				
2					
3				LSD	Omega
4	Treatment	Treatment	Difference	Alpha = 0.05	Alpha = 0.05
5	*Bumper 1*	*Bumper 2*	-105.9	100.99	133.45
6		*Bumper 3*	-103.8	100.99	133.45
7		*Bumper 4*	31.8	100.99	133.45
8	*Bumper 2*	*Bumper 3*	2.1	100.99	133.45
9		*Bumper 4*	137.7	100.99	133.45
10	*Bumper 3*	*Bumper 4*	135.6	100.99	133.45

Tukey and Fisher's LSD with the Bonferroni Adjustment ($\alpha = .05/6 = .0083$)

	A	B	C	D	E
1	**Multiple Comparisons**				
2					
3				LSD	Omega
4	Treatment	Treatment	Difference	Alpha = 0.0083	Alpha = 0.05
5	*Bumper 1*	*Bumper 2*	-105.9	139.11	133.45
6		*Bumper 3*	-103.8	139.11	133.45
7		*Bumper 4*	31.8	139.11	133.45
8	*Bumper 2*	*Bumper 3*	2.1	139.11	133.45
9		*Bumper 4*	137.7	139.11	133.45
10	*Bumper 3*	*Bumper 4*	135.6	139.11	133.45

The printout includes ω (Tukey's method), the differences between sample means for each combination of populations, and Fisher's LSD. (The Bonferroni adjustment is made by specifying another value for α.)

INSTRUCTIONS

1. Type or import the data into adjacent columns. (Open Xm14-02.)
2. Click **Add-Ins, Data Analysis Plus** and **Multiple Comparisons.**
3. Specify the **Input Range** (A1:D11). Type the value of α. To use the Bonferroni adjustment, divide α by $C = k(k - 1)/2$. For Tukey, Excel computes ϖ only for $\alpha = .05$.

MINITAB

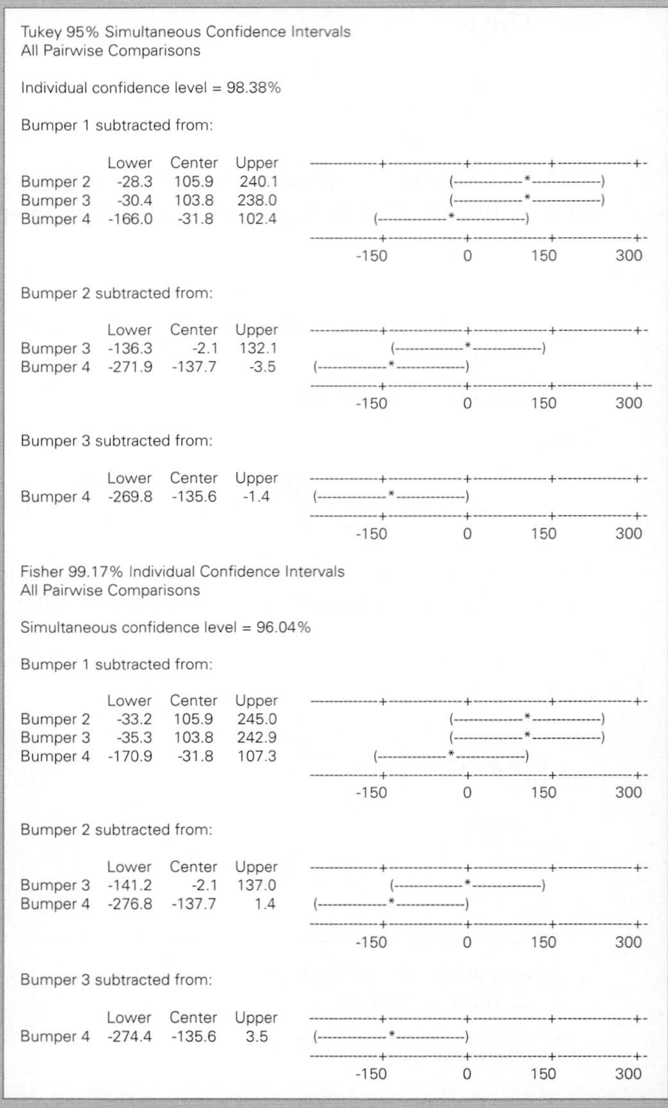

```
Tukey 95% Simultaneous Confidence Intervals
All Pairwise Comparisons

Individual confidence level = 98.38%

Bumper 1 subtracted from:

          Lower   Center   Upper   ------------+------------+------------+------------+-
Bumper 2  -28.3    105.9   240.1                     (-----------*-----------)
Bumper 3  -30.4    103.8   238.0                     (-----------*-----------)
Bumper 4  -166.0   -31.8   102.4        (-----------*-----------)
                                   ------------+------------+------------+------------+-
                                             -150           0          150          300

Bumper 2 subtracted from:

          Lower   Center   Upper   ------------+------------+------------+------------+-
Bumper 3  -136.3    -2.1   132.1             (-----------*-----------)
Bumper 4  -271.9  -137.7    -3.5   (-----------*-----------)
                                   ------------+------------+------------+------------+--
                                             -150           0          150          300

Bumper 3 subtracted from:

          Lower   Center   Upper   ------------+------------+------------+------------+-
Bumper 4  -269.8  -135.6    -1.4   (-----------*-----------)
                                   ------------+------------+------------+------------+-
                                             -150           0          150          300

Fisher 99.17% Individual Confidence Intervals
All Pairwise Comparisons

Simultaneous confidence level = 96.04%

Bumper 1 subtracted from:

          Lower   Center   Upper   ------------+------------+------------+------------+-
Bumper 2  -33.2    105.9   245.0                     (-----------*-----------)
Bumper 3  -35.3    103.8   242.9                     (-----------*-----------)
Bumper 4  -170.9   -31.8   107.3        (-----------*-----------)
                                   ------------+------------+------------+------------+-
                                             -150           0          150          300

Bumper 2 subtracted from:

          Lower   Center   Upper   ------------+------------+------------+------------+-
Bumper 3  -141.2    -2.1   137.0             (-----------*-----------)
Bumper 4  -276.8  -137.7     1.4   (-----------*-----------)
                                   ------------+------------+------------+------------+-
                                             -150           0          150          300

Bumper 3 subtracted from:

          Lower   Center   Upper   ------------+------------+------------+------------+-
Bumper 4  -274.4  -135.6     3.5   (-----------*-----------)
                                   ------------+------------+------------+------------+-
                                             -150           0          150          300
```

Minitab reports the results of Tukey's multiple comparisons by printing interval estimates of the differences between each pair of means. The estimates are computed by calculating the pairwise difference between sample means minus ϖ for the lower limit and plus ϖ for the upper limit. The calculations are described in the following table.

Tukey's Method

Pair of Population Means Compared	Difference	Lower Limit	Upper Limit
Bumper 2–Bumper 1	105.9	−28.3	240.1
Bumper 3–Bumper 1	103.8	−30.4	238.0
Bumper 4–Bumper 1	−31.8	−166.0	102.4
Bumper 3–Bumper 2	−2.1	−136.3	132.1
Bumper 4–Bumper 2	−137.7	−271.9	−3.5
Bumper 4–Bumper 3	−135.6	−269.8	−1.4

A similar calculation is performed for Fisher's method replacing ϖ by LSD.

Fisher's Method

Pair of Population Means Compared	Difference	Lower Limit	Upper Limit
Bumper 2–Bumper 1	105.9	−33.2	245.0
Bumper 3–Bumper 1	103.8	−35.3	242.9
Bumper 4–Bumper 1	−31.8	−170.9	107.3
Bumper 3–Bumper 2	−2.1	−141.2	137.0
Bumper 4–Bumper 2	−137.7	−276.8	1.4
Bumper 4–Bumper 3	−135.6	−274.7	3.5

We interpret the test results in the following way. If the interval includes 0, there is not enough evidence to infer that the pair of means differ. If the entire interval is above or the entire interval is below 0, we conclude that the pair of means differ.

INSTRUCTIONS

1. Type or import the data either in stacked or unstacked format. (Open Xm14-02.)
2. Click **Stat, ANOVA**, and **Oneway (Unstacked) . . . .**
3. Type or **Select** the variables in the **Responses (in separate columns)** box (Bumper 1, Bumper 2, Bumper 3, Bumper 4).
4. Click **Comparisons . . . .** Select Tukey's method and specify α. Select Fisher's method and specify α. For the Bonferroni adjustment divide α by $C = k(k - 1)/2$.

INTERPRET

Using the Bonferroni adjustment of Fisher's LSD method, we discover that none of the bumpers differ. Tukey's method tells us that Bumper 4 differs from both Bumpers 2 and 3. Based on this sample, Bumper 4 appears to have the lowest cost of repair. Because there was not enough evidence to conclude that Bumpers 1 and 4 differ, we would consider using Bumper 1 if there are advantages that it has over Bumper 4.

Which Multiple Comparison Method to Use

Unfortunately, no one procedure works best in all types of problems. Most statisticians agree with the following guidelines:

> If you have identified two or three pairwise comparisons that you wish to make before conducting the analysis of variance, use the Bonferroni method. This means that if in a problem there are 10 populations but you're particularly interested in comparing, say populations 3 and 7 and populations 5 and 9, use Bonferroni with $C = 2$.
>
> If you plan to compare all possible combinations, use Tukey.
>
> When do we use Fisher's LSD? If the purpose of the analysis is to point to areas that should be investigated further, Fisher's LSD method is indicated.

Incidentally, to employ Fisher's LSD or the Bonferroni adjustment, you must perform the analysis of variance first. Tukey's method can be employed instead of the analysis of variance.

EXERCISES

Developing an Understanding of Statistical Concepts

14.21 a. Use Fisher's LSD method with $\alpha = .05$ to determine which population means differ in the following problem:

$k = 3$ $n_1 = 10$ $n_2 = 10$ $n_3 = 10$

$\text{MSE} = 700$ $\bar{x}_1 = 128.7$ $\bar{x}_2 = 101.4$ $\bar{x}_3 = 133.7$

b. Repeat Part a, using the Bonferroni adjustment.

c. Repeat Part a, using Tukey's multiple comparison method.

14.22 a. Use Fisher's LSD procedure with $\alpha = .05$ to determine which population means differ given the following statistics:

$k = 5$ $n_1 = 5$ $n_2 = 5$ $n_3 = 5$

$\text{MSE} = 125$ $\bar{x}_1 = 227$ $\bar{x}_2 = 205$ $\bar{x}_3 = 219$

$n_4 = 5$ $n_5 = 5$

$\bar{x}_4 = 248$ $\bar{x}_5 = 202$

b. Repeat Part a, using the Bonferroni adjustment.

c. Repeat Part a, using Tukey's multiple comparison method.

Applications

Unless specified otherwise use a 5% significance level.

14.23 Apply Tukey's method to determine which brands differ in Exercise 14.5.

14.24 Refer to Exercise 14.6.

a. Employ Fisher's LSD method to determine which degrees differ (use $\alpha = .10$).

b. Repeat Part a, using the Bonferroni adjustment.

The following exercises require the use of a computer and software. The answers may be calculated manually. See Appendix A for the sample statistics.

14.25 Xr14-09 a. Apply Fisher's LSD method with the Bonferroni adjustment to determine which schools differ in Exercise 14.9.

b. Repeat Part a, applying Tukey's method instead.

14.26 Xr14-10 a. Apply Tukey's multiple comparison method to determine which forms differ in Exercise 14.10.

b. Repeat Part a, applying the Bonferroni adjustment.

14.27 Xr14-27 Police cars, ambulances, and other emergency vehicles are required to carry road flares. One of the most important features of flares is their burning times. To help decide which of four brands on the market to use, a police laboratory technician measured the burning time for a random sample of 10 flares of each brand. The results were recorded to the nearest minute.

a. Can we conclude that differences exist between the burning times of the four brands of flares?

b. Apply Fisher's LSD method with the Bonferroni adjustment to determine which flares are better.

c. Repeat Part b, using Tukey's method.

14.28 Xr14-12 Refer to Exercise 14.12.

a. Apply Fisher's LSD method with the Bonferroni adjustment to determine which lacquers differ.

b. Repeat Part a, applying Tukey's method instead.

14.29 Xr14-29 An engineering student who is about to graduate decided to survey various firms in Silicon Valley to see which offered the best chance for early promotion and career advancement. He surveyed 30 small firms (size level is based on gross revenues), 30 medium-size firms, and 30 large firms and determined how much time must elapse before an average engineer can receive a promotion.

a. Can the engineering student conclude that speed of promotion varies between the three sizes of engineering firms?

b. If differences exist, which of the following is true? Use Tukey's method.

i. Small firms differ from the other two.

ii. Medium-size firms differ from the other two.

iii. Large firms differ from the other two.

iv. All three firms differ from one another.

v. Small firms differ from large firms.

14.30 Xr14-14 a. Apply Tukey's multiple comparison method to determine which fertilizers differ in Exercise 14.14.

b. Repeat Part a, applying the Bonferroni adjustment.

14.3 Analysis of Variance Experimental Designs

Since we introduced the matched pairs experiment in Section 13.3, the experimental design has been one of the factors that determines which technique we use. Statistics practitioners often design experiments to help extract the information they need to assist them in making decisions. The one-way analysis of variance introduced in Section 14.1 is only one of many different experimental designs of the analysis of variance. For each type of experiment, we can describe the behavior of the response variable using a mathematical expression or model. Although we will not exhibit the mathematical expressions in this chapter (we introduce models in Chapter 16), we think it is useful for you to be aware of the elements that distinguish one experimental design or model from another. In this section, we present some of these elements, and in so doing, we introduce two of the experimental designs that will be presented later in this chapter.

Single-Factor and Multifactor Experimental Designs

As we pointed out in Section 14.1, the criterion by which we identify populations is called a *factor*. The experiment described in Section 14.1 is a single-factor analysis of variance because it addresses the problem of comparing two or more populations defined on the basis of only one factor. A **multifactor experiment** is one where there are two or more factors that define the treatments. The experiment described in Example 14.1 is a single-factor design because we had one treatment, age of the head of the household. That is, the factor is the age, and the four age categories were the levels of this factor.

Suppose that in another study, we can also look at the gender of the household head. We would then develop a two-factor analysis of variance where the first factor, age, has four levels and the second factor, gender has two levels. We will discuss two-factor experiments in Section 14.5.

Independent Samples and Blocks

In Section 13.3, we introduced statistical techniques where the data were gathered from a matched pairs experiment. This type of experimental design reduces the variation within the samples, making it easier to detect differences between the two populations. When the problem objective is to compare more than two populations, the experimental design that is the counterpart of the matched pairs experiment is called the **randomized block design.** The term *block* refers to a matched group of observations from each population. Suppose that in Examples 13.4 and 13.5 we had wanted to compare the salary offers for finance, marketing, accounting, and operations management majors. To redo Example 13.5 we would conduct a randomized block experiment where the blocks are the 25 GPA groups and the treatments are the four MBA majors. Once again the experimental design should reduce the variation in each treatment to make it easier to detect differences.

We can also perform a blocked experiment by using the same subject (person, plant, and store) for each treatment. For example, we can determine whether sleeping pills are effective by giving three brands of pills to the same group of people to measure the effects. Such experiments are called **repeated measures** designs. Technically, this is a different design than the randomized block. However, the data are analyzed in the same way for both designs. Hence, we will treat repeated measures designs as randomized block designs.

The randomized block experiment is also called the **two-way analysis of variance.** In Section 14.4, we introduce the technique used to calculate the test statistic for this type of experiment.

Fixed and Random Effects

If our analysis includes all possible levels of a factor, the technique is called a **fixed-effects analysis of variance.** If the levels included in the study represent a random sample of all the levels that exist, the technique is called a **random-effects analysis of variance.** In Example 14.2, there were only four possible bumpers. Consequently, the study is a fixed-effects experiment. However, if there were other bumpers besides the four described in the example, and we wanted to know whether there were differences in repair costs between all bumpers, the application would be a random-effects experiment. Here's another example.

To determine whether there is a difference in the number of units produced by the machines in a large factory, four machines out of 50 in the plant are randomly selected for study. The number of units each produces per day for 10 days will be recorded. This experiment is a random-effects experiment because we selected a random sample of four machines, and therefore the statistical results will allow us to determine whether there are differences between the 50 machines.

In some experimental designs, there are no differences in calculations of the test statistic between fixed and random effects. However, in others, including the two-factor experiment presented in Section 14.5, the calculations are different.

14.4 Randomized Block (Two-Way) Analysis of Variance

The purpose of designing a randomized block experiment is to reduce the within-treatments variation to more easily detect differences between the treatment means. In the one-way analysis of variance, we partitioned the total variation into the between-treatments and the within-treatments variation. That is

$$SS(\text{Total}) = SST + SSE$$

In the randomized block design of the analysis of variance, we partition the total variation into three sources of variation.

$$SS(\text{Total}) = SST + SSB + SSE$$

where SSB, the **sum of squares for blocks,** measures the variation between the blocks. When the variation associated with the blocks is removed, SSE is reduced, making it easier to determine whether differences exist between the treatment means.

At this point in our presentation of statistical inference, we will deviate from our usual procedure of solving examples in three ways: manually, using Excel, and using Minitab. The calculations for this experimental design and for the experiment presented in the next section are so time-consuming that solving them by hand adds little to your understanding of the technique. Consequently, although we will continue to present the concepts by discussing how the statistics are calculated, we will solve the problems only by computer.

To help you understand the formulas, we will use the following notation:

$\bar{x}[T]_j$ = Mean of the observations in the jth treatment ($j = 1, 2, \ldots, k$)

$\bar{x}[B]_i$ = Mean of the observations in the ith block ($i = 1, 2, \ldots, b$)

b = Number of blocks

Table 14.4 summarizes the notation we use in this experimental design.

TABLE **14.4** Notation for the Randomized Block Analysis of Variance

	Treatment				
Block	**1**	**2**		k	**Block Mean**
1	x_{11}	x_{12}	. . .	x_{1k}	$\bar{x}[B]_1$
2	x_{21}	x_{22}	. . .	x_{2k}	$\bar{x}[B]_2$
	⋮	⋮		⋮	⋮
b	x_{b1}	x_{b2}	. . .	x_{bk}	$\bar{x}[B]_b$
Treatment mean	$\bar{x}[T]_1$	$\bar{x}[T]_2$	. . .	$\bar{x}[T]_k$	

The definitions of SS(Total) and SST in the randomized block design are identical to those in the independent samples design. SSE in the independent samples design is equal to the sum of SSB and SSE in the randomized block design.

Sums of Squares in the Randomized Block Experiment

$$\text{SS(Total)} = \sum_{j=1}^{k}\sum_{i=1}^{b}(x_{ij} - \bar{\bar{x}})^2$$

$$\text{SST} = \sum_{j=1}^{k} b(\bar{x}[T]_j - \bar{\bar{x}})^2$$

$$\text{SSB} = \sum_{i=1}^{b} k(\bar{x}[B]_i - \bar{\bar{x}})^2$$

$$\text{SSE} = \sum_{j=1}^{k}\sum_{i=1}^{b}(x_{ij} - \bar{x}[T]_j - \bar{x}[B]_i + \bar{\bar{x}})^2$$

The test is conducted by determining the mean squares, which are computed by dividing the sums of squares by their respective degrees of freedom.

Mean Squares for the Randomized Block Experiment

$$\text{MST} = \frac{\text{SST}}{k - 1}$$

$$\text{MSB} = \frac{\text{SSB}}{b - 1}$$

$$\text{MSE} = \frac{\text{SSE}}{n - k - b + 1}$$

Finally, the test statistic is the ratio of mean squares, as described in the box.

Test Statistic for the Randomized Block Experiment

$$F = \frac{\text{MST}}{\text{MSE}}$$

which is F-distributed with $\nu_1 = k - 1$ and $\nu_2 = n - k - b + 1$ degrees of freedom.

An interesting, and sometimes useful, by-product of the test of the treatment means is that we can also test to determine whether the block means differ. This will allow us to determine whether the experiment should have been conducted as a randomized block design. (If there are no differences between the blocks, the randomized block design is less likely to detect real differences between the treatment means.) Such a discovery could be useful in future similar experiments. The test of the block means is almost identical to that of the treatment means except the test statistic is

$$F = \frac{\text{MSB}}{\text{MSE}}$$

which is F-distributed with $\nu_1 = b - 1$ and $\nu_2 = n - k - b + 1$ degrees of freedom.

As with the one-way experiment, the statistics generated in the randomized block experiment are summarized in an ANOVA table, whose general form is exhibited in Table 14.5.

TABLE **14.5** ANOVA Table for the Randomized Block Analysis of Variance

Source of Variation	Degrees of Freedom	Sums of Squares	Mean Squares	F-Statistic
Treatments	$k - 1$	SST	MST = SST/$(k - 1)$	F = MST/MSE
Blocks	$b - 1$	SSB	MSB = SSB/$(b - 1)$	F = MSB/MSE
Error	$n - k - b + 1$	SSE	MSE = SSE/$(n - k - b + 1)$	
Total	$n - 1$	SS(Total)		

EXAMPLE 14.3

DATA Xm14-03

Comparing Cholesterol-Lowering Drugs

Many North Americans suffer from high levels of cholesterol, which can lead to heart attacks. For those with very high levels (over 280), doctors prescribe drugs to reduce cholesterol levels. A pharmaceutical company has recently developed four such drugs. To determine whether any differences exist in their benefits, an experiment was organized. The company selected 25 groups of four men, each of whom had cholesterol levels in excess of 280. In each group, the men were matched according to age and weight. The drugs were administered over a 2-month period, and the reduction in cholesterol was recorded. Do these results allow the company to conclude that differences exist between the four new drugs?

Group	Drug 1	Drug 2	Drug 3	Drug 4
1	6.6	12.6	2.7	8.7
2	7.1	3.5	2.4	9.3
3	7.5	4.4	6.5	10
4	9.9	7.5	16.2	12.6
5	13.8	6.4	8.3	10.6
6	13.9	13.5	5.4	15.4
7	15.9	16.9	15.4	16.3
8	14.3	11.4	17.1	18.9
9	16	16.9	7.7	13.7
10	16.3	14.8	16.1	19.4
11	14.6	18.6	9	18.5
12	18.7	21.2	24.3	21.1
13	17.3	10	9.3	19.3
14	19.6	17	19.2	21.9
15	20.7	21	18.7	22.1
16	18.4	27.2	18.9	19.4
17	21.5	26.8	7.9	25.4
18	20.4	28	23.8	26.5
19	21.9	31.7	8.8	22.2
20	22.5	11.9	26.7	23.5
21	21.5	28.7	25.2	19.6
22	25.2	29.5	27.3	30.1
23	23	22.2	17.6	26.6
24	23.7	19.5	25.6	24.5
25	28.4	31.2	26.1	27.4

SOLUTION

IDENTIFY

The problem objective is to compare four populations, and the data are interval. Because the researchers recorded the cholesterol reduction for each drug for each member of the similar groups of men, we identify the experimental design as randomized block. The response variable is the cholesterol reduction, the treatments are the drugs, and the blocks are the 25 similar groups of men. The hypotheses to be tested are as follows:

$$H_0: \quad \mu_1 = \mu_2 = \mu_3 = \mu_4$$

H_1: At least two means differ

COMPUTE

EXCEL

	A	B	C	D	E	F	G
36	ANOVA						
37	*Source of Variation*	*SS*	*df*	*MS*	*F*	*P-value*	*F crit*
38	Rows	3848.66	24	160.36	10.11	9.70E-15	1.6695
39	Columns	195.95	3	65.32	4.12	0.0094	2.7318
40	Error	1142.56	72	15.87			
41							
42	Total	5187.17	99				

(continued)

Note the use of scientific notation for one of the *p*-values. The number 9.70E-15 (E stands for *exponent*) is 9.70 multiplied by 10 raised to the power −15, that is, 9.70×10^{-15}. You can increase or decrease the number of decimal places and you can convert the number into a regular number, but you would need many decimal places, which is why Excel uses scientific notation when the number is very small. (Excel also uses scientific notation for very large numbers as well.)

The output includes block and treatment statistics (sums, averages, and variances, which are not shown here), and the ANOVA table. The *F*-statistic to determine whether differences exist between the four drugs (**Columns**) is 4.12. Its *p*-value is .0094. The other *F*-statistic, 10.11 (p-value $= 9.70 \times 10^{-15}$ = virtually 0) indicates that there are differences between the groups of men (**Rows**).

INSTRUCTIONS

1. Type or import the data into adjacent columns. (Open Xm14-03.)
2. Click **Data, Data Analysis . . . ,** and **Anova: Two-Factor Without Replication.**
3. Specify the **Input Range** (A1:E26). Click **Labels** if applicable. If you do, both the treatments and blocks must be labeled (as in Xm14-03). Specify the value of α (.05).

MINITAB

Two-way ANOVA: Reduction versus Group, Drug

Analysis of Variance for Reduction

Source	DF	SS	MS	F	P
Group	24	3848.7	160.4	10.11	0.000
Drug	3	196.0	65.3	4.12	0.009
Error	72	1142.6	15.9		
Total	99	5187.2			

The *F*-statistic for **Drug** is 4.12 with a *p*-value of .009. The *F*-statistic for the blocks (**Group**) is 10.11, with a *p*-value of 0.

INSTRUCTIONS

The data must be in stacked format in three columns. One column contains the responses, another contains codes for the levels of the blocks, and a third column contains codes for the levels of the treatments.

1. Click **Stat, ANOVA,** and **Twoway . . . .**
2. Specify the **Responses, Row factor,** and **Column factor.**

INTERPRET

A Type I error occurs when you conclude that differences exist when, in fact, they do not. A Type II error is committed when the test reveals no difference when at least two means differ. It would appear that both errors are equally costly. Accordingly, we judge the *p*-value against a standard of 5%. Because the *p*-value = .0094, we conclude that there is sufficient evidence to infer that at least two of the drugs differ. An examination reveals that cholesterol reduction is greatest using drugs 2 and 4. Further testing is recommended to determine which is better.

Checking the Required Conditions

The F-test of the randomized block design of the analysis of variance has the same requirements as the independent samples design. That is, the random variable must be normally distributed and the population variances must be equal. The histograms (not shown) appear to support the validity of our results; the reductions appear to be normal. The equality of variances requirement also appears to be met.

Violation of the Required Conditions

When the response is not normally distributed, we can replace the randomized block analysis of variance with the Friedman test, which is introduced in Section 19.4.

Criteria for Blocking

In Section 13.3, we listed the advantages and disadvantages of performing a matched pairs experiment. The same comments are valid when we discuss performing a blocked experiment. The purpose of blocking is to reduce the variation caused by differences between the experimental units. By grouping the experimental units into homogeneous blocks with respect to the response variable, the statistics practitioner increases the chances of detecting actual differences between the treatment means. Hence, we need to find criteria for blocking that significantly affect the response variable. For example, suppose that a statistics professor wants to determine which of four methods of teaching statistics is best. In a one-way experiment he might take four samples of 10 students, teach each sample by a different method, grade the students at the end of the course, and perform an F-test to determine whether differences exist. However, it is likely that there are very large differences between the students within each class that may hide differences between classes. To reduce this variation, the statistics professor must identify variables that are linked to a student's grade in statistics. For example, overall ability of the student, completion of mathematics courses, and exposure to other statistics courses are all related to performance in a statistics course.

The experiment could be performed in the following way. The statistics professor selects four students at random whose average grade before statistics is 95–100. He then randomly assigns the students to one of the four classes. He repeats the process with students whose average is 90–95, 85–90, . . . , and 50–55. The final grades would be used to test for differences between the classes.

Any characteristics that are related to the experimental units are potential blocking criteria. For example, if the experimental units are people, we may block according to age, gender, income, work experience, intelligence, residence (country, county, or city), weight, or height. If the experimental unit is a factory and we're measuring number of units produced hourly, blocking criteria include workforce experience, age of the plant, and quality of suppliers.

Developing an Understanding of Statistical Concepts

As we explained previously, the randomized block experiment is an extension of the matched pairs experiment discussed in Section 13.3. In the matched pairs experiment, we simply remove the effect of the variation caused by differences between the experimental units. The effect of this removal is seen in the decrease in the value of the standard error (compared to the standard error in the test statistic produced from independent samples) and the increase in the value of the t-statistic. In the randomized block experiment of the analysis of variance, we actually measure the variation between the blocks by computing SSB. The sum of squares for error is reduced by SSB, making it easier to detect differences between the treatments. Additionally, we can test to determine whether the blocks differ—a procedure we were unable to perform in the matched pairs experiment.

To illustrate, let's return to Examples 13.4 and 13.5, which were experiments to determine whether there was a difference in starting salaries offered to finance and marketing MBA majors. (In fact, we tested to determine whether finance majors draw higher salary offers than do marketing majors. However, the analysis of variance can test only for differences.) In Example 13.4 (independent samples), there was insufficient evidence to infer a difference between the two types of majors. In Example 13.5 (matched pairs experiment), there was enough evidence to infer a difference. As we pointed out in Section 13.3, matching by grade point average allowed the statistics practitioner to more easily discern a difference between the two types of majors. If we repeat Examples 13.4 and 13.5 using the analysis of variance, we come to the same conclusion. The Excel outputs are shown here. (Minitab's printouts are similar.)

Excel Analysis of Variance Output for Example 13.4

	A	B	C	D	E	F	G
1	Anova: Single Factor						
2							
3	SUMMARY						
4	*Groups*	*Count*	*Sum*	*Average*	*Variance*		
5	Finance	25	1,640,595	65,624	360,433,294		
6	Marketing	25	1,510,570	60,423	262,228,559		
7							
8							
9	ANOVA						
10	*Source of Variation*	*SS*	*df*	*MS*	*F*	*P-value*	*F crit*
11	Between Groups	338,130,013	1	338,130,013	1.09	0.3026	4.0427
12	Within Groups	14,943,884,470	48	311,330,926			
13							
14	Total	15,282,014,483	49				

Excel Analysis of Variance Output for Example 13.5

	A	B	C	D	E	F	G
34	ANOVA						
35	*Source of Variation*	*SS*	*df*	*MS*	*F*	*P-value*	*F crit*
36	Rows	21,415,991,654	24	892,332,986	40.39	4.17E-14	1.9838
37	Columns	320,617,035	1	320,617,035	14.51	0.0009	4.2597
38	Error	530,174,605	24	22,090,609			
39							
40	Total	22,266,783,295	49				

In Example 13.4, we partition the total sum of squares [SS(Total) = 15,282,014,483] into two sources of variation: SST = 338,130,013 and SSE = 14,943,884,470. In Example 13.5, the total sum of squares is SS(Total) = 22,266,783,295, SST (sum of squares for majors) = 320,617,035, SSB (sum of squares for GPA) = 21,415,991,654, and SSE = 530,174,605. As you can see, the sums of squares for treatments are approximately equal (338,130,013 and 320,617,035). However, the two calculations differ in the sums of squares for error. SSE in Example 13.5 is much smaller than SSE in Example 13.4 because the randomized block experiment allows us to measure and remove the effect of the variation between MBA students with the same majors. The sum of squares for blocks (sum of squares for GPA groups) is 21,415,991,654, a statistic that measures how much variation exists between the salary offers within majors. As a result of removing this variation, SSE is small. Thus, we conclude in Example 13.5 that the salary offers differ between majors, whereas there was not enough evidence in Example 13.4 to draw the same conclusion.

Notice that in both examples the t-statistic squared equals the F-statistic. That is in Example 13.4, $t = 1.04$, which when squared equals 1.09, which is the F-statistic (rounded). In Example 13.5, $t = 3.81$, which when squared equals 14.51, the F-statistic for the test of the treatment means. Moreover the p-values are also the same.

We now complete this section by listing the factors that we need to recognize to use this experiment of the analysis of variance.

Factors That Identify the Randomized Block of the Analysis of Variance

1. **Problem objective:** Compare two or more populations
2. **Data type:** Interval
3. **Experimental design:** Blocked samples

EXERCISES

Developing an Understanding of Statistical Concepts

14.31 The following statistics were generated from a randomized block experiment with $k = 3$ and $b = 7$:

SST = 100 SSB = 50 SSE = 25

a. Test to determine whether the treatment means differ. (Use $\alpha = .05$.)
b. Test to determine whether the block means differ. (Use $\alpha = .05$.)

14.32 A randomized block experiment produced the following statistics:

$k = 5$ b = 12 SST = 1,500 SSB = 1,000 SS(Total) = 3,500

a. Test to determine whether the treatment means differ. (Use $\alpha = .01$.)
b. Test to determine whether the block means differ. (Use $\alpha = .01$.)

14.33 Suppose the following statistics were calculated from data gathered from a randomized block experiment with $k = 4$ and $b = 10$:

SS(Total) = 1,210 SST = 275 SSB = 625

a. Can we conclude from these statistics that the treatment means differ? (Use $\alpha = .01$.)
b. Can we conclude from these statistics that the block means differ? (Use $\alpha = .01$.)

14.34 A randomized block experiment produced the following statistics:

$k = 3$ $b = 8$ SST = 1,500 SS(Total) = 3,500

a. Test at the 5% significance level to determine whether the treatment means differ given that SSB = 500.
b. Repeat Part a with SSB = 1,000.
c. Repeat Part a with SSB = 1,500.
d. Describe what happens to the test statistic as SSB increases.

14.35 a. Assuming that the data shown here were generated from a randomized block experiment, calculate SS(Total), SST, SSB, and SSE.
b. Assuming that the data below were generated from a one-way (independent samples) experiment, calculate SS(Total), SST, and SSE.
c. Why does SS(Total) remain the same for both experimental designs?
d. Why does SST remain the same for both experimental designs?
e. Why does SSB + SSE in Part a equal SSE in Part b?

Treatment		
1	2	3
7	12	8
10	8	9
12	16	13
9	13	6
12	10	11

14.36 a. Calculate SS(Total), SST, SSB, and SSE, assuming that the accompanying data were generated from a randomized block experiment.
b. Calculate SS(Total), SST, and SSE, assuming that the data below were generated from a one-way (independent samples) experiment.
c. Explain why SS(Total) remains the same for both experimental designs.
d. Explain why SST remains the same for both experimental designs.

e. Explain why SSB + SSE in Part a equals SSE in Part b.

Treatment			
1	2	3	4
6	5	4	4
8	5	5	6
7	6	5	6

Applications

14.37 Xr14-37 As an experiment to understand measurement error, a statistics professor asks four students to measure the height of the professor, a male student, and a female student. The differences (in centimeters) between the correct dimension and the ones produced by the students are listed here. Can we infer that there are differences in the errors between the subjects being measured? (use $\alpha = .05$.)

	Errors in Measuring Heights of		
Student	Professor	Male Student	Female Student
1	1.4	1.5	1.3
2	3.1	2.6	2.4
3	2.8	2.1	1.5
4	3.4	3.6	2.9

14.38 Xr14-38 How well do diets work. In a preliminary study, 20 people who were more than 50 pounds overweight were recruited to compare four diets. The people were matched by age. The oldest four became block 1, the next oldest four became block 2, and so on. The number of pounds that each person lost are listed in the following table. Can we infer at the 1% significance level that there are differences between the four diets?

	Diet			
Block	1	2	3	4
1	5	2	6	8
2	4	7	8	10
3	6	12	9	2
4	7	11	16	7
5	9	8	15	14

The following exercises require the use of a computer and software. The answers may be calculated manually. See Appendix A for the sample statistics. ***Use a 5% significance level unless specified otherwise.***

14.39 Xr14-39 In recent years, lack of confidence in the Postal Service has led many companies to send all of their correspondence by private courier. A large company is in the process of selecting one of three possible couriers to act as its sole delivery method. To help in making the decision, an experiment was performed whereby letters were sent using each of the three couriers at 12 different times of the day to a delivery point across town. The number of minutes required for delivery was recorded.

a. Can we conclude that there are differences in delivery times between the three couriers?
b. Did the statistics practitioner choose the correct design? Explain.

14.40 Xr14-40 Refer to Exercise 14.14. Despite failing to show that differences in the three types of fertilizer exist, the scientist continued to believe that there were differences, and that the differences were masked by the variation between the plots of land. Accordingly, he conducted another experiment. In the second experiment he found 20 3-acre plots of land scattered across the county. He divided each into three plots and applied the three types of fertilizer on each of the 1-acre plots. The crop yields were recorded

a. Can the scientist infer that there are differences between the three types of fertilizer?
b. What do these test results reveal about the variation between the plots?

14.41 Xr14-41 A recruiter for a computer company would like to determine whether there are differences in sales ability between business, arts, and science graduates. She takes a random sample of 20 business graduates who have been working for the company for the past 2 years. Each is then matched with an arts graduate and a science graduate with similar educational and working experience. The commission earned by each (in $1,000s) in the last year was recorded.

a. Is there sufficient evidence to allow the recruiter to conclude that there are differences in sales ability between the holders of the three types of degrees?
b. Conduct a test to determine whether an independent samples design would have been a better choice.
c. What are the required conditions for the test in Part a?
d. Are the required conditions satisfied?

14.42 Xr14-42 Exercise 14.10 described an experiment that involved comparing the completion times associated with four different income tax forms. Suppose the experiment is redone in the following way. Thirty people are asked to fill out all four forms. The completion times (in minutes) are recorded.

a. Is there sufficient evidence at the 1% significance level to infer that differences in the completion times exist between the four forms?
b. Comment on the suitability of this experimental design in this problem.

14.43 Xr14-43 The advertising revenues commanded by a radio station depend on the number of listeners it has. The manager of a station that plays mostly hard rock music wants to learn more about its listeners—mostly teenagers and young adults. In particular, he wants to know whether the amount of time they spend listening to radio music varies by the day of the week. If the manager discovers that the mean time per day is about the same, he will schedule the most popular music evenly throughout the week. Otherwise, the top hits will be played mostly on the days that attract the greatest audience. An opinion survey company is hired, and it randomly selects 200 teenagers and asks them to record the amount of time spent listening to music on the radio for each day of the previous week. What can the manager conclude from these data?

14.44 Xr14-44 Do medical specialists differ in the amount of time they devote to patient care? To answer this question a statistics practitioner organized a study. The numbers of hours of patient care per week were recorded for five specialists. The experimental design was randomized blocks. The physicians were blocked by age. (Adapted from the *Statistical Abstract of the United States*, 2000, Table 190)

a. Can we infer that there are differences in the amount of patient care between medical specialties?
b. Can we infer that blocking by age was appropriate?

14.45 Xr14-45 Refer to Exercise 14.9. Another study was conducted in the following way. Students from each of the high schools who were admitted to the business program were matched according to their high school averages. The average grades in the first year were recorded. Can the university admissions officer conclude that there are differences in grading standards between the four high schools?

14.5 Two-Factor Analysis of Variance

In Section 14.1, we addressed problems where the data were generated from single-factor experiments. In Example 14.1, the treatments were the four age categories. Thus, there were four levels of a single factor. In this section, we address the problem where the experiment features two factors. The general term for such data-gathering procedures is **factorial experiment.** In factorial experiments, we can examine the effect on the response variable of two or more factors, although in this book we address the problem of only two factors. We can use the analysis of variance to determine whether the levels of each factor are different from one another.

We will present the technique for fixed effects only. That means we will address problems where all the levels of the factors are included in the experiment. As was the case with the randomized block design, calculation of the test statistic in this type of experiment is quite time-consuming. As a result, we will use Excel and Minitab to produce our statistics.

EXAMPLE 14.4*

DATA Xm14-04

Comparing the Lifetime Number of Jobs by Educational Level

One measure of the health of a nation's economy is how quickly it creates jobs. One aspect of this issue is the number of jobs individuals hold. As part of a study on job tenure, a survey was conducted wherein Americans aged between 37 and 45 were asked how many jobs they have held in their lifetimes. Also recorded were gender and educational attainment. The categories are

Less than high school (E1)
High school (E2)

*Adapted from the *Statistical Abstract of the United States*, 2006, Table 598.

Some college/university but no degree (E3)
At least one university degree (E4)

The data are shown for each of the eight categories of gender and education. Can we infer that differences exist between genders and educational levels?

Male E1	Male E2	Male E3	Male E4	Female E1	Female E2	Female E3	Female E4
10	12	15	8	7	7	5	7
9	11	8	9	13	12	13	9
12	9	7	5	14	6	12	3
16	14	7	11	6	15	3	7
14	12	7	13	11	10	13	9
17	16	9	8	14	13	11	6
13	10	14	7	13	9	15	10
9	10	15	11	11	15	5	15
11	5	11	10	14	12	9	4
15	11	13	8	12	13	8	11

SOLUTION

IDENTIFY

We begin by treating this example as a one-way analysis of variance. Notice that there are eight treatments. However, the treatments are defined by two different factors. One factor is gender, which has two levels. The second factor is educational attainment, which has four levels.

We can proceed to solve this problem in the same way we did in Section 14.1; that is, we test the following hypotheses:

H_0: $\mu_1 = \mu_2 = \mu_3 = \mu_4 = \mu_5 = \mu_6 = \mu_7 = \mu_8$

H_1: At least two means differ

COMPUTE

EXCEL

	A	B	C	D	E	F	G
1	Anova: Single Factor						
2							
3	SUMMARY						
4	*Groups*	*Count*	*Sum*	*Average*	*Variance*		
5	Male E1	10	126	12.60	8.27		
6	Male E2	10	110	11.00	8.67		
7	Male E3	10	106	10.60	11.60		
8	Male E4	10	90	9.00	5.33		
9	Female E1	10	115	11.50	8.28		
10	Female E2	10	112	11.20	9.73		
11	Female E3	10	94	9.40	16.49		
12	Female E4	10	81	8.10	12.32		
13							
14							
15	ANOVA						
16	*Source of Variation*	*SS*	*df*	*MS*	*F*	*P-value*	*F crit*
17	Between Groups	153.35	7	21.91	2.17	0.0467	2.1397
18	Within Groups	726.20	72	10.09			
19							
20	Total	879.55	79				

MINITAB

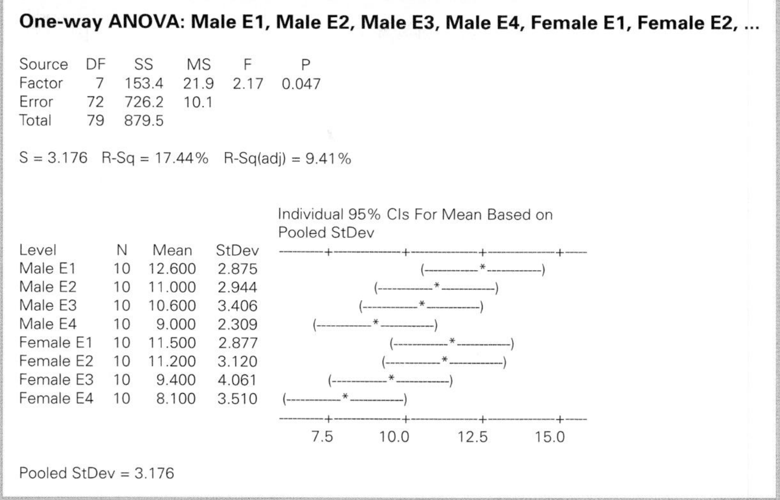

One-way ANOVA: Male E1, Male E2, Male E3, Male E4, Female E1, Female E2, ...

Source	DF	SS	MS	F	P
Factor	7	153.4	21.9	2.17	0.047
Error	72	726.2	10.1		
Total	79	879.5			

S = 3.176 R-Sq = 17.44% R-Sq(adj) = 9.41%

Individual 95% CIs For Mean Based on Pooled StDev

Level	N	Mean	StDev
Male E1	10	12.600	2.875
Male E2	10	11.000	2.944
Male E3	10	10.600	3.406
Male E4	10	9.000	2.309
Female E1	10	11.500	2.877
Female E2	10	11.200	3.120
Female E3	10	9.400	4.061
Female E4	10	8.100	3.510

Pooled StDev = 3.176

INTERPRET

The value of the test statistic is $F = 2.17$ with a p-value of .0467. We conclude that there are differences in the number of jobs between the eight treatments.

This statistical result raises more questions. Namely, can we conclude that the differences in the mean number of jobs are caused by differences between males and females? Or are they caused by differences between educational levels? Or, perhaps, are there combinations called **interactions** of gender and education that result in especially high or low numbers? To show how we test for each type of difference, we need to develop some terminology

A **complete factorial experiment** is an experiment in which the data for all possible combinations of the levels of the factors are gathered. That means that in Example 14.4 we measured the number of jobs for all eight combinations. This experiment is called a complete 2×4 factorial experiment.

In general, we will refer to one of the factors as factor A (arbitrarily chosen). The number of levels of this factor will be denoted by a. The other factor is called factor B, and its number of levels is denoted by b. This terminology becomes clearer when we present the data from Example 14.4 in another format. Table 14.6 depicts the layout for a two-way classification, which is another name for the complete factorial experiment. The number of observations for each combination is called a **replicate.** The number of replicates is denoted by r. In this book, we address only problems in which the number of replicates is the same for each treatment. Such a design is called **balanced.**

Thus, we use a complete factorial experiment where the number of treatments is ab with r replicates per treatment. In Example 14.4, $a = 2, b = 4$, and $r = 10$. As a result, we have 10 observations for each of the eight treatments.

If you examine the ANOVA table, you can see that the total variation is SS(Total) = 879.55, the sum of squares for treatments is SST = 153.35, and the sum of squares for error is SSE = 726.20. The variation caused by the treatments is measured by SST. In order to determine whether the differences are due to factor A, factor B, or some

TABLE **14.6** Two-Way Classification for Example 14.4

	Male	Female
Less than high school	10	7
	9	13
	12	14
	16	6
	14	11
	17	14
	13	13
	9	11
	11	14
	15	12
High School	12	7
	11	12
	9	6
	14	15
	12	10
	16	13
	10	9
	10	15
	5	12
	11	13
Less than bachelor's degree	15	5
	8	13
	7	12
	7	3
	7	13
	9	11
	14	15
	15	5
	11	9
	13	8
At least one bachelor's degree	8	7
	9	9
	5	3
	11	7
	13	9
	8	6
	7	10
	11	15
	10	4
	8	11

interaction between the two factors, we need to partition SST into three sources. These are SS(A), SS(B), and SS(AB).

For those whose mathematical confidence is high, we have provided an explanation of the notation as well as the definitions of the sums of squares. Learning how the sums of squares are calculated is useful but hardly essential to your ability to conduct the tests. Uninterested readers should jump to the box on page 555 where we describe the individual F-tests.

How the Sums of Squares for Factors A and B and Interaction Are Computed

To help you understand the formulas, we will use the following notation:

x_{ijk} = kth observation in the ijth treatment

$\overline{x}[\mathrm{AB}]_{ij}$ = Mean of the response variable in the ijth treatment (mean of the treatment when the factor A level is i and the factor B level is j)

$\overline{x}[\mathrm{A}]_i$ = Mean of the observations when the factor A level is i

$\overline{x}[\mathrm{B}]_j$ = Mean of the observations when the factor B level is j

$\overline{\overline{x}}$ = Mean of all the observations

a = Number of factor A levels

b = Number of factor B levels

r = Number of replicates

In this notation, $\overline{x}[\mathrm{AB}]_{11}$ is the mean of the responses for factor A level 1 and factor B level 1. The mean of the responses for factor A level 1 is $\overline{x}[\mathrm{A}]_1$. The mean of the responses for factor B level 1 is $\overline{x}[\mathrm{B}]_1$.

Table 14.7 describes the notation for the two-factor analysis of variance.

TABLE 14.7 Notation for Two-Factor Analysis of Variance

	Factor A							
Factor B	1		2		. . .	a		
1	x_{111} x_{112} . . . x_{11r}	$\overline{x}[\mathrm{AB}]_{11}$	x_{211} x_{212} . . . x_{21r}	$\overline{x}[\mathrm{AB}]_{21}$		x_{a11} x_{a12} . . . x_{a1r}	$\overline{x}[\mathrm{AB}]_{a1}$	$\overline{x}[\mathrm{B}]_1$
2	x_{121} x_{122} . . . x_{12r}	$\overline{x}[\mathrm{AB}]_{12}$	x_{221} x_{222} . . . x_{22r}	$\overline{x}[\mathrm{AB}]_{22}$		x_{a21} x_{a22} . . . x_{a2r}	$\overline{x}[\mathrm{AB}]_{a2}$	$\overline{x}[\mathrm{B}]_2$
. . .								
b	x_{1b1} x_{1b2} . . . x_{1br}	$\overline{x}[\mathrm{AB}]_{1b}$	x_{2b1} x_{2b2} . . . x_{2br}	$\overline{x}[\mathrm{AB}]_{2b}$		x_{ab1} x_{ab2} . . . x_{abr}	$\overline{x}[\mathrm{AB}]_{ab}$	$\overline{x}[\mathrm{B}]_b$
	$\overline{x}[\mathrm{A}]_1$		$\overline{x}[\mathrm{A}]_2$			$\overline{x}[\mathrm{A}]_a$		$\overline{\overline{x}}$

The sums of squares are defined as follows.

Sums of Squares in the Two-Factor Analysis of Variance

$$SS(\text{Total}) = \sum_{i=1}^{a}\sum_{j=1}^{b}\sum_{k=1}^{r}(x_{ijk} - \bar{\bar{x}})^2$$

$$SS(A) = rb\sum_{i=1}^{a}(\bar{x}[A]_i - \bar{\bar{x}})^2$$

$$SS(B) = ra\sum_{j=1}^{b}(\bar{x}[B]_j - \bar{\bar{x}})^2$$

$$SS(AB) = r\sum_{i=1}^{a}\sum_{j=1}^{b}(\bar{x}[AB]_{ij} - \bar{x}[A]_i - \bar{x}[B]_j + \bar{\bar{x}})^2$$

$$SSE = \sum_{i=1}^{a}\sum_{j=1}^{b}\sum_{k=1}^{r}(x_{ijk} - \bar{x}[AB]_{ij})^2$$

To compute SS(A), we calculate the sum of the squared differences between the factor A level means, which are denoted $\bar{x}[A]_i$, and the grand mean, $\bar{\bar{x}}$. The sum of squares for factor B, SS(B), is defined similarly. The interaction sum of squares, SS(AB), is calculated by taking each treatment mean (a treatment consists of a combination of a level of factor A and a level of factor B), subtracting the factor A level mean, subtracting the factor B level mean, adding the grand mean, squaring this quantity, and adding. The sum of squares for error, SSE, is calculated by subtracting the treatment means from the observations, squaring, and adding.

To test for each possibility, we conduct several F-tests similar to the one performed in Section 14.1. Figure 14.4 illustrates the partitioning of the total sum of squares that leads to the F-tests. We've included in this figure the partitioning used in the one-way study. When the one-way analysis of variance allows us to infer that differences between the treatment means exist, we continue our analysis by partitioning the treatment sum of squares into three sources of variation. The first is sum of squares for factor A, which we label SS(A), which measures the variation between the levels of factor A. Its degrees of freedom are $a - 1$. The second is the sum of squares for factor B, whose degrees of freedom are $b - 1$. SS(B) is the variation between the levels of factor B. The interaction sum of squares is labeled SS(AB), which is a measure of the amount of variation between

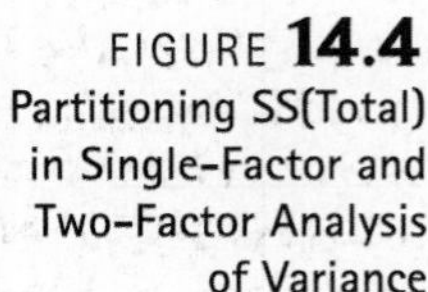

FIGURE **14.4** Partitioning SS(Total) in Single-Factor and Two-Factor Analysis of Variance

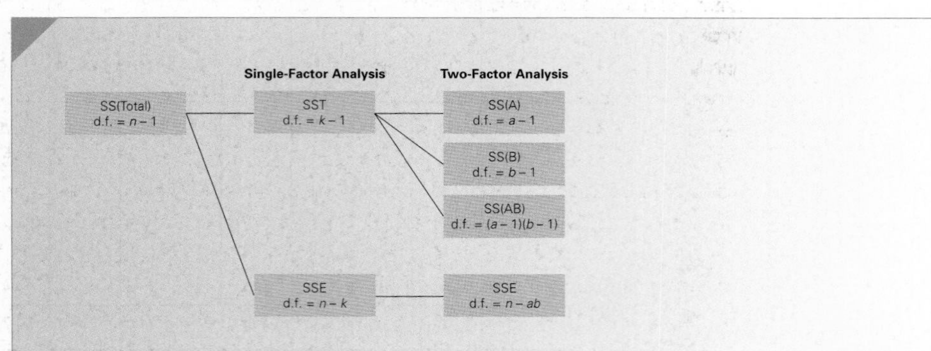

the combinations of factors A and B; its degrees of freedom are $(a - 1) \times (b - 1)$. The sum of squares for error is SSE, and its degrees of freedom are $n - ab$. (Recall that n is the total sample size, which in this experiment is $n = abr$.) Notice that SSE and its number of degrees of freedom are identical in both partitions. As in the previous experiment, SSE is the variation within the treatments.

F-Tests Conducted in Two-Factor Analysis of Variance

Test for Differences between the Levels of Factor A

H_0: The means of the a levels of factor A are equal

H_1: At least two means differ

Test statistic: $F = \frac{\text{MS(A)}}{\text{MSE}}$

Test for Differences between the Levels of Factor B

H_0: The means of the b levels of factor B are equal

H_1: At least two means differ

Test statistic: $F = \frac{\text{MS(B)}}{\text{MSE}}$

Test for Interaction between Factors A and B

H_0: Factors A and B do not interact to affect the mean responses

H_1: Factors A and B do interact to affect the mean responses

Test statistic: $F = \frac{\text{MS(AB)}}{\text{MSE}}$

Required Conditions

1. The distribution of the response is normally distributed.
2. The variance for each treatment is identical.
3. The samples are independent.

As in the two previous experimental designs of the analysis of variance, we summarize the results in an ANOVA table. Table 14.8 depicts the general form of the table for the complete factorial experiment.

TABLE **14.8** ANOVA Table for the Two-Factor Experiment

Source of Variation	Degrees of Freedom	Sums of Squares	Mean Squares	*F*-Statistic
Factor A	$a - 1$	SS(A)	MS(A) = SS(A)/$(a - 1)$	F = MS(A)/MSE
Factor B	$b - 1$	SS(B)	MS(B) = SS(B)/$(b - 1)$	F = MS(B)/MSE
Interaction	$(a - 1)(b - 1)$	SS(AB)	MS(AB) = SS(AB)/$[(a - 1)(b - 1)]$	F = MS(AB)/MSE
Error	$n - ab$	SSE	MSE = SSE/$(n - ab)$	
Total	$n - 1$	SS(Total)		

We'll illustrate the techniques using the data in Example 14.4. All calculations will be performed by Excel and Minitab.

EXCEL

	A	B	C	D	E	F	G
1	Anova: Two-Factor with Replication						
2							
3	SUMMARY	Male	Female	Total			
4	*Less than HS*						
5	Count	10	10	20			
6	Sum	126	115	241			
7	Average	12.6	11.5	12.1			
8	Variance	8.27	8.28	8.16			
9							
10	*High School*						
11	Count	10	10	20			
12	Sum	110	112	222			
13	Average	11.0	11.2	11.1			
14	Variance	8.67	9.73	8.73			
15							
16	*Less than Bachelor's*						
17	Count	10	10	20			
18	Sum	106	94	200			
19	Average	10.6	9.4	10.0			
20	Variance	11.6	16.49	13.68			
21							
22	*Bachelor's or more*						
23	Count	10	10	20			
24	Sum	90	81	171			
25	Average	9.0	8.1	8.6			
26	Variance	5.33	12.32	8.58			
27							
28	*Total*						
29	Count	40	40				
30	Sum	432	402				
31	Average	10.8	10.1				
32	Variance	9.50	12.77				
33							
34	ANOVA						
35	*Source of Variation*	*SS*	*df*	*MS*	*F*	*P-value*	*F crit*
36	Sample	135.85	3	45.28	4.49	0.0060	2.7318
37	Columns	11.25	1	11.25	1.12	0.2944	3.9739
38	Interaction	6.25	3	2.08	0.21	0.8915	2.7318
39	Within	726.20	72	10.09			
40							
41	Total	879.55	79				

In the ANOVA table, **Sample** refers to factor B (educational level) and **Columns** refers to factor A (gender). Thus, MS(B) = 45.28, MS(A) = 11.25, MS(AB) = 2.08 and MSE = 10.09. The F-statistics are 4.49 (educational level), 1.12 (gender), and .21 (interaction).

INSTRUCTIONS

1. Type or import the data using the same format as Xm14-04a. (You must label the rows and columns as we did.)
2. Click **Data, Data Analysis,** and **Anova: Two-Factor with Replication.**
3. Specify the **Input Range** (A1:C41). Type the number of replications in the **Rows per sample** box (10).
4. Specify a value for α (.05).

MINITAB

Two-way ANOVA: Jobs versus Gender, Education

```
Source       DF      SS       MS      F      P
Gender        1   11.25  11.2500   1.12  0.294
Education     3  135.85  45.2833   4.49  0.006
Interaction   3    6.25   2.0833   0.21  0.892
Error        72  726.20  10.0861
Total        79  879.55

S = 3.176  R-Sq = 17.44%  R-Sq(adj) = 9.41%

                  Individual 95% CIs For Mean Based on Pooled StDev
Gender   Mean    -+--------------+--------------+--------------+------------
1        10.80              (---------------------*---------------------)
2        10.05    (---------------------*--------------------)
                 -+--------------+--------------+--------------+------------
                 9.10         9.80          10.50          11.20

                     Individual 95% CIs For Mean Based on Pooled StDev
Education   Mean    --------+--------------+--------------+--------------+------
1          12.05                                (-------------*-------------)
2          11.10                            (------------*------------)
3          10.00                   (-------------*------------)
4           8.55    (-----------*------------)
                    --------+--------------+--------------+--------------+------
                           8.0            9.6           11.2           12.8
```

INSTRUCTIONS

1. Type or import the data in stacked format in three columns. One column contains the responses, another contains codes for the levels of factor A, and a third column contains codes for the levels of factor B. (Open Xm14-04b.)
2. Click **Stat, ANOVA,** and **Twoway . . . .**
3. Specify the **Responses,** (Jobs) **Row factor** (Gender), and **Column factor** (Education).
4. To produce the graphics check **Display means.**

Test for Differences in Number of Jobs between Men and Women

H_0: The means of the two levels of factor A are equal

H_1: At least two means differ

Test statistic: $F = \dfrac{\text{MS(A)}}{\text{MSE}}$

Value of the test statistic: From the computer output, we have MS(A) = 11.25, MSE = 10.09, and F = 11.25/10.09 = 1.12 (p-value = .2944)

There is not evidence at the 5% significance level to infer that differences in the number of jobs exist between men and women.

Test for Differences in Number of Jobs between Education Levels

H_0: The means of the four levels of factor B are equal

H_1: At least two means differ

Test statistic: $F = \dfrac{\text{MS(B)}}{\text{MSE}}$

Value of the test statistic: From the computer output, we find MS(B) = 45.28 and MSE = 10.09. Thus, F = 45.28/10.09 = 4.49 (p-value = .0060)

There is sufficient evidence at the 5% significance level to infer that differences in the number of jobs exist between educational levels.

Test for Interaction between Factors A and B

H_0: Factors A and B do not interact to affect the mean number of jobs

H_1: Factors A and B do interact to affect the mean number of jobs

Test statistic: $F = \frac{\text{MS(AB)}}{\text{MSE}}$

Value of the test statistic: From the printouts, MS(AB) = 2.08, MSE = 10.09, and $F = 2.08/10.09 = .21$ (p-value = .8915)

There is not enough evidence to conclude that there is an interaction between gender and education.

INTERPRET

Figure 14.5 is a graph of the mean responses for each of the eight treatments. As you can see, there are small (not significant) differences between males and females. There are significant differences between men and women with different educational backgrounds. Finally, there is no interaction.

FIGURE **14.5**
Mean Responses for Example 14.4

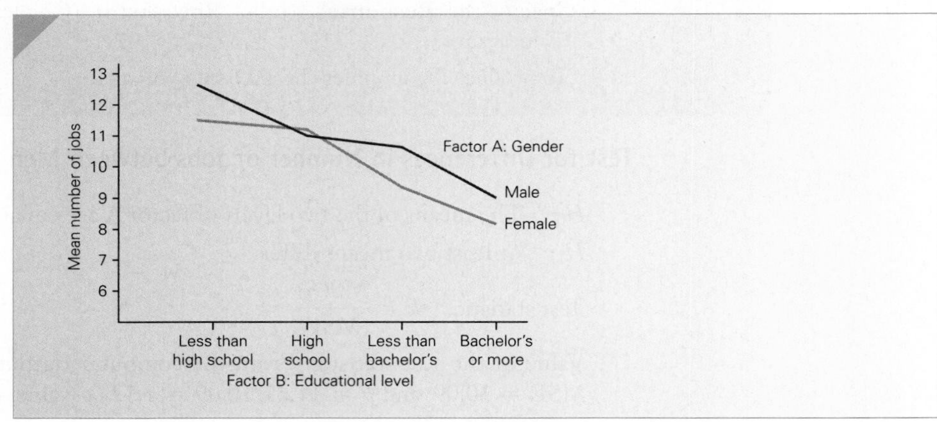

What Is Interaction?

To more fully understand interaction, we have changed the sample associated with men who have not finished high school (Treatment 1). We subtracted 6 from the original numbers so that the sample in treatment 1 is

4 3 6 10 8 11 7 3 5 9

The new data are stored in Xm14-04c (Excel format) and Xm14-04d (Minitab format). The mean is 6.6. Here are the Excel and Minitab ANOVA tables.

EXCEL

	A	B	C	D	E	F	G
35	ANOVA						
36	*Source of Variation*	*SS*	*df*	*MS*	*F*	*P-value*	*F crit*
37	Sample	75.85	3	25.28	2.51	0.0657	2.7318
38	Columns	11.25	1	11.25	1.12	0.2944	3.9739
39	Interaction	120.25	3	40.08	3.97	0.0112	2.7318
40	Within	726.20	72	10.09			
41							
42	Total	933.55	79				

MINITAB

Two-way ANOVA: Jobs versus Gender, Education

Source	DF	SS	MS	F	P
Gender	1	11.25	11.2500	1.12	0.294
Education	3	75.85	25.2833	2.51	0.066
Interaction	3	120.25	40.0833	3.97	0.011
Error	72	726.20	10.0861		
Total	79	933.55			

INTERPRET

In this example there is not enough evidence (at the 5% significance level) to infer that there are differences between men and women and between the educational levels. However, there is sufficient evidence to conclude that there is interaction between gender and education.

The means for each of the eight treatments are

	Male	Female
Less than high school	6.6	11.5
High school	11.0	11.2
Less than bachelor's	10.6	9.4
Bachelor's or more	9.0	8.1

The graph of the treatment means in Figure 14.6 depicts the results.

FIGURE **14.6** Mean Responses for Example 14.4a

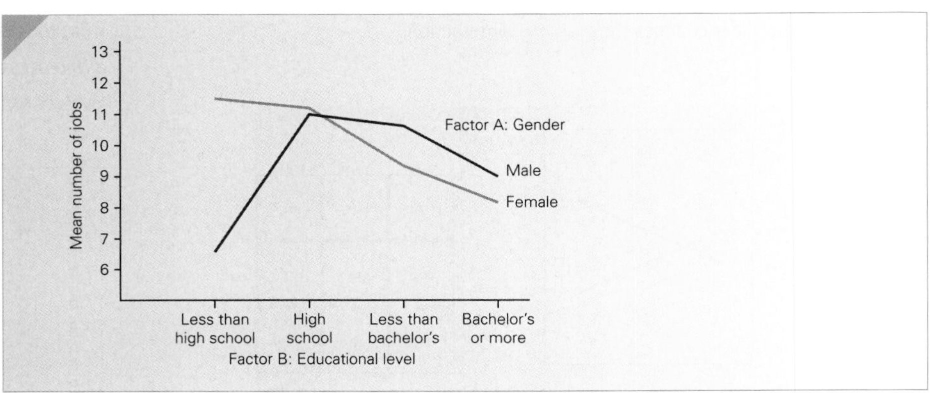

Compare Figures 14.5 and 14.6. In Figure 14.5, the lines joining the response means for males and females are quite similar. In particular, we see that the lines are almost parallel. However, in Figure 14.6 the lines are no longer almost parallel. It is apparent that the mean of treatment 1 is smaller; the pattern is different. For whatever reason, in this case men with less than high school have a smaller number of jobs.

Conducting the Analysis of Variance for the Complete Factorial Experiment

In addressing the problem outlined in Example 14.4, we began by conducting a one-way analysis of variance to determine whether differences existed between the eight treatment means. This was done primarily for pedagogical reasons to enable you to see that when the treatment means differ, we need to analyze the reasons for the differences. However, in practice, we generally do not conduct this test in the complete factorial experiment (although it should be noted that some statistics practitioners prefer this "two-stage" strategy). We recommend that you proceed directly to the two-factor analysis of variance.

In the two versions of Example 14.4, we conducted the tests of each factor and then the test for interaction. However, if there is evidence of interaction, the tests of the factors are irrelevant. There may or may not be differences between the levels of factor A and of the levels of factor B. Accordingly, we change the order of conducting the *F*-Tests.

Order of Testing in the Two-Factor Analysis of Variance

Test for interaction first. If there is enough evidence to infer that there is interaction, do not conduct the other tests.

If there is not enough evidence to conclude that there is interaction, proceed to conduct the *F*-tests for factors A and B.

SEEING STATISTICS

Applet 17: Plots of Two-Way ANOVA Effects

This applet provides a graph similar to those in Figures 14.5 and 14.6. There are three sliders, one for rows, one for columns, and one for interaction. Moving the top slider changes the difference between the row means. The second slider changes the difference between the column means. The third slider allows us to see the effects of interaction.

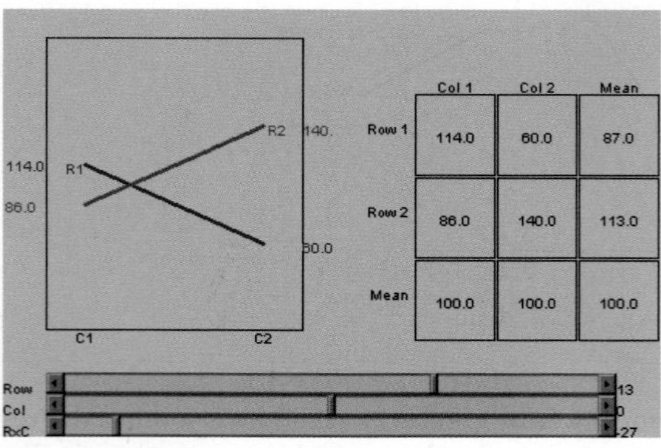

Applet Exercises

Label the columns factor A and the rows factor B. Move the sliders to arrange for each of the following differences. Describe what the resulting figure tells you about differences between levels of factor A, levels of factor B, and interaction.

	Row	Col	R × C
17.1	−30	0	0
17.2	0	25	0
17.3	0	0	−20
17.4	25	−30	0
17.5	30	0	30
17.6	30	0	−30
17.7	0	20	20
17.8	0	20	−20
17.9	30	30	30
17.10	30	30	−30

Developing an Understanding of Statistical Concepts

You may have noticed that there are similarities between the two-factor experiment and the randomized block experiment. In fact, when the number of replicates is one, the calculations are identical. (Minitab uses the same command.) This raises the question, What is the difference between a factor in a multifactor study and a block in a randomized block experiment? In general, the difference between the two experimental designs is that in the randomized block experiment, blocking is performed specifically to reduce variation, whereas in the two-factor model the effect of the factors on the response variable is of interest to the statistics practitioner. The criteria that define the blocks are always characteristics of the experimental units. Consequently, factors that are characteristics of the experimental units will be treated not as factors in a multifactor study, but as blocks in a randomized block experiment.

Let's review how we recognize the need to use the procedure described in this section.

Factors That Identify the Independent Samples Two-Factor Analysis of Variance

1. **Problem objective:** Compare two or more populations (populations are defined as the combinations of the levels of two factors)
2. **Data type:** Interval
3. **Experimental design:** Independent samples

EXERCISES

14.46 A two-factor analysis of variance experiment was performed with $a = 3$, $b = 4$, and $r = 20$. The following sums of squares were computed:

SS(Total) = 42,450 SS(A) = 1,560
SS(B) = 2,880 SS(AB) = 7,605

a. Determine the one-way ANOVA table.
b. Test at the 1% significance level to determine whether differences exist between the 12 treatments.
c. Conduct whatever test you deem necessary at the 1% significance level to determine whether there are differences between the levels of factor A, the levels of factor B, or interaction between factors A and B.

14.47 A statistics practitioner conducted a two-factor analysis of variance experiment with $a = 4$, $b = 3$, and $r = 8$. The sums of squares are listed here:

SS(Total) = 9420 SS(A) = 203 SS(B) = 859
SS(AB) = 513

a. Test at the 5% significance level to determine whether factors A and B interact.
b. Test at the 5% significance level to determine whether differences exist between the levels of factor A.
c. Test at the 5% significance level to determine whether differences exist between the levels of factor B.

14.48 Xr14-48 The following data were generated from a 2 × 2 factorial experiment with 3 replicates:

	Factor B	
Factor A	**1**	**2**
1	6	12
	9	10
	7	11
2	9	15
	10	14
	5	10

a. Test at the 5% significance level to determine whether factors A and B interact.
b. Test at the 5% significance level to determine whether differences exist between the levels of factor A.
c. Test at the 5% significance level to determine whether differences exist between the levels of factor B.

14.49 Xr14-49 The data shown here were taken from a 2×3 factorial experiment with 4 replicates:

Factor A	Factor B 1	2
1	23	20
	18	17
	17	16
	20	19
2	27	29
	23	23
	21	27
	28	25
3	23	27
	21	19
	24	20
	16	22

a. Test at the 5% significance level to determine whether factors A and B interact.
b. Test at the 5% significance level to determine whether differences exist between the levels of factor A.
c. Test at the 5% significance level to determine whether differences exist between the levels of factor B.

14.50 Xr14-50 Refer to Example 14.4. We've revised the data by adding 2 to each of the numbers of the men. What do these data tell you?

14.51 Xr14-51 Refer to Example 14.4. We've altered the data by subtracting 4 from the numbers of treatment 8. What do these data tell you?

Applications

The following exercises require the use of a computer and software

14.52 Xr14-52 Refer to Exercise 14.10. Suppose that the experiment is redone in the following way. Thirty taxpayers fill out each of the four forms. However, 10 taxpayers in each group are in the lowest income bracket, 10 are in the next income bracket, and the remaining 10 are in the highest bracket. The amount of time needed to complete the returns is recorded.

Column 1: Group number
Column 2: times to complete form 1 (first 10 rows = low income, next 10 rows = next income bracket, and last 10 rows = highest bracket)
Column 3: times to complete form 2 (same format as column 2)
Column 4: times to complete form 3 (same format as column 2)
Column 5: times to complete form 4 (same format as column 2)

a. How many treatments are there in this experiment?
b. How many factors are there? What are they?
c. What are the levels of each factor?
d. Is there evidence at the 5% significance level of interaction between the two factors?
e. Can we conclude at the 5% significance level that differences exist between the four forms?
f. Can we conclude at the 5% significance level that taxpayers in different brackets require different amounts of time to complete their tax forms?

14.53 Xr14-53 Detergent manufacturers frequently make claims about the effectiveness of their products. A consumer-protection service decided to test the five best selling brands of detergent, where each manufacturer claims that its product produces the "whitest whites" in all water temperatures. The experiment was conducted in the following way. One hundred fifty white sheets were equally soiled. Thirty sheets were washed in each brand—10 with cold water, 10 with warm water, and 10 with hot water. After washing, the "whiteness" scores for each sheet were measured with laser equipment.

Column 1: Water temperature code
Column 2: Scores for detergent 1 (first 10 rows = cold water, middle 10 rows = warm, and last 10 rows = hot)
Column 2: Scores for detergent 2 (same format as column 2)
Column 3: Scores for detergent 3 (same format as column 2)
Column 4: Scores for detergent 4 (same format as column 2)
Column 5: Scores for detergent 5 (same format as column 2)

a. What are the factors in this experiment?
b. What is the response variable?
c. Identify the levels of each factor.
d. Perform a statistical analysis using a 5% significance level to determine whether there is sufficient statistical evidence to infer that there are differences in whiteness scores between the five detergents, differences in whiteness scores between the three water temperatures, or interaction between detergents and temperatures.

14.54 Xr14-54 Headaches are one of the most common, but least understood, ailments. Most people get headaches several times per month; over-the-counter medication is usually sufficient to eliminate their pain. However, for a significant proportion of people, headaches are debilitating and make their lives almost unbearable. Many such people have investigated a wide spectrum of possible treatments, including narcotic drugs, hypnosis, biofeedback, and acupuncture, with little or no success. In the last few years, a promising new treatment has been developed. Simply described, the treatment involves a series of injections of a local anesthetic to the occipital nerve (located in the back of the neck). The current treatment procedure is to schedule the injections once a week for 4 weeks. However, it has been suggested that another procedure may be better—one that features one injection every other day for a total of four injections. Additionally, some physicians recommend other combinations of drugs that may increase the effectiveness of the injections. To analyze the problem, an experiment was organized. It was decided to test for a difference between the two schedules of injection and to determine whether there are differences between four drug mixtures. Because of the possibility of an interaction between the schedule and the drug, a complete factorial experiment was chosen. Five headache patients were randomly selected for each combination of schedule and drug. Forty patients were treated and each was asked to report the frequency, duration, and severity of his or her headache prior to treatment and for the 30 days following the last injection. An index ranging from 0 to 100 was constructed for each patient, where 0 indicates no headache pain and 100 specifies the worst headache pain. The improvement in the headache index for each patient was recorded and reproduced in the accompanying table. (A negative value indicates a worsening condition.) (The author is grateful to Dr. Lorne Greenspan for his help in writing this example.)

a. What are the factors in this experiment?
b. What is the response variable?
c. Identify the levels of each factor.
d. Analyze the data and conduct whichever tests you deem necessary at the 5% significance level to determine whether there is sufficient statistical evidence to infer that there are differences in the improvement in the headache index between the two schedules, differences in the improvement in the headache index between the four drug mixtures, or interaction between schedules and drug mixtures.

Improvement in Headache Index

	Drug Mixture			
Schedule	**1**	**2**	**3**	**4**
One Injection	17	24	14	10
Every Week	6	15	9	−1
(4 Weeks)	10	10	12	0
	12	16	0	3
	14	14	6	−1
One Injection	18	−2	20	−2
Every 2 Days	9	0	16	7
(4 Days)	17	17	12	10
	21	2	17	6
	15	6	18	7

14.55 Xr14-55 Most college instructors prefer to have their students participate actively in class. Ideally, students will ask their professor questions and answer their professor's questions, making the classroom experience more interesting and useful. Many professors seek ways to encourage their students to participate in class. A statistics professor at a community college in upper New York state believes that there are a number of external factors that affect student participation. He believes that the time of day and the configuration of seats are two such factors. Consequently, he organized the following experiment. Six classes of about 60 students each were scheduled for one semester. Two classes were scheduled at 9:00 A.M., two at 1:00 P.M., and two at 4:00 P.M. At each of the three times, one of the classes was assigned to a room where the seats were arranged in rows of 10 seats. The other class was a U-shaped, tiered room, where students not only face the instructor, but face their fellow students as well. In each of the six classrooms, over 5 days, student participation was measured by counting the number of times students asked and answered questions. These data are displayed in the accompanying table.

a. How many factors are there in this experiment? What are they?
b. What is the response variable?
c. Identify the levels of each factor.
d. What conclusions can the professor draw from these data?

Class	Time		
Configuration	**9:00 A.M.**	**1:00 P.M.**	**4:00 P.M.**
Rows	10	9	7
	7	12	12
	9	12	9
	6	14	20
	8	8	7
U-Shape	15	4	7
	18	4	4
	11	7	9
	13	4	8
	13	6	7

14.6 (Optional) Applications in Operations Management: Finding and Reducing Variation

In the introduction to Example 12.3, we pointed out that variation in the size, weight, or volume of a product's components causes the product to fail or not function properly. Unfortunately, it is impossible to eliminate all variation. Designers of products and the processes that make the products understand this phenomenon. Consequently, when they specify the length, weight, or some other measurable characteristic of the product they allow for some variation, called the tolerance. For example, the diameters of the piston rings of a car are supposed to be .826 millimeter (mm) with a tolerance of .006 mm. That is, the product will function provided that the diameter is between $.826 - .006 = .820$ and $.826 + .006 = .832$ mm. These quantities are called the lower and upper specification limits (LSL and USL), respectively.

Suppose that the diameter of the piston rings is actually a random variable that is normally distributed with a mean of .826 and a standard deviation of .003 mm. We can compute the probability that a piston ring's diameter is between the specification limits. Thus,

$$
\begin{aligned}
P(.820 < X < .832) &= P\left(\frac{.820 - .826}{.003} < \frac{X - \mu}{\sigma} < \frac{.832 - .826}{.003}\right) \\
&= P(-2.0 < Z < 2.0) \\
&= .9772 - .0228 \\
&= .9544
\end{aligned}
$$

The probability that the diameter does not meet specifications is $1 - .9544 = .0456$. This probability is a measure of the process capability.

If we can decrease the standard deviation, a greater proportion of piston rings will have diameters that meet specification. Suppose that the operations manager has decreased the diameter's standard deviation to .002. The proportion of piston rings that do not meet specifications is .0026. When the probabilities are quite low, we express the probabilities as the number of defective units per million or per billion. Thus, if the standard deviation is .002, the number of defective piston rings is expected to be 2,600 per million. The goal of many firms is to reduce the standard deviation so that the lower specification and upper specification limits are at least 6 standard deviations away from the mean. If the standard deviation is .001, the proportion of nonconforming piston rings is $1 - P(-6 < Z < 6)$, which is 2 per billion. (Incidentally, this figure is often erroneously quoted as 3.4 per million.) The goal is called six sigma. Figure 14.7 depicts the proportion of conforming and nonconforming piston rings for $\sigma = .003, .002$, and .001.

Another way to measure how well the process works is the process capability index, denoted by C_p, which is defined as

$$C_p = \frac{\text{USL} - \text{LSL}}{6\sigma}$$

Thus, in the illustration, USL = .832 and LSL = .820. If the standard deviation is .002, then

$$C_p = \frac{\text{USL} - \text{LSL}}{6\sigma} = \frac{.832 - .820}{6(.002)} = 1.0$$

The larger the process capability index, the more capable is the process in meeting specifications. A value of 1.0 describes a production process where the specification limits are equal to 3 standard deviations above and below the mean. A process capability index

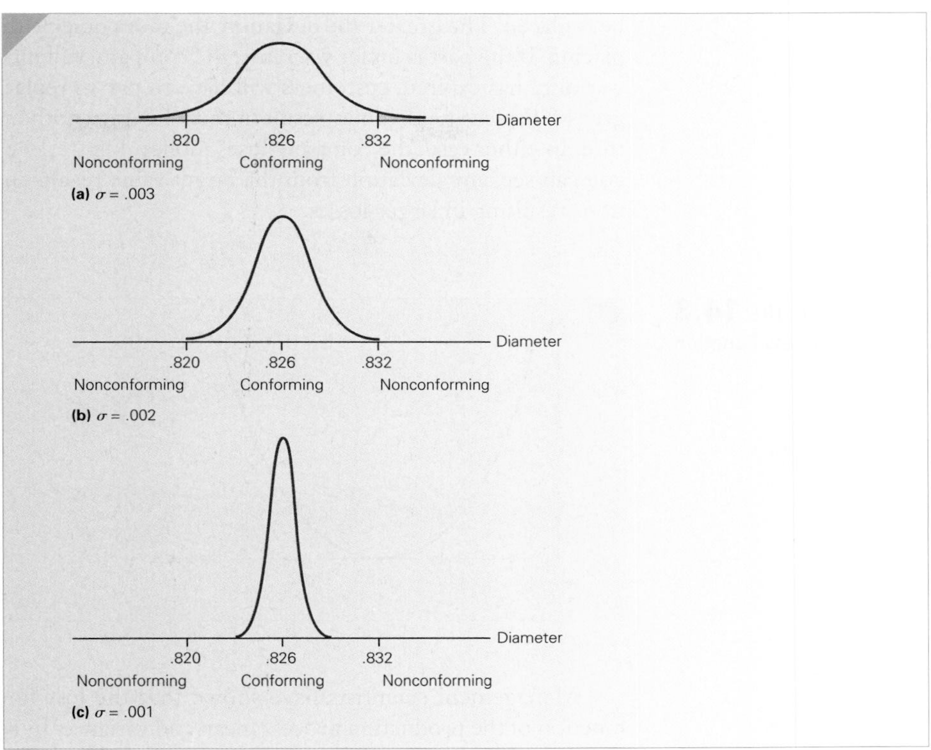

FIGURE **14.7** Proportion of Conforming and Nonconforming Piston Rings

of 2.0 means that the upper and lower limits are 6 standard deviations above and below the mean. This is the goal for many firms.

In practice the standard deviation must be estimated from the data. We will address this issue again in Chapter 21.

Taguchi Loss Function

Historically, operations managers applied the "goalpost" philosophy, a name derived from the game of football. If the ball is kicked *anywhere* between the goalposts, the kick is equally as successful as one that is in the center of the goalposts. Under this philosophy, a piston ring that has a diameter of .821 works as well as one that is exactly .826. That is, the company sustains a loss only when the product falls outside the goalposts. Products that lie between the goalposts suffer no financial loss. For many firms, this philosophy has now been replaced by the Taguchi loss function (named for Genichi Taguchi, a Japanese statistician whose ideas and techniques permeate any discussion of statistical applications in quality management).

Products whose length or weight fall within the tolerances of their specifications do not all function in exactly the same way. There is a difference between a product that barely falls between the goalposts and one that is in the exact center. The Taguchi loss function recognizes that any deviation from the target value results in a financial loss. In addition, the farther the product's variable is from the target value, the greater the loss. The piston ring described previously is specified to have a diameter of exactly .826 mm, an amount specified by the manufacturer to work at the optimum level. Any deviation will cause that part and perhaps other parts to wear out prematurely. Although customers will not know the reason for the problem, they will know that the unit had to be replaced. The greater the deviation, the more quickly the part will wear and need replacing. If the part is under warranty, the company will incur a loss in replacing it. If the warranty has expired, customers will have to pay to replace the unit, causing some degree of displeasure that may result in them buying another company's product in the future. In either case, the company loses money. Figure 14.8 depicts the loss function. As you can see, any deviation from the target value results in some loss, with large deviations resulting in larger losses.

FIGURE **14.8**
Taguchi Loss Function

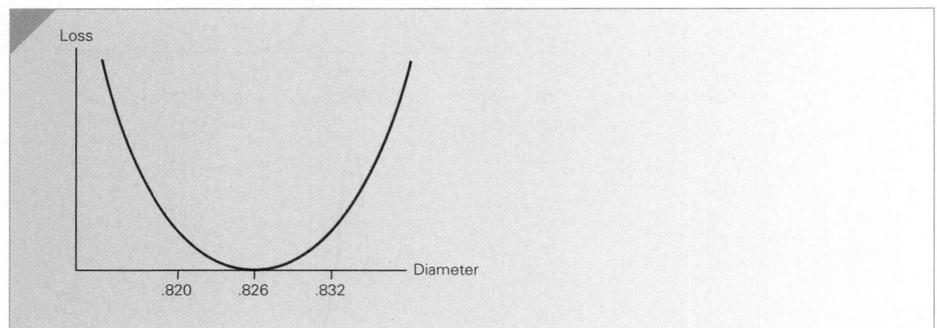

Management scientists have shown that the loss function can be expressed as a function of the production process mean and variance. In Figure 14.9 we describe a normal distribution of the diameter of the machined part with a target value of .826 mm. When the mean of the distribution is .826, any loss is caused by the variance. The statistical techniques introduced in Chapter 21 are usually employed to center the distribution on the target value. However, reducing the variance is considerably more difficult. To reduce variation, it is necessary to first find the sources of variation. We do so by conducting experiments. The principles are quite straightforward, drawing on the concepts developed in the previous section.

FIGURE **14.9**
Taguchi Loss Function and the Distribution of Piston Rings

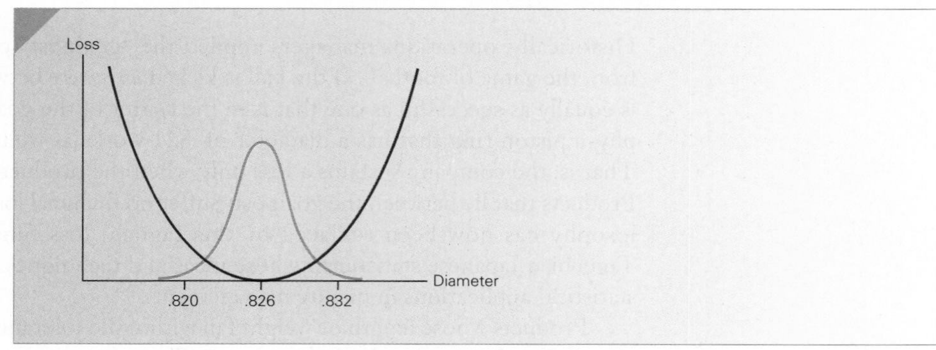

An important function of operations management is production design wherein decisions are made about how a product is manufactured. The objective is to produce the highest quality product at a reasonable cost. This objective is achieved by choosing the machines, materials, methods, and "manpower" (personnel), the so-called 4 M's. By altering some or all of these elements, the operations manager can alter the size, weight, or volume and ultimately, the quality of the product.

The causes of variation example that opened this chapter illustrates this strategy. Because we have limited our discussion to the two-factor model, the example features this experimental design. It should be understood, however, that more complicated models are needed to fully investigate sources of variation.

Causes of Variation: Solution

© Peter Poulides/Stone/Getty Images

IDENTIFY

The response variable is the length of the rods. The two factors are the operators and the machines. There are three levels of operators and four levels of machines. The model we employ is the two-factor model with interaction. The computer output is shown here.

COMPUTE

EXCEL

	A	B	C	D	E	F	G
1	Anova: Two-Factor With Replication						
2							
3	ANOVA						
4	*Source of Variation*	*SS*	*df*	*MS*	*F*	*P-value*	*F crit*
5	Sample	0.0151	2	0.0076	6.98	0.0022	3.1907
6	Columns	0.0034	3	0.0011	1.04	0.3856	2.7981
7	Interaction	0.0046	6	0.0008	0.71	0.6394	2.2946
8	Within	0.0520	48	0.0011			
9							
10	Total	0.0751	59				

MINITAB

Xm14-00a stores the data in Minitab format.

Two-way Analysis of Variance

Analysis of Variance for Rods

Source	DF	SS	MS	F	P
Machines	3	3363	1121	1.04	0.386
Operator	2	15133	7566	6.98	0.002
Interaction	6	4646	774	0.71	0.639
Error	48	51995	1083		
Total	59	75137			

INTERPRET

The test for interaction yields $F = .71$ and a p-value of .6394. There is not enough evidence to infer that the two factors interact. The F-statistic for the operator factor (Sample) is 6.98 (p-value $= .0022$). The F-statistic for the machine factor (Columns) is 1.04 (p-value $= .3856$). We conclude that there are differences only between the levels of the operators. Thus, the only source of variation here is the different operators. The operations manager can now focus on reducing or eliminating this variation. For example, the manager may use only one operator in the future or investigate why the operators differ.

Design of Experiments and Taguchi Methods

In the example just discussed, the experiment used only two factors. In practice, there are frequently many more factors. The problem is that the total number of treatments or combinations can be quite high, making any experimentation both time-consuming and expensive. For example, if there are 10 factors each with 2 levels, the number of treatments is $2^{10} = 1{,}024$. If we measure each treatment with 10 replicates the number of observations, 10,240, makes this experiment prohibitive. Fortunately, it is possible to reduce this number considerably. Through the use of *orthogonal arrays*, we can conduct *fractional factorial experiments* that can produce useful results at a small fraction of the cost. The experimental designs and statistical analyses are beyond the level of this book. Interested readers can find a variety of books at different levels of mathematical and statistical sophistication to learn more about this application.

EXERCISES

Applications

The following exercises require the use of a computer and software. ***Use a 5% significance level.***

14.56 Xr14-56 The headrests on a car's front seats are designed to protect the driver and front-seat passenger from whiplash when the car is hit from behind. The frame of the headrest is made from metal rods. A machine is used to bend the rod into a U-shape exactly 440 millimeters wide. The width is critical; too wide or too narrow and it won't fit into the holes drilled into the car seat frame. The company has experimented with several different metal alloys in the hope of finding a material that will result in more headrest frames that fit. Another possible source of variation is the machines used. To learn more about the process, the operations manager conducts an experiment. Both of the machines are used to produce 10 headrests from each of the five metal alloys now being used. Each frame is measured and the data (in millimeters) are recorded using the format shown here. Analyze the data to determine whether the alloys, machines, or both are sources of variation.

Column 1: Machine 1, rows 1 to 10 alloy A, rows 11 to 20, alloy B
Column 2: Machine 2, rows 1 to 10 alloy A, rows 11 to 20, alloy B

14.57 Xr14-57 A paint manufacturer is attempting to improve the process that fills the 1-gallon containers. The foreperson has suggested that the nozzle can be made from several different alloys. Furthermore, the way that the process "knows" when to stop the flow of paint can be accomplished in two ways: by setting a predetermined amount or by measuring the amount of paint already in the can. To determine what factors lead to variation, an experiment is conducted. For each of the four alloys that could be used to make the nozzles and the two measuring devices, five cans are filled. The amount of paint in each container is precisely measured. The data in liters were recorded in the following way:

Column 1: Device 1, rows 1 to 5 alloy A, rows 6 to 10 alloy B, etc.
Column 2: Device 2, rows 1 to 5 alloy A, rows 6 to 10 alloy B, etc.

Can we infer that the alloys, the measuring devices, or both are sources of variation?

14.58 Xr14-58 The marketing department of a firm that manufactures office furniture has ascertained that there is a growing market for a specialized desk that houses the various parts of a computer system. The operations manager is summoned to put together a plan that will produce high-quality desks at low cost. The characteristics of the desk have been dictated by the marketing department, which has specified the material that the desk will be made from and the machines used to produce the parts. However, there are three methods that can be utilized. Moreover, because of the complexity of the operation, the manager realizes that it is possible that different skill levels of the workers can yield different results. Accordingly, he organized an experiment. Workers from each of three skill levels were chosen. These groups were further divided into two subgroups. Each subgroup assembled the desks using methods A and B. The amount of time taken to assemble each of eight desks was recorded as follows. Columns 1 and 2 contain the times for methods A and B; rows 1 to 8, 9 to 16, and 17 to 24 store the times for the three skill levels. What can we infer from these data?

Chapter Summary

The analysis of variance allows us to test for differences between populations when the data are interval. The analyses of the results of three different experimental designs were presented in this chapter. The one-way analysis of variance defines the populations as the levels of one factor. The second experimental design also defines the treatments on the basis of one factor. However, the randomized block design uses data gathered by observing the results of a matched or blocked experiment (two-way analysis of variance). The third design is the two-factor experiment wherein the treatments are defined as the combinations of the levels of two factors. All the analyses of variance are based on partitioning the total sum of squares into sources of variation from which the mean squares and F-statistics are computed.

Additionally, we introduced three multiple comparison methods, which allow us to determine which means differ in the one-way analysis of variance.

Finally, we described an important application in operations management that employs the analysis of variance.

Important Terms

Analysis of variance 514
Treatment means 514
One-way analysis of variance 514
Response variable 516
Responses 516
Experimental unit 516
Factor 516
Level 516
Between-treatments variation 516
Sum of squares for treatments (SST) 516
Within-treatments variation 517
Sum of squares for error (SSE) 517
Mean squares 518
Mean square for treatments 518
Mean square for error 518
F-Statistic 519
Analysis of variance (ANOVA) table 520
Total variation 520
SS(Total) 520
Completely randomized design 522
Multiple comparisons 530
Least Significance Difference 532
Bonferroni adjustment 533
Tukey's multiple comparison method 534
Multifactor experiment 539
Randomized block design 539
Repeated measures 539
Two-way analysis of variance 540
Fixed effects analysis of variance 540
Random effects analysis of variance 540
Sum of squares for blocks 540
Factorial experiment 549
Interactions 551
Complete factorial experiment 551
Replicate 551
Balanced 551

Symbols

Symbol	Pronounced	Represents
$\bar{\bar{x}}$	*x-double-bar*	Overall or grand mean
q		Studentized range
ω	*Omega*	Critical value of Tukey's multiple comparison method
$q_\alpha(k, \nu)$	*q-sub-alpha-k-v*	Critical value of the Studentized range
n_g		Number of observations in each of k samples
$\bar{x}[T]_j$	*x-bar-T-sub-j*	Mean of the jth treatment
$\bar{x}[B]_i$	*x-bar-B-sub-i*	Mean of the ith block
$\bar{x}[AB]_{ij}$	*x-bar-A-B-sub-i-j*	Mean of the ijth treatment
$\bar{x}[A]_i$	*x-bar-A-sub-i*	Mean of the observations when the factor A level is i
$\bar{x}[B]_j$	*x-bar-B-sub-j*	Mean of the observations when the factor B level is j

FORMULAS

One-way analysis of variance

$$\text{SST} = \sum_{j=1}^{k} n_j(\bar{x}_j - \bar{\bar{x}})^2$$

$$\text{SSE} = \sum_{j=1}^{k}\sum_{i=1}^{n_j}(x_{ij} - \bar{x}_j)^2$$

$$\text{MST} = \frac{\text{SST}}{k-1}$$

$$\text{MSE} = \frac{\text{SSE}}{n-k}$$

$$F = \frac{\text{MST}}{\text{MSE}}$$

Least significant difference comparison method

$$\text{LSD} = t_{\alpha/2}\sqrt{\text{MSE}\left(\frac{1}{n_i} + \frac{1}{n_j}\right)}$$

Tukey's multiple comparison method

$$\omega = q_\alpha(k, \nu)\sqrt{\frac{\text{MSE}}{n_g}}$$

Two-way analysis of variance (randomized block design of experiment)

$$\text{SS(Total)} = \sum_{j=1}^{k}\sum_{i=1}^{b}(x_{ij} - \bar{\bar{x}})^2$$

$$\text{SST} = \sum_{i=1}^{k} b(\bar{x}[T]_j - \bar{\bar{x}})^2$$

$$\text{SSB} = \sum_{i=1}^{b} k(\bar{x}[B]_i - \bar{\bar{x}})^2$$

$$\text{SSE} = \sum_{j=1}^{k}\sum_{i=1}^{b}(x_{ij} - \bar{x}[T]_j - \bar{x}[B]_i + \bar{\bar{x}})^2$$

$$\text{MST} = \frac{\text{SST}}{k-1}$$

$$\text{MSB} = \frac{\text{SSB}}{b-1}$$

$$\text{MSE} = \frac{\text{SSE}}{n-k-b+1}$$

$$F = \frac{\text{MST}}{\text{MSE}}$$

$$F = \frac{\text{MSB}}{\text{MSE}}$$

Two-factor analysis of variance

$$\text{SS(Total)} = \sum_{i=1}^{a}\sum_{j=1}^{b}\sum_{k=1}^{r}(x_{ijk} - \bar{\bar{x}})^2$$

$$\text{SS(A)} = rb\sum_{i=1}^{a}(\bar{x}[A]_i - \bar{\bar{x}})^2$$

$$\text{SS(B)} = ra\sum_{j=1}^{b}(\bar{x}[B]_j - \bar{\bar{x}})^2$$

$$\text{SS(AB)} = r\sum_{i=1}^{a}\sum_{j=1}^{b}(\bar{x}[AB]_{ij} - \bar{x}[A]_i - \bar{x}[B]_j + \bar{\bar{x}})^2$$

$$\text{SSE} = \sum_{i=1}^{a}\sum_{j=1}^{b}\sum_{k=1}^{r}(x_{ijk} - \bar{x}[AB]_{ij})^2$$

$$\text{MS(A)} = \frac{\text{SS(A)}}{a-1}$$

$$\text{MS(B)} = \frac{\text{SS(B)}}{b-1}$$

$$\text{MS(AB)} = \frac{\text{SS(AB)}}{(a-1)(b-1)}$$

$$F = \frac{\text{MS(A)}}{\text{MSE}}$$

$$F = \frac{\text{MS(B)}}{\text{MSE}}$$

$$F = \frac{\text{MS(AB)}}{\text{MSE}}$$

COMPUTER OUTPUT AND INSTRUCTIONS

Technique	Excel	Minitab
One-Way ANOVA	521	521
Multiple comparisons (LSD, Bonferroni adjustment, and Tukey)	535	536
Two-Way (randomized block) ANOVA	543	544
Two-factor ANOVA	556	557

CHAPTER EXERCISES

The following exercises require the use of a computer and software. ***Use a 5% significance level.***

14.59 Xr14-59 Each year billions of dollars are lost because of worker injuries on the job. Costs can be decreased if injured workers can be rehabilitated quickly. As part of an analysis of the amount of time taken for workers to return to work, a sample of male blue-collar workers aged 35 to 45 who suffered a common wrist fracture was taken. The researchers believed that the mental and physical condition of the individual affects recovery time. Each man was given a questionnaire to complete, which measured whether he tended to be optimistic or pessimistic. Their physical condition was also evaluated and categorized as very physically fit, average, or in poor condition. The number of days until the wrist returned to full function was measured for each individual. These data were recorded in the following way:

Column 1: Time to recover for optimists (columns 1–10) = very fit, rows 11–20 = in average condition, rows 21– 30 = poor condition)

Column 2: Time to recover for pessimists (same format as column 1)

a. What are the factors in this experiment? What are the levels of each factor?
b. Can we conclude that pessimists and optimists differ in their recovery times?
c. Can we conclude that physical condition affects recovery times?

14.60 Xr14-60 In the past decade, American companies have spent nearly $1 trillion on computer systems. However, productivity gains have been quite small. During the 1980s, productivity in U.S. service industries (where most computers are used) grew by only .7% annually. In the 1990s, this figure rose to 1.5%. (*Source: New York Times Service*, February 22, 1995). The problem of small productivity increases may be caused by employee difficulty in learning how to use the computer. Suppose that in an experiment to examine the problem, 100 firms were studied. Each company had bought a new computer system 5 years ago. The companies reported their increase in productivity over the 5-year period and were also classified as offering extensive employee training, some employee training, little employee training, or no formal employee training in the use of computers. (There were 25 firms in each group.)

a. Can we conclude that differences in productivity gain exist between the four groups of companies?
b. If there are differences, what are they?

14.61 Xr14-61 The possible imposition of a residential property tax has been a sensitive political issue in a large city that consists of five boroughs. Currently, property tax is based on an assessment system that dates back to 1950. This system has produced numerous inequities whereby newer homes tend to be assessed at higher values than older homes. A new system based on the market value of the house has been proposed. Opponents of the plan argue that residents of some boroughs would have to pay considerably more on the average, while residents of other boroughs would pay less. As part of a study examining this issue, several homes in each borough were assessed under both plans. The percentage increase (a decrease is represented by a negative increase) in each case was recorded.

a. Can we conclude that there are differences in the effect the new assessment system would have on the five boroughs?
b. If differences exist, which boroughs differ? Use Tukey's multiple comparison method.
c. What are the required conditions for your conclusions to be valid?
d. Are the required conditions satisfied?

14.62 Xr14-62 The editor of the student newspaper was in the process of making some major changes in the newspaper's layout. He was also contemplating changing the typeface of the print used. To help himself make a decision, he set up an experiment in which 20 individuals were asked to read four newspaper pages, with each page printed in a different typeface. If the reading speed differed, then the typeface that was read fastest would be used. However, if there was not enough evidence to allow the editor to conclude that such differences existed, the current typeface would be continued. The times (in seconds) to completely read one page were recorded. What should the editor do?

14.63 Xr14-63 In marketing children's products, it is extremely important to produce television commercials that hold the attention of the children who view them. A psychologist hired by a marketing research firm wants to determine whether differences in attention span exist between children watching advertisements for different types of products. One hundred fifty children under 10 years of age were recruited for an experiment. One-third watched a 60-second commercial for a new computer game, one-third watched a commercial for a breakfast cereal, and one-third watched a commercial for children's clothes. Their attention spans (in seconds) were measured and recorded. Do these data provide enough evidence to conclude that there are differences in attention span between the three products advertised?

14.64 Xr14-64 Upon reconsidering the experiment in Exercise 14.63, the psychologist decides that the age of the child may influence the attention span. Consequently, the experiment is redone in the following way. Three 10-year-olds, three 9-year-olds, three 8-year-olds, three 7-year-olds, three 6-year-olds, three 5-year-olds, and three 4-year-olds are randomly assigned to watch one of the commercials, and their attention spans are measured. Do the results indicate that there are differences in the abilities of the products advertised to hold children's attention?

14.65 Xr14-65 It is important for salespeople to be knowledgeable about how people shop for certain products. Suppose that a new car salesman believes that the age and gender of a car shopper affect the way he or she makes an offer on a car. He records the initial offers made by a group of men and women shoppers on a $20,000 Mercury Sable. Besides the gender of the shopper, the salesman also notes the age category. The amount of money below the asking price that each person offered initially for the car was recorded using the following format: Column 1 contains the data for the under 30 group; the first 25 rows store the results for female shoppers, and the last 25 rows are the male shoppers. Columns 2 and 3 store the data for the 30–45 age category and over-45 category, respectively. What can we conclude from these data?

14.66 Xr14-66 Many of you reading this page probably learned how to read using the whole-language method. This strategy maintains that the natural and effective way is to be exposed to whole words in context. Students learn how to read by recognizing words they have seen before. In the past generation this has been the dominant teaching strategy throughout North America. It replaced phonics, wherein children were taught to sound out the letters to form words. The whole language method was instituted with little or no research and has been severely criticized in the past. A recent study may have resolved the question of which method should be employed. Barbara Foorman, an educational psychologist at the University of Houston, described the experiment at the annual meeting of the American Association for the Advancement of Science. The subjects were 375 low-achieving, poor, first-grade students in Houston schools. The students were divided into three groups. One was educated according to the whole-language philosophy, a second group was taught using a pure phonics strategy, and the third was taught employing a mixed or embedded phonics technique. At the end of the term, students were asked to read a list of 50 words. The number of words each child could read was recorded.

a. Can we infer that differences exist between the effects of the three teaching strategies?
b. If differences exist, identify which method appears to be best.

14.67 Xr14-67 Are babies who are exposed to music before their birth smarter than those who are not? And, if so, what kind of music is best? Researchers at the University of Wisconsin conducted an experiment with rats. The researchers selected a random sample of pregnant rats and divided the sample into three groups. Mozart's works were played to one group, a second group was exposed to white noise (a steady hum with no musical elements), and the third group listened to Philip Glass's music (very simple compositions). The researchers then trained the young rats to run a maze in search of food. The amount of time for the rats to complete the maze was measured for all three groups.

a. Can we infer from these data that there are differences between the three groups?
b. If there are differences, determine which group is best.

14.68 Xr14-68 Increasing tuition has resulted in some students being saddled with large debts upon graduation. To examine this issue, a random sample of recent graduates was asked to report whether they had student loans, and if so, how much was the debt at graduation. Each person who reported that they owed money was also asked to whether their degree was a B.A., B.Sc., B.B.A., or other. Can we conclude that debt levels differ between the four types of degree?

14.69 Xr14-69 Studies indicate that single male investors tend to take the most risk, whereas married female investors tend to be conservative. This raises the question, Which does best? The risk-adjusted returns for single and married men, and for single and married women were recorded. Can we infer that differences exist between the four groups of investors?

14.70 Xr14-70 Like all other fine restaurants, Ye Olde Steak House in Windsor, Ontario, attempts to have three "seatings" on weekend nights. Three seatings means that each table gets three different sets of customers. Obviously, any group that lingers over dessert and coffee may result in the loss of one seating and profit for the restaurant. In an effort to determine which types of groups tend to linger, a random sample of 150 groups was drawn. For each group, the number of members and the length of time that the group stayed were recorded in the following way:

Column A: Length of time for 2 people
Column B: Length of time for 3 people
Column C: Length of time for 4 people
Column D: Length of time for more than 4 people

Do these data allow us to infer that the length of time in the restaurant depends on the size of the party?

14.71 Xr14-71 When the stock market has a large 1-day decline, does it bounce back the next day or does the bad news endure? To answer this question, an economist examined a random sample of daily changes to the Toronto Stock Index (TSE). He recorded the percent change. He classified declines as

Down by less than 0.5%
Down by 0.5% to 1.5%
Down by 1.5% to 2.5%
Down by more than 2.5%

For each of these days he recorded the percent loss the following day. Do these data allow us to infer that there are differences in changes to the TSE depending on the loss the previous day? (This exercise is based on a study undertaken by Tim Whitehead, an economist for Left Bank Economics, a consulting firm near Paris, Ontario.)

14.72 Xr14-72 Stock market investors are always seeking the "Holy Grail," a sign that tells them the market has bottomed out or achieved its highest level. There are several indicators. One is the buy signal developed by Gerald Appel, who believed that a bottom has been reached when the difference between the weekly close of the New York Stock Exchange (NYSE) index and the 10-week moving average (see Chapter 20) is −4.0 points or more. Another bottom indicator is based on identifying a certain pattern in the line chart of the stock market index. As an experiment, a financial analyst randomly selected 100 weeks. For each week he determined whether there was an Appel buy, a chart buy, or no indication. For each type of week he recorded the percentage change over the next 4 weeks. Can we infer that the two buy indicators are not useful?

The following exercises use data files associated with three exercises seen previously in this book.

14.73 Xr12-92* In Exercise 12.92 marketing managers for the JCPenney department store chain segmented the market for women's apparel on the basis of personal and family values. The segments are Conservative, Traditional, and Contemporary. Recall that the classification was done on the basis of questionnaires. Suppose that in addition to identifying the segment, the questionnaire also asked each woman to report family income (in $1,000s). Do these data allow us to infer that family incomes differ between the three market segments?

14.74 Xr13-15* Exercise 13.15 addressed the problem of determining whether the distances young (under 25) males and females drive annually differ. Included in the data is also the number of accidents that each person was involved in the past 2 years. Responses are 0, 1, or 2 or more. Do the data allow us to infer that the distances driven differ between the drivers who have had 0, 1, or 2 or more accidents?

14.75 Xr13-87* The objective in Exercise 13.87 was to determine whether various market segments were more likely to use the Quik Lube service. Included with the data is also the age (in months) of the car. Do the data allow us to conclude that there are differences in the age between the four market segments?

CASE 14.1 Comparing Three Methods of Treating Childhood Ear Infections*

© Elizabeth Hathon/Corbis

DATA C14-01

Acute otitis media, an infection of the middle ear, is a common childhood illness. There are various ways to treat the problem. To help determine the best way, researchers conducted an experiment. One hundred and eighty children between 10 months and 2 years with recurrent acute otitis media were divided into three equal groups. Group 1 was treated by surgically removing the adenoids (adenoidectomy), the second using the drug Sulfafurazole, and the third with a placebo. Follow-up lasted for two years. Each child was tracked for 2 years, during which time all symptoms and episodes of acute otitis media were recorded. The data were recorded in the following way:

Column 1: ID number
Column 2: Group number
Column 3: Number of episodes of the illness
Column 4: Number of visits to a physician because of any infection
Column 5: Number of prescriptions
Column 6: Number of days with symptoms of respiratory infection

a. Are there differences between the three groups with respect to the number of episodes, number of physician visits, number of prescriptions, and number of days with symptoms of respiratory infection?

b. Assume that you are working for the company that makes the drug Sulfafurazole. Write a report to the company's executives discussing your results.

*This case is adapted from the *British Medical Journal*, February 2004.

APPENDIX 14 / REVIEW OF CHAPTERS 12 TO 14

The number of techniques introduced in Chapters 12 to 14 is up to 23. As we did in Appendix 13, we provide a list of the techniques in Table A14.1, a flowchart to help you identify the correct technique in Figure A14.1, and 18 exercises to give you practice in how to choose the appropriate method. The table and the flowchart have been amended to include the three analysis of variance techniques introduced in this chapter and the three multiple comparison methods.

TABLE **A14.1** Summary of Statistical Techniques in Chapters 12 to 14

t-test of μ
Estimator of μ (including small population estimator of μ and large and small population estimators of $N\mu$)
χ^2-test of σ^2
Estimator of σ^2
z-test of p
Estimator of p (including small population estimator of p and large and small population estimators of Np)
Equal-variances t-test of $\mu_1 - \mu_2$
Equal-variances estimator of $\mu_1 - \mu_2$
Unequal-variances t-test of $\mu_1 - \mu_2$
Unequal-variances estimator of $\mu_1 - \mu_2$
t-test of μ_D
Estimator of μ_D
F-test of σ_1^2/σ_2^2
Estimator of σ_1^2/σ_2^2
z-test of $p_1 - p_2$ (Case 1)
z-test of $p_1 - p_2$ (Case 2)
Estimator of $p_1 - p_2$
One-way analysis of variance (including multiple comparisons)
Two-way (randomized blocks) analysis of variance
Two-factor analysis of variance

FIGURE **A14.1** Flowchart of Techniques in Chapters 12 to 14

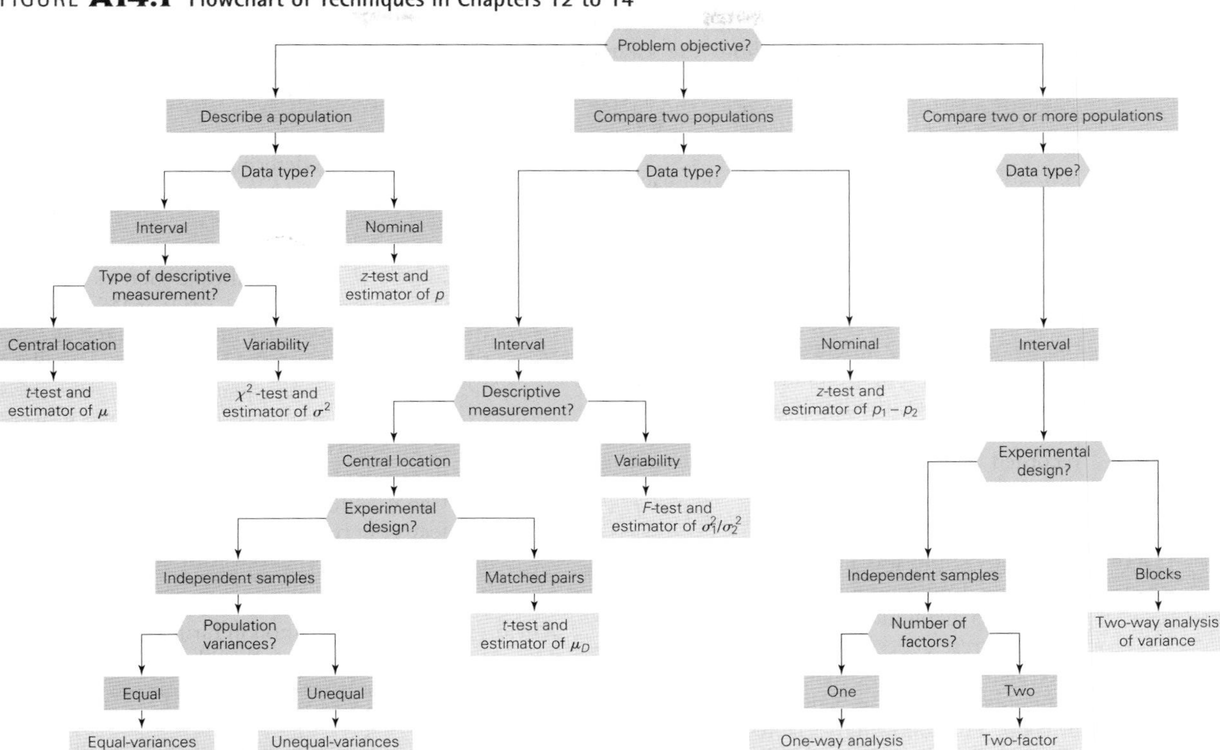

EXERCISES

Note that as we did in Appendix 13, we do not specify a significance level in exercises requiring a test of hypothesis. We leave this decision to you. After analyzing the issues raised in the exercise, use your own judgment to determine whether the p-value is small enough to reject the null hypothesis.

A14.1 XrA14-01 Sales of a product may depend on its placement in a store. Candy manufacturers frequently offer discounts to retailers who display their products more prominently than competing brands. To examine this phenomenon more carefully, a candy manufacturer (with the assistance of a national chain of restaurants) planned the following experiment. In 20 restaurants, the manufacturer's brand was displayed behind the cashier's counter with all the other brands (this was called position 1). In another 20 restaurants, the brand was placed separately, but close to the other brands (position 2). In a third group of 20 restaurants, the candy was placed in a special display next to the cash register (position 3). The number of packages sold during 1 week at each restaurant was recorded. Is there sufficient evidence to infer that sales of candy differ according to placement?

A14.2 XrA14-02 Advertising is critical in the residential real estate industry. Agents are always seeking ways to increase sales through improved advertising methods. A particular agent believes that he can increase the number of inquiries (and thus the probability of making a sale) by describing the house for sale without indicating its asking price. To support his belief, he conducted an experiment in which 100 houses for sale were advertised in two ways—with and without the asking price. The number of inquiries for each house was recorded as well as whether the customer saw the ad with or without the asking price shown. Do these data allow the real estate agent to infer that ads with no price shown are more effective in generating interest in a house?

A14.3 XrA14-03 A professor of statistics hands back his graded midterms in class by calling out the name of each student and personally handing the exam over to its owner. At the end of the process, he notes that there are several exams left over, the result of students missing that class. He forms the theory that the absence is caused by a poor performance by those students on the test. If the theory is correct, the leftover papers will have lower marks than those papers handed back. He recorded the marks (out of 100) for the leftover papers and the marks of the returned papers. Do the data support the professor's theory?

A14.4 XrA14-04 A study was undertaken to determine whether a drug commonly used to treat epilepsy could help alcoholics to overcome their addiction. The researchers took a sample of 103 hard-core alcoholics. Fifty-five drinkers were given Topiramate and the remaining 48 were given a placebo. The following variables were recorded after 6 months:

Column 1: Identification number
Column 2: 1 = Topiramate and 2 = placebo
Column 3: Abstain from alcohol for one month (1 = no, 2 = yes)
Column 4: Did not binge in final month (1 = no, 2 = yes)

Do these data provide sufficient evidence to infer that Topiramate is effective in

a. causing abstinence for the first month?
b. causing alcoholics to refrain from binge drinking in the final month?

A14.5 XrA14-05 Health care costs in the United States and Canada are concerns for citizens and politicians. The question is, How can we devise a system wherein people's medical bills are covered, but individuals attempt to reduce costs. An American company has come up with a possible solution. Golden Rule is an insurance company in Indiana with 1,300 employees. The company offered its employees a choice of programs. One choice was a medical savings account (MSA) plan. Here's how it works. To ensure that a major illness or accident does not financially destroy an employee, Golden Rule offers catastrophic insurance—a policy that covers all expenses above $2,000 per year. At the beginning of the year, the company deposits $1,000 (for a single employee) and $2,000 (for an employee with a family) into the MSA. For minor expenses, the employee pays from his or her MSA. As an incentive for the employee to spend wisely, any money left in the MSA at the end of the year can be withdrawn by the employee. To determine how well it works, a random sample of employees who opted for the medical savings account plan was compared to employees who chose the regular plan. At the end of the year the medical expenses for each employee were recorded. Critics of MSA say that the plan leads to poorer health care, and as a result employees are less likely to be in excellent health. To address this issue, each employee was examined. The results of the examination were recorded where 1 = excellent health and 2 = not in excellent health.

a. Can we infer from these data that MSA is effective in reducing costs?
b. Can we infer that the critics of MSA are correct?

A14.6 XrA14-06 Discrimination in hiring has been illegal for a number of years. It is illegal to discriminate against any person on the basis of race, gender, or religion. It is also illegal to discriminate because of a person's handicap if it in no way prevents that person from performing that job. In recent years, the definition of "handicap" has widened. Several people have successfully sued companies because they were denied employment for no other reason than that the applicant was overweight. A study was conducted to examine attitudes toward overweight people. The experiment involved showing a number of subjects video of an applicant being interviewed for a job. Prior to the interview, the subject was given a description of the job. Following the interview, the subject was asked to score the applicant in terms of how well the applicant was suited for the job. The score was out of 100, where higher scores described greater suitability. (The scores are interval data.) The same procedure was repeated for each subject. However, the gender and weight (average and overweight) of the applicant varied. The results were recorded using the following format:

Column 1: Score for average weight males
Column 2: Score for overweight males
Column 3: Score for average weight females
Column 4: Score for overweight females

a. Can we infer that the scores of the four groups of applicants differ?
b. Are the differences detected in Part a due to weight, gender, or some interaction?

A14.7 XrA14-07 Most automobile repair shops now charge according to a schedule that is claimed to be based on average times. This means that instead of determining the actual time to make a repair and multiplying this value by their hourly rate, repair shops determine the cost from a schedule that is calculated from average times. A critic of this policy is examining how closely this schedule adheres to the actual time to complete a job. He randomly selects five jobs. According to the schedule, these jobs should take 45 minutes, 60 minutes, 80 minutes, 100 minutes, and 125 minutes, respectively. The critic then takes a random sample of repair shops and records the actual times for each of 20 cars for each job. For each job, can we infer that the time specified by the schedule is greater than the actual time?

A14.8 XrA14-08 Automobile insurance appraisers examine cars that have been involved in accidental collisions and estimate the cost of repairs. An insurance executive claims that there are significant differences in the estimates from different appraisers. To support his claim, he takes a random sample of 25 cars that have recently been damaged in accidents. Three appraisers then estimated the repair costs of each car. The estimates were recorded for each appraiser. From the data can we conclude that the executive's claim is true?

A14.9 XrA14-09 The widespread use of salt on roads in Canada and the northern United States during the winter and acid precipitation throughout the year combine to cause rust on cars. Car manufacturers and other companies offer rust-proofing services to help purchasers preserve the value of their cars. A consumer protection agency decides to determine whether there are any differences between the rust protection provided by automobile manufacturers and that provided by two competing types of rust-proofing services. As an experiment, 60 identical new cars are selected. Of these, 20 are rust-proofed by the manufacturer. Another 20 are rust-proofed using a method that applies a liquid to critical areas of the car. The liquid hardens, forming a (supposedly) lifetime bond with the metal. The last 20 are treated with oil and are retreated every 12 months. The cars are then driven under similar conditions in a Minnesota city. The number of months until the first rust appears was recorded. Is there sufficient evidence to conclude that at least one rust-proofing method is different from the others?

A14.10 XrA14-10 One of the ways in which advertisers measure the value of television commercials is by telephone surveys conducted shortly after commercials are aired. Respondents who watched a certain television station at a given time period, during which the commercial appeared, are asked whether they can recall the name of the product in the commercial. Suppose an advertiser wants to compare the recall proportions of two commercials. The first commercial is relatively inexpensive. A second commercial shown a week later is quite expensive to produce. The advertiser decides that the second commercial is viable only if its recall proportion is more than 15% higher than the recall proportion of the first commercial. Two surveys of 500 television viewers each were conducted after each commercial was aired. Each person was asked whether he or she remembered the product name. The results are stored in columns 1 (commercial 1) and 2 (commercial 2) (2 = remembered the product name, 1 = did not remember the product name). Can we infer that the second commercial is viable?

A14.11 XrA14-11 In the door-to-door selling of vacuum cleaners, various factors influence sales. The Birk Vacuum Cleaner Company considers its sales pitch and overall package to be extremely important. As a result, it often thinks of new ways to sell its product. Because the company's management develops so many new sales pitches each year, there is a two-stage testing process. In stage 1, a new plan is tested with a relatively small sample. If there is sufficient evidence that the plan increases sales, a second, considerably larger, test is undertaken. In a stage 1 test to determine whether the inclusion of a "free" 10-year service contract increases sales, 100 sales representatives were selected at random from the company's list of several thousand. The monthly sales of these representatives were recorded for 1 month prior to use of the new sales pitch and for 1 month after its introduction. Should the company proceed to stage 2?

A14.12 XrA14-12 The cost of workplace injuries is high for the individual worker, for the company, and or society. It is in everyone's interest to rehabilitate the injured worker as quickly as possible. A statistician working for an insurance company has investigated the problem. He believes that a major determinant in how quickly a worker returns to his or her job after sustaining an injury is the physical condition. To help determine whether he is on the right track, he organized an experiment. He took a random sample of male and female workers who were injured in the last year. He recorded their gender, their physical condition and the number of working days until they returned to their job. These data were recorded in the following way. Columns 1 and 2 store the number of working days until return to work for men and women, respectively. In each column the first 25 observations relate to those who are physically fit, the next 25 rows relate to individuals who are moderately fit, and the last 25 observations are for those who are in poor physical shape. Can we infer that the six groups differ? If differences exist, determine whether the differences are due to gender, physical fitness, or some combination of gender and physical fitness.

A14.13 XrA14-13 Does driving an ABS-equipped car change the behavior of drivers? To help answer this question, the following experiment was undertaken. A random sample of 200 drivers who currently operate cars without ABS was selected. Each person was given an identical car to drive for 1 year. Half the sample were given cars that had ABS, and the other half were given cars with standard-equipment brakes. Computers on the cars recorded the average speed (in miles per hour) during the year. Can we infer that operating an ABS-equipped car changes the behavior of the driver?

A14.14 XrA14-14 We expect the demand for a product depends on its price: the higher the price, the lower the demand. However, this may not be entirely true. In an experiment conducted by professors at Northwestern University and M.I.T., a mail-order dress was available at the prices $34, $39, and $44. The number of dresses sold weekly was recorded over a 20-week period. The prices were randomized over 60 weeks. Conduct a test to determine whether demand differed and if so, which price elicited the highest sales.

A14.15 XrA14-15 Researchers at the University of Washington conducted an experiment to determine whether the herbal remedy Echinacea is effective in treating children's colds and other respiratory infection (*National Post*, December 3, 2003). A sample of 524 children was recruited. Half the sample treated their colds with Echinacea and the other half was given a placebo. For each infection the duration of the colds (in days) was measured and recorded. Can we conclude that Echinacea is effective?

A14.16 XrA14-16 The marketing manager of a large ski resort wants to advertise that his ski resort has the shortest lift lines of any resort in the area. To avoid the possibility of a false advertising liability suit, he collects data on the times skiers wait in line at his resort and at each of two competing resorts on each of 14 days.

a. Can he conclude that there are differences in waiting times between the three resorts?
b. What are the required conditions for these techniques?
c. How would you check to determine that the required conditions are satisfied?

A14.17 XrA14-17 A popularly held belief about university professors is that they don't work very hard, and that the higher their rank, the less work they do. A statistics student decided to determine whether the belief is true. She took a random sample of 20 university instructors in each of the faculties of business, engineering, arts, and sciences. In each sample of 20, five were instructors, five were assistant professors, five were associate professors, and five were full professors. Each professor was surveyed and asked to report confidentially the number of weekly hours of work. These data were recorded in the following way:

Column 1: hours of work for business professors (first 5 rows = instructors, next 5 rows = assistant professors, next 5 rows = associate professors, and last 5 rows = full professors)
Column 2: hours of work for engineering professors (same formal as column 1)
Column 3: hours of work for arts professors (same format as column 1)
Column 4: hours of work for science professors (same format as column 1)

a. If we conduct the test under the single-factor analysis of variance, how many levels are there? What are they?
b. Test to determine whether differences exist using a single-factor analysis of variance.
c. If we conduct tests using the two-factor analysis of variance, what are the factors? What are their levels?
d. Is there evidence of interaction?
e. Are there differences between the four ranks of instructor?
f. Are there differences between the four faculties?

A14.18 XrA14-18 Billions of dollars are spent annually by Americans for the care and feeding of pets. A survey conducted by the American Veterinary Medical Association drew a random sample of 1,328 American households and asked whether they owned a pet and if so, the type of animal.

Additionally, each was asked to report the veterinary expenditures for the previous 12 months. Column 1 contains the expenditures for dogs and column 2 stores the expenditures for cats. The results are that 474 households reported that they owned at least one dog and 419 owned at least one cat. The latest census indicates that there are 112 million households in the United States. (*Source: Statistical Abstract of the United States*, 2006, Table 1232)

a. Estimate with 95% confidence the total number of households owning at least one dog.
b. Repeat Part a for cats.
c. If we assume that there are 40 million households with at least one dog, estimate with 95% confidence the total amount spent on veterinary expenditures for dogs.
d. If we assume that there are 35 million households with at least on cat, estimate with 95% confidence the total amount spent on veterinary expenditures for cats.

© Bob Elsdale/Workbook Stock/Jupiterimages

15 CHI-SQUARED TESTS

Exit Polls in Ohio

DATA Xm15-00

After the polls close on Election Day, networks compete to be the first to predict which candidate will win. The predictions are based on counts in certain precincts and on exit polls. Exit polls are conducted by asking random samples of voters who have just exited the polling booth for which candidate they voted. In addition to asking for whom they voted, respondents are asked a variety of other questions that provide information to politicians, journalists, and other citizens. The responses to questions about gender, age, education, and the vote cast (Gore or Bush) in the state of Ohio during the 2000 elections were collected and the tables were created. What do these results tell about the vote in Ohio?

© David Butow/Corbis Saba

See page 594 for the answer.

INTRODUCTION

We have seen a variety of statistical techniques that are used when the data are nominal. In Chapter 2 we introduced bar and pie charts, both graphical techniques to describe a set of nominal data. Later in Chapter 2 we showed how to describe the relationship between two sets of nominal data by producing a frequency table and a bar chart. However, these techniques simply describe the data, which may represent a sample or a population. In this chapter we deal with similar problems, but the goal is to use statistical techniques to make inferences about populations from sample data.

This chapter develops two statistical techniques that involve nominal data. The first is a *goodness-of-fit test* applied to data produced by a *multinomial experiment,* a generalization of a binomial experiment. The second uses data arranged in a table (called a *contingency table*) to determine whether two classifications of a population of nominal data are statistically independent; this test can also be interpreted as a comparison of two or more populations. The sampling distribution of the test statistics in both tests is the chi-squared distribution introduced in Chapter 8.

15.1 CHI-SQUARED GOODNESS-OF-FIT TEST

This section presents another test designed to describe a population of nominal data. The first such test was introduced in Section 12.3, where we discussed the statistical procedure employed to test hypotheses about a population proportion. In that case, the nominal variable could assume one of only two possible values, success or failure. Our tests dealt with hypotheses about the proportion of successes in the entire population. Recall that the experiment that produces the data is called a binomial experiment. In this section, we introduce the **multinomial experiment,** which is an extension of the binomial experiment, wherein there are two or more possible outcomes per trial.

Multinomial Experiment

A multinomial experiment is one possessing the following properties.

1. The experiment consists of a fixed number n of trials.
2. The outcome of each trial can be classified into one of k categories, called cells.
3. The probability p_i that the outcome will fall into cell i remains constant for each trial. Moreover, $p_1 + p_2 + \cdots + p_k = 1$.
4. Each trial of the experiment is independent of the other trials.

When $k = 2$, the multinomial experiment is identical to the binomial experiment. Just as we count the number of successes (recall that we label the number of successes x) and failures in a binomial experiment, we count the number of outcomes falling into each of the k cells in a multinomial experiment. In this way, we obtain a set of observed frequencies $f_1, f_2, \ldots, f_k$ where f_i is the observed frequency of outcomes falling into cell i, for $i = 1, 2, \ldots, k$. Because the experiment consists of n trials and an outcome must fall into some cell,

$$f_1 + f_2 + \cdots + f_k = n$$

Just as we used the number of successes x (by calculating the sample proportion $\hat{p}$, which is equal to x/n) to draw inferences about p, so do we use the observed frequencies to draw inferences about the cell probabilities. We'll proceed in what by now has become a standard procedure. We will set up the hypotheses and develop the test statistic and its sampling distribution. We'll demonstrate the process with the following example.

EXAMPLE 15.1 Testing Market Shares

Company A has recently conducted aggressive advertising campaigns to maintain and possibly increase its share of the market for fabric softener. Their main competitor, Company B, has 40% of the market and a number of other competitors account for the remaining 15%. To determine whether the market shares changed after the advertising campaign, the marketing manager for Company A solicited the preferences of a random sample of 200 customers of fabric softener. Of the 200 customers, 102 indicated a preference for Company A's product, 82 preferred Company B's fabric softener, and the remaining 16 preferred the products of one of the competitors. Can the analyst infer at the 5% significance level that customer preferences have changed from their levels before the advertising campaigns were launched?

SOLUTION

The population in question is composed of the brand preferences of the fabric softener customers. The data are nominal because each respondent will choose one of three possible answers: product A, product B, or other. If there were only two categories, or if we were interested only in the proportion of one company's customers (which we would label as successes and label the others as failures), we would identify the technique as the z-test of p. However, in this problem we're interested in the proportions of all three categories. We recognize this experiment as a multinomial experiment, and we identify the technique as the **chi-squared goodness-of-fit test.**

Because we want to know whether the market shares have changed, we specify those precampaign market shares in the null hypothesis.

$$H_0: \quad p_1 = .45, p_2 = .40, p_3 = .15$$

The alternative hypothesis attempts to answer our question, Have the proportions changed? Thus,

$$H_1: \quad \text{At least one } p_i \text{ is not equal to its specified value}$$

Test Statistic

If the null hypothesis is true, we would expect the number of customers selecting brand A, brand B, and other to be 200 times the proportions specified under the null hypothesis. That is,

$$e_1 = 200(.45) = 90$$
$$e_2 = 200(.40) = 80$$
$$e_3 = 200(.15) = 30$$

In general, the **expected frequency** for each cell is given by

$$e_i = np_i$$

This expression is derived from the formula for the expected value of a binomial random variable, introduced in Section 7.4.

Figure 15.1 is a bar chart (created by Excel) showing the comparison of actual and expected frequencies.

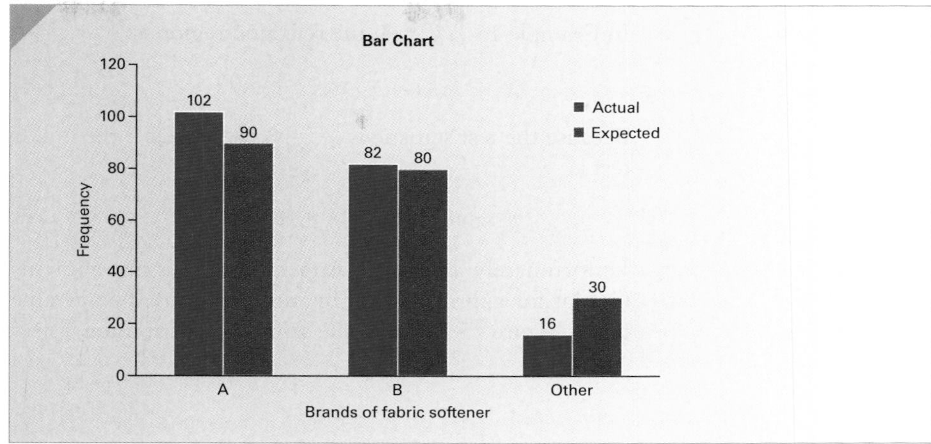

FIGURE 15.1 Bar Chart for Example 15.1

If the expected frequencies e_i and the **observed frequencies** f_i are quite different, we would conclude that the null hypothesis is false, and we would reject it. However, if the expected and observed frequencies are similar, we would not reject the null hypothesis. The test statistic defined in the box measures the similarity of the expected and observed frequencies.

Chi-Squared Goodness-of-Fit Test Statistic

$$\chi^2 = \sum_{i=1}^{k} \frac{(f_i - e_i)^2}{e_i}$$

The sampling distribution of the test statistic is approximately chi-squared distributed with $\nu = k - 1$ degrees of freedom, provided that the sample size is large. We will discuss this required condition later. (The chi-squared distribution was introduced in Section 8.4.)

The following table demonstrates the calculation of the test statistic. Thus, the value $\chi^2 = 8.18$. As usual, we judge the size of this test statistic by specifying the rejection region or by determining the p-value.

Company	Observed Frequency f_i	Expected Frequency e_i	$(f_i - e_i)$	$\frac{(f_i - e_i)^2}{e_i}$
A	102	90	12	1.60
B	82	80	2	0.05
Other	16	30	−14	6.53
Total	200	200		$\chi^2 = 8.18$

When the null hypothesis is true, the observed and expected frequencies should be similar, in which case the test statistic will be small. Thus, a small test statistic supports the null hypothesis. If the null hypothesis is untrue, some of the observed and expected frequencies will differ and the test statistic will be large. Consequently, we want to reject the null hypothesis when χ^2 is greater than $\chi^2_{\alpha,k-1}$. That is, the rejection region is

$$\chi^2 > \chi^2_{\alpha,k-1}$$

In Example 15.1, $k = 3$; the rejection region is

$$\chi^2 > \chi^2_{\alpha,k-1} = \chi^2_{.05,2} = 5.99$$

Because the test statistic is $\chi^2 = 8.18$, we reject the null hypothesis. The p-value of the test is

$$p\text{-value} = P(\chi^2 > 8.18)$$

Unfortunately Table 5 in Appendix B does not allow us to perform this calculation (except for approximation by interpolation). The p-value must be produced by computer. Figure 15.2 depicts the sampling distribution, rejection region, and p-value.

FIGURE **15.2**
Sampling Distribution for Example 15.1

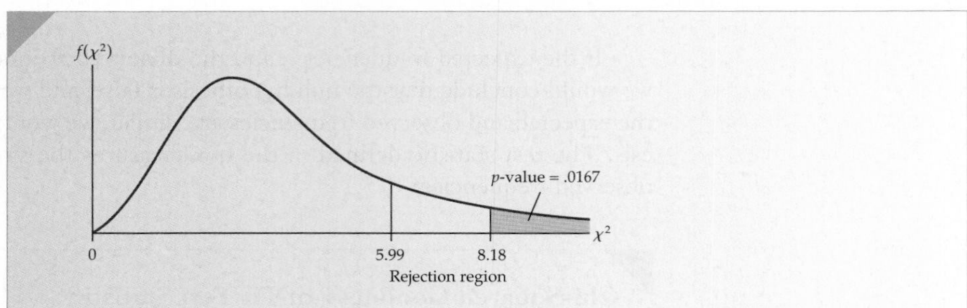

EXCEL

The output from the commands listed here is the p-value of the test. It is .0167.

INSTRUCTIONS

1. Type the observed values into one column and the expected values into another column. (If you wish you can type the cell probabilities specified in the null hypothesis and let Excel convert these into expected values by multiplying by the sample size.)
2. Activate an empty cell and type

   =CHITEST([Actual_range], [Expected_range])

   where the ranges are the cells containing the actual observations and the expected values.

You can also perform what-if analyses to determine for yourself the effect of changing some of the observed values and the sample size.

If we have the raw data representing the nominal responses, we must first determine the frequency of each category (the observed values) using the **COUNTIF** function described on page 20.

MINITAB*

Chi-Square Goodness-of-Fit Test for Observed Counts in Variable: C1

Category	Observed	Test Proportion	Expected	Contribution to Chi-Sq
1	102	0.45	90	1.60000
2	82	0.40	80	0.05000
3	16	0.15	30	6.53333

N	DF	Chi-Sq	P-Value
200	2	8.18333	0.017

INSTRUCTIONS

1. Click **Stat, Tables**, and **Chi-square Goodness-of-Fit Test (One Variable). . . .**
2. Type the observed values into the **Observed counts:** box (102 82 16). If you have a column of data, click **Categorical data:** and specify the column or variable name.
3. Click **Proportions specified by historical counts** and **Input constants.** Type the values of the proportions under the null hypothesis (.45 .40 .15)

INTERPRET

There is sufficient evidence at the 5% significance level to infer that the proportions have changed since the advertising campaigns were implemented. If the sampling was conducted properly, we can be quite confident in our conclusion. This technique has only one required condition, which is satisfied. (See the next subsection.) It is probably a worthwhile exercise to determine the nature and causes of the changes. The results of this analysis will determine the design and timing of other advertising campaigns.

Required Condition

The actual sampling distribution of the test statistic defined previously is discrete, but it can be approximated by the chi-squared distribution provided that the sample size is large. This requirement is similar to the one we imposed when we used the normal approximation to the binomial in the sampling distribution of a proportion. In that approximation we needed np and $n(1 - p)$ to be 5 or more. A similar rule is imposed for the chi-squared test statistic. It is called the *rule of five*, which states that the sample size must be large enough so that the expected value for each cell must be 5 or more. Where necessary, cells should be combined to satisfy this condition. We discuss this required condition and provide more details on its application in CD Appendix W.

Factors That Identify the Chi-Squared Goodness-of-Fit Test

1. **Problem objective:** Describe a single population
2. **Data type:** Nominal
3. **Number of categories:** 2 or more

*Earlier versions of Minitab do not conduct this procedure. CD Appendix V, "Minitab (version 14) Instructions for the Chi-Squared Goodness-of-Fit Test and the Test for Normality," covers this technique.

EXERCISES

Developing an Understanding of Statistical Concepts

*Exercises 15.1–15.6 are "what-if analyses" designed to determine what happens to the test statistic of the goodness-of-fit test when elements of the statistical inference change. These problems can be solved manually or using Excel's **CHITEST**.*

15.1 Consider a multinomial experiment involving $n = 300$ trials and $k = 5$ cells. The observed frequencies resulting from the experiment are shown in the accompanying table, and the null hypothesis to be tested is as follows:

$$H_0: \ p_1 = .1, p_2 = .2, p_3 = .3, p_4 = .2, p_5 = .2$$

Test the hypothesis at the 1% significance level.

Cell	1	2	3	4	5
Frequency	24	64	84	72	56

15.2 Repeat Exercise 15.1 with the following frequencies:

Cell	1	2	3	4	5
Frequency	12	32	42	36	28

15.3 Repeat Exercise 15.1 with the following frequencies:

Cell	1	2	3	4	5
Frequency	6	16	21	18	14

15.4 Review the results of Exercises 15.1–15.3. What is the effect of decreasing the sample size?

15.5 Consider a multinomial experiment involving $n = 150$ trials and $k = 4$ cells. The observed frequencies resulting from the experiment are shown in the accompanying table, and the null hypothesis to be tested is as follows:

$$H_0: \ p_1 = .3, p_2 = .3, p_3 = .2, p_4 = .2$$

Cell	1	2	3	4
Frequency	38	50	38	24

Test the hypotheses, using $\alpha = .05$.

15.6 For Exercise 15.5, retest the hypotheses, assuming that the experiment involved twice as many trials ($n = 300$) and that the observed frequencies were twice as high as before, as shown here.

Cell	1	2	3	4
Frequency	76	100	76	48

*The following exercises require the use of a computer and software. The answers may be calculated manually. See Appendix A for the sample statistics. **Use a 5% significance level unless otherwise directed.***

15.7 Xr15-07 The results of a multinomial experiment with $k = 5$ were recorded. Each outcome is identified by the numbers 1 to 5. Test to determine whether there is enough evidence to infer that the proportion of outcomes differ.

15.8 Xr15-08 A multinomial experiment was conducted with $k = 4$. Each outcome is stored as an integer from 1 to 4 and the results of a survey were recorded. Test the following hypotheses.

$$H_0: \ p_1 = .15, p_2 = .40, p_3 = .35, p_4 = .10$$

H_1: At least one p_i is not equal to its specified value

15.9 Xr15-09 To determine whether a single die is balanced, or fair, the die was rolled 600 times. Is there sufficient evidence to allow you to conclude that the die is not fair?

Applications

15.10 Xr15-10 Grades assigned by an economics instructor have historically followed a symmetrical distribution: 5% A's, 25% B's, 40% C's, 25% D's, and 5% F's. This year, a sample of 150 grades was drawn and the grades (1 = A, 2 = B, 3 = C, 4 = D, and 5 = F) were recorded. Can you conclude, at the 10% level of significance, that this year's grades are distributed differently from grades in the past?

15.11 Xr15-11 Pat Statsdud is about to write a multiple-choice exam but as usual knows absolutely nothing. Pat plans to guess one of the five choices. Pat has been given one of the professor's previous exams with the correct answers marked. The correct choices were recorded where 1 = (a), 2 = (b), 3 = (c), 4 = (d), and 5 = (e). Help Pat determine whether this professor does not randomly distribute the correct answer over the five choices? If this is true, how does it affect Pat's strategy?

15.12 Xr15-12 Financial managers are interested in the speed with which customers who make purchases on credit pay their bills. In addition to calculating the average number of days that unpaid bills (called accounts receivable) remain outstanding, they often prepare an aging schedule. An aging schedule classifies outstanding accounts receivable according to the time that has elapsed since billing and records the proportion of accounts receivable belonging to each classification. A large firm has determined its aging schedule for the past 5 years. These results are shown in the accompanying table. During the past few months, however, the economy has taken a downturn. The company would like to know whether the recession has affected the aging schedule. A random sample of 250 accounts receivable was drawn and each account was classified as follows:

1 = 0–14 days outstanding

2 = 15–29 days outstanding

3 = 30–59 days outstanding

4 = 60 or more days outstanding

Number of Days Outstanding	Proportion of Accounts Receivable Past 5 Years
0–14	.72
15–29	.15
30–59	.10
60 and over	.03

Determine whether the aging schedule has changed.

15.13 Xr15-13 License records in a county reveal that 15% of cars are subcompacts (1), 25% are compacts (2), 40% are midsize (3), and the rest are an assortment of other styles and models (4). A random sample of accidents involving cars licensed in the county was drawn. The type of car was recorded using the codes in parentheses. Can we infer that certain sizes of cars are involved in a higher than expected percentage of accidents?

15.14 Xr15-14 In an election held last year that was contested by three parties, party A captured 31% of the vote, party B garnered 51%, and party C received the remaining votes. A survey of 1,200 voters asked each to identify the party that they would vote for in the next election. These results were recorded where 1 = party A, 2 = party B, and 3 = party C. Can we infer at the 10% significance level that voter support has changed since the election?

15.15 Xr15-15 In a number of pharmaceutical studies volunteers who take placebos (but are told they have taken a cold remedy) report the following side effects:

Headache (1)	5%
Drowsiness (2)	7%
Stomach upset (3)	4%
No side effect (4)	84%

A random sample of 250 people who were given a placebo (but who thought they had taken an antiinflammatory) reported whether they had experienced each of the side effects. These responses were recorded using the codes in parentheses. Do these data provide enough evidence to infer that the reported side effects of the placebo for an antiinflammatory differ from that of a cold remedy?

APPLICATIONS in MARKETING

© Paul Thomas/ The Image Bank/Getty Images

Market Segmentation

Market segmentation was introduced in Section 12.4, where a statistical technique was used to estimate the size of a segment. In Chapters 13 and 14 statistical procedures were applied to determine whether market segments differ in their purchases of products and services. Exercise 15.16 requires you to apply the chi-squared goodness-of-fit test to determine whether the relative sizes of segments have changed.

15.16 Xr12-91* Refer to Exercise 12.91 where the statistics practitioner estimated the size of market segments based on education among California adults. Suppose that census figures from 10 years ago showed the education levels and the proportions of California adults, as follows:

Level	Proportion
1. Did not complete high school	.23
2. Completed high school only	.40
3. Some college or university	.15
4. College or university graduate	.22

Determine whether there has been a change in these proportions.

15.2 CHI-SQUARED TEST OF A CONTINGENCY TABLE

In Chapter 2 we developed the **cross-classification table** as a first step in graphing the relationship between two nominal variables (see page 57). Our goal was to determine whether the two variables were related. In this section we extend the technique to statistical inference. We introduce another chi-squared test, this one designed to satisfy two different problem objectives. The **chi-squared test of a contingency table** is used to determine whether there is enough evidence to infer that two nominal variables are related and to infer that differences exist between two or more populations of nominal variables. Completing both objectives entails classifying items according to two different criteria. To see how this is done, consider the following example.

EXAMPLE 15.2

Relationship between Undergraduate Degree and MBA Major

DATA
Xm15-02

The MBA program was experiencing problems scheduling their courses. The demand for the program's optional courses and majors was quite variable from one year to the next. In one year students seem to want marketing courses and in other years accounting or finance are the rage. In desperation, the dean of the business school turned to a statistics professor for assistance. The statistics professor believed that the problem may be the variability in the academic background of the students and that the undergraduate degree affects the choice of major. As a start, he took a random sample of last year's MBA students and recorded the undergraduate degree and the major selected in the graduate program. The undergraduate degrees were B.A., B.Eng., B.B.A., and several others. There are three possible majors for the MBA students; accounting, finance, and marketing. The results were summarized in a cross-classification table (introduced in Chapter 2) shown here. Can the statistician conclude that the undergraduate degree affects the choice of major?

	MBA Major			
Undergraduate Degree	Accounting	Finance	Marketing	Total
B.A.	31	13	16	60
B.Eng.	8	16	7	31
B.B.A.	12	10	17	39
Other	10	5	7	22
Total	61	44	47	152

SOLUTION

One way to solve the problem is to consider that there are two variables; undergraduate degree and MBA major. Both are nominal. The values of the undergraduate degree are B.A., B.Eng., B.B.A., and other. The values of MBA major are accounting, finance, and marketing. The problem objective is to analyze the relationship between the two variables. Specifically, we want to know whether one variable is related to the other.

Another way of addressing the problem is to determine whether differences exist between B.A.'s, B.Eng.'s, B.B.A.'s, and others. That is, we treat the holders of each undergraduate degree as a separate population. Each population has three possible values represented by the MBA major. The problem objective is to compare four populations. (We can also answer the question by treating the MBA majors as populations and the undergraduate degrees as the values of the random variable.)

As you will shortly discover, both objectives lead to the same test. Consequently, we address both objectives at the same time.

The null hypothesis will specify that there is no relationship between the two variables. We state this in the following way:

H_0: The two variables are independent

The alternative hypothesis specifies one variable affects the other, expressed as

H_1: The two variables are dependent

Graphical Technique

Figure 15.3 depicts the graphical technique introduced in Chapter 2 to show the relationship (if any) between the two nominal variables.

FIGURE **15.3** Bar Chart for Example 15.2

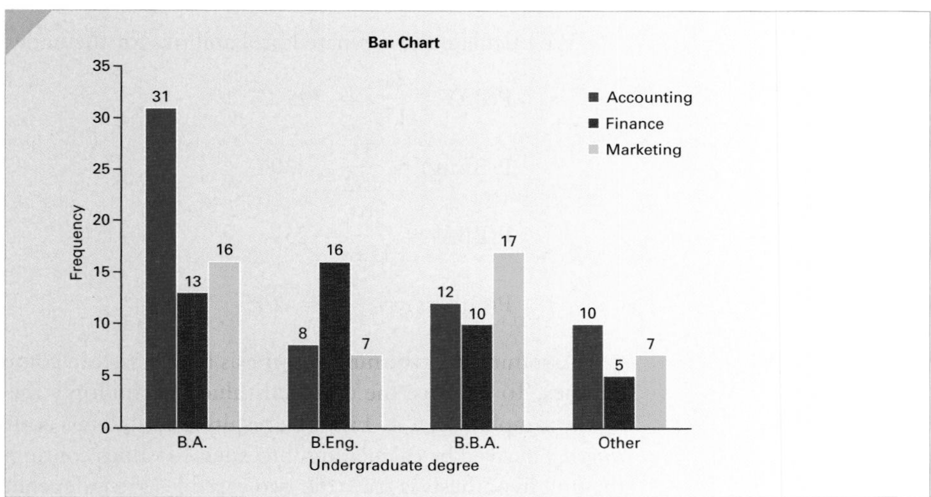

The bar chart displays the data from the sample. It does appear that there is a relationship between the two nominal variables in the sample. However, to draw inferences about the population of MBA students, we need to apply an inferential technique.

Test Statistic

The test statistic is the same as the one used to test proportions in the goodness-of-fit-test. That is, the test statistic is

$$\chi^2 = \sum_{i=1}^{k} \frac{(f_i - e_i)^2}{e_i}$$

where k is the number of cells in the cross-classification table. If you examine the null hypothesis described in the goodness-of-fit test and the one described above, you will discover a major difference. In the goodness-of-fit test, the null hypothesis lists values for the probabilities p_i. The null hypothesis for the chi-squared test of a contingency table only states that the two variables are independent. However, we need the probabilities to compute the expected values e_i, which in turn are needed to calculate the value of the test statistic. (The entries in the table are the observed values f_i.) The question immediately arises, From where do we get the probabilities? The answer is that they must come from the data after we assume that the null hypothesis is true.

In Chapter 6 we introduced independent events and showed that if two events A and B are independent, the joint probability P(A and B) is equal to the product of P(A) and P(B). That is,

$$P(\text{A and B}) = P(\text{A}) \times P(\text{B})$$

The events in this example are the values each of the two nominal variables can assume. Unfortunately we do not have the probabilities of A and B. However, these probabilities can be estimated from the data. Using relative frequencies, we calculate the estimated probabilities for the MBA major.

$$P(\text{Accounting}) = \frac{61}{152} = .401$$

$$P(\text{Finance}) = \frac{44}{152} = .289$$

$$P(\text{Marketing}) = \frac{47}{152} = .309$$

We calculate the estimated probabilities for the undergraduate degree.

$$P(\text{BA}) = \frac{60}{152} = .395$$

$$P(\text{BEng}) = \frac{31}{152} = .204$$

$$P(\text{BBA}) = \frac{39}{152} = .257$$

$$P(\text{Others}) = \frac{22}{152} = .145$$

Assuming that the null hypothesis is true, we can compute the estimated joint probabilities. To produce the expected values, we multiply the estimated joint probabilities by the sample size, $n = 152$. The results are listed in a **contingency table,** the word *contingency* derived by calculating the expected values contingent upon the assumption that the null hypothesis is true (the two variables are independent).

Undergraduate Degree	MBA Major: Accounting	Finance	Marketing	Total
B.A.	$152 \times \frac{60}{152} \times \frac{61}{152} = 24.08$	$152 \times \frac{60}{152} \times \frac{44}{152} = 17.37$	$152 \times \frac{60}{152} \times \frac{47}{152} = 18.55$	60
B.Eng.	$152 \times \frac{31}{152} \times \frac{61}{152} = 12.44$	$152 \times \frac{31}{152} \times \frac{44}{152} = 8.97$	$152 \times \frac{31}{152} \times \frac{47}{152} = 9.59$	31
B.B.A.	$152 \times \frac{39}{152} \times \frac{61}{152} = 15.65$	$152 \times \frac{39}{152} \times \frac{44}{152} = 11.29$	$152 \times \frac{39}{152} \times \frac{47}{152} = 12.06$	39
Others	$152 \times \frac{22}{152} \times \frac{61}{152} = 8.83$	$152 \times \frac{22}{152} \times \frac{44}{152} = 6.37$	$152 \times \frac{22}{152} \times \frac{47}{152} = 6.80$	22
Total	61	44	47	152

As you can see, the expected value for each cell is computed by multiplying the row total by the column total and dividing by the sample size. For example, the B.A. and Accounting cell expected value is

$$152 \times \frac{60}{152} \times \frac{61}{152} = \frac{60 \times 61}{152} = 24.08$$

All the other expected values would be determined similarly.

Expected Frequencies for a Contingency Table

The expected frequency of the cell in row i and column j is

$$e_{ij} = \frac{\text{Row } i \text{ total} \times \text{Column } j \text{ total}}{\text{Sample size}}$$

The expected cell frequencies are shown in parentheses in the following table. As in the case of the goodness-of-fit test, the expected cell frequencies should satisfy the rule of five.

	MBA Major		
Undergraduate Degree	Accounting	Finance	Marketing
B.A.	31 (24.08)	13 (17.37)	16 (18.55)
B.Eng.	8 (12.44)	16 (8.97)	7 (9.59)
B.B.A.	12 (15.65)	10 (11.29)	17 (12.06)
Other	10 (8.83)	5 (6.37)	7 (6.80)

We can now calculate the value of the test statistic:

$$\begin{aligned}
\chi^2 = \sum_{i=1}^{k} \frac{(f_i - e_i)^2}{e_i} &= \frac{(31 - 24.08)^2}{24.08} + \frac{(13 - 17.37)^2}{17.37} + \frac{(16 - 18.55)^2}{18.55} \\
&+ \frac{(8 - 12.44)^2}{12.44} + \frac{(16 - 8.97)^2}{8.97} + \frac{(7 - 9.59)^2}{9.59} + \frac{(12 - 15.65)^2}{15.65} \\
&+ \frac{(10 - 11.29)^2}{11.29} + \frac{(17 - 12.06)^2}{12.06} + \frac{(10 - 8.33)^2}{8.33} \\
&+ \frac{(5 - 6.37)^2}{6.37} + \frac{(7 - 6.80)^2}{6.80} \\
&= 14.70
\end{aligned}$$

Notice that we continue to use a single subscript in the formula of the test statistic when we should use two subscripts, one for the rows and one for the columns. We believe that it is clear that, for each cell, we must calculate the squared difference between the observed and expected frequencies divided by the expected frequency. We don't believe that the satisfaction of using the mathematically correct notation overcomes the unnecessary complication.

Rejection Region and *p*-Value

To determine the rejection region, we must know the number of degrees of freedom associated with the chi-squared statistic. The number of degrees of freedom for a contingency table with r rows and c columns is $\nu = (r - 1)(c - 1)$. For this example, the number of degrees of freedom is $v = (r - 1)(c - 1) = (4 - 1)(3 - 1) = 6$.

If we employ a 5% significance level, the rejection region is

$$\chi^2 > \chi^2_{\alpha,\nu} = \chi^2_{.05,6} = 12.6$$

Because $\chi^2 = 14.70$, we reject the null hypothesis and conclude that there is evidence of a relationship between undergraduate degree and MBA major.

The p-value of the test statistic is

$$P(\chi^2 > 14.70)$$

Unfortunately, we cannot determine the p-value manually.

Using the Computer

Excel and Minitab can produce the chi-squared statistic either from a cross-classification table whose frequencies have already been calculated or from raw data. The respective printouts are almost identical.

File Xm15-02 contains the raw data using the following codes:

Column1 (Undergraduate Degree)	Column 2 (MBA Major)
1 = B.A.	1 = Accounting
2 = B.Eng.	2 = Finance
3 = B.B.A.	3 = Marketing
4 = Other	

EXCEL

	A	B	C	D	E	F
1	Contingency Table					
2						
3		*Degree*				
4	*MBA Major*		1	2	3	TOTAL
5		1	31	13	16	60
6		2	8	16	7	31
7		3	12	10	17	39
8		4	10	5	7	22
9		TOTAL	61	44	47	152
10						
11						
12		chi-squared Stat			14.70	
13		df			6	
14		p-value			0.0227	
15		chi-squared Critical			12.5916	

INSTRUCTIONS (RAW DATA)

1. Type or import the data into two adjacent columns. (Open Xm15-02.) The codes must be positive integers greater than 0.
2. Click **Add-Ins, Data Analysis Plus,** and **Contingency Table (Raw Data).**
3. Specify the **Input Range** (A1:B153) and specify the value of α (.05).

INSTRUCTIONS (COMPLETED TABLE)

1. Type the frequencies into adjacent columns.
2. Click **Add-Ins, Data Analysis Plus,** and **Contingency Table.**
3. Specify the **Input Range.** Click **Labels** if the first row and first column of the input range contain the names of the categories. Specify the value for α.

MINITAB

Tabulated statistics: Degree, MBA Major

Rows: Degree Columns: MBA Major

	1	2	3	All
1	31	13	16	60
2	8	16	7	31
3	12	10	17	39
4	10	5	7	22
All	61	44	47	152

Cell Contents: Count

Pearson Chi-Square = 14.702, DF = 6, P-Value = 0.023
Likelihood Ratio Chi-Square = 13.781, DF = 6, P-Value = 0.032

INSTRUCTIONS (RAW DATA)

1. Type or import the data into two columns. (Open Xm15-02.)
2. Click **Stat, Tables,** and **Cross Tabulation and Chi-Square . . . .**
3. In the **Categorical variables** box, select or type the variables **For rows** (Degree) and **For columns** (MBA Major). Click **Chi-Square . . . .**
4. Under **Display,** click **Chi-Square analysis.** Specify **Chi-Square analysis.**

INSTRUCTIONS (COMPLETED TABLE)

1. Type the observed frequencies into adjacent columns.
2. Click **Stat, Tables,** and **Chi-Square Test (Table in Worksheet) . . . .**
3. Select or type the names of the variables representing the columns.

INTERPRET

There is strong evidence to infer that the undergraduate degree and MBA major are related. This suggests that the dean can predict the number of optional courses by counting the number of MBA students with each type of undergraduate degree. We can see that B.A.'s favor marketing courses, B.Eng.'s prefer finance, B.B.A.'s drift to accounting, and others show no particular preference.

If the null hypothesis is true, undergraduate degree and MBA major are independent of one another. This means that whether an MBA student earned a B.A., B.Eng., B.B.A., or other degree does not affect his or her choice of major program in the MBA. Consequently, there is no difference in major choice among the graduates of the undergraduate programs. If the alternative hypothesis is true, undergraduate degree does affect the choice of MBA major. Thus, there are differences between the four undergraduate degree categories.

Rule of Five

In the previous section, we pointed out that the expected values should be at least 5 to ensure that the chi-squared distribution provides an adequate approximation of the sampling distribution. In a contingency table where one or more cells have expected values of less than 5, we need to combine rows or columns to satisfy the rule of five. This subject is discussed in CD Appendix W.

Exit Polls in Ohio: Solution

© David Butow/Corbis Saba

For each of these tables, we conduct the chi-squared test of a contingency table. The Excel printouts appear here. (Minitab's printouts are similar.)

	A	B	C	D
1	**Contingency Table**			
2				
3		*Gore*	*Bush*	TOTAL
4	*Men*	109	138	247
5	*Women*	152	112	264
6	TOTAL	261	250	511
7				
8	chi-squared Stat			9.23
9	df			1
10	p-value			0.0024
11	chi-squared Critical			3.8415

	A	B	C	D
1	**Contingency Table**			
2				
3		*Gore*	*Bush*	TOTAL
4	*18-29*	44	120	164
5	*30-44*	86	128	214
6	*45-59*	73	128	201
7	*60+*	61	123	184
8	TOTAL	264	499	763
9				
10	chi-squared Stat			7.76
11	df			3
12	p-value			0.0512
13	chi-squared Critical			7.8147

	A	B	C	D
1	**Contingency Table**			
2				
3		*Gore*	*Bush*	TOTAL
4	*No HS*	16	99	115
5	*High School*	55	128	183
6	*Some college*	78	133	211
7	*College/University degree*	59	133	192
8	*Post-graduate degree*	51	115	166
9	TOTAL	259	608	867
10				
11	chi-squared Stat			19.18
12	df			4
13	p-value			0.0007
14	chi-squared Critical			9.4877

INTERPRET

There is enough evidence at the 5% significance level to infer that gender and education affect the way people voted in the presidential election of 2000. However, there is only weak evidence to infer that age and presidential vote are related.

Here is a summary of the factors that tell us when to apply the chi-squared test of a contingency table. Note that there are two problem objectives satisfied by this statistical procedure.

Factors That Identify the Chi-Squared Test of a Contingency Table

1. **Problem objectives:** Analyze the relationship between two variables and compare two or more populations
2. **Data type:** Nominal

EXERCISES

Developing an Understanding of Statistical Concepts

15.17 Conduct a test to determine whether the two classifications L and M are independent, using the data in the accompanying cross classification table. (Use $\alpha = .05$.)

	M_1	M_2
L_1	28	68
L_2	56	36

15.18 Repeat Exercise 15.17 using the following table:

	M_1	M_2
L_1	14	34
L_2	28	18

15.19 Repeat Exercise 15.17 using the following table:

	M_1	M_2
L_1	7	17
L_2	14	9

15.20 Review the results of Exercises 15.17–15.19. What is the effect of decreasing the sample size?

15.21 Conduct a test to determine whether the two classifications *R* and *C* are independent, using the data in the accompanying cross classification table. (Use $\alpha = .10$.)

	C_1	C_1	C_3
R_1	40	32	48
R_2	30	48	52

Applications

Use a 5% significance level unless specified otherwise.

15.22 The trustee of a company's pension plan has solicited the opinions of a sample of the company's employees about a proposed revision of the plan. A breakdown of the responses is shown in the accompanying table. Is there enough evidence to infer that the responses differ between the three groups of employees?

Responses	Blue-Collar Workers	White-Collar Workers	Managers
For	67	32	11
Against	63	18	9

15.23 The operations manager of a company that manufactures shirts wants to determine whether there are differences in the quality of workmanship among the three daily shifts. She randomly selects 600 recently made shirts and carefully inspects them. Each shirt is classified as either perfect or flawed, and the shift that produced it is also recorded. The accompanying table summarizes the number of shirts that fell into each cell. Do these data provide sufficient evidence to infer that there are differences in quality between the three shifts?

	Shift		
Shirt Condition	1	2	3
Perfect	240	191	139
Flawed	10	9	11

15.24 One of the issues that came up in a recent national election (and is likely to arise in many future elections) is how to deal with a sluggish economy. Specifically, should governments cut spending, raise taxes, inflate the economy (by printing more money), or do none of the above and let the deficit rise? And as with most other issues, politicians need to know which parts of the electorate support these options. Suppose that a random sample of 1,000 people was asked which option they support and their political affiliations. The possible responses to the question about political affiliation were Democrat, Republican, and Independent (which included a variety of political persuasions). The responses are summarized in the accompanying table. Do these results allow us to conclude at the 1% significance level that political affiliation affects support for the economic options?

Economic	Political Affiliation		
Options	Democrat	Republican	Independent
Cut spending	101	282	61
Raise taxes	38	67	25
Inflate the economy	131	88	31
Let deficit increase	61	90	25

15.25 Econetics Research Corporation, a well-known Montreal-based consulting firm, wants to test how it can influence the proportion of questionnaires returned from surveys. In the belief that the inclusion of an inducement to respond may be important, the firm sends out 1,000 questionnaires: Two hundred promise to send respondents a summary of the survey results, 300 indicate that 20 respondents (selected by lottery) will be awarded gifts, and 500 are accompanied by no inducements. Of these, 80 questionnaires promising a summary, 100 questionnaires offering gifts, and 120 questionnaires offering no inducements are returned. What can you conclude from these results?

The following exercises require the use of a computer and software. The answers may be calculated manually. See Appendix A for the sample statistics. ***Use a 5% significance level unless specified otherwise.***

15.26 Xm02-10 (Example 2.10 revisited) In a major North American city there are four competing newspapers: the *Globe and Mail* (*G&M*), *Post*, *Sun*, and *Star*. To help design advertising campaigns, the advertising managers of the newspapers need to know which segments of the newspaper market are reading their papers. A survey was conducted to analyze the relationship between newspapers read and occupation. A sample of newspaper readers was asked to report which newspaper they read: *Globe and Mail* (1) *Post* (2), *Star* (3), *Sun* (4), and to indicate whether they were a blue-collar worker (1), white-collar worker (2), or professional (3). Can we infer that occupation and newspaper are related?

15.27 Xr15-27 An investor who can correctly forecast the direction and size of changes in foreign currency exchange rates is able to reap huge profits in the international currency markets. A knowledgeable reader of *The Wall Street Journal* (in particular, of the currency futures market quotations) can determine the direction of change in various exchange rates that is predicted by all investors, viewed collectively. Predictions from 216 investors, together with the subsequent actual directions of change, were recorded in the following way: Column 1: predicted change where 1 = positive and 2 = negative; column 2: actual change where 1 = positive and 2 = negative.

a. Can we infer at the 10% significance level that a relationship exists between the predicted and actual directions of change?
b. To what extent would you make use of these predictions in formulating your forecasts of future exchange rate changes?

15.28 Xr02-80 (Exercise 2.80 revisited) Is there brand loyalty among car owners in their purchases of gasoline? To help answer the question, a random sample of car owners was asked to record the brand of gasoline in their last two purchases: 1 = Exxon, 2 = Amoco, 3 = Texaco, 4 = Other. Can we conclude that there is brand loyalty in gasoline purchases?

15.29 Xr15-29 During the past decade, many cigarette smokers have attempted to quit. Unfortunately, nicotine is highly addictive. Smokers use a large number of different methods to help them quit. These include nicotine patches, hypnosis, and various forms of therapy. A researcher for the Addiction Research Council wanted to determine why some people quit while others attempted to quit successfully but failed. He surveyed 1,000 people who planned to quit smoking. He determined their educational level and whether they continued to smoke 1 year later. Educational level was recorded in the following way:

1 = Did not finish high school
2 = High school graduate
3 = University or college graduate
4 = Completed a postgraduate degree

A continuing smoker was recorded as 1; a quitter was recorded as 2. Can we infer that the amount of education is a factor in determining whether a smoker will quit?

15.30 Xr15-30 Because television audiences of newscasts tend to be older (and because older people suffer from a variety of medical ailments), pharmaceutical companies' advertising often appears on national news on the three networks (ABC, CBS, and NBC). To determine how effective the ads are, a survey was undertaken. Adults over 50 were asked about their primary sources of news. The responses are

1. ABC news 2. CBS news 3. NBC news
4. Newspapers 5. Radio 6. None of the above

Each person was also asked whether they suffer from heartburn, and if so, what remedy they take. The answers were recorded as follows:

1. Do not suffer from heartburn
2. Suffer from heartburn but take no remedy
3. Suffer from heartburn and take an over-the-counter-remedy (Tums, Gavoscol, . . .)
4. Suffer from heartburn and take a prescription pill (Nexium, . . .)

Is there a relationship between an adult's source of news and his or her heartburn condition?

15.31 Xr02-79 (Exercise 2.79 revisited.) The associate dean of a business school was looking for ways to improve the quality of the applicants to its MBA program. In particular, she wanted to know whether the undergraduate degree of applicants differed among her school and the three nearby universities with MBA programs. She sampled 100 applicants of her program and an equal number from each of the other universities. She recorded their undergraduate degree (1 = B.A.,

2 = B.Eng., 3 = B.B.A., 4 = other) as well the university (codes 1, 2, 3, and 4). Do these data provide sufficient evidence to infer that undergraduate degree and the university each person applied are related?

15.32 Xr15-52 The relationship between drug companies and medical researchers is under scrutiny because of possible conflict of interest. The issue that started the controversy was a 1995 case control study that suggested that the use of calcium-channel blockers to treat hypertension led to an increase risk of heart disease. This led to an intense debate both in technical journals and in the press. Researchers writing in the *New England Journal of Medicine* ("Conflict of Interest in the Debate over Calcium Channel Antagonists," January 8, 1998, p. 101) looked at the 70 reports that appeared during 1996–1997, classifying them as favorable, neutral, or critical toward the drugs. The researchers then contacted the authors of the reports and questioned them about financial ties to drug companies. The results were recorded in the following way:

Column 1: Results of the scientific study; 1 = favorable, 2 = neutral, 3 = critical

Column 2: 1 = financial ties to drug companies, 2 = no ties to drug companies

Do these data allow us to infer that the research findings for calcium-channel blockers are affected by whether the research is funded by drug companies?

15.33 Xr15-33 After a thorough analysis of the market, a publisher of business and economics statistics books has divided the market into three general approaches to teach applied statistics. These are (1) use of a computer and statistical software with no manual calculations; (2) traditional teaching of concepts and solution of problems by hand; (3) mathematical approach with emphasis on derivations and proofs. The publisher wanted to know whether this market could be segmented on the basis of the educational background of the instructor. As a result, the statistics editor organized a survey that asked 195 professors of business and economics statistics to report their approach to teaching and which one of the following categories represents their highest degree:

1. Business (MBA or Ph.D. in business)
2. Economics
3. Mathematics or engineering
4. Other

a. Can the editor infer that there are differences in type of degree among the three teaching approaches? If so, how can the editor use this information?

b. Suppose that you work in the marketing department of a textbook publisher. Prepare a report for the editor that describes this analysis.

15.3 Summary of Tests on Nominal Data

At this point in the textbook, we've described four tests that are used when the data are nominal:

z-test of p (Section 12.3)

z-test of $p_1 - p_2$ (Section 13.5)

Chi-squared goodness-of-fit test (Section 15.1)

Chi-squared test of a contingency table (Section 15.2)

In the process of presenting these techniques, it was necessary to concentrate on one technique at a time and focus on the kinds of problems each addresses. However, this approach tends to conflict somewhat with our promised goal of emphasizing the "when" of statistical inference. In this section, we summarize the statistical tests on nominal data to ensure that you are capable of selecting the correct method.

There are two critical factors in identifying the technique used when the data are nominal. The first, of course, is the problem objective. The second is the number of categories that the nominal variable can assume. Table 15.1 provides a guide to help select the correct technique.

TABLE 15.1 Statistical Techniques for Nominal Data

Problem Objective	Number of Categories	Statistical Technique
Describe a population	2	z-test of p or the chi-squared goodness-of-fit test
Describe a population	More than 2	Chi-squared goodness-of-fit test
Compare two populations	2	z-test of $p_1 - p_2$ or Chi-squared test of a contingency table
Compare two populations	More than 2	Chi-squared test of a contingency table
Compare two or more populations	2 or more	Chi-squared test of a contingency table
Analyze the relationship between two variables	2 or more	Chi-squared test of a contingency table

Notice that when we describe a population of nominal data with exactly two categories, we can use either of two techniques. We can employ the z-test of p or the chi-squared goodness-of-fit test. These two tests are equivalent because if there are only two categories, the multinomial experiment is actually a binomial experiment (one of the categorical outcomes is labeled success and the other is labeled failure). Mathematical statisticians have established that if we square the value of z, the test statistic for the test of p, we produce the χ^2-statistic. That is, $z^2 = \chi^2$. Thus, if we want to conduct a two-tail test of a population proportion, we can employ either technique. However, the chi-squared goodness-of-fit test can test only to determine whether the hypothesized values of p_1 (which we can label p) and p_2, (which we call $1 - p$) are not equal to their specified values. Consequently, to perform a one-tail test of a population proportion, we must use the z-test of p. (This issue was discussed in Chapter 14 when we pointed out that we can use either the t-test of $\mu_1 - \mu_2$ or the analysis of variance to conduct a test to determine whether two population means differ.)

When we test for differences between two populations of nominal data with two categories, we can also use either of two techniques: the z-test of $p_1 - p_2$ (Case 1) or the chi-squared test of a contingency table. Once again, we can use either technique to perform a two-tail test about $p_1 - p_2$. (Squaring the value of the z-statistic yields the value of χ^2-statistic.) However, one-tail tests must be conducted by the z-test of $p_1 - p_2$. The rest of the table is quite straightforward. Notice that when we want to compare two populations when there are more than two categories, we use the chi-squared test of a contingency table.

Figure 15.4 offers another summary of the tests that deal with nominal data introduced in this book. There are two groups of tests: those that test hypotheses about single populations and those that test either for differences or for independence. In the first set, we have the z-test of p, which can be replaced by the chi-squared test of a multinomial experiment. The latter test is employed when there are more than two categories.

To test for differences between two proportions, we apply the z-test of $p_1 - p_2$. Instead, we can use the chi-squared test of a contingency table, which can be applied to a variety of other problems.

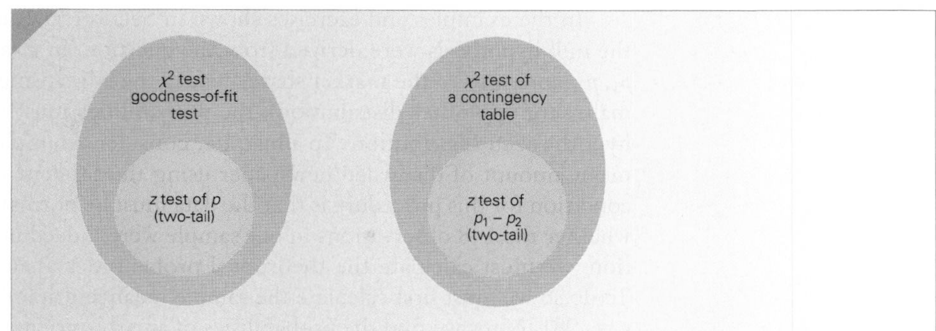

FIGURE **15.4**
Tests on Nominal Data

Data Format

As you have already seen, we need the observed frequencies in order to calculate the chi-squared test statistic either manually or using a computer. Both Excel and Minitab can convert raw data (where the data are the codes representing the categories) into the frequencies or directly into the test statistic. For the chi-squared good-of-fit test, the data must be in one column. For the chi-squared test of a contingency table, the data must be in two columns where one column stores the codes for one variable and the second stores the codes for the second variable. It may be necessary to stack the data in order to calculate the chi-squared test statistic. For example, suppose that we wish to compare three populations where the samples are stored in three columns. To use Excel or Minitab to calculate the test statistic, we need to stack the data so that one column contains the codes for the variable of interest (for example, marital status) and the second contains the codes representing the samples. CD Appendix R shows how to stack data.

Notice how different this is from the data format when we draw inferences about the difference between two proportions. To compute the z-test and estimator of $p_1 - p_2$, the data must be stored in two columns where one column represents the codes for one sample and the second column contains the codes for the second sample. If necessary you may have to unstack the data.

Developing an Understanding of Statistical Concepts

Table 15.1 and Figure 15.4 summarize how we deal with nominal data. We determine the frequency of each category and use these frequencies to compute test statistics. We can then compute proportions to calculate z-statistics or use the frequencies to calculate χ^2-statistics. Because squaring a standard normal random variable produces a chi-squared variable, we can employ either statistic to test for differences. As a consequence, when you encounter nominal data in the problems described in this book (and other introductory applied statistics books), the most logical starting point in selecting the appropriate technique will be either a z-statistic or a χ^2-statistic. However, you should know that there are other statistical procedures that can be applied to nominal data, techniques that are not included in this book.

15.4 (Optional) Chi-Squared Test for Normality

We can use the goodness-of-fit test presented in Section 15.2 in another way. We can test to determine whether data were drawn from any distribution. The most common application of this procedure is a test of normality.

In the examples and exercises shown in Section 15.1, the probabilities specified in the null hypothesis were derived from the question. In Example 15.1, the probabilities p_1, p_2, and p_3 were the market shares before the advertising campaign. To test for normality (or any other distribution), the probabilities must first be calculated using the hypothesized distribution. To illustrate, consider Example 12.1 where we tested the mean amount of discarded newspaper using the Student t distribution. The required condition for this procedure is that the data must be normally distributed. To determine whether the 148 observations in our sample were indeed taken from a normal distribution, we must calculate the theoretical probabilities assuming a normal distribution. To do so we must first calculate the sample mean and standard deviation: $\bar{x} = 2.18$ and $s = .981$. Next, we find the probabilities of an arbitrary number of intervals. For example, we can find the probabilities of the following intervals:

Interval 1: $X \le .709$

Interval 2: $.709 < X \le 1.69$

Interval 3: $1.69 < X \le 2.67$

Interval 4: $2.67 < X \le 3.65$

Interval 5: $X > 3.65$

We will discuss the reasons for our choices of intervals later.

The probabilities are computed using the normal distribution and the values of $\bar{x}$ and s as estimators of μ and σ. We calculated the sample mean and standard deviation as $\bar{x} = 2.18$ and $s = .981$. Thus,

$$P(X \le .709) = P\left(\frac{X-\mu}{\sigma} \le \frac{.709 - 2.18}{.981}\right) = P(Z \le -1.5) = .0668$$

$$P(.709 < X \le 1.69) = P\left(\frac{.709 - 2.18}{.981} < \frac{X-\mu}{\sigma} \le \frac{1.69 - 2.18}{.981}\right)$$
$$= P(-1.5 < Z \le -.5) = .2417$$

$$P(1.69 < X \le 2.67) = P\left(\frac{1.69 - 2.18}{.981} < \frac{X-\mu}{\sigma} \le \frac{2.67 - 2.18}{.981}\right)$$
$$= P(-.5 < Z \le .5) = .3829$$

$$P(2.67 < X \le 3.65) = P\left(\frac{2.67 - 2.18}{.981} < \frac{X-\mu}{\sigma} \le \frac{3.65 - 2.18}{.981}\right)$$
$$= P(.5 < Z \le 1.5) = .2417$$

$$P(X > 3.65) = P\left(\frac{X-\mu}{\sigma} > \frac{3.65 - 2.18}{.981}\right) = P(Z > 1.5) = .0668$$

To test for normality is to test the following hypotheses:

H_0: $p_1 = .0668, p_2 = .2417, p_3 = .3829, p_4 = .2417, p_5 = .0668$

H_1: At least two proportions differ from their specified values

We complete the test as we did in Section 15.1, except that the number of degrees of freedom associated with the chi-squared statistic is the number of intervals minus 1 minus the number of parameters estimated, which in this illustration is two. (We estimated the population mean μ and the population standard deviation σ.) Thus, in this case the number of degrees of freedom is $5 - 1 - 2 = 2$.

The expected values are

$$e_1 = np_1 = 148(.0668) = 9.89$$
$$e_2 = np_2 = 148(.2417) = 35.78$$
$$e_3 = np_3 = 148(.3829) = 56.67$$
$$e_4 = np_4 = 148(.2417) = 35.78$$
$$e_5 = np_5 = 148(.0668) = 9.89$$

The observed values are determined manually by counting the number of values in each interval. Thus,

$$f_1 = 10$$
$$f_2 = 36$$
$$f_3 = 54$$
$$f_4 = 39$$
$$f_5 = 9$$

The chi-squared statistic is

$$\chi^2 = \sum_{i=1}^{k} \frac{(f_i - e_i)^2}{e_i} = \frac{(10 - 9.89)^2}{9.89} + \frac{(36 - 35.78)^2}{35.78} + \frac{(54 - 56.67)^2}{56.67} + \frac{(39 - 35.78)^2}{35.78} + \frac{(9 - 9.89)^2}{9.89} = .50$$

The rejection region is

$$\chi^2 > \chi^2_{\alpha,k-3} = \chi^2_{.05,2} = 5.99$$

There is not enough evidence to conclude that these data are not normally distributed.

Class Intervals

In practice you can use any intervals you like. We chose the intervals we did to facilitate the calculation of the normal probabilities. The number of intervals was chosen to comply with the rule of five, which requires that all expected values be at least equal to 5. Because the number of degrees of freedom is $k - 3$, the minimum number of intervals is $k = 4$.

Using the Computer

EXCEL

	A	B	C	D
1	Chi-Squared Test of Normality			
2				
3		*Newspaper*		
4	Mean	2.18		
5	Standard deviation	0.981		
6	Observations	148		
7				
8	Intervals	Probability	Expected	Observed
9	(z <= -1.5)	0.0668	9.89	10
10	(-1.5 < z <= -0.5)	0.2417	35.78	36
11	(-0.5 < z <= 1.5)	0.3829	56.67	54
12	(0.5 < z <=1.5)	0.2417	35.78	39
13	(z > 1.5)	0.0668	9.89	9
14				
15				
16	chi-squared Stat	0.50		
17	df	2		
18	p-value	0.7792		
19	chi-squared Critical	5.9915		

We programmed Excel to calculate the value of the test statistic so that the expected values are at least 5 (where possible) and the minimum number of intervals is 4. Hence if the number of observations is more than 220, the intervals and probabilities are

Interval	Probability
$Z \le -2$	.0228
$-2 < Z \le -1$	.1359
$-1 < Z \le 0$	.3413
$0 < Z \le 1$	.3413
$1 < Z \le 2$	.1359
$Z > 2$	.0228

If the sample size is less than or equal to 220 and greater than 80, the intervals are

Interval	Probability
$Z \le -1.5$	.0668
$-1.5 < Z \le -0.5$	.2417
$-0.5 < Z \le 0.5$	.3829
$0.5 < Z \le 1.5$	.2417
$Z > 1.5$	.0668

If the sample size is less than or equal to 80, we employ the minimum number of intervals 4. When the sample size is less than 32, at least one expected value will be less than 5. The intervals are

Interval	Probability
$Z \le -1$	.1587
$-1 < Z \le 0$	.3413
$0 < Z \le 1$	.3413
$Z > 1$	.1587

INSTRUCTIONS

1. Type or import the data into one column. (Open Xm12-01.)
2. Click **Add-Ins, Data Analysis Plus,** and **Chi-Squared Test of Normality.**
3. Specify the **Input Range** (A1:A149) and the value of α (.05).

MINITAB

Minitab does not conduct this procedure. However, you can use Minitab to perform several parts of the statistical procedure. See CD Appendix V.

Interpreting the Results of a Chi-Squared Test for Normality

In the example above we found that there was little evidence to conclude that the weight of discarded newspaper is not normally distributed. However, had we found evidence of nonnormality, this would not necessarily invalidate the t-test we conducted in Example 12.1. As we pointed out in Chapter 12, the t-test of a mean is a robust procedure, which means that only if the variable is extremely nonnormal and the sample size is small, is the conclusion of the technique suspect. The problem here is that if the sample size is large and the variable is only slightly nonnormal, the chi-squared test for normality will, in many cases, conclude that the variable is not normally distributed. However, if the variable is even quite nonnormal and the sample size is large, the t-test will still be valid. Although there are situations where we need to know whether a variable is nonnormal, we continue to advocate that the way to decide if the normality requirement for almost all statistical techniques applied to interval data is satisfied is to draw histograms and look for shapes that are far from bell shaped (e.g., highly skewed or bimodal). We will use this approach in Chapter 19 where we introduce nonparametric techniques that are used when interval data are nonnormal.

EXERCISES

15.34 Suppose that a random sample of 100 observations was drawn from a population. After calculating the mean and standard deviation, each observation was standardized and the number of observations in each of the following intervals was counted. Can we infer at the 5% significance level that the data were not drawn from a normal population?

Interval	Frequency
$Z \le -1.5$	10
$-1.5 < Z \le -0.5$	18
$-0.5 < Z \le 0.5$	48
$0.5 < Z \le 1.5$	16
$Z > 1.5$	8

15.35 A random sample of 50 observations yielded the following frequencies for the standardized intervals:

Interval	Frequency
$Z \le -1$	6
$-1 < Z \le 0$	27
$0 < Z \le 1$	14
$Z > 1$	3

Can we infer that the data are not normal? (Use $\alpha = .10$.)

The following exercises require the use of a computer and software.

15.36 Xr12-29 Refer to Exercise 12.29. Test at the 10% significance level to determine whether the amount of time spent working at part-time jobs is normally distributed. If there is evidence of nonnormality, is the t-test invalid?

15.37 Xr12-35 The t-test in Exercise 12.35 requires that the costs of prescriptions is normally distributed. Conduct a test with $\alpha = .05$ to determine whether the required condition is unsatisfied. If there is enough evidence to conclude that the requirement is not satisfied, does this indicate that the t-test is invalid?

15.38 Xr13-19 Exercise 13.19 required you to conduct a t-test of the difference between two means. Each sample's productivity data are required to be normally distributed. Is that required condition violated? Test with $\alpha = .05$.

15.39 Xr13-20 Exercise 13.20 asked you to conduct a t-test of the difference between two means (reaction times). Test to determine whether there is enough evidence to infer that the reaction times are not normally distributed. A 5% significance level is judged to be suitable.

15.40 Xr13-47 In Exercise 13.47, you performed a test of the mean matched pairs difference. The test result depends on the requirement that the differences are normally distributed. Test with a 10% significance level to determine whether the requirement is violated.

CHAPTER SUMMARY

This chapter introduced three statistical techniques. The first is the chi-squared goodness-of-fit test, which is applied when the problem objective is to describe a single population of nominal data with two or more categories. The second is the chi-squared test of a contingency table. This test has two objectives: to analyze the relationship between two nominal variables and to compare two or more populations of nominal data. The last procedure is designed to test for normality.

IMPORTANT TERMS

Multinomial experiment 581
Chi-squared goodness-of-fit test 582
Expected frequency 582
Observed frequencies 583
Cross-classification table 588
Chi-squared test of a contingency table 588
Contingency table 520

SYMBOLS

Symbol	Pronounced	Represents
f_i	*f-sub-i*	Frequency of the *i*th category
e_i	*e-sub-i*	Expected value of the *i*th category
χ^2	*Chi-squared*	Test statistic

FORMULA

Test statistic for all procedures

$$\chi^2 = \sum_{i=1}^{k} \frac{(f_i - e_i)^2}{e_i}$$

COMPUTER OUTPUT AND INSTRUCTIONS

Technique	Excel	Minitab
Chi-squared goodness-of-fit test	584	585
Chi-squared test of a contingency table (raw data)	592	593
Chi-squared test of a contingency table	592	593
Chi-squared test of normality	602	

CHAPTER EXERCISES

Use a 5% significance level unless specified otherwise.

15.41 An organization dedicated to ensuring fairness in television game shows is investigating *Wheel of Fortune*. In this show, three contestants are required to solve puzzles by selecting letters. Each contestant gets to select the first letter and continues selecting until he or she chooses a letter that is not in the hidden word, phrase, or name. The order of contestants is random. However, contestant 1 gets to start game 1, contestant 2 starts game 2, and so on. The contestant who wins the most money is declared the winner and he or she is given an opportunity to win a grand prize. Usually, more than three games are played per show, and as a result it appears that contestant 1 has an advantage: contestant 1 will start two games, whereas contestant 3 will usually start only one game. To see whether this is the case, a random sample of 30 shows was taken and the starting position of the winning contestant for each show was recorded. These are shown in the following table:

Starting position	1	2	3
Number of wins	14	10	6

Do the tabulated results allow us to conclude that the game is unfair?

15.42 It has been estimated that employee absenteeism costs North American companies more than $100 billion per year. As a first step in addressing the rising cost of absenteeism, the personnel department of a large corporation recorded the weekdays during which individuals in a sample of 362 absentees were away over the past several months. Do these data suggest that absenteeism is higher on some days of the week than on others?

Day of the Week	Monday	Tuesday	Wednesday	Thursday	Friday
Number absent	87	62	71	68	74

15.43 Suppose that the personnel department in Exercise 15.42 continued its investigation by categorizing absentees according to the shift on which they worked, as shown in the accompanying table. Is there sufficient evidence at the 10% significance level of a relationship between the days on which employees are absent and the shift on which the employees work?

Shift	Monday	Tuesday	Wednesday	Thursday	Friday
Day	52	28	37	31	33
Evening	35	34	34	37	41

15.44 A management behavior analyst has been studying the relationship between male/female supervisory structures in the workplace and the level of employees' job satisfaction. The results of a recent survey are shown in the accompanying table. Is there sufficient evidence to infer that the level of job satisfaction depends on the boss/employee gender relationship?

	Boss/Employee			
Level of Satisfaction	Female/ Male	Female/ Female	Male/ Male	Male/ Female
Satisfied	21	25	54	71
Neutral	39	49	50	38
Dissatisfied	31	48	10	11

The following exercises require the use of a computer and software. The answers may be calculated manually. See Appendix A for the sample statistics. ***Use a 5% significance level unless specified otherwise.***

15.45 Xr15-45 Stress is a serious medical problem that costs businesses and government billions of dollars annually. As a result, it is important to determine the causes and possible cures. It would be helpful to know whether the causes are universal or do they vary from country to country. In a survey, American and Canadian adults were asked to report the primary source of stress in their lives. The responses are

1. Job 2. Finances 3. Health
4. Family life 5. Other

The data were recorded using the codes above plus 1 = American and 2 = Canadian. Do these data provide sufficient evidence to conclude that Americans and Canadians differ in their sources of stress?

15.46 Xr15-46 According to NBC News (March 11, 1994) more than 3,000 Americans quit smoking each day. (Unfortunately, more than 3,000 Americans start smoking each day.) Because nicotine is one of the most addictive drugs, quitting smoking is a difficult and frustrating task. It usually takes several tries before success is achieved. There are various methods, including cold turkey, nicotine patch, hypnosis, and group therapy sessions. In an experiment to determine how these methods differ, a random sample of smokers who have decided to quit is selected. Each smoker has chosen one of the methods listed above. After one year, the respondents report whether they have quit (1 = yes and 2 = no) and which method they used (1 = cold turkey; 2 = nicotine patch; 3 = hypnosis; 4 = group therapy sessions). Is there sufficient evidence to conclude that the four methods differ in their success?

15.47 Xr15-47 A newspaper publisher, trying to pinpoint his market's characteristics, wondered whether the way people read a newspaper is related to the reader's educational level. A survey asked adult readers which section of the paper they read first and asked to report their highest educational level. These data were recorded (Column 1 = first section read where 1 = front page, 2 = sports, 3 = editorial, and 4 = other) and column 2 = educational level where 1 = did not complete high school, 2 = high school graduate, 3 = university or college graduate, and 4 = postgraduate degree). What do these data tell the publisher about how educational level affects the way adults read the newspaper?

15.48 Xr15-48 Every week the Florida Lottery draws six numbers between 1 and 49. Lottery ticket buyers are naturally interested in whether certain numbers are drawn more frequently than others. To assist players, the *Sun-Sentinel* publishes the number of times each of the 49 numbers has been drawn in the past 52 weeks. The numbers and the frequency with which each occurred were recorded. (These data are from the Sunday, January 5, 1997, edition.)

a. If the numbers are drawn from a uniform distribution, what is the expected frequency for each number?
b. Can we infer that the data were not generated from a uniform distribution?

15.49 Xr15-49 Canadians have the option of investing income in registered retirement savings plans (RRSPs). Subject to limits calculated on the basis of income, employer retirement plans, and previous RRSPs, money invested in RRSPs is not taxable. (Money withdrawn from retirement plans is taxable.) Critics argue that RRSPs are a tax loophole for the rich because only wealthier people are in a position to take advantage of the tax provisions. In a study to determine who uses RRSPs, the Caledon Institute of Social Policy randomly sampled a variety of Canadians in different tax brackets. (Survey results were published in the *Globe and Mail*, February 5, 1994.) For each respondent, the researchers recorded the tax bracket (1 = less than $20,000; 2 = $20,000– $39,999; 3 = $40,000– $59,999; 4 = $60,000– $100,000; 5 = over $100,000), and whether they invested in an RRSP this year (2 = yes; 1 = no). Can we infer from these data that there are differences in RRSP positions among the income groups?

In Section 15.4 we showed how to test for normality. However, we can use the same process to test for any other distribution.

15.50 Xr15-50 A scientist believes that the gender of a child is a binomial random variable with probability = .5 for a boy and .5 for a girl. To help test her belief, she randomly samples 100 families with five children. She records the number of boys. Can the scientist infer that the number of boys in families with five children is not a binomial random variable with $p = .5$? (*Hint:* Find the probability of $X = 0, 1, 2, 3, 4,$ and 5 from a binomial distribution with $n = 5$ and $p = .5$.

15.51 Xr15-51 Given the high cost of medical care, research that points the way to avoid illness is welcome. Previously performed research tells us that stress affects the immune system. Two scientists at Carnegie Mellon Hospital in Pittsburgh asked 114 healthy adults about their social circles; they were asked to list every group they had contact with at least once every 2 weeks—family, co-workers, neighbors, friends, religious, and community groups. Participants also reported negative life events over the past year, events such as death of a friend or relative, divorce, or job-related problems. The participants were divided into four groups:

Group 1: Highly social and highly stressed
Group 2: Not highly social and highly stressed
Group 3: Highly social and not highly stressed
Group 4: Not highly social and not highly stressed

Each individual was classified in this way. In addition, whether each person contracted a cold over the next 12 weeks was recorded (1 = cold, 2 = no cold). Can we infer that there are differences between the four groups in terms of contracting a cold?

The following exercises employ data files associated with examples and exercises seen earlier in this book.

15.52 Xr12-76* Exercise 12.76 described the problem of a looming shortage of professors, possibly made worse by professors desiring to retire before the age of 65. A survey asked a random sample of professors whether he or she intended to retire before 65. The responses are "no" (1) "yes" (2). In addition, the survey asked in which faculty did the professor belong (1 = Arts, 2 = Science, 3 = Business, 4 = Engineering, 5 = other). Do these provide sufficient evidence to infer that whether a professor wishes to retire is related to the faculty?

15.53 Xr12-81* Refer to Exercise 12.81. Determine whether there is enough evidence to infer that there are differences in the choice of Christmas tree between the three age categories.

15.54 Xr12-82* Exercise 12.82 described a study to determine whether viewers (older than 50) of the network news had contacted their physician to ask about one of the prescription drugs advertised during the newscast. The responses (1 = no and 2 = yes) were recorded. Also recorded were which of the three networks they normally watch (1 = ABC, 2 = CBS, 3 = NBC). Can we conclude that there are differences in responses between the three network news shows?

15.55 Xr12-86* In addition to asking a sample of Toronto newspaper readers whether they would subscribe to a new proposed newspaper (1 = no and 2 = yes) in Exercise 12.86, the statistics practitioner asked which of the current papers they currently read (1 = *Globe and Mail*, 2 = *Star*, 3 = *Sun*). Do these data allow the statistics practitioner to infer that the response to subscribing to the new paper differs between the readers of the current newspapers?

15.56 Xr13-86* Exercise 13.86 described a survey of adults wherein on the basis of several probing questions each was classified as either a member of the health conscious group (code = 1) or not (code = 2) and whether he or she buys Special X (1 = no, 2 = yes). Additionally, his or her educational attainment (1 = did not finish high school, 2 = finished high school, 3 = finished college or university, 4 = postgraduate degree) was recorded.

a. Do the data allow the surveyor to conclude that there differences in educational attainment between those who do and those who do not belong to the health conscious group?
b. Can we infer that there is a relationship between the four educational groups and whether or not a person buys Special X?

15.57 Xm12-00* The Chapter 12 opening example describes a survey of 5,000 television viewers at 9:00 to 9:30 P.M. on Sunday, April 1, 2007, where each respondent was classified in one of the following categories:

Network	Show	Code
ABC	*Desperate Housewives*	1
CBS	*The Amazing Race 11*	2
NBC	*Deal or No Deal*	3
Fox	*Family Guy*	4
Television turned off or watched some other channel		5

Respondents were also asked to indicate their household income group (1 = under \$25,000, 2 = \$25,000 to \$49,999, 3 = \$50,000 to \$75,000, 4 = over \$75,000). Do the data provide enough evidence to conclude that what people watch at 9:00 P.M. is related to income?

15.58 Xm12-05* Example 12.5 described exit polls wherein people are asked whether they voted for the Democrat or Republican candidate for president. The surveyors also record gender (1 = female, 2 = male), educational attainment (1 = did not finish high school, 2 = completed high school, 3 = completed college or university, 4 = postgraduate degree), and income level (1 = under \$25,000, 2 = \$25,000 to \$49,999, 3 = \$50,000 to \$75,000), 4 = over \$75,000).

a. Is there sufficient evidence to infer that voting and gender are related?
b. Do the data allow the conclusion that voting and educational level are related?
c. Can we infer that voting and income are related?

APPLICATIONS in MARKETING

© Richard Brown/ Taxi/Getty Images

Market Segmentation

In Section 12.4 and in Chapters 13 and 14 we described how marketing managers use statistical analyses to estimate the size of market segments and to determine whether there are differences between segments.

The following exercises require the application of the chi-squared test of a contingency table to determine whether market segments differ with respect to some nominal variable.

15.59 Xr12-92* Exercise 12.92 described the market segments defined by JC Penney. Another question included in the questionnaire that classified the women surveyed was asked whether each worked outside the home. The responses were

1. No
2. Part-time job
3. Full-time job

These data plus the classifications (1 = conservative, 2 = traditional, and 3 = contemporary) were recorded. Can we infer from these data that there are differences in employment status between the three market segments?

15.60 Xr12-92* Refer to Exercise 12.92. The women in the survey were also asked to define value by identifying what they considered to be the most important attribute of value. The responses are

1. Price
2. Quality
3. Fashion

The responses and the classifications of segments (1 = conservative, 2 = traditional, and 3 = contemporary) were recorded. Do these data allow us to infer that there are differences in the definition of value between the three market segments?

15.61 Xm12-06* Refer to Example 12.6. In segmenting the breakfast cereal market, a food manufacturer uses health and diet consciousness as the segmentation variable. Four segments are developed:

1. Concerned about eating healthy foods
2. Concerned primarily about weight
3. Concerned about health because of illness
4. Unconcerned

A survey was undertaken and each person was asked how often they ate a healthy breakfast (defined as cereal and/or fruit). The responses are

1. Never
2. Seldom
3. Often
4. Always

The responses and the market segments of each respondent were recorded. Can we infer that there are differences in frequency of healthy breakfasts between the market segments?

CASE 15.1 Predicting the Outcomes of Basketball, Baseball, Football, and Hockey Games from Intermediate Results*

© Dave Sanford/ Allsport/Getty Images

DATA [C15-01]

Some basketball fans generally believe that it doesn't pay to watch an entire game because the outcome is determined in the last few minutes (some say the last 2 minutes) of the game. Is this really true, and, if so, is basketball different in this respect from other professional sports played in North America? For example, is it true that the team that leads a baseball game after seven innings almost always wins the game? To address these questions, three researchers tracked basketball, baseball, football, and hockey games. The results (whether the early-game leader and whether the late-game leader won) of games during the 1990 season (for baseball and football) and during the 1990–1991 season (for basketball and hockey) were recorded. Early-game leaders are defined as the teams that are ahead after one quarter of basketball and football, one period of hockey, or three innings of baseball. Late-game leaders are defined as the teams that are ahead after three quarters of basketball and football, two periods of hockey, or seven innings of baseball.

The data were recorded in the following way:

Column 1: Results of games where 2 = early-game leader wins and 1 = early-game leader loses

Column 2: Early-game leader game where 1 = basketball, 2 = baseball, 3 = football, and 4 = hockey games

Column 3: Results of games where 2 = late-game leader wins and 1 = late-game leader loses

Column 4: Late-game leader game where 1 = basketball, 2 = baseball, 3 = football, and 4 = hockey games

Can we infer from these data that all four professional sports experience the same proportion of early-game leaders winning the game?

Can we infer from these data that all four professional sports experience the same proportion of late-game leaders winning the game?

*Adapted from H. Cooper, K. M. DeNeve, and E. Mosteller, "Predicting Professional Sports Game Outcomes from Intermediate Game Scores," *Chance* 5, No. 3–4 (1992): 18–22.

APPENDIX 15 / REVIEW OF CHAPTERS 12 TO 15

Here are the updated list of statistical techniques (Table A15.1) and the flowchart (Figure A15.1) for Chapters 12 to 15. Counting the two techniques of chi-squared tests introduced here (we do not include the chi-squared test for normality), we have covered 25 statistical methods.

TABLE **A15.1** **Summary of Statistical Techniques in Chapters 12 to 15**

t-test of μ
Estimator of μ (including small population estimator of μ and large and small population estimators of $N\mu$)
χ^2-test of σ^2
Estimator of σ^2
z-test of p
Estimator of p (including small population estimator of p and large and small population estimators of Np)
Equal-variances t-test of $\mu_1 - \mu_2$
Equal-variances estimator of $\mu_1 - \mu_2$
Unequal-variances t-test of $\mu_1 - \mu_2$
Unequal-variances estimator of $\mu_1 - \mu_2$
t-test of μ_D
Estimator of μ_D
F-test of σ_1^2/σ_2^2
Estimator of σ_1^2/σ_2^2
z-test of $p_1 - p_2$ (Case 1)
z-test of $p_1 - p_2$ (Case 2)
Estimator of $p_1 - p_2$
One-way analysis of variance (including multiple comparisons)
Two-way (randomized blocks) analysis of variance
Two-factor analysis of variance
χ^2-goodness-of-fit test
χ^2-test of a contingency table

FIGURE **A15.1** Flowchart of Techniques in Chapters 12 to 15

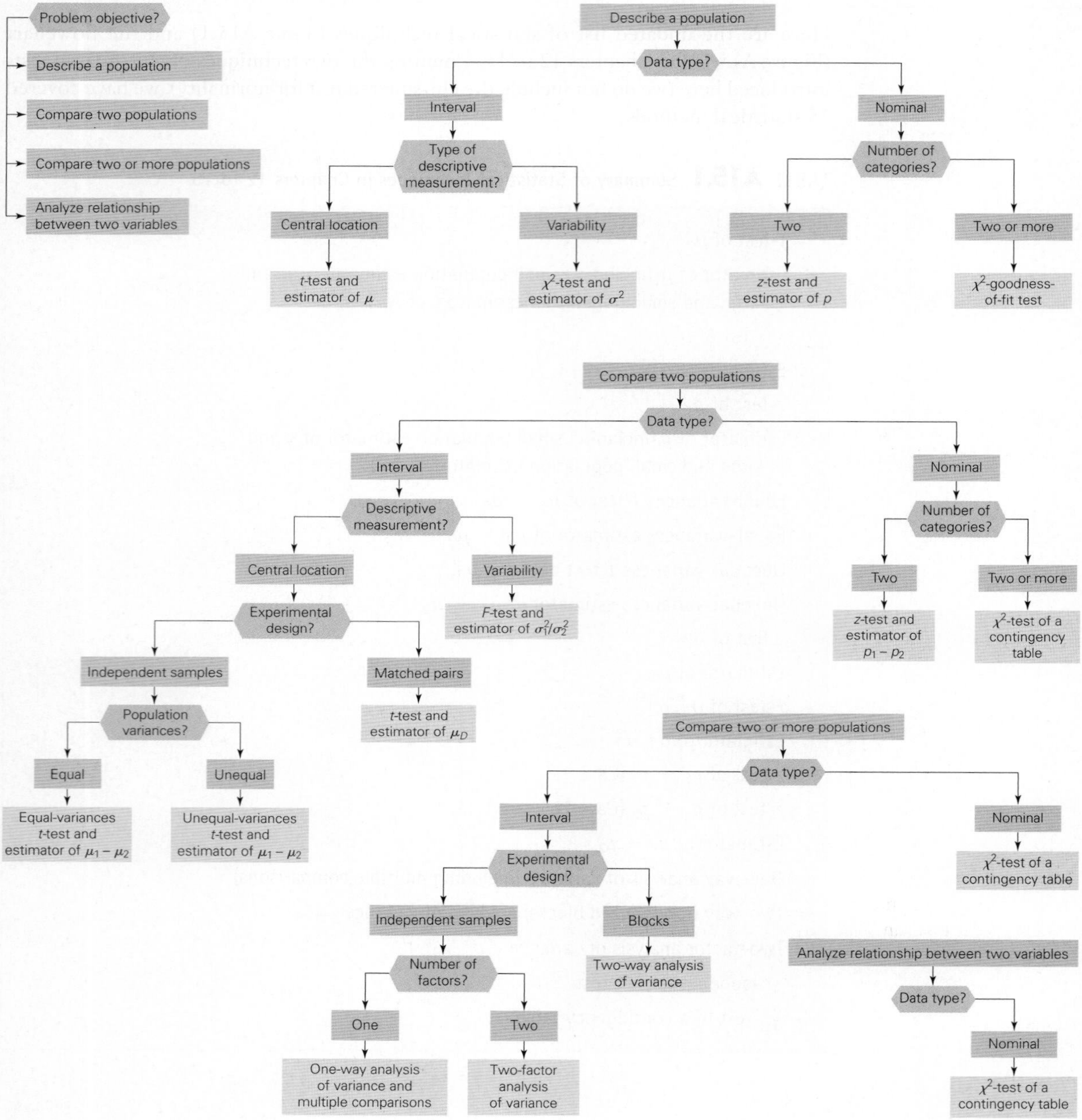

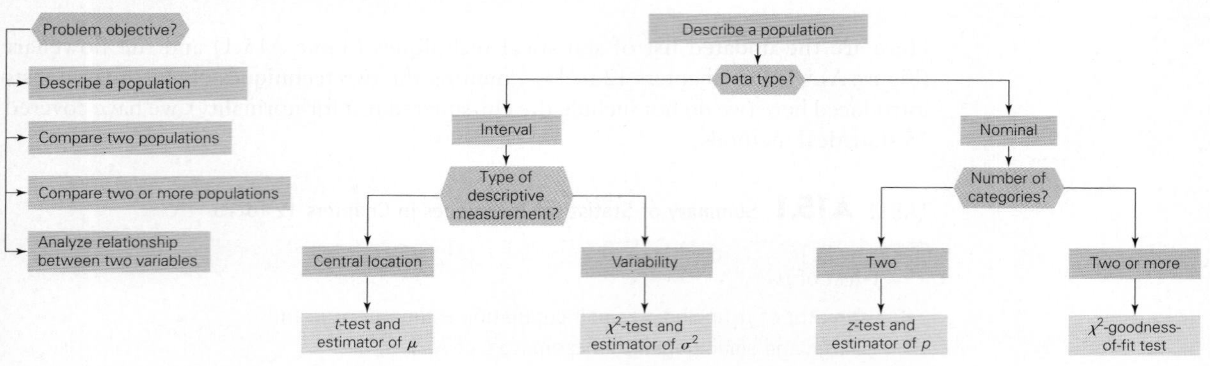

EXERCISES

We remind you that we do not specify significance levels in the exercise that follow. Choose your own.

A15.1 XrA15-01 An analysis of the applicants of all MBA programs in North America reveals that the proportions of each type of undergraduate degree are as follows:

Undergraduate Degree	Proportion
B.A. (1)	50%
B.B.A. (2)	20%
B.Sc. (3)	15%
B.Eng. (4)	10%
Other (5)	5%

The director of Wilfrid Laurier University's (WLU) MBA program recorded the undergraduate degree of the applicants for this year using the codes in parentheses. Do these data indicate that applicants to WLU's MBA program are different in terms of their undergraduate degrees from the population of MBA applicants?

A15.2 XrA15-02 The experiment to determine the effect of taking a preparatory course to improve SAT scores in Exercise A13.16 was criticized by other statisticians. They argued that the first test would provide a valuable learning experience that would produce a higher test score from the second exam even without the preparatory course. Consequently, another experiment was performed. Forty students wrote the SAT without taking any preparatory course. At the next scheduled exam (3 months later), these same students took the exam again (again with no preparatory course). The scores for both exams were recorded in columns 1 (first test scores) and 2 (second test scores). Can we infer that repeating the SAT produces higher exam scores even without the preparatory course?

A15.3 XrA15-03 How does dieting affect the brain? This question was addressed by researchers in Australia. The experiment used 40 middle-age women in Adelaide, Australia, half of whom were on a diet and the other half of whom were not. (*National Post*, December 1, 2003). The mental arithmetic part of the experiment required the participants to add two three-digit numbers. The amount of time taken to solve the 48 problems was recorded. The participants were given another test that required them to repeat a string of five letters they had been told 10 seconds earlier. They were asked to repeat the test with five words told to them 10 seconds earlier. The data were recorded in the following way:

Column 1: Identification number

Column 2: 1 = dieting, 2 = not dieting

Column 3: Time to solve 48 problems (seconds)

Column 4: Repeat string of 5 letters (1 = no, 2 = yes)

Column 5: Repeat string of 5 words (1 = no, 2 = yes)

Is there sufficient evidence to infer that dieting adversely affects the brain?

A15.4 XrA15-04 A small but important part of a university library's budget is the amount collected in fines on overdue books. Last year, a library collected $75,652.75 in fine payments; however, the head librarian suspects that some employees are not bothering to collect the fines on overdue books. In an effort to learn more about the situation, she asked a sample of 400 students (out of a total student population of 50,000) how many books they had returned late to the library in the previous 12 months. They were also asked how many days overdue the books had been. The results indicated that the total number of days overdue ranged from 0 to 55 days. The number of days overdue was recorded.

a. Estimate with 95% confidence the average number of days overdue for all 50,000 students at the university.

b. If the fine is 25 cents per day, estimate the amount that should be collected annually. Should the librarian conclude that not all the fines were collected?

A15.5 XrA15-05 An apple juice manufacturer has developed a new product—a liquid concentrate that, when mixed with water, produces 1 liter of apple juice. The product has several attractive features. First, it is more convenient than bottled apple juice, which is the way apple juice is currently sold. Second, because the apple juice that is sold in bottles is actually made from concentrate, the quality of the new product is at least as high as that of bottled apple juice. Third, the cost of the new product is slightly lower than that of bottled apple juice. The marketing manager has to decide how to market the new product. She can create advertising that emphasizes convenience, quality, or price. To facilitate a decision, she conducts an experiment in three different small cities. In one city, she launches the product with advertising stressing the convenience of the liquid concentrate (e.g., easy to carry from store to home and takes up less room in the freezer). In the second city, the advertisements emphasize the quality of the product ("average" shoppers are depicted discussing how good the apple juice tastes). Advertising that highlights the relatively low cost of the liquid concentrate is used in the third city. The number of packages sold weekly is recorded for the 20 weeks following the beginning of the campaign. The marketing manager wants to know whether differences in sales exist between the three advertising strategies. (We will assume that except for the type of advertising, the three cities are identical.)

A15.6 XrA15-06 Mutual funds are a popular way of investing in the stock market. A financial analyst wanted to determine the effect income had on ownership of mutual funds and whether the relationship had changed from 4 years earlier. She took a random sample of adults 25 years of age and older and asked each person whether he or she owned mutual funds (No = 1 and Yes = 2) and to report the annual household income. The categories are

1. Less than $25,000
2. $25,000 to $34,999
3. $35,000 to $49,999
4. $50,000 to $74,999
5. $75,000 to $100,000
6. Over $100,000

Can we infer from the data that household income and ownership of mutual funds are related? (Adapted from the *Statistical Abstract of the United States*, 2006, Table 1200)

A15.7 XrA15-07 Refer to Exercise A15.5. Suppose that in addition to varying the marketing strategy, the manufacturer also decided to advertise in one of the two media that are available: television and newspapers. As a consequence, the experiment was repeated in the following way. Six different small cities were selected. In city 1, the marketing emphasized convenience, and all the advertising was conducted on television. In city 2, marketing also emphasized convenience, but all the advertising was conducted in the daily newspaper. Quality was emphasized in cities 3 and 4. City 3 learned about the product from television commercials, and city 4 saw newspaper advertising. Price was the marketing emphasis in cities 5 and 6. City 5 saw television commercials, and city 6 saw newspaper advertisements. In each city, the weekly sales for each of 10 weeks were recorded. What conclusions can be drawn from these data?

A15.8 XrA15-08 After a recent study, researchers reported on the effects of folic acid on the occurrence of spina bifida—a birth defect in which there is incomplete formation of the spine. A sample of 2,000 women who gave birth to children with spina bifida and who were planning another pregnancy was recruited. Before attempting to get pregnant again, half the sample was given regular doses of folic acid, and the other half was given a placebo. After 18 months, researchers recorded the result for each woman: 1 = birth to normal baby, 2 = birth to baby with spina bifida, 3 = not pregnant or no baby yet delivered. Can we infer that folic acid reduces the incidence of spina bifida in newborn babies?

A15.9 XrA15-09 Slow play of golfers is a serious problem for golf clubs. Slow play results in fewer rounds of golf and less profits for public course owners. To examine this problem, a random sample of British and American golf courses was selected. The amount of time taken (in minutes) was recorded for a random sample of British and American golfers. Can we conclude that British golfers play golf in less time than do American golfers? (*Source: Golf Magazine*, July 2001)

A15.10 XrA15-10 The United States and Canada (among others) are countries in which a significant proportion of citizens are immigrants. Many arrive in North America with few assets but quickly adapt to a changed economic environment. The question often arises, How quickly do immigrants increase their standard of living? A study initiated by Statistics Canada surveyed three different types of families:

1. Immigrants who arrived before 1976
2. Immigrants who came to Canada after 1986
3. Canadian-born families

The survey measured family wealth, which includes houses, cars, income, and savings, and recorded the results ($1,000s). Can we infer that differences exist between the three groups? If so, what are those differences?

A15.11 XrA15-11 During the decade of the 1980s, professional baseball thrived in North America. However, in the 1990s attendance dropped and the number of television viewers also decreased. To examine the popularity of baseball relative to other sports, surveys were performed. In 1985 and again in 1992, a Harris Poll asked a random sample of 500 people to name their favorite sport. The results, which were published in *The Wall Street Journal* (July 6, 1993), were recorded in the following way: favorite sport (1 = professional football, 2 = baseball, 3 = professional basketball, 4 = college basketball, 5 = college football, 6 = golf, 7 = auto racing, 8 = tennis, and 9 = other); year (1 = 1985, 2 = 1992). Do these results indicate that North Americans changed their favorite sport between 1985 and 1992?

A15.12 XrA15-12 In an attempt to learn more about traffic congestion in a large North American city, the number of cars passing through intersections was determined. (*National Post*, October 18, 2006). The number of cars was counted in 5-minute samples throughout several days. The counts for one busy intersection were recorded. Estimate with 95% confidence the mean number of cars in 5-minutes. Use this estimate to estimate the counts for a 24-hour day.

A15.13 XrA15-13 Organizations that sponsor various leisure activities need to know the number of people who wish to participate. Bureaucrats need to know the number because many organizations apply for government grants to pay the costs. The U.S. National Endowment for the Arts conducts surveys of American adults to acquire this type of information. One part of the survey asked a random sample of adults whether they participated in exercise programs. The responses (1 = yes and 2 = no) were recorded. A recent census reveals that there are 205.9 million adults in the United States. Estimate with 95% confidence the number of American adults who participate in exercise programs. (Adapted from the *Statistical Abstract of the United States*, 2006, Table 1227)

A15.14 XrA15-14 Low back pain is a common medical problem that sometimes results in disability and absence from work. Any method of treatment that decreases absence would be welcome by individuals and insurance companies. A randomized control study (published in *Annals of Internal Medicine*, January 2004) was undertaken to determine whether an alternate form of treatment is effective. The study examined 134 workers who were absent from work because of low back pain. Half the sample was assigned to graded activity, a physical exercise program designed to stimulate rapid return to work. The other half was assigned to the usual care, which involves mostly rest. For each worker the number of days absent from work because of low back pain in the following 6 months was recorded. Do these data provide sufficient evidence to infer that the graded activity is effective?

A15.15 XrA15-15 Clinical depression is a serious and sometimes debilitating disease. It is often treated by antidepressants such as Prozac and Zoloft. Recent studies may indicate another possible remedy. Researchers took a random sample of people who are clinically depressed and divided them into three groups. The first group was treated with antidepressants and light therapy, the second was treated with a placebo and light therapy and the third group treated with a placebo. Whether the patient showed improvement (code = 1) or not (code = 2) and the group number were recorded. Can we infer that there are differences between the three groups?

A15.16 How well do airlines keep to their schedules. To help answer this question, an economist conducted a survey of 780 takeoffs in the United States and determined that 77.4% of them departed on time (defined as a departure that is within 15 minutes of its scheduled time). There were 7,140,596 flight departures in the United States in 2005. Estimate with 95% confidence the total number of on-time departures.

CASE A15.1 Which Diets Work?

DATA CA15-01

Every year millions of people start new diets. There is a bewildering array of diets to choose from. The question for many people is which ones work. Researchers at Tufts University in Boston made an attempt to point dieters in the right direction. Four diets were used:

1. Atkins low carbohydrate diet
2. Zone high-protein moderate carbohydrate diet
3. Weight Watchers diet
4. Dr. Ornish's low-fat diet

The study recruited 160 overweight people and randomly assigned 40 to each diet. The average weight before dieting was 220 pounds and all needed to lose between 30 and 80 pounds. All volunteers agreed to follow their diets for 2 months. No exercise or regular meetings were required. The following variables were recorded for each dieter using the format shown here:

Column 1: Identification number
Column 2: Diet
Column 3: Percent weight loss
Column 4: percent low-density lipoprotein (LDL)—"bad" cholesterol decrease
Column 5: percent high-density lipoprotein (HDL)—"good" cholesterol increase
Column 6: Quit after 2 months?
 1 = yes, 2 = no
Column 7: Quit after 1 year?
 1 = yes, 2 = no

Is there enough evidence to conclude that there are differences between the diets with respect to

a. percent weight loss?
b. percent LDL decrease?
c. percent HDL increase?
d. proportion quitting within 2 months?
e. proportion quitting after 1 year?

© Mark Joseph/Stone/Getty Images

16

SIMPLE LINEAR REGRESSION AND CORRELATION

Foreign Index Funds*

DATA Xm16-00

Most Americans who invest in the stock market buy stocks that are listed on the New York Stock Exchange or the NASDAQ. Canadians tend to invest in the Toronto Stock Exchange as well as the two American exchanges. However, restricting equity purchases in these ways likely limits potential profits. It also increases risk because investors are not taking advantage of a potential source of diversification (see Section 7.3).

A certain investor prefers index mutual funds, which are constructed by buying a wide assortment of stocks so that the fund more or less mirrors the entire exchange. Thus, for example, if an investor believes that the NASDAQ will increase rapidly over the next 2 years but is not confident that he can pick winners among individual stocks, he may instead buy a NASDAQ index fund.

© Terryvine/Stone/Getty Images

For the answer, turn to page 645.

*The author is grateful to Ariel Aminof for gathering the data. The data were adapted from Morgan Stanley Dean Witter.

The investor has determined that a foreign index fund is beneficial to him if it is weakly correlated with an American index fund that he owns. He examines a Japanese index constructed by Morgan Stanley Dean Witter, a well-respected financial Institution in the United States. The monthly returns on his U.S. index and the Japanese index were computed over a 59-month period. He decides that if there is evidence of a linear relationship between the returns on the U.S. and Japanese indexes, he will not buy the Japanese index. What should he do?

INTRODUCTION

Regression analysis is used to predict the value of one variable on the basis of other variables. This technique may be the most commonly used statistical procedure because, as you can easily appreciate, almost all companies and government institutions forecast variables such as product demand interest rates, inflation rates, prices of raw materials, and labor costs.

The technique involves developing a mathematical equation or model that describes the relationship between the variable to be forecast, which is called the **dependent variable,** and variables that the statistics practitioner believes are related to the dependent variable. The dependent variable is denoted Y, whereas the related variables are called **independent variables** and are denoted $X_1, X_2, \ldots, X_k$ (where k is the number of independent variables).

If we are interested only in determining whether a relationship exists, we employ correlation analysis, a technique that we have already introduced. In Chapter 2, we presented the graphical method to describe the association between two interval variables—the scatter diagram. We introduced the coefficient of correlation and covariance in Chapter 4.

Because regression analysis involves a number of new techniques and concepts, we divided the presentation into three chapters. In this chapter, we present techniques that allow us to determine the relationship between only two variables. In Chapter 17, we expand our discussion to more than two variables, and in Chapter 18, we discuss how to build regression models.

Here are three illustrations of the use of regression analysis.

Illustration 1 The product manager in charge of a particular brand of children's breakfast cereal would like to predict the demand for the cereal during the next year. To use regression analysis, she and her staff list the following variables as likely to affect sales:

Price of the product

Number of children 5 to 12 years of age (the target market)

Price of competitors' products

Effectiveness of advertising (as measured by advertising exposure)

Annual sales this year

Annual sales in previous years

Illustration 2 A gold speculator is considering a major purchase of gold bullion. He would like to forecast the price of gold 2 years from now (his planning horizon), using regression analysis. In preparation, he produces the following list of independent variables:

Interest rates

Inflation rate

Price of oil

Demand for gold jewelry

Demand for industrial and commercial gold

Dow Jones Industrial Average

Illustration 3 A real estate agent wants to predict the selling price of houses more accurately. She believes that the following variables affect the price of a house:

Size of the house (number of square feet)

Number of bedrooms

Frontage of the lot

Condition

Location

In each of these illustrations, the primary motive for using regression analysis is forecasting. Nonetheless, analyzing the relationship among variables can also be quite useful in managerial decision making. For instance, in the first application, the product manager may want to know how price is related to product demand so that a decision about a prospective change in pricing can be made.

Regardless of why regression analysis is performed, the next step in the technique is to develop a mathematical equation or model that accurately describes the nature of the relationship that exists between the dependent variable and the independent variables. This stage—which is only a small part of the total process—is described in the next section. In the ensuing sections of this chapter (and in Chapter 17), we will spend considerable time assessing and testing how well the model fits the actual data. Only when we're satisfied with the model do we use it to estimate and forecast.

16.1 MODEL

The job of developing a mathematical equation can be quite complex, because we need to have some idea about the nature of the relationship between each of the independent variables and the dependent variable. The number of different mathematical models that could be proposed is virtually infinite. Here is an example from Chapter 4.

$$\text{Profit} = (\text{Price per unit} - \text{variable cost per unit}) \times (\text{Number of units sold}) - \text{Fixed costs}$$

You may encounter the next example in a finance course:

$$F = P(1 + i)^n$$

where F = future value of an investment

P = principle or present value

i = interest rate per period

n = number of periods

These are all examples of **deterministic models,** so named because such equations allow us to determine the value of the dependent variable (on the left side of the equation) from the values of the independent variables. In many practical applications of interest to us, deterministic models are unrealistic. For example, is it reasonable to believe that we can determine the selling price of a house solely on the basis of its size? Unquestionably, the size of a house affects its price, but many other variables (some of which may not be measurable) also influence price. What must be included in most practical models is a method to represent the randomness that is part of a real-life process. Such a model is called a **probabilistic model.**

To create a probabilistic model, we start with a deterministic model that approximates the relationship we want to model. We then add a term that measures the random error of the deterministic component.

Suppose that in Illustration 3, the real estate agent knows that the cost of building a new house is about $100 per square foot and that most lots sell for about $100,000. The approximate selling price would be

$$y = 100{,}000 + 100x$$

where $y =$ selling price and $x =$ size of the house in square feet. A house of 2,000 square feet would therefore be estimated to sell for

$$y = 100{,}000 + 100(2{,}000) = 300{,}000$$

We know, however, that the selling price is not likely to be exactly $300,000. Prices may actually range from $200,000 to $400,000. In other words, the deterministic model is not really suitable. To represent this situation properly, we should use the probabilistic model

$$y = 100{,}000 + 100x + \varepsilon$$

where ε (the Greek letter epsilon) represents the **error variable**—the difference between the actual selling price and the estimated price based on the size of the house. The error thus accounts for all the variables, measurable and immeasurable, that are not part of the model. The value of ε will vary from one sale to the next, even if x remains constant. That is, houses of exactly the same size will sell for different prices because of differences in location and number of bedrooms and bathrooms, as well as other variables.

In the three chapters devoted to regression analysis, we will present only probabilistic models. In this chapter we describe only the straight-line model with one independent variable. This model is called the **first-order linear model**—sometimes called the **simple linear regression model.***

First-Order Linear Model

$$y = \beta_0 + \beta_1 x + \varepsilon$$

where

$y =$ dependent variable

$x =$ independent variable

$\beta_0 = y$-intercept

$\beta_1 =$ slope of the line (defined as rise/run)

$\varepsilon =$ error variable

The problem objective addressed by the model is to analyze the relationship between two variables, x and y, both of which must be interval. To define the relationship between x and y, we need to know the value of the coefficients β_0 and β_1. However, these coefficients are population parameters, which are almost always unknown. In the next section, we discuss how these parameters are estimated.

---

*We use the term *linear* in two ways. The "linear" in linear regression refers to the form of the model wherein the terms form a linear combination of the coefficients β_0 and β_1. Thus, for example, the model $y = \beta_0 + \beta_1 x^2 + \varepsilon$ is a linear combination whereas $y = \beta_0 + \beta_1^2 x + \varepsilon$ is not. The simple linear regression model $y = \beta_0 + \beta_1 x + \varepsilon$ describes a straight-line or linear relationship between the dependent variable and one independent variable. In this book we use the linear regression technique only. Hence, when we use the word *linear* we will be referring to the straight-line relationship between the variables.

16.2 ESTIMATING THE COEFFICIENTS

We estimate the parameters β_0 and β_1 in a way similar to the methods used to estimate all the other parameters discussed in this book. We draw a random sample from the population of interest and calculate the sample statistics we need. However, because β_0 and β_1 represent the coefficients of a straight line, their estimators are based on drawing a straight line through the sample data. The straight line that we wish to use to estimate β_0 and β_1 is the "best" straight line, best in the sense that it comes closest to the sample data points. This best straight line, called the least squares line, is derived from calculus and is represented by the following equation:

$$\hat{y} = b_0 + b_1 x$$

Here b_0 is the y-intercept, b_1 is the slope, and $\hat{y}$ is the predicted or fitted value of y. In Chapter 4 we introduced the **least squares method,** which produces a straight line that minimizes the sum of the squared differences between the points and the line. The coefficients b_0 and b_1 are calculated so that the sum of squared deviations

$$\sum_{i=1}^{n} (y_i - \hat{y}_i)^2$$

is minimized. That is, the values of $\hat{y}$ on average come closest to the observed values of y. The calculus derivation is available in CD Appendix X, which shows how the following formulas, first shown in Chapter 4, were produced.

Least Squares Line Coefficients

$$b_1 = \frac{s_{xy}}{s_x^2}$$

$$b_0 = \bar{y} - b_1 \bar{x}$$

where

$$s_{xy} = \frac{\sum_{i=1}^{n} (x_i - \bar{x})(y_i - \bar{y})}{n - 1}$$

$$s_x^2 = \frac{\sum_{i=1}^{n} (x_i - \bar{x})^2}{n - 1}$$

$$\bar{x} = \frac{\sum_{i=1}^{n} x_i}{n}$$

$$\bar{y} = \frac{\sum_{i=1}^{n} y_i}{n}$$

In Chapter 4 we provided shortcut formulas for the sample variance (page 109) and the sample covariance (page 124). Combining them provides a shortcut method to manually calculate the slope coefficient.

Shortcut Formula for b_1

$$b_1 = \frac{s_{xy}}{s_x^2}$$

$$s_{xy} = \frac{1}{n-1}\left[\sum_{i=1}^{n} x_i y_i - \frac{\sum_{i=1}^{n} x_i \sum_{i=1}^{n} y_i}{n}\right]$$

$$s_x^2 = \frac{1}{n-1}\left[\sum_{i=1}^{n} x_i^2 - \frac{\left(\sum_{i=1}^{n} x_i\right)^2}{n}\right]$$

Statisticians have shown that b_0 and b_1 are unbiased estimators of β_0 and β_1, respectively.

Although the calculations are straightforward, we would rarely compute the regression line manually because the work is time-consuming. However, we illustrate the manual calculations for a very small sample.

EXAMPLE 16.1

DATA Xm16-01

Annual Bonus and Years of Experience

The annual bonuses ($1,000s) of six employees with different years of experience were recorded as follows. We wish to determine the straight-line relationship between annual bonus and years of experience.

Years of Experience x	1	2	3	4	5	6
Annual Bonus y	6	1	9	5	17	12

SOLUTION

To apply the shortcut formula, we need to compute four summations. Using a calculator, we find

$$\sum_{i=1}^{n} x_i = 21$$

$$\sum_{i=1}^{n} y_i = 50$$

$$\sum_{i=1}^{n} x_i y_i = 212$$

$$\sum_{i=1}^{n} x_i^2 = 91$$

The covariance and the variance of x can now be computed:

$$s_{xy} = \frac{1}{n-1}\left[\sum_{i=1}^{n} x_i y_i - \frac{\sum_{i=1}^{n} x_i \sum_{i=1}^{n} y_i}{n}\right] = \frac{1}{6-1}\left[212 - \frac{(21)(50)}{6}\right] = 7.4$$

$$s_x^2 = \frac{1}{n-1}\left[\sum_{i=1}^{n} x_i^2 - \frac{\left(\sum_{i=1}^{n} x_i\right)^2}{n}\right] = \frac{1}{6-1}\left[91 - \frac{(21)^2}{6}\right] = 3.5$$

The sample slope coefficient is calculated next:

$$b_1 = \frac{s_{xy}}{s_x^2} = \frac{7.4}{3.5} = 2.114$$

The y-intercept is computed as follows:

$$\bar{x} = \frac{\sum x_i}{n} = \frac{21}{6} = 3.5$$

$$\bar{y} = \frac{\sum y_i}{n} = \frac{50}{6} = 8.333$$

$$b_0 = \bar{y} - b_1\bar{x} = 8.333 - (2.114)(3.5) = .934$$

Thus, the least squares line is

$$\hat{y} = .934 + 2.114x$$

Figure 16.1 depicts the least squares (or regression) line. As you can see, the line fits the data reasonably well. We can measure how well by computing the value of the minimized sum of squared deviations. The deviations between the actual data points and the line are called **residuals,** denoted e_i. That is,

$$e_i = y_i - \hat{y}_i$$

The residuals are observations of the error variable. Consequently, the minimized sum of squared deviations is called the **sum of squares for error,** denoted SSE.

FIGURE **16.1** **Scatter Diagram with Regression Line for Example 16.1**

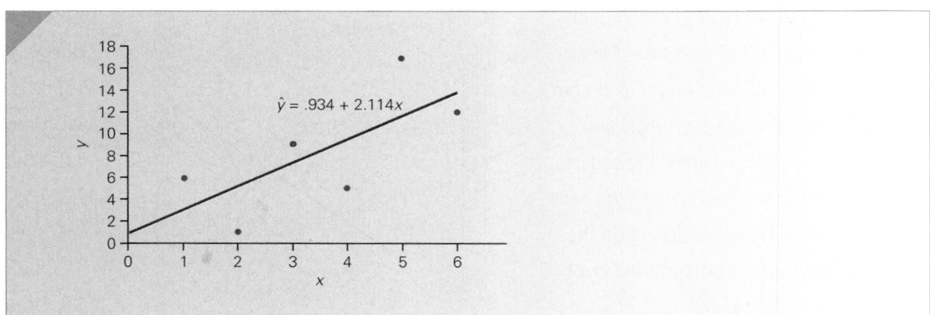

The calculation of the residuals in this example is shown in Figure 16.2. Notice that we compute $\hat{y}_i$ by substituting x_i into the formula of the regression line. The residuals are the differences between the observed values of y_i and the fitted or predicted values of $\hat{y}_i$. Table 16.1 describes these calculations.

Thus, SSE = 81.104. No other straight line will produce a sum of squared deviations as small as 81.104. In that sense, the regression line fits the data best. The sum of squares for error is an important statistic because it is the basis for other statistics that assess how well the linear model fits the data. We will introduce these statistics in Section 16.4.

FIGURE **16.2** Calculation of Residuals in Example 16.1

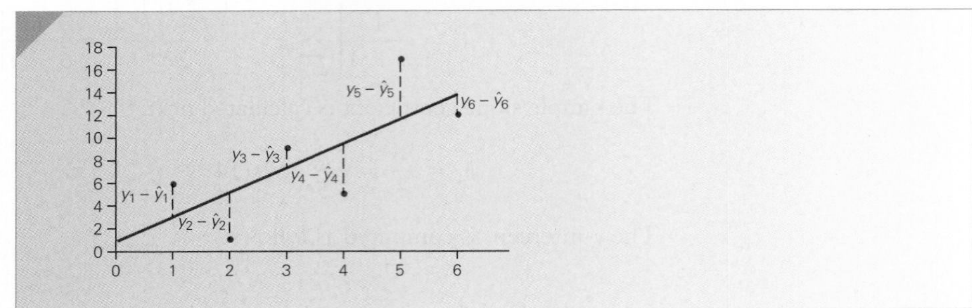

TABLE **16.1** Calculation of Residuals in Example 16.1

x_i	y_i	$\hat{y}_i = .934 + 2.114x_i$	$y_i - \hat{y}_i$	$(y_i - \hat{y}_i)^2$
1	6	3.048	2.952	8.714
2	1	5.162	−4.162	17.322
3	9	7.276	1.724	2.972
4	5	9.390	−4.390	19.272
5	17	11.504	5.496	30.206
6	12	13.618	−1.618	2.618
				$\sum(y_i - \hat{y}_i)^2 = 81.104$

SEEING STATISTICS

Applet 18: Fitting the Regression Line

This applet allows you to experiment with the data in Example 16.1. Click or drag the mouse in the graph to change the slope of the line. The errors are measured by the red lines. The squares represent the squared errors. (You can hide or show them by clicking on the **Hide/Show Errors button.**) The error meter on the left keeps track of your progress. The amount of the error that turns green is the proportion of the squared error you eliminate by finding a better regression line. The sum of squared errors is shown at the bottom. The coefficient of correlation squared (which is the coefficient of determination, explained in Section 16.4) is shown at the top. Change the slope until the sum of squares for error as indicated in the error meter is minimized. If you need help, click the **Find Best Model** button.

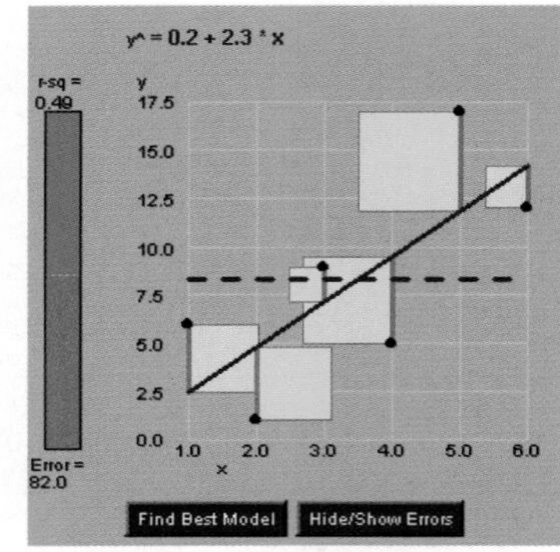

Applet Exercises

Change the slope (if necessary) so that the line is horizontal.

17.1 What is the slope of this line?

17.2 What is the *y*-intercept?

17.3 The *y*-intercept is equal to $\bar{y}$. What does this tell you about predicting the value of *y*?

17.4 Drag the mouse to change the slope to 1. What is the sum of squared errors?

17.5 Drag the mouse to change the slope to .5. What is the sum of squared errors?

17.6 Experiment with different lines. What point is common to all the lines?

EXAMPLE 16.2

DATA Xm16-02*

Odometer Reading and Prices of Used Toyota Camrys, Part 1

Car dealers across North America use the "Blue Book" to help them determine the value of used cars that their customers trade in when purchasing new cars. The book, which is published monthly, lists the trade-in values for all basic models of cars. It provides alternative values for each car model according to its condition and optional features. The values are determined on the basis of the average paid at recent used-car auctions, the source of supply for many used-car dealers. However, the Blue Book does not indicate the value determined by the odometer reading, despite the fact that a critical factor for used-car buyers is how far the car has been driven. To examine this issue, a used-car dealer randomly selected 100 3-year-old Toyota Camrys that were sold at auction during the past month. Each car was in top condition and equipped with all the features that come standard with this car. The dealer recorded the price ($1,000) and the number of miles (thousands) on the odometer. Some of these data are listed here. The dealer wants to find the regression line.

Car	Price ($1,000)	Odometer (1,000 mi)
1	14.6	37.4
2	14.1	44.8
3	14.0	45.8
⋮	⋮	⋮
98	14.5	33.2
99	14.7	39.2
100	14.3	36.4

SOLUTION

IDENTIFY

Notice that the problem objective is to analyze the relationship between two interval variables. Because we believe that the odometer reading affects the selling price, we identify the former as the independent variable, which we label x, and the latter as the dependent variable, which we label y.

COMPUTE

MANUALLY

From the data set, we find

$$\sum_{i=1}^{n} x_i = 3{,}601.1$$

$$\sum_{i=1}^{n} y_i = 1{,}484.1$$

$$\sum_{i=1}^{n} x_i y_i = 53{,}155.93$$

$$\sum_{i=1}^{n} x_i^2 = 133{,}986.59$$

Next we calculate the covariance and the variance of the independent variable x:

$$s_{xy} = \frac{1}{n-1}\left[\sum_{i=1}^{n} x_i y_i - \frac{\sum_{i=1}^{n} x_i \sum_{i=1}^{n} y_i}{n}\right]$$

$$= \frac{1}{100-1}\left[53{,}155.93 - \frac{(3{,}601.1)(1{,}484.1)}{100}\right] = -2.909$$

$$s_x^2 = \frac{1}{n-1}\left[\sum_{i=1}^{n} x_i^2 - \frac{\left(\sum_{i=1}^{n} x_i\right)^2}{n}\right]$$

$$= \frac{1}{100-1}\left[133{,}986.59 - \frac{(3{,}601.1)^2}{100}\right] = 43.509$$

The sample slope coefficient is calculated next:

$$b_1 = \frac{s_{xy}}{s_x^2} = \frac{-2.909}{43.509} = -.0669$$

The y-intercept is computed as follows:

$$\bar{x} = \frac{\sum x_i}{n} = \frac{3{,}601.1}{100} = 36.011$$

$$\bar{y} = \frac{\sum y_i}{n} = \frac{1{,}484.1}{100} = 14.841$$

$$b_0 = \bar{y} - b_1\bar{x} = 14.841 - (-.0669)(36.011) = 17.25$$

The sample regression line is

$$\hat{y} = 17.25 - 0.0669x$$

EXCEL

	A	B	C	D	E	F
1	SUMMARY OUTPUT					
2						
3	*Regression Statistics*					
4	Multiple R	0.8052				
5	R Square	0.6483				
6	Adjusted R Square	0.6447				
7	Standard Error	0.3265				
8	Observations	100				
9						
10	ANOVA					
11		*df*	*SS*	*MS*	*F*	*Significance F*
12	Regression	1	19.26	19.26	180.64	5.75E-24
13	Residual	98	10.45	0.11		
14	Total	99	29.70			
15						
16		*Coefficients*	*Standard Error*	*t Stat*	*P-value*	
17	Intercept	17.25	0.182	94.73	3.57E-98	
18	Odometer	-0.0669	0.0050	-13.44	5.75E-24	

INSTRUCTIONS

1. Type or import data into two columns, one storing the dependent variable and the other the independent variable. (Open Xm16-02.)
2. Click **Data, Data Analysis,** and **Regression.**
3. Specify the **Input Y Range** (A1:A101) and the **Input X Range** (B1:B101).

To draw the scatter diagram, follow the instructions provided in Chapter 2 on page 67.

MINITAB

Regression Analysis: Price versus Odometer

The regression equation is
Price = 17.2 - 0.0669 Odometer

Predictor	Coef	SE Coef	T	P
Constant	17.2487	0.1821	94.73	0.000
Odometer	-0.066861	0.004975	-13.44	0.000

S = 0.326489 R-Sq = 64.8% R-Sq(adj) = 64.5%

Analysis of Variance

Source	DF	SS	MS	F	P
Regression	1	19.256	19.256	180.64	0.000
Residual Error	98	10.446	0.107		
Total	99	29.702			

INSTRUCTIONS

1. Type or import the data into two columns. (Open Xm16-02.)
2. Click **Stat, Regression,** and **Regression . . . .**
3. Type the name of the dependent variable in the **Response** box (Price) and the name of the independent variable in the **Predictors** box (Odometer).

To draw the scatter diagram, click **Stat, Regression,** and **Fitted Line Plot.** Alternatively, follow the instructions provided in Chapter 2 on page 67.

The printouts include more statistics than we need right now. However, we will be discussing the rest of the printouts later.

INTERPRET

The slope coefficient b_1 is -0.0669, which means that for each additional 1,000 miles on the odometer, the price decreases by an average of \$.0669 thousand. Expressed more simply, the slope tells us that for each additional mile on the odometer, the price decreases on average by \$.0669 or 6.69 cents.

The intercept is $b_0 = 17.25$. Technically, the intercept is the point at which the regression line and the y-axis intersect. This means that when $x = 0$ (i.e., the car was not driven at all), the selling price is \$17.25 thousand or \$17,250. We might be tempted to interpret this number as the price of cars that have not been driven. However, in this case, the intercept is probably meaningless. Because our sample did not include any cars with zero miles on the odometer, we have no basis for interpreting b_0. As a general rule, we cannot determine the value of $\hat{y}$ for a value of x that is far outside the range of the sample values of x. In this example, the smallest and largest values of x are 19.1 and 49.2, respectively. Because $x = 0$ is not in this interval, we cannot safely interpret the value of $\hat{y}$ when $x = 0$.

It is important to bear in mind that the interpretation of the coefficients pertains only to the sample, which consists of 100 observations. To infer information about the population, we need statistical inference techniques, which are described subsequently.

In the sections that follow, we will return to this problem and the computer output to introduce other statistics associated with regression analysis.

EXERCISES

16.1 The term *regression* was originally used in 1885 by Sir Francis Galton in his analysis of the relationship between the heights of children and parents. He formulated the "law of universal regression," which specifies that "each peculiarity in a man is shared by his kinsmen, but on average in a less degree." (Evidently, people spoke this way in 1885.) In 1903, two statisticians, K. Pearson and A. Lee, took a random sample of 1,078 father-son pairs to examine Galton's law ("On the Laws of Inheritance in Man, I. Inheritance of Physical Characteristics," *Biometrika* 2:457–462). Their sample regression line was Son's height = 33.73 + .516 × Father's height

a. Interpret the coefficients.
b. What does the regression line tell you about the heights of sons of tall fathers?
c. What does the regression line tell you about the heights of sons of short fathers?

16.2 Xr16-02 Attempting to analyze the relationship between advertising and sales, the owner of a furniture store recorded the monthly advertising budget ($thousands) and the sales ($millions) for a sample of 12 months. The data are listed here:

Advertising	23	46	60	54	28	33
Sales	9.6	11.3	12.8	9.8	8.9	12.5

Advertising	25	31	36	88	90	99
Sales	12.0	11.4	12.6	13.7	14.4	15.9

a. Draw a scatter diagram. Does it appear that advertising and sales are linearly related?
b. Calculate the least squares line and interpret the coefficients.

16.3 Xr16-03 To determine how the number of housing starts is affected by mortgage rates, an economist recorded the average mortgage rate and the number of housing starts in a large county for the past 10 years. These data are listed here.

Rate	8.5	7.8	7.6	7.5	8.0
Starts	115	111	185	201	206

Rate	8.4	8.8	8.9	8.5	8.0
Starts	167	155	117	133	150

a. Determine the regression line.
b. What do the coefficients of the regression line tell you about the relationship between mortgage rates and housing starts?

16.4 Xr16-04 Critics of television often refer to the detrimental effects that all the violence shown on television has on children. However, there may be another problem. It may be that watching television also reduces the amount of physical exercise, causing weight gains. A sample of 15 10-year-old children was taken. The number of pounds each child was overweight was recorded (a negative number indicates the child is underweight). Additionally, the number of hours of television viewing per week was also recorded. These data are listed here.

Television	42	34	25	35	37	38	31	33
Overweight	18	6	0	−1	13	14	7	7

Television	19	29	38	28	29	36	18
Overweight	−9	8	8	5	3	14	−7

a. Draw the scatter diagram.
b. Calculate the sample regression line and describe what the coefficients tell you about the relationship between the two variables.

16.5 Xr16-05 To help determine how many beers to stock, the concession manager at Yankee Stadium wanted to know how the temperature affected beer sales. Accordingly, she took a sample of 10 games and recorded the number of beers sold and the temperature in the middle of the game.

Temperature	80	68	78	79	87
Number of beers	20,533	1,439	13,829	21,286	30,985
Temperature	74	86	92	77	84
Number of beers	17,187	30,240	37,596	9,610	28,742

a. Compute the coefficients of the regression line.
b. Interpret the coefficients.

The exercises that follow were created to allow you to see how regression analysis is used to solve realistic problems. As a result, most feature a large number of observations. We anticipate that most students will solve these problems using a computer and statistical software. However, for students without these resources, we have computed the means, variances, and covariances that will permit them to complete the calculations manually. (See Appendix A.)

16.6 Xr16-06* In television's early years, most commercials were 60 seconds long. Now, however, commercials can be any length. The objective of commercials remains the same—to have as many viewers as possible remember the product in a favorable way and eventually buy it. In an experiment to determine how the length of a commercial is related to people's memory of it, 60 randomly selected people were asked to watch a 1-hour television program. In the middle of the show, a commercial advertising a brand of toothpaste appeared. Some viewers watched a commercial that lasted for 20 seconds, others watched one that lasted for 24 seconds, 28 seconds, . . . , 60 seconds. The essential content of the commercials was the same. After the show, each person was given a test to measure how much he or she remembered about the product. The commercial times and test scores (on a 30-point test) were recorded.

a. Draw a scatter diagram of the data to determine whether a linear model appears to be appropriate.
b. Determine the least squares line.
c. Interpret the coefficients.

16.7 Xr16-07 Florida condominiums are popular winter retreats for many North Americans. In recent years the price has steadily increased. A real estate agent wanted to know why prices of similar-size apartments in the same building vary. A possible answer lies in the floor. It may be that the higher the floor, the greater the sale price of the apartment. He recorded the price (in $1,000s) of 1,200 sq. ft. condominiums in several buildings in the same location that have sold recently and the floor number of the condominium.

a. Determine the regression line.
b. What do the coefficients tell you about the relationship between the two variables?

16.8 Xr16-08 Is the value of an education worth its cost? Education costs are easy to calculate. They include tuition, books, and living expenses as well as the cost of foregone income that a job would have provided. In an effort to learn more about the value of education, a university vice-president gathered data from a sample of 150 30-year-old men and women. Each was asked how many years of formal education he or she had completed and his or her income ($1,000s) for the previous 12 months.

a. Draw a chart that describes the relationship between years of education and income.
b. Calculate the coefficients of the regression line.
c. Interpret the coefficients.

APPLICATIONS in HUMAN RESOURCES MANAGEMENT

Retaining Workers

Human resource managers are responsible for a variety of tasks within organizations. Personnel/human resource managers are involved with recruiting new workers, determining which applicants are most suitable to hire, and in various aspects of monitoring the workforce, including absenteeism and worker turnover. For many firms, worker turnover is a costly problem. First, there is the cost of recruiting and attracting qualified workers. The firm must advertise vacant positions and make certain that applicants are judged properly. Second, the cost of training hirees can be high, particularly in technical areas. Third, new employees are often not as productive and efficient as experienced employees. Consequently, it is in the interests of the firm to attract and keep the best workers. Any information that the personnel manager can obtain is likely to be useful.

16.9 Xr16-09 The human resource manager of a telemarketing firm is concerned about the rapid turnover of the firm's telemarketers. It appears that many telemarketers do not work very long before quitting. There may be a number of reasons, including relatively low pay, personal unsuitability for the work, and the low probability of advancement. Because of the high cost of hiring and training new workers, the manager decided to examine the factors that influence workers to quit. He reviewed the work history of a random sample of workers who have quit in the last year and recorded the number of weeks on the job before quitting and the age of each worker when originally hired.

a. Use regression analysis to describe how the work period and age are related.
b. Briefly discuss what the coefficients tell you.

16.10 Xr16-10 Besides the known long-term effects of smoking, do cigarettes also cause short-term illnesses such as colds? To help answer this question, a sample of smokers was drawn. Each person was asked to report the average number of cigarettes smoked per day and the number of days absent from work due to colds last year.

a. Determine the regression line.
b. What do the coefficients tell you about the relationship between smoking cigarettes and sick days due to colds?

16.11 Xr16-11 Fire damage in the United States amounts to billions of dollars, much of it insured. The time taken to arrive at the fire is critical. This raises the question, Should insurance companies lower premiums if the home to be insured is close to a fire station? To help make a decision, a study was undertaken wherein a number of fires were investigated. The distance to the nearest fire station (in kilometers) and the percentage of fire damage were recorded. Determine the least squares line and interpret the coefficients.

16.12 Xr16-12* A real estate agent specializing in commercial real estate wanted a more precise method of judging the likely selling price (in $1,000s) of apartment buildings. As a first effort, she recorded the price of a number of apartment buildings sold recently and the number of square feet (in 1,000s) in the building.

a. Calculate the regression line.
b. What do the coefficients tell you about the relationship between price and square footage?

16.13 Xr16-13 There are millions of boats registered in the United States. As is the case with automobiles, there is an active used-boat market. Many of the boats purchased require bank financing, and, as a result, it is important for financial institutions to be capable of accurately estimating the price of boats. One of the variables that affects the price is the number of hours the engine has been run. To determine the effect of the hours on the price, a financial analyst recorded the price (in $1,000s) of a sample of 2003 24-foot Sea Ray cruisers (one of the most popular boats) and the number of hours it had been run. Determine the least squares line and explain what the coefficients tell you.

16.14 (Exercise 2.90 revisited) Xr02-90 In an attempt to determine the factors that affect the amount of energy used, 200 households were analyzed. In each, the number of occupants and the amount of electricity used were measured. Determine the regression line.

16.15 Xr16-15 An economist for the federal government is attempting to produce a better measure of poverty than is currently in use. To help acquire information, she recorded the annual household income (in $1,000s) and the amount of money spent on food during one week for a random sample of households. Determine the regression line and interpret the coefficients.

16.16 Xr16-16* An economist wanted to investigate the relationship between office rents (the dependent variable) and vacancy rates. Accordingly, he took a random sample of monthly office rents and the percentage of vacant office space in 30 different cities.

a. Determine the regression line.
b. Interpret the coefficients.

16.17 (Exercise 2.92 revisited) Xr02-92 One general belief held by observers of the business world is that taller men earn more money than shorter men. In a University of Pittsburgh study, 250 MBA graduates, all about 30 years old, were polled and asked to report their height (in inches) and their annual income (to the nearest $1,000).

a. Determine the regression line.
b. What do the coefficients tell you?

APPLICATIONS in HUMAN RESOURCES MANAGEMENT

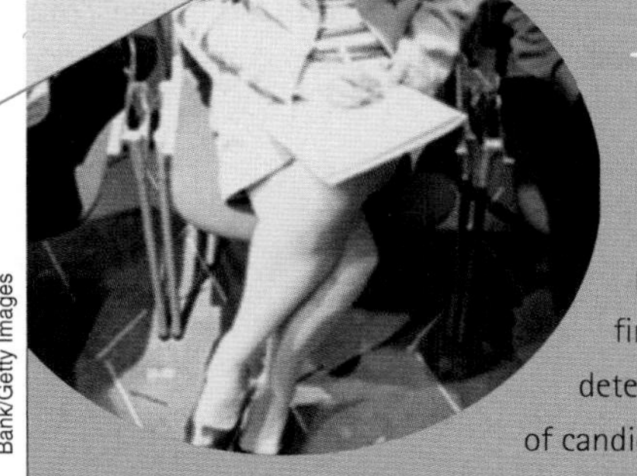
© Romilly Lockyer/The Image Bank/Getty Images

Testing Job Applicants

The recruitment process at many firms involves tests to determine the suitability of candidates. The tests may be written to determine whether the applicant has sufficient knowledge in his or her area of expertise to perform well on the job. There may be oral tests to determine whether the applicant's personality matches the needs of the job. Manual or technical skills can be tested through a variety of physical tests. The test results contribute to the decision to hire. In some cases, the test result is the only criterion to hire. Consequently, it is vital to ensure that the test is a reliable predictor of job performance. If the tests are poor predictors, they should be discontinued. Statistical analyses allow personnel managers to examine the link between the test results and job performance.

16.18 Xr16-18 Although a large number of tasks in the computer industry are robotic, a number of operations require human workers. Some jobs require a great deal of dexterity to properly position components into place. A large North American computer maker routinely tests applicants for these jobs by giving a dexterity test that involves a number of intricate finger and hand movements. The tests are scored on a 100-point scale. Only those who have scored above 70 are hired. To determine whether the tests are valid predictors of job performance, the personnel manager drew a random sample of 45 workers who were hired 2 months ago. He recorded their test scores and the percentage of nondefective computers they produced in the last week. Determine the regression line and interpret the coefficients.

16.3 ERROR VARIABLE: REQUIRED CONDITIONS

In the previous section, we used the least squares method to estimate the coefficients of the linear regression model. A critical part of this model is the error variable ε. In the next section, we will present an inferential method that determines whether there is a relationship between the dependent and independent variables. Later we will show how we use the regression equation to estimate and predict. For these methods to be valid, however, four requirements involving the probability distribution of the error variable must be satisfied.

Required Conditions for the Error Variable

1. The probability distribution of ε is normal.
2. The mean of the distribution is 0; that is, $E(\varepsilon) = 0$.
3. The standard deviation of ε is σ_ε, which is a constant regardless of the value of x.
4. The value of ε associated with any particular value of y is independent of ε associated with any other value of y.

Requirements 1, 2, and 3 can be interpreted in another way: For each value of x, y is a normally distributed random variable whose mean is

$$E(y) = \beta_0 + \beta_1 x$$

and whose standard deviation is σ_ε. Notice that the mean depends on x. The standard deviation, however, is not influenced by x, because it is a constant over all values of x. Figure 16.3 depicts this interpretation. Notice that for each value of x, $E(y)$ changes, but the shape of the distribution of y remains the same. That is, for each x, y is normally distributed with the same standard deviation.

FIGURE 16.3
Distribution of *y* Given *x*

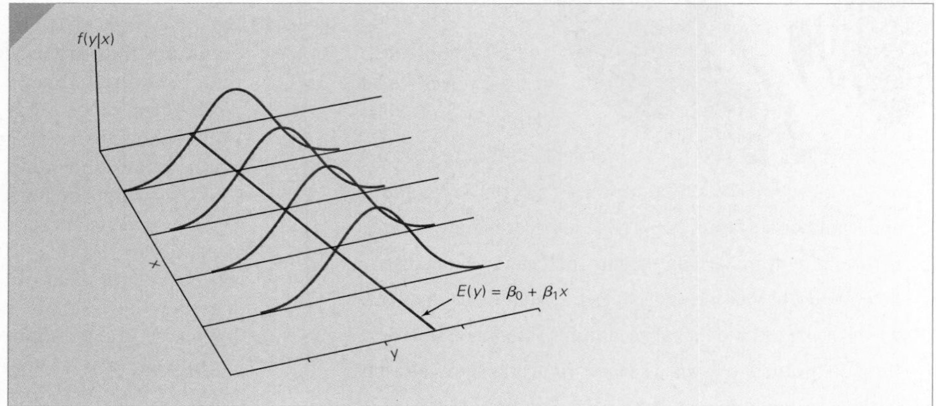

In Section 16.6, we will discuss how departures from these required conditions affect the regression analysis and how they are identified.

Observational and Experimental Data

In Chapter 2 and again in Chapter 13 we described the difference between observational and experimental data. We pointed out that statistics practitioners often design controlled experiments to enable them to interpret the results of their analyses more clearly than would be the case after conducting an observational study. Example 16.2 is an illustration of observational data. In that example, we merely observed the odometer reading and auction selling price of 100 randomly selected cars.

If you examine Exercise 16.6, you will see experimental data gathered through a controlled experiment. To determine the effect of the length of a television commercial on its viewers' memories of the product advertised, the statistics practitioner arranged for 60 television viewers to watch a commercial of differing lengths and then tested their memories of that commercial. Each viewer was randomly assigned a commercial length. The values of x ranged from 20 to 60 and were set by the statistics practitioner as part of the experiment. For each value of x, the distribution of the memory test scores is assumed to be normally distributed with a constant variance.

We can summarize the difference between the experiment described in Example 16.2 and the one described in Exercise 16.6. In Example 16.2 both the odometer reading and the auction selling price are random variables. We hypothesize that for each possible odometer reading, there is a theoretical population of auction selling prices that are normally distributed with a mean that is a linear function of the odometer reading and a variance that is constant. In Exercise 16.6 the length of the commercial is not a random variable, but a series of values selected by the statistics practitioner. For each commercial length, the memory test scores are required to be normally distributed with a constant variance.

Regression analysis can be applied to data generated from either observational or controlled experiments. In both cases our objective is to determine how the independent variable is related to the dependent variable. However, observational data can be analyzed in another way. When the data are observational, both variables are random variables. We need not specify that one variable is independent and the other is dependent. We can simply determine *whether* the two variables are related. The equivalent of the required conditions described in the previous box is that the two variables are bivariate normally distributed. (Recall that in Section 7.2 we introduced the bivariate distribution which describes the joint probability of two variables.) A bivariate normal distribution is described in Figure 16.4. As you can see, it is a three-dimensional bell-shaped curve. The dimensions are the variables x, y, and the joint density function $f(x, y)$.

FIGURE **16.4** Bivariate Normal Distribution

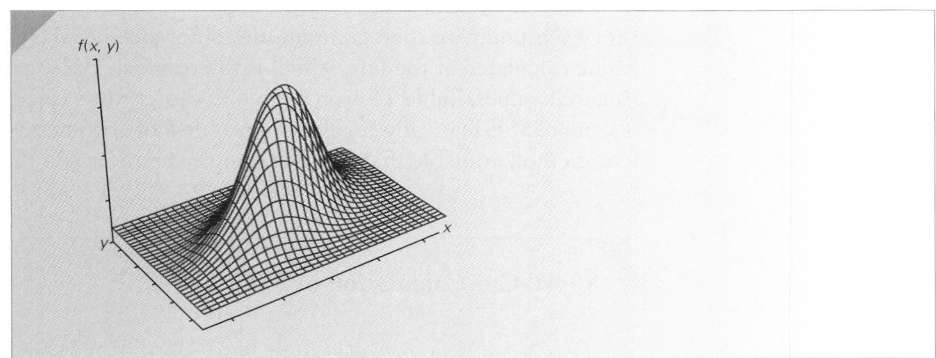

In Section 16.4 we will discuss the statistical technique that is used when both x and y are random variables and they are bivariate normally distributed. In Chapter 19 we will introduce a procedure applied when the normality requirement is not satisfied.

EXERCISES

16.19 Describe what the required conditions mean in Exercise 16.6. If the conditions are satisfied, what can you say about the distribution of memory test scores?

16.20 What are the required conditions for Exercise 16.8? Do these seem reasonable?

16.21 Assuming that the required conditions are satisfied in Exercise 16.13, what does this tell you about the distribution of used boat prices?

16.4 ASSESSING THE MODEL

The least squares method produces the best straight line. However, there may in fact be no relationship or perhaps a nonlinear relationship between the two variables. If so, a straight line model is likely to be impractical. Consequently, it is important for us to assess how well the linear model fits the data. If the fit is poor, we should discard the linear model and seek another one.

Several methods are used to evaluate the model. In this section, we present two statistics and one test procedure to determine whether a linear model should be employed. They are the **standard error of estimate,** the t-test of the slope, and the **coefficient of determination.** All these methods are based on the sum of squares for error.

Sum of Squares for Error

The least squares method determines the coefficients that minimize the sum of squared deviations between the points and the line defined by the coefficients. Recall from Section 16.2 that the minimized sum of squared deviations is called the sum of squares for error, denoted SSE. In that section we demonstrated the direct method of calculating SSE. For each value of x, we compute the value of $\hat{y}$. That is, for $i = 1$ to n, we compute

$$\hat{y}_i = b_0 + b_1 x_i$$

For each point we then compute the difference between the actual value of y and the value calculated at the line, which is the residual. We square each residual and sum the squared values. Table 16.1 on page 622 shows these calculations for Example 16.1. To calculate SSE manually requires a great deal of arithmetic. Fortunately, there is a shortcut method available that uses the sample variances and the covariance.

Short-Cut Calculation of SSE

$$\text{SSE} = \sum_{i=1}^{n}(y_i - \hat{y}_i)^2 = (n-1)\left(s_y^2 - \frac{s_{xy}^2}{s_x^2}\right)$$

where s_y^2 is the sample variance of the dependent variable.

Standard Error of Estimate

In Section 16.3, we pointed out that the error variable ε is normally distributed with mean 0 and standard deviation σ_ε. If σ_ε is large, some of the errors will be large, which implies that the model's fit is poor. If σ_ε is small, the errors tend to be close to the mean (which is 0), and, as a result, the model fits well. Hence, we could use σ_ε to measure the suitability of using a linear model. Unfortunately, σ_ε is a population parameter and, like most other parameters, is unknown. We can, however, estimate σ_ε from the data. The estimate is based on SSE. The unbiased estimator of the variance of the error variable σ_ε^2 is

$$s_\varepsilon^2 = \frac{\text{SSE}}{n-2}$$

The square root of s_ε^2 is called the *standard error of estimate*.

Standard Error of Estimate

$$s_\varepsilon = \sqrt{\frac{\text{SSE}}{n-2}}$$

EXAMPLE 16.3

Odometer Reading and Prices of Used Toyota Camrys, Part 2

Find the standard error of estimate for Example 16.2 and describe what it tells you about the model's fit.

SOLUTION

COMPUTE

MANUALLY

To compute the standard error of estimate, we must compute SSE, which is calculated from the sample variances and the covariance. We have already determined the covariance and the variance of x. They are -2.909 and 43.509, respectively. The sample variance of y (applying the shortcut method) is

$$s_y^2 = \frac{1}{n-1}\left[\sum_{i=1}^{n} y_i^2 - \frac{\left(\sum_{i=1}^{n} y_i\right)^2}{n}\right]$$

$$= \frac{1}{100-1}\left[22{,}055.23 - \frac{(1{,}484.1)^2}{100}\right]$$

$$= .300$$

$$\text{SSE} = (n-1)\left(s_y^2 - \frac{s_{xy}^2}{s_x^2}\right)$$

$$= (100-1)\left(.300 - \frac{[-2.909]^2}{43.509}\right)$$

$$= 10.445$$

The standard error of estimate follows:

$$s_\varepsilon = \sqrt{\frac{\text{SSE}}{n-2}} = \sqrt{\frac{10.445}{98}} = .3265$$

EXCEL

	A	B
7	Standard Error	0.3265

This part of the Excel printout was copied from the complete printout on page 624.

MINITAB

S = 0.326489

This part of the Minitab printout was copied from the complete printout on page 625.

INTERPRET

The smallest value that s_ε can assume is 0, which occurs when SSE $= 0$, that is, when all the points fall on the regression line. Thus, when s_ε is small, the fit is excellent, and the linear model is likely to be an effective analytical and forecasting tool. If s_ε is large, the model is a poor one, and the statistics practitioner should improve it or discard it.

We judge the value of s_ε by comparing it to the values of the dependent variable y or more specifically to the sample mean $\bar{y}$. In this example, because $s_\varepsilon = .3265$ and $\bar{y} = 14.841$, it does appear that the standard error of estimate is small. However, because there is no predefined upper limit on s_ε, it is often difficult to assess the model in this way. In general, the standard error of estimate cannot be used as an absolute measure of the model's utility.

Nonetheless, s_ε is useful in comparing models. If the statistics practitioner has several models from which to choose, the one with the smallest value of s_ε should generally be the one used. As you'll see, s_ε is also an important statistic in other procedures associated with regression analysis.

Testing the Slope

To understand this method of assessing the linear model, consider the consequences of applying the regression technique to two variables that are not at all linearly related. If we could observe the entire population and draw the regression line, we would observe the scatter diagram shown in Figure 16.5. The line is horizontal, which means that no matter what value of x is used, we would estimate the same value for $\hat{y}$, thus y is not linearly related to x. Recall that a horizontal straight line has a slope of 0, that is, $\beta_1 = 0$.

FIGURE **16.5**
Scatter Diagram of Entire Population with $\beta_1 = 0$

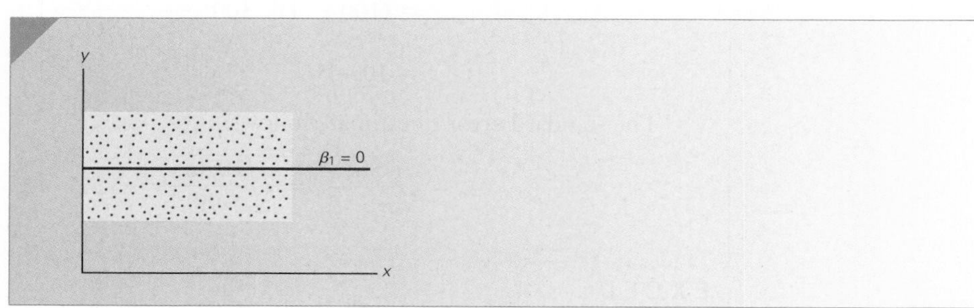

Because we rarely examine complete populations, the parameters are unknown. However; we can draw inferences about the population slope β_1 from the sample slope b_1.

The process of testing hypotheses about β_1 is identical to the process of testing any other parameter. We begin with the hypotheses. The null hypothesis specifies that there is no linear relationship, which means that the slope is 0. Thus, we specify

$$H_0: \quad \beta_1 = 0$$

It must be noted that if the null hypothesis is true, it does not necessarily mean that no relationship exists. For example, there may be a quadratic relationship described in Figure 16.6 may exist where $\beta_1 = 0$.

FIGURE **16.6**
Quadratic Relationship

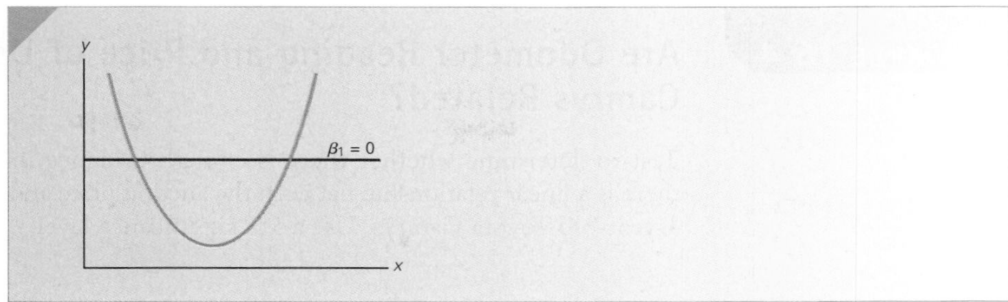

We can conduct one- or two-tail tests of β_1. Most often we perform a two-tail test to determine whether there is sufficient evidence to infer that a linear relationship exists.* We test the alternative hypothesis

$$H_1: \quad \beta_1 \neq 0$$

Estimator and Sampling Distribution

In Section 16.2 we pointed out that b_1 is an unbiased estimator of β_1. That is,

$$E(b_1) = \beta_1$$

The estimated standard error of b_1 is

$$s_{b_1} = \frac{s_\varepsilon}{\sqrt{(n-1)s_x^2}}$$

where s_ε is the standard error of estimate and s_x^2 is the sample variance of the independent variable. If the required conditions outlined in Section 16.3 are satisfied, the sampling distribution of the t-statistic

$$t = \frac{b_1 - \beta_1}{s_{b_1}}$$

is Student t with degrees of freedom $\nu = n - 2$. Notice that the standard error of b_1 decreases when the sample size increases (which makes b_1 a consistent estimator of β_1) and/or the variance of the independent variable increases.

Thus, the test statistic and confidence interval estimator are as follows.

Test Statistic for β_1

$$t = \frac{b_1 - \beta_1}{s_{b_1}} \qquad \nu = n - 2$$

Confidence Interval Estimator of β_1

$$b_1 \pm t_{\alpha/2} s_{b_1} \qquad \nu = n - 2$$

*If the alternative hypothesis is true, it may be that a linear relationship exists or that a nonlinear relationship exists, but that the relationship can be approximated by a straight line.

EXAMPLE 16.4 Are Odometer Reading and Price of Used Toyota Camrys Related?

Test to determine whether there is enough evidence in Example 16.2 to infer that there is a linear relationship between the auction price and the odometer reading for all 3-year-old Toyota Camrys. Use a 5% significance level.

SOLUTION

We test the hypotheses

$$H_0: \quad \beta_1 = 0$$
$$H_1: \quad \beta_1 \neq 0$$

If the null hypothesis is true, no linear relationship exists. If the alternative hypothesis is true, some linear relationship exists.

COMPUTE

MANUALLY

To compute the value of the test statistic we need b_1 and s_{b_1}. In Example 16.2 we found

$$b_1 = -.0669$$

and

$$s_x^2 = 43.509$$

Thus,

$$s_{b_1} = \frac{s_\varepsilon}{\sqrt{(n-1)s_x^2}} = \frac{.3265}{\sqrt{(99)(43.509)}} = .00497$$

The value of the test statistic is

$$t = \frac{b_1 - \beta_1}{s_{b_1}} = \frac{-.0669 - 0}{.00497} = -13.46$$

The rejection region is

$$t < -t_{\alpha/2,\nu} = -t_{.025,98} \approx -1.984 \qquad \text{or} \qquad t > t_{\alpha/2,\nu} = t_{.025,98} \approx 1.984$$

EXCEL

	A	B	C	D	E
16		Coefficients	Standard Error	t Stat	P-value
17	Intercept	17.25	0.182	94.73	3.57E-98
18	Odometer	-0.0669	0.0050	-13.44	5.75E-24

MINITAB

Predictor	Coef	SE Coef	T	P
Constant	17.2487	0.1821	94.73	0.000
Odometer	-0.066861	0.004975	-13.44	0.000

INTERPRET

The value of the test statistic is $t = -13.44$, with a p-value of 0. Excel uses scientific notation when the p-value is very small. In this example Excel's p-value is equal to 5.75×10^{-24}, which is approximately 0. There is overwhelming evidence to infer that a linear relationship exists. What this means is that the odometer reading may affect the auction selling price of the cars. (See the subsection on cause and effect relationship on page 641.)

As was the case when we interpreted the y-intercept, the conclusion we draw here is valid only over the range of the values of the independent variable. That is, we can infer that there is a relationship between odometer reading and auction price for the 3-year-old Toyota Camrys whose odometer readings lie between 19.1 (thousand) and 49.2 (thousand) miles (the minimum and maximum values of x in the sample). Because we have no observations outside this range, we do not know how or even whether, the two variables are related.

Notice that the printout includes a test for β_0. However, as we pointed out before, interpreting the value of the y-intercept can lead to erroneous, if not ridiculous, conclusions. Consequently, we generally ignore the test of β_0.

We can also acquire information about the relationship by estimating the slope coefficient. In this example the 95% confidence interval estimate (approximating $t_{.025}$ with 98 degrees of freedom with $t_{.025}$ with 100 degrees of freedom) is

$$b_1 \pm t_{\alpha/2} s_{b_1} = -.0669 \pm 1.984(.00497) = -.0669 \pm .0099$$

We estimate that the slope coefficient lies between $-.0768$ and $-.0570$.

One-Tail Tests

If we wish to test for positive or negative linear relationships, we conduct one-tail tests. To illustrate, suppose that in Example 16.2 we wanted to know whether there is evidence of a negative linear relationship between odometer reading and auction selling price. We would specify the hypotheses as

$$H_0: \quad \beta_1 = 0$$
$$H_1: \quad \beta_1 < 0$$

The value of the test statistic would be exactly as computed previously (Example 16.4). However, in this case the p-value would be the two-tail p-value divided by 2, which, using Excel's p-value, would be $(5.75 \times 10^{-24})/2 = 2.875 \times 10^{-24}$, which is again, approximately 0.

Coefficient of Determination

The test of β_1 addresses only the question of whether there is enough evidence to infer that a linear relationship exists. In many cases, however, it is also useful to measure the strength of that linear relationship, particularly when we want to compare several different models. The statistic that performs this function is the coefficient of determination, which is denoted R^2. Statistics practitioners often refer to this statistic as the "*R*-square." Recall that we introduced the coefficient of determination in Chapter 4 where we pointed out that this statistic is a measure of the amount of variation in the dependent variable that is explained by the variation in the independent variable. However, we did not describe why we interpret the *R*-square in this way.

Coefficient of Determination

$$R^2 = \frac{s_{xy}^2}{s_x^2 s_y^2}$$

With a little algebra, statisticians can show that

$$R^2 = 1 - \frac{\text{SSE}}{\sum (y_i - \bar{y})^2}$$

We'll return to Example 16.1 to learn more about how to interpret the coefficient of determination. In Chapter 14, we partitioned the total sum of squares into two sources of variation. We do so here as well. We begin by adding and subtracting $\hat{y}_i$ from the deviation between y_i from the mean $\bar{y}$. That is,

$$(y_i - \bar{y}) = (y_i - \bar{y}) + \hat{y}_i - \hat{y}_i$$

We observe that by rearranging the terms, the deviation between y_i and $\bar{y}$ can be decomposed into two parts. That is,

$$(y_i - \bar{y}) = (y_i - \hat{y}_i) + (\hat{y}_i - \bar{y})$$

This equation is represented graphically (for $i = 5$) in Figure 16.7.

FIGURE **16.7** Partitioning the Deviation for $i = 5$

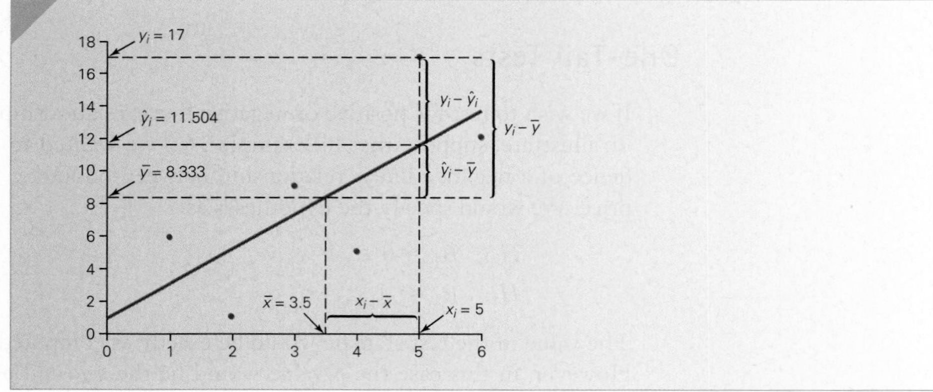

Now we ask why the values of *y* are different from one another. From Figure 16.7, we see that part of the difference between y_i and $\bar{y}$ is the difference between $\hat{y}_i$ and $\bar{y}$, which is accounted for by the difference between x_i and $\bar{x}$. That is, some of the variation in *y* is explained by the changes to *x*. The other part of the difference between y_i and $\bar{y}$, however, is accounted for by the difference between y_i and $\hat{y}_i$. This difference is the residual, which represents variables not otherwise represented by the model. As a result, we say that this part of the difference is *unexplained* by the variation in x.

If we now square both sides of the equation, sum over all sample points, and perform some algebra, we produce

$$\sum(y_i - \bar{y})^2 = \sum(y_i - \hat{y}_i)^2 + \sum(\hat{y}_i - \bar{y})^2$$

The quantity on the left side of this equation is a measure of the variation in the dependent variable y. The first quantity on the right side of the equation is SSE, and the second term is denoted SSR, for sum of squares for regression. We can rewrite the equation as

$$\text{Variation in } y = \text{SSE} + \text{SSR}$$

As we did in the analysis of variance, we partition the variation of y into two parts: SSE, which measures the amount of variation in y that remains unexplained, and SSR, which measures the amount of variation in y that is explained by the variation in the independent variable x. We can incorporate this analysis into the definition of R^2.

Coefficient of Determination

$$R^2 = 1 - \frac{\text{SSE}}{\sum(y_i - \bar{y})^2} = \frac{\sum(y_i - \bar{y})^2 - \text{SSE}}{\sum(y_i - \bar{y})^2} = \frac{\text{Explained variation}}{\text{Variation in } y}$$

It follows that R^2 measures the proportion of the variation in y that can be explained by the variation in x.

SEEING STATISTICS

Applet 19: Analysis of Regression Deviations

This applet provides another way to understand the coefficient of determination.

Move the regression line to reduce the sum of squared errors. The vertical line from each point to the horizontal line depicts the deviation from the mean. In regression this is divided into two parts—the green part, which is the deviation that is eliminated by using the regression line, and the red part, which is the deviation remaining. Note that for some points the deviations become larger.

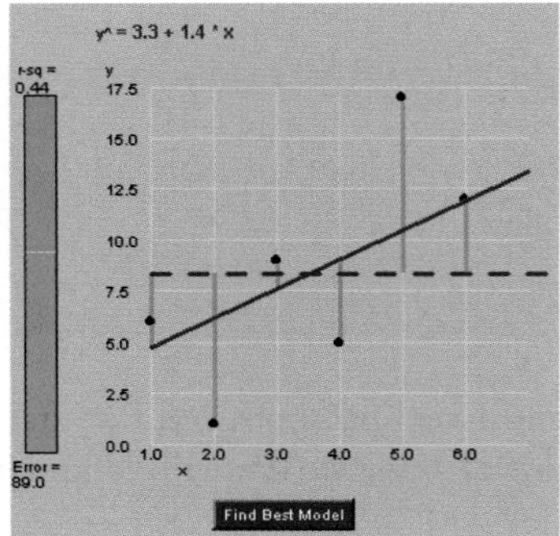

Applet Exercises

Change the slope (if necessary) so that the line is horizontal.

19.1 How much of the variation in y is explained by the variation in x? Why is this so?

Move the line so that it goes through the sixth point ($x = 6$).

19.2 What is the value of R^2?

19.3 How much of the variation between y_6 and $\bar{y}$ is explained by the variation between x_6 and $\bar{x}$? Why is this so?

Produce the least squares line. (Click the ***Find Best Model*** *button.)*

19.4 How much of the variation in y is explained by the variation in x?

EXAMPLE 16.5 Measuring the Strength of the Linear Relationship between Odometer Reading and Price of Used Toyota Camrys

Find the coefficient of determination for Example 16.2 and describe what this statistic tells you about the regression model.

SOLUTION

COMPUTE

MANUALLY

We have already calculated all the necessary components of this statistic. In Example 16.2 we found

$$s_{xy} = -2.909$$

$$s_x^2 = 43.509$$

and from Example 16.3

$$s_y^2 = .300$$

Thus,

$$R^2 = \frac{s_{xy}^2}{s_x^2 s_y^2} = \frac{(-2.909)^2}{(43.509)(.300)} = .6483$$

EXCEL

	A	B
5	R Square	0.6483

MINITAB

R-Sq = 64.8%

Both Minitab and Excel print a second R^2 statistic called the *coefficient of determination adjusted for degrees of freedom*. We will define and describe this statistic in Chapter 17.

INTERPRET

We found that R^2 is equal to .6483. This statistic tells us that 64.83% of the variation in the auction selling prices is explained by the variation in the odometer readings. The remaining 35.17% is unexplained. Unlike the value of a test statistic, the coefficient of determination does not have a critical value that enables us to draw conclusions.

In general, the higher the value of R^2, the better the model fits the data. From the t-test of β_1, we already know that there is evidence of a linear relationship. The coefficient of determination merely supplies us with a measure of the strength of that relationship. As you will discover in the next chapter, when we improve the model, the value of R^2 increases.

Other Parts of the Computer Printout

The last part of the printout shown on pages 624 and 625 relates to our discussion of the interpretation of the value of R^2, when its meaning is derived from the partitioning of the variation in y. The values of SSR and SSE are shown in an analysis of variance table similar to the tables introduced in Chapter 14. The general form of the table is shown in Table 16.2. The F-test performed in the ANOVA table will be explained in Chapter 17.

TABLE **16.2** General Form of the ANOVA Table in the Simple Linear Regression Model

Source	d.f.	Sums of Squares	Mean Squares	F-Statistic
Regression	1	SSR	MSR = SSR/1	F = MSR/MSE
Error	$n - 2$	SSE	MSE = SSE/$(n - 2)$	
Total	$n - 1$	Variation in y		

Note: Excel uses the word "Residual" to refer to the second source of variation, which we called "Error."

Developing an Understanding of Statistical Concepts

Once again, we encounter the concept of explained variation. We first discussed the concept in Chapter 13 when we introduced the matched pairs experiment, where the experiment was designed to reduce the variation among experimental units. This concept was extended in the analysis of variance, where we partitioned the total variation into two or more sources (depending on the experimental design). And now in regression analysis, we use the concept to measure how the dependent variable is related to the independent variable. We partition the variation of the dependent variable into the sources: the variation explained by the variation in the independent variable and the unexplained variation. The greater the explained variation, the better the model is. We often refer to the coefficient of determination as a measure of the explanatory power of the model.

Cause-and-Effect Relationship

A common mistake is made by many students when they attempt to interpret the results of a regression analysis when there is evidence of a linear relationship. They imply that changes in the independent variable cause changes in the dependent variable. It must be emphasized that we cannot infer a causal relationship from statistics alone. Any inference about the cause of the changes in the dependent variable must be justified by a reasonable theoretical relationship. For example, statistical tests established that the more one smoked, the greater the probability of developing lung cancer. However, this analysis did not prove that smoking causes lung cancer. It only demonstrated that smoking and lung cancer were somehow related. Only when medical investigations established the connection were scientists able to confidently declare that smoking causes lung cancer.

As another illustration, consider Example 16.2 where we showed that the odometer reading is linearly related to the auction price. Although it seems reasonable to conclude that decreasing the odometer reading would cause the auction price to rise, the conclusion may not be entirely true. It is theoretically possible that the price is determined by the overall condition of the car and that the condition generally worsens when the car is driven longer. Another analysis would be needed to establish the veracity of this conclusion.

Be cautious about the use of the terms *explained variation* and *explanatory power of the model.* Do not interpret the word *explained* to mean *caused.* We say that the coefficient of determination measures the amount of variation in y that is explained (not caused) by the variation in x. Thus, regression analysis can only show that a statistical relationship exists. We cannot infer that one variable causes another.

Recall that we first pointed this out in Chapter 2 using the following sentence:

Correlation is not causation.

Testing the Coefficient of Correlation

When we introduced the coefficient of correlation (also called the **Pearson coefficient of correlation**) in Chapter 4, we observed that it is used to measure the strength of association between two variables. However, the coefficient of correlation can be useful in another way. We can use it to test for a linear relationship between two variables.

When we are interested in determining *how* the independent variable is related to the dependent variable, we estimate and test the linear regression model. The t-test of the slope presented previously allows us to determine whether a linear relationship actually exists. As we pointed out in Section 16.3, the statistical test requires that for each value of x, there exists a population of values of y that are normally distributed with a constant variance. This condition is required whether the data are experimental or observational.

In many circumstances we're interested in determining only *whether* a linear relationship exists and not the form of the relationship. When the data are observational and the two variables are bivariate normally distributed (see Section 16.3.), we can calculate the coefficient of correlation and use it to test for linear association.

As we noted in Chapter 4, the population coefficient of correlation is denoted ρ (the Greek letter *rho*). Because ρ is a population parameter (which is almost always unknown), we must estimate its value from the sample data. Recall that the sample coefficient of correlation is defined as follows.

Sample Coefficient of Correlation

$$r = \frac{s_{xy}}{s_x s_y}$$

When there is no linear relationship between the two variables, $\rho = 0$. To determine whether we can infer that ρ is 0, we test the hypotheses

$$H_0: \quad \rho = 0$$
$$H_1: \quad \rho \neq 0$$

The test statistic is defined in the following way.

Test Statistic for Testing $\rho = 0$

$$t = r\sqrt{\frac{n-2}{1-r^2}}$$

which is Student t distributed with $\nu = n - 2$ degrees of freedom provided that the variables are bivariate normally distributed.

EXAMPLE 16.6

Are Odometer Reading and Price of Used Toyota Camrys Linearly Related? Testing the Coefficient of Correlation

Conduct the t-test of the coefficient of correlation to determine whether odometer reading and auction selling price are linearly related in Example 16.2. Assume that the two variables are bivariate normally distributed.

SOLUTION

COMPUTE

MANUALLY

The hypotheses to be tested are

$$H_0: \quad \rho = 0$$
$$H_1: \quad \rho \neq 0$$

In Example 16.2, we found $s_{xy} = -2.909$ and $s_x^2 = 43.509$. In Example 16.5, we determined that $s_y^2 = .300$. Thus,

$$s_x = \sqrt{43.509} = 6.596$$
$$s_y = \sqrt{.300} = .5477$$

The coefficient of correlation is

$$r = \frac{s_{xy}}{s_x s_y} = \frac{-2.909}{(6.596)(.5477)} = -.8052$$

The value of the test statistic is

$$t = r\sqrt{\frac{n-2}{1-r^2}} = -.8052\sqrt{\frac{100-2}{1-(-.8052)^2}} = -13.44$$

Notice that this is the same value we produced in the t-test of the slope in Example 16.4. Because both sampling distributions are Student t with 98 degrees of freedom, the p-value and conclusion are also identical.

EXCEL

	A	B
1	**Correlation**	
2		
3	*Price and Odometer*	
4	Pearson Coefficient of Correlation	-0.8052
5	t Stat	-13.44
6	df	98
7	P(T<=t) one tail	0
8	t Critical one tail	1.6606
9	P(T<=t) two tail	0
10	t Critical two tail	1.9845

INSTRUCTIONS

1. Type or import the data into two adjacent columns. (Open Xm16-02.)
2. Click **Add-ins, Data Analysis Plus,** and **Correlation (Pearson).**
3. Specify the **Variable 1 Input Range** (A1:A101), **Variable 2 Input Range** (B1:B101), and α (.05).

MINITAB

Correlations: Odometer, Price

Pearson correlation of Price and Odometer = -0.805
P-Value = 0.000

INSTRUCTIONS

1. Type or import the data into two adjacent columns. (Open Xm16-02.)
2. Click **Stat, Basic Statistics,** and **Correlation.**
3. Type the names of the variables in the **Variables** box (Odometer Price).

Notice that the t-test of ρ and the t-test of β_1 in Example 16.4 produced identical results. This should not be surprising because both tests are conducted to determine whether there is evidence of a linear relationship. The decision about which test to use is based on the type of experiment and the information we seek from the statistical analysis. If we're interested in discovering the relationship between two variables, or if we've conducted an experiment where we controlled the values of the independent variable (as in Exercise 16.6), the t-test of β_1 should be applied. If we're interested only in determining *whether* two random variables that are bivariate normally distributed are linearly related, the t-test of ρ should be applied.

As is the case with the t-test of the slope, we can also conduct one-tail tests. We can test for a positive or a negative linear relationship.

Foreign Index Funds: Solution

© Terryvine/Stone/Getty Images

IDENTIFY

The problem objective is to analyze the relationship between two interval variables. Because we're not interested in the form of the linear relationship but only *whether* a linear relationship exists between the two variables and the data are observational, the parameter of interest is the coefficient of correlation. We test the following hypotheses:

$$H_0: \quad \rho = 0$$

$$H_1: \quad \rho \neq 0$$

The test statistic is

$$t = r\sqrt{\frac{n-2}{1-r^2}}$$

COMPUTE

MANUALLY

The rejection region of the test is

$$t < -t_{\alpha/2,\nu} = -t_{.025,57} \approx -2.004 \qquad \text{or} \qquad t > t_{\alpha/2,\nu} = t_{.025,57} \approx 2.004$$

The value of r is calculated from the covariance and two standard deviations:

$$s_{xy} = .001279$$

$$s_x = .0509$$

$$s_y = .0512$$

$$r = \frac{s_{xy}}{s_x s_y} = \frac{.001279}{(.0509)(.0512)} = .491$$

The value of the test statistic is

$$t = r\sqrt{\frac{n-2}{1-r^2}} = (.491)\sqrt{\frac{59-2}{1-(.491)^2}} = 4.26$$

EXCEL

	A	B	C	D
1	**Correlation**			
2				
3	*US Index and Japanese Index*			
4	Pearson Coefficient of Correlation			0.4911
5	t Stat			4.26
6	df			57
7	P(T<=t) one tail			0
8	t Critical one tail			1.672
9	P(T<=t) two tail			0
10	t Critical two tail			2.0025

MINITAB

Correlations: US Index, Japanese Index

Pearson correlation of US Index and Japanese Index = 0.491
P-Value = 0.000

INTERPRET

The coefficient of correlation is $r = .4911$, and the value of the test statistic is $t = 4.26$, which has a p-value of 0. There is overwhelming evidence of a linear relationship between the two indexes. The investor should not buy the Japanese index.

Violation of the Required Condition

When the normality requirement is unsatisfied, we can use a nonparametric technique—Spearman rank correlation coefficient (Chapter 19*) to replace the t-test of ρ.

*Instructors who wish to teach the use of the Spearman rank correlation coefficient here can use CD Appendix Y, "Spearman Rank Correlation Coefficient and Test."

EXERCISES

Use a 5% significance level for all tests of hypotheses.

16.22 You have been given the following data:

x	1	3	4	6	9	8	10
y	1	8	15	33	75	70	95

a. Draw the scatter diagram. Does it appear that x and y are related? If so, how?
b. Test to determine whether there is evidence of a linear relationship.

16.23 Suppose that you have the following data:

x	3	5	2	6	1	4
y	25	110	9	250	3	71

a. Draw the scatter diagram. Does it appear that x and y are related? If so, how?
b. Test to determine whether there is evidence of a linear relationship.

16.24 Refer to Exercise 16.2.
a. Determine the standard error of estimate.
b. Is there evidence of a linear relationship between advertising and sales?
c. Estimate β_1 with 95% confidence.
d. Compute the coefficient of determination and interpret this value.
e. Briefly summarize what you have learned in Parts a, b, c, and d.

16.25 Calculate the coefficient of determination and conduct a test to determine whether a linear relationship exists between housing starts and mortgage interest in Exercise 16.3.

16.26 Is there evidence of a linear relationship between the number of hours of television viewing and how overweight the child is in Exercise 16.4?

16.27 Determine whether there is evidence of a negative linear relationship between temperature and the number of beers sold at Yankee Stadium in Exercise 16.5.

The following exercises require the use of a computer and software. The answers may be calculated manually. See Appendix A for the sample statistics.

16.28 Refer to Exercise 16.6.
a. What is the standard error of estimate? Interpret its value.
b. Describe how well the memory test scores and length of television commercial are linearly related.

c. Are the memory test scores and length of commercial linearly related? Test using a 5% significance level.
d. Estimate the slope coefficient with 90% confidence.

16.29 Refer to Exercise 16.7. Apply the three methods of assessing the model to determine how well the linear model fits.

16.30 Is there enough evidence to infer that education and income are linearly related in Exercise 16.8?

16.31 Refer to Exercise 16.9. Use two statistics to measure the strength of the linear association. What do these statistics tell you?

16.32 Is there evidence of a linear relationship between number of cigarettes smoked and number of sick days in Exercise 16.10?

16.33 Refer to Exercise 16.11.
a. Test to determine whether there is evidence of a linear relationship between distance to the nearest fire station and percentage of damage.
b. Estimate the slope coefficient with 95% confidence.
c. Determine the coefficient of determination. What does this statistic tell you about the relationship?

16.34 Refer to Exercise 16.12.
a. Determine the standard error of estimate, and describe what this statistic tells you about the regression line.
b. Can we conclude that the size and price of the apartment building are linearly related?
c. Determine the coefficient of determination and discuss what its value tells you about the two variables.

16.35 Is there enough evidence to infer that as the number of hours of engine use increases, the price decreases in Exercise 16.13?

16.36 Assess fit of the regression line in Exercise 16.14.

16.37 Refer to Exercise 16.15.
a. Determine the coefficient of determination and describe what it tells you.
b. Conduct a test to determine whether there is evidence of a linear relationship between household income and food budget.

16.38 Can we infer that office rents and vacancy rates are linearly related in Exercise 16.16?

16.39 Are height and income in Exercise 16.17 positively linearly related?

16.40 Refer to Exercise 16.18.
a. Compute the coefficient of determination and describe what it tells you.
b. Can we infer that aptitude test scores and percentages of nondefectives are linearly related?

16.41 Repeat Exercise 16.13 using the *t*-test of the coefficient of correlation to determine whether there is a negative linear relationship between the number of hours of engine use and the selling price of the used boats.

16.42 Repeat Exercise 16.6 using the *t*-test of the coefficient of correlation. Is this result identical to the one you produced in Exercise 16.6?

16.43 Are food budget and household income in Exercise 16.15 linearly related? Employ the *t*-test of the coefficient of correlation to answer the question.

16.44 Refer to Exercise 16.10. Use the *t*-test of the coefficient of correlation to determine whether there is evidence of a positive linear relationship between number of cigarettes smoked and the number of sick days.

16.5 USING THE REGRESSION EQUATION

Using the techniques in Section 16.4, we can assess how well the linear model fits the data. If the model fits satisfactorily, we can use it to forecast and estimate values of the dependent variable. To illustrate, suppose that in Example 16.2, the used-car dealer wanted to predict the selling price of a 3-year-old Toyota Camry with 40 (thousand) miles on the odometer. Using the regression equation, with $x = 40$, we get

$$\hat{y} = 17.25 - .0669x = 17.25 - 0.0669(40) = 14.574$$

We call this value the point prediction and $\hat{y}$ is the point estimate or predicted value for y when $x = 40$. Thus, the dealer would predict that the car would sell for $14,574.

By itself, however, the point prediction does not provide any information about how closely the value will match the true selling price. To discover that information, we must use an interval. In fact, we can use one of two intervals: the prediction interval of a particular value of *y* or the confidence interval estimator of the expected value of *y*.

Predicting the Particular Value of *y* for a Given *x*

The first confidence interval we present is used whenever we want to predict a one-time occurrence for a particular value of the dependent variable when the independent variable is a given value x_g. This interval, often called the **prediction interval,** is calculated in the usual way (point estimator ± bound on the error of estimation). Here the point estimate for *y* is $\hat{y}$ and the bound on the error of estimation is shown below.

Prediction Interval

$$\hat{y} \pm t_{\alpha/2,n-2}s_\varepsilon\sqrt{1 + \frac{1}{n} + \frac{(x_g - \bar{x})^2}{(n-1)s_x^2}}$$

where x_g is the given value of *x* and

$$\hat{y} = b_0 + b_1 x_g$$

Estimating the Expected Value of *y* for a Given *x*

The conditions described in Section 16.3 imply that, for a given value of *x*, there is a population of values of *y* whose mean is

$$E(y) = \beta_0 + \beta_1 x$$

To estimate the mean of *y* or long-run average value of *y*, we would use the following interval referred to simply as the confidence interval. Again the point estimator is $\hat{y}$, but the bound on the error of estimation is different from the prediction interval shown below.

Confidence Interval Estimator of the Expected Value of *y*

$$\hat{y} \pm t_{\alpha/2,n-2}s_\varepsilon\sqrt{\frac{1}{n} + \frac{(x_g - \bar{x})^2}{(n-1)s_x^2}}$$

Unlike the formula for the prediction interval, this formula does not include the 1 under the square-root sign. As a result, the **confidence interval estimate of the expected value of *y*** will be narrower than the prediction interval for the same given value of *x* and confidence level. This is because there is less error in estimating a mean value as opposed to predicting an individual value.

EXAMPLE 16.7

Predicting the Price and Estimating the Mean Price of Used Toyota Camrys

a. A used-car dealer is about to bid on a 3-year-old Toyota Camry equipped with all the standard features, and with 40,000 ($x_g = 40$) miles on the odometer. To help him decide how much to bid, he needs to predict the selling price.
b. The used-car dealer mentioned in Part a has an opportunity to bid on a lot of cars offered by a rental company. The rental company has 250 Toyota Camrys all equipped with standard features. All the cars in this lot have about 40,000 ($x_g = 40$) miles on the odometer. The dealer would like an estimate of the selling price of all the cars in the lot.

SOLUTION

IDENTIFY

a. The dealer would like to predict the selling price of a single car. Thus, he must employ the prediction interval

$$\hat{y} \pm t_{\alpha/2,n-2}s_{\varepsilon}\sqrt{1 + \frac{1}{n} + \frac{(x_g - \bar{x})^2}{(n - 1)s_x^2}}$$

b. The dealer wants to determine the mean price of a large lot of cars, so he needs to calculate the confidence interval estimator of the expected value:

$$\hat{y} \pm t_{\alpha/2,n-2}s_{\varepsilon}\sqrt{\frac{1}{n} + \frac{(x_g - \bar{x})^2}{(n - 1)s_x^2}}$$

Technically, this formula is used for infinitely large populations. However, we can interpret our problem as attempting to determine the average selling price of all Toyota Camrys equipped as described above, all with 40,000 miles on the odometer. The crucial factor in Part b is the need to estimate the mean price of a number of cars. We arbitrarily select a 95% confidence level.

COMPUTE

MANUALLY

From previous calculations, we have the following:

$$\hat{y} = 17.25 - .0669(40) = 14.574$$

$$s_{\varepsilon} = .3265$$

$$s_x^2 = 43.509$$

$$\bar{x} = 36.011$$

From Table 4 in Appendix B, we find

$$t_{\alpha/2} = t_{.025,98} \approx t_{.025,100} = 1.984$$

a. The 95% prediction interval is

$$\hat{y} \pm t_{\alpha/2,n-2} s_\varepsilon \sqrt{1 + \frac{1}{n} + \frac{(x_g - \bar{x})^2}{(n-1)s_x^2}}$$

$$= 14.574 \pm 1.984 \times .3265 \sqrt{1 + \frac{1}{100} + \frac{(40 - 36.011)^2}{(100-1)(43.509)}}$$

$$= 14.574 \pm .652$$

The lower and upper limits of the prediction interval are $13,922 and $15,226, respectively.

b. The 95% confidence interval estimator of the mean price is

$$\hat{y} \pm t_{\alpha/2,n-2} s_\varepsilon \sqrt{\frac{1}{n} + \frac{(x_g - \bar{x})^2}{(n-1)s_x^2}}$$

$$= 14.574 \pm 1.984 \times .3265 \sqrt{\frac{1}{100} + \frac{(40 - 36.011)^2}{(100-1)(43.509)}}$$

$$= 14.574 \pm .076$$

The lower and upper limits of the confidence interval estimate of the expected value are $14,498 and 14,650, respectively.

EXCEL

	A	B	C
1	**Prediction Interval**		
2			
3			Price
4			
5	Predicted value		14.574
6			
7	Prediction Interval		
8	Lower limit		13.922
9	Upper limit		15.227
10			
11	Interval Estimate of Expected Value		
12	Lower limit		14.498
13	Upper limit		14.650

INSTRUCTIONS

1. Type or import the data into two columns. (Open Xm16-02.)
2. Type the given value of x into any cell. We suggest the next available row in the column containing the independent variable.
3. Click **Add-Ins, Data Analysis Plus,** and **Prediction Interval.**
4. Specify the **Input Y Range** (A1:A101), the **Input X Range** (B1:B101), the **Given X Range** (B102), and the **Confidence Level** (.95).

MINITAB

```
Predicted Values for New Observations

New
Obs    Fit      SE Fit        95% CI                95% PI
  1  14.5743   0.0382   (14.4985, 14.6501)   (13.9220, 15.2266)


Values of Predictors for New Observations

New
Obs   Odometer
  1      40.0
```

The output includes the predicted value $\hat{y}$ (**Fit**), the standard deviation of $\hat{y}$ (**SE Fit**), the 95% confidence interval estimate of the expected value of y (**CI**), and the 95% prediction interval (**PI**).

INSTRUCTIONS

1. Proceed through the three steps of regression analysis described on page 625. Do not click **OK.** Click **Options . . . .**
2. Specify the given value of x in the **Prediction intervals for new observations** box (40).
3. Specify the confidence level (.95).

INTERPRET

We predict that one car will sell for between \$13,925 and \$15,226. The average selling price of the population of 3-year-old Toyota Camrys is estimated to lie between \$14,498 and \$14,650. Because predicting the selling price of one car is more difficult than estimating the mean selling price of all similar cars, the prediction interval is wider than the interval estimate of the expected value.

Effect of the Given Value of x on the Intervals

Calculating the two intervals for various values of x results in the graph in Figure 16.8. Notice that both intervals are represented by curved lines. This is because the farther the given value of x is from $\bar{x}$, the greater the estimated error becomes.

FIGURE **16.8**
Interval Estimates and Prediction Intervals

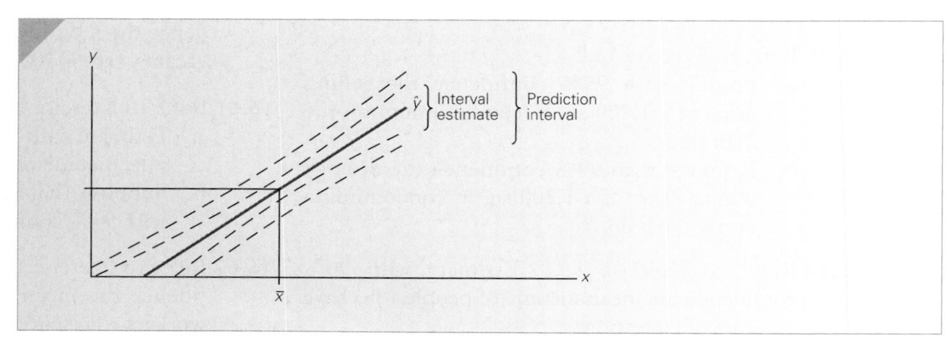

This part of the estimated error is measured by

$$\frac{(x_g - \bar{x})^2}{(n - 1)s_x^2}$$

which appears in both the prediction interval and the interval estimate of the expected value.

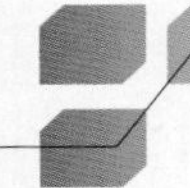

EXERCISES

16.45 Briefly describe the difference between predicting a value of y and estimating the expected value of y.

16.46 Use the regression equation in Exercise 16.2 to predict with 90% confidence the sales when the advertising budget is $90,000.

16.47 Estimate with 90% confidence the mean monthly number of housing starts when the mortgage interest rate is 8% in Exercise 16.3.

16.48 Refer to Exercise 16.4.

a. Predict with 90% confidence the number of pounds overweight for a child who watches 30 hours of television per week.

b. Estimate with 90% confidence the mean number of pounds overweight for children who watch 30 hours of television per week.

16.49 Refer to Exercise 16.5. Predict with 90% confidence the number of beers to be sold when the temperature is 80 degrees.

The following exercises require the use of a computer and software. The answers may be calculated manually. See Appendix A for the sample statistics.

16.50 Refer to Exercise 16.6.

a. Predict with 95% confidence the memory test score of a viewer who watches a 36-second commercial.

b. Estimate with 95% confidence the mean memory test score of people who watch 36-second commercials.

16.51 Refer to Exercise 16.7.

a. Predict with 95% confidence the selling price of a 1,200 sq. ft. condominium on the 25th floor.

b. Estimate with 99% confidence the average selling price of a 1,200 sq. ft. condominium on the 12th floor.

16.52 Refer to Exercise 16.8. Estimate with 90% confidence the mean income of people who have 15 years of education.

16.53 Refer to Exercise 16.9. The company has just hired a 25-year-old telemarketer. Predict with 95% confidence how long he will stay with the company.

16.54 Refer to Exercise 16.10. Predict with 95% confidence the number of sick days for individuals who smoke on average 30 cigarettes per day.

16.55 Refer to Exercise 16.11.

a. Predict with 95% confidence the percentage loss due to fire for a house that is 5 miles away from the nearest fire station.

b. Estimate with 95% confidence the average percentage loss due to fire for houses that are 2 miles away from the nearest fire station.

16.56 Refer to Exercise 16.12. Estimate with 95% confidence the mean price of 50,000 sq. ft. apartment buildings.

16.57 Refer to Exercise 16.13. Predict with 99% confidence the price of a 1999 24-foot Sea Ray cruiser with 500 hours of engine use.

16.58 Refer to Exercise 16.14. Estimate with 90% confidence the mean electricity consumption for households with 5 occupants.

16.59 Refer to Exercise 16.15. Predict the food budget of a family whose household income is $50,000. Use a 90% confidence level.

16.60 Refer to Exercise 16.16. Predict with 95% confidence the monthly office rent in a city when the vacancy rate is 10%.

16.61 Refer to Exercise 16.17.

a. Estimate with 95% confidence the mean annual income of 6'-tall men.

b. Suppose that an individual is 5'6". Predict with 95% confidence his annual income.

16.62 Refer to Exercise 16.18. Estimate with 95% confidence the mean percentage of defectives for workers who score 75 on the dexterity test.

16.6 Regression Diagnostics—I

In Section 16.3, we described the required conditions for the validity of regression analysis. Simply put, the error variable must be normally distributed with a constant variance, and the errors must be independent of each other. In this section, we show how to diagnose violations. Additionally, we discuss how to deal with observations that are unusually large or small. Such observations must be investigated to determine whether an error was made in recording them.

Residual Analysis

Most departures from required conditions can be diagnosed by examining the residuals, which we discussed in Section 16.4. Most computer packages allow you to output the values of the residuals and apply various graphical and statistical techniques to this variable.

We can also compute the standardized residuals. We standardize residuals in the same way we standardize all variables, by subtracting the mean and dividing by the standard deviation. The mean of the residuals is 0 and because the standard deviation σ_ε is unknown, we must estimate its value. The simplest estimate is the standard error of estimate s_ε. Thus

$$\text{Standardized residuals for point } i = \frac{e_i}{s_\varepsilon}$$

EXCEL

Excel calculates the standardized residuals by dividing the residuals by the standard deviation of the residuals. (The difference between the standard error of estimate and the standard deviation of the residuals is that in the formula of the former, the denominator is $n - 2$, whereas in the formula for the latter, the denominator is $n - 1$.)

Part of the printout (we show only the first five and last five values.) for Example 16.2 follows.

	A	B	C	D
1	RESIDUAL OUTPUT			
2				
3	*Observation*	*Predicted Price*	*Residuals*	*Standard Residuals*
4	1	14.748	-0.148	-0.456
5	2	14.253	-0.153	-0.472
6	3	14.186	-0.186	-0.574
7	4	15.183	0.417	1.285
8	5	15.129	0.471	1.449
9				
10				
11				
12	96	14.828	-0.028	-0.087
13	97	14.962	-0.362	-1.115
14	98	15.029	-0.529	-1.628
15	99	14.628	0.072	0.222
16	100	14.815	-0.515	-1.585

INSTRUCTIONS

Proceed with the three steps of regression analysis described on page 625. Before clicking **OK,** select **Residuals** and **Standardized Residuals.** The predicted values, residuals, and standardized residuals will be printed.

We can also standardize by computing the standard deviation of each residual. Statisticians have determined that the standard deviation of the residual for observation i is defined as follows.

Standard Deviation of the ith Residual

$$s_{r_i} = s_\varepsilon \sqrt{1 - h_i}$$

where

$$h_i = \frac{1}{n} + \frac{(x_i - \bar{x})^2}{(n-1)s_x^2}$$

The quantity h_i should look familiar; it was used in the formula for the prediction interval and confidence interval estimate of the expected value of y in Section 16.6. Minitab computes this version of the standardized residuals. Part of the printout (we show only the first five and last five values) for Example 16.2 is shown below.

MINITAB

Obs	Odometer	Price	Fit	SE Fit	Residual	St Resid
1	37.4	14.6000	14.7481	0.0334	-0.1481	-0.46
2	44.8	14.1000	14.2534	0.0546	-0.1534	-0.48
3	45.8	14.0000	14.1865	0.0586	-0.1865	-0.58
4	30.9	15.6000	15.1827	0.0414	0.4173	1.29
5	31.7	15.6000	15.1292	0.0391	0.4708	1.45
96	36.2	14.8000	14.8284	0.0327	-0.0284	-0.09
97	34.2	14.6000	14.9621	0.0339	-0.3621	-1.12
98	33.2	14.5000	15.0289	0.0355	-0.5289	-1.63
99	39.2	14.7000	14.6278	0.0363	0.0722	0.22
100	36.4	14.3000	14.8150	0.0327	-0.5150	-1.59

INSTRUCTIONS

Proceed with the three steps of regression analysis as described on page 625. After specifying the **Response** and **Predictors,** click **Results . . . , In addition, the full table of fits and residuals.** The predicted values, residuals, and standardized residuals will be printed.

An analysis of the residuals will allow us to determine whether the error variable is nonnormal, whether the error variance is constant, and whether the errors are independent. We begin with nonnormality.

Nonnormality

As we've done throughout this book, we check for normality by drawing the histogram of the residuals. Figure 16.9 is Excel's version (Minitab's is similar). As you can see, the histogram is bell shaped, leading us to believe that the error is normally distributed.

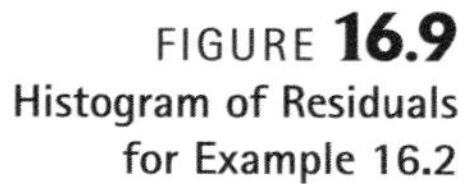

FIGURE **16.9**
Histogram of Residuals for Example 16.2

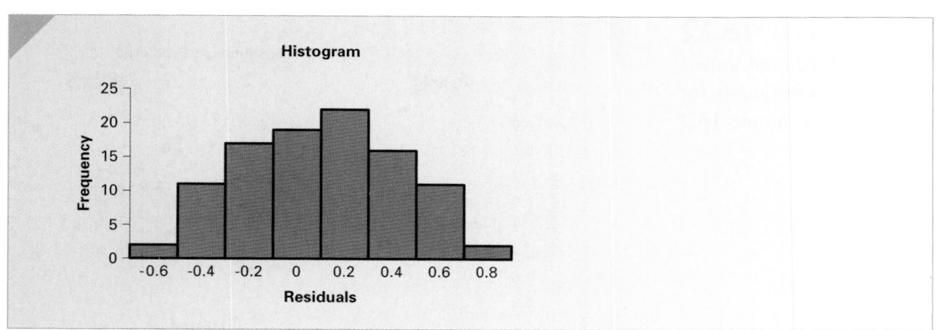

Heteroscedasticity

The variance of the error variable σ_ε^2 is required to be constant. When this requirement is violated, the condition is called **heteroscedasticity.** (You can impress friends and relatives by using this term. If you can't pronounce it, try **homoscedasticity,** which refers to the condition where the requirement is satisfied.) One method of diagnosing heteroscedasticity is to plot the residuals against the predicted values of y. We then look for a change in the spread of the plotted points.* Figure 16.10 describes such a situation. Notice that in this illustration, σ_ε^2 appears to be small when $\hat{y}$ is small and large when $\hat{y}$ is large. Of course, many other patterns could be used to depict this problem.

FIGURE **16.10**
Plot of Residuals Depicting Heteroscedasticity

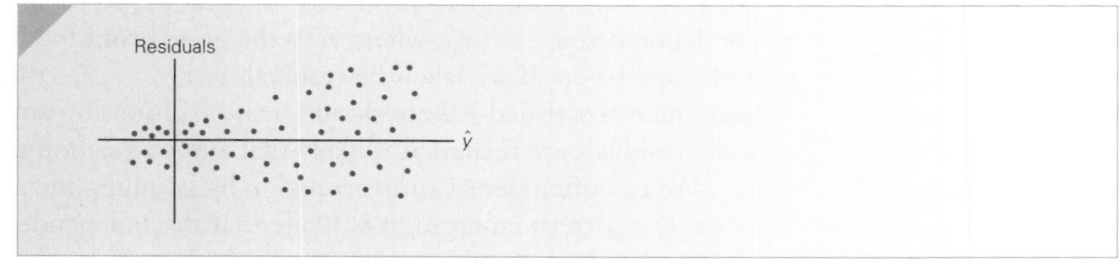

Figure 16.11 illustrates a case in which σ_ε^2 is constant. As a result, there is no apparent change in the variation of the residuals.

FIGURE **16.11**
Plot of Residuals Depicting Homoscedasticity

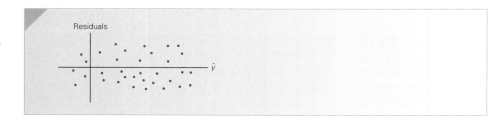

Excel's plot of the residuals versus the predicted values of y for Example 16.2 is shown in Figure 16.12. There is no sign of heteroscedasticity.

*CD Appendix Z describes Szroeter's test for heteroscedasticity.

FIGURE **16.12**
Plot of Predicted Values versus Residuals for Example 16.2

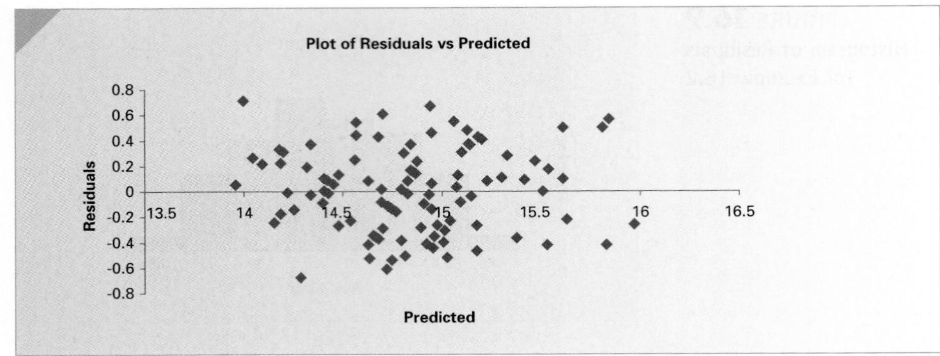

Nonindependence of the Error Variable

In Chapter 2, we briefly described the difference between cross-sectional and time series data. Cross-sectional data are observations made at approximately the same time, whereas a time series is a set of observations taken at successive points of time. The data in Example 16.2 are cross sectional because all of the prices and odometer readings were taken at about the same time. If we were to observe the auction price of cars every week for, say, a year, that would constitute a time series.

Condition 4 states that the values of the error variable are independent. When the data are time series, the errors often are correlated. Error terms that are correlated over time are said to be **autocorrelated** or **serially correlated.** For example, suppose that, in an analysis of the relationship between annual gross profits and some independent variable, we observe the gross profits for the years 1988 to 2007. The observed values of y are denoted $y_1, y_2, \ldots, y_{20}$, where y_1 is the gross profit for 1988, y_2 is the gross profit for 1989, and so on. If we label the residuals $e_1, e_2, \ldots, e_{20}$, then—if the independence requirement is satisfied—there should be no relationship among the residuals. However, if the residuals are related, it is likely that autocorrelation exists.

We can often detect autocorrelation by graphing the residuals against the time periods. If a pattern emerges, it is likely that the independence requirement is violated. Figures 16.13 (alternating positive and negative residuals) and 16.14 (increasing residuals) exhibit patterns indicating autocorrelation. (Notice that we joined the points to make it easier to see the patterns.) Figure 16.15 shows no pattern (the residuals appear to be randomly distributed over the time periods) and thus likely represent the occurrence of independent errors.

In Chapter 17, we introduce the Durbin-Watson test, which is another statistical test to determine whether one form of autocorrelation is present.

FIGURE **16.13**
Plot of Residuals versus Time Indicating Autocorrelation (Alternating)

FIGURE **16.14**
Plot of Residuals versus Time Indicating Autocorrelation (Increasing)

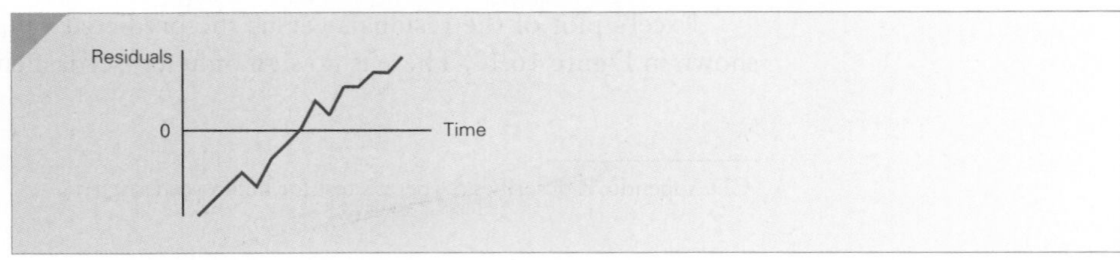

FIGURE **16.15**
Plot of Residuals versus Time Indicating Independence

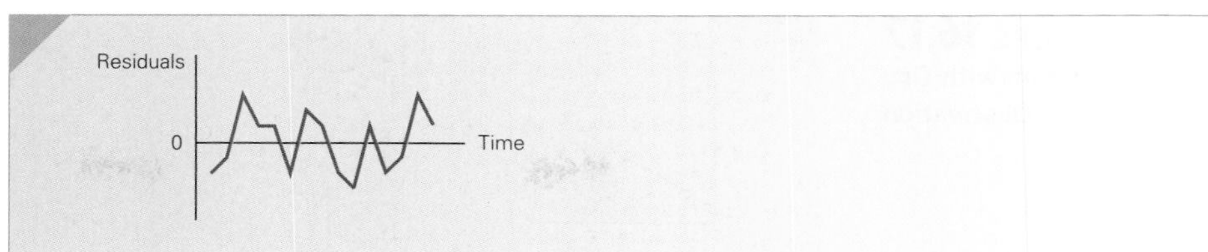

Outliers

An outlier is an observation that is unusually small or unusually large. To illustrate, consider Example 16.2, where the range of odometer readings was 19.1 to 49.2 thousand miles. If we had observed a value of 5,000 miles, we would identify that point as an outlier. We need to investigate several possibilities.

1. There was an error in recording the value. To detect an error, we would check the point or points in question. In Example 16.2, we could check the car's odometer to determine whether a mistake was made. If so, we would correct it before proceeding with the regression analysis.

2. The point should not have been included in the sample. Occasionally, measurements are taken from experimental units that do not belong with the sample. We can check to ensure that the car with the 5,000-mile odometer reading was actually 3 years old. We should also investigate the possibility that the odometer was rolled back. In either case, the outlier should be discarded.

3. The observation was simply an unusually large or small value that belongs to the sample and that was recorded properly. In this case, we would do nothing to the outlier. It would be judged to be valid.

Outliers can be identified from the scatter diagram. Figure 16.16 depicts a scatter diagram with one outlier. The statistics practitioner should check to determine whether the measurement was recorded accurately and whether the experimental unit should be included in the sample.

FIGURE **16.16**
Scatter Diagram with One Outlier

The standardized residuals also can be helpful in identifying outliers. Large absolute values of the standardized residuals should be thoroughly investigated. Minitab automatically reports standardized residuals that are less than -2 and greater than 2.

Influential Observations

Occasionally, in a regression analysis, one or more observations have a large influence on the statistics. Figure 16.17 describes such an observation and the resulting least squares line. If the point had not been included, the least squares line in Figure 16.18 would have been produced. Obviously, one point has had an enormous influence on the results. Influential points can be identified by the scatter diagram. The point may be an outlier and as such must be investigated thoroughly. Minitab also identities influential observations.

FIGURE **16.17**
Scatter Diagram with One Influential Observation

FIGURE **16.18**
Scatter Diagram without the Influential Observation

Procedure for Regression Diagnostics

The order of the material presented in this chapter is dictated by pedagogical requirements. Consequently, we presented the least squares methods of assessing the model's fit, predicting and estimating using the regression equation, coefficient of correlation, and finally, the regression diagnostics. In a practical application, the regression diagnostics would be conducted earlier in the process. It is appropriate to investigate violations of the required conditions when the model is assessed and before using the regression equation to predict and estimate. The following steps describe the entire process. (In Chapter 18, we will discuss model building, for which the following steps represent only a part of the entire procedure.)

1. Develop a model that has a theoretical basis. That is, for the dependent variable in question, find an independent variable that you believe is linearly related to it.
2. Gather data for the two variables. Ideally, conduct a controlled experiment. If that is not possible, collect observational data.
3. Draw the scatter diagram to determine whether a linear model appears to be appropriate. Identify possible outliers.
4. Determine the regression equation.
5. Calculate the residuals and check the required conditions:

   Is the error variable nonnormal?
   Is the variance constant?
   Are the errors independent?
   Check the outliers and influential observations.

6. Assess the model's fit.

   Compute the standard error of estimate.
   Test to determine whether there is a linear relationship. (Test β_1 or ρ.)
   Compute the coefficient of determination.

7. If the model fits the data, use the regression equation to predict a particular value of the dependent variable and/or estimate its mean.

EXERCISES

16.63 You are given the following six points:

x	−5	−2	0	3	4	7
y	15	9	7	6	4	1

a. Determine the regression equation.
b. Use the regression equation to determine the predicted values of y.
c. Use the predicted and actual values of y to calculate the residuals.
d. Compute the standardized residuals.
e. Identify possible outliers.

16.64 Refer to Exercise 16.2. Calculate the residuals and the predicted values of y.

16.65 Calculate the residuals and predicted values of y in Exercise 16.3.

16.66 Refer to Exercise 16.4.
a. Calculate the residuals.
b. Calculate the predicted values of y.
c. Plot the residuals (on the vertical axis) and the predicted values of y.

16.67 Calculate and plot the residuals and predicted values of y for Exercise 16.5.

The following exercises require the use of a computer and software.

16.68 Refer to Exercise 16.6.
a. Determine the residuals and the standardized residuals.
b. Draw the histogram of the residuals. Does it appear that the errors are normally distributed? Explain.
c. Identify possible outliers.
d. Plot the residuals versus the predicted values of y. Does it appear that heteroscedasticity is a problem? Explain.

16.69 Refer to Exercise 16.7.
a. Does it appear that the errors are normally distributed? Explain.
b. Does it appear that heteroscedasticity is a problem? Explain.

16.70 Are the required conditions satisfied in Exercise 16.8?

16.71 Refer to Exercise 16.9.
a. Determine the residuals and the standardized residuals.
b. Draw the histogram of the residuals. Does it appear that the errors are normally distributed? Explain.
c. Identify possible outliers.
d. Plot the residuals versus the predicted values of y. Does it appear that heteroscedasticity is a problem? Explain.

16.72 Refer to Exercise 16.10. Are the required conditions satisfied?

16.73 Refer to Exercise 16.11.
a. Determine the residuals and the standardized residuals.
b. Draw the histogram of the residuals. Does it appear that the errors are normally distributed? Explain.
c. Identify possible outliers.
d. Plot the residuals versus the predicted values of y. Does it appear that heteroscedasticity is a problem? Explain.

16.74 Check the required conditions for Exercise 16.12.

16.75 Refer to Exercise 16.13. Are the required conditions satisfied?

16.76 Refer to Exercise 16.14.
a. Determine the residuals and the standardized residuals.
b. Draw the histogram of the residuals. Does it appear that the errors are normally distributed? Explain.
c. Identify possible outliers.
d. Plot the residuals versus the predicted values of y. Does it appear that heteroscedasticity is a problem? Explain.

16.77 Are the required conditions satisfied for Exercise 16.15?

16.78 Check to ensure that the required conditions for Exercise 16.16 are satisfied.

16.79 Are the required conditions satisfied for Exercise 16.17?

16.80 Perform a complete diagnostic analysis for Exercise 16.18 to determine whether the required conditions are satisfied.

CHAPTER SUMMARY

Simple linear regression and correlation are techniques for analyzing the relationship between two interval variables. Regression analysis assumes that the two variables are linearly related. The least squares method produces estimates of the intercept and the slope of the regression line. Considerable effort is expended in assessing how well the linear model fits the data. We calculate the standard error of estimate, which is an estimate of the standard deviation of the error variable. We test the slope to determine whether there is sufficient evidence of a linear relationship. The strength of the linear association is measured by the coefficient of determination. When the model provides a good fit, we can use it to predict the particular value and to estimate the expected value of the dependent variable. We can also use the Pearson correlation coefficient to measure and test the relationship between two bivariate normally distributed variables. We completed this chapter with a discussion of how to diagnose violations of the required conditions.

IMPORTANT TERMS

Regression analysis 616
Dependent variable 616
Independent variable 616
Deterministic model 617
Probabilistic model 617
Error variable 618
First-order linear model 618
Simple linear regression model 618
Least squares method 619
Residuals 621
Sum of squares for error 621
Standard error of estimate 631
Coefficient of determination 631
Pearson coefficient of correlation 642
Prediction interval 648
Confidence interval estimate of the expected value of *y* 648
Heteroscedasticity 655
Homoscedasticity 655
Autocorrelated 656
Serially correlated 656

SYMBOLS

Symbol	Pronounced	Represents
β_0	*Beta-sub-zero* or *beta-zero*	*y*-intercept coefficient
β_1	*Beta-sub-one* or *beta-one*	Slope coefficient
ε	*Epsilon*	Error variable
$\hat{y}$	*y-hat*	Fitted or calculated value of *y*
b_0	*b-sub-zero* or *b-zero*	Sample *y*-intercept coefficient
b_1	*b-sub-one* or *b-one*	Sample slope coefficient
σ_ε	*Sigma-sub-epsilon* or *sigma-epsilon*	Standard deviation of error variable
s_ε	*s-sub-epsilon* or *s-epsilon*	Standard error of estimate
s_{b_1}	*s-sub-b-sub-one* or *s-b-one*	Standard error of b_1
R^2	*R-squared*	Coefficient of determination
x_g	*x-sub-g* or *x-g*	Given value of *x*
ρ	*Rho*	Pearson coefficient of correlation
r		Sample coefficient of correlation
e_i	*e-sub-i* or *e-i*	Residual of *i*th point

FORMULAS

Sample slope

$$b_1 = \frac{s_{xy}}{s_x^2}$$

Sample *y*-intercept

$$b_0 = \bar{y} - b_1\bar{x}$$

Sum of squares for error

$$\text{SSE} = \sum_{i=1}^{n}(y_i - \hat{y}_i)^2$$

Standard error of estimate

$$s_\varepsilon = \sqrt{\frac{\text{SSE}}{n-2}}$$

Test statistic for the slope

$$t = \frac{b_1 - \beta_1}{s_{b_1}}$$

Standard error of b_1

$$s_{b_1} = \frac{s_\varepsilon}{\sqrt{(n-1)s_x^2}}$$

Coefficient of determination

$$R^2 = \frac{s_{xy}^2}{s_x^2 s_y^2} = 1 - \frac{\text{SSE}}{\sum(y_i - \bar{y})^2}$$

Prediction interval

$$\hat{y} \pm t_{\alpha/2,n-2}s_\varepsilon\sqrt{1 + \frac{1}{n} + \frac{(x_g - \bar{x})^2}{(n-1)s_x^2}}$$

Confidence interval estimator of the expected value of y

$$\hat{y} \pm t_{\alpha/2,n-2}s_\varepsilon\sqrt{\frac{1}{n} + \frac{(x_g - \bar{x})^2}{(n-1)s_x^2}}$$

Sample coefficient of correlation

$$r = \frac{s_{xy}}{s_x s_y}$$

Test statistic for testing $\rho = 0$

$$t = r\sqrt{\frac{n-2}{1-r^2}}$$

COMPUTER OUTPUT AND INSTRUCTIONS

Technique	Excel	Minitab
Regression	624	625
Correlation	644	644
Prediction interval	650	651
Regression diagnostics	653	654

CHAPTER EXERCISES

The following exercises require the use of a computer and software. The answers to some of the questions may be calculated manually. See Appendix A for the sample statistics. ***Conduct all tests of hypotheses at the 5% significance level.***

16.81 Xr16-81 The manager of Colonial Furniture has been reviewing weekly advertising expenditures. During the past 6 months, all advertisements for the store have appeared in the local newspaper. The number of ads per week has varied from one to seven. The store's sales staff has been tracking the number of customers who enter the store each week. The number of ads and the number of customers per week for the past 26 weeks were recorded.

a. Determine the sample regression line.
b. Interpret the coefficients.
c. Can the manager infer that the larger the number of ads, the larger the number of customers?
d. Find and interpret the coefficient of determination.
e. In your opinion, is it a worthwhile exercise to use the regression equation to predict the number of customers who will enter the store, given that Colonial intends to advertise five times in the newspaper? If so, find a 95% prediction interval. If not, explain why not.

16.82 Xr16-82 The president of a company that manufactures car seats has been concerned about the number and cost of machine breakdowns. The problem is that the machines are old and becoming quite unreliable. However, the cost of replacing them is quite high, and the president is not certain that the cost can be made up in today's slow economy. To help make a decision about replacement, he gathered data about last month's costs for repairs and the ages (in months) of the plant's 20 welding machines.

a. Find the sample regression line.
b. Interpret the coefficients.
c. Determine the coefficient of determination, and discuss what this statistic tells you.
d. Conduct a test to determine whether the age of a machine and its monthly cost of repair are linearly related.
e. Is the fit of the simple linear model good enough to allow the president to predict the monthly repair cost of a welding machine that is 120 months old? If so, find a 95% prediction interval. If not, explain why not.

16.83 Xr16-83 An agronomist wanted to investigate the factors that determine crop yield. Accordingly, she undertook an experiment wherein a farm was divided into 30 1-acre plots. The amount of fertilizer applied to each plot was varied. Corn was then planted, and the amount of corn harvested at the end of the season was recorded.

a. Find the sample regression line, and interpret the coefficients.
b. Can the agronomist conclude that there is a linear relationship between the amount of fertilizer and the crop yield?
c. Find the coefficient of determination, and interpret its value.
d. Does the simple linear model appear to be a useful tool in predicting crop yield from the amount of fertilizer applied? If so, produce a 95% prediction interval of the crop yield when 300 pounds of fertilizer are applied. If not, explain why not.

16.84 Xr16-84 Every year the United States Federal Trade Commission rates cigarette brands according to their levels of tar and nicotine, substances that are hazardous to smokers' health. Additionally, the commission includes the amount of carbon monoxide, which is a by-product of burning tobacco that seriously affects the heart. A random sample of 25 brands was taken.

a. Are the levels of tar and nicotine linearly related?
b. Are the levels of nicotine and carbon monoxide linearly related?

16.85 Xr16-85 Some critics of television complain that the amount of violence shown on television contributes to violence in our society. Others point out that television also contributes to the high level of obesity among children. We may have to add financial problems to the list. A sociologist theorized that people who watch television frequently are exposed to many commercials, which in turn leads them to buy more, finally resulting in increasing debt. To test this belief, a sample of 430 families was drawn. For each, the total debt and the number of hours the television is turned on per week were recorded. Perform a statistical procedure to help test the theory.

16.86 Xr16-86 The analysis the human resources manager performed in Exercise 16.18 indicated that the dexterity test is not a predictor of job performance. However, before discontinuing the test he decided that the problem is that the statistical analysis was flawed because it examined the relationship between test score and job performance only for those who scored well in the test. (Recall that only those who scored above 70 were hired; applicants who achieved scores below 70 were not hired.) The manager decided to perform another statistical analysis. A sample of 50 job applicants who scored above 50 were hired, and as before, the workers' performance was measured. The test scores and percentages of nondefective computers produced were recorded. On the basis of these data should the manager discontinue the dexterity tests?

16.87 Xr16-87 Mutual funds minimize risks by diversifying the investments they make. There are mutual funds that specialize in particular types of investments. For example, the TD Precious Metal Mutual Fund buys shares in gold mining companies. The value of this mutual fund depends on a number of factors related to the companies in which the fund invests as well as on the price of gold. To investigate the relationship between the value of the fund and the price of gold, an MBA student gathered the daily fund price and the daily price of gold for a 28-day period. Can we infer from these data that there is a positive linear relationship between the value of the fund and the price of gold? (The authors are grateful to Jim Wheat for writing this exercise.)

16.88 (Exercise 2.95 revisited) Xr02-95 A very large contribution to profits for a movie theater is the sale of popcorn, soft drinks, and candy. A movie theater manager speculated that the longer the time between showings of a movie, the greater the sales of concessions. To acquire more information, the manager conducted an experiment. For a month he varied the amount of time between movie showings and calculated the sales. Can the manager conclude that when the times between movies increase, so do sales?

The following exercises employ data files associated with two previous exercises.

16.89 Xr12-29* In addition to the data recorded for Exercises 12.29 and 13.114, we recorded the grade point average of the students who held down part-time jobs. Determine whether there is evidence of a linear relationship between the hours spent at part-time jobs and the grade point averages.

16.90 Xr13-13* Exercise 13.13 described a survey that asked people between 18 and 34 years of age and 35 to 50 years of age how much time they spent listening to FM radio each day. Also recorded were the amounts spent on music throughout the year. Can we infer that a linear relationship exists between listening times and amounts spent on music?

CASE 16.1 Insurance Compensation for Lost Revenues*

DATA C16-01

© Spencer Grant/PhotoEdit

In July 1990, a rock-and-roll museum opened in Atlanta, Georgia. The museum was located in a large city block containing a variety of stores. In late July 1992, a fire that started in one of these stores burned the entire block, including the museum. Fortunately, the museum had taken out insurance to cover the cost of rebuilding as well as lost revenue. As a general rule, insurance companies base their payment on how well the company performed in the past. However, the owners of the museum argued that the revenues were increasing, and hence they were entitled to more money under their insurance plan. The argument was based on the revenues and attendance figures of an amusement park featuring rides and other similar attractions that had opened nearby. The amusement park opened in December 1991. The two entertainment facilities were operating jointly during the last 4 weeks of 1991 and the first 28 weeks of 1992 (the point at which the fire destroyed the museum). In April 1995, the museum reopened with considerably more features than the original one.

The attendance figures for both facilities for December 1991 to October 1995 are listed in columns 1 (museum) and 2 (amusement park). During the period when the museum was closed, the data show zero attendance.

The owners of the museum argued that the weekly attendance from the 29th week of 1992 to the 16th week of 1995 should be estimated using the most current data (17th to 42nd week of 1995). The insurance company argued that the estimates should be based on the 4 weeks of 1991 and the 28 weeks of 1992, when both facilities were operating and before the museum reopened with more features than the original museum.

a. Estimate the coefficients of the simple regression model based on the insurance company's argument. That is, use the attendance figures for the last 4 weeks in 1991 and the next 28 weeks in 1992 to estimate the coefficients. Then, use the model to calculate point predictions for the museum's weekly attendance figures when the museum was closed. Calculate the predicted total attendance.

b. Repeat Part a using the museum's argument. That is, use the attendance figures after the reopening in 1995 to estimate the regression coefficients and use the equation to predict the weekly attendance when the museum was closed. Calculate the total attendance that was lost because of the fire.

c. Write a report to the insurance company discussing this analysis and include your recommendation about how much the insurance company should award the museum?

*The case and the data are real. The names have been changed to preserve anonymity. The authors wish to thank Dr. Kevin Leonard for supplying the problem and the data.

CASE 16.2 Predicting University Grades from High School Grades*

© Purestock/Jupiterimages

DATA C16-02

Ontario high school students must complete a minimum of six Ontario Academic Credits (OACs) to gain admission to a university in the province. Most students take more than six OACs because universities take the average of the best six in deciding which students to admit. Most programs at universities require high school students to select certain courses. For example, science programs require two courses of chemistry, biology, and physics. Students applying to engineering must complete at least two mathematics OACs as well as physics. In recent years, one business program began an examination of all aspects of its program including the criteria used to admit students. Students are required to take English and calculus OACs, and the minimum high school average is about 85%. Strangely enough, even though students are required to complete English and calculus, the marks in these subjects are not included in the average unless they are in the top six courses in a student's transcript. To examine the issue, the registrar took a random sample of students who recently graduated with the BBA (Bachelor of Business Administration) degree. He recorded the university GPA (range 0 to 12), the high school average based on the best six courses, and the high school average using English and calculus and the four next best marks.

a. Is there a relationship between university grades and high school average using the best six OACs?
b. Is there a relationship between university grades and high school average using the best four OACs plus calculus and English?
c. Write a report to the university's academic vice president describing your statistical analysis and your recommendations.

---

*The authors are grateful to Leslie Grauer for her help in gathering the data for this case.

APPENDIX 16 / REVIEW OF CHAPTERS 12 TO 16

We have now presented more than two dozen inferential techniques. Undoubtedly, the task of choosing the appropriate technique is growing more difficult. Table A16.1 lists all the statistical inference methods covered since Chapter 12. Figure A16.1 is a flowchart to help you choose the correct technique.

TABLE **A16.1** **Summary of Statistical Techniques in Chapters 12 to 16**

t-test of μ

Estimator of μ (including small population estimator of μ and large and small population estimators of $N\mu$)

χ^2-test of σ^2

Estimator of σ^2

z-test of p

Estimator of p (including small population estimator of p and large and small population estimators of Np)

Equal-variances t-test of $\mu_1 - \mu_2$

Equal-variances estimator of $\mu_1 - \mu_2$

Unequal-variances t-test of $\mu_1 - \mu_2$

Unequal-variances estimator of $\mu_1 - \mu_2$

t-test of μ_D

Estimator of μ_D

F-test of σ_1^2/σ_2^2

Estimator of σ_1^2/σ_2^2

z-test of $p_1 - p_2$ (Case 1)

z-test of $p_1 - p_2$ (Case 2)

Estimator of $p_1 - p_2$

One-way analysis of variance (including multiple comparisons)

Two-way (randomized blocks) analysis of variance

Two-factor analysis of variance

χ^2-goodness-of-fit test

χ^2-test of a contingency table

Simple linear regression and correlation (including t-tests of β_1 and ρ, and prediction and confidence intervals)

FIGURE **A16.1** Flowchart of Techniques in Chapters 12 to 16

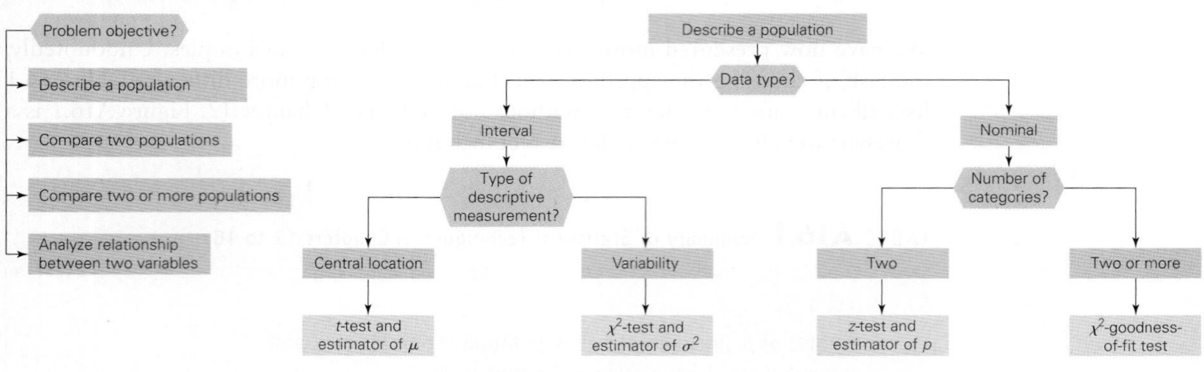

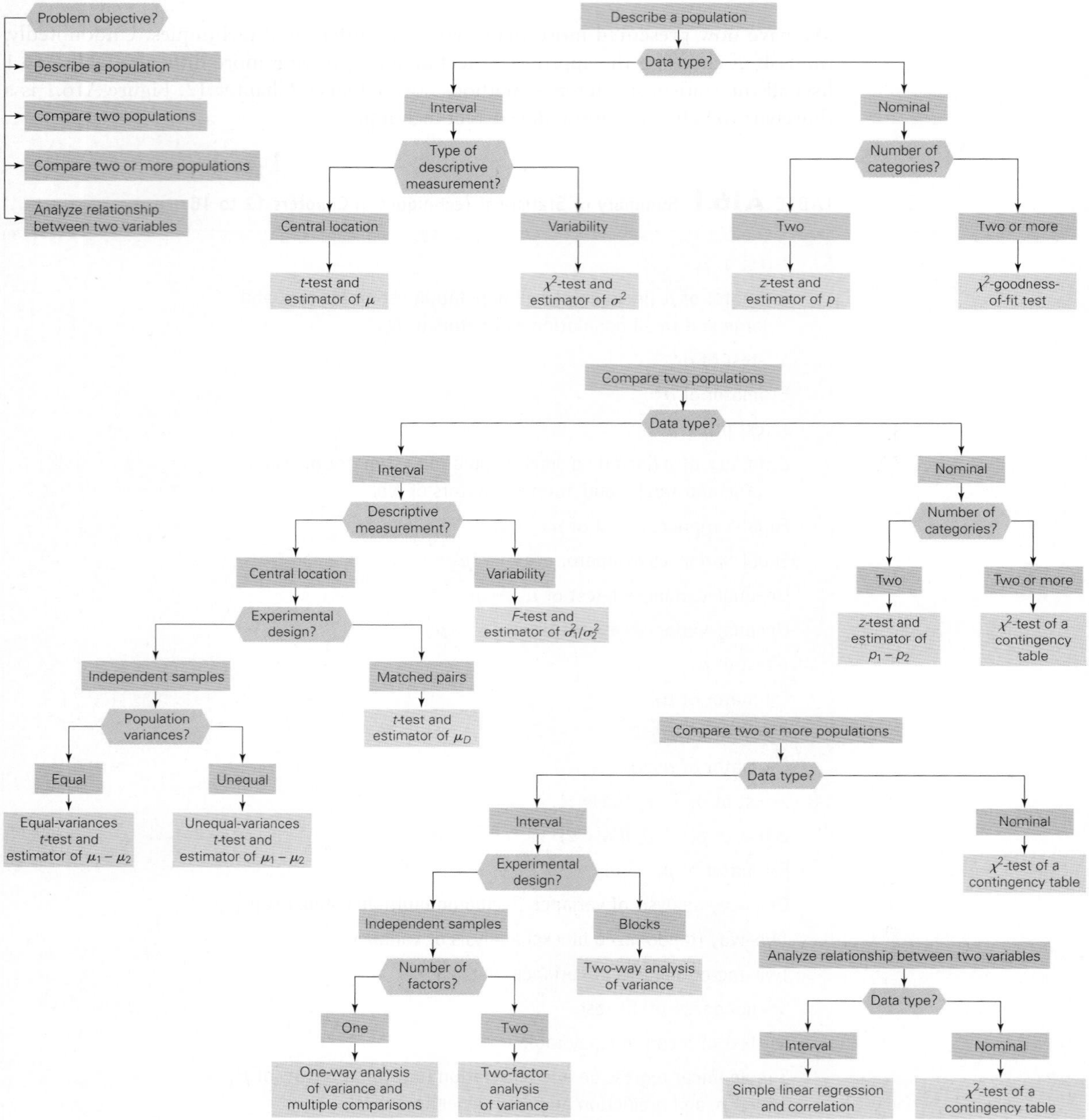

EXERCISES

A16.1 XrA16-01 In the last decade, society in general and the judicial system in particular have altered their opinions on the seriousness of drunken driving. In most jurisdictions, driving an automobile with a blood alcohol level in excess of .08 is a felony. Because of a number of factors, it is difficult to provide guidelines on when it is safe for someone who has consumed alcohol to drive a car. In an experiment to examine the relationship between blood alcohol level and the weight of a drinker,

50 men of varying weights were each given three beers to drink, and one hour later their blood alcohol level was measured. If we assume that the two variables are normally distributed, can we conclude that blood alcohol level and weight are related?

A16.2 XrA16-02 An article in the journal *Appetite* (Vol. 41, No. 3, December 2003) described an experiment to determine the effect that breakfast meals have on school children. A sample of 29 children was tested on four successive days, having a different breakfast each day. The breakfast meals were

1. Cereal (Cheerios)
2. Cereal (Shreddies)
3. A glucose drink
4. No breakfast

The order of breakfast meals was randomly assigned. A computerized test of working memory was conducted prior to breakfast and again two hours later. The decrease in scores was recorded. Do these data allow us to infer that there are differences in the decrease depending on the type of breakfast?

A16.3 Do cell phones cause cancer? This is a multibillion dollar question. There are currently dozens of lawsuits pending that claim cell phone use has caused cancer. To help shed light on the issue, several scientific research projects have been undertaken. One such project was conducted by Danish researchers (*Source*: *Journal of the National Cancer Institute*, 2001). The 13-year study examined 420,000 Danish cell phone users. The scientists determined the number of Danes who would be expected to contract various forms of cancer. The expected number and the actual number of cell phone users who developed each type of cancer are listed here.

Cancer	Expected Number	Actual Number
Brain and nervous system	143	135
Salivary glands	9	7
Leukemia	80	77
Pharynx	52	32
Esophagus	57	42
Eye	12	8
Thyroid	13	13

a. Can we infer from these data that there is a relationship between cell phone use and cancer?
b. Discuss the results, including whether the data are observational or experimental. Provide several interpretations of the statistics. In particular, indicate whether you can infer that cell phone use causes cancer.

A16.4 XrA16-04 A new antiflu vaccine designed to reduce the duration of symptoms has been developed. However, the effect of the drug varies from person to person. To examine the effect of age on the effectiveness of the drug, a sample of 140 flu sufferers was drawn. Each person reported how long the symptoms of the flu persisted and his or her age. Do these data provide sufficient evidence to infer that the older the patient, the longer it takes for the symptoms to disappear?

A16.5 XrA16-05 Several years ago we heard about the "Mommy Track," the phenomenon of women being underpaid in the corporate world because of what is seen as their divided loyalties between home and office. There may also be a "Daddy Differential," which refers to the situation where men whose wives stay at home earn more than men whose wives work. It is argued that the differential occurs because bosses reward their male employees if they come from "traditional families." Linda Stroh of Loyola University of Chicago studied a random sample of 348 male managers employed by 20 *Fortune 500* companies. Each manager reported whether his wife stayed at home to care for their children or worked outside the home, and his annual income. The incomes (in thousands of dollars) were recorded. The incomes of the managers whose wives stay at home are stored in column 1. Column 2 contains the incomes of managers whose wives work outside the home.

a. Can we conclude that men whose wives stay at home earn more than men whose wives work outside the home?
b. If your answer in Part a is affirmative, does this establish a case for discrimination? Can you think of another cause-and-effect scenario? Explain.

A16.6 XrA16-06 There are enormous differences between health care systems in the United States and Canada. In a study to examine one dimension of these differences, 300 heart attack victims in each country were randomly selected. (Results of the study conducted by Dr. Daniel Mark of Duke University Medical Center, Dr. David Naylor of Sunnybrook Hospital in Toronto, and Dr. Paul Armstrong of the University of Alberta were published in the *Toronto Sun*, October 27, 1994.) Each patient was asked the following questions regarding the effect of his or her treatment:

1. How many days did it take you to return to work?
2. Do you still have chest pain? (This question was asked 1 month, 6 months, and 12 months after the patients' heart attacks.)

The responses were recorded in the following way:

Column 1: Code representing nationality: 1 = U.S.; 2 = Canada
Column 2: Responses to question 1
Column 3: Responses to question 2—1 month after heart attack: 2 = yes; 1 = no
Column 4: Responses to question 2—6 months after heart attack: 2 = yes; 1 = no
Column 5: Responses to question 2—12 months after heart attack: 2 = yes; 1 = no

Can we conclude that recovery is faster in the United States?

A16.7 XrA16-07 Betting on the results of National Football League games is a popular North American activity. In many states and provinces, it is legal to do so provided that wagers are made through the government-authorized betting organization. In the province of Ontario, Pro-Line serves that function. Bettors can choose any team on which to wager, and Pro-Line sets the odds, which determine the winning payoffs. It is also possible to bet that in any game a tie will be the result. (A tie is defined as a game in which the winning margin is 3 or fewer points. A win occurs when the winning margin is greater than 3.) To assist bettors, Pro-Line lists the favorite for each game and predicts the point spread between the two teams. To judge how well Pro-Line predicts outcomes, the Creative Statistics Company tracked the results of a recent season. It recorded whether a team was favored by 3 or fewer points (1); 3.5 to 7 points (2); 7.5 to 11 points (3); or 11.5 or more points (4). It also recorded whether the favored team won (1); lost (2); or tied (3). These data are recorded in columns 1 (Pro-Line's predictions) and 2 (game results). Can we conclude that Pro-Line's forecasts are useful for bettors?

A16.8 XrA16-08 As all baseball fans know, first base is the only base that the base runner may overrun. At second and third base, the runner may be tagged out if he runs past them. Consequently, on close plays at second and third base, the runner will slide, enabling him to stop at the base. In recent years, however, several players have chosen to slide headfirst when approaching first base, claiming that this is faster than simply running over the base. In an experiment to test this claim, the 25 players on one National League team were recruited. Each player ran to first base with and without sliding, and the times to reach the base were recorded. Can we conclude that sliding is slower than not sliding?

A16.9 XrA16-09 How does mental outlook affect a person's health? The answer to this question may allow physicians to care more effectively for their patients. In an experiment to examine the relationship between attitude and physical health, Dr. Daniel Mark, a heart specialist at Duke University, studied 1,719 men and women who had recently undergone a heart catheterization, a procedure that checks for clogged arteries. Patients undergo this procedure when heart disease results in chest pain. All of the patients in the experiment were in about the same condition. In interviews, 14% of the patients doubted that they would recover sufficiently to resume their daily routines. Dr. Mark identified these individuals as pessimists; the others were (by default) optimists. After one year, Dr. Mark recorded how many patients were still alive. The data are stored in columns 1 (1 = optimist, 2 = pessimist) and 2 (2 = alive, 1 = dead). Do these data allow us to infer that pessimists are less likely to survive than optimists with similar physical ailments?

A16.10 XrA16-10 Physicians have been recommending more exercise for their patients, particularly those who are overweight. One benefit of regular exercise appears to be a reduction in cholesterol, a substance associated with heart disease. To study the relationship more carefully, a physician took a random sample of 50 patients who do not exercise and measured their cholesterol levels. He then started them on regular exercise programs. After 4 months, he asked each patient how many minutes per week (on average) he or she exercised and also measured their cholesterol levels. (column 1 = weekly exercise in minutes; column 2 = cholesterol level before exercise program; and column 3 = cholesterol level after exercise program).

a. Do these data allow us to infer that the amount of exercise and the reduction in cholesterol levels are related?
b. Produce a 95% interval of the amount of cholesterol reduction for someone who exercises for 100 minutes per week.
c. Produce a 95% interval for the average cholesterol reduction for people who exercise for 120 minutes per week.

A16.11 XrA16-11 An economist working for a state university wanted to acquire information about salaries in publicly funded and private colleges and universities. She conducted a survey of 926 faculty members, asking each to report his or her rank (instructor, assistant professor, associate professor, and professor) and current salary. For each rank, determine whether there is enough evidence to infer that the private college and university salaries are higher than those at publicly funded ones. (Adapted from American Association of University Professors, *AAUP Annual Report on the Economic Status of the Profession*) [*Statistical Abstract of the United States*, 2006, Table 282]

A16.12 XrA16-12 Millions of people suffer from migraine headaches. The costs in work days lost, medication, and treatment are measured in the billions of dollars. A study reported in the *Journal of the American Medical Association* (2005, 203: 2118–2125) described an experiment that examined whether acupuncture is an effective procedure in treating migraines. A random sample of 302 migraine patients was selected and divided into three groups. Group 1 was treated with acupuncture; group 2 was treated with sham acupuncture (patients believed that they were being treated with acupuncture, but were not); group 3 was not treated at all. The number of headache days per month was recorded for each patient before the treatments began. The number of headache days per month after treatment was also measured.

a. Conduct a test to determine whether there are differences in the number of headache days before treatment between the three groups of patients.
b. Test to determine whether differences exist after treatment. If so, what are the differences?
c. Why was the test in Part a conducted?

A16.13 XrA16-13 The battle between customers and car dealerships is often intense. Customers want the lowest price and dealers want to extract as much money as possible. One source of conflict is the trade-in car. Most dealers will offer a relatively low trade-in in anticipation of negotiating the final package. In an effort to determine how dealers operate, a consumer organization undertook an experiment. Seventy-two individuals were recruited. Each solicited an offer on their 5-year-old Toyota Camry. The exact same car was used throughout the experiment. The only variables were the age and gender of the "owner." The ages were categorized as (1) young, (2) middle, and (3) senior. The cash offers are stored in columns 1 and 2. Column 1 stores the data for female owners and column 2 contains the offers made to male owners. The first 12 rows in both columns represent the offers made to young people, the next 12 rows represent the middle group, and the last 12 rows represent the elderly owners.

a. Can we infer that differences exist between the six groups?
b. If differences exist, determine whether the differences are due to gender, age, or some interaction.

A16.14 XrA16-14 In the presidential elections in 2000 and 2004, the vote in the state of Florida was crucial. It is important for the political parties to track party affiliation. The numbers of Democrats, Republicans, and other parties were recorded for both counties and for all 4 years. Surveys in Broward County and in Miami-Dade were conducted in 1990, 1996, 2000, and 2004. Test each of the following.

a. Party affiliation changed over the four surveys in Broward.
b. Party affiliation changed over the four surveys in Miami-Dade.
c. There were differences between Broward and Miami-Dade in 2004.

A16.15 XrA16-15 Auto manufacturers are required to test their vehicles for a variety of pollutants in the exhaust. The amount of pollutant varies even among identical vehicles, so that several vehicles must be tested. The engineer in charge of testing has collected data (in grams per kilometer driven) on the amounts of two pollutants, carbon monoxide and nitrous oxide, for 50 identical vehicles. The engineer believes the company can save money by testing for only one of the pollutants because the two pollutants are closely linked. That is, if a car is emitting a large amount of carbon monoxide, it will also emit a large amount of nitrous oxide. Do the data support the engineer's belief?

A16.16 In 2003, there were 129,142,000 workers in the United States (*Source:* U.S. Census Bureau). The general manager for a public transportation company wanted to learn more about how workers commute to work and how long it takes them. A random sample of workers was interviewed. Each reported how they typically get to work and how long it took them. Estimate with 95% confidence the total amount of time spent commuting. (Data for this exercise were adapted from the *Statistical Abstract of the United States*, 2006, Table 1083)

CASE A16.1 Do Banks Discriminate against Women Business Owners? Part 2

DATA CA16-01

To help explain the apparent discrimination against women documented in Case 13.1, researchers performed further analyses. In the original study, the following pieces of information were gathered for each company:

1. Form of business
   a. Proprietorship
   b. Partnership
   c. Corporation
2. Annual gross sales
3. Age of the firm

These data, together with the data from Case 13.1, were stored in the following way.

Column 1: Rates above prime paid by women
Column 2: Type of women's business (1 = proprietorship: 2 = partnership; 3 = corporation)
Column 3: Annual gross sales of women's businesses (in thousands of dollars)
Column 4: Age of women's businesses
Column 5: Rates above prime paid by men
Column 6: Type of men's business (1 = proprietorship; 2 = partnership; 3 = corporation)
Column 7: Annual gross sales of men's businesses (in thousands of dollars)
Column 8: Age of men's businesses

What do these data tell you about the alleged discrimination against women by banks?

CASE A16.2 Nutrition Education Programs*

DATA CA16-02

Nutrition education programs, which teach their clients how to lose weight or reduce cholesterol levels through better eating patterns, have been growing in popularity. The nurse in charge of one such program at a local hospital wanted to know whether the programs actually work. A random sample of 33 clients who attended a nutrition education program for those with elevated cholesterol levels was drawn. The study recorded the weight, cholesterol levels, total dietary fat intake per average day, total dietary cholesterol intake per average day, and percent of daily calories from fat. These data were gathered both before and 3 months after the program. The researchers also determined the gender, age, and height of the clients. The data are stored in the following way:

Column 1: Gender (1 = female; 2 = male)
Column 2: Age
Column 3: Height (in meters)
Columns 4 and 5: Weight, before and after (in kilograms)
Columns 6 and 7: Cholesterol level, before and after
Columns 8 and 9: Total dietary fat intake per average day, before and after (in grams)
Columns 10 and 11: Dietary cholesterol intake per average day, before and after (in milligrams)
Columns 12 and 13: Percent daily calories from fat, before and after

The nurse would like the following information:

a. In terms of each of weight, cholesterol level, fat intake, cholesterol intake, and calories from fat, is the program a success?
b. Does gender affect the amount of reduction in each of weight, cholesterol level, fat intake, cholesterol intake, and calories from fat?
c. Does age affect the amount of reduction in weight, cholesterol level, fat intake, cholesterol intake, and calories from fat cholesterol?

*The authors would like to thank Karen Cavrag for writing this case.

© Comstock Images/Jupiterimages

17 MULTIPLE REGRESSION

MBA Program Admissions Policy

DATA Xm17-00*

The MBA program at a large university is facing a pleasant problem—too many applicants. The current admissions policy requires students to have completed at least 3 years of work experience and an undergraduate degree with a B− average or better. Until 3 years ago, the school admitted any applicant who met these requirements. However, because the program recently converted from a 2-year program (four semesters) to a 1-year program (three semesters), the number of applicants has increased substantially. The dean, who teaches statistics courses, wants to raise the admissions standards by developing a method that more accurately predicts how well an applicant will perform in the MBA program. She believes that

© Eyewire Collection/Getty Images

Our answer appears on page 686.

the primary determinants of success are the following:

Undergraduate grade point average (GPA)

Graduate Management Admissions Test (GMAT) score

Number of years of work experience

She randomly sampled students who completed the MBA and recorded their MBA program GPA, as well as the three variables listed here. Develop a plan to decide which applicants to admit.

INTRODUCTION

In the previous chapter, we employed the simple linear regression model to analyze how one variable (the dependent variable y) is related to another interval variable (the independent variable x). The restriction of using only one independent variable was motivated by the need to simplify the introduction to regression analysis. Although there are a number of applications where we purposely develop a model with only one independent variable (see Section 4.6, for example), in general we prefer to include as many independent variables as are believed to affect the dependent variable. Arbitrarily limiting the number of independent variables also limits the usefulness of the model.

In this chapter, we allow for any number of independent variables. In so doing, we expect to develop models that fit the data better than would a simple linear regression model. We begin by describing the multiple regression model and listing the required conditions. We let the computer produce the required statistics and use them to assess the model's fit and diagnose violations of the required conditions. We use the model by interpreting the coefficients, predicting the particular value of the dependent variable, and estimating its expected value.

17.1 MODEL AND REQUIRED CONDITIONS

We now assume that k independent variables are potentially related to the dependent variable. Thus, the model is represented by the following equation:

$$y = \beta_0 + \beta_1 x_1 + \beta_2 x_2 + \cdots + \beta_k x_k + \varepsilon$$

where y is the dependent variable, $x_1, x_2, \ldots, x_k$ are the independent variables, $\beta_0, \beta_1, \ldots, \beta_k$ are the coefficients, and ε is the error variable. The independent variables may actually be functions of other variables. For example, we might define some of the independent variables as follows:

$$x_2 = x_1^2$$

$$x_5 = x_3 x_4$$

$$x_7 = \log(x_6)$$

In Chapter 18, we will discuss how and under what circumstances such functions can be used in regression analysis.

The error variable is retained because, even though we have included additional independent variables, deviations between predicted values of y and actual values of y will still occur. Incidentally, when there is more than one independent variable in the regression model, we refer to the graphical depiction of the equation as a **response surface** rather than as a straight line. Figure 17.1 depicts a scatter diagram of a response surface with $k = 2$. (When $k = 2$, the regression equation creates a plane.) Of course, whenever k is greater than 2, we can only imagine the response surface; we cannot draw it.

FIGURE **17.1** Scatter Diagram and Response Surface with $k = 2$

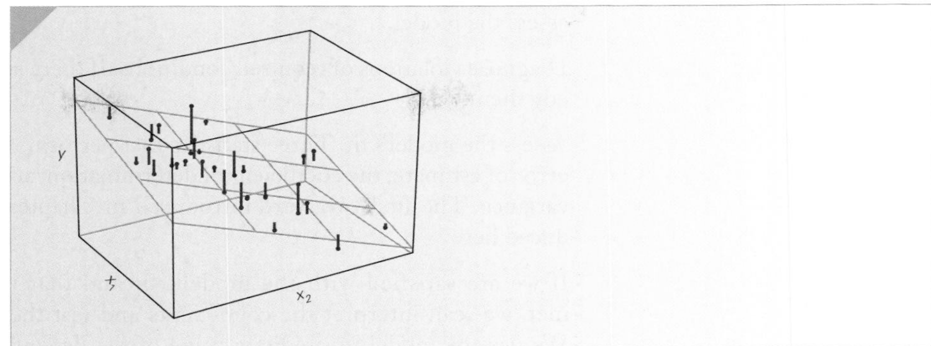

An important part of the regression analysis comprises several statistical techniques that evaluate how well the model fits the data. These techniques require the following conditions, which we introduced in the previous chapter.

Required Conditions for Error Variable

1. The probability distribution of the error variable ε is normal.
2. The mean of the error variable is 0.
3. The standard deviation of ε is σ_ε, which is a constant.
4. The errors are independent.

In Section 16.6, we discussed how to recognize when the requirements are unsatisfied. Those same procedures can be used to detect violations of required conditions in the multiple regression model. We now proceed as we did in Chapter 16. We discuss how the model's coefficients are estimated and how we assess the model's fit. However, there is one major difference between Chapters 16 and 17. In Chapter 16, we allowed for the possibility that some students will perform the calculations manually. The multiple regression model involves so many computations that it is virtually impossible to conduct the analysis without a computer. All analyses in this chapter will be performed by Excel and Minitab. Your job will be to interpret the output.

17.2 Estimating the Coefficients and Assessing the Model

The multiple regression equation is expressed similarly to the simple regression equation. The general form is

$$\hat{y} = b_0 + b_1x_1 + b_2x_2 + \cdots + b_kx_k$$

where k is the number of independent variables.

The procedures introduced in Chapter 16 are extended to the multiple regression model. However, in Chapter 16, we first discussed how to interpret the coefficients and then discussed how to assess the model's fit. In practice, we reverse the process. That is, the first step is to determine how well the model fits. If the model's fit is poor, there is no point in a further analysis of the coefficients of that model. A much higher priority is assigned to the task of improving the model. We will discuss the art and science of model building in Chapter 18. In this chapter, we show how a regression analysis is performed. The steps we use are as follows:

1. Use a computer and software to generate the coefficients and the statistics used to assess the model.
2. Diagnose violations of required conditions. If there are problems, attempt to remedy them.
3. Assess the model's fit. Three statistics that perform this function are the standard error of estimate, the coefficient of determination, and the F-test of the analysis of variance. The first two were introduced in Chapter 16; the third will be introduced here.
4. If we are satisfied with the model's fit and that the required conditions are met, we scan interpret the coefficients and test them as we did in Chapter 16. We use the model to predict a value of the dependent variable or estimate the expected value of the dependent variable.

We illustrate these techniques with the following example.

APPLICATIONS in **OPERATIONS MANAGEMENT**

Location Analysis

Location analysis is one function of operations management function. Deciding where to locate a plant, warehouse, or retail outlet is a critical decision for any organization. A large number of variables must be considered in this decision problem. For example, a production facility must be located close to suppliers of raw resources and supplies, skilled labor, and transportation to customers. Retail outlets must consider the type and number of potential customers. In the next example we describe an application of regression analysis to find profitable locations for a motel chain.

EXAMPLE 17.1*

DATA Xm17-01

Selecting Sites for La Quinta Inns

La Quinta Motor Inns is a moderately priced chain of motor inns located across the United States. Its market is the frequent business traveler. The chain recently launched a campaign to increase market share by building new inns. The management of the chain is aware of the difficulty in choosing locations for new motels. Moreover, making decisions without adequate information often results in poor decisions. Consequently the chain's management acquired data on 100 randomly selected inns belonging to La Quinta. The objective was to predict which sites are likely to be profitable.

To measure profitability, La Quinta used *operating margin*, which is the ratio of the sum of profit, depreciation, and interest expenses divided by total revenue. (Although occupancy is often used as a measure of a motel's success, the company statistician concluded that occupancy was too unstable, especially during economic turbulence.) The higher the operating margin, the greater the success of the inn. La Quinta defines profitable inns as those with an operating margin in excess of 50% and unprofitable inns with margins of

*Adapted from Sheryl E. Kimes and James A. Fitzsimmons, "Selecting Profitable Hotel Sites at La Quinta Motor Inns," *Interfaces* 20 (March–April, 1990): 12–20.

less than 30%. After a discussion with a number of experienced managers, La Quinta decided to select one or two independent variables from each of the categories: competition, market awareness, demand generators, demographics, and physical. To measure the degree of competition, they determined the total number of motel and hotel rooms within 3 miles of each La Quinta Inn. Market awareness was measured by the number of miles to the closest competing motel. Two variables that represent sources of customers were chosen. The amount of office space and college and university enrollment in the surrounding community are demand generators. Both of these are measures of economic activity. A demographic variable that describes the community is the median household income. Finally, as a measure of the physical qualities of the location, La Quinta chose the distance to the downtown core. These data are stored using the following format:

Column 1: y = operating margin, in percent

Column 2: x_1 = Total number of motel and hotel rooms within 3 miles of La Quinta Inn

Column 3: x_2 = Number of miles to closest competition

Column 4: x_3 = Office space in thousands of square feet in surrounding community

Column 5: x_4 = College and university enrollment (in thousands) in nearby university and/or college

Column 6: x_5 = Median household income (in $thousands) in surrounding community

Column 7: x_6 = Distance (in miles) to the downtown core.

Some of these data are shown here. Conduct a regression analysis and analyze the results.

Margin	Number	Nearest	Office Space	Enrollment	Income	Distance
55.5	3203	4.2	549	8	37	2.7
33.8	2810	2.8	496	17.5	35	14.4
49.0	2890	2.4	254	20	35	2.6
. . .	. . .	. . .	. . .	. . .	. . .	. . .
40.0	3397	1.6	855	19.5	32	3.1
39.8	3823	3.6	202	17	38	4.8
35.2	3251	1.7	275	13	35	4.3

SOLUTION

EXCEL

	A	B	C	D	E	F
1	SUMMARY OUTPUT					
2						
3	*Regression Statistics*					
4	Multiple R	0.7246				
5	R Square	0.5251				
6	Adjusted R Square	0.4944				
7	Standard Error	5.51				
8	Observations	100				
9						
10	ANOVA					
11		*df*	*SS*	*MS*	*F*	*Significance F*
12	Regression	6	3123.8	520.64	17.14	3.03E-13
13	Residual	93	2825.6	30.38		
14	Total	99	5949.5			
15						
16		*Coefficients*	*Standard Error*	*t Stat*	*P-value*	
17	Intercept	38.14	6.99	5.45	4.04E-07	
18	Number	-0.0076	0.0013	-6.07	2.77E-08	
19	Nearest	1.65	0.63	2.60	0.0108	
20	Office Space	0.020	0.0034	5.80	9.24E-08	
21	Enrollment	0.21	0.13	1.59	0.1159	
22	Income	0.41	0.14	2.96	0.0039	
23	Distance	-0.23	0.18	-1.26	0.2107	

INSTRUCTIONS

1. Type or import the data so that the independent variables are in adjacent columns. (Open Xm17-01.)
2. Click **Data, Data Analysis,** and **Regression.**
3. Specify the **Input Y Range** (A1:A101), the **Input X Range** (B1:G101), and a value for α (.05).

MINITAB

Regression Analysis: Margin versus Number, Nearest, ...

The regression equation is
Margin = 38.1 - 0.00762 Number + 1.65 Nearest + 0.0198 Office Space + 0.212 Enrollment + 0.413 Income - 0.225 Distance

Predictor	Coef	SE Coef	T	P
Constant	38.139	6.993	5.45	0.000
Number	-0.007618	0.001255	-6.07	0.000
Nearest	1.6462	0.6328	2.60	0.011
Office Space	0.019766	0.003410	5.80	0.000
Enrollment	0.2118	0.1334	1.59	0.116
Income	0.4131	0.1396	2.96	0.004
Distance	-0.2253	0.1787	-1.26	0.211

S = 5.512 R-Sq = 52.5% R-Sq(adj) = 49.4%

Analysis of Variance

Source	DF	SS	MS	F	P
Regression	6	3123.83	520.64	17.14	0.000
Residual Error	93	2825.63	30.38		
Total	99	5949.46			

INSTRUCTIONS

1. Click **Stat, Regression,** and **Regression . . . .**
2. Specify the dependent variable in the **Response** box (Margin) and the independent variables in the **Predictors** box (Number, Nearest, Office Space, Enrollment, Income, Distance).

INTERPRET

The regression model is estimated by

$$\hat{y} = 38.14 - .0076x_1 + 1.65x_2 + .020x_3 + .21x_4 + .41x_5 - .23x_6$$

We assess the model in three ways: the standard error of estimate, the coefficient of determination (both introduced in Chapter 16), and the F-test of the analysis of variance (presented subsequently).

Standard Error of Estimate

Recall that σ_ε is the standard deviation of the error variable ε and that, because σ_ε is a population parameter, it is necessary to estimate its value by using s_ε. In multiple regression, the standard error of estimate is defined as follows.

Standard Error of Estimate

$$s_\varepsilon = \sqrt{\frac{\text{SSE}}{n - k - 1}}$$

where n is the sample size and k is the number of independent variables in the model.

As we noted in Chapter 16, each of our software packages reports the standard error of estimate in a different way.

EXCEL

	A	B
7	Standard Error	5.51

MINITAB

S = 5.512

INTERPRET

Recall that we judge the magnitude of the standard error of estimate relative to the values of the dependent variable, and particularly to the mean of y. In this example, $\bar{y} = 45.739$ (not shown in printouts). It appears that the standard error of estimate is not particularly small.

Coefficient of Determination

Recall from Chapter 16 that the coefficient of determination is defined as

$$R^2 = 1 - \frac{\text{SSE}}{\sum(y_i - \bar{y})^2}$$

EXCEL

	A	B
5	R Square	0.5251

MINITAB

R-Sq = 52.5%

INTERPRET

This means that 52.51% of the total variation in operating margin is explained by the six independent variables, whereas 47.49% remains unexplained.

Notice that Excel and Minitab print a second R^2 statistic, called the **coefficient of determination adjusted for degrees of freedom,** which has been adjusted to take into account the sample size and the number of independent variables. The rationale for this statistic is that, if the number of independent variables k is large relative to the sample size n, the unadjusted R^2 value may be unrealistically high. To understand this point, consider what would happen if the sample size is 2 in a simple linear regression model. The line would fit the data perfectly, resulting in $R^2 = 1$ when, in fact, there may be no linear relationship. To avoid creating a false impression, the adjusted R^2 is often calculated. Its formula follows.

Coefficient of Determination Adjusted for Degrees of Freedom

$$\text{Adjusted } R^2 = 1 - \frac{\text{SSE}/(n-k-1)}{\sum(y_i - \bar{y})^2/(n-1)} = 1 - \frac{\text{MSE}}{s_y^2}$$

If n is considerably larger than k, the unadjusted and adjusted R^2 values will be similar. But if SSE is quite different from 0 and k is large compared to n, the unadjusted and adjusted values of R^2 will differ substantially. If such differences exist, the analyst should be alerted to a potential problem in interpreting the coefficient of determination. In Example 17.1, the adjusted coefficient of determination is 49.44%, indicating that, no matter how we measure the coefficient of determination, the model's fit is moderately good.

Testing the Validity of the Model

In the simple linear regression model, we tested the slope coefficient to determine whether sufficient evidence existed to allow us to conclude that there was a linear relationship between the independent variable and the dependent variable. However, because there is only one independent variable in that model, that same t-test also tested to determine whether that model is valid. When there is more than one independent variable, we need another method to test the overall validity of the model. The technique is a version of the analysis of variance, which we introduced in Chapter 14.

To test the validity of the regression model, we specify the following hypotheses:

$$H_0: \beta_1 = \beta_2 = \cdots = \beta_k = 0$$

$$H_1: \text{At least one } \beta_i \text{ is not equal to } 0$$

If the null hypothesis is true, none of the independent variables $x_1, x_2, \ldots, x_k$ is linearly related to y, and therefore the model is invalid. If at least one β_i is not equal to 0, the model does have some validity.

When we discussed the coefficient of determination in Chapter 16, we noted that the total variation in the dependent variable [measured by $\sum(y_i - \bar{y})^2$] can be decomposed into two parts: the explained variation (measured by SSR) and the unexplained variation (measured by SSE). That is,

$$\text{Total Variation in } y = \text{SSR} + \text{SSE}$$

Furthermore, we established that, if SSR is large relative to SSE, the coefficient of determination will be high—signifying a good model. On the other hand, if SSE is large, most of the variation will be unexplained, which indicates that the model provides a poor fit and consequently has little validity.

The test statistic is the same one we encountered in Section 14.1, where we tested for the equivalence of two or more population means. To judge whether SSR is large enough relative to SSE to allow us to infer that at least one coefficient is not equal to 0, we compute the ratio of the two mean squares. (Recall that the mean square is the sum of squares divided by its degrees of freedom; recall, too, that the ratio of two mean squares is F distributed as long as the underlying population is normal—a required condition for this application.) The calculation of the test statistic is summarized in an analysis of variance (ANOVA) table, whose general form appears in Table 17.1. The Excel and Minitab ANOVA tables are shown next.

TABLE **17.1**
Analysis of Variance Table for Regression Analysis

Source of Variation	Degrees of Freedom	Sums of Squares	Mean Squares	F-Statistic
Regression	k	SSR	MSR = SSR/k	F = MSR/MSE
Residual	$n - k - 1$	SSE	MSE = SSE/$(n - k - 1)$	
Total	$n - 1$	$\sum(y_i - \bar{y})^2$		

EXCEL

	A	B	C	D	E	F
10	ANOVA					
11		df	SS	MS	F	Significance F
12	Regression	6	3123.8	520.64	17.14	3.03E-13
13	Residual	93	2825.6	30.38		
14	Total	99	5949.5			

MINITAB

Analysis of Variance

Source	DF	SS	MS	F	P
Regression	6	3123.83	520.64	17.14	0.000
Residual Error	93	2825.63	30.38		
Total	99	5949.46			

A large value of F indicates that most of the variation in y is explained by the regression equation and that the model is valid. A small value of F indicates that most of the variation in y is unexplained. The rejection region allows us to determine whether F is large enough to justify rejecting the null hypothesis. For this test, the rejection region is

$$F > F_{\alpha,k,n-k-1}$$

In Example 17.1, the rejection region (assuming $\alpha = .05$) is

$$F > F_{\alpha,k,n-k-1} = F_{.05,6,93} \approx 2.20$$

As you can see from the printout, $F = 17.14$. The printout also includes the p-value of the test, which is 0. Obviously, there is a great deal of evidence to infer that the model is valid.

Although each assessment measurement offers a different perspective, all agree in their assessment of how well the model fits the data, because all are based on the sum of squares for error, SSE. The standard error of estimate is

$$s_\varepsilon = \sqrt{\frac{\text{SSE}}{n - k - 1}}$$

and the coefficient of determination is

$$R^2 = 1 - \frac{\text{SSE}}{\sum(y_i - \bar{y})^2}$$

When the response surface hits every single point, SSE $= 0$. Hence $s_\varepsilon = 0$ and $R^2 = 1$.

If the model provides a poor fit, we know that SSE will be large [its maximum value is $\sum(y_i - \bar{y})^2$], s_ε will be large, and [since SSE is close to $\sum(y_i - \bar{y})^2$] R^2 will be close to 0.

The F-statistic also depends on SSE. Specifically,

$$F = \frac{\text{MSR}}{\text{MSE}} = \frac{\left(\sum(y_i - \bar{y})^2 - \text{SSE}\right)/k}{\text{SSE}/(n - k - 1)}$$

When SSE $= 0$,

$$F = \frac{\sum(y_i - \bar{y})^2/k}{0/(n - k - 1)}$$

which is infinitely large. When SSE is large, SSE is close to $\sum(y_i - \bar{y})^2$ and F is quite small.

The relationship among s_ε, R^2, and F are summarized in Table 17.2.

TABLE 17.2 **Relationship among SSE, s_ε, R^2, and F**

SSE	s_ε	R^2	F	Assessment of Model
0	0	1	∞	Perfect
Small	Small	Close to 1	Large	Good
Large	Large	Close to 0	Small	Poor
$\sum(y_i - \bar{y})^2$	$\sqrt{\dfrac{\sum(y_i - \bar{y})^2}{n - k - 1}}$*	0	0	Useless

*When n is large and k is small, this quantity is approximately equal to the standard deviation of y.

If we're satisfied that the model fits the data as well as possible, and that the required conditions are satisfied, we can interpret and test the individual coefficients and use the model to predict and estimate.

Interpreting the Coefficients

The coefficients $b_0, b_1, \ldots, b_k$ describe the relationship between each of the independent variables and the dependent variable in the sample. We need to use inferential methods (described below) to draw conclusions about the population. In Example 17.1, the sample consists of the 100 observations. The population is composed of all La Quinta inns.

Intercept

The intercept is $b_0 = 38.14$. This is the average operating margin when all the independent variables are zero. As we observed in Chapter 16, it is often misleading to try to interpret this value, particularly if 0 is outside the range of the values of the independent variables (as is the case here).

Number of Motel and Hotel Rooms

The relationship between operating margin and the number of motel and hotel rooms within 3 miles is described by $b_1 = -.0076$. From this number, we learn that in this model, for each additional room within 3 miles of the La Quinta Inn, the operating margin decreases on average by .0076, assuming that the other independent variables in this model are held constant. Changing the units we can interpret b_1 to say that for each additional 1,000 rooms, the margin decreases by 7.6%.

Distance to Nearest Competitor

The coefficient $b_2 = 1.65$ specifies that for each additional mile that the nearest competitor is to a La Quinta Inn, the average operating margin increases by 1.65%, assuming the constancy of the other independent variables.

The nature of the relationship between operating margin and the number of motel and hotel rooms and between operating margin and the distance to the nearest competitor was expected. The closer the competition, the lower the operating margin becomes.

Office Space

The relationship between office space and operating margin is expressed by $b_3 = .020$. Because office space is measured in thousands of square feet, we interpret this number as the average increase in operating margin for each additional thousand square feet of office space, keeping the other independent variables fixed. So, for every extra 100,000 square feet of office space, the operating margin increases on average by 2.0%.

College and University Enrollment

The relationship between operating margin and college and university enrollment is described by $b_4 = .21$, which we interpret to mean that for each additional thousand students the average operating margin increases by .21% when the other variables are constant.

Both office space and enrollment produced positive coefficients, indicating that these measures of economic activity are positively related to the operating margin.

Median Household Income

The relationship between operating margin and median household income is described by $b_5 = .41$. For each additional thousand dollar increase in median household income, the average operating margin increases by .41%, holding all other variables constant. This statistic suggests that motels in more affluent communities have higher operating margins.

Distance to Downtown Core

The last variable in the model is distance to the downtown core. Its relationship with operating margin is described by $b_6 = -.23$. This tells us that for each additional mile to the downtown center, the operating margin decreases on average by .23%, keeping the other independent variables constant. It may be that people prefer to stay at motels that are closer to town centers.

Testing the Coefficients

In Chapter 16, we described how to test to determine whether there is sufficient evidence to infer that, in the simple linear regression model, x and y are linearly related. The null and alternative hypotheses were

$$H_0: \quad \beta_1 = 0$$
$$H_1: \quad \beta_1 \neq 0$$

The test statistic was

$$t = \frac{b_1 - \beta_1}{s_{b_1}}$$

which is Student t distributed with $\nu = n - 2$ degrees of freedom.

In the multiple regression model, we have more than one independent variable. For each such variable, we can test to determine whether there is enough evidence of a linear relationship between it and the dependent variable for the entire population when the other independent variables are included in the model.

Testing the Coefficients

$$H_0: \quad \beta_i = 0$$
$$H_1: \quad \beta_i \neq 0$$

(for $i = 1, 2, \ldots, k$); the test statistic is

$$t = \frac{b_i - \beta_i}{s_{b_i}}$$

which is Student t distributed with $\nu = n - k - 1$ degrees of freedom.

To illustrate, we test each of the coefficients in the multiple regression model in Example 17.1. The tests that follow are performed just as all other tests in this book have been performed. We set up the null and alternative hypotheses, identify the test statistic, and use the computer to calculate the value of the test statistic and its p-value. For each independent variable, we test ($i = 1, 2, 3, 4, 5, 6$)

$$H_0: \quad \beta_i = 0$$
$$H_1: \quad \beta_i \neq 0$$

Refer to pages 675 and 676 and examine the computer output for Example 17.1. The output includes the t-tests of β_i. The results of these tests pertain to the entire population of La Quinta inns. It is also important to add that these test results were determined when the other independent variables were included in the model. We add this statement because a simple linear regression will very likely result in different values of the test statistics and possibly the conclusion.

Test of β_1 (Coefficient of the number of motel and hotel rooms)

Value of the test statistic: $t = -6.07$; p-value $= 0$

There is overwhelming evidence to infer that the number of motel and hotel rooms within 3 miles of the La Quinta Inn and the operating margin are linearly related.

Test of β_2 (Coefficient of the distance to nearest competitor)

Value of the test statistic: $t = 2.60$; p-value $= .0108$

There is evidence to conclude that the distance to the nearest motel and the operating margin of the La Quinta Inn are linearly related.

Test of β_3 (Coefficient of office space)

Value of the test statistic: $t = 5.80$; p-value $= 0$

This test allows us to infer that there is a linear relationship between the operating margin and the amount of office space around the inn.

Test of β_4 (Coefficient of college and university enrollment)

Value of the test statistic: $t = 1.59$; p-value $= .1159$

There is no evidence to conclude that the coefficient of college enrollment is not equal to 0. This may mean that there is no evidence of a linear relationship between college enrollment in the community around the inn and operating margin. However, it may also mean that there is a linear relationship between the two variables, but because of a condition called *multicollinearity*, the t-test of β_4 revealed no linear relationship. We will discuss multicollinearity in Section 17.3.

Test of β_5 (Coefficient of median household income)

Value of the test statistic: $t = 2.96$; p-value $= .0039$

There is sufficient statistical evidence to indicate that the operating margin and the median household income are linearly related.

Test of β_6 (Coefficient of distance to downtown)

Value of the test statistic: $t = -1.26$; p-value $= .2107$

There is not enough evidence to infer the existence of a linear relationship between the distance to the downtown center and the operating margin of the La Quinta Inn in the presence of the other independent variables. Once again the result may be the result of multicollinearity.

INTERPRET

We have discovered that in this model the number of hotel and motel rooms, distance to the nearest motel, amount of office space, and median household income are linearly related to the operating margin. Moreover, in this model we found little evidence to infer that college enrollment and distance to the downtown center are linearly related to operating margin. The t-tests tell La Quinta's management that in choosing the site of a new motel, they should look for locations where there are few other motels nearby, where there is a great deal of office space, and where the surrounding households are relatively affluent.

A Cautionary Note about Interpreting the Results

Care should be taken when interpreting the results of this and other regression analyses. We might find that in one model there is enough evidence to conclude that a particular independent variable is linearly related to the dependent variable, but that in another model no such evidence exists. Consequently, whenever a particular t-test is *not* significant, we state that there is not enough evidence to infer that the independent and dependent variable are linearly related *in this model*. The implication is that another model may yield different conclusions.

Furthermore, if one or more of the required conditions are violated, the results may be invalid. In Section 16.6 we introduced the procedures that allow the statistics practitioner to examine the model's requirements. We will add to this discussion in Section 17.3. We also remind you that it is dangerous to extrapolate far outside the range of the observed values of the independent variables.

t-Tests and the Analysis of Variance

The t-tests of the individual coefficients allow us to determine whether $\beta_i \neq 0$ (for $i = 1, 2, \ldots, k$), which tells us whether a linear relationship exists between x_i and y. There is a t-test for each independent variable. Consequently, the computer automatically performs k t-tests. (It actually conducts $k + 1$ t-tests, including the one for the intercept β_0, which we usually ignore.) The F-test in the analysis of variance combines these t-tests into a single test. That is, we test all the β_i at one time to determine whether at least one of them is not equal to 0. The question naturally arises, Why do we need the F-test if it is nothing more than the combination of the previously performed t-tests? Recall that we addressed this issue before. In Chapter 14, we pointed out that we can replace the analysis of variance by a series of t-tests of the difference between two means. However, by doing so we increase the probability of making a Type I error. That means that even when there is no linear relationship between each of the independent variables and the dependent variable, multiple t-tests will likely show some are significant. As a result, you will conclude erroneously that, since at least one β_i is not equal to 0, the model is valid. The F-test, on the other hand, is performed only once. Because the probability that a Type I error will occur in a single trial is equal to α, the chance of erroneously concluding that the model is valid is substantially less with the F-test than with multiple t-tests.

There is another reason that the F-test is superior to multiple t-tests. Because of a commonly occurring problem called *multicollinearity*, the t-tests may indicate that some independent variables are not linearly related to the dependent variable, when in fact they are. The problem of multicollinearity does not affect the F-test, nor does it inhibit us from developing a model that fits the data well. Multicollinearity is discussed in Section 17.3.

The F-Test and the t-Test in the Simple Linear Regression Model

It is useful for you to know that we can use the F-test to test the validity of the simple linear regression model. However, this test is identical to the t-test of β_1. The t-test of β_1 in the simple linear regression model tells us whether that independent variable is linearly related to the dependent variable. However, because there is only one independent variable, the t-test of β_1 also tells us whether the model is valid, which is the purpose of the F-test.

The relationship between the t-test of β_1 and the F-test can be explained mathematically. Statisticians can show that if we square a t-statistic with ν degrees of freedom, we produce an F-statistic with 1 and ν degrees of freedom. (We briefly discussed this relationship in Chapter 14.) To illustrate, consider Example 16.2 on page 623. We found the t-test of β_1 to be -13.44, with degrees of freedom equal to 98. The p-value was 5.75×10^{-24}. The output included the analysis of variance table where $F = 180.64$ and p-value $= 5.75 \times 10^{-24}$. The t-statistic squared is $t^2 = (-13.44)^2 = 180.63$. (The difference is due to rounding errors.) Notice that the degrees of freedom of the F-statistic are 1 and 98. Thus, we can use either test to test the validity of the simple linear regression model.

Using the Regression Equation

As was the case with simple linear regression, we can use the multiple regression equation in two ways: We can produce the prediction interval for a particular value of y, and we can produce the confidence interval estimate of the expected value of y. Like the other calculations associated with multiple regression, we call on the computer to do the work.

Suppose that in Example 17.1 a manager investigated a potential site for a La Quinta Inn and found the following characteristics. There are 3,815 rooms within 3 miles of the site ($x_1 = 3{,}815$) and the closest other hotel or motel is .9 miles away ($x_2 = .9$). The amount of office space is 476,000 square feet ($x_3 = 476$). There is one college and one university nearby with a total enrollment of 24,500 students ($x_4 = 24.5$). From the census, the manager learns that the median household income in the area (rounded to the nearest thousand) is \$35,000 ($x_5 = 35$). Finally, the distance to the downtown center has been measured at 11.2 miles ($x_6 = 11.2$). The manager wants to predict the operating margin if and when the inn is built. As you discovered in the previous chapter, both Excel and Minitab output the prediction interval for one inn and interval estimate of the expected (average) operating margin for all sites with the given variables.

EXCEL

	A	B	C
1	**Prediction Interval**		
2			
3			Margin
4			
5	Predicted value		37.1
6			
7	Prediction Interval		
8	Lower limit		25.4
9	Upper limit		48.8
10			
11	Interval Estimate of Expected Value		
12	Lower limit		33.0
13	Upper limit		41.2

INSTRUCTIONS

See the instructions on page 650. In cells B102 to G102 we input the values **3815, .9, 476, 24.5, 35, 11.2,** respectively. We specified 95% confidence.

MINITAB

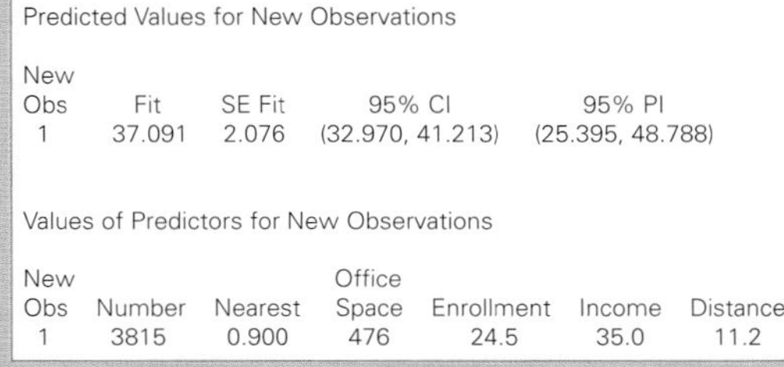

Predicted Values for New Observations

New Obs	Fit	SE Fit	95% CI	95% PI
1	37.091	2.076	(32.970, 41.213)	(25.395, 48.788)

Values of Predictors for New Observations

New Obs	Number	Nearest	Office Space	Enrollment	Income	Distance
1	3815	0.900	476	24.5	35.0	11.2

INSTRUCTIONS

See the instructions on page 651. We input the values **3815 .9 476 24.5 35 11.2.** We specified 95% confidence.

INTERPRET

As you can see, we predict that the operating margin will fall between 25.4 and 48.8. This interval is quite wide, confirming the need to have extremely well-fitting models to make accurate predictions. However, management defines a profitable inn as one with an operating margin greater than 50% and an unprofitable inn as one with an operating margin below 30%. Because the entire prediction interval is below 50 and part of it is below 30, the management of La Quinta will pass on this site.

The expected operating margin of all sites that fit this category is estimated to be between 33.0 and 41.2. We interpret this to mean that if we built inns on an infinite number of sites that fit the category described above, the mean operating margin would fall between 33.0 and 41.2. In other words, the average inn would not be profitable.

MBA Program Admissions Policy: Solution

© Eyewire Collection/Getty Images

The model we need to estimate and analyze is

$$y = \beta_0 + \beta_1 x_1 + \beta_2 x_2 + \beta_3 x_3 + \varepsilon$$

where

y = MBA program GPA

x_1 = undergraduate GPA (range: 0 to 12)

x_2 = GMAT score (range: 200 to 800)

x_3 = number of years of work experience (minimum = 3)

EXCEL

	A	B	C	D	E	F
1	SUMMARY OUTPUT					
2						
3	*Regression Statistics*					
4	Multiple R	0.6808				
5	R Square	0.4635				
6	Adjusted R Square	0.4446				
7	Standard Error	0.7879				
8	Observations	89				
9						
10	ANOVA					
11		*df*	*SS*	*MS*	*F*	*Significance F*
12	Regression	3	45.60	15.20	24.48	1.64E-11
13	Residual	85	52.77	0.62		
14	Total	88	98.37			
15						
16		*Coefficients*	*Standard Error*	*t Stat*	*P-value*	
17	Intercept	0.47	1.51	0.31	0.7576	
18	UnderGPA	0.0628	0.120	0.52	0.6017	
19	GMAT	0.0113	0.0014	8.16	2.71E-12	
20	Work	0.0926	0.0309	3.00	0.0036	

MINITAB

```
The regression equation is
MBA GPA = 0.47 + 0.063 UnderGPA + 0.0113 GMAT + 0.0926 Work

Predictor     Coef      SE Coef     T      P
Constant      0.466     1.506       0.31   0.758
UnderGPA      0.0628    0.1199      0.52   0.602
GMAT          0.011281  0.001383    8.16   0.000
Work          0.09259   0.03091     3.00   0.004

S = 0.7879   R-Sq = 46.4%   R-Sq(adj) = 44.5%

Analysis of Variance

Source          DF   SS       MS       F      P
Regression       3   45.597   15.199   24.48  0.000
Residual Error  85   52.772   0.621
Total           88   98.369
```

INTERPRET

The coefficient of determination is .4635, which tells the dean that 46.35% of the variation in MBA program GPA is explained by the variation in the independent variables. The test for validity produces a test statistic of $F = 24.48$ with a p-value of 0. The t-tests indicate that GMAT score and years of work experience are linearly related to MBA program GPA, but there is no evidence of a linear relationship between undergraduate GPA and MBA GPA. These results suggest that admissions should be based on the GMAT score as well as the number of years of work experience.

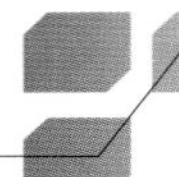

EXERCISES

The following exercises require the use of a computer and statistical software. Exercises 17.1–17.4 can be solved manually. See Appendix A for the sample statistics. ***Use a 5% significance level.***

17.1 Xr17-01 A developer who specializes in summer cottage properties is considering purchasing a large tract of land adjoining a lake. The current owner of the tract has already subdivided the land into separate building lots and has prepared the lots by removing some of the trees. The developer wants to forecast the value of each lot. From previous experience, she knows that the most important factors affecting the price of the lot are size, number of mature trees, and distance to the lake. From a nearby area, she gathers the relevant data for 60 recently sold lots.

a. Find the regression equation.
b. What is the standard error of estimate? Interpret its value.
c. What is the coefficient of determination? What does this statistic tell you?
d. What is the coefficient of determination, adjusted for degrees of freedom? Why does this value differ from the coefficient of determination? What does this tell you about the model?
e. Test the validity of the model. What does the p-value of the test statistic tell you?
f. Interpret each of the coefficients.
g. Test to determine whether each of the independent variables is linearly related to the price of the lot in this model.
h. Predict with 90% confidence the selling price of a 40,000-square-foot lot that has 50 mature trees and is 25 feet from the lake.
i. Estimate with 90% confidence the average selling price of 50,000-square-foot lots that have 10 mature trees and are 75 feet from the lake.

17.2 Xr17-02 Pat Statsdud, a student ranking near the bottom of the statistics class, decided that a certain amount of studying could actually improve final grades. However, too much studying would not be warranted, since Pat's ambition (if that's what one could call it) was to ultimately graduate with the absolute minimum level of work. Pat was registered in a statistics course, which had only 3 weeks to go before the final exam, and where the final grade was determined in the following way:

Total mark = 20% (Assignment)
+ 30% (Midterm test)
+ 50% (Final exam)

To determine how much work to do in the remaining 3 weeks, Pat needed to be able to predict the final exam mark on the basis of the assignment mark (worth 20 points) and the midterm mark (worth 30 points). Pat's marks on these were 12/20 and 14/30, respectively. Accordingly, Pat undertook the following analysis. The final exam mark, assignment mark, and midterm test mark for 30 students who took the statistics course last year were collected.

a. Determine the regression equation.
b. What is the standard error of estimate? Briefly describe how you interpret this statistic.
c. What is the coefficient of determination? What does this statistic tell you?
d. Test the validity of the model.
e. Interpret each of the coefficients.
f. Can Pat infer that the assignment mark is linearly related to the final grade in this model?
g. Can Pat infer that the midterm mark is linearly related to the final grade in this model?
h. Predict Pat's final exam mark with 95% confidence.
i. Predict Pat's final grade with 95% confidence.

17.3 Xr17-03 The president of a company that manufactures drywall wants to analyze the variables that affect demand for his product. Drywall is used to construct walls in houses and offices. Consequently, the president decides to develop a regression model in which the dependent variable is monthly sales of drywall (in hundreds of 4 × 8 sheets) and the independent variables are

Number of building permits issued in the county
Five-year mortgage rates (in percentage points)
Vacancy rate in apartments (in percentage points)
Vacancy rate in office buildings (in percentage points)

To estimate a multiple regression model, he took monthly observations from the past 2 years.

a. Analyze the data using multiple regression.
b. What is the standard error of estimate? Can you use this statistic to assess the model's fit? If so, how?
c. What is the coefficient of determination, and what does it tell you about the regression model?
d. Test the overall validity of the model.
e. Interpret each of the coefficients.
f. Test to determine whether each of the independent variables is linearly related to drywall demand in this model.
g. Predict next month's drywall sales with 95% confidence if the number of building permits is 50, the 5-year mortgage rate is 9.0%, and the vacancy rates are 3.6% in apartments and 14.3% in office buildings.

17.4 Xr17-04 The general manager of the Cleveland Indians baseball team is in the process of determining which minor-league players to draft. He is aware that his team needs home-run hitters and would like to find a way to predict the number of home runs a player will hit. Being an astute statistician, he gathers a random sample of players and records the number of home runs each player hit in his first two full years as a major-league player, the number of home runs he hit in his last full year in the minor leagues, his age, and the number of years of professional baseball.

a. Develop a regression model, and use a software package to produce the statistics.
b. Interpret each of the coefficients.
c. How well does the model fit?
d. Test the model's validity.
e. Do each of the independent variables belong in the model?
f. Calculate the 95% interval of the number of home runs in the first two years of a player who is 25 years old, has played professional baseball for 7 years, and hit 22 home runs in his last year in the minor leagues.
g. Calculate the 95% interval of the expected number of home runs in the first two years of players who are 27 years old, have played professional baseball for 5 years, and hit 18 home runs in their last year in the minors.

APPLICATIONS in HUMAN RESOURCES MANAGEMENT

Severance Pay

In most firms the entire issue of compensation falls into the domain of the human resources manager. The manager must ensure that the method used to determine compensation contributes to the firm's objectives. Moreover, the firm needs to ensure that discrimination or bias of any kind is not a factor. Another function of the personnel manager is to develop severance packages for employees whose services are no longer needed because of downsizing or merger. The size and nature of severance is rarely part of any working agreement and must be determined by a variety of factors. Regression analysis is often useful in this area.

17.5 Xr17-05 When one company buys another company, it is not unusual that some workers are terminated. The severance benefits offered to the laid-off workers are often the subject of dispute. Suppose that the Laurier Company recently bought the Western Company and subsequently terminated 20 of Western's employees. As part of the buyout agreement, it was promised that the severance packages offered to the former Western employees would be equivalent to those offered to Laurier employees who had been terminated in the past year. Thirty-six-year-old Bill Smith, a Western employee for the past 10 years, earning \$32,000 per year, was one of those let go. His severance package included an offer of 5 weeks' severance pay. Bill complained that this offer was less than that offered to Laurier's employees when they were laid off, in contravention of the buyout agreement. A statistician was called in to settle the dispute. The statistician was told that severance is determined by three factors: age, length of service with the company, and pay. To determine how generous the severance package had been, a random sample of 50 Laurier ex-employees was taken. For each, the following variables were recorded:

Number of weeks of severance pay

Age of employee

Number of years with the company

Annual pay (in thousands of dollars)

a. Determine the regression equation.
b. Comment on how well the model fits the data.
c. Do all the independent variables belong in the equation? Explain.
d. Perform an analysis to determine whether Bill is correct in his assessment of the severance package.

17.6 Xr17-06 The admissions officer of a university is trying to develop a formal system of deciding which students to admit to the university. She believes that determinants of success include the standard variables—high school grades and SAT scores. However, she also believes that students who have participated in extracurricular activities are more likely to succeed than those who have not. To investigate the issue, she randomly sampled 100 fourth-year students and recorded the following variables:

GPA for the first 3 years at the university (range: 0 to 12)

GPA from high school (range: 0 to 12)

SAT score (range: 400 to 1600)

Number of hours on average spent per week in organized extracurricular activities in the last year of high school

a. Develop a model that helps the admissions officer decide which students to admit, and use the computer to generate the usual statistics.
b. What is the coefficient of determination? Interpret its value.
c. Test the overall validity of the model.
d. Test to determine whether each of the independent variables is linearly related to the dependent variable in this model.
e. Determine the 95% interval of the GPA for the first 3 years of university for a student whose high school GPA is 10, whose SAT score is 1200, and who worked an average of 2 hours per week on organized extracurricular activities in the last year of high school.
f. Find the 90% interval of the mean GPA for the first 3 years of university for all students whose high school GPA is 8, whose SAT

score is 1100, and who worked an average of 10 hours per week on organized extracurricular activities in the last year of high school.

17.7 Xr17-07 The marketing manager for a chain of hardware stores needed more information about the effectiveness of the three types of advertising that the chain used. These are localized direct mailing (in which flyers describing sales and featured products are distributed to homes in the area surrounding a store), newspaper advertising, and local television advertisements. To determine which type is most effective, the manager collected 1 week's data from 100 randomly selected stores. For each store, the following variables were recorded:

Weekly gross sales

Weekly expenditures on direct mailing

Weekly expenditures on newspaper advertising

Weekly expenditures on television commercials

All variables were recorded in thousands of dollars.

a. Find the regression equation.
b. What are the coefficient of determination and the coefficient of determination, adjusted for degrees of freedom? What do these statistics tell you about the regression equation?
c. What does the standard error of estimate tell you about the regression model?
d. Test the validity of the model.
e. Which independent variables are linearly related to weekly gross sales in this model? Explain.
f. Compute the 95% interval of the week's gross sales if a local store spent \$800 on direct mailing, \$1,200 on newspaper advertisements, and \$2,000 on television commercials.
g. Calculate the 95% interval of the mean weekly gross sales for all stores that spend \$800 on direct mailing, \$1,200 on newspaper advertising, and \$2,000 on television commercials.
h. Discuss the difference between the two intervals found in Parts f and g.

17.8 Xr17-08 For many cities around the world, garbage is an increasing problem. Many North American cities have virtually run out of space to dump the garbage. A consultant for a large American city decided to gather data about the problem. She took a random sample of houses and determined the following:

Y = the amount of garbage per average week (pounds)

X_1 = Size of the house (square feet)

X_2 = Number of children

X_3 = Number of adults who are usually home during the day

a. Conduct a regression analysis.
b. Is the model valid?
c. Interpret each of the coefficients.
d. Test to determine whether each of the independent variables is linearly related to the dependent variable.

17.9 Xr17-09 The administrator of a school board in a large county was analyzing the average mathematics test scores in the schools under her control. She noticed that there were dramatic differences in scores among the schools. In an attempt to improve the scores of all the schools, she attempted to determine the factors that account for the differences. Accordingly, she took a random sample of 40 schools across the county and, for each, determined the mean test score last year, the percentage of teachers in each school who have at least one university degree in mathematics, the mean age, and the mean annual income (in \$thousands) of the mathematics teachers.

a. Conduct a regression analysis to develop the equation.
b. Is the model valid?
c. Interpret and test the coefficients.
d. Predict with 95% confidence the test score at a school where 50% of the mathematics teachers have mathematics degrees, the mean age is 43, and the mean annual income is \$48,300.

17.10 Xr17-10* Life insurance companies are keenly interested in predicting how long their customers will live, because their premiums and profitability depend on such numbers. An actuary for one insurance company gathered data from 100 recently deceased male customers. He recorded the age at death of the customer plus the ages at death of his mother and father, the mean ages at death of his grandmothers, and the mean ages at death of his grandfathers.

a. Perform a multiple regression analysis on these data.
b. Is the model valid?
c. Interpret and test the coefficients.
d. Determine the 95% interval of the longevity of a man whose parents lived to the age of 70, whose grandmothers averaged 80 years, and whose grandfathers averaged 75 years.
e. Find the 95% interval of the mean longevity of men whose mothers lived to 75 years, whose fathers lived to 65 years, whose grandmothers averaged 85 years, and whose grandfathers averaged 75 years.

17.11 Xr17-11 University students often complain that universities reward professors for research, but not for teaching, and argue that professors react to this situation by devoting more time and energy to the publication of their findings and less time and energy to classroom activities. Professors counter that research and teaching go hand in hand: More research makes better teachers. A student organization at one university decided to investigate the issue. They randomly selected 50 economics professors who are employed by a multicampus university. The students recorded the salaries (in $thousands) of the professors, their average teaching evaluations (on a 10-point scale), and the total number of journal articles published in their careers. Perform a complete analysis (produce the regression equation, assess it, and report your findings).

17.12 Xr17-12* One of the critical factors that determine the success of a catalog store chain is the availability of products that consumers want to buy. If a store is sold out, future sales to that customer are less likely. Accordingly, delivery trucks operating from a central warehouse regularly re-supply stores. In an analysis of a chain's operations, the general manager wanted to determine the factors that are related to how long it takes to unload delivery trucks. A random sample of 50 deliveries to one store was observed. The times (in minutes) to unload the truck, the total number of boxes, and the total weight (in hundreds of pounds) of the boxes were recorded.

a. Determine the multiple regression equation.
b. How well does the model fit the data? Explain.
c. Interpret and test the coefficients.
d. Produce a 95% interval of the amount of time needed to unload a truck with 100 boxes weighing 5,000 pounds.
e. Produce a 95% interval of the average amount of time needed to unload trucks with 100 boxes weighing 5,000 pounds.

17.13 Xr17-13 Lotteries have become important sources of revenue for governments. Many people have criticized lotteries, however, referring to them as a tax on the poor and uneducated. In an examination of the issue, a random sample of 100 adults was asked how much they spend on lottery tickets and was interviewed about various socioeconomic variables. The purpose of this study is to test the following beliefs:

1. Relatively uneducated people spend more on lotteries than do relatively educated people.
2. Older people buy more lottery tickets than younger people.
3. People with more children spend more on lotteries than people with fewer children.
4. Relatively poor people spend a greater proportion of their income on lotteries than relatively rich people.

The following data were recorded:

Amount spent on lottery tickets as a percentage of total household income

Number of years of education

Age

Number of children

Personal income (in thousands of dollars)

a. Develop the multiple regression equation.
b. Is the model valid?
c. Test each of the beliefs. What conclusions can you draw?

17.3 Regression Diagnostics—II

In Section 16.6, we discussed how to determine whether the required conditions are unsatisfied. The same procedures can be used to diagnose problems in the multiple regression model. Here is a brief summary of the diagnostic procedure we described in Chapter 16. Calculate the residuals and check the following:

1. *Is the error variable nonnormal?* Draw the histogram of the residuals.
2. *Is the error variance constant?* Plot the residuals versus the predicted values of y.
3. *Are the errors independent (time-series data)?* Plot the residuals versus the time periods.
4. *Are there observations that are inaccurate or do not belong to the target population?*

Double-check the accuracy of outliers and influential observations.

If the error is nonnormal and/or the variance is not a constant, several remedies can be attempted. These are described at the end of this section.

Outliers and influential observations are checked by examining the data in question to ensure accuracy.

Nonindependence of a time series can sometimes be detected by graphing the residuals and the time periods and looking for evidence of autocorrelation. In Section 17.4, we introduce the Durbin-Watson test, which tests for one form of autocorrelation. We will offer a corrective measure for nonindependence.

There is another problem that is applicable to multiple regression models only. *Multicollinearity* is a condition wherein the independent variables are highly correlated. Multicollinearity distorts the t-tests of the coefficients, making it difficult to determine whether any of the independent variables are linearly related to the dependent variable. It also makes interpreting the coefficients problematic. We will discuss this condition and its remedy next.

Multicollinearity

Multicollinearity (also called *collinearity* and *intercorrelation*) is a condition that exists when the independent variables are correlated with one another. The adverse effect of multicollinearity is that the estimated regression coefficients of the independent variables that are correlated tend to have large sampling errors. There are two consequences of multicollinearity. First, because the variability of the coefficients is large, the sample coefficient may be far from the actual population parameter, including the possibility that the statistic and parameter may have opposite signs. Second, when the coefficients are tested, the t-statistics will be small, which leads to the inference that there is no linear relationship between the affected independent variables and the dependent variable. In some cases, this inference will be wrong. Fortunately, multicollinearity does not affect the F-test of the analysis of variance. We will illustrate the effects and remedy with the following example.

EXAMPLE 17.2 Predicting the Selling Prices of Houses

DATA Xm17-02*

A real estate agent wanted to develop a model to predict the selling price of a home. The agent believed that the most important variables in determining the price of a house are its size, number of bedrooms, and lot size. Accordingly, he took a random sample of 100 homes that recently sold and recorded the selling price (y), the number of bedrooms (x_1), the size in square feet (x_2), and the lot size in square feet (x_3). Analyze the relationship among the four variables.

The proposed multiple regression model is

$$y = \beta_0 + \beta_1 x_1 + \beta_2 x_2 + \beta_3 x_3 + \varepsilon$$

SOLUTION

EXCEL

	A	B	C	D	E	F
1	SUMMARY OUTPUT					
2						
3	*Regression Statistics*					
4	Multiple R	0.7483				
5	R Square	0.5600				
6	Adjusted R Square	0.5462				
7	Standard Error	25023				
8	Observations	100				
9						
10	ANOVA					
11		*df*	*SS*	*MS*	*F*	*Significance F*
12	Regression	3	76501718347	25500572782	40.73	4.57E-17
13	Residual	96	60109046053	626135896		
14	Total	99	136610764400			
15						
16		*Coefficients*	*Standard Error*	*t Stat*	*P-value*	
17	Intercept	37718	14177	2.66	0.0091	
18	Bedrooms	2306	6994	0.33	0.7423	
19	House Size	74.30	52.98	1.40	0.1640	
20	Lot Size	-4.36	17.02	-0.26	0.7982	

MINITAB

Regression Analysis: Price versus Bedrooms, House Size, Lot Size

The regression equation is
Price = 37718 + 2306 Bedrooms + 74.3 House Size - 4.4 Lot Size

Predictor	Coef	SE Coef	T	P
Constant	37718	14177	2.66	0.009
Bedrooms	2306	6994	0.33	0.742
House Size	74.30	52.98	1.40	0.164
Lot Size	-4.36	17.02	-0.26	0.798

S = 25022.7   R-Sq = 56.0%   R-Sq(adj) = 54.6%

Analysis of Variance

Source	DF	SS	MS	F	P
Regression	3	76501718347	25500572782	40.73	0.000
Residual Error	96	60109046053	626135896		
Total	99	1.36611E+11			

INTERPRET

The regression output reveals that none of the independent variables is significantly related to the selling price. (The p-values of the t-tests are .7423, .1640, and .7982, respectively.) However, the F-test ($F = 40.73$ and p-value $= 0$) indicates that the complete model is valid. Moreover, the coefficient of determination is 56.0%, which tells us that the model's fit is good. How can the model be valid and fit well, when none of the independent variables that make up the model are linearly related to price? To answer this question we perform a t-test of the coefficient of correlation between each of the independent variables and the dependent variable.

EXCEL

	A	B
1	**Correlation**	
2		
3	*Price and Bedrooms*	
4	Pearson Coefficient of Correlation	0.6454
5	t Stat	8.36
6	df	98
7	P(T<=t) one tail	0
8	t Critical one tail	1.6606
9	P(T<=t) two tail	0
10	t Critical two tail	1.9845

	A	B
1	**Correlation**	
2		
3	*Price and House Size*	
4	Pearson Coefficient of Correlation	0.7478
5	t Stat	11.15
6	df	98
7	P(T<=t) one tail	0
8	t Critical one tail	1.6606
9	P(T<=t) two tail	0
10	t Critical two tail	1.9845

	A	B
1	**Correlation**	
2		
3	*Price and Lot Size*	
4	Pearson Coefficient of Correlation	0.7409
5	t Stat	10.92
6	df	98
7	P(T<=t) one tail	0
8	t Critical one tail	1.6606
9	P(T<=t) two tail	0
10	t Critical two tail	1.9845

MINITAB

Correlations: Price, Bedrooms

Pearson correlation of Price and Bedrooms = 0.645
P-Value = 0.000

Correlations: Price, House Size

Pearson correlation of Price and House Size = 0.748
P-Value = 0.000

Correlations: Price, Lot Size

Pearson correlation of Price and Lot Size = 0.741
P-Value = 0.000

INTERPRET

The t-tests in the multiple regression model lead to the inference that no independent variable is a factor in determining the selling price. The three t-tests of the correlation coefficients contradict this conclusion. They tell us that the number of bedrooms, the house size, and the lot size are all linearly related to the price. How do we account for this contradiction? The answer is that the three independent variables are correlated with each other. It is reasonable to believe that larger houses have more bedrooms and are situated on larger lots, and that smaller houses have fewer bedrooms and are located on smaller lots. To confirm this belief, we computed the correlations between the three independent variables.

EXCEL

	A	B	C	D
1		*Bedrooms*	*House Size*	*Lot Size*
2	Bedrooms	1		
3	House Size	0.8465	1	
4	Lot Size	0.8374	0.9936	1

MINITAB

Correlations: Bedrooms, House Size, Lot Size

	Bedrooms	House Size
House Size	0.846	
	0.000	
Lot Size	0.837	0.994
	0.000	0.000

Cell Contents: Pearson correlation
P-Value

INTERPRET

The coefficient of correlation between number of bedrooms and house size is .846; the correlation between number of bedrooms and lot size is .837; the correlation between house size and lot size is .994. In the multiple regression model, multicollinearity affected the *t*-tests so that they implied that none of the independent variables is linearly related to price when, in fact, all are.

Another problem caused by multicollinearity is the interpretation of the coefficients. We interpret the coefficients as measuring the change in the dependent variable when the corresponding independent variable increases by 1 unit while all the other independent variables are held constant. This interpretation may be impossible when the independent variables are highly correlated, because when the independent variable increases by 1 unit, some or all of the other independent variables will change. In the multiple regression model in this example, the coefficient of the number of bedrooms is 2,306. Without multicollinearity, we would interpret this coefficient to mean that for each additional bedroom the average price increases by $2,306, provided that the other variables are held constant. However, because the number of bedrooms is correlated with house size and lot size, it is impossible to increase the number of bedrooms by 1 and hold the other variables constant.

This raises two important questions for the statistics practitioner. First, how do we recognize the problem of multicollinearity when it occurs and second, how do we avoid or correct it?

Multicollinearity exists in virtually all multiple regression models. In fact, finding two completely uncorrelated variables is rare. The problem becomes serious, however, only when two or more independent variables are highly correlated. Unfortunately, we do not have a critical value that indicates when the correlation between two independent variables is large enough to cause problems. To complicate the issue, multicollinearity also occurs when a combination of several independent variables is correlated with another independent variable or with a combination of other independent variables. Consequently, even with access to all the correlation coefficients, determining when the multicollinearity problem has reached the serious stage may be extremely difficult. A good indicator of the problem is a large F-statistic, but small t-statistics.

Minimizing the effect of multicollinearity is often easier than correcting it. The statistics practitioner must try to include independent variables that are independent of each other. For example, the real estate agent wanted to include house size, the number of bedrooms, and the lot size, three variables that are clearly related. Rather than developing a model that uses all such variables, the statistics practitioner may choose to include only house size, plus several other variables that measure other aspects of a house's value.

Another alternative is to use a stepwise regression package. *Forward stepwise regression* brings independent variables into the equation one at a time. Only if an independent variable improves the model's fit is it included. If two variables are strongly correlated, the inclusion of one of them in the model makes the second one unnecessary. *Backward stepwise regression* starts with all the independent variables included in the equation and removes variables if they are not strongly related to the dependent variable. Because the stepwise technique excludes redundant variables, it minimizes multicollinearity. Stepwise regression is presented in Chapter 18.

The most commonly used method to remedy nonnormality or heteroscedasticity is to transform the dependent variable. This process is discussed in CD Appendix AA.

EXERCISES

The following exercises require a computer and software.

17.14 Compute the residuals and the predicted values for the regression analysis in Exercise 17.1.

a. Is the normality requirement violated? Explain.
b. Is the variance of the error variable constant? Explain.

17.15 Calculate the coefficients of correlation for each pair of independent variables in Exercise 17.1. What do these statistics tell you about the independent variables and the t-tests of the coefficients?

17.16 Refer to Exercise 17.2.

a. Determine the residuals and predicted values.
b. Does it appear that the normality requirement is violated? Explain.
c. Is the variance of the error variable constant? Explain.
d. Determine the coefficient of correlation between the assignment mark and the midterm mark. What does this statistic tell you about the t-tests of the coefficients?

17.17 Compute the residuals and predicted values for the regression analysis in Exercise 17.3.

a. Does it appear that the error variable is not normally distributed?
b. Is the variance of the error variable constant?
c. Is multicollinearity a problem?

17.18 Refer to Exercise 17.4. Find the coefficients of correlation of the independent variables.

a. What do these correlations tell you about the independent variables?
b. What do they say about the t-tests of the coefficients?

17.19 Calculate the residuals and predicted values for the regression analysis in Exercise 17.5.
a. Does the error variable appear to be normally distributed?
b. Is the variance of the error variable constant?
c. Is multicollinearity a problem?

17.20 Are the required conditions satisfied in Exercise 17.6?

17.21 Refer to Exercise 17.7.
a. Conduct an analysis of the residuals to determine whether any of the required conditions are violated.
b. Does it appear that multicollinearity is a problem?
c. Identify any observations that should be checked for accuracy.

17.22 Are the required conditions satisfied for the regression analysis in Exercise 17.8?

17.23 Determine whether the required conditions are satisfied in Exercise 17.9.

17.24 Refer to Exercise 17.10. Calculate the residuals and predicted values.
a. Is the normality requirement satisfied?
b. Is the variance of the error variable constant?
c. Is multicollinearity a problem?

17.25 Determine whether there are violations of the required conditions in the regression model used in Exercise 17.11.

17.26 Determine whether the required conditions are satisfied in Exercise 17.12.

17.27 Refer to Exercise 17.13.
a. Are the required conditions satisfied?
b. Is multicollinearity a problem? If so, explain the consequences.

17.28 Refer to the MBA Program Admissions Policy example. Are the required conditions satisfied?

17.4 Regression Diagnostics-III (Time Series)

In Chapter 16, we pointed out that, in general, we check to see whether the errors are independent when the data constitute a *times series*—data gathered sequentially over a series of time periods. In Section 16.6, we described the graphical procedure for determining whether the required condition that the errors be independent is violated. We plot the residuals versus the time periods and look for patterns. In this section, we augment that procedure with the **Durbin-Watson test.**

Durbin-Watson Test

The Durbin-Watson test allows the statistics practitioner to determine whether there is evidence of **first-order autocorrelation**—a condition in which a relationship exists between consecutive residuals e_i and e_{i-1}, where i is the time period. The Durbin-Watson statistic is defined as

$$d = \frac{\sum_{i=2}^{n}(e_i - e_{i-1})^2}{\sum_{i=1}^{n} e_i^2}$$

The range of the values of d is

$$0 \le d \le 4$$

where small values of d ($d < 2$) indicate a positive first-order autocorrelation and large values of d ($d > 2$) imply a negative first-order autocorrelation. Positive first-order autocorrelation is a common occurrence in business and economic time series. It occurs when consecutive residuals tend to be similar. In that case, $(e_i - e_{i-1})^2$ will be small, producing a small value for d. Negative first-order autocorrelation occurs when consecutive residuals differ widely. For example, if positive and negative residuals generally alternate, $(e_i - e_{i-1})^2$ will be large, and as a result, d will be greater than 2. Figures 17.2 and 17.3 depict positive first-order autocorrelation, whereas Figure 17.4 illustrates negative autocorrelation. Notice that in Figure 17.2 the first residual is a small number; the second residual, also a small number, is somewhat larger; and that trend continues. In Figure 17.3, the first residual is large, and in general, succeeding residuals decrease. In both figures, consecutive residuals are similar. In Figure 17.4, the first residual is a positive number and is followed by a negative residual. The remaining residuals follow this pattern (with some exceptions). Consecutive residuals are quite different.

FIGURE 17.2 Positive First-Order Autocorrelation

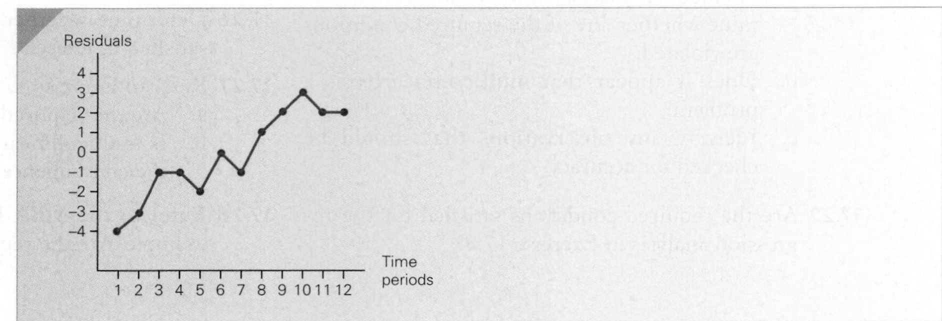

FIGURE 17.3 Positive First-Order Autocorrelation

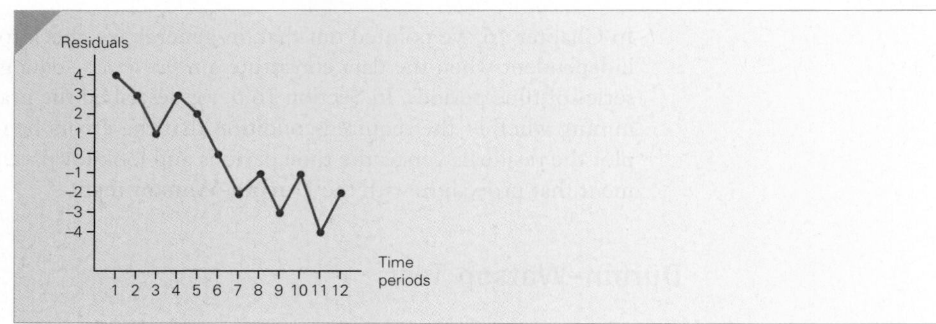

FIGURE 17.4 Negative First-Order Autocorrelation

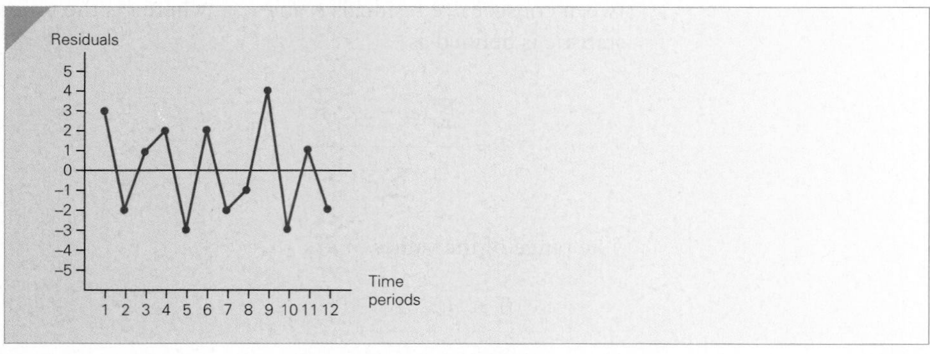

Table 8 in Appendix B is designed to test for positive first-order autocorrelation by providing values of d_L and d_U for a variety of values of n and k and for $\alpha = .01$ and .05.

The decision is made in the following way. If $d < d_L$, we conclude that there is enough evidence to show that positive first-order autocorrelation exists. If $d > d_U$, we conclude that there is not enough evidence to show that positive first-order autocorrelation exists. And if $d_L \leq d \leq d_U$, the test is inconclusive. The recommended course of action when the test is inconclusive is to continue testing with more data until a conclusive decision can be made.

For example, to test for positive first-order autocorrelation with $n = 20$, $k = 3$, and $\alpha = .05$, we test the following hypotheses:

H_0: There is no first-order autocorrelation.

H_1: There is positive first-order autocorrelation.

The decision is made as follows:

If $d < d_L = 1.00$, reject the null hypothesis in favor of the alternative hypothesis.

If $d > d_U = 1.68$, do not reject the null hypothesis.

If $1.00 \leq d \leq 1.68$, the test is inconclusive.

To test for negative first-order autocorrelation, we change the critical values. If $d > 4 - d_L$, we conclude that negative first-order autocorrelation exists. If $d < 4 - d_U$, we conclude that there is not enough evidence to show that negative first-order autocorrelation exists. If $4 - d_U \leq d \leq 4 - d_L$, the test is inconclusive. We can also test simply for first-order autocorrelation by combining the two one-tail tests. If $d < d_L$ or $d > 4 - d_L$, we conclude that autocorrelation exists. If $d_U \leq d \leq 4 - d_U$ we conclude that there is no evidence of autocorrelation. If $d_L \leq d \leq d_U$ or $4 - d_U \leq d \leq 4 - d_L$, the test is inconclusive. The significance level will be 2α (where α is the one-tail significance level). Figure 17.5 describes the range of values of d and the conclusion for each interval.

FIGURE **17.5**
Durbin–Watson Test

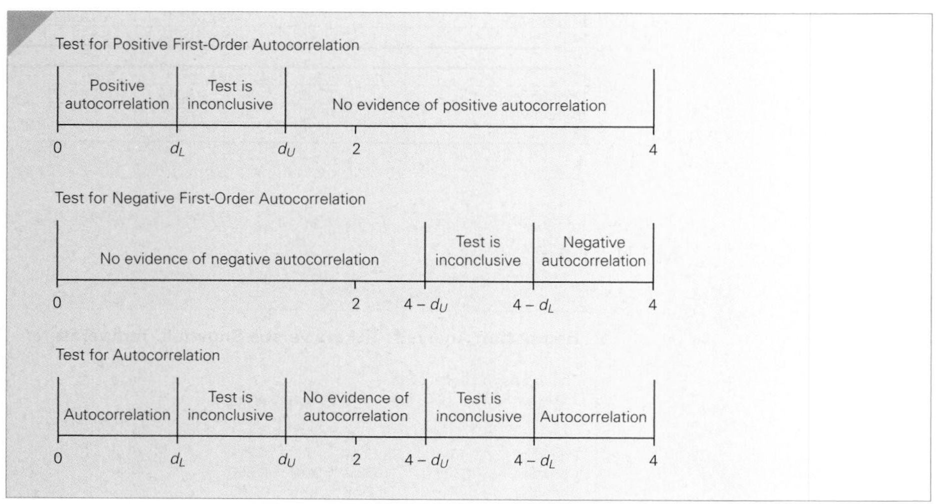

For time-series data, we add the Durbin-Watson test to our list of regression diagnostics. That is, we determine whether the error variable is normally distributed with constant variance (as we did in Section 17.3), we identify outliers and (if our software allows it) influential observations that should be verified, and we conduct the Durbin-Watson test.

EXAMPLE 17.3 Christmas Week Ski Lift Sales

DATA Xm17-03

Christmas week is a critical period for most ski resorts. Because many students and adults are free from other obligations, they are able to spend several days indulging in their favorite pastime, skiing. A large proportion of gross revenue is earned during this period. A ski resort in Vermont wanted to determine the effect that weather had on its sales of lift tickets. The manager of the resort collected data on the number of lift tickets sold during Christmas week (y), the total snowfall in inches (x_1), and the average temperature in degrees Fahrenheit (x_2) for the past 20 years. Develop the multiple regression model, and diagnose any violations of the required conditions.

SOLUTION

The model is

$$y = \beta_0 + \beta_1 x_1 + \beta_2 x_2 + \varepsilon$$

EXCEL

	A	B	C	D	E	F
1	SUMMARY OUTPUT					
2						
3	*Regression Statistics*					
4	Multiple R	0.3465				
5	R Square	0.1200				
6	Adjusted R Square	0.0165				
7	Standard Error	1712				
8	Observations	20				
9						
10	ANOVA					
11		*df*	*SS*	*MS*	*F*	*Significance F*
12	Regression	2	6793798	3396899	1.16	0.3373
13	Residual	17	49807214	2929836		
14	Total	19	56601012			
15						
16		*Coefficients*	*Standard Error*	*t Stat*	*P-value*	
17	Intercept	8308	904	9.19	5.24E-08	
18	Snowfall	74.59	51.57	1.45	0.1663	
19	Temperature	-8.75	19.70	-0.44	0.6625	

MINITAB

Regression Analysis: Tickets versus Snowfall, Temperature

The regression equation is
Tickets = 8308 + 74.6 Snowfall - 8.8 Temperature

Predictor	Coef	SE Coef	T	P
Constant	8308.0	903.7	9.19	0.000
Snowfall	74.59	51.57	1.45	0.166
Temperature	-8.75	19.70	-0.44	0.662

S = 1712 R-Sq = 12.0% R-Sq(adj) = 1.7%

Analysis of Variance

Source	DF	SS	MS	F	P
Regression	2	6793798	3396899	1.16	0.337
Residual Error	17	49807214	2929836		
Total	19	56601012			

INTERPRET

As you can see the coefficient of determination is small ($R^2 = 12\%$) and the p-value of the F-test is .3373, both of which indicate that the model is poor. We used Excel to draw the histogram (Figure 17.6) of the residuals and plot the predicted values of y versus the residuals in Figure 17.7. Because the observations constitute a time series, we also used Excel to plot the time periods (years) versus the residuals (Figure 17.8).

FIGURE **17.6** Histogram of Residuals in Example 17.3

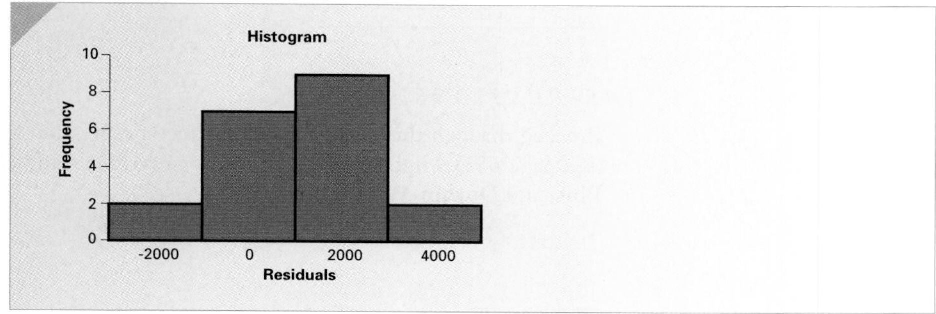

The histogram reveals that the error may be normally distributed.

FIGURE **17.7** Plot of Predicted Values versus Residuals in Example 17.3

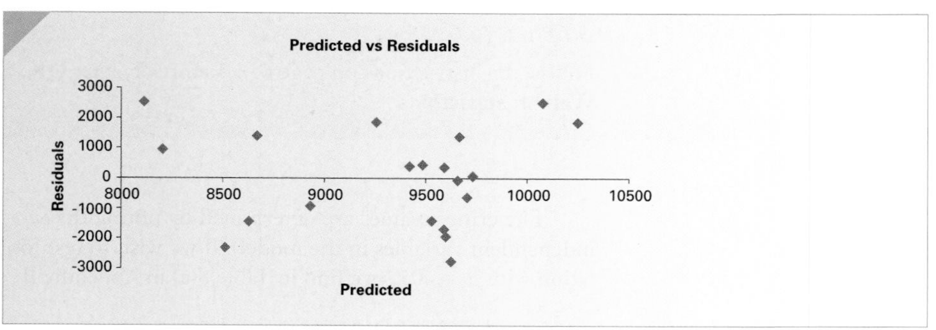

There does not appear to be any evidence of heteroscedasticity.

FIGURE **17.8** Plot of Time Periods versus Residuals in Example 17.3

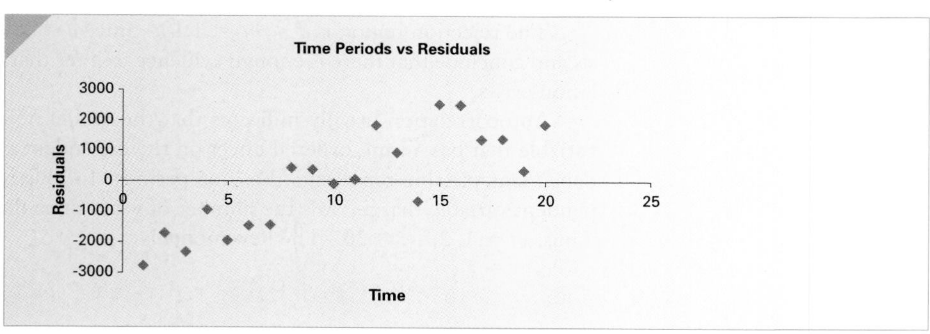

This graph reveals a serious problem. There is a strong relationship between consecutive values of the residuals, which indicates that the requirement that the errors are independent has been violated. To confirm this diagnosis, we instructed Excel and Minitab to calculate the Durbin-Watson statistic.

EXCEL

	A	B	C
1	Durbin-Watson Statistic		
2			
3	d = 0.5931		

INSTRUCTIONS

Proceed through the usual steps to conduct a regression analysis and print the residuals (see page 653). Highlight the entire list of residuals and click **Add-Ins, Data Analysis Plus,** and **Durbin-Watson Statistic.**

MINITAB

Durbin-Watson statistic = 0.593140

INSTRUCTIONS

Follow the instructions on page 654. Before clicking **OK,** click **Options . . .** and **Durbin-Watson statistic.**

The critical values are determined by noting that $n = 20$ and $k = 2$ (there are two independent variables in the model). If we wish to test for positive first-order autocorrelation with $\alpha = .05$, we find in Table 8(a) in Appendix B

$$d_L = 1.10 \text{ and } d_U = 1.54$$

The null and alternative hypotheses are

H_0: There is no first-order autocorrelation.

H_1: There is positive first-order autocorrelation.

The rejection region is $d < d_L = 1.10$. Since $d = .59$, we reject the null hypothesis and conclude that there is enough evidence to infer that positive first-order autocorrelation exists.

Autocorrelation usually indicates that the model needs to include an independent variable that has a time-ordered effect on the dependent variable. The simplest such independent variable represents the time periods. To illustrate, we included a third independent variable that records the number of years since the year the data were gathered. Thus, $x_3 = 1, 2, \ldots, 20$. The new model is

$$y = \beta_0 + \beta_1 x_1 + \beta_2 x_2 + \beta_3 x_3 + \varepsilon$$

EXCEL

	A	B	C	D	E	F
1	SUMMARY OUTPUT					
2						
3	*Regression Statistics*					
4	Multiple R	0.8608				
5	R Square	0.7410				
6	Adjusted R Square	0.6924				
7	Standard Error	957				
8	Observations	20				
9						
10	ANOVA					
11		*df*	*SS*	*MS*	*F*	*Significance F*
12	Regression	3	41940217	13980072	15.26	0.0001
13	Residual	16	14660795	916300		
14	Total	19	56601012			
15						
16		*Coefficients*	*Standard Error*	*t Stat*	*P-value*	
17	Intercept	5966	631.3	9.45	6.00E-08	
18	Snowfall	70.18	28.85	2.43	0.0271	
19	Temperature	-9.23	11.02	-0.84	0.4145	
20	Time	230.0	37.13	6.19	1.29E-05	

MINITAB

Regression Analysis: Tickets versus Snowfall, Temperature, Time

The regression equation is
Tickets = 5966 + 70.2 Snowfall - 9.2 Temperature + 230 Time

Predictor	Coef	SE Coef	T	P
Constant	5965.6	631.3	9.45	0.000
Snowfall	70.18	28.85	2.43	0.027
Temperature	-9.23	11.02	-0.84	0.414
Time	229.97	37.13	6.19	0.000

S = 957.2 R-Sq = 74.1% R-Sq(adj) = 69.2%

Analysis of Variance

Source	DF	SS	MS	F	P
Regression	3	41940217	13980072	15.26	0.000
Residual Error	16	14660795	916300		
Total	19	56601012			

As we did before, we calculate the residuals and conduct regression diagnostics using Excel. The results are shown in Figures 17.9–17.11.

FIGURE **17.9** Histogram of Residuals in Example 17.3 (Time Variable Included)

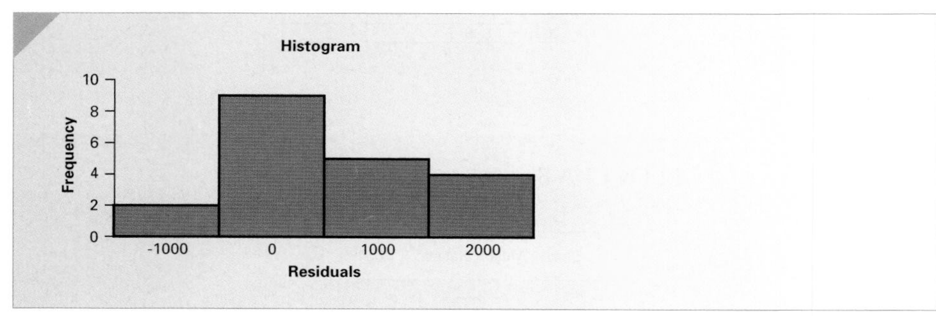

The histogram reveals that the error may be normally distributed.

FIGURE **17.10** Plot of Predicted Values versus Residuals in Example 17.3 (Time Variable Included)

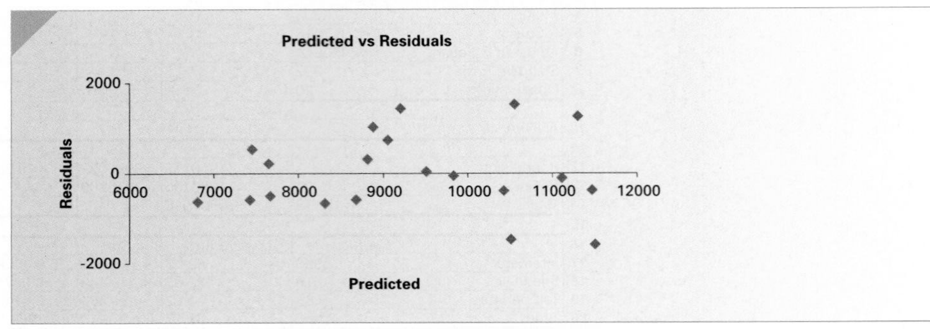

The error variable variance appears to be constant.

FIGURE **17.11** Plot of Time Periods versus Residuals in Example 17.3 (Time Variable Included)

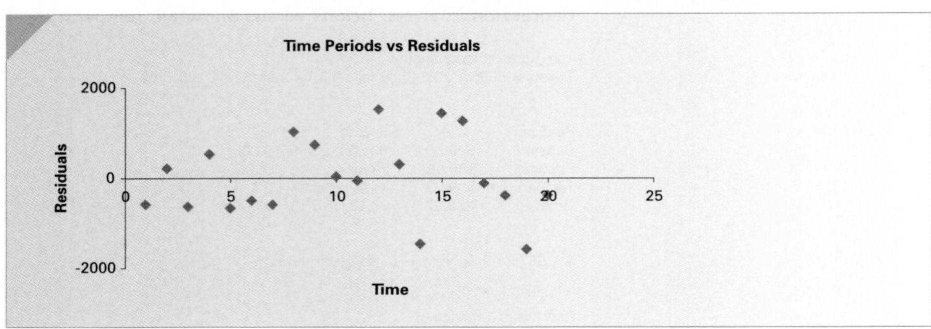

There is no sign of autocorrelation. To confirm our diagnosis, we conducted the Durbin-Watson test.

EXCEL

	A	B	C
1	**Durbin-Watson Statistic**		
2			
3	d = 1.885		

MINITAB

Durbin-Watson statistic = 1.88499

From Table 8(a) in Appendix B, we find the critical values of the Durbin-Watson test. With $k = 3$ and $n = 20$, we find

$$d_L = 1.00 \text{ and } d_U = 1.68$$

Since $d > 1.68$, we conclude that there is not enough evidence to infer the presence of positive first-order autocorrelation.

Notice that the model is improved dramatically. The F-test tells us that the model is valid. The t-tests tell us that both the amount of snowfall and time are significantly linearly related to the number of lift tickets. This information could prove useful in advertising for the resort. For example, if there has been a recent snowfall, the resort could emphasize that in its advertising. If no new snow has fallen, the resort may emphasize its snow-making facilities.

Developing an Understanding of Statistical Concepts

Notice that the addition of the time variable explained a large proportion of the variation in the number of lift tickets sold. That is, the resort experienced a relatively steady increase in sales over the past 20 years. Once this variable was included in the model, the amount of snowfall became significant because it was able to explain some of the remaining variation in lift ticket sales. Without the time variable, the amount of snowfall and the temperature were unable to explain a significant proportion of the variation in ticket sales. The graph of the residuals versus the time periods and the Durbin-Watson test enabled us to identify the problem and correct it. In overcoming the autocorrelation problem, we improved the model so that we identified the amount of snowfall as an important variable in determining ticket sales. This result is quite common. Correcting a violation of a required condition will frequently improve the model.

EXERCISES

17.29 Perform the Durbin-Watson test at the 5% significance level to determine whether positive first-order autocorrelation exists when $d = 1.10$, $n = 25$, and $k = 3$.

17.30 Determine whether negative first-order autocorrelation exists when $d = 2.85$, $n = 50$, and $k = 5$. (Use a 1% significance level.)

17.31 Given the following information, perform the Durbin-Watson test to determine whether first-order autocorrelation exists.

$$n = 25 \quad k = 5 \quad \alpha = .10 \quad d = .90$$

17.32 Test the following hypotheses with $\alpha = .05$.

H_0: There is no first-order autocorrelation
H_1: There is positive first-order autocorrelation
$n = 50, k = 2, d = 1.38$

17.33 Test tile following hypotheses with $\alpha = .02$.

H_0: There is no first-order autocorrelation
H_1: There is first-order autocorrelation
$n = 90, k = 5, d = 1.60$

17.34 Test the following hypotheses with $\alpha = .05$.

H_0: There is no first-order autocorrelation
H_1: There is negative first-order autocorrelation
$n = 33, k = 4, d = 2.25$

The following exercises require a computer and software.

17.35 Xr17-35 Observations of variables y, x_1, and x_2 were taken over 100 consecutive time periods.

a. Conduct a regression analysis of these data.
b. Plot the residuals versus the time periods. Describe the graph.
c. Perform the Durbin-Watson test. Is there evidence of autocorrelation? Use $\alpha = .10$.
d. If autocorrelation was detected in Part c, propose an alternative regression model to remedy the problem. Use the computer to generate the statistics associated with this model.
e. Redo Parts b and c. Compare the two models.

17.36 Xr17-36 Weekly sales of a company's product (y) and those of its main competitor (x) were recorded for one year.

a. Conduct a regression analysis of these data.
b. Plot the residuals versus the time periods. Does there appear to be autocorrelation?
c. Perform the Durbin-Watson test. Is there evidence of autocorrelation? Use $\alpha = .10$.
d. If autocorrelation was detected in Part c, propose an alternative regression model to remedy the problem. Use the computer to generate the statistics associated with this model.
e. Redo Parts b and c. Compare the two models.

17.37 Refer to Exercise 17.3. Is there evidence of positive first-order autocorrelation?

17.38 Refer to Exercise 16.81. Determine whether there is evidence of first-order autocorrelation.

17.39 Xr17-39 The manager of a tire store in Minneapolis has been concerned with the high cost of inventory. The current policy is to stock all the snow tires that are predicted to sell over the entire winter at the beginning of the season (end of October). The manager can reduce inventory costs by having suppliers deliver snow tires regularly from October to February. However, he needs to be able to predict weekly sales to avoid stockouts that will ultimately lose sales. To help develop a forecasting model, he records the number of snow tires sold weekly during the last winter and the amount of snowfall (in inches) in each week.

a. Develop a regression model and use a software package to produce the statistics.
b. Perform a complete diagnostic analysis to determine whether the required conditions are satisfied.
c. If one or more conditions are unsatisfied, attempt to remedy the problem.
d. Use whatever procedures you wish to assess how well the new model fits the data.
e. Interpret and test each of the coefficients.

CHAPTER SUMMARY

The multiple regression model extends the model introduced in Chapter 16. The statistical concepts and techniques are similar to those presented in simple linear regression. We assess the model in three ways: standard error of estimate, the coefficient of determination (and the coefficient of determination adjusted for degrees of freedom), and the F-test of the analysis of variance. We can use the t-tests of the coefficients to determine whether each of the independent variables is linearly related to the dependent variable. As we did in Chapter 16, we showed how to diagnose violations of the required conditions and to identify other problems. We introduced multicollinearity and demonstrated its effect and its remedy. Finally, we presented the Durbin-Watson test to detect first-order autocorrelation.

IMPORTANT TERMS

Response surface 672
Coefficient of determination adjusted for degrees of freedom 678
Multicollinearity 692
Durbin-Watson test 697
First-order autocorrelation 697

SYMBOLS

Symbol	Pronounced	Represents
β_i	*Beta-sub-i* or *beta-i*	Coefficient of ith independent variable
b_i	*b-sub-i* or *b-i*	Sample coefficient

FORMULAS

Standard error of estimate

$$s_\varepsilon = \sqrt{\frac{\text{SSE}}{n-k-1}}$$

Test statistic for β_i

$$t = \frac{b_i - \beta_i}{s_{b_i}}$$

Coefficient of determination

$$R^2 = \frac{s_{xy}^2}{s_x^2 s_y^2} = 1 - \frac{\text{SSE}}{\sum (y_i - \bar{y})^2}$$

Adjusted coefficient of determination

$$\text{Adjusted } R^2 = 1 - \frac{\text{SSE}/(n - k - 1)}{\sum (y_i - \bar{y})^2/(n - 1)}$$

Mean square for error

$$\text{MSE} = \text{SSE}/k$$

Mean square for regression

$$\text{MSR} = \text{SSR}/(n - k - 1)$$

F-statistic

$$F = \text{MSR}/\text{MSE}$$

Durbin-Watson statistic

$$d = \frac{\sum_{i=2}^{n} (e_i - e_{i-1})^2}{\sum_{i=1}^{n} e_i^2}$$

COMPUTER OUTPUT AND INSTRUCTIONS

Technique	Excel	Minitab
Regression	675	676
Prediction interval	685	685
Durbin-Watson statistic	702	702

CHAPTER EXERCISES

The following exercises require the use of a computer and statistical software. ***Use a 5% significance level.***

17.40 Xr17-40 The agronomist referred to in Exercise 16.83 believed that the amount of rainfall as well as the amount of fertilizer used would affect the crop yield. She redid the experiment in the following way. Thirty greenhouses were rented. In each, the amount of fertilizer and the amount of water were varied. At the end of the growing season, the amount of corn was recorded.

a. Determine the sample regression line, and interpret the coefficients.
b. Do these data allow us to infer that there is a linear relationship between the amount of fertilizer and the crop yield?
c. Do these data allow us to infer that there is a linear relationship between the amount of water and the crop yield?
d. What can you say about the fit of the multiple regression model?
e. Is it reasonable to believe that the error variable is normally distributed with constant variance?
f. Predict the crop yield when 100 kilograms of fertilizer and 1,000 liters of water are applied. Use a confidence level of 95%.

17.41 Xr16-12* Exercise 16.12 addressed the problem of determining the relationship between the price of apartment buildings and number of square feet. Hoping to improve the predictive capability of the model, the real estate agent also recorded the number of apartments, the age, and the number of floors.

a. Calculate the regression equation.
b. Is the model valid?
c. Compare your answer with that of Exercise 16.12.

17.42 Xr16-16* In Exercise 16.16, a statistics practitioner examined the relationship between office rents and the city's office vacancy rate. The model appears to be quite poor. It was decided to add another variable that measures the state of the economy. The city's unemployment rate was chosen for this purpose.

a. Determine the regression equation.
b. Determine the coefficient of determination and describe what this value means.
c. Test the model's validity in explaining office rent.
d. Determine which of the two independent variables is linearly related to rents.
e. Determine whether the error is normally distributed with a constant variance.
f. Determine whether there is evidence of autocorrelation.
g. Predict with 95% confidence the office rent in a city whose vacancy rate is 10% and whose unemployment rate is 7%.

CASE 17.1 An Analysis of Mutual Fund Managers, Part 1*

© Ed Pritchard/Stone/Getty Images

DATA [C17-01]

There are thousands of mutual funds available (see page 175 for a brief introduction to mutual funds). There is no shortage of sources of information about them. Newspapers regularly report the value of each unit, mutual fund companies and brokers advertise extensively, and there are books on the subject. Many of the advertisements imply that individuals should invest in the advertiser's mutual fund because it has performed well in the past. Unfortunately, there is little evidence to infer that past performance is a predictor of the future. However, it may be possible to acquire useful information by examining the managers of mutual funds. Several researchers have studied the issue. One project gathered data concerning the performance of 2,029 funds.

The performance of each fund was measured by its ***risk-adjusted excess return,*** which is the difference between the return on investment of the fund and a return that is considered a standard. The standard is based on a variety of variables including the risk-free rate.

There are four variables that describe the fund manager. They are age, tenure (how many years the manager has been in charge), whether the manager had an MBA (1 = yes, 0 = no), and a measure of the quality of the manager's education [the average Scholastic Achievement Test (SAT) score of students at the university where the manager received his or her undergraduate degree].

Conduct an analysis of the data. Discuss how the average SAT score of the manager's alma mater, whether he or she has an MBA, and his or her age and tenure are related to the performance of the fund.

*This case is based on Judith Chevalier and Glenn Ellison, "Are Some Mutual Fund Managers Better Than Others? Cross-Sectional Patterns in Behavior and Performance," Working Paper 5852, National Bureau of Economic Research.

CASE 17.2 An Analysis of Mutual Fund Managers, Part 2

© BananaStock/Jupiterimages

DATA [C17-02a] [C17-02b]

In addition to analyzing the relationship between the managers' characteristic and the performance of the fund, researchers wanted to determine whether the same characteristics are related to the behavior of the fund. In particular, they wanted to know whether the risk of the fund and its management expense ratio (MER) were related to the managers' age, tenure, university SAT score, and whether he or she had an MBA.

In Section 4.6 we introduced the market model wherein we measure the systematic risk of stocks by the stock's beta. The beta of a portfolio is the average of the betas of the stocks that make up the portfolio. File C17-02a stores the same managers' characteristics as those in file C17-01. However, the first column contains the betas of the mutual funds.

To analyze the management expense ratios, it was decided to include a measure of the size of the fund. The logarithm of the funds' assets (in $millions) was recorded with the MER. These data are stored in file C17-02b.

Analyze both sets of data and write a brief report of your findings.

APPENDIX 17 / REVIEW OF CHAPTERS 12 TO 17

Table A17.1 presents a list of inferential methods presented thus far and Figure A17.1 depicts a flowchart designed to help students identify the correct statistical technique.

TABLE **A17.1** Summary of Statistical Techniques in Chapters 12 to 17

- t-test of μ
- Estimator of μ (including small population estimator of μ and large and small population estimators of $N\mu$)
- χ^2-test of σ^2
- Estimator of σ^2
- z-test of p
- Estimator of p (including small population estimator of p and large and small population estimators of Np)
- Equal-variances t-test of $\mu_1 - \mu_2$
- Equal-variances estimator of $\mu_1 - \mu_2$
- Unequal-variances t-test of $\mu_1 - \mu_2$
- Unequal-variances estimator of $\mu_1 - \mu_2$
- t-test of μ_D
- Estimator of μ_D
- F-test of σ_1^2/σ_2^2
- Estimator of σ_1^2/σ_2^2
- z-test of $p_1 - p_2$ (Case 1)
- z-test of $p_1 - p_2$ (Case 2)
- Estimator of $p_1 - p_2$
- One-way analysis of variance (including multiple comparisons)
- Two-way (randomized blocks) analysis of variance
- Two-factor analysis of variance
- χ^2-goodness-of-fit test
- χ^2-test of a contingency table
- Simple linear regression and correlation (including t-tests of β_1 and ρ, and prediction and confidence intervals)
- Multiple regression (including t-tests of β_i, F-test, and prediction and confidence intervals)

FIGURE A17.1 Flowchart of Techniques in Chapters 12 to 17

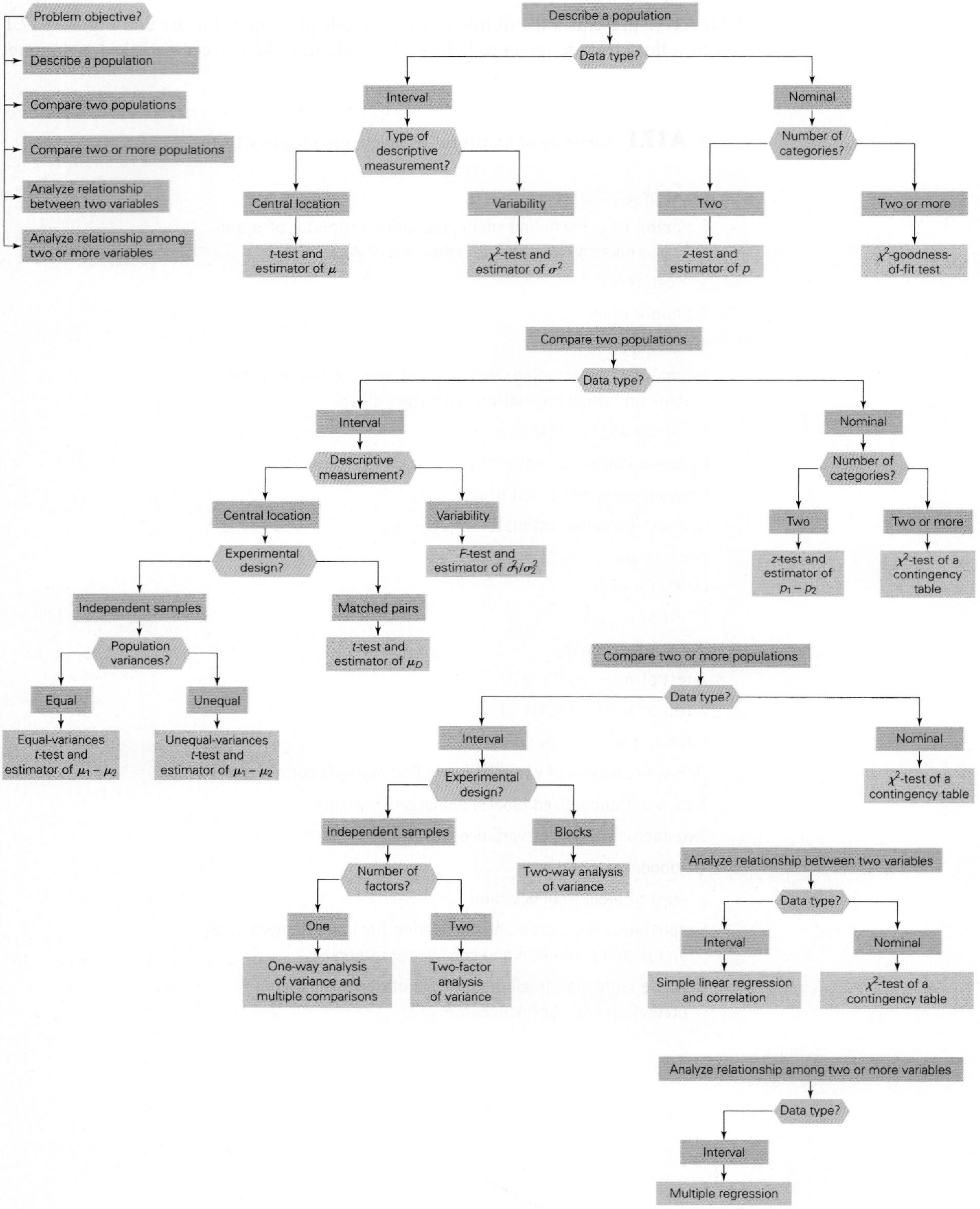

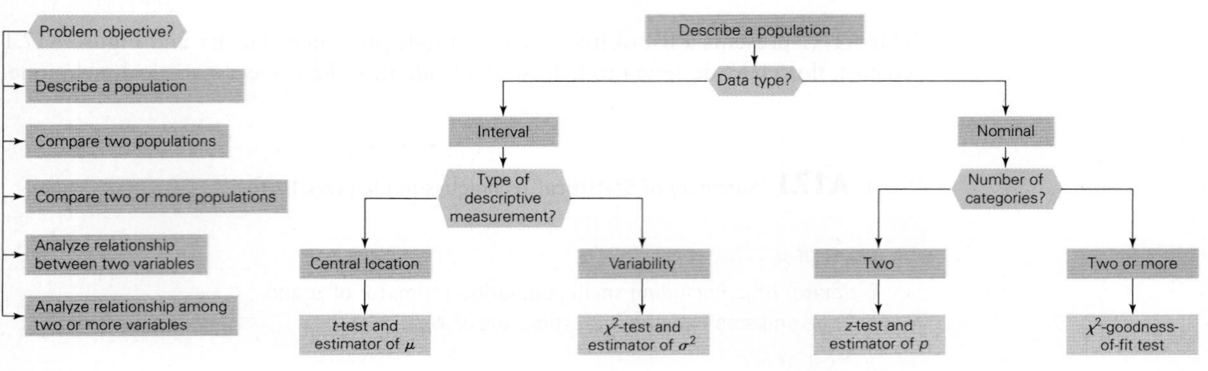

EXERCISES

A17.1 XrA17-01 Garlic has long been considered a remedy to ward off the common cold. A British researcher organized an experiment to see if this generally held belief is true. A random sample of 146 volunteers was recruited. Half the sample took one capsule of an allicin-containing garlic supplement each day. The others took a placebo. The results for each volunteer after the winter months were recorded in the following way:

Column

1. Identification number
2. 1 = allicin-containing capsule; 2 = placebo
3. Suffered a cold (1 = no, 2 = yes)
4. If individual caught a cold, the number of days until recovery (999 was recorded if no cold)

a. Can the researcher conclude that garlic does help prevent colds?
b. Does garlic reduce the number of days until recovery if a cold was caught?

A17.2 XrA17-02 Since shelf space is a limited resource of a retail store, product selection, shelf space allocation, and shelf space placement decisions must be made according to a careful analysis of profitability and inventory turnover. The manager of a chain of variety stores wishes to see whether shelf location affects the sales of canned soup. She believes that placing the product at eye level will result in greater sales than will placing the product on a lower shelf. She observed the number of sales of the product in 40 different stores. Sales were observed over 2 weeks, with product placement at eye level one week and on a lower shelf the other week. Can we conclude that placement of the product at eye level significantly increases sales?

A17.3 XrA17-03 In an effort to explain the results of Exercise A15.9, a researcher recorded the distances for the random sample of British and American courses. Can we infer that British courses are shorter than American courses?

A17.4 XrA17-04 It is generally assumed that alcohol consumption tends to make drinkers more impulsive. However, a recent study in the journal *Alcohol and Alcoholism* may contradict this assumption. The study took a random sample of 76 male undergraduate students and divided them into three groups. One group remained sober, a second group was given flavored drinks with not enough alcohol to intoxicate, and the students in third group were intoxicated. Each student was offered a chance of receiving \$15 at the end of the session or double that amount later. The results were recorded using the following format:

Column 1: Group number

Column 2: Code 1 = chose \$15, 2 = chose \$30 later

Do the data allow us to infer that there is a relationship between the choices students make and their level of intoxication?

A17.5 XrA17-05 Refer to Exercise 13.29. The executive did a further analysis by taking another random sample. This time she tracked the number of customers who have had an accident in the last 5 years. For each she recorded the total amount of repairs and the credit score. Do these data allow the executive to conclude that the higher the credit score, the lower the cost of repairs will be?

A17.6 XrA17-06 The U.S. National Endowment for the Arts conducts surveys of American adults to determine, among other things, their participation in various arts activities. A recent survey asked a random sample of American adults whether they participate in photography. The responses are 1 = yes and 2 = no. There were 205.8 million American adults. Estimate with 95% confidence the number of American adults who participate in photography. (*Source:* Adapted from the *Statistical Abstract of the United States*, 2006, Table 1228)

A17.7 XrA17-07 Mouth-to-mouth resuscitation has long been considered better than chest compression for people who have suffered a heart attack. To determine if this indeed is the better way, Japanese researchers analyzed 4,068 adult patients who had cardiac arrest witnessed by bystanders. Of those, 439 received only chest compressions from bystanders and 712 received conventional CPR-compressions and breaths. The results for each group was recorded where 1 = did not survive with good neurological function and 2 = did survive with good neurological function. What conclusions can be drawn from these data?

A17.8 XrA17-08 Refer to Exercise A15.6. The Financial analyst undertook another project wherein respondents were also asked the age of the head of the household. The choices are

1. Under 25
2. 25 to 34
3. 35 to 44
4. 45 to 54
5. 55 to 64
6. 65 and over

The responses to questions about ownership of mutual funds is No = 1 and Yes = 2. Do these data allow us to infer that the age of the head of the household is related to whether he or she owns mutual funds? (*Source:* Adapted from the *Statistical Abstract of the United States*, 2006, Table 1200)

A17.9 XrA17-09 Over the last decade (1995–2005), the number of hip and knee replacement surgeries has increased by 87%. Because the costs of hip and knee replacements are so expensive, private health insurance and government-operated health cares plans have become more concerned. To get more information, random samples of people who have had hip replacements in 1995 and in 2005 were drawn. From the files, the ages of the patients were recorded. Is there enough evidence to infer that the ages of people who require hip replacements are getting smaller? (*Source:* Canadian Joint Replacement Registry)

A17.10 XrA17-10 Refer to Exercise A17.9. A major factor that determines whether a person will need a hip or knee replacement and, if so, at what age is weight. To learn more about the topic, a medical researcher randomly sampled individuals who had hip replacement (code = 1) and knee replacement (code = 2) and one of the following categories:

1. Underweight 2. Normal range
3. Overweight but not obese 4. Obese.

Do the data allow the researcher to conclude that weight and the joint needing replacement are related?

A17.11 XrA17-11 Television shows with large amounts of sex or violence tend to attract more viewers. Advertisers want large audiences, but they also want viewers to remember the brand names of their products. To determine the effect that shows with sex and violence have on their viewers, a study was undertaken. A random sample of 328 adults was divided into three groups. Group 1 watched violent programs, group 2 watched sexually explicit shows, and group 3 watched neutral shows. The researchers spliced nine 30-second commercials for a wide range of products. After the show, the subjects were quizzed to see if they could recall the brand name of the products. They were also asked to name the brands 24 hours later. The number of correct answers was recorded. Conduct a test to determine whether differences exist between the three groups of viewers and which type of program does best in brand recall. Results were published in the *Journal of Applied Psychology* (*National Post* August 16, 2004).

A17.12 XrA17-12 In an effort to explain to customers why their electricity bills have been so high lately, and how, specifically, they could save money by reducing the thermostat settings on both space heaters and water heaters, a public utility commission has collected total kilowatt consumption figures for last year's winter months, as well as thermostat settings on space and water heaters, for 100 homes.

a. Determine the regression equation.
b. Determine the coefficient of determination, and describe what it tells you.
c. Test the validity of the model.
d. Find the 95% interval of the electricity consumption of a house whose space heater thermostat is set at 70 and whose water heater thermostat is set at 130.
e. Calculate the 95% interval of the average electricity consumption for houses whose space heater thermostat is set at 70 and whose water heater thermostat is set at 130.

A17.13 XrA17-13 An economist wanted to learn more about total compensation packages. She conducted a survey of 858 workers and asked each to report their hourly wages and salaries, their total benefits, and whether the company they worked for produced goods or services. Determine whether differences exist between goods-producing and services-producing firms in terms of hourly wages and total benefits. (Adapted from the *Statistical Abstract of the United States*, 2006, Table 637)

A17.14 XrA17-14 Professional athletes in North America are paid very well for their ability to play games that amateurs play for fun. To determine the factors that influence a team to pay a hockey player's salary, an MBA student randomly selected 50 hockey players who played in the 1992–1993 and 1993–1994 seasons. He recorded their salaries at the end of the 1993–1994 season as well as a number of performance measures in the previous two seasons. The following data were recorded:

Columns 1 and 2: Games played in 1992–1993 and 1993–1994

Columns 3 and 4: Goals scored in 1992–1993 and 1993–1994
Columns 5 and 6: Assists recorded in 1992–1993 and 1993–1994
Columns 7 and 8: Plus/minus score in 1992–1993 and 1993–1994
Columns 9 and 10: Penalty minutes served in 1992–1993 and 1993–1994
Column 11: Salary in U.S. dollars

(Plus/minus is the number of goals scored by his team minus the number of goals scored by the opposing team while the player is on the ice.) Develop a model that analyzes the relationship between salary and the performance measures. Describe your findings. (The author wishes to thank Gordon Barnett for writing this exercise.)

A17.15 XrA17-15 The risks associated with smoking are well known. Virtually all physicals recommend that their patients quit. This raises the question, What are the risks for people who quit smoking compared to continuing smokers and those who have never smoked? In a study described in the *Journal of Internal Medicine* (February 2004, 255(2): 266–272) researchers took samples of each of the following groups:

Group 1: Never smokers
Group 2: Continuing smokers
Group 3: Smokers who quit

At the beginning of the 10-year research project, there were 238 people who had never smoked and 155 smokers. Over the span of a year, 39 smokers quit. The weight gain, increase in systolic (SBP) blood pressure, and the increase in diastolic (DBP) blood pressure was measured and recorded. Determine whether differences exist between the three groups in terms of weight gain, systolic blood pressure increase, diastolic blood pressure increase, and which groups differ.

A17.16 XrA17-16 A survey was conducted among Canadian farmers. Each farmer was asked to report the number of acres in his or her farm. There were a total of 229,373 farms in Canada in 2006 (*Source:* Statistics Canada). Estimate with 95% confidence the total amount of area (in acres) that was farmed in Canada in 2006.

CASE A17.1 Testing a More Effective Device to Keep Arteries Open

DATA
CA17-01

A stent is a metal mesh cylinder that holds a coronary artery open after a blockage has been removed. However, in many patients, the stents, which are made out of bare metal, become blocked as well. One cause of the reoccurrence of blockages is the body's rejection of the foreign object. In a study published in the *New England Journal of Medicine* (January 2004), a new stent was tested. The new stents are polymer based and after insertion slowly release a drug (paclitaxel) to prevent the rejection problem. A sample of 1,314 patients who were receiving a stent in a single, previously untreated coronary artery blockage was recruited. A total of 652 were randomly assigned to receive a bare-metal stent, and 662 to receive an identical-looking polymer drug releasing stent. The results were recorded in the following way:

Column 1: Patient identification number
Column 2: Stent type (1 = bare-metal, 2 = polymer based)
Column 3: Reference-vessel diameter (the diameter of the artery that is blocked, in millimeters)
Column 4: Lesion length (the length of the blockage, in millimeters)

Reference-vessel diameters and lesion lengths were measured before the stents were inserted.

The following data were recorded 12 months after the stents were inserted:

Column 5: Blockage reoccurrence after 9 months (2 = yes, 1 = no)
Column 6: Blockage that needed to be reopened (2 = yes, 1 = no)
Column 7: Death from cardiac causes (2 = yes, 1 = no)
Column 8: Stroke caused by stent (2 = yes, 1 = no)

a. Using the variables stored in columns 3 through 8, determine whether there is enough evidence to infer that the polymer-based stent is superior to the bare metal stent?

b. As a laboratory researcher in the pharmaceutical company, write a report that describes this experiment and the results.

DATA
CA17-02

CASE A17.2 Automobile Crashes and the Ages of Drivers*

Setting premiums for insurance is a complex task. If the premium is too high, the insurance company will lose customers; if it is too low, the company will lose money. Statistics plays a critical role in almost all aspects of the insurance business. As part of a statistical analysis, an insurance company in Florida studied the relationship between the severity of car crashes and the ages of the drivers. A random sample of crashes in 2002 in the state of Florida was drawn. For each crash, the age category of the driver was recorded as well as whether the driver was injured or killed. The data were stored as follows:

Column 1 Crash number
Column 2 Age category

1. 15 to 34
2. 35 to 44
3. 45 to 54
4. 55 to 64
5. 65 and over

Column 3 Medical status of driver

1 = Uninjured
2 = Injured (but not killed)
3 = Killed

a. Is there enough evidence to conclude that age and medical status of the driver in car crashes are related?

b. Estimate with 95% confidence the proportion of all Florida drivers in crashes in 2002 who were uninjured.

*Adapted from Florida Department of Highway Safety and Vehicles as reported in the *Miami Herald*, January 1, 2004, p. 2B.

18

© Digital Vision/Alamy

MODEL BUILDING

MBA Program Admissions Policy, Part 2

DATA Xm17-00*

After considering the results of the initial study (see Chapter 17 opening example), the dean realized that she may have omitted an important variable, the type of undergraduate degree. She returned to her sample of students and recorded the type of undergraduate degree using the following codes:

1 = BA
2 = BBA (including similar business or management degrees)
3 = BEng or BSc
4 = Other (including no undergraduate degree)

These data were included with the data from the original example. Can the dean conclude that the undergraduate degree is a factor in determining how well a student performs in the MBA program?

© Comstock Images/ Royalty-free/Getty Images

On page 729 we provide a copy of this chart.

INTRODUCTION

Chapters 16 and 17 introduced the techniques and concepts of regression analysis. We discussed how the model is developed, interpreted, assessed, and diagnosed for violations of required conditions. However, there is more to regression analysis. In this chapter, we demonstrate why this procedure is one of the most powerful and commonly used techniques in statistics. Regression analysis allows the statistics practitioner to use mathematical models to realistically describe relationships between the dependent variable and independent variables.

In Section 18.1, we introduce models in which the relationship between the dependent variable and the independent variables may not be linear. Section 18.2 introduces indicator variables, which allow us to use nominal independent variables. We describe pay equity, an important human resources management application that employs nominal independent variables in Section 18.3. Section 18.4 presents logistic regression wherein the dependent variable is nominal. In Section 18.5, we introduce stepwise regression, which enables the statistics practitioner to include the independent variables that yield the best fitting models. Finally, Section 18.6 discusses how to properly use regression analysis in building models.

18.1 POLYNOMIAL MODELS

Chapter 17 introduced the multiple regression model:

$$y = \beta_0 + \beta_1 x_1 + \beta_2 x_2 + \cdots + \beta_k x_k + \varepsilon$$

We included variables $x_1, x_2, \ldots, x_k$ because we believed that these variables were each linearly related to the dependent variable. In this section, we discuss models where the independent variables may be functions of a smaller number of predictor variables. The simplest form of the **polynomial model** is described in the box.

Polynomial Model with One Predictor Variable

$$y = \beta_0 + \beta_1 x + \beta_2 x^2 + \cdots + \beta_p x^p + \varepsilon$$

Technically, this is a multiple regression model with p independent variables. However, all independent variables are based on only one variable, which we label the **predictor variable.** That is, $x_1 = x$, $x_2 = x^2, \ldots, x_p = x^p$. In this model, p is the **order** of the equation. For reasons that we discuss later, we rarely propose a model whose order is greater than 3. However, it is worthwhile to devote individual attention to situations where $p = 1, 2,$ and 3.

First-Order Model

When $p = 1$, we have the now-familiar simple linear regression model introduced in Chapter 16. It is also called the **first-order** polynomial model.

$$y = \beta_0 + \beta_1 x + \varepsilon$$

Obviously, this model is chosen when the statistics practitioner believes that there is a straight-line relationship between the dependent and independent variables over the range of the values of x.

Second-Order Model

With $p = 2$, the polynomial model is

$$y = \beta_0 + \beta_1 x + \beta_2 x^2 + \varepsilon$$

When we plot x versus y, the graph is shaped like a parabola, as shown in Figures 18.1 and 18.2. The coefficient β_0 represents the intercept where the response surface strikes the y-axis. The signs of β_1 and β_2 control the position of the parabola relative to the y-axis. If $\beta_1 = 0$, for example, the parabola is symmetric and centered around $x = 0$. If β_1 and β_2 have the same sign, the parabola shifts to the left. If β_1 and β_2 have opposite signs, the parabola shifts to the right. The coefficient β_2 describes the curvature. If $\beta_2 = 0$, there is no curvature. If β_2 is negative, the graph is concave (as in Figure 18.1). If β_2 is positive, the graph is convex (as in Figure 18.2). The greater the absolute value of β_2, the greater the rate of curvature, as can be seen in Figure 18.3.

FIGURE 18.1 Second-Order Model with $\beta_2 < 0$

FIGURE 18.2 Second-Order Model with $\beta_2 > 0$

FIGURE 18.3 Second-Order Model with Various Values of β_2

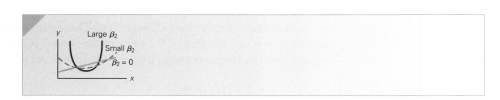

Third-Order Model

By setting $p = 3$, we produce the third-order model

$$y = \beta_0 + \beta_1 x + \beta_2 x^2 + \beta_3 x^3 + \varepsilon$$

Figures 18.4 and 18.5 depict this equation, whose curvature can change twice.

FIGURE **18.4**
Third-Order Model with $\beta_3 < 0$

FIGURE **18.5**
Third-Order Model with $\beta_3 > 0$

As you can see, when β_3 is negative, y is decreasing over the range of x, and when β_3 is positive, y increases. The other coefficients determine the position of the curvature changes and the point at which the curve intersects the y-axis.

The number of real-life applications of this model is quite small. Statistics practitioners rarely encounter problems involving more than one curvature reversal. Therefore, we will not discuss any higher order models.

Polynomial Models with Two Predictor Variables

If we believe that two predictor variables influence the dependent variable, we can use one of the following polynomial models. The general form of this model is rather cumbersome, so we will not show it. Instead we discuss several specific examples.

First-Order Model

The first-order model is represented by

$$y = \beta_0 + \beta_1 x_1 + \beta_2 x_2 + \varepsilon$$

This model is used whenever the statistics practitioner believes that, on average, y is linearly related to each of x_1 and x_2, and the predictor variables do not interact. (Recall that we introduced interaction in Chapter 14.) This means that the effect of one predictor variable on y is independent of the value of the second predictor variable. For example, suppose that the sample regression line of the first-order model is

$$\hat{y} = 5 + 3x_1 + 4x_2$$

If we examine the relationship between y and x_1 for several values of x_2 (say, $x_2 = 1$, 2, and 3), we produce the following equations:

x_2	$\hat{y} = 5 + 3x_1 + 4x_2$
1	$\hat{y} = 9 + 3x_1$
2	$\hat{y} = 13 + 3x_1$
3	$\hat{y} = 17 + 3x_1$

The only difference in the three equations is the intercept. (See Figure 18.6.) The coefficient of x_1 remains the same, which means that the effect of x_1 on y remains the same no matter what the value of x_2. (We could also have shown that the effect of x_2 on y remains the same no matter what the value of x_1.) As you can see from Figure 18.6, the first-order model with no interaction produces parallel straight lines.

FIGURE **18.6** First-Order Model with Two Independent Variables: No Interaction

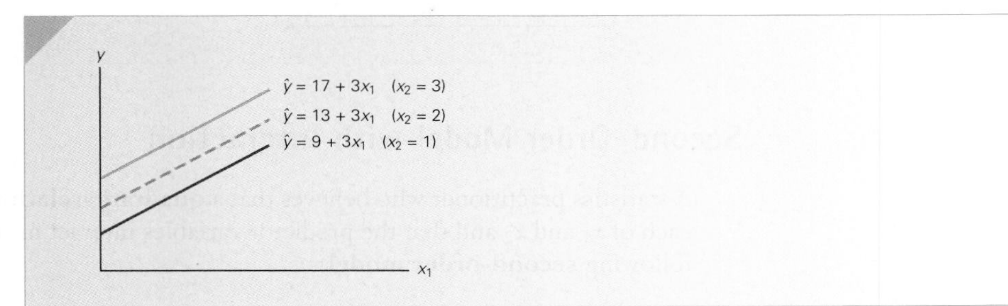

A statistics practitioner who thinks that the effect of one predictor variable on y is influenced by the other predictor variable can use the model described next.

First-Order Model with Two Predictor Variables and Interaction

Interaction means that the effect of x_1 on y is influenced by the value of x_2. (It also means that the effect of x_2 on y is influenced by x_1.)

First-Order Model with Interaction

$$y = \beta_0 + \beta_1 x_1 + \beta_2 x_2 + \beta_3 x_1 x_2 + \varepsilon$$

Suppose that the sample regression line is

$$\hat{y} = 5 + 3x_1 + 4x_2 - 2x_1 x_2$$

If we examine the relationship between y and x_1 for $x_2 = 1, 2,$ and 3, we produce the following table of equations:

x_2	$\hat{y} = 5 + 3x_1 + 4x_2 - 2x_1x_2$
1	$\hat{y} = 9 + x_1$
2	$\hat{y} = 13 - x_1$
3	$\hat{y} = 17 - 3x_1$

As you can see, not only is the intercept different, but the coefficient of x_1 also varies. Obviously, the effect of x_1 on y is influenced by the value of x_2. Figure 18.7 depicts these equations. The straight lines are clearly not parallel.

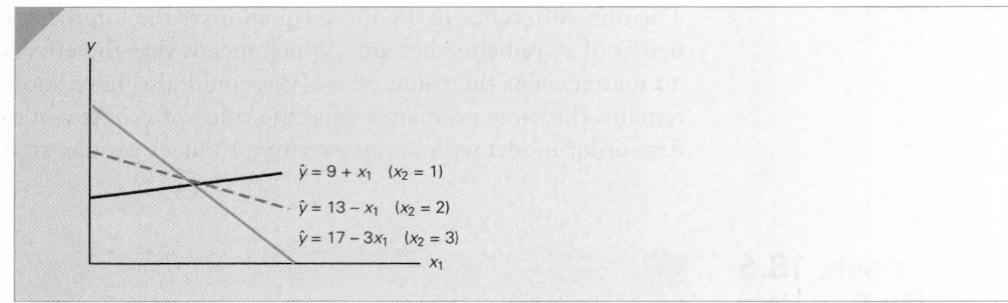

FIGURE 18.7 First-Order Model with Interaction

Second-Order Model with Interaction

A statistics practitioner who believes that a **quadratic relationship** exists between y and each of x_1 and x_2 and that the predictor variables interact in their effect on y can use the following **second-order model.**

Second-Order Model with Interaction

$$y = \beta_0 + \beta_1 x_1 + \beta_2 x_2 + \beta_3 x_1^2 + \beta_4 x_2^2 + \beta_5 x_1 x_2 + \varepsilon$$

Figures 18.8 and 18.9, respectively, depict this model without and with the interaction term.

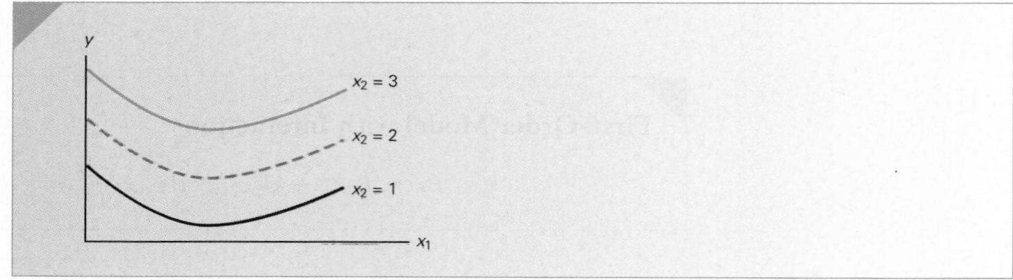

FIGURE 18.8 Second-Order Model without Interaction

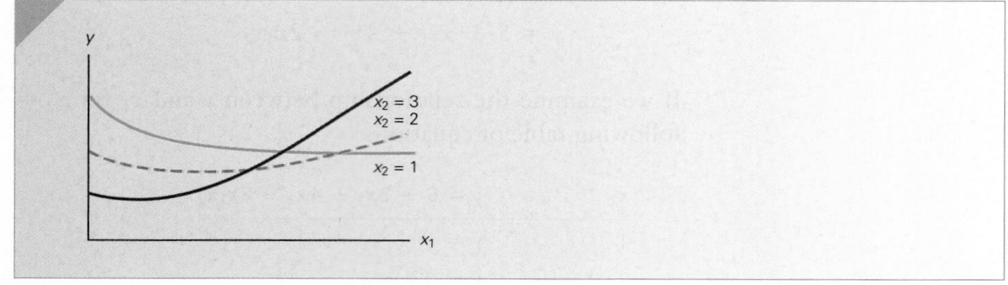

FIGURE 18.9 Second-Order Model with Interaction

Now that we've introduced several different models, how do we know which model to use? The answer is that we employ a model based on our knowledge of the variables involved and then test that model using the statistical techniques presented in this and the preceding chapters.

EXAMPLE 18.1

Selecting Sites for a Fast-Food Restaurant, Part 1

In trying to find new locations for their restaurants, fast-food restaurant chains like McDonald's and Wendy's usually consider a number of factors. Suppose that an analyst working for a fast-food restaurant chain has been asked to construct a regression model that will help identify new locations that are likely to be profitable. The analyst knows that this type of restaurant has, as its primary market, middle-income adults and their children, particularly those between the ages of 5 and 12. Which model should the analyst propose?

SOLUTION

The dependent variable is gross revenue or net profit. The predictor variables will be mean annual household income and the mean age of children in the restaurant's neighborhood.

The relationship between the dependent variable and each predictor variable is probably quadratic. That is, members of relatively poor or relatively affluent households are less likely to eat at this chain's restaurants, since the restaurants attract mostly middle-income customers. Figure 18.10 depicts the hypothesized relationship.

FIGURE **18.10** **Relationship between Annual Gross Revenue and Mean Household Income**

A similar relationship can be proposed for revenue and age. Neighborhoods where the mean age of children is either quite low or quite high will probably produce lower revenues than in similar areas where the mean age lies in the middle of the 5-to-12 range.

The question of whether to include the interaction term is more difficult to answer. When in doubt, it is probably best to include it. Thus, the model to be tested is

$$y = \beta_0 + \beta_1 x_1 + \beta_2 x_2 + \beta_3 x_1^2 + \beta_4 x_2^2 + \beta_5 x_1 x_2 + \varepsilon$$

where

y = annual gross sales

x_1 = mean annual household income in the neighborhood

x_2 = mean age of children in the neighborhood

EXAMPLE 18.2

DATA Xm18-02

Selecting Sites for a Fast-Food Restaurant, Part 2

To determine whether the second-order model with interaction is appropriate, the analyst in Example 18.1 selected 25 areas at random. Each area consists of approximately 5,000 households, as well as one of her employer's restaurants and three competing fast-food restaurants. The previous year's annual gross sales, the mean annual household income, and the mean age of children (the latter two figures are available from the latest census) were recorded; some of these data are listed here (the file also contains x_1^2, x_2^2, and x_1x_2). What conclusions can be drawn from these data?

Area	Annual Gross Revenue ($ thousands) y	Mean Annual Household ($ thousands) x_1	Mean Age of Children x_2
1	$1,128	$23.5	10.5
2	1,005	17.6	7.2
3	1,212	26.3	7.6
⋮			
25	950	17.8	6.1

SOLUTION

Both Excel and Minitab were employed to produce the regression analysis shown here.

EXCEL

	A	B	C	D	E	F
1	SUMMARY OUTPUT					
2						
3	*Regression Statistics*					
4	Multiple R	0.9521				
5	R Square	0.9065				
6	Adjusted R Square	0.8819				
7	Standard Error	44.70				
8	Observations	25				
9						
10	ANOVA					
11		*df*	*SS*	*MS*	*F*	*Significance F*
12	Regression	5	368140	73628	36.86	3.86E-09
13	Residual	19	37956	1998		
14	Total	24	406096			
15						
16		*Coefficients*	*Standard Error*	*t Stat*	*P-value*	
17	Intercept	-1134.0	320.0	-3.54	0.0022	
18	Income	173.20	28.20	6.14	6.66E-06	
19	Age	23.55	32.23	0.73	0.4739	
20	Income sq	-3.726	0.542	-6.87	1.48E-06	
21	Age sq	-3.869	1.179	-3.28	0.0039	
22	(Income)(Age)	1.967	0.944	2.08	0.0509	

MINITAB

Regression Analysis: Revenue Versus Income, Age, Inc-sq, Age-sq, Inc-Age

The regression equation is
Revenue = - 1134 + 173 Income + 23.5 Age - 3.73 Inc-sq - 3.87 Age-sq
+ 1.97 Inc-Age

Predictor	Coef	SE Coef	T	P
Constant	-1134.0	320.0	-3.54	0.002
Income	173.20	28.20	6.14	0.000
Age	23.55	32.23	0.73	0.474
Inc-sq	-3.7261	0.5422	-6.87	0.000
Age-sq	-3.869	1.179	-3.28	0.004
Inc-Age	1.9673	0.9441	2.08	0.051

S = 44.70 R-Sq = 90.7% R-Sq(adj) = 88.2%

Analysis of Variance

Source	DF	SS	MS	F	P
Regression	5	368140	73628	36.86	0.000
Residual Error	19	37956	1998		
Total	24	406096			

INTERPRET

From the computer output, we determine that the value of the coefficient of determination (R^2) is 90.65%, which tells us that the model fits the data quite well. The value of the F-statistic is 36.86, which has a p-value of approximately 0. This confirms that the model is valid.

Care must be taken when interpreting the t-tests of the coefficients in this type of model. Not surprisingly, each variable will be correlated with its square, and the interaction variable will be correlated with both of its components. As a consequence, multicollinearity distorts the t-tests of the coefficients in some cases, making it appear that some of the components should be eliminated from the model. In fact, in Example 18.2, Minitab warns (not shown) the analyst that multicollinearity is a problem. However, in most such applications, the objective is to forecast the dependent variable and multicollinearity does not affect the model's fit or forecasting capability.

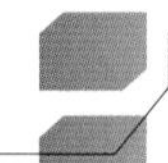

EXERCISES

18.1 Graph y versus x_1 for $x_2 = 1, 2$, and 3, for each of the following equations.

a. $y = 1 + 2x_1 + 4x_2$

b. $y = 1 + 2x_1 + 4x_2 - x_1x_2$

18.2 Graph y versus x_1 for $x_2 = 2, 4$, and 5 for each of the following equations.

a. $y = 0.5 + 1x_1 - 0.7x_2 - 1.2x_1^2 + 1.5x_2^2$

b. $y = 0.5 + 1x_1 - 0.7x_2 - 1.2x_1^2 + 1.5x_2^2 + 2x_1x_2$

The following exercises require the use of a computer and software. Exercises 18.3–18.5 can be solved manually. See Appendix A.

18.3 Xr18-03 The general manager of a supermarket chain believes that sales of a product are influenced by the amount of space the product is allotted on shelves. If true, this would have great significance, because the more profitable items could be given more shelf space. The manager realizes that sales volume would likely increase with more space only up to a certain point. Beyond that point, sales would likely flatten and perhaps decrease (because customers are often dismayed by very large displays). To test his belief, the manager records the number of boxes of detergent sold during 1 week in 25 stores in the chain. For each store, he records the shelf space (in inches) allotted to the detergent.

a. Write the equation that represents the model.

b. Discuss how well the model fits.

APPLICATIONS in ECONOMICS

© Royalty-free/Corbis

Demand Curve

The law of supply and demand states that other things being equal, the higher the price of a product or service, the lower is the quantity demanded. The relationship between quantity and price is called a *demand curve*. Generally, such a curve is modeled by a quadratic equation. To estimate the demand curve, we measure the demand at several different prices and then employ regression analysis to calculate the coefficients of the model.

18.4 Xr18-04 A fast-food restaurant chain whose menu features hamburgers and chicken sandwiches is about to add a fish sandwich to its menu. There was considerable debate among the executives about the likely demand and what the appropriate price should be. A recently hired economics graduate observed that the demand curve would reveal a great deal about the relationship between price and demand. She convinced the executives to conduct an experiment. A random sample of 20 restaurants was drawn. The restaurants were almost identical in terms of sales and in the demographics of the surrounding area. At each restaurant, the fish sandwich was sold at a different price. The number of sandwiches sold over a 7-day period and the price were recorded. A first-order model and a second-order model were proposed.

a. Write the equation for each model.
b. Use regression analysis to estimate the coefficients and other statistics for each model.
c. Which model seems to fit better? Explain.
d. Use the better model to calculate the point prediction for weekly sales when the price is $2.95.

APPLICATIONS in OPERATIONS MANAGEMENT

© Royalty-free/Corbis

Learning Curve

A well-established phenomenon in operations management is the *learning curve*, which describes how quickly new workers learn to do their jobs. A number of mathematical models are used to describe the relationship between time on the job and productivity. Regression analysis allows the operations manager to select the appropriate model and use it to predict when workers achieve their highest level of productivity.

18.5 Xr18-05 A person starting a new job always takes a certain amount of time to adjust fully. In repetitive-task situations, such as on an assembly line, significant productivity gains can occur within a few days. In an experiment to study this phenomenon, the average amount of time required for a new employee to install electronic components in a computer was measured for her first 10 days. These data are shown here.

Day	1	2	3	4	5	6	7	8	9	10
Mean times (minutes)	40	41	36	38	33	32	30	32	29	30

A first-order model and a second-order model were proposed.

a. Write the equation for each model.
b. Analyze both models. Determine whether they are valid.
c. Which model fits better? Explain.

18.6 Xm17-00* Refer to the MBA Program Admissions Policy example introduced in Chapter 17 (chapter-opening example). The dean of the school of business wanted to improve the regression model, which was developed to describe the relationship between MBA program GPA and undergraduate GPA, GMAT score, and years of work experience. The dean now believes that an interaction effect may exist between undergraduate GPA and the GMAT test score.

a. Write the equation that describes the model.
b. Use a computer to generate the regression statistics. Use whatever statistics you deem necessary to assess the model's fit. Is this model valid?
c. Compare your results with those achieved in the original example.

18.7 Xr18-07 The manager of the food concession at a major league baseball stadium wanted to be able to predict the attendance of a game 24 hours in advance to prepare the correct amount of food for sale. He believed that the two most important factors were the home team's winning percentage and the visiting team's winning percentage. In order to examine his beliefs, he collected the attendance figures, the home team's winning percentage, and the visiting team's winning percentage for 40 randomly selected games from all the teams in the league.

a. Conduct a regression analysis using a first-order model with interaction.
b. Do these results indicate that your model is valid? Explain.

18.8 Xr18-08 The manager of a large hotel on the Riviera in southern France wanted to forecast the monthly vacancy rate (as a percentage) during the peak season. After considering a long list of potential variables, she identified two variables that she believed were most closely related to the vacancy rate: the average daily temperature and the value of the currency in American dollars. She collected data for 25 months.

a. Perform a regression analysis using a first-order model with interaction.
b. Perform a regression analysis using a second-order model with interaction.
c. Which model fits better? Explain.

18.9 Xr18-09 The coach and the general manager of a team in the National Hockey League are trying to decide what kinds of players to draft. To help in making their decision, they need to know which variables are most closely related to the goals differential—the difference between the number of goals their team scores and the number of goals scored by their team's opponents. (A positive differential means that their team wins, and a negative differential is a loss.) After some consideration, they decide that there are two important variables: the percentage of face-offs won and the penalty-minutes differential. The latter variable is the difference between the number of penalty minutes assessed against their team and the number of penalty minutes assessed against their team's opponents. The data from 100 games were recorded.

a. Perform a regression analysis using a first-order model with interaction.
b. Is this model valid?
c. Should the interaction term be included?

18.10 Xr18-10 The production manager of a chemical plant wants to determine the roles that temperature and pressure play in the yield of a particular chemical produced at the plant. From past experience, she believes that when pressure is held constant, lower and higher temperatures tend to reduce the yield. When temperature is held constant, higher and lower pressures tend to increase the yield. She does not have any idea about how the yield is affected by various combinations of pressure and temperature. She observes 80 batches of the chemical in which the pressure and temperature were allowed to vary.

a. Which model should be used? Explain.
b. Conduct a regression analysis using the model you specified in Part a.
c. Assess how well the model fits the data.

18.2 Nominal Independent Variables

When we introduced regression analysis, we pointed out that all the variables must be interval. But in many real-life cases, one or more independent variables are nominal. For example, suppose that the used-car dealer in Example 16.2 believed that the color of a car is a factor in determining its auction price. Color is clearly a nominal variable. If we assign numbers to each possible color, these numbers will be completely arbitrary, and using them in a regression model will usually be pointless. For example, suppose the dealer believes the colors that are most popular, white and silver, are likely to lead to different prices than other colors. Accordingly, he assigns a code of 1 to white cars, a code of 2 to silver cars, and a code of 3 to all other colors. If we now conduct a multiple regression analysis using odometer reading and color as independent variables, the following results would be obtained. (File Xm16-02* contains these data. Interested readers can produce the following regression equation.)

$$\hat{y} = 17.342 - .0671x_1 - .0434x_2$$

Aside from the inclusion of the variable x_2, this equation is very similar to the one we produced in the simple regression model ($\hat{y} = 17.25 - .0669x$). The t-test of color (t-statistic $= -1.11$, and p-value $= .2694$) indicates that there is not enough evidence to infer that color is not linearly related to price. There are two possible explanations for this result. First, there is no relationship between color and price. Second, color is a factor in determining the car's price, but the way in which the dealer assigned the codes to the colors made detection of that fact impossible. That is, the dealer treated the nominal variable, color, as an interval variable. To further understand why we cannot use nominal data in regression analysis, try to interpret the coefficient of color. Such an effort is similar to attempting to interpret the mean of a sample of nominal data. It is futile. Even though this effort failed, it is possible to include nominal variables in the regression model. This is accomplished through the use of *indicator variables.*

An **indicator variable** (also called a **dummy variable**) is a variable that can assume either one of only two values (usually 0 and 1), where 1 represents the existence of a certain condition and 0 indicates that the condition does not hold. In this illustration we would create two indicator variables to represent the color of the car:

$$I_1 = \begin{cases} 1 & \text{(if color is white)} \\ 0 & \text{(if color is not white)} \end{cases}$$

and

$$I_2 = \begin{cases} 1 & \text{(if color is silver)} \\ 0 & \text{(if color is not silver)} \end{cases}$$

Notice that we need only two indicator variables to represent the three categories. A white car is represented by $I_1 = 1$ and $I_2 = 0$. A silver car is represented by $I_1 = 0$ and $I_2 = 1$. Because cars that are painted some other color are neither white nor silver, they are represented by $I_1 = 0$ and $I_2 = 0$. It should be apparent that we cannot have $I_1 = 1$ and $I_2 = 1$, as long as we assume that no Toyota Camry is two-toned.

The effect of using these two indicator variables is to create three equations, one for each of the three colors. As you're about to discover, we can use the equations to determine how the car's color relates to its auction selling price.

In general, to represent a nominal variable with m categories, we must create $m - 1$ indicator variables. The last category represented by $I_1 = I_2 = \cdots = I_{m-1} = 0$ is called the **omitted category.**

Interpreting and Testing the Coefficients of Indicator Variables

In file Xm16-02a, we stored the values of I_1 and I_2. We then performed a multiple regression analysis using the variables odometer reading (x), I_1, and I_2.

EXCEL

	A	B	C	D	E	F
1	SUMMARY OUTPUT					
2						
3	*Regression Statistics*					
4	Multiple R	0.8371				
5	R Square	0.7008				
6	Adjusted R Square	0.6914				
7	Standard Error	0.3043				
8	Observations	100				
9						
10	ANOVA					
11		*df*	*SS*	*MS*	*F*	*Significance F*
12	Regression	3	20.81	6.94	74.95	4.65E-25
13	Residual	96	8.89	0.0926		
14	Total	99	29.70			
15						
16		*Coefficients*	*Standard Error*	*t Stat*	*P-value*	
17	Intercept	16.84	0.197	85.42	2.28E-92	
18	Odometer	-0.0591	0.0051	-11.67	4.04E-20	
19	I-1	0.0911	0.0729	1.25	0.2143	
20	I-2	0.3304	0.0816	4.05	0.0001	

MINITAB

Regression Analysis: Price Versus Odometer, I-1, I-2

The regression equation is
Price = 16.8 - 0.0591 Odometer + 0.0911 I-1 + 0.330 I-2

Predictor	Coef	SE Coef	T	P
Constant	16.8372	0.1971	85.42	0.000
Odometer	-0.059123	0.005065	-11.67	0.000
I-1	0.09113	0.07289	1.25	0.214
I-2	0.33037	0.08165	4.05	0.000

S = 0.304258 R-Sq = 70.1% R-Sq(adj) = 69.1%

Analysis of Variance

Source	DF	SS	MS	F	P
Regression	3	20.8149	6.9383	74.95	0.000
Residual Error	96	8.8870	0.0926		
Total	99	29.7019			

INTERPRET

The regression equation is

$$\hat{y} = 16.837 - .0591x_1 + .0911I_1 + .3304I_2$$

The intercept and the coefficient of odometer reading are interpreted in the usual manner. The intercept ($b_0 = 16.837$) is meaningless in the context of this problem. The coefficient of the odometer reading ($b_1 = -.0591$) tells us that for each additional mile on the odometer, the auction price decreases, on average, by 5.91 cents holding the color constant. Now examine the remaining two coefficients:

$$b_2 = .0911$$
$$b_3 = .3304$$

Recall that we interpret the coefficients in a multiple regression model by holding the other variables constant. In this example we interpret the coefficient of I_1 as follows. In this sample a white Camry sells for .0911 thousand or \$91.10 on average more than other colors (nonwhite, nonsilver) with the same odometer reading. A silver car sells for \$330.40 on average more than other colors with the same odometer reading. The reason both comparisons are made with other colors is that such cars are represented by $I_1 = I_2 = 0$. Thus, for a nonwhite and nonsilver car, the equation becomes

$$\hat{y} = 16.837 - .0591x + .0911(0) + .3304(0)$$

which is

$$\hat{y} = 16.837 - .0591x$$

For a white car ($I_1 = 1$ and $I_2 = 0$), the regression equation is

$$\hat{y} = 16.837 - .0591x + .0911(1) + .3304(0)$$

which is

$$\hat{y} = 16.928 - .0591x$$

Finally, for a silver car ($I_1 = 0$ and $I_2 = 1$), the regression equation is

$$\hat{y} = 16.837 - .0591x + .0911(0) + .3304(1)$$

which simplifies to

$$\hat{y} = 17.167 - .0591x$$

Figure 18.11 depicts the graph of price versus odometer reading for the three different color categories. Notice that the three lines are parallel (with slope = .0591) while the intercepts differ.

FIGURE **18.11**
Price Versus Odometer Reading for Three Colors

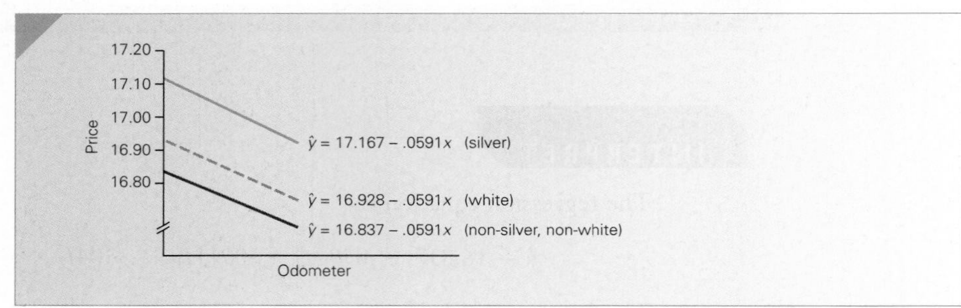

We can also perform *t*-tests on the coefficients of I_1 and I_2. However, because the variables I_1 and I_2 represent different groups (the three color categories), these *t*-tests allow us to draw inferences about the differences in auction selling prices between the groups for the entire population of similar 3-year-old Toyota Camrys.

The test of the coefficient of I_1, which is β_2, is conducted as follows:

H_0: $\beta_2 = 0$
H_1: $\beta_2 \neq 0$
Test statistic: $t = 1.25$ p-value $= .2143$

There is insufficient evidence to infer that white Camrys have a different mean selling price than do Camrys in the omitted color category in the population of 3-year-old Camrys with the same odometer reading.

To determine whether silver-colored Camrys sell for a different price than Camrys in the other color category, we test

H_0: $\beta_3 = 0$
H_1: $\beta_3 \neq 0$
Test statistic: $t = 4.05$ p-value $= .0001$

We can conclude that there are differences in the mean auction selling prices between all 3-year-old, silver-colored Camrys and the omitted color category with the same odometer readings.

MBA Program Admissions Policy, Part 2: Solution

© Comstock Images/ Royalty-free/Getty Images

Because the dean is a statistician, she knows that the codes representing the type of degree are nominal and cannot be used in regression analysis. She creates the following indicator variables (All the data are stored in Xm18-00):

$$I_1 = \begin{cases} 1 & \text{if BA} \\ 0 & \text{otherwise} \end{cases}$$

$$I_2 = \begin{cases} 1 & \text{if BBA} \\ 0 & \text{otherwise} \end{cases}$$

$$I_3 = \begin{cases} 1 & \text{if BSc or BEng} \\ 0 & \text{otherwise} \end{cases}$$

EXCEL

	A	B	C	D	E	F
1	SUMMARY OUTPUT					
2						
3	*Regression Statistics*					
4	Multiple R	0.7461				
5	R Square	0.5566				
6	Adjusted R Square	0.5242				
7	Standard Error	0.7293				
8	Observations	89				
9						
10	ANOVA					
11		*df*	*SS*	*MS*	*F*	*Significance F*
12	Regression	6	54.75	9.13	17.16	9.59E-13
13	Residual	82	43.62	0.532		
14	Total	88	98.37			
15						
16		*Coefficients*	*Standard Error*	*t Stat*	*P-value*	
17	Intercept	0.190	1.41	0.13	0.8930	
18	UnderGPA	-0.0061	0.114	-0.05	0.9577	
19	GMAT	0.0128	0.0014	9.43	9.92E-15	
20	Work	0.0982	0.0303	3.24	0.0017	
21	I-1	-0.345	0.224	-1.54	0.1269	
22	I-2	0.706	0.241	2.93	0.0043	
23	I-3	0.0348	0.209	0.17	0.8684	

MINITAB

Regression Analysis: MBA GPA Versus UnderGPA, GMAT, Work, I-1, I-2, I-3

The regression equation is
MBA GPA = 0.19 - 0.006 UnderGPA + 0.0128 GMAT + 0.0982 Work - 0.345 I-1
+ 0.706 I-2 + 0.035 I-3

Predictor	Coef	SE Coef	T	P
Constant	0.190	1.407	0.13	0.893
UnderGPA	-0.0061	0.1140	-0.05	0.958
GMAT	0.012793	0.001356	9.43	0.000
Work	0.09818	0.03032	3.24	0.002
I-1	-0.3450	0.2237	-1.54	0.127
I-2	0.7057	0.2405	2.93	0.004
I-3	0.0348	0.2094	0.17	0.868

S = 0.7293 R-Sq = 55.7% R-Sq(adj) = 52.4%

Analysis of Variance

Source	DF	SS	MS	F	P
Regression	6	54.7518	9.1253	17.16	0.000
Residual Error	82	43.6174	0.5319		
Total	88	98.3692			

INTERPRET

The model has improved somewhat; the coefficient of determination has increased to .5566. The GMAT score and years of work experience continue to be significant variables. The coefficients of the indicator variables are interpreted as follows.

Variable I_1, coefficient $b_4 = -.345$: In this sample the MBA program GPAs of students with a B.A. degree are on average .345 lower than the other group, assuming the same undergraduate GPA, GMAT, and number of years of work experience. The *t*-test of β_4 ($t = -1.54$, *p*-value = .1269) indicates that there is no evidence of a difference between otherwise identical students with a B.A. degree and the other group.

Variable I_2, coefficient $b_5 = .706$: In this sample the MBA program GPAs of students with a B.B.A. degree are on average .706 higher than the other group, comparing students with the same undergraduate GPA, GMAT score, and years of work experience. The *t*-test of β_5 ($t = 2.93$, *p*-value = .0043) indicates that there is evidence of a difference between otherwise identical students with a B.B.A. degree and the other group.

Variable I_3, coefficient $b_6 = .0348$: In this sample the MBA program GPAs of students with a B.Sc. or B.Eng. degree are on average .0348 higher than the other group, once again assuming that the other variables are identical. The *t*-test of β_6 ($t = .17$, *p*-value = .8684) indicates that there is no evidence of a difference between otherwise identical students with a B.Sc. or B.Eng. degree and the other group.

Because of the difference between B.B.A. degree holders and the other group, we can infer that the type of degree is a determinant of success in the MBA program.

EXERCISES

18.11 How many indicator variables must be created to represent a nominal independent variable that has 5 categories?

18.12 Create and identify indicator variables to represent the following nominal variables.

a. Religious affiliation (Catholic, Protestant, and others)

b. Working shift (8:00 A.M. to 4:00 P.M., 4:00 P.M. to 12:00 midnight, and 12:00 midnight to 8:00 A.M.)

c. Supervisor (Jack Jones, Mary Brown, George Fosse, and Elaine Smith)

18.13 In a study of computer applications, a survey asked which microcomputer a number of companies used. The following indicator variables were created.

$$I_1 = \begin{cases} 1 & \text{(if IBM)} \\ 0 & \text{(if not)} \end{cases} \qquad I_2 = \begin{cases} 1 & \text{(if Macintosh)} \\ 0 & \text{(if not)} \end{cases}$$

Which computer is being referred to by each of the following pairs of values?

a. $I_1 = 0; I_2 = 1$

b. $I_1 = 1; I_2 = 0$

c. $I_1 = 0; I_2 = 0$

The following exercises require the use of a computer and software. Exercises 18.15 and 18.16 can be solved manually. See Appendix A.

18.14 Refer to the MBA Program Admissions Policy, Part 2 example.

a. Predict with 95% confidence the MBA program GPA of a B.Eng. whose undergraduate GPA was 9.0, whose GMAT score was 700, and who has had 10 years of work experience.

b. Repeat Part a for a B.A. student.

18.15 Xr17-10* Refer to Exercise 17.10, where a multiple regression analysis was performed to predict men's longevity based on the parents' and grandparents' longevity. Suppose that in addition to these data, the actuary also recorded whether the man was a smoker (1 = yes and 0 = no).

a. Use regression analysis to produce the statistics for this model.

b. Compare the equation produced below to that produced in Exercise 17.10. Describe the differences.

c. Are smoking and length of life related? Explain.

18.16 Xr18-16 The manager of an amusement park would like to be able to predict daily attendance, in order to develop more accurate plans about how much food to order and how many ride operators to hire. After some consideration, he decided that the following three factors are critical:

Yesterday's attendance

Weekday or weekend

Predicted weather

He then took a random sample of 40 days. For each day, he recorded the attendance, the previous day's attendance, day of the week, and weather forecast. The first independent variable is interval, but the other two are nominal. Accordingly, he created the following sets of indicator variables:

$$I_1 = \begin{cases} 1 & \text{(if weekend)} \\ 0 & \text{(if not)} \end{cases}$$

$$I_2 = \begin{cases} 1 & \text{(if mostly sunny is predicted)} \\ 0 & \text{(if not)} \end{cases}$$

$$I_3 = \begin{cases} 1 & \text{(if rain is predicted)} \\ 0 & \text{(of not)} \end{cases}$$

a. Conduct a regression analysis.
b. Is this model valid? Explain.
c. Can we conclude that weather is a factor in determining attendance?
d. Do these results provide sufficient evidence that weekend attendance is, on average, larger than weekday attendance?

18.17 Xm17-02* The real estate agent described in Example 17.2 has become so fascinated by the multiple regression technique that he decided to improve the model. After some consideration, he decided that structure of the house was also a factor. There are four structures: two story, side split, back split, and ranch. Each house was classified. Three indicator variables were created as follows:

$$I_1 = \begin{cases} 1 & \text{(if two story)} \\ 0 & \text{(if not)} \end{cases}$$

$$I_2 = \begin{cases} 1 & \text{(if side split)} \\ 0 & \text{(if not)} \end{cases}$$

$$I_3 = \begin{cases} 1 & \text{(if back split)} \\ 0 & \text{(if not)} \end{cases}$$

a. Perform a multiple regression analysis, and compare your results with Example 17.2.
b. Interpret and test the coefficients of the indicator variables.

18.18 Xr16-06* Recall Exercise 16.6 where a statistics practitioner analyzed the relationship between the length of a commercial and viewers' memory of the commercial's product. However, in the experiment, not only was the length varied, but also the type of commercial. There were three types: humorous (1), musical (2), and serious (3). The memory test scores, lengths, and type of commercial (using the codes in parentheses) were recorded.

a. Perform a regression analysis using the codes provided in the data file.
b. Can we infer that the memory test score is related to the type of commercial? Test with $\alpha = .05$.
c. Create indicator variables to describe the type of commercial and perform another regression analysis.
d. Repeat Part b using the second model.
e. Discuss the reasons for the differences between Parts b and d.

18.19 Xr17-12* Refer to Exercise 17.12 where the amount of time to unload a truck was analyzed. The manager realized that another variable, the time of day, may affect unloading time. He recorded the following codes: 1 = morning, 2 = early afternoon, and 3 = late afternoon.

a. Run a regression using the codes for time of day.
b. Create indicator variables to represent time of day. Perform a regression analysis with these new variables.
c. Which model fits better? Explain.
d. Is time of day related to time to unload?

18.20 Xr18-20 Profitable banks are ones that make good decisions on loan applications. *Credit scoring* is the statistical technique that helps banks make that decision. However, many branches overturn credit scoring recommendations, whereas other banks do not use the technique. In an attempt to determine the factors that affect loan decisions, a statistics practitioner surveyed 100 banks and recorded the percentage of bad loans (any loan that is not completely repaid), the average size of the loan, and whether a scorecard is used, and if so, whether scorecard recommendations are overturned more than 10% of the time. These results are stored in columns 1 (percentage good loans); 2 (average loan); and 3 (code 1 = no scorecard; 2 = scorecard overturned more than 10% of the time; 3 = scorecard overturned less than 10% of the time).

a. Create indicator variables to represent the codes.
b. Perform a regression analysis.
c. How well does the model fit the data?
d. Is multicollinearity a problem?

e. Interpret and test the coefficients. What does this tell you?

f. Predict with 95% confidence the percentage of bad loans for a bank whose average loan is $10,000 and that does not use a scorecard.

18.21 Xr18-21 Refer to Exercise 16.82, where a simple linear regression model was used to analyze the relationship between welding machine breakdowns and the age of the machine. The analysis proved to be so useful to company management that they decided to expand the model to include other machines. Data were gathered for two other machines. These data as well as the original data were recorded in the following way:

Column 1: Cost of repairs

Column 2: Age of machine

Column 3: Machine (1 = welding machine; 2 = lathe; 3 = stamping machine)

a. Develop a multiple regression model.

b. Interpret the coefficients.

c. Can we conclude that welding machines cost less to repair than other machines?

APPLICATIONS in HUMAN RESOURCES MANAGEMENT

© Reza Estakhrian/Stone/ Getty Images

Performance Measurement

Most aspects of workers' performance fall into the domain of the human resources/ personnel department. An important performance measurement is the attendance record of each worker. Personnel managers need to know what factors are likely to influence a worker to be absent more frequently than the norm. This can enable the manager to determine whether someone should be hired in the first place. Once hired, the manager needs to be able to influence workers' attitudes and performance.

18.22 Xr18-22 Absenteeism is a serious employment problem in most countries. It is estimated that absenteeism reduces potential output by more than 10%. Two economists launched a research project to learn more about the problem. They randomly selected 100 organizations to participate in a 1-year study. For each organization, they recorded the average number of days absent per employee and several variables thought to affect absenteeism. The following data were recorded:

Column 1: Average number of days absent per employee

Column 2: Average employee wage

Column 3: Percentage of part-time employees

Column 4: Percentage of unionized employees

Column 5: Availability of shiftwork (1 = yes; 0 = no)

Column 6: Union-management relationship (1 = good; 0 = not good)

a. Conduct a regression analysis.

b. Can we infer at the 5% significance level that the availability of shiftwork is related to absenteeism?

c. Is there enough evidence at the 5% significance level to infer that in organizations where the union-management relationship is good, absenteeism is lower?

(The authors are grateful to James Fong and Diana Mansour for developing this exercise. The data are based on M. Chadhury and I. Ng, "Absenteeism Predictors," *Canadian Journal of Economics*, August 1992.)

18.3 (OPTIONAL) APPLICATIONS IN HUMAN RESOURCES MANAGEMENT: PAY EQUITY

In the history of North America there are many examples of racial, ethnic, and gender discrimination. In the last three decades there have been a number of endeavors designed to eliminate discriminatory practices and to right past wrongs. One of these efforts is pay equity, a program that attempts to correct discrimination in the way workers are paid. Our goal in this section is to describe the statistical component of the pay equity issue.

There are two forms of pay equity. The first is "equal pay for equal work." This form is relatively straightforward, arguing that if two individuals do the same job with similar qualifications and experience, they should be paid the same. In many jurisdictions it is illegal to violate "equal pay for equal work." The second form is "equal pay for work of equal value." This form is controversial for several reasons, including the use of subjectively assigned measures of qualifications and working conditions.

Regression analysis is used extensively in pay-equity cases. However, the methodology used in equal pay for equal work cases differs from that used for equal pay for work of equal value cases. The following example illustrates how statistical analyses can be utilized for the former.

EXAMPLE 18.3 Testing for Pay Equity, Equal Pay for Equal Work

DATA
Xm18-03

A large firm employing tens of thousands of workers has been accused of discriminating against its female managers. The accusation is based on a random sample of 100 managers. The mean annual salary of the 38 female managers is $76,189, whereas the mean annual salary of the 62 male managers is $97,832. A statistical analysis reveals that the t-test of the difference between two means yields a p-value of less than 1%, which provides overwhelming evidence that male managers are paid more than female managers. In rebuttal, the president of the firm points out that the company has a strict policy of equal pay for equal work and that the difference may be due to other variables. Accordingly, he found and recorded the number of years of education and the number of years of experience for each of the 100 managers in the sample. Also recorded are the salary and gender (0 = female, 1 = male). The president wanted to know whether a regression analysis would shed some light on the issue.

SOLUTION

Using salary as the dependent variable, a multiple regression analysis was performed with the results shown here.

EXCEL

	A	B	C	D	E	F
1	SUMMARY OUTPUT					
2						
3	*Regression Statistics*					
4	Multiple R	0.8326				
5	R Square	0.6932				
6	Adjusted R Square	0.6836				
7	Standard Error	16274				
8	Observations	100				
9						
10	ANOVA					
11		*df*	*SS*	*MS*	*F*	*Significance F*
12	Regression	3	57434095083	19144698361	72.29	1.55E-24
13	Residual	96	25424794888	264841613		
14	Total	99	82858889971			
15						
16		*Coefficients*	*Standard Error*	*t Stat*	*P-value*	
17	Intercept	-5835	16083	-0.36	0.7175	
18	Education	2119	1018	2.08	0.0401	
19	Experience	4099	317	12.92	9.89E-23	
20	Gender	1851	3703	0.50	0.6183	

MINITAB

Regression Analysis:

The regression equation is
Salary = - 5835 + 2119 Education + 4099 Experience + 1851 Gender

Predictor	Coef	StDev	T	P
Constant	-5835	16083	-0.36	0.718
Education	2119	1018	2.08	0.040
Experience	4099.3	317.2	12.92	0.000
Gender	1851	3703	0.50	0.618

S = 16274 R-Sq = 69.3% R-Sq(adj) = 68.4%

Analysis of Variance

Source	DF	SS	MS	F	P
Regression	3	57434095083	19144698361	72.29	0.000
Residual Error	96	25424794888	264841613		
Total	99	82858889971			

INTERPRET

The model fits quite well. The coefficient of determination is .6932, which tells the president that 69.32% of the variation in salaries is explained by the model. The F-statistic is 72.29, which has a p-value of 0. There is overwhelming evidence to allow us to infer that the model is valid.

The p-values of the t-tests to determine whether there is evidence of a linear relationship between salary and each of education, experience, and gender are .0401, 0, and .6183, respectively. Both the years of education and the years of experience are linearly related to salary. However, the t-test of the slope for gender tells us that there is not enough evidence to infer that the mean salaries of all the firm's male and female managers with the same amount of education and experience differ. That is, on average, the female managers in this firm have less education and experience than their male counterpart, which explains their lower mean salary. Before the regression analysis we calculated the difference in sample mean salaries to be $97,832 − $76,189 = $21,643. After removing the effects of education and experience in this sample that difference was reduced to $1851, which is statistically insignificant.

Regression Analysis for Equal-Pay-for-Work-of-Equal-Value Cases

Cases involving the issue of equal pay for work of equal value are much more difficult. The issue generally revolves around female-dominated and male-dominated jobs. The former refers to jobs that are generally held by women (e.g., secretaries) and the latter refers to jobs generally held by men (e.g., maintenance workers). Women's groups claim that male-dominated jobs are more highly paid. Here the issue is not underpaying women doing exactly the same job performed by men. Instead the issue is that women's jobs are undervalued. Thus, it is necessary to evaluate jobs.

Several jurisdictions have enacted laws requiring pay equity for work of equal value. One such jurisdiction is the province of Manitoba. The Manitoba Pay Equity Act is mandatory in the province's civil service, crown corporations, hospitals, and universities. The act defines gender-dominated job classes as ones with at least 10 workers where at least 70% are of the same gender. The act requires that all such jobs be evaluated to determine whether female-dominated jobs are undervalued and underpaid compared to male-dominated jobs.

Although regression analysis is employed, there are major differences between the technique described in Example 18.3 and the one used in this case. Rather than estimate a regression model that explains how several related variables affect pay, we need to develop a job evaluation system (JES). The system is used to assign a score to each job, which is then used as an independent variable in regression where pay is again the dependent variable. The regression analysis can be conducted in several ways. The simple linear regression equation can be estimated using the male-dominated jobs only. The coefficients are then used to calculate the "correct" female-dominated job pay rates. The difference between the "correct" and the actual pay rates represents the degree of underpayment. Alternatively, a regression analysis with both male- and female-dominated jobs can be employed. An indicator variable representing gender is included. The value of the indicator variable's coefficient represents the difference between male- and female-dominated jobs and the degree of underpayment. The following example illustrates the latter type of analysis, which was adapted from the province of Manitoba Pay Equity Act manuals that describe the law and how it is to be administered.

EXAMPLE 18.4

Testing for Pay Equity, Equal Pay for Work of Equal Value

In a university a total of eight jobs are identified as gender-dominated. The female-dominated jobs are cleaner, secretary, and workers in the bookstore and cafeteria. The male-dominated jobs are maintenance worker, security guard, gardener, and technician. Perform a pay-equity analysis to determine whether and to what degree female-dominated jobs are undervalued and underpaid.

SOLUTION

The hourly pay rates are as follows:

Job Categories	Pay Rate
Maintenance	13.55
Security	15.65
Gardener	13.80
Technician	19.90
Cleaner	11.85
Secretary	14.75
Bookstore	18.90
Cafeteria	13.30

After some consideration the following factors were selected as part of the job evaluation system:

Knowledge and training
Responsibility
Mental effort
Physical effort
Working conditions

Each factor is assigned a weight that reflects its importance. The weights (which must sum to 1) are 25%, 23%, 22%, 15%, and 15%, respectively.

A score for each job is determined by assigning a value between 1 and 10 for each of the five factors and then multiplying by the weight. Smaller values represent less demanding requirements or better conditions. The male-dominated jobs are evaluated as follows:

Factors	Weight	Maintenance	Security	Gardener	Technician
Knowledge and training	.25	1	2	3	9
Responsibility	.23	2	7	1	7
Mental effort	.22	2	3	1	8
Physical effort	.15	7	1	6	4
Working conditions	.15	7	4	8	1
Total score		3.25	3.52	3.30	6.37

As you can see, the scores assigned to the maintenance workers and gardeners reflect relatively small demands on knowledge, training, and mental effort, but high demands on physical effort and poor working conditions. The technician, on the other hand, has excellent working conditions but requires a high level of knowledge and training.

The evaluations of the female dominated jobs are as follows:

Factors	Weight	Cleaner	Secretary	Bookstore	Cafeteria
Knowledge and training	.25	1	6	4	2
Responsibility	.23	2	7	7	2
Mental effort	.22	2	6	7	2
Physical effort	.15	7	3	2	5
Working conditions	.15	5	1	1	6
Total score		2.95	5.03	4.60	3.05

As was the case with the male-dominated jobs, the scores for the female-dominated jobs are based on a subjective assessment of the requirements and work that the jobs entail.

The score and an indicator variable are used as independent variables in a regression analysis with pay as the dependent variable. The following data are used in the regression analysis:

Job Categories	Pay Rate	Score	Gender
Maintenance	13.55	3.25	1
Security	15.65	3.52	1
Gardener	13.80	3.30	1
Technician	19.90	6.37	1
Cleaner	11.85	2.95	0
Secretary	14.75	5.03	0
Bookstore	18.90	4.60	0
Cafeteria	13.30	3.05	0

where

$$\text{Gender} = \begin{cases} 1 & \text{if male-dominated job} \\ 0 & \text{if female-dominated job} \end{cases}$$

The results of the regression are shown below.

EXCEL

	A	B	C	D	E	F
1	SUMMARY OUTPUT					
2						
3	*Regression Statistics*					
4	Multiple R	0.8515				
5	R Square	0.7251				
6	Adjusted R Square	0.6152				
7	Standard Error	1.75				
8	Observations	8				
9						
10	ANOVA					
11		*df*	*SS*	*MS*	*F*	*Significance F*
12	Regression	2	40.39	20.19	6.59	0.0396
13	Residual	5	15.31	3.06		
14	Total	7	55.70			
15						
16		*Coefficients*	*Standard Error*	*t Stat*	*P-value*	
17	Intercept	7.15	2.31	3.10	0.0270	
18	Score	1.93	0.547	3.54	0.0166	
19	Gender	0.633	1.242	0.51	0.6318	

MINITAB

Regression Analysis:

The regression equation is
Pay rate = 7.15 + 1.93 Score + 0.63 Gender

Predictor	Coef	StDev	T	P
Constant	7.145	2.309	3.10	0.027
Score	1.9334	0.5468	3.54	0.017
Gender	0.633	1.242	0.51	0.632

S = 1.7   R-Sq = 72.5%   R-Sq(adj) = 61.5%

Analysis of Variance

Source	DF	SS	MS	F	P
Regression	2	40.388	20.194	6.59	0.040
Residual Error	5	15.310	3.062		
Total	7	55.699			

INTERPRET

We cannot apply the usual statistical inference because the eight observations represent the entire population under consideration. Instead, we simply use the coefficients of interest. In this case we discover that male-dominated jobs are paid an average of .63 more than female-dominated jobs after adjusting for the value of each job. If we accept the validity of this analysis (see Exercises 18.24 and 18.25), we conclude that the holders of female-dominated jobs need to have their pay rates increased by 63 cents per hour.

EXERCISES

The following exercises require a computer and software.

18.23 Xr18-23 Pay equity for men and women has been an ongoing source of conflict for a number of years in North America. Suppose that a statistics practitioner is investigating the factors that affect salary differences between male and female university professors. He believes that the following variables have some impact on a professor's salary:

Number of years since first degree

$$\text{Highest degree} = \begin{cases} 1 & \text{if highest degree is a Ph.D.} \\ 0 & \text{if highest degree is not a Ph.D.} \end{cases}$$

Average score on teaching evaluations

Number of articles published in refereed journals

$$\text{Gender} = \begin{cases} 1 & \text{if professor is male} \\ 0 & \text{if professor is female} \end{cases}$$

A random sample of 100 university professors was taken and the following data were recorded:

Column 1: Annual salary
Column 2: Number of years since first degree
Column 3: Highest degree
Column 4: Mean score on teaching evaluation
Column 5: Number of articles published
Column 6: Gender

a. Can the statistics practitioner conclude that the model is valid?
b. Can the statistics practitioner conclude at the 5% significance level that there is gender discrimination?

An Excel spreadsheet, ***Pay Equity*** *(stored in the* ***Excel Workbooks*** *folder) was created to perform the analysis described in Example 18.4. The jobs, pay rates, job scores, and the values of the indicator variable are shown at the bottom of the sheet. These data were used as inputs in the regression analysis. The worksheet is set up so that any change in the factor scores and/or weights automatically changes the job scores at the bottom of the page.*

18.24 Redo Example 18.4. Change the weights for knowledge and training to 15% and for working conditions to 25%. What effect does this have on the conclusion? Briefly explain why the result was predictable.

18.25 Redo Example 18.4 by assigning your own values to each factor and to the weights. What conclusion did you reach?

18.26 Discuss how the factor values and weights affect the final result. Explain the strengths and weaknesses of the statistical analysis.

18.4 (OPTIONAL) LOGISTIC REGRESSION

One of the requirements of regression analysis is that the dependent variable must be interval. However, there are numerous applications where the dependent variable is nominal and where there are only two possible values. For example, financial analysts would like to predict whether a company will become bankrupt in the next year, using independent variables such as company sales, product costs, market share, and profits. Because the dependent variable is nominal (values are bankrupt and solvent), the least squares technique is not appropriate. Instead a technique called **logistic regression** is employed.

Logistic regression is used to estimate the probability that a particular outcome will occur. The dependent variable in logistic regression is the odds ratio, which is another way to express probability.

Odds Ratio and Probability

$$\text{Odds ratio} = \frac{\text{Probability of event}}{1 - \text{Probability of event}}$$

For example, if the probability of an event is .5 we find

$$\text{Odds ratio} = \frac{.5}{1 - .5} = 1$$

which we express as 1 to 1.

If the probability of the event is 2/3, the odds ratio is

$$\text{Odds ratio} = \frac{2/3}{1 - 2/3} = \frac{2/3}{1/3} = 2$$

In this case, we express the probability that the odds of the event occurring is 2 to 1. (*Note:* This definition refers to the odds ratio *for* the event. Gamblers use the term *odds* to refer to the odds *against* the event occurring.)

We can also calculate the probability of an event given the odds ratio.

Probability and Odds Ratio

$$\text{Probability} = \frac{\text{Odds ratio}}{\text{Odds ratio} + 1}$$

For example, if we're told that the odds ratio of a student achieving a grade of A+ in a statistics course is 5 (also expressed as 5 to 1), the probability that the student will achieve a grade of A+ is

$$\text{Probability} = \frac{\text{Odds ratio}}{\text{Odds ratio} + 1} = \frac{5}{5 + 1} = \frac{5}{6} = .833$$

The logistic regression model is similar to linear regression except for the dependent variable.

Logistic Regression Model

$$\ln(y) = \beta_0 + \beta_1 x_1 + \beta_2 x_2 + \cdots + \beta_k x_k + \varepsilon$$

where

y = odds ratio

$\ln(y)$ = natural logarithm of the odds ratio

$x_1, x_2, \ldots, x_k$ are the independent variables

$\beta_0, \beta_1, \ldots, \beta_k$ are the coefficients of the independent variables

ε = the error variable

The coefficients are estimated using a statistical technique called maximum likelihood estimation. The result is the logistic regression equation.

Logistic Regression Equation

$$\ln(\hat{y}) = b_0 + b_1x_1 + b_2x_2 + \cdots + b_kx_k$$

The value of $\hat{y}$ is the estimated odds ratio, which can be calculated from $\ln(\hat{y})$. That is,

$$\hat{y} = e^{\ln(\hat{y})}$$

Finally, the estimated probability of the event can be calculated from the estimated odds ratio:

$$\text{Estimated probability of the event} = \frac{\hat{y}}{\hat{y} + 1}$$

Several statistical software packages, including Minitab (Excel does not conduct logistic regression analysis) perform the required calculations. We will not show how the calculations are completed. Instead, we will provide the coefficients and use them to estimate probabilities.

EXAMPLE 18.5

Estimating the Probability of a Heart Attack among Diabetics

Travelers frequently buy insurance, which pays for medical emergencies while traveling. The premiums are determined primarily on the basis of age. However, additional variables are often considered. Foremost among these are continuing medical problems such as cancer and previous heart attacks. To help refine the calculation of premiums, an actuary was in the process of determining the probabilities of various outcomes. An area of interest is individuals who have diabetes. It is known that diabetics suffer a greater incidence of heart attacks than nondiabetics. After consulting medical specialists, the actuary learned that diabetics who smoke, have high cholesterol levels, and are overweight have a much higher probability of heart attacks. Additionally, age and gender also affect the probability in virtually all populations. To more precisely evaluate the risks, she took a random sample of diabetics and used the following logistic regression model:

$$\ln(y) = \beta_0 + \beta_1x_1 + \beta_2x_2 + \beta_3x_3 + \beta_4x_4 + \beta_5x_5 + \varepsilon$$

where

y = odds ratio of suffering a heart attack in the next 5 years

x_1 = average number of cigarettes smoked per day

x_2 = cholesterol level

x_3 = number of pounds overweight

x_4 = age

x_5 = gender (1 = female and 0 = male)

Notice that logistic regression allows us to use nominal independent variables in the same way that we included nominal independent variables in linear regression.

The coefficients of the logistic regression equation are

$$b_0 = -2.15$$
$$b_1 = .00847$$
$$b_2 = .00214$$
$$b_3 = .00539$$
$$b_4 = .00989$$
$$b_5 = -.288$$

The actuary would like to estimate the probability of a heart attack in the next 2 years for each of the following individuals.

Individual	Average Number of Cigarettes per Day	Cholesterol Level	Number of Pounds Overweight	Age	Gender
1	0	200	20	48	1
2	40	230	80	41	0
3	15	210	35	62	0
4	0	165	0	54	1
5	60	320	150	66	0

SOLUTION

The logistic regression equation is

$$\ln(\hat{y}) = -2.15 + .00847x_1 + .00214x_2 + .00539x_3 + .00989x_4 - .288x_5$$

Individual 1

$$\ln(\hat{y}) = -2.15 + .00847(0) + .00214(200) + .00539(20) + .00989(48) - .288(1) = -1.427$$

$$\hat{y} = e^{\ln(\hat{y})} = e^{-1.427} = .2399$$

$$\text{Probability of heart attack} = \frac{\hat{y}}{\hat{y}+1} = \frac{.2399}{.2399+1} = .1935$$

Individual 2

$$\ln(\hat{y}) = -2.15 + .00847(40) + .00214(230) + .00539(80) + .00989(41) - .288(0) = -.4823$$

$$\hat{y} = e^{\ln(\hat{y})} = e^{-.4823} = .6174$$

$$\text{Probability of heart attack} = \frac{\hat{y}}{\hat{y}+1} = \frac{.6174}{.6174+1} = .3817$$

Individual 3

$$\ln(\hat{y}) = -2.15 + .00847(15) + .00214(210) + .00539(35) + .00989(62) - .288(0) = -.7717$$

$$\hat{y} = e^{\ln(\hat{y})} = e^{-.7717} = .4622$$

$$\text{Probability of heart attack} = \frac{\hat{y}}{\hat{y}+1} = \frac{.4622}{.4622+1} = .3161$$

Individual 4

$$\ln(\hat{y}) = -2.15 + .00847(0) + .00214(165) + .00539(0) + .00989(54) - .288(1) = -1.5508$$

$$\hat{y} = e^{\ln(\hat{y})} = e^{-1.5508} = .2121$$

$$\text{Probability of heart attack} = \frac{\hat{y}}{\hat{y}+1} = \frac{.2121}{.2121+1} = .1750$$

Individual 5

$$\ln(\hat{y}) = 2.15 + .00847(60) + .00214(320) + .00539(150) + .00989(66) - .288(0) = .5042$$

$$\hat{y} = e^{\ln(\hat{y})} = e^{.5042} = 1.6557$$

$$\text{Probability of heart attack} = \frac{\hat{y}}{\hat{y}+1} = \frac{1.6557}{1.6557+1} = .6235$$

Interpreting the Coefficients

In linear regression the coefficients describe the relationship between each of the independent variables and the dependent variable. A positive coefficient of an independent variable means that when that variable increases by 1 unit, on average the dependent variable will increase by the amount of the coefficient (holding all the other variables constant). If the coefficient is negative, an increase of 1 unit in the independent variable will result in a decrease by the amount equal to the coefficient in the dependent variable (holding all the other variables constant).

Interpreting the coefficients of the logistic regression is somewhat more complex. Rather than interpreting the magnitude of the coefficient, we will instead describe the effect of the sign.

If a coefficient is positive, an increase in that independent variable will result in an increase in the probability of the event. Notice that in Example 18.5 the coefficients of the number of cigarettes, cholesterol level, number of pounds overweight, and age are positive. This tells us that older smokers who are overweight, and who have high cholesterol levels have a higher probability of suffering a heart attack in the next 5 years. The coefficient of gender is negative and because females are designated by the indicator variable equaling 1 (males are represented by 0), females have a lower probability than men with exactly the same characteristics.

APPLICATIONS in **BANKING**

Credit Scoring

On page 49 we introduced credit scorecards, which produce a number that tells lending institutions whether to approve loan applications. In fact, the purpose of credit scoring is to produce a probability that an applicant will repay his or her loan. The statistical method that produces these results is logistic regression. There are two methods of scoring. The first method is called application scoring wherein the information contained in a loan application is converted into a probability that the applicant will repay. The independent variables include the applicant's age, income, the number of years at his or her current job, and the number of years at the current residence. (See Exercise 18.34 for another form of credit scoring called bureau scoring.) The following example (and several exercises) was provided by SCORE, a well-known statistical consulting company. The author is grateful to Dr. Kevin Leonard, president of SCORE.

EXAMPLE 18.6 Estimating the Probability of Loan Repayment

Financial institutions build application scorecards by recording the data derived from thousands of applications and whether the loans were fully repaid. The data are stored in the same way as in multiple regression. The result of the loan is coded as 1 = loan repaid and 0 = defaulted. The following is an example from SCORE. The coefficients were slightly altered to protect the property of SCORE and the financial institution that paid for it.

The logistic regression model is

$$\ln(y) = \beta_0 + \beta_1 x_1 + \beta_2 x_2 + \beta_3 x_3 + \beta_4 x_4 + \varepsilon$$

where

x_1 = age

x_2 = income (\$1,000s)

x_3 = time at current job (years)

x_4 = time at current address (years)

The coefficients of the logistic regression equation are

$b_0 = .1524$

$b_1 = .0281$

$b_2 = .0223$

$b_3 = .0152$

$b_4 = .0114$

Suppose that the bank has received the applications from three people whose application forms produced the data in the accompanying table. The bank would like to estimate the probability that each applicant will fully repay the loan.

Applicant	Age	Income ($1,000s)	Time at Current Job	Time at Current Address
1	27	55	6	5
2	48	78	3	12
3	37	39	12	10

Use the logistic regression equation to produce the information the bank needs.

SOLUTION

The logistic regression equation is

$$\ln(\hat{y}) = .1524 + .0281x_1 + .0223x_2 + .0152x_3 + .0114x_4$$

Applicant 1

$$\ln(\hat{y}) = .1524 + .0281(27) + .0223(55) + .0152(6) + .0114(5) = 2.2858$$

$$\hat{y} = e^{\ln(\hat{y})} = e^{2.2858} = 9.8335$$

$$\text{Probability of repaying loan} = \frac{\hat{y}}{\hat{y}+1} = \frac{9.8335}{9.8335+1} = .9077$$

Applicant 2

$$\ln(\hat{y}) = .1524 + .0281(48) + .0223(78) + .0152(3) + .0114(12) = 3.4230$$

$$\hat{y} = e^{\ln(\hat{y})} = e^{3.4230} = 30.6613$$

$$\text{Probability of repaying loan} = \frac{\hat{y}}{\hat{y}+1} = \frac{30.6613}{30.6613+1} = .9684$$

Applicant 3

$$\ln(\hat{y}) = .1524 + .0281(37) + .0223(39) + .0152(12) + .0114(10) = 2.3582$$

$$\hat{y} = e^{\ln(\hat{y})} = e^{2.3582} = 10.5719$$

$$\text{Probability of repaying loan} = \frac{\hat{y}}{\hat{y}+1} = \frac{10.5719}{10.5719+1} = .9136$$

The decision to grant the loan would be made on the basis of these probabilities and the bank's risk policy. Most banks would insist on probabilities above 95%.

EXERCISES

Exercises 18.27–18.30 refer to Example 18.5.

18.27 Calculate the probability of a heart attack in the next 5 years for the following individual who suffers from diabetes.

Average number of cigarettes per day: 20

Cholesterol level: 200

Number of pounds overweight: 25

Age: 50

Gender: male

18.28 Recalculate the probability of a heart attack if the individual in Exercise 18.27 is able to quit smoking.

18.29 Refer to Exercise 18.27. What is the probability of a heart attack if the individual is able to reduce his cholesterol level to 150?

18.30 Refer to Exercise 18.27. Recalculate the probability of a heart attack if the individual loses 25 pounds.

Exercises 18.31–18.33 refer to Example 18.6.

18.31 If applicants want to improve the chances of having their loan applications approved by increasing the value of one variable, what variable should they select? Explain.

18.32 Redo the original example increasing the age by 10 years. What happened to the probability? Was this result expected?

18.33 Redo the original example increasing the income by 10 ($10,000). What happened to the probability? Was this result expected?

18.34 Another method of credit scoring is based on bureau reports. Bureaus such as Equifax and Trans Union keep records of the credit history of millions of North Americans. Among other bits of information, they record the following information.

Percent utilization: This is the proportion of available credit that is used. For example, if one's credit limit (sum of all credit cards and lines of credit) is $50,000 and that individual owes $10,000, the utilization is .20.

Worst rating: This records whether an individual has missed payments on loans or credit cards. For example, suppose that someone was 2 months late paying one credit card and 4 months late paying another. The worst rating would be recorded as 4. (This variable is sometimes recorded as the number of days late.)

Number of open trades: This statistic is the number of credit cards and lines of credit that the individual has applied for in the past 2 years.

The following logistic regression analysis was supplied by SCORE. Once again the coefficients were slightly altered. The model is

$$\ln(y) = \beta_0 + \beta_1 x_1 + \beta_2 x_2 + \beta_3 x_3 + \varepsilon$$

where

x_1 = utilization

x_2 = worst rating

x_3 = number of open trades

The coefficients of the logistic regression equation are

$b_0 = 5.8687$

$b_1 = -5.1546$

$b_2 = -.3725$

$b_3 = -.3213$

The bank has an application from an individual. A check with a credit bureau indicates that he currently has a credit limit of \$60,000 of which \$15,000 is owed, at one time he was 3 months late in paying a credit card, and he has applied for 1 more credit card in the last 2 years. Use the logistic regression equation to determine the probability that the applicant will repay the loan.

18.35 Refer to Exercise 18.34. Describe what each coefficient tells us about the probability.

18.36 Recent years have seen the growth of a new crime, identity theft. The theft of an identity is often accomplished by stealing or otherwise acquiring the mail of the intended victim. The criminal determines a number of critical facts about his victim including name, address, social security number, and which credit cards are held. With this information, the criminal creates credit cards with the correct name and number. In the past, financial institutions attempted to prevent credit card fraud by developing a model that describes how the owner of the card actually uses it. Variations from this pattern would alert the bank to the possibility of fraud. However, these models were rarely successful, missing many frauds and falsely identifying illegal use by the card's owner (what statisticians would call Type I and Type II errors). SCORE has pioneered the use of logistic regression to help identify the pattern of the fraudulent use of credit cards. The relevant independent variables are as follows.

Number of purchases or cash advances within a 3-hour period (a statistic called the velocity) (x_1)

Amount of credit card purchases or cash advances (x_2)

Code representing products purchased or transactions where 1 = jewelry or cash advances and 0 = other (x_3)

The model is designed to estimate the probability that the use is legal. The following regression coefficients were computed:

$b_0 = 6.1889$

$b_1 = -.6462$

$b_2 = -.0009254$

$b_3 = -.8721$

The bank's computer has just received the following information from one credit card. There have been four purchases or transactions over the last 3 hours, the total amount was \$1,855, and at least one transaction was a cash advance. Calculate the probability that the transactions are fraudulent. How should the estimated probabilities be employed? Discuss various issues associated with any policy you propose.

18.37 One of the questions in the course evaluation completed by students at the end of their university and college course asks, Would you recommend this course to a friend? To determine the factors that influence the decision, a statistics practitioner performed a logistic regression analysis with the following independent variables:

Midterm mark (out of 100)

Expected final mark (out of 100)

Grade point average (range: 0 to 12)

The following regression coefficients were computed:

$b_0 = -2.583$

$b_1 = .0277$

$b_2 = .0335$

$b_3 = .0041$

Determine the estimated probability that a student whose midterm mark was 72, who expects a final mark of 65, and whose grade point average is 7.5 will recommend the course to a friend.

18.5 (Optional) Stepwise Regression

In Section 17.3 we introduced multicollinearity and described the problems it causes by distorting the t-tests of the coefficients. If one of the objectives of the regression analysis is to determine whether and how each independent variable is related to the dependent variable, it is necessary to reduce the extent of multicollinearity.

As we discussed in Section 17.3, one of the ways to reduce multicollinearity is to include independent variables that appear to be uncorrelated with each other. A correlation matrix is usually produced to determine the correlation coefficients for each pair of variables. In many cases the correlation matrix will not be able to identify whether multicollinearity is a serious problem because there are many ways for variables to be related. For example, one variable may be a function of several other variables. Consequently, a correlation matrix may not reveal the problem. In this section, we introduce *stepwise regression*, a procedure that eliminates correlated independent variables.

Stepwise regression is an iterative procedure that adds and deletes one independent variable at a time. The decision to add or delete a variable is made on the basis of whether that variable improves the model. Minitab features this and related procedures. Excel does not. However, we created a macro that performs stepwise regression. It should be noted that this macro is different from the ones you have encountered thus far in this book. Specific instructions for its use are provided here.

Stepwise Regression Procedure

The procedure begins by computing the simple regression model for each independent variable. The independent variable with the largest F-statistic (which in a simple regression model is the t-statistic squared) or equally, with the smallest p-value, is chosen as the first entering variable. (Minitab uses the F-statistic, and the Excel macro can use either the F-statistic or the p-value.) The standard is usually set at $F = 4.0$, chosen because the significance level is about 5%. The standard may be changed in both Minitab and Excel. The standard is called the *F-to-enter*. If no independent variable exceeds the F-to-enter, the procedure ceases with no regression model produced. If at least one variable exceeds the standard, the procedure continues. It then considers whether the model would be improved by adding a second independent variable. It examines all such models to determine which is best and whether the F-statistic of the second variable (with the first variable already in the equation) is greater than the F-to-enter.

If two independent variables are highly correlated, only one of them will enter the equation. Once the first variable is included, the added explanatory power of the second variable will be minimal and its F-statistic will not be large enough to enter the model. In this way multicollinearity is reduced.

The procedure continues by deciding whether to add another independent variable at each step. The computer also checks to see whether the inclusion of previously added variables is warranted. At each step the p-values of all variables are computed and compared to the *F-to-remove*. If a variable's F-statistic falls below this standard, it is removed from the equation. These steps are repeated until no more variables are added or removed.

To illustrate, we'll return to Example 17.2 wherein the real estate agent wanted to determine a method to predict the selling prices of houses on the basis of the size of the house, the number of bedrooms, and the lot size. However, the regression model produced contradictory results because of multicollinearity. The model's fit was excellent, the coefficient of determination was high, and yet the t-tests of the coefficients revealed that there was no evidence of a linear relationship between each of the independent variables and the dependent variable. Here is the stepwise regression output.

EXCEL

	A	B	C	D	E	F	G
1	**Results of stepwise regression**						
2							
3	**Step 1 - Entering variable: House_Size**						
4							
5	Summary measures						
6		Multiple R	0.7478				
7		R-Square	0.5591				
8		Adj R-Square	0.5547				
9		StErr of Est	24,790				
10							
11	ANOVA Table						
12		Source	df	SS	MS	F	p-value
13		Explained	1	76,385,711,728	76,385,711,728	124.30	0.0000
14		Unexplained	98	60,225,052,672	614,541,354		
15							
16	Regression coefficients						
17			Coefficient	Std Err	t-value	p-value	
18		Constant	40,066	10521	3.81	0.0002	
19		House_Size	64.20	5.759	11.15	0.0000	

INSTRUCTIONS

1. Type or import the data according to the following instructions.
2. Click **Add-Ins, Data Analysis Plus,** and **Stepwise Regression.**
3. Specify the dependent variable (**response variable**) and the independent variables (**explanatory variables**). Highlight the entire list.
4. Select the **Significance option.** Click **OK** to choose **p-values** as the criterion. Click **OK** to accept the default values for **p-to-enter** and **p-to-leave.**

Special Instructions for Stepwise Regression

Before beginning, place the cursor somewhere in the data set. If you don't, you will be forced to close the macro and repeat.

The macro will ask about missing values. The data sets that are stored on the CD do not have missing values. Simply accept the default and proceed. However, if you have created your own data set that has missing values, click the box to erase the check mark.

Instead of specifying the input range of the data, you simply specify variable names for the dependent and the independent variables. The names must start with a letter or underscore. Do not use names that are the same as cell addresses (e.g., A2), R or C, or names consisting only of numbers (e.g., 55). Do not use symbols other than letters, numbers, and underscores.

The dependent variable must be interval. It cannot be an indicator (dummy) variable. You may use independent indicator variables.

You will be asked to specify the significance options, which refers to the way the stepwise regression decides which variables to enter and which variables to leave. Until you learn more about these choices, we suggest that you accept the defaults.

The macro will draw several different plots. Specify the ones you want.

The output will be placed on the same worksheet as the data or on a different worksheet. If you choose the latter, the macro will ask you to name the worksheet.

MINITAB

Stepwise Regression: Price Versus Bedrooms, House Size, Lot Size

Forward selection. Alpha-to-Enter: 0.25

Response is Price on 3 predictors, with N = 100

```
Step              1
Constant      40066

House Size     64.2
T-Value       11.15
P-Value       0.000

S             24790
R-Sq          55.91
R-Sq(adj)     55.47
```

INSTRUCTIONS

1. Click **Stat, Regression,** and **Stepwise.**
2. Specify the dependent variable (**Response**) and the independent variables (**Predictors**).
3. Click **Methods** and **Forward selection.** Specify **Alpha to enter.** (We recommend .05.)

INTERPRET

The regression equation is

$$\hat{y} = 40{,}066 + 64.20x_1$$

where x_1 is the house size. The coefficient of determination is .5591, which is only slightly less than the one obtained in the regression equation in Example 17.2 where $R^2 = .5600$. Note that the variable most strongly correlated with price in Example 17.2 was house size. As a result, the one and only variable included in the stepwise regression model is house size. Once house size is included in the equation, the other two variables contributed so little to explaining the variation in price that they were excluded from the model.

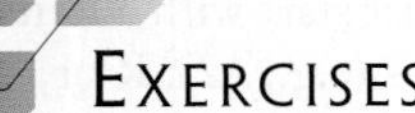

EXERCISES

The following exercises require the use of a computer and statistical software.

18.38 Refer to the chapter-opening example.

a. Use stepwise regression to compute the regression equation.
b. Explain why these results differ from those produced previously.

18.39 Refer to Exercise 17.9.

a. Using stepwise regression determine the regression line.
b. What are the differences between this equation and the one produced in Exercise 18.9?

c. Explain why the variable age was included in the stepwise regression equation whereas it was not statistically significant in Exercise 17.9.

18.40 Refer to Exercise 17.4.
a. Determine the regression equation using stepwise regression.
b. What differences are there between these statistics and the ones produced in Exercise 17.4?
c. Explain why the two printouts differ.

18.41 Refer to Exercise 17.13.
a. Use the stepwise regression method to estimate the coefficients of the regression equation.
b. Discuss any differences between these statistics and the statistics computed in Exercise 17.13.

18.6 Model Building

At this point, we have described several different regression models. You now have the use of nominal predictor variables and the tools to describe a variety of nonlinear relationships. In this section, we describe how the statistics practitioner builds a model.

Regression analysis is used either to determine how one or more predictor variables are related to a dependent variable or to predict the value of the dependent variable and estimate its expected value. Although the process differs between the two objectives, there are many similarities in the approach.

Here is the procedure that is employed in the building of a model.

Procedure for Building a Model

1. *Identify the dependent variable.* Clearly define the variable that you wish to analyze or predict. For example, if you want to forecast sales, decide whether it is to be the number of units sold, gross revenue, or perhaps net profits. Additionally, decide whether to forecast weekly, monthly, or annual figures.

2. *List potential predictors.* Using your knowledge of the dependent variable, produce a list of predictors that may be related to the dependent variable. Although we cannot establish a causal relationship, we should attempt to include predictor variables that cause changes in the dependent variable. Bear in mind the problems caused by multicollinearity and the cost of gathering, storing, and processing data. Be selective in your choices. It is best to use the fewest independent variables that produce a satisfactory model.

3. *Gather the required observations for the potential models.* A general rule is that there should be at least six observations for each independent variable used in the equation.

4. *Identify several possible models.* Once again, use your knowledge of the dependent variable and predictor variables to formulate a model. For example, if you believe that a predictor variable affects the dependent variable, but you are uncertain about the form of the relationship, formulate first-order and second-order models with and without interaction. It may be helpful to draw a scatter diagram of the dependent variable and each predictor variable to discover the nature of the relationship.

5. *Use statistical software to estimate the models.* Use one or more of the variable selection methods described in the previous section to determine which variables to include in the model. If the objective is to determine which predictor variables are related to the dependent variable, you will need to ensure that multicollinearity is not a problem. If it is, attempt to reduce the number of independent variables.

6. *Determine whether the required conditions are satisfied.* If not, attempt to correct the problem. At this point, you may have several "equal" models from which to choose.

7. *Use your judgment and the statistical output to select the best model.* This may be the most difficult part of the process. There may be a model that fits best, but another one may be a better predictor, and yet another may feature fewer variables and, thus, be easier to work with. Experience with regression helps. Taking another statistics course is likely your best strategy.

CHAPTER SUMMARY

This chapter completes our discussion of the regression technique, which began in Chapter 17. We presented several additional models for predicting the value of one variable on the basis of other variables. Polynomial models with one and two independent variables were presented. We discussed how indicator variables allow us to use nominal variables and we described how indicator variables are used in pay equity discussions. Logistic regression was introduced to address the problem of a nominal dependent variable. To help choose the model that is best for our purposes, we introduced stepwise regression. We completed the chapter by providing some advice on how statisticians build models.

IMPORTANT TERMS

Polynomial model 716
Predictor variable 716
Order 716
First-order model 716
Interaction 719
Quadratic relationship 720
Second-order model 720
Indicator variable 726
Dummy variable 726
Omitted category 726
Logistic regression 738
Stepwise regression 746

SYMBOLS

Symbol	Pronounced	Represents
I_i	*I-sub-i* or *I-i*	Indicator variable

COMPUTER OUTPUT AND INSTRUCTIONS

Technique	Excel	Minitab
Stepwise regression	747	748

CHAPTER EXERCISES

The following exercises require the use of a computer and statistical software. ***Use a 5% significance level.***

18.42 Xr18-42 Car designers have been experimenting with ways to improve gas mileage for many years. An important element in this research is the way in which a car's speed affects how quickly fuel is burned. Competitions whose objective is to drive the farthest on the smallest amount of gas have determined that low speeds and high speeds are inefficient. Designers would like to know which speed burns gas most efficiently. As an experiment, 50 identical cars are driven at different speeds and the gas mileage measured.

a. Write the equation of the model you think is appropriate.
b. Perform a regression analysis using your model.
c. How well does it fit?

18.43 Xr18-43 The number of car accidents on a particular stretch of highway seems to be related to the number of vehicles that travel over it and the speed at which they are traveling. A city alderman has decided to ask the county sheriff to provide him with statistics covering the last few years, with the intention of examining these data statistically so that he can (if possible) introduce new speed laws that will reduce traffic accidents. Using the number of accidents as the dependent variable, he obtains estimates of the number of cars passing along a stretch of road and their average speeds (in miles per hour). The observations for 60 randomly selected days were recorded.

a. Which model should the alderman use? Explain.
b. Conduct a regression analysis using a first-order model with interaction.
c. Is the model valid?

18.44 Refer to Exercise 18.43.

a. Estimate a second-order model with interaction.
b. Is this model valid in predicting the number of accidents? Test at the 10% significance level.

18.45 Xr18-45 After analyzing whether the number of ads is related to the number of customers, the manager in Exercise 16.81 decided to determine whether the advertising made any difference. As a result, he reorganized the experiment. Each week he advertised several times per week, but in only one of the advertising media. He again recorded the weekly number of customers, the number of ads, and the media of that week's advertisement (1 = newspaper, 2 = radio, 3 = television).

a. Create indicator variables to describe the advertising medium.
b. Conduct a regression analysis. Test to determine whether the model is valid.
c. Does the advertising medium make a difference? Explain.

18.46 Xr18-46 A baseball fan has been collecting data from a newspaper on the various American League teams. She wants to explain each team's winning percentage as a function of its batting average and its earned run average plus an indicator variable for whether the team fired its manager within the last 12 months (1 = fired manager, and 0 = did not fire manager). The data for 50 randomly selected teams over the last five seasons were recorded.

a. Perform a regression analysis using a first-order model (no interaction).
b. Do these data provide sufficient evidence that a team that fired its manager within the last 12 months wins less frequently than a team that did not fire its manager?

18.47 Xr18-47 A growing segment of the textile industry in the United States is based on piecework, wherein workers are paid for each unit they produce, instead of receiving an hourly wage. The manager of one such company has observed that inexperienced workers perform quite poorly, but they usually improve quickly. However, very experienced workers do not perform as well as expected. Analysts attribute this phenomenon to boredom. More experienced workers grow weary of the monotonous work and become less productive. In an attempt to learn more about piecework labor, a statistics practitioner took a random sample of workers with varying years of experience and counted the number of units each produced in 8 hours.

a. Write the equation of the model you think would fit.
b. Perform a regression analysis using your model.
c. Describe how well the model fits.

18.48 Xr18-48 The maintenance of swimming pools is quite costly because of all the chlorine that is needed to keep the water clear and relatively free of germs. A chain of hotels (all with outdoor pools) seeking to reduce costs decided to analyze the factors that determine how much chlorine is needed. They commissioned a chemist to conduct an analysis. It is believed that the speed at which chlorine in a pool is depleted is dependent on the temperature of the water (higher temperature uses chlorine faster); pH level, which is a measure of the acidity of the water (pH ranges from 0 to 14, where 0 is very acidic and 14 is very alkaline; levels around 7.5 use the least chlorine); and weather (sunshine uses up chlorine). The chemist conducted the following experiment. The percentage of chlorine depletion during 8-hour days was recorded under varying conditions of pH level, water temperature, and weather conditions. These data were recorded in the following way:

Column 1: Percentage of chlorine depletion over 8 hours

Column 2: Temperature (degrees Fahrenheit)

Column 3: pH level

Column 4: 1 = mainly cloudy, 2 = sunny, 3 = partly sunny

a. Write the equation of the model that you would suggest.
b. Use regression analysis to estimate the model's coefficients.
c. Test to determine whether the model is valid.
d. Can we infer that higher temperatures deplete chlorine more quickly?
e. Is there evidence to infer that the belief about the relationship between chlorine depletion and pH level is correct?
f. Can we infer that weather is a factor in chlorine depletion?

© Stephen Webster/Photonica/Getty Images

19 NONPARAMETRIC STATISTICS

Retaining Workers

DATA Xm19-00

Because of the high cost of hiring and training new employees, employers would like to ensure that they retain highly qualified workers. To help develop a hiring program, the human resources manager of a large company wanted to compare how long business and nonbusiness university graduates worked for the company before quitting to accept a position elsewhere. The manager selected a random sample of 25 business and 20 nonbusiness graduates who had been hired 5 years ago. The number of months each had worked for the company was recorded. (Those who had not quit were recorded as having worked for 60 months.) The data are listed below. Can the human resources manager conclude at the 5% significance level that a difference in duration of employment exists between business and nonbusiness graduates?

© White Packert/The Image Bank/ Getty Images

See page 763 for our answer.

Duration of Employment (Months)

Business Graduates												
60	11	18	19	5	25	60	7	8	17	37	4	8
28	27	11	60	25	5	13	22	11	17	9	4	

Nonbusiness Graduates										
25	60	22	24	23	36	39	15	35	16	28
9	60	29	16	22	60	17	60	32		

INTRODUCTION

Throughout this book we have presented statistical techniques that are used when the data are either interval or nominal. In this chapter, we introduce statistical techniques that deal with ordinal data. We will introduce three methods that compare two populations, two procedures used to compare two or more populations, and a technique to analyze the relationship between two variables. As you've seen when we compare two or more populations of interval data, we measure the difference between means. However, as we discussed in Chapter 2, when the data are ordinal, the mean is not an appropriate measure of location. As a result, the methods in this chapter do not enable us to test the difference in population means; instead, we will test characteristics of populations without referring to specific parameters. For this reason, these techniques are called **nonparametric techniques.** Rather than testing to determine whether the population means differ, we will test to determine whether the *population locations* differ.

Although nonparametric methods are designed to test ordinal data, they have another area of application. The statistical tests described in Sections 13.1 and 13.3 and in Chapter 14 require that the populations be normally distributed. If the data are extremely nonnormal, the t-tests and F-test are invalid. Fortunately, nonparametric techniques can be used instead. For this reason, nonparametric procedures are often (perhaps more accurately) called **distribution-free statistics.** The techniques presented here can be used when the data are interval and the required condition of normality is unsatisfied. In such circumstances we will treat the interval data as if they were ordinal. For this reason, even when the data are interval and the mean is the appropriate measure of location, we will choose instead to test population locations.

Figure 19.1 depicts the distributions of two populations when their locations are the same. Notice that, since we don't know (or care) anything about the shape of the

FIGURE **19.1** Population Locations Are the Same

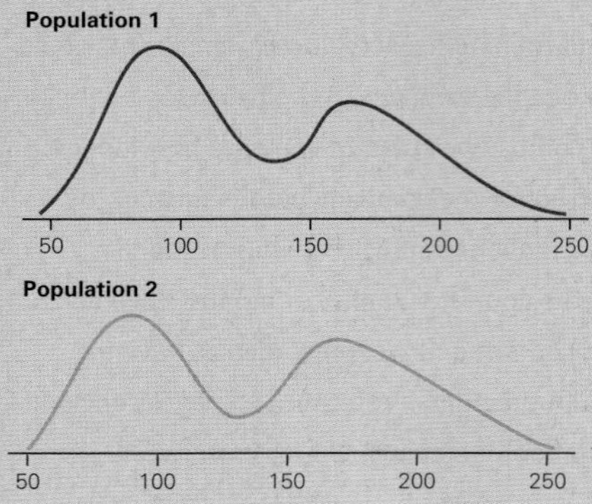

FIGURE **19.2** Location of Population 1 Is to the Right of the Location of Population 2

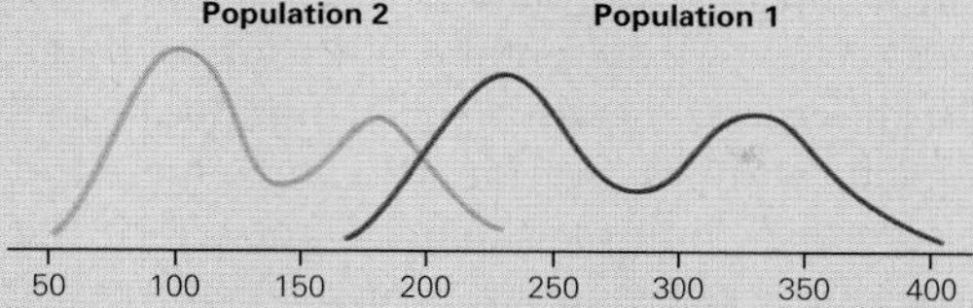

FIGURE **19.3** Location of Population 1 Is to the Left of the Location of Population 2

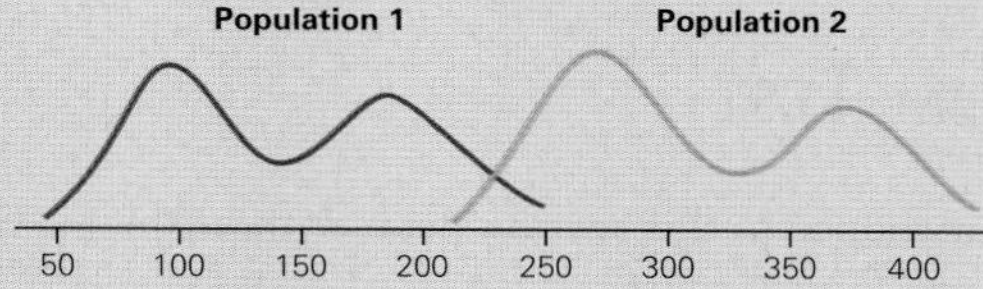

distributions, we represent them as nonnormal. Figure 19.2 describes a circumstance when the location of population 1 is to the right of the location of population 2. The location of population 1 is to the left of the location of population 2 in Figure 19.3.

When the problem objective is to compare two populations, the null hypothesis will state

H_0: The two population locations are the same

The alternative hypothesis can take on any one of the following three forms.

1. If we want to know whether there is sufficient evidence to infer that there is a difference between the two populations, the alternative hypothesis is

H_1: The location of population 1 is different from the location of population 2

2. If we want to know whether we can conclude that the random variable in population 1 is larger in general than the random variable in population 2 (see Figure 19.2), the alternative hypothesis is

H_1: The location of population 1 is to the right of the location of population 2

3. If we want to know whether we can conclude that the random variable in population 1 is smaller in general than the random variable in population 2 (see Figure 19.3), the alternative hypothesis is

H_1: The location of population 1 is to the left of the location of population 2

As you will see, nonparametric tests utilize a ranking procedure as an integral part of the calculations. You've actually dealt with such a process already in this book. In

Chapter 4, we introduced the median as a measure of central location. The median is computed by placing the observations in order and selecting the observation that falls in the middle. Thus, the appropriate measure of central location of ordinal data is the median, a statistic that is the product of a ranking process.

In the next section, we present the Wilcoxon rank sum test employed when we wish to test for the differences between population locations when the data are generated from independent samples. Section 19.2 introduces the sign test and the Wilcoxon signed rank sum test, both of which are applied to the matched pairs experiment. Sections 19.3 and 19.4 introduce the Kruskal-Wallis test and the Friedman test, respectively, procedures that are employed when the objective is to compare two or more populations. The Spearman rank correlation coefficient, which analyzes the relationship between two variables is presented in Section 19.5.

19.1 Wilcoxon Rank Sum Test

The test we introduce in this section deals with problems with the following characteristics:

1. The problem objective is to compare two populations.
2. The data are either ordinal or interval where the normality requirement necessary to perform the equal-variances t-test of $\mu_1 - \mu_2$ is unsatisfied.
3. The samples are independent.

To illustrate how to compute the test statistic for the **Wilcoxon rank sum test,** we offer the following example.

EXAMPLE 19.1

Wilcoxon Rank Sum Test

Suppose that we want to determine whether the following observations drawn from two populations allow us to conclude at the 5% significance level that the location of population 1 is to the left of the location of population 2.

Sample 1: 22 23 20

Sample 2: 18 27 26

We want to test the following hypotheses:

H_0: The two population locations are the same

H_1: The location of population 1 is to the left of the location of population 2

Test Statistic

The first step is to rank all six observations with rank 1 assigned to the smallest observation and rank 6 to the largest.

Sample 1	Rank	Sample 2	Rank
22	3	18	1
23	4	27	6
20	2	26	5
	$T_1 = 9$		$T_2 = 12$

Observe that 18 is the smallest number, so it receives a rank of 1; 20 is the second-smallest number, and it receives a rank of 2. We continue until rank 6 is assigned to 27, which is the largest of the observations. In case of ties, we average the ranks of the tied observations. The second step is to calculate the sum of the ranks of each sample. The rank sum of sample 1, denoted T_1, is 9. The rank sum of sample 2, denoted T_2, is 12. (Note that T_1 plus T_2 must equal the sum of the integers from 1 to 6, which is 21.) We can use either rank sum as the test statistic. We arbitrarily select T_1 as the test statistic and label it T. The value of the test statistic in this example is $T = T_1 = 9$.

Sampling Distribution of the Test Statistic

A small value of T indicates that most of the smaller observations are in sample 1 and that most of the larger observations are in sample 2. This would imply that the location of population 1 is to the left of the location of population 2. Therefore, in order for us to conclude statistically that this is the case, we need to show that T is small. The definition of "small" comes from the sampling distribution of T. As we did in Section 9.1 when we derived the sampling distribution of the sample mean, we can derive the sampling distribution of T by listing all possible values of T. In Table 19.1 we show all possible rankings of two samples of size 3.

TABLE **19.1** **All Possible Ranks and Rank Sums of Two Samples of Size 3**

Ranks of Sample 1	Rank Sum	Ranks of Sample 2	Rank Sum
1, 2, 3	6	4, 5, 6	15
1, 2, 4	7	3, 5, 6	14
1, 2, 5	8	3, 4, 6	13
1, 2, 6	9	3, 4, 5	12
1, 3, 4	8	2, 5, 6	13
1, 3, 5	9	2, 4, 6	12
1, 3, 6	10	2, 4, 5	11
1, 4, 5	10	2, 3, 6	11
1, 4, 6	11	2, 3, 5	10
1, 5, 6	12	2, 3, 4	9
2, 3, 4	9	1, 5, 6	12
2, 3, 5	10	1, 4, 6	11
2, 3, 6	11	1, 4, 5	10
2, 4, 5	11	1, 3, 6	10
2, 4, 6	12	1, 3, 5	9
2, 5, 6	13	1, 3, 4	8
3, 4, 5	12	1, 2, 6	9
3, 4, 6	13	1, 2, 5	8
3, 5, 6	14	1, 2, 4	7
4, 5, 6	15	1, 2, 3	6

If the null hypothesis is true and the two population locations are identical, then it follows that each possible ranking is equally likely. Because there are 20 different possibilities, each value of T has the same probability, namely, 1/20, Notice that there is one value of 6, one value of 7, two values of 8, and so on. Table 19.2 summarizes the values of T and their probabilities, and Figure 19.4 depicts this sampling distribution.

TABLE **19.2** **Sampling Distribution of T with Two Samples of Size 3**

T	$P(T)$
6	1/20
7	1/20
8	2/20
9	3/20
10	3/20
11	3/20
12	3/20
13	2/20
14	1/20
15	1/20
Total	1

FIGURE **19.4** **Sampling Distribution of T with Two Samples of Size 3**

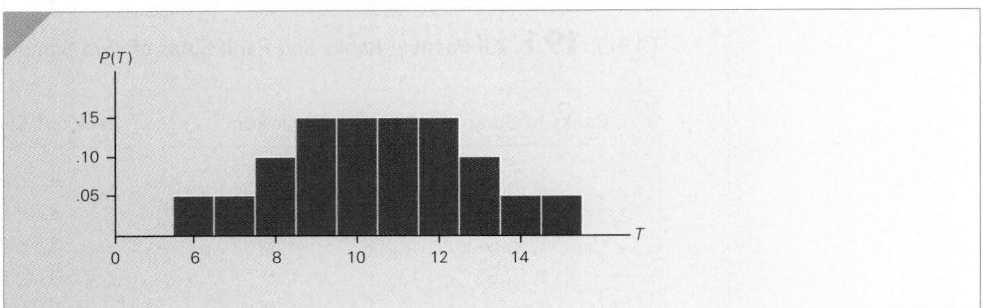

From this sampling distribution we can see that $P(T \leq 6) = P(T = 6) = 1/20 = .05$. Because we're trying to determine whether the value of the test statistic is small enough for us to reject the null hypothesis at the 5% significance level, we specify the rejection region as $T \leq 6$. Since $T = 9$, we cannot reject the null hypothesis.

Statisticians have generated the sampling distribution of T for various combinations of sample sizes. The critical values are provided in Table 9 in Appendix B and reproduced here as Table 19.3. Table 19.3 provides values of T_L and T_U for sample sizes between 3 and 10 (n_1 is the size of sample 1 and n_2 is the size of sample 2). The values of T_L and T_U in Part a of the table are such that

$$P(T \leq T_L) = P(T \geq T_U) = .025$$

The values of T_L and T_U in Part b of the table are such that

$$P(T \leq T_L) = P(T \geq T_U) = .05$$

TABLE **19.3** Critical Values of the Wilcoxon Rank Sum Test

(a) $\alpha = .025$ one-tail; $\alpha = .05$ two-tail

$n_2 \backslash n_1$	3		4		5		6		7		8		9		10	
	T_L	T_U	T_L	T_U	T_L	T_U	T_L	T_U	T_L	T_U	T_L	T_U	T_L	T_U	T_L	T_U
4	6	18	11	25	17	33	23	43	31	53	40	64	50	76	61	89
5	6	21	12	28	18	37	25	47	33	58	42	70	52	83	64	96
6	7	23	12	32	19	41	26	52	35	63	44	76	55	89	66	104
7	7	26	13	35	20	45	28	56	37	68	47	81	58	95	70	110
8	8	28	14	38	21	49	29	61	39	73	49	87	60	102	73	117
9	8	31	15	41	22	53	31	65	41	78	51	93	63	108	76	124
10	9	33	16	44	24	56	32	70	43	83	54	98	66	114	79	131

(b) $\alpha = .05$ one-tail; $\alpha = .10$ two-tail

$n_2 \backslash n_1$	3		4		5		6		7		8		9		10	
	T_L	T_U	T_L	T_U	T_L	T_U	T_L	T_U	T_L	T_U	T_L	T_U	T_L	T_U	T_L	T_U
3	6	15	11	21	16	29	23	37	31	46	39	57	49	68	60	80
4	7	17	12	24	18	32	25	41	33	51	42	62	52	74	63	87
5	7	20	13	27	19	36	26	46	35	56	45	67	55	80	66	94
6	8	22	14	30	20	40	28	50	37	61	47	73	57	87	69	101
7	9	24	15	33	22	43	30	54	39	66	49	79	60	93	73	107
8	9	27	16	36	24	46	32	58	41	71	52	84	63	99	76	114
9	10	29	17	39	25	50	33	63	43	76	54	90	66	105	79	121
10	22	31	18	42	26	54	35	67	46	80	57	95	69	111	83	127

Source: From F. Wilcoxon and R.A. Wilcox, "Some Rapid Approximate Statistical Procedures" (1964), p. 28. Reproduced with the permission of American Cyanamid Company.

Part a is used either in a two-tail test with $\alpha = .05$ or in a one-tail test with $\alpha = .025$. Part b is employed either in a two-tail test with $\alpha = .10$ or in a one-tail test with $\alpha = .05$. Because no other values are provided, we are restricted to those values of α.

Although it is possible to derive the sampling distribution of the test statistic for any other sample sizes, the process can be quite tedious. Fortunately it is also unnecessary. Statisticians have shown that when the sample sizes are larger than 10, the test statistic is approximately normally distributed with mean $E(T)$ and standard deviation σ_T where

$$E(T) = \frac{n_1(n_1 + n_2 + 1)}{2}$$

and

$$\sigma_T = \sqrt{\frac{n_1 n_2(n_1 + n_2 + 1)}{12}}$$

Thus, the standardized test statistic is

$$z = \frac{T - E(T)}{\sigma_T}$$

EXAMPLE 19.2

Comparing Pharmaceutical Painkillers

DATA Xm19-02

A pharmaceutical company is planning to introduce a new painkiller. In a preliminary experiment to determine its effectiveness, 30 people were randomly selected, of whom 15 were given the new painkiller and 15 were given aspirin. All 30 were told to use the drug when headaches or other minor pains occurred and to indicate which of the following statements most accurately represented the effectiveness of the drug they took:

5 = The drug was extremely effective.
4 = The drug was quite effective.
3 = The drug was somewhat effective.
2 = The drug was slightly effective.
1 = The drug was not at all effective.

The responses are listed here using the codes. Can we conclude at the 5% significance level that the new painkiller is perceived to be more effective?

New painkiller: 3, 5, 4, 3, 2, 5, 1, 4, 5, 3, 3, 5, 5, 5, 4
Aspirin: 4, 1, 3, 2, 4, 1, 3, 4, 2, 2, 2, 4, 3, 4, 5

SOLUTION

IDENTIFY

The objective is to compare two populations: the perceived effectiveness of the new painkiller and of aspirin. We recognize that the data are ordinal, because, except for the order of the codes, the numbers used to record the results are arbitrary. Finally, the samples are independent. These factors tell us that the appropriate technique is the Wilcoxon rank sum test. We denote the effectiveness scores of the new painkiller as sample 1 and the effectiveness scores of aspirin as sample 2. Because we want to know whether the new painkiller is better than aspirin, the alternative hypothesis is

H_1: The location of population 1 is to the right of the location of population 2

We specify the null hypothesis as

H_0: The two population locations are the same

COMPUTE

MANUALLY

If the alternative hypothesis is true, the location of population 1 will be located to the right of the location of population 2. It follows that T and z would be large. Our job is to determine whether z is large enough to reject the null hypothesis in favor of the alternative hypothesis. Thus, the rejection region is

$$z > z_\alpha = z_{.05} = 1.645$$

We compute the test statistic by ranking all the observations.

New Pain Killer	Rank	Aspirin	Rank
3	12	4	19.5
5	27	1	2
4	19.5	3	12
3	12	2	6
2	6	4	19.5
5	27	1	2
1	2	3	12
4	19.5	4	19.5
5	27	2	6
3	12	2	6
3	12	2	6
5	27	4	19.5
5	27	3	12
5	27	4	19.5
4	19.5	5	27
	$T_1 = 276.5$		$T_2 = 188.5$

Notice that there are three "ones" that occupy ranks 1, 2, and 3. The average is 2. Thus, each "one" is assigned a rank of 2. There are five "twos" whose ranks are 4, 5, 6, 7, and 8, the average of which is 6. We continue until all the observations have been similarly ranked. The rank sums are computed with $T_1 = 276.5$ and $T_2 = 188.5$. The unstandardized test statistic is $T = T_1 = 276.5$. To standardize, we determine $E(T)$ and σ_T as follows.

$$E(T) = \frac{n_1(n_1 + n_2 + 1)}{2} = \frac{15(31)}{2} = 232.5$$

$$\sigma_T = \sqrt{\frac{n_1 n_2(n_1 + n_2 + 1)}{12}} = \sqrt{\frac{(15)(15)(31)}{12}} = 24.1$$

The standardized test statistic is calculated next:

$$z = \frac{T - E(T)}{\sigma_T} = \frac{276.5 - 232.5}{24.1} = 1.83$$

The p-value of the test is

$$p\text{-value} = P(Z > 1.83) = 1 - .9664 = .0336$$

EXCEL

	A	B	C	D	E
1	**Wilcoxon Rank Sum Test**				
2					
3			Rank Sum	Observations	
4	*New*		276.5	15	
5	*Aspirin*		188.5	15	
6	z Stat		1.83		
7	P(Z<=z) one-tail		0.0340		
8	z Critical one-tail		1.6449		
9	P(Z<=z) two-tail		0.0680		
10	z Critical two-tail		1.96		

INSTRUCTIONS

1. Type or import the data into two adjacent columns. (Open Xm19-02.)
2. Click **Add-Ins, Data Analysis Plus,** and **Wilcoxon Rank Sum Test.**
3. Specify the **Variable 1 Range** (A1:A16), the **Variable 2 Range** (B1:B16), and the value of α (.05).

MINITAB

Mann-Whitney Test and CI: New, Aspirin

```
         N  Median
New     15   4.000
Aspirin 15   3.000


Point estimate for ETA1-ETA2 is 1.000
95.4 Percent CI for ETA1-ETA2 is (0.001,2.000)
W = 276.5
Test of ETA1 = ETA2 vs ETA1 > ETA2 is significant at 0.0356
The test is significant at 0.0321 (adjusted for ties)
```

Minitab performs the **Mann-Whitney** test rather than the Wilcoxon test. However, the tests are equivalent. In the output, **ETA** represents the population median. The test statistic $W = 276.5$, is the value of the Wilcoxon rank sum statistic. That is, $T = W = 276.5$. The output includes the p-value (0.0356) and another p-value calculated by adjusting for tied observations (0.0321). We will report the first p-value only.

INSTRUCTIONS

1. Type or import the data into two adjacent columns. (Open Xm19-02.)
2. Click **Stat, Nonparametrics,** and **Mann-Whitney . . . .**
3. Type the variable name for the **First Sample** (New), the variable name of the **Second Sample** (Aspirin), and click one of **less than, not equal to,** or **greater than** in the **Alternative** box (greater than).

INTERPRET

The data provide sufficient evidence to infer that the new painkiller is perceived to be more effective than aspirin. We note that the data were generated from a controlled experiment. That is, the subjects were assigned to take either the new painkiller or aspirin. (When subjects decide for themselves which medication to take, the data are observational.) This factor helps support the claim that the new painkiller is indeed more effective than aspirin. Factors that weaken the argument are small sample sizes and the inexactness of the responses. There may be methods to measure the effectiveness less subjectively. Additionally, a double-blind experiment should have been conducted.

As we pointed out in the introduction to this chapter, the Wilcoxon rank sum test is used to compare two populations when the data are either ordinal or interval. Example 19.2 illustrated the use of the Wilcoxon rank sum test when the data are ordinal. In the example that opened this chapter, we demonstrate its use when the data are interval.

Retaining Workers: Solution

IDENTIFY

The problem objective is to compare two populations whose data are interval. The samples are independent. Thus, the appropriate parametric technique is the t-test of $\mu_1 - \mu_2$, which requires that the populations be normally distributed. However, when the histograms are drawn (see Figures 19.5 and 19.6), it becomes clear that this requirement is unsatisfied. It follows that the correct statistical procedure is the Wilcoxon rank sum test. The null and alternative hypotheses are

H_0: The two population locations are the same

H_0: The location of population I (business graduates) is different from the location of population 2 (nonbusiness graduates)

FIGURE **19.5** **Histogram of Length of Employment of Business Graduates in Retaining Workers Example**

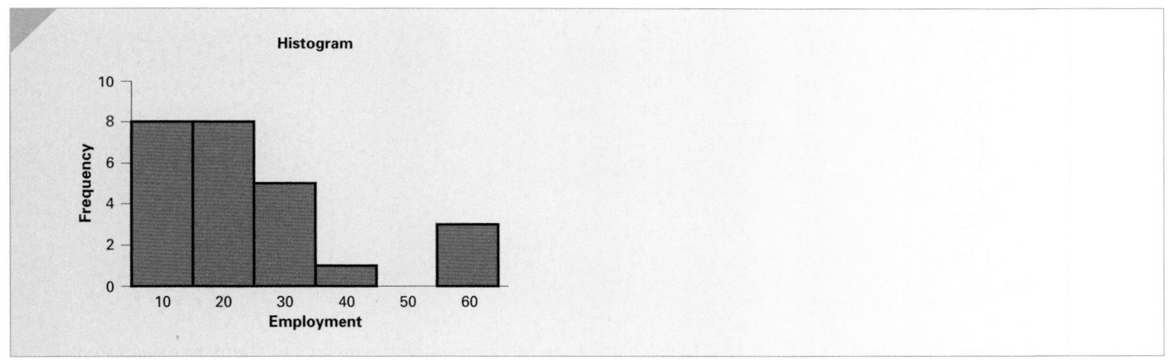

FIGURE **19.6** **Histogram of Length of Employment of Nonbusiness Graduates in Retaining Workers Example**

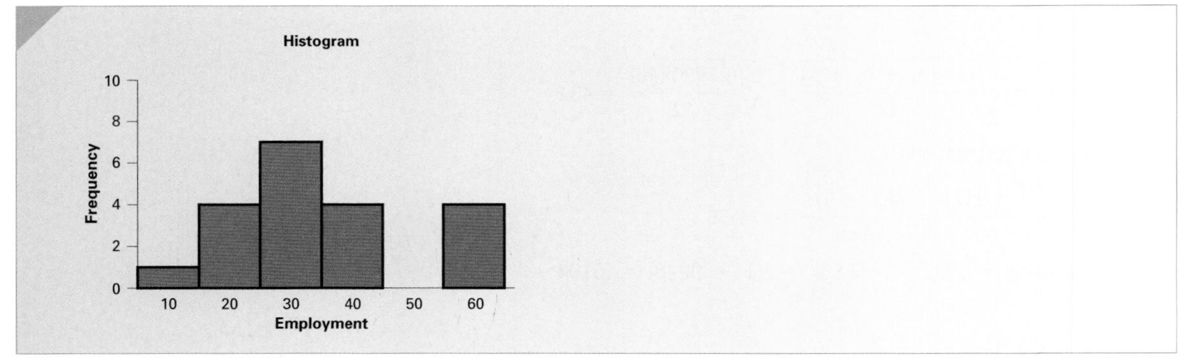

COMPUTE

MANUALLY

The rejection region is

$$z < -z_{\alpha/2} = -z_{.025} = -1.96 \qquad \text{or} \qquad z > z_{\alpha/2} = z_{.025} = 1.96$$

We calculate the value of the test statistic in the following way.

Business	Rank	Nonbusiness	Rank
60	42	25	28
11	11	60	42
18	20	22	23
19	21	24	26
5	3.5	23	25
25	28	36	36
60	42	39	38
7	5	15	14
8	6.5	35	35
17	18	16	15.5
37	37	28	31.5
4	1.5	9	8.5
8	6.5	60	42
28	31.5	29	33
27	30	16	15.5
11	11	22	23
60	42	60	42
25	28	17	18
5	3.5	60	42
13	13	32	34
22	23	$T_2 = 572$	
11	11		
17	18		
9	8.5		
4	1.5		
$T_1 = 463$			

The unstandardized test statistic is $T = T_1 = 463$. To calculate z, we first determine the mean and standard deviation of T. Note that $n_1 = 25$ and $n_2 = 20$.

$$E(T) = \frac{n_1(n_1 + n_2 + 1)}{2} = \frac{25(46)}{2} = 575$$

$$\sigma_T = \sqrt{\frac{n_1 n_2(n_1 + n_2 + 1)}{12}} = \sqrt{\frac{(25)(20)(46)}{12}} = 43.8$$

The standardized test statistic is

$$z = \frac{T - E(T)}{\sigma_T} = \frac{463 - 575}{43.8} = -2.56$$

$$p\text{-value} = 2P(Z < -2.56) = 2(1 - .9948) = .0104$$

EXCEL

	A	B	C	D	E
1	**Wilcoxon Rank Sum Test**				
2					
3			Rank Sum	Observations	
4	*Business*		463	25	
5	*Non-Bus*		572	20	
6	z Stat		-2.56		
7	P(Z<=z) one-tail		0.0053		
8	z Critical one-tail		1.6449		
9	P(Z<=z) two-tail		0.0106		
10	z Critical two-tail		1.96		

MINITAB

Mann-Whitney Test and CI: Business, Non-Bus

```
          N   Median
Business  25   17.00
Non-Bus   20   26.50


Point estimate for ETA1-ETA2 is -11.00
95.2 Percent CI for ETA1-ETA2 is (-19.00,-3.00)
W = 463.0
Test of ETA1 = ETA2 vs ETA1 not = ETA2 is significant at 0.0109
The test is significant at 0.0107 (adjusted for ties)
```

INTERPRET

There is strong evidence to infer that the duration of employment is different for business and nonbusiness graduates. The data cannot tell us the cause of this conclusion. For example, we don't know whether business graduates are in greater demand, making it more likely that such employees will leave for better jobs, or whether nonbusiness graduates are more satisfied with their jobs and thus remain longer. Moreover, we don't know what the results would have been had we surveyed employees 10 years after they were employed.

Required Conditions

The Wilcoxon rank sum test (like the other nonparametric tests presented in this chapter) actually tests to determine whether the population *distributions* are identical. This means that it tests not only for identical locations, but for identical spreads (variances) and shapes (distributions) as well. Unfortunately this means that the rejection of the null hypothesis may not necessarily signify a difference in population locations. The rejection of the null hypothesis may be due instead to a difference in distribution shapes and/or spreads. To avoid this problem, we will require that the two probability distributions be identical except with respect to location, which then becomes the sole focus of the test. This requirement is made for the tests introduced in the next three sections (sign test and Wilcoxon signed rank sum test, Kruskal-Wallis test, and Friedman test).

Both histograms (Figures 19.5 and 19.6) are approximately bimodal. Although there are differences between them, it would appear that the required condition for the use of the Wilcoxon rank sum test is roughly satisfied in the example about retaining workers.

Developing an Understanding of Statistical Concepts

When applying nonparametric techniques, we do not perform any calculations using the original data. Instead, we perform computations only on the ranks. (We determine the rank sums and use them to make our decision.) As a result, we do not care about the actual distribution of the data (hence the name *distribution-free techniques*), and we do not specify parameters in the hypotheses (hence the name nonparametric techniques). Although there are other techniques that do not specify parameters in the hypotheses, we use the term *nonparametric* for procedures that feature these concepts.

Here is a summary of how to identify the Wilcoxon rank sum test.

Factors That Identify the Wilcoxon Rank Sum

1. **Problem objective:** Compare two populations
2. **Data type:** Ordinal or interval but nonnormal
3. **Experimental design:** Independent samples

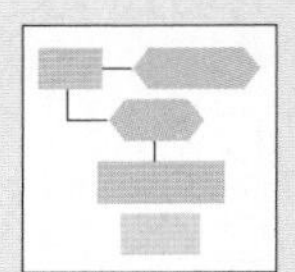

EXERCISES

Developing an Understanding of Statistical Concepts

Exercises 19.1–19.2 are "what-if analyses" designed to determine what happens to the test statistics and p-values when elements of the statistical inference change. These problems can be solved manually or using the ***Wilcoxon Rank Sum Test*** *worksheet in the* ***Test Statistics*** *workbook.*

19.1 a. Given the following statistics, calculate the value of the test statistic to determine whether the population locations differ.

$T_1 = 250$ $n_1 = 15$

$T_2 = 215$ $n_2 = 15$

b. Repeat Part a with $T_1 = 275$ and $T_2 = 190$.

c. Describe the effect on the test statistic of increasing T_1 to 275.

19.2 a. From the following statistics, test (with $\alpha = .05$) to determine whether the location of population 1 is to the right of the location of population 2.

$T_1 = 1{,}205$ $n_1 = 30$

$T_2 = 1{,}280$ $n_2 = 40$

b. Repeat Part a with $T_1 = 1{,}065$.

c. Discuss the effect on the test statistic and p-value of decreasing T_1 to 1,065.

19.3 Use the Wilcoxon rank sum test on the following data to determine whether the location of population 1 is to the left of the location of population 2. (Use $\alpha = .05$.)

Sample 1: 75 60 73 66 81

Sample 2: 90 72 103 82 78

19.4 Xr19-04 Use the Wilcoxon rank sum test on the following data to determine whether the two population locations differ. (Use a 10% significance level.)

Sample 1: 15 7 22 20 32 18 26 17 23 30

Sample 2: 8 27 17 25 20 16 21 17 10 18

Exercises 19.5 to 19.16 require the use of a computer and software. ***Conduct tests of hypotheses at the 5% significance level.***

19.5 a. Xr19-05a In a taste test of a new beer, 25 people rated the new beer and another 25 rated the leading brand on the market. The possible ratings were Poor, Fair, Good, Very good, and Excellent. The responses for the new beer and the leading beer were stored using a 1–2–3–4–5 coding system. Can we infer that the new beer is less highly rated than the leading brand?

b. Xr19-05b The responses were recoded so that 3 = Poor, 8 = Fair, 22 = Good, 37 = Very good, and 55 = Excellent. Can we infer that the new beer is less highly rated than the leading brand?

c. What does this exercise tell you about ordinal data?

19.6 a. Xr19-06a To determine whether the satisfaction rating of an airline differs between business class and economy class, a survey was performed. Random samples of both groups were asked to rate their satisfaction with the quality of service using the following responses:

Very satisfied
Quite satisfied
Somewhat satisfied
Neither satisfied nor dissatisfied
Somewhat dissatisfied
Quite dissatisfied
Very dissatisfied

Using a 7–6–5–4–3–2–1 coding system, the results were recorded. Can we infer that business and economy class differ in their degree of satisfaction with the service?

b. Xr19-06b The responses were recoded using the values 88–67–39–36–25–21–18. Can we infer that business and economy class differ in their degree of satisfaction with the airline?

c. What is the effect of changing the codes? Why was this expected?

19.7 a. Xr19-07 Refer to Example 19.2. Suppose that the responses were coded as follows:

100 = The drug was extremely effective.
60 = The drug was quite effective.
40 = The drug was somewhat effective.
35 = The drug was slightly effective.
10 = The drug was not at all effective.

Determine whether we can infer that the new painkiller is more effective than aspirin.

b. Why are the results of Example 19.2 and Part a identical?

Applications

Exercises 19.8–19.16 may be solved manually. See Appendix A for the sample statistics.

19.8 Xr19-08 A survey of statistics professors asked them to rate the importance of teaching nonparametric techniques. The possible responses are

Very important
Quite important
Somewhat important
Not too important
Not important at all

The professors were classified as either a member of the Mathematics Department or a member of some other department. The responses were coded (codes 5, 4, 3, 2, and 1, respectively) and recorded. Can we infer that members of the Mathematics Department rate nonparametric techniques as more important than do members of other departments?

19.9 Xr19-09 In recent years, insurance companies offering medical coverage have given discounts to companies that are committed to improving the health of their employees. To help determine whether this policy is reasonable, the general manager of one large insurance company organized a study of a random sample of 30 workers who regularly participate in their company's lunchtime exercise program and 30 workers who do not. Over a 2-year period he observed the total dollar amount of medical expenses for each individual. Can the manager conclude that companies that provide exercise programs should be given discounts?

19.10 Xr19-10 Feminist organizations often use the issue of who does the housework in two-career families as a gauge of equality. Suppose that a study was undertaken and a random sample of 125 two-career families was taken. The wives were asked to report the number of hours of housework they performed the previous week. The results, together with the responses from a survey performed last year (with a different sample of two-career families), were recorded. Can we conclude that women are doing less housework today than last year?

19.11 Xr19-11 The American public's support for the space program is important for the program's continuation and for the financial health of the aerospace industry. In a poll conducted by the Gallup organization last year, a random sample of 100 Americans was asked, "Should the amount of money being spent on the space program be increased or kept at current levels (3), decreased (2), or ended altogether (1)?" The survey was conducted again this year. The results were recorded using the codes in parentheses. Can we conclude that public support decreased between this year and last year?

19.12 Xr19-12 Certain drugs differ in their side effects depending on the gender of the patient. In a study to determine whether men or women suffer more serious side effects when taking a powerful penicillin substitute, 50 men and 50 women were given the drug. Each was asked to evaluate the level of stomach upset on a 4-point scale, where 4 = extremely upset, 3 = somewhat upset, 2 = not too upset, and 1 = not upset at all. Can we conclude that men and women experience different levels of stomach upset from the drug?

19.13 Xr19-13 The president of Tastee Inc., a baby-food producer, claims that her company's product is superior to that of her leading competitor because babies gain weight faster with her product. As an experiment, 40 healthy newborn infants are randomly selected. For two months, 15 of the babies are fed Tastee baby food and the other 25 are fed the competitor's product. Each baby's weight gain (in ounces) was recorded. If we use weight gain as our criterion, can we conclude that Tastee baby food is indeed superior? (This exercise is identical to Exercise 13.11 except for the data).

19.14 Xr19-14 Do the ways that women dress influence the ways that other women judge them? This question was addressed by a researcher at Ohio State University (*Working Mother*, April 1992). The experiment consisted of asking women to rate how professional two women looked. One woman wore a size 6 dress and the other wore a size 14. Suppose that the researcher asked 20 women to rate the woman wearing the size 6 dress and another 20 to rate the woman wearing the size 14 dress. The ratings were as follows:

4 = highly professional
3 = somewhat professional

2 = not very professional

1 = not at all professional

Do these data provide sufficient evidence to infer that women perceive another woman wearing a size 6 dress as more professional than one wearing a size 14 dress?

19.15 Xr19-15 The image of the lowly prune is not very good. It is perceived as a product used by seniors to help avoid constipation. However, in reality it is a nutritious and (for many) a tasty treat. To help improve the image of the prune, a company that produces the product decided to see the effect of changing its name to dried plums (which is what a prune is). To gauge the effect, a random sample of shoppers was asked how likely it was that they would purchase the product. Half the sample was shown a package that identified its contents as prunes. The other half was shown packages labeled dried plums. The responses are

Highly unlikely (1)

Somewhat unlikely (2)

Somewhat likely (3)

Highly likely (4)

a. Can we infer from these data that changing the name of prunes to dried plums increases the likelihood that shoppers will buy the product?

b. Write a report to the marketing manager of the company and describe your findings.

19.16 Xr19-16 Burger King Restaurants regularly survey customers to determine how well they are doing. Suppose that a survey asked customers to rate (among other things) the speed of service. The responses are

1 = Poor

2 = Good

3 = Very good

4 = Excellent

The responses for the day shift and night shift were recorded. Can we infer that night shift customers rate the service differently than the day shift?

19.2 Sign Test and Wilcoxon Signed Rank Sum Test

In the preceding section, we discussed the nonparametric technique for comparing two populations of data that are either ordinal or interval (nonnormal) and where the data are independently drawn. In this section, the problem objective and data type remain as they were in Section 19.1, but we will be working with data generated from a matched pairs experiment. We have dealt with this type of experiment before. In Section 13.3, we dealt with the mean of the paired differences represented by the parameter μ_D. In this section, we introduce two nonparametric techniques that test hypotheses in problems with the following characteristics:

1. The problem objective is to compare two populations.
2. The data are either ordinal or interval (where the normality requirement necessary to perform the parametric test is unsatisfied).
3. The samples are matched pairs.

To extract all the potential information from a matched pairs experiment, we must create the matched pair differences. Recall that we did so when conducting the t-test and estimate of μ_D. We then calculated the mean and standard deviation of these differences and determined the test statistic and confidence interval estimator. The first step in both nonparametric methods presented here is the same: Compute the differences for each pair of observations. However, if the data are ordinal, we cannot perform any calculations on those differences because differences between ordinal values have no meaning.

To understand this point, consider comparing two populations of responses of people rating a product or service. The responses are "excellent," "good," "fair," and "poor." Recall that we can assign any numbering system as long as the order is maintained. The simplest system is 4–3–2–1. However, any other system such as

66–38–25–11 (or another set of numbers of decreasing order) is equally valid. Now suppose that in one matched pair the sample 1 response was "excellent" and the sample 2 response was "good." Calculating the matched pairs difference under the 4–3–2–1 system gives a difference of $4 - 3 = 1$. Using the 66–38–25–11 system gives a difference of $66 - 38 = 28$. If we treat this and other differences as real numbers, we are likely to produce different results depending on which numbering system we used. Thus, we cannot use any method that uses the actual differences. However, we can use the sign of the differences. In fact, when the data are ordinal, that is the only method that is valid. In other words, no matter what numbering system is used, we know that "excellent" is better than "good." In the 4–3–2–1 system the difference between "excellent" and "good" is +1. In the 66–38–25–11 system the difference is +28. If we ignore the magnitude of the number and record only the sign, the two numbering systems (and all other systems where the rank order is maintained) will produce exactly the same result.

As you will shortly discover, the sign test uses only the sign of the differences. That's why it's called the sign test.

When the data are interval, however, differences have real meaning. Although we can use the sign test when the data are interval, doing so results in a loss of potentially useful information. For example, knowing that the difference in sales between two matched used-car salespeople is 25 cars is much more informative than simply knowing that the first salesperson sold more cars than the second salesperson. As a result, when the data are interval, but not normal, we will use the *Wilcoxon signed rank sum test*, which incorporates not only the sign of the difference (hence the name), but the magnitude as well.

Sign Test

The **sign test** is employed in the following situations:

1. The problem objective is to compare two populations.
2. The data are ordinal.
3. The experimental design is matched pairs.

Test Statistic and Sampling Distribution

The sign test is quite simple. For each matched pair, we calculate the difference between the observation in sample 1 and the related observation in sample 2. We then count the number of positive differences and the number of negative differences. If the null hypothesis is true, we expect the number of positive differences to be approximately equal to the number of negative differences. Expressed another way, we expect the number of positive differences and the number of negative differences each to be approximately equal to half the total sample size. If either number is too large or too small, we reject the null hypothesis. By now you know that the determination of what is too large or too small comes from the sampling distribution of the test statistic. We will arbitrarily choose the test statistic to be the number of positive differences, which we denote x. The test statistic x is a binomial random variable, and under the null hypothesis, the binomial proportion is $p = .5$. Thus, the sign test is none other than the z-test of p introduced in Section 12.3.

Recall from Sections 7.4 and 9.2 that x is binomially distributed and that, for sufficiently large n, x is approximately normally distributed with mean $\mu = np$ and standard deviation $\sqrt{np(1-p)}$. Thus, the standardized test statistic is

$$z = \frac{x - np}{\sqrt{np(1-p)}}$$

The null hypothesis

H_0: The two population locations are the same

is equivalent to testing

H_0: $p = .5$

Therefore, the test statistic, assuming that the null hypothesis is true, becomes

$$z = \frac{x - np}{\sqrt{np(1-p)}} = \frac{x - .5n}{\sqrt{n(.5)(.5)}} = \frac{x - .5n}{.5\sqrt{n}}$$

The normal approximation of the binomial distribution is valid when $np \geq 5$ and $n(1-p) \geq 5$. When $p = .5$,

$$np = n(.5) \geq 5$$

and

$$n(1-p) = n(1-.5) = n(.5) \geq 5$$

implies that n must be greater than or equal to 10. Thus, this is one of the required conditions of the sign test. However, the quality of the inference with so small a sample size is poor. Larger sample sizes are recommended and will be used in the examples and exercises that follow.

It is common practice in this type of test to eliminate the matched pairs of observations when the differences equal 0. Consequently, n equals the number of nonzero differences in the sample.

EXAMPLE 19.3

DATA
Xm19-03

Comparing the Comfort of Two Midsize Cars

In an experiment to determine which of two cars is perceived to have the more comfortable ride, 25 people rode (separately) in the back seat of an expensive European model and also in the back seat of a North American midsize car. Each of the 25 people was asked to rate the ride on the following 5-point scale:

1 = Ride is very uncomfortable.
2 = Ride is quite uncomfortable.
3 = Ride is neither uncomfortable nor comfortable.
4 = Ride is quite comfortable.
5 = Ride is very comfortable.

The results are shown here. Do these data allow us to conclude at the 5% significance level that the European car is perceived to be more comfortable than the North American car?

Respondent	Comfort Ratings	
	European Car	North American Car
1	3	4
2	2	1
3	5	4
4	3	2
5	2	1
6	5	3
7	2	3
8	4	2
9	4	2
10	2	2
11	2	1
12	3	4
13	2	1
14	3	4
15	2	1
16	4	3
17	5	4
18	2	3
19	5	4
20	3	1
21	4	2
22	3	3
23	3	4
24	5	2
25	5	3

SOLUTION

IDENTIFY

The problem objective is to compare two populations of ordinal data. Because the same 25 people rated both cars, we recognize the experimental design as matched pairs. The sign test is applied, with the following hypotheses:

H_0: The two population locations are the same

H_1: The location of population 1 (European car rating) is to the right of the location of population 2 (North American car rating)

COMPUTE

MANUALLY

The rejection region is

$$z > z_\alpha = z_{.05} = 1.645$$

To calculate the value of the test statistic, we calculate the paired differences and count the number of positive, negative, and zero differences. The matched pairs differences are

−1 1 1 1 1 2 −1 2 2 0 1 −1 1

−1 1 1 1 −1 1 2 2 0 −1 3 2

There are 17 positive, 6 negative, and 2 zero differences. Thus, $x = 17$ and $n = 23$. The value of the test statistic is

$$z = \frac{x - .5n}{.5\sqrt{n}} = \frac{17 - .5(23)}{.5\sqrt{23}} = 2.29$$

Because the test statistic is normally distributed, we can calculate the p-value of the test:

$$p\text{-value} = P(Z > 2.29) = 1 - .9890 = .0110$$

EXCEL

	A	B	C	D	E
1	**Sign Test**				
2					
3	Difference			*European - American*	
4					
5	Positive Differences			17	
6	Negative Differences			6	
7	Zero Differences			2	
8	z Stat			2.29	
9	P(Z<=z) one-tail			0.0109	
10	z Critical one-tail			1.6449	
11	P(Z<=z) two-tail			0.0218	
12	z Critical two-tail			1.96	

INSTRUCTIONS

1. Type or import the data into two adjacent columns. (Open Xm19-03.)
2. Click **Add-Ins, Data Analysis Plus,** and **Sign Test.**
3. Specify the **Variable 1 Range** (A1:A26), the **Variable 2 range** (B1:B26), and the value of α (.05).

MINITAB

Sign Test for Median: Differences

Sign test of median = 0.00000 versus > 0.00000

	N	Below	Equal	Above	P	Median
Differences	25	6	2	17	0.0173	1.000

Minitab prints the number of differences that are negative (Below), the number of zero differences (Equal), and the number of positive differences (Above). The p-value .0173 is based on the actual distribution of the number of positive differences, which is binomial.

INSTRUCTIONS

1. Type or import the data into two columns. (Open Xm19-03.)
2. Create a new variable, the paired difference.
3. Click **Stat, Nonparametrics,** and **1-Sample Sign . . . .**
4. Type or select the new variable in the **Variables** box, select **Test median** and type **0**, and specify the alternative hypothesis in the **Alternative** box (greater than).

INTERPRET

There is relatively strong evidence to indicate that people perceive the European car to provide a more comfortable ride than the North American car. There are, however, two aspects of the experiment that may detract from the conclusion that European cars provide a more comfortable ride. First, did the respondents know in which car they were riding? If so, they may have answered on their preconceived bias that European cars are more expensive and therefore better. If the subjects were blindfolded, we would be more secure in our conclusion. Second, was the order in which each subject rode the two cars varied? If all the subjects rode in the North American car first and the European car second, that may have influenced their ratings. The experiment should have been conducted so that the car each subject rode in first was randomly determined.

Checking the Required Conditions

As we noted in Section 19.1, the sign test requires that the populations be identical in shape and spread. The histogram of the ratings for the European car (Figure 19.7) suggests that the ratings may be uniformly distributed between 2 and 5. The histogram of the ratings for the North American car (Figure 19.8) seems to indicate that the ratings are uniformly distributed between 1 and 4. Thus both sets of ratings have the same shape and spread, but their locations differ. The other condition is that the sample size exceeds 10.

FIGURE **19.7** Histogram of Ratings of European Car in Example 19.3

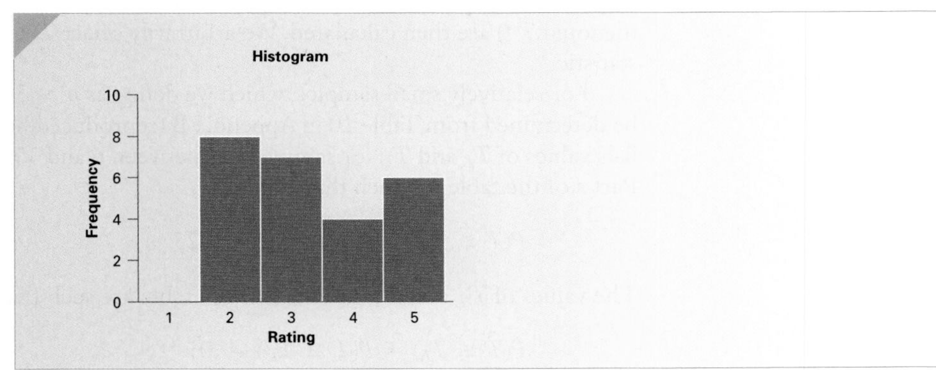

FIGURE **19.8** Histogram of Ratings of North American Car in Example 19.3

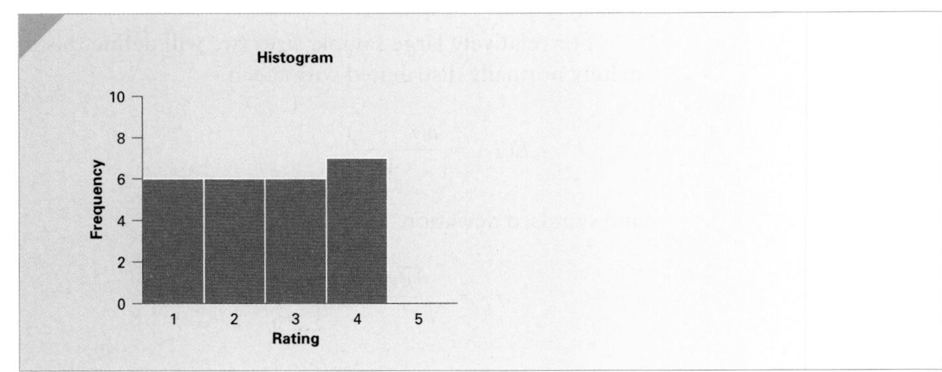

Wilcoxon Signed Rank Sum Test

The **Wilcoxon signed rank sum test** is used under the following circumstances:

1. The problem objective is to compare two populations.
2. The data (matched pairs differences) are interval, but not normally distributed.
3. The samples are matched pairs.

The Wilcoxon signed rank sum test is the nonparametric counterpart of the t-test of μ_D. Because the data are interval, we can refer to the Wilcoxon signed rank sum test as a test of μ_D. However, to be consistent with the other nonparametric techniques and to avoid confusion, we will express the hypotheses to be tested in the same way as we did in Section 19.1.

Test Statistic and Sampling Distribution

We begin by computing the paired differences. As we did in the sign test, we eliminate all differences that are equal to 0. Next we rank the absolute values of the nonzero differences where 1 = smallest value and n = largest value, with n = number of nonzero differences. (We average the ranks of tied observations.) The sum of the ranks of the positive differences (denoted T^+) and the sum of the ranks of the negative differences (denoted T^-) are then calculated. We arbitrarily select T^+, which we label T, as our test statistic.

For relatively small samples, which we define as $n \leq 30$, the critical values of T can be determined from Table 10 in Appendix B (reproduced here as Table 19.4). This table lists values of T_L and T_U for sample sizes between 6 and 30. The values of T_L and T_U in Part a of the table are such that

$$P(T \leq T_L) = P(T \geq T_U) = .025$$

The values of T_L and T_U in Part b of the table are such that

$$P(T \leq T_L) = P(T \geq T_U) = .05$$

Part a is used either in a two-tail test with $\alpha = .05$ or in a one-tail test with $\alpha = .025$. Part b is employed either in a two-tail test with $\alpha = .10$ or in a one-tail test with $\alpha = .05$.

For relatively large sample sizes (we will define this to mean $n > 30$), T is approximately normally distributed with mean

$$E(T) = \frac{n(n+1)}{4}$$

and standard deviation

$$\sigma_T = \sqrt{\frac{n(n+1)(2n+1)}{24}}$$

Thus, the standardized test statistic is

$$z = \frac{T - E(T)}{\sigma_T}$$

TABLE **19.4** Critical Values of the Wilcoxon Signed Rank Sum Test

	(a) $\alpha = .025$ One-Tail $\alpha = .05$ Two-Tail		(b) $\alpha = .05$ One-Tail $\alpha = .10$ Two-Tail	
n	T_L	T_U	T_L	T_U
6	1	20	2	19
7	2	26	4	24
8	4	32	6	30
9	6	39	8	37
10	8	47	11	44
11	11	55	14	52
12	14	64	17	61
13	17	74	21	70
14	21	84	26	79
15	25	95	30	90
16	30	106	36	100
17	35	118	41	112
18	40	131	47	124
19	46	144	54	136
20	52	158	60	150
21	59	172	68	163
22	66	187	75	178
23	73	203	83	193
24	81	219	92	208
25	90	235	101	224
26	98	253	110	241
27	107	271	120	258
28	117	289	130	276
29	127	308	141	294
30	137	328	152	313

EXAMPLE 19.4

DATA Xm19-04

Comparing Flextime and Fixed Time Schedules

Traffic congestion on roads and highways costs industry billions of dollars annually as workers struggle to get to and from work. Several suggestions have been made about how to improve this situation, one of which is called *flextime*, which involves allowing workers to determine their own schedules (provided they work a full shift). Such workers will likely choose an arrival and departure time to avoid rush-hour traffic. In a preliminary experiment designed to investigate such a program, the general manager of a large company wanted to compare the times it took workers to travel from their homes to work at 8:00 A.M. with travel time under the flextime program. A random sample of 32 workers was selected. The employees recorded the time (in minutes) it took to arrive at work at 8:00 A.M. on Wednesday of one week.

The following week, the same employees arrived at work at times of their own choosing. The travel time on Wednesday of that week was recorded. These results are listed in the following table. Can we conclude at the 5% significance level that travel times under the flextime program are different from travel times to arrive at work at 8:00 A.M.?

	Travel Time	
Worker	Arrival at 8:00 A.M.	Flextime Program
1	34	31
2	35	31
3	43	44
4	46	44
5	16	15
6	26	28
7	68	63
8	38	39
9	61	63
10	52	54
11	68	65
12	13	12
13	69	71
14	18	13
15	53	55
16	18	19
17	41	38
18	25	23
19	17	14
20	26	21
21	44	40
22	30	33
23	19	18
24	48	51
25	29	33
26	24	21
27	51	50
28	40	38
29	26	22
30	20	19
31	19	21
32	42	38

SOLUTION

IDENTIFY

The objective is to compare two populations; the data are interval and were produced from a matched pairs experiment. If matched pairs differences are normally distributed, we should apply the t-test of μ_D. To judge whether the data are normal, we computed the paired differences and drew the histogram (actually Excel did). Figure 19.9 depicts this histogram. Apparently, the normality requirement is not satisfied, indicating that we should employ the Wilcoxon signed rank sum test.

FIGURE **19.9** Histogram of the Differences for Example 19.4

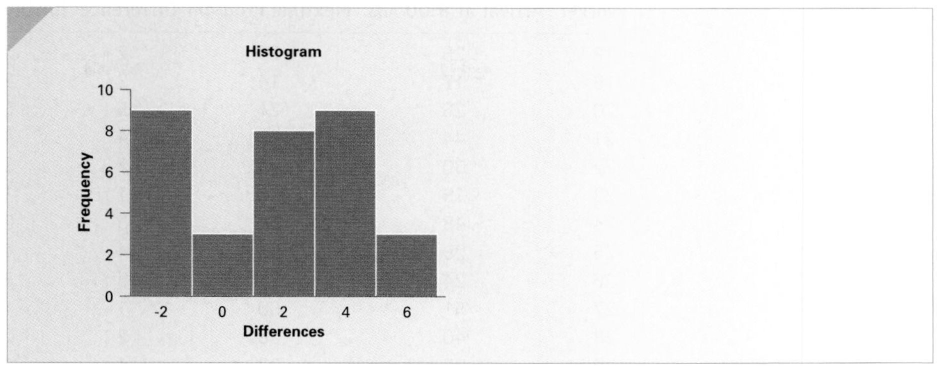

Because we want to know whether the two groups of times differ, we perform a two-tail test whose hypotheses are

H_0: The two population locations are the same

H_1: The location of population 1 (travel times for current work schedule) is different from the location of population 2 (travel times for flextime program)

COMPUTE

MANUALLY

For each worker, we compute the difference between travel time with arrival at 8:00 A.M. and travel time under flextime.

	Travel Time					
Worker	Arrival at 8:00 A.M.	Flextime Program	Difference	|Difference|	Rank	|Rank|
1	34	31	3	3	21.0	
2	35	31	4	4	27.0	
3	43	44	−1	1		4.5
4	46	44	2	2	13.0	
5	16	15	1	1	4.5	
6	26	28	−2	2		13.0
7	68	63	5	5	31.0	
8	38	39	−1	1		4.5
9	61	63	−2	2		13.0
10	52	54	−2	2		13.0
11	68	65	3	3	21.0	
12	13	12	1	1	4.5	
13	69	71	−2	2		13.0
14	18	13	5	5	31.0	
15	53	55	−2	2		13.0
16	18	19	−1	1		4.5
17	41	38	3	3	21.0	

(continued)

	Travel Time					
Worker	Arrival at 8:00 A.M.	Flextime Program	Difference	|Difference|	Rank	|Rank|
18	25	23	2	2	13.0	
19	17	14	3	3	21.0	
20	26	21	5	5	31.0	
21	44	40	4	4	27.0	
22	30	33	−3	3		21.0
23	19	18	1	1	4.5	
24	48	51	−3	3		21.0
25	29	33	−4	4		27.0
26	24	21	3	3	21.0	
27	51	50	1	1	4.5	
28	40	38	2	2	13.0	
29	26	22	4	4	27.0	
30	20	19	1	1	4.5	
31	19	21	−2	2		13.0
32	42	38	4	4	27.0	
					$T^+ = 367.5$	$T^- = 160.5$

The differences and the absolute values of the differences are calculated. We rank the absolute differences. (If there were any zero differences, we would eliminate them before ranking the absolute differences.) Ties are resolved by calculating the averages. The ranks of the negative differences are offset to facilitate the summing of the ranks. The rank sums of the positive and negative differences are

$$T^+ = 367.5 \quad \text{and} \quad T^- = 160.5$$

The test statistic is

$$z = \frac{T - E(T)}{\sigma_T}$$

where

$$T = T^+ = 367.5$$

$$E(T) = \frac{n(n+1)}{4} = \frac{32(33)}{4} = 264$$

$$\sigma_T = \sqrt{\frac{n(n+1)(2n+1)}{24}} = \sqrt{\frac{32(33)(65)}{24}} = 53.48$$

Thus,

$$z = \frac{T - E(T)}{\sigma_T} = \frac{367.5 - 264}{53.48} = 1.94$$

The rejection region is

$$z < -z_{\alpha/2} = -z_{.025} = -1.96 \text{ or } z > z_{\alpha/2} = z_{.025} = 1.96$$

The p-value is $2P(Z > 1.94) = 2(1 - .9738) = .0524$

EXCEL

	A	B	C	D
1	**Wilcoxon Signed Rank Sum Test**			
2				
3	Difference		*8:00-Arr - Flextime*	
4				
5	T+		367.5	
6	T-		160.5	
7	Observations (for test)		32	
8	z Stat		1.94	
9	P(Z<=z) one-tail		0.0265	
10	z Critical one-tail		1.6449	
11	P(Z<=z) two-tail		0.0530	
12	z Critical two-tail		1.96	

INSTRUCTIONS

1. Type or import the data into two adjacent columns. (Open Xm19-04.)
2. Click **Add-Ins**, **Data Analysis Plus,** and **Wilcoxon Signed Rank Sum Test.**
3. Specify the **Variable 1 Range** (A1:A33), the **Variable 2 Range** (B1:B33), and the value of α (.05).

MINITAB

Wilcoxon Signed Rank Test: Difference

Test of median = 0.000000 versus median not = 0.000000

	N	N for Test	Wilcoxon Statistic	P	Estimated Median
Difference	32	32	367.5	0.054	1.000

The output includes the original sample size, the number of nonzero differences (N for Test), the value of T^+ (Wilcoxon Statistic 367.5), and the p-value (0.054).

INSTRUCTIONS

1. Type or import the data into two columns. (Open Xm19-04.)
2. Follow the same steps as in the sign test. At step 2, click **Stat, Nonparametrics,** and **1-Sample Wilcoxon . . . .**

INTERPRET

There is not enough evidence to infer that flextime commutes are different from the commuting times under the current schedule. This conclusion may be due primarily to the way in which this experiment was performed. All of the drivers recorded their travel time with 8:00 A.M. arrival on the first Wednesday and their flextime travel time on the second Wednesday. If the second day's traffic was heavier than usual, that may account for the conclusion reached. As we pointed out in Example 19.4, the order of schedules should have been randomly determined for each employee. In this way, the effect of varying traffic conditions could have been minimized.

Here is how we recognize when to use the two techniques introduced in this section.

Factors That Identify the Sign Test

1. **Problem objective:** Compare two populations
2. **Data type:** Ordinal
3. **Experimental design:** Matched pairs

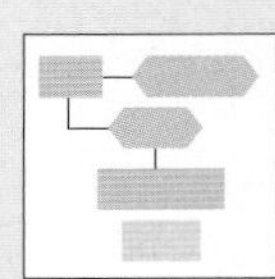

Factors That Identify the Wilcoxon Signed Rank Sum Test

1. **Problem objective:** Compare two populations
2. **Data type:** Interval
3. **Distribution of differences:** Nonnormal
4. **Experimental design:** Matched pairs

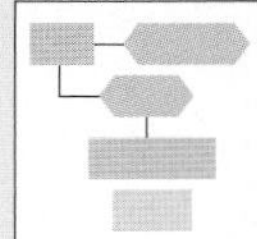

EXERCISES

19.17 In a matched pairs experiment, if we find 30 negative, 5 zero, and 15 positive differences, perform the sign test to determine whether the two population locations differ. (Use a 5% significance level.)

19.18 Suppose that in a matched pairs experiment we find 28 positive differences, 7 zero differences, and 41 negative differences. Can we infer at the 10% significance level that the location of population 1 is to the left of the location of population 2?

19.19 A matched pairs experiment yielded the following results:

Positive differences: 18

Zero differences: 0

Negative differences: 12

Can we infer at the 5% significance level that the location of population 1 is to the right of the location of population 2?

19.20 Xr19-20 Use the sign test on the following data to determine whether the location of population 1 is to the right of the location of population 2. (Use $\alpha = .05$.)

Pair	1	2	3	4	5	6	7	8	9	10	11	12	13	14	15	16
Sample 1	5	3	4	2	3	4	3	5	4	3	4	5	4	5	3	2
Sample 2	3	2	4	3	3	1	3	4	2	5	1	2	2	3	1	2

19.21 Given the following statistics from a matched pairs experiment, perform the Wilcoxon signed rank sum test to determine whether we can infer at the 5% significance level that the two population locations differ.

$T^+ = 660 \quad T^- = 880 \quad n = 55$

19.22 A matched pairs experiment produced the following statistics. Conduct a Wilcoxon signed rank sum test to determine whether the location of population 1 is to the right of the location of population 2. (Use $\alpha = .01$.)

$T^+ = 3{,}457 \quad T^- = 2{,}429 \quad n = 108$

19.23 Perform the Wilcoxon signed rank sum test for the following matched pairs to determine whether the two population locations differ. (Use $\alpha = .10$.)

Pair	1	2	3	4	5	6
Sample 1	9	12	13	8	7	10
Sample 2	5	10	11	9	3	9

19.24 Xr19-24 Perform the Wilcoxon signed rank sum test to determine whether the location of population 1 differs from the location of population 2 given the data shown here. (Use $\alpha = .05$.)

Pair	1	2	3	4	5	6	7	8	9	10	11	12
Sample 1	18.2	14.1	24.5	11.9	9.5	12.1	10.9	16.7	19.6	8.4	21.7	23.4
Sample 2	18.2	14.1	23.6	12.1	9.5	11.3	9.7	17.6	19.4	8.1	21.9	21.6

Exercises 19.25–19.39 require the use of a computer and software. ***Use a 5% significance level unless specified otherwise.***

Developing an Understanding of Statistical Concepts

19.25 a. Xr19-25a In a taste test of a new beer, 100 people rated the new beer and the leading brand on the market. The possible ratings were Poor, Fair, Good, Very good, and Excellent. The responses for the new beer and the leading beer were recorded using a 1–2–3–4–5 coding system. Can we infer that the new beer is more highly rated than the leading brand?

b. Xr19-25b The responses were recoded so that 3 = Poor, 8 = Fair, 22 = Good, 37 = Very good, and 55 = Excellent. Can we infer that the new beer is more highly rated than the leading brand?

c. Why are the answers to Parts a and b identical?

19.26 a. Xr19-26a A random sample of 50 people was asked to rate two brands of ice cream using the following responses:

Delicious

OK

Not bad

Terrible

The responses were converted to codes 4, 3, 2, and 1, respectively. Can we infer that Brand A is preferred?

b. Xr19-26b The responses were recoded using the values 28–25–16–3. Can we infer that Brand A is preferred?

c. Compare your answers for Parts a and b. Are they identical? Explain why?

19.27 Xr19-27 Refer to Example 19.3. Suppose that the responses have been recoded in the following way:

6 = Ride is very uncomfortable.

24 = Ride is quite uncomfortable.

28 = Ride is neither uncomfortable nor comfortable.

53 = Ride is quite comfortable.

95 = Ride is very comfortable.

a. Do these data allow us to conclude that the European car is perceived to be more comfortable than the North American car?

b. Compare your answer with that obtained in Example 19.3. Explain why the results are identical

19.28 a. Xr19-28 Data from a matched pairs experiment were recorded. Use the sign test to determine whether the population locations differ.

b. Repeat Part a using the Wilcoxon signed rank sum test.

c. Why do the answers to Parts a and b differ?

19.29 a. Data from a matched pairs experiment were recorded. Use the sign test to determine whether the population locations differ.

b. Repeat Part a using the Wilcoxon signed rank sum test.

c. Why do the results of Parts a and b differ?

Applications

Exercises 19.30–19.39 may be solved manually. See Appendix A for the sample statistics.

19.30 Xr19-30 Research scientists at a pharmaceutical company have recently developed a new nonprescription sleeping pill. They decide to test its effectiveness by measuring the time it takes for people to fall asleep after taking the pill. Preliminary analysis indicates that the time to fall asleep varies considerably from one person to another. Consequently, they organize the experiment in the following way. A random sample of 100 volunteers who regularly suffer from insomnia is chosen. Each person is given one pill containing the newly developed drug and one placebo. (A placebo is a pill that contains absolutely no medication.) Participants are told to take one pill one night and the second pill one night a week later. (They do not know whether the pill they are taking is the placebo or the new drug, and the order of use is random.) Each participant is fitted with a device that measures the time until sleep occurs. Can we conclude that the new drug is effective? (This exercise is identical to Exercise 13.92 except for the data.)

19.31 Xr19-31 Suppose the housework study referred to in Exercise 19.10 was repeated with some changes. In the revised experiment, 60 women were asked last year and again this year how many hours of housework they perform weekly. Can we conclude at the 1% significance level that women as a group are doing less housework now than last year?

19.32 Xr19-32 At the height of the energy shortage during the 1970s, governments were actively seeking ways to persuade consumers to reduce their energy consumption. Among other efforts undertaken, several advertising campaigns were launched. To provide input on how to design effective advertising messages, a poll was taken in which people were asked how concerned they were about shortages of gasoline and electricity. There were four possible responses to the questions:

Not concerned at all (1)

Not too concerned (2)

Somewhat concerned (3)

Very concerned (4)

A poll of 150 individuals was undertaken. Do these data provide enough evidence to allow us to infer that concern about a gasoline shortage exceeded concern about an electricity shortage?

19.33 Xr19-33 A locksmith is in the process of selecting a new key-cutting machine. If there is a difference in key-cutting speed between the two machines under consideration, he will purchase the faster one. If there is no difference, he will purchase the cheaper machine. The times (in seconds) required to cut each of the 35 most common types of keys were recorded. What should he do?

19.34 Xr19-34 A large sporting-goods store located in Florida is planning a renovation that will result in an increase in the floor space for one department. The manager of the store has narrowed her choice about which department's floor space to increase to two possibilities: the tennis-equipment department or the swimming-accessories department. The manager would like to enlarge the tennis-equipment department because she believes that this department improves the overall image of the store. She decides, however, that if the swimming-accessories department can be shown to have higher gross sales, she will choose that department. She has collected each of the two departments' weekly gross sales data for the past 32 weeks. Which department should be enlarged?

19.35 Xr19-35 Does the brand name of an ice cream affect consumers' perceptions of it? The marketing manager of a major dairy pondered this question. She decided to ask 60 randomly selected people to taste the same flavor of ice cream in two different dishes. The dishes contained exactly the same ice cream but were labeled differently. One was given a name that suggested that its maker was European and sophisticated; the other was given a name that implied that the product was domestic and inexpensive. The tasters were asked to rate each ice cream on a 5-point scale, where 1 = poor, 2 = fair, 3 = good, 4 = very good, 5 = excellent. Do the results allow the manager to conclude at the 10% significance level that the European brand is preferred?

19.36 Xr19-36 Do children feel less pain than adults? That question was addressed by nursing professors at the University of Alberta and the University of Saskatchewan. Suppose that in a preliminary study, 50 8-year-old children and their mothers were subjected to moderately painful pressure on their hands. Each was asked to rate the level of pain as very severe (4), severe (3), moderate (2), or weak (1). The data were recorded using the codes in parentheses. Can we conclude at the 1% significance level that children feel less pain than adults?

19.37 Xr19-37 In a study to determine whether gender affects salary offers for graduating MBA students, 45 pairs of students were selected. Each pair consisted of a male and a female student who had almost identical grade-point averages, courses taken, ages, and previous work experience. The highest salary offered to each student upon graduation was recorded. Is there sufficient evidence to allow us to conclude that the salary offers differ between men and women? (This exercise is identical to Exercise 13.97 except for the data.)

19.38 Xr19-38 Admissions officers at universities and colleges face the problem of comparing grades achieved at different high schools. As a step toward developing a more informed interpretation of such grades, an admissions officer at a large state university conducts the following experiment. The records of 100 students from the same local high school (high school 1) who just completed their first year at the university were selected. Each of these students was paired (according to average grade in the last year of high school) with a student from another local high school (high school 2) who also just completed the first year at the university. For each matched pair, the average letter grades (4 = A, 3 = B, 2 = C, 1 = D, or 0 = F) in the first year of university study were recorded. Do these results allow us to conclude that, in comparing two students with the same high-school average (one from high school 1 and the other from high school 2), preference in admissions should be given to the student from high school 1?

19.39 Xr19-39 Some movie studios believe that by adding sexually explicit scenes to the home video version of a movie, they can increase the movie's appeal and profitability (*Wall Street Journal*, October 14, 1988). A studio executive decided to test this belief. She organized a study that involved 40 movies that were rated PG-13. Versions of each movie were created by adding scenes that changed the rating to R. The two versions of the movies were then made available to rental shops. For each of the 40 pairs of movies, the total number of rentals in one major city during a 1-week period was recorded.

a. Do these data provide enough evidence to support the belief?
b. As an analyst for a movie studio, write a report detailing the statistical analysis.

19.3 KRUSKAL-WALLIS TEST

In this section we introduce the first of two statistical procedures designed to compare two or more populations. The **Kruskal-Wallis test** is applied to problems with the following characteristics:

1. The problem objective is to compare two or more populations.
2. The data are either ordinal or interval, but nonnormal.
3. The samples are independent.

When the data are interval and normal, we use the analysis of variance F-test presented in Chapter 14 to determine whether differences exist. When the data are not normal, we will treat the data as if they were ordinal and employ the Kruskal-Wallis test.

The null and alternative hypotheses for this test are similar to those we specified in the analysis of variance. Because the data are ordinal or are treated as ordinal, however, we test population locations instead of population means. In all applications of the Kruskal-Wallis test, the null and alternative hypotheses are

H_0: The locations of all k populations are the same

H_1: At least two population locations differ

Here, k represents the number of populations to be compared.

Test Statistic

The test statistic is calculated in a way that closely resembles the way in which the Wilcoxon rank sum test was calculated. The first step is to rank all the observations. As before, 1 = smallest observation and n = largest observation, where $n = n_1 + n_2 + \cdots + n_k$. In case of ties, average the ranks.

If the null hypothesis is true, the ranks should be evenly distributed among the k samples. The degree to which this is true is judged by calculating the rank sums (labeled $T_1, T_2, \ldots, T_k$). The last step is to calculate the test statistic, which is denoted H.

Test Statistic for Kruskal-Wallis Test

$$H = \left[\frac{12}{n(n+1)} \sum_{j=1}^{k} \frac{T_j^2}{n_j}\right] - 3(n+1)$$

Although it is impossible to see from this formula, if the rank sums are similar, the test statistic will be small. As a result, a small value of H supports the null hypothesis. Conversely, if considerable differences exist between the rank sums, the test statistic will be large. To judge the value of H, we need to know its sampling distribution.

Sampling Distribution of the Test Statistic

The distribution of the test statistic can be derived in the same way we derived the sampling distribution of the test statistic in the Wilcoxon rank sum test. That is, we can list all possible combinations of ranks and their probabilities to yield the sampling distribution. A table of critical values can then be determined However, this is necessary only for small sample sizes. For sample sizes greater than or equal to 5, the test statistic H is approximately chi-squared distributed with $k - 1$ degrees of freedom. Recall that we introduced the chi-squared distribution in Section 8.4.

Rejection Region and *p*-Value

As we noted previously, large values of H are associated with different population locations. Consequently, we want to reject the null hypothesis if H is sufficiently large. Thus, the rejection region is

$$H > \chi^2_{\alpha,k-1}$$

and the p-value is

$$P(\chi^2 > H)$$

Figure 19.10 describes this sampling distribution, the rejection region, and the p-value.

FIGURE **19.10** Sampling Distribution of *H*

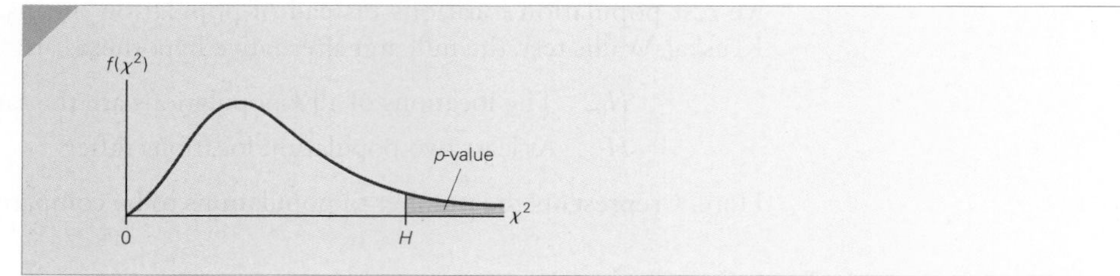

EXAMPLE 19.5

DATA Xm19-05

Comparing Quality in Three Shifts

The management of fast-food restaurants is extremely interested in knowing how their customers rate the quality of food and service and the cleanliness of the restaurants. Customers are given the opportunity to fill out customer comment cards. Suppose that one franchise wanted to compare how customers rate the three shifts (4:00 P.M. to midnight, midnight to 8:00 A.M., and 8:00 A.M. to 4:00 P.M.). In a preliminary study, 10 customer cards were randomly selected from each shift. The responses to the question concerning speed of service were recorded where 4 = excellent, 3 = good, 2 = fair, and 1 = poor and are listed here. Do these data provide sufficient evidence at the 5% significance level to indicate whether customers perceive the speed of service to be different between the three shifts?

4:00 P.M. to Midnight	Midnight to 8:00 P.M.	8:00 A.M. to 4:00 P.M.
4	3	3
4	4	1
3	2	3
4	2	2
3	3	1
3	4	3
3	3	4
3	3	2
2	2	4
3	3	1

SOLUTION

IDENTIFY

The problem objective is to compare three populations of ordinal data (the ratings of the three shifts), and the samples are independent. These factors are sufficient to justify the use of the Kruskal-Wallis test. The null and alternative hypotheses are

H_0: The locations of all three populations are the same

H_1: At least two population locations differ

COMPUTE

MANUALLY

The data are ranked in the same way as we ranked the data in the Wilcoxon rank sum test. Ties are resolved by assigning the average rank. When all the data have been ranked, the rank sums are computed.

Sample 1	Rank	Sample 2	Rank	Sample 3	Rank
4	27.0	3	16.5	3	16.5
4	27.0	4	27.0	1	2.0
3	16.5	2	6.5	3	16.5
4	27.0	2	6.5	2	6.5
3	16.5	3	16.5	1	2.0
3	16.5	4	27.0	3	16.5
3	16.5	3	16.5	4	27.0
3	16.5	3	16.5	2	6.5
2	6.5	2	6.5	4	27.0
3	16.5	3	16.5	1	2.0
	$T_1 = 186.5$		$T_2 = 156.0$		$T_3 = 122.5$

The value of the test statistic is

$$H = \left[\frac{12}{n(n+1)}\sum_{j=1}^{k}\frac{T_j^2}{n_j}\right] - 3(n+1)$$

$$= \frac{12}{n(n+1)}\left(\frac{T_1^2}{n_1} + \frac{T_2^2}{n_2} + \frac{T_3^2}{n_3}\right) - 3(n+1)$$

$$= \frac{12}{30(30+1)}\left(\frac{186.5^2}{10} + \frac{156^2}{10} + \frac{122.5^2}{10}\right) - 3(30+1)$$

$$= 2.64$$

The rejection region is

$$H > \chi^2_{\alpha,k-1} = \chi^2_{.05,2} = 5.99$$

EXCEL

	A	B	C	D
1	**Kruskal-Wallis Test**			
2				
3	*Group*	Rank Sum	Observations	
4	*4:00-mid*	186.5	10	
5	*Mid-8:00*	156	10	
6	*8:00-4:00*	122.5	10	
7				
8	H Stat		2.64	
9	df		2	
10	p-value		0.2665	
11	chi-squared Critical		5.9915	

INSTRUCTIONS

1. Type or import the data into adjacent columns. (Open Xm19-05.)
2. Click **Add-Ins, Data Analysis Plus,** and **Kruskal Wallis Test.**
3. Specify the **Input Range** (A1:C11) and the value of α (.05).

MINITAB

Kruskal-Wallis Test: Ratings Versus Shift

Kruskal-Wallis Test on Ratings

Shift	N	Median	Ave Rank	Z
1	10	3.000	18.7	1.39
2	10	3.000	15.6	0.04
3	10	2.500	12.3	-1.43
Overall	30		15.5	

H = 2.64 DF = 2 P = 0.267
H = 3.01 DF = 2 P = 0.222 (adjusted for ties)

INSTRUCTIONS

1. The data must be stacked so that the responses are in one column and the codes identifying the shift are in a second column.
2. Click **Stat, Nonparametrics,** and **Kruskal-Wallis . . . .**
3. Select the name of the **Response** variable and the name of **Factor** variable.

INTERPRET

There is not enough evidence to infer that a difference in speed of service exists between the three shifts. Management should assume that all three of the shifts are equally rated, and any action to improve service should be applied to all three shifts. Management should also bear in mind that the data were generated from a self-selected sample. (See the story of the *Literary Digest* presented in Chapter 5.)

Kruskal-Wallis Test and the Wilcoxon Rank Sum Test

When the Kruskal-Wallis test is used to test for a difference between two populations, it will produce the same outcome as the two-tail Wilcoxon rank sum test. However, the Kruskal-Wallis test can determine only whether a difference exists. To determine, for example, if one population is located to the right of another, we must apply the Wilcoxon rank sum test.

We complete this section with a review of how to recognize the use of the Kruskal-Wallis test.

Factors That Identify the Kruskal-Wallis Test

1. **Problem objective:** Compare two or more populations
2. **Data type:** Ordinal or interval but not normal
3. **Experimental design:** Independent samples

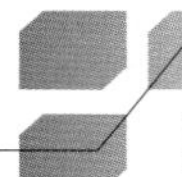

EXERCISES

19.40 Conduct the Kruskal-Wallis test on the following statistics. Use a 5% significance level.

$T_1 = 984$ $n_1 = 23$
$T_2 = 1{,}502$ $n_2 = 36$
$T_3 = 1{,}430$ $n_3 = 29$

19.41 From the following statistics, test (with $\alpha = .01$) to determine whether the population locations differ.

$T_1 = 1{,}207$ $n_1 = 25$
$T_2 = 1{,}088$ $n_2 = 25$
$T_3 = 1{,}310$ $n_3 = 25$
$T_4 = 1{,}445$ $n_4 = 25$

19.42 Use the following statistics to determine whether there is enough statistical evidence at the 10% significance level to infer that the population locations differ.

$T_1 = 3{,}741$ $n_1 = 47$
$T_2 = 1{,}610$ $n_2 = 29$
$T_3 = 4{,}945$ $n_3 = 67$

19.43 Use the Kruskal Wallis test on the following data to determine whether the population locations differ. (Use $\alpha = .05$.)

Sample 1:	27	33	18	29	41	52	75
Sample 2:	37	12	17	22	30		
Sample 3:	19	12	33	41	28	18	

19.44 Using the Kruskal-Wallis test, determine whether there is enough evidence provided by the accompanying data to enable us to infer that at least two population locations differ. (Use $\alpha = .05$.)

Sample 1:	25	15	20	22	23
Sample 2:	19	21	23	22	28
Sample 3:	27	25	22	29	28

19.45 Apply the Kruskal-Wallis test to determine whether there is enough evidence at the 5% significance level to infer that at least two population locations differ, given the following data.

Sample 1:	39	8	7	21	14	40	42	15
Sample 2:	31	43	50	57	54	46	26	29
Sample 3:	13	58	16	26	44	17	15	19
Sample 4:	50	51	55	28	57	37	41	37

Exercises 19.46–19.55 require the use of a computer and software. ***Use a 5% significance level.***

Developing an Understanding of Statistical Concepts

19.46 a. Xr19-46a Four random samples of 50 people each were asked to rate four different computer printers in terms of their ease of use. The responses are

Very easy to use
Easy to use
Difficult to use
Very difficult to use

The responses were coded using a 4–3–2–1 system. Do these data yield enough evidence to infer that differences in ratings exist between the four printers?

b. Xr19-46b The responses were recoded using a 25–22–5–2 system. Do these data yield enough evidence to infer that differences in ratings exist among the four printers?

c. Why are the results of Parts a and b identical?

19.47 a. Xr19-47 Refer to Example 19.5. The responses were recoded using a 100–90–55–25 system. Do these data yield enough evidence to infer that differences in ratings exist between the three shifts?

b. Why are the answers to Example 19.5 and Part a identical?

Applications

Exercises 19.48–19.55 may be solved manually. See Appendix A for the sample statistics.

19.48 Xr19-48 In an effort to determine whether differences exist between three methods of teaching statistics, a professor of business taught his course differently in each of three large sections. In the first section, he taught by lecturing; in the second, he taught by the case method; and in the third, he used a computer software package extensively. At the end of the semester, each student was asked to evaluate the course on a 7-point scale, where 1 = atrocious, 2 = poor, 3 = fair, 4 = average, 5 = good, 6 = very good, and 7 = excellent. From each section, the professor chose 25 evaluations at random. Is there evidence that differences in student satisfaction exist with respect to at least two of the three teaching methods?

19.49 Xr19-49 Applicants to MBA programs must take the Graduate Management Admission Test (GMAT). There are several companies that offer assistance in preparing for the test. To determine whether they work, and if so, which one is best, an experiment was conducted. Several hundred MBA applicants were surveyed and asked to report their GMAT score and which, if any, GMAT preparation course they took. The responses are course A, course B, course C, or no preparatory course. Do these data allow us to infer that there are differences between the four groups of GMAT scores?

19.50 Xr19-50 Because there are no national or regional standards, it is difficult for university admission committees to compare graduates of different high schools. University administrators have noted that an 80% average at a high school with low standards may be equivalent to a 70% average at another school with higher standards of grading. In an effort to more equitably compare applications, a pilot study was initiated. Random samples of students who were admitted the previous year from four local high schools were drawn. All the students entered the business program with averages between 70% and 80%. Their average grades in the first year at the university were computed. Can the university admissions officer conclude that there are differences in grading standards between the four high schools? (This exercise is identical to Exercise 14.9 except for the data.)

19.51 Xr19-51 A consumer testing service compared the effectiveness of four different brands of drain cleaner. The experiment consisted of using each product on 50 different clogged sinks and measuring the amount of time that elapsed until each drain became unclogged. The recorded times were measured in minutes.

a. Which techniques should be considered as possible procedures to apply to determine whether differences exist? What are the required conditions? How do you decide?

b. If a statistical analysis has shown that the times are not normally distributed, can the service conclude that differences exist between the speeds at which the four brands perform?

19.52 Xr19-52 During the last presidential campaign, the Gallup organization surveyed a random sample of 30 registered Democrats in January, another 30 in February, and yet another 30 in March. All 90 Democrats were asked to "rate the chances of the Democrats winning the presidential race in your state." The responses and their numerical codes were excellent (4), good (3), fair (2), and poor (1). Do these data allow us to infer that Democrats' ratings of their chances of winning the presidency changed over the 3-month period?

19.53 Xr19-53 It is common practice in the advertising business to create several different advertisements and then ask a random sample of potential customers to rate the ads on several different dimensions. Suppose that an advertising firm developed four different ads for a new breakfast cereal and asked a sample of 400 shoppers to rate the believability of the advertisements. One hundred people viewed ad 1, another 100 viewed ad 2, another 100 saw ad 3, and another 100 saw ad 4. The ratings were very believable (4), quite believable (3), somewhat believable (2), and not believable at all (1). Can the firm's management conclude that differences exist in believability between the four ads?

19.54 Xr19-54 Do university students become more supportive of their varsity teams as they progress through their 4-year stint? To help answer this question, a sample of students was drawn. Each was asked their class standing (freshman, sophomore, junior or senior) and to what extent they supported the university's football team, the Hawks. The responses to the latter question are

Wildly fanatic

Support the Hawks wholeheartedly

Support the Hawks, but not that enthusiastically

Who are the Hawks?

The responses were coded using a 4–3–2–1 numbering system. Can we conclude that the four levels of students differ in their support for the Hawks?

19.55 Xr19-55 In anticipation of buying a new scanner, a student turned to a web site that reported the results of surveys of users of the different scanners. A sample of 133 responses was listed, showing the ease of use of five different brands. The survey responses were

Very easy

Easy

Not easy

Difficult

Very difficult

The responses were assigned numbers from 1 to 5. Can we infer that there are differences in perceived ease of use between the five brands of scanners?

19.4 Friedman Test

This section introduces another statistical technique whose objective is to compare two or more populations of ordinal or interval data. In Section 15.3, we presented the randomized block experimental design of the analysis of variance. In this section, we present its nonparametric counterpart. The **Friedman test** is applied to problems with the following characteristics:

1. The problem objective is to compare two or more populations.
2. The data are either ordinal or interval, but not normal.
3. The data are generated from a randomized block experiment.

The null and alternative hypotheses are identical to the ones tested in the Kruskal-Wallis test:

H_0: The locations of all k populations are the same

H_1: At least two population locations differ

Test Statistic

To calculate the test statistic, we first rank each observation within each block, where $1 =$ smallest observation and $k =$ largest observation, averaging the ranks of ties. Then we compute the rank sums, which we label $T_1, T_2, \ldots, T_k$. The test statistic is defined as follows. (Recall that $b =$ number of blocks.)

Test Statistic for the Friedman Test

$$F_r = \left[\frac{12}{b(k)(k+1)}\sum_{j=1}^{k} T_j^2\right] - 3b(k+1)$$

Sampling Distribution of the Test Statistic

The test statistic is approximately chi-squared distributed with $k - 1$ degrees of freedom, provided that either k or b is greater than or equal to 5. As was the case with the Kruskal-Wallis test, we reject the null hypothesis when the test statistic is large. Hence, the rejection region is

$$F_r > \chi^2_{\alpha,k-1}$$

and the p-value is

$$P(\chi^2 > F_r)$$

Figure 19.11 depicts the sampling distribution, rejection region, and p-value.

FIGURE **19.11** Sampling Distribution of F_r

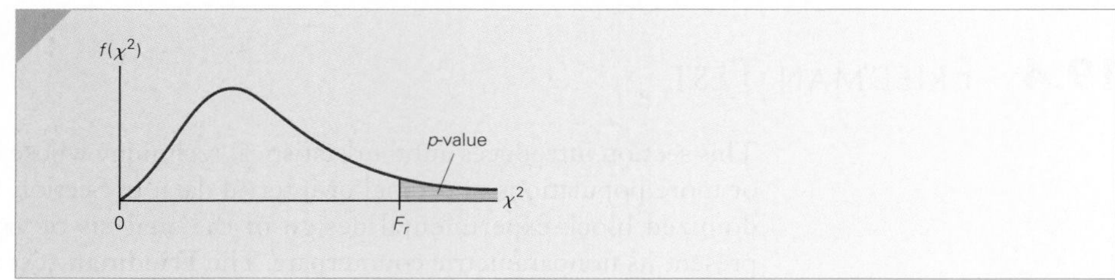

This test, like all the other nonparametric tests, requires that the populations being compared be identical in shape and spread.

EXAMPLE 19.6 Comparing Managers' Evaluations of Job Applicants

DATA
Xm19-06

The personnel manager of a national accounting firm has been receiving complaints from senior managers about the quality of recent hirings. All new accountants are hired through a process whereby four managers interview the candidate and rate her or him on several dimensions, including academic credentials, previous work experience, and personal suitability. Each manager then summarizes the results and produces an evaluation of the candidate. There are five possibilities:

1. The candidate is in the top 5% of applicants.
2. The candidate is in the top 10% of applicants, but not in the top 5%.
3. The candidate is in the top 25% of applicants, but not in the top 10%.
4. The candidate is in the top 50% of applicants, but not in the top 25%.
5. The candidate is in the bottom 50% of applicants.

The evaluations are then combined in making the final decision. The personnel manager believes that the quality problem is caused by the evaluation system. However, she needs to know whether there is general agreement or disagreement between the interviewing managers in their evaluations. To test for differences between the managers, she takes a random sample of the evaluations of eight applicants. The results are shown below. What conclusions can the personnel manager draw from these data? Employ a 5% significance level.

	Manager			
Applicant	1	2	3	4
1	2	1	2	2
2	4	2	3	2
3	2	2	2	3
4	3	1	3	2
5	3	2	3	5
6	2	2	3	4
7	4	1	5	5
8	3	2	5	3

SOLUTION

IDENTIFY

The problem objective is to compare the four populations of managers' evaluations, which we can see are ordinal data. This experiment is identified as a randomized block design because the eight applicants were evaluated by all four managers. (The treatments are the managers, and the blocks are the applicants.) The appropriate statistical technique is the Friedman test. The null and alternative hypotheses are as follows:

H_0: The locations of all four populations are the same

H_1: At least two population locations differ

COMPUTE

MANUALLY

The rejection region is

$$F_r > \chi^2_{\alpha,k-1} = \chi^2_{.05,3} = 7.81$$

The following table demonstrates how the ranks are assigned and the rank sums calculated. Notice how the ranks are assigned by moving across the rows (blocks) and the rank sums computed by adding down the columns (treatments).

	Manager			
Applicant	1 (Rank)	2 (Rank)	3 (Rank)	4 (Rank)
1	2 (3)	1 (1)	2 (3)	2 (3)
2	4 (4)	2 (1.5)	3 (3)	2 (1.5)
3	2 (2)	2 (2)	2 (2)	3 (4)
4	3 (3.5)	1 (1)	3 (3.5)	2 (2)
5	3 (2.5)	2 (1)	3 (2.5)	5 (4)
6	2 (1.5)	2 (1.5)	3 (3)	4 (4)
7	4 (2)	1 (1)	5 (3.5)	5 (3.5)
8	3 (2.5)	2 (1)	5 (4)	3 (2.5)
	$T_1 = 21$	$T_2 = 10$	$T_3 = 24.5$	$T_4 = 24.5$

The value of the test statistic is

$$F_r = \left[\frac{12}{b(k)(k+1)}\sum_{j=1}^{k} T_j^2\right] - 3b(k+1)$$

$$= \left[\frac{12}{(8)(4)(5)}(21^2 + 10^2 + 24.5^2 + 24.5^2)\right] - 3(8)(5)$$

$$= 10.61$$

EXCEL

	A	B	C
1	**Friedman Test**		
2			
3	Group		Rank Sum
4	*Manager 1*		21
5	*Manager 2*		10
6	*Manager 3*		24.5
7	*Manager 4*		24.5
8			
9	Fr Stat		10.61
10	df		3
11	p-value		0.0140
12	chi-squared Critical		7.8147

INSTRUCTIONS

1. Type or import the data into adjacent columns. (Open Xm19-06.)
2. Click **Add-Ins, Data Analysis Plus,** and **Friedman Test.**
3. Specify the **Input Range** (A1:D9) and the value of α (.05).

MINITAB

Friedman Test: Ratings Versus Manager Blocked by Applicant

S = 10.61 DF = 3 P = 0.014
S = 12.86 DF = 3 P = 0.005 (adjusted for ties)

Manager	N	Est Median	Sum of Ranks
1	8	2.8750	21.0
2	8	2.0000	10.0
3	8	3.0000	24.5
4	8	3.1250	24.5

Grand median = 2.7500

INSTRUCTIONS

1. Type or import the data in stacked format. The responses are stored in one column, the treatment codes are stored in another column, and the block codes are stored in a third column.
2. Click **Stat, Nonparametrics,** and **Friedman . . . .**
3. Select the **Response** variable, the **Treatment** variable, and the **Blocks** variable.

INTERPRET

There appears to be sufficient evidence to indicate that the managers' evaluations differ. The personnel manager should attempt to determine why the evaluations differ. Is there a problem with the way in which the assessments are conducted, or are some managers are using different criteria? If it is the latter, those managers may need additional training.

The Friedman Test and the Sign Test

The relationship between the Friedman and sign tests is the same as the relationship between the Kruskal-Wallis and Wilcoxon rank sum tests. That is, we can use the Friedman test to determine whether two populations differ. The conclusion will be the same as that produced from the sign test. However, we can use the Friedman test to determine only whether a difference exists. If we want to determine whether one population is, for example, to the left of another population, we must use the sign test.

Here is a list of the factors that tell us when to use the Friedman test.

Factors That Identify the Friedman Test

1. **Problem objective:** Compare two or more populations
2. **Data type:** Ordinal or interval but not normal
3. **Experimental design:** Randomized blocks

EXERCISES

19.56 Xr19-56 Apply the Friedman test to the accompanying table of data to determine whether we can conclude that at least two population locations differ. (Use α = .10.)

	Treatment			
Block	1	2	3	4
1	10	12	15	9
2	8	10	11	6
3	13	14	16	11
4	9	9	12	13
5	7	8	14	10

19.57 Xr19-57 The following data were generated from a blocked experiment. Conduct a Friedman test to determine whether at least two population locations differ. (Use α = .05.)

	Treatment		
Block	1	2	3
1	7.3	6.9	8.4
2	8.2	7.0	7.3
3	5.7	6.0	8.1
4	6.1	6.5	9.1
5	5.9	6.1	8.0

19.58 Xr19-58 Ten judges were asked to test the quality of four different brands of orange juice. The judges assigned scores using a 5-point scale where 1 = bad, 2 = poor, 3 = average, 4 = good, and 5 = excellent. The results are shown here. Can we conclude at the 5% significance level that there are differences in sensory quality between the four brands of orange juice?

	Orange Juice Brand			
Judge	1	2	3	4
1	3	5	4	3
2	2	3	5	4
3	4	4	3	4
4	3	4	5	2
5	2	4	4	3
6	4	5	5	3
7	3	3	4	4
8	2	3	3	3
9	4	3	5	4
10	2	4	5	3

19.59 Xr19-59 The manager of a personnel company is in the process of examining her company's advertising programs. Currently, the company advertises in each of the three local newspapers for a wide variety of positions, including computer programmers, secretaries, and receptionists. The manager has decided that only one newspaper will be used if it can be determined that there are differences between the newspapers in the number of inquiries. The following experiment was performed. For 1 week (6 days), six different jobs were advertised in each of the three newspapers. The number of inquiries was counted, and the results appear in the accompanying table.

	Newspaper		
Job Advertised	1	2	3
Receptionist	14	17	12
Systems analyst	8	9	6
Junior secretary	25	20	23
Computer programmer	12	15	10
Legal secretary	7	10	5
Office manager	5	9	4

a. What techniques should be considered to apply in reaching a decision? What are the required conditions? How do we determine whether the conditions are satisfied?

b. Assuming that the data are not normally distributed, can we conclude at the 5% significance level that differences exist between the newspapers' abilities to attract potential employees?

Exercises 19.60–19.65 require the use of a computer and software. ***Use a 5% significance level.***

Developing an Understanding of Statistical Concepts

19.60 a. Xr19-60a A random sample of 30 people was asked to rate each of four different premium brands of coffee. The ratings are

Excellent

Good

Fair

Poor

The responses were assigned numbers 1 through 4, respectively. Can we infer that differences exist between the ratings of the four brands of coffee?

b. Xr19-60b Suppose that the codes were 12, 31, 66, and 72, respectively. Can we infer that differences exist between the ratings of the four brands of coffee?

c. Compare your answers in Parts a and b. Why are they identical?

19.61 a. Xr19-61 Refer to Example 19.6. Suppose that the responses were recoded so that the numbers equaled the midpoint of the range of percentiles. That is,

97.5 = The candidate is in the top 5% of applicants.

92.5 = The candidate is in the top 10% of applicants, but not in the top 5%.

82.5 = The candidate is in the top 25% of applicants, but not in the top 10%.

62.5 = The candidate is in the top 50% of applicants, but not in the top 25%.

25 = The candidate is in the bottom 50% of applicants.

Can we conclude that differences exist between the ratings assigned by the four professors?

b. Compare your answer in Part a with the one obtained in Example 19.6. Are they the same? Explain why.

Applications

Exercises 19.62–19.65 may be solved manually. See Appendix A for the sample statistics.

19.62 Xr19-62 A well-known soft-drink manufacturer has used the same secret recipe for its product since its introduction over 100 years ago. In response to a decreasing market share, however, the president of the company is contemplating changing the recipe. He has developed two alternative recipes. In a preliminary study, he asked 20 people to taste the original recipe and the two new recipes. He asked each to evaluate the taste of the product on a 5-point scale, where 1 = awful, 2 = poor, 3 = fair, 4 = good, and 5 = wonderful. The president decides that unless significant differences exist between evaluations of the products, he will not make any changes. Can we conclude that there are differences in the ratings of the three recipes?

19.63 Xr19-63 The manager of a chain of electronic-products retailers is trying to decide on a location for its newest store. After a thorough analysis, the choice has been narrowed to three possibilities. An important factor in the decision is the number of people passing each location. The number of people passing each location per day was counted during 30 days.

a. Which techniques should be considered to determine whether the locations differ? What are the required conditions? How do you select a technique?

b. Can management conclude that there are differences in the numbers of people passing the three locations if the number of people passing each location is not normally distributed?

19.64 Xr19-64 In recent years, lack of confidence in the U.S. Postal Service has led many companies to send all their correspondence by private courier. A large company is in the process of selecting one of three possible couriers to act as its sole delivery method. To help make the decision, an experiment was performed whereby letters were sent using each of the three couriers at 12 different times of the day to a delivery point across town. The number of minutes required for delivery was recorded. Can we conclude that there are differences in delivery times between the three couriers? (This exercise is identical to Exercise 14.39 except for the data.)

19.65 Xr19-65 Many North Americans suffer from high levels of cholesterol, which can lead to heart attacks. For those with very high levels (over 280), doctors prescribe drugs to reduce cholesterol levels. A pharmaceutical company has recently developed four such drugs. To determine whether any differences exist in their benefits, an experiment was organized. The company selected 25 groups of four men, each of whom had cholesterol levels in excess of 280. In each group, the men were matched according to age and weight. The drugs were administered over a 2-month period, and the reduction in cholesterol was recorded. Do these results allow the company to conclude differences exist between the four new drugs? (This exercise is identical to Example 14.3 except for the data.)

19.5 SPEARMAN RANK CORRELATION COEFFICIENT

In Section 16.4 we introduced the test of the coefficient of correlation, which allows us to determine whether there is evidence of a linear relationship between two interval variables. Recall that the required condition for the t-test of ρ is that the variables are bivariate normally distributed. In many situations, however, one or both variables may be ordinal; or if both variables are interval, the normality requirement may not be satisfied. In such cases, we measure and test to determine whether a relationship exists by employing a nonparametric technique, the **Spearman rank correlation coefficient.**

The Spearman rank correlation coefficient is calculated like all of the other previously introduced nonparametric methods by first ranking the data. We then calculate the *Pearson correlation coefficient* of the ranks. The population Spearman correlation coefficient is labeled ρ_s, and the sample statistic used to estimate its value is labeled r_s.

Sample Spearman Rank Correlation Coefficient

$$r_s = \frac{s_{ab}}{s_a s_b}$$

where a and b are the ranks of x and y, respectively, s_{ab} is the covariance of the values of a and b, s_a is the standard deviation of the values of a, and s_b is the standard deviation of the values of b.

We can test to determine whether a relationship exists between the two variables. The hypotheses to be tested are

$$H_0: \quad \rho_s = 0$$
$$H_1: \quad \rho_s \neq 0$$

(We also can conduct one-tail tests.) The test statistic is the absolute value of r_s. To determine whether the value of r_s is large enough to reject the null hypothesis, we refer to Table 11 in Appendix B, reproduced here as Table 19.5, which lists the critical values of the test statistic for one-tail tests. To conduct a two-tail test, the value of α must be doubled. The table lists critical values for $\alpha = .01, .025$, and $.05$ and for $n = 5$ to 30. When n is greater than 30, r_s is approximately normally distributed with mean 0 and standard deviation $1/\sqrt{n-1}$. Thus, for $n > 30$, the test statistic is as shown in the box.

Test Statistic for Testing $\rho_s = 0$ When $n > 30$

$$z = \frac{r_s - 0}{1/\sqrt{n-1}} = r_s\sqrt{n-1}$$

which is standard normally distributed

TABLE **19.5** Critical Values for the Spearman Rank Correlation Coefficient

The α values correspond to a one-tail test of H_0: $\rho_s = 0$. The value should be doubled for two-tail tests.

n	$\alpha = .05$	$\alpha = .025$	$\alpha = .01$
5	.900	—	—
6	.829	.886	.943
7	.714	.786	.893
8	.643	.738	.833
9	.600	.683	.783
10	.564	.648	.745
11	.523	.623	.736
12	.497	.591	.703
13	.475	.566	.673
14	.457	.545	.646
15	.441	.525	.623
16	.425	.507	.601
17	.412	.490	.582
18	.399	.476	.564
19	.388	.462	.549
20	.377	.450	.534
21	.368	.438	.521
22	.359	.428	.508
23	.351	.418	.496
24	.343	.409	.485
25	.336	.400	.475
26	.329	.392	.465
27	.323	.385	.456
28	.317	.377	.448
29	.311	.370	.440
30	.305	.364	.432

EXAMPLE 19.7

DATA Xm19-07

Testing the Relationship between Aptitude Tests and Performance

The production manager of a firm wants to examine the relationship between aptitude test scores given prior to hiring production-line workers and performance ratings received by the employees 3 months after starting work. The results of the study would allow the firm to decide how much weight to give to these aptitude tests relative to other work-history information obtained, including references. The aptitude test results range from 0 to 100. The performance ratings are as follows:

1 = Employee has performed well below average.

2 = Employee has performed somewhat below average.

3 = Employee has performed at the average level.

4 = Employee has performed somewhat above average.

5 = Employee has performed well above average.

A random sample of 40 production workers yielded the results listed here. Can the firm's manager infer at the 5% significance level that aptitude test scores are correlated with performance rating?

Employee	Aptitude	Performance
1	59	3
2	47	2
3	58	4
4	66	3
5	77	2
6	57	4
7	62	3
8	68	3
9	69	5
10	36	1
11	48	3
12	65	3
13	51	2
14	61	3
15	40	3
16	67	4
17	60	2
18	56	3
19	76	3
20	71	2
21	52	3
22	62	5
23	54	2
24	50	3
25	57	1
26	59	5
27	66	4
28	84	5
29	56	2
30	61	1
31	53	4
32	76	3
33	42	4
34	59	4
35	58	2
36	66	4
37	58	2
38	53	1
39	63	5
40	85	3

SOLUTION

IDENTIFY

The problem objective is to analyze the relationship between two variables. The aptitude test score is interval, but the performance rating is ordinal. We will treat the aptitude test score as if it were ordinal and calculate the Spearman rank correlation coefficient. To answer the question, we specify the hypotheses as

$$H_0: \quad \rho_s = 0$$
$$H_1: \quad \rho_s \neq 0$$

COMPUTE

MANUALLY

We rank each of the variables separately, averaging any ties that we encounter. The original data and ranks are as follows.

Employee	Aptitude	Rank *a*	Performance	Rank *b*
1	59	20	3	20.5
2	47	4	2	9
3	58	17	4	31.5
4	66	30	3	20.5
5	77	38	2	9
6	57	14.5	4	31.5
7	62	25.5	3	20.5
8	68	33	3	20.5
9	69	34	5	38
10	36	1	1	2.5
11	48	5	3	20.5
12	65	28	3	20.5
13	51	7	2	9
14	61	23.5	3	20.5
15	40	2	3	20.5
16	67	32	4	31.5
17	60	22	2	9
18	56	12.5	3	20.5
19	76	36	3	20.5
20	71	35	2	9
21	52	8	3	20.5
22	62	25.5	5	38
23	54	11	2	9
24	50	6	3	20.5
25	57	14.5	1	2.5
26	59	20	5	38
27	66	30	4	31.5
28	84	39	5	38
29	56	12.5	2	9
30	61	23.5	1	2.5
31	53	9.5	4	31.5
32	76	37	3	20.5
33	42	3	4	31.5
34	59	20	4	31.5

(continued)

Employee	Aptitude	Rank *a*	Performance	Rank *b*
35	58	17	2	9
36	66	30	4	31.5
37	58	17	2	9
38	53	9.5	1	2.5
39	63	27	5	38
40	85	40	3	20.5

The next step is to calculate the following sums:

$$\sum a_i b_i = 18{,}319$$
$$\sum a_i = \sum b_i = 820$$
$$\sum a_i^2 = 22{,}131.5$$
$$\sum b_i^2 = 21{,}795.5$$

Using the short-cut calculation on page 124, we determine that the covariance of the ranks is

$$s_{ab} = \frac{1}{n-1}\left[\sum a_i b_i - \frac{\sum a_i \sum b_i}{n}\right] = \frac{1}{40-1}\left[18{,}319 - \frac{(820)(820)}{40}\right] = 38.69$$

The sample variances of the ranks (using the short-cut formula on page 109) are

$$s_a^2 = \frac{1}{n-1}\left[\sum a_i^2 - \frac{\left(\sum a_i\right)^2}{n}\right] = \frac{1}{40-1}\left[22{,}131.5 - \frac{(820)^2}{40}\right] = 136.45$$

$$s_b^2 = \frac{1}{n-1}\left[\sum b_i^2 - \frac{\left(\sum b_i\right)^2}{n}\right] = \frac{1}{40-1}\left[21{,}795.5 - \frac{(820)^2}{40}\right] = 127.83$$

The standard deviations are

$$s_a = \sqrt{s_a^2} = \sqrt{136.45} = 11.68$$
$$s_b = \sqrt{s_b^2} = \sqrt{127.83} = 11.31$$

Thus,

$$r_s = \frac{s_{ab}}{s_a s_b} = \frac{38.69}{(11.68)(11.31)} = .2929$$

The value of the test statistic is

$$z = r_s\sqrt{n-1} = .2929\sqrt{40-1} = 1.83$$

$$p\text{-value} = 2P(Z > 1.83) = 2(1 - .9664) = .0672$$

EXCEL

	A	B	C	D
1	**Spearman Rank Correlation**			
2				
3	*Aptitude and Performance*			
4	Spearman Rank Correlation			0.2930
5	z Stat			1.83
6	P(Z<=z) one tail			0.0337
7	z Critical one tail			1.6449
8	P(Z<=z) two tail			0.0674
9	z Critical two tail			1.96

INSTRUCTIONS

1. Type or import the data into two adjacent columns. (Open Xm19-07.)
2. Click **Add-Ins, Data Analysis Plus,** and **Correlation (Spearman).**
3. Specify the **Input Range** (A1:B41) and the value of α (.05).

MINITAB

Correlations: Rank Aptitude, Rank Performance

Pearson correlation of Rank Aptitude and Rank Performance = 0.293
P-Value = 0.067

INSTRUCTIONS

1. Click **Data** and **Rank . . .** to rank each variable.
2. Click **Stat, Basic Statistics,** and **Correlation.** Select the variables representing the ranks.

INTERPRET

There is not enough evidence to believe that the aptitude test scores and performance ratings are related. This conclusion suggests that the aptitude test should be improved to better measure the knowledge and skill required by a production-line worker. If this proves impossible, the aptitude test should be discarded.

EXERCISES

19.66 Test the following hypotheses:

H_0: $\rho_s = 0$

H_1: $\rho_s \neq 0$

$n = 50 \quad r_s = .23 \quad \alpha = .05$

19.67 Is there sufficient evidence at the 5% significance level to infer that there is a positive relationship between two ordinal variables given that $r_s = .15$ and $n = 12$?

19.68 Xr19-68 A statistics student asked seven first-year economics students to report their grades in the required mathematics and economics courses. The results (where 1 = F, 2 = D, 3 = C, 4 = B, 5 = A) are as follows:

Mathematics	4	2	5	4	2	2	1
Economics	5	2	3	5	3	3	2

Calculate the Spearman rank correlation coefficient, and test to determine whether we can infer that a relationship exists between the grades in the two courses. (Use $\alpha = .05$.)

19.69 Xr19-69 Does the number of commercials shown during a half-hour television program affect how viewers rate the show? In a preliminary study, eight people were asked to watch a pilot for a situation comedy and rate the show (1 = terrible, 2 = bad, 3 = OK, 4 = good, 5 = very good). Each person was shown a different number of 30-second commercials. The data are shown here. Calculate the Spearman rank correlation coefficient and test with a 10% significance level to determine whether there is a relationship between the two variables.

Number of Commercials	1	2	3	4	5	6	7	8
Rating	4	5	3	3	3	2	3	1

19.70 Xr19-70 The weekly returns of two stocks for a 13-week period were recorded and are listed here. Assuming that the returns are not normally distributed, can we infer at the 5% significance level that the stock returns are correlated?

Stock 1	−7	−4	−7	−3	2	−10	−10
Stock 2	6	6	−4	9	3	−3	7

Stock 1	5	1	−4	2	6	−13
Stock 2	−3	4	7	9	5	−7

19.71 Xr19-71 The general manager of an engineering firm wants to know whether a draftsman's experience influences the quality of his work. She selects 24 draftsmen at random and records their years of work experience and their quality rating (as assessed by their supervisors where 5 = excellent, 4 = very good, 3 = average, 2 = fair, and 1 = poor). The data are listed here. Can we infer from these data that years of work experience is a factor in determining the quality of work performed? Use $\alpha = .05$.

Draftsman	Experience	Rating	Draftsman	Experience	Rating
1	1	1	13	8	2
2	17	4	14	20	5
3	20	4	15	21	3
4	9	5	16	19	2
5	2	2	17	1	1
6	13	4	18	22	3
7	9	3	19	20	4
8	23	5	20	11	3
9	7	2	21	18	5
10	10	5	22	14	4
11	12	5	23	21	3
12	24	2	24	21	1

The following exercises require the use of a computer and software. ***Use a 5% significance level.***

19.72 Xr16-02 Refer to Example 16.2. If the required condition is not satisfied, conduct another more appropriate test to determine whether odometer reading and price are related.

19.73 Xr19-73 At the completion of most courses in universities and colleges a course evaluation is undertaken. Some professors believe that the way in which students fill out the evaluations is based on how well the student is doing in the course. To test this theory, a random sample of course evaluations was selected. Two answers were recorded. The questions and answers are

a. How would you rate the course?

1. Poor 2. Fair 3. Good 4. Very good 5. Excellent

b. What grade do you expect in this course?

1. F 2. D 3. C 4. B 5. A

Is there enough evidence to conclude that the theory is correct?

19.74 Xr19-74 Many people suffer from heartburn. It appears, however, that the problem may increase with age. A researcher for a pharmaceutical company wanted to determine whether age and the incidence and extent of heartburn are related. A random sample of 325 adults was drawn. Each person was asked to give his or her age and to rate the severity of heartburn (1 = low, 2 = moderate, 3 = high, 4 = very high). Do these data provide sufficient evidence to indicate that older people suffer more severe heartburn?

19.75 Xr16-06 Refer to Exercise 16.6. Assume that the conditions for the test conducted in Exercise 16.6 are not met. Do the data allow us to conclude that the longer the commercial, the higher the memory test score will be?

19.76 Xr16-07 Assume that the normality requirement in Exercise 16.7 is not met. Test to determine whether the price of a condominium and floor number are positively related.

19.77 Xr19-77 Many people who quit smoking gain weight. Many explain that after they quit smoking, food tastes better. To examine the relationship between smoking and taste, a researcher randomly sample 280 smokers. Each was asked how many cigarettes they smoked on an average day. In addition, each person was asked to taste and rate some vanilla ice cream. The responses are 5 = excellent, 4 = very good, 3 = good, 2 = fair, 1 = poor. Can the researcher infer that the more a person smokes, the less taste sensation he or she has?

19.78 Xr19-78 Gambling on sports is big business in the United States and Canada. A television executive wants to know whether the amount of money wagered on a professional football game affects the enjoyment of viewers. A random sample of 200 men who regularly watch football Sunday afternoons and wager on the outcomes was drawn. Each was asked to report the amount wagered on the game they watched and to rate the enjoyment (where 1 = not enjoyable, 2 = somewhat enjoyable, 3 = moderately enjoyable 4 = very enjoyable). Do these data provide enough evidence to conclude that the greater the wager the more enjoyable the game is for the viewer?

CHAPTER SUMMARY

Nonparametric statistical tests are applied to problems where the data are either ordinal or interval but not normal. The Wilcoxon rank sum test is used to compare two populations of ordinal or interval data when the data are generated from independent samples. The sign test is used to compare two populations of ordinal data drawn from a matched pairs experiment. The Wilcoxon signed rank sum test is employed to compare two populations of nonnormal interval data taken from a matched pairs experiment. When the objective is to compare two or more populations of independently sampled ordinal or interval nonnormal data, the Kruskal Wallis test is employed. The Friedman test is used instead of the Kruskal Wallis test when the samples are blocked. To determine whether two variables are related, we employ the test of the Spearman rank correlation coefficient.

IMPORTANT TERMS

Nonparametric techniques 754
Distribution-free statistics 754
Wilcoxon rank sum test 756
Sign test 769
Wilcoxon signed rank sum test 774
Kruskal-Wallis test 783
Friedman test 789
Spearman rank correlation coefficient 795

SYMBOLS

Symbol	Pronounced	Represents
T_i	*T-sub-i* or *T-i*	Rank sum of sample i ($i = 1, 2, \ldots, k$)
T^+	*T-plus*	Rank sum of positive differences
T^-	*T-minus*	Rank sum of negative differences
σ_T	*sigma-sub-T* or *sigma-T*	Standard deviation of the sampling distribution of T
ρ_S	*Rho-sub S* or *rho-S*	Spearman rank correlation coeficient

FORMULAS

Wilcoxon rank sum test

$$T = T_1$$

$$E(T) = \frac{n_1(n_1 + n_2 + 1)}{2}$$

$$\sigma_T = \sqrt{\frac{n_1 n_2(n_1 + n_2 + 1)}{12}}$$

$$z = \frac{T - E(T)}{\sigma_T}$$

Sign test

x = number of positive differences

$$z = \frac{x - .5n}{.5\sqrt{n}}$$

Wilcoxon signed rank sum test

$$T = T^+$$

$$E(T) = \frac{n(n + 1)}{4}$$

$$\sigma_T = \sqrt{\frac{n(n + 1)(2n + 1)}{24}}$$

$$z = \frac{T - E(T)}{\sigma_T}$$

Kruskal-Wallis test

$$H = \left[\frac{12}{n(n + 1)} \sum_{j=1}^{k} \frac{T_j^2}{n_j}\right] - 3(n + 1)$$

Friedman test

$$F_r = \left[\frac{12}{b(k)(k + 1)} \sum_{j=1}^{k} T_j^2\right] - 3b(k + 1)$$

Spearman rank correlation coefficient

$$r_S = \frac{s_{ab}}{s_a s_b}$$

Spearman test statistic for $n > 30$

$$z = r_S\sqrt{n - 1}$$

COMPUTER OUTPUT AND INSTRUCTIONS

Technique	Excel	Minitab
Wilcoxon rank sum test	761	762
Sign test	772	772
Wilcoxon signed rank sum test	779	779
Kruskal-Wallis test	786	786
Friedman test	792	792
Spearman rank correlation	799	800

CHAPTER EXERCISES

The following exercises require the use of a computer and software. ***Use a 5% significance level.***

19.79 Xr19-79 Are education and income related? To answer this question, a random sample of people was selected and each was asked to indicate into which of the following categories of education they belonged:

1. Less than high school
2. High school graduate
3. Some college or university but no degree
4. University degree
5. Postgraduate degree

Additionally respondents were asked for their annual income group from the following choices:

1. Under $25,000
2. $25,000 up to but not including $40,000
3. $40,000 up to but not including $60,000
4. $60,000 up to $100,000
5. Greater than $100,000

Conduct a test to determine whether more education and higher incomes are linked.

19.80 Xr19-80 In a study to determine which of two teaching methods is perceived to be better, two sections of an introductory marketing course were taught in different ways by the same professor. At the course's completion, each student rated the course on a boring/stimulating spectrum, with 1 = very boring, 2 = somewhat boring, 3 = a little boring, 4 = neither boring nor stimulating, 5 = a little stimulating, 6 = somewhat stimulating, and 7 = very stimulating. Can we conclude that the ratings of the two teaching methods differ?

19.81 Xr19-81 The researchers at a large carpet manufacturer have been experimenting with a new dyeing process in hopes of reducing the streakiness that frequently occurs with the current process. As an experiment, 15 carpets are dyed using the new process, and another 15 are dyed using the existing method. Each carpet is rated on a 5-point scale of streakiness, where 5 is extremely streaky, 4 is quite streaky, 3 is somewhat streaky, 2 is a little streaky, and 1 is not streaky at all. Is there enough evidence to infer that the new method is better?

19.82 Xr19-82 The editor of the student newspaper was in the process of making some major changes in the newspaper's layout. He was also contemplating changing the typeface of the print used. To help make a decision, he set up an experiment in which 20 individuals were asked to read four newspaper pages, with each page printed in a different typeface. If the reading speed differed, the typeface that was read fastest would be used. However, if there was not enough evidence to allow the editor to conclude that such differences exist, the current typeface would be continued. The times (in seconds) to completely read one page were recorded. We have determined that the times are not normally distributed. Determine the course of action the editor should follow. (This exercise is identical to Exercise 14.62, except in this exercise, the data are not normally distributed.)

19.83 Xr19-83 Large potential profits for pharmaceutical companies exist in the area of hair growth drugs. The head chemist for a large pharmaceutical company is conducting experiments to determine which of two new drugs is more effective in growing hair among balding men. One experiment was conducted as follows. A total of 30 pairs of men—each pair of which was matched according to their degree of baldness—was selected. One man used drug A, and the other used drug B. After 10 weeks, the men's new hair growth was examined, and the new growth was judged using the following ratings:

0 = No growth
1 = Some growth
2 = Moderate growth

Do these data provide sufficient evidence that drug B is more effective?

19.84 Xr19-84 Suppose that a precise measuring device for new hair growth has been developed and is used in the experiment described in Exercise 19.83. The percentages of new hair growth for the 30 pairs of men involved in the experiment were recorded. Do these data allow the chemist to conclude that drug B is more effective?

19.85 Xr19-85 The printing department of a publishing company wants to determine whether there are differences in durability between three types of book bindings. Twenty-five books with each type of binding were selected and placed in machines that continually opened and closed them. The numbers of openings and closings until the pages separated from the binding were recorded.

a. What techniques should be considered to determine whether differences exist between the types of bindings? What are the required conditions? How do you decide which technique to use?
b. If we know that the number of openings and closings is not normally distributed, test to determine whether differences exist between the types of bindings.

19.86 Xr19-86 In recent years, consumers have become more safety conscious, particularly about children's products. A manufacturer of children's pajamas is looking for material that is as nonflammable as possible. In an experiment to compare a new fabric with the kind now being used, 50 pieces of each kind were exposed to an open flame, and the number of seconds until the fabric burst into flames was recorded. Because the new material is much more expensive than the current material, the manufacturer will switch only if the new material can be shown to be better. On the basis of these data, what should the manufacturer do?

19.87 Xr19-87 Samuel's is a chain of family restaurants. Like many other service companies, Samuel's surveys its customers on a regular basis to monitor their opinions. Two questions (among others) asked in the survey are as follows:

a. While you were at Samuel's, did you find the service slow (1), moderate (2), or fast (3)?
b. What day was your visit to Samuel's?

The responses of a random sample of 269 customers were recorded. Can the manager infer that there are differences in customer perceptions of the speed of service between the days of the week?

19.88 Xr19-88 An advertising firm wants to determine the relative effectiveness of two recently produced commercials for a car dealership. An important attribute of such commercials is their believability. To judge this aspect of the commercials, 60 people were randomly selected. Each watched both commercials and then rated them on a 5-point scale, where 1 = not believable, 2 = somewhat believable, 3 = moderately believable, 4 = quite believable, and 5 = very believable. Do these data provide sufficient evidence to indicate that there are differences in believability between the two commercials?

19.89 Xr19-89 Researchers at the U.S. National Institute of Aging in Bethesda, Maryland, have been studying hearing loss. They have hypothesized that as men age they will lose their hearing faster than comparably aged women because many more men than women have worked at jobs where noise levels have been excessive. To test their beliefs, the researchers randomly selected one man and one woman aged $45, 46, 47, \ldots, 78, 79, 80$ and measured the percentage hearing loss for each person. What conclusions can be drawn from these data?

19.90 Xr19-90 In a Gallup poll this year, 200 people were asked, "Do you feel that the newspaper you read most does a good job of presenting the news?" The same question was asked of another 200 people 10 years ago. The possible responses were as follows:

3 = Good job

2 = Fair job

1 = Not a good job

Do these data provide enough evidence to infer that people perceive newspapers as doing a better job 10 years ago than today?

19.91 Xr19-91 It is common practice in many MBA programs to require applicants to arrange for a letter of reference. Some universities have their own forms in which referees assess the applicant using the following categories:

5 The candidate is in the top 5% of applicants.

4 The candidate is in the top 10% of applicants, but not in the top 5%.

3 The candidate is in the top 25% of applicants, but not in the top 10%.

2 The candidate is in the top 50% of applicants, but not in the top 25%.

1 The candidate is in the bottom 50% of applicants.

However, the question arises, Are the referees' ratings related to how well the applicant performs in the MBA program? To answer the question, a random sample of recently graduated MBAs was drawn. For each, the rating of the referee and the MBA grade-point average (GPA) were recorded. Do these data present sufficient evidence to infer that the letter of reference and the MBA GPA are related?

19.92 Xr19-92 The increasing number of traveling businesswomen represents a large potential clientele for the hotel industry. Many hotel chains have made changes designed to attract more women. To help direct these changes, a hotel chain commissioned a study to determine whether major differences exist between male and female business travelers. A total of 100 male and 100 female executives were questioned on a variety of topics, one of which was the number of trips they had taken in the previous 12 months. We would like to know whether these data provide enough evidence to allow us to conclude that businesswomen and businessmen differ in the number of business trips taken per year.

19.93 Xr19-93 To examine the effect that a tough midterm test has on student evaluations of professors, a statistics professor had her class evaluate her teaching effectiveness before the midterm test. The questionnaire asked for opinions on a number of dimensions, but the last question is considered the most important. It is, "How would you rate the overall performance of the instructor?" The possible responses are 1 = poor, 2 = fair, 3 = good, and 4 = excellent. After a difficult test, the evaluation was redone. The evaluation scores before and after the test for each of the 40 students in the class were recorded. Do the data allow the professor to conclude that the results of the midterm negatively influence student opinion?

19.94 Xr19-94 The town of Stratford, Ontario, is very much dependent upon the Shakespearean Festival it holds every summer for its financial well-being. Thousands of people visit Stratford to attend one or more Shakespearean plays and spend money in hotels, restaurants, and gift shops. As a consequence, any sign that the number of visitors will decrease in the future is cause for concern. Two years ago, a survey of 100 visitors asked how likely it was that they would return within the next 2 years. This year the survey was repeated with another 100 visitors. The likelihood of returning within 2 years was measured as

4 = Very likely
3 = Somewhat likely
2 = Somewhat unlikely
1 = Very unlikely

Conduct whichever statistical procedures you deem necessary to determine whether the citizens of Stratford should be concerned about the results of the two surveys.

19.95 Xr19-95 Scientists have been studying the effects of lead in children's blood, bones, and tissue for a number of years. It is known that lead reduces intelligence and can cause a variety of other problems. A study directed by Dr. Herman Needleman, a psychiatrist at the University of Pittsburgh Medical Center, examined some of these problems. Two hundred boys attending public schools in Pittsburgh were recruited. Each boy was categorized as having low or high levels of lead in their bones. Each boy was then assessed by his teachers on a 4-point scale (1 = low, 2 = moderate, 3 = high, and 4 = extreme) on degrees of aggression. Is there evidence to infer that boys with high levels of lead are more aggressive than boys with low levels of lead?

19.96 Xr19-96 How does gender affect teaching evaluations? Several researchers addressed this question during the past decade. (See Peter Seldin, "The Use and Abuse of Student Ratings of Professors," *Chronicle of Higher Education*, July 2, 1993.) In one study several female and male professors in the same department with similar backgrounds were selected. A random sample of 100 female students was drawn. Each student evaluated a female professor and a male professor. A sample of 100 male students was drawn and each also evaluated a female professor and a male professor. The ratings were based on a 4-point scale where 1 = poor, 2 = fair, 3 = good, and 4 = excellent. The evaluations were recorded in the following way:

Column 1 = Female student
Column 2 = Female professor rating
Column 3 = Male professor rating
Column 4 = Male student
Column 5 = Female professor rating
Column 6 = Male professor rating

a. Can we infer that female students rate female professors higher than they rate male professors?
b. Can we infer that male students rate male professors higher than they rate female professors?

19.97 Xr19-97 It is an unfortunate fact of life that the characteristics that one is born with play a critical role in later life. For example, race is a critical factor in almost all aspects of North American life. Height and weight also determine how friends, teachers, employers, and customers will treat you. And now we may add physical attractiveness to this list. A recent study conducted by economists Jeff Biddle of Michigan State University and Daniel Hamermesh from the University of Texas followed the careers of students from a prestigious U.S. law school. A panel of independent raters examined the graduation yearbook photos of the students and rated their appearance as unattractive, neither attractive nor unattractive, or attractive. The annual incomes in thousands of dollars 5 years after graduation were recorded. Assuming that incomes are not normally distributed, can we infer that incomes of lawyers are affected by physical attractiveness?

19.98 Xr19-98 According to a CNN news report broadcast on November 26, 1995, 9% of full-time workers telecommute. This means that they do not work in their employers' offices but instead perform their work at home using a computer and modem. To ascertain whether such workers are more satisfied than their nontelecommuting counterparts, a study was undertaken. A random sample of telecommuters and regular office workers was taken. Each was asked how satisfied they were with their current employment. The responses are 1 = very unsatisfied; 2 = somewhat unsatisfied; 3 = somewhat satisfied; and 4 = very satisfied. What conclusions can we draw from these data?

19.99 Xr19-99 Can you become addicted to exercise? In a study conducted at the University of Wisconsin at Madison, a random sample of dedicated exercisers who usually work out every day was drawn. Each completed a questionnaire that gauged their mood on a 5-point scale, where 5 = very relaxed and happy, 4 = somewhat relaxed and happy, 3 = neutral feeling, 2 = tense and anxious, and 1 = very tense and anxious. The group was then instructed to abstain from all workouts for the next 3 days. Moreover, they were told to be as physically inactive as possible. Each day their mood was measured using the same questionnaire. Column 1 stores the code identifying the respondent and columns 2 through 5 store the measures of mood for the day before the experiment began and for the 3 days of the experiment, respectively.

a. Can we infer that for each day the exercisers abstained from physical activity they were less happy than when they were exercising?
b. Do the data indicate that by the third day moods were improving?
c. Draw two possible conclusions from your findings.

19.100 Xr19-100 How does alcohol affect judgment? To provide some insight, an experiment was conducted. A random sample of customers of an Ohio club was selected. Each respondent was asked to assess the attractiveness of members of the opposite sex who were in the club at the time. The assessment was to be made on a 5-point scale (1 = very unattractive, 2 = unattractive, 3 = neither attractive nor unattractive, 4 = attractive, and 5 = very attractive). The survey was conducted 3 hours before closing and again just before closing using another group of respondents. Can we conclude that the assessments made just before closing are higher than those made 3 hours earlier? If so, what does this imply about the effects of alcohol on judgment? (The survey results were reported in the September 1997 edition of the *Report on Business*.)

CASE 19.1 Customer Ratings of an Automobile Service Center

© Digital Vision/ Royalty-free/Getty Images

DATA C19-01

A number of retailers regularly survey their customers to determine among other things, whether they were happy with their purchase or service and whether they intended to return. A chain of hardware stores/automobile service centers is one such company. At the completion of repair work, customers are asked to fill out the following form:

Tell us what you think.

ARE YOU SATISFIED?	VERY GOOD	GOOD	FAIR	POOR
1. Quality of work performed				
2. Fairness of price				
3. Explanation of work and guarantee				
4. Checkout process				
5. Will return in future				

Comments? YES NO

A random sample of 134 responses was drawn. The responses to questions 1 through 4 (1 = poor; 2 = fair; 3 = good, 4 = very good) are stored in columns 1 through 4, respectively. Responses to question 5 (2 = yes; 1 = no) are stored in column 5. Column 6 stores a 1 if a positive comment was made, 2 if a negative comment was made, and 3 if no comment was made.

a. Can we infer that those who say they will return assess each category higher than those who will not return?
b. Is there sufficient evidence to infer that those who make positive comments, negative comments, and no comments differ in their assessment of each category?
c. Prepare a presentation for the company's executives describing your analysis.

APPENDIX 19 Review of Statistical Inference, Chapters 12 to 19

Although there are four more chapters to go in this book, we have completed our presentation of statistical inference. (The remaining techniques, times-series analysis and forecasting, statistical process control, and decision analysis address different kinds of problems, which tend to be easy to identify.) The list of statistical techniques in Table A19.1 and the flowchart in Figure A19.1 now contain all the statistical inference methods presented in this book. Use them to determine how each of the exercises and cases is to be addressed. Because these exercises and cases were drawn from a wide variety of applications and collectively require the use of all the techniques introduced in this book, they provide the same kind of challenge faced by real statistics practitioners. By attempting to solve these problems, you will be getting realistic exposure to statistical applications. Incidentally, this also provides practice in the approach required to succeed in a statistics course examination.

TABLE **A19.1**
Summary of Statistical Techniques in Chapters 12 to 19

Problem objective: Describe a population

Data type: Interval

Descriptive measurement: Central location

Parameter: μ

Test statistic: $t = \dfrac{\bar{x} - \mu}{s/\sqrt{n}}$

Interval estimator: $\bar{x} \pm t_{\alpha/2}\dfrac{s}{\sqrt{n}}$

Required condition: Population is normal.

Descriptive measurement: Variability

Parameter: σ^2

Test statistic: $\chi^2 = \dfrac{(n-1)s^2}{\sigma^2}$

Interval estimator: LCL $= \dfrac{(n-1)s^2}{\chi^2_{\alpha/2}}$ UCL $= \dfrac{(n-1)s^2}{\chi^2_{1-\alpha/2}}$

Required condition: Population is normal.

Data type: Nominal

Number of categories: Two

Parameter: p

Test statistic: $z = \dfrac{\hat{p} - p}{\sqrt{p(1-p)/n}}$

Interval estimator: $\hat{p} \pm z_{\alpha/2}\sqrt{\hat{p}(1-\hat{p})/n}$

Required condition: $np \geq 5$ and $n(1-p) \geq 5$ (for test)

$n\hat{p} \geq 5$ and $n(1-\hat{p}) \geq 5$ for estimate

(continued)

TABLE **A19.1**
(continued)

Number of categories: Two or more

Parameters: $p_1, p_2, \ldots, p_k$

Statistical technique: Chi-squared goodness-of-fit

Test statistic: $\chi^2 = \sum \frac{(f_i - e_i)^2}{e_i}$

Required condition: $e_i \geq 5$

Problem objective: Compare two populations

Data type: Interval

Descriptive measurement: Central location

Experimental design: Independent samples

Population variances: $\sigma_1^2 = \sigma_2^2$

Parameter: $\mu_1 - \mu_2$

Test statistic: $t = \dfrac{(\bar{x}_1 - \bar{x}_2) - (\mu_1 - \mu_2)}{\sqrt{s_p^2\left(\dfrac{1}{n_1} + \dfrac{1}{n_2}\right)}}$

Interval estimator: $(\bar{x}_1 - \bar{x}_2) \pm t_{\alpha/2}\sqrt{s_p^2\left(\dfrac{1}{n_1} + \dfrac{1}{n_2}\right)}$

Required condition: Populations are normal.

If populations are nonnormal, apply the Wilcoxon rank sum test.

Population variances: $\sigma_1^2 \neq \sigma_2^2$

Parameter: $\mu_1 - \mu_2$

Test statistic: $t = \dfrac{(\bar{x}_1 - \bar{x}_2) - (\mu_1 - \mu_2)}{\sqrt{\left(\dfrac{s_1^2}{n_1} + \dfrac{s_2^2}{n_2}\right)}}$

Interval estimator: $(\bar{x}_1 - \bar{x}_2) \pm t_{\alpha/2}\sqrt{\left(\dfrac{s_1^2}{n_1} + \dfrac{s_2^2}{n_2}\right)}$

Required condition: Populations are normal.

Experimental design: Matched pairs

Parameter: μ_D

Test statistic: $t = \dfrac{\bar{x}_D - \mu_D}{s_D/\sqrt{n_D}}$

Interval estimator: $\bar{x}_D \pm t_{\alpha/2}\dfrac{s_D}{\sqrt{n_D}}$

Required condition: Differences are normal.

If differences are nonnormal, apply Wilcoxon signed rank sum test.

Nonparametric technique: Wilcoxon signed rank sum test

Test statistic: $z = \dfrac{T - E(T)}{\sigma_T}$

Required condition: Populations are identical in shape and spread.

TABLE **A19.1**
(continued)

Descriptive measurement: Variability

Parameter: σ_1^2/σ_2^2

Test statistic: $F = \dfrac{s_1^2}{s_2^2}$

Interval estimator: $\text{LCL} = \left(\dfrac{s_1^2}{s_2^2}\right)\dfrac{1}{F_{\alpha/2,\nu_1,\nu_2}}$ $\quad \text{UCL} = \left(\dfrac{s_1^2}{s_2^2}\right)F_{\alpha/2,\nu_2,\nu_1}$

Required condition: Populations are normal.

Data type: Ordinal

Experimental design: Independent samples

Nonparametric technique: Wilcoxon rank sum test

Test statistic: $z = \dfrac{T - E(T)}{\sigma_T}$

Required condition: Populations are identical in shape and spread.

Experimental design: Matched pairs

Nonparametric technique: Sign test

Test statistic: $z = \dfrac{x - .5n}{.5\sqrt{n}}$

Required condition: Populations are identical in shape and spread.

Data type: Nominal

Number of categories: Two

Parameter: $p_1 - p_2$

Test statistic:

Case 1: H_0: $p_1 - p_2 = 0$ $\quad z = \dfrac{(\hat{p}_1 - \hat{p}_2)}{\sqrt{\hat{p}(1 - \hat{p})\left(\dfrac{1}{n_1} + \dfrac{1}{n_2}\right)}}$

Case 2: H_0: $p_1 - p_2 = D$ $\quad (D \neq 0)$ $\quad z = \dfrac{(\hat{p}_1 - \hat{p}_2) - (p_1 - p_2)}{\sqrt{\dfrac{\hat{p}_1(1 - \hat{p}_1)}{n_1} + \dfrac{\hat{p}_2(1 - \hat{p}_2)}{n_2}}}$

Interval estimator: $(\hat{p}_1 - \hat{p}_2) \pm z_{\alpha/2}\sqrt{\dfrac{\hat{p}_1(1 - \hat{p}_1)}{n_1} + \dfrac{\hat{p}_2(1 - \hat{p}_2)}{n_2}}$

Required condition: $n_1\hat{p}_1$, $n_1(1 - \hat{p}_1)$, $n_2\hat{p}_2$, and $n_2(1 - \hat{p}_2) \geq 5$

Number of categories: Two or more

Statistical technique: Chi-squared test of a contingency table

Test statistic: $\chi^2 = \sum \dfrac{(f_i - e_i)^2}{e_i}$

Required condition: $e_i \geq 5$

(continued)

TABLE **A19.1**
(continued)

Problem objective: Compare two or more populations

Data type: Interval

Experimental design: Independent samples

Number of factors: One

Parameters: $\mu_1, \mu_2, \ldots, \mu_k$

Statistical technique: (One-way analysis of variance)

Test statistic: $F = \frac{\text{MST}}{\text{MSE}}$

Statistical technique: Multiple comparisons:

Fisher and Bonferroni adjustment: $\text{LSD} = t_{\alpha/2}\sqrt{\text{MSE}\left(\frac{1}{n_i} + \frac{1}{n_j}\right)}$

Tukey: $\omega = q_\alpha(k, \nu)\sqrt{\frac{\text{MSE}}{n_g}}$

Required conditions: Populations are normal with equal variances.

Number of factors: Two

Parameters: $\mu_1, \mu_2, \ldots, \mu_k$

Statistical technique: (Two-factor analysis of variance)

Test statistics: $F = \frac{\text{MS(AB)}}{\text{MSE}} \quad F = \frac{\text{MS(A)}}{\text{MSE}} \quad F = \frac{\text{MS(B)}}{\text{MSE}}$

Required conditions: Populations are normal with equal variances.

Experimental design: Randomized blocks

Parameters: $\mu_1, \mu_2, \ldots, \mu_k$

Statistical technique: (Two-way analysis of variance)

Test statistics: $F = \frac{\text{MST}}{\text{MSE}} \quad F = \frac{\text{MSB}}{\text{MSE}}$

Required conditions: Populations are normal with equal variances.

Data type: Nominal

Number of categories: Two or more

Statistical technique: Chi-squared test of a contingency table

Test statistic: $\chi^2 = \sum \frac{(f_i - e_i)^2}{e_i}$

Required condition: $e_i \geq 5$

TABLE **A19.1** *(continued)*

Problem objective: Analyze the relationship between two variables

Data type: Interval

Parameters: β_0, β_1, ρ

Statistical technique: Simple linear regression and correlation

Test statistic: $t = \frac{b_1 - \beta_1}{s_{b_1}}$; $t = r\sqrt{\frac{n-2}{1-r^2}}$

Prediction interval: $\hat{y} \pm t_{\alpha/2,n-2}s_\varepsilon\sqrt{1 + \frac{1}{n} + \frac{(x_g - \bar{x})^2}{(n-1)s_x^2}}$

Interval estimator of expected value: $\hat{y} \pm t_{\alpha/2,n-2}s_\varepsilon\sqrt{\frac{1}{n} + \frac{(x_g - \bar{x})^2}{(n-1)s_x^2}}$

Required conditions: ε is normally distributed with mean 0 and standard deviation σ_ε; ε values are independent.

Data type: Nominal

Statistical technique: Chi-squared test of a contingency table

Test statistic: $\chi^2 = \sum \frac{(f_i - e_i)^2}{e_i}$

Required condition: $e_i \geq 5$

Problem objective: Analyze the relationship among two or more variables

Data type: Interval

Parameters: $\beta_0, \beta_1, \beta_2, \ldots, \beta_k$

Statistical technique: multiple regression

Test statistics: $t = \frac{b_i - \beta_i}{s_{b_i}}$ $(i = 1, 2, \ldots, k)$; $F = \frac{\text{MSR}}{\text{MSE}}$

Required conditions: ε is normally distributed with mean 0 and standard deviation σ_ε; ε values are independent.

FIGURE **A19.1**
Flowchart of All Statistical Inference Techniques

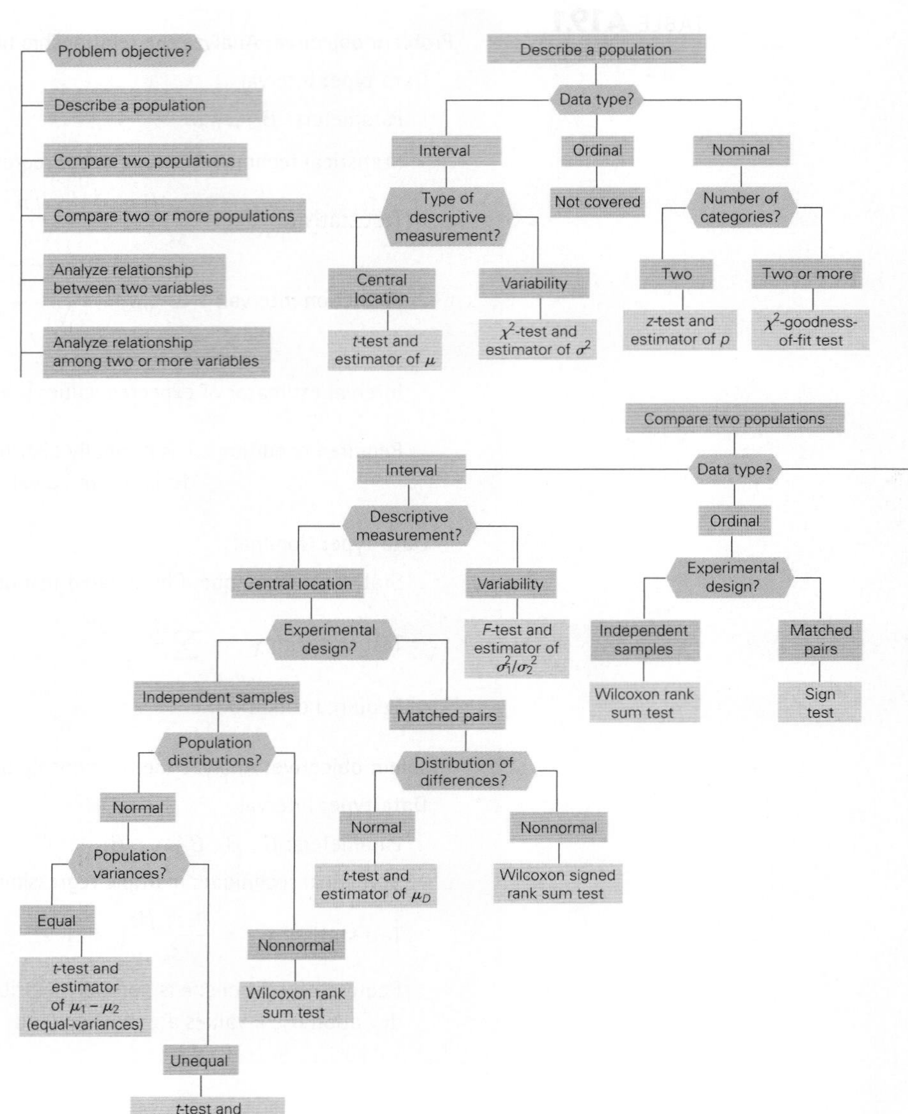

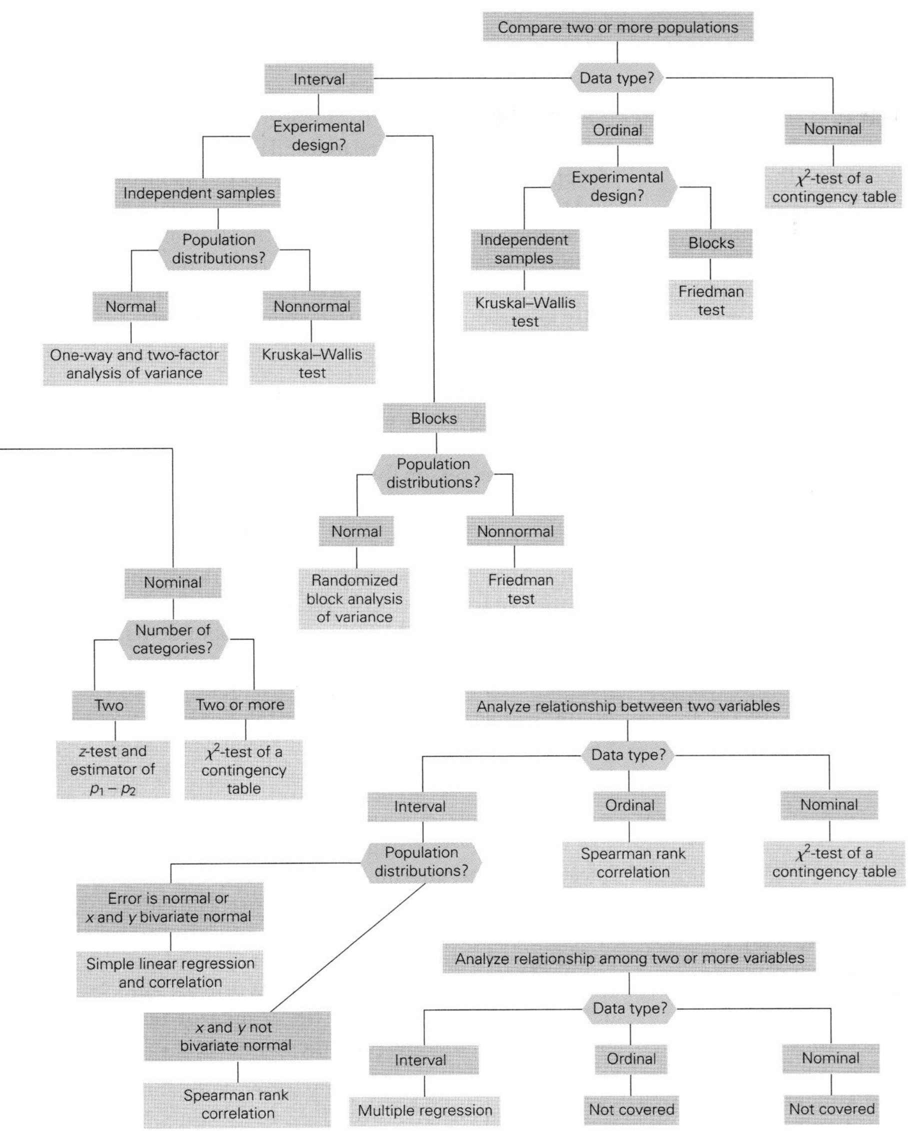

Compare two or more populations
Data type?
Interval
Experimental design?
Independent samples
Population distributions?
Normal
Nonnormal
One-way and two-factor analysis of variance
Kruskal–Wallis test
Blocks
Population distributions?
Normal
Nonnormal
Randomized block analysis of variance
Friedman test
Ordinal
Experimental design?
Independent samples
Kruskal–Wallis test
Blocks
Friedman test
Nominal
χ^2-test of a contingency table
Nominal
Number of categories?
Two
Two or more
z-test and estimator of $p_1 - p_2$
χ^2-test of a contingency table
Analyze relationship between two variables
Data type?
Interval
Population distributions?
Error is normal or x and y bivariate normal
Simple linear regression and correlation
x and y not bivariate normal
Spearman rank correlation
Ordinal
Spearman rank correlation
Nominal
χ^2-test of a contingency table
Analyze relationship among two or more variables
Data type?
Interval
Multiple regression
Ordinal
Not covered
Nominal
Not covered

EXERCISES

A19.1 XrA19-01 Most supermarkets load groceries into plastic bags. However, plastic bags take many years to decompose in garbage dumps. To ascertain the problem, a random sample of American households asked each to determine the number of plastic bags they use and discard in a week. The latest census reveals that there are 112 million households in the United States. Estimate with 95% confidence the total number of plastic bags discarded per week. (*Source:* NBC News May 7, 2007)

A19.2 XrA19-02 Some customers spend a great deal of time researching before choosing a particular brand of a product, particularly an expensive product. Does this result in a more satisfied customer? To shed light on this question, a random sample of people who purchased a new car within the last two years was drawn. Each was asked to report the amount of time spent researching (reading newspaper reports, brochures, Internet) in hours and their level of satisfaction measured in the following way:

1. Extremely dissatisfied
2. Somewhat dissatisfied
3. Neither satisfied or dissatisfied
4. Somewhat satisfied
5. Extremely satisfied

Do the data allow us to infer that those who do more research are more satisfied with their choice?

A19.3 Xr13-30* Refer to Exercise 13.30 where respondents were asked to taste the same wine in two different bottles. The first bottle was capped using a cork and the second had a metal screw cap. Respondents were also asked to taste the wine and rate it using the following categories:

Poor (1), fair (2), good (3), very good (4), excellent (5)

Do these data provide sufficient evidence to indicate that wine bottled with a screw cap is perceived to be inferior?

A19.4 XrA19-04 In an effort to analyze the results of Exercise A15.9, the researcher recorded the total travel length of the course. This variable measures the total distance golfers must walk to play a round of golf. It is the sum of the golf course playing distance plus the distance golfers must walk from the green to the next tee. What can you conclude from these data?

A19.5 XrA19-05* In city after city, downtown cores have become less and less populated and poorer because shoppers have taken their money to the suburbs and to shopping malls. One reason often given for the decline in downtown shops is the difficulty in parking. To shed more light on the issue, a random sample of 197 adults was asked to rate the difficulty in parking using the following responses:

Poor (1), Acceptable (2), Good (3), Very good (4), Excellent (5)

They also asked how often the subject shopped at a downtown store in a typical month. Do these data allow us to infer that the problem of parking is one reason for the decline in downtown shopping?

A19.6 XrA19-05* Refer to A19.5. To acquire some information about who is shopping downtown, the statistics practitioner also recorded the annual household income ($1,000) of the respondents. Is there enough evidence to infer that affluent people shop downtown more frequently than poorer people? (The author is grateful to Patricia Gafoor-Darlington and Michael Kirby-MacLean for writing these exercises.)

A19.7 XrA19-07 Why do some students do well in university while others struggle? To help answer this question a random sample of first-year students at four universities was selected. Those who had a grade point average of more than 3.0 (group 1) and those who had a grade point average of less than 2.0 (group 2) were surveyed. For each student, researchers recorded the results of tests (score 0 to 10) that measure the following:

Interpersonal skills (strong social skills, ability to interact effectively)

Stress management (being able to work well under pressure or resist and delay an impulse)

(*Source: National Post* August 16, 2004.) Do these data provide sufficient evidence to infer that students whose GPA is more than 3.0 score higher in interpersonal skills and stress management than do students whose GPA is less than 2.0?

A19.8 XrA19-08 The issue of immigration, legal and illegal, has political and economic ramifications. An important component of the issue is how well immigrants integrate into the American economy. A University of Florida study attempted to answer this question. They randomly surveyed U.S. born Americans, immigrants who arrived in the United States before 1980, and immigrants who arrived after 1980 in Miami-Dade, the state of Florida, and in the United States. Each respondent had full-time employment and was asked to report their annual earnings. Conduct tests to determine whether differences exist between the three groups in each of the three geographic regions.

A19.9 XrA19-09 The high price of gasoline is likely to lead to less travel. An economist specializing in energy uses wanted to learn more about driving habits and, in particular, distances traveled by cars, buses, and vans, pickups and SUVs. In the latest year available (2003), there were 136 million cars, 776,000 buses, and 87 million vans, pickups, and SUVs. (*Source:* U.S. Federal Highway Administration, *Highway Statistics*) A survey of each type of vehicle was undertaken and the number of miles driven (1,000) was recorded. For each type of vehicle, estimate with 95% confidence the mean number of miles driven and the total number of miles driven. (Adapted from the *Statistical Abstract of the United States* 2006, Table 1084)

A19.10 XrA19-10 Refer to Exercise A19.9 The economist also wanted to know whether there are differences in miles driven between cars, buses, and vans, pickups, and SUVs.

A19.11 XrA19-11 An important measure of the health of a nation's economy is total debt. A Canadian survey asked a random sample of households how much money ($1,000) they owed. This includes mortgages, loans, and credit card debt. Assuming that there are 10 million households in Canada, estimate with 95% confidence the total debt in Canada.

A19.12 XrA19-12 In Chapter 7 we showed that diversification reduces the risk associated with a portfolio of investments. (Most experts advise their clients that the portfolios should contain between 20 and 30 stocks scattered in different industries.) Do investors understand this concept? To help answer this question, a random sample of investors' portfolios was sampled. This survey was a duplicate of the one done 5 years ago. The number of stocks in each sampled portfolio was recorded. Do these data allow us to infer that investors' portfolios are becoming more diverse? (Adapted from W. Goetzman and A. Kumar, "Equity Portfolio Diversification," NBER Paper 8686)

A19.13 XrA19-13 The cost of taking an extra year to earn an MBA is quite high. To determine whether it is worthwhile, a BBA graduate surveyed 200 people who had either a BBA or an MBA and recorded their annual salary ($1,000) after 5 years of work. The student determined that the added cost of the MBA is warranted only if the mean income of MBAs is more than $5,000 greater than that of BBA's. Can we conclude that acquiring an MBA is worthwhile?

A19.14 XrA19-14 Flonase is a nasal allergy treatment and, like all other drugs, has side effects. Before approving a drug, the company (GlaxoSmithKline) performs a number of experiments to determine the side effects. In one such experiment, a random sample of 1707 volunteers was drawn. Of these, 167 were given a 100 mcg dose once a day (1), 782 were given 200 mcg daily (2), and the remaining 758 were given a placebo spray once a day (3). Each person reported whether they had any side effects, and if so, what was the most serious. The following data were recorded:

1. Headache
2. Pharyngitis (sore throat)
3. Epistaxis (nosebleed)
4. Other side effect
5. No side effect

Do these data allow us to infer that there are differences in side effects between the three groups of volunteers?

A19.15 XrA19-15 Simco Inc. is a manufacturer that purchased a new piece of equipment designed to reduce costs. After several months of operation, the results were quite unsatisfactory. The operations manager believes that the problem lies with the machine's operators, who were unable to master the required skills. It was decided to establish a training program to upgrade the skills of those workers with the greatest likelihood of success. To do so, the company needed to know which skills are most needed to run the machine. Experts identified six such skills. They are dexterity, attention to detail, teamwork skills, mathematical ability, problem-solving skills, and technical knowledge. To examine the issue, a random sample of workers was drawn. Workers were measured on each of the six skills through a series of paper-and-pencil tests and through supervisor ratings. Additionally, each worker received a score on the quality of his or her actual work on the machine. These data are stored in columns 1 through 7. (Column 1 stores the quality-of-work scores and columns 2 to 7 are the scores on the skill tests. All data are interval.) Identify the skills that affect the quality of work. (We are grateful to Scott Bergen for writing this exercise.)

A19.16 XrA19-16 Obesity among children in North America is said to be at near-epidemic proportions. Some experts blame television for the problem, citing the statistic that children watch on average about 26 hours per week. During this time, children are not engaged in any physical activity, which results in weight gain. However, the problem may be compounded by a reduction in metabolic rate. In an experiment to address this issue (the study results were published in the February 1993 issue of the medical journal *Pediatrics*), scientists from Memphis State University and the University of Tennessee at Memphis took a random sample of 223 children aged 8 to 12, of whom 41 were obese. Each child's metabolic rate (the amount of calories burned per hour) was measured while at rest and also measured while the child watched a television program (*The Wonder Years*). The differences between the two rates were recorded where column 1 contains the numbers representing the decrease in metabolic rate and column 2 codes the children as 1 = obese and 2 = nonobese.

a. Do these data allow us to conclude that there is a decrease in metabolism when children watch television?
b. Can we conclude that the decrease in metabolism while watching television is greater among obese children?

A19.17 XrA19-17 The game of Scrabble is one of the oldest and most popular board games. It is played all over the world and there is even an annual world championship competition. The game is played by forming words and placing them on the board to obtain the maximum number of points. It is generally believed that a large vocabulary is the only skill required to be successful. However, there is a strategic element to the game that suggests that mathematical skills are just as necessary. To determine which skills are most in demand, a statistician recruited a random sample of fourth-year university English and mathematics majors and asked them to play the game. A total of 500 games was played by different pairs of English and mathematics majors. The scores in each game were recorded. (The authors would like to thank Scott Bergen for his assistance in writing this exercise.)

a. Can we conclude that mathematics majors win more frequently than do English majors?
b. Do these data allow us to infer that the average score obtained by English majors is greater than that for mathematics majors?
c. Why are the results of Parts a and b not the same?

A19.18 XrA19-18 Since germs have been discovered, parents have been telling their children to wash their hands. Common sense tells us that this should help minimize the spread of infectious diseases and lead to better health. A study reported in the *University of California at Berkeley Wellness Letter* (Volume 13, Issue 6, March 1997) may confirm the advice our parents gave us. A study in Michigan tracked a random sample of children, some of whom washed their hands four or more times during the school day. The number of sick days due to colds and flu and the number of sick days due to stomach illness were recorded for the past year. Column 1 contains a code representing whether the child washed his or her hands four or more times per school day (1) or not (2). Column 2 stores the number of sick days due to cold and flu and column 3 contains the number of sick days due to stomach illness.

a. Do these data allow us to infer that a child who washed his or her hands four or more times during the school day will have fewer sick days due to cold and flu than other children?
b. Repeat Part a for sick days due to stomach illness.

A19.19 XrA19-19 Under the rules of Canada's Employment Insurance (EI) plan, some workers can use EI repeatedly after working only a short time. The amount of time needed to qualify for EI varies by region and by occupation. In a study undertaken by researchers, regular users of EI were surveyed and asked, among other questions, how frequently they used EI and how satisfied they were with their employment situation. The responses are

1. Very unsatisfied
2. Somewhat unsatisfied
3. Neither unsatisfied or satisfied
4. Somewhat satisfied
5. Very satisfied

Do the data allow us to conclude that workers who use EI more often are more satisfied with their employment situation? (*Source: National Post Business*, July 2001)

A19.20 XrA19-20 Winter is the influenza season in North America. Each winter thousands of elderly and sick people die from the flu and its attendant complications. Consequently, many elderly people receive flu shots in the fall. It has generally been accepted that young healthy North Americans need not receive flu shots because, although many contract the disease, few die from it. However, there are economic consequences. Sick days cost both the employee and employer. A study published in the *New England Journal of Medicine* reported the results of an experiment to determine whether it is useful for young healthy people to take flu shots. A random sample of working adults was selected. Half received a flu shot in November; the other half received a placebo. The numbers of sick days over the next 6-month period were recorded in columns 1 (flu shot) and 2 (placebo). Columns 3 (flu shot) and 4 (placebo) contain the number of visits to the doctor.

a. Can we conclude that the number of sick days is less for those who take flu shots?
b. Can we conclude that those who take flu shots visit their doctors less frequently?

A19.21 XrA19-21 The high cost of medical care makes it imperative that hospitals operate efficiently and effectively. As part of a larger study, patients leaving a hospital were surveyed. They were asked how satisfied they were with the treatment they received. The responses were recorded with a measure of the degree of severity of their illness (as determined by the admitting physician) and the length of stay. These data are recorded in the following way:

Column 1: Satisfaction level (1 = very unsatisfied; 2 = somewhat unsatisfied; 3 = neither satisfied nor dissatisfied; 4 = somewhat satisfied; 5 = very satisfied)

Column 2: Severity of illness (1 = least severe and 10 = most severe)

Column 3: Number of days in hospital

a. Is the satisfaction level affected by the severity of illness?
b. Is the satisfaction level higher for patients who stay for shorter periods of time?

A19.22 XrA19-22 What should be the priority of paramedics who respond to accidents? Should they treat the patients with their limited facilities or should they rush the victims to the nearest hospital (an approach known as "scoop and run")? A research project begun in 1993 may provide the answer. Researchers looked at the care of 1,846 trauma patients—those with life-threatening injuries in Montreal (1), Toronto (2), and Quebec City (3). Montreal use physicians to provide advanced life support (ALS) at the scene of the accident. Toronto uses paramedics to provide ALS and Quebec City uses emergency medical services who apply only basic life support. The outcomes (survived = 1, died = 2) and city were recorded. Determine whether there are differences in the death rate between the three cities. What recommendation would you make? (Adapted from *Annals of Surgery*, 2002. The author is grateful to Jie Hunag for creating this exercise.)

A19.23 XrA19-23 How many golfers are there in the United States? A survey of American adults (age 18 and above) asked each whether they had played golf at least once a month during the summer. The responses were 2 = yes and 1 = no. The survey also asked respondents to indicate which of the following household income category they fell into:

1. Under $15,000
2. $15,000 to $24,999
3. $25,000 to $34,999
4. $35,000 to $49,999
5. $50,000 to $75,000
6. Over $75,000

The latest census reveals that the number of American households in each of the income categories is as follows:

1. 75.7 million
2. 36.9 million
3. 28.3 million
4. 27.8 million
5. 21.9 million
6. 17.1 million

(*Source*: *Statistical Abstract of the United States*, 2006, Tables 685 and 1238)

a. Estimate with 95% confidence the total number of golfers.
b. Estimate with 95% confidence the number of golfers who earn at least $75,000.
c. Test to determine whether income is a determinant in who plays golf.

A19.24 XrA19-24 One of the arguments put forth by advocates of lower tuition fees is that children of low- or moderate-income families will not be able to pay for a child's university education. To examine this issue, a random sample of families whose children were at least 20 years old was drawn. Each family was asked to specify which of the following household income categories they fell into and whether at least one child had attended university (2 = yes and 1 = no). (*Source: Globe and Mail*, October 15, 2003)

1. Under $25,000
2. $50,000 to $75,000
3. Over $100,000

Do these data allow researchers to conclude that family income affects whether children attend university?

A19.25 XrA19-25 Because of the high cost of hospital stays, anything that can reduce the length of stays and the costs of medication would be appreciated by insurance companies, hospitals, and patients. A physician-researcher was searching for ways to reduce costs and decided to investigate the effect of the room the patient stayed in. She gathered data on the length of stay and the amount of pain medication (measured in morphine equivalents). Also recorded was whether the room was sunny or dim. Do these data allow the researcher to conclude that the length of stay and the amount of pain medication is lower in bright rooms than in dim ones? (*Source: USA Today*, March 3, 2004)

A19.26 Repeat Cases 13.1 (page 504) and A16.1 (page 670), assuming that interest rates, annual gross sales, and ages of businesses are not normally distributed.

CASE A19.1 Type A, B, and C Personalities and Job Satisfaction and Performance

DATA CA19-01

Some psychologists believe that there are at least three different personality types: Type A is the aggressive workaholic; type B is the relaxed underachiever; and type C displays various characteristics of types A and B. The personnel manager of a large insurance company believes that, among life insurance salespersons, there are equal numbers of all three personality types and that their degree of satisfaction and sales ability are not influenced by personality type. In a survey of 150 randomly selected salespersons, he determined the type of personality, measured their job satisfaction on a 7-point scale (where 1 = very dissatisfied and 7 = very satisfied), and determined the total amount of life insurance sold by each during the previous year. The results were recorded using the following format:

Column 1: Personality type (A = 1; B = 2; C = 3)
Column 2: Job satisfaction
Column 3: Total amount of life insurance sold (in hundreds of thousands of dollars)

Test all three of the personnel manager's beliefs. That is, test to determine whether there is enough evidence to justify the following conclusions:

a. The proportions of each personality type are different.
b. The job satisfaction measures for each personality type are different.
c. The life insurance sales for each personality type are different.

CASE A19.2 Do Banks Discriminate against Women Business Owners? Part 3

DATA CA19-02

A statistician made a final effort to determine whether banks discriminate against women business owners because of their gender (see Cases 13.1 and A16.1). For each of the women business owners who had received a loan, he attempted to find a male business owner whose characteristics closely matched. The matching was done on the basis of type of business (proprietorship, partnership, or corporation), gross sales, and age of company. A match was made when the type of business was the same, the gross sales were within $10,000 of each other, and the ages were within 1 year of each other. The interest rates (points above prime) for each pair were recorded in columns 2 (women's rates) and 3 (men's rates); column 1 stores the pair number. What do these data tell you? Is this analysis definitive? Explain.

© Walter Hodges/Brand X Pictures/Jupiterimages

20 TIME-SERIES ANALYSIS AND FORECASTING

Housing Starts

DATA Xm20-00

At the end of 2005, a major builder of residential houses in the northeastern United States wanted to predict the number of housing units to be started in 2006. This information would be extremely useful in determining a variety of variables, including housing demand, availability of labor, and the price of building materials. To help develop an accurate forecasting model, an economist collected data on the number of housing starts (in thousands) for the previous 60 months (2001–2005). Forecast the number of housing starts for the 12 months of 2006. (*Source:* Standard & Poor's Industry Surveys)

© John Lund/Drew Kelly/Blend Images/Jupiterimages

See page 844 for the answer.

Introduction

Any variable that is measured over time in sequential order is called a **time series.** We introduced times series in Chapter 2 and demonstrated how we use a line chart to graphically display the data. Our objective in this chapter is to analyze time series in order to detect patterns that will enable us to forecast future values of the time series. There is an almost unlimited number of such applications in management and economics. Some examples follow.

1. Governments want to know future values of interest rates, unemployment rates, and percentage increases in the cost of living.
2. Housing industry economists must forecast mortgage interest rates, demand for housing, and the cost of building materials.
3. Many companies attempt to predicts the demand for their products and their share of the market.
4. Universities and colleges often try to forecast the number of students who will be applying for acceptance at postsecondary-school institutions.

Forecasting is a common practice among managers and government decision makers. This chapter will focus on time-series forecasting, which is forecasting that uses historical time series data to predict future values of variables such as sales or unemployment rates. This entire chapter is an application tool for both economists and managers in all functional areas of business because forecasting is such a vital factor in decision making in these areas.

For example, the starting point for aggregate production planning by operations managers is to forecast demand for the company's products. These forecasts will make use of economists' forecasts of macroeconomic variables (such as gross domestic product, disposable income, and housing starts) as well as the marketing managers' internal forecasts of their customers' future needs. Not only are these sales forecasts critical to production planning, but they are also the key to accurate pro forma (i.e., forecasted) financial statements, which are produced by the accounting and financial managers to assist in their planning for future financial needs such as borrowing. Likewise, the human resources department will find such forecasts of a company's growth prospects to be invaluable in their planning for future manpower requirements.

There are many different forecasting techniques. Some are based on developing a model that attempts to analyze the relationship between a dependent variable and one or more independent variables. We presented some of these methods in the chapters on regression analysis (Chapters 16, 17, and 18). The forecasting methods to be discussed in this chapter are all based on time series, which we discuss in the next section. In Sections 20.2 and 20.3, we deal with methods for detecting and measuring which time-series components exist. After we uncover this information, we can develop forecasting tools. We will only scratch the surface of this topic. Our objective is to expose you to the concepts of forecasting and to introduce some of the simpler techniques. The level of this text precludes the investigation of more complicated methods.

A related subject, index numbers, is presented in CD Appendix AC.

20.1 Time Series Components

A time series can consist of four different components as described in the box.

Time-Series Components

1. Long-term trend
2. Cyclical variation
3. Seasonal variation
4. Random variation

A **trend** (also known as a secular trend) is a long-term, relatively smooth pattern or direction exhibited by a series. Its duration is more than 1 year. For example, the population of the United States exhibited a trend of relatively steady growth from 157 million in 1950 to 299 million in 2006. (The data are stored in Ch20:\Fig20-01.) Figure 20.1 exhibits the line chart.

FIGURE **20.1**
U.S. Population (millions) 1950–2006

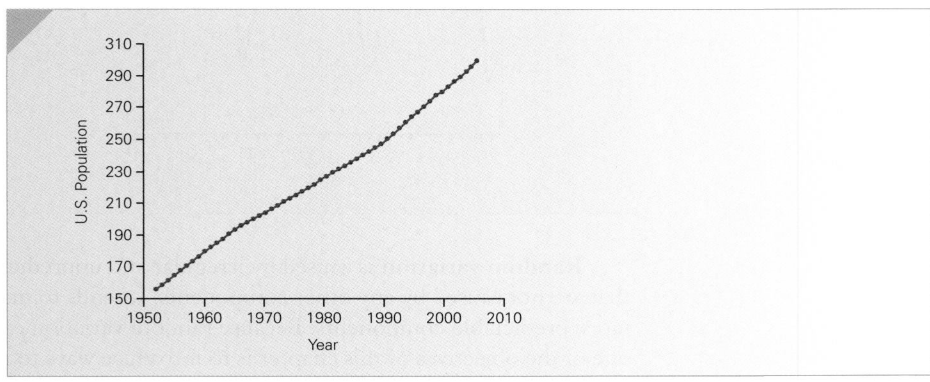

The trend of a time series is not always linear. For example, Figure 20.2 describes U.S. per capita beer consumption from 1970 to 2003. As you can see from the line chart in Figure 20.2, per capita consumption grew between 1970 to 1980 (data are stored in Ch20:\Fig20-02) and then more or less steadily decreased.

FIGURE **20.2**
U.S. Beer Consumption per Capita 1970–2004

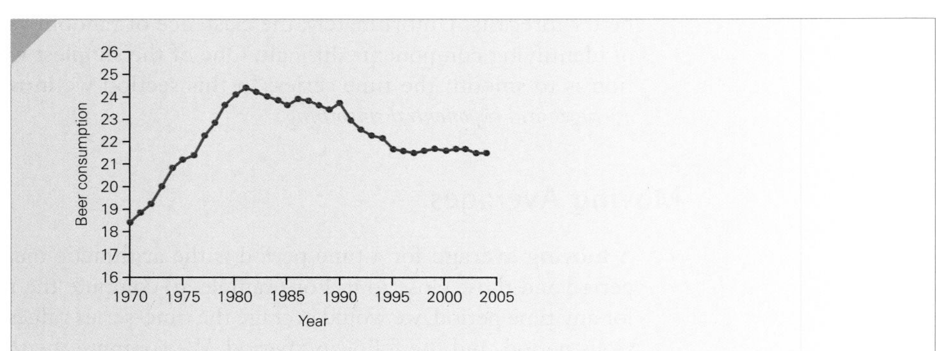

Cyclical variation is a wavelike pattern describing a long-term trend that is generally apparent over a number of years, resulting in a cyclical effect. By definition, it has duration of more than 1 year. Examples include business cycles that record periods of economic recession and inflation, long-term product-demand cycles, and cycles in monetary and financial sectors. However, cyclical patterns that are consistent and predictable are quite rare. For practical purposes, we will ignore this type of variation.

Seasonal variation refers to cycles that occur over short repetitive calendar periods and, by definition, have a duration of less than 1 year. The term *seasonal variation* may refer to the four traditional seasons or to systematic patterns that occur during a month, a week, or even 1 day. Demand for restaurants feature "seasonal" variation throughout the day.

An illustration of seasonal variation is provided in Figure 20.3, which graphs monthly U.S. traffic volume (in billions of miles and where period 1 is January). (Data are in Ch20:\Fig20-03.) It is obvious from the graph that Americans drive more during the summer months than during the winter months.

FIGURE **20.3** Traffic Volume (billions of miles)

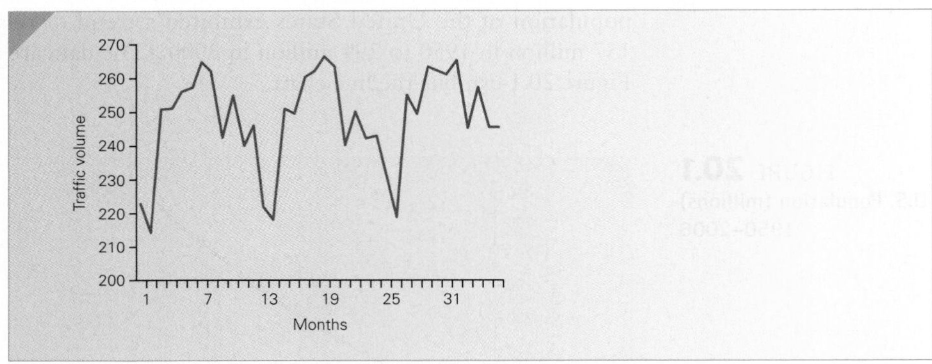

Random variation is caused by irregular and unpredictable changes in a time series that are not caused by any other components. It tends to mask the existence of the other more predictable components. Because random variation exists in almost all time series, one of the objectives of this chapter is to introduce ways to reduce the random variation, which will enable statistics practitioners to describe and measure the other components. By doing so, we hope to be able to make accurate predictions of the time series.

20.2 Smoothing Techniques

If we can determine which components actually exist in a time series, we can develop better forecasts. Unfortunately, the existence of random variation often makes the task of identifying components difficult. One of the simplest ways to reduce random variation is to smooth the time series. In this section we introduce two methods: *moving averages* and *exponential smoothing*.

Moving Averages

A **moving average** for a time period is the arithmetic mean of the values in that time period and those close to it. For example, to compute the three-period moving average for any time period, we would average the time-series values in that time period, the previous period, and the following period. We compute the three-period moving averages for all time periods except the first and the last. To calculate the five-period moving average, we average the value in that time period, the values in the two preceding periods, and the values in the two following time periods. We can choose any number of periods with which to calculate the moving averages.

EXAMPLE 20.1

Gasoline Sales, Part 1

DATA
Xm20-01

As part of an effort to forecast futures sales, an operator of five independent gas stations recorded the quarterly gasoline sales (in thousands of gallons) for the past 4 years. These data are shown below. Calculate the three-quarter and five-quarter moving averages. Draw graphs of the time series and the moving averages.

Time Period	Year	Quarter	Gasoline Sales (thousands of gallons)
1	1	1	39
2		2	37
3		3	61
4		4	58
5	2	1	18
6		2	56
7		3	82
8		4	27
9	3	1	41
10		2	69
11		3	49
12		4	66
13	4	1	54
14		2	42
15		3	90
16		4	66

SOLUTION

COMPUTE

MANUALLY

To compute the first three-quarter moving average, we group the gasoline sales in periods 1, 2, and 3, and then average them. Thus, the first moving average is

$$\frac{39 + 37 + 61}{3} = \frac{137}{3} = 45.7$$

The second moving average is calculated by dropping the first period's sales (39), adding the fourth period's sales (58), and then computing the new average. Thus, the second moving average is

$$\frac{37 + 61 + 58}{3} = \frac{156}{3} = 52.0$$

The process continues as shown in the following table. Similar calculations are made to produce the five-quarter moving averages (also shown in the table).

Time Period	Gasoline Sales	Three-Quarter Moving Average	Five-Quarter Moving Average
1	39	—	—
2	37	45.7	—
3	61	52.0	42.6
4	58	45.7	46.0
5	18	44.0	55.0
6	56	52.0	48.2
7	82	55.0	44.8
8	27	50.0	55.0
9	41	45.7	53.6
10	69	53.0	50.4
11	49	61.3	55.8
12	66	56.3	56.0
13	54	54.0	60.2
14	42	62.0	63.6
15	90	66.0	—
16	66	—	—

Notice that we place the moving averages in the center of the group of values being averaged. It is for this reason that we prefer to use an odd number of periods in the moving averages. Later in this section, we discuss how to deal with an even number of periods.

Figure 20.4 displays the line chart for gasoline sales and Figure 20.5 shows the three-period and five-period moving average.

FIGURE **20.4** Quarterly Gasoline Sales

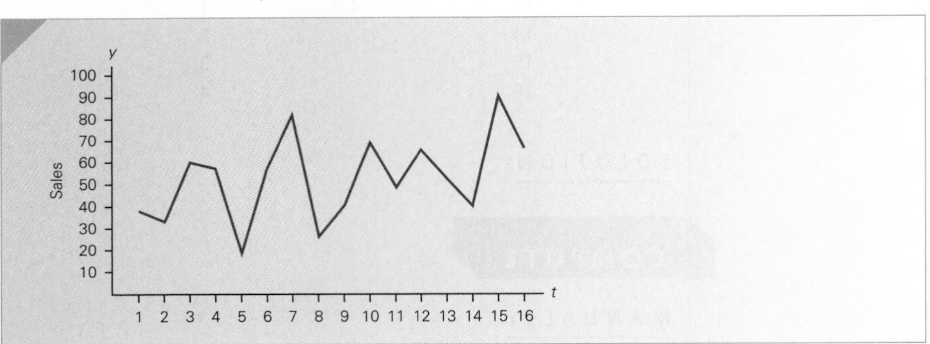

FIGURE **20.5** Quarterly Gasoline Sales and Three-Quarter and Five-Quarter Moving Averages

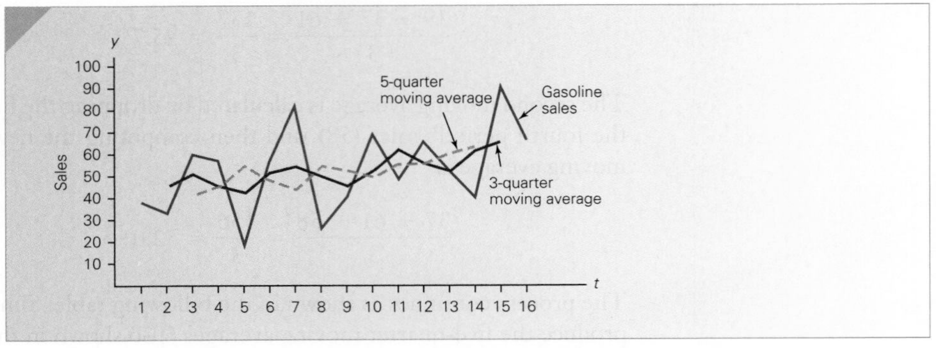

EXCEL

	A	B
1	Gas Sales	Moving Average
2	39	
3	37	45.7
4	61	52.0
5	58	45.7
6	18	44.0
7	56	52.0
8	82	55.0
9	27	50.0
10	41	45.7
11	69	53.0
12	49	61.3
13	66	56.3
14	54	54.0
15	42	62.0
16	90	66.0
17	66	

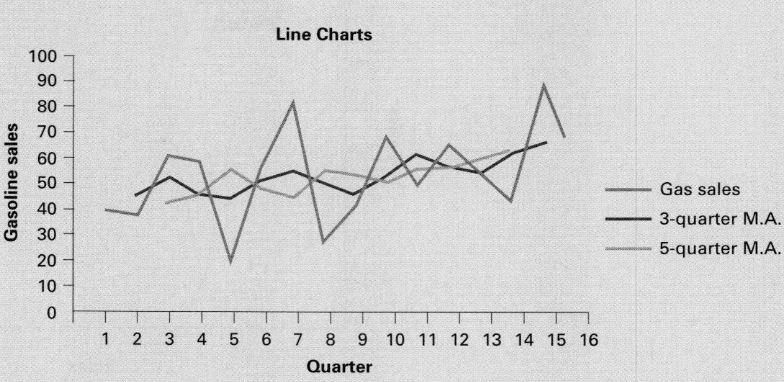

INSTRUCTIONS

1. Type or import the data into one column. (Open Xm20-01.)
2. Click **Data, Data Analysis,** and **Moving Average.**
3. Specify the **Input Range** (A1:A17). Specify the number of periods (3), and the **Output Range** (B1).
4. Delete the cells containing N/A.
5. To draw the line charts, follow the instructions on page 51.

MINITAB

Gas Sales	AVER1	AVER2
39	*	*
37	45.6667	*
61	52.0000	42.6
58	45.6667	46.0
18	44.0000	55.0
56	52.0000	48.2
82	55.0000	44.8
27	50.0000	55.0
41	45.6667	53.6
69	53.0000	50.4
49	61.3333	55.8
66	56.3333	56.0
54	54.0000	60.2
42	62.0000	63.6
90	66.0000	
66		

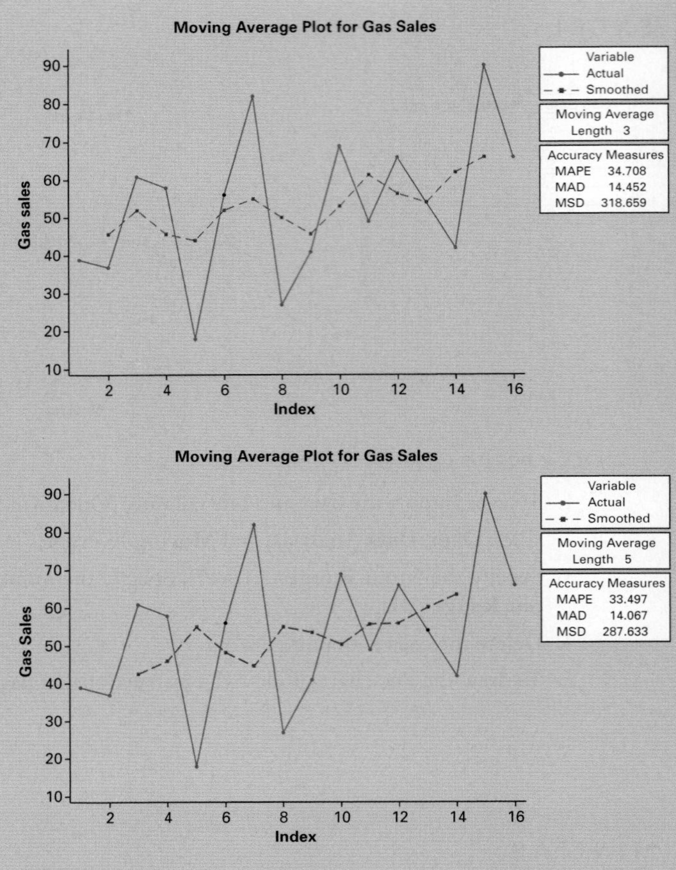

INSTRUCTIONS

1. Type or import the data into one column. (Open Xm20-01.)
2. Click **Stat, Time Series,** and **Moving Average . . . .**
3. Type or select the variable in the **Variable** box (Gas sales) and type the number of periods in the **MA length** box (3). Click **Center the moving averages, Storage . . . .** and **Moving averages.**
4. To draw the graph, click **Graphs . . .** and **Plot smoothed vs. actual.**

INTERPRET

To see how the moving averages remove some of the random variation, examine Figures 20.4 and 20.5. Figure 20.4 depicts the quarterly gasoline sales. Discerning any of the time-series components is difficult because of the large amount of random variation. Now consider the three-quarter moving average in Figure 20.5. You should be able to detect the seasonal pattern that exhibits peaks in the third quarter of each year (periods 3, 7, 11, and 15) and valleys in the first quarter of the year (periods 5, 9, and 13). There is also a small but discernible long-term trend of increasing sales.

Notice also in Figure 20.5 that the five-quarter moving average produces more smoothing than the three-quarter moving average. In general, the longer the time period over which we average, the smoother the series becomes. Unfortunately, in this case we've smoothed too much—the seasonal pattern is no longer apparent in the five-quarter moving average. All we can see is the long-term trend. It is important to realize that our objective is to smooth the time series sufficiently to remove the random variation and to reveal the other components (trend, cycle, and/or season) present. With too little smoothing, the random variation disguises the real pattern. With too much smoothing, however, some or all of the other effects may be eliminated along with the random variation.

Centered Moving Averages

Using an even number of periods to calculate the moving averages presents a problem about where to place the moving averages in a graph or table. For example, suppose that we calculate the four-period moving average of the following time series:

Period	Time Series
1	15
2	27
3	20
4	14
5	25
6	11

The first moving average is

$$\frac{15 + 27 + 20 + 14}{4} = 19.0$$

However, because this value represents time periods 1, 2, 3, and 4, we must place it between periods 2 and 3. The next moving average is

$$\frac{27 + 20 + 14 + 25}{4} = 21.5$$

and it must be placed between periods 3 and 4. The moving average that falls between periods 4 and 5 is

$$\frac{20 + 14 + 25 + 11}{4} = 17.5$$

There are several problems that result from placing the moving averages between time periods, including graphing difficulties. Centering the moving average corrects the problem. We do this by computing the two-period moving average of the four-period moving average. Thus the centered moving average for period 3 is

$$\frac{19.0 + 21.5}{2} = 20.25$$

The centered moving average for period 4 is

$$\frac{21.5 + 17.5}{2} = 19.50$$

The following table summarizes these results.

Period	Time Series	Four-Period Moving Average	Four-Period Centered Moving Average
1	15		–
		–	
2	27		–
		19.0	
3	20		20.25
		21.5	
4	14		19.50
		17.5	
5	25		–
		–	
6	11		–

Minitab centers the moving averages on command. Excel does not center the moving averages.

Exponential Smoothing

Two drawbacks are associated with the moving average method of smoothing time series. First, we do not have moving averages for the first and last sets of time periods. If the time series has few observations, the missing values can represent an important loss of information. Second, the moving average "forgets" most of the previous time-series values. For example, in the five-quarter moving average described in Example 20.1, the average for quarter 4 reflects quarters 2, 3, 4, 5, and 6 but is not affected by quarter 1. Similarly, the moving average for quarter 5 forgets quarters 1 and 2. Both of these problems are addressed by **exponential smoothing.**

Exponentially Smoothed Time Series

$$S_t = wy_t + (1 - w)S_{t-1} \quad \text{for } t \geq 2$$

where

S_t = exponentially smoothed time series at time period t

y_t = time series at time period t

S_{t-1} = exponentially smoothed time series at time period $t - 1$

w = smoothing constant, where $0 \leq w \leq 1$

We begin by setting

$$S_1 = y_1$$

Then

$$\begin{aligned} S_2 &= wy_2 + (1 - w)S_1 \\ &= wy_2 + (1 - w)y_1 \\ S_3 &= wy_3 + (1 - w)S_2 \\ &= wy_3 + (1 - w)[wy_2 + (1 - w)y_1] \\ &= wy_3 + w(1 - w)y_2 + (1 - w)^2 y_1 \end{aligned}$$

and so on. In general, we have

$$S_t = wy_t + w(1 - w)y_{t-1} + w(1 - w)^2 y_{t-2} + \cdots + (1 - w)^{t-1} y_1$$

This formula states that the smoothed time series in period t depends on all the previous observations of the time series.

The smoothing constant w is chosen on the basis of how much smoothing is required. A small value of w produces a great deal of smoothing. A large value of w results in very little smoothing. Figure 20.6 depicts a time series and two exponentially smoothed series with $w = .1$ and $w = .5$.

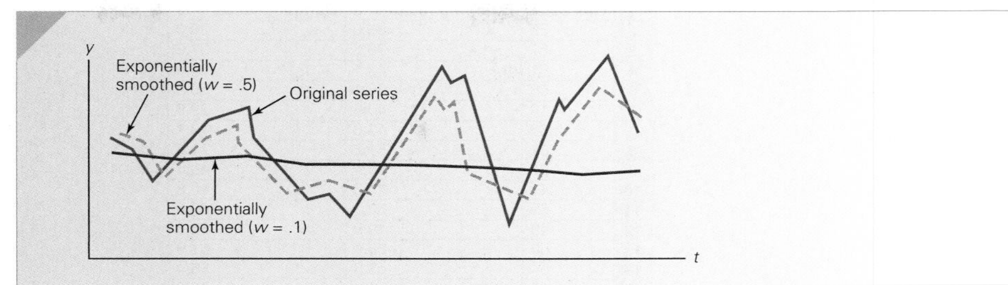

FIGURE **20.6** Time Series and Two Exponentially Smoothed Series

EXAMPLE 20.2

Gasoline Sales, Part 2

Apply the exponential smoothing technique with $w = .2$ and $w = .7$ to the data in Example 20.1, and graph the results.

SOLUTION

COMPUTE

MANUALLY

The exponentially smoothed values are calculated from the formula

$$S_t = wy_t + (1 - w)S_{t-1}$$

The results with $w = .2$ and $w = .7$ are shown in the following table.

Time Period	Gasoline Sales	Exponentially Smoothed with $w = .2$	Exponentially Smoothed with $w = .7$
1	39	39.0	39.0
2	37	38.6	37.6
3	61	43.1	54.0
4	58	46.1	56.8
5	18	40.5	29.6
6	56	43.6	48.1
7	82	51.2	71.8
8	27	46.4	40.4
9	41	45.3	40.8
10	69	50.1	60.6
11	49	49.8	52.5
12	66	53.1	61.9
13	54	53.3	56.4
14	42	51.0	46.3
15	90	58.8	76.9
16	66	60.2	69.3

Figure 20.7 shows the exponentially smoothed time series.

EXCEL

	A	B	C
1	Gas Sales	Damping factor = .8	Damping factor = .3
2	39	39.0	39.0
3	37	38.6	37.6
4	61	43.1	54.0
5	58	46.1	56.8
6	18	40.5	29.6
7	56	43.6	48.1
8	82	51.2	71.8
9	27	46.4	40.4
10	41	45.3	40.8
11	69	50.1	60.6
12	49	49.8	52.5
13	66	53.1	61.9
14	54	53.3	56.4
15	42	51.0	46.3
16	90	58.8	76.9
17	66	60.2	69.3

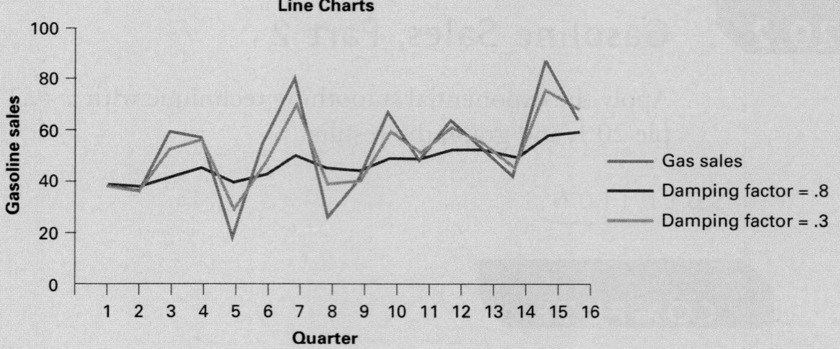

INSTRUCTIONS

1. Type or import the data into one column (Open Xm20-01.)
2. Click **Data, Data Analysis,** and **Exponential Smoothing.**
3. Specify the **Input Range** (A1:A17). Type the **Damping factor,** which is $1 - w$ (.8). Specify the **Output Range** (B1). To calculate the second exponentially smoothed times series, specify $1 - w$ (.3) **Output Range** (C1).

To modify the table so that the smoothed values appear the way we calculated manually, click the cell containing the last smoothed value displayed here (58.8) and drag it to the cell below to reveal the final smoothed value (60.2 and 69.3).

MINITAB

Gas Sales	SMOO1	SMOO2
39	39.0000	39.0000
37	38.6000	37.6000
61	43.0800	53.9800
58	46.0640	56.7940
18	40.4512	29.6382
56	43.5610	48.0915
82	51.2488	71.8274
27	46.3990	40.4482
41	45.3192	40.8345
69	50.0554	60.5503
49	49.8443	52.4651

(continued)

Gas Sales	SMOO1	SMOO2
66	53.0754	61.9395
54	53.2603	56.3819
42	51.0083	46.3146
90	58.8066	76.8944
66	60.2453	69.2683

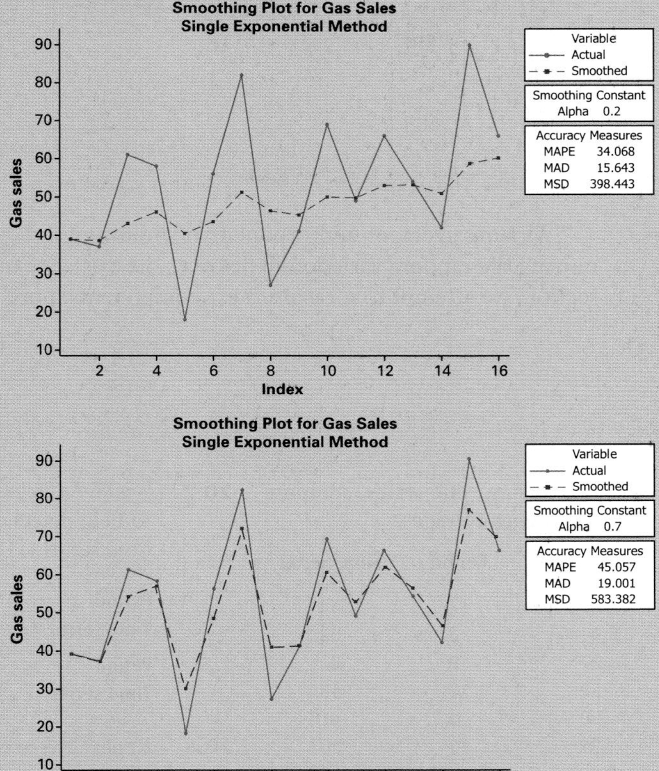

INSTRUCTIONS

1. Type or import the data into one column. (Open Xm20-01.)
2. Click **Stat, Time Series,** and **Single Exp Smoothing . . . .**
3. Type or select the time variable in the **Variable** box (Gas Sales). Click **Use** under **Weight to use in Smoothing** and type the value of w (.2) (.7). Click **Options . . . .**
4. Type 1 in the **Use average of first k observations K** = box.
5. Click **Storage . . .** and **Smoothed data.**

To draw the graph, click **Graphs . . .** and **Plot smoothed vs. actual.**

INTERPRET

Figure 20.7 depicts the graph of the original time series and the exponentially smoothed series. As you can see, $w = .7$ results in very little smoothing, whereas $w = .2$ results in perhaps too much smoothing. In both smoothed time series, it is difficult to discern the seasonal pattern that we detected by using moving averages. A different value of w (perhaps $w = .5$) would be likely to produce more satisfactory results.

FIGURE **20.7** Quarterly Gasoline Sales and Exponentially Smoothed Sales with $w = .2$ and $w = .7$

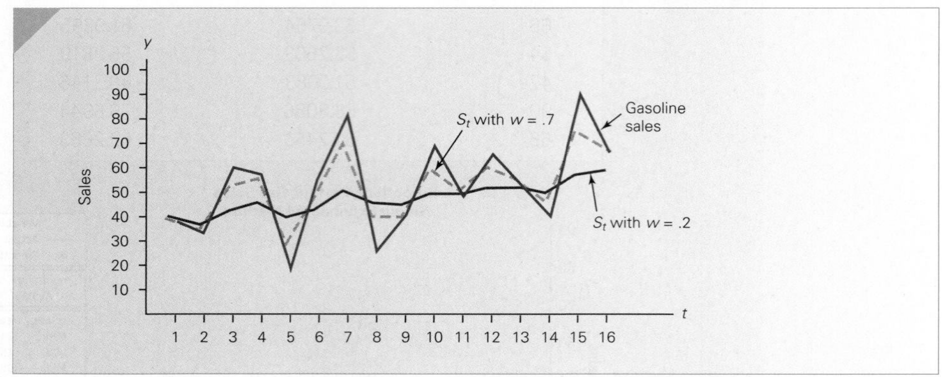

Moving averages and exponential smoothing are relatively crude methods of removing the random variation to discover the existence of other components. In the next section, we attempt to measure these components more precisely.

EXERCISES

20.1 Xr20-01 For the following time series, compute the three-period moving averages.

Period	Time Series	Period	Time Series
1	48	7	43
2	41	8	52
3	37	9	60
4	32	10	48
5	36	11	41
6	31	12	30

20.2 Compute the five-period moving averages for the time series in Exercise 20.1.

20.3 For Exercises 20.1 and 20.2, graph the time series and the two moving averages.

20.4 Xr20-04 For the following time series, compute the three-period moving averages.

Period	Time Series	Period	Time Series
1	16	7	24
2	22	8	29
3	19	9	21
4	24	10	23
5	30	11	19
6	26	12	15

20.5 For Exercise 20.4, compute the five-period moving averages.

20.6 For Exercises 20.4 and 20.5, graph the time series and the two moving averages.

20.7 Xr20-07 Apply exponential smoothing with $w = .1$ to help detect the components of the following time series.

Period	1	2	3	4	5
Time series	12	18	16	24	17
Period	6	7	8	9	10
Time series	16	25	21	23	14

20.8 Repeat Exercise 20.7 with $w = .8$.

20.9 For Exercises 20.7 and 20.8, draw the time series and the two sets of exponentially smoothed values. Does there appear to be a trend component in the times series?

20.10 Xr20-10 Apply exponential smoothing with $w = .1$ to help detect the components of the following time series.

Period	1	2	3	4	5
Time series	38	43	42	45	46
Period	6	7	8	9	10
Time series	48	50	49	46	45

20.11 Repeat Exercise 20.10 with $w = .8$.

20.12 For Exercises 20.10 and 20.11, draw the time series and the two sets of exponentially smoothed values. Does there appear to be a trend component in the times series?

20.13 Xr20-12 The following daily sales figures have been recorded in a medium-size merchandising firm:

Day	Week 1	2	3	4
Monday	43	51	40	64
Tuesday	45	41	57	58
Wednesday	22	37	30	33
Thursday	25	22	33	38
Friday	31	25	37	25

a. Compute the three-day moving averages.
b. Plot the series and the moving averages on a graph.
c. Does there appear to be a seasonal (weekly) pattern?

20.14 For Exercise 20.13, compute the five-day moving averages, and superimpose these on the same graph. Does this help you answer Part c of Exercise 20.13?

20.15 Xr20-15 The following quarterly sales of a department store chain were recorded for the years 1997–2000:

Quarter	Year 2001	2002	2003	2004
1	18	33	25	41
2	22	20	36	33
3	27	38	44	52
4	31	26	29	45

a. Calculate the four-quarter centered moving averages.
b. Graph the time series and the moving averages.
c. What can you conclude from your time-series smoothing?

20.16 Repeat Exercise 20.15, using exponential smoothing with $w = .4$.

20.17 Repeat Exercise 20.15, using exponential smoothing with $w = .8$.

20.3 Trend and Seasonal Effects

In the previous section we described how smoothing a time series can give us a clearer picture of which components are present. In order to forecast, however, we often need more precise measurements of the time-series components.

Trend Analysis

A trend can be linear or nonlinear and, indeed can take on a whole host of functional forms. The easiest way of measuring the long-term trend is by regression analysis, where the independent variable is time. If we believe that the long-term trend is approximately linear, we will use the linear model introduced in Chapter 16:

$$y = \beta_0 + \beta_1 t + \varepsilon$$

If we believe that the trend is nonlinear, we can use one of the polynomial models described in Chapter 18. For example, the quadratic model is

$$y = \beta_0 + \beta_1 t + \beta_2 t^2 + \varepsilon$$

In most realistic applications, the linear model is used. We will demonstrate how the long-term trend is measured and applied later in this section.

Seasonal Analysis

Seasonal variation may occur within a year or within shorter intervals, such as a month, week, or day. To measure the seasonal effect, we compute seasonal indexes, which gauge the degree to which the seasons differ from one another. One requirement necessary to calculate seasonal indexes is a time series sufficiently long enough to allow us to observe the variable over several seasons. For example, if the seasons are defined as the quarters of a year, we need to observe the time series for at least 4 years. The **seasonal indexes** are computed in the following way.

Procedure for Computing Seasonal Indexes

1. Remove the effect of seasonal and random variation by regression analysis. That is, compute the sample regression line

$$\hat{y}_t = b_0 + b_1 t$$

2. For each time period, compute the ratio

$$\frac{y_t}{\hat{y}_t}$$

This ratio removes most of the trend variation.

3. For each type of season, compute the average of the ratios in step 2. This procedure removes most (but seldom all) of the random variation leaving a measure of seasonality.

4. Adjust the averages in step 3 so that the average of all the seasons is 1 (if necessary).

Hotel Quarterly Occupancy Rates

DATA Xm20-03

The tourist industry is subject to seasonal variation. In most resorts the spring and summer seasons are considered the "high" seasons. Fall and winter (except for Christmas and New Years) are "low" seasons. A hotel in Bermuda has recorded the occupancy rate for each quarter for the past 5 years. These data are shown here. Measure the seasonal variation by computing the seasonal indexes.

Year	Quarter	Occupancy Rate
2003	1	.561
	2	.702
	3	.800
	4	.568
2004	1	.575
	2	.738
	3	.868
	4	.605
2005	1	.594
	2	.738
	3	.729
	4	.600
2006	1	.622
	2	.708
	3	.806
	4	.632
2007	1	.665
	2	.835
	3	.873
	4	.670

SOLUTION

COMPUTE

MANUALLY

We performed a regression analysis with y = occupancy rate and t = time period 1, 2, . . . , 20. The regression equation is

$$\hat{y} = .639368 + .005246t$$

For each time period, we computed the ratio

$$\frac{y_t}{\hat{y}_t}$$

In the next step we collected the ratios associated with each quarter and computed the average. We then computed the seasonal indexes by adjusting the average ratios so that they summed to 4.0, if necessary. In this example, it was not necessary.

Year	Quarter	t	y_t	$\hat{y} = .639368 + .005246t$	Ratio $\frac{y_t}{\hat{y}_t}$
2003	1	1	.561	.645	.870
	2	2	.702	.650	1.080
	3	3	.800	.655	1.221
	4	4	.568	.660	.860
2004	1	5	.575	.666	.864
	2	6	.738	.671	1.100
	3	7	.868	.676	1.284
	4	8	.605	.681	.888
2005	1	9	.594	.687	.865
	2	10	.738	.692	1.067
	3	11	.729	.697	1.046
	4	12	.600	.702	.854
2006	1	13	.622	.708	.879
	2	14	.708	.713	.993
	3	15	.806	.718	1.122
	4	16	.632	.723	.874
2007	1	17	.665	.729	.913
	2	18	.835	.734	1.138
	3	19	.873	.739	1.181
	4	20	.670	.744	.900

	Quarter			
Year	1	2	3	4
2003	.870	1.080	1.221	.860
2004	.864	1.100	1.284	.888
2005	.865	1.067	1.046	.854
2006	.879	.993	1.122	.874
2007	.913	1.138	1.181	.900
Average	.878	1.076	1.171	.875
Index	.878	1.076	1.171	.875

EXCEL

	A	B
1	Seasonal Indexes	
2		
3	Season	Index
4	1	0.8782
5	2	1.0756
6	3	1.1709
7	4	0.8753

INSTRUCTIONS

1. Type or import the time series in chronological order into one column and the codes representing the seasons into the next adjacent column. (Open Xm20-03.)
2. Click **Tools, Data Analysis Plus**, and **Seasonal Indexes.**
3. Specify the **Input Range** (B1:C21).

MINITAB

Time Series Decomposition for Rate

Fitted Trend Equation

Yt = 0.638747 + 0.00535513*t

Seasonal Indices

Period	Index
1	0.87684
2	1.07939
3	1.17764
4	0.86613

Note: Some output has been omitted.

INSTRUCTIONS

1. Type or import the data into one column. (Open file Xm20-03.)
2. Click **Stat, Time Series,** and **Decomposition . . . .**
3. Select the **variable** (Rate), specify the **Seasonal length** (4), and click **Multiplicative** under **Model Type** and **Trend plus seasonal** under **Model Components.** Click **Options . . . .**
4. Type 1 in the **First obs. is in period box.**
5. Click **Storage . . .** and **Seasonals.**

INTERPRET

Note that Minitab's results differ from those computed manually and by Excel. (Minitab uses a technique other than ordinary least squares to produce the trend line.) In this example the differences are small. However, in other cases there can be substantial differences. The discussion that follows uses Excel's figures.

The seasonal indexes tell us that, on average, the occupancy rates in the first and fourth quarters are below the annual average, and the occupancy rates in the second and third quarters are above the annual average. Using Excel's figures (the manually calculated seasonal indexes are the same, but Minitab's indexes differ slightly), we expect the occupancy rate in the first quarter to be 12.2% (100% − 87.8%) below the annual rate. The second and third quarters' rates are expected to be 7.6% and 17.1%, respectively, above the annual rate. The fourth quarter's rate is 12.5% below the annual rate.

Figure 20.8 depicts the time series and the regression trend line.

FIGURE **20.8** Time Series and Trend for Example 20.3

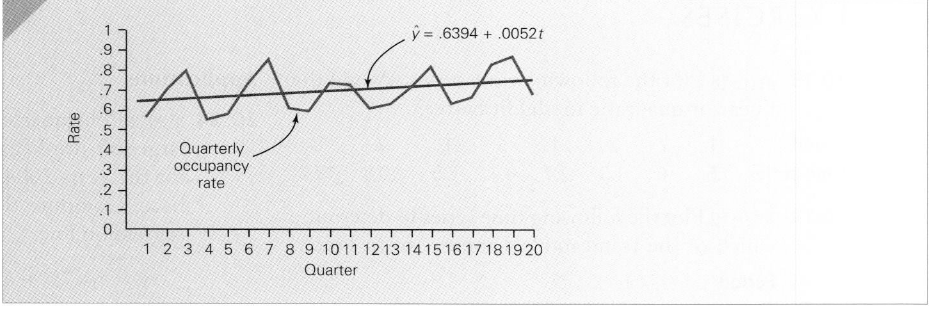

Deseasonalizing a Time Series

One application of seasonal indexes is to remove the seasonal variation in a time series. The process is called **deseasonalizing,** and the result is called a **seasonally adjusted time series.** Often this allows the statistics practitioner to more easily compare the time series across seasons. For example, the unemployment rate varies according to the season. During the winter months unemployment usually rises; it falls in the spring and summer. The seasonally adjusted unemployment rate allows economists to determine whether unemployment has increased or decreased over the previous months. The process is easy: Simply divide the time series by the seasonal indexes. To illustrate, we have deseasonalized the occupancy rates in Example 20.3 (using the seasonal indexes produced by Excel). The results are shown here.

Year	Quarter	Occupancy Rate y_t	Seasonal Index	Seasonally Adjusted Occupancy Rate
2003	1	.561	.878	.639
	2	.702	1.076	.652
	3	.800	1.171	.683
	4	.568	.875	.649
2004	1	.575	.878	.655
	2	.738	1.076	.686
	3	.868	1.171	.741
	4	.605	.875	.691
2005	1	.594	.878	.677
	2	.738	1.076	.686
	3	.729	1.171	.623
	4	.600	.875	.686
2006	1	.622	.878	.708
	2	.708	1.076	.658
	3	.806	1.171	.688
	4	.632	.875	.722
2007	1	.665	.878	.757
	2	.835	1.076	.776
	3	.873	1.171	.746
	4	.670	.875	.766

By removing the seasonality, we can see when there has been a "real" increase or decrease in the occupancy rate. This enables the statistics practitioner to examine the factors that produced the rate change. We can more easily see that there has been an increase in the occupancy rate over the 5-year period.

In the next section we show how to forecast with seasonal indexes.

EXERCISES

20.18 Xr20-18 Plot the following time series. Would the linear or quadratic model fit better?

Period	1	2	3	4	5	6	7	8
Time series	.5	.6	1.3	2.7	4.1	6.9	10.8	19.2

20.19 Xr20-19 Plot the following time series to determine which of the trend models appears to fit better.

Period	1	2	3	4	5
Time series	55	57	53	49	47
Period	6	7	8	9	10
Time series	39	41	33	28	20

20.20 Refer to Exercise 20.18. Use regression analysis to calculate the linear and quadratic trends. Which line fits better?

20.21 Refer to Exercise 20.19. Use regression analysis to calculate the linear and quadratic trends. Which line fits better?

20.22 Xr20-22 For the following time series compute the seasonal (daily) indexes.

The regression line is

$$\hat{y} = 16.8 + .366t \quad (t = 1, 2, \ldots, 20)$$

	Week			
Day	**1**	**2**	**3**	**4**
Monday	12	11	14	17
Tuesday	18	17	16	21
Wednesday	16	19	16	20
Thursday	25	24	28	24
Friday	31	27	25	32

20.23 Xr20-23 Given the following time series, compute the seasonal indexes.

The regression equation is

$$\hat{y} = 47.7 - 1.06t \quad (t = 1, 2, \ldots, 20)$$

	Year				
Quarter	**1**	**2**	**3**	**4**	**5**
1	55	41	43	36	50
2	44	38	39	32	25
3	46	37	39	30	24
4	39	30	35	25	22

Applications

20.24 Xr20-24 The quarterly earnings (in $millions) of a large soft-drink manufacturer have been recorded for the years 2004 to 2007. These data are listed here. Compute the seasonal indexes given the regression line

$$\hat{y} = 61.75 + 1.18t \quad (t = 1, 2, \ldots, 16)$$

	Year			
Quarter	**2004**	**2005**	**2006**	**2007**
1	52	57	60	66
2	67	75	77	82
3	85	90	94	98
4	54	61	63	67

The following exercises require a computer and software.

20.25 Xr20-25 College and university enrollment increased sharply during the 1970s and 1980s. However, since then the rate of growth has slowed. To help forecast future enrollments, an economist recorded the total U.S. college and university enrollment from 1993 to 2003. These data (in thousands) are listed here.

Year	1993	1994	1995	1996	1997	1998
Enrollment	13,898	15,022	14,715	15,226	15,436	15,546
Year	1999	2000	2001	2002	2003	
Enrollment	15,203	15,314	15,873	16,497	16,638	

Source: Statistical Abstract of the United States, 2006, Table 268.

a. Plot the time series
b. Use regression analysis to determine the trend.

20.26 Xr20-26 Foreign trade is important to the United States. No country exports and imports more. However, there has been a large trade imbalance in many sectors. To measure the extent of the problem, an economist recorded the difference between exports and imports of merchandise for the years 1980 to 2004 (latest available). *Source: Statistical Abstract of the United States*, 2006, Table 1295.

a. Plot the trade balance.
b. Apply regression analysis to measure the trend.

20.27 Xr20-27 The number of cable television subscribers has increased over the past 5 years. The marketing manager for a cable company has recorded the numbers of subscribers for the past 24 quarters.

a. Plot the numbers.
b. Compute the seasonal (quarterly) indexes.

20.28 Xr20-28 The owner of a pizzeria wants to forecast the number of pizzas she will sell each day. She recorded the numbers sold daily for the past 4 weeks. Calculate the seasonal (daily) indexes.

20.29 Xr20-29 A manufacturer of ski equipment is in the process of reviewing his accounts receivable. He noticed that there appears to be a seasonal pattern with the accounts receivable increasing in the winter months and decreasing during the summer. The quarterly accounts receivable (in $millions) were recorded. Compute the seasonal (quarterly) indexes.

20.4 Introduction to Forecasting

Many different forecasting methods are available for the statistics practitioner. One of the factors to be considered in choosing among them is the type of component that makes up the time series. Even then, however, we have several different methods from which to choose. One way of deciding which method to apply is to select the technique that achieves the greatest forecast accuracy. The most commonly used measures of forecast accuracy are **mean absolute deviation (MAD)** and the **sum of squares for forecast errors (SSE).**

Mean Absolute Deviation

$$\text{MAD} = \frac{\sum_{i=1}^{n} |y_t - F_t|}{n}$$

where

y_t = actual value of the time series at time period t

F_t = forecasted value of the time series at time period t

n = number of time periods

Sum of Squares for Forecast Error

$$\text{SSE} = \sum_{i=1}^{n} (y_t - F_t)^2$$

MAD averages the absolute differences between the actual and forecast values; SSE is the sum of the squared differences. Which measure to use in judging forecast accuracy depends on the circumstances. If avoiding large errors is important, SSE should be used because it penalizes large deviations more heavily than does MAD. Otherwise use MAD.

It is probably best to use some of the observations of the time series to develop several competing forecasting models and then forecast for the remaining time periods. Afterward, compute MAD or SSE for the forecasts. For example, if we have 5 years of monthly observations, use the first 4 years to develop the forecasting models and then use them to forecast the fifth year. Since we know the actual values in the fifth year, we can choose the technique that results in the most accurate forecast using either MAD or SSE.

EXAMPLE 20.4 Comparing Forecasting Models

Annual data from 1974 to 2003 were used to develop three different forecasting models. Each model was used to forecast the time series for 2004, 2005, 2006, and 2007. The forecasted and actual values for these years are shown here. Use MAD and SSE to determine which model performed best.

		Forecast Using Method		
Year	Actual Time Series	1	2	3
2004	129	136	118	130
2005	142	148	141	146
2006	156	150	158	170
2007	183	175	163	180

SOLUTION

For model 1, we have

$$\text{MAD} = \frac{|129 - 136| + |142 - 148| + |156 - 150| + |183 - 175|}{4}$$

$$= \frac{7 + 6 + 6 + 8}{4} = 6.75$$

$$\text{SSE} = (129 - 136)^2 + (142 - 148)^2 + (156 - 150)^2 + (183 - 175)^2$$

$$= 49 + 36 + 36 + 64 = 185$$

For model 2, we compute

$$\text{MAD} = \frac{|129 - 118| + |142 - 141| + |156 - 158| + |183 - 163|}{4}$$

$$= \frac{11 + 1 + 2 + 20}{4} = 8.5$$

$$\text{SSE} = (129 - 118)^2 + (142 - 141)^2 + (156 - 158)^2 + (183 - 163)^2$$

$$= 121 + 1 + 4 + 400 = 526$$

The measures of forecast accuracy for model 3 are

$$\text{MAD} = \frac{|129 - 130| + |142 - 146| + |156 - 170| + |183 - 180|}{4}$$

$$= \frac{1 + 4 + 14 + 3}{4} = 5.5$$

$$\text{SSE} = (129 - 130)^2 + (142 - 146)^2 + (156 - 170)^2 + (183 - 180)^2$$

$$= 1 + 16 + 196 + 9 = 222$$

Model 2 is inferior to both models 1 and 3, no matter how we measure forecast accuracy. Using MAD, model 3 is best, but using SSE, model 1 is most accurate. The choice between model 1 and model 3 should be made on the basis of whether we prefer a model that consistently produces moderately accurate forecasts (model 1) or one whose forecasts come quite close to most actual values but miss badly in a small number of time periods (model 3).

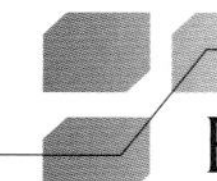

EXERCISES

20.30 For the actual and forecast values of a time series shown here, calculate MAD and SSE

Period	1	2	3	4	5
Forecast	173	186	192	211	223
Actual value	166	179	195	214	220

20.31 Two forecasting models were used to predict the future values of a time series. These are shown here together with the actual values. Compute MAD and SSE for each model to determine which was more accurate.

Period	1	2	3	4
Forecast (Model 1)	7.5	6.3	5.4	8.2
Forecast (Model 2)	6.3	6.7	7.1	7.5
Actual	6.0	6.6	7.3	9.4

20.32 Calculate MAD and SSE for the forecasts that follow.

Period	1	2	3	4	5
Forecast	63	72	86	71	60
Actual	57	60	70	75	70

20.33 Three forecasting techniques were used to predict the values of a time series. These values are given in the following table. Compute MAD and SSE for each technique to determine which was most accurate.

Period	1	2	3	4	5
Forecast (Model 1)	21	27	29	31	35
Forecast (Model 2)	22	24	26	28	30
Forecast (Model 3)	17	20	25	31	39
Actual	19	24	28	32	38

20.5 FORECASTING MODELS

There is a large number of different forecasting techniques available to statistics practitioners. However, many are beyond the level of this book. In this section we present three models. Similar to the method of choosing the correct statistical inference technique in Chapters 12 to 19, the choice of model depends on the time-series components.

Forecasting with Exponential Smoothing

If the time series displays a gradual or no trend and no evidence of seasonal variation, exponential smoothing can be effective as a forecasting method. Suppose that t represents the most recent time period and we've computed the exponentially smoothed value S_t. This value is then the forecasted value at time $t + 1$. That is,

$$F_{t+1} = S_t$$

If we wish, we can forecast two or three or any number of periods into the future:

$$F_{t+2} = S_t \qquad \text{or} \qquad F_{t+3} = S_t$$

It must be understood that the accuracy of the forecast decreases rapidly for predictions more than one time period into the future. However, as long as we're dealing with time series with no cyclical or seasonal variation, we can produce reasonably accurate predictions for the next time period.

Forecasting with Seasonal Indexes

If the time series is composed of seasonal variation and long-term trend, we can use seasonal indexes and the regression equation to forecast.

Forecast of Trend and Seasonality

The forecast for time period t is

$$F_t = [b_0 + b_1 t] \times SI_t$$

where

$$F_t = \text{forecast for period } t$$
$$b_0 + b_1 t = \text{regression equation}$$
$$SI_t = \text{seasonal index for period } t$$

EXAMPLE 20.5

Forecasting Hotel Occupancy Rates

Forecast hotel occupancy rates for next year in Example 20.3.

SOLUTION

In the process of computing the seasonal indexes, we computed the trend line. It is

$$\hat{y} = .639 + .00525t$$

For $t = 21, 22, 23$, and 24 we calculate the forecasted trend values.

Quarter	t	$\hat{y} = .639 + .00525t$
1	21	$.639 + .00525(21) = .749$
2	22	$.639 + .00525(22) = .755$
3	23	$.639 + .00525(23) = .760$
4	24	$.639 + .00525(24) = .765$

We now multiply the forecasted trend values by the seasonal indexes calculated in Example 20.3. The seasonalized forecasts are as follow:

Quarter	t	Trend Value $\hat{y}_t$	Seasonal Index	Forecast $F_t = \hat{y}_t \times SI_t$
1	21	.749	.878	$.749 \times .878 = .658$
2	22	.755	1.076	$.755 \times 1.076 = .812$
3	23	.760	1.171	$.760 \times 1.171 = .890$
4	24	.765	.875	$.765 \times .875 = .670$

INTERPRET

We forecast that the quarterly occupancy rates during the next year will be .658, .812, .890, and .670.

Autoregressive Model

In Chapter 17 we discussed autocorrelation wherein the errors are not independent of one another. The existence of strong autocorrelation is an indication that the model has been misspecified, which usually means that until we improve the regression model, it will not provide an adequate fit. However, autocorrelation also provides us with an opportunity to develop another forecasting technique. If there is no obvious trend or seasonality and we believe that there is a correlation between consecutive residuals, the **autoregressive model** may be most effective.

Autoregressive Forecasting Model

$$y_t = \beta_0 + \beta_1 y_{t-1} + \varepsilon$$

The model specifies that consecutive values of the time series are correlated. We estimate the coefficient in the usual way. The estimated regression line is defined as

$$\hat{y}_t = b_0 + b_1 y_{t-1}$$

EXAMPLE 20.6

DATA Xm20-06

Forecasting Changes to the Consumer Price Index

The consumer price index (CPI) is used as a general measure of inflation. It is an important measure because a high rate of inflation often influences governments to take corrective measures. The table below lists the consumer price index from 1978 to 2006 and the annual percentage increases in the CPI. Forecast next year's change in the CPI.

Year	CPI	% Change	Year	CPI	% Change
1978	65.2		1993	144.5	3.0%
1979	72.6	11.3%	1994	148.2	2.6%
1980	82.4	13.5%	1995	152.4	2.8%
1981	90.9	10.4%	1996	156.9	2.9%
1982	96.5	6.2%	1997	160.5	2.3%
1983	99.6	3.2%	1998	163.0	1.5%
1984	103.9	4.4%	1999	166.6	2.2%
1985	107.6	3.5%	2000	172.2	3.4%
1986	109.7	1.9%	2001	177.0	2.8%
1987	113.6	3.6%	2002	179.9	1.6%
1988	118.3	4.1%	2003	184.0	2.3%
1989	123.9	4.8%	2004	188.9	2.7%
1990	130.7	5.4%	2005	195.3	3.4%
1991	136.2	4.2%	2006	201.6	3.2%
1992	140.3	3.0%			

Source: Statistical Abstract of the United States, 2003, Table 713.

SOLUTION

Notice that we included the CPI for 1978 because we wanted to determine the percentage change for 1979. We will use the percentage changes for 1979 to 2005 as the independent variable and the percentage change from 1980 to 2006 as the dependent variable. File Xm20-06 stores the data in the format necessary to determine the autoregressive model.

EXCEL

	A	B	C	D	E
16		Coefficients	Standard Error	t Stat	P-value
17	Intercept	0.0070	0.0044	1.58	0.12681
18	% Change X	0.761	0.0867	8.78	4.12E-09

MINITAB

Regression Analysis: % Change (t) Versus % Change (t-1)

The regression equation is
% Change (t) = 0.00698 + 0.761 % Change (t-1)

INTERPRET

The regression line is

$$\hat{y}_t = .0070 + .761y_{t-1}$$

Because the last CPI change is 3.2%, our forecast for 2007 is

$$\hat{y}_{2007} = .0070 + .761y_{2006}$$
$$= .0070 + .761(3.2) = 2.44$$

The autoregressive model forecasts a 2.44% increase in the CPI for the year 2007.

Housing Starts: Solution

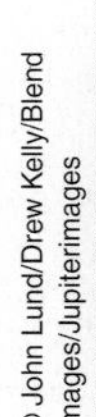

A preliminary examination of the data reveals that there is a very small upward trend over the 5-year period. Moreover the number of housing starts varies by month. The presence of these components suggests that we determine the linear trend and seasonal (monthly) indexes.

EXCEL

With housing starts as the dependent variable and the month as the independent variable, Excel yielded the following regression line:

$$\hat{y} = 11.46 + .0808t \qquad t = 1, 2, \ldots, 60$$

The seasonal indexes were computed as follows. (See the instructions on page 836.)

	A	B
1	**Seasonal Indexes**	
2		
3	**Season**	**Index**
4	1	0.5974
5	2	0.6548
6	3	0.9800
7	4	1.0697
8	5	1.1110
9	6	1.1917
10	7	1.2050
11	8	1.2276
12	9	1.0960
13	10	1.0226
14	11	0.9960
15	12	0.8483

The regression equation was used again to predict the number of housing starts based on the linear trend:

$$\hat{y} = 11.46 + .0808t \qquad t = 61, 62, \ldots, 72$$

These figures were multiplied by the seasonal indexes, which resulted in the following forecasts.

Period	Month	$\hat{y} = 11.46 + .0808t$	Seasonal Index	Forecasts
61	January	16.39	.5974	9.79
62	February	16.47	.6548	10.79
63	March	16.55	.9800	16.22
64	April	16.63	1.0697	17.79
65	May	16.71	1.1110	18.57
66	June	16.79	1.1917	20.01
67	July	16.87	1.2050	20.33
68	August	16.95	1.2276	20.81
69	September	17.04	1.0960	18.67
70	October	17.12	1.0226	17.50
71	November	17.20	.9960	17.13
72	December	17.28	.8483	14.66

This table displays the actual and forecasted housing starts for 2006 (the latest figures available). Figure 20.9 depicts the time series, trend line, and forecasts.

Period	Month	Forecasts	Actual
61	January	9.79	13.3
62	February	10.79	10.1
63	March	16.22	12.9
64	April	17.79	16.0
65	May	18.57	18.8
66	June	20.01	16.1
67	July	20.33	13.7
68	August	20.81	15.6
69	September	18.67	12.3
70	October	17.50	13.3
71	November	17.13	12.2
72	December	14.66	12.9

The size of the error was measured by MAD and SSE. They are

$$\text{MAD} = 43.55$$

$$\text{SSE} = 199.13$$

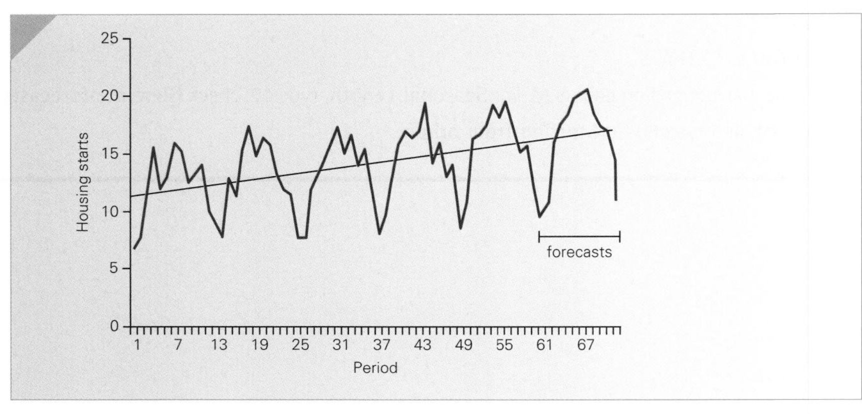

FIGURE **20.9** **Time Series, Trend, and Forecasts of Housing Starts**

MINITAB

Minitab produces the regression equation, seasonal indexes, and forecasts in one step. We remind readers that the outputs for Minitab and Excel's outputs often differ.

Time Series Decomposition for Starts

Multiplicative Model

Data	Starts
Length	60
NMissing	0

Fitted Trend Equation

Yt = 11.921 + 0.0658*t

Seasonal Indices

Period	Index
1	0.57912
2	0.63392
3	0.95450
4	1.03338
5	1.13799
6	1.19369
7	1.21443
8	1.23142
9	1.09832
10	1.04138
11	0.00315
12	0.87869

Accuracy Measures

MAPE	6.98073
MAD	0.95337
MSD	1.32163

Forecasts

Period	Forecast
61	9.2293
62	10.1445
63	15.3375
64	16.6730
65	18.4357
66	19.4166
67	19.8341
68	20.1926
69	18.0823
70	17.2135
71	16.6477
72	14.6400

INSTRUCTIONS

Follow the instructions on page 836. For **Seasonal Length,** type 12, check **Generate forecasts**, type 12 for the **Number of forecast**, and type 60 for **Starting from origin.**

EXERCISES

20.34 The following trend line and seasonal indexes were computed from 10 years of quarterly observations. Forecast the next year's time series.

$$\hat{y} = 150 + 3t \qquad t = 1, 2, \ldots, 40$$

Quarter	Seasonal Index
1	.7
2	1.2
3	1.5
4	.6

20.35 The following trend line and seasonal indexes were computed from 4 weeks of daily observations. Forecast the seven values for next week.

$$\hat{y} = 120 + 2.3t \qquad t = 1, 2, \ldots, 28$$

Day	Seasonal Index
Sunday	1.5
Monday	.4
Tuesday	.5
Wednesday	.6
Thursday	.7
Friday	1.4
Saturday	1.9

20.36 Use the following autoregressive equation to forecast the next value of the time series if the last observed value is 65.

$$\hat{y}_t = 625 - 1.3y_{t-1}$$

20.37 The following autoregressive equation was developed. Forecast the next value if the last observed value was 11.

$$\hat{y}_t = 155 + 21y_{t-1}$$

20.38 Apply exponential smoothing with $w = .4$ to forecast the next four quarters in Exercise 20.15.

20.39 Use the seasonal indexes and trend line to forecast the time series for the next 5 days in Exercise 20.22.

20.40 Refer to Exercise 20.23. Use the seasonal indexes and the trend line to forecast the time series for the next four quarters.

Applications

20.41 Use the seasonal indexes and trend line to forecast the quarterly earnings for the years 2005 and 2006 in Exercise 20.24.

20.42 Refer to Exercise 20.25 Forecast next year's enrollment using the following methods.
a. Autoregressive forecasting model.
b. Exponential smoothing method with $w = .5$.

20.43 Refer to Exercise 20.26. Forecast next year's merchandise trade balance using the following methods.
a. Autoregressive forecasting model.
b. Exponential smoothing method with $w = .7$.

20.44 Use the seasonal indexes and trend line from Exercise 20.27 to forecast the number of cable subscribers for the next four quarters.

20.45 Refer to Exercise 20.28. Use the seasonal indexes and trend line to forecast the number of pizzas to be sold for each of the next 7 days.

20.46 Apply the trend line and seasonal indexes from Exercise 20.29 to forecast accounts receivable for the next four quarters.

Exercises 20.47–20.51 are based on the following problem.

Xr20-47 The revenues (in $millions) of a chain of ice cream stores are listed for each quarter during the previous 5 years.

	Year				
Quarter	2003	2004	2005	2006	2007
1	16	14	17	18	21
2	25	27	31	29	30
3	31	32	40	45	52
4	24	23	27	24	32

20.47 Plot the time series.

20.48 Discuss why exponential smoothing is not recommended as a forecasting tool in this problem.

20.49 Use regression analysis to determine the trend line.

20.50 Determine the seasonal indexes.

20.51 Using the seasonal indexes and trend line, forecast revenues for the next four quarters.

Chapter Summary

In this chapter we discussed the classical time series and its decomposition into trend, seasonal, and random variation. Moving averages and exponential smoothing were used to remove some of the random variation, making it easier to detect trend and seasonality. The long-term trend was measured by regression analysis. Seasonal variation was measured by computing the seasonal indexes. Three forecasting techniques were described in this chapter: exponential smoothing, forecasting with seasonal indexes, and the autoregressive model.

IMPORTANT TERMS

Time series 820
Trend 821
Cyclical variation 822
Seasonal variation 822
Random variation 822
Moving average 822
Exponential smoothing 828
Seasonal indexes 834
Deseasonalizing 837
Seasonally adjusted time-series 837
Mean absolute deviation (MAD) 839
Sum of squares for forecast error (SSE) 839
Autoregressive model 843

SYMBOLS

Symbol	Represents
y_t	Time series
S_t	Exponentially smoothed time series
w	Smoothing constant
F_t	Forecasted time series

FORMULAS

Exponential smoothing

$$S_t = wy_t + (1 - w)S_{t-1}$$

Mean absolute deviation

$$\text{MAD} = \frac{\sum_{i=1}^{n} |y_t - F_t|}{n}$$

Sum of squares for error

$$\text{SSE} = \sum_{i=1}^{n} (y_t - F_t)^2$$

Forecast of trend and seasonality

$$F_t = [b_0 + b_1 t] \times SI_t$$

Autoregressive model

$$y_t = \beta_0 + \beta_1 y_{t-1} + \varepsilon$$

COMPUTER INSTRUCTIONS

Technique	Excel	Minitab
Moving averages	825	825
Exponential smoothing	830	830
Seasonal indexes	836	836

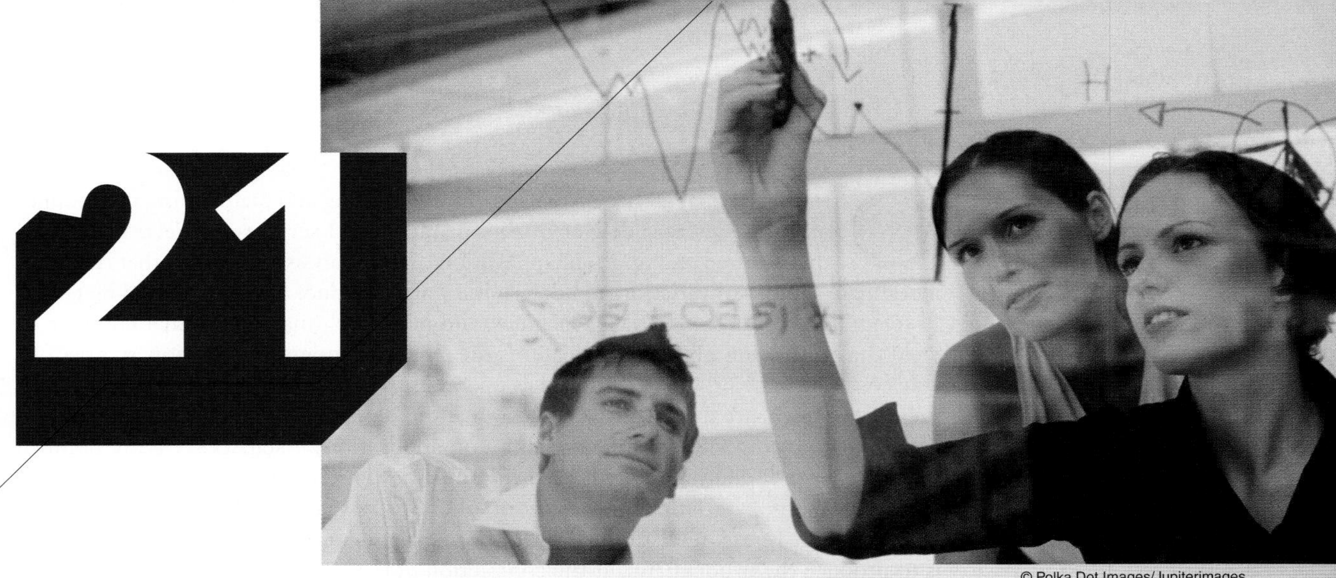

© Polka Dot Images/Jupiterimages

21 STATISTICAL PROCESS CONTROL

Detecting the Source of Defective Disks

DATA Xm21-00

A company that produces compact disks (CDs) has been receiving complaints from its customers about the large number of disks that will not store data properly. Company management has decided to institute statistical process control to remedy the problem. Every hour, a random sample of 200 disks is taken, and each disk is tested to determine whether it is defective. The number of defective disks in the samples of size 200 for the first 40 hours is shown here (in chronological order). Using these data, draw a *p* chart to monitor the production process. Was the process out of control when the sample results were generated?

19	5	16	20	6	12	18	6	13	15	10	6	7	10	18	20	13	6	8	3
8	7	4	19	3	19	9	10	10	18	15	16	5	14	3	10	19	13	19	9

© Comstock Images/Royalty-free/Getty Images

See page 873 for the answer.

INTRODUCTION

Operations managers are responsible for developing and maintaining the production processes that deliver quality products and services. In Section 14.6 we demonstrated an important application of the analysis of variance that is used to investigate sources of variation and determine ways to reduce that variation. The goal is to select the methods, materials, machines, and personnel (manpower) that combine to yield the production process that features the smallest amount of variation at a reasonable cost. Once the production process is operating, it is necessary to constantly monitor the process to ensure that it functions the way it was designed. The statistical methods we are about to introduce are the most common application of statistics. At any point in time, there are literally thousands of firms applying these methods. This chapter deals with the subject of **statistical process control** or **SPC** (formerly called **quality control**).

There are two general approaches to the management of quality. The first approach is to produce the product and, at the completion of the production process, inspect the unit to determine whether it conforms to specifications; if it doesn't, it is either discarded or repaired. This approach has several drawbacks. Foremost among them is that it is costly to produce substandard products that are later discarded or fixed. In recent years, this approach has been employed by a decreasing number of companies. Instead, many firms have adopted the **prevention approach.** Using the concepts of hypothesis testing, statistics practitioners concentrate on the production process. Rather than inspect the product, they inspect the process to determine when the process starts producing units that do not conform to specifications. This allows them to correct the production process before it creates a large number of defective products.

In the next section, we discuss the problem of process variation and why it is often the key to the management of quality. In Section 21.2 we also introduce the concept and logic of control charts and show why they work. In the rest of the chapter, we introduce three specific control charts.

21.1 PROCESS VARIATION

All production processes result in variation; that is, no product is exactly the same as another. You can see for yourself that this is true by weighing, for example, two boxes of breakfast cereal that are supposed to weigh 16 ounces each. Not only will they not weigh exactly 16 ounces, they will also not even have equal weights. All products exhibit some degree of variation. There are two sources of variation: *chance* and *assignable variation*. **Chance** or **common variation** is caused by a number of randomly occurring events that are part of the production process and that in general cannot be eliminated without changing the process. In effect, chance variation was built into the product when the production process was first set up, perhaps as a result of a statistical analysis that attempted to minimize but not necessarily eliminate such variation. In Section 14.6 we discussed statistical techniques that allow firms to experiment to search for sources of variation, and in so doing, reduce the variation.

Assignable or **special variation** is caused by specific events or factors that are frequently temporary and that can usually be identified and eliminated. To illustrate, consider a paint company that produces and sells paint in 1-gallon cans. The cans are filled by an automatic valve that regulates the amount of paint in each can. The designers of the valve acknowledge that there will be some variation in the amount of paint even when the valve is working as it was designed to work. This is chance variation. Occasionally, the valve will malfunction, causing the variation in the amount delivered to each can to increase. This increase is the assignable variation.

Perhaps the best way to understand what is happening is to consider the volume of paint in each can as a random variable. If the only sources of variation are caused by chance, then each can's volume is drawn from identical distributions. That is, each distribution has the same shape, mean, and standard deviation. Under such circumstances the production process is said to be **under control.** In recognition of the fact that variation in output will occur even when the process is under control and operating properly, most processes are designed so that their products will fall within designated **specification limits** or "specs." For example, the process that fills the paint cans may be designed so that the cans contain between .99 and 1.01 gallons. Inevitably, some event or combination of factors in a production process will cause the process distribution to change. When it does, the process is said to be **out of control.** There are several possible ways for the process to go out of control. Here is a list of the most commonly occurring possibilities and their likely assignable causes.

1. **Level shift** This is a change in the mean of the process distribution. Assignable causes include machine breakdown, new machine and/or operator, or a change in the environment. In the paint-can illustration, a temperature or humidity change may affect the density of the paint, resulting in less paint in each can.

2. **Instability** This is the name we apply to the process when the standard deviation increases. (As we discuss later, a decrease in the standard deviation is desirable.) This may be caused by a machine in need of repair, defective materials, wear of tools, or a poorly trained operator. Suppose, for example, that a part in the valve that controls the amount of paint wears down, causing greater variation than normal.

3. **Trend** When there is a slow steady shift (either up or down) in the process distribution mean, the result is a trend. This is frequently the result of less-than-regular maintenance, operator fatigue, residue or dirt buildup, or gradual loss of lubricant. If the paint-control valve becomes increasingly clogged, we would expect to see a steady decrease in the amount of paint delivered.

4. **Cycle** This is a repeated series of small observations followed by large observations. Likely assignable causes include environmental changes, worn parts, or operator fatigue. If there are changes in the voltage in the electricity that runs the machines in the paint-can example, we might see series of overfilled cans and series of underfilled cans.

The key to quality is to detect when the process goes out of control so that we can correct the malfunction and restore control of the process. The control chart is the statistical method that we use to detect problems.

EXERCISES

21.1 What is meant by chance variation?

21.2 Provide two examples of production processes and their associated chance variation.

21.3 What is meant by special variation?

21.4 Your education as a statistics practitioner can be considered a production process overseen by the course instructor. The variable we measure is the grade achieved by each student.

a. Discuss chance variation, that is, describe the sources of variation that the instructor has no control over.

b. Discuss special variation.

21.2 CONTROL CHARTS

A **control chart** is a plot of statistics over time. For example, an $\bar{x}$ **chart** plots a series of sample means taken over a period of time. Each control chart contains a **centerline** and *control limits*. The control limit above the centerline is called the **upper control limit** and that below the centerline is called the **lower control limit.** If, when the sample statistics are plotted, all points are randomly distributed between the control limits, we conclude that the process is under control. If the points are not randomly distributed between the control limits, we conclude that the process is out of control.

To illustrate the logic of control charts, let us suppose that in the paint can example described previously, we want to determine whether the central location of the distribution has changed from one period to another. We will draw our conclusion from an $\bar{x}$ chart. For the moment, let us assume that we know the mean μ and standard deviation σ of the process when it is under control. We can construct the $\bar{x}$ chart, as shown in Figure 21.1. The chart is drawn so that the vertical axis plots the values of $\bar{x}$ that will be calculated and the horizontal axis tracks the samples in the order in which they are drawn. The centerline is the value of μ. The control limits are set at three standard errors from the centerline. Recall that the standard error of $\bar{x}$ is $\sigma/\sqrt{n}$. Hence, we define the control limits as follows:

$$\text{Lower control limit} = \mu - 3\frac{\sigma}{\sqrt{n}}$$

$$\text{Upper control limit} = \mu + 3\frac{\sigma}{\sqrt{n}}$$

FIGURE **21.1**
$\bar{x}$ Chart: μ and σ Known

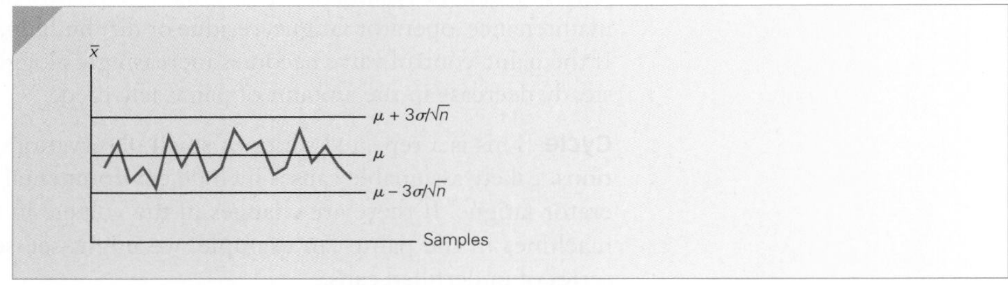

After we've constructed the chart by drawing the centerline and control limits, we use it to plot the sample means, which are joined to make it easier to interpret. The principles underlying control charts are identical to the principles of hypothesis testing. The null and alternative hypotheses are

H_0: The process is under control
H_1: The process is out of control

For an $\bar{x}$ chart, the test statistic is the sample mean $\bar{x}$. However, because we're dealing with a dynamic process rather than a fixed population, we test a series of sample means. That is, we compute the mean for each of a continuing series of samples taken over time. For each series of samples, we want to determine whether there is sufficient evidence to infer that the process mean has changed. We reject the null hypothesis if at any time the sample mean falls outside the control limits. It is logical to ask why we use 3 standard errors and not 2 or 1.96 or 1.645, as we did when we tested hypotheses about a population mean in Chapter 11. The answer lies in the way in which all tests are conducted. Because test conclusions are based on sample data, there are two possible errors. In statistical process control, a Type I error occurs if we conclude that the process is out of control when in fact it is not. This error can be quite expensive, because the production process must be stopped and the causes of the variation found and repaired. Consequently, we want the probability of a Type I error to be small. With control limits set at 3 standard errors from the mean, the probability of a Type I error for each sample is

$$\alpha = P(|z| > 3) = .0026$$

Recall that a small value of α results in a relatively large value of the probability of a Type II error. A Type II error occurs when at any sample we do not reject a false null hypothesis. This means that, for each sample, we are less likely to recognize when the process goes out of control. However, because we will be performing a series of tests (one for each sample), we will eventually discover that the process is out of control and take steps to rectify the problem.

Suppose that in order to test the production process that fills 1-gallon paint cans, we choose to take a sample of size 4 every hour. Let us also assume that we know the mean and standard deviation of the process distribution of the amount of paint when the process is under control, say, $\mu = 1.001$ and $\sigma = .006$. (This means that when the valve is working the way it was designed, the amount of paint put into each can is a random variable whose mean is 1.001 gallons and whose standard deviation is .006 gallon.) Thus,

$$\text{Centerline} = \mu = 1.001$$

$$\text{Lower control limit} = \mu - 3\frac{\sigma}{\sqrt{n}} = 1.001 - 3\frac{.006}{\sqrt{4}} = 1.001 - .009 = .992$$

$$\text{Upper control limit} = \mu + 3\frac{\sigma}{\sqrt{n}} = 1.001 + 3\frac{.006}{\sqrt{4}} = 1.001 + .009 = 1.010$$

Figure 21.2 depicts a situation in which the first 15 samples were taken when the process was under control. However, after the 15th sample was drawn, the process went out of control and produced sample means outside the control limits. We conclude that the process distribution has changed, because the data display variability beyond that predicted for a process with the specified mean and standard deviation. This means that the variation is assignable and that the cause must be identified and corrected.

FIGURE **21.2** $\bar{x}$ Chart: Process Out of Control

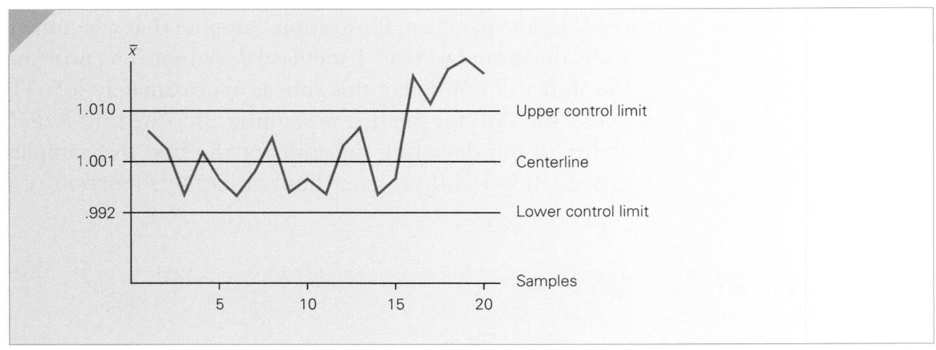

As we stated previously, SPC is a slightly different form of hypothesis testing. The concept is the same, but there are differences that you should be aware of. The most important difference is that when we tested means and proportions in Chapters 11 and 12, we were dealing with fixed but unknown parameters of populations. For instance, in Example 11.1 the population we dealt with was the account balances of the department store customers. The population mean balance was a constant value that we simply did not know. The purpose of the test was to determine whether there was enough statistical evidence to allow us to infer that the mean balance was greater than \$170. So we took one sample and based the decision on the sample mean. When dealing with a production process, it's important to realize that the process distribution itself is variable. That is, at any time, the process distribution of the amount of paint fill may change if the valve malfunctions. Consequently, we do not simply take one sample and make the decision. Instead, we plot a number of statistics over time in the control chart. Simply put, in Chapters 11 through 18, we assumed static population distributions with fixed but unknown parameters whereas in this chapter we assume a dynamic process distribution with parameters subject to possible shifts.

Sample Size and Sampling Frequency

In designing a control chart, the statistics practitioner must select a sample size and a sampling frequency. These decisions are based on several factors, including the costs of making Type I and Type II errors, the length of the production run, and the typical change in the process distribution when the process goes out of control. A useful aid in making the decision is the **operating characteristic (OC) curve.**

Operating Characteristic Curve

Recall that in Chapter 11 we drew the operating characteristic (OC) curve that plotted the probabilities of Type II errors and population means. Here is how the OC curve for the $\bar{x}$ chart is drawn.

Suppose that when the production process is under control, the mean and standard deviation of the process variable are μ_0 and σ, respectively. For specific values of α and n, we can compute the probability of a Type II error when the process mean changes to $\mu_1 = \mu_0 + k\sigma$. A Type II error occurs when a sample mean falls between the control limits when the process is out of control. In other words the probability of a Type II error is the probability that the $\bar{x}$ chart will be unable to detect a shift of $k\sigma$ in the process mean on the first sample after the shift has occurred. Figure 21.3 depicts the OC curve for $n = 2, 3, 4$, and 5. Figure 21.4 is the OC curve for $n = 10, 15, 20$, and 25. (We drew two sets of curves because one alone would not provide the precision we need.) We can use the OC curves to help determine the sample size we should use.

Figure 21.3 tells us that for small shifts in the mean of 1 standard deviation or less, samples of size 2 to 5 produce probabilities of not detecting shifts that range between .8 and .95 (approximately). To appreciate the effect of large probabilities of Type II errors, consider the paint can illustration. Suppose that when the process goes out of control, it shifts the mean by about 1 standard deviation. The probability that the first sample after the shift will not detect this shift is approximately .85. The probability that it will not detect the shift for the first m samples after the shift is $.85^m$. Thus for $m = 5$, the probability of not detecting the shift for the first five samples after the shift is .44. If the process fills 1,000 cans per hour, a large proportion of the 5,000 cans filled will be

FIGURE 21.3 Operating Characteristic Curve for $n = 2, 3, 4,$ and 5

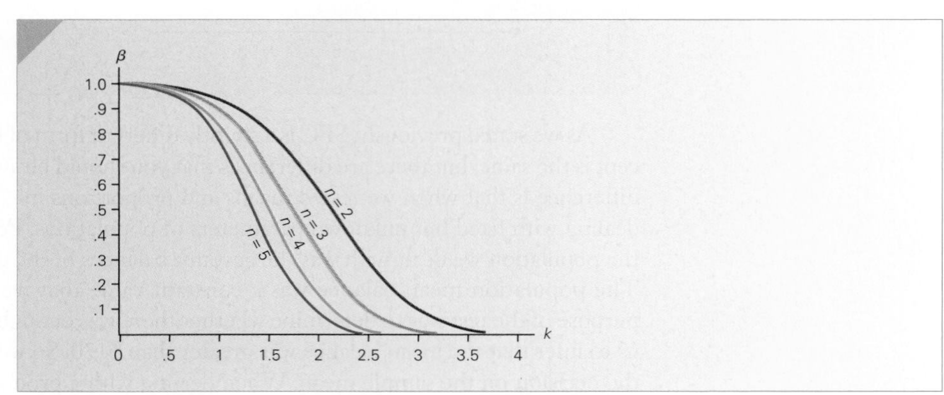

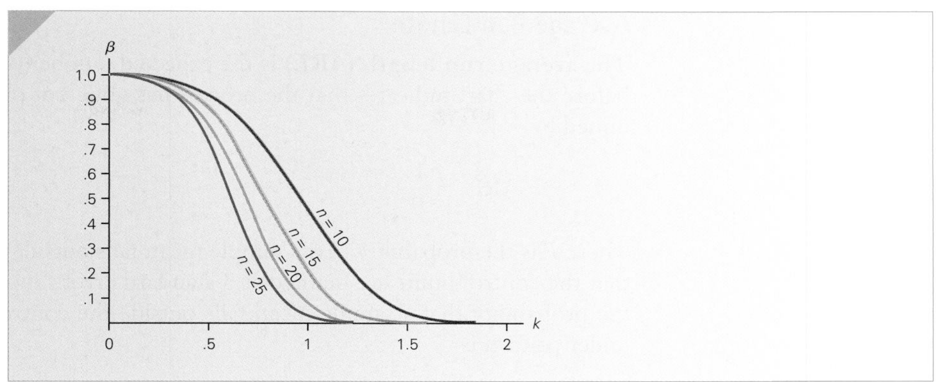

FIGURE **21.4**
Operating Characteristic Curve for $n = 10, 15, 20,$ and 25

overfilled or underfilled (depending on the direction of the shift). Figure 21.4 suggests that when the shift moves the process mean by 1 standard deviation, samples of size 15 or 20 are recommended. For $n = 15$, the probability that a shift of 1 standard deviation will not be detected by the first sample is approximately .2.

If the typical shift is 2 or more standard deviations, samples of size 4 or 5 will likely suffice. For $n = 5$, the probability of a Type II error is about .07.

Using the Computer

EXCEL

As we did in Chapter 11, we use can the computer to determine the probability of a Type II error. Here is the printout for $k = 1.5$ and $n = 4$. (See page 372 for instructions.)

	A	B	C	D
1	**Type II Error**			
2				
3	**H0: MU**	0	**Critical values**	**-1.51**
4	**SIGMA**	1		**1.51**
5	**Sample size**	4	**Prob(Type II error)**	**0.5046**
6	**ALPHA**	0.0026	**Power of the test**	**0.4954**
7	**H1: MU**	1.5		

MINITAB

As we did in Chapter 11, we use can the computer to determine the probability of a Type II error. Here is the printout for $k = 1.5$ and $n = 4$. (See page 373 for instructions.)

Power and Sample Size

```
1-Sample Z Test

Testing mean = null (versus not = null)
Calculating power for mean = null + difference
Alpha = 0.0026  Assumed standard deviation = 1


              Sample
Difference     Size     Power
   1.5           4     0.4954
```

Minitab prints the power of the test. Recall that the power of the test is $1 - \beta$.

Average Run Length

The **average run length (ARL)** is the expected number of samples that must be taken before the chart indicates that the process has gone out of control. The ARL is determined by

$$\text{ARL} = \frac{1}{P}$$

where P is the probability that a sample mean falls outside the control limits. Assuming that the control limits are defined as 3 standard errors above and below the centerline, the probability that a sample mean falls outside the control limits when the process is under control is

$$P = P(|z| > 3) = .0026$$

Thus,

$$\text{ARL} = \frac{1}{.0026} = 385$$

This means that when the process is under control, the $\bar{x}$ chart will erroneously conclude that it is out of control once every 385 samples on average. If the sampling plan calls for samples to be taken every hour, on average there will be a *false alarm* once every 385 hours.

We can use the OC curve to determine the average run length until the $\bar{x}$ chart detects a process that is out of control. Suppose that when the process goes out of control, it typically shifts the process mean 1.5 standard deviations to the right or left. From Figure 21.3, we can see that for $n = 4$ and $k = 1.5$, the probability of a Type II error is approximately .5. That is, the probability that a mean falls between the control limits, which indicates that the process is under control when there has been a shift of 1.5 standard deviations, is .5. The probability that the sample mean falls outside the control limits is $P = 1 - .5 = .5$. Thus, the average run length is

$$\text{ARL} = \frac{1}{.5} = 2$$

This means that the control chart will require two samples on average to detect a shift of 1.5 standard deviations. Suppose that a shift of this magnitude results in an unacceptably high number of nonconforming cans. We can reduce that number in two ways: by sampling more frequently or increasing the sample size. For example, if we take samples of size 4 every half hour, then on average it will take 1 hour to detect the shift and make repairs. If we take samples of size 10 every hour, Figure 21.4 indicates that the probability of a Type II error when the shift is 1.5 standard deviations is about .05. Thus, $P = 1 - .05 = .95$ and

$$\text{ARL} = \frac{1}{.95} = 1.05$$

This tells us that a sample size of 10 will allow the statistics practitioner to detect a shift of 1.5 standard deviations about twice as quickly as a sample of size 4.

Changing the Control Limits

Another way to decrease the probability of a Type II error is to increase the probability of making a Type I error. Thus, we may define the control limits so that they are 2 standard errors above and below the centerline. In order to judge whether this is advisable, it is necessary to draw the operating characteristic curve for this plan.

In our demonstration of the logic of control charts, we resorted to traditional methods of presenting inferential methods; we assumed that the process parameters were known. When the parameters are unknown, we estimate their values from the sample data. In the next two sections, we discuss how to construct and use control charts in more realistic situations. In Section 21.3, we present control charts when the data are interval. In the context of statistical process control, we call these **control charts for variables.** Section 21.4 demonstrates the use of control charts that record whether a unit is defective or nondefective. These are called **control charts for attributes.**

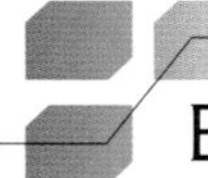

EXERCISES

21.5 If the control limits of an $\bar{x}$ chart are set at 2.5 standard errors from the centerline, what is the probability that on any sample the control chart will indicate that the process is out of control when it is under control?

21.6 Refer to Exercise 21.5. What is the average run length until the $\bar{x}$ chart signals that the process is out of control when it is under control?

21.7 The control limits of an $\bar{x}$ chart are set at 2 standard errors from the centerline. Calculate the probability that on any sample the control chart will indicate that the process is out of control when it is under control.

21.8 Refer to Exercise 21.7. Determine the ARL until the $\bar{x}$ chart signals that the process is out of control when it is under control.

Exercises 21.9– 21.15 are based on the following scenario.

A production facility produces 100 units per hour and uses an $\bar{x}$ chart to monitor its quality. The control limits are set at 3 standard errors from the mean. When the process goes out of control, it usually shifts the mean by 1.5 standard deviations. Sampling is conducted once per hour with a sample size of 3.

21.9 On average how many units will be produced until the control chart signals that the process is out of control when it is under control?

21.10 Refer to Exercise 21.9.
a. Find the probability that the $\bar{x}$ chart does not detect a shift of 1.5 standard deviations on the first sample after the shift occurs.
b. Compute the probability that the $\bar{x}$ chart will not detect the shift for the first 8 samples after the shift.

21.11 Refer to Exercise 21.10. Find the average run length to detect the shift.

21.12 The operations manager is unsatisfied with the current sampling plan. He changes it to samples of size 2 every half hour. What is the average number of units produced until the chart indicates that the process is out of control when it is not?

21.13 Refer to Exercise 21.12.
a. Find the probability that the $\bar{x}$ chart does not detect a shift of 1.5 standard deviations on the first sample after the shift occurs.
b. Compute the probability that the $\bar{x}$ chart will not detect the shift for the first 8 samples after the shift.

21.14 Refer to Exercise 21.13. What is the average run length to detect the shift.

21.15 Write a brief report comparing the sampling plans described in Exercises 21.9 and 21.12. Discuss the relative costs of the two plans and the frequency of Type I and Type II errors.

Exercises 21.16 to 21.22 are based on the following scenario.

A firm that manufactures notebook computers uses statistical process control to monitor all its production processes. For one component, the company draws samples of size 10 every 30 minutes. The company makes 4,000 of these components per hour. The control limits of the $\bar{x}$ chart are set at 3 standard errors from the mean. When the process goes out of control, it usually shifts the mean by .75 standard deviation.

21.16 On average how many units will be produced until the control chart signals that the process is out of control when it is under control?

21.17 Refer to Exercise 21.16.
a. Find the probability that the $\bar{x}$ chart does not detect a shift of .75 standard deviation on the first sample after the shift occurs.
b. Compute the probability that the $\bar{x}$ chart will not detect the shift for the first 4 samples after the shift.

21.18 Refer to Exercise 21.17. Find the average run length to detect the shift.

21.19 The company is considering changing the sampling plan so that 20 components are sampled every hour. What is the average number of units produced until the chart indicates that the process is out of control when it is not?

21.20 Refer to Exercise 21.19.

a. Find the probability that the $\bar{x}$ chart does not detect a shift of .75 standard deviations on the first sample after the shift occurs.

b. Compute the probability that the $\bar{x}$ chart will not detect the shift for the first 4 samples after the shift.

21.21 Refer to Exercise 21.20. What is the average run length to detect the shift?

21.22 Write a brief report comparing the sampling plans described in Exercises 21.16 and 21.19. Discuss the relative costs of the two plans and the frequency of Type I and Type II errors.

21.3 Control Charts for Variables: $\bar{X}$ and S Charts

There are several ways to judge whether a change in the process distribution has occurred when the data are interval. To determine whether the distribution means have changed, we employ the $\bar{x}$ chart. To determine whether the process distribution standard deviation has changed, we can use the S (which stands for standard deviation) chart or the R (which stands for range) chart.

Throughout this textbook we have used the sample standard deviation to estimate the population standard deviation. However, for a variety of reasons, SPC frequently employs the range instead of the standard deviation. This is primarily because computing the range is simpler than computing the standard deviation. Because many practitioners conducting SPC perform calculations by hand (with the assistance of a calculator), they select the computationally simple range as the method to estimate the process standard deviation. In this section, we will introduce control charts that feature the sample standard deviation. In CD Appendix AD we employ the sample range to construct our charts.

$\bar{x}$ Chart

In Section 21.2 we determined the centerline and control limits of an $\bar{x}$ chart using the mean and standard deviation of the process distribution. However, it is unrealistic to believe that the mean and standard deviation of the process distribution are known. Thus, to construct the $\bar{x}$ chart, we need to estimate the relevant parameters from the data.

We begin by drawing samples when we have determined that the process is under control. The sample size must lie between 2 and 25. We discuss later how to determine that the process is under control. For each sample, we compute the mean and the standard deviation. The estimator of the mean of the distribution is the mean of the sample means (denoted $\bar{\bar{x}}$):

$$\bar{\bar{x}} = \frac{\sum_{j=1}^{k} \bar{x}_j}{k}$$

where $\bar{x}_j$ is the mean of the jth sample and there are k samples. (Note that $\bar{\bar{x}}$ is simply the average of all nk observations.)

To estimate the standard deviation of the process distribution, we calculate the sample variance s_j^2 for each sample. We then compute the pooled standard deviation,* which we denote S and define as

$$S = \sqrt{\frac{\sum_{j=1}^{k} s_j^2}{k}}$$

In the previous section, where we assumed that the process distribution mean and variance were known, the centerline and control limits were defined as

$$\text{Centerline} = \mu$$

$$\text{Lower control limit} = \mu - 3\frac{\sigma}{\sqrt{n}}$$

$$\text{Upper control limit} = \mu + 3\frac{\sigma}{\sqrt{n}}$$

Because the values of μ and σ are unknown, we must use the sample data to estimate them. The estimator of μ is $\bar{\bar{x}}$ and the estimator of σ is S. Therefore, the centerline and control limits are as shown in the box.

Centerline and Control Limits for $\bar{x}$ Chart

$$\text{Centerline} = \bar{\bar{x}}$$

$$\text{Lower control limit} = \bar{\bar{x}} - 3\frac{S}{\sqrt{n}}$$

$$\text{Upper control limit} = \bar{\bar{x}} + 3\frac{S}{\sqrt{n}}$$

EXAMPLE 21.1†

DATA
Xm21-01

Statistical Process Control at Lear Seating, Part 1

Lear Seating of Kitchener, Ontario, manufactures seats for Chrysler, Ford, and General Motors cars. Several years ago, Lear instituted statistical process control, which has resulted in improved quality and lower costs. One of the components of a front-seat cushion is a wire spring, produced from 4-mm (millimeter) steel wire. A machine is used to bend the wire so that the spring's length is 500 mm. If the springs are longer than 500 mm, they will loosen and eventually fall out. If they are too short, they won't easily fit into position. (In fact, in the past, when there were a relatively large number of short springs, workers incurred arm and hand injuries when attempting to install the springs.) To determine whether the process is under control, random samples of four springs are taken every hour. The last 25 samples are shown here. Construct an $\bar{x}$ chart from these data.

*This formula requires that the sample size be the same for all samples, a condition that is imposed throughout this chapter.

†The authors are grateful to Pat Bourke, Barry Cress, Kevin Lewis, and Brian Riehl of Lear Seating Ltd. for their assistance in writing this example and several exercises.

Sample				
1	501.02	501.65	504.34	501.10
2	499.80	498.89	499.47	497.90
3	497.12	498.35	500.34	499.33
4	500.68	501.39	499.74	500.41
5	495.87	500.92	498.00	499.44
6	497.89	499.22	502.10	500.03
7	497.24	501.04	498.74	503.51
8	501.22	504.53	499.06	505.37
9	499.15	501.11	497.96	502.39
10	498.90	505.99	500.05	499.33
11	497.38	497.80	497.57	500.72
12	499.70	500.99	501.35	496.48
13	501.44	500.46	502.07	500.50
14	498.26	495.54	495.21	501.27
15	497.57	497.00	500.32	501.22
16	500.95	502.07	500.60	500.44
17	499.70	500.56	501.18	502.36
18	501.57	502.09	501.18	504.98
19	504.20	500.92	500.02	501.71
20	498.61	499.63	498.68	501.84
21	499.05	501.82	500.67	497.36
22	497.85	494.08	501.79	501.95
23	501.08	503.12	503.06	503.56
24	500.75	501.18	501.09	502.88
25	502.03	501.44	498.76	499.39

SOLUTION

COMPUTE

MANUALLY

The means and standard deviations for each sample were computed and are listed in Table 21.1. We then calculated the mean of the means (which is also the mean of all 100 numbers) and the pooled standard deviation:

$$\bar{\bar{x}} = 500.296$$
$$S = 1.971$$

Thus, the centerline and control limits are

$$\text{Centerline} = \bar{\bar{x}} = 500.296$$

$$\text{Lower control limit} = \bar{\bar{x}} - 3\frac{S}{\sqrt{n}} = 500.296 - 3\frac{1.971}{\sqrt{4}} = 497.340$$

$$\text{Upper control limit} = \bar{\bar{x}} + 3\frac{S}{\sqrt{n}} = 500.296 + 3\frac{1.971}{\sqrt{4}} = 503.253$$

The centerline and control limits are drawn and the sample means plotted in the order in which they occurred. The manually drawn chart is identical to the Excel and Minitab versions shown here.

TABLE **21.1** Means and Standard Deviations of Samples in Example 21.1

Sample					$\bar{x}_j$	s_j
1	501.02	501.65	504.34	501.10	502.03	1.567
2	499.80	498.89	499.47	497.90	499.02	0.833
3	497.12	498.35	500.34	499.33	498.79	1.376
4	500.68	501.39	499.74	500.41	500.56	0.683
5	495.87	500.92	498.00	499.44	498.56	2.152
6	497.89	499.22	502.10	500.03	499.81	1.763
7	497.24	501.04	498.74	503.51	500.13	2.741
8	501.22	504.53	499.06	505.37	502.55	2.934
9	499.15	501.11	497.96	502.39	500.15	1.978
10	498.90	505.99	500.05	499.33	501.07	3.316
11	497.38	497.80	497.57	500.72	498.37	1.578
12	499.70	500.99	501.35	496.48	499.63	2.216
13	501.44	500.46	502.07	500.50	501.12	0.780
14	498.26	495.54	495.21	501.27	497.57	2.820
15	497.57	497.00	500.32	501.22	499.03	2.059
16	500.95	502.07	500.60	500.44	501.02	0.735
17	499.70	500.56	501.18	502.36	500.95	1.119
18	501.57	502.09	501.18	504.98	502.46	1.724
19	504.20	500.92	500.02	501.71	501.71	1.796
20	498.61	499.63	498.68	501.84	499.69	1.507
21	499.05	501.82	500.67	497.36	499.73	1.943
22	497.85	494.08	501.79	501.95	498.92	3.741
23	501.08	503.12	503.06	503.56	502.71	1.106
24	500.75	501.18	501.09	502.88	501.48	0.955
25	502.03	501.44	498.76	499.39	500.41	1.576

EXCEL

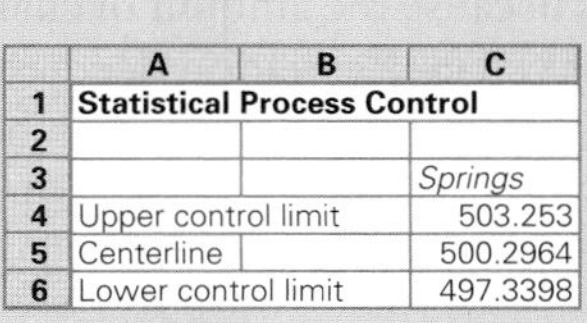

	A	B	C
1	Statistical Process Control		
2			
3			*Springs*
4	Upper control limit		503.253
5	Centerline		500.2964
6	Lower control limit		497.3398

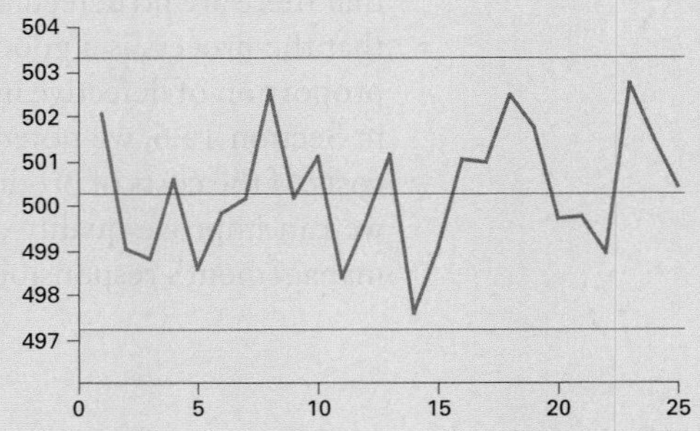

INSTRUCTIONS

1. Type or import the data into one column. (Open Xm21-01.)
2. Click **Add-Ins, Data Analysis Plus,** and **Statistical Process Control.**
3. Specify the **Input Range** (A1:101) and the **Sample Size** (4). Click **XBAR (Using S).**

MINITAB

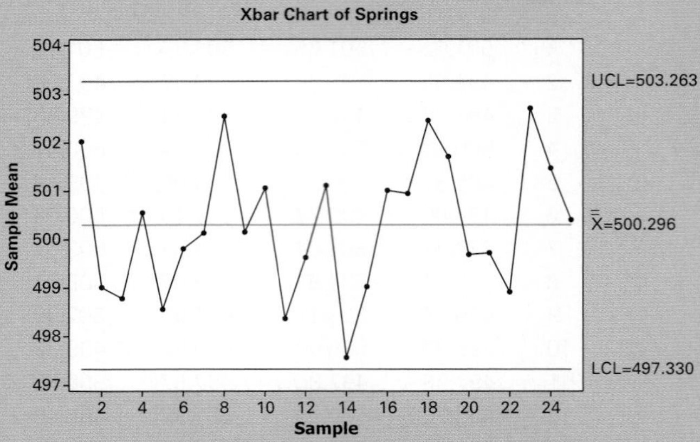

INSTRUCTIONS

1. Type or import the data into one column. (Open Xm21-01.)
2. Click **Stat, Control Charts, Variable charts for subgroups,** and **Xbar . . . .**
3. Specify **All observations for a chart are in one column.** Type or select the variable (Springs). Type the **Subgroup sizes** (4). Click **Xbar Options . . . .**
4. Click **Estimate . . . .** Under **Method for estimating standard deviation,** click **Pooled standard deviation.**

INTERPRET

As you can see, no point lies outside the control limits. We conclude that the variation in the lengths of the springs is caused by chance. That is, there is not enough evidence to infer that the process is out of control. No remedial action by the operator is called for.

We stress that statistical process control allows us to detect assignable variation only. In Example 21.1, we determined that the process is under control, which means that there are no detectable sources of assignable variation. However, this does not mean that the process is a good one. It may well be that the production process yields a large proportion of defective units because the amount of chance variation is large. Recall that in Section 14.6, we noted that chance variation decreases product quality and increases costs. If the costs of producing defective units are high because of large chance variation, we can improve quality and reduce costs only by changing the process itself, which is management's responsibility.

Pattern Tests to Determine When the Process Is Out of Control

When we tested hypotheses in the other parts of this book, we used only one sample statistic to make a decision. However, in statistical process control, the decision is made from a series of sample statistics. In the $\bar{x}$ chart we make the decision after plotting at least 25 sample means. As a result, we can develop tests that are based on the pattern the sample means make when plotted. To describe them, we need to divide the $\bar{x}$ chart between the control limits into six zones, as shown in Figure 21.5. The C zones represent the area within one standard error of the centerline. The B zones are the regions between one and two standard errors from the centerline. The spaces between two and three standard errors from the centerline are defined as A zones.

FIGURE **21.5**
Zones of $\bar{x}$ Chart

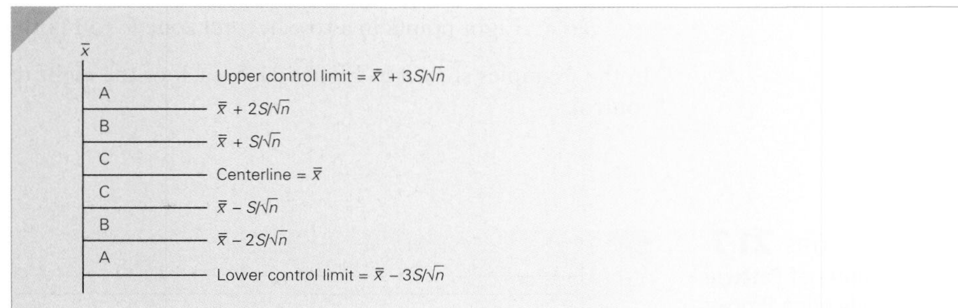

The width of the zones is one standard error of $\bar{x}$ $(S/\sqrt{n})$. If the calculations were performed manually, the value of S will be known. However, if a computer was used, the centerline and control limits are the only statistics printed. We can calculate $S/\sqrt{n}$ by finding the difference between the upper and lower control limits and dividing the difference by 6. That is,

$$S/\sqrt{n} = \frac{(\bar{\bar{x}} + 3S/\sqrt{n}) - (\bar{\bar{x}} - 3S/\sqrt{n})}{6} = \frac{503.253 - 497.340}{6} = .9855$$

Figure 21.6 describes the centerline, control limits, and zones for Example 21.1.

FIGURE **21.6**
Zones of $\bar{x}$ Chart: Example 21.1

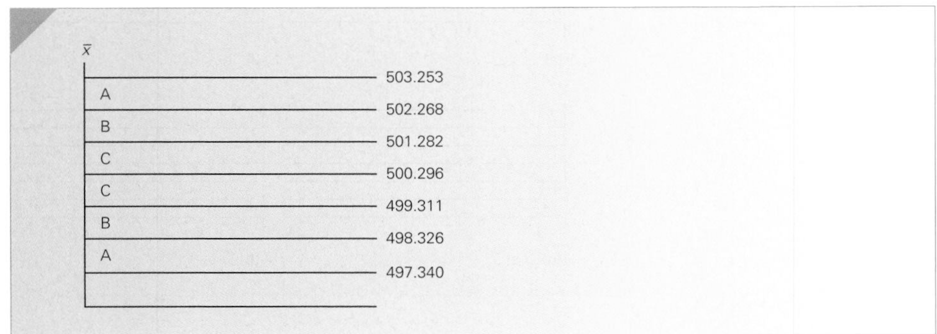

Several pattern tests can be applied. We list eight tests that are conducted by Minitab and by Data Analysis Plus.

Test 1: One point beyond zone A. This is the method discussed previously, where we conclude that the process is out of control if any point is outside the control limits.

Test 2: Nine points in a row in zone C or beyond (on the same side of the centerline).

Test 3: Six increasing or six decreasing points in a row.

Test 4: Fourteen points in a row alternating up and down.

Test 5: Two out of three points in a row in zone A or beyond (on the same side of the centerline).

Test 6: Four out of five points in a row in zone B or beyond (on the same side of the centerline).

Test 7: Fifteen points in a row in zone C (on both sides of the centerline).

Test 8: Eight points in a row beyond zone C (on both sides of the centerline).

In the examples shown in Figure 21.7, each of the eight tests indicates a process out of control.

FIGURE **21.7** Examples of Patterns Indicating Process Out of Control

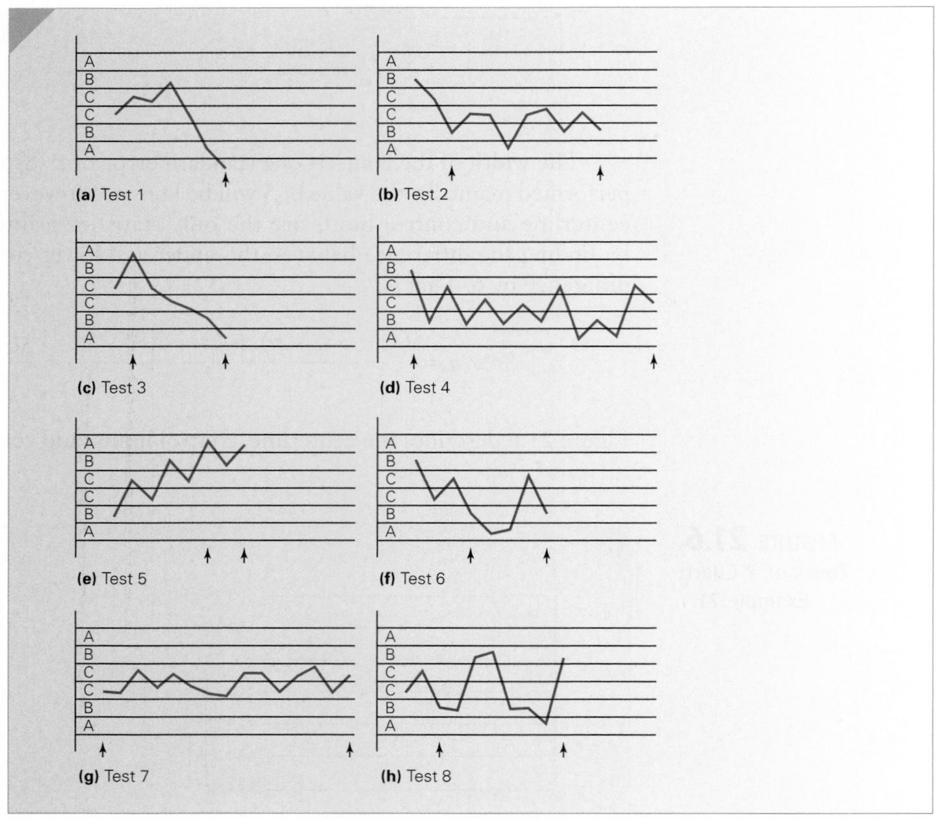

All eight tests are based on the same concepts used to test hypotheses throughout this book. That is, each of these patterns is a rare event, unlikely to occur when a process is under control. Thus, when any one of these patterns is recognized, the statistics practitioner has reason to believe that the process is out of control. In fact, it is often possible to identify the cause of the problem from the pattern in the control chart.

Figure 21.8 depicts the zones and the means for Example 21.1. After checking each of the eight pattern tests, we conclude that the process is under control.

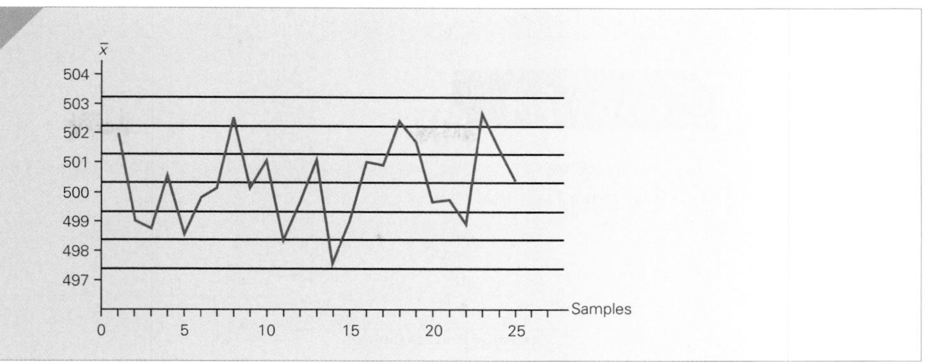

FIGURE **21.8** $\bar{x}$ Chart with Zones: Example 21.1

EXCEL

Excel automatically performs all eight tests. Any that fail will be reported on the spreadsheet.

MINITAB

Minitab will perform any or all of the pattern tests and report any that fail.

INSTRUCTIONS

After clicking **Xbar Options . . .** click **Tests** and click **Perform all tests for special causes.** To draw the zones, click **S Limits . . .** and type **1 2 3** in the **Display control limits at** box.

Pattern Tests in Practice

There appears to be a great deal of disagreement among statisticians with regard to pattern tests. Some authors and statistical software packages apply eight tests whereas others employ a different number. In addition, some statisticians apply pattern tests to $\bar{x}$ charts, but not to other charts. Rather than joining the debate with our own opinions, we will follow Minitab's rules. There are eight pattern tests for $\bar{x}$ charts, no pattern tests for S and R charts, and four pattern tests for the chart presented in Section 21.4 (p charts). The same rules apply to Data Analysis Plus.

S Charts

The ***S* chart** graphs sample standard deviations to determine whether the process distribution standard deviation has changed. The format is similar to that of the $\bar{x}$ chart: The S chart will display a centerline and control limits. However, the formulas for the centerline and control limits are more complicated than those for the $\bar{x}$ chart. Consequently, we will not display the formulas; instead we will let the computer do all the work.

EXAMPLE 21.2

Statistical Process Control at Lear Seating, Part 2

Using the data provided in Example 21.1, determine whether there is evidence to indicate that the process distribution standard deviation has changed over the period when the samples were taken.

SOLUTION

COMPUTE

EXCEL

	A	B	C
1	**Statistical Process Control**		
2			
3			*Springs*
4	Upper control limit		4.1288
5	Centerline		1.822
6	Lower control limit		0

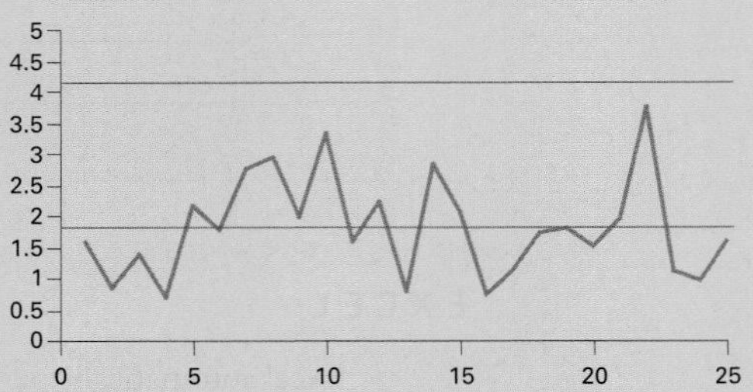

INSTRUCTIONS

1. Type or import the data into one column. (Open Xm21-01.)
2. Click **Add-Ins, Data Analysis Plus,** and **Statistical Process Control.**
3. Specify the **Input Range** (A1:101) and the **Sample Size** (4). Click **S.**

MINITAB

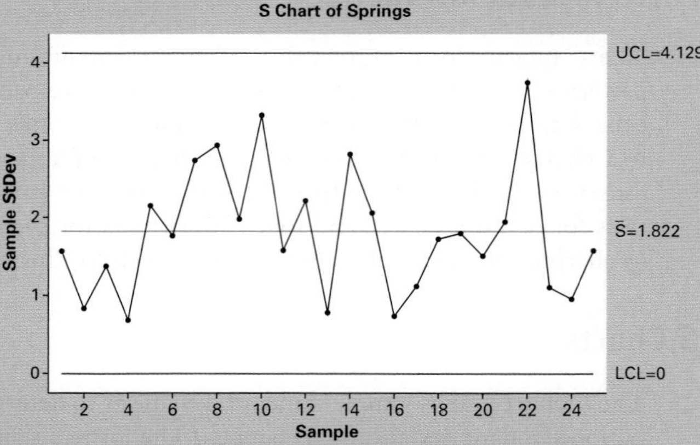

INSTRUCTIONS

1. Type or import the data into one column. (Open Xm21-01.)
2. Click **Stat, Control Charts, Variable charts for subgroups,** and **S . . . .**
3. Specify **All observations for a chart are in one column.** Type or select the variable (Springs). Type the **Subgroup sizes** (4). Click **S Options . . . .**
4. Click **Estimate . . . .** Under **Method for estimating standard deviation,** click **Pooled standard deviation.**

INTERPRET

There are no points outside the control limits. Because we do not apply any of the pattern tests, we conclude that there is no evidence to believe that the standard deviation has changed over this period.

Good News and Bad News about *S* Charts

In analyzing *S* charts, we would conclude that the process distribution has changed if we observe points outside the control limits. Obviously, points above the upper control limit indicate that the process standard deviation has increased—an undesirable situation. Points below the lower control limit also indicate that the process standard deviation has changed. However, cases in which the standard deviation has decreased are welcome occurrences because reducing the variation generally leads to improvements in quality. The operations manager should investigate cases where the sample standard deviations or ranges are small to determine the factors that produced such results. The objective is to determine whether permanent improvements in the production process can be made. Care must be exercised in cases where the *S* chart reveals a decrease in the standard deviation, because this is often caused by improper sampling.

Using the $\bar{x}$ and *S* Charts

In this section, we have introduced $\bar{x}$ and *S* charts as separate procedures. In actual practice, however, the two charts must be drawn and assessed together. The reason for this is that the $\bar{x}$ chart uses *S* to calculate the control limits and zone boundaries. Consequently, if the *S* chart indicates that the process is out of control, the value of *S* will not lead to an accurate estimate of the standard deviation of the process distribution. The usual procedure is to draw the *S* chart first. If it indicates that the process is under control, we then draw the $\bar{x}$ chart. If the $\bar{x}$ chart also indicates that the process is under control, we are then in a position to use both charts to maintain control. If either chart shows that the process was out of control at some time during the creation of the charts, we can detect and fix the problem and then redraw the charts with new data.

Monitoring the Production Process

When the process is under control, we can use the control chart limits and centerline to monitor the process in the future. We do so by plotting all future statistics on the control chart.

EXAMPLE 21.3

Statistical Process Control at Lear Seating, Part 3

After determining that the process is under control, the company in Example 21.1 began using the statistics generated in the creation of the $\bar{x}$ and *S* charts to monitor the production process. The sampling plan calls for samples of size 4 every hour. The following table lists the lengths of the springs taken during the first 6 hours.

Sample				
1	502.653	498.354	502.209	500.080
2	501.212	494.454	500.918	501.855
3	500.086	500.826	496.426	503.591
4	502.994	500.481	502.996	503.113
5	500.549	498.780	502.480	499.836
6	500.441	502.666	502.569	503.248

SOLUTION

After each sample is taken, the mean and standard deviation are computed. The standard deviations are plotted on the S chart using the previously determined control limits when the process variation was deemed to be in control. The sample means are plotted on the $\bar{x}$ chart, again using the zone limits determined when the process was deemed to be in control and the pattern tests are checked after each point is plotted. The first six samples are shown in Figure 21.9. After the standard deviation and mean of the sixth sample are plotted, the technician would stop the production process. Although the process variation still appears to be in control, the fourth and sixth means on the $\bar{x}$ chart combine to indicate that test 5 has failed; there are two out of three points in a row that are in zone A or beyond. Thus, it appears that the process mean has shifted upward. Technicians need to find the source of the problem and make repairs. After repairs are completed, production resumes and new control charts and their centerlines and control limits are recalculated.

FIGURE **21.9** *S* and $\bar{x}$ Charts for Example 21.3

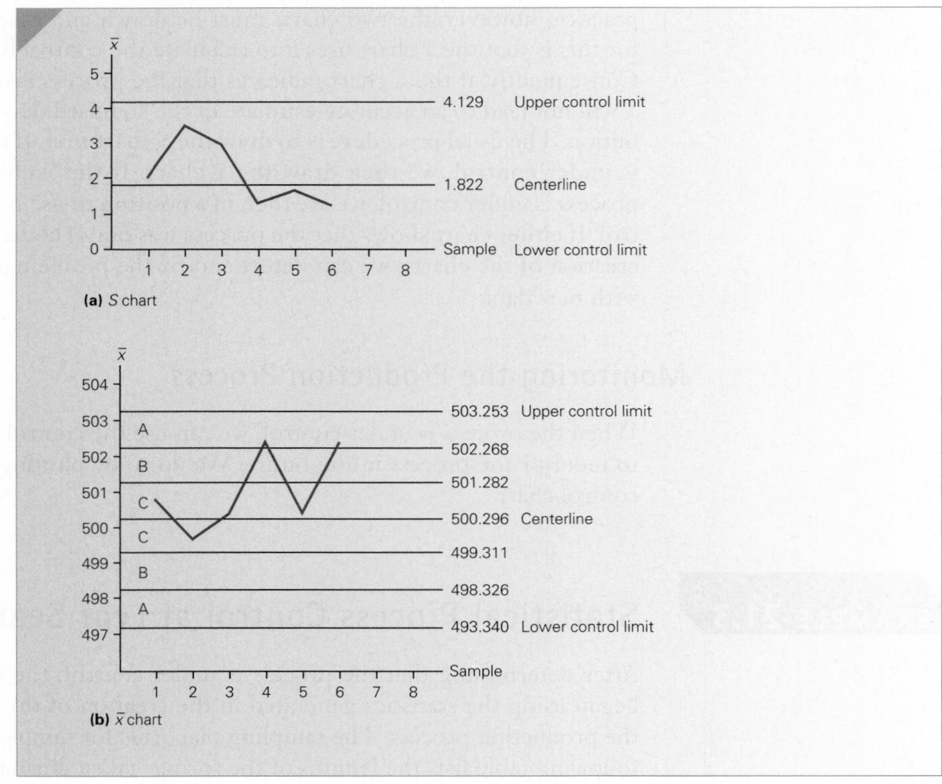

Process Capability Index

In Section 14.6 we discussed the process capability index, which measures the capability of the process to produce units whose dimensions fall within the specifications. We defined the index as

$$C_p = \frac{\text{USL} - \text{LSL}}{6\sigma}$$

where USL and LSL are the upper and lower specification limits, respectively. To compute the process capability index, we need to know these limits and the process standard deviation. The standard deviation is a population parameter that is generally unknown. Thus, C_p measures the theoretical or potential process capability. To produce a measure of the process's actual capability, we must use statistics computed in the construction of the control chart. Suppose that, in Example 21.1, the operations manager determined that the springs will fit provided that their lengths fall between the lower specification limit LSL = 493 and the upper specification limit USL = 507. In Example 21.1 we found $\bar{\bar{x}} = 500.296$ and $S = 1.971$.

We define the following:

$$\text{CPL} = \frac{\bar{\bar{x}} - \text{LSL}}{3S} = \frac{500.296 - 493}{3(1.971)} = 1.23$$

$$\text{CPU} = \frac{\text{USL} - \bar{\bar{x}}}{3S} = \frac{507 - 500.296}{3(1.971)} = 1.13$$

We define the process capability index as the smaller of these two indexes. That is,

$$C_{pk} = \text{Min}(\text{CPL}, \text{CPU}) = 1.13$$

As is the case with C_p, the larger C_{pk} is, the better the production process meets specifications. By determining this value, operations managers can measure improvements in the production process.

Reducing the Process Variance

In Section 14.6 we described how experimentation with the four M's allows statistics practitioners to discover sources of variation and ultimately reduce that variation. Control charts may also contribute to this effort. By examining the results of control charts, we may be able to find additional sources of variation.

Developing an Understanding of Statistical Concepts

The concepts that underlie statistical process control are the same as the fundamental principles of hypothesis testing. That is, statistics that are not consistent with the null hypothesis lead us to reject the null hypothesis. However, there are two critical differences between SPC and hypothesis testing. First, in SPC we test processes rather than parameters of populations. That is, we test to determine whether there is evidence that the process distribution has changed. Second, in SPC we test a series of statistics taken over time. From a pedagogical point of view, there is another fundamental difference. Many students of statistics have difficulty identifying the correct hypothesis-testing procedure to employ. However, SPC applications tend to be rather uncomplicated. We use control charts for variables to determine whether the process is under control when the product produced must be measured quantitatively. Identifying the correct technique is seldom difficult and thus does not require the technique-identification skills developed throughout this book.

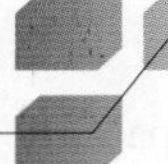

EXERCISES

21.23 Given the following statistics drawn from 30 samples of size 4, calculate the centerline and control limits for the $\bar{x}$ chart.

$\bar{\bar{x}} = 453.6 \qquad S = 12.5$

21.24 The mean of the sample means and the pooled standard deviation of 40 samples of size 9 taken from a production process under control are shown here. Compute the centerline, control limits, and zone boundaries for the $\bar{x}$ chart.

$\bar{\bar{x}} = 181.1 \qquad S = 11.0$

21.25 Twenty-five samples of size 4 were taken from a production process. The sample means are listed in chronological order below. The mean of the sample means and the pooled standard deviation are $\bar{\bar{x}} = 13.3$ and $S = 3.8$, respectively.

14.5	10.3	17.0	9.4	13.2	9.3	17.1
5.5	5.3	16.3	10.5	11.5	8.8	12.6
10.5	16.3	8.7	9.4	11.4	17.6	20.5
21.1	16.3	18.5	20.9			

a. Find the centerline and control limits for the $\bar{x}$ chart.
b. Plot the sample means on the $\bar{x}$ chart.
c. Is the process under control? Explain.

The following exercises require a computer and statistical software.

21.26 Xr21-26 Thirty samples of size 4 were drawn from a production process.
a. Construct an S chart.
b. Construct an $\bar{x}$ chart.
c. Do the charts allow you to conclude that the process is under control?
d. If the process went out of control, which of the following is the likely cause: level shift, instability, trend, or cycle?

21.27 Xr21-27 The fence of a saw is set so that it automatically cuts 2-by-4's into 96-inch lengths needed to produce prefabricated homes. To ensure that the lumber is cut properly, three pieces of wood are measured after each 100 cuts are made. The measurements in inches for the last 40 samples were recorded.
a. Do these data indicate that the process is out of control?
b. If so, when did it go out of control? What is the likely cause: level shift, instability, trend, or cycle?
c. Speculate on how the problem could be corrected.

21.28 Xr21-28 An Arc Extinguishing Unit (AEU) is used in the high-voltage electrical industry to eliminate the occurrence of electrical flash from one live 25,000-volt switch contact to another. A small but important component of an AEU is a nonconductive sliding bearing called a (ST-90811) pin guide. The dimensional accuracy of this pin guide is critical to the overall operation of the AEU. If any one of its dimensions is "out of spec" (specification), the part will bind within the AEU, causing failure. This would cause the complete destruction of both the AEU and the 25,000 volt-switch contacts, resulting in a power blackout. A pin guide has a square shape with a circular hole in the center, as shown below with its specified dimensions. The specification limits are LSL = .4335 and USL = .4435.

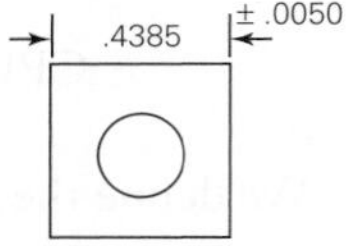

Due to the critical nature of the dimensions of the pin guide, statistical process control is used during long production runs to check that the production process is under control. Suppose that samples of five pin guides are drawn every hour. The results of the last 25 samples were recorded. Do these data allow the technician to conclude that the process is out of control?

21.29 Refer to Exercise 21.28. Find the process capability index C_{pk}.

21.30 Xr21-30 KW Paints is a company that manufactures various kinds of paints and sells them in 1- and 4-liter cans. The cans are filled on an assembly line with an automatic valve regulating the amount of paint. If the cans are overfilled, paint and money will be wasted. If the cans are underfilled, customers will complain. To ensure that the proper amount of paint goes into each can, statistical process control is used. Every hour five cans are opened, and the volume of paint is measured. The results from the last 30 hours from the l-liter production line were recorded. To avoid rounding errors, we recorded the volumes in cubic centimeters (cc) after subtracting 1,000. Thus the file contains the amounts of overfill and underfill. Draw the $\bar{x}$ and S charts to determine whether the process is under control.

21.31 Refer to Exercise 21.30. If the lower and upper specification limits are 995 cc and 1,005 cc, respectively, what is C_{pk}?

21.32 Xr21-32 Lear Seating of Kitchener, Ontario, produces seats for Cadillacs and other GM cars and trucks. The Cadillac seat includes a part called the EK headrest. The frame of the headrest is made from steel rods. A machine is used to bend the rod into a U-shape described as shown. The width is critical; if it is too wide or too narrow, it will not fit into the holes drilled into the seat frame. The process is checked by drawing samples of size 3 every 2 hours. The last 20 samples were recorded.

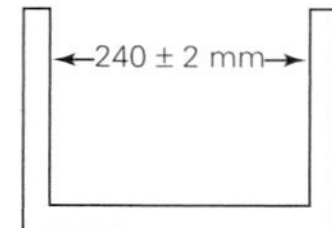

a. What do these data tell you about the process?
b. If it went out of control, at what sample did this occur?
c. What is the likely assignable cause?

21.33 Xr21-33 The degree to which nuts and bolts are tightened in numerous places on a car is often important. For example, in Toyota cars, a nut holds the rear signal light. If the nut is not tightened sufficiently, it will loosen and fall off; if it is too tight, the light may break. The nut is tightened with a torque wrench with a set clutch. The target torque is 8 kgf/cm (kilogram-force per centimeter) with specification limits LSL = 7kgf/cm and USL = 9 kgf/cm. Statistical process control is employed to constantly check the process. Random samples of size 4 are drawn after every 200 nuts are tightened. The data from the last 25 samples were recorded. (The authors are grateful to Ted Couves for contributing this exercise.)
a. Determine whether the process is under control.
b. If it is out of control, identify when this occurred and the likely cause.

21.34 Xr21-34 The seats for the F-150 series Ford trucks are manufactured by Lear Seating. The frames must be 1,496 mm wide with specification limits LSL = 1,486 mm and USL = 1,506 mm. Frames that are wider than 1,506 mm or narrower than 1,486 mm result in assembly problems, because seat cushions and/or other parts won't fit. The process is tested by drawing random samples of five frames every 2 hours. The last 25 samples were recorded. What can we conclude from these data?

21.35 Xr21-35 Long Manufacturing produces heat exchangers, primarily for the automotive industry. One such product, a transmission oil cooler, is used in the cooling of bus transmissions. It is composed of a series of copper tubes that are soldered into a header. The header must have a diameter of 4.984 inches with specification limits LSL = 4.978 inches and USL = 4.990 inches. Oversize headers result in fluid mixing and possible failure of the device. For every 100 headers produced, the operations manager draws a sample of size 4. The data from the last 25 samples were recorded. What can we conclude from these data?

21.36 Find the process capability index for Exercise 21.35.

21.37 Xr21-37 Refer to Exercise 21.35. Nuts and bolts are used in the assembly of the transmission oil coolers. They are supposed to be tightened by a torque wrench to 7 foot-pounds with specification limits LSL = 6 foot-pounds and USL = 8 foot-pounds. To test the process, three nuts are tested every 3 hours. The results for the last 75 hours were recorded. Does it appear that the process is under control?

21.38 Xr21-38 Motor oil is packaged and sold in plastic bottles. The bottles are often handled quite roughly, either in delivery to the stores (bottles are packed in boxes, which are stacked to conserve truck space), in the stores themselves, or by the consumer. The bottles must be hardy enough to withstand this treatment without leaking. Before leaving the plant, the bottles undergo statistical process control procedures. Five out of every 10,000 bottles are sampled. The burst strength (the pressure required to burst the bottle) is measured in pounds per square inch (psi). The process is designed to produce bottles that can withstand up to 800 psi. The burst strengths of the last 30 samples were recorded.
a. Draw the appropriate control chart(s).
b. Does it appear that the process went out of control? If so, when did this happen, and what are the likely causes and remedies?

21.39 Xr21-39 Almost all computer hardware and software producers offer a toll-free telephone number to solve problems associated with their product. The ability to work quickly to resolve difficulties is critical. One software maker's policy is that all calls must be answered by a software consultant within 120 seconds. (All calls are initially answered by computer and the caller is put on hold until a consultant attends to the caller.) To help maintain the quality of the service, four calls per day are monitored. The amount of time before the consultant responds to the calls was recorded for the last 30 days.
a. Draw the appropriate control chart(s).
b. Does it appear that the process went out of control? If so, when did this happen, and what are the likely causes and remedies?

21.40 Xr21-40 Plastic pipe is used for plumbing in almost all new homes. If the pipes are too narrow or too wide, they will not connect properly with other parts of the plumbing system. A manufacturer of 3-inch diameter pipes uses statistical process control to maintain the quality of its products. The sampling plan is to draw samples of three 10-foot long pipes every hour and measure the diameters. Twenty hours ago, the production process was shut down for repairs. The results of the first 20 samples taken since were recorded. Does it appear that the production process is under control?

21.41 If the specification limits for the plastic pipes in Exercise 21.40 are LSL = 2.9 inches and USL = 3.1 inches, determine the process capability index C_{pk}.

21.42 Calculate the process capability index for Exercise 21.34. Does the value of this index indicate that the production process is poor? Explain.

21.4 CONTROL CHARTS FOR ATTRIBUTES: *P* CHART

In this section, we introduce a control chart that is used to monitor a process whose results are categorized as either defective or nondefective. We construct a ***p* chart** to track the proportion of defective units in a series of samples.

p Chart

We draw the p chart in a way similar to the construction of the $\bar{x}$ chart. We draw samples of size n from the process at a minimum of 25 time periods. For each sample, we calculate the sample proportion of defective units, which we label $\hat{p}_j$. We then compute the mean of the sample proportions, which is labeled $\bar{p}$. That is,

$$\bar{p} = \frac{\sum_{j=1}^{k} \hat{p}_j}{k}$$

The centerline and control limits are as follows.

Centerline and Control Limits for the *p* Chart

$$\text{Centerline} = \bar{p}$$

$$\text{Lower control limit} = \bar{p} - 3\sqrt{\frac{\bar{p}(1-\bar{p})}{n}}$$

$$\text{Upper control limit} = \bar{p} + 3\sqrt{\frac{\bar{p}(1-\bar{p})}{n}}$$

If the lower limit is negative, set it equal to 0.

Pattern Tests

As we did in the previous section, we use Minitab's pattern tests. Minitab performs only tests 1 through 4, which are as follows:

Test 1: One point beyond zone A.
Test 2: Nine points in a row in zone C or beyond (on the same side of the centerline).
Test 3: Six increasing or six decreasing points in a row.
Test 4: Fourteen points in a row alternating up and down.

We'll demonstrate this technique using the chapter opening example.

Detecting the Source of Defective Disks: Solution

© Comstock Images/ Royalty-free/Getty Images

For each sample, we compute the proportion of defective disks and calculate the mean sample proportion, which is $\bar{p} = .05762$. Thus,

$$\text{Centerline} = \bar{p} = .05762$$

$$\begin{aligned}\text{Lower control limit} &= \bar{p} - 3\sqrt{\frac{\bar{p}(1-\bar{p})}{n}} \\ &= .05762 - 3\sqrt{\frac{(.05762)(1-.05762)}{200}} \\ &= .008188\end{aligned}$$

$$\begin{aligned}\text{Upper control limit} &= \bar{p} + 3\sqrt{\frac{\bar{p}(1-\bar{p})}{n}} \\ &= .05762 + 3\sqrt{\frac{(.05762)(1-.05762)}{200}} \\ &= .1071\end{aligned}$$

Because

$$\sqrt{\frac{\bar{p}(1-\bar{p})}{n}} = \sqrt{\frac{(.05762)(1-.05762)}{200}} = .01648$$

the boundaries of the zones are as follows:

Zone C: $.05762 \pm .01648 = (.04114, .0741)$

Zone B: $.05762 \pm 2(.01648) = (.02467, .09057)$

Zone A: $.05762 \pm 3(.01648) = (.008188, .1071)$

The following output exhibits this p chart.

EXCEL

	A	B	C
1	**Statistical Process Control**		
2			
3			*Disks*
4	Upper control limit		0.1071
5	Centerline		0.0576
6	Lower control limit		0.0082

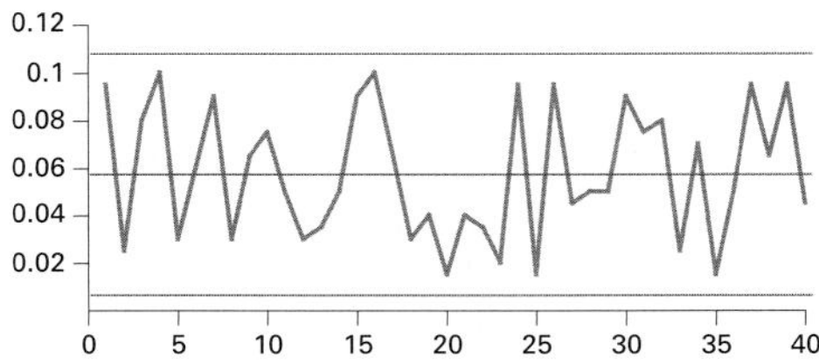

INSTRUCTIONS

1. Type or import the data into one column. (Open Xm21-00.)
2. Click **Add-Ins, Data Analysis Plus,** and **Statistical Process Control.**
3. Specify the **Input Range** (A1:41) and the **Sample Size** (200). Click **P.**

MINITAB

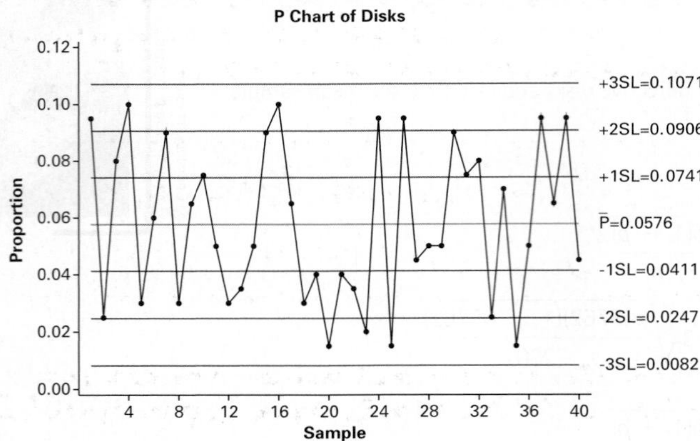

INSTRUCTIONS

1. Type or import the data into one column. (Open Xm21-00.)
2. Click **Stat, Control Charts, Attribute Charts,** and **P . . . .**
3. Type or select the variable (Disks). Type the sample size in the **Subgroup sizes** box (200). Click **P Chart Options . . . .**
4. Click **S Limits** and type **1 2 3.**
5. Click **Tests** and **Perform all four tests.**

INTERPRET

None of the points lies outside the control limits (test 1), and the other test results are negative. There is no evidence to infer that the process is out of control. However, this does not mean that 5.76% is an acceptable proportion of defects. Management should continually improve the process to reduce the defective rate and to improve the process.

The comment we made about *S* charts is also valid for *p* charts. That is, sample proportions that are less than the lower control limit indicate a change in the process that we would like to make permanent. We need to investigate the reasons for such a change just as vigorously as we investigate the causes of large proportions of defects.

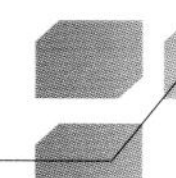

EXERCISES

21.43 To ensure that a manufacturing process is under control, 40 samples of size 1,000 were drawn and the number of defectives in each sample was counted. The mean sample proportion was .035. Compute the centerline and control limits for the p chart.

21.44 Xr21-44 Random samples of 200 copier machines were taken on an assembly line every hour for the past 25 hours. The number of defective machines is shown here. Are there any points beyond the control limits? If so, what do they tell you about the production process?

3 5 3 2 2 11 12 6 7 5 0 7 8
2 10 6 4 2 10 5 4 11 10 13 14

21.45 Xr21-45 Raytheon of Canada Limited produces printed circuit boards (PCBs), which involve a number of soldering operations. At the end of the process, the PCBs are tested to determine whether they work properly. There are several causes of PCB failure including bad flux, improper heating, and impurities. A reject rate of less than .80% is considered acceptable. Statistical process control is used by Raytheon to constantly check quality. Every hour, 500 PCBs are tested. The number of defective PCBs for the past 25 hours is shown here. Draw a p chart and apply the pattern tests to determine whether the process is under control.

3 1 2 2 1 2 3 3 3 2 3 0 0
0 2 0 0 2 4 1 1 1 4 1 3

21.46 Xr21-46 A plant produces 1,000 cordless telephones daily. A random sample of 100 telephones is inspected each day. After 30 days, the following number of defectives were found. Construct a p chart to determine whether the process is out of control.

5 0 4 3 0 3 1 1 5 0 2 1 6 0 3
0 5 5 8 5 0 1 9 6 11 6 6 4 5 10

21.47 Xr21-47 The Woodsworth Publishing Company produces millions of books containing hundred of millions of pages each year. To ensure the quality of the printed page, Woodsworth uses statistical process control. In each production run, 1,000 pages are randomly inspected. The examiners look for print clarity and whether the material is centered on the page properly. The numbers of defective pages in the last 40 production runs are listed here. Draw the p chart. Using the pattern tests, can we conclude that the production process is under control?

11 9 17 19 15 15 18 21 18 6 27 14 7 18
18 19 17 15 7 16 17 22 12 12 12 16 12
9 21 17 20 17 17 18 23 29 24 27 23 21

The following exercises require the use of a computer and statistical software.

21.48 Xr21-48 A company that manufactures batteries employs statistical process control to ensure that its product functions properly. The sampling plan for the D-cell batteries calls for samples of 500 batteries to be taken and tested. The numbers of defective batteries in the last 30 samples were recorded. Determine whether the process is under control.

21.49 Xr21-49 A courier delivery company advertises that it guarantees delivery by noon the following day. The statistical process control plan calls for sampling 2,000 deliveries each day to ensure that the advertisement is reasonable. The numbers of late deliveries for the last 30 days were recorded. What can we conclude from these data?

21.50 Xr21-50 Optical scanners are used in all supermarkets to speed the checkout process. Whenever the scanner fails to read the bar code on the product, the cashier is required to manually punch the code into the register. Obviously, unreadable bar codes slow the checkout process. Statistical process control is used to determine whether the scanner is working properly. Once a day at each checkout counter, a sample of 500 scans is taken and the number of times the scanner is unable to read the bar code is determined. (The sampling process is performed automatically by the cash register.) The results for one checkout counter for the past 25 days were recorded.

a. Draw the appropriate control chart(s).
b. Does it appear that the process went out of control? If so, identify when this happened and suggest several possible explanations for the cause.

CHAPTER SUMMARY

In this chapter, we introduced statistical process control and explained how it contributes to the maintenance of quality. We discussed how control charts detect changes in the process distribution and introduced the $\bar{x}$ chart, S chart, and p chart.

IMPORTANT TERMS

SYMBOLS

Symbol	Pronounced	Represents
S		Pooled standard deviation
s_j	*s-sub-j*	Standard deviation of the jth sample
$\hat{p}_j$	*p-hat-sub-j*	Proportion of defectives in jth sample
$\bar{p}$	*p-bar*	Mean proportion of defectives

FORMULAS

Centerline and control limits for $\bar{x}$ chart using S

$$\text{Centerline} = \bar{\bar{x}}$$

$$\text{Lower control limit} = \bar{\bar{x}} - 3\frac{S}{\sqrt{n}}$$

$$\text{Upper control limit} = \bar{\bar{x}} + 3\frac{S}{\sqrt{n}}$$

Centerline and control limits for the p chart

$$\text{Centerline} = \bar{p}$$

$$\text{Lower control limit} = \bar{p} - 3\sqrt{\frac{\bar{p}(1-\bar{p})}{n}}$$

$$\text{Upper control limit} = \bar{p} + 3\sqrt{\frac{\bar{p}(1-\bar{p})}{n}}$$

COMPUTER OUTPUT AND INSTRUCTIONS

Technique	Excel	Minitab
$\bar{x}$ chart using S	861	862
S chart	866	866
p chart	873	874

22

© Mike Kemp/Rubberball/Jupiterimages

DECISION ANALYSIS

Acceptance Sampling

A factory produces a small but important component used in computers. The factory manufactures the component in 1,000-unit lots. Because of the relatively advanced technology, the manufacturing process results in a large proportion of defective units. In fact, the operations manager has observed that the percentage of defective units per lot has been either 15% or 35%. In the past year, 60% of the lots have had 15% defectives and 40% have had 35% defectives. The present policy of the company is to send the lot to the customer, replace all defectives, and pay any additional costs. The total cost of replacing a defective unit that has been sent to the customer is $10/unit. Because of the high costs, the company management is

© Image Source Pink/ Image Source/Getty Images

See pages 890–891 for the answer.

considering inspecting all units and replacing the defective units before shipment. The sampling cost is $2/unit, and the replacement cost is $.50/unit. Each unit sells for $5.

a. Based on the history of the past year, should the company adopt the 100% inspection plan?
b. Is it worthwhile to take a sample of size 2 from the lot before deciding whether to inspect 100%?

INTRODUCTION

In previous chapters, we dealt with techniques for summarizing data in order to make decisions about population parameters and population characteristics. Our focus in this chapter is also on decision making, but the types of problems we deal with here differ in several ways. First, the technique for hypothesis testing concludes with either rejecting or not rejecting some hypothesis concerning a dimension of a population. In decision analysis, we deal with the problem of selecting one alternative from a list of several possible decisions. Second, in hypothesis testing the decision is based on the statistical evidence available. In decision analysis, there may be no statistical data, or if there are data, the decision may depend only partly on them. Third, costs (and profits) are only indirectly considered (in the selection of a significance level or in interpreting the *p*-value) in the formulation of a hypothesis test. Decision analysis directly involves profits and losses. Because of these major differences, the only topics covered previously in the text that are required for an understanding of decision analysis are probability (including Bayes's Law) and expected value.

22.1 DECISION PROBLEM

You would think that, by this point in the text, we would already have introduced all the necessary concepts and terminology. Unfortunately, because decision analysis is so radically different from statistical inference, several more terms must be defined. They will be introduced in the following example.

EXAMPLE 22.1 An Investment Decision

A man wants to invest $1 million for 1 year. After analyzing and eliminating numerous possibilities, he has narrowed his choice to one of three alternatives. These alternatives are referred to as **acts** and are denoted a_i:

a_1: Invest in a guaranteed income certificate paying 10%.

a_2: Invest in a bond with a coupon value of 8%.

a_3: Invest in a well-diversified portfolio of stocks.

He believes that the payoffs associated with the last two acts depend on a number of factors, foremost among which is interest rates. He concludes that there are three possible **states of nature,** denoted s_j:

s_1: Interest rates increase.

s_2: Interest rates stay the same.

s_3: Interest rates decrease.

After further analysis, he determines the amount of profit he will make for each possible combination of an act and a state of nature. Of course, the payoff for the guaranteed income certificate will be \$100,000 no matter which state of nature occurs. The profits from each alternative investment are summarized in Table 22.1, in what is called a **payoff table.** Notice that, for example, when the decision is a_2 and the state of nature is s_1, the investor would suffer a \$50,000 loss, which is represented by a −\$50,000 payoff.

TABLE **22.1** Payoff Table for Example 22.1

States of Nature	a_1 (GIC)	a_2 (bond)	a_3 (stocks)
s_1 (interest rates increase)	\$100,000	−\$50,000	\$150,000
s_2 (interest rates stay the same)	100,000	80,000	90,000
s_3 (interest rates decrease)	100,000	180,000	40,000

Another way of expressing the consequence of an act involves measuring the opportunity loss associated with each combination of an act and a state of nature. An **opportunity loss** is the difference between what the decision maker's profit for an act is and what the profit could have been had the best decision been made. For example, consider the first row of Table 22.1. If s_1 is the state of nature that occurs and the investor chooses act a_1, he makes a profit of \$100,000. However, had he chosen act a_3, he would have made a profit of \$150,000. The difference between what he could have made (\$150,000) and what he actually made (\$100,000) is the opportunity loss. Thus, given that s_1 is the state of nature, the opportunity loss of act a_1 is \$50,000. The opportunity loss of act a_2 is \$200,000, which is the difference between \$150,000 and −\$50,000. The opportunity loss of act a_3 is 0, because there is no opportunity loss when the best alternative is chosen. In a similar manner, we can compute the remaining opportunity losses for this example (see Table 22.2). Notice that we can never experience a negative opportunity loss.

TABLE **22.2** Opportunity Loss Table for Example 22.1

States of Nature	a_1 (GIC)	a_2 (bond)	a_3 (stocks)
s_1 (interest rates increase)	\$50,000	\$200,000	0
s_2 (interest rates stay the same)	0	20,000	10,000
s_3 (interest rates decrease)	80,000	0	140,000

Decision Trees

Most problems involving a simple choice of alternatives can readily be resolved by using the payoff table (or the opportunity loss table). In other situations, however, the decision maker must choose between sequences of acts. In Section 22.2, we introduce one form of such situations. In these cases, a payoff table will not suffice to determine the best alternative; instead, we require a **decision tree.**

In Chapter 6, we suggested the probability tree as a useful device for computing probabilities. In this type of tree, all the branches represent stages of events. In a decision tree, however, the branches represent both acts and events (states of nature). We distinguish between them in the following way: A square node represents a point where

FIGURE **22.1** Decision Tree for Example 22.1

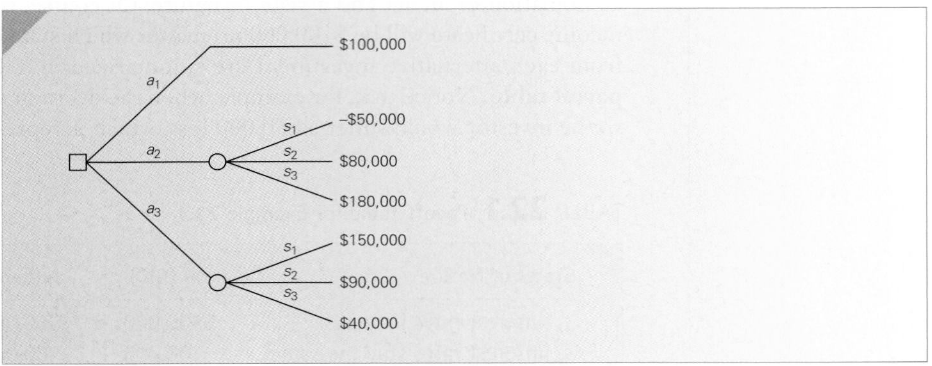

a decision is to be made; a point where a state of nature occurs is represented by a round node. Figure 22.1 depicts the decision tree for Example 22.1.

The tree begins with a square node; that is, we begin by making a choice among a_1, a_2, and a_3. The branches emanating from the square node represent these alternatives. At the ends of branches a_2 and a_3, we reach round nodes representing the occurrence of some state of nature. These are depicted as branches representing s_1, s_2, and s_3. At the end of branch a_1, we don't really have a state of nature, because the payoff is fixed at $100,000 no matter what happens to interest rates.

At the ends of the branches, the payoffs are shown (alternatively, we could have worked with opportunity losses instead of with payoffs). These are, of course, the same values that appear in Table 22.1.

Up to this point, all we have done is set up the problem; we have not made any attempt to determine the decision. It should be noted that in many real-life problems, determining the payoff table or decision tree can be a formidable task in itself. Many managers, however, have observed that this task is often extremely helpful in decision making.

Expected Monetary Value Decision

In many decision problems, it is possible to assign probabilities to the states of nature. For example, if the decision involves trying to decide whether to draw to an inside straight in the game of poker, the probability of succeeding can easily be determined by the use of simple rules of probability. If we must decide whether to replace a machine that has broken down frequently in the past, we can assign probabilities on the basis of the relative frequency of the breakdowns. In many other instances, however, formal rules and techniques of probability cannot be applied. In Example 22.1, the historic relative frequencies of the ups and downs of interest rates will supply scant useful information to help the investor assign probabilities to the behavior of interest rates during the coming year. In such cases, probabilities must be assigned subjectively. In other words, the determination of the probabilities must be based on the experience, knowledge, and (perhaps) guesswork of the decision maker.

If in Example 22.1 the investor has some knowledge about a number of economic variables, he might have a reasonable guess about what will happen to interest rates in the next year. Suppose, for example, that our investor believes that future interest rates are most likely to remain essentially the same as they are today and that (of the remaining two states of nature) rates are more likely to decrease than to increase. He might then guess the following probabilities:

$$P(s_1) = .2 \qquad P(s_2) = .5 \qquad P(s_3) = .3$$

Because the probabilities are subjective, we would expect another decision maker to produce a completely different set of probabilities. In fact, if this were not true, we would rarely have buyers and sellers of stocks (or any other investment), because everyone would be a buyer (and there would be no sellers) or everyone would be a seller (with no buyers).

After determining the probabilities of the states of nature, we can address the *expected monetary value decision*. We now calculate what we expect will happen for each decision. Because we generally measure the consequences of each decision in monetary terms, we compute the **expected monetary value (EMV)** of each act. Recall from Section 7.1 that we calculate expected values by multiplying the values of the random variables by their respective probabilities and then summing the products. Thus, in our example, the expected monetary value of alternative a_1 is

$$\text{EMV}(a_1) = .2(100{,}000) + .5(100{,}000) + .3(100{,}000) = \$100{,}000$$

The expected values of the other decisions are found in the same way:

$$\text{EMV}(a_2) = .2(-50{,}000) + .5(80{,}000) + .3(180.000) = \$84{,}000$$
$$\text{EMV}(a_3) = .2(150{,}000) + .5(90{,}000) + .3(40{,}000) = \$87{,}000$$

We choose the decision with the largest expected monetary value, which is a_1, and label its expected value EMV*. Hence, EMV* = \$100,000.

In general, the expected monetary values do not represent possible payoffs. For example, the expected monetary value of act a_2 is \$84,000, yet the payoff table indicates that the only possible payoffs from choosing a_2 are −\$50,000, \$80,000, and \$180,000. Of course, the expected monetary value of act a_1 (\$100,000) is possible, because that is the only payoff of the act.

What, then, does the expected monetary value represent? If the investment is made a large number of times, with exactly the same payoffs and probabilities, the expected monetary value is the average payoff per investment. That is, if the investment is repeated an infinite number of times with act a_2, 20% of the investments will result in a \$50,000 loss, 50% will result in an \$80,000 profit, and 30% will result in a \$180,000 profit. The average of all these investments is the expected monetary value, \$84,000. If act a_3 is chosen, the average payoff in the long run will be \$87,000.

An important point is raised by the question of how many investments are going to be made. The answer is one. Even if the investor intends to make the same type of investment annually, the payoffs and the probabilities of the states of nature will undoubtedly change from year to year. Hence, we are faced with having determined the expected monetary value decision on the basis of an infinite number of investments, when there will be only one investment. We can rationalize this apparent contradiction in two ways. First, the expected value decision is the only method that allows us to combine the two most important factors in the decision process—the payoffs and their probabilities. It seems inconceivable that, where both factors are known, the investor would want to ignore either one. (There are processes that make decisions on the basis of the payoffs alone; however, these processes assume no knowledge of the probabilities, which is not the case with our example.) Second, typical decision makers make a large number of decisions over their lifetimes. By using the expected value decision, the decision maker should perform at least as well as anyone else. Thus, despite the problem of interpretation, we advocate the expected monetary value decision.

Expected Opportunity Loss Decision

We can also calculate the **expected opportunity loss (EOL)** of each act. From the opportunity loss table (Table 22.2), we get the following values:

$$\text{EOL}(a_1) = .2(50{,}000) + .5(0) + .3(80{,}000) = \$34{,}000$$
$$\text{EOL}(a_2) = .2(200{,}000) + .5(20{,}000) + .3(0) = \$50{,}000$$
$$\text{EOL}(a_3) = .2(0) + .5(10{,}000) + .3(140{,}000) = \$47{,}000$$

Because we want to minimize losses, we choose the act that produces the smallest expected opportunity loss, which is a_1. We label its expected value EOL*. Observe that the EMV decision is the same as the EOL decision. This is not a coincidence—the opportunity loss table was produced directly from the payoff table.

Rollback Technique for Decision Trees

Figure 22.2 presents the decision tree for Example 22.1, with the probabilities of the states of nature included. The process of determining the EMV decision is called the rollback technique; it operates as follows. Beginning at the end of the tree (right-hand side), we calculate the expected monetary value at each round node. The numbers above the round nodes in Figure 22.2 specify these expected monetary values.

At each square node, we make a decision by choosing the branch with the largest EMV. In our example, there is only one square node. Our optimal decision is, of course, a_1.

FIGURE **22.2** Rollback Technique for Example 22.1

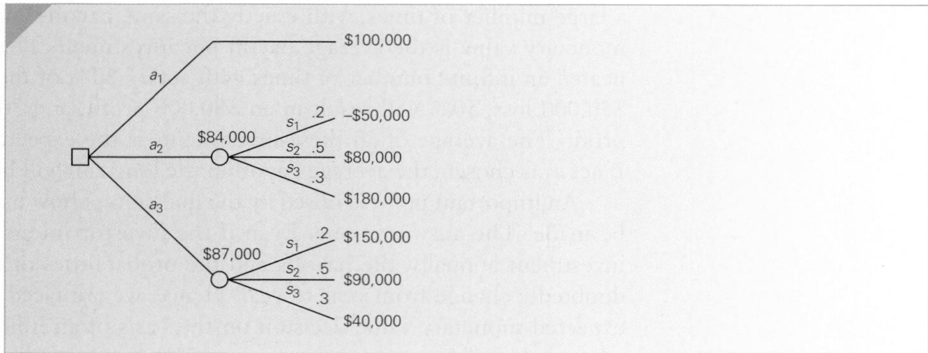

Using the Computer

There is yet another Excel add-in called Treeplan that enables Excel to draw the decision tree and produce the optimal solution. CD Appendix AE provides a brief introduction to get you started.

EXERCISES

22.1 Set up the opportunity loss table from the following payoff table.

	a_1	a_2
s_1	55	26
s_2	43	38
s_3	29	43
s_4	15	51

22.2 Draw the decision tree for Exercise 22.1.

22.3 If we assign the following probabilities to the states of nature in Exercise 22.1, determine the EMV decision.

$P(s_1) = .4 \quad P(s_2) = .1 \quad P(s_3) = .3 \quad P(s_4) = .2$

22.4 Given the following payoff table, draw the decision tree.

	a_1	a_2	a_3
s_1	20	5	−1
s_2	8	5	4
s_3	−10	5	10

22.5 Refer to Exercise 22.4. Set up the opportunity loss table.

22.6 If we assign the following probabilities to the states of nature in Exercise 22.5, determine the EOL decision.

$P(s_1) = .2 \quad P(s_2) = .6 \quad P(s_3) = .2$

Applications

22.7 A baker must decide how many specialty cakes to bake each morning. From past experience, she knows that the daily demand for cakes ranges from 0 to 3. Each cake costs $3.00 to produce and sells for $8.00 and any unsold cakes are thrown into the garbage at the end of the day.

a. Set up a payoff table to help the baker decide how many cakes to bake.
b. Set up the opportunity loss table.
c. Draw the decision tree.

22.8 Refer to Exercise 22.7. Assume that the probability of each value of demand is the same for all possible demands.

a. Determine the EMV decision.
b. Determine the EOL decision.

22.9 The manager of a large shopping center in Buffalo is in the process of deciding on the type of snow-clearing service to hire for his parking lot. Two services are available. The White Christmas Company will clear all snowfalls for a flat fee of $40,000 for the entire winter season. The Weplowem Company charges $18,000 for each snowfall it clears. Set up the payoff table to help the manager decide, assuming that the number of snowfalls per winter season ranges from 0 to 4.

22.10 Refer to Exercise 22.9. Using subjective assessments, the manager has assigned the following probabilities to the number of snowfalls. Determine the optimal decision.

$P(0) = .05 \quad P(1) = .15 \quad P(2) = .30 \quad P(3) = .40$

$P(4) = .10$

22.11 The owner of a clothing store must decide how many men's shirts to order for the new season. For a particular type of shirt, she must order in quantities of 100 shirts. If she orders 100 shirts, her cost is $10 per shirt; if she orders 200 shirts, her cost is $9 per shirt; and if she orders 300 or more shirts, her cost is $8.50 per shirt. Her selling price for the shirt is $12, but any shirts that remain unsold at the end of the season are sold at her famous "half-price, end-of-season sale." For the sake of simplicity, she is willing to assume that the demand for this type of shirt will be 100, 150, 200, or 250 shirts. Of course, she cannot sell more shirts than she stocks. She is also willing to assume that she will suffer no loss of goodwill among her customers if she understocks and the customers cannot buy all the shirts they want. Furthermore, she must place her order today for the entire season; she cannot wait to see how the demand is running for this type of shirt.

a. Construct the payoff table to help the owner decide how many shirts to order.
b. Set up the opportunity loss table.
c. Draw the decision tree.

22.12 Refer to Exercise 22.11. The owner has assigned the following probabilities.

$P(\text{Demand} = 100) = .20 \quad P(\text{Demand} = 150) = .25$

$P(\text{Demand} = 200) = .40 \quad P(\text{Demand} = 250) = .15$

Find the EMV decision.

22.13 A building contractor must decide how many mountain cabins to build in the ski resort area of Chick-oh-pee. He builds each cabin at a cost of $26,000 and sells each for $33,000. All cabins unsold after 10 months will be sold to a local investor for $20,000. The contractor believes that the demand for cabins follows a Poisson distribution, with a mean of .5. He assumes that any probability less than .01 can be treated as 0. Construct the payoff table and the opportunity loss table for this decision problem.

22.14 The electric company is in the process of building a new power plant. There is some uncertainty regarding the size of the plant to be built. If the community that the plant will service attracts a large number of industries, the demand for electricity will be high. If commercial establishments (offices and retail stores) are attracted, demand will be moderate. If neither industries nor commercial stores locate in the community, the electricity demand will be low. The company can build a small, medium, or large plant, but if the plant is too small, the company will incur extra costs. The total costs (in $millions) of all options are shown in the accompanying table.

	Size of Plant		
Electricity Demand	**Small**	**Medium**	**Large**
Low	220	300	350
Moderate	330	320	350
High	440	390	350

The following probabilities are assigned to the electricity demand.

Demand	*P*(Demand)
Low	.15
Moderate	.55
High	.30

a. Determine the act with the largest expected monetary value. (*Caution:* All the values in the table are costs.)
b. Draw up an opportunity loss table.
c. Calculate the expected opportunity loss for each decision, and determine the optimal decision.

22.15 A retailer buys bushels of mushrooms for $2 each and sells them for $5 each. The quality of the mushrooms begins to decline after the first day they are offered for sale; therefore, to sell the mushrooms for $5/bushel, he must sell them on the first day. Bushels not sold on the first day can be sold to a wholesaler who buys day-old mushrooms at the following rates.

Amount purchased (bushels)	1	2	3	4 or more
Price per bushel	$2.00	$1.75	$1.50	$1.25

A 90-day observation of past demand yields the following information.

Daily demand (bushels)	10	11	12	13
Number of days	9	18	36	27

a. Set up a payoff table that could be used by the retailer to decide how many bushels to buy.
b. Find the optimal number of bushels the retailer should buy to maximize profit.

22.16 An international manufacturer of electronic products is contemplating introducing a new type of compact disk player. After some analysis of the market, the president of the company concludes that, within 2 years, the new product will have a market share of 5%, 10%, or 15%. She assesses the probabilities of these events as .15, .45, and .40, respectively. The vice president of finance informs her that if the product captures only a 5% market share, the company will lose $28 million. A 10% market share will produce a $2-million profit, and a 15% market share will produce an $8-million profit. If the company decides not to begin production of the new compact disk player, there will be no profit or loss. Based on the expected value decision, what should the company do?

22.2 Acquiring, Using, and Evaluating Additional Information

In this section we discuss methods of introducing and incorporating additional information into the decision process. Such information generally has value, but it also has attendant costs; that is, we can acquire useful information from consultants, surveys, or other experiments, but we usually must pay for this information. We can calculate the maximum price that a decision maker should be willing to pay for any information by determining the value of perfect information. We begin by calculating the expected payoff with perfect information (EPPI).

If we knew in advance which state of nature would occur, we would certainly make our decisions accordingly. For instance, if the investor in Example 22.1 knew before investing his money what interest rates would do, he would choose the best act to suit that case. Referring to Table 22.1, if he knew that s_1 was going to occur, he would choose act a_3; if s_2 were certain to occur, he'd choose a_1, and if s_3 were certain, he'd choose a_2. Thus, in the long run, his expected payoff from perfect information would be

$$\text{EPPI} = .2(150{,}000) + .5(100{,}000) + .3(180{,}000) = \$134{,}000$$

Notice that we compute EPPI by multiplying the probability of each state of nature by the largest payoff associated with that state of nature and then summing the products.

This figure, however, does not represent the maximum amount he'd be willing to pay for perfect information. Because the investor could make an expected profit of EMV* = $100,000 without perfect information, we subtract EMV* from EPPI to determine the **expected value of perfect information (EVPI).** That is,

$$\text{EVPI} = \text{EPPI} - \text{EMV}^* = \$134{,}000 - \$100{,}000 = \$34{,}000$$

This means that, if perfect information were available, the investor should be willing to pay up to $34,000 to acquire it.

You may have noticed that the expected value of perfect information (EVPI) equals the smallest expected opportunity loss (EOL*). Again, this is not a coincidence—it will always be the case. In future questions, if the opportunity loss table has been determined, you need only calculate EOL* in order to know EVPI.

Decision Making with Additional Information

Suppose the investor in our continuing example wants to improve his decision-making capabilities. He learns about Investment Management Consultants (IMC), who, for a fee of $5,000, will analyze the economic conditions and forecast the behavior of interest rates over the next 12 months. The investor, who is quite shrewd (after all, he does have $1 million to invest), asks for some measure of IMC's past successes. IMC has been forecasting interest rates for many years and so provides him with various conditional probabilities (referred to as **likelihood probabilities**), as shown in Table 22.3. Table 22.3 uses the following notation:

I_1: IMC predicts that interest rates will increase.

I_2: IMC predicts that interest rates will stay the same.

I_3: IMC predicts that interest rates will decrease.

TABLE **22.3** Likelihood Probabilities $P(I_i \mid s_j)$

	I_1 (predict s_1)	I_2 (predict s_2)	I_3 (predict s_3)
s_1	$P(I_1 \mid s_1) = .60$	$P(I_2 \mid s_1) = .30$	$P(I_3 \mid s_1) = .10$
s_2	$P(I_1 \mid s_2) = .10$	$P(I_2 \mid s_2) = .80$	$P(I_3 \mid s_2) = .10$
s_3	$P(I_1 \mid s_3) = .10$	$P(I_2 \mid s_3) = .20$	$P(I_3 \mid s_3) = .70$

The I_i terms are referred to as experimental outcomes, and the process by which we gather additional information is called the experiment.

Examine the first line of Table 22.3. When s_1 actually did occur in the past, IMC correctly predicted s_1 60% of the time; 30% of the time, it predicted s_2; and 10% of the time, it predicted s_3. The second row gives the conditional probabilities of I_1, I_2, and I_3

when s_2 actually occurred. The third row shows the conditional probabilities of I_1, I_2, and I_3 when s_3 actually occurred.

The following question now arises: How is the investor going to use the forecast that IMC produces? One approach is simply to assume that whatever it forecasts will actually take place and to choose the act accordingly. There are several drawbacks to this approach. Foremost among them is that it puts the investor in the position of ignoring whatever knowledge (in the form of subjective probabilities) he had concerning the issue. Instead the decision maker should use this information to modify his initial assessment of the probabilities of the states of nature. To incorporate the investor's subjective probabilities with the consultant's forecast requires the use of Bayes's Law, which we introduced in Section 6.4. We'll review Bayes's Law in the context of our example.

Suppose that the investor pays IMC the $5,000 fee and IMC forecasts that s_1 will occur. We want to revise our estimates for the probabilities of the states of nature, given that I_1 is the outcome of the experiment, that is, we want $P(s_1 | I_1)$, $P(s_2 | I_1)$, and $P(s_3 | I_1)$. Before proceeding, let's develop some terminology. Recall from Section 6.4 that the original probabilities, $P(s_1)$, $P(s_2)$, and $P(s_3)$, are called **prior probabilities,** because they were determined prior to the acquisition of any additional information. In this example, they were based on the investor's experience. The set of probabilities we want to compute—$P(s_1 | I_1)$, $P(s_2 | I_1)$, and $P(s_3 | I_1)$—are called posterior or revised probabilities.

Now we will calculate the **posterior probabilities,** first by using a probability tree and then by applying a less time-consuming method. Figure 22.3 depicts the probability tree. We begin with the branches of the prior probabilities, which are followed by the likelihood probabilities.

FIGURE **22.3** Probability Tree to Compute Posterior Probabilities

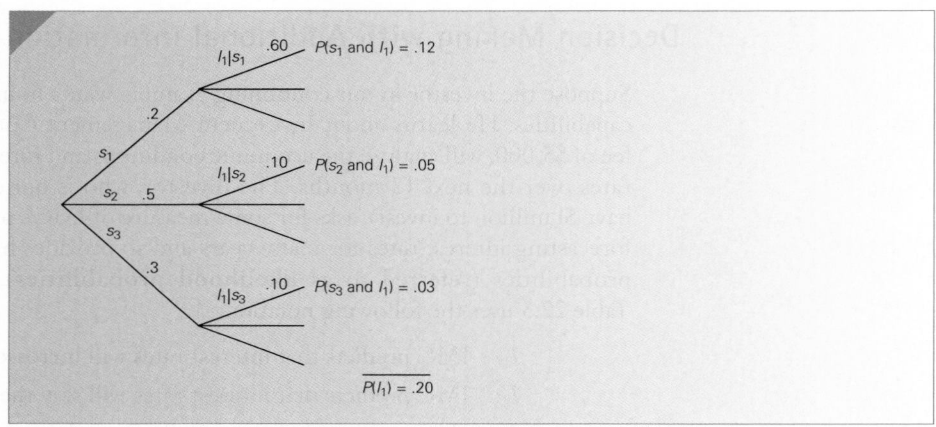

Notice that we label only $P(I_1 | s_1)$, $P(I_1 | s_2)$, and $P(I_1 | s_3)$ because (at this point) we are assuming that I_1 is the experimental outcome. Now recall that conditional probability is defined as

$$P(A|B) = \frac{P(A \text{ and } B)}{P(B)}$$

At the ends of each branch, we have the joint probability $P(s_j \text{ and } I_1)$. By summing the joint probabilities $P(s_j \text{ and } I_1)$ for $j = 1, 2,$ and 3, we calculate $P(I_1)$. Finally,

$$P(s_j|I_1) = \frac{P(s_j \text{ and } I_1)}{P(I_1)}$$

Table 22.4 performs exactly the same calculations as the probability tree except without the tree. So for example, our revised probability for s_3, which was initially .3 is now .15.

TABLE **22.4**
Posterior Probabilities for I_1

s_j	$P(s_j)$	$P(I_1 \mid s_j)$	$P(s_j \text{ and } I_1)$	$P(s_j \mid I_1)$
s_1	.2	.60	(.2)(.60) = .12	.12/.20 = .60
s_2	.5	.10	(.5)(.10) = .05	.05/.20 = .25
s_3	.3	.10	(.3)(.10) = .03	.03/.20 = .15
			$P(I_1)$ = .20	

After the probabilities have been revised, we can use them in exactly the same way we used the prior probabilities. That is, we can calculate the expected monetary value of each act:

$$\text{EMV}(a_1) = .60(100{,}000) + .25(100{,}000) + .15(100{,}000) = \$100{,}000$$
$$\text{EMV}(a_2) = .60(-50{,}000) + .25(80{,}000) + .15(180{,}000) = \$17{,}000$$
$$\text{EMV}(a_3) = .60(150{,}000) + .25(90{,}000) + .15(40{,}000) = \$118{,}500$$

Thus, if IMC forecasts s_1, the optimal act is a_3, and the expected monetary value of the decision is $118,500.

As a further illustration, we now repeat the process for I_2 and I_3 in Tables 22.5 and 22.6, respectively.

TABLE **22.5**
Posterior Probabilities for I_2

s_j	$P(s_j)$	$P(I_2 \mid s_j)$	$P(s_j \text{ and } I_2)$	$P(s_j \mid I_2)$
s_1	.2	.30	(.2)(.30) = .06	.06/.52 = .115
s_2	.5	.80	(.5)(.80) = .40	.40/.52 = .770
s_3	.3	.20	(.3)(.20) = .06	.06/.52 = .115
			$P(I_2)$ = .52	

Applying the posterior probabilities for I_2 from Table 22.5 to the payoff table, we find the following:

$$\text{EMV}(a_1) = .115(100{,}000) + .770(100{,}000) + .115(100{,}000) = \$100{,}000$$
$$\text{EMV}(a_2) = .115(-50{,}000) + .770(80{,}000) + .115(180{,}000) = \$76{,}550$$
$$\text{EMV}(a_3) = .115(150{,}000) + .770(90{,}000) + .115(40{,}000) = \$91{,}150$$

As you can see, if IMC predicts that s_2 will occur, the optimal act is a_1, with an expected monetary value of $100,000.

TABLE **22.6**
Posterior Probabilities for I_3

s_j	$P(s_j)$	$P(I_3 \mid s_j)$	$P(s_j \text{ and } I_3)$	$P(s_j \mid I_3)$
s_1	.2	.10	(.2)(.10) = .02	.02/.28 = .071
s_2	.5	.10	(.5)(.10) = .05	.05/.28 = .179
s_3	.3	.70	(.3)(.70) = .21	.21/.28 = .750
			$P(I_3)$ = .28	

With the set of posterior probabilities for I_3 from Table 22.6, the expected monetary values are as follows:

$$\text{EMV}(a_1) = .071(100{,}000) + .179(100{,}000) + .750(100{,}000) = \$100{,}000$$
$$\text{EMV}(a_2) = .071(-50{,}000) + .179(80{,}000) + .750(180{,}000) = \$145{,}770$$
$$\text{EMV}(a_3) = .071(150{,}000) + .179(90{,}000) + .750(40{,}000) = \$56{,}760$$

If IMC predicts that s_3 will occur, the optimal act is a_2, with an expected monetary value of \$145,770.

At this point, we know the following:

If IMC predicts s_1, then the optimal act is a_3.
If IMC predicts s_2, then the optimal act is a_1.
If IMC predicts s_3, then the optimal act is a_2.

Thus, even before IMC makes its forecast, the investor knows which act is optimal for each of the three possible IMC forecasts. Of course, all these calculations can be performed before paying IMC its \$5,000 fee. This leads to an extremely important calculation. By performing the computations just described, the investor can determine *whether* he should hire IMC; that is, he can determine whether the value of IMC's forecast exceeds the cost of its information. Such a determination is called a **preposterior analysis.**

Preposterior Analysis

The objective of a preposterior analysis is to determine whether the value of the prediction is greater or less than the cost of the information. *Posterior* refers to the revision of the probabilities, and the *pre* indicates that this calculation is performed before paying the fee.

We begin by finding the expected monetary value of using the additional information. This value is denoted EMV′, which for our example is determined on the basis of the following analysis:

If IMC predicts s_1, then the optimal act is a_3, and the expected payoff is \$118,500.
If IMC predicts s_2, then the optimal act is a_1, and the expected payoff is \$100,000.
If IMC predicts s_3, then the optimal act is a_2, and the expected payoff is \$145,770.

A useful by-product of calculating the posterior probabilities is the set of probabilities of I_1, I_2, and I_3:

$$P(I_1) = .20 \qquad P(I_2) = .52 \qquad P(I_3) = .28$$

(Notice that these probabilities sum to 1.) Now imagine that the investor seeks the advice of IMC an infinite number of times. (This is the basis for the expected value decision.) The set of probabilities of I_1, I_2, and I_3 indicates the following outcome distribution: 20% of the time, IMC will predict s_1 and the expected monetary value will be \$118,500; 52% of the time, IMC will predict s_2 and the expected monetary value will be \$100,000; and 28% of the time, IMC will predict s_3 and the expected monetary value will be \$145,770.

The expected monetary value with additional information is the weighted average of the expected monetary values, where the weights are $P(I_1)$, $P(I_2)$, and $P(I_3)$. Hence,

$$\text{EMV}' = .20(118{,}500) + .52(100{,}000) + .28(145{,}770) = \$116{,}516$$

The value of IMC's forecast is the difference between the expected monetary value with additional information (EMV′) and the expected monetary value without additional information (EMV*). This difference is called the **expected value of sample information** and is denoted EVSI. Thus,

$$\text{EVSI} = \text{EMV}' - \text{EMV}^* = \$116{,}516 - \$100{,}000 = \$16{,}516$$

By using IMC's forecast, the investor can make an average additional profit of $16,516 in the long run. Because the cost of the forecast is only $5,000, the investor is advised to hire IMC.

If you review this problem, you'll see that the investor had to make two decisions. The first (chronologically) was whether to hire IMC, and the second was which type of investment to make. A decision tree is quite helpful in describing the acts and states of nature in this question. Figure 22.4 provides the complete tree diagram.

FIGURE **22.4** **Complete Decision Tree for Example 22.1**

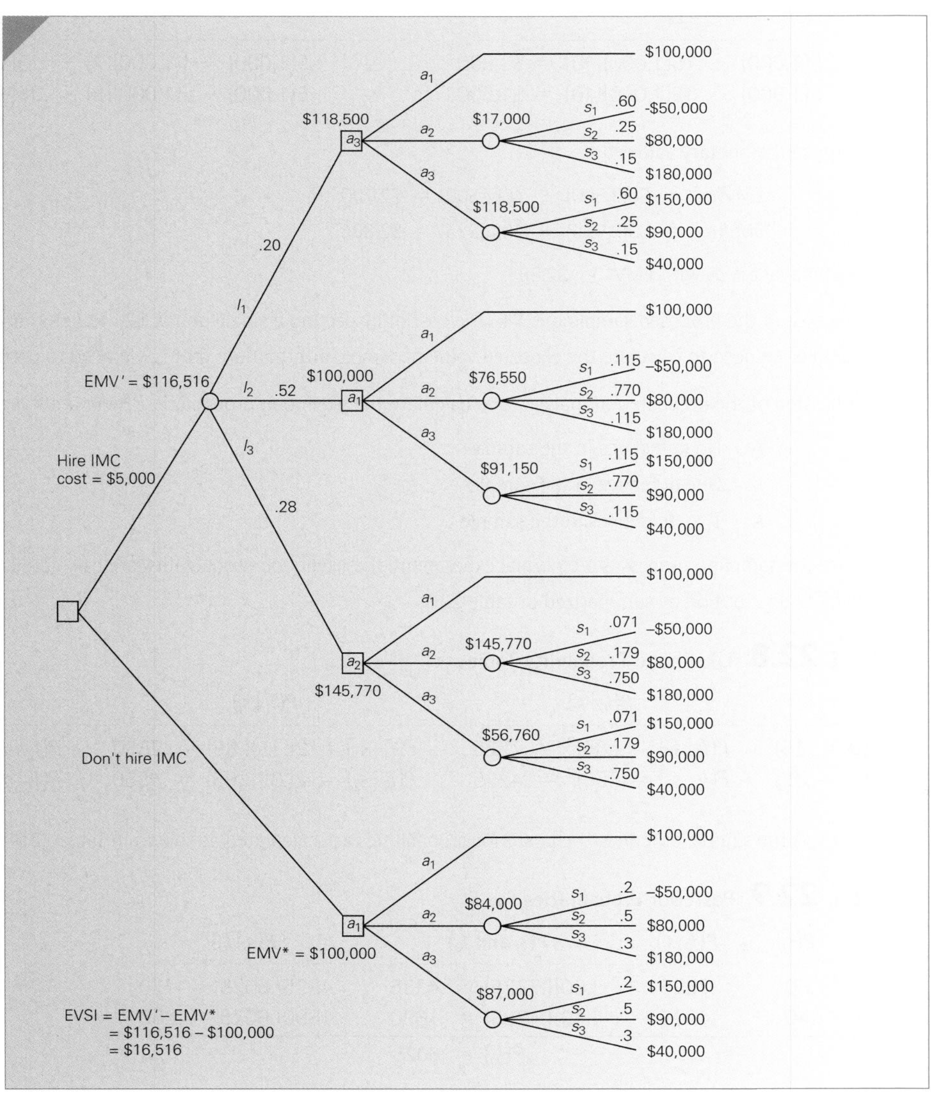

Acceptance Sampling: Solution

© Image Source Pink/ Image Source/Getty Images

a. The two alternatives are

a_1: No inspection (the current policy)

a_2: 100% inspection

The two states of nature are

s_1: The lot contains 15% defectives

s_2: The lot contains 35% defectives

Based on the past year's historical record,

$$P(s_1) = .60 \quad \text{and} \quad P(s_2) = .40$$

The payoff table is constructed as shown in Table 22.7.

TABLE **22.7** **Payoff Table**

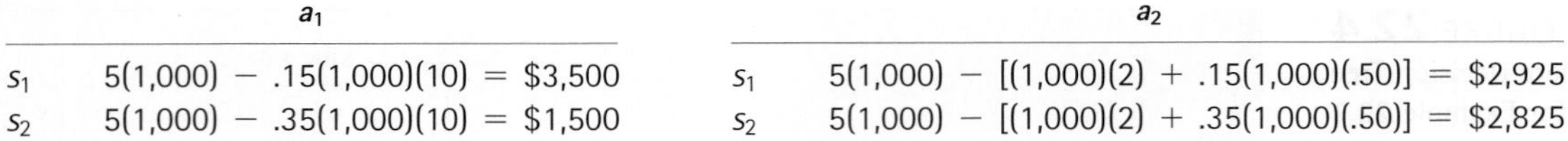

	a_1		a_2
s_1	$5(1,000) - .15(1,000)(10) = \$3,500$	s_1	$5(1,000) - [(1,000)(2) + .15(1,000)(.50)] = \$2,925$
s_2	$5(1,000) - .35(1,000)(10) = \$1,500$	s_2	$5(1,000) - [(1,000)(2) + .35(1,000)(.50)] = \$2,825$

The expected monetary values are

$$\text{EMV}(a_1) = .60(3,500) + .40(1,500) = \$2,700$$

$$\text{EMV}(a_2) = .60(2,925) + .40(2,825) = \$2,885$$

The optimal act is a_2 with EMV* = $2,885.

b. The cost of the proposed sampling is $4. (The cost of inspecting a single unit is $2.) To determine whether we should sample, we need to calculate the expected value of sample information; that is, we need to perform a preposterior analysis.

The first step of the preposterior analysis is to calculate the likelihood probabilities. There are three possible sample outcomes:

I_0: No defectives in the sample

I_1: One defective in the sample

I_2: Two defectives in the sample

Because the sampling process is a binomial experiment, the likelihood probabilities are calculated by using the binomial probability distribution as summarized in Table 22.8.

TABLE **22.8** **Likelihood Probability Table**

	$P(I_0 \mid s_j)$	$P(I_1 \mid s_j)$	$P(I_2 \mid s_j)$
s_1 ($p = .15$)	$P(I_0 \mid s_1) = (.85)^2 = .7225$	$P(I_1 \mid s_1) = 2(.15)(.85) = .2550$	$P(I_2 \mid s_1) = (.15)^2 = .0225$
s_2 ($p = .35$)	$P(I_0 \mid s_2) = (.65)^2 = .4225$	$P(I_1 \mid s_2) = 2(.35)(.65) = .4550$	$P(I_2 \mid s_2) = (.35)^2 = .1225$

If I_0 is the sample outcome, the posterior probabilities are calculated as shown in Table 22.9.

TABLE **22.9** **Posterior Probabilities for I_0**

s_j	$P(s_j)$	$P(I_0 \mid s_j)$	$P(s_j \text{ and } I_0)$	$P(s_j \mid I_0)$
s_1	.60	.7225	(.60)(.7225) = .4335	.4335/.6025 = .720
s_2	.40	.4225	(.40)(.4225) = .1690	.1690/.6025 = .280
			$P(I_0) = .6025$	

The expected monetary values if the sample outcome is I_0 are

$$EMV(a_1) = .720(3{,}500) + .280(1{,}500) = \$2{,}940$$

$$EMV(a_2) = .720(2{,}925) + .280(2{,}825) = \$2{,}897$$

Therefore, the optimal act is a_1.

If I_1 is the sample outcome, the posterior probabilities are calculated as shown in Table 22.10.

TABLE **22.10** Posterior Probabilities for I_1

s_j	$P(s_j)$	$P(I_1 \mid s_j)$	$P(s_j \text{ and } I_1)$	$P(s_j \mid I_1)$
s_1	.60	.2550	(.60)(.2550) = .1530	.1530/.3350 = .457
s_2	.40	.4550	(.40)(.4550) = .1820	.1820/.3350 = .543
			$P(I_1)$ = .3350	

The expected monetary values if the sample outcome is I_1 are

$$EMV(a_1) = .457(3{,}500) + .543(1{,}500) = \$2{,}414$$

$$EMV(a_2) = .457(2{,}925) + .543(2{,}825) = \$2{,}871$$

Therefore, the optimal act is a_2.

If I_2 is the sample outcome, the posterior probabilities are calculated as shown in Table 22.11.

TABLE **22.11** Posterior Probabilities for I_2

s_j	$P(s_j)$	$P(I_2 \mid s_j)$	$P(s_j \text{ and } I_2)$	$P(s_j \mid I_2)$
s_1	.60	.0225	(.60)(.0225) = .0135	.0135/.0625 = .216
s_2	.40	.1225	(.40)(.1225) = .0490	.0490/.0625 = .784
			$P(I_2)$ = .0625	

The expected monetary values if the sample outcome is I_2 are

$$EMV(a_1) = .216(3{,}500) + .784(1{,}500) = \$1{,}932$$

$$EMV(a_2) = .216(2{,}925) + .784(2{,}825) = \$2{,}847$$

Therefore, the optimal act is a_2.

We can now summarize these results, as shown in Table 22.12.

TABLE **22.12** Summary of Optimal Acts

Sample Outcome	Probability	Optimal Act	Expected Monetary Value
I_0	.6025	a_1	\$2,940
I_1	.3350	a_2	\$2,871
I_2	.0625	a_2	\$2,847

The expected monetary value with additional information is

$$EMV' = .6025(2{,}940) + .3350(2{,}871) + .0625(2{,}847) = \$2{,}911$$

The expected value of sample information is

$$EVSI = EMV' - EMV^* = 2{,}911 - 2{,}885 = \$26$$

Because the expected value of sample information is \$26 and the sampling cost is \$4, the company should take a sample of 2 units before deciding whether to inspect 100%. The optimal sequence is as follows:

1. Take a sample of 2 units.
2. If there are no defective units in the sample, continue the current policy of no inspection. If either one or two of the sample units are defective, perform a complete inspection of the lot.

Bayesian Statistics

In Chapters 10–19 we dealt with inference about unknown parameters. In Chapter 10 we pointed out that when interpreting the confidence interval estimate, we cannot make probability statements about parameters because they are not variables. However, Bayesian statistics specifies that parameters are variables and we can assume various probability distributions. The acceptance sampling example illustrates this concept. The parameter was the proportion p of defective units in the 1,000-unit batch. The example was unrealistic because we allowed the parameter to assume one of only two values, 15% and 35%. We assigned prior probabilities using the relative frequency approach. That is, based on historic records we had

$$P(p = 15\%) = .60 \quad \text{and} \quad P(p = 35\%) = .40$$

To make the problem more realistic, we let p be a continuous random variable rather than a discrete one. That is, p can take on any value between 0 and 100%. We assign a density function also based on historic records. We can express the payoffs as a linear function of p. Then, using calculus, we can determine the optimum decision. We can also revise the prior probabilities based on the outcome of the sampling of 2 units. The technique requires some calculus, but the concept is the same as the one developed in this chapter. It should be noted that there is a parallel universe of Bayesian statistics that more or less matches the material in the inference part of this book. Interested readers can learn more about Bayesian statistics from additional courses dedicated to the subject.

EXERCISES

22.17 Find EPPI, EMV*, and EVPI for the accompanying payoff table and probabilities.

	a_1	a_2	a_3
s_1	60	110	75
s_2	40	110	150
s_3	220	120	85
s_4	250	120	130

$P(s_1) = .10 \quad P(s_2) = .25 \quad P(s_3) = .50 \quad P(s_4) = .15$

22.18 For Exercise 22.17, determine the opportunity loss table and compute EOL*. Confirm that EOL* = EVPI.

22.19 Given the following payoff table and probabilities, determine EVPI.

	a_1	a_2	a_3	a_4
s_1	65	20	45	30
s_2	70	110	80	95

$P(s_1) = .5 \quad P(s_2) = .5$

22.20 Redo Exercise 22.19, changing the probabilities to the following values.

a. $P(s_1) = .75 \quad P(s_2) = .25$
b. $P(s_1) = .95 \quad P(s_2) = .05$

22.21 What conclusion can you draw about the effect of the probabilities on EVPI from Exercises 22.19 and 22.20?

22.22 Determine the posterior probabilities, given the following prior and likelihood probabilities.

Prior Probabilities

$P(s_1) = .25 \quad P(s_2) = .40 \quad P(s_3) = .35$

Likelihood Probabilities

	l_1	l_2	l_3	l_4
s_1	.40	.30	.20	.10
s_2	.25	.25	.25	.25
s_3	0	.30	.40	.30

22.23 Calculate the posterior probabilities from the prior and likelihood probabilities that follow.

Prior Probabilities

$P(s_1) = .5 \quad P(s_2) = .5$

Likelihood Probabilities

	l_1	l_2
s_1	.98	.02
s_2	.05	.95

22.24 With the accompanying payoff table and the prior and posterior probabilities computed in Exercise 22.23, calculate the following.

a. The optimal act for each experimental outcome
b. The expected value of sample information

Payoff Table

	a_1	a_2	a_3
s_1	10	18	23
s_2	22	19	15

22.25 Given the following payoff table, prior probabilities, and likelihood probabilities, find the expected value of sample information.

Payoff Table

	a_1	a_2
s_1	60	90
s_2	90	90
s_3	150	90

Prior Probabilities

$P(s_1) = \frac{1}{3}$ $\quad P(s_2) = \frac{1}{3}$ $\quad P(s_3) = \frac{1}{3}$

Likelihood Probabilities

	I_1	I_2
s_1	.7	.3
s_2	.5	.5
s_3	.2	.8

22.26 Repeat Exercise 22.25 with the following prior probabilities.

$P(s_1) = .5$ $\quad P(s_2) = .4$ $\quad P(s_3) = .1$

22.27 Repeat Exercise 22.25 with the following prior probabilities.

$P(s_1) = .90$ $\quad P(s_2) = .05$ $\quad P(s_3) = .05$

22.28 What conclusions can you draw about the effect of the prior probabilities on EVSI from Exercises 22.25–22.27?

Applications

22.29 A sporting-goods storeowner has the opportunity to purchase a lot of 50,000 footballs for $100,000. He believes that he can sell some or all by taking out mail-order advertisements in a magazine. Each football will be sold for $6.00. The advertising cost is $25,000, and the mailing cost per football is $1.00. He believes that the demand distribution is as follows.

Demand	P(Demand)
10,000	.2
30,000	.5
50,000	.3

What is the maximum price the owner should pay for additional information about demand?

22.30 What is the maximum price the electronics product manufacturer should be willing to pay for perfect information regarding the market share in Exercise 22.16?

22.31 To improve her decision-making capability, the electronics products manufacturer in Exercise 22.16 performs a survey of potential buyers of compact disk players. She describes the product to 25 individuals, 3 of whom say they would buy it. Using this additional information together with the prior probabilities, determine whether the new product should be produced.

22.32 A radio station currently directing its programming toward middle-age listeners is contemplating switching to rock-and-roll music. After analyzing advertising revenues and operating costs, the owner concludes that, for each percentage point of market share, revenues increase by $100,000 per year. Fixed annual operating costs are $700,000. The owner believes that, with the change, the station will get a 5%, 10%, or 20% market share, with probabilities .4, .4, and .2, respectively. The current annual profit is $285,000.

a. Set up the payoff table.
b. Determine the optimal act.
c. What is the most the owner should be willing to pay to acquire additional information about the market share?

22.33 There is a garbage crisis in North America—too much garbage and no place to put it. As a consequence, the idea of recycling has become quite popular. A waste-management company in a large city is willing to begin recycling newspapers, aluminum cans, and plastic containers. However, it is profitable to do so only if a sufficiently large proportion of households is willing to participate. In this city, 1 million households are potential recyclers. After some analysis, it was determined that, for every 1,000 households that participate in the program, the contribution to profit is $500. It was also discovered that fixed costs are $55,000 per year. It is believed that 50,000, 100,000, 200,000, or 300,000 households will participate, with probabilities .5, .3, .1, and .1, respectively, A preliminary survey was performed wherein 25 households were asked whether they would be willing to be part of this recycling program. Suppose only 3 of the 25 respond affirmatively; incorporate this information into the decision-making process to decide whether the waste-management company should proceed with the recycling venture.

22.34 Repeat Exercise 22.33, given that 12 out of 100 households respond affirmatively.

22.35 Suppose that in Exercise 22.14 a consultant offers to analyze the problem and predict the amount of electricity required by the new community. In order to induce the electric company to hire her, the consultant provides the set of likelihood probabilities given here. Perform a preposterior analysis to determine the expected value of the consultant's sample information.

	I_1 (predict low demand)	I_2 (predict moderate demand)	I_3 (predict high demand)
s_1	.5	.3	.2
s_2	.3	.6	.1
s_3	.2	.2	.6

22.36 In Exercise 22.32, suppose that it is possible to survey radio listeners to determine whether they would tune in to the station if the format changed to rock and roll. What would a survey of size 2 be worth?

22.37 Suppose that in Exercise 22.32 a random sample of 25 radio listeners revealed that 2 people would be regular listeners of the station. What is the optimal decision now?

22.38 The president of an automobile battery company must decide which one of three new types of batteries to produce. The fixed and variable costs of each battery are shown in the accompanying table.

Battery	Fixed Cost	Variable Cost (per unit)
1	$900,000	$20
2	1,150,000	17
3	1,400,000	15

The president believes that demand will be 50,000, 100,000, or 150,000 batteries, with probabilities .3, .3, and .4, respectively. The selling price of the battery will be $40.

a. Determine the payoff table.
b. Determine the opportunity loss table.
c. Find the expected monetary value for each act, and select the optimal one.
d. What is the most the president should be willing to pay for additional information about demand?

22.39 Credibility is often the most effective feature of an advertising campaign. Suppose that, for a particular advertisement, 32% of people surveyed currently believe what the ad claims. A marketing manager believes that for each 1-point increase in that percentage, annual sales will increase by $1 million. For each 1-point decrease, annual sales will decrease by $1 million. The manager believes that a change in the advertising approach can influence the ad's credibility. The probability distribution of the potential percentage changes is listed here.

Percentage Change	Probability
−2	.1
−1	.1
0	.2
+1	.3
+2	.3

If for each dollar of sales the profit contribution is 10 cents and the overall cost of changing the ad is $58,000, should the ad be changed?

22.40 Suppose that in Exercise 22.39 it is possible to perform a survey to determine the percentage of people who believe the ad. What would a sample of size 1 be worth?

22.41 Suppose that in Exercise 22.39 a sample of size 5 showed that only one person believes the new ad. In light of this additional information, what should the manager do?

22.42 Max the Bookie is trying to decide how many telephones to install in his new bookmaking operation. Because of heavy police activity, he cannot increase or decrease the number of telephones once he sets up his operation. He has narrowed the possible choices to three. He can install 25, 50, or 100 telephones. His profit for 1 year (the usual length of time he can remain in business before the police close him down) depends on the average number of calls he receives. The number of calls is Poisson distributed. After some deliberation, he concludes that the average number of calls per minute can be .5. 1.0, or 1.5, with probabilities of .50, .25, and .25, respectively. Max then produces the payoffs given in the accompanying table.

Max's assistant, Lefty (who attended a business school for 2 years), points out that Max may be able to get more information by observing a competitor's similar operation. However, he will be able to watch for only 10 minutes, and doing so will cost him $4,000. Max determines that if he counts fewer than 8 calls, that would be a low number; at least 8 but fewer than 17 would be a medium number; and at least 17 would be a large number of calls. Max also decides that, if the experiment is run, he will record only whether there is a small, medium, or large number of calls. Help Max by performing a preposterior analysis to determine whether the sample should be taken. Conclude by specifying clearly what the optimal strategy is.

Payoff Table

	25 telephones	50 telephones	100 telephones
s_1 ($\mu = .5$)	\$50,000	\$30,000	\$20,000
s_2 ($\mu = 1.0$)	50,000	60,000	40,000
s_3 ($\mu = 1.5$)	50,000	60,000	80,000

22.43 The Megabuck Computer Company is thinking of introducing two new products. The first, Model 101, is a small computer designed specifically for children between the ages of 8 and 16. The second, Model 202, is a medium-size computer suitable for managers. Because of limited production capacity, Megabuck has decided to produce only one of the products.

The profitability of each model depends on the proportion of the potential market that would actually buy the computer. For Model 101, the size of the market is estimated at 10 million, whereas for Model 202, the estimate is 3 million.

After careful analysis, the management of Megabuck has concluded that the percentage of buyers of Model 101 is 5%, 10%, or 15%. The respective profits are given here.

Percentage Who Buy Model 101	Net Profits (\$millions)
5%	20
10	100
15	210

An expert in probability from the local university estimated the probability of the percentages as $P(5\%) = .2$, $P(10\%) = .4$, and $P(15\%) = .4$.

A similar analysis for Model 202 produced the following table.

Percentage Who Buy Model 202	Net Profits (\$millions)
30%	70
40	100
50	150

For this model, the expert estimated the probabilities as $P(30\%) = .1$, $P(40\%) = .4$, and $P(50\%) = .5$

a. Based on this information, and with the objective of maximizing expected profit, which model should Megabuck produce?

b. To make a better decision, Megabuck sampled 10 potential buyers of Model 101 and 20 potential buyers of Model 202. Only 1 of the 10 wished to purchase the Model 101, whereas 9 of the 20 indicated that they would buy Model 202. Given this information, revise the prior probabilities and determine which model should be produced.

22.44 A major movie studio has just completed its latest epic, a musical comedy about the life of Attila the Hun. Because the movie is different (no sex or violence), the studio is uncertain about how to distribute it. The studio executives must decide whether to release the movie to North American audiences or to sell it to a European distributor and realize a profit of \$12 million. If the movie is shown in North America, the studio profit depends on its level of success, which can be classified as excellent, good, or fair. The payoffs and the prior subjective probabilities of the success levels are shown in the accompanying table.

Success Level	Payoff	Probability
Excellent	\$33 million	.5
Good	12 million	.3
Fair	−15 million	.2

Another possibility is to have the movie shown to a random sample of North Americans and use their collective judgment to help the studio make a decision. These judgments are categorized as "rave review," "lukewarm response," or "poor response." The cost of the sample is \$100,000. The sampling process has been used several times in the past. The likelihood probabilities describing the audience judgments and the movie's success level are shown next. Perform a preposterior analysis to determine what the studio executives should do.

	Judgment		
Success Level	Rave Review	Lukewarm Response	Poor Response
Excellent	.8	.1	.1
Good	.5	.3	.2
Fair	.4	.3	.3

CHAPTER SUMMARY

The objective of decision analysis is to select the optimal act from a list of alternative acts. We define as optimal the act with the largest expected monetary value or smallest expected opportunity loss. The expected values are calculated after assigning prior probabilities to the states of nature. The acts, states of nature, and their consequences may be presented in a payoff table, an opportunity loss table, or a decision tree. We also discussed a method by which additional information in the form of an experiment can be incorporated in the analysis. This method involves combining prior and likelihood probabilities to produce posterior probabilities. The preposterior analysis allows us to decide whether to pay for and acquire the experimental outcome. That decision is based on the expected value of sample information and on the sampling cost.

IMPORTANT TERMS

SYMBOLS

Symbol	Represents
a_i	Acts
s_j	States of nature
I_i	Experimental outcomes
$P(s_j)$	Prior probability
$P(I_i \mid s_j)$	Likelihood probability
$P(s_j \text{ and } I_i)$	Joint probability
$P(s_j \mid I_i)$	Posterior probability

FORMULAS

Expected value of perfect information

$$\text{EVPI} = \text{EPPI} - \text{EMV}^*$$

Expected value of sample information

$$\text{EVSI} = \text{EMV}' - \text{EMV}^*$$

© Sandra Baker/Stone/Getty Images

CONCLUSION

We have come to the end of the journey that began with the words "Statistics is a way to get information from data." You will shortly write the final examination in your statistics course. (We assume that readers of this book are taking a statistics course and not just reading it for fun.) If you believe that this event will be the point where you and statistics part company, you could not be more wrong. In the world into which you are about to graduate, the potential applications of statistical techniques are virtually limitless. However, if you are unable or unwilling to employ statistics, you cannot consider yourself to be competent. Can you imagine a marketing manager who does not fully understand marketing concepts and techniques? Can an accountant who knows little about accounting principles do his or her job? Similarly, you cannot be a competent decision maker without a comprehension of statistical concepts and techniques.

In our experience, we have come across far too many people who display an astonishing ignorance of probability and statistics. In some cases, this is displayed in the way they gamble. (Talk to people in a casino in Las Vegas or Atlantic City and discover how many understand probability; see how many of them lose money.) We have seen managers who regularly make decisions involving millions of dollars who don't understand the fundamental principles that should govern the way decisions are made. The worst may be the managers who have access to vast amounts of information no further away than the nearest computer but don't know how to get it or even that it is there.

This raises the question, What statistical concepts and techniques will you need for your life after the final exam? We don't expect students to remember the formulas (or computer commands) that calculate the confidence interval estimates or test statistics. (Statistics reference books are available for that purpose.) However, you must know what you can and cannot do with statistical techniques. You must remember a number of important principles that were covered in this book. To assist you, we have selected the 12 most important concepts and listed them. They are drawn from the "Developing an Understanding of Statistical Concepts" subsections that are scattered throughout the book. We hope that they prove useful to you.

Twelve Statistical Concepts You Need for Life after the Statistics Final Exam

1. Statistical techniques are processes that convert data into information. Descriptive techniques describe and summarize; inferential techniques allow us to make estimates and draw conclusions about populations from samples.

2. We need a large number of techniques because there are numerous objectives and types of data. There are three types of data: interval (real numbers), nominal (categories), and ordinal (ratings). Each combination of data type and objective requires specific techniques.

3. We gather data by various sampling plans. However, the validity of any statistical outcome is dependent on the validity of the sampling. "Garbage in-garbage out" very much applies in statistics.

4. The sampling distribution is the source of statistical inference. The confidence interval estimator and the test statistic are derived directly from the sampling distribution. All inferences are actually probability statements based on the sampling distribution.

5. All tests of hypotheses are conducted similarly. We assume that the null hypothesis is true. We then compute the value of the test statistic. If the difference between what we have observed (and calculated) and what we expect to observe is too large, we reject the null hypothesis. The standard that decides what is "too large" is determined by the probability of a Type I error.

6. In any test of hypothesis (and in most decisions) there are two possible errors, Type I and Type II errors. The relationship between the probabilities of these errors helps us decide where to set the standard. If we set the standard so high that the probability of a Type I error is very small, we increase the probability of a Type II error. A procedure designed to decrease the probability of a Type II error must have a relatively large probability of a Type I error.

7. We can improve the exactitude of a confidence interval estimator or decrease the probability of a Type II error by increasing the sample size. More data mean more information, which results in narrower intervals or lower probabilities of making mistakes, which in turn leads to better decisions.

8. The sampling distributions that are used for interval data are the Student *t* and the *F*. These distributions are related so that the various techniques for interval data are themselves related. We can use the analysis of variance in place of the *t*-test of two means. We can use regression analysis with indicator variables in place of the analysis of variance. We often build a model to represent relationships among interval variables, including indicator variables.

9. In analyzing interval data, we attempt to explain as much of the variation as possible. By doing so, we can learn a great deal about whether populations differ and what variables affect the response (dependent) variable.

10. The techniques used on nominal data require that we count the number of times each category occurs. The counts are then used to compute statistics. The sampling distributions we use for nominal data are the standard normal and the chi-squared. These distributions are related, as are the techniques.

11. The techniques used on ordinal data are based on a ranking procedure. We call these techniques nonparametric. Because the requirements for the use of nonparametric techniques are less stringent than those for a parametric procedure, we often use nonparametric techniques in place of parametric ones when the required conditions for the parametric test are not satisfied. To ensure the validity of a statistical technique, we must check the required conditions.

12. We can obtain data through experimentation or by observation. Observational data lend themselves to several conflicting interpretations. Data gathered by an experiment are more likely to lead to a definitive interpretation. In addition to designing experiments, statistics practitioners can also select particular sample sizes to produce the accuracy and confidence they desire.

APPENDIX A

Data File Sample Statistics

Chapter 10

10.30 $\bar{x} = 252.38$
10.31 $\bar{x} = 1810.16$
10.32 $\bar{x} = 12.10$
10.33 $\bar{x} = 10.21$
10.34 $\bar{x} = .510$
10.35 $\bar{x} = 26.81$
10.36 $\bar{x} = 19.28$
10.37 $\bar{x} = 15.00$
10.38 $\bar{x} = 585{,}063$
10.39 $\bar{x} = 14.98$
10.40 $\bar{x} = 27.19$

Chapter 11

11.35 $\bar{x} = 5065$
11.36 $\bar{x} = 29{,}120$
11.37 $\bar{x} = 569$
11.38 $\bar{x} = 19.13$
11.39 $\bar{x} = -1.20$
11.40 $\bar{x} = 55.8$
11.41 $\bar{x} = 5.04$
11.42 $\bar{x} = 19.39$
11.43 $\bar{x} = 105.7$
11.44 $\bar{x} = 4.84$
11.45 $\bar{x} = 5.64$
11.46 $\bar{x} = 29.92$
11.47 $\bar{x} = 231.56$

Chapter 12

12.29 $\bar{x} = 7.15, s = 1.65, n = 200$
12.30 $\bar{x} = 4.66, s = 2.37, n = 240$
12.31 $\bar{x} = 63.70, s = 18.94, n = 162$
12.32 $\bar{x} = 15{,}137, s = 5{,}263, n = 306$
12.33 $\bar{x} = 59.04, s = 20.62, n = 122$
12.34 $\bar{x} = 2.67, s = 2.50, n = 188$
12.35 $\bar{x} = 29.69, s = 27.53, n = 900$
12.36 $\bar{x} = 422.36, s = 122.77, n = 176$
12.37 $\bar{x} = 13.94, s = 2.16, n = 212$
12.38 $\bar{x} = 15.27, s = 5.72, n = 116$
12.39 $\bar{x} = 3.44, s = 3.33, n = 471$
12.40 $\bar{x} = 89.27, s = 17.30, n = 85$
12.41 $\bar{x} = 15.02, s = 8.31, n = 84$
12.42 $\bar{x} = 460.38, s = 38.83, n = 50$
12.50 $s^2 = 270.58, n = 25$
12.51 $s^2 = 22.56, n = 245$
12.52 $s^2 = 4.72, n = 90$
12.53 $s^2 = 174.47, n = 100$
12.54 $s^2 = 19.68, n = 25$
12.76 $n(1) = 466, n(2) = 55$
12.78 $n(1) = 137, n(2) = 430$
12.79 $n(1) = 153, n(2) = 24$
12.80 $n(1) = 92, n(2) = 28$
12.81 $n(1) = 603, n(2) = 905$
12.82 $n(1) = 92, n(2) = 334$
12.83 $n(1) = 57, n(2) = 35, n(3) = 4, n(4) = 4$
12.85 $n(1) = 303, n(2) = 518, n(3) = 418, n(4) = 397, n(5) = 3364$
12.86 $n(1) = 335, n(2) = 65$
12.87 $n(1) = 155, n(2) = 655$
12.88 $n(1) = 518, n(2) = 132$
12.89 $n(1) = 48, n(2) = 31, n(3) = 45, n(4) = 269, n(5) = 1984$
12.90 $n(1) = 81, n(2) = 47, n(3) = 167, n(4) = 146, n(5) = 34$
12.91 $n(1) = 63, n(2) = 125, n(3) = 45, n(4) = 87$
12.92 $n(1) = 418, n(2) = 536, n(3) = 882$
12.93 $n(1) = 290, n(2) = 35$
12.94 $n(1) = 72, n(2) = 77, n(3) = 37, n(4) = 50, n(5) = 176$
12.95 $n(1) = 289, n(2) = 51$
12.98 $\bar{x} = 229.18, s = 67.36, n = 500$
12.101 $\bar{x} = 313.47, s = 55.53, n = 100$
12.103 $\bar{x} = 12{,}940, s = 4{,}139.18, n = 188$

Chapter 13

13.11 Tastee: $\bar{x}_1 = 36.93, s_1 = 4.23, n_1 = 15$; Competitor: $\bar{x}_2 = 31.36, s_2 = 3.35, n_2 = 25$
13.12 Oat bran: $\bar{x}_1 = 10.01, s_1 = 4.43, n_1 = 120$; Other: $\bar{x}_2 = 9.12, s_2 = 4.45, n_2 = 120$
13.13 18-to-34: $\bar{x}_1 = 58.99, s_1 = 30.77, n_1 = 250$; 35-to-50: $\bar{x}_2 = 52.96, s_2 = 43.32, n_2 = 250$
13.14 2 yrs ago: $\bar{x}_1 = 59.81, s_1 = 7.02, n_1 = 125$; This year: $\bar{x}_2 = 57.40, s_2 = 6.99, n_2 = 159$
13.15 Male: $\bar{x}_1 = 10.23, s_1 = 2.87, n_1 = 100$; Female: $\bar{x}_2 = 9.66, s_2 = 2.90, n_2 = 100$
13.16 A: $\bar{x}_1 = 115.50, s_1 = 21.69, n_1 = 30$; B: $\bar{x}_2 = 110.20, s_2 = 21.93, n_2 = 30$
13.17 Men: $\bar{x}_1 = 5.56, s_1 = 5.36, n_1 = 306$; Women: $\bar{x}_2 = 5.49, s_2 = 5.58, n_2 = 290$
13.18 A: $\bar{x}_1 = 70.42, s_1 = 20.54, n_1 = 24$; B: $\bar{x}_2 = 56.44, s_2 = 9.03, n_2 = 16$
13.19 Successful: $\bar{x}_1 = 5.02, s_1 = 1.39, n_1 = 200$; Unsuccessful: $\bar{x}_2 = 7.80, s_2 = 3.09, n_2 = 200$
13.20 Phone: $\bar{x}_1 = .646, s_1 = .045, n_1 = 125$; Not: $\bar{x}_2 = .601, s_2 = .053, n_2 = 145$
13.21 Chitchat: $\bar{x}_1 = .654, s_1 = .048, n_1 = 95$; Political: $\bar{x}_2 = .662, s_2 = .045, n_2 = 90$
13.22 Planner: $\bar{x}_1 = 6.18, s_1 = 1.59, n_1 = 64$; Broker: $\bar{x}_2 = 5.94, s_2 = 1.61, n_2 = 81$
13.23 Textbook: $\bar{x}_1 = 63.71, s_1 = 5.90, n_1 = 173$; No book: $\bar{x}_2 = 66.80, s_2 = 6.85, n_2 = 202$
13.24 Wendy's: $\bar{x}_1 = 149.85, s_1 = 21.82, n_1 = 213$; McDonald's: $\bar{x}_2 = 154.43, s_2 = 23.64, n_2 = 202$
13.25 General: $\bar{x}_1 = 53.05, s_1 = 3.06, n_1 = 79$; Pediatrics: $\bar{x}_2 = 51.67, s_2 = 3.64, n_2 = 91$
13.26 Applied: $\bar{x}_1 = 130.93, s_1 = 31.99, n_1 = 100$; Contacted: $\bar{x}_2 = 126.14, s_2 = 26.00, n_2 = 100$
13.27 New: $\bar{x}_1 = 73.60, s_1 = 15.60, n_1 = 20$; Existing: $\bar{x}_2 = 69.20, s_2 = 15.06, n_2 = 20$
13.28 Fixed: $\bar{x}_1 = 60{,}245, s_1 = 10{,}506, n_1 = 90$; Commission: $\bar{x}_2 = 63{,}563, s_2 = 10{,}755, n_2 = 90$
13.29 Accident: $\bar{x}_1 = 633.97, s_1 = 49.45, n_1 = 93$; No accident: $\bar{x}_2 = 661.86, s_2 = 52.69, n_2 = 338$
13.30 Cork: $\bar{x}_1 = 14.20, s_1 = 2.84, n_1 = 130$; Metal: $\bar{x}_2 = 11.27, s_2 = 4.42, n_2 = 130$
13.45 $D = X$[This year] $- X$[5 years ago]: $\bar{x}_D = 12.4, s_D = 99.1, n_D = 150$
13.46 $D = X$[Waiter] $- X$[Waitress]: $\bar{x}_D = -1.16, s_D = 2.22, n_D = 50$
13.47 $D = X$[This year] $- X$[Last year]: $\bar{x}_D = 19.75, s_D = 30.63, n_D = 40$
13.48 $D = X$[Uninsulated] $- X$[Insulated]: $\bar{x}_D = 57.40, s_D = 13.14, n_D = 15$
13.49 $D = X$[Men] $- X$[Women]: $\bar{x}_D = -42.94, s_D = 317.16, n_D = 45$
13.50 $D = X$[Last year] $- X$[Previous year]: $\bar{x}_D = -183.35, s_D = 1568.94, n_D = 170$
13.51 $D = X$[This year] $- X$[Last year]: $\bar{x}_D = .0422, s_D = .1634, n_D = 38$
13.52 $D = X$[Company 1] $- X$[Company 2]: $\bar{x}_D = 520.85, s_D = 1854.92, n_D = 55$
13.53 $D = X$[New] $- X$[Existing]: $\bar{x}_D = 4.55, s_D = 7.22, n_D = 20$
13.55 $D = X$[Finance] $- X$[Marketing]: $\bar{x}_D = 4{,}587, s_D = 22{,}851, n_D = 25$
13.57 a. $D = X$[After] $- X$[Before]: $\bar{x}_D = -.10, s_D = 1.95, n_D = 42$
b. $D = X$[After] $- X$[Before]: $\bar{x}_D = 1.24, s_D = 2.83, n_D = 98$
13.63 Week 1: $s_1^2 = 19.38, n_1 = 100$; Week 2: $s_2^2 = 12.70, n_2 = 100$
13.64 A: $s_1^2 = 41{,}309, n_1 = 100$; B: $s_2^2 = 19{,}850, n_2 = 100$
13.65 Portfolio 1: $s_1^2 = .0261, n_1 = 52$; Portfolio 2: $s_2^2 = .0875, n_2 = 52$
13.66 Teller 1: $s_1^2 = 3.35, n_1 = 100$; Teller 2: $s_2^2 = 10.95, n_2 = 100$
13.81 Cadillac: $n(1) = 33, n(2) = 317$; Lincoln: $n(1) = 33, n(2) = 261$
13.82 Smokers: $n_1(1) = 28, n_1(2) = 10$; Nonsmokers: $n_2(1) = 150, n_2(2) = 12$
13.83 Last year: $n_1(1) = 68, n_1(2) = 332$; This year: $n_2(1) = 74, n_2(2) = 426$

13.84 This year: $n_1(1) = 306$, $n_1(2) = 171$; 10 years ago: $n_2(1) = 304$, $n_2(2) = 158$

13.85 A: $n_1(1) = 189$, $n_1(2) = 11$; B: $n_2(1) = 178$, $n_2(2) = 22$

13.86 Health conscious: $n_1(1) = 199$, $n_1(2) = 32$; Not health conscious: $n_2(1) = 563$, $n_2(2) = 56$

13.87 Segment 1: $n(1) = 68$, $n(2) = 95$; Segment 2: $n(1) = 20$, $n(2) = 34$; Segment 3: $n(1) = 10$, $n(2) = 13$; Segment 4: $n(1) = 29$, $n(2) = 79$

13.88 Source 1: $n_1(1) = 344$, $n_1(2) = 38$; Source 2: $n_2(1) = 275$, $n_2(2) = 41$

Chapter 14

14.9

Sample	$\bar{x}_i$	s_i^2	n_i
1	68.83	52.28	20
2	65.08	37.38	26
3	62.01	63.46	16
4	64.64	56.88	19

14.10

Sample	$\bar{x}_i$	s_i^2	n_i
1	90.17	991.5	30
2	95.77	900.9	30
3	106.8	928.7	30
4	111.2	1023	30

14.11

Sample	$\bar{x}_i$	s_i^2	n_i
1	196.8	914.1	41
2	207.8	861.1	73
3	223.4	1195	86
4	232.7	1080	79

14.12

Sample	$\bar{x}_i$	s_i^2	n_i
1	164.6	1164	25
2	185.6	1719	25
3	154.8	1113	25
4	182.6	1657	25
5	178.9	841.8	25

14.13

Sample	$\bar{x}_i$	s_i^2	n_i
1	22.21	121.6	39
2	18.46	90.39	114
3	15.49	85.25	81
4	9.31	65.40	67

14.14

Sample	$\bar{x}_i$	s_i^2	n_i
1	551.5	2742	20
2	576.8	2641	20
3	559.5	3129	20

14.15

Sample	$\bar{x}_i$	s_i^2	n_i
1	5.81	6.22	100
2	5.30	4.05	100
3	5.33	3.90	100

14.16

Sample	$\bar{x}_i$	s_i^2	n_i
1	74.10	250.0	30
2	75.67	184.2	30
3	78.50	233.4	30
4	81.30	242.9	30

14.17

Size

Sample	$\bar{x}_i$	s_i^2	n_i
1	24.97	48.23	50
2	21.65	54.54	50
3	17.84	33.85	50

Nicotine

Sample	$\bar{x}_i$	s_i^2	n_i
1	15.52	3.72	50
2	13.39	3.59	50
3	10.08	3.83	50

14.18a.

Sample	$\bar{x}_i$	s_i^2	n_i
1	31.30	28.34	63
2	34.42	23.20	81
3	37.38	31.16	40
4	39.93	72.03	111

b.

Sample	$\bar{x}_i$	s_i^2	n_i
1	37.22	39.82	63
2	38.91	40.85	81
3	41.48	61.38	40
4	41.75	46.59	111

c.

Sample	$\bar{x}_i$	s_i^2	n_i
1	11.75	3.93	63
2	12.41	3.39	81
3	11.73	4.26	40
4	11.89	4.30	111

14.19

Sample	$\bar{x}_i$	s_i^2	n_i
1	153.6	654.3	20
2	151.5	924.0	20
3	133.3	626.8	20

14.20

Sample	$\bar{x}_i$	s_i^2	n_i
1	18.54	178.0	61
2	19.34	171.4	83
3	20.29	297.5	91

14.27

Sample	$\bar{x}_i$	s_i^2	n_i
1	61.60	80.49	10
2	57.30	70.46	10
3	61.80	22.18	10
4	51.80	75.29	10

14.29

Sample	$\bar{x}_i$	s_i^2	n_i
1	53.17	194.6	30
2	49.37	152.6	30
3	44.33	129.9	30

14.39 $k = 3$, $b = 12$, SST = 204.2, SSB = 1150.2, SSE = 495.1

14.40 $k = 3$, $b = 20$, SST = 7131, SSB = 177,465, SSE = 1098

14.41 $k = 3$, $b = 20$, SST = 10.26, SSB = 3020.30, SSE = 226.71

14.42 $k = 4$, $b = 30$, SST = 4206, SSB = 126,843, SSE = 5764

14.43 $k = 7$, $b = 200$, SST = 28,674, SSB = 209,835, SSE = 479,125

14.44 $k = 5$, $b = 36$, SST = 1406.4, SSB = 7309.7, SSE = 4593.9

14.45 $k = 4$, $b = 21$, SST = 563.82, SSB = 1327.33, SSE = 748.70

Chapter 15

15.7 $n(1) = 28$, $n(2) = 17$, $n(3) = 19$, $n(4) = 17$, $n(5) = 19$

15.8 $n(1) = 41$, $n(2) = 107$, $n(3) = 66$, $n(4) = 19$

15.9 $n(1) = 114$, $n(2) = 92$, $n(3) = 84$, $n(4) = 101$, $n(5) = 107$, $n(6) = 102$

15.10 $n(1) = 11$, $n(2) = 32$, $n(3) = 62$, $n(4) = 29$, $n(5) = 16$

15.11 $n(1) = 8$, $n(2) = 4$, $n(3) = 3$, $n(4) = 8$, $n(5) = 2$

15.12 $n(1) = 159$, $n(2) = 28$, $n(3) = 47$, $n(4) = 16$

15.13 $n(1) = 36$, $n(2) = 58$, $n(3) = 74$, $n(4) = 29$

15.14 $n(1) = 408$, $n(2) = 571$, $n(3) = 221$

15.15 $n(1) = 19$, $n(2) = 23$, $n(3) = 14$, $n(4) = 194$

15.16 $n(1) = 63$, $n(2) = 125$, $n(3) = 45$, $n(4) = 87$

15.26

	Newspaper			
Occupation	G&M	Post	Star	Sun
Blue collar	27	18	38	37
White collar	29	43	21	15
Professional	33	51	22	20

15.27

	Actual	
Predicted	Positive	Negative
Positive	65	64
Negative	39	48

15.28

	Last			
Second-last	1	2	3	4
1	39	36	51	23
2	36	32	46	20
3	54	46	65	29
4	24	20	28	10

15.29

Education	Continuing	Quitter
1	34	23
2	251	212
3	159	248
4	16	57

15.30

	Heartburn Condition			
Source	1	2	3	4
ABC	60	23	13	25
CBS	65	19	14	28
NBC	73	26	9	24
Newspaper	67	11	10	7
Radio	57	16	9	14
None	47	21	10	10

15.31

	Degree			
University	B.A.	B.Eng.	B.B.A.	Other
1	44	11	34	11
2	52	14	27	7
3	31	27	18	24
4	40	12	42	6

15.32

	Financial Ties	
Results	Yes	No
Favorable	29	1
Neutral	10	7
Critical	9	14

15.33

	Degree			
Approach	1	2	3	4
1	51	8	5	11
2	24	14	12	8
3	26	9	19	8

Chapter 16

16.6 Lengths: $\bar{x} = 38.00$, $s_x^2 = 193.90$,
Test: $\bar{y} = 13.80$, $s_y^2 = 47.96$;
$n = 60$, $s_{xy} = 51.86$

16.7 Floors: $\bar{x} = 13.68$ $s_x^2 = 59.32$,
Price: $\bar{y} = 210.42$, $s_y^2 = 496.41$;
$n = 50$, $s_{xy} = 86.93$

16.8 Education: $\bar{x} = 13.17$, $s_x^2 = 11.12$,
Income: $\bar{y} = 78.13$, $s_y^2 = 437.90$;
$n = 150$, $s_{xy} = 46.02$

16.9 Age: $\bar{x} = 37.28$, $s_x^2 = 55.11$,
Employment: $\bar{y} = 26.28$, $s_y^2 = 4.00$;
$n = 80$, $s_{xy} = -6.44$

16.10 Cigarettes: $\bar{x} = 37.64$, $s_x^2 = 108.3$,
Days: $\bar{y} = 14.43$, $s_y^2 = 19.80$;
$n = 231$, $s_{xy} = 20.55$

16.11 Distance: $\bar{x} = 4.88$, $s_x^2 = 4.27$,
Percent: $\bar{y} = 49.22$, $s_y^2 = 243.94$;
$n = 85$, $s_{xy} = 22.83$

16.12 Size: $\bar{x} = 53.93$, $s_x^2 = 688.18$,
Price: $\bar{y} = 6{,}465$, $s_y^2 = 11{,}918{,}489$;
$n = 40$, $s_{xy} = 30{,}945$

16.13 Hours: $\bar{x} = 1199$, $s_x^2 = 59{,}153$,
Price: $\bar{y} = 27.73$, $s_y^2 = 3.62$;
$n = 60$, $s_{xy} = -81.78$

16.14 Occupants: $\bar{x} = 4.75$, $s_x^2 = 4.84$,
Electricity: $\bar{y} = 762.6$, $s_y^2 = 56{,}725$;
$n = 200$, $s_{xy} = 310.0$

16.15 Income: $\bar{x} = 59.42$, $s_x^2 = 115.24$,
Food: $\bar{y} = 270.3$, $s_y^2 = 1{,}797.25$;
$n = 150$, $s_{xy} = 225.66$

16.16 Vacancy: $\bar{x} = 11.33$, $s_x^2 = 35.47$,
Rent: $\bar{y} = 17.20$, $s_y^2 = 11.24$;
$n = 30$, $s_{xy} = -10.78$

16.17 Height: $\bar{x} = 68.95$, $s_x^2 = 9.966$,
Income: $\bar{y} = 59.59$, $s_y^2 = 71.95$;
$n = 250$, $s_{xy} = 6.020$

16.18 Test: $\bar{x} = 79.47$, $s_x^2 = 16.07$,
Nondefective: $\bar{y} = 93.89$, $s_y^2 = 1.28$;
$n = 45$, $s_{xy} = .83$

16.81 Ads: $\bar{x} = 4.12$, $s_x^2 = 3.47$,
Customers: $\bar{y} = 384.81$, $s_y^2 = 18{,}552$;
$n = 26$, $s_{xy} = 74.02$

16.82 Age: $\bar{x} = 113.35$, $s_x^2 = 378.77$,
Repairs: $\bar{y} = 395.21$, $s_y^2 = 4{,}094.79$;
$n = 20$, $s_{xy} = 936.82$

16.83 Fertilizer: $\bar{x} = 300$, $s_x^2 = 20{,}690$,
Yield: $\bar{y} = 318.60$, $s_y^2 = 5{,}230$;
$n = 30$, $s_{xy} = 2538$

16.85 Television: $\bar{x} = 30.43$, $s_x^2 = 99.11$,
Debt: $\bar{y} = 126{,}604$, $s_y^2 = 2{,}152{,}602{,}614$;
$n = 430$, $s_{xy} = 255{,}877$

16.86 Test: $\bar{x} = 71.92$, $s_x^2 = 90.97$,
Nondefective: $\bar{y} = 94.44$, $s_y^2 = 11.84$;
$n = 50$, $s_{xy} = 13.08$

Chapter 17

17.1 $R^2 = .2425$, $R^2\,(adjusted) = .2019$,
$s_\varepsilon = 40.24$, $F = 5.97$, p-value $= .0013$

	Coefficients	Standard Error	t-Statistic	p-Value
Intercept	51.39	23.52	2.19	.0331
Lot size	.700	.559	1.25	.2156
Trees	.679	.229	2.96	.0045
Distance	−.378	.195	−1.94	.0577

17.2 $R^2 = .7629$, $R^2\,(adjusted) = .7453$,
$s_\varepsilon = 3.75$, $F = 43.43$, p-value $= 0$

	Coefficients	Standard Error	t-Statistic	p-Value
Intercept	13.01	3.53	3.69	.0010
Assignment	.194	.200	.97	.3417
Midterm	1.11	.122	9.12	0

17.3 $R^2 = .8935$, $R^2\,(adjusted) = .8711$,
$s_\varepsilon = 40.13$, $F = 39.86$, p-value $= 0$

	Coefficients	Standard Error	t-Statistic	p-Value
Intercept	−111.83	134.34	−.83	.4155
Permits	4.76	.395	12.06	0
Mortgage	16.99	15.16	1.12	.2764
Apartment vacancy	−10.53	6.39	−1.65	.1161
Office vacancy	1.31	2.79	.47	.6446

17.4 $R^2 = .3511$, $R^2\,(adjusted) = .3352$,
$s_\varepsilon = 6.99$, $F = 22.01$, p-value $= 0$

	Coefficients	Standard Error	t-Statistic	p-Value
Intercept	−1.97	9.55	−.21	.8369
Minor HR	.666	.087	7.64	0
Age	.136	.524	.26	.7961
Years pro	1.18	.671	1.75	.0819

Chapter 18

18.3 $R^2 = .4068$, $R^2\,(adjusted) = .3528$,
$s_\varepsilon = 41.15$, $F = 7.54$, p-value $= .0032$

	Coefficients	Standard Error	t-Statistic	p-Value
Intercept	−108.99	97.24	−1.12	.2744
Space	33.09	8.59	3.85	.0009
Space^2	−.666	.177	−3.75	.0011

18.4 First-order model
$R^2 = .8553$, $R^2\,(adjusted) = .8473$,
$s_\varepsilon = 13.29$, $F = 106.44$, p-value $= 0$

	Coefficients	Standard Error	t-Statistic	p-Value
Intercept	453.6	15.18	29.87	0
Price	−68.91	6.68	−10.32	0

Second-order model
$R^2 = .9726$, $R^2\,(adjusted) = .9693$, $s_\varepsilon = 5.96$,
$F = 301.15$, p-value $= 0$

	Coefficients	Standard Error	t-Statistic	p-Value
Intercept	766.9	37.40	20.50	0
Price	−359.1	34.19	−10.50	0
Price^2	64.55	7.58	8.52	0

18.5 First-order model
$R^2 = .8504$, $R^2\,(adjusted) = .8317$,
$s_\varepsilon = 1.79$, $F = 45.48$, p-value $= .0001$

	Coefficients	Standard Error	t-Statistic	p-Value
Intercept	41.4	1.22	33.90	0
Day	−1.33	.197	−6.74	.0001

Second-order model
$R^2 = .8852$, $R^2\,(adjusted) = .8524$, $s_\varepsilon = 1.67$,
$F = 26.98$, p-value $= .0005$

	Coefficients	Standard Error	t-Statistic	p-Value
Intercept	43.73	1.97	22.21	0
Day	−2.49	.822	−3.03	.0191
Day^2	.106	.073	1.46	.1889

18.15 $R^2 = .8051$, $R^2\,(adjusted) = .7947$,
$s_\varepsilon = 2.32$, $F = 77.66$, p-value $= 0$

	Coefficients	Standard Error	t-Statistic	p-Value
Intercept	23.57	5.98	3.94	.0002
Mother	.306	.054	5.65	0
Father	.303	.048	6.37	0
Gmothers	.032	.058	.55	.5853
Gfathers	.078	.057	1.36	.1777
Smoker	−3.72	.669	−5.56	0

18.16 $R^2 = .7002$, $R^2\,(adjusted) = .6659$,
$s_\varepsilon = 810.8$, $F = 20.43$, p-value $= 0$

	Coefficients	Standard Error	t-Statistic	p-Value
Intercept	3490	469.2	7.44	0
Yest Att	.369	.078	4.73	0
I1	1623	492.6	3.30	.0023
I2	733.5	394.4	1.86	.0713
I3	−766.5	484.7	−1.58	.1232

Chapter 19

19.8 $T_1 = 6{,}807$, $n_1 = 82$, $T_2 = 5{,}596$,
$n_2 = 75$

19.9 $T_1 = 797$, $n_1 = 30$, $T_2 = 1{,}033$, $n_2 = 30$

19.10 $T_1 = 14{,}873.5$, $n_1 = 125$, $T_2 = 16{,}501.5$,
$n_2 = 125$

19.11 $T_1 = 10{,}691$, $n_1 = 100$, $T_2 = 9{,}409$,
$n_2 = 100$

19.12 $T_1 = 2{,}810$, $n_1 = 50$, $T_2 = 2{,}240$,
$n_2 = 50$

19.13 $T_1 = 383.5$, $n_1 = 15$, $T_2 = 436.5$,
$n_2 = 25$

19.14 $T_1 = 439.5$, $n_1 = 20$, $T_2 = 380.5$,
$n_2 = 20$

19.15 $T_1 = 13{,}078$, $n_1 = 125$, $T_2 = 18{,}297$, $n_2 = 125$
19.16 $T_1 = 32{,}225.5$, $n_1 = 182$, $T_2 = 27{,}459.5$, $n_2 = 163$
19.30 $T^+ = 378.5$, $T^- = 2{,}249.5$, $n = 72$
19.31 $T^+ = 62$, $T^- = 758$, $n = 40$
19.32 n(positive) = 60, n(negative) = 38
19.33 $T^+ = 40.5$, $T^- = 235.5$, $n = 23$
19.34 $T^+ = 111$, $T^- = 240$, $n = 26$
19.35 n(positive) = 30, n(negative) = 8
19.36 n(positive) = 5, n(negative) = 15
19.37 $T^+ = 190$, $T^- = 135$, $n = 25$
19.38 n(positive) = 32, n(negative) = 21
19.39 $T^+ = 48$, $T^- = 732$, $n = 39$
19.48 $T_1 = 767.5$, $n_1 = 25$, $T_2 = 917$, $n_2 = 25$, $T_3 = 1165.5$, $n_3 = 25$
19.49 $T_1 = 17{,}116.5$, $n_1 = 80$, $T_2 = 16{,}816.5$, $n_2 = 90$, $T_3 = 17{,}277$, $n_3 = 77$, $T_4 = 29{,}391$, $n_4 = 154$
19.50 $T_1 = 2195$, $n_1 = 33$, $T_2 = 1650.5$, $n_2 = 34$, $T_3 = 2830$, $n_3 = 34$, $T_4 = 2102.5$, $n_4 = 31$
19.51 $T_1 = 4180$, $n_1 = 50$, $T_2 = 5262$, $n_2 = 50$, $T_3 = 5653$, $n_3 = 50$, $T_4 = 5005$, $n_4 = 50$
19.52 $T_1 = 1565$, $n_1 = 30$, $T_2 = 1358.5$, $n_2 = 30$, $T_3 = 1171.5$, $n_3 = 30$
19.53 $T_1 = 21{,}246$, $n_1 = 100$, $T_2 = 19{,}784$, $n_2 = 100$, $T_3 = 20{,}976$, $n_3 = 100$, $T_4 = 18{,}194$, $n_4 = 100$
19.54 $T_1 = 28{,}304$, $n_1 = 123$, $T_2 = 21{,}285$, $n_2 = 109$, $T_3 = 21{,}796$, $n_3 = 102$, $T_4 = 20{,}421$, $n_4 = 94$
19.55 $T_1 = 638.5$, $n_1 = 18$, $T_2 = 1233.5$, $n_2 = 14$, $T_3 = 1814.5$, $n_3 = 26$, $T_4 = 3159.5$, $n_4 = 42$, $T_5 = 2065$, $n_5 = 33$
19.62 $T_1 = 33$, $T_2 = 39.5$, $T_3 = 47.5$
19.63 $T_1 = 46$, $T_2 = 72$, $T_3 = 62$
19.64 $T_1 = 28.5$, $T_2 = 22.5$, $T_3 = 21$
19.65 $T_1 = 59.5$, $T_2 = 63.5$, $T_3 = 64$, $T_4 = 63$

APPENDIX B

TABLES

TABLE 1 Binomial Probabilities

Tabulated values are $P(X \leq k) = \sum_{x=0}^{k} p(xi)$. (Values are rounded to four decimal places.)

$n = 5$

	p														
k	**0.01**	**0.05**	**0.10**	**0.20**	**0.25**	**0.30**	**0.40**	**0.50**	**0.60**	**0.70**	**0.75**	**0.80**	**0.90**	**0.95**	**0.99**
0	0.9510	0.7738	0.5905	0.3277	0.2373	0.1681	0.0778	0.0313	0.0102	0.0024	0.0010	0.0003	0.0000	0.0000	0.0000
1	0.9990	0.9774	0.9185	0.7373	0.6328	0.5282	0.3370	0.1875	0.0870	0.0308	0.0156	0.0067	0.0005	0.0000	0.0000
2	1.0000	0.9988	0.9914	0.9421	0.8965	0.8369	0.6826	0.5000	0.3174	0.1631	0.1035	0.0579	0.0086	0.0012	0.0000
3	1.0000	1.0000	0.9995	0.9933	0.9844	0.9692	0.9130	0.8125	0.6630	0.4718	0.3672	0.2627	0.0815	0.0226	0.0010
4	1.0000	1.0000	1.0000	0.9997	0.9990	0.9976	0.9898	0.9688	0.9222	0.8319	0.7627	0.6723	0.4095	0.2262	0.0490

$n = 6$

	p														
k	**0.01**	**0.05**	**0.10**	**0.20**	**0.25**	**0.30**	**0.40**	**0.50**	**0.60**	**0.70**	**0.75**	**0.80**	**0.90**	**0.95**	**0.99**
0	0.9415	0.7351	0.5314	0.2621	0.1780	0.1176	0.0467	0.0156	0.0041	0.0007	0.0002	0.0001	0.0000	0.0000	0.0000
1	0.9985	0.9672	0.8857	0.6554	0.5339	0.4202	0.2333	0.1094	0.0410	0.0109	0.0046	0.0016	0.0001	0.0000	0.0000
2	1.0000	0.9978	0.9842	0.9011	0.8306	0.7443	0.5443	0.3438	0.1792	0.0705	0.0376	0.0170	0.0013	0.0001	0.0000
3	1.0000	0.9999	0.9987	0.9830	0.9624	0.9295	0.8208	0.6563	0.4557	0.2557	0.1694	0.0989	0.0159	0.0022	0.0000
4	1.0000	1.0000	0.9999	0.9984	0.9954	0.9891	0.9590	0.8906	0.7667	0.5798	0.4661	0.3446	0.1143	0.0328	0.0015
5	1.0000	1.0000	1.0000	0.9999	0.9998	0.9993	0.9959	0.9844	0.9533	0.8824	0.8220	0.7379	0.4686	0.2649	0.0585

$n = 7$

	p														
k	**0.01**	**0.05**	**0.10**	**0.20**	**0.25**	**0.30**	**0.40**	**0.50**	**0.60**	**0.70**	**0.75**	**0.80**	**0.90**	**0.95**	**0.99**
0	0.9321	0.6983	0.4783	0.2097	0.1335	0.0824	0.0280	0.0078	0.0016	0.0002	0.0001	0.0000	0.0000	0.0000	0.0000
1	0.9980	0.9556	0.8503	0.5767	0.4449	0.3294	0.1586	0.0625	0.0188	0.0038	0.0013	0.0004	0.0000	0.0000	0.0000
2	1.0000	0.9962	0.9743	0.8520	0.7564	0.6471	0.4199	0.2266	0.0963	0.0288	0.0129	0.0047	0.0002	0.0000	0.0000
3	1.0000	0.9998	0.9973	0.9667	0.9294	0.8740	0.7102	0.5000	0.2898	0.1260	0.0706	0.0333	0.0027	0.0002	0.0000
4	1.0000	1.0000	0.9998	0.9953	0.9871	0.9712	0.9037	0.7734	0.5801	0.3529	0.2436	0.1480	0.0257	0.0038	0.0000
5	1.0000	1.0000	1.0000	0.9996	0.9987	0.9962	0.9812	0.9375	0.8414	0.6706	0.5551	0.4233	0.1497	0.0444	0.0020
6	1.0000	1.0000	1.0000	1.0000	0.9999	0.9998	0.9984	0.9922	0.9720	0.9176	0.8665	0.7903	0.5217	0.3017	0.0679

TABLE **1** (*Continued*)

$n = 8$

	p														
k	**0.01**	**0.05**	**0.10**	**0.20**	**0.25**	**0.30**	**0.40**	**0.50**	**0.60**	**0.70**	**0.75**	**0.80**	**0.90**	**0.95**	**0.99**
0	0.9227	0.6634	0.4305	0.1678	0.1001	0.0576	0.0168	0.0039	0.0007	0.0001	0.0000	0.0000	0.0000	0.0000	0.0000
1	0.9973	0.9428	0.8131	0.5033	0.3671	0.2553	0.1064	0.0352	0.0085	0.0013	0.0004	0.0001	0.0000	0.0000	0.0000
2	0.9999	0.9942	0.9619	0.7969	0.6785	0.5518	0.3154	0.1445	0.0498	0.0113	0.0042	0.0012	0.0000	0.0000	0.0000
3	1.0000	0.9996	0.9950	0.9437	0.8862	0.8059	0.5941	0.3633	0.1737	0.0580	0.0273	0.0104	0.0004	0.0000	0.0000
4	1.0000	1.0000	0.9996	0.9896	0.9727	0.9420	0.8263	0.6367	0.4059	0.1941	0.1138	0.0563	0.0050	0.0004	0.0000
5	1.0000	1.0000	1.0000	0.9988	0.9958	0.9887	0.9502	0.8555	0.6846	0.4482	0.3215	0.2031	0.0381	0.0058	0.0001
6	1.0000	1.0000	1.0000	0.9999	0.9996	0.9987	0.9915	0.9648	0.8936	0.7447	0.6329	0.4967	0.1869	0.0572	0.0027
7	1.0000	1.0000	1.0000	1.0000	1.0000	0.9999	0.9993	0.9961	0.9832	0.9424	0.8999	0.8322	0.5695	0.3366	0.0773

$n = 9$

	p														
k	**0.01**	**0.05**	**0.10**	**0.20**	**0.25**	**0.30**	**0.40**	**0.50**	**0.60**	**0.70**	**0.75**	**0.80**	**0.90**	**0.95**	**0.99**
0	0.9135	0.6302	0.3874	0.1342	0.0751	0.0404	0.0101	0.0020	0.0003	0.0000	0.0000	0.0000	0.0000	0.0000	0.0000
1	0.9966	0.9288	0.7748	0.4362	0.3003	0.1960	0.0705	0.0195	0.0038	0.0004	0.0001	0.0000	0.0000	0.0000	0.0000
2	0.9999	0.9916	0.9470	0.7382	0.6007	0.4628	0.2318	0.0898	0.0250	0.0043	0.0013	0.0003	0.0000	0.0000	0.0000
3	1.0000	0.9994	0.9917	0.9144	0.8343	0.7297	0.4826	0.2539	0.0994	0.0253	0.0100	0.0031	0.0001	0.0000	0.0000
4	1.0000	1.0000	0.9991	0.9804	0.9511	0.9012	0.7334	0.5000	0.2666	0.0988	0.0489	0.0196	0.0009	0.0000	0.0000
5	1.0000	1.0000	0.9999	0.9969	0.9900	0.9747	0.9006	0.7461	0.5174	0.2703	0.1657	0.0856	0.0083	0.0006	0.0000
6	1.0000	1.0000	1.0000	0.9997	0.9987	0.9957	0.9750	0.9102	0.7682	0.5372	0.3993	0.2618	0.0530	0.0084	0.0001
7	1.0000	1.0000	1.0000	1.0000	0.9999	0.9996	0.9962	0.9805	0.9295	0.8040	0.6997	0.5638	0.2252	0.0712	0.0034
8	1.0000	1.0000	1.0000	1.0000	1.0000	1.0000	0.9997	0.9980	0.9899	0.9596	0.9249	0.8658	0.6126	0.3698	0.0865

TABLE 1 (*Continued*)

$n = 10$

	p														
k	**0.01**	**0.05**	**0.10**	**0.20**	**0.25**	**0.30**	**0.40**	**0.50**	**0.60**	**0.70**	**0.75**	**0.80**	**0.90**	**0.95**	**0.99**
0	0.9044	0.5987	0.3487	0.1074	0.0563	0.0282	0.0060	0.0010	0.0001	0.0000	0.0000	0.0000	0.0000	0.0000	0.0000
1	0.9957	0.9139	0.7361	0.3758	0.2440	0.1493	0.0464	0.0107	0.0017	0.0001	0.0000	0.0000	0.0000	0.0000	0.0000
2	0.9999	0.9885	0.9298	0.6778	0.5256	0.3828	0.1673	0.0547	0.0123	0.0016	0.0004	0.0001	0.0000	0.0000	0.0000
3	1.0000	0.9990	0.9872	0.8791	0.7759	0.6496	0.3823	0.1719	0.0548	0.0106	0.0035	0.0009	0.0000	0.0000	0.0000
4	1.0000	0.9999	0.9984	0.9672	0.9219	0.8497	0.6331	0.3770	0.1662	0.0473	0.0197	0.0064	0.0001	0.0000	0.0000
5	1.0000	1.0000	0.9999	0.9936	0.9803	0.9527	0.8338	0.6230	0.3669	0.1503	0.0781	0.0328	0.0016	0.0001	0.0000
6	1.0000	1.0000	1.0000	0.9991	0.9965	0.9894	0.9452	0.8281	0.6177	0.3504	0.2241	0.1209	0.0128	0.0010	0.0000
7	1.0000	1.0000	1.0000	0.9999	0.9996	0.9984	0.9877	0.9453	0.8327	0.6172	0.4744	0.3222	0.0702	0.0115	0.0001
8	1.0000	1.0000	1.0000	1.0000	1.0000	0.9999	0.9983	0.9893	0.9536	0.8507	0.7560	0.6242	0.2639	0.0861	0.0043
9	1.0000	1.0000	1.0000	1.0000	1.0000	1.0000	0.9999	0.9990	0.9940	0.9718	0.9437	0.8926	0.6513	0.4013	0.0956

$n = 15$

	p														
k	**0.01**	**0.05**	**0.10**	**0.20**	**0.25**	**0.30**	**0.40**	**0.50**	**0.60**	**0.70**	**0.75**	**0.80**	**0.90**	**0.95**	**0.99**
0	0.8601	0.4633	0.2059	0.0352	0.0134	0.0047	0.0005	0.0000	0.0000	0.0000	0.0000	0.0000	0.0000	0.0000	0.0000
1	0.9904	0.8290	0.5490	0.1671	0.0802	0.0353	0.0052	0.0005	0.0000	0.0000	0.0000	0.0000	0.0000	0.0000	0.0000
2	0.9996	0.9638	0.8159	0.3980	0.2361	0.1268	0.0271	0.0037	0.0003	0.0000	0.0000	0.0000	0.0000	0.0000	0.0000
3	1.0000	0.9945	0.9444	0.6482	0.4613	0.2969	0.0905	0.0176	0.0019	0.0001	0.0000	0.0000	0.0000	0.0000	0.0000
4	1.0000	0.9994	0.9873	0.8358	0.6865	0.5155	0.2173	0.0592	0.0093	0.0007	0.0001	0.0000	0.0000	0.0000	0.0000
5	1.0000	0.9999	0.9978	0.9389	0.8516	0.7216	0.4032	0.1509	0.0338	0.0037	0.0008	0.0001	0.0000	0.0000	0.0000
6	1.0000	1.0000	0.9997	0.9819	0.9434	0.8689	0.6098	0.3036	0.0950	0.0152	0.0042	0.0008	0.0000	0.0000	0.0000
7	1.0000	1.0000	1.0000	0.9958	0.9827	0.9500	0.7869	0.5000	0.2131	0.0500	0.0173	0.0042	0.0000	0.0000	0.0000
8	1.0000	1.0000	1.0000	0.9992	0.9958	0.9848	0.9050	0.6964	0.3902	0.1311	0.0566	0.0181	0.0003	0.0000	0.0000
9	1.0000	1.0000	1.0000	0.9999	0.9992	0.9963	0.9662	0.8491	0.5968	0.2784	0.1484	0.0611	0.0022	0.0001	0.0000
10	1.0000	1.0000	1.0000	1.0000	0.9999	0.9993	0.9907	0.9408	0.7827	0.4845	0.3135	0.1642	0.0127	0.0006	0.0000
11	1.0000	1.0000	1.0000	1.0000	1.0000	0.9999	0.9981	0.9824	0.9095	0.7031	0.5387	0.3518	0.0556	0.0055	0.0000
12	1.0000	1.0000	1.0000	1.0000	1.0000	1.0000	0.9997	0.9963	0.9729	0.8732	0.7639	0.6020	0.1841	0.0362	0.0004
13	1.0000	1.0000	1.0000	1.0000	1.0000	1.0000	1.0000	0.9995	0.9948	0.9647	0.9198	0.8329	0.4510	0.1710	0.0096
14	1.0000	1.0000	1.0000	1.0000	1.0000	1.0000	1.0000	1.0000	0.9995	0.9953	0.9866	0.9648	0.7941	0.5367	0.1399

TABLE 1 *(Continued)*

n = 20

	p														
k	**0.01**	**0.05**	**0.10**	**0.20**	**0.25**	**0.30**	**0.40**	**0.50**	**0.60**	**0.70**	**0.75**	**0.80**	**0.90**	**0.95**	**0.99**
0	0.8179	0.3585	0.1216	0.0115	0.0032	0.0008	0.0000	0.0000	0.0000	0.0000	0.0000	0.0000	0.0000	0.0000	0.0000
1	0.9831	0.7358	0.3917	0.0692	0.0243	0.0076	0.0005	0.0000	0.0000	0.0000	0.0000	0.0000	0.0000	0.0000	0.0000
2	0.9990	0.9245	0.6769	0.2061	0.0913	0.0355	0.0036	0.0002	0.0000	0.0000	0.0000	0.0000	0.0000	0.0000	0.0000
3	1.0000	0.9841	0.8670	0.4114	0.2252	0.1071	0.0160	0.0013	0.0000	0.0000	0.0000	0.0000	0.0000	0.0000	0.0000
4	1.0000	0.9974	0.9568	0.6296	0.4148	0.2375	0.0510	0.0059	0.0003	0.0000	0.0000	0.0000	0.0000	0.0000	0.0000
5	1.0000	0.9997	0.9887	0.8042	0.6172	0.4164	0.1256	0.0207	0.0016	0.0000	0.0000	0.0000	0.0000	0.0000	0.0000
6	1.0000	1.0000	0.9976	0.9133	0.7858	0.6080	0.2500	0.0577	0.0065	0.0003	0.0000	0.0000	0.0000	0.0000	0.0000
7	1.0000	1.0000	0.9996	0.9679	0.8982	0.7723	0.4159	0.1316	0.0210	0.0013	0.0002	0.0000	0.0000	0.0000	0.0000
8	1.0000	1.0000	0.9999	0.9900	0.9591	0.8867	0.5956	0.2517	0.0565	0.0051	0.0009	0.0001	0.0000	0.0000	0.0000
9	1.0000	1.0000	1.0000	0.9974	0.9861	0.9520	0.7553	0.4119	0.1275	0.0171	0.0039	0.0006	0.0000	0.0000	0.0000
10	1.0000	1.0000	1.0000	0.9994	0.9961	0.9829	0.8725	0.5881	0.2447	0.0480	0.0139	0.0026	0.0000	0.0000	0.0000
11	1.0000	1.0000	1.0000	0.9999	0.9991	0.9949	0.9435	0.7483	0.4044	0.1133	0.0409	0.0100	0.0001	0.0000	0.0000
12	1.0000	1.0000	1.0000	1.0000	0.9998	0.9987	0.9790	0.8684	0.5841	0.2277	0.1018	0.0321	0.0004	0.0000	0.0000
13	1.0000	1.0000	1.0000	1.0000	1.0000	0.9997	0.9935	0.9423	0.7500	0.3920	0.2142	0.0867	0.0024	0.0000	0.0000
14	1.0000	1.0000	1.0000	1.0000	1.0000	1.0000	0.9984	0.9793	0.8744	0.5836	0.3828	0.1958	0.0113	0.0003	0.0000
15	1.0000	1.0000	1.0000	1.0000	1.0000	1.0000	0.9997	0.9941	0.9490	0.7625	0.5852	0.3704	0.0432	0.0026	0.0000
16	1.0000	1.0000	1.0000	1.0000	1.0000	1.0000	1.0000	0.9987	0.9840	0.8929	0.7748	0.5886	0.1330	0.0159	0.0000
17	1.0000	1.0000	1.0000	1.0000	1.0000	1.0000	1.0000	0.9998	0.9964	0.9645	0.9087	0.7939	0.3231	0.0755	0.0010
18	1.0000	1.0000	1.0000	1.0000	1.0000	1.0000	1.0000	1.0000	0.9995	0.9924	0.9757	0.9308	0.6083	0.2642	0.0169
19	1.0000	1.0000	1.0000	1.0000	1.0000	1.0000	1.0000	1.0000	1.0000	0.9992	0.9968	0.9885	0.8784	0.6415	0.1821

TABLE **1** (*Continued*)

$n = 25$

	p														
k	**0.01**	**0.05**	**0.10**	**0.20**	**0.25**	**0.30**	**0.40**	**0.50**	**0.60**	**0.70**	**0.75**	**0.80**	**0.90**	**0.95**	**0.99**
0	0.7778	0.2774	0.0718	0.0038	0.0008	0.0001	0.0000	0.0000	0.0000	0.0000	0.0000	0.0000	0.0000	0.0000	0.0000
1	0.9742	0.6424	0.2712	0.0274	0.0070	0.0016	0.0001	0.0000	0.0000	0.0000	0.0000	0.0000	0.0000	0.0000	0.0000
2	0.9980	0.8729	0.5371	0.0982	0.0321	0.0090	0.0004	0.0000	0.0000	0.0000	0.0000	0.0000	0.0000	0.0000	0.0000
3	0.9999	0.9659	0.7636	0.2340	0.0962	0.0332	0.0024	0.0001	0.0000	0.0000	0.0000	0.0000	0.0000	0.0000	0.0000
4	1.0000	0.9928	0.9020	0.4207	0.2137	0.0905	0.0095	0.0005	0.0000	0.0000	0.0000	0.0000	0.0000	0.0000	0.0000
5	1.0000	0.9988	0.9666	0.6167	0.3783	0.1935	0.0294	0.0020	0.0001	0.0000	0.0000	0.0000	0.0000	0.0000	0.0000
6	1.0000	0.9998	0.9905	0.7800	0.5611	0.3407	0.0736	0.0073	0.0003	0.0000	0.0000	0.0000	0.0000	0.0000	0.0000
7	1.0000	1.0000	0.9977	0.8909	0.7265	0.5118	0.1536	0.0216	0.0012	0.0000	0.0000	0.0000	0.0000	0.0000	0.0000
8	1.0000	1.0000	0.9995	0.9532	0.8506	0.6769	0.2735	0.0539	0.0043	0.0001	0.0000	0.0000	0.0000	0.0000	0.0000
9	1.0000	1.0000	0.9999	0.9827	0.9287	0.8106	0.4246	0.1148	0.0132	0.0005	0.0000	0.0000	0.0000	0.0000	0.0000
10	1.0000	1.0000	1.0000	0.9944	0.9703	0.9022	0.5858	0.2122	0.0344	0.0018	0.0002	0.0000	0.0000	0.0000	0.0000
11	1.0000	1.0000	1.0000	0.9985	0.9893	0.9558	0.7323	0.3450	0.0778	0.0060	0.0009	0.0001	0.0000	0.0000	0.0000
12	1.0000	1.0000	1.0000	0.9996	0.9966	0.9825	0.8462	0.5000	0.1538	0.0175	0.0034	0.0004	0.0000	0.0000	0.0000
13	1.0000	1.0000	1.0000	0.9999	0.9991	0.9940	0.9222	0.6550	0.2677	0.0442	0.0107	0.0015	0.0000	0.0000	0.0000
14	1.0000	1.0000	1.0000	1.0000	0.9998	0.9982	0.9656	0.7878	0.4142	0.0978	0.0297	0.0056	0.0000	0.0000	0.0000
15	1.0000	1.0000	1.0000	1.0000	1.0000	0.9995	0.9868	0.8852	0.5754	0.1894	0.0713	0.0173	0.0001	0.0000	0.0000
16	1.0000	1.0000	1.0000	1.0000	1.0000	0.9999	0.9957	0.9461	0.7265	0.3231	0.1494	0.0468	0.0005	0.0000	0.0000
17	1.0000	1.0000	1.0000	1.0000	1.0000	1.0000	0.9988	0.9784	0.8464	0.4882	0.2735	0.1091	0.0023	0.0000	0.0000
18	1.0000	1.0000	1.0000	1.0000	1.0000	1.0000	0.9997	0.9927	0.9264	0.6593	0.4389	0.2200	0.0095	0.0002	0.0000
19	1.0000	1.0000	1.0000	1.0000	1.0000	1.0000	0.9999	0.9980	0.9706	0.8065	0.6217	0.3833	0.0334	0.0012	0.0000
20	1.0000	1.0000	1.0000	1.0000	1.0000	1.0000	1.0000	0.9995	0.9905	0.9095	0.7863	0.5793	0.0980	0.0072	0.0000
21	1.0000	1.0000	1.0000	1.0000	1.0000	1.0000	1.0000	0.9999	0.9976	0.9668	0.9038	0.7660	0.2364	0.0341	0.0001
22	1.0000	1.0000	1.0000	1.0000	1.0000	1.0000	1.0000	1.0000	0.9996	0.9910	0.9679	0.9018	0.4629	0.1271	0.0020
23	1.0000	1.0000	1.0000	1.0000	1.0000	1.0000	1.0000	1.0000	0.9999	0.9984	0.9930	0.9726	0.7288	0.3576	0.0258
24	1.0000	1.0000	1.0000	1.0000	1.0000	1.0000	1.0000	1.0000	1.0000	0.9999	0.9992	0.9962	0.9282	0.7226	0.2222

TABLE 2 Poisson Probabilities

Tabulated values are $P(X \leq k) = \sum_{x=0}^{k} p(xi)$. (Values are rounded to four decimal places.)

	μ															
k	**0.10**	**0.20**	**0.30**	**0.40**	**0.50**	**1.0**	**1.5**	**2.0**	**2.5**	**3.0**	**3.5**	**4.0**	**4.5**	**5.0**	**5.5**	**6.0**
0	0.9048	0.8187	0.7408	0.6703	0.6065	0.3679	0.2231	0.1353	0.0821	0.0498	0.0302	0.0183	0.0111	0.0067	0.0041	0.0025
1	0.9953	0.9825	0.9631	0.9384	0.9098	0.7358	0.5578	0.4060	0.2873	0.1991	0.1359	0.0916	0.0611	0.0404	0.0266	0.0174
2	0.9998	0.9989	0.9964	0.9921	0.9856	0.9197	0.8088	0.6767	0.5438	0.4232	0.3208	0.2381	0.1736	0.1247	0.0884	0.0620
3	1.0000	0.9999	0.9997	0.9992	0.9982	0.9810	0.9344	0.8571	0.7576	0.6472	0.5366	0.4335	0.3423	0.2650	0.2017	0.1512
4		1.0000	1.0000	0.9999	0.9998	0.9963	0.9814	0.9473	0.8912	0.8153	0.7254	0.6288	0.5321	0.4405	0.3575	0.2851
5				1.0000	1.0000	0.9994	0.9955	0.9834	0.9580	0.9161	0.8576	0.7851	0.7029	0.6160	0.5289	0.4457
6						0.9999	0.9991	0.9955	0.9858	0.9665	0.9347	0.8893	0.8311	0.7622	0.6860	0.6063
7						1.0000	0.9998	0.9989	0.9958	0.9881	0.9733	0.9489	0.9134	0.8666	0.8095	0.7440
8							1.0000	0.9998	0.9989	0.9962	0.9901	0.9786	0.9597	0.9319	0.8944	0.8472
9								1.0000	0.9997	0.9989	0.9967	0.9919	0.9829	0.9682	0.9462	0.9161
10									0.9999	0.9997	0.9990	0.9972	0.9933	0.9863	0.9747	0.9574
11									1.0000	0.9999	0.9997	0.9991	0.9976	0.9945	0.9890	0.9799
12										1.0000	0.9999	0.9997	0.9992	0.9980	0.9955	0.9912
13											1.0000	0.9999	0.9997	0.9993	0.9983	0.9964
14												1.0000	0.9999	0.9998	0.9994	0.9986
15													1.0000	0.9999	0.9998	0.9995
16														1.0000	0.9999	0.9998
17															1.0000	0.9999
18																1.0000
19																
20																

TABLE **2** (*Continued*)

	μ												
k	**6.50**	**7.00**	**7.50**	**8.00**	**8.50**	**9.00**	**9.50**	**10**	**11**	**12**	**13**	**14**	**15**
0	0.0015	0.0009	0.0006	0.0003	0.0002	0.0001	0.0001	0.0000	0.0000	0.0000	0.0000	0.0000	0.0000
1	0.0113	0.0073	0.0047	0.0030	0.0019	0.0012	0.0008	0.0005	0.0002	0.0001	0.0000	0.0000	0.0000
2	0.0430	0.0296	0.0203	0.0138	0.0093	0.0062	0.0042	0.0028	0.0012	0.0005	0.0002	0.0001	0.0000
3	0.1118	0.0818	0.0591	0.0424	0.0301	0.0212	0.0149	0.0103	0.0049	0.0023	0.0011	0.0005	0.0002
4	0.2237	0.1730	0.1321	0.0996	0.0744	0.0550	0.0403	0.0293	0.0151	0.0076	0.0037	0.0018	0.0009
5	0.3690	0.3007	0.2414	0.1912	0.1496	0.1157	0.0885	0.0671	0.0375	0.0203	0.0107	0.0055	0.0028
6	0.5265	0.4497	0.3782	0.3134	0.2562	0.2068	0.1649	0.1301	0.0786	0.0458	0.0259	0.0142	0.0076
7	0.6728	0.5987	0.5246	0.4530	0.3856	0.3239	0.2687	0.2202	0.1432	0.0895	0.0540	0.0316	0.0180
8	0.7916	0.7291	0.6620	0.5925	0.5231	0.4557	0.3918	0.3328	0.2320	0.1550	0.0998	0.0621	0.0374
9	0.8774	0.8305	0.7764	0.7166	0.6530	0.5874	0.5218	0.4579	0.3405	0.2424	0.1658	0.1094	0.0699
10	0.9332	0.9015	0.8622	0.8159	0.7634	0.7060	0.6453	0.5830	0.4599	0.3472	0.2517	0.1757	0.1185
11	0.9661	0.9467	0.9208	0.8881	0.8487	0.8030	0.7520	0.6968	0.5793	0.4616	0.3532	0.2600	0.1848
12	0.9840	0.9730	0.9573	0.9362	0.9091	0.8758	0.8364	0.7916	0.6887	0.5760	0.4631	0.3585	0.2676
13	0.9929	0.9872	0.9784	0.9658	0.9486	0.9261	0.8981	0.8645	0.7813	0.6815	0.5730	0.4644	0.3632
14	0.9970	0.9943	0.9897	0.9827	0.9726	0.9585	0.9400	0.9165	0.8540	0.7720	0.6751	0.5704	0.4657
15	0.9988	0.9976	0.9954	0.9918	0.9862	0.9780	0.9665	0.9513	0.9074	0.8444	0.7636	0.6694	0.5681
16	0.9996	0.9990	0.9980	0.9963	0.9934	0.9889	0.9823	0.9730	0.9441	0.8987	0.8355	0.7559	0.6641
17	0.9998	0.9996	0.9992	0.9984	0.9970	0.9947	0.9911	0.9857	0.9678	0.9370	0.8905	0.8272	0.7489
18	0.9999	0.9999	0.9997	0.9993	0.9987	0.9976	0.9957	0.9928	0.9823	0.9626	0.9302	0.8826	0.8195
19	1.0000	1.0000	0.9999	0.9997	0.9995	0.9989	0.9980	0.9965	0.9907	0.9787	0.9573	0.9235	0.8752
20			1.0000	0.9999	0.9998	0.9996	0.9991	0.9984	0.9953	0.9884	0.9750	0.9521	0.9170
21				1.0000	0.9999	0.9998	0.9996	0.9993	0.9977	0.9939	0.9859	0.9712	0.9469
22					1.0000	0.9999	0.9999	0.9997	0.9990	0.9970	0.9924	0.9833	0.9673
23						1.0000	0.9999	0.9999	0.9995	0.9985	0.9960	0.9907	0.9805
24							1.0000	1.0000	0.9998	0.9993	0.9980	0.9950	0.9888
25									0.9999	0.9997	0.9990	0.9974	0.9938
26									1.0000	0.9999	0.9995	0.9987	0.9967
27										0.9999	0.9998	0.9994	0.9983
28										1.0000	0.9999	0.9997	0.9991
29											1.0000	0.9999	0.9996
30												0.9999	0.9998
31												1.0000	0.9999
32													1.0000

TABLE 3 Cumulative Standardized Normal Probabilities

$P(-\infty < Z < z)$

Z	0.00	0.01	0.02	0.03	0.04	0.05	0.06	0.07	0.08	0.09
−3.0	0.0013	0.0013	0.0013	0.0012	0.0012	0.0011	0.0011	0.0011	0.0010	0.0010
−2.9	0.0019	0.0018	0.0018	0.0017	0.0016	0.0016	0.0015	0.0015	0.0014	0.0014
−2.8	0.0026	0.0025	0.0024	0.0023	0.0023	0.0022	0.0021	0.0021	0.0020	0.0019
−2.7	0.0035	0.0034	0.0033	0.0032	0.0031	0.0030	0.0029	0.0028	0.0027	0.0026
−2.6	0.0047	0.0045	0.0044	0.0043	0.0041	0.0040	0.0039	0.0038	0.0037	0.0036
−2.5	0.0062	0.0060	0.0059	0.0057	0.0055	0.0054	0.0052	0.0051	0.0049	0.0048
−2.4	0.0082	0.0080	0.0078	0.0075	0.0073	0.0071	0.0069	0.0068	0.0066	0.0064
−2.3	0.0107	0.0104	0.0102	0.0099	0.0096	0.0094	0.0091	0.0089	0.0087	0.0084
−2.2	0.0139	0.0136	0.0132	0.0129	0.0125	0.0122	0.0119	0.0116	0.0113	0.0110
−2.1	0.0179	0.0174	0.0170	0.0166	0.0162	0.0158	0.0154	0.0150	0.0146	0.0143
−2.0	0.0228	0.0222	0.0217	0.0212	0.0207	0.0202	0.0197	0.0192	0.0188	0.0183
−1.9	0.0287	0.0281	0.0274	0.0268	0.0262	0.0256	0.0250	0.0244	0.0239	0.0233
−1.8	0.0359	0.0351	0.0344	0.0336	0.0329	0.0322	0.0314	0.0307	0.0301	0.0294
−1.7	0.0446	0.0436	0.0427	0.0418	0.0409	0.0401	0.0392	0.0384	0.0375	0.0367
−1.6	0.0548	0.0537	0.0526	0.0516	0.0505	0.0495	0.0485	0.0475	0.0465	0.0455
−1.5	0.0668	0.0655	0.0643	0.0630	0.0618	0.0606	0.0594	0.0582	0.0571	0.0559
−1.4	0.0808	0.0793	0.0778	0.0764	0.0749	0.0735	0.0721	0.0708	0.0694	0.0681
−1.3	0.0968	0.0951	0.0934	0.0918	0.0901	0.0885	0.0869	0.0853	0.0838	0.0823
−1.2	0.1151	0.1131	0.1112	0.1093	0.1075	0.1056	0.1038	0.1020	0.1003	0.0985
−1.1	0.1357	0.1335	0.1314	0.1292	0.1271	0.1251	0.1230	0.1210	0.1190	0.1170
−1.0	0.1587	0.1562	0.1539	0.1515	0.1492	0.1469	0.1446	0.1423	0.1401	0.1379
−0.9	0.1841	0.1814	0.1788	0.1762	0.1736	0.1711	0.1685	0.1660	0.1635	0.1611
−0.8	0.2119	0.2090	0.2061	0.2033	0.2005	0.1977	0.1949	0.1922	0.1894	0.1867
−0.7	0.2420	0.2389	0.2358	0.2327	0.2296	0.2266	0.2236	0.2206	0.2177	0.2148
−0.6	0.2743	0.2709	0.2676	0.2643	0.2611	0.2578	0.2546	0.2514	0.2483	0.2451
−0.5	0.3085	0.3050	0.3015	0.2981	0.2946	0.2912	0.2877	0.2843	0.2810	0.2776
−0.4	0.3446	0.3409	0.3372	0.3336	0.3300	0.3264	0.3228	0.3192	0.3156	0.3121
−0.3	0.3821	0.3783	0.3745	0.3707	0.3669	0.3632	0.3594	0.3557	0.3520	0.3483
−0.2	0.4207	0.4168	0.4129	0.4090	0.4052	0.4013	0.3974	0.3936	0.3897	0.3859
−0.1	0.4602	0.4562	0.4522	0.4483	0.4443	0.4404	0.4364	0.4325	0.4286	0.4247
−0.0	0.5000	0.4960	0.4920	0.4880	0.4840	0.4801	0.4761	0.4721	0.4681	0.4641

TABLE **3** (*Continued*)

$P(-\infty < Z < z)$.

Z	0.00	0.01	0.02	0.03	0.04	0.05	0.06	0.07	0.08	0.09
0.0	0.5000	0.5040	0.5080	0.5120	0.5160	0.5199	0.5239	0.5279	0.5319	0.5359
0.1	0.5398	0.5438	0.5478	0.5517	0.5557	0.5596	0.5636	0.5675	0.5714	0.5753
0.2	0.5793	0.5832	0.5871	0.5910	0.5948	0.5987	0.6026	0.6064	0.6103	0.6141
0.3	0.6179	0.6217	0.6255	0.6293	0.6331	0.6368	0.6406	0.6443	0.6480	0.6517
0.4	0.6554	0.6591	0.6628	0.6664	0.6700	0.6736	0.6772	0.6808	0.6844	0.6879
0.5	0.6915	0.6950	0.6985	0.7019	0.7054	0.7088	0.7123	0.7157	0.7190	0.7224
0.6	0.7257	0.7291	0.7324	0.7357	0.7389	0.7422	0.7454	0.7486	0.7517	0.7549
0.7	0.7580	0.7611	0.7642	0.7673	0.7704	0.7734	0.7764	0.7794	0.7823	0.7852
0.8	0.7881	0.7910	0.7939	0.7967	0.7995	0.8023	0.8051	0.8078	0.8106	0.8133
0.9	0.8159	0.8186	0.8212	0.8238	0.8264	0.8289	0.8315	0.8340	0.8365	0.8389
1.0	0.8413	0.8438	0.8461	0.8485	0.8508	0.8531	0.8554	0.8577	0.8599	0.8621
1.1	0.8643	0.8665	0.8686	0.8708	0.8729	0.8749	0.8770	0.8790	0.8810	0.8830
1.2	0.8849	0.8869	0.8888	0.8907	0.8925	0.8944	0.8962	0.8980	0.8997	0.9015
1.3	0.9032	0.9049	0.9066	0.9082	0.9099	0.9115	0.9131	0.9147	0.9162	0.9177
1.4	0.9192	0.9207	0.9222	0.9236	0.9251	0.9265	0.9279	0.9292	0.9306	0.9319
1.5	0.9332	0.9345	0.9357	0.9370	0.9382	0.9394	0.9406	0.9418	0.9429	0.9441
1.6	0.9452	0.9463	0.9474	0.9484	0.9495	0.9505	0.9515	0.9525	0.9535	0.9545
1.7	0.9554	0.9564	0.9573	0.9582	0.9591	0.9599	0.9608	0.9616	0.9625	0.9633
1.8	0.9641	0.9649	0.9656	0.9664	0.9671	0.9678	0.9686	0.9693	0.9699	0.9706
1.9	0.9713	0.9719	0.9726	0.9732	0.9738	0.9744	0.9750	0.9756	0.9761	0.9767
2.0	0.9772	0.9778	0.9783	0.9788	0.9793	0.9798	0.9803	0.9808	0.9812	0.9817
2.1	0.9821	0.9826	0.9830	0.9834	0.9838	0.9842	0.9846	0.9850	0.9854	0.9857
2.2	0.9861	0.9864	0.9868	0.9871	0.9875	0.9878	0.9881	0.9884	0.9887	0.9890
2.3	0.9893	0.9896	0.9898	0.9901	0.9904	0.9906	0.9909	0.9911	0.9913	0.9916
2.4	0.9918	0.9920	0.9922	0.9925	0.9927	0.9929	0.9931	0.9932	0.9934	0.9936
2.5	0.9938	0.9940	0.9941	0.9943	0.9945	0.9946	0.9948	0.9949	0.9951	0.9952
2.6	0.9953	0.9955	0.9956	0.9957	0.9959	0.9960	0.9961	0.9962	0.9963	0.9964
2.7	0.9965	0.9966	0.9967	0.9968	0.9969	0.9970	0.9971	0.9972	0.9973	0.9974
2.8	0.9974	0.9975	0.9976	0.9977	0.9977	0.9978	0.9979	0.9979	0.9980	0.9981
2.9	0.9981	0.9982	0.9982	0.9983	0.9984	0.9984	0.9985	0.9985	0.9986	0.9986
3.0	0.9987	0.9987	0.9987	0.9988	0.9988	0.9989	0.9989	0.9989	0.9990	0.9990

TABLE 4
Critical Values of the Student *t* Distribution

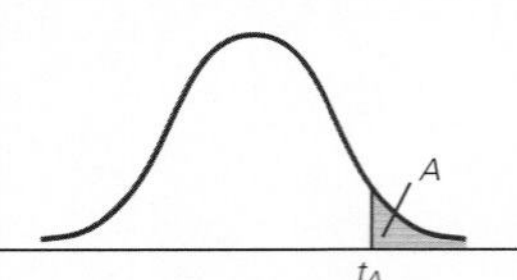

Degrees of Freedom	$t_{.100}$	$t_{.050}$	$t_{.025}$	$t_{.010}$	$t_{.005}$
1	3.078	6.314	12.706	31.821	63.657
2	1.886	2.920	4.303	6.965	9.925
3	1.638	2.353	3.182	4.541	5.841
4	1.533	2.132	2.776	3.747	4.604
5	1.476	2.015	2.571	3.365	4.032
6	1.440	1.943	2.447	3.143	3.707
7	1.415	1.895	2.365	2.998	3.499
8	1.397	1.860	2.306	2.896	3.355
9	1.383	1.833	2.262	2.821	3.250
10	1.372	1.812	2.228	2.764	3.169
11	1.363	1.796	2.201	2.718	3.106
12	1.356	1.782	2.179	2.681	3.055
13	1.350	1.771	2.160	2.650	3.012
14	1.345	1.761	2.145	2.624	2.977
15	1.341	1.753	2.131	2.602	2.947
16	1.337	1.746	2.120	2.583	2.921
17	1.333	1.740	2.110	2.567	2.898
18	1.330	1.734	2.101	2.552	2.878
19	1.328	1.729	2.093	2.539	2.861
20	1.325	1.725	2.086	2.528	2.845
21	1.323	1.721	2.080	2.518	2.831
22	1.321	1.717	2.074	2.508	2.819
23	1.319	1.714	2.069	2.500	2.807
24	1.318	1.711	2.064	2.492	2.797
25	1.316	1.708	2.060	2.485	2.787
26	1.315	1.706	2.056	2.479	2.779
27	1.314	1.703	2.052	2.473	2.771
28	1.313	1.701	2.048	2.467	2.763
29	1.311	1.699	2.045	2.462	2.756
30	1.310	1.697	2.042	2.457	2.750
35	1.306	1.690	2.030	2.438	2.724
40	1.303	1.684	2.021	2.423	2.704
45	1.301	1.679	2.014	2.412	2.690
50	1.299	1.676	2.009	2.403	2.678
55	1.297	1.673	2.004	2.396	2.668
60	1.296	1.671	2.000	2.390	2.660
65	1.295	1.669	1.997	2.385	2.654
70	1.294	1.667	1.994	2.381	2.648
75	1.293	1.665	1.992	2.377	2.643
80	1.292	1.664	1.990	2.374	2.639
85	1.292	1.663	1.988	2.371	2.635
90	1.291	1.662	1.987	2.368	2.632
95	1.291	1.661	1.985	2.366	2.629
100	1.290	1.660	1.984	2.364	2.626
110	1.289	1.659	1.982	2.361	2.621
120	1.289	1.658	1.980	2.358	2.617
130	1.288	1.657	1.978	2.355	2.614
140	1.288	1.656	1.977	2.353	2.611
150	1.287	1.655	1.976	2.351	2.609
160	1.287	1.654	1.975	2.350	2.607
170	1.287	1.654	1.974	2.348	2.605
180	1.286	1.653	1.973	2.347	2.603
190	1.286	1.653	1.973	2.346	2.602
200	1.286	1.653	1.972	2.345	2.601
∞	1.282	1.645	1.960	2.326	2.576

TABLE 5 Critical Values of the χ^2 Distribution

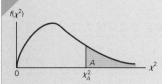

Degrees of Freedom	$\chi^2_{.995}$	$\chi^2_{.990}$	$\chi^2_{.975}$	$\chi^2_{.950}$	$\chi^2_{.900}$	$\chi^2_{.100}$	$\chi^2_{.050}$	$\chi^2_{.025}$	$\chi^2_{.010}$	$\chi^2_{.005}$
1	0.000039	0.000157	0.000982	0.00393	0.0158	2.71	3.84	5.02	6.63	7.88
2	0.0100	0.0201	0.0506	0.103	0.211	4.61	5.99	7.38	9.21	10.6
3	0.072	0.115	0.216	0.352	0.584	6.25	7.81	9.35	11.3	12.8
4	0.207	0.297	0.484	0.711	1.06	7.78	9.49	11.1	13.3	14.9
5	0.412	0.554	0.831	1.15	1.61	9.24	11.1	12.8	15.1	16.7
6	0.676	0.872	1.24	1.64	2.20	10.6	12.6	14.4	16.8	18.5
7	0.989	1.24	1.69	2.17	2.83	12.0	14.1	16.0	18.5	20.3
8	1.34	1.65	2.18	2.73	3.49	13.4	15.5	17.5	20.1	22.0
9	1.73	2.09	2.70	3.33	4.17	14.7	16.9	19.0	21.7	23.6
10	2.16	2.56	3.25	3.94	4.87	16.0	18.3	20.5	23.2	25.2
11	2.60	3.05	3.82	4.57	5.58	17.3	19.7	21.9	24.7	26.8
12	3.07	3.57	4.40	5.23	6.30	18.5	21.0	23.3	26.2	28.3
13	3.57	4.11	5.01	5.89	7.04	19.8	22.4	24.7	27.7	29.8
14	4.07	4.66	5.63	6.57	7.79	21.1	23.7	26.1	29.1	31.3
15	4.60	5.23	6.26	7.26	8.55	22.3	25.0	27.5	30.6	32.8
16	5.14	5.81	6.91	7.96	9.31	23.5	26.3	28.8	32.0	34.3
17	5.70	6.41	7.56	8.67	10.1	24.8	27.6	30.2	33.4	35.7
18	6.26	7.01	8.23	9.39	10.9	26.0	28.9	31.5	34.8	37.2
19	6.84	7.63	8.91	10.1	11.7	27.2	30.1	32.9	36.2	38.6
20	7.43	8.26	9.59	10.9	12.4	28.4	31.4	34.2	37.6	40.0
21	8.03	8.90	10.3	11.6	13.2	29.6	32.7	35.5	38.9	41.4
22	8.64	9.54	11.0	12.3	14.0	30.8	33.9	36.8	40.3	42.8
23	9.26	10.2	11.7	13.1	14.8	32.0	35.2	38.1	41.6	44.2
24	9.89	10.9	12.4	13.8	15.7	33.2	36.4	39.4	43.0	45.6
25	10.5	11.5	13.1	14.6	16.5	34.4	37.7	40.6	44.3	46.9
26	11.2	12.2	13.8	15.4	17.3	35.6	38.9	41.9	45.6	48.3
27	11.8	12.9	14.6	16.2	18.1	36.7	40.1	43.2	47.0	49.6
28	12.5	13.6	15.3	16.9	18.9	37.9	41.3	44.5	48.3	51.0
29	13.1	14.3	16.0	17.7	19.8	39.1	42.6	45.7	49.6	52.3
30	13.8	15.0	16.8	18.5	20.6	40.3	43.8	47.0	50.9	53.7
40	20.7	22.2	24.4	26.5	29.1	51.8	55.8	59.3	63.7	66.8
50	28.0	29.7	32.4	34.8	37.7	63.2	67.5	71.4	76.2	79.5
60	35.5	37.5	40.5	43.2	46.5	74.4	79.1	83.3	88.4	92.0
70	43.3	45.4	48.8	51.7	55.3	85.5	90.5	95.0	100	104
80	51.2	53.5	57.2	60.4	64.3	96.6	102	107	112	116
90	59.2	61.8	65.6	69.1	73.3	108	113	118	124	128
100	67.3	70.1	74.2	77.9	82.4	118	124	130	136	140

TABLE **6(a)** Critical Values of the *F*-Distribution: $A = .05$

f(F) A F 0 F_A

ν_2 \ ν_1 (Denominator degrees of freedom \ Numerator degrees of freedom)	1	2	3	4	5	6	7	8	9	10	11	12	13	14	15	16	17	18	19	20
1	161	199	216	225	230	234	237	239	241	242	243	244	245	245	246	246	247	247	248	248
2	18.5	19.0	19.2	19.2	19.3	19.3	19.4	19.4	19.4	19.4	19.4	19.4	19.4	19.4	19.4	19.4	19.4	19.4	19.4	19.4
3	10.1	9.55	9.28	9.12	9.01	8.94	8.89	8.85	8.81	8.79	8.76	8.74	8.73	8.71	8.70	8.69	8.68	8.67	8.67	8.66
4	7.71	6.94	6.59	6.39	6.26	6.16	6.09	6.04	6.00	5.96	5.94	5.91	5.89	5.87	5.86	5.84	5.83	5.82	5.81	5.80
5	6.61	5.79	5.41	5.19	5.05	4.95	4.88	4.82	4.77	4.74	4.70	4.68	4.66	4.64	4.62	4.60	4.59	4.58	4.57	4.56
6	5.99	5.14	4.76	4.53	4.39	4.28	4.21	4.15	4.10	4.06	4.03	4.00	3.98	3.96	3.94	3.92	3.91	3.90	3.88	3.87
7	5.59	4.74	4.35	4.12	3.97	3.87	3.79	3.73	3.68	3.64	3.60	3.57	3.55	3.53	3.51	3.49	3.48	3.47	3.46	3.44
8	5.32	4.46	4.07	3.84	3.69	3.58	3.50	3.44	3.39	3.35	3.31	3.28	3.26	3.24	3.22	3.20	3.19	3.17	3.16	3.15
9	5.12	4.26	3.86	3.63	3.48	3.37	3.29	3.23	3.18	3.14	3.10	3.07	3.05	3.03	3.01	2.99	2.97	2.96	2.95	2.94
10	4.96	4.10	3.71	3.48	3.33	3.22	3.14	3.07	3.02	2.98	2.94	2.91	2.89	2.86	2.85	2.83	2.81	2.80	2.79	2.77
11	4.84	3.98	3.59	3.36	3.20	3.09	3.01	2.95	2.90	2.85	2.82	2.79	2.76	2.74	2.72	2.70	2.69	2.67	2.66	2.65
12	4.75	3.89	3.49	3.26	3.11	3.00	2.91	2.85	2.80	2.75	2.72	2.69	2.66	2.64	2.62	2.60	2.58	2.57	2.56	2.54
13	4.67	3.81	3.41	3.18	3.03	2.92	2.83	2.77	2.71	2.67	2.63	2.60	2.58	2.55	2.53	2.51	2.50	2.48	2.47	2.46
14	4.60	3.74	3.34	3.11	2.96	2.85	2.76	2.70	2.65	2.60	2.57	2.53	2.51	2.48	2.46	2.44	2.43	2.41	2.40	2.39
15	4.54	3.68	3.29	3.06	2.90	2.79	2.71	2.64	2.59	2.54	2.51	2.48	2.45	2.42	2.40	2.38	2.37	2.35	2.34	2.33
16	4.49	3.63	3.24	3.01	2.85	2.74	2.66	2.59	2.54	2.49	2.46	2.42	2.40	2.37	2.35	2.33	2.32	2.30	2.29	2.28
17	4.45	3.59	3.20	2.96	2.81	2.70	2.61	2.55	2.49	2.45	2.41	2.38	2.35	2.33	2.31	2.29	2.27	2.26	2.24	2.23
18	4.41	3.55	3.16	2.93	2.77	2.66	2.58	2.51	2.46	2.41	2.37	2.34	2.31	2.29	2.27	2.25	2.23	2.22	2.20	2.19
19	4.38	3.52	3.13	2.90	2.74	2.63	2.54	2.48	2.42	2.38	2.34	2.31	2.28	2.26	2.23	2.21	2.20	2.18	2.17	2.16
20	4.35	3.49	3.10	2.87	2.71	2.60	2.51	2.45	2.39	2.35	2.31	2.28	2.25	2.22	2.20	2.18	2.17	2.15	2.14	2.12
22	4.30	3.44	3.05	2.82	2.66	2.55	2.46	2.40	2.34	2.30	2.26	2.23	2.20	2.17	2.15	2.13	2.11	2.10	2.08	2.07
24	4.26	3.40	3.01	2.78	2.62	2.51	2.42	2.36	2.30	2.25	2.22	2.18	2.15	2.13	2.11	2.09	2.07	2.05	2.04	2.03
26	4.23	3.37	2.98	2.74	2.59	2.47	2.39	2.32	2.27	2.22	2.18	2.15	2.12	2.09	2.07	2.05	2.03	2.02	2.00	1.99
28	4.20	3.34	2.95	2.71	2.56	2.45	2.36	2.29	2.24	2.19	2.15	2.12	2.09	2.06	2.04	2.02	2.00	1.99	1.97	1.96
30	4.17	3.32	2.92	2.69	2.53	2.42	2.33	2.27	2.21	2.16	2.13	2.09	2.06	2.04	2.01	1.99	1.98	1.96	1.95	1.93
35	4.12	3.27	2.87	2.64	2.49	2.37	2.29	2.22	2.16	2.11	2.07	2.04	2.01	1.99	1.96	1.94	1.92	1.91	1.89	1.88
40	4.08	3.23	2.84	2.61	2.45	2.34	2.25	2.18	2.12	2.08	2.04	2.00	1.97	1.95	1.92	1.90	1.89	1.87	1.85	1.84
45	4.06	3.20	2.81	2.58	2.42	2.31	2.22	2.15	2.10	2.05	2.01	1.97	1.94	1.92	1.89	1.87	1.86	1.84	1.82	1.81
50	4.03	3.18	2.79	2.56	2.40	2.29	2.20	2.13	2.07	2.03	1.99	1.95	1.92	1.89	1.87	1.85	1.83	1.81	1.80	1.78
60	4.00	3.15	2.76	2.53	2.37	2.25	2.17	2.10	2.04	1.99	1.95	1.92	1.89	1.86	1.84	1.82	1.80	1.78	1.76	1.75
70	3.98	3.13	2.74	2.50	2.35	2.23	2.14	2.07	2.02	1.97	1.93	1.89	1.86	1.84	1.81	1.79	1.77	1.75	1.74	1.72
80	3.96	3.11	2.72	2.49	2.33	2.21	2.13	2.06	2.00	1.95	1.91	1.88	1.84	1.82	1.79	1.77	1.75	1.73	1.72	1.70
90	3.95	3.10	2.71	2.47	2.32	2.20	2.11	2.04	1.99	1.94	1.90	1.86	1.83	1.80	1.78	1.76	1.74	1.72	1.70	1.69
100	3.94	3.09	2.70	2.46	2.31	2.19	2.10	2.03	1.97	1.93	1.89	1.85	1.82	1.79	1.77	1.75	1.73	1.71	1.69	1.68
120	3.92	3.07	2.68	2.45	2.29	2.18	2.09	2.02	1.96	1.91	1.87	1.83	1.80	1.78	1.75	1.73	1.71	1.69	1.67	1.66
140	3.91	3.06	2.67	2.44	2.28	2.16	2.08	2.01	1.95	1.90	1.86	1.82	1.79	1.76	1.74	1.72	1.70	1.68	1.66	1.65
160	3.90	3.05	2.66	2.43	2.27	2.16	2.07	2.00	1.94	1.89	1.85	1.81	1.78	1.75	1.73	1.71	1.69	1.67	1.65	1.64
180	3.89	3.05	2.65	2.42	2.26	2.15	2.06	1.99	1.93	1.88	1.84	1.81	1.77	1.75	1.72	1.70	1.68	1.66	1.64	1.63
200	3.89	3.04	2.65	2.42	2.26	2.14	2.06	1.98	1.93	1.88	1.84	1.80	1.77	1.74	1.72	1.69	1.67	1.66	1.64	1.62
∞	3.84	3.00	2.61	2.37	2.21	2.10	2.01	1.94	1.88	1.83	1.79	1.75	1.72	1.69	1.67	1.64	1.62	1.60	1.59	1.57

ν_2 \ ν_1	NUMERATOR DEGREES OF FREEDOM 22	24	26	28	30	35	40	45	50	60	70	80	90	100	120	140	160	180	200	∞
DENOMINATOR DEGREES OF FREEDOM 1	249	249	249	250	250	251	251	251	252	252	252	253	253	253	253	253	254	254	254	254
2	19.5	19.5	19.5	19.5	19.5	19.5	19.5	19.5	19.5	19.5	19.5	19.5	19.5	19.5	19.5	19.5	19.5	19.5	19.5	19.5
3	8.65	8.64	8.63	8.62	8.62	8.60	8.59	8.59	8.58	8.57	8.57	8.56	8.56	8.55	8.55	8.55	8.54	8.54	8.54	8.53
4	5.79	5.77	5.76	5.75	5.75	5.73	5.72	5.71	5.70	5.69	5.68	5.67	5.67	5.66	5.66	5.65	5.65	5.65	5.65	5.63
5	4.54	4.53	4.52	4.50	4.50	4.48	4.46	4.45	4.44	4.43	4.42	4.41	4.41	4.41	4.40	4.39	4.39	4.39	4.39	4.37
6	3.86	3.84	3.83	3.82	3.81	3.79	3.77	3.76	3.75	3.74	3.73	3.72	3.72	3.71	3.70	3.70	3.70	3.69	3.69	3.67
7	3.43	3.41	3.40	3.39	3.38	3.36	3.34	3.33	3.32	3.30	3.29	3.29	3.28	3.27	3.27	3.26	3.26	3.25	3.25	3.23
8	3.13	3.12	3.10	3.09	3.08	3.06	3.04	3.03	3.02	3.01	2.99	2.99	2.98	2.97	2.97	2.96	2.96	2.95	2.95	2.93
9	2.92	2.90	2.89	2.87	2.86	2.84	2.83	2.81	2.80	2.79	2.78	2.77	2.76	2.76	2.75	2.74	2.74	2.73	2.73	2.71
10	2.75	2.74	2.72	2.71	2.70	2.68	2.66	2.65	2.64	2.62	2.61	2.60	2.59	2.59	2.58	2.57	2.57	2.57	2.56	2.54
11	2.63	2.61	2.59	2.58	2.57	2.55	2.53	2.52	2.51	2.49	2.48	2.47	2.46	2.46	2.45	2.44	2.44	2.43	2.43	2.41
12	2.52	2.51	2.49	2.48	2.47	2.44	2.43	2.41	2.40	2.38	2.37	2.36	2.36	2.35	2.34	2.33	2.33	2.33	2.32	2.30
13	2.44	2.42	2.41	2.39	2.38	2.36	2.34	2.33	2.31	2.30	2.28	2.27	2.27	2.26	2.25	2.25	2.24	2.24	2.23	2.21
14	2.37	2.35	2.33	2.32	2.31	2.28	2.27	2.25	2.24	2.22	2.21	2.20	2.19	2.19	2.18	2.17	2.17	2.16	2.16	2.13
15	2.31	2.29	2.27	2.26	2.25	2.22	2.20	2.19	2.18	2.16	2.15	2.14	2.13	2.12	2.11	2.11	2.10	2.10	2.10	2.07
16	2.25	2.24	2.22	2.21	2.19	2.17	2.15	2.14	2.12	2.11	2.09	2.08	2.07	2.07	2.06	2.05	2.05	2.04	2.04	2.01
17	2.21	2.19	2.17	2.16	2.15	2.12	2.10	2.09	2.08	2.06	2.05	2.03	2.03	2.02	2.01	2.00	2.00	1.99	1.99	1.96
18	2.17	2.15	2.13	2.12	2.11	2.08	2.06	2.05	2.04	2.02	2.00	1.99	1.98	1.98	1.97	1.96	1.96	1.95	1.95	1.92
19	2.13	2.11	2.10	2.08	2.07	2.05	2.03	2.01	2.00	1.98	1.97	1.96	1.95	1.94	1.93	1.92	1.92	1.91	1.91	1.88
20	2.10	2.08	2.07	2.05	2.04	2.01	1.99	1.98	1.97	1.95	1.93	1.92	1.91	1.91	1.90	1.89	1.88	1.88	1.88	1.84
22	2.05	2.03	2.01	2.00	1.98	1.96	1.94	1.92	1.91	1.89	1.88	1.86	1.86	1.85	1.84	1.83	1.82	1.82	1.82	1.78
24	2.00	1.98	1.97	1.95	1.94	1.91	1.89	1.88	1.86	1.84	1.83	1.82	1.81	1.80	1.79	1.78	1.78	1.77	1.77	1.73
26	1.97	1.95	1.93	1.91	1.90	1.87	1.85	1.84	1.82	1.80	1.79	1.78	1.77	1.76	1.75	1.74	1.73	1.73	1.73	1.69
28	1.93	1.91	1.90	1.88	1.87	1.84	1.82	1.80	1.79	1.77	1.75	1.74	1.73	1.73	1.71	1.71	1.70	1.69	1.69	1.65
30	1.91	1.89	1.87	1.85	1.84	1.81	1.79	1.77	1.76	1.74	1.72	1.71	1.70	1.70	1.68	1.68	1.67	1.66	1.66	1.62
35	1.85	1.83	1.82	1.80	1.79	1.76	1.74	1.72	1.70	1.68	1.66	1.65	1.64	1.63	1.62	1.61	1.61	1.60	1.60	1.56
40	1.81	1.79	1.77	1.76	1.74	1.72	1.69	1.67	1.66	1.64	1.62	1.61	1.60	1.59	1.58	1.57	1.56	1.55	1.55	1.51
45	1.78	1.76	1.74	1.73	1.71	1.68	1.66	1.64	1.63	1.60	1.59	1.57	1.56	1.55	1.54	1.53	1.52	1.52	1.51	1.47
50	1.76	1.74	1.72	1.70	1.69	1.66	1.63	1.61	1.60	1.58	1.56	1.54	1.53	1.52	1.51	1.50	1.49	1.49	1.48	1.44
60	1.72	1.70	1.68	1.66	1.65	1.62	1.59	1.57	1.56	1.53	1.52	1.50	1.49	1.48	1.47	1.46	1.45	1.44	1.44	1.39
70	1.70	1.67	1.65	1.64	1.62	1.59	1.57	1.55	1.53	1.50	1.49	1.47	1.46	1.45	1.44	1.42	1.42	1.41	1.40	1.35
80	1.68	1.65	1.63	1.62	1.60	1.57	1.54	1.52	1.51	1.48	1.46	1.45	1.44	1.43	1.41	1.40	1.39	1.38	1.38	1.33
90	1.66	1.64	1.62	1.60	1.59	1.55	1.53	1.51	1.49	1.46	1.44	1.43	1.42	1.41	1.39	1.38	1.37	1.36	1.36	1.30
100	1.65	1.63	1.61	1.59	1.57	1.54	1.52	1.49	1.48	1.45	1.43	1.41	1.40	1.39	1.38	1.36	1.35	1.35	1.34	1.28
120	1.63	1.61	1.59	1.57	1.55	1.52	1.50	1.47	1.46	1.43	1.41	1.39	1.38	1.37	1.35	1.34	1.33	1.32	1.32	1.26
140	1.62	1.60	1.57	1.56	1.54	1.51	1.48	1.46	1.44	1.41	1.39	1.38	1.36	1.35	1.33	1.32	1.31	1.30	1.30	1.23
160	1.61	1.59	1.57	1.55	1.53	1.50	1.47	1.45	1.43	1.40	1.38	1.36	1.35	1.34	1.32	1.31	1.30	1.29	1.28	1.22
180	1.60	1.58	1.56	1.54	1.52	1.49	1.46	1.44	1.42	1.39	1.37	1.35	1.34	1.33	1.31	1.30	1.29	1.28	1.27	1.20
200	1.60	1.57	1.55	1.53	1.52	1.48	1.46	1.43	1.41	1.39	1.36	1.35	1.33	1.32	1.30	1.29	1.28	1.27	1.26	1.19
∞	1.54	1.52	1.50	1.48	1.46	1.42	1.40	1.37	1.35	1.32	1.29	1.28	1.26	1.25	1.22	1.21	1.19	1.18	1.17	1.00

TABLE 6(b) Values of the F-Distribution: $A = .025$

ν_2 \ ν_1	1	2	3	4	5	6	7	8	9	10	11	12	13	14	15	16	17	18	19	20
	NUMERATOR DEGREES OF FREEDOM																			
1	648	799	864	900	922	937	948	957	963	969	973	977	980	983	985	987	989	990	992	993
2	38.5	39.0	39.2	39.2	39.3	39.3	39.4	39.4	39.4	39.4	39.4	39.4	39.4	39.4	39.4	39.4	39.4	39.4	39.4	39.4
3	17.4	16.0	15.4	15.1	14.9	14.7	14.6	14.5	14.5	14.4	14.4	14.3	14.3	14.3	14.3	14.2	14.2	14.2	14.2	14.2
4	12.2	10.6	10.0	9.60	9.36	9.20	9.07	8.98	8.90	8.84	8.79	8.75	8.71	8.68	8.66	8.63	8.61	8.59	8.58	8.56
5	10.0	8.43	7.76	7.39	7.15	6.98	6.85	6.76	6.68	6.62	6.57	6.52	6.49	6.46	6.43	6.40	6.38	6.36	6.34	6.33
6	8.81	7.26	6.60	6.23	5.99	5.82	5.70	5.60	5.52	5.46	5.41	5.37	5.33	5.30	5.27	5.24	5.22	5.20	5.18	5.17
7	8.07	6.54	5.89	5.52	5.29	5.12	4.99	4.90	4.82	4.76	4.71	4.67	4.63	4.60	4.57	4.54	4.52	4.50	4.48	4.47
8	7.57	6.06	5.42	5.05	4.82	4.65	4.53	4.43	4.36	4.30	4.24	4.20	4.16	4.13	4.10	4.08	4.05	4.03	4.02	4.00
9	7.21	5.71	5.08	4.72	4.48	4.32	4.20	4.10	4.03	3.96	3.91	3.87	3.83	3.80	3.77	3.74	3.72	3.70	3.68	3.67
10	6.94	5.46	4.83	4.47	4.24	4.07	3.95	3.85	3.78	3.72	3.66	3.62	3.58	3.55	3.52	3.50	3.47	3.45	3.44	3.42
11	6.72	5.26	4.63	4.28	4.04	3.88	3.76	3.66	3.59	3.53	3.47	3.43	3.39	3.36	3.33	3.30	3.28	3.26	3.24	3.23
12	6.55	5.10	4.47	4.12	3.89	3.73	3.61	3.51	3.44	3.37	3.32	3.28	3.24	3.21	3.18	3.15	3.13	3.11	3.09	3.07
13	6.41	4.97	4.35	4.00	3.77	3.60	3.48	3.39	3.31	3.25	3.20	3.15	3.12	3.08	3.05	3.03	3.00	2.98	2.96	2.95
14	6.30	4.86	4.24	3.89	3.66	3.50	3.38	3.29	3.21	3.15	3.09	3.05	3.01	2.98	2.95	2.92	2.90	2.88	2.86	2.84
15	6.20	4.77	4.15	3.80	3.58	3.41	3.29	3.20	3.12	3.06	3.01	2.96	2.92	2.89	2.86	2.84	2.81	2.79	2.77	2.76
16	6.12	4.69	4.08	3.73	3.50	3.34	3.22	3.12	3.05	2.99	2.93	2.89	2.85	2.82	2.79	2.76	2.74	2.72	2.70	2.68
17	6.04	4.62	4.01	3.66	3.44	3.28	3.16	3.06	2.98	2.92	2.87	2.82	2.79	2.75	2.72	2.70	2.67	2.65	2.63	2.62
18	5.98	4.56	3.95	3.61	3.38	3.22	3.10	3.01	2.93	2.87	2.81	2.77	2.73	2.70	2.67	2.64	2.62	2.60	2.58	2.56
19	5.92	4.51	3.90	3.56	3.33	3.17	3.05	2.96	2.88	2.82	2.76	2.72	2.68	2.65	2.62	2.59	2.57	2.55	2.53	2.51
20	5.87	4.46	3.86	3.51	3.29	3.13	3.01	2.91	2.84	2.77	2.72	2.68	2.64	2.60	2.57	2.55	2.52	2.50	2.48	2.46
22	5.79	4.38	3.78	3.44	3.22	3.05	2.93	2.84	2.76	2.70	2.65	2.60	2.56	2.53	2.50	2.47	2.45	2.43	2.41	2.39
24	5.72	4.32	3.72	3.38	3.15	2.99	2.87	2.78	2.70	2.64	2.59	2.54	2.50	2.47	2.44	2.41	2.39	2.36	2.35	2.33
26	5.66	4.27	3.67	3.33	3.10	2.94	2.82	2.73	2.65	2.59	2.54	2.49	2.45	2.42	2.39	2.36	2.34	2.31	2.29	2.28
28	5.61	4.22	3.63	3.29	3.06	2.90	2.78	2.69	2.61	2.55	2.49	2.45	2.41	2.37	2.34	2.32	2.29	2.27	2.25	2.23
30	5.57	4.18	3.59	3.25	3.03	2.87	2.75	2.65	2.57	2.51	2.46	2.41	2.37	2.34	2.31	2.28	2.26	2.23	2.21	2.20
35	5.48	4.11	3.52	3.18	2.96	2.80	2.68	2.58	2.50	2.44	2.39	2.34	2.30	2.27	2.23	2.21	2.18	2.16	2.14	2.12
40	5.42	4.05	3.46	3.13	2.90	2.74	2.62	2.53	2.45	2.39	2.33	2.29	2.25	2.21	2.18	2.15	2.13	2.11	2.09	2.07
45	5.38	4.01	3.42	3.09	2.86	2.70	2.58	2.49	2.41	2.35	2.29	2.25	2.21	2.17	2.14	2.11	2.09	2.07	2.04	2.03
50	5.34	3.97	3.39	3.05	2.83	2.67	2.55	2.46	2.38	2.32	2.26	2.22	2.18	2.14	2.11	2.08	2.06	2.03	2.01	1.99
60	5.29	3.93	3.34	3.01	2.79	2.63	2.51	2.41	2.33	2.27	2.22	2.17	2.13	2.09	2.06	2.03	2.01	1.98	1.96	1.94
70	5.25	3.89	3.31	2.97	2.75	2.59	2.47	2.38	2.30	2.24	2.18	2.14	2.10	2.06	2.03	2.00	1.97	1.95	1.93	1.91
80	5.22	3.86	3.28	2.95	2.73	2.57	2.45	2.35	2.28	2.21	2.16	2.11	2.07	2.03	2.00	1.97	1.95	1.92	1.90	1.88
90	5.20	3.84	3.26	2.93	2.71	2.55	2.43	2.34	2.26	2.19	2.14	2.09	2.05	2.02	1.98	1.95	1.93	1.91	1.88	1.86
100	5.18	3.83	3.25	2.92	2.70	2.54	2.42	2.32	2.24	2.18	2.12	2.08	2.04	2.00	1.97	1.94	1.91	1.89	1.87	1.85
120	5.15	3.80	3.23	2.89	2.67	2.52	2.39	2.30	2.22	2.16	2.10	2.05	2.01	1.98	1.94	1.92	1.89	1.87	1.84	1.82
140	5.13	3.79	3.21	2.88	2.66	2.50	2.38	2.28	2.21	2.14	2.09	2.04	2.00	1.96	1.93	1.90	1.87	1.85	1.83	1.81
160	5.12	3.78	3.20	2.87	2.65	2.49	2.37	2.27	2.19	2.13	2.07	2.03	1.99	1.95	1.92	1.89	1.86	1.84	1.82	1.80
180	5.11	3.77	3.19	2.86	2.64	2.48	2.36	2.26	2.19	2.12	2.07	2.02	1.98	1.94	1.91	1.88	1.85	1.83	1.81	1.79
200	5.10	3.76	3.18	2.85	2.63	2.47	2.35	2.26	2.18	2.11	2.06	2.01	1.97	1.93	1.90	1.87	1.84	1.82	1.80	1.78
∞	5.03	3.69	3.12	2.79	2.57	2.41	2.29	2.19	2.11	2.05	1.99	1.95	1.90	1.87	1.83	1.80	1.78	1.75	1.73	1.71

DENOMINATOR DEGREES OF FREEDOM

ν_2 \ ν_1	NUMERATOR DEGREES OF FREEDOM																			
DENOMINATOR DEGREES OF FREEDOM	22	24	26	28	30	35	40	45	50	60	70	80	90	100	120	140	160	180	200	∞
1	995	997	999	1000	1001	1004	1006	1007	1008	1010	1011	1012	1013	1013	1014	1015	1015	1015	1016	1018
2	39.5	39.5	39.5	39.5	39.5	39.5	39.5	39.5	39.5	39.5	39.5	39.5	39.5	39.5	39.5	39.5	39.5	39.5	39.5	39.5
3	14.1	14.1	14.1	14.1	14.1	14.1	14.0	14.0	14.0	14.0	14.0	14.0	14.0	14.0	13.9	13.9	13.9	13.9	13.9	13.9
4	8.53	8.51	8.49	8.48	8.46	8.43	8.41	8.39	8.38	8.36	8.35	8.33	8.33	8.32	8.31	8.30	8.30	8.29	8.29	8.26
5	6.30	6.28	6.26	6.24	6.23	6.20	6.18	6.16	6.14	6.12	6.11	6.10	6.09	6.08	6.07	6.06	6.06	6.05	6.05	6.02
6	5.14	5.12	5.10	5.08	5.07	5.04	5.01	4.99	4.98	4.96	4.94	4.93	4.92	4.92	4.90	4.90	4.89	4.89	4.88	4.85
7	4.44	4.41	4.39	4.38	4.36	4.33	4.31	4.29	4.28	4.25	4.24	4.23	4.22	4.21	4.20	4.19	4.18	4.18	4.18	4.14
8	3.97	3.95	3.93	3.91	3.89	3.86	3.84	3.82	3.81	3.78	3.77	3.76	3.75	3.74	3.73	3.72	3.71	3.71	3.70	3.67
9	3.64	3.61	3.59	3.58	3.56	3.53	3.51	3.49	3.47	3.45	3.43	3.42	3.41	3.40	3.39	3.38	3.38	3.37	3.37	3.33
10	3.39	3.37	3.34	3.33	3.31	3.28	3.26	3.24	3.22	3.20	3.18	3.17	3.16	3.15	3.14	3.13	3.13	3.12	3.12	3.08
11	3.20	3.17	3.15	3.13	3.12	3.09	3.06	3.04	3.03	3.00	2.99	2.97	2.96	2.96	2.94	2.94	2.93	2.92	2.92	2.88
12	3.04	3.02	3.00	2.98	2.96	2.93	2.91	2.89	2.87	2.85	2.83	2.82	2.81	2.80	2.79	2.78	2.77	2.77	2.76	2.73
13	2.92	2.89	2.87	2.85	2.84	2.80	2.78	2.76	2.74	2.72	2.70	2.69	2.68	2.67	2.66	2.65	2.64	2.64	2.63	2.60
14	2.81	2.79	2.77	2.75	2.73	2.70	2.67	2.65	2.64	2.61	2.60	2.58	2.57	2.56	2.55	2.54	2.54	2.53	2.53	2.49
15	2.73	2.70	2.68	2.66	2.64	2.61	2.59	2.56	2.55	2.52	2.51	2.49	2.48	2.47	2.46	2.45	2.44	2.44	2.44	2.40
16	2.65	2.63	2.60	2.58	2.57	2.53	2.51	2.49	2.47	2.45	2.43	2.42	2.40	2.40	2.38	2.37	2.37	2.36	2.36	2.32
17	2.59	2.56	2.54	2.52	2.50	2.47	2.44	2.42	2.41	2.38	2.36	2.35	2.34	2.33	2.32	2.31	2.30	2.29	2.29	2.25
18	2.53	2.50	2.48	2.46	2.44	2.41	2.38	2.36	2.35	2.32	2.30	2.29	2.28	2.27	2.26	2.25	2.24	2.23	2.23	2.19
19	2.48	2.45	2.43	2.41	2.39	2.36	2.33	2.31	2.30	2.27	2.25	2.24	2.23	2.22	2.20	2.19	2.19	2.18	2.18	2.13
20	2.43	2.41	2.39	2.37	2.35	2.31	2.29	2.27	2.25	2.22	2.20	2.19	2.18	2.17	2.16	2.15	2.14	2.13	2.13	2.09
22	2.36	2.33	2.31	2.29	2.27	2.24	2.21	2.19	2.17	2.14	2.13	2.11	2.10	2.09	2.08	2.07	2.06	2.05	2.05	2.00
24	2.30	2.27	2.25	2.23	2.21	2.17	2.15	2.12	2.11	2.08	2.06	2.05	2.03	2.02	2.01	2.00	1.99	1.99	1.98	1.94
26	2.24	2.22	2.19	2.17	2.16	2.12	2.09	2.07	2.05	2.03	2.01	1.99	1.98	1.97	1.95	1.94	1.94	1.93	1.92	1.88
28	2.20	2.17	2.15	2.13	2.11	2.08	2.05	2.03	2.01	1.98	1.96	1.94	1.93	1.92	1.91	1.90	1.89	1.88	1.88	1.83
30	2.16	2.14	2.11	2.09	2.07	2.04	2.01	1.99	1.97	1.94	1.92	1.90	1.89	1.88	1.87	1.86	1.85	1.84	1.84	1.79
35	2.09	2.06	2.04	2.02	2.00	1.96	1.93	1.91	1.89	1.86	1.84	1.82	1.81	1.80	1.79	1.77	1.77	1.76	1.75	1.70
40	2.03	2.01	1.98	1.96	1.94	1.90	1.88	1.85	1.83	1.80	1.78	1.76	1.75	1.74	1.72	1.71	1.70	1.70	1.69	1.64
45	1.99	1.96	1.94	1.92	1.90	1.86	1.83	1.81	1.79	1.76	1.74	1.72	1.70	1.69	1.68	1.66	1.66	1.65	1.64	1.59
50	1.96	1.93	1.91	1.89	1.87	1.83	1.80	1.77	1.75	1.72	1.70	1.68	1.67	1.66	1.64	1.63	1.62	1.61	1.60	1.55
60	1.91	1.88	1.86	1.83	1.82	1.78	1.74	1.72	1.70	1.67	1.64	1.63	1.61	1.60	1.58	1.57	1.56	1.55	1.54	1.48
70	1.88	1.85	1.82	1.80	1.78	1.74	1.71	1.68	1.66	1.63	1.60	1.59	1.57	1.56	1.54	1.53	1.52	1.51	1.50	1.44
80	1.85	1.82	1.79	1.77	1.75	1.71	1.68	1.65	1.63	1.60	1.57	1.55	1.54	1.53	1.51	1.49	1.48	1.47	1.47	1.40
90	1.83	1.80	1.77	1.75	1.73	1.69	1.66	1.63	1.61	1.58	1.55	1.53	1.52	1.50	1.48	1.47	1.46	1.45	1.44	1.37
100	1.81	1.78	1.76	1.74	1.71	1.67	1.64	1.61	1.59	1.56	1.53	1.51	1.50	1.48	1.46	1.45	1.44	1.43	1.42	1.35
120	1.79	1.76	1.73	1.71	1.69	1.65	1.61	1.59	1.56	1.53	1.50	1.48	1.47	1.45	1.43	1.42	1.41	1.40	1.39	1.31
140	1.77	1.74	1.72	1.69	1.67	1.63	1.60	1.57	1.55	1.51	1.48	1.46	1.45	1.43	1.41	1.39	1.38	1.37	1.36	1.28
160	1.76	1.73	1.70	1.68	1.66	1.62	1.58	1.55	1.53	1.50	1.47	1.45	1.43	1.42	1.39	1.38	1.36	1.35	1.35	1.26
180	1.75	1.72	1.69	1.67	1.65	1.61	1.57	1.54	1.52	1.48	1.46	1.43	1.42	1.40	1.38	1.36	1.35	1.34	1.33	1.25
200	1.74	1.71	1.68	1.66	1.64	1.60	1.56	1.53	1.51	1.47	1.45	1.42	1.41	1.39	1.37	1.35	1.34	1.33	1.32	1.23
∞	1.67	1.64	1.61	1.59	1.57	1.52	1.49	1.46	1.43	1.39	1.36	1.33	1.31	1.30	1.27	1.25	1.23	1.22	1.21	1.00

TABLE **6(c)** Values of the F-Distribution: $A = .01$

ν_2 \ ν_1	1	2	3	4	5	6	7	8	9	10	11	12	13	14	15	16	17	18	19	20
DENOMINATOR DEGREES OF FREEDOM	NUMERATOR DEGREES OF FREEDOM																			
1	4052	4999	5403	5625	5764	5859	5928	5981	6022	6056	6083	6106	6126	6143	6157	6170	6181	6192	6201	6209
2	98.5	99.0	99.2	99.2	99.3	99.3	99.4	99.4	99.4	99.4	99.4	99.4	99.4	99.4	99.4	99.4	99.4	99.4	99.4	99.4
3	34.1	30.8	29.5	28.7	28.2	27.9	27.7	27.5	27.3	27.2	27.1	27.1	27.0	26.9	26.9	26.8	26.8	26.8	26.7	26.7
4	21.2	18.0	16.7	16.0	15.5	15.2	15.0	14.8	14.7	14.5	14.5	14.4	14.3	14.2	14.2	14.2	14.1	14.1	14.0	14.0
5	16.3	13.3	12.1	11.4	11.0	10.7	10.5	10.3	10.2	10.1	9.96	9.89	9.82	9.77	9.72	9.68	9.64	9.61	9.58	9.55
6	13.7	10.9	9.78	9.15	8.75	8.47	8.26	8.10	7.98	7.87	7.79	7.72	7.66	7.60	7.56	7.52	7.48	7.45	7.42	7.40
7	12.2	9.55	8.45	7.85	7.46	7.19	6.99	6.84	6.72	6.62	6.54	6.47	6.41	6.36	6.31	6.28	6.24	6.21	6.18	6.16
8	11.3	8.65	7.59	7.01	6.63	6.37	6.18	6.03	5.91	5.81	5.73	5.67	5.61	5.56	5.52	5.48	5.44	5.41	5.38	5.36
9	10.6	8.02	6.99	6.42	6.06	5.80	5.61	5.47	5.35	5.26	5.18	5.11	5.05	5.01	4.96	4.92	4.89	4.86	4.83	4.81
10	10.0	7.56	6.55	5.99	5.64	5.39	5.20	5.06	4.94	4.85	4.77	4.71	4.65	4.60	4.56	4.52	4.49	4.46	4.43	4.41
11	9.65	7.21	6.22	5.67	5.32	5.07	4.89	4.74	4.63	4.54	4.46	4.40	4.34	4.29	4.25	4.21	4.18	4.15	4.12	4.10
12	9.33	6.93	5.95	5.41	5.06	4.82	4.64	4.50	4.39	4.30	4.22	4.16	4.10	4.05	4.01	3.97	3.94	3.91	3.88	3.86
13	9.07	6.70	5.74	5.21	4.86	4.62	4.44	4.30	4.19	4.10	4.02	3.96	3.91	3.86	3.82	3.78	3.75	3.72	3.69	3.66
14	8.86	6.51	5.56	5.04	4.69	4.46	4.28	4.14	4.03	3.94	3.86	3.80	3.75	3.70	3.66	3.62	3.59	3.56	3.53	3.51
15	8.68	6.36	5.42	4.89	4.56	4.32	4.14	4.00	3.89	3.80	3.73	3.67	3.61	3.56	3.52	3.49	3.45	3.42	3.40	3.37
16	8.53	6.23	5.29	4.77	4.44	4.20	4.03	3.89	3.78	3.69	3.62	3.55	3.50	3.45	3.41	3.37	3.34	3.31	3.28	3.26
17	8.40	6.11	5.18	4.67	4.34	4.10	3.93	3.79	3.68	3.59	3.52	3.46	3.40	3.35	3.31	3.27	3.24	3.21	3.19	3.16
18	8.29	6.01	5.09	4.58	4.25	4.01	3.84	3.71	3.60	3.51	3.43	3.37	3.32	3.27	3.23	3.19	3.16	3.13	3.10	3.08
19	8.18	5.93	5.01	4.50	4.17	3.94	3.77	3.63	3.52	3.43	3.36	3.30	3.24	3.19	3.15	3.12	3.08	3.05	3.03	3.00
20	8.10	5.85	4.94	4.43	4.10	3.87	3.70	3.56	3.46	3.37	3.29	3.23	3.18	3.13	3.09	3.05	3.02	2.99	2.96	2.94
22	7.95	5.72	4.82	4.31	3.99	3.76	3.59	3.45	3.35	3.26	3.18	3.12	3.07	3.02	2.98	2.94	2.91	2.88	2.85	2.83
24	7.82	5.61	4.72	4.22	3.90	3.67	3.50	3.36	3.26	3.17	3.09	3.03	2.98	2.93	2.89	2.85	2.82	2.79	2.76	2.74
26	7.72	5.53	4.64	4.14	3.82	3.59	3.42	3.29	3.18	3.09	3.02	2.96	2.90	2.86	2.81	2.78	2.75	2.72	2.69	2.66
28	7.64	5.45	4.57	4.07	3.75	3.53	3.36	3.23	3.12	3.03	2.96	2.90	2.84	2.79	2.75	2.72	2.68	2.65	2.63	2.60
30	7.56	5.39	4.51	4.02	3.70	3.47	3.30	3.17	3.07	2.98	2.91	2.84	2.79	2.74	2.70	2.66	2.63	2.60	2.57	2.55
35	7.42	5.27	4.40	3.91	3.59	3.37	3.20	3.07	2.96	2.88	2.80	2.74	2.69	2.64	2.60	2.56	2.53	2.50	2.47	2.44
40	7.31	5.18	4.31	3.83	3.51	3.29	3.12	2.99	2.89	2.80	2.73	2.66	2.61	2.56	2.52	2.48	2.45	2.42	2.39	2.37
45	7.23	5.11	4.25	3.77	3.45	3.23	3.07	2.94	2.83	2.74	2.67	2.61	2.55	2.51	2.46	2.43	2.39	2.36	2.34	2.31
50	7.17	5.06	4.20	3.72	3.41	3.19	3.02	2.89	2.78	2.70	2.63	2.56	2.51	2.46	2.42	2.38	2.35	2.32	2.29	2.27
60	7.08	4.98	4.13	3.65	3.34	3.12	2.95	2.82	2.72	2.63	2.56	2.50	2.44	2.39	2.35	2.31	2.28	2.25	2.22	2.20
70	7.01	4.92	4.07	3.60	3.29	3.07	2.91	2.78	2.67	2.59	2.51	2.45	2.40	2.35	2.31	2.27	2.23	2.20	2.18	2.15
80	6.96	4.88	4.04	3.56	3.26	3.04	2.87	2.74	2.64	2.55	2.48	2.42	2.36	2.31	2.27	2.23	2.20	2.17	2.14	2.12
90	6.93	4.85	4.01	3.53	3.23	3.01	2.84	2.72	2.61	2.52	2.45	2.39	2.33	2.29	2.24	2.21	2.17	2.14	2.11	2.09
100	6.90	4.82	3.98	3.51	3.21	2.99	2.82	2.69	2.59	2.50	2.43	2.37	2.31	2.27	2.22	2.19	2.15	2.12	2.09	2.07
120	6.85	4.79	3.95	3.48	3.17	2.96	2.79	2.66	2.56	2.47	2.40	2.34	2.28	2.23	2.19	2.15	2.12	2.09	2.06	2.03
140	6.82	4.76	3.92	3.46	3.15	2.93	2.77	2.64	2.54	2.45	2.38	2.31	2.26	2.21	2.17	2.13	2.10	2.07	2.04	2.01
160	6.80	4.74	3.91	3.44	3.13	2.92	2.75	2.62	2.52	2.43	2.36	2.30	2.24	2.20	2.15	2.11	2.08	2.05	2.02	1.99
180	6.78	4.73	3.89	3.43	3.12	2.90	2.74	2.61	2.51	2.42	2.35	2.28	2.23	2.18	2.14	2.10	2.07	2.04	2.01	1.98
200	6.76	4.71	3.88	3.41	3.11	2.89	2.73	2.60	2.50	2.41	2.34	2.27	2.22	2.17	2.13	2.09	2.06	2.03	2.00	1.97
∞	6.64	4.61	3.78	3.32	3.02	2.80	2.64	2.51	2.41	2.32	2.25	2.19	2.13	2.08	2.04	2.00	1.97	1.94	1.91	1.88

ν_2 \ ν_1	NUMERATOR DEGREES OF FREEDOM																			
DENOMINATOR DEGREES OF FREEDOM	22	24	26	28	30	35	40	45	50	60	70	80	90	100	120	140	160	180	200	∞
1	6223	6235	6245	6253	6261	6276	6287	6296	6303	6313	6321	6326	6331	6334	6339	6343	6346	6348	6350	6366
2	99.5	99.5	99.5	99.5	99.5	99.5	99.5	99.5	99.5	99.5	99.5	99.5	99.5	99.5	99.5	99.5	99.5	99.5	99.5	99.5
3	26.6	26.6	26.6	26.5	26.5	26.5	26.4	26.4	26.4	26.3	26.3	26.3	26.3	26.2	26.2	26.2	26.2	26.2	26.2	26.1
4	14.0	13.9	13.9	13.9	13.8	13.8	13.7	13.7	13.7	13.7	13.6	13.6	13.6	13.6	13.6	13.5	13.5	13.5	13.5	13.5
5	9.51	9.47	9.43	9.40	9.38	9.33	9.29	9.26	9.24	9.20	9.18	9.16	9.14	9.13	9.11	9.10	9.09	9.08	9.08	9.02
6	7.35	7.31	7.28	7.25	7.23	7.18	7.14	7.11	7.09	7.06	7.03	7.01	7.00	6.99	6.97	6.96	6.95	6.94	6.93	6.88
7	6.11	6.07	6.04	6.02	5.99	5.94	5.91	5.88	5.86	5.82	5.80	5.78	5.77	5.75	5.74	5.72	5.72	5.71	5.70	5.65
8	5.32	5.28	5.25	5.22	5.20	5.15	5.12	5.09	5.07	5.03	5.01	4.99	4.97	4.96	4.95	4.93	4.92	4.92	4.91	4.86
9	4.77	4.73	4.70	4.67	4.65	4.60	4.57	4.54	4.52	4.48	4.46	4.44	4.43	4.41	4.40	4.39	4.38	4.37	4.36	4.31
10	4.36	4.33	4.30	4.27	4.25	4.20	4.17	4.14	4.12	4.08	4.06	4.04	4.03	4.01	4.00	3.98	3.97	3.97	3.96	3.91
11	4.06	4.02	3.99	3.96	3.94	3.89	3.86	3.83	3.81	3.78	3.75	3.73	3.72	3.71	3.69	3.68	3.67	3.66	3.66	3.60
12	3.82	3.78	3.75	3.72	3.70	3.65	3.62	3.59	3.57	3.54	3.51	3.49	3.48	3.47	3.45	3.44	3.43	3.42	3.41	3.36
13	3.62	3.59	3.56	3.53	3.51	3.46	3.43	3.40	3.38	3.34	3.32	3.30	3.28	3.27	3.25	3.24	3.23	3.23	3.22	3.17
14	3.46	3.43	3.40	3.37	3.35	3.30	3.27	3.24	3.22	3.18	3.16	3.14	3.12	3.11	3.09	3.08	3.07	3.06	3.06	3.01
15	3.33	3.29	3.26	3.24	3.21	3.17	3.13	3.10	3.08	3.05	3.02	3.00	2.99	2.98	2.96	2.95	2.94	2.93	2.92	2.87
16	3.22	3.18	3.15	3.12	3.10	3.05	3.02	2.99	2.97	2.93	2.91	2.89	2.87	2.86	2.84	2.83	2.82	2.81	2.81	2.75
17	3.12	3.08	3.05	3.03	3.00	2.96	2.92	2.89	2.87	2.83	2.81	2.79	2.78	2.76	2.75	2.73	2.72	2.72	2.71	2.65
18	3.03	3.00	2.97	2.94	2.92	2.87	2.84	2.81	2.78	2.75	2.72	2.70	2.69	2.68	2.66	2.65	2.64	2.63	2.62	2.57
19	2.96	2.92	2.89	2.87	2.84	2.80	2.76	2.73	2.71	2.67	2.65	2.63	2.61	2.60	2.58	2.57	2.56	2.55	2.55	2.49
20	2.90	2.86	2.83	2.80	2.78	2.73	2.69	2.67	2.64	2.61	2.58	2.56	2.55	2.54	2.52	2.50	2.49	2.49	2.48	2.42
22	2.78	2.75	2.72	2.69	2.67	2.62	2.58	2.55	2.53	2.50	2.47	2.45	2.43	2.42	2.40	2.39	2.38	2.37	2.36	2.31
24	2.70	2.66	2.63	2.60	2.58	2.53	2.49	2.46	2.44	2.40	2.38	2.36	2.34	2.33	2.31	2.30	2.29	2.28	2.27	2.21
26	2.62	2.58	2.55	2.53	2.50	2.45	2.42	2.39	2.36	2.33	2.30	2.28	2.26	2.25	2.23	2.22	2.21	2.20	2.19	2.13
28	2.56	2.52	2.49	2.46	2.44	2.39	2.35	2.32	2.30	2.26	2.24	2.22	2.20	2.19	2.17	2.15	2.14	2.13	2.13	2.07
30	2.51	2.47	2.44	2.41	2.39	2.34	2.30	2.27	2.25	2.21	2.18	2.16	2.14	2.13	2.11	2.10	2.09	2.08	2.07	2.01
35	2.40	2.36	2.33	2.30	2.28	2.23	2.19	2.16	2.14	2.10	2.07	2.05	2.03	2.02	2.00	1.98	1.97	1.96	1.96	1.89
40	2.33	2.29	2.26	2.23	2.20	2.15	2.11	2.08	2.06	2.02	1.99	1.97	1.95	1.94	1.92	1.90	1.89	1.88	1.87	1.81
45	2.27	2.23	2.20	2.17	2.14	2.09	2.05	2.02	2.00	1.96	1.93	1.91	1.89	1.88	1.85	1.84	1.83	1.82	1.81	1.74
50	2.22	2.18	2.15	2.12	2.10	2.05	2.01	1.97	1.95	1.91	1.88	1.86	1.84	1.82	1.80	1.79	1.77	1.76	1.76	1.68
60	2.15	2.12	2.08	2.05	2.03	1.98	1.94	1.90	1.88	1.84	1.81	1.78	1.76	1.75	1.73	1.71	1.70	1.69	1.68	1.60
70	2.11	2.07	2.03	2.01	1.98	1.93	1.89	1.85	1.83	1.78	1.75	1.73	1.71	1.70	1.67	1.65	1.64	1.63	1.62	1.54
80	2.07	2.03	2.00	1.97	1.94	1.89	1.85	1.82	1.79	1.75	1.71	1.69	1.67	1.65	1.63	1.61	1.60	1.59	1.58	1.50
90	2.04	2.00	1.97	1.94	1.92	1.86	1.82	1.79	1.76	1.72	1.68	1.66	1.64	1.62	1.60	1.58	1.57	1.55	1.55	1.46
100	2.02	1.98	1.95	1.92	1.89	1.84	1.80	1.76	1.74	1.69	1.66	1.63	1.61	1.60	1.57	1.55	1.54	1.53	1.52	1.43
120	1.99	1.95	1.92	1.89	1.86	1.81	1.76	1.73	1.70	1.66	1.62	1.60	1.58	1.56	1.53	1.51	1.50	1.49	1.48	1.38
140	1.97	1.93	1.89	1.86	1.84	1.78	1.74	1.70	1.67	1.63	1.60	1.57	1.55	1.53	1.50	1.48	1.47	1.46	1.45	1.35
160	1.95	1.91	1.88	1.85	1.82	1.76	1.72	1.68	1.66	1.61	1.58	1.55	1.53	1.51	1.48	1.46	1.45	1.43	1.42	1.32
180	1.94	1.90	1.86	1.83	1.81	1.75	1.71	1.67	1.64	1.60	1.56	1.53	1.51	1.49	1.47	1.45	1.43	1.42	1.41	1.30
200	1.93	1.89	1.85	1.82	1.79	1.74	1.69	1.66	1.63	1.58	1.55	1.52	1.50	1.48	1.45	1.43	1.42	1.40	1.39	1.28
∞	1.83	1.79	1.76	1.73	1.70	1.64	1.59	1.56	1.53	1.48	1.44	1.41	1.38	1.36	1.33	1.30	1.28	1.26	1.25	1.00

TABLE 6(d) Values of the F-Distribution: $A = .005$

ν_2 \ ν_1 (NUMERATOR DEGREES OF FREEDOM)	1	2	3	4	5	6	7	8	9	10	11	12	13	14	15	16	17	18	19	20
1	16211	19999	21615	22500	23056	23437	23715	23925	24091	24224	24334	24426	24505	24572	24630	24681	24727	24767	24803	24836
2	199	199	199	199	199	199	199	199	199	199	199	199	199	199	199	199	199	199	199	199
3	55.6	49.8	47.5	46.2	45.4	44.8	44.4	44.1	43.9	43.7	43.5	43.4	43.3	43.2	43.1	43.0	42.9	42.9	42.8	42.8
4	31.3	26.3	24.3	23.2	22.5	22.0	21.6	21.4	21.1	21.0	20.8	20.7	20.6	20.5	20.4	20.4	20.3	20.3	20.2	20.2
5	22.8	18.3	16.5	15.6	14.9	14.5	14.2	14.0	13.8	13.6	13.5	13.4	13.3	13.2	13.1	13.1	13.0	13.0	12.9	12.9
6	18.6	14.5	12.9	12.0	11.5	11.1	10.8	10.6	10.4	10.3	10.1	10.0	9.95	9.88	9.81	9.76	9.71	9.66	9.62	9.59
7	16.2	12.4	10.9	10.1	9.52	9.16	8.89	8.68	8.51	8.38	8.27	8.18	8.10	8.03	7.97	7.91	7.87	7.83	7.79	7.75
8	14.7	11.0	9.60	8.81	8.30	7.95	7.69	7.50	7.34	7.21	7.10	7.01	6.94	6.87	6.81	6.76	6.72	6.68	6.64	6.61
9	13.6	10.1	8.72	7.96	7.47	7.13	6.88	6.69	6.54	6.42	6.31	6.23	6.15	6.09	6.03	5.98	5.94	5.90	5.86	5.83
10	12.8	9.43	8.08	7.34	6.87	6.54	6.30	6.12	5.97	5.85	5.75	5.66	5.59	5.53	5.47	5.42	5.38	5.34	5.31	5.27
11	12.2	8.91	7.60	6.88	6.42	6.10	5.86	5.68	5.54	5.42	5.32	5.24	5.16	5.10	5.05	5.00	4.96	4.92	4.89	4.86
12	11.8	8.51	7.23	6.52	6.07	5.76	5.52	5.35	5.20	5.09	4.99	4.91	4.84	4.77	4.72	4.67	4.63	4.59	4.56	4.53
13	11.4	8.19	6.93	6.23	5.79	5.48	5.25	5.08	4.94	4.82	4.72	4.64	4.57	4.51	4.46	4.41	4.37	4.33	4.30	4.27
14	11.1	7.92	6.68	6.00	5.56	5.26	5.03	4.86	4.72	4.60	4.51	4.43	4.36	4.30	4.25	4.20	4.16	4.12	4.09	4.06
15	10.8	7.70	6.48	5.80	5.37	5.07	4.85	4.67	4.54	4.42	4.33	4.25	4.18	4.12	4.07	4.02	3.98	3.95	3.91	3.88
16	10.6	7.51	6.30	5.64	5.21	4.91	4.69	4.52	4.38	4.27	4.18	4.10	4.03	3.97	3.92	3.87	3.83	3.80	3.76	3.73
17	10.4	7.35	6.16	5.50	5.07	4.78	4.56	4.39	4.25	4.14	4.05	3.97	3.90	3.84	3.79	3.75	3.71	3.67	3.64	3.61
18	10.2	7.21	6.03	5.37	4.96	4.66	4.44	4.28	4.14	4.03	3.94	3.86	3.79	3.73	3.68	3.64	3.60	3.56	3.53	3.50
19	10.1	7.09	5.92	5.27	4.85	4.56	4.34	4.18	4.04	3.93	3.84	3.76	3.70	3.64	3.59	3.54	3.50	3.46	3.43	3.40
20	9.94	6.99	5.82	5.17	4.76	4.47	4.26	4.09	3.96	3.85	3.76	3.68	3.61	3.55	3.50	3.46	3.42	3.38	3.35	3.32
22	9.73	6.81	5.65	5.02	4.61	4.32	4.11	3.94	3.81	3.70	3.61	3.54	3.47	3.41	3.36	3.31	3.27	3.24	3.21	3.18
24	9.55	6.66	5.52	4.89	4.49	4.20	3.99	3.83	3.69	3.59	3.50	3.42	3.35	3.30	3.25	3.20	3.16	3.12	3.09	3.06
26	9.41	6.54	5.41	4.79	4.38	4.10	3.89	3.73	3.60	3.49	3.40	3.33	3.26	3.20	3.15	3.11	3.07	3.03	3.00	2.97
28	9.28	6.44	5.32	4.70	4.30	4.02	3.81	3.65	3.52	3.41	3.32	3.25	3.18	3.12	3.07	3.03	2.99	2.95	2.92	2.89
30	9.18	6.35	5.24	4.62	4.23	3.95	3.74	3.58	3.45	3.34	3.25	3.18	3.11	3.06	3.01	2.96	2.92	2.89	2.85	2.82
35	8.98	6.19	5.09	4.48	4.09	3.81	3.61	3.45	3.32	3.21	3.12	3.05	2.98	2.93	2.88	2.83	2.79	2.76	2.72	2.69
40	8.83	6.07	4.98	4.37	3.99	3.71	3.51	3.35	3.22	3.12	3.03	2.95	2.89	2.83	2.78	2.74	2.70	2.66	2.63	2.60
45	8.71	5.97	4.89	4.29	3.91	3.64	3.43	3.28	3.15	3.04	2.96	2.88	2.82	2.76	2.71	2.66	2.62	2.59	2.56	2.53
50	8.63	5.90	4.83	4.23	3.85	3.58	3.38	3.22	3.09	2.99	2.90	2.82	2.76	2.70	2.65	2.61	2.57	2.53	2.50	2.47
60	8.49	5.79	4.73	4.14	3.76	3.49	3.29	3.13	3.01	2.90	2.82	2.74	2.68	2.62	2.57	2.53	2.49	2.45	2.42	2.39
70	8.40	5.72	4.66	4.08	3.70	3.43	3.23	3.08	2.95	2.85	2.76	2.68	2.62	2.56	2.51	2.47	2.43	2.39	2.36	2.33
80	8.33	5.67	4.61	4.03	3.65	3.39	3.19	3.03	2.91	2.80	2.72	2.64	2.58	2.52	2.47	2.43	2.39	2.35	2.32	2.29
90	8.28	5.62	4.57	3.99	3.62	3.35	3.15	3.00	2.87	2.77	2.68	2.61	2.54	2.49	2.44	2.39	2.35	2.32	2.28	2.25
100	8.24	5.59	4.54	3.96	3.59	3.33	3.13	2.97	2.85	2.74	2.66	2.58	2.52	2.46	2.41	2.37	2.33	2.29	2.26	2.23
120	8.18	5.54	4.50	3.92	3.55	3.28	3.09	2.93	2.81	2.71	2.62	2.54	2.48	2.42	2.37	2.33	2.29	2.25	2.22	2.19
140	8.14	5.50	4.47	3.89	3.52	3.26	3.06	2.91	2.78	2.68	2.59	2.52	2.45	2.40	2.35	2.30	2.26	2.22	2.19	2.16
160	8.10	5.48	4.44	3.87	3.50	3.24	3.04	2.88	2.76	2.66	2.57	2.50	2.43	2.38	2.33	2.28	2.24	2.20	2.17	2.14
180	8.08	5.46	4.42	3.85	3.48	3.22	3.02	2.87	2.74	2.64	2.56	2.48	2.42	2.36	2.31	2.26	2.22	2.19	2.15	2.12
200	8.06	5.44	4.41	3.84	3.47	3.21	3.01	2.86	2.73	2.63	2.54	2.47	2.40	2.35	2.30	2.25	2.21	2.18	2.14	2.11
∞	7.88	5.30	4.28	3.72	3.35	3.09	2.90	2.75	2.62	2.52	2.43	2.36	2.30	2.24	2.19	2.14	2.10	2.07	2.03	2.00

DENOMINATOR DEGREES OF FREEDOM (ν_2)

ν_2 \ ν_1	NUMERATOR DEGREES OF FREEDOM																			
DENOMINATOR DEGREES OF FREEDOM	22	24	26	28	30	35	40	45	50	60	70	80	90	100	120	140	160	180	200	∞
1	24892	24940	24980	25014	25044	25103	25148	25183	25211	25253	25283	25306	25323	25337	25359	25374	25385	25394	25401	25464
2	199	199	199	199	199	199	199	199	199	199	199	199	199	199	199	199	199	199	199	199
3	42.7	42.6	42.6	42.5	42.5	42.4	42.3	42.3	42.2	42.1	42.1	42.1	42.0	42.0	42.0	42.0	41.9	41.9	41.9	41.8
4	20.1	20.0	20.0	19.9	19.9	19.8	19.8	19.7	19.7	19.6	19.6	19.5	19.5	19.5	19.5	19.4	19.4	19.4	19.4	19.3
5	12.8	12.8	12.7	12.7	12.7	12.6	12.5	12.5	12.5	12.4	12.4	12.3	12.3	12.3	12.3	12.3	12.2	12.2	12.2	12.1
6	9.53	9.47	9.43	9.39	9.36	9.29	9.24	9.20	9.17	9.12	9.09	9.06	9.04	9.03	9.00	8.98	8.97	8.96	8.95	8.88
7	7.69	7.64	7.60	7.57	7.53	7.47	7.42	7.38	7.35	7.31	7.28	7.25	7.23	7.22	7.19	7.18	7.16	7.15	7.15	7.08
8	6.55	6.50	6.46	6.43	6.40	6.33	6.29	6.25	6.22	6.18	6.15	6.12	6.10	6.09	6.06	6.05	6.04	6.03	6.02	5.95
9	5.78	5.73	5.69	5.65	5.62	5.56	5.52	5.48	5.45	5.41	5.38	5.36	5.34	5.32	5.30	5.28	5.27	5.26	5.26	5.19
10	5.22	5.17	5.13	5.10	5.07	5.01	4.97	4.93	4.90	4.86	4.83	4.80	4.79	4.77	4.75	4.73	4.72	4.71	4.71	4.64
11	4.80	4.76	4.72	4.68	4.65	4.60	4.55	4.52	4.49	4.45	4.41	4.39	4.37	4.36	4.34	4.32	4.31	4.30	4.29	4.23
12	4.48	4.43	4.39	4.36	4.33	4.27	4.23	4.19	4.17	4.12	4.09	4.07	4.05	4.04	4.01	4.00	3.99	3.98	3.97	3.91
13	4.22	4.17	4.13	4.10	4.07	4.01	3.97	3.94	3.91	3.87	3.84	3.81	3.79	3.78	3.76	3.74	3.73	3.72	3.71	3.65
14	4.01	3.96	3.92	3.89	3.86	3.80	3.76	3.73	3.70	3.66	3.62	3.60	3.58	3.57	3.55	3.53	3.52	3.51	3.50	3.44
15	3.83	3.79	3.75	3.72	3.69	3.63	3.58	3.55	3.52	3.48	3.45	3.43	3.41	3.39	3.37	3.36	3.34	3.34	3.33	3.26
16	3.68	3.64	3.60	3.57	3.54	3.48	3.44	3.40	3.37	3.33	3.30	3.28	3.26	3.25	3.22	3.21	3.20	3.19	3.18	3.11
17	3.56	3.51	3.47	3.44	3.41	3.35	3.31	3.28	3.25	3.21	3.18	3.15	3.13	3.12	3.10	3.08	3.07	3.06	3.05	2.99
18	3.45	3.40	3.36	3.33	3.30	3.25	3.20	3.17	3.14	3.10	3.07	3.04	3.02	3.01	2.99	2.97	2.96	2.95	2.94	2.87
19	3.35	3.31	3.27	3.24	3.21	3.15	3.11	3.07	3.04	3.00	2.97	2.95	2.93	2.91	2.89	2.87	2.86	2.85	2.85	2.78
20	3.27	3.22	3.18	3.15	3.12	3.07	3.02	2.99	2.96	2.92	2.88	2.86	2.84	2.83	2.81	2.79	2.78	2.77	2.76	2.69
22	3.12	3.08	3.04	3.01	2.98	2.92	2.88	2.84	2.82	2.77	2.74	2.72	2.70	2.69	2.66	2.65	2.63	2.62	2.62	2.55
24	3.01	2.97	2.93	2.90	2.87	2.81	2.77	2.73	2.70	2.66	2.63	2.60	2.58	2.57	2.55	2.53	2.52	2.51	2.50	2.43
26	2.92	2.87	2.84	2.80	2.77	2.72	2.67	2.64	2.61	2.56	2.53	2.51	2.49	2.47	2.45	2.43	2.42	2.41	2.40	2.33
28	2.84	2.79	2.76	2.72	2.69	2.64	2.59	2.56	2.53	2.48	2.45	2.43	2.41	2.39	2.37	2.35	2.34	2.33	2.32	2.25
30	2.77	2.73	2.69	2.66	2.63	2.57	2.52	2.49	2.46	2.42	2.38	2.36	2.34	2.32	2.30	2.28	2.27	2.26	2.25	2.18
35	2.64	2.60	2.56	2.53	2.50	2.44	2.39	2.36	2.33	2.28	2.25	2.22	2.20	2.19	2.16	2.15	2.13	2.12	2.11	2.04
40	2.55	2.50	2.46	2.43	2.40	2.34	2.30	2.26	2.23	2.18	2.15	2.12	2.10	2.09	2.06	2.05	2.03	2.02	2.01	1.93
45	2.47	2.43	2.39	2.36	2.33	2.27	2.22	2.19	2.16	2.11	2.08	2.05	2.03	2.01	1.99	1.97	1.95	1.94	1.93	1.85
50	2.42	2.37	2.33	2.30	2.27	2.21	2.16	2.13	2.10	2.05	2.02	1.99	1.97	1.95	1.93	1.91	1.89	1.88	1.87	1.79
60	2.33	2.29	2.25	2.22	2.19	2.13	2.08	2.04	2.01	1.96	1.93	1.90	1.88	1.86	1.83	1.81	1.80	1.79	1.78	1.69
70	2.28	2.23	2.19	2.16	2.13	2.07	2.02	1.98	1.95	1.90	1.86	1.84	1.81	1.80	1.77	1.75	1.73	1.72	1.71	1.62
80	2.23	2.19	2.15	2.11	2.08	2.02	1.97	1.94	1.90	1.85	1.82	1.79	1.77	1.75	1.72	1.70	1.68	1.67	1.66	1.57
90	2.20	2.15	2.12	2.08	2.05	1.99	1.94	1.90	1.87	1.82	1.78	1.75	1.73	1.71	1.68	1.66	1.64	1.63	1.62	1.52
100	2.17	2.13	2.09	2.05	2.02	1.96	1.91	1.87	1.84	1.79	1.75	1.72	1.70	1.68	1.65	1.63	1.61	1.60	1.59	1.49
120	2.13	2.09	2.05	2.01	1.98	1.92	1.87	1.83	1.80	1.75	1.71	1.68	1.66	1.64	1.61	1.58	1.57	1.55	1.54	1.43
140	2.11	2.06	2.02	1.99	1.96	1.89	1.84	1.80	1.77	1.72	1.68	1.65	1.62	1.60	1.57	1.55	1.53	1.52	1.51	1.39
160	2.09	2.04	2.00	1.97	1.93	1.87	1.82	1.78	1.75	1.69	1.65	1.62	1.60	1.58	1.55	1.52	1.51	1.49	1.48	1.36
180	2.07	2.02	1.98	1.95	1.92	1.85	1.80	1.76	1.73	1.68	1.64	1.61	1.58	1.56	1.53	1.50	1.49	1.47	1.46	1.34
200	2.06	2.01	1.97	1.94	1.91	1.84	1.79	1.75	1.71	1.66	1.62	1.59	1.56	1.54	1.51	1.49	1.47	1.45	1.44	1.32
∞	1.95	1.90	1.86	1.82	1.79	1.72	1.67	1.63	1.59	1.54	1.49	1.46	1.43	1.40	1.37	1.34	1.31	1.30	1.28	1.00

TABLE **7(a)** Critical Values of the Studentized Range, $\alpha = .05$

ν	k: 2	3	4	5	6	7	8	9	10	11	12	13	14	15	16	17	18	19	20
1	18.0	27.0	32.8	37.1	40.4	43.1	45.4	47.4	49.1	50.6	52.0	53.2	54.3	55.4	56.3	57.2	58.0	58.8	59.6
2	6.08	8.33	9.80	10.9	11.7	12.4	13.0	13.5	14.0	14.4	14.7	15.1	15.4	15.7	15.9	16.1	16.4	16.6	16.8
3	4.50	5.91	6.82	7.50	8.04	8.48	8.85	9.18	9.46	9.72	9.95	10.2	10.3	10.5	10.7	10.8	11.0	11.1	11.2
4	3.93	5.04	5.76	6.29	6.71	7.05	7.35	7.60	7.83	8.03	8.21	8.37	8.52	8.66	8.79	8.91	9.03	9.13	9.23
5	3.64	4.60	5.22	5.67	6.03	6.33	6.58	6.80	6.99	7.17	7.32	7.47	7.60	7.72	7.83	7.93	8.03	8.12	8.21
6	3.46	4.34	4.90	5.30	5.63	5.90	6.12	6.32	6.49	6.65	6.79	6.92	7.03	7.14	7.24	7.34	7.43	7.51	7.59
7	3.34	4.16	4.68	5.06	5.36	5.61	5.82	6.00	6.16	6.30	6.43	6.55	6.66	6.76	6.85	6.94	7.02	7.10	7.17
8	3.26	4.04	4.53	4.89	5.17	5.40	5.60	5.77	5.92	6.05	6.18	6.29	6.39	6.48	6.57	6.65	6.73	6.80	6.87
9	3.20	3.95	4.41	4.76	5.02	5.24	5.43	5.59	5.74	5.87	5.98	6.09	6.19	6.28	6.36	6.44	6.51	6.58	6.64
10	3.15	3.88	4.33	4.65	4.91	5.12	5.30	5.46	5.60	5.72	5.83	5.93	6.03	6.11	6.19	6.27	6.34	6.40	6.47
11	3.11	3.82	4.26	4.57	4.82	5.03	5.20	5.35	5.49	5.61	5.71	5.81	5.90	5.98	6.06	6.13	6.20	6.27	6.33
12	3.08	3.77	4.20	4.51	4.75	4.95	5.12	5.27	5.39	5.51	5.61	5.71	5.80	5.88	5.95	6.02	6.09	6.15	6.21
13	3.06	3.73	4.15	4.45	4.69	4.88	5.05	5.19	5.32	5.43	5.53	5.63	5.71	5.79	5.86	5.93	5.99	6.05	6.11
14	3.03	3.70	4.11	4.41	4.64	4.83	4.99	5.13	5.25	5.36	5.46	5.55	5.64	5.71	5.79	5.85	5.91	5.97	6.03
15	3.01	3.67	4.08	4.37	4.59	4.78	4.94	5.08	5.20	5.31	5.40	5.49	5.57	5.65	5.72	5.78	5.85	5.90	5.96
16	3.00	3.65	4.05	4.33	4.56	4.74	4.90	5.03	5.15	5.26	5.35	5.44	5.52	5.59	5.66	5.73	5.79	5.84	5.90
17	2.98	3.63	4.02	4.30	4.52	4.70	4.86	4.99	5.11	5.21	5.31	5.39	5.47	5.54	5.61	5.67	5.73	5.79	5.84
18	2.97	3.61	4.00	4.28	4.49	4.67	4.82	4.96	5.07	5.17	5.27	5.35	5.43	5.50	5.57	5.63	5.69	5.74	5.79
19	2.96	3.59	3.98	4.25	4.47	4.65	4.79	4.92	5.04	5.14	5.23	5.31	5.39	5.46	5.53	5.59	5.65	5.70	5.75
20	2.95	3.58	3.96	4.23	4.45	4.62	4.77	4.90	5.01	5.11	5.20	5.28	5.36	5.43	5.49	5.55	5.61	5.66	5.71
24	2.92	3.53	3.90	4.17	4.37	4.54	4.68	4.81	4.92	5.01	5.10	5.18	5.25	5.32	5.38	5.44	5.49	5.55	5.59
30	2.89	3.49	3.85	4.10	4.30	4.46	4.60	4.72	4.82	4.92	5.00	5.08	5.15	5.21	5.27	5.33	5.38	5.43	5.47
40	2.86	3.44	3.79	4.04	4.23	4.39	4.52	4.63	4.73	4.82	4.90	4.98	5.04	5.11	5.16	5.22	5.27	5.31	5.36
60	2.83	3.40	3.74	3.98	4.16	4.31	4.44	4.55	4.65	4.73	4.81	4.88	4.94	5.00	5.06	5.11	5.15	5.20	5.24
120	2.80	3.36	3.68	3.92	4.10	4.24	4.36	4.47	4.56	4.64	4.71	4.78	4.84	4.90	4.95	5.00	5.04	5.09	5.13
∞	2.77	3.31	3.63	3.86	4.03	4.17	4.29	4.39	4.47	4.55	4.62	4.68	4.74	4.80	4.85	4.89	4.93	4.97	5.01

TABLE **7(b)** Critical Values of the Studentized Range, $\alpha = .01$

ν	k																		
	2	3	4	5	6	7	8	9	10	11	12	13	14	15	16	17	18	19	20
1	90.0	135	164	186	202	216	227	237	246	253	260	266	272	277	282	286	290	294	298
2	14.0	19.0	22.3	24.7	26.6	28.2	29.5	30.7	31.7	32.6	33.4	34.1	34.8	35.4	36.0	36.5	37.0	37.5	37.9
3	8.26	10.6	12.2	13.3	14.2	15.0	15.6	16.2	16.7	17.1	17.5	17.9	18.2	18.5	18.8	19.1	19.3	19.5	19.8
4	6.51	8.12	9.17	9.96	10.6	11.1	11.5	11.9	12.3	12.6	12.8	13.1	13.3	13.5	13.7	13.9	14.1	14.2	14.4
5	5.70	6.97	7.80	8.42	8.91	9.32	9.67	9.97	10.2	10.5	10.7	10.9	11.1	11.2	11.4	11.6	11.7	11.8	11.9
6	5.24	6.33	7.03	7.56	7.97	8.32	8.61	8.87	9.10	9.30	9.49	9.65	9.81	9.95	10.1	10.2	10.3	10.4	10.5
7	4.95	5.92	6.54	7.01	7.37	7.68	7.94	8.17	8.37	8.55	8.71	8.86	9.00	9.12	9.24	9.35	9.46	9.55	9.65
8	4.74	5.63	6.20	6.63	6.96	7.24	7.47	7.68	7.87	8.03	8.18	8.31	8.44	8.55	8.66	8.76	8.85	8.94	9.03
9	4.60	5.43	5.96	6.35	6.66	6.91	7.13	7.32	7.49	7.65	7.78	7.91	8.03	8.13	8.23	8.32	8.41	8.49	8.57
10	4.48	5.27	5.77	6.14	6.43	6.67	6.87	7.05	7.21	7.36	7.48	7.60	7.71	7.81	7.91	7.99	8.07	8.15	8.22
11	4.39	5.14	5.62	5.97	6.25	6.48	6.67	6.84	6.99	7.13	7.25	7.36	7.46	7.56	7.65	7.73	7.81	7.88	7.95
12	4.32	5.04	5.50	5.84	6.10	6.32	6.51	6.67	6.81	6.94	7.06	7.17	7.26	7.36	7.44	7.52	7.59	7.66	7.73
13	4.26	4.96	5.40	5.73	5.98	6.19	6.37	6.53	6.67	6.79	6.90	7.01	7.10	7.19	7.27	7.34	7.42	7.48	7.55
14	4.21	4.89	5.32	5.63	5.88	6.08	6.26	6.41	6.54	6.66	6.77	6.87	6.96	7.05	7.12	7.20	7.27	7.33	7.39
15	4.17	4.83	5.25	5.56	5.80	5.99	6.16	6.31	6.44	6.55	6.66	6.76	6.84	6.93	7.00	7.07	7.14	7.20	7.26
16	4.13	4.78	5.19	5.49	5.72	5.92	6.08	6.22	6.35	6.46	6.56	6.66	6.74	6.82	6.90	6.97	7.03	7.09	7.15
17	4.10	4.74	5.14	5.43	5.66	5.85	6.01	6.15	6.27	6.38	6.48	6.57	6.66	6.73	6.80	6.87	6.94	7.00	7.05
18	4.07	4.70	5.09	5.38	5.60	5.79	5.94	6.08	6.20	6.31	6.41	6.50	6.58	6.65	6.72	6.79	6.85	6.91	6.96
19	4.05	4.67	5.05	5.33	5.55	5.73	5.89	6.02	6.14	6.25	6.34	6.43	6.51	6.58	6.65	6.72	6.78	6.84	6.89
20	4.02	4.64	5.02	5.29	5.51	5.69	5.84	5.97	6.09	6.19	6.29	6.37	6.45	6.52	6.59	6.65	6.71	6.76	6.82
24	3.96	4.54	4.91	5.17	5.37	5.54	5.69	5.81	5.92	6.02	6.11	6.19	6.26	6.33	6.39	6.45	6.51	6.56	6.61
30	3.89	4.45	4.80	5.05	5.24	5.40	5.54	5.65	5.76	5.85	5.93	6.01	6.08	6.14	6.20	6.26	6.31	6.36	6.41
40	3.82	4.37	4.70	4.93	5.11	5.27	5.39	5.50	5.60	5.69	5.77	5.84	5.90	5.96	6.02	6.07	6.12	6.17	6.21
60	3.76	4.28	4.60	4.82	4.99	5.13	5.25	5.36	5.45	5.53	5.60	5.67	5.73	5.79	5.84	5.89	5.93	5.98	6.02
120	3.70	4.20	4.50	4.71	4.87	5.01	5.12	5.21	5.30	5.38	5.44	5.51	5.56	5.61	5.66	5.71	5.75	5.79	5.83
∞	3.64	4.12	4.40	4.60	4.76	4.88	4.99	5.08	5.16	5.23	5.29	5.35	5.40	5.45	5.49	5.54	5.57	5.61	5.65

SOURCE: From E. S. Pearson and H. O. Hartley, *Biometrika Tables for Statisticians*, 1: 176–77. Reproduced by permission of the Biometrika Trustees.

TABLE **8(a)** Critical Values for the Durbin-Watson Statistic, $\alpha = .05$

	$k = 1$		$k = 2$		$k = 3$		$k = 4$		$k = 5$	
n	d_L	d_U	d_L	d_U	d_L	d_U	d_L	d_U	d_L	d_U
15	1.08	1.36	.95	1.54	.82	1.75	.69	1.97	.56	2.21
16	1.10	1.37	.98	1.54	.86	1.73	.74	1.93	.62	2.15
17	1.13	1.38	1.02	1.54	.90	1.71	.78	1.90	.67	2.10
18	1.16	1.39	1.05	1.53	.93	1.69	.82	1.87	.71	2.06
19	1.18	1.40	1.08	1.53	.97	1.68	.86	1.85	.75	2.02
20	1.20	1.41	1.10	1.54	1.00	1.68	.90	1.83	.79	1.99
21	1.22	1.42	1.13	1.54	1.03	1.67	.93	1.81	.83	1.96
22	1.24	1.43	1.15	1.54	1.05	1.66	.96	1.80	.86	1.94
23	1.26	1.44	1.17	1.54	1.08	1.66	.99	1.79	.90	1.92
24	1.27	1.45	1.19	1.55	1.10	1.66	1.01	1.78	.93	1.90
25	1.29	1.45	1.21	1.55	1.12	1.66	1.04	1.77	.95	1.89
26	1.30	1.46	1.22	1.55	1.14	1.65	1.06	1.76	.98	1.88
27	1.32	1.47	1.24	1.56	1.16	1.65	1.08	1.76	1.01	1.86
28	1.33	1.48	1.26	1.56	1.18	1.65	1.10	1.75	1.03	1.85
29	1.34	1.48	1.27	1.56	1.20	1.65	1.12	1.74	1.05	1.84
30	1.35	1.49	1.28	1.57	1.21	1.65	1.14	1.74	1.07	1.83
31	1.36	1.50	1.30	1.57	1.23	1.65	1.16	1.74	1.09	1.83
32	1.37	1.50	1.31	1.57	1.24	1.65	1.18	1.73	1.11	1.82
33	1.38	1.51	1.32	1.58	1.26	1.65	1.19	1.73	1.13	1.81
34	1.39	1.51	1.33	1.58	1.27	1.65	1.21	1.73	1.15	1.81
35	1.40	1.52	1.34	1.58	1.28	1.65	1.22	1.73	1.16	1.80
36	1.41	1.52	1.35	1.59	1.29	1.65	1.24	1.73	1.18	1.80
37	1.42	1.53	1.36	1.59	1.31	1.66	1.25	1.72	1.19	1.80
38	1.43	1.54	1.37	1.59	1.32	1.66	1.26	1.72	1.21	1.79
39	1.43	1.54	1.38	1.60	1.33	1.66	1.27	1.72	1.22	1.79
40	1.44	1.54	1.39	1.60	1.34	1.66	1.29	1.72	1.23	1.79
45	1.48	1.57	1.43	1.62	1.38	1.67	1.34	1.72	1.29	1.78
50	1.50	1.59	1.46	1.63	1.42	1.67	1.38	1.72	1.34	1.77
55	1.53	1.60	1.49	1.64	1.45	1.68	1.41	1.72	1.38	1.77
60	1.55	1.62	1.51	1.65	1.48	1.69	1.44	1.73	1.41	1.77
65	1.57	1.63	1.54	1.66	1.50	1.70	1.47	1.73	1.44	1.77
70	1.58	1.64	1.55	1.67	1.52	1.70	1.49	1.74	1.46	1.77
75	1.60	1.65	1.57	1.68	1.54	1.71	1.51	1.74	1.49	1.77
80	1.61	1.66	1.59	1.69	1.56	1.72	1.53	1.74	1.51	1.77
85	1.62	1.67	1.60	1.70	1.57	1.72	1.55	1.75	1.52	1.77
90	1.63	1.68	1.61	1.70	1.59	1.73	1.57	1.75	1.54	1.78
95	1.64	1.69	1.62	1.71	1.60	1.73	1.58	1.75	1.56	1.78
100	1.65	1.69	1.63	1.72	1.61	1.74	1.59	1.76	1.57	1.78

SOURCE: From J. Durbin and G. S. Watson, "Testing for Serial Correlation in Least Squares Regression, II," *Biometrika* 30 (1951): 159–78. Reproduced by permission of the Biometrika Trustees.

TABLE **8(b)** Critical Values for the Durbin-Watson Statistic, $\alpha = .01$

	$k = 1$		$k = 2$		$k = 3$		$k = 4$		$k = 5$	
n	d_L	d_U	d_L	d_U	d_L	d_U	d_L	d_U	d_L	d_U
15	.81	1.07	.70	1.25	.59	1.46	.49	1.70	.39	1.96
16	.84	1.09	.74	1.25	.63	1.44	.53	1.66	.44	1.90
17	.87	1.10	.77	1.25	.67	1.43	.57	1.63	.48	1.85
18	.90	1.12	.80	1.26	.71	1.42	.61	1.60	.52	1.80
19	.93	1.13	.83	1.26	.74	1.41	.65	1.58	.56	1.77
20	.95	1.15	.86	1.27	.77	1.41	.68	1.57	.60	1.74
21	.97	1.16	.89	1.27	.80	1.41	.72	1.55	.63	1.71
22	1.00	1.17	.91	1.28	.83	1.40	.75	1.54	.66	1.69
23	1.02	1.19	.94	1.29	.86	1.40	.77	1.53	.70	1.67
24	1.04	1.20	.96	1.30	.88	1.41	.80	1.53	.72	1.66
25	1.05	1.21	.98	1.30	.90	1.41	.83	1.52	.75	1.65
26	1.07	1.22	1.00	1.31	.93	1.41	.85	1.52	.78	1.64
27	1.09	1.23	1.02	1.32	.95	1.41	.88	1.51	.81	1.63
28	1.10	1.24	1.04	1.32	.97	1.41	.90	1.51	.83	1.62
29	1.12	1.25	1.05	1.33	.99	1.42	.92	1.51	.85	1.61
30	1.13	1.26	1.07	1.34	1.01	1.42	.94	1.51	.88	1.61
31	1.15	1.27	1.08	1.34	1.02	1.42	.96	1.51	.90	1.60
32	1.16	1.28	1.10	1.35	1.04	1.43	.98	1.51	.92	1.60
33	1.17	1.29	1.11	1.36	1.05	1.43	1.00	1.51	.94	1.59
34	1.18	1.30	1.13	1.36	1.07	1.43	1.01	1.51	.95	1.59
35	1.19	1.31	1.14	1.37	1.08	1.44	1.03	1.51	.97	1.59
36	1.21	1.32	1.15	1.38	1.10	1.44	1.04	1.51	.99	1.59
37	1.22	1.32	1.16	1.38	1.11	1.45	1.06	1.51	1.00	1.59
38	1.23	1.33	1.18	1.39	1.12	1.45	1.07	1.52	1.02	1.58
39	1.24	1.34	1.19	1.39	1.14	1.45	1.09	1.52	1.03	1.58
40	1.25	1.34	1.20	1.40	1.15	1.46	1.10	1.52	1.05	1.58
45	1.29	1.38	1.24	1.42	1.20	1.48	1.16	1.53	1.11	1.58
50	1.32	1.40	1.28	1.45	1.24	1.49	1.20	1.54	1.16	1.59
55	1.36	1.43	1.32	1.47	1.28	1.51	1.25	1.55	1.21	1.59
60	1.38	1.45	1.35	1.48	1.32	1.52	1.28	1.56	1.25	1.60
65	1.41	1.47	1.38	1.50	1.35	1.53	1.31	1.57	1.28	1.61
70	1.43	1.49	1.40	1.52	1.37	1.55	1.34	1.58	1.31	1.61
75	1.45	1.50	1.42	1.53	1.39	1.56	1.37	1.59	1.34	1.62
80	1.47	1.52	1.44	1.54	1.42	1.57	1.39	1.60	1.36	1.62
85	1.48	1.53	1.46	1.55	1.43	1.58	1.41	1.60	1.39	1.63
90	1.50	1.54	1.47	1.56	1.45	1.59	1.43	1.61	1.41	1.64
95	1.51	1.55	1.49	1.57	1.47	1.60	1.45	1.62	1.42	1.64
100	1.52	1.56	1.50	1.58	1.48	1.60	1.46	1.63	1.44	1.65

SOURCE: From J. Durbin and G. S. Watson, "Testing for Serial Correlation in Least Squares Regression, II," *Biometrika* 30 (1951): 159–78. Reproduced by permission of the Biometrika Trustees.

TABLE 9 Critical Values for the Wilcoxon Rank Sum Test

(a) $\alpha = .025$ one-tail; $\alpha = .05$ two-tail

n_2 \ n_1	3		4		5		6		7		8		9		10	
	T_L	T_U	T_L	T_U	T_L	T_U	T_L	T_U	T_L	T_U	T_L	T_U	T_L	T_U	T_L	T_U
4	6	18	11	25	17	33	23	43	31	53	40	64	50	76	61	89
5	6	11	12	28	18	37	25	47	33	58	42	70	52	83	64	96
6	7	23	12	32	19	41	26	52	35	63	44	76	55	89	66	104
7	7	26	13	35	20	45	28	56	37	68	47	81	58	95	70	110
8	8	28	14	38	21	49	29	61	39	63	49	87	60	102	73	117
9	8	31	15	41	22	53	31	65	41	78	51	93	63	108	76	124
10	9	33	16	44	24	56	32	70	43	83	54	98	66	114	79	131

(b) $\alpha = .05$ one-tail; $\alpha = .10$ two-tail

n_2 \ n_1	3		4		5		6		7		8		9		10	
	T_L	T_U	T_L	T_U	T_L	T_U	T_L	T_U	T_L	T_U	T_L	T_U	T_L	T_U	T_L	T_U
3	6	15	11	21	16	29	23	37	31	46	39	57	49	68	60	80
4	7	17	12	24	18	32	25	41	33	51	42	62	52	74	63	87
5	7	20	13	27	19	37	26	46	35	56	45	67	55	80	66	94
6	8	22	14	30	20	40	28	50	37	61	47	73	57	87	69	101
7	9	24	15	33	22	43	30	54	39	66	49	79	60	93	73	107
8	9	27	16	36	24	46	32	58	41	71	52	84	63	99	76	114
9	10	29	17	39	25	50	33	63	43	76	54	90	66	105	79	121
10	11	31	18	42	26	54	35	67	46	80	57	95	69	111	83	127

Source: From F. Wilcoxon and R. A. Wilcox, "Some Rapid Approximate Statistical Procedures" (1964), p. 28. Reproduced with the permission of American Cyanamid Company.

TABLE **10**
Critical Values for the Wilcoxon Signed Rank Sum Test

	(a) α = .025 one-tail; α = .05 two-tail		(b) α = .05 one-tail; α = .10 two-tail	
n	T_L	T_U	T_L	T_U
6	1	20	2	19
7	2	26	4	24
8	4	32	6	30
9	6	39	8	37
10	8	47	11	44
11	11	55	14	52
12	14	64	17	61
13	17	74	21	70
14	21	84	26	79
15	25	95	30	90
16	30	106	36	100
17	35	118	41	112
18	40	131	47	124
19	46	144	54	136
20	52	158	60	150
21	59	172	68	163
22	66	187	75	178
23	73	203	83	193
24	81	219	92	208
25	90	235	101	224
26	98	253	110	241
27	107	271	120	258
28	117	289	130	276
29	127	308	141	294
30	137	328	152	313

Source: From F. Wilcoxon and R. A. Wilcox, "Some Rapid Approximate Statistical Procedures" (1964), p. 28. Reproduced with the permission of American Cyanamid Company.

TABLE **11** Critical Values for the Spearman Rank Correlation Coefficient

The α values correspond to a one-tail test of H_0: $\rho_s = 0$. The value should be doubled for two-tail tests.

n	$\alpha = .05$	$\alpha = .025$	$\alpha = .01$
5	.900	—	—
6	.829	.886	.943
7	.714	.786	.893
8	.643	.738	.833
9	.600	.683	.783
10	.564	.648	.745
11	.523	.623	.736
12	.497	.591	.703
13	.475	.566	.673
14	.457	.545	.646
15	.441	.525	.623
16	.425	.507	.601
17	.412	.490	.582
18	.399	.476	.564
19	.388	.462	.549
20	.377	.450	.534
21	.368	.438	.521
22	.359	.428	.508
23	.351	.418	.496
24	.343	.409	.485
25	.336	.400	.475
26	.329	.392	.465
27	.323	.385	.456
28	.317	.377	.448
29	.311	.370	.440
30	.305	.364	.432

Source: From E. G. Olds, "Distribution of Sums of Squares of Rank Differences for Small Samples," *Annals of Mathematical Statistics* 9 (1938). Reproduced with the permission of the Institute of Mathematical Statistics.

TABLE **12** Control Chart Constants

SAMPLE SIZE n	A_2	d_2	d_3	D_3	D_4
2	1.880	1.128	.853	.000	3.267
3	1.023	1.693	.888	.000	2.575
4	.729	2.059	.880	.000	2.282
5	.577	2.326	.864	.000	2.115
6	.483	2.534	.848	.000	2.004
7	.419	2.704	.833	.076	1.924
8	.373	2.847	.820	.136	1.864
9	.337	2.970	.808	.184	1.816
10	.308	3.078	.797	.223	1.777
11	.285	3.173	.787	.256	1.744
12	.266	3.258	.778	.284	1.716
13	.249	3.336	.770	.308	1.692
14	.235	3.407	.762	.329	1.671
15	.223	3.472	.755	.348	1.652
16	.212	3.532	.749	.364	1.636
17	.203	3.588	.743	.379	1.621
18	.194	3.640	.738	.392	1.608
19	.187	3.689	.733	.404	1.596
20	.180	3.735	.729	.414	1.586
21	.173	3.778	.724	.425	1.575
22	.167	3.819	.720	.434	1.566
23	.162	3.858	.716	.443	1.557
24	.157	3.895	.712	.452	1.548
25	.153	3.931	.709	.459	1.541

SOURCE: From E. S. Pearson, "The Percentage Limits for the Distribution of Range in Samples from a Normal Population," *Biometrika* 24 (1932): 416. Reproduced by permission of the Biometrika Trustees.

APPENDIX C

ANSWERS TO SELECTED EXERCISES

All answers have been double-checked for accuracy. However, we cannot be absolutely certain that there are no errors. Students should not automatically assume that answers that don't match ours are wrong. When and if we discover mistakes we will post corrected answers on our web page. (See page 8 for instructions.) If you find any errors, please e-mail the author (address on web page). We will be happy to acknowledge you with the discovery.

Chapter 1

1.3 a. The complete production run
b. 1,000 chips **c.** Proportion defective
d. Proportion of sample chips that are defective (7.5%) **e.** Parameter **f.** Statistic
g. Because the sample proportion is less than 10%, we can conclude that the claim is true.

1.4 Descriptive statistics summarizes a set of data. Inferential statistics makes inferences about populations from samples.

1.6 a. Fuel mileage of all the taxis in the fleet.
b. Mean mileage. **c.** The 50 observations.
d. Mean of the 50 observations.
e. The statistic would be used to estimate the parameter from which the owner can calculate total costs. We computed the sample mean to be 19.8 mpg.

1.7 a. Flip the coin 100 times and count the number of heads and tails **b.** Outcomes of flips **c.** Outcomes of the 100 flips **d.** Proportion of heads **e.** Proportion of heads in the 100 flips

Chapter 2

2.2 a. Interval **b.** Interval **c.** Nominal
d. Ordinal

2.4 a. Nominal **b.** Interval **c.** Nominal
d. Interval **e.** Ordinal

2.6 a. Interval **b.** Interval **c.** Nominal
d. Ordinal **e.** Interval

2.8 a. Interval **b.** Ordinal **c.** Nominal
d. Ordinal

2.10 a. Ordinal **b.** Ordinal **c.** Ordinal

2.32 10 or 11

2.34 a. 7 to 9 **b.** Upper limits: 5.25, 5.40, 5.55, 5.70, 5.85, 6.00, 6.15

2.44 d. The histogram is positively skewed, unimodal, and not bell shaped. We learn that most prices lie between \$500,000 and \$1,100,000 with a small number of houses selling for more than \$1,100,000.

2.46 a. 9 or 10 **c.** Positively skewed. **d.** No

2.54 d. This scorecard is a much better predictor.

2.80 There does not appear to be any brand loyalty.

2.82 There are differences between men and women in terms of the reason for unemployment.

2.84 There does not appear to be a linear relationship between the two variables.

2.86 b. There is a positive linear relationship between calculus and statistics marks.

2.88 b. There is a linear relationship.

2.90 b. There is a moderately strong positive linear relationship.

2.92 b. There is a very weak positive linear relationship.

2.94 There is a moderately strong positive linear relationship.

2.103 b. The slope is positive.
c. There is a moderately strong linear relationship.

2.110 There is a moderately strong negative linear relationship.
2.115 There is a weak positive linear relationship.

Chapter 4

4.2 $\bar{x} = 6$, median = 5, mode = 5

4.4 a. $\bar{x} = 39.3$, median = 38, mode = all

4.6 $R_g = .19$

4.8 a. $\bar{x} = .106$, median = .10 **b.** $R_g = .102$
c. Geometric mean

4.10 a. .20, 0, .25, .33 **b.** $\bar{x} = .195$, median = .225 **c.** $R_g = .188$
d. Geometric mean

4.12 a. $\bar{x} = 24{,}329$, median = 24,461

4.14 a. $\bar{x} = 128.07$, median = 136.00

4.16 a. $\bar{x} = 30.53$, median = 31

4.18 a. $\bar{x} = 519.20$, median = 523.00

4.20 $s^2 = 1.14$

4.22 $s^2 = 15.12$, $s = 3.89$

4.24 a. $s^2 = 51.5$ **b.** $s^2 = 6.5$
c. $s^2 = 174.5$

4.26 6, 6, 6, 6, 6

4.28 a. 16% **b.** 97.5% **c.** 16%

4.30 a. nothing **b.** at least 75% lie between 60 and 180 **c.** at least 88.9% lie between 30 and 210

4.32 $s^2 = 40.73$ mph^2, and $s = 6.38$ mph; at least 75% of the speeds lie within 12.76 mph of the mean; at least 88.9% of the speeds lie within 19.14 mph of the mean

4.34 $s^2 = .0858$ cm^2, and $s = .2929$cm; at least 75% of the lengths lie within .5858 of the mean; at least 88.9% of the rods will lie within .8787 cm of the mean.

4.36 a. $s = 15.01$

4.38 22.3, 30.8

4.40 13.05, 14.7, 15.6

4.42 IQR = 2.55

4.44 IQR = 9.25

4.50 The starting salaries of B.A. and other are the lowest and least variable. Starting salaries for B.B.A. and B.Sc. are higher.

4.52 b. Time taken to complete rounds on the public course are larger and more variable.

4.54 697.19, 804.90, 909.38

4.56 $r = -.7813$; there is a moderately strong negative linear relationship.
b. $R^2 = .6104$; 61.04% of the variation in y is explained by the variation in x.

4.58 a. 98.52 **b.** .8811 **c.** .7763
d. $\hat{y} = 5.917 + 1.705x$

4.60 .4009

4.62 a. .112

4.64 $R^2 = .0069$; the relationship is very weak.

4.66 Estimated fixed costs = \$263.40, estimated variable costs = \$71.65

4.68 a. $R^2 = .017$; there is a very weak relationship between the two variables.
b. The away attendance increases on average by 31.28 for each win.

4.70 a. \$19.231 million. **b.** $R^2 = .024$

4.76 $R^2 = .4603$

4.78 $R^2 = .050$; $\hat{y} = 17.93 + .604x$

4.80 a. $\hat{y} = 105.3 + 7.034x$

4.87 $R^2 = .369$; $\hat{y} = 89.54 + .128x$

4.91 $R^2 = .412$; $\hat{y} = -8.289 + 3.146x$

4.92 $R^2 = .548$; $\hat{y} = 49{,}337 - 553.7x$

Chapter 6

6.2 a. Subjective approach **b.** If all the teams in major league baseball have exactly the same players, the New York Yankees will win 25% of all World Series.

6.4 a. Subjective approach **b.** The Dow Jones Industrial Index will increase on 60% of the days if economic conditions remain unchanged.

6.6 {Adams wins, Brown wins, Collins wins, Dalton wins}

6.8 a. {0, 1, 2, 3, 4, 5} **b.** {4, 5} **c.** .10
d. .65 **e.** 0

6.10 2/6, 3/6, 1/6

6.12 a. .40 **b.** .90

6.14 a. P(single) = .15, P(married) = .50, P(divorced) = .25, P(widowed) = .10
b. Relative frequency approach

6.16 $P(A_1) = .3$, $P(A_2) = .4$, $P(A_3) = .3$. $P(B_1) = .6$, $P(B_2) = .4$.

6.18 a. .57 **b.** .43 **c.** It is not a coincidence.

6.20 The events are not independent.

6.22 The events are independent.

6.24 $P(A_1) = .40$, $P(A_2) = .45$, $P(A_3) = .15$. $P(B_1) = .45$, $P(B_2) = .55$.
6.26 **a.** .85. **b.** .75 **c.** .50
6.28 **a.** .36 **b.** .49 **c.** .83
6.30 **a.** .31 **b.** .85 **c.** .387 **d.** .044
6.32 **a.** .390 **b.** .66 **c.** No
6.34 **a.** .11 **b.** .043 **c.** .091 **d.** .909
6.36 **a.** .33 **b.** .30 **c.** Yes, the events are dependent.
6.38 **a.** .623 **b.** .650 **c.** .598
6.40 **a.** .591 **b.** .638 **c.** .561
6.42 **a.** .8309 **b.** .5998 **c.** .0751
6.44 No, the events are independent.
6.46 **a.** .1898 **b.** .2131 **c.** .3538 **d.** .6462
6.52 **a.** .81 **b.** .01 **c.** .18 **d.** .99
6.54 **b.** .8091 **c.** .0091 **d.** .1818 **e.** .9909
6.56 **a.** .28 **b.** .30 **c.** .42
6.58 .038
6.60 .335
6.62 .698
6.64 .2520
6.66 .033
6.68 .00000001
6.70 .6125
6.72 **a.** .696 **b.** .304 **c.** .889 **d.** .111
6.74 .526
6.76 .327
6.78 .661
6.80 .593
6.82 .843
6.84 .920, .973, .1460, .9996
6.86 .3333
6.87 **a.** .3285 **b.** .2403
6.91 .825
6.92 .2214
6.96 2/3
6.97 .295
6.99 **a.** .19 **b.** .517 **c.** No
6.100 .9710
6.103 **a.** .290 **b.** .290 **c.** Yes

Chapter 7

7.2 **a.** Any value between 0 and several hundred miles **b.** No **c.** No **d.** Continuous
7.4 **a.** 0, 1, 2, . . . , 100 **b.** Yes **c.** Yes, 101 values **d.** Discrete.
7.6 $P(x) = 1/6$, for $x = 1, 2, \ldots, 6$
7.8 **a.** .950 .020 .680 **b.** 3.066 **c.** 1.085
7.10 **a.** .8 **b.** .8 **c.** .8 **d.** .3
7.12 .0156
7.14 **a.** .25 **b.** .25 **c.** .25 **d.** .25
7.18 **a.** 1.40, 17.04 **c.** 7.00, 426.00 **d.** 7.00, 426.00
7.20 **a.** .6 **b.** 1.7, .81
7.22 **a.** .40 **b.** .95
7.24 1.025, .168
7.26 **a.** .06 **b.** 0 **c.** .35 **d.** .65
7.28 **a.** .21 **b.** .31 **c.** .26
7.30 2.76, 1.517
7.32 3.86, 2.60
7.34 E(value of coin) = \$460; take the \$500
7.36 \$18
7.38 4.00, 2.40
7.40 1.85
7.42 3,409
7.44 .14, .58
7.46 **b.** 2.8, .76
7.48 0, 0
7.50 **b.** 2.9, .45 **c.** Yes
7.54 **c.** 1.07, .505 **d.** .93, .605 **e.** −.045, −.081
7.56 **a.** .412 **b.** .286 **c.** .148
7.58 145, 5.57
7.60 168, 24.0
7.62 **a.** .211, .1081 **b.** .211, .1064 **c.** .211, .1052
7.64 .1060, .1456
7.68 .00005, .03981
7.70 .0056, .0562
7.72 .0051, .0545
7.74 .0232 .1046
7.78 .0139 .0578
7.84 **a.** .2668 **b.** .1029 **c.** .0014
7.86 **a.** .26683 **b.** .10292 **c.** .00145
7.88 **a.** .2457 **b.** .0819 **c.** .0015
7.90 **a.** .1711 **b.** .0916 **c.** .9095 **d.** .8106
7.92 **a.** .4219 **b.** .3114 **c.** .25810
7.94 **a.** .0646 **b.** 9666 **c.** .9282 **d.** 22.5
7.96 .0081
7.98 .1244
7.100 .00317
7.102 **a.** .3369 **b.** .75763
7.104 **a.** .2990 **b.** .91967
7.106 **a.** .69185 **b.** .12519 **c.** .44069
7.108 **a.** .05692 **b.** .47015
7.110 **a.** .1353 **b.** .1804 **c.** .0361
7.112 **a.** .0302 **b.** .2746 **c.** .3033
7.114 **a.** .1353 **b.** .0663
7.116 **a.** .20269 **b.** .26761
7.118 .6703
7.120 **a.** .4422 **b.** .1512
7.122 **a.** .2231 **b.** .7029 **c.** .5768
7.123 .95099, .04803, .00097, .00001, 0, 0
7.126 .0064
7.127 .08755
7.130 .0473
7.131 **a.** 1.46, 1.49 **b.** 2.22, 1.45
7.134 **a.** .0993 **b.** .8088 **c.** .8881
7.135 **a.** .1612 **b.** .0095 **c.** .0132
7.138 **a.** .8 **b.** .4457
7.139 **a.** .00793 **b.** 56 **c.** 4.10

Chapter 8

8.2 **a.** .1200 **b.** .3333 **c.** .6667 **d.** .1867
8.4 **b.** 0 **c.** .25 **d.** .005
8.6 **a.** .1667 **b.** .3333 **c.** 0
8.8 57 minutes
8.10 123 tons
8.12 **b.** .5 **c.** .25
8.14 **b.** .25 **c.** .33
8.16 .9345
8.18 .0559
8.20 .0107
8.22 .9251
8.24 .0475
8.26 .1196
8.28 .0010
8.30 0
8.32 1.70
8.34 .0122
8.36 .4435
8.38 **a.** .6759 **b.** .3745 **c.** .1469
8.40. .6915
8.42 **a.** .2023 **b.** .3372
8.44 **a.** .1056 **b.** .1056 **c.** .8882
8.46 Top 5%: 34.4675, Bottom 5%: 29.5325
8.48 .1151
8.50 **a.** .1170 **b.** .3559 **c.** .0162 **d.** 4.05 hours
8.52 9,636 pages
8.54 **a.** .3336 **b.** .0314 **c.** .0436 **d.** \$32.88
8.56 **a.** .0099 **b.** \$12.88
8.58 132.80 (rounded to 133)
8.60 .5948
8.62 .0465
8.64 171
8.66 872.5 (rounded to 873)
8.68 .8159
8.70 **a.** .2327 **b.** .2578
8.71 .4857
8.72 .1353
8.75 .8647
8.76 **a.** .5488 **b.** .6988 **c.** .1920 **d.** 0
8.82 .1889
8.84 **a.** 2.724 **b.** 1.282 **c.** 2.132 **d.** 2.528
8.86 **a.** 1.6556 **b.** 2.6810 **c.** 1.9600 **d.** 1.6602
8.88 **a.** .1744 **b.** .0231 **c.** .0251 **d.** .0267
8.90 **a.** 17.3 **b.** 50.9 **c.** 2.71 **d.** 53.5
8.92 **a.** 33.5705 **b.** 866.911 **c.** 24.3976 **d.** 261.058
8.94 **a.** .4881 **b.** .9158 **c.** .9988 **d.** .9077
8.96 **a.** 2.84 **b.** 1.93 **c.** 3.60 **d.** 3.37
8.98 **a.** 1.5204 **b.** 1.5943 **c.** 2.8397 **d.** 1.1670
8.100 **a.** .1050 **b.** .1576 **c.** .0001 **d.** .0044

Chapter 9

9.2 **a.** 1/36 **b.** 1/36
9.4 The variance of $\bar{X}$ is smaller than the variance of X.

9.6 No, because the sample mean is approximately normally distributed.

9.8 a. .1056 b. .1587 c. .0062

9.10 a. .4435 b. .7333 c. .8185

9.12 a. .1191 b. .2347 c. .2902

9.14 a. 15.00 b. 21.80 c. 49.75

9.18 a. .0918 b. .0104 c. .00077

9.20 a. .3085 b. 0

9.22 a. .0038 b. It appears to be false.

9.26 .1170

9.28 .9319

9.30 a. 0 b. .0409 c. .5

9.32 .1056

9.34 .0035

9.36 a. .1151 b. .0287

9.38 .0096; the commercial is dishonest.

9.40 a. .0071 b. The claim appears to be false.

9.42 .0066

9.44 The claim appears to be false.

9.47 .3050

9.48 1

9.49 .8413

9.51 .8413

9.53 .0033

Chapter 10

10.10 a. 200 ± 19.60 b. 200 ± 9.80 c. 200 ± 3.92 d. The interval narrows.

10.12 a. 500 ± 3.95 b. 500 ± 3.33 c. 500 ± 2.79 d. The interval narrows.

10.14 a. 10 ± .82 b. 10 ± 1.64 c. 10 ± 2.60 d. The interval widens.

10.16 a. 400 ± 1.29 b. 200 ± 1.29 c. 100 ± 1.29 d. The width of the interval is unchanged.

10.18 Yes, because the variance decreases as the sample size increases.

10.20 a. 500 ± 3.50

10.22 LCL = 36.82, UCL = 50.68

10.24 LCL = 6.91, UCL = 12.79

10.26 LCL = 12.83, UCL = 20.97

10.28 LCL = 10.41, UCL = 15.89

10.30 LCL = 249.44, UCL = 255.32

10.32 LCL = 11.86, UCL = 12.34

10.34 LCL = .494, UCL = .526

10.36 LCL = 18.66, UCL = 19.90

10.38 LCL = 579,545, UCL = 590,581

10.40 LCL = 25.62, UCL = 28.76

10.41 217

10.43 1,083

10.45 2,149

10.46 a. 1,537 b. 500 ± 10

Chapter 11

11.2 H_0: I will complete the Ph.D.
H_1: I will not be able to complete the Ph.D.

11.4 H_0: Risky investment is more successful
H_1: Risky investment is not more successful

11.6 O. J. Simpson

All p-values and probabilities of Type II errors were calculated manually using Table 3 in Appendix B.

11.8 $z = .60$; rejection region: $z > 1.88$; p-value = .2743; not enough evidence that $\mu > 50$.

11.10 $z = 0$; rejection region: $z < -1.96$ or $z > 1.96$; p-value = 1.0; not enough evidence that $\mu \neq 100$.

11.12 $z = -1.33$; rejection region: $z < -1.645$; p-value = .0918; not enough evidence that $\mu < 50$

11.14 a. .2743 b. .1587 c. .0013 d. The test statistic decreases and the p-value decreases.

11.16 a. .2112 b. .3768 c. .5764 d. The test statistic increases and the p-value increases.

11.18 a. .0013 b. .0028 c. .1587 d. The test statistic decreases and the p-value increases.

11.20 a. $z = 4.57$, p-value = 0 b. $z = 1.60$, p-value = .0548.

11.22 a. $z = -.62$, p-value = .2676 b. $z = -1.38$, p-value = .0838

11.24 p-values: .5, .3121, .1611, .0694, .0239, .0062, .0015, 0, 0

11.26 a. $z = 2.30$, p-value = .0214 b. $z = .46$, p-value = .6456

11.28 $z = 2.11$, p-value = .0174; yes

11.30 $z = -1.29$, p-value = .0985; yes

11.32 $z = .95$, p-value = .1711; no

11.34 $z = 1.85$, p-value = .0322; no

11.36 $z = -2.06$, p-value = .0197; yes

11.38 $z = 1.65$, p-value = .0495; yes

11.40 $z = 2.26$, p-value = .0119; no

11.42 $z = -1.22$, p-value = .1112; no

11.44 $z = 3.33$, p-value = 0; yes

11.46 $z = -2.73$, p-value = .0032; yes

11.49 p-value = .9931; no evidence that the new system will not be cost effective.

11.50 a. .6103 b. .8554 c. β increases.

11.56 .6480

11.57 .1492

11.59 a. .4404 b. .6736 c. β increases.

11.65 .1170

11.66 .1635 (with $\alpha = .05$)

The answers for the exercises in Chapters 12 through 19 were produced in the following way. In exercises where the statistics are provided in the question or in Appendix A, the solutions were produced manually. The solutions to exercises requiring the use of a computer were produced using Excel. When the test result is calculated manually and the test statistic is normally distributed (z statistic) the p-value was computed manually using the normal table (Table 3 in Appendix B). The p-value for all other test statistics was determined using Excel.

Chapter 12

12.2 a. 1500 ± 59.52 b. 1500 ± 39.68 c. 1500 ± 19.84 d. Interval narrows

12.4 a. 10 ± .20 b. 10 ± .79 c. 10 ± 1.98 d. Interval widens

12.6 a. 63 ± 1.77 b. 63 ± 2.00 c. 63 ± 2.71 d. Interval widens

12.8 a. $t = -3.21$, p-value = .0015 b. $t = -1.57$, p-value = .1177 c. $t = -1.18$, p-value = .2400 d. t decreases and p-value increases

12.10 a. $t = .67$, p-value = .5113 b. $t = .52$, p-value = .6136 c. $t = .30$, p-value = .7804 d. t decreases and p-value increases

12.12 a. $t = 1.71$, p-value = .0448 b. $t = 2.40$, p-value = .0091 c. $t = 4.00$, p-value = .0001 d. t increases and p-value decreases

12.14 a. 175 ± 28.60 b. 175 ± 22.07 c. Because the distribution of Z is narrower than that of the Student t

12.16 a. 350 ± 11.56 b. 350 ± 11.52 c. When n is large, the distribution of Z is virtually identical to that of the Student t.

12.18 a. $t = -1.30$, p-value = .1126; no b. $z = -1.30$, p-value = .0964; yes c. Because the distribution of Z is narrower than that of the Student t

12.20 a. $t = 1.58$, p-value = .0569 b. $z = 1.58$, p-value = .0571 c. When n is large, the distribution of Z is virtually identical to that of the Student t.

12.22 LCL = 14,422, UCL = 33,680

12.24 $t = -4.49$, p-value = .0002; yes

12.26 LCL = 18.11, UCL = 35.23

12.28 $t = -2.45$, p-value = .0185; yes

12.30 LCL = 427 million, UCL = 505 million

12.32 LCL = $727,350 million, UCL = $786,350 million

12.34 LCL = 2.31, UCL = 3.03

12.36 LCL = $51,725 million, UCL = $56,399 million

12.38 $t = .51$, p-value = .3061; no

12.40 $t = 2.28$, p-value = .0127; yes

12.42 $t = 1.89$, p-value = .0323; yes

12.44 a. $\chi^2 = 39.20$, p-value = .1596; no b. $\chi^2 = 79.20$, p-value = .0714; no c. χ^2 increases and the p-value decreases.

12.46 LCL = .00046, UCL = .00300

12.48 $\chi^2 = 5.02$, p-value = .6854; no

12.50 a. $\chi^2 = 25.98$, p-value = .7088; no b. Demand is required to be normally distributed. c. The histogram is approximately bell shaped.

12.52 LCL = 3.72, UCL = 6.08

12.54 LCL = 11.99, UCL = 38.09

12.56 a. .50 ± .0490 b. .33 ± .0461 c. .10 ± .0294 d. The interval narrows.

12.58 a. $z = .65$, p-value = .2578 b. $z = .44$, p-value = .3300 c. $z = .22$, p-value = .4129 d. The z statistic decreases and the p-value increases.

12.60 **a.** .5 ± .03 **b.** Yes, because the sample size was chosen to produce this interval.

12.62 n = 564

12.64 **a.** .92 ± .0188 **b.** The interval is narrower. **c.** Yes, because the interval estimate is better than specified.

12.66 LCL = .6431, UCL = .7369

12.68 LCL = .6057, UCL = .6943

12.70 LCL = .2326, UCL = .3274

12.72 LCL = $399,600, UCL = $620,400

12.74 LCL = .0035, UCL = .0229

12.76 LCL = .0792, UCL = .1320

12.78 LCL = .7232, UCL = .7936

12.80 LCL = .1576, UCL = .3090

12.82 LCL = .7449, UCL = .8231

12.84 z = 2.00, p-value = .0228; yes

12.86 z = 2.62, p-value = .0044; yes

12.88 **a.** LCL = .861 million, UCL = 1.17 million

12.90 LCL = 13,195,985, UCL = 14,720,803

12.92 **a.** LCL = .2711, UCL = .3127 **b.** LCL = 29,060,293, UCL = 33,519,564

12.94 LCL = 26.928 million, UCL = 38.447 million

12.96 LCL = 5.4 (rounded to 5), UCL = 239.9 (rounded to 240)

12.98 LCL = $16,419,433, UCL = $17,290,194

12.100 LCL = 33, UCL = 186

12.102 LCL = 192, UCL = 482

12.104 LCL = 8,613, UCL = 15,946

12.106 t = −1.64, p-value = .0569; no

12.108 LCL = .304, UCL = .412

12.110 LCL = 108.19, UCL = 126.89

12.112 **a.** LCL = 6.84, UCL = 6.98 **b.** The histogram is bell shaped. **c.** t = −3.48, p-value = .0004; yes

12.114 LCL = 5.11, UCL = 6.47

12.116 χ^2 = 161.25, p-value = .0001; yes

12.118 LCL = .94, UCL = 1.26

12.120 n = 4144

12.122 Number of cars: t = .46, p-value = .3351; there is not enough evidence to infer that the employee is stealing by lying about the number of cars.
Amount of time: t = 7.00, p-value = 0; there is enough evidence to infer that the employee is stealing by lying about the amount of time.

For all exercises in Chapter 13 and all chapter appendixes we employed the F-test of two variances at the 5% significance level to decide which one of the equal-variances or unequal-variances t-test and estimator of the difference between two means to use to solve the problem. Additionally, for exercises that compare two populations and are accompanied by data files, our answers were derived by defining the sample from population 1 as the data stored in the first column (often column A). The data stored in the second column represent the sample from population 2. Paired differences were defined as the difference between the variable in the first column minus the variable in the second column.

Chapter 13

13.2 **a.** t = .43, p-value = .6703; no **b.** t = .04, p-value = .9716; no **c.** The t-statistic decreases and the p-value increases. **d.** t = 1.53, p-value = .1282; no **e.** The t-statistic increases and the p-value decreases. **f.** t = .72, p-value = .4796; no **g.** The t-statistic increases and the p-value decreases.

13.4 **a.** t = .62, p-value = .2689; no **b.** t = 2.46, p-value = .0074; yes **c.** The t-statistic increases and the p-value decreases.

13.8 t = −2.04, p-value = .0283; yes

13.10 t = −1.59, p-value = .1368; no

13.12 t = 1.55, p-value = .1204; no

13.14 **a.** t = 2.88, p-value = .0021; yes **b.** LCL = .24, UCL = 4.58

13.16 t = .94, p-value = .1753; switch to supplier B.

13.18 **a.** t = 2.94, p-value = .0060; yes **b.** LCL = 4.31, UCL = 23.65 **c.** The times are required to be normally distributed. **d.** conditions are satisfied.

13.20 t = 7.54, p-value = 0; yes

13.22 t = .90, p-value = .1858; no

13.24 t = −2.05, p-value = .0412; yes

13.26 t = 1.16, p-value = .2467; no

13.28 t = −2.09, p-value = .0189; yes

13.30 t = 6.28, p-value = 0; yes

13.32 Experimental

13.34 **a.** Observational **b.** Randomly assign students to a software package. **c.** Better students choose software B.

13.40 t = −3.22, p-value = .0073; yes

13.42 t = 1.98, p-value = .0473; yes

13.44 **a.** t = 1.82, p-value = .0484; yes **b.** LCL = −.66, UCL = 6.82

13.46 t = −3.70, p-value = .0006

13.48 **a.** t = 16.92, p-value = 0; yes **b.** LCL = 50.12, UCL = 64.48 **c.** Differences are required to be normally distributed.

13.50 t = −1.52, p-value = .0647; no

13.52 t = 2.08, p-value = .0210; yes

13.58 **a.** F = .50, p-value = .0669; yes **b.** F = .50, p-value = .2071; no **c.** The value of the test statistic is unchanged, but the conclusion did change.

13.60 F = .50, p-value = .3179; no

13.62 F = 3.23, p-value = .0784; no

13.64 F = 2.08, p-value = .0003; yes

13.66 F = .31, p-value = 0; yes

13.68 **a.** z = 1.07, p-value = .2846 **b.** z = 2.01, p-value = .0444 **c.** The p-value decreases.

13.70 z = 1.70, p-value = .0446; yes

13.72 z = 1.74, p-value = .0409; yes

13.74 z = −2.85, p-value = .0022; yes

13.76 **a.** z = −4.04, p-value = 0; yes

13.78 z = 2.00, p-value = .0228; yes

13.80 z = −1.19, p-value = .1170; no

13.82 **a.** z = 3.35, p-value = 0; yes **b.** LCL = .0668, UCL = .3114

13.84 z = .53, p-value = .2981; no

13.86 z = 2.04, p-value = .0207; yes

13.88 z = −1.25, p-value = .2112; no

13.90 t = .88, p-value = .1931; no

13.92 t = −6.09, p-value = 0; yes

13.94 z = −2.30, p-value = .0106; yes

13.96 **a.** t = −1.06, p-value = .2980; no **b.** t = −2.87, p-value = .0040; yes

13.98 z = 2.26, p-value = .0119; yes

13.100 z = −4.28, p-value = 0; yes

13.102 **a.** t = −2.62, p-value = .0059; yes **b.** t = −4.08, p-value = .0001; yes

13.104 **a.** t = 4.14, p-value = .0001; yes **b.** LCL = 1.84, UCL = 5.36

13.106 t = −2.40, p-value = .0100; yes

13.108 z = 1.20, p-value = .1141; no

13.110 t = 14.07, p-value = 0; yes

13.112 t = −2.40, p-value = .0092; yes

13.114 t = 1.08, p-value = .2820; no

13.116 t = 2.28, p-value = .0115; yes

13.118 t = 4.15, p-value = 0; yes

13.120 t = −2.09, p-value = .0184; yes

Chapter 14

14.4 F = 4.82, p-value = .0377; yes

14.6 F = 3.91, p-value = .0493; yes

14.8 F = .81, p-value = .5224; no

14.10 **a.** F = 2.94, p-value = .0363; evidence of differences

14.12 F = 3.32, p-value = .0129; no

14.14 F = 1.17, p-value = .3162; no

14.16 F = 1.33, p-value = .2675; no

14.18 **a.** F = 25.60, p-value = 0; yes **b.** Incomes: F = 7.37, p-value = .0001; yes **c.** Education: F = 1.82, p-value = .1428; no

14.20 F = .26, p-value = .7730; no

14.22 **a.** μ_1 and μ_2, μ_1 and μ_4, μ_1 and μ_5, μ_2 and μ_4, μ_3 and μ_4, μ_3 and μ_5, and μ_4 and μ_5 differ. **b.** μ_1 and μ_5, μ_2 and μ_4, μ_3 and μ_4, and μ_4 and μ_5 differ. **c.** μ_1 and μ_2, μ_1 and μ_5, μ_2 and μ_4, μ_3 and μ_4, and μ_4 and μ_5 differ.

14.24 **a.** B.A. and B.B.A. differ. **b.** B.A. and B.B.A. differ.

14.26 **a.** Forms 1 and 4 differ. **b.** No forms differ.

14.28 **a.** Lacquers 2 and 3 differ. **b.** Lacquers 2 and 3 differ.

14.30 No fertilizers differ.

14.32 **a.** F = 16.50, p-value = 0; treatment means differ **b.** F = 4.00, p-value = .0005; block means differ

14.34 **a.** F = 7.00, p-value = .0078; treatment means differ **b.** F = 10.50, p-value = .0016; treatment means differ **c.** F = 21.00, p-value = .0001; treatment means differ **d.** F-statistic increases and p-value decreases.

14.36 a. SS(Total) 14.9, SST = 8.9, SSB = 4.2, SSE = 1.8 b. SS(Total) 14.9, SST = 8.9, SSE = 6.0

14.38 F = 1.65, p-value = .2296; no

14.40 a. F = 123.360, p-value = 0; yes
b. F = 323.16, p-value = 0; plots vary.

14.42 a. F = 21.16, p-value = 0; yes
b. F = 66.02, p-value = 0; randomized block design is best

14.44 a. F = 10.72, p-value = 0; yes
b. F = 6.36, p-value = 0; yes

14.46 a. F = 9.53. There is enough evidence to infer that factors A and B interact, which makes the tests in Parts b and c irrelevant.

14.48 a. F = .31, p-value = .5943; no evidence that factors A and B interact.
b. F = 1.23, p-value = .2995; no evidence of differences between the levels of factor A.
c. F = 13.00, p-value = .0069; evidence that differences between the levels of factor B.

14.50 F = .21, p-value = .8915; no evidence that educational level and gender interact.
F = 4.49, p-value = .0060; evidence of differences between educational levels.
F = 15.00, p-value = .0002; evidence of a difference between men and women.

14.52 a. There are 12 treatments.
b. Two factors—tax form and income group
c. There are a = 4 forms and b = 3 income groups. d. F = 1.04, p-value = .4030; no evidence of interaction.
e. F = 2.56, p-value = .0586; no
f. F = 4.11, p-value = .0190; yes

14.54 a. Factor A is the drug and factor B is the schedule. b. The improvement index
c. There are a = 4 drugs and b = 2 schedules.
d. F = 7.27, p-value = .0007. There is evidence of interaction, which makes the tests in Parts e and f irrelevant.

14.56 Both machines and alloys are sources of variation.

14.58 The only source of variation is skill level.

14.60 a. F = 7.67, p-value = .0001; yes

14.62 F = 13.79, p-value = 0; use the typeface that was read the fastest.

14.64 F = 7.72, p-value = .0070; yes

14.66 a. F = 136.58, p-value = 0; yes
b. All three means differ from one another. Pure method is best.

14.68 F = 14.47, p-value = 0; yes

14.70 F = 13.84, p-value = 0; yes

14.72 F = 1.62, p-value = .2022; no

14.74 F = 3.24, p-value = .0414; yes

Chapter 15

15.2 χ^2 = 2.26, p-value = .6868; no evidence that at least one p_i is not equal to its specified value.

15.6 χ^2 = 9.96, p-value = .0189; evidence that at least one p_i is not equal to its specified value.

15.8 χ^2 = 6.85, p-value = .0769; not enough evidence that at least one p_i is not equal to its specified value.

15.10 χ^2 = 14.07, p-value = .0071; yes

15.12 χ^2 = 33.85, p-value = 0; yes

15.14 χ^2 = 6.35, p-value = .0419; yes

15.16 χ^2 = 5.70, p-value = .1272; no

15.18 χ^2 = 9.56, p-value = .0020; evidence of a relationship

15.22 χ^2 = 2.27, p-value = .3221; no

15.24 χ^2 = 70.675, p-value = 0; yes

15.26 χ^2 = 32.57, p-value = 0; yes

15.28 χ^2 = .67, p-value = .9999; no

15.30 χ^2 = 22.36, p-value = .0988; no

15.32 χ^2 = 21.00, p-value = 0; yes

15.34 χ^2 = 8.71, p-value = .0128; yes

15.36 χ^2 = 16.62, p-value = .0002; yes

15.38 Successful firms: χ^2 = 3.03, p-value = .2199; no. Unsuccessful firms: χ^2 = 1.13, p-value = .5670; no

15.40 χ^2 = .055, p-value = .8140; no

15.42 χ^2 = 4.77, p-value = .3119; no

15.44 χ^2 = 74.47, p-value = 0; yes

15.46 χ^2 = .58, p-value = .9009; no

15.48 a. χ^2 = 38.22, p-value = .8427; no

15.50 χ^2 = 4.13, p-value = .5310; no

15.52 χ^2 = 9.73, p-value = .0452; yes

15.54 χ^2 = 4.57, p-value = .1016; no

15.56 a. χ^2 = 26.71, p-value = 0; yes
b. χ^2 = 2.94, p-value = .4007; no

15.58 a. χ^2 = .65, p-value = .4207; no
b. χ^2 = 7.72, p-value = .0521; no
c. χ^2 = 23.11, p-value = 0; yes

15.60 χ^2 = 4.51, p-value = .3411; no

Chapter 16

16.2 $\hat{y}$ = 9.107 + .0582x

16.4 $\hat{y}$ = −24.72 + .9675x

16.6 b. $\hat{y}$ = 3.635 + .2675x

16.8 b. $\hat{y}$ = 23.63 + 4.138x

16.10 $\hat{y}$ = 7.286 + .1898x

16.12 $\hat{y}$ = 4,040 + 44.97x

16.14 $\hat{y}$ = 458.4 + 64.05x

16.16 $\hat{y}$ = 20.64 − .3039x

16.18 $\hat{y}$ = 89.81 + .0514x

16.22 b. t = 10.09, p-value = 0; evidence of linear relationship

16.24 a. 1.347 b. t = 3.93, p-value = .0028; yes c. LCL = .0252, UCL = .0912
d. 6067

16.26 t = 6.55, p-value = 0; yes

16.28 a. 5.888 b. .2892 c. t = 4.86, p-value = 0; yes d. LCL = .1756, UCL = .3594

16.30 t = 10.67, p-value = 0; yes

16.32 t = 7.50, p-value = 0; yes

16.34 a. 3,287 b. t = 2.24, p-value = .0309; yes c. .1167

16.36 s_ε = 191.1; R^2 = .3500; t = 10.39, p-value = 0

16.38 t = −3.39, p-value = .0021; yes

16.40 a. .0331 b. t = 1.22, p-value = .2319; no

16.42 t = 4.86, p-value = 0; yes

16.44 t = 7.49, p-value = 0; yes

16.46 11.60, 17.10

16.48 a. −2.70, 11.31 b. 2.51, 6.10

16.50 a. 1.39, 25.15 b. 11.73, 14.81

16.52 83.26, 88.14

16.54 5.12, 20.84

16.56 5,227, 7,351

16.58 756.17, 801.13

16.60 11.62, 23.58

16.62 93.13, 94.15

16.82 a. $\hat{y}$ = 115.24 + 2.47x c. .5659
d. t = 4.84, p-value = .0001; yes
e. Lower prediction limit = 318.1, upper prediction limit = 505.2

16.84 a. t = 21.78, p-value = 0; yes
b. t = 11.76, p-value = 0; yes

16.86 t = 3.01, p-value = .0042; yes

16.88 t = 1.67, p-value = .0522; no

16.90 t = 29.51, p-value = 0; yes

Chapter 17

17.2 a. $\hat{y}$ = 13.01 + 194x_1 + 1.11x_2
b. 3.75 c. .7629 d. F = 43.43, p-value = 0; evidence that the model is valid.
f. t = .97, p-value = .3417; no g. t = 9.12, p-value = 0; yes h. 23,39 i. 49,65

17.4 c. s_ε = 6.99, R^2 = .3511; model is not very good d. F = 22.01, p-value = 0; evidence that the model is valid. e. Minor league home runs: t = 7.64, p-value = 0; Age: t = .26, p-value = .7961; Years professional: t = 1.75, p-value = .0819. Only the number of minor league home runs is linearly related to the number of major league home runs. f. 9.86 (rounded to 10), 38.76 (rounded to 39) g. 14.66, 24.47

17.6 b. .2882 c. F = 12.96, p-value = 0; evidence that the model is valid. d. High school GPA: t = 6.06, p-value = 0;
SAT: t = .94, p-value = .3485;
Activities: t = .72, p-value = .4720
e. 4.45, 12.00 (actual value = 12.65; 12 is the maximum) f. 6.90, 8.22

17.8 b. F = 29.80, p-value = 0; evidence to conclude that the model is valid.
d. House size: t = 3.21, p-value = .0014;
Number of children: t = 7.84, p-value = 0
Number of adults at home: t = 4.48, p-value = 0

17.10 b. F = 67.97, p-value = 0; evidence that the model is valid. d. 65.54, 77.31
e. 68.75, 74.66

17.12 a. $\hat{y}$ = −28.43 + .604x_1 + .374x_2
b. s_ε = 7.07 and R^2 = .8072; the model fits well. d. 35.16, 66.24 e. 44.43, 56.96

17.30 d_L = 1.16, d_U = 1.59; 4 − d_L = 2.84, 4 − d_U = 2.41; evidence of negative first-order autocorrelation

17.32 d_L = 1.46, d_U = 1.63. There is evidence of positive first-order autocorrelation.

17.34 4 − d_U = 2.27, 4 − d_L = 2.81. There is no evidence of negative first-order autocorrelation.

17.36 **a.** The regression equation is $\hat{y}$ = 2260 + .423x **c.** d = .7859. There is evidence of first-order autocorrelation.

17.38 d = 2.2003; d_L = 1.30, d_U = 1.46, $4 - d_U$ = 2.70, $4 - d_L$ = 2.54. There is no evidence of first-order autocorrelation.

17.40 **a.** $\hat{y}$ = 164.01 + .140x_1 + .0313x_2
b. t = 1.72, p-value = .0974; no
c. t = 4.64, p-value = .0001; yes
d. s_ε = 63.08 and R^2 = .4752; the model fits moderately well. **f.** 69.2, 349.3

17.42 **a.** $\hat{y}$ = 29.60 − .309x_1 − 1.11x_2
b. R^2 = .6123; the model fits moderately well.
c. F = 21.32, p-value = 0; evidence to conclude that the model is valid.
d. Vacancy rate: t = −4.58, p-value = .0001; Unemployment rate: t = −4.73, p-value = .0001 **e.** The error is approximately normally distributed with a constant variance.
f. d = 2.07; no evidence of first-order autocorrelation. **g.** \$14.18, \$23.27

Chapter 18

18.4 **a.** First-order model: Demand = β_0 + β_1 Price + ε; Second-order model: Demand = β_0 + β_1 Price + β_2 Price^2 + ε **c.** The second order model fits better because its standard error of estimate is 5.96, whereas that of the first-order models is 13.29. **d.** 269.3

18.6 **a.** MBA GPA = β_0 + β_1 UnderGPA + β_2 GMAT + β_3 Work + β_4 UnderGPA × GMAT + ε
b. F = 18.43, p-value = 0; s_ε = .790 and R^2 = .4674. The model is valid, but the fit is relatively poor.

18.8 **c.** Both models fit equally well. The standard errors of estimate and coefficients of determination are quite similar.

18.10 **a.** Yield = β_0 + β_1Pressure + β_2 Temperature + β_3 Pressure^2 + β_4 Temperature^2 + β_5 Pressure Temperature + ε
c. s_ε = 512 and R^2 = .6872. The model's fit is good.

18.14 **a.** 8.55, 11.67 **b.** 8.15, 11.31

18.16 **b.** F = 20.43, p-value = 0; evidence that the model is valid. **c.** I_2: t = 1.86, p-value = .0713; I_3: t = −1.58, p-value = .1232
d. t = 3.30, p-value = .0012; yes

18.18 **b.** t = −1.43, p-value = .1589; no
d. I_1: t = 1.61, p-value = .1130; I_2: t = 3.01, p-value = .0039

18.20 **c.** s_ε = 4.20 and R^2 = .5327
f. 1.39, 18.49

18.22 **b.** t = 3.11, p-value = .0025; yes
c. t = −5.36, p-value = 0; yes

18.24 Male-dominated jobs are paid on average \$.039 (3.9 cents) *less* than female-dominated jobs after adjusting for the value of each job.

18.28 .2511

18.30 .2577

18.32 .9287, .9760, .9333

18.34 .9586

18.36 .7341

18.38 **a.** $\hat{y}$ = −.0629 + .0129 GMAT + .7891 I_2 + .1047 Work

18.40 **a.** $\hat{y}$ = .4525 + .668 Minor HR + 1.305 Years Pro

18.42 **a.** Mileage = β_0 + β_1 Speed + β_2 Speed^2 + ε **c.** s_ε = 3.86 and R^2 = .7102. The model fits moderately well.

18.44 **b.** F = 32.65, p-value = 0. There is enough evidence to infer that the model is valid.

18.46 **b.** t = −8.61, p-value = 0; yes

18.48 **a.** Depletion = β_0 + β_1 Temperature + β_2 PH-level + β_3 PH-level^2 + $\beta_4 I_4$ + $\beta_5 I_5$ + ε
where I_1 = 1 if mainly cloudy
I_1 = 0 otherwise
I_2 = 1 if sunny
I_2 = 0 otherwise
c. F = 77.00, p-value = 0; evidence to infer that the model is valid. **d.** t = 6.78, p-value = 0; yes **e.** t = 18.07, p-value = 0; yes **f.** I_1: t = −1.53, p-value = .128; I_2: t = 1.65, p-value = .0997; weather is not a factor in chlorine depletion.

Chapter 19

19.2 **a.** z = 1.66, p-value = .0485; no
b. z = 0, p-value = .5; no

19.4 T = 118; no

19.6 **a.** z = 3.61, p-value = .0004; yes
b. z = 3.61, p-value = .0004; yes

19.8 z = 1.16, p-value = .1230; no

19.10 z = −1.42, p-value = .0778; no

19.12 z = 1.964, p-value = .05; yes

19.14 z = .80, p-value .2119; no

19.16 z = .80, p-value = .4238; no

19.18 z = −1.57; .0582; yes

19.20 z = 2.31, .0104; yes

19.22 z = 1.58, p-value = .0571; no

19.24 T = 34.5; no

19.26 **a.** z = 1.00, p-value = .1587; no
b. z = 1.00, p-value = .1587; no
c. Identical results

19.28 **a.** z = −2.06, p-value = .0396; yes
b. z = −.52, p-value = .6032; no
c. The sign test ignores the magnitudes of the differences.

19.30 z = −5.25, p-value = 0; yes

19.32 z = 2.22, p-value = .0132; yes

19.34 T = 111; no

19.36 z = −2.24, p-value = .0125; no

19.38 z = 1.51, p-value = .0655; no

19.40 H = 1.56; not enough evidence that the population locations differ.

19.42 H = 6.30; evidence that the population locations differ.

19.44 H = 4.46, p-value = .1075; not enough evidence that the population locations differ.

19.46 **a.** H = .899, p-value = .8257; no
b. H = .899, p-value = .8257; no
c. All codes that preserve the order produce the same results.

19.48 H = 6.81, p-value = .0333; yes

19.50 H = 14.04, p-value = .0029; yes

19.52 H = 3.78, p-value = .1507; no

19.54 H = 4.64, p-value = .1999; no

19.56 F_r = 7.74, p-value = .0517; yes

19.58 F_r = 9.42, p-value = .0242; yes

19.60 **a.** F_r = 8.00, p-value = .0460; yes
b. F_r = 8.00, p-value = .0460; yes
c. All codes that preserve the order produce the same results.

19.62 F_r = 5.28, p-value = .0715; no

19.64 F_r = 2.63, p-value = .2691; no

19.66 z = 1.61, p-value = .1074; do not reject the null hypothesis

19.68 r_s = .6931; no

19.70 r_s = .2250; no

19.72 z = −.20, p-value = .8412; no

19.74 z = .54, p-value = .2931; no

19.76 z = 3.87, p-value = .0001; yes

19.78 z = 5.52, p-value = 0; yes

19.80 z = 2.65, p-value = .0082; yes

19.82 F_r = 12.615, p-value = .0055; yes

19.84 T = 36; yes

19.86 z = 1.53, p-value = .0630; no

19.88 z = −1.00, p-value = .3174; no

19.90 z = −2.23, p-value = .0130; yes

19.92 z = .70, p-value = .4840; no

19.94 z = 1.80, p-value = .0360; yes

19.96 **a.** z = 5.27, p-value = 0; yes
b. z = −1.39, p-value = .0828; no

19.98 z = 2.16, p-value = .0153; yes

19.100 z = −10.06, p-value = 0; yes

Chapter 20

20.2 38.8, 35.4, 35.8, 38.8, 44.4, 46.8, 48.8, 46.2

20.4 19.00, 21.67, 24.33, 26.67, 26.67, 26.33, 24.67, 24.33, 21.00, 19.00

20.8 12.00, 16.80, 16.16, 22.43, 18.09, 16.42, 23.28, 21.46, 22.69, 15.74

20.10 38.00, 38.50, 38.85, 39.47, 40.12, 40.91, 41.82, 42.53, 42.88, 43.09; There appears to be a gradual upward trend.

20.14 33.20, 34.80, 34.00, 37.00, 36.40, 35.20, 33.00, 36.20, 34.80, 37.00, 39.40, 44.20, 44.40, 45.00, 46.00, 43.60

20.16 18.00, 19.60, 22.56, 25.94, 28.76, 25.26, 30.35, 28.61, 27.17, 30.70, 36.02, 33.21, 36.33, 35.00, 41.80, 43.08

20.18 The quadratic model would appear to be the best model.

20.20 Linear Model: $\hat{y}$ = −4.96 + 2.38t (R^2 = .81)
Quadratic model: $\hat{y}$ = 3.14 − 2.48t + .54t^2 (R^2 = .98)
The quadratic trend line fits better.

20.22 .675, .892, .864, 1.212, 1.357

20.24 .839, 1.058, 1.270, .833

20.26 **b.** $\hat{y}$ = 39,854 − 20.10 Year

20.28 1.404, .517, .515, .621, .675, 1.145, 2.123

20.30 MAD = 4.60, SSE = 125

20.32 MAD = 9.6, SSE = 552

20.34 191.1, 331.2, 418.5, 169.2

20.36 540.5

20.38 43.08

20.40 30.71, 23.38, 22.67, 19.21

20.42 **a.** 16,494 **b.** 16,340

20.44 349.20, 323.16, 318.28, 367.60

20.46 136.47, 121.44, 88.60, 164.81

20.50 .646, 1.045, 1.405, .904

Chapter 21

21.6 81

21.8 22

21.10 **a.** .6603 **b.** .0361

2.12 19,250

21.14 5.36

21.16 770,000

21.18 3.83

21.20 **a.** .3659 **b.** .0179

21.24 Zone boundaries: 170.10, 173.77, 177.44, 181.10, 184.77, 188.43, 192.10

21.26 **c.** The process is out of control at samples 29 and 30. **d.** A level shift occurred.

21.28 The process is under control.

21.30 The process is under control.

21.32 **a.** The process is out of control.
b. The process is out of control at sample 19.
c. The width became too small.

21.34 The process is out of control at sample 23 (S chart).

21.36 .92

21.38 The process went out of control at sample 29.

21.40 The process is under control.

21.42 Index = .53. The value of the index is low because the statistics used to calculate the control limits and centerline were taken when the process was out of control.

21.44 Centerline = .0324; control limits: 0, .06996. The process is out of control at sample 25.

21.46 Centerline = .0383; control limits: 0, .0959. The process is out of control at samples 25 and 30.

21.48 Centerline = .0257; control limits: .0045, .047. The process is out of control at sample 28.

21.50 Centerline = .0126; control limits: 0, .0275. The process is out of control at sample 24.

Chapter 22

22.6 a_1

22.8 **a. and b.** Bake 2 cakes.

22.10 Pay flat fee.

22.12 Order 200 shirts.

22.14 **a.** Build a medium size plant.
b. Build a medium size plant.

22.16 Don't produce.

22.18 32.5

22.20 **a.** 10 **b.** 2

22.22 I_1: .50, .50, 0
I_2: .268, .357, .375
I_3: .172, .345, .483
I_4: .109, .435, .456

22.24 **a.** I_1: a_3; I_2: a_1 **b.** 3.191

22.26 .30

22.28 As the prior probabilities become more diverse, EVSI decreases.

22.30 4.1 million

22.32 **b.** Switch to rock and roll.
c. 194,000

22.34 Don't proceed with recycling venture.

22.36 38,104

22.38 **c.** Produce Battery 2. **d.** $50,000

22.40 315

22.42 Don't sample; Small number: 25 telephones.
Medium number: 50 telephones.
Large number: 100 telephones.

22.44 Rave review: Release in North America.
Lukewarm or poor response: Sell to European distributor.

INDEX